MONOGRAPHS ON NUMERICAL ANALYSIS

General Editors

L. FOX, J. WALSH

To my family.

THE NUMERICAL TREATMENT OF INTEGRAL EQUATIONS

By

CHRISTOPHER T. H. BAKER M.A., D.Phil.

Reader in Mathematics,
University of Manchester

CLARENDON PRESS · OXFORD

Oxford University Press, Walton Street, Oxford OX2 6DP

OXFORD LONDON GLASGOW NEW YORK
TORONTO MELBOURNE WELLINGTON CAPE TOWN
IBADAN NAIROBI DAR ES SALAAM LUSAKA ADDIS ABABA
KUALA LUMPUR SINGAPORE JAKARTA HONG KONG TOKYO
DELHI BOMBAY CALCUTTA MADRAS KARACHI

ISBN 0 19 853406 X

First published 1977
Reprinted with corrections 1978

Printed in Great Britain by
Thomson Litho Ltd., East Kilbride, Scotland.

PREFACE

'... of making many books there is no end;
and much study is a weariness of the flesh.'
Ecclesiastes XI v. 12.

At the time of writing, few books are devoted to the numerical solution of integral equations. The book by Bückner (1952) is written in German, and that of Anselone (1971) is devoted principally to an abstract theory and its applications; the proceedings edited by Delves and Walsh (1974), which provides a useful theory, is, however, limited by space. Finally, the works of Atkinson (1976) and Ivanov (1976) will have been published by the time that this work appears; each relates to a restricted class of equation. It is therefore my hope that this book will go some way to fill what is, ostensibly, a gap in the English literature.

Hamming (1962) states that the purpose of computing is insight, not numbers. My own view can be similarly but more moderately expressed. In practice, any insight derived from computing follows from accurate (or controlled) computation. I assume, here, that the reader is interested in practical numerical methods for the approximate solution of integral equations, and insight into the extent of their accuracy and any limitations they may possess. By careful selection of their reading, those with varying requirements should find material on this subject to suit their individual tastes; I hope that this book will be helpful to those whose interest lies in the solution of a particular equation as well as those with wider mathematical interests.

In keeping with my general philosophy, I have isolated some of the more abstract theory, in Chapter 5. Unfortunately, this type of theory appears to be required to analyse the treatment of the more difficult problems which can occur in practice. The reader who prefers a succinct and fairly abstract treatment of such a theory may refer to Anselone (1971). Ivanov (1976) treats the singular equations mentioned in Chapter 5 in some detail.

The majority of the numerical methods discussed in this book are illustrated by simple test calculations, which were performed, in general, using the Atlas and ICL/CDC system provided for use of members of the University of Manchester.

The writing of this book has doubtless placed a strain upon the forebearance of both my colleagues and my family, and on the secretarial staff of the Department of Mathematics at Manchester and (whilst I was there on leave at the invitation of Professor Thos. E. Hull) of the Computer Science Department at the University of Toronto. In particular, the secretaries at Manchester have borne, ably, the brunt of typing the initial manuscript. My wife Helen assisted with the checking of the manuscript, proof-reading and helping in the preparation of the references and index. I wish to thank all those mentioned, and also all who have offered constructive criticism of the book, in particular (in alphabetical order) L. Fox, I. Gladwell, E.T. Goodwin, M.S. Keech, D. Kershaw, G.F. Miller, M.R. O'Donohoe, H.H. Robertson, A. Spence and K. Wright, all of whom have made a substantial contribution to relieving the burden on the author. To Professor L. Fox and Dr. E.T. Goodwin, in their rôles as editors of the series at the time of writing, go my thanks for their encouragement, whilst thanks also go to the staff of the Clarendon Press for their assistance.

Christopher T.H. Baker

September, 1975.
Department of Mathematics
The Victoria University of Manchester.

September, 1978.

The occasion of a second printing provides an opportunity to eliminate a number of misprints and similar minor errors. I am greatly indebted to Dr. Derek Arthur of Edinburgh University, who, by his meticulous reading of the book, assisted in this task.

C.T.H.B.

CONTENTS

INTRODUCTION xiii

1. INTEGRAL EQUATIONS 1

1.1. General remarks 2

1.2. Preliminary classification of linear integral equations 3

1.3. Linear equations of various types 6

1.4. The solvability of linear integral equations 10

1.5. Linear operators on a function space 17

1.6. Theoretical discussion of the eigenproblem 28

1.7. Equations of the second kind 33

1.8. Equations of the first kind 40

1.9. Elementary functional analysis 44

1.10. Differential equations and integral equations 56

1.11. Smoothness of kernels 68

1.12. Non-linear integral equations 77

2. NUMERICAL ANALYSIS 85

2.1. Introduction 86

2.2. Interpolation 87

2.3. Error in polynomial interpolation 92

2.4. Least-squares fitting 95

2.5. Fourier series 97

2.6. Error bounds 99

2.7. Generalized Fourier series 100

2.8. Closure and completeness 101

2.9. Discrete weighted least-squares approximations 103

2.10. Approximation in more than one dimension 104

2.11. Numerical integration 105

2.12. Repeated quadrature rules 109

2.13. Modifications of the trapezium rule 113

2.14. Classification of quadrature rules 123

2.15. Convergence 124

2.16. Precise error bounds 128
2.17. Integration of products 131
2.18. Integration in more than one dimension 144
2.19. Asymptotic upper and lower bounds 147
2.20. Numerical methods 152

3. EIGENVALUE PROBLEMS 167
3.1. Introduction 168
3.2. The quadrature method 169
3.3. A modification of the quadrature method 193
3.4. Product-integration methods 198
3.5. Hämmerlin's approximate kernel method 202
3.6. Expansion methods 205
3.7. The collocation method 206
3.8. The Galerkin method 214
3.9. A least-squares method 219
3.10. The Rayleigh quotient 220
3.11. The Rayleigh-Ritz equations 222
3.12. Further remarks 225
3.13. Theoretical results 227
3.14. The quadrature method for degenerate kernels 231
3.15. The quadrature method for general kernels 234
3.16. Error bounds for computed eigenvalues of Hermitian kernels 269
3.17. A posteriori error bounds dependent on the local error 278
3.18. A dual approach: Hermitian kernels 287
3.19. The rôle of Peano theory 289
3.20. Asymptotic error bounds for simple eigenvalues 293
3.21. Asymptotic expansions for the error in a simple eigenvalue and the corresponding eigenfunction 297
3.22. Extensions 314
3.23. Expansion methods 316
3.24. The Rayleigh-Ritz method for Hermitian kernels 316

3.25. Error bounds for the Rayleigh-Ritz method for Hermitian kernels 322
3.26. Convergence of the eigenvalues and eigenfunctions 325
3.27. A further error bound 329
3.28. The Rayleigh quotient 330
3.29. Galerkin's method and the Rayleigh-Ritz method 330
3.30. Convergence of the Galerkin method 336
3.31. Convergence of the collocation method 340
3.32. A product-integration method 346
3.33. A generalized Rayleigh quotient and a least-squares solution 349

4. LINEAR EQUATIONS OF THE SECOND KIND 352
4.1. Introductory remarks 353
4.2. The Fredholm equation of the second kind 353
4.3. The quadrature method 356
4.4. A modification to the quadrature method 375
4.5. A further modification 379
4.6. A product-integration method 380
4.7. Reduction to a degenerate kernel 385
4.8. Bateman's method 388
4.9. Expansion methods and quadrature methods 389
4.10. An infinite system of equations 391
4.11. The collocation method 396
4.12. Ritz-Galerkin type methods and minimization methods 406
4.13. Linear programming methods 416
4.14. Further remarks; conditioning 421
4.15. The theory of methods for Fredholm equations of the second kind 423
4.16. Analysis of the quadrature method 432
4.17. The Anselone-Brakhage-Mysovskih approach 457
4.18. Asymptotic expansions 466
4.19. Deferred correction 473
4.20. The modified quadrature method 477

4.21. A product-integration method 479

4.22. Expansion methods 481

4.23. Variational formulation of the Rayleigh-Ritz method 482

4.24. Collocation 487

4.25. Ritz-Galerkin-type methods 496

4.26. The classical Galerkin method 500

4.27. The Galerkin methods using Chebyshev polynomials 505

4.28. The method of least squares and the method of moments 507

4.29. Projection methods 509

5. FURTHER DISCUSSION OF THE TREATMENT OF FREDHOLM EQUATIONS 519

5.1. Introduction 520

5.2. Linear equations 521

5.3. 'Mild' and weak singularities in linear equations 526

5.4. The modified quadrature method applied to equations with weakly singular kernels 534

5.5. Simple product-integration formulae for weakly singular kernels 539

5.6. Practical product-integration methods applied to weakly singular kernels 546

5.7. Expansion methods for weakly singular kernels 556

5.8. The eigenvalue problem for a weakly singular kernel 563

5.9. Equations of the second kind which are not uniquely solvable 570

5.10. Singular integral equations 577

5.11. Theory of methods for continuous and weakly singular kernels 589

5.12. An abstract theory 596

5.13. Convergence of the modified quadrature method 603

5.14. Convergence of the general product-integration method 616

5.15. Fredholm equations of the first kind 635

5.16. Quadrature methods and expansion methods 645

5.17. A least-squares approach and regularization methods 655
5.18. Non-linear integral equations 685
5.19. The quadrature methods 686
5.20. Expansion methods 699
5.21. Collocation 701
5.22. Galerkin methods 707
5.23. Variational formulation 709
5.24. Variational methods 717
5.25. Theoretical study of methods for non-linear integral equations 719

6. VOLTERRA INTEGRAL EQUATIONS 755
6.1. Introduction 755
6.2. The linear equation of the second kind and Volterra type 759
6.3. Stability and ill-conditioning 788
6.4. Quadrature methods for non-linear integral equations 825
6.5. Starting values 835
6.6. Runge-Kutta-type methods for Volterra equations 849
6.7. A block-by-block method 864
6.8. Product-integration methods 885
6.9. Product-integration techniques for singular linear and non-linear equations 892
6.10. Non-singular Volterra equations of the first kind 896
6.11. Product-integration methods for equations of the first kind with continuous or weakly singular kernels 915
6.12. Convergence properties of numerical methods for equations of the second kind 922
6.13. Convergence of certain block-by-block methods 952
6.14. Convergence of product-integration methods 959
6.15. Convergence of methods for non-singular equations of the first kind 963

6.16. Block-by-block methods for non-singular equations of the first kind 974

6.17. Theory of product-integration methods for first kind equations 979

REFERENCES 985

INDEX 1031

INTRODUCTION

This book is concerned with the numerical analysis of integral equations. We are not principally concerned with the abstract theory of integral equations, nor with applications of mathematics where integral equations arise, but the first chapter is devoted to a review of the theory of integral equations. The survey of certain aspects of numerical analysis in Chapter 2 is intended to emphasize various topics which are of relevance in the study of numerical methods for integral equations.

In practice, mathematical and physical insight is of value in any attempt to solve a particular equation. It would doubtless have proved instructive had I selected a particular equation and investigated the problem of approximating its solution, taking advantage of all that is known about the equation and its origin. Instead, I have chosen to write in general terms about various classes of equation.

I have generally tried to separate the exposition of numerical methods from the underlying theory of the methods. This has been done in order that the reader may pursue his own interests. However, practice not substantiated by theory is risky; equally, theory not motivated by practice is something of an affectation. The principal results of the theory are accessible in the statements of theorems. Theorems relating to convergence and order of convergence provide helpful insight into the behaviour of numerical methods, and may lead to rigorous error bounds or well-founded error estimates (both of which are desirable in practice as well as in theory).

Some theorems, of the type referred to above, may be viewed with a natural suspicion. In practical computation on a digital computer, limiting processes cannot take place and rounding errors occur which obscure the behaviour predicted by 'convergence theory'. A classic example occurs in the numerical solution of Volterra equations of the second kind, where a class of methods can be shown to be 'convergent' but not all are satisfactory in practice.

Whilst each of Chapters 3, 4, 5, and 6 is clearly divided into sections on practice and on theory, there are some areas where such a division cannot be properly maintained. This is most apparent in Chapter 5.

The book can also be divided into three parts, containing respectively Chapters 1 and 2 (which give background material), Chapters 3 and 4 (in which we consider certain types of *well-behaved* equations), and Chapters 5 and 6 (where the discussion includes coverage of some more difficult problems). Each chapter is prefaced by a summary of its contents, since it often happens that a reader consulting a book has a particular problem in mind.

Although a number of results and proofs which have not previously appeared in print are presented here, it has been my aim to give (within the constraints, and where possible) a discussion in terms of fairly fundamental concepts of analysis.

A word on the system of notation applied throughout the book is appropriate here. The length of the book and the range of topics covered do not permit a complete standardization of the notation, but certain principles are followed closely. Thus, real or complex scalars are denoted by *italic* or Greek symbols, as are functions assuming such values, whilst matrices are denoted by upper-case elite or Greek letters with wavy underlines (as in $\underset{\sim}{\mathrm{A}}$ and $\underset{\sim}{\Sigma}$) and lower-case elite or Greek letters having wavy underlines, as in $\underset{\sim}{\mathrm{a}}$ and $\underset{\sim}{\alpha}$, denote column vectors. Amongst functions, we reserve $f(x)$, if necessary with embellisments as in $f_r^{\pm}(x)$, to denote the solution of an integral equation. Computed approximations are denoted, in general, by adding a tilde as in $\tilde{f}(x)$; it is sometimes necessary to consider more than one type of approximation, however, and we then employ such symbols as $\hat{f}(x)$, $\mathring{f}(x)$, or $\tilde{\mathrm{f}}(x)$. Since the notation $\underset{\sim}{\mathrm{f}}$, $\underset{\sim}{\tilde{\mathrm{f}}}$, etc., is used to denote a vector whose components are values of the functions $f(x)$, $\tilde{f}(x)$, etc., consistency requires that functions denoted by $\tilde{f}(x)$ and by $\tilde{\mathrm{f}}(x)$ give rise to identical vectors $\underset{\sim}{\tilde{\mathrm{f}}}$.

These comments are intended to serve as an introduction to the system of notation to which we hope the reader will grow accustomed. It will be noted that certain Greek symbols serve to denote mathematical symbols (as Σ for summation) in addition to their normal rôles. Where I considered confusion possible, the symbol ε for membership is replaced by ϵ. Finally, it should be noted that the house style employed involves the repetition of an arithmetic operation at a break in a mathematical expression. The symbol * in the text concludes an Example.

1

INTEGRAL EQUATIONS

In Chapter 1 the reader is introduced to the theory of integral equations. We shall content ourselves with an outline of this theory, and a brief survey of some related mathematical material. This chapter is not intended to provide a complete theory, and in many cases results which are standard in the literature are quoted without proof. We shall cover in more detail various results which assume some importance in later parts of the book.

In this chapter, and elsewhere throughout the book, more space is devoted to non-singular linear integral equations than to non-linear or singular equations, although the latter equations may be more common in practice. We have three reasons for this: First, more is known about the linear theory; secondly, we can often gain some insight into the general case by considering simple examples; and thirdly, we can sometimes treat equations which are difficult to solve by considering related, but simpler, equations. Some indication of such techniques is given below; and I have taken pains to cover the types of equation which most commonly arise in practice.

It could be argued that the pedagogically correct technique for discussing integral equations is within an abstract framework of functional analysis. I have foresaken such an approach in the interest of simplicity, but have devoted sections to an introduction to the type of analysis used in the abstract discussion of integral equations.

The reader is asked to be selective in his study. A core of the material necessary for a discussion of numerical methods is given in sections 1.1-1.4 and 1.10-1.12, and the remainder of the chapter could be treated as a reference section for later use.

Should an extensive, systematic or rigorous development of the theory of integral equations be required, I would refer the reader (in particular) to the work of Smithies (1962), of Cochran (1972), and of Zabreyko, Koshelev, Krasnosel'skii, Mikhlin, Rakovshchik and Stet'senko (1975).

1.1. *General remarks*

A functional equation in which the unknown function appears under an integral sign is called an integral equation; see, however, Zabreyko *et al.* (1975, p.1). For example, the quations

$$\frac{1}{f_0(x)} + \tfrac{1}{2}\int_0^1 \frac{xf_0(y)}{x+y}\,dy = 1 \qquad (0 \leq x \leq 1)\,, \tag{1.1}$$

$$f_1(x) - \int_0^x (x-y)\,\{f_1(y)\}^2\,dy = e^{-x} \qquad (x \geq 0)\,, \tag{1.2}$$

$$\int_{-\infty}^{\infty} \exp\,\{-(x-y)^2\}\,f_2(y)\,dy = e^{x} \qquad (-\infty < x < \infty)\,, \tag{1.3}$$

and

$$\lambda\int_{-1}^{1}\int_{-1}^{1} |xu + yv|\,f_3(u,v)\,du\,dv = f_3(x,y) \quad (-1 < x,y < 1) \tag{1.4}$$

are all integral equations. The functions to be found are $f_0(x)$, $f_1(x)$, $f_2(x)$, and $f_3(x,y)$ respectively. If the derivative of the solution appears in the integral equation, as in

$$\{f''(x)\}^2 - \int_{-1}^{1} \sin xy\; f(y)\,dy = \cos x \quad (-1 < x < 1)\,, \tag{1.5}$$

the equation is usually known as an integro-differential equation, and additional boundary conditions are required to determine the solution $f(x)$.

Integral equations occur in the mathematical theory for a number of branches of science. In particular, they occur in the study of acoustics, optics and laser theory, potential theory, radiative transfer theory, cardiology, and in fluid mechanics and statistics.

The solutions occur non-linearly in eqns (1.1), (1.2), and (1.5), and these equations are therefore said to be non-linear equations. The other examples are linear equations (the unknown function appears linearly). In the early parts of this book we treat linear equations (in which the solution is a function of a single variable, as in eqn (1.3)). Later in the book we shall consider more general types of integral equation.

One of our first tasks is to classify the different types of linear integral equation.

1.2. *Preliminary classification of linear integral equations*

A general form of linear integral equation is given by the equation

$$\alpha(x)\ f(x) - \lambda \int_a^b K(x,y)\ f(y)\mathrm{d}y = g(x)\ , \tag{1.6}$$

where the functions involved may be supposed to be complex-valued functions of real variables. We shall suppose that eqn (1.6) is valid for $a \leq x \leq b$, and we shall suppose, in general, that a and b are finite. The functions $\alpha(x)$, $g(x)$, and $K(x,y)$ are known for $a \leq x,\ y \leq b$, and λ is a constant (which, when its value is known, is sometimes absorbed in $K(x,y)$) . The function $K(x,y)$ is called the *kernel* of the integral equation.

We shall consider some particular cases of (1.6). The first of these is

$$\int_a^b K(x,y)\ f(y)\mathrm{d}y = g(x) \qquad (c \leq x \leq d)\ , \tag{1.7}$$

which is called an *equation of the first kind*. We can ensure, by a change of variable, that $c=a$ and $d=b$, given that $|abcd|$ is finite.

Example 1.1

The equation

$$\int_0^1 (x^2 + y^2)^{\frac{1}{2}} f(y)\mathrm{d}y = \frac{1}{3}\ \{(1 + x^2)^{\frac{3}{2}} - x^3\}\ (0 \leq x \leq 1)$$

has a solution $f(x) = x$. *

The equation

$$f(x) - \lambda\int_a^b K(x,y)\, f(y)\mathrm{d}y = g(x) \qquad (a \le x \le b)\ , \tag{1.8}$$

where λ is a *known* constant, is called an *equation of the second kind.*

Example 1.2

The equation

$$f(x) - \int_0^1 \mathrm{e}^{xy}\, f(y)\mathrm{d}y = 1 - \frac{1}{x}(\mathrm{e}^x - 1) \quad (0 \le x \le 1)$$

has the unique solution $f(x) = 1$. *

Example 1.3

The equation

$$f(x) - 3\int_0^1 xy\, f(y)\mathrm{d}y = x^2 \qquad (0 \le x \le 1)$$

has no solution. *

Connected with eqn (1.8) is an *eigenvalue problem* associated with the equation

$$f(x) = \lambda\int_a^b K(x,y)\, f(y)\mathrm{d}y \quad . \quad (a \le x \le b) \tag{1.9}$$

and eqn (1.10) below. In solving (1.9), we seek values of the parameter λ for which (1.9) has a non-null solution $f(x)$. Such a value λ is called a *characteristic value.*† Since $f(x)$ is non-null it is clear that $\lambda \neq 0$ and we may write $\kappa = 1/\lambda$, so that eqn (1.9) becomes

† Not all authors use our terminology. See, for example, Tricomi (1957).

$$\kappa f(x) = \int_a^b K(x,y)\, f(y)\mathrm{d}y \qquad (a \le x \le b)\,. \qquad (1.10)$$

A value of κ for which eqn (1.10) has a non-null solution will be called an *eigenvalue*¶ of the kernel $K(x,y)$ (the range $a \le x,y \le b$ will be implied, but not stated, unless there is a possibility of confusion). The corresponding function $f(x)$ will be called an *eigenfunction* (some authors call it a *characteristic function*). A given kernel may have more than one eigenvalue, or it may have none. We refer to the problem of determining the solutions of (1.10) as the eigenvalue problem or *eigenproblem* of the kernel.

The set of eigenvalues of a kernel, together with zero, is called the *spectrum* of $K(x,y)$. (Another definition of the term spectrum is given in section 1.9.)

Example 1.4

The equation

$$\int_0^{2\pi} \sin x \cos y\, f(y)\mathrm{d}y = \kappa\, f(x) \qquad (0 \le x \le 2\pi)$$

has no non-zero eigenvalues. Any non-null function $f(x)$ for which

$$\int_0^{2\pi} \cos y\, f(y)\mathrm{d}y = 0$$

is an eigenfunction corresponding to the eigenvalue zero. *

Example 1.5

Consider the eigenvalue problem

$$\int_0^{2\pi} H(x - y)\, f(y)\mathrm{d}y = \kappa f(x)\,, \qquad (0 \le x \le 2\pi)$$

where $H(z) = \frac{1}{2}\alpha_0 + \sum_{r=1}^{\infty} \alpha_r \cos rz$ (this series being convergent).

¶ Some authors require κ to be non-zero; we do not.

The spectrum may be shown to be the set of eigenvalues:

$$\{\pi\alpha_0, \pi\alpha_1 \text{ (twice)}, \pi\alpha_2 \text{ (twice)}, \ldots\},$$

together with zero (which, as $\alpha_r \to 0$, is a limit point of the eigenvalues). One eigenfunction corresponding to the eigenvalue $\pi\alpha_0$ is the function $f(x) \equiv 1$. Any other function which is constant and non-zero is also an eigenfunction corresponding to this eigenvalue, and, more generally, the eigenfunctions of any integral equation are only determined to within an arbitrary multiplicative constant.

Corresponding to the eigenvalue $\pi\alpha_r$, for $r \geq 1$, there are two linearly independent eigenfunctions, for example, $f(x) = \cos rx$ and $f(x) = \sin rx$. Any linear combination of these two eigenfunctions, of the form $f(x) = \alpha \cos rx + \beta \sin rx$, is also an eigenfunction corresponding to the same eigenvalue. *

1.3. *Linear equations of various types*

The classification of section 1.2 is used for a wide class of equations, including those in which the limits of integration a,b are infinite, and/or those in which the kernel $K(x,y)$ is so badly behaved that the integrals in eqns (1.6)-(1.9) must be interpreted as Cauchy principal values. We shall call such equations singular equations, and we defer their consideration until later; unless otherwise stated we suppose that the integrals have a meaning as Riemann integrals.

The theoretical treatment of eqns (1.7)-(1.9) usually covers (Smithies 1962) the case where

$$\int_a^b |\, g(t)\,|^2\, dt < \infty \quad \text{and}^\dagger \quad \int_a^b \int_a^b |\, K(x,y)\,|^2\, dx dy < \infty\,,$$

---

† Where we refer to a square-integrable kernel, we imply that the kernel $K(x,y)$ is an L^2-kernel in the sense of Smithies (1962), so that

$$\int_a^b |K(x,y)|^2 dy\,, \quad \int_a^b |K(x,y)|^2 dx\,, \text{ and } \int_a^b \int_a^b |K(x,y)|^2 dx\, dy$$

are finite.

and these integrals exist as Lebesgue integrals. We shall impose rather more restrictive conditions. Unless we state otherwise, we shall suppose that the kernel $K(x,y)$ is, at worst, piecewise-continuous (and bounded), and that any discontinuities occur on a finite number of continuous curves $y = \phi_i(x)$, and we assume that $g(x)$ is bounded and piecewise-continuous.ǂ We shall also suppose that the range of integration $[a,b]$ is finite; as a consequence we can presume, where convenient, that the range has been normalized by a change of variable so that $a = 0$ and $b = 1$. It may also be possible to transform an infinite range of integration into $[0,1]$, but the transformation is likely to produce a kernel $K(x,y)$ which does not satisfy the rather restrictive conditions which we will impose in our general discussion.

One type of kernel which satisfies our conditions of piecewise-continuity occurs when $K(x,y)$ is continuous for $0 \le y \le x$, and $K(x,y) = 0$ for $y > x$. In such a case the integral equation

$$f(x) - \lambda \int_0^1 K(x,y)\, f(y)\, dy = g(x) \tag{1.11}$$

may be written as

$$f(x) - \lambda \int_0^x K(x,y)\, f(y) dy = g(x). \tag{1.12}$$

Whereas eqn (1.11) may be valid for $0 \le x \le 1$, we often encounter equations of the form (1.12) which are valid for all x, or for $0 \le x < \infty$, or for $-\infty < x \le 0$. Irrespective of the range of values of x for which (1.12) is valid, we say that eqn (1.12) is an equation of *Volterra type*, whereas an integral equation in which the limits of integration appear as fixed constants is said to be formulated as an equation

ǂ The term piecewise-continuous will be used, throughout, to denote continuity except at a finite number of points where finite jump discontinuities can occur, and the term bounded is only added for emphasis.

of *Fredholm type*. (Thus eqns (1.7), (1.8), and (1.9) are equations of Fredholm type.) We use the terms 'equation of the first kind' and 'equation of the second kind' for Volterra equations just as for Fredholm equations. We say that a kernel $K(x,y)$ is a *Volterra kernel* if $K(x,y) = 0$ when $y > x$, and we call it a *continuous Volterra kernel* if $K(x,y)$ is continuous when $y < x$.

Equations of the first kind may sometimes be transformed into equations of the second kind. This is well known for equations of Volterra type (Tricomi 1957, p.15). Suppose that we seek a continuous solution of the equation

$$\int_a^x K(x,y)\, f(y)\mathrm{d}y = g(x) \qquad (-\infty < x < \infty)\,, \tag{1.13}$$

where $K(x,y)$ is a continuous Volterra kernel. This is a Volterra equation of the first kind, and for a solution we require that $g(a) = 0$. If $K_x(x,y) = (\partial/\partial x)\, K(x,y)$ is continuous for $y \le x$ and $g'(x)$ is continuous, we may differentiate eqn (1.13) and obtain

$$K(x,x)\, f(x) + \int_a^x K_x(x,y)\, f(y)\mathrm{d}y = g'(x)\,,$$

and, if $K(x,x)$ is non-vanishing, we may write

$$f(x) + \int_a^x H_0(x,y)\, f(y)\mathrm{d}y = g'(x)/K(x,x)\,,$$

where $H_0(x,y) = K_x(x,y)/K(x,x)$. We thus obtain an equation of the second kind and of the Volterra type.

On the other hand, suppose $K_y(x,y) = (\partial/\partial y)K(x,y)$ is continuous for $y \le x$. We may write

$$F(x) = \int_a^x f(y)\mathrm{d}y\,.$$

This is a Volterra equation of the first kind for $f(x)$ in terms of $F(x)$, and its solution is $f(x) = F'(x)$. If the integral in (1.13) is rewritten, using integration by parts for the left-hand term, we obtain

$$K(x,x)\ F(x) - \int_a^x K_y(x,y)\ F(y)dy = g(x)\ .$$

If $K(x,x)$ is non-vanishing, we obtain a Volterra equation of the second kind for $F(x)$:

$$F(x) - \int_a^x H_1(x,y)\ F(y)dy = g(x)/K(x,x)\quad ,$$

where $H_1(x,y) = K_y(x,y)/K(x,x)$. The function $f(x)$ is obtained by differentiating the solution $F(x)$ of this equation.

The equations

$$f(x) - \lambda \int_{\phi_0(x)}^{\phi_1(x)} K(x,y)\ f(y)dy = g(x)$$

and

$$\int_{\phi_0(x)}^{\phi_1(x)} K(x,y)\ f(y)dy = g(x)\ ,$$

where $\phi_i(x)$ is continuous for $i = 0,1$ and $K(x,y)$ is continuous for $\phi_0(x) \le y \le \phi_1(x)$, are sometimes similar to Volterra equations. Such equations may, in special circumstances, be recast as Volterra equations or as systems of Volterra equations.

Example 1.6

Suppose that

$$f(x) - \int_{-x}^{x} K(x,y)\ f(y)dy = g(x) \qquad (- \infty < x < \infty)\ .$$

We will write $\phi(x) = f(x)$ for $x \ge 0$, $\phi(x) = 0$ for $x < 0$ and $\psi(x) = f(-x)$ for $x \ge 0$, $\psi(x) = 0$ for $x < 0$. Now, for $x \ge 0$,

$$\phi(x) - \int_0^x K(x,y)\ \phi(y)dy - \int_0^x K(x,-y)\ \psi(y)dy = g(x)$$

and (also for $x \geq 0$)

$$\psi(x) + \int_0^x K(-x,-y)\ \psi(y)dy + \int_0^x K(-x,y)\ \phi(y)dy = g(-x)\ .$$

We thus have a system of coupled Volterra equations.

If we determine $\phi(x)$ and $\psi(x)$, we may then set $f(x) = \phi(x)$ for $x \geq 0$, and $f(x) = \psi(-x)$ for $x \leq 0$. *

1.4. *The solvability of linear integral equations*

The reader may feel that the classification of equations outlined in sections 1.2 and 1.3 is somewhat arbitrary or inconclusive. The terminology introduced above is useful, however, for discussing the solvability of linear integral equations. We shall also see, later, that the numerical and theoretical treatment of an integral equation is determined, in part, by the kind (whether first or second) and type (Fredholm or Volterra) of the equation.

Perhaps I should emphasize that I do not aim to give, in this book, a complete and rigorous treatment of the theory of integral equations, but merely to supply the background material which is relevant to the numerical solution of the equations we consider. The first question to answer is whether or not a solution exists.

Recalling that our functions are generally supposed to be piecewise-continuous and bounded, we may summarize the existence and uniqueness properties of solutions to certain types of linear integral equations in sections 1.4.1, 1.4.2, and 1.4.3. We return to a discussion of the underlying theory in sections 1.6, 1.7, and 1.8. At this point, however, we shall state some well-known points which arise in the solution of systems of linear algebraic equations, since there are a number of similarities with our study of linear integral equations.

We suppose that $\underset{\sim}{A}$ is a matrix of order n. In the eigenvalue problem for the matrix $\underset{\sim}{A}$, we seek values μ_i which correspond to non-zero vectors $\underset{\sim}{x}_i$ such that $\underset{\sim}{A}\,\underset{\sim}{x}_i = \mu_i \underset{\sim}{x}_i$. The problem can be recast as follows: we seek values of μ which permit the equation $(\underset{\sim}{A} - \mu\underset{\sim}{I})\underset{\sim}{x} = \underset{\sim}{0}$ to have a non-trivial solution ($\underset{\sim}{x} = \underset{\sim}{0}$ is always a solution). Now, given a matrix $\underset{\sim}{B}$ of order n, and a vector $\underset{\sim}{b}$, the equation $\underset{\sim}{B}\,\underset{\sim}{x} = \underset{\sim}{b}$ has a unique solution if and only if $\det(\underset{\sim}{B}) \neq 0$. If $\det(\underset{\sim}{B}) = 0$, there will be a solution provided[§] that $\underset{\sim}{b}$ is a vector such that $\underset{\sim}{y}^*\underset{\sim}{b} = 0$ for every vector $\underset{\sim}{y}$ which satisfies the equation $\underset{\sim}{B}^*\underset{\sim}{y} = 0$. (Indeed, there will then be an infinite family of solutions $\underset{\sim}{x}$.) Now, we may apply this result to the eigenvalue problem by setting $\underset{\sim}{B} = \underset{\sim}{A} - \mu\underset{\sim}{I}, \underset{\sim}{b}=\underset{\sim}{0}$, and we see that the eigenvalues of $\underset{\sim}{A}$ are the zeros of the polynomial $\det(\underset{\sim}{A} - \mu\underset{\sim}{I})$. There is an infinite number of corresponding eigenvectors, since if $\underset{\sim}{x}$ is one eigenvector and α is non-zero, $\alpha\underset{\sim}{x}$ is another eigenvector corresponding to the same eigenvalue. However, there is the possibility (when μ_i is a multiple zero of $\det(\underset{\sim}{A} - \mu\underset{\sim}{I})$) that we can find p linearly independent eigenvectors $\underset{\sim}{x}_{i_j}$ $(j=1,2,3,\ldots,p)$ such that $\underset{\sim}{A}\,\underset{\sim}{x}_{i_j} = \mu_i \underset{\sim}{x}_{i_j}$ (where p is less than or equal to the multiplicity of μ_i as a zero of the characterisic polynomial). Similarly, if we consider $\underset{\sim}{B}\,\underset{\sim}{x} = \underset{\sim}{b}$, where $\det(\underset{\sim}{B}) = 0$, there are p linearly independent solutions $\underset{\sim}{x}_i$ if we can find p linearly independent vectors $\underset{\sim}{y}_i$ $(i=1,2,\ldots,p)$ such that $\underset{\sim}{B}^*\underset{\sim}{y}_i = 0$. Lovitt (1924) compares this situation with the corresponding situation for integral equations.

We now return to our study of linear integral equations.

1.4.1.

In the eigenvalue problem we seek an eigenvalue κ and a corresponding non-null eigenfunction $f(x)$ such that

$$\kappa f(x) = \int_a^b K(x,y)\, f(y)\,dy \; .$$

[§] We write $\underset{\sim}{y}^T$ as $\underset{\sim}{y}^*$ and $\underset{\sim}{\bar{B}}^T$ as $\underset{\sim}{B}^*$.

In general we cannot guarantee the existence of any solution $\kappa \neq 0$ to this problem. Thus, a continuous Volterra kernel (for which $K(x,y) = 0$ if $y>x$) has no continuous eigenfunctions for $\kappa \neq 0$; for another case see Example 1.4. It is sometimes a difficult task to establish that a given kernel does have non-zero eigenvalues, but there are cases where the existence of eigenvalues is easily shown.

In particular, suppose that $K(x,y)$ is not identically zero and either it is real and $K(x,y) = K(y,x)$ or, more generally, that $K(x,y) = \overline{K(y,x)}$. (In the first case $K(x,y)$ is said to be real and *symmetric* and in the second case it is said to be *Hermitian*.) In either case, there is at least one non-zero eigenvalue and all the eigenvalues are real. Moreover, if $f_1(x)$ and $f_2(x)$ are eigenfunctions corresponding to two different eigenvalues, they are *orthogonal*, that is,

$$\int_a^b f_1(t)\,\overline{f_2(t)}\,dt = 0 \quad \text{(see, for example, Smithies 1962).}$$

In general, the set of non-zero eigenvalues of a kernel $K(x,y)$ is countable, and there is no non-zero limit point (or cluster point) for the eigenvalues. This last statement presupposes some method of counting the multiplicity of an eigenvalue, and we shall discuss this point in section 1.6. Our statement implies that each non-zero eigenvalue has a finite multiplicity.

The eigenvalues of a Hermitian kernel are real, but the eigenvalues of a general kernel may be complex. If the kernel is real, then any complex eigenvalues must occur in complex conjugate pairs.

Remark. It is useful as an exercise to obtain some bounds on any eigenvalues which a continuous kernel may possess. If $f(x)$ is continuous,† non-null, and

$$\kappa f(x) = \int_a^b K(x,y)\, f(y)\,dy \,,$$

---

† Any square-integrable eigenfunction of a continuous kernel, corresponding to a non-zero eigenvalue, is continuous.

then

$$|\kappa|\ |f(x)| \le (b-a) \sup_{a\le y\le b} \{|K(x,y)\ f(y)|\}$$

$$\le (b-a) \sup_{a\le y\le b} |K(x,y)| \sup_{a\le y\le b} |f(y)| ,$$

if $x \in [a,b]$. We take the supremum over values of x to obtain (since $\sup_{a\le x\le b} |f(x)| \ne 0$)

$$|\kappa| \le (b-a) \sup_{a\le x,y\le b} |K(x,y)| . \tag{1.14a}$$

A stronger inequality which we may obtain using the Cauchy-Schwartz inequality is

$$|\kappa| \le \{\int_a^b \int_a^b |K(x,y)|^2 \,dx dy\}^{\frac{1}{2}} . \tag{1.14b}$$

The maximum of the moduli of the eigenvalues of $K(x,y)$ is called the *spectral radius* of K, and we shall denote it by $\rho(K)$.

The degree of smoothness of any eigenfunctions is of interest to us. We suppose that $f(x)$ is an eigenfunction corresponding to a non-zero eigenvalue κ . We first consider the function

$$F(x) = \int_a^b K(x,y)\ \phi(y)dy \qquad (a \le x \le b) ,$$

where $\phi(x)$ is 'absolutely integrable', that is,

$$\int_a^b |\phi(y)|dy < \infty .$$

We suppose that $K(x,y)$ is continuous for $a \le x, y \le b$, so that given $\varepsilon > 0$ there is a $\delta > 0$, independent of y, such that $|K(x_1,y) - K(x_2,y)| < \varepsilon$ if $|x_1 - x_2| < \delta$ and $a \le x_1, x_2 \le b$. When x_1 and x_2 satisfy these inequalities, we may show that

$$|F(x_1) - F(x_2)| \le \sup_{a \le y \le b} |K(x_1,y) - K(x_2,y)| \int_a^b |\phi(y)|\,dy$$

$$\le \varepsilon \int_a^b |\phi(y)|\,dy = \varepsilon_1 .$$

and ε_1 may be made as small as required by taking δ sufficiently small. Thus $F(x)$ is continuous. If we now set $\phi(x) = (1/\kappa)\, f(x)$, then $F(x)$ becomes the eigenfunction $f(x)$, and $f(x)$ is continuous if it is absolutely integrable. (A function which is square-integrable over a finite range is also absolutely integrable.)

If we proceed further, we may show that $f'(x)$ exists if the eigenfunction $f(x)$ is absolutely integrable and $\kappa \ne 0$, whenever $K_x(x,y) \equiv (\partial/\partial x)K(x,y)$ is continuous.

Sometimes, $K(x,y)$ has a finite jump discontinuity at $x = y$, whilst $(\partial/\partial x)\, K(x,y)$ is continuous for $a \le x < y$ and for $y < x \le b$. We then write

$$\kappa\, f(x) = \int_a^x K(x,y)\, f(y)\,dy + \int_x^b K(x,y)\, f(y)\,dy .$$

If $\kappa \ne 0$, we may show that if $f(x)$ is square-integrable then it is continuous. Further,

$$\kappa\, f'(x) = -\Delta(x)\, f(x) + \int_a^x K_x(x,y)\, f(y)\, dy + \int_x^b K_x(x,y)\, f(y)\, dy ,$$

where $\Delta(x) = K(x,x+) - K(x,x-)$ is the jump in $K(x,y)$ at $y = x$. Thus $f'(x)$ exists.

It is sometimes possible to investigate the higher differentiability of eigenfunctions of kernels in a similar fashion.

1.4.2.

A value of λ such that the equation

$$f(x) - \lambda\int_a^b K(x,y)\, f(y)dy = g(x) \qquad (a \le x \le b) \tag{1.15}$$

is *uniquely* solvable (when $g(x)$ is piecewise-continuous but otherwise arbitrary) is known as a *regular value*. If λ is a characteristic value and $\psi(x)$ a corresponding eigenfunction then to any solution $f(x)$ of eqn (1.15) there corresponds another solution $f(x) + \alpha\psi(x)$, where α is arbitrary. Thus if λ is a characteristic value it cannot be a regular value. Moreover, if λ is not a characteristic value it can be shown that eqn (1.15) has a unique solution, for arbitrary $g(x)$, and hence that λ is a regular value. Thus either λ is a regular value or λ is a characteristic value; this result is known as *Fredholm's alternative*.

Remark. A number of results are associated with the name of Fredholm (see, for example, Mikhlin 1964 *a*). Amongst these results is the statement that eqn (1.15) has a solution if and only if

$$\int_a^b \phi(t)\, \overline{g(t)}\, dt = 0$$

for every function $\phi(x)$ which satisfies the equation

$$\phi(x) = \bar{\lambda}\int_a^b \overline{K(y,x)}\, \phi(y)dy \ .$$

We can show that such (non-null) functions $\phi(x)$ exists only if λ is a characteristic value of $K(x,y)$. If λ is a characteristic value and $g(x)$ satisfies the required conditions, there is a solution to eqn (1.15). However, this solution is not unique.

Example 1.7

Every value λ satisfying the inequality

$$(1/|\lambda|) \ge (b - a) \times \sup_{a\le x,y\le b} |K(x,y)|$$

is a regular value. (We use (1.14a), and the relation between eigenvalues and characteristic values, to establish this result.) *

Example 1.8

Since a continuous Volterra kernel has no finite characteristic values, the corresponding equation

$$f(x) - \lambda\int_a^x K(x,y)\, f(y)\mathrm{d}y = g(x) \quad (a \leq x \leq X < \infty)$$

has a unique solution for all finite λ . *

The differentiability of solutions of equations of the second kind may be discussed in a manner similar to our treatment of eigenfunctions.

Example 1.9

If $f(x) - \int_0^1 \mathrm{e}^{xy}\, f(y)\mathrm{d}y = \mathrm{e}^{2x}$

then $f(x)$ is infinitely differentiable, and for $r \geq 0$

$$f^{(r)}(x) - \int_0^1 (y^r\, \mathrm{e}^{xy})f(y)\mathrm{d}y = 2^r\mathrm{e}^{2x}. \; *$$

Example 1.10

Suppose $f(x) - \int_0^1 K(x,y)\, f(y)\mathrm{d}y = \cos x$,

where $K(x,y) = x$ for $x \leq y$ and $K(x,y) = y$ for $x \geq y$. Then $f(x)$ is infinitely differentiable. If the term $\cos x$ is replaced by $x^{\frac{1}{2}}$, then $f'(x)$ is unbounded at $x = 0$. *

1.4.3.

The solution of equations of the first kind is, in general, much more difficult than the solution of equations of the second kind. For this reason, it is sometimes helpful to rewrite Volterra equations of the first kind as Volterra equations of the second kind, as we indicated in section 1.3.

It is possible to discuss the solvability of equations of the first kind in terms of 'singular functions' and 'singular values' of the kernel $K(x,y)$, but we defer the investigation of this subject until section 1.8. We shall give some examples which indicate some of the problems involved.

If we suppose that $K(x,y)$ has a zero eigenvalue with corresponding eigenfunction $\psi(x)$, then (for $a \leq x \leq b$)

$$\int_a^b K(x,y)\ \psi(y)\mathrm{d}y = 0 \ .$$

(We say that $\psi(x)$ is annihilated by the integral transform.) In such circumstances, any solution

$$\int_a^b K(x,y)\ f(y)\mathrm{d}y = g(x) \qquad (x \in [a,b]) \tag{1.16}$$

to the equation of first kind is not unique, since $f(x) + \psi(x)$ is another solution.

Now suppose that $K(x,y) = X(x)\ Y(y)$. Then

$$\int_a^b K(x,y)\ f(y)\mathrm{d}y = \gamma X(x) \ ,$$

where
$$\gamma = \int_a^b f(y)\ Y(y)\ \mathrm{d}y \ .$$

Thus, in this case, $g(x)$ must have the form $\gamma X(x)$ if eqn (1.16) is to have a solution. It will also be clear that a perturbation of $g(x)$, changing it to $g(x) + \varepsilon(x)$, say, may mean that a solution no longer exists, though one may have existed previously. This indicates that the approximate solution of equations of the first kind is difficult.

1.5. *Linear operators on a function space*

Associated with a kernel $K(x,y)$ and an interval of integration $[a,b]$ is a linear *integral operator* K which transforms a given function

ϕ (we† shall write $\phi(x)$) into a new function $\psi = K\phi$ (we shall write $\psi(x) = (K\phi)(x)$). The function $\psi(x)$ is to be defined by the relation

$$\psi(x) = \int_a^b K(x,y)\ \phi(y)dy\ .$$

Clearly, $\psi(x)$ exists if $\phi(x)$ is integrable, since we assume $K(x,y)$ is piecewise-continuous.

Now the set $C[a,b]$ of complex-valued functions which are continuous on $[a,b]$ forms a linear space (or vector space) when we define $(\phi_1 + \phi_2)(x) = \phi_1(x) + \phi_2(x)$ and $(\alpha\phi_1)(x) = \alpha\phi_1(x)$, where α is a complex number. If the complex-valued function $K(x,y)$ is continuous for $a \leq x,y \leq b$, then $(K\phi)(x)$ is a member of $C[a,b]$ when $\phi(x)$ is continuous (see section 1.3). The integral operator then 'maps' $C[a,b]$ into itself, and the operator K is said to be linear because $(K(\alpha_1\phi_1 + \alpha_2\phi_2))(x) = \alpha_1(K\phi_1)(x) + \alpha_2(K\phi_2)(x)$, where α_1, α_2 are complex values. Of course, the set $C[a,b]$ may be replaced by another set of functions if different conditions are imposed on $K(x,y)$.

The elements of $C[a,b]$ are functions, and with each continuous function $\phi(x)$ we may associate a positive real number, known as a *norm* of $\phi(x)$ and denoted by $||\phi(x)||$. A norm is a measurement of the size of $\phi(x)$. An example is $||\phi(x)|| = \{\int_a^b |\phi(t)|^2 dt\}^{\frac{1}{2}}$, which we write $||\phi(x)||_2$.

There are different ways of defining $||\phi(x)||$ (and different spaces of functions may have different norms) but the definition must be consistent with the following properties:

(i) $||\phi(x)|| = 0$ if and only if $\phi(x) = 0$,

---

† The notation $\phi(x)$ should be interpreted according to the context, either as the function, or as the numerical value of the function when the argument assumes the value x . In the former rôle, our choice of notation $\phi(x)$ indicates the number and type of the arguments of the function; we refer to Collatz (1966 *b*) for an apologia for this convention.

(ii) $||\alpha\phi(x)|| = |\alpha| \; || \phi(x) ||$,

(iii) $||(\phi_1 + \phi_2)(x) || \leq || \phi_1(x) || + || \phi_2(x) ||$.

A linear space for which a norm is defined is called a *normed linear space*.

Example 1.11

(a) The set $R[a,b]$ of functions which are properly Riemann-integrable on $[a,b]$, with the *uniform norm* $||\phi(x)||_\infty = \sup_{a\leq x\leq b} |\phi(x)|$, forms a normed linear space. The elements of $R[a,b]$ are those functions which are bounded and continuous 'almost everywhere'. Further examples of normed linear spaces are (b) for $r = 0, 1, \ldots,$ the set $C^r[a,b]$ of functions which have a continuous rth derivative on $[a,b]$, the norm being $\max_{0\leq s\leq r} \sup_{a\leq x\leq b} |\phi^{(s)}(x)|$, also (c) the same set with the norm $\sum_{s=0}^{r} \sup_{a\leq x\leq b} |\phi^{(s)}(x)|$, and (d) the spaces $L^p[a,b]$ (Taylor 1958, p.16).

In future we shall not distinguish in our notation between the set $C[a,b]$ and the space $C[a,b]$ equipped with a norm, usually $|| \phi(x) ||_\infty$.*

If we have a sequence of functions $\{\phi_n(x)\} (n = 0, 1, \ldots)$ and $||\phi(x) - \phi_n(x)|| \to 0$ as $n \to \infty$, we say that $\phi_n(x)$ *converges in norm* to $\phi(x)$. We talk of *uniform convergence* if the norm is $||\cdot||_\infty$ (see Example 1.11) and we speak of *convergence in mean* if the norm is $||\cdot||_2$ (see above, and Example 1.13 below). We shall say that $\phi_n(x)$ *converges essentially* to $\phi(x)$ if there is a sequence $\{\alpha_n\}$ of numbers with modulus unity such that $|| \phi(x) - \alpha_n \phi_n(x)|| \to 0$, in some specified norm.

Remark. There is another type of convergence which we encounter, called *relatively uniform convergence*. We say that $\phi_n(x)$ converges *relatively uniformly* to $\phi(x)$ on $[a,b]$ if there is some function $\psi(x)$ (which is usually assumed to be square-integrable) such that $|\phi_n(x) - \phi(x)| \leq \varepsilon_n\psi(x)$ for $a \leq x \leq b$, where $\lim_{n\to\infty} \varepsilon_n = 0$.

The set $C[a,b]$ of continuous functions measured with the uniform norm has a useful property known as *completeness*. To motivate the discussion of this property, let us recall the condition that a sequence $\{\alpha_n\}$ of complex numbers has a limit $\alpha = \lim_{n\to\infty} \alpha_n$. Such a limit α exists if and only if $\{\alpha_n\}$ is a Cauchy sequence, that is, if and only if, given $\varepsilon > 0$, we may find an integer $N(\varepsilon) = N$ such that $|\alpha_n - \alpha_m| < \varepsilon$ whenever $n,m > N$. Now suppose that $\{\phi_n(x)\}$ is a sequence of continuous functions defined on $[a,b]$. Suppose that when we are given any $\varepsilon > 0$ we may find an integer N depending on ε but independent of x, such that $|\phi_n(x) - \phi_m(x)| < \varepsilon$ for *all* x in $[a,b]$, when $n,m > N$. (It follows that when $n,m > N$, $||\phi_n(x) - \phi_m(x)||_\infty < \varepsilon$.) Then clearly $\lim_{n\to\infty} \phi_n(x)$ exists for each x. However, if we define $\phi(x) = \lim_{n\to\infty} \phi_n(x)$, it turns out that $\phi(x)$ is continuous. This property of the continuous functions $C[a,b]$ is expressed by the statement that $C[a,b]$ is *complete in the uniform norm* (see also section 1.9).

Example 1.12

The set of continuous functions of two variables, having the form $F(x,y)$, forms a normed linear space with the norm $||F(x,y)||_\infty = \max_{a\le x\le y\le b} |F(x,y)|$. The terminology used for convergence of functions of one variable is also used for functions of two or more variables. *

The set $C[a,b]$ possesses another structure, associated with the *inner product* of two functions (which may indeed be square integrable rather than continuous) in the form

$$(\phi,\psi) = \int_a^b \phi(t)\ \overline{\psi(t)}\ dt\ . \tag{1.17}$$

Unless otherwise indicated, we suppose that (ϕ,ψ) has the meaning of (1.17), but, more generally, an inner product is a complex-valued functional of any two functions, $\phi(x)$ and $\psi(x)$ satisfying

(i) $(\phi,\psi) = \overline{(\psi, \phi)}$,

(ii) $(\alpha\phi,\psi) = \alpha(\phi,\psi)$,

(iii) $(\phi + \psi, \chi) = (\phi,\chi) + (\psi, \chi)$,

(iv) $(\phi,\phi) \geq 0$

with equality if and only if $\phi(x) = 0$. It is easily seen that, given that $\omega(x)$ vanishes only on a set of measure zero,

$$<\phi,\psi> = \int_a^b \omega(x)\, \phi(x)\, \overline{\psi(x)} dx \tag{1.18}$$

is an inner product on $C[a,b]$ if the *weight function* $\omega(x)$ is integrable and $\omega(x) \geq 0$ for $a \leq x \leq b$. We may define a norm in terms of an inner product if we set $||\phi(x)||_\omega = \{<\phi,\phi>\}^{\frac{1}{2}}$.

<u>Example 1.13</u>

The expression $\{\int_a^b |\phi(t)|^p dt\}^{1/p}$ defines the L_p-norm $||\phi(x)||_p$ for every real p , on the space of functions for which it exists (regarding functions as equivalent if the L_p-norm of their difference vanishes). Do not confuse $||\phi(x)||_p$ with $||\phi(x)||_\omega$ above!

We have a norm with $||\phi(x)||_2 = \{(\phi,\phi)\}^{\frac{1}{2}}$. The angle θ between two functions $\phi(x)$ and $\psi(x)$ may be defined by the relation¶ $\cos\theta = |(\phi,\psi)|/\{\ ||\phi(x)||_2\, ||\psi(x)||_2\}$; see Davis (1965, p.199). *

We have illustrated methods of defining a norm or an inner product on a *function space*, S, (that is, a linear space whose elements are functions). The existence of such a norm or inner product permits the development of various properties of linear operators acting on the space, and, in particular, of integral operators. Thus, suppose that T is any linear operator on S ; that is, T acts on a normed linear function space S and maps it into itself. Thus $(T\phi)(x)$ is in S , and the function norm of S induces a *subordinate operator norm* which may be defined in terms of any of the following definitions, which are equivalent, in (1.19) below.

¶ The Cauchy-Schwarz inequality yields $|\cos\theta| \leq 1$. (Taylor (1958) gives various basic inequalities which we shall use subsequently.)

TABLE 1.1

Function norm; choice of space	Operator norm; integral operator $(K\phi)(x) = \int_a^b K(x,y)\phi(y)\mathrm{d}y$
$K(x,y)$ is presumed to have properties which ensure that K maps the chosen space into itself.	
$\Vert\phi(x)\Vert_1 = \int_a^b \vert\phi(t)\vert\mathrm{d}t$; $C[a,b]$, $R[a,b]$, $L^1[a,b]$	$\Vert K\Vert_1 = \sup_{a\le y\le b}\int_a^b \vert K(x,y)\vert\mathrm{d}x \le \vert b-a\vert \sup_{x,y}\vert K(x,y)\vert$
$\Vert\phi(x)\Vert_2 = \{\int_a^b \vert\phi(t)\vert^2\mathrm{d}t\}^{\frac{1}{2}}$; $C[a,b]$, $R[a,b]$, $L^2[a,b]$	$\Vert K\Vert_2 = \{\rho(K^*K)\}^{\frac{1}{2}}$ = the square-root of the largest eigenvalue of the kernel $\int_a^b \overline{K(z,x)}K(z,y)\mathrm{d}z$, with eigenfunction in the given space $\le \{\int_a^b\int_a^b \vert K(x,y)\vert^2\mathrm{d}x\,\mathrm{d}y\}^{\frac{1}{2}}$
$\Vert\phi(x)\Vert_\infty = \sup_{a\le t\le b}\vert\phi(t)\vert$; $C[a,b]$, $R[a,b]$	$\Vert K\Vert_\infty = \sup_{a\le x\le b}\int_a^b \vert K(x,y)\vert\mathrm{d}y \le \vert b-a\vert \sup_{x,y}\vert K(x,y)\vert$.

Vector norm;	Operator norm;
$\underset{\sim}{x} = [x_0, \ldots, x_n]^T$	matrix $\underset{\sim}{A} = [A_{ij}]$
$\lVert \underset{\sim}{x} \rVert_1 = \sum_{i=0}^{n} \lvert x_i \rvert$	$\lVert \underset{\sim}{A} \rVert_1 = \max_j \sum_{i=0}^{n} \lvert A_{ij} \rvert$
$\lVert \underset{\sim}{x} \rVert_2 = \{\sum_{i=0}^{n} \lvert x_i \rvert^2\}^{\frac{1}{2}}$	$\lVert \underset{\sim}{A} \rVert_2 = \{\rho(\underset{\sim}{A}^*\underset{\sim}{A})\}^{\frac{1}{2}}$ = the square-root of the largest eigenvalue of $\underset{\sim}{A}^*\underset{\sim}{A} \leq \{\sum_{i=0}^{n} \sum_{j=0}^{n} \lvert A_{ij} \rvert^2\}^{\frac{1}{2}}$.
$\lVert \underset{\sim}{x} \rVert_\infty = \max_i \lvert x_i \rvert$	$\lVert \underset{\sim}{A} \rVert_\infty = \max_i \sum_{j=0}^{n} \lvert A_{ij} \rvert$.

$$\|T\| = \sup_{\|\phi(x)\| \neq 0} \frac{\|(T\phi)(x)\|}{\|\phi(x)\|},$$

$$\|T\| = \sup_{\|\phi(x)\| = 1} \|(T\phi)(x)\| , \qquad (1.19)$$

$$\|T\| = \inf \{M \mid \|(T\phi)(x)\| \leq M \|\phi(x)\|\} .$$

We deduce that $\|(T\phi)(x)\| \leq \|T\| \; \|\phi(x)\|$. If $\|T\| < \infty$ we refer to T as a bounded linear operator, and the set of bounded linear operators on S (that is from S into S) forms a normed linear space when we define

$$(\alpha_1 T_1 + \alpha_2 T_2)\phi(x) = \alpha T_1\phi(x) + \alpha_2 T_2\phi(x) .$$

Notice that we distinguish between $\|K\|$ and $\|K(x,y)\|$ (see Example 1.12), in the case of an integral operator K with kernel $K(x,y)$. If the function norm has a suffix we write the operator norm with the same suffix. $\|K\|_\infty$ is defined by eqn (1.19) in terms of $\|\phi(x)\|_\infty$, for example.

The operator norm $\|K\|_\infty$ depends upon the underlying space of functions. This may be taken to be $R[a,b]$ (see Example 1.11(a)) unless otherwise stated. The choice $R[a,b]$ is made to admit the possibility that $K(x,y)$ be piecewise-continuous. (If the kernel is continuous, the choice $C[a,b]$ is equally satisfactory.)

The reader may recognize a similarity between linear operators on function spaces and matrices of order n acting on an n-dimensional normed vector space. The similarity is reinforced in Table 1.1, where we give expressions and bounds for some subordinate norms of an integral operator and for some subordinate norms of a matrix. Observe that if K_1 and K_2 are integral operators then so is the product operator K_1K_2; its kernel is $\int_a^b K_1(x,z)K_2(z,y)\,dz$. We can therefore define $K*K$ for use in Table 1.1, and, for later use, the powers or *iterates* K^n ($n = 2,3,\ldots$) whose kernels $K^n(x,y)$ are given by eqn (1.23). (The function $K^n(x,y)$ is known as the nth iterated kernel of the kernel $K(x,y)$.)

We may readily show that $\|T_1 T_2\| \leq \|T_1\| \, \|T_2\|$. The operator norm also satisfies the relations

(i) $\|T\| > 0$ if $T \neq 0$ and $\|T\| = 0$ if $T = 0$,

(ii) $\|T_1 + T_2\| \leq \|T_1\| + \|T_2\|$,

(iii) $\|\alpha T\| = |\alpha| \, \|T\|$, where αT is the operator which maps $f(x)$ into $\alpha(Tf)(x)$.

An operator with a finite norm is said to be *bounded*.

In particular we may replace T by K or by $(\mu I - K)$, where I is the identity operator and we obtain bounded operators. It is not difficult to show that $\| I \| = 1$, and we may then show that $\|(K - \mu I)\| \leq \|K\| + |\mu|$.

Remark. Suppose that $K(x,y)$ is continuous or, more generally, that the associated integral operator K maps the space $C[a,b]$ (with the uniform norm, say) into itself. The *spectral radius* of the operator K acting on the given space is $\rho(K) = \sup |\kappa_i|$, where $\kappa_i f_i(x) = (Kf_i)(x)$, and the supremum is taken over all of those eigenfunctions which are in $C[a,b]$. We have $|\kappa_i| \, \|f_i(x)\| \leq \|K\| \, \| f_i(x)\|$ and hence $|\kappa_i| \leq \|K\|$, and thus $\rho(K) \leq \|K\|$. However, it is clear that the choice of norm of $C[a,b]$ is not relevant in this argument, and any norm of $C[a,b]$ gives a corresponding operator norm satisfying $\| K \| \geq \rho(K)$. We can therefore generalize (1.14).

It is also clear that, if $K(x,y)$ is continuous, every square-integrable eigenfunction is continuous, and vice versa. Thus, in the choice of space, $R[a,b]$, $L^2[a,b]$, and $C[a,b]$ are equally satisfactory, and $\rho(K)$ is independent of this choice. The same conclusion is true if some iterated kernel of $K(x,y)$ is continuous. For, if $\kappa_i \neq 0$ and $\kappa_i f_i(x) = Kf_i(x)$, then it follows that $\kappa_i^n f_i(x) = K^n f_i(x)$. If $K^n(x,y)$ is continuous, then $f_i(x)$ is continuous when it is square-integrable.

If $K(x,y)$ is bounded and piecewise-continuous, then square-integrable eigenfunctions are piecewise-continuous and integrable but not necessarily continuous.

In view of the above remarks, we make the following definition. Suppose that T is a linear operator from a given normed linear space X into itself. Then the spectral radius of T, $\rho(T)$ is defined to be $\sup |\kappa_i|$, where $\kappa_i f_i = Tf_i$ and $f_i \in X$ is non-null. Here, it becomes clear that the definition of a characteristic value or of an eigenvalue (and hence of the spectral radius of a linear operator) generally depends upon the space on which the operator acts. We can similarly define a *regular value* of a linear operator T on X to be a value λ such that whenever $g \in X$, $(I - \lambda T)f = g$ has a unique solution $f \in X$. For an integral operator K, the space X may be taken to be the space $L^2[a,b]$ with the mean-square norm. Unless otherwise stated, a regular value or characteristic value of $K(x,y)$ will be interpreted as a regular value or characteristic value of the associated operator K, acting on this space. What has been indicated, in the comments above, is that the space $C[a,b]$ could be taken in place of $L^2[a,b]$, when defining the eigenvalues or characteristic values, provided that $K^n(x,y)$ is continuous for some n. To establish a similar result for the definition of the regular values, we require an abstract formulation of Fredholm alternative (see section 1.9).

We have seen a connection between $\rho(K)$ and $\|K\|_2$, $\|K\|_1$, $\|K\|_\infty$, where $K^n(x,y)$ is sufficiently well behaved, for some n. In Table 1.1 we give some bounds for $\|K\|$, where the operator K is defined on the space of functions for which the corresponding norm exists. The corresponding matrix results are given for comparison.

The inner product (ϕ,ψ) permits us to define the *adjoint* of an operator T on a space X with this inner product. The adjoint of T is denoted by T^* and is the operator which satisfies, for $\phi,\psi \in X$,

$$(T\phi,\psi) = (\phi,T^*\psi) . \qquad (1.20)$$

It is easily shown that when K is an integral operator, with kernel $K(x,y)$, on $C[a,b]$, $R[a,b]$, or $L^2[a,b]$, with the inner product (1.17), then K^* is an integral operator whose kernel $K^*(x,y)$ is the function $\overline{K(y,x)}$. Thus, $K^*\phi(x) = \int_a^b \overline{K(y,x)}\, \phi(y)\, dy$.

Example 1.14

A Hermitian kernel is *self-adjoint*, that is, $K = K^*$, since $K^*(x,y) = \overline{K(y,x)}$. The adjoint of the operator $T = I - \lambda K$ is $T^* = I - \overline{\lambda} K^*$, where I is the identity operator : $(I\, \phi)(x) = \phi(x)$. *

We may define the product of two integral operators K_1 and K_2 in this order. The product is itself an integral operator, $K_1 K_2$, and its kernel is

$$(K_1 K_2)(x,y) = \int_a^b K_1(x,z)\, K_2(z,y) dz \ . \qquad (1.21)$$

If $K_1 = K_2$ we can define the iterated kernel

$$K^2(x,y) = \int_a^b K(x,z)\, K(z,y) dz \qquad (1.22)$$

and we can define the *nth iterated kernel*

$$K^n(x,y) = \int_a^b K(x,z)\, K^{n-1}(z,y) dz \ , \qquad (1.23)$$

which is the kernel of the nth power of the operator K .

The product of any two operators T_1, T_2 acting on the same space is defined similarly: $(T_1 T_2 \phi)(x) = (T_1(T_2\phi))(x)$ as indicated above§.

Example 1.15

Denote the identity operator by I . The operator $(K - \mu I)^\rho$ may be written in terms of I and the powers K^r $(1 \leq r \leq \rho)$ by expanding formally by the binomial theorem. (We assume ρ is an integer.) *

Example 1.16

A kernel $K(x,y)$ and the corresponding operator K are said to be *normal* if $K^*K = KK^*$. A Hermitian kernel is normal. *

---

§ If the linear operator P is idempotent $(P^2 = P)$ it is a *projection*. In the sequel, projections are generally bounded.

1.6. *Theoretical discussion of the eigenproblem*

In principle, the solution of a linear integral equation of the first or second kind may be reduced to the solution of a suitable eigenvalue problem for the determination of eigenvalues and eigenfunctions. This is not always a practical approach, and we shall not pursue it far. However, we have already seen, in section 1.4.2, the intimate connection between eigenvalues and the equation of second kind as expressed in the Fredholm alternative theorem.

The eigenvalue problem and the inhomogeneous problem are intertwined, and we shall defer, until section 1.7, a few of our remarks on the eigenvalue problem.

Recall that we seek κ and $f(x)$ such that $f(x)$ is non-null and

$$\int_a^b K(x,y)f(y)\mathrm{d}y = \kappa f(x) \ .$$

It may happen that there are a number of different eigenfunctions $f_r(x)$ $(r = 1, 2, 3, \ldots, m)$, which correspond to the same eigenvalue κ . More precisely, suppose that

$$\int_a^b K(x,y)f_r(y)\mathrm{d}y = \kappa f_r(x) \ (r = 1, 2, 3, \ldots, m) \ ,$$

and that the m functions $f_r(x)$ are linearly independent. (A set of functions is said to be *linearly independent* if no one of them can be expressed as a non-trivial linear combination of the others.) It is clear that κ must be regarded, in some sense, as a *multiple eigenvalue* if $m > 1$. If m is the largest possible integer which we can associate with κ in this fashion we may regard it as one measure of the multiplicity of κ; it is called the *geometric multiplicity* (or the *rank*) of the eigenvalue κ .

There is another way of describing the multiplicity of an eigenvalue κ , which results in a definition of the (*algebraic*) *multiplicity* of κ . The natural way to define the algebraic multiplicity is in terms of the multiplicity of κ as a root of an equation, and we indicate this in section 1.7. For the present, we shall content ourselves with the following description of the algebraic multiplicity of κ, which we can take as its definition.

Suppose that $\mathcal{F}$ is the set of functions $f(x)$ which satisfy the relation $(K - \kappa I)^p f(x) = 0$, for any $p \geq 1$. Clearly $\mathcal{F}$ contains the eigenfunctions corresponding to κ, since we may set p equal to 1. The other functions in $\mathcal{F}$ are called *generalized eigenfunctions*. The maximum number of linearly independent functions in $\mathcal{F}$ (which is the *dimension* of $\mathcal{F}$) is the algebraic multiplicity of κ.

In the case of a Hermitian kernel the algebraic and geometric multiplicities of any particular eigenvalue are equal to each other. More generally, the geometric multiplicity of an eigenvalue is less than or equal to its algebraic multiplicity (for the set $\mathcal{F}$ contains the eigenfunctions).

Example 1.17

Each of the eigenvalues $\pi\alpha_r$ $(r \geq 0)$ occurring in Example 1.4 have algebraic and geometric multiplicities which are equal. *

Generalized eigenfunctions are sometimes called principal functions and, in the analogous case for matrices, the generalized eigenvectors are often called principal vectors (Forsythe and Moler 1967; Wilkinson 1965).

Example 1.18

Consider the kernel

$$K(x,y) = \sum_{i=1}^{N} X_i(x)\, Y_i(y) \quad (0 \leq x,\ y \leq 1),$$

where $X_r(x) = \sin\ r\pi x$ and $Y_r(x) = 2\{\kappa \sin\ r\pi x + \sin\ (r-1)\pi x\}$. This kernel has an eigenvalue κ whose geometric multiplicity is 1 and whose algebraic multiplicity is N. (This example reduces to a matrix problem, see below.) *

If a kernel $K(x,y)$ possesses an eigenvalue κ, then (Smithies 1962, p.103) the adjoint kernel $K^*(x,y)$ possesses an eigenvalue $\bar{\kappa}$. We then have $\int_a^b K(x,y)\, f(y)dy = \kappa f(x)$ and $\int_a^b \overline{K(y,x)}\, \psi(y)dy = \bar{\kappa}\, \psi(x)$, where $\psi(x)$ is an eigenfunction of $K^*(x,y)$. If we set $\phi(x) = \overline{\psi(x)}$ then

$$\int_a^b \phi(y)\, K(y,x)dy = \kappa\phi(x) ,$$

so that κ is an eigenvalue of the *transposed kernel* $K^T(x,y) = K(y,x)$ and $\phi(x)$ is a corresponding eigenfunction. We sometimes say that $\phi(x)$ is a *left eigenfunction* of $K(x,y)$ corresponding to κ, whilst the eigenfunction $f(x)$ is then known as a *right eigenfunction*.

Example 1.19

Let $(Kf_1)(x) = \kappa_1 f_1(x)$ and $(K^*f_2)(x) = \kappa_2 f_2(x)$. Then $(f_1, f_2) = 0$ if $\kappa_1 \neq \kappa_2$. Thus, in particular, a pair of eigenfunctions of a Hermitian kernel are orthogonal if they correspond to different eigenvalues. *

If $K(x,y) = \overline{K(y,x)}$, the eigenvalues are real and the algebraic and geometric multiplicities of each eigenvalue are equal, so that the term 'multiplicity' is unambiguous. It is sometimes convenient to order the eigenvalues of a Hermitian kernel, and we may index them according to their ordering on the real line. We shall denote the non-negative eigenvalues by κ_r^+ and the negative eigenvalues by κ_r^-, and we suppose them to be indexed so that

$$\kappa_1^+ \geq \kappa_2^+ \geq \ldots \geq 0 , \quad \kappa_1^- \leq \kappa_2^- \leq \kappa_3^- \leq \ldots < 0 ,$$

where every eigenvalue is repeated according to its multiplicity. With every eigenvalue κ we associate m linearly independent eigenfunctions where m is the multiplicity of κ. These functions may be chosen to be normalized, and orthogonal to each other if $m > 1$. An eigenfunction corresponding to $\kappa_r^{\pm}$ will be denoted $f_r^{\pm}(x)$. The totality of eigenfunctions of the Hermitian kernel $K(x,y)$, chosen in the way described, forms an *orthonormal* set; in other words, the eigenfunctions are mutually orthogonal and normalized.

In some instances, it is convenient to ignore the distinction between positive and negative eigenvalues and suppose the eigenvalues to be labelled κ_r and to be ordered so that $|\kappa_r| \leq |\kappa_s|$ if $r \geq s$. The total set of eigenvalues and corresponding eigenfunctions, represented according to their multiplicity as described above, may now be represented as an *eigensystem* $\{\kappa_r, f_r(x)\}$ of $K(x,y)$.

Remark. The eigenvalues of a Hermitian kernel may be characterized by the relations

$$\kappa_r^+ = \sup\{(K\phi,\phi) \,|\, \|\phi(x)\|_2 = 1;\ (\phi,f_s^+) = 0,\ s = 1, 2, 3,\ldots, (r-1)\}$$

if κ_r^+ exists and

$$\kappa_r^- = \inf\{(K\phi,\phi) \,|\, \|\phi(x)\|_2 = 1;\ (\phi,f_s^-) = 0,\ s = 1, 2, 3,\ldots, (r-1)\}$$

if κ_r^- exists (see Smithies 1962). The following alternative characterization will be of some use to us in section 3.24. If κ_r^+ exists, and $K(x,y)$ is Hermitian, then

$$\kappa_r^+ = \sup_{\substack{\phi_1(x),\ldots,\phi_r(x) \\ (\phi_i,\phi_j)=\delta_{ij}}} \ \inf_{\substack{\phi(x)=\sum_1^r a_i\phi_i(x) \\ \|\phi(x)\|_2=1}} (K\phi,\phi)$$

This result is discussed by Cochran (1972, p.122).

The corresponding result for κ_r^- is obtained by considering the eigenvalues of the Hermitian kernel $-K(x,y)$.

In certain instances, the eigenvalue problem for a kernel is equivalent to the eigenvalue problem for a matrix. This is true when the kernel $K(x,y)$ has the form $K(x,y) = \sum_{i=0}^{n} X_i(x)\, Y_i(y)$. (Such a kernel is said to be *degenerate*. The kernel in Example 1.18 is degenerate.) We shall suppose that $K(x,y)$ has this form, where the $(n+1)$ functions $X_i(x)$ and the $(n+1)$ functions $Y_i(x)$ are each linearly independent. A degenerate kernel may always be written in such a form.

The eigenvalue problem (1.10) now becomes

$$\sum_{j=0}^{n} X_j(x) \int_a^b Y_j(y)\, f(y)\mathrm{d}y = \kappa\, f(x)\,. \tag{1.24}$$

Thus, if $\kappa \neq 0$, the corresponding eigenfunction has the form

$$(1/\kappa) \sum_{i=0}^{n} a_i\, X_i(x)\,,$$

where

$$a_i = \int_a^b Y_i(y)\, f(y)\mathrm{d}y\,.$$

Eqn (1.24) then has the form

$$\sum_{j=0}^{n} a_j\, X_j(x) = \kappa\, f(x)\,, \tag{1.25}$$

and if we multiply eqn (1.25) by $Y_i(x)$ and integrate we find

$$\sum_{j=0}^{n} \int_a^b Y_i(x)\, X_j(x)\, \mathrm{d}x\, a_j = \kappa\, a_i$$

or

$$\sum_{j=0}^{n} A_{ij}\, a_j = \kappa\, a_i\,, \tag{1.26}$$

where $A_{ij} = \int_a^b Y_i(x)\, X_j(x)\mathrm{d}x$. Thus, κ is an eigenvalue of the matrix $\underset{\sim}{A}$ with elements A_{ij} $(i,\ j = 0,\ 1,\ 2,\ldots,\ n)$. If $\kappa \neq 0$ and the corresponding eigenvector is $\underset{\sim}{a} = [a_0,\ldots,\ a_n]^T$, then an unscaled eigenfunction $f(x)$ corresponding to κ is

$$\sum_{i=0}^{n} a_i\, X_i\,(x)\ .$$

If there are m linearly independent eigenvectors, there are precisely m independent eigenfunctions which correspond to these vectors because the functions $\{X_i(x)\}$ have been assumed independent. Thus the geometric multiplicity of a non-zero value κ as an eigenvalue of $\underset{\sim}{A}$ is the same as its geometric multiplicity as an eigenvalue of $K(x,y)$. The algebraic multiplicities also agree.

1.7. *Equations of the second kind*

If the kernel $K(x,y)$ is degenerate, we may reduce the equation of the second kind,

$$f(x) - \lambda \int_a^b K(x,y)\ f(y)\mathrm{d}y = g(x)\ ,$$

to the solution of a system of linear equations.

We suppose that $K(x,y)$ has the form

$$\sum_{i=0}^{n} X_i(x)\ Y_i(y)\ ,$$

as in section 1.6. The kernel is degenerate if it has this form. The equation of the second kind then becomes

$$f(x) - \lambda \sum_{j=0}^{n} X_j(x) \int_a^b Y_j(y)\ f(y)\mathrm{d}y = g(x)\ ,$$

so that

$$f(x) = g(x) + \lambda \sum_{j=0}^{n} a_j\ X_j(x)\ , \tag{1.27}$$

where $a_i = \int_a^b Y_i(y)\, f(y)\mathrm{d}y$. If we multiply eqn (1.27) by $Y_i(x)$, and integrate, we obtain

$$a_i = b_i + \lambda \sum_{j=0}^{n} A_{ij}\, a_j \,, \tag{1.28}$$

where $A_{ij} = \int_a^b X_j\,(x)\, Y_i\,(x)\,\mathrm{d}x$ and $b_i = \int_a^b g(x)\, Y_i(x)\mathrm{d}x$. The system of equations may be written $(\underset{\sim}{I} - \lambda\underset{\sim}{A})\, \underset{\sim}{a} = \underset{\sim}{b}$ in matrix notation, where $\underset{\sim}{a} = [a_0, \ldots, a_n]^T$, $\underset{\sim}{b} = [b_0, \ldots, b_n]^T$, and $\underset{\sim}{A} = [A_{ij}]$ is the matrix occurring in section 1.6. If we find a solution $[a_0, \ldots, a_n]^T$ then substitution of the a_i in eqn (1.27) yields the solution $f(x)$. A solution $\underset{\sim}{a}$ exists if $\det(\underset{\sim}{I} - \lambda\underset{\sim}{A}) \neq 0$, that is, if λ is not a characteristic value of $K(x,y)$, and it is not difficult to verify the Fredholm alternative theorem for a degenerate kernel.

We now consider a more general form of kernel. A simple iterative scheme, for the solution of the equation

$$f(x) - \lambda \int_a^b K(x,y)\, f(y)\mathrm{d}y = g(x) \tag{1.29}$$

is defined by

$$f_{n+1}(x) = g(x) + \lambda \int_a^b K(x,y)\, f_n(y)\mathrm{d}y \,. \tag{1.30}$$

We may take $f_0(x) = g(x)$, and form $f_n(x)$ $(n = 1, 2, 3, \ldots)$. The iteration is sometimes successful in the sense that $f_n(x)$ converges to the true solution, but it may fail: it is necessary to assume rather more than the solvability of eqn (1.29) in order to guarantee success. It is possible to show that, when $|\lambda|$ is smaller than the modulus of the smallest characteristic value of $K(x,y)$, the iteration will 'converge' to the solution of eqn (1.29), which is unique since λ must be a regular value.

To gain insight, we obtain a condition for convergence. If we write $\delta_n(x) = f(x) - f_n(x)$, we obtain from (1.29) and (1.30) the equation

$$\delta_{n+1}(x) = \lambda\int_a^b K(x,y)\ \delta_n(y)\mathrm{d}y\ . \tag{1.31}$$

Thus, $\|\delta_{n+1}(x)\| \leq |\lambda|\ \|K\|\ \|\delta_n(x)\|$, and if $|\lambda|\ \|K\| < 1$ it follows that $\|\delta_n(x)\| \to 0$ as $n \to \infty$, so that $f_n(x)$ converges in norm to $f(x)$. (The analysis is independent of the choice of $f_0(x)$.) In particular, we obtain uniform convergence if $f(x)$ is continuous and $|\lambda|\ \|K\|_\infty < 1$, and the latter condition is satisfied (see Table 1.1) when $|\lambda(b-a)|\ \|K(x,y)\|_\infty < 1$.

We obtain a less restrictive result if we develop the analysis more carefully. From eqn (1.31), we obtain (setting $n = 0, 1, 2,\ldots, (N-1)$) the equation

$$\delta_N(x) = \lambda^N \int_a^b K^N(x,y)\ \delta_0(y)\mathrm{d}y\ ,$$

and consequently

$$\|\delta_N(x)\| \leq |\lambda|^N\ \|K^N\|\ \|\delta_0(x)\|\ .$$

Thus $\|\delta_N(x)\| \to 0$ if $|\lambda|^N \|K^N\| \to 0$. Now it may be shown that $\rho(K) = \lim_{N\to\infty}\{\|K^N\|_2\}^{1/N}$ (Riesz and Sz. Nagy 1965) so that we obtain mean-square convergence in the iteration when $|\lambda|\ \rho(K) < 1$. This convergence is actually uniform if $g(x)$ and $K(x,y)$ are continuous and $f_0(x) = g(x)$.

Example 1.20

If $g(x)$ is continuous and $K(x,y)$ is a continuous Volterra kernel, the iteration

$$f_{n+1}(x) = \lambda\int_a^x K(x,y)\ f_n(y)\mathrm{d}y + g(x)$$

converges uniformly to the solution

$$f(x) = \lambda\int_a^x K(x,y)\, f(y)dy + g(x)$$

on any finite interval. *

Eqn (1.30) is a recurrence relation, and we may use it to obtain an explicit expression for $f_n(x)$ in terms of $f_0(x) = g(x)$. Clearly

$$f_1(x) = \lambda \int_a^b K(x,y)\, g(y)dy + g(x)\ , \quad f_2(x) = \lambda^2 \int_a^b K^2(x,y)\, g(y)dy + $$

$$+ \lambda \int_a^b K(x,y)\, g(y)dy + g(x)\ ,$$

where

$$K^2(x,y) = \int_a^b K(x,z)\, K(z,y)dz\ ,$$

and, in general,

$$f_n(x) = \lambda \int_a^b R_n(x,y;\lambda)g(y)dy + g(x)\ , \tag{1.32}$$

where

$$R_n(x,y;\lambda) = \sum_{j=1}^{n} \lambda^{j-1}K^j\,(x,y)\ ,$$

and $K^p(x,y)$ is the pth iterated kernel (see eqn (1.23)) of $K(x,y)$. When the iteration converges, the solution $f(x)$ is obtained formally by setting $n = \infty$ in eqn (1.32). The expression

$$R_\infty(x,y;\lambda) = \sum_{j=1}^{\infty} \lambda^{j-1}\, K^j(x,y)$$

will be known as the *Neumann series* for the *resolvent kernel* of $K(x,y)$. Smithies (1962, p.27) establishes the 'relatively uniform absolute convergence' of the Neumann series for the resolvent kernel when $|\lambda|\ \|K(x,y)\|_2 < 1$.

We see, formally, that $R_\infty(x,y;\lambda)$ satisfies the equation

$$R(x,y;\lambda) + \lambda \int_a^b K(x,z)R(z,y;\lambda)dz = K(x,y) . \tag{1.33}$$

More generally, when λ is a regular value we can find a kernel $R(x,y;\lambda)$, which depends on λ and is known as the resolvent kernel, satisfying (1.33). The solution of eqn (1.29) is then given as

$$f(x) = g(x) + \lambda \int_a^b R(x,y;\lambda)\, g(y)dy , \tag{1.34}$$

which is clearly an extension of eqn (1.32). The determination of the resolvent is somewhat analogous to finding the inverse of a matrix. We may write (1.34) in the form $f(x) = (I + \lambda R_\lambda)g(x)$. The operator $(I + \lambda R_\lambda)$ is the inverse of $(I - \lambda K)$ and is known as the *resolvent operator*.

We shall now return to the solution of the equation of the second kind. It is not too difficult to treat the general equation (1.29) by writing an arbitrary kernel $K(x,y)$ as the sum $K_n(x,y) + k(x,y)$, where

$$K_n(x,y) = \sum_{i=0}^{n} X_i(x)\, Y_i(y)$$

is degenerate and is chosen so that $|\lambda|\ \|k(x,y)\|_2 < 1$, say. Here, $K_n(x,y)$ is an approximation to $K(x,y)$. Weierstrass's theorem guarantees that we may choose $K_n(x,y)$ arbitrarily close to $K(x,y)$ when the latter is continuous,† and the condition on $k(x,y)$ ensures the convergence of the Neumann series for its resolvent kernel.

Eqn (1.29) may now be rewritten

$$f(x) - \lambda \int_a^b K_n(x,y)\, f(y)dy - \lambda \int_a^b k(x,y)f(y)dy = g(x) \tag{1.35}$$

† Smithies (1962) discusses the case where $K(x,y)$ is square-integrable.

Thus

$$f(x) - \lambda \int_a^b k(x,y)f(y)\mathrm{d}y = g(x) + \lambda \int_a^b K_n(x,y)f(y)\mathrm{d}y \quad .$$

If $r(x,y;\lambda)$ is the Neumann series for the resolvent kernel of $k(x,y)$, we obtain

$$f(x) = g(x) + \lambda \int_a^b K_n(x,y)f(y)\mathrm{d}y + \lambda \int_a^b r(x,y;\lambda)g(y)\mathrm{d}y + \\ + \lambda^2 \int_a^b \int_a^b r(x,z;\lambda)\, K_n(z,y)f(y)\mathrm{d}z\mathrm{d}y \ . \tag{1.36a}$$

Using operator notation, this equation becomes

$$f(x) = g(x) + \lambda(K_n f)(x) + \lambda(r_\lambda g)(x) + \lambda^2(r_\lambda K_n f)(x) \quad ,$$

or (on rearranging)

$$f(x) - \lambda(K_n + \lambda\, r_\lambda K_n)f(x) = g(x) + \lambda(r_\lambda g)(x) \ . \tag{1.36b}$$

Now the kernel of the operator $K_n + \lambda r_\lambda K_n$ in eqn (1.36b) is a degenerate kernel of the form $\sum_{i=0}^{n} Z_i(x)\ Y_i(y)$, where

$$Z_i(x) = X_i(x) + \lambda \int_a^b r(x,z;\lambda)X_i(z)\mathrm{d}z \ .$$

In theory, therefore, eqn (1.36b) may be solved by reducing it to a system of linear equations in $(n + 1)$ unknowns.

There will be no unique solution to these equations if $D(\lambda) = \det(\underset{\sim}{I} - \lambda\underset{\sim}{A}(\lambda)) = 0$, where $\underset{\sim}{A}(\lambda)$ is the matrix whose (i,j)th element is $\int_a^b Z_i(t)\ X_j(t)\mathrm{d}t$ (note that $Z_i(x)$ depends on λ). The values of λ which make $D(\lambda)$ zero are the characteristic values of $K(x,y)$. The algebraic multiplicity of a characteristic value λ_r is its multiplicity as a zero of $D(\lambda)$, and this is equal to the algebraic multiplicity of the corresponding eigenvalue $\kappa_r = {}^1/\lambda_r$.

Remark. When $K(x,y)$ is Hermitian, we may obtain a solution to an equation of the second kind when the total eigensystem $\{\kappa_r, f_r(x)\}$ of $K(x,y)$ is known. The solution of the equation

$$f(x) - \lambda \int_a^b K(x,y)\, f(y)\mathrm{d}y = g(x)$$

is given by

$$f(x) = g(x) + \lambda \sum_{n=1}^{\infty} \frac{\kappa_n (g,f_n)\, f_n(x)}{(1 - \lambda\kappa_n)} \tag{1.37}$$

(Smithies 1962), where the right-hand side converges in the mean to $f(x)$ if λ is not a characteristic value.

It is interesting to note that eqn (1.29), of the second kind, is equivalent to the equation

$$f(x) - \int_a^b S_\lambda(x,y)\, f(y)\mathrm{d}y = d_\lambda(x) , \tag{1.38}$$

where

$$S_\lambda(x,y) = \lambda K(x,y) + \bar{\lambda} K^*(x,y) - \lambda\bar{\lambda} \int_a^b K^*(x,z)\, K(z,y)\mathrm{d}z$$

and

$$d_\lambda(x) = g(x) - \bar{\lambda} \int_a^b K^*(x,z) g(z)\mathrm{d}z .$$

The kernel $S_\lambda(x,y)$ is Hermitian so that eqn (1.38) may be solved as above. As a practical procedure, however, this approach may suffer some disadvantages.

1.8. *Equations of the first kind*

We have indicated earlier that the equation of the first kind,[§]

$$\int_a^b K(x,y)\, f(y)\mathrm{d}y = g(x) \tag{1.39}$$

presents some difficulties.

Let us consider an example. If we differentiate the equation of the first kind

$$\int_0^x f(y)\mathrm{d}y = g(x)$$

we see that the solution is $f(x) = g'(x)$ provided that $g(0) = 0$ and $g'(x)$ exists. In this simple example we see that the equation has a solution only when $g(x)$ is a special type of function. If we suppose that $g'(x)$ does exist, and we wish to find the solution $f(x)$ we must somehow determine $g'(x)$, given $g(x)$. This is not always easy, and in practice we will have exchanged one type of difficulty for another. In the case of the equation

$$\int_0^x (x - y)^r f(y)\mathrm{d}y = g(x) \ ,$$

any solution depends on $g^{(r)}(x)$. Similar phenomena can arise with Fredholm equations of the first kind.

Example 1.21

Consider eqn (1.39) with $a = 0$, $b = 1$ and $g(0) = g'(1) = 0$,

$$K(x,y) = x \qquad (0 \le x \le y \le 1) \ ;$$

$$K(x,y) = y \qquad (0 \le y \le x \le 1) \ .$$

---

§ In the theoretical discussion (Smithies 1962) we aim to satisfy eqn (1.39) 'almost everywhere' on $[a,b]$; see section 1.9 for this term.

The solution is $f(x) = -g''(x)$ when $g''(x)$ exists. (Differentiate the equation twice.) In the case $g(x) = \frac{4}{3}x^{\frac{3}{2}} - 2x$ the solution is $-x^{-\frac{1}{2}}$, which is not continuous, is unbounded, and is not square-integrable. *

Suppose now that $K(x,y)$ is Hermitian and that $g(x)$ has, formally, an expansion

$$g(x) = \sum_{r=0}^{\infty} \gamma_r f_r(x) , \qquad (1.40)$$

where $\{\kappa_r, f_r(x)\}$ is the total eigensystem of $K(x,y)$. We shall seek a solution of (1.39) with the form

$$f(x) = \sum_{r=0}^{\infty} \alpha_r f_r(x) . \qquad (1.41)$$

Now

$$\int_a^b K(x,y) f_r(y) dy = \kappa_r f_r(x)$$

so that, formally, the series (1.41) is transformed by K into the series

$$\sum_{r=0}^{\infty} \alpha_r (Kf_r)(x) = \sum_{r=0}^{\infty} \alpha_r \kappa_r f_r(x) . \qquad (1.42)$$

The right-hand side of eqn (1.42) is supposed to agree with (1.40), and, by a purely formal argument, we see that this is true when

$$\gamma_r = \alpha_r \kappa_r .$$

If we obtain the coefficients γ_r in (1.40), we may deduce the coefficients α_r in (1.41) using the relation $\alpha_r = \gamma_r / \kappa_r$. This motivates, in a heuristic way, the solution of (1.39) in the case where $K(x,y)$ is Hermitian.

Now the kernel in Example 1.21 is Hermitian, but in that Example a solution exists only when $g''(x)$ exists. It is clear, therefore, that the rigorous details of the foregoing argument are of some importance. We shall content ourselves with a statement of some key theorem and refer to Smithies (1962) and Tricomi (1957) for a full discussion.

The following results may be established.

Let $\{\kappa_n, f_n(x)\}$ be the eigensystem of the Hermitian kernel $K(x,y)$. If $f(x)$ is square integrable and

$$g(x) = \int_a^b K(x,y)\, f(y)\mathrm{d}y ,$$

then the series $\sum_{r=0}^{\infty} (g,f_r)\, f_r(x)$ converges in mean (section 1.5) to $g(x)$, and $(g,f_r) = \kappa_r(f,f_r)$. Thus, if eqn (1.39) is to have a solution, the function $g(x)$ must have an appropriate series expansion. Clearly, $g(x)$ must be orthogonal to any function $\psi(x)$ which is annihilated by K (set $f_r(x)=\psi(x)$, $\kappa_r=0$).

Now, given a suitable function $g(x)$, one question is whether (1.41) converges when $\alpha_r = \gamma_r/\kappa_r$. We state the following result.¶

Eqn (1.39) has a square-integrable solution $f(x)$ if and only if

(i) $$\sum_{n=0}^{\infty} \kappa_n^{-2}\left|(g,f_n)\right|^2 < \infty$$

and (ii) $(g, \psi) = 0$ for every function $\psi(x)$ annihilated by K. Condition (i) guarantees that the right-hand side of eqn (1.41) converges in the mean to a square-integrable function $f(x)$ (which need not be unique, continuous, or even bounded) given $\alpha_r=(g,f_r)/\kappa_r$.

Example 1.22

The eigensystem $\{\kappa_n,f_n(x)\}$ of the kernel in Example 1.21 is given by the relations

$$\kappa_n = 4/\{(2n+1)\pi\}^2 \quad \text{and} \quad f_n(x) = \sqrt{2}\left[\sin\{\tfrac{1}{2}(2n+1)\pi x\}\right](n = 0, 1, 2,\ldots).$$

¶ The case where $K(x,y)$ is not Hermitian is discussed below.

Recall that the solution of the equation in this Example is $-g''(x)$. If $g(x)$ has the expansion

$$\sum_{n=0}^{\infty} \gamma_n f_n(x)$$

then, formally, $g''(x)$ has the expansion

$$\sum_{n=0}^{\infty} \gamma_n f''_n(x) = -\sum_{n=0}^{\infty} (\gamma_n/\kappa_n) f_n(x) ,$$

using the expression for $f_n(x)$ above. The series converges in the mean if

$$\sum_{n=0}^{\infty} |\gamma_n \kappa_n|^2 < \infty .$$

On the other hand, if $g''(x)$ is square-integrable it has the expansion $\sum_{n=0}^{\infty} (g'',f_n)f_n(x)$. Now $g(0) = g'(1) = 0$ and

$$(g'',f_n) = \sqrt{2} \int_0^1 g''(t) \sin \{(2n + 1)\pi t/2\} \, dt$$

and this integral may be evaluated, using repeated integration by parts, to give $(g'',f_n) = -(g,f_n)/\kappa_n$. *

In the case of a non-Hermitian kernel the role of the eigenfunctions is assumed by the *singular functions* of a kernel.

We shall use the terminology employed by Smithies. The *singular system* of a kernel $K(x,y)$ (on an interval $[a,b]$) is the complete set of ordered triples $\{u_n(x), v_n(x); \mu_n\}$ satisfying, for $n= 0,1,2,\ldots$,

$$\mu_n > 0,$$

$$u_n(x) = \mu_n \int_a^b K(x,y) \, v_n(y) \mathrm{d}y ,$$

and

$$v_n(x) = \mu_n \int_a^b \overline{K(y,x)} \, u_n(y) \mathrm{d}y .$$

The functions $u_n(x)$ and $v_n(x)$ are called *singular functions*. We suppose that the *singular values* μ_n are ordered so that $0 < \mu_0 \leq \mu_1 \leq \mu_2 \leq \ldots$. It is not difficult to see that $u_n(x)$ is an eigenfunction of $KK^*(x,y)$ corresponding to the eigenvalue $1/\mu_n^2$ (and hence corresponding to the characteristic value μ_n^2). The function $v_n(x)$ is, similarly, an eigenfunction of $K^*K(x,y)$.

A solution to eqn (1.39) exists§ if and only if

(i) $\sum_{n=0}^{\infty} \mu_n^2 |(g,u_n)|^2 < \infty$ and (ii) $(g,\psi) = 0$ for all functions $\psi(x)$ annihiliated by K^*. A solution is then given by the series

$$\sum_{n=0}^{\infty} \mu_n (g,u_n)\, v_n(x),$$

which converges in the mean.

1.9. *Elementary functional analysis*

We have touched on a few of the basic ideas of analysis in section 1.5. The reader who wishes to pursue some of the more abstract work on the theoretical or numerical solution of integral equations will require some acquaintance with the elementary ideas of functional analysis, and some of these ideas will be introduced in the present section. The advanced reader may find Collatz (1966 *b*) a useful reference.

A *metric* $\rho(\phi_1,\phi_2)$ defined on a (function) space provides a measure of the distance between any two elements $(\phi_1(x)$ and $\phi_2(x))$. The definition of $\rho(\ ,\ )$ must satisfy the following: (i) $\rho(\phi_1,\phi_1) = 0$; (ii) $\rho(\phi_1,\phi_2) = 0$ implies $\phi_1(x) = \phi_2(x)$; (iii) $\rho(\phi_1,\phi_2) = \rho(\phi_2,\phi_1)$; and (iv) $\rho(\phi_1,\phi_3) \leq \rho(\phi_1,\phi_2) + \rho(\phi_2,\phi_3)$. It is easily seen that we obtain a metric on a normed linear function space if we define $\rho(\phi_1,\phi_2) = \|\phi_1(x)-\phi_2(x)\|$. Such a metric defined on a normed linear space will be called the natural or *induced metric*. Any linear space with a metric $\rho(\ ,\ )$ satisfying (i) - (iv) is called

---

§ Recall the footnote on p. 40.

a *metric space*.¶ Some of the material of Collatz is presented in terms of *pseudometric* spaces (Collatz 1966 *b*, p.51).

A sequence $\{\phi_n(x)\}$ in a space is a *fundamental sequence* if it satisfies the *Cauchy criterion* $\rho(\phi_n,\phi_m) \leq \varepsilon$ for all $n, m \geq N(\varepsilon)$, where $\varepsilon > 0$. If $\rho(\phi,\phi_n) \to 0$ as $n \to \infty$ we say that $\phi_n(x)$ converges to $\phi(x)$. A space is said to be *complete* if every fundamental sequence converges to an element in the space. If the set of limits of all fundamental sequences in a space is added to the space, the new space is called the *completion* of the original one.

Example 1.23

If

$$||\phi(x)|| = \sup_{a\leq x\leq b} |\phi(x)|$$

the induced metric for the space $C[a,b]$ of continuous functions is

$$\rho(\phi_1,\phi_2) = \sup_{a\leq x\leq b} |\phi_1(x) - \phi_2(x)|$$

and the space is complete. Convergence with respect to the metric is 'uniform convergence'. *

We say that a normed linear space is complete if it is complete in the metric induced by the norm. It is, of course, simpler to express the Cauchy criterion directly in terms of the norm. A complete normed linear space is a *Banach space*.

Example 1.24

The set of functions which are continuous on $[a,b]$ is not complete in the norm $||\phi(x)||_2 = \{\int_a^b |\phi(x)|^2 dt\}^{\frac{1}{2}}$, a sequence of continuous functions may converge in the mean to a discontinuous function. *

¶ A general metric space is a set with a metric ρ defined on it; the set need not have the structure of a linear space.

The space $L^2[a,b]$ of functions which are square-integrable, in the sense of Lebesgue, provides an example of a Banach space. We can find orthonormal functions $\phi_0(x), \phi_1(x), \phi_2(x), \ldots$ in $L^2[a,b]$ such that any square-integrable function $\phi(x)$ can be written in the form

$\phi(x) = \sum_{i=0}^{\infty} \alpha_i \phi_i(x)$, where $\alpha_i = (\phi, \phi_i)$ and the series on the right

converges in the mean to the element $\phi(x)$. (The Legendre polynomials can be used to give a set of such functions $\phi_i(x)$.) Thus, the functions $\{\phi_i(x)\}$ form what we regard as a countable 'basis' for the space $L^2[a,b]$. It is clear that if $\psi(x)$ and $\phi(x)$ differ only at a single point then $(\psi, \phi_i) = (\phi, \phi_i)$, so that 'differing' functions give rise to the same coefficients α_i. However, if $(\psi, \phi_i) = (\phi, \phi_i)$ $(i = 0, 1, 2, \ldots)$ then $\psi(x) = \phi(x)$ 'almost everywhere',† and, since $\|\psi(x) - \phi(x)\|_2 = 0$, we regard $\psi(x)$ and $\phi(x)$ as equivalent; some authors write $\psi(x) \overset{\circ}{=} \phi(x)$. If $\psi(x)$ and $\phi(x)$ are both continuous then $\psi(x) \equiv \phi(x)$.

In the above discussion, we have a sequence of functions $\phi_i(x)$ such that if $(\phi,\phi_i) = 0$ $(i = 0, 1, 2, \ldots)$, then $\phi(x) \overset{\circ}{=} 0$. Such a *sequence* of functions is said to be *complete* (in $L^2[a,b]$). The sequence of functions $\phi_i(x)$ can be chosen to be continuous, and they are then also complete 'in $C[a,b]$'.¶

If the sequence of functions $\phi_0(x), \phi_1(x), \phi_2(x), \ldots$ is complete in $L^2[a,b]$, then

$$\left\| \sum_{i=0}^{n} \alpha_i \phi_i(x) - \phi(x) \right\|_2$$

can be made arbitrarily small by taking n sufficiently large, and choosing $\alpha_0, \alpha_1, \ldots, \alpha_n$ appropriately. More generally, suppose that $\phi_0(x), \phi_1(x), \phi_2(x), \ldots$ is a sequence of functions in a normed

---

† For a rigorous discussion see Taylor (1958); see also Smithies (1962, p.6).

¶ At this point I should warn the reader that the use of the term 'complete' is not standard in the literature. Terms such as 'complete', 'compact' and 'closed' (see below) should be interpreted by the context.

linear space, with a norm $\|\ \|$, and every element $\phi(x)$ can be approximated arbitrarily closely by finite linear combinations of $\phi_0(x), \phi_1(x), \phi_2(x)\ldots$. Then this sequence $\{\phi_i(x)\}$ is called *closed*.

Remark. If $\{\phi_i(x)\}$ is a sequence which is complete in $C[a,b]$, in the sense above, it does not follow that the sequence is closed in the space $C[a,b]$ with the *uniform* norm (sometimes called the L_∞-norm). There is, we note , a different definition of completeness which is equivalent to closure in any given normed linear space (see P.J. Davis 1965).

Remark. A normed linear space is called separable if it contains a countable sequence which is closed. Most spaces considered here are separable.

Example 1.25

The following spaces are Banach spaces.

(i) $R[a,b]$, with the norm $\|\phi(x)\|_\infty = \sup_{a\le x\le b} |\phi(x)|$;

(ii) $C^r[a,b]$, with the norm

$$\|\phi(x)\| = \max_{0\le s\le r} \|\phi^{(s)}(x)\|_\infty \quad (r = 0, 1, 2, \ldots) \;;$$

(iii) $C^r[a,b]$, with the norm

$$\|\phi(x)\| = \sum_{s=0}^{r} \|\phi^{(s)}(x)\|_\infty \quad (r = 0, 1, 2, \ldots)$$

(see also Example 1.11). The space (i) is not separable. *

Some further terminology is needed. A subset S of a metric space X is said to be *compact*† in X if every sequence of elements in S contains a subsequence which converges, with respect to the metric, to an element of X . Thus, S is compact in itself if it is closed and compact in X. The metric space may be a normed linear space with induced metric.

† The term (relatively) sequentially compact is used by some authors. We sometimes write 'sequentially compact' in X.

Example 1.26

Consider the set S of functions $\{\phi(x)\}$ which are continuous and differentiable on $[a,b]$, and which satisfy $\|\phi(x)\|_\infty < 1$, $\|\phi'(x)\|_\infty < 1$. The set S may be shown to be compact in $C[a,b]$. *

An operator is *continuous* if, briefly, it maps every convergent sequence into a corresponding convergent sequence. Thus, a linear operator of finite norm is continuous. A continuous operator T which maps a normed space S_1 into a normed linear space S_2 is said to be *completely continuous* (sometimes a *compact operator*†) if the image of an arbitrary bounded set $B \subset S_1$ is compact in S_2. (The *image* of B is the set $\{T\phi \mid \phi \in B\}$. If $B = S_1$ the image is called the *range* of T.) In the case of a function space, a *bounded* set B is a set of functions which are *uniformly bounded* in the norm; that is, there exists some constant M independent of $\phi(x)$ such that $\|\phi(x)\| \leq M$ for each $\phi(x)$ in B.

If $K(x,y)$ is bounded for $a \leq x, y \leq b$, and continuous, except possibly on a finite number of continuous curves $y = \phi_i(x)$, then the corresponding integral operator K is completely continuous on the space $C[a,b]$, with the uniform norm (Kolmogorov and Fomin 1957).

Example 1.27

Suppose that the operator T acting on a space S is linear and bounded. If the range of T is finite-dimensional then T is completely continuous (see Riesz and Sz. Nagy 1955, p.179).

An operator H acting on the space $C[a,b]$ may be defined by setting

$$(H\phi)(x) = \sum_{j=0}^{n} w_j K(x,y_j)\phi(y_j),$$

where $K(x,y)$ is continuous for $a \leq x, y \leq b$, and $a \leq y_j \leq b$ $(j = 0, 1, 2, \ldots, n)$. This operator H is compact since any function in the range is a linear combination of the functions

† We shall treat these terms as synonymous, though this is not universal practice.

$$\psi_j(x) = K(x,y_j) \ (j = 0, 1, 2,\ldots, n) \ . \ *$$

One reason for our interest in completely continuous operators is the validity of the Fredholm alternative. If T is a completely continuous linear operator acting on a Banach space X, then[†] either λ is a regular value of T, and the equation $(I - \lambda T)f = g$ has a unique solution $f \varepsilon X$ for every $g \varepsilon X$, or λ is a characteristic value of T. The same conclusion holds if T^m is completely continuous for some m, even if T is not itself completely continuous.

Further, suppose that the norm of the Banach space is an inner-product norm, with $\| \phi \| = (\phi, \phi)^{\frac{1}{2}}$ for $\phi \varepsilon X$ (such a Banach space is called a *Hilbert space*). If T is completely continuous, then $(I - \lambda T) f = g$ has a solution if $(g,\psi) = 0$ for any ψ such that $\psi - \overline{\lambda}T^*\psi = 0$. The solution is unique if λ is not a characteristic value. The set of values of μ for which $(\mu I - T)^{-1}$ exists is called the *resolvent set* of T. The other values of μ form the *spectrum* of T. The spectrum of a completely continuous operator is precisely the set of eigenvalues, along with zero if X is infinite-dimensional. The non-zero spectrum is discrete and its only possible limit point is zero.

A set S of functions defined on $[a,b]$ is said to be *equicontinuous* if, given $\varepsilon > 0$, there exists a $\delta = \delta(\varepsilon)$ such that for any x_1,x_2 in $[a,b]$ and satisfying $|x_1 - x_2| < \delta$ we have $|f(x_1) - f(x_2)| < \varepsilon$ for all $f(x)$ in S. (The number δ is to be independent of the particular $f(x)$ in S.)

Example 1.28

Suppose that $K(x,y)$ is continuous on $a \leq x, y \leq b$ and that S_1 is a set of uniformly bounded piecewise-continuous functions. Then the set S_2 of transformed functions of the form $K\phi(x)$, where $\phi(x) \varepsilon S_1$ is equicontinuous (Courant and Hilbert 1953) (see also below). *

The above concepts are used in the Arzela-Ascoli theorem (Courant and Hilbert 1953; Taylor 1958; Simmons 1963). This states that a

---

† Liusternik and Sobolev (1961, p.141), Kolmogorov and Fomin (1957, p.117).

§ For the result where T is completely continuous and X is not complete see Taylor (1958).

necessary and sufficient condition that a set of functions which are continuous on $[a,b]$ should be compact in $C[a,b]$ (with uniform norm) is that the set should be (i) equicontinuous and (ii) uniformly bounded.

The Arzela-Ascoli theorem permits us to investigate whether or not an operator is completely continuous on the space $C[a,b]$. In particular, if

$$K\phi(x) = \int_a^b K(x,y)\ \phi(y)dy\ ,$$

where

$$\sup_{a\leq x\leq b} \int_a^b |K(x,y)|dy < \infty\ ,$$

then $\|K\|_\infty < \infty$, so that K is a bounded linear operator

and if

$$\lim_{\delta\to 0}\ \sup_{a\leq x\leq b} \int_a^b |K(x+\delta,y) - K(x,y)|dy = 0$$

in addition, then it can be shown that K is compact on the space $C[a,b]$, with the uniform norm. (Write $\psi(x) = K\phi(x)$. Then

$$|\psi(x+\delta) - \psi(x)| \leq \int_a^b |K(x+\delta,y) - K(x,y)|dy \times \|\phi(x)\|_\infty$$

and the result follows from the Arzela-Ascoli theorem.)

The following properties of functions are useful. If a set $\{\delta_i(x)\}$ of continuous functions on $[a,b]$ is equicontinuous, and $\lim_{n\to\infty} \|\delta_n(x)\|_2 = 0$ then $\lim_{n\to\infty} \|\delta_n(x)\|_\infty = 0$, so that in this case mean square convergence implies uniform convergence. If the sequence $\{\phi_n(x)\}$ is equicontinuous and converges in the mean to $\phi(x)$, it converges uniformly to $\phi(x)$. Courant and Hilbert (1953) discuss these results.

Remark. The preceding comments may be applied to explain why the functions $f_n(x)$ given by the iteration

$$f_0(x) = g(x) ,$$

$$f_{n+1}(x) = \lambda \int_a^b K(x,y)\, f_n(y)\,dy + g(x) \quad (n \geq 0)$$

converge uniformly to the solution of eqn (1.29) when $g(x)$ and $K(x,y)$ are continuous and $|\lambda| \; \|K\|_2 < 1$. (It may be that $|\lambda| \; \|K\|_\infty \geq 1$.) With $\delta_n(x) = f(x) - f_n(x)$ as before, we know that $\|\delta_n(x)\|_2 \to 0$, and thus $f_n(x)$ converges in the mean to $f(x)$. But $\delta_1(x)$ is an integral transform of the continuous function $\delta_0(x)$; by (1.31), $\delta_2(x)$ is an integral transform of $\delta_1(x)$, etc., and by the result in Example 1.28 the set $\{\delta_1(x), \delta_2(x), \delta_3(x), \ldots\}$ is equicontinuous. Thus $\lim \|\delta_n(x)\|_2 = 0$ implies $\lim \|\delta_n(x)\|_\infty = 0$ and we actually obtain uniform convergence of $f_n(x)$ to $f(x)$. There is a substantial difference in practice between mere mean-square convergence and uniform convergence.

We shall, in our analysis of numerical methods, frequently appeal to a well-known theorem which we now state. The result is linked with our previous discussion.

THEOREM 1.1. *GEOMETRIC SERIES THEOREM. Suppose* X *is a normed linear space, and let* G *be a bounded linear operator mapping* X *into itself, with the subordinate norm* $\|G\| < 1$. *Suppose either* (a) *that* X *is complete (a Banach space) or* (b) *that* G *is a compact (completely continuous) operator on* X . *Then,* $(I-G)^{-1}$ *exists as a bounded linear operator mapping* X *into itself, with* $(I-G)^{-1} = \sum_{r=0}^{\infty} G^r$ *and* $\|(I-G)^{-1}\| \leq (1 - \|G\|)^{-1}$.

Remark. The operator G provides an example of a *contraction mapping*, that is, G is an operator such that $\|G\phi - G\psi\| \leq \rho\|\phi-\psi\|$, $\rho < 1$.

<u>Proof</u> (see also Lonseth 1947). Consider the equation $(I-G)f = g$ for arbitrary $g \in X$. Given g, any possible solution f is unique. (For, if f^0, f^1 are two possible solutions, $f^0 - f^1 = G(f^0 - f^1)$ so that $\|f^0 - f^1\| \le \|G\| \; \|f^0 - f^1\|$, where $\|G\| < 1$.)

Consider the iteration defined by $f_0 = g$, and $f_n = Gf_{n-1} + g$ for $n \ge 1$. Thus $f_n = g + Gg + \ldots + G^n g$. If a solution f exists then

$$(I-G^{n+1})\, f = (\sum_{r=0}^{n} G^r)(I-G)f = \sum_{r=0}^{n} G^r g = f_n ,$$

and $\|f - f_n\| < \|G\|^{n+1}\|f\| \to 0$ as $n \to \infty$. We consider $\{f_n\}$, without assuming the existence of f. Now

$$f_{n+p} - f_n = \sum_{r=n+1}^{n+p} G^r g ,$$

so that

$$\|f_{n+p} - f_n\| \le \{ \sum_{r=n+1}^{n+p} \|G^r\| \} \; \|g\| \le$$

$$\|G\|^n \; \|g\| / \{1 - \|G\|\} .$$

Thus, $\{f_n\}$ satisfies the Cauchy criterion and if X is complete it follows that $\lim_{n\to\infty} f_n$ exists in X. Suppose, on the other hand, that G is a compact operator. For $n \ge 0$,

$$f_n = \sum_{r=0}^{n} G^r g ,$$

so that $\|f_n\| \le \sum_{r=0}^{\infty} \|G\|^r \; \|g\|$ for all n, and the set $\{f_n\}$ is uniformly bounded. Furthermore, G is compact and $f_n - g = Gf_{n-1}$ so that the uniformly bounded set $\{f_n\}$ is transformed into a set compact *in* X. That is to say, the sequence of functions $\{f_n - g\}$

has a convergent subsequence $\{f_{n_k}-g\}$ with a limit point in X. On the other hand, $\|f_{n_k}-f_n\| \to 0$ as $n_k \to \infty$ and $n \to \infty$. (Set $n_k = n + p_k$ for some p_k.) Thus $\lim_{n\to\infty} f_n$ exists in X.

Now $f_n = Gf_{n-1} + g\ (n \geq 1)$ and it follows that $\lim f_n = \lim Gf_{n-1}+g$. Since G is a bounded linear transformation it is continuous, hence $\lim f_n = G \lim f_{n-1}+g$, and if $f = \lim f_n$, then $f = Gf+g$. We can write $f = \sum_{r=0}^{\infty} G^r g$. Thus a required solution f exists for each g. We have seen that it is unique.

Now $\|f_n\| \leq \sum_{r=0}^{\infty} \|G\|^r \|g\|$ for all n, so that

$$\|f\| \leq \sum_{r=0}^{\infty} \|G\|^r \|g\| = (1 - \|G\|)^{-1} \|g\| .$$

Hence $\|(I-G)^{-1}\| \leq (1 - \|G\|)^{-1}$.

Observe that we can write $(I-G)^{-1} = \sum_{r=0}^{\infty} G^r$, since $f = \sum_{r=0}^{\infty} G^r g$ for each $g \in X$. This completes our proof.

We shall appeal to the following results later in this book.

LEMMA 1.1 *The sequence* $\{f_\gamma(x)\}$ *converges to* $f(x)$ *as* $\gamma \to 0$ *(in norm) if and only if for every infinite subsequence* $\{f_{\gamma_r}(x)\}$ *with* $\gamma_r \to 0$ *there exists a convergent subsequence* $\{f_{\gamma_{r_s}}(x)\}$ *converging (in norm) to* $f(x)$.

THEOREM 1.2. *Suppose that the functions* $\{f_\gamma(x)\}$, *defined for* $a \leq x \leq b$, *are uniformly bounded* ($\|f_\gamma(x)\|_\infty \leq M$ *for all* γ, *say) and equicontinuous, and suppose that the functions* $v_i(x)$ $(i = 0,1,2,\ldots)$

form a complete orthonormal basis for $L^2[a,b]$. *If*
$\lim_{\gamma\to 0} (f_\gamma, v_i) = (f, v_i)$ $(i = 0,1,2, \ldots)$, *where* $f(x) \in C[a,b]$, *then*
$\lim_{\gamma\to 0} \|f_\gamma(x) - f(x)\|_\infty = 0$.

<u>Proof</u>. By the Arzela-Ascoli theorem, the functions $\{f_{\gamma_r}(x)\}$ contain a subsequence $\{f_{\gamma_{r_s}}(x)\}$ converging uniformly to some function.
Suppose for some such subsequence that $g(x)$ is the corresponding limit, then $\lim \|f_{\gamma_{r_s}}(x) - g(x)\|_\infty = 0$, and, *a fortiori*,
$\lim \|f_{\gamma_{r_s}}(x) - g(x)\|_2 = 0$.

Now

$$|(f_{\gamma_{r_s}} - g, v_i)| \leq \|f_{\gamma_r}(x) - g(x)\|_2 \, \|v_i(x)\|_2$$

so

$$\lim (f_{\gamma_{r_s}}, v_i) = (g, v_i) \quad (i = 0,1,2,\ldots) .$$

But $\lim_{\gamma\to 0} (f_\gamma, v_i) = (f, v_i)$, so $(g, v_i) = (f, v_i)$ $(i = 0,1,2,\ldots)$ and hence $g(x) = f(x)$ 'almost everywhere'. Since both $g(x)$ and $f(x)$ are continuous, $f(x) \equiv g(x)$.

Thus every subsequence $\{f_{\gamma_r}(x)\}$ has a convergent subsequence $\{f_{\gamma_{r_s}}(x)\}$ converging uniformly to $f(x)$ and, by Lemma 1.1,
$\lim_{\gamma\to 0} \|f_\gamma(x) - f(x)\|_\infty = 0$.

The following sequence of results also have applications later.

THEOREM 1.3. *Let* X, Y *be Banach spaces and let* T, T_n $(n = 1, 2, \ldots)$ *be bounded linear operators mapping* X *into* Y. *Let* E *be a dense subset of* X *(so that for any* $\phi \in X$ *and for every* $\varepsilon > 0$, *there exists a corresponding* $\tilde{\phi} \in X$ *with* $\|\phi - \tilde{\phi}\| < \varepsilon$). *In order that* $\|T_n\phi - T\phi\| \to 0$ *for each* $\phi \in X$ *it is necessary and sufficient that*

(i) $\|T_n\tilde{\phi} - T\tilde{\phi}\| \to 0$ *for all* $\tilde{\phi} \in E$, *and*

(ii) $\sup_n \|T_n\| < \infty$.

The necessity of condition (ii) is usually referred to as the 'principle of uniform boundedness'. For a discussion of the proof of the theorem see, for example, the treatment of Collatz (1966 *b*, pp.104-6). The application of this result to the convergence of quadrature rules will be given in Chapter 2. See also Krylov (1962, p.61) and Anselone (1971, p.7).

Remark. The preceding result is related to the *Banach-Steinhaus* theorem (Zaanen 1964, p.135; Cheney 1966, p.211). This states that if $\{T_\alpha\}$ is a set of linear operators on a Banach space X and $\|T\phi\|$ is bounded for $T \in \{T_\alpha\}$, separately for each $\phi \in X$, then there exists a number $M < \infty$ such that $\|T\| \leq M$ for $T \in \{T_\alpha\}$.

The following theorem is a corollary of Theorem 1.3 (see Anselone 1971, p.7). We shall use Theorem 1.4 in our analysis of methods for Fredholm equations.

THEOREM 1.4. *Let* X, Y *be Banach spaces and* $\{T_n\}$ *a sequence of bounded linear operators mapping* X *into* Y. *If* $T_n\phi$ *converges (in* Y*) as* $n \to \infty$, *for any* $\phi \in X$, *this convergence is uniform on compact subsets of* X.

Proof. By Theorem 1.3, there exists a finite M such that $\|T_n\| \leq M$ for all n. Thus $\|T_n\phi - T_n\psi\| = \|T_n(\phi - \psi)\| \leq M\|\phi - \psi\|$ for all n, that is, the sequence $\{T_n\}$ is 'equicontinuous'. The convergence of a uniformly bounded and equicontinuous family of operators is uniform on compact sets (*cf.* real analysis) and the result follows.

1.10. *Differential equations and integral equations*

It is quite well known that there is a strong connection between some types of integral and differential equations. The theoretical treatment of differential equations sometimes involves recasting the problem as an integral equation. In the case of the numerical solution of differential equations, the transformation of a differential equation into an integral equation is not always to be recommended, and the converse procedure may sometimes be advisable. Each type of equation should be examined on its merits, in the light of the methods of solution which are proposed.

We shall indicate some of the ways in which differential equations may be re-formulated as integral equations. Fuller details may be found in, for example, Courant and Hilbert (1953), Hildebrand (1965), Lovitt (1924), and Tricomi (1957).

The initial-value problem defined by the differential equation

$$\phi^{(n)}(x) + \sum_{r=0}^{n-1} a_{n-r}(x)\, \phi^{(r)}(x) = \gamma(x) \qquad (x \geq 0) ,$$

with continuous coefficients, and the initial conditions $\phi^{(r)}(0) = \alpha_r$ $(r = 0, 1, 2, \ldots, (n-1))$, may be re-formulated as a Volterra equation of the second kind. We obtain

$$f(x) + \int_0^x K(x,y)\, f(y)\mathrm{d}y = g(x) ,$$

where $g(x) = \gamma(x) - \sum_{r=1}^{n} \sum_{s=n-r}^{n-1} \alpha_s x^{s-n+r} a_r(x)/(s-n+r)!$,

$$K(x,y) = \sum_{r=1}^{n} a_r(x)\, (x-y)^{r-1}/(r-1)! ,$$

and $f(x) = \phi^{(n)}(x)$ (see Tricomi 1957, p.18). We note that the kernel $K(x,y)$ is a function $k(x - y)$ say, of the difference $(x - y)$. Such a kernel will be called a *convolution* kernel. The solution of a Volterra equation with a convolution kernel may be obtained, in theory, by using Laplace transforms (Hildebrand 1965, p.275); but the technique is not particularly appropriate to a numerical method of solution.

Example 1.29

The problem $f''(x) + f(x) = \gamma(x)$, $x \geq 0$, with $f(0) = 1$, $f'(0) = 0$ is equivalent to the equation

$$f(x) + \int_0^x (x - y)\, f(y)\mathrm{d}y = 1 + \int_0^x (x - y)\, \gamma(y)\mathrm{d}y \quad \text{for } x \geq 0 . \quad *$$

We obtain a Volterra equation from the initial value problem above, and we obtain a Fredholm equation from certain types of boundary value problem.

We consider the problem solving

$$L\{f\}(x) \equiv \sum_{r=0}^{n} a_{n-r}(x) f^{(r)}(x) = -\gamma(x) \qquad (a \leq x \leq b) \qquad (1.43)$$

subject to homogeneous boundary conditions at $x = a$ and $x = b$. The boundary conditions have the form

$$\sum_{r=0}^{n} \alpha_{n-r,s}\, f^{(r)}(a) = 0 \quad (s = 1, 2, 3, \ldots n'),$$

$$\sum_{r=0}^{n} \alpha_{n-r,s}\, f^{(r)}(b) = 0 \qquad (s = (n' + 1), (n' + 2), \ldots, n). \qquad (1.44)$$

Example 1.30

An example frequently discussed is that of the second-order equation

$$\frac{\mathrm{d}}{\mathrm{d}x}\{p(x)f'(x)\} + q(x)\, f(x) = -\, \gamma(x) \qquad (p(x) \neq 0)$$

with boundary conditions of the form $\alpha_0 f(a) + \beta_0 f'(a) = 0$, $\alpha_1 f(b) + \beta_1 f'(b) = 0$. *

Associated with the general problem is a kernel known as the *Green's function* $G(x,y)$, which satisfies the following conditions.

(i) If y is held fixed, then $G(x,y)$ is a function of x which satisfies the differential equation for $x < y$ and for $x > y$.

(ii) If y is held fixed, $G(x,y)$ satisfies the homogeneous boundary conditions at $x = a$ and at $x = b$.

(iii) $G(x,y)$ and its first $(n - 2)$ x-derivatives are continuous for $a \leq x,y \leq b$.

(iv) The $(n - 1)$th x-derivative $G(x,y)$ has a jump of magnitude $- 1/a_0(x)$ as x increases through y .

If $G(x,y)$ can be determined, then

$$f(x) = \int_a^b G(x,y)\ \gamma(y)dy\ . \tag{1.45}$$

Now, we obtain an eigenvalue problem in the differential equation if we set $\gamma(x) = \lambda f(x)$ and look for values of λ which correspond to non-null solutions. Such a value λ will be a characteristic value of $G(x,y)$, and $f(x)$ will be a corresponding eigenfunction. An inhomogeneous equation is obtained if we replace $\gamma(x)$ by $\lambda r(x)\ f(x) - g(x)$, say. We then have an equation of the second kind:

$$f(x) - \lambda \int_a^b G(x,y)\ r(y)\ f(y)dy = d(x)\ ,$$

in which the kernel is $K(x,y) = G(x,y)r(y)$, and $d(x)$ $= \int_a^b G(x,y)\ g(y)dy$.

Example 1.31

The Green's function associated with $L\{f\} = f^{iv}(x)$, $f(0) = f'(0) = f(1) = f'(1) = 0$ is the Hermitian kernel such that

$$G(x,y) = \frac{1}{6}x^2(1-y)^2 \cdot (2xy + x - 3y)$$

when $x \leq y$, and $G(y, x) = G(x,y)$. *

In a number of cases it is not difficult to determine the Green's function. Some examples are given in Table 1.2.

An integral equation corresponding to a boundary-value problem in which there are linear inhomogeneous boundary conditions may be obtained using the Green's function for the corresponding homogeneous boundary conditions.

It may be that there is no suitable function satisfying the requirements (i) - (iv) specified for the Green's function. This is the case when there is a non-trivial function satisfying the boundary conditions and also satisfying $L\{f\}(x) = 0$; we may then introduce a *generalized Green's function* (Hildebrand 1965, p.300; Courant and Hilbert 1953, p.356) which assumes to some extent the rôle of $G(x, y)$. Let us indicate the rôle of the *generalized* Green's function by considering the case where $L\{f\}(x) = -d(x)$, with

$$L\{f\}(x) = \frac{d}{dx}\{p(x)\,f'(x)\} - q(x)f(x)$$

for $a \leq x \leq b$, where $p(x) > 0$ for $x \in [a,b]$, and $d(x), p(x), p'(x)$, $q(x) \in C[a,b]$ are given (real-valued) functions. In our boundary-value problem we require $f(x)$ to satisfy, in addition, certain homogeneous boundary conditions. These, we suppose, are also satisfied by a function $u_0(x)$, where $L\{u_0\}(x) = 0$. Clearly if $f(x)$ is a solution of our boundary-value problem, $f(x) + \alpha u_0(x)$ is another solution. On the other hand, a solution of the boundary problem exists only if

$$(d,u_0) = \int_a^b d(x)u_0(x)\,dx = 0 ,$$

TABLE 1.2

Green's functions and generalized Green's functions

Differential expression	Boundary conditions	Green's function $G(x,y)$ for $x \le y$ (For $y > x$, $G(x,y)=G(y,x)$.)
$L\{f\}(x) = f''(x)$	$f(0) = f(1) = 0$	$G(x,y) = x(1-y)$ $(0 \le x \le y \le 1)$
$L\{f\}(x) = f''(x)$	$f(-1) = f(1) = 0$	$G(x,y) = -\frac{1}{2}\{xy-1+\lvert x-y\rvert\}$ $(-1 \le x \le y \le 1)$
$L\{f\}(x) = f''(x)$	$f(0) = f'(1) = 0$	$G(x,y) = x$ $(0 \le x \le y \le 1)$
$L\{f\}(x) = f''(x)$	$f'(0) = f'(1) = 0$	$G(x,y) = \frac{1}{3}+\frac{1}{2}(x^2+y^2) - y$ $(0 \le x \le y \le 1)$
$L\{f\}(x) = \frac{\mathrm{d}}{\mathrm{d}x}\{a_0(x)\, f'(x)\}$ $(a_0(x) \neq 0)$	$f(0) = f(1) = 0$	$G(x,y) = A_0(x)\{1-A_0(y)/A_0(1)\}$, where $A_0(x) = \int_0^x \frac{1}{a_0(t)}\,\mathrm{d}t$ $(0 \le x \le y \le 1)$

$L\{f\}(x) = \frac{d}{dx}\{a_0(x)\, f'(x)\}$	$f(0) = f'(1) = 0$	$G(x,y) = A_0(x)$, (see above) $(0 \le x \le y \le 1)$
$L\{f\}(x) = xf''(x) + f'(x)$	$\lvert f(0)\rvert < \infty$, $f(1) = 0$	$G(x,y) = -\ln y$ $(0 \le x \le y \le 1)$
$L\{f\}(x) = xf''(x) + f'(x) - \frac{n^2 f(x)}{x}$	$\lvert f(0)\rvert < \infty$, $f(1) = 0$	$G(x,y) = \frac{x^n}{2ny^n}(1-y^{2n})$ $(0 \le x < y \le 1)$
$L\{f\}(x) = f^{iv}(x)$	$f(0) = f'(0) = f''(1) = f'''(1) = 0$	$G(x,y) = \frac{1}{6}(3y-x)x^2$, $(0 \le x < y \le 1)$
$L\{f\}(x) = f^{iv}(x)$	$f(0) = f''(0) = f(1) = f''(1) = 0$	$G(x,y) = \frac{1}{6}x(y-1)(x^2+y^2-2y)$ $(0 \le x \le y \le 1)$

Differing sign conventions prevail in the definition of Green's functions, and it is a wise precaution to check the sign when determining expressions from tables.

and the generalized Green's function $G(x,y)$ then gives a solution

$$f(x) = \int_a^b G(x,y)d(y)\mathrm{d}y$$

which is orthogonal to $u_0(x)$. Observe (see Courant and Hilbert 1953, p.356) that the generalized Green's function $G(x,y)$ is defined so that

$$\int_a^b G(x,y)\ u_0(y)\mathrm{d}y = 0\ ,$$

and if $(d,u_0) \neq 0$ then the function $f(x) = \int_a^b G(x,y)\ d(y)\mathrm{d}y$ satisfies the boundary conditions and the equation

$$L\{f\}(x) = -d(x) + (d,u_0)\ u_0(x)\ .$$

For a more detailed discussion we refer to Courant and Hilbert (1953). The above material is sufficient for use in the following Example.

Example 1.32

Consider the equation $f''(x) = -d(x)$ $(0 \leq x \leq 1)$ with boundary conditions $f'(0) = f'(1) = 0$. The only solutions to the homogeneous boundary-value problem (corresponding to the case $d(x) \equiv 0$) have the form $f_0(x) = \alpha$, where α is a constant. A solution of the given problem exists if $\int_a^b d(x)\mathrm{d}x = 0$. Using the (generalized) Green's function supplied in Table 1.2, where

$$G(x,y) = \frac{1}{3} + \tfrac{1}{2}(x^2 + y^2) - y \quad (0 \leq x \leq y \leq 1)$$

and $G(x,y) = G(y,x)$, the general solution of the boundary-value problem is then

$$f(x) = \int_0^1 G(x,y)d(y)\mathrm{d}y + \alpha \quad *$$

Remark. Some authors define the Green's function as the negative of the function defined above.

The Green's function can sometimes be employed to relate an integro-differential equation to an integral equation. Thus if the solution of the second-order equation $L\{f\}(x) = -d(x)\ (a \le x \le b)$ satisfying the boundary conditions,

$$\alpha_0 f(a) + \beta_0 f'(a) = 0\ ,\quad \alpha_1 f(b) + \beta_1 f'(b) = 0$$

is obtained by writing

$$f(x) = \int_a^b G(x,y)\ d(y)\,dy$$

then we can obtain the solution of the corresponding integro-differential equation

$$L\{\phi\} = \int_a^b M(x,y)\ \phi(y)\,dy - d(x)$$

subject to the boundary conditions

$$\alpha_0\phi(a) + \beta_0\phi'(a) = 0\ ,\ \alpha_1\phi(b) + \beta_1\phi'(b) = 0.$$

It is sufficient to consider the equivalent formulation

$$\phi(x) + \int_a^b K(x,y)\phi(y)\,dy = g(x)\ ,$$

where $K(x,y) = \int_a^b G(x,z)\ M(z,y)\,dz$ and $g(x) = \int_a^b G(x,z)\,d(z)\,dz$. A corresponding analysis exists for the case where no Green's function exists in the ordinary sense and $G(x,y)$ is taken to be the generalized Green's function. This case has applications in the study of regularization methods for Fredholm equations of the first kind, and the analysis given in the following Example is therefore instructive, and will be referred to later (in Chapter 5).

Example 1.33

We consider the problem of finding the function $\phi(x)$ satisfying the equation

$$\mu\,\phi''(x) - \int_0^1 M(x,y)\phi(y)\,dy = d(x) \qquad (0 \le x \le 1) \tag{1.46}$$

and the boundary conditions

$$\phi'(0) = \phi'(1) = 0\ . \tag{1.47}$$

We shall suppose that $\mu > 0$, and that $M(x,y) = \int_0^1 K^*(x,z)\,K(z,y)\,dz$ and $d(x) = \int_0^1 K^*(x,z)g(z)\,dz$ for some kernel $K(x,y)$. We assume that there is no non-null function $\psi(x)$ such that

$$(K\psi)(x) \equiv \int_0^1 K(x,y)\psi(y)\,dy$$

vanishes.

Let us first establish that any solution of the integro-differential equation which satisfies the boundary conditions is unique. For, suppose that $\phi_0(x)$ and $\phi_1(x)$ are two solutions. Then if $\varepsilon(x) = \phi_0(x) - \phi_1(x)$, it is clear that

$$\mu\,\varepsilon''(x) - \int_0^1 M(x,y)\varepsilon(y)\,dy = 0$$

and

$$\varepsilon'(0) = \varepsilon'(1) = 0.$$

Thus $\mu \int_0^1 \varepsilon''(x)\overline{\varepsilon(x)}\,dx = \int_0^1 M(x,y)\varepsilon(y)\overline{\varepsilon(x)}\,dx\,dy = \|(K\varepsilon)(x)\|_2^2 \ge 0$.

But

$$\mu \int_0^1 \varepsilon''(x)\overline{\varepsilon(x)}dx = \mu\Big[\overline{\varepsilon(x)}\varepsilon'(x)\Big]_{x=0}^{x=1} - \mu \int_0^1 |\varepsilon'(x)|^2 dx \leq 0 ,$$

on substituting $\varepsilon'(0) = \varepsilon'(1) = 0$. Consequently $||(K\varepsilon)(x)||_2 = 0$. By our hypothesis on $K(x,y), \varepsilon(x) = 0$, and it follows that $\phi_1(x) = \phi_2(x)$. (Observe that the essential feature of $M(x,y)$ is that $(M\varepsilon,\varepsilon) > 0$ if $\varepsilon(x) \not\equiv 0$.)

Let us now establish the existence of a solution $\phi(x)$ of the given boundary-value problem. The generalized Green's function for $L\{\phi\} = \phi''(x)$, and the boundary conditions $\phi'(0) = \phi'(1) = 0$ were given in Table 1.1; we shall denote the function by $G(x,y)$. Now from our remarks in the preceding Example, the function $\phi(x) = \phi_0(x) + \alpha$ satisfies the boundary conditions $\phi'(a) = \phi'(b) = 0$ and the equation

$$\mu\phi''(x) = \int_0^1 M(x,y)\ \phi(y)dy\ - d(x) - \beta , \tag{1.48}$$

where

$$\beta = \int_a^b \{(M\phi)(x) - d(x)\}dx ,$$

provided that

$$\mu\phi_0(x) + \int_a^b \{\int_a^b G(x,z)\ M(z,y)dx\}\ \phi_0(y)dy = \int_a^b \{G(x,z)\ d(z)\}dz . \tag{1.49}$$

The choice of β ensures that the right-hand side of (1.48) is orthogonal to the function $u_0(x) = e(x)$ where $e(x) \equiv 1$. This condition is necessary for the solvability of eqn (1.48).

We shall establish that eqn (1.49) has a (unique) solution $\phi_0(x)$ and that α can be chosen (uniquely) so that $\beta = 0$. Then $\phi(x)$ satisfies the integro-differential equation (1.46) and the associated boundary conditions (1.47).

Since $\phi_0(x)$ is defined by the integral equation (1.48) we must establish that $\mu > 0$ is not an eigenvalue of the kernel

$$-\int_0^1 G(x,z)\, M(z,y)\mathrm{d}z \;.$$

Suppose the contrary holds, and $\psi(x)$ is a non-null eigenfunction so that, in compact notation

$$\mu\, \psi(x) = -(GK^*K)\, \psi(x) \;.$$

Then it follows that, if $e(x) \equiv 1$, then $\psi'(0) = \psi'(1) = 0$, and $\mu\psi''(x) = K^*K\psi(x) - \gamma e(x)$, where γ is chosen so that this boundary-value problem for $\psi(x)$ has a solution. We require $\gamma = (K^*K\psi, e)$. We observe that $\mu(\psi,e) = (GK^*K\psi,e) = (K^*K\psi,Ge)$ from the symmetry of $G(x,y)$. Since $Ge(x) \equiv 0$ and $\mu > 0$, it follows that $(\psi,e) = 0$. Then $\mu(\psi,\psi'') = (\psi, K^*K\psi) - \gamma(\psi,e) = (\psi, K^*K\psi)$ and $\psi'(0) = \psi'(1) = 0$. Using these boundary conditions and integrating by parts, $\mu(\psi,\psi'') \le 0$, but $(\psi, K^*K\psi) = \|K\psi(x)\|_2^2 \ge 0$ with equality only if $\psi(x) \equiv 0$. It follows that $\psi(x) \equiv 0$, and hence μ is not an eigenvalue.

We now know that $\phi_0(x)$ exists, and is defined uniquely by eqn (1.49). Consequently, if $\phi(x) = \phi_0(x) + \alpha$, the value of β in (1.48) is

$$\int_a^b \{M\phi_0(x) - d(x)\}\mathrm{d}x + \alpha \int_a^b Me(x)\mathrm{d}x \;.$$

The coefficient of α in this expression is

$$\int_a^b Me(x)\mathrm{d}x = \int_a^b\!\!\int_a^b M(x,y)\; e(y)\mathrm{d}y\; \mathrm{d}x = (Me,e) = \|Ke(x)\|_2^2 \ne 0$$

and thus α can be chosen (in a unique way) to ensure that $\beta = 0$. It follows that the corresponding function $\phi(x)$ satisfies (1.46) and (1.47) and is the unique solution of this problem. *

In the Example discussed above, any scalar-valued function satisfies the boundary conditions and is annihilated by the differential operator. The analysis extends immediately to the more general case where an arbitrary multiple of a function $u_0(x)$ satisfies prescribed boundary conditions and is annihilated by an associated differential operator. We can also treat the case where linearly independent functions $u_0(x), u_1(x), \ldots, u_m(x)$ satisfy the boundary conditions and are annihilated by the differential operator.

The notion of a Green's function can be extended to partial differential equations, and the resulting integral equations relate functions of more than one variable. A number of examples of Green's functions for partial differential equations appear in Morse and Feshbach (1953).

Example 1.34

If $F(r,\theta)$ satisfies $\nabla^2 F(r,\theta) + \lambda F(r,\theta) = 0$ in the interior of the circle $|r| = 1$, and $F(r,\theta)$ vanishes when $|r| = 1$, then

$$F(r,\theta) = \int_0^1 \int_0^{2\pi} G(r,\theta;\, \rho,\phi)\, \lambda F(\rho,\phi) d\rho\, d\phi ,$$

where

$$G(r,\theta;\, \rho,\phi) = \frac{1}{4\pi} \ln \frac{1 - 2\rho r \cos(\theta-\phi) + \rho^2 r^2}{\rho^2 - 2\rho \cos(\theta-\phi) + r^2} . \quad *$$

Green's function for a practical problem may be unknown, but it may be possible to obtain it by the use of conformal mapping to relate the given boundary to some standard one for which the Green's function is known. Some methods of conformal mapping employ integral equation techniques.

Example 1.35 is illustrative of the fact that some partial differential equations for functions in n variables may be re-formulated as integral equations for functions in $(n - 1)$ variables. This is one of the advantages claimed for an integral equation approach.

Example 1.35

The plane interior Dirichlet problem associated with a smooth closed contour L consists in finding $F(x,y)$ satisfying Laplace's equation $\nabla^2 F(x,y) = 0$ in the interior of L with prescribed values of $F(x,y)$ on the boundary L. The problem may be reduced to the solution of a linear integral equation for a function of a single variable (Tricomi 1957; Mikhlin 1964) with a continuous kernel. When L is the ellipse $(^x/a)^2 + (^y/b)^2 = 1$ with eccentricity e we may reduce the problem to the solution of the equation

$$\mu(\tau) + \frac{b}{2\pi a}\int_0^{2\pi} \frac{\mu(t)}{1 - e^2\cos^2\frac{(t+\tau)}{2}}\, dt = g(\tau) ,$$

where $g(\tau)$ involves prescribed values of F on the boundary. *

1.11. Smoothness of kernels

A number of Green's functions suffer discontinuities along $x = y$, possibly in some low-order derivative.

We may also encounter kernels which become unbounded at $x = y$, for example, the kernel $K(x,y) = H(x,y)/|x - y|^\alpha$, where $H(x,y)$ is continuous†, $H(x,x) \neq 0$ and $0 < \alpha < 1$. This kernel is said to have a *weak singularity* (such a value of α is called the *exponent* of the kernel). Permitting $H(x,x) = 0$ gives a general class of weak singularity¶.

It is my purpose in this section to consider such kernels and to investigate what action may be taken to restate our problem as an equation with a smoother kernel. Such a course of action is appropriate for eigenvalue problems and equations of the second kind. However, it is interesting to note the commonly held view that an equation of the first kind where $K(x,y)$ is badly behaved at $x = y$ is better posed than a similar equation where $K(x,y)$ is continuous. There is some truth in this, and we would regard it as unwise to convert one of the

---

† In the case of a Volterra equation, continuity of $H(x,y)$ is required only when $y \leq x$. Some authors impose less restrictive conditions.

¶ Thus the kernel $\ln|x-y|$ has a 'logarithmic weak singularity'.

first class of equations into one of the second class. However, difficulty is likely to be encountered with any equations $Kf = g$, where K is a completely continuous operator (section 1.9) on a space X (with $g \in X$), and we seek a solution $f \in X$. (In particular, as we note later, weakly singular kernels give rise to integral operators which are compact on $L^2[a,b]$.)

We shall give some examples of linear equations with badly behaved kernels.

Example 1.36

The generalized Abel equation

$$\int_0^x \frac{f(y)}{(x-y)^\alpha} dy = g(x) \quad (0 < \alpha < 1),$$

has the solution $f(x) = \dfrac{\sin \alpha\pi}{\pi} \dfrac{d}{dx} \displaystyle\int_0^x \frac{g(y)}{(x-y)^{1-\alpha}} dy$,

under appropriate restrictions on $g(x)$. *

Example 1.37

Consider the equation of the first kind

$$\frac{1}{2\pi} \fint_{-1}^{1} \frac{f(y)}{x-y} dy = g(x),$$

in which the integral exists as a Cauchy principal value (Collatz 1966 *a*, p.505) and $g(x)$ is continuous. A solution of this equation is obtained if we solve

$$\sqrt{(1-x^2)}\pi f(x) + \int_{-1}^{1} f(y) dy = 2 \fint_{-1}^{1} \frac{g(y)\sqrt{(1-y^2)}}{y-x} dy$$

(Collatz 1966 *a*, p.510; see also Mikhlin 1964 *a*). *

Example 1.38

The equation

$$f(x) - \int_0^{\pi} \ln|\cos(x) - \cos(y)| f(y) dy = 1 \quad (0 \le x \le \pi)$$

has a kernel $K(x,y)$ with a 'logarithmic singularity', and $K^2(x,y)$ is bounded. *

If we suppose that $K(x,y)$ is a badly behaved kernel, we can usually derive a smoother kernel by forming an iterated kernel $K^p(x,y)$. However, there are kernels which cannot be made non-singular by iteration, and Morse and Feshbach (1953, Vol. 1, p.924) label such kernels *intrinsically singular*. In the case where $0 \le \alpha < 1$ and $H(x,y)$ is continuous we have the general *weakly singular* kernel

$$K(x,y) = H(x,y)/|x-y|^{\alpha} .$$

(On the other hand, the kernel $1/(x-y)$ is an *intrinsically singular* kernel (Example 1.37).) If $0<\alpha<\frac{1}{2}$ above then the kernel $K^2(x,y)$ is bounded, and no longer weakly singular. More generally, if $0 < \alpha < 1$, the kernel $K^p(x,y)$ has an exponent which is at most $p\alpha - (p-1)$, so that $K^p(x,y)$ is bounded if p is larger than $1/(1-\alpha)$.

Example 1.39

If $K(x,y) = \ln|x - y|$, $0 \le x, y \le 1$, then $K^2(x,y)$ is bounded. *

Suppose, then, that $K(x,y)$ is a badly behaved kernel appearing in a linear integral equation, and $K^p(x,y)$ is continuous for some p (In particular, we may suppose that $K(x,y)$ is weakly singular.) We shall investigate how, in such circumstances, the eigenvalue problem and the equation of the second kind can be transformed into similar equations with continuous kernels. In simple cases, such a technique can be implemented to simplify a practical problem.

Remark. The technique of transforming an equation of the second kind with a weakly singular kernel into one with a continuous kernel enables us to establish, as we indicate below, the validity of the Fredholm theory in the weakly singular case. The validity of the theory can also be established by considering the abstract theory associated with completely continuous operators, as indicated in section 1.9.

In the case of an eigenvalue problem

$$\int_a^b K(x,y)\ f(y)\mathrm{d}y = \kappa\ f(x)\ ,$$

we find

$$\int_a^b K^p(x,y)\ f(y)\mathrm{d}y = \kappa \int_a^b K^{p-1}(x,y)\ f(y)\mathrm{d}y = \ldots$$
$$= \kappa^p\ f(x)\ ,$$

so that the kernel $K^p(x,y)$ has an eigenvalue κ^p and the corresponding eigenfunction $f(x)$ whenever κ and $f(x)$ are corresponding eigenvalue and eigenfunction for $K(x,y)$. The converse problem is of some interest. We would hope to solve the eigenproblem for $K(x,y)$ by considering the eigenvalue problem for the kernel $K^p(x,y)$. Suppose

$$\int_a^b K^p(x,y)\ \psi(y)\mathrm{d}y = \nu\ \psi(x)\ .$$

We must then ask whether $\psi(x)$ is an eigenfunction for $K(x,y)$ corresponding to an eigenvalue $\kappa = \nu^{1/p}$. The difficulty is that there are p roots of ν, and they need not all be eigenvalues of $K(x,y)$. For example, if we consider a kernel $K(x,y)$ which is Hermitian and has only positive eigenvalues, then $K^2(x,y)$ has positive real eigenvalues $\{\nu_r\}$ and we must exclude the negative square-roots of these ν_r. Some independent test of each of the roots would therefore seem to be called for.

See also Smithies (1962, pp. 121-2) and Goursat (1964, pp. 77-8) for further comments.

Example 1.40

Consider the eigenvalue problem

$$\int_0^x \{f(y)/(x^2-y^2)^{\frac{1}{2}}\}dy = \kappa\, f(x)$$

($0 \leq x \leq 1$, say). Well-behaved Volterra kernels have no eigenvalues corresponding to square-integrable eigenfunctions, but in this example the kernel is not even weakly singular. A solution to the eigenvalue problem is given, for each $\mu \geq 0$, by setting $f_\mu(x) = x^\mu$ and

$\kappa_\mu = \int_0^{\frac{1}{2}\pi} (\sin t)^\mu \, dt$. Thus, every point $\kappa \in (0, \frac{1}{2}\pi]$ is an eigenvalue;

we say that the spectrum is not discrete. Observe that the kernel is not square-integrable. Determine the success or failure of the construction of the iterated kernels to solve the eigenvalue problem, noting that if an operator is compact, it must have a discrete spectrum.*

In the case of an equation of the second kind with a weakly singular kernel, the situation is perhaps less complicated. For the equation

$$f(x) - \lambda \int_a^b K(x,y)\, f(y)dy = g(x)$$

(where λ is not a characteristic value), we write

$$f(x) - g(x) = \lambda \int_a^b K(x,y)\, \{g(y) + \lambda \int_a^b K(y,z)\, f(z)\, dz\}dy$$

or, on rearranging,

$$f(x) - \lambda^2 \int_a^b K^2(x,z)\, f(z)dz = g_\lambda(x) ,$$

where

$$g_\lambda(x) = g(x) + \lambda \int_a^b K(x,y)\, g(y)dy .$$

If $K^2(x,y)$ is continuous, the last equation is solvable if λ^2 is a regular value of $K^2(x,y)$, in particular if neither λ nor $-\lambda$ is a characteristic value of $K(x,y)$. If $K^2(x,y)$ is weakly singular the process may be repeated and we can consider the equation

$$f(x) - \lambda^3\int_a^b K^3(x,y)f(y)dy = g_\lambda(x) + \lambda^2\int_a^b K^2(x,y)g(y)dy ,$$

and so on.

The preceding analysis is useful if both λ and $-\lambda$ are known to be regular values of $K(x,y)$. If λ is known to be a regular value of the weakly singular kernel $K(x,y)$, then it can be shown that, for some m, $K^m(x,y)$ is continuous and, with $\omega = e^{2\pi i/m}$ an mth root of unity, $\lambda\omega, \lambda\omega^2, \ldots, \lambda\omega^{m-1}$ are regular values of $K(x,y)$. The equation $f(x) - \lambda Kf(x) = g(x)$ can then be shown to be equivalent to the equation

$$f(x) - \lambda^m K^m f(x) = (I - \lambda\omega K)(I - \lambda\omega^2 K) \ldots (I - \lambda\omega^{m-1}K)g(x) ,$$

and the kernel of the operator K^m is continuous. This approach is adopted by Mikhlin (1964 *a*, pp.65-6) but it appears to be a theoretical device (of more theoretical than practical interest) unless some information about the eigenvalues of $K(x,y)$ is known. Obvious cases where the latter condition is satisfied arise in the case of a Volterra equation, or an equation with a Hermitian kernel (see Example 1.41).

Remark. We leave it to the reader to investigate the usefulness of the preceding techniques when $K(x,y)$ is not weakly singular but does suffer a discontinuity in a derivative at $x = y$.

Example 1.41

Suppose that $K(x,y)$ is Hermitian, and weakly singular, and that $K^2(x,y)$ is continuous. Given that λ is real and not a characteristic value, the above analysis applies with $m = 3$, since $\lambda e^{\pi i/3}$ and

$\lambda e^{2\pi i/3}$ cannot be eigenvalues of $K(x,y)$. These values are complex, and the eigenvalues of $K(x,y)$ are real. *

Example 1.42

Consider the Volterra kernel $K(x,y) = x - y$ if $y \leq x$, 0 if $y > x$. $K^2(x,y)$ is also a Volterra kernel and $(\partial/\partial x)\, K^2(x,y)$ is continuous at $x = y$, whereas $(\partial/\partial x)\, K(x,y)$ is 1 if $y \leq x$ and 0 otherwise. If $K(x,y)$ is a Volterra kernel, weakly singular or continuous, its iterated kernels are also Volterra kernels. If $g(x)$ is continuous, say, a continuous solution to a linear Volterra equation with a weakly singular kernel may always be found by transforming the equation as indicated above. *

In view of the practical importance of the case where $K(x,y)$ is weakly singular we shall make some further remarks on this subject. We have seen that equations with a weakly singular kernel can sometimes be recast as equations where the kernel is well-behaved. A direct treatment is often necessary or convenient, however, and in some numerical methods it is important to know the behaviour of $(K\phi)(x)$ given the behaviour of $\phi(x)$. In the case where $H(x,y)$ is continuous, and

$$K(x,y) = H(x,y)/|x - y|^{\alpha}$$

with $0 < \alpha < \frac{1}{2}$, it follows that

$$\int_a^b \int_a^b |K(x,y)|^2 \, dx dy < \infty ;$$

this case fits into the general theory covered by Smithies (1962). The theory of sections 1.6-1.8 is applicable in this case.

More generally, suppose that $0 < \alpha < 1$. Then $\psi(x) = (K\phi)(x)$ is square-integrable (in the sense of Lebesgue) when $\phi(x)$ is square-integrable. It does not follow that $\psi(x)$ is continuous or even bounded, as can be seen by considering the case

$$\alpha \varepsilon \, [\tfrac{1}{2},1), \; [a,b] = [0,1], \; H(x,y) \equiv 1, \; \phi(x) = x^{-\beta}, \quad \beta = 1 - \alpha \, .$$

On the other hand, if $\phi(x)$ is continuous on $[a,b]$, then $\psi(x)$ is continuous on $[a,b]$. It is important to note that the operator K corresponding to a weakly singular kernel maps $L^2[a,b]$ into itself and, moreover, (Mikhlin 1964 *a*, p.160; Zaanen 1964, p.329) it is compact on this space. As a consequence the Fredholm alternative, and the associated theorems of Fredholm, apply. If λ is a regular value the equation of the second kind is solvable when $g(x) \,\varepsilon\, L^2[a,b]$. We may then ask, when $g(x) \,\varepsilon\, C[a,b]$, if the solution is in $C[a,b]$. The answer is affirmative, since, indeed, K is compact† on the space $C[a,b]$ (with uniform norm) and the Fredholm alternative again applies; the regular values are in each case the same.

Remark. We also note in passing that K is compact on the space $L^p[a,b]$ (where $1 \leq p \leq \infty$) of functions whose norm

$$\{\int_a^b |f(x)|^p dx\}^{1/p}$$

exists (see Zaanen 1964, p.326; $L_p[a,b]$ and $L^p[a,b]$ are the same).

Example 1.43

Suppose that $K(x,y) = |x - y|^{-\alpha}$ and $||K||_2$ is the corresponding operator norm on $L^2[0,1]$. Then $||K||_2 \leq 2^{\alpha}/(1 - \alpha)$ for $0 < \alpha < 1$, although $\int_0^1 \int_0^1 |K(x,y)|^2 dx\, dy = \infty$ if $\tfrac{1}{2} \leq \alpha < 1$. (Zaanen 1964, p.236). *

We now proceed to a brief mention of integral equations which are truly singular and in which the integral is defined as a Cauchy principal value. Anything approaching a full discussion is beyond our scope. We shall refer, instead, to Muskhelishvili (1953) for a systematic treatment.

---

† $K(x,y)$ satisfies the conditions given in section 1.9 for the compactness of K on $C[a,b]$ (see also Mikhlin 1964 *a*, p.162).

The book of Pogorzelski (1966) provides a readable introduction to our topic. Let us observe, in passing, that the techniques employed to reduce an equation with a weakly singular kernel to an equation with a continuous kernel can sometimes be employed to convert an equation with an intrinsically singular kernel to one with a weakly singular kernel. Such a procedure is associated with the name of Vekua (Pogorzelski 1966, p.512).

We content ourselves here with a few remarks concerning a particular equation. Consider the equation

$$(Af)(t) \equiv a(t)\, f(t) + \frac{b(t)}{\pi i}\int_\Gamma \frac{f(\tau)\mathrm{d}\tau}{\tau - t} + \int_\Gamma K(t,\tau)f(\tau)\mathrm{d}t = g(t) \ ,$$

where $a(t)$, $b(t)$, $K(t,\tau)$, and $f(t)$ are Lipschitz-continuous, and Γ is a simple smooth closed curve with a continuously turning tangent. The integrals occurring in this equation are line integrals, and the first integral is to be interpreted as a Cauchy principal value. We suppose that

$$\min_{t\in\Gamma}|a^2(t) - b^2(t)| > 0 \ .$$

The number of solutions to this equation depends on the *index*

$$m = \frac{1}{2\pi i}\left[\ln \frac{a(t) - b(t)}{a(t) + b(t)}\right]_\Gamma$$

where $[\ \ ]_\Gamma$ indicates the increment in the expression as one traverses Γ in the positive direction. If $m \geq 0$ the equation is solvable for any $g(t)$; if $m < 0$, $g(t)$ must satisfy certain orthogonality relations. The first two terms in $(Af)(t)$ define the *dominant part* A^0 of A ,

$$(A^0 f)(t) = a(t)\, f(t) + \frac{b(t)}{\pi i}\int_\Gamma \frac{f(\tau)}{\tau - t}\,\mathrm{d}\tau \ .$$

If $m \geq 0$ we can express the solution of the equation

$$(A^0 f)(t) = \phi(t)$$

in the form

$$f(t) = a^*(t)\,\phi(t) - \frac{b^*(t)\,z(t)}{\pi i}\int_\Gamma \frac{\phi(\tau)\,d\tau}{z(\tau)(\tau - t)},$$

where $a^*(t)$, $b^*(t)$, and $z(t)$ may be obtained explicitly. If we now substitute

$$g(t) - \int_\Gamma K(t,\tau)\,f(\tau)d\tau$$

for $\phi(t)$ and interchange an order of integration, we obtain a Fredholm equation of the second kind in which the kernel is Lipschitz-continuous. If we parametrize Γ, the integral is over an interval and the problem is now tractable (for fuller details see Chapter 6 of Muskhelishvili (1953)).

The idea we have pursued in this section has been the equivalence of badly behaved equations with well-behaved equations. There are examples, however, where such ideas fail us either because the implementation of a technique is too complicated or because the techniques themselves fail.

Example 1.44

If $K(x,y) = (y/x)^{\frac{1}{2}}$ $(0 \leq x, y \leq 1)$ then $K^p(x,y) = K(x,y)$, and all the iterated kernels are unbounded (see Goursat 1964, pp. 36-7). Notice that $K(x,y)$ is a degenerate kernel. *

1.12. Non-linear integral equations

The theory in section 1.10 provides some motivation for studying certain types of non-linear equations. If we replace $\gamma(x)$ in eqn (1.43) by the function $M(x,f(x))$, the resulting integral equation (1.45) assumes the form

$$f(x) - \int_a^b K(x,y)\, M(y,f(y))dy = 0 \ . \tag{1.50}$$

This is a non-linear equation of *Hammerstein type*. Suppose that $K(x,y)$ is real, symmetric, and continuous, and all its eigenvalues are negative. It is possible to show that (1.50) has at least one continuous solution if $M(u,v)$ is a continuous function such that

$$|M(u,v)| \leq c_1|v| + c_2 \ ,$$

where c_1 and c_2 are positive and $^1/c_1 \geq \rho(K)$. The solution is unique if $M(u,v)$ is non-decreasing in v , when u is fixed. Tricomi (1957) discusses these types of result, with some variations on the conditions (see also Anselone 1964).

Eqn (1.50) is an example of a more general type of non-linear equation of the form

$$f(x) - \lambda \int_a^b F(x,y;\, f(y))dy = g(x) \ , \tag{1.51}$$

where $F(x,y;v)$ is a function which is non-linear in v . Eqn (1.51) may be called a Fredholm equation because the limits of integration, a and b , are constants. An example of the corresponding Volterra equation arises when $K(x,y)$ is a Volterra kernel in eqn (1.50).

A constructive proof of the existence of a solution to a non-linear integral equation (1.51) may sometimes be obtained by considering the iteration

$$f_{n+1}(x) = \lambda \int_a^b F(x,y;\, f_n(y))dy + g(x) \ . \tag{1.52}$$

We assume at least the continuity of $F(x,y;v)$ in a suitable domain.

The iteration is clearly an extension of (1.30). The convergence of (1.30) was discussed using the assumption of a unique solution $f(x)$. We shall now indicate how the existence of a solution of eqn (1.51) may sometimes be established by considering the convergence of (1.52).

We shall write $\varepsilon_n(x) = f_{n+1}(x) - f_n(x)$, and we suppose that $(\partial/\partial v)\, F(x,y;v)$ exists. Then, from (1.52),

$$\varepsilon_n(x) = \lambda \int_a^b F(x,y;\, f_n(y))\mathrm{d}y - \lambda \int_a^b F(x,y;\, f_{n-1}(y))\mathrm{d}y$$

$$= \lambda \int_a^b \varepsilon_{n-1}(y)\, (\partial/\partial v)\, F(x,y;\, v_n(y))\mathrm{d}y\ ,$$

by a mean-value theorem. (Here, $v_n(y)$ depends on $f_n(y)$ and $f_{n-1}(y)$, but is unknown.) Now let us suppose that

$$|\lambda(b-a)|\left|(\partial/\partial v)\, F(x,y;\, v)\right| \le \Lambda < 1$$

for all possible values of x,y, and v. Then

$$\|\varepsilon_n(x)\|_\infty \le \Lambda\, \|\varepsilon_{n-1}(x)\|_\infty\ ,$$

and hence

$$\|\varepsilon_n(x)\|_\infty \le \Lambda^n\, \|\varepsilon_0(x)\|_\infty\ .$$

By the triangle inequality, if $n > m$,

$$\|f_n(x) - f_m(x)\|_\infty \le \sum_{r=m+1}^{n} \|f_r(x) - f_{r-1}(x)\|_\infty$$

$$= \sum_{r=m+1}^{n} \|\varepsilon_{r-1}(x)\|_\infty$$

$$\le \{\sum_{r=m+1}^{n} \Lambda^{r-1}\}\, \|\varepsilon_0(x)\|_\infty$$

by the result above. Thus

$$\|f_n(x) - f_m(x)\|_\infty \le \frac{\Lambda^m(1 - \Lambda^{n-m})}{1 - \Lambda} \|\varepsilon_0(x)\|_\infty .$$

Since $n > m$, $\|f_n(x) - f_m(x)\|_\infty \to 0$ as $m \to \infty$ and the sequence $\{f_n(x)\}$ is a Cauchy sequence in the uniform norm. Now if each function $f_n(x)$ is continuous, the completeness of $C[a,b]$ assures us that $\lim_{n\to\infty} f_n(x) = f(x)$, where $f(x)$ is continuous. We can then verify that this limit function $f(x)$ is indeed a solution of eqn (1.52). Some extensions of this type of analysis are possible. In particular, we can establish the uniqueness of a solution to eqn (1.51) under the conditions we have assumed For, if $f(x)$ and $\phi(x)$ are two solutions, we can show that

$$\|f(x) - \phi(x)\|_\infty \le \Lambda \|f(x) - \phi(x)\|_\infty ,$$

which implies that $\|f(x) - \phi(x)\|_\infty = 0$, or $f(x) \equiv \phi(x)$, if $\Lambda < 1$.

<u>Example 1.45</u>

The Volterra equation

$$f(x) - \int_0^x K(x,y;f(y))dy = g(x) \qquad (0 \le x \le 1)$$

has a unique continuous solution if $g(x)$ is continuous, $K(x,y;v)$ is uniformly continuous, and $|K(x,y;v_1) - K(x,y;v_2)| \le L|v_1 - v_2|$

for all v_1, v_2 and for $0 \le y \le x \le 1$. *

The iteration in eqn (1.52) may not yield a convergent sequence, even with mild restrictions on the integral equation. A more promising iteration is provided by applying 'Newton's method' to the integral equation (1.51). This method is not always successful, but, if $F(x,y;v)$ is sufficiently differentiable and $f_0(x)$ is sufficiently close to a solution of the equation, the successive iterates converge to this solution provided λ is not a characteristic value of the kernel

$$K(x,y) = \left[(\partial/\partial v)\ F(x,y;v)\right]_{v=f(y)} .$$

Newton's method for the integral equation (1.51) is defined by the equations

$$r_k(x) = f_k(x) - \lambda\int_a^b F(x,y;\ f_k(y))dy - g(x) , \quad (1.53)$$

$$\varepsilon_k(x) - \lambda\int_a^b K_k(x,y)\ \varepsilon_k(y)dy = r_k(x) , \quad (1.54)$$

where

$$K_k(x,y) = \left[(\partial/\partial v)\ F(x,y;v)\right]_{v=f_k(y)} , \quad (1.55)$$

and

$$f_{k+1}(x) = f_k(x)-\varepsilon_k(x) . \quad (1.56)$$

The iteration is defined by a choice of $f_0(x)$, provided λ is a regular value of $K_q(x,y)$ $(q = 0, 1, 2, \ldots)$. To motivate the method, we suppose that $(\partial^2/\partial v^2)\ F(x,y,v)$ is continuous for $a \le x, y \le b$, $|v| < \infty$, and suppose that the continuous function $\tilde{f}_0(x)$ is an approximate solution of (1.51). Unless $\tilde{f}_0(x)$ is an exact solution, the residual

$$r_0(x) = \tilde{f}_0(x)-\lambda\int_a^b F(x,y;\ \tilde{f}_0(y))dy - g(x)$$

is non-zero and the iteration continues.

Now suppose that $f(x)$ is an exact solution and

$$f_0(x) = f(x) + e_0(x) ,$$

where $\|e_0(x)\|_\infty$ is small. Then

$$\int_a^b F(x,y;\ f(y))dy$$
$$= \int_a^b F(x,y;\ f_0(y))dy - \int_a^b \frac{\partial}{\partial v} F(x,y;\ f_0(y))\ e_0(y)dy + O(\|e_0(x)\|_\infty^2) .$$

Thus

$$r_0(x) = f(x) + e_0(x) - \lambda\int_a^b F(x,y;\ f(y))dy -$$
$$- \lambda\int_a^b K_0(x,y)\ e_0(y)dy + O(\|e_0(x)\|_\infty^2) - g(x)$$
$$= e_0(x) - \lambda\int_a^b K_0(x,y)\ e_0(y)dy + O(\|e_0(x)\|_\infty^2) ,$$

where

$$K_0(x,y) = \frac{\partial}{\partial v} F(x,y;v)\Big|_{v=f_0(y)} .$$

If the order term is sufficiently small to be negligible, we may write

$$e_0(x) - \lambda\int_a^b K_0(x,y)\ e_0(y)dy \simeq r_0(x) .$$

If the equation

$$\varepsilon_0(x) - \lambda\int_a^b K_0(x,y)\ \varepsilon_0(y)dy = r_0(x)$$

has a (unique) solution, we can form the function $f_1(x) = f_0(x) - \varepsilon_0(x)$, and obtain a new approximation to the solution $f(x)$. We thus have a basis of an iterative method characterized by eqns (1.53) - (1.56).

As we remarked, $\varepsilon_k(x)$ $(k = 0, 1, 2, \ldots)$ is defined if λ is a regular value of $K_k(x,y)$. However, we seek conditions which ensure that the sequence $\{f_k(x)\}$ is uniformly convergent. Conditions for a very general application of Newton's method to operator equations can be applied to the present situation. We then obtain the following result.

THEOREM 1.5. *Suppose that* $f_0(x) \in C[a,b]$ *and* $\rho_0 > 0$ *can be found such that*

(i) *for* $\|\phi(x) - f_0(x)\|_\infty \leq \rho_0$, *the function*

$$(\partial/\partial v)^2 F(x,y;\ \phi(y)) \equiv (\partial/\partial v)^2\ F(x,y;v)|_{v = \phi(y)}$$

is continuous for $a \leq x,\ y \leq b$ *and*

$$|\lambda| \int_a^b |(\partial/\partial v)^2\ F(x,y;\ \phi(y))|dy \leq C < \infty \quad ;$$

and

(ii) $$\|(I - \lambda K_0)^{-1}\|_\infty \leq \beta_0 \quad ,$$

$$\|(I - \lambda K_0)^{-1} r_0(x)\|_\infty \leq \eta_0$$

(*where* K_0 *is the integral operator with kernel* $K_0(x,y)$ $= (\partial/\partial v)\ F(x,y;\ f_0(y))$) ; *and*

(iii) $\gamma_0 \equiv \beta_0 \eta_0 C \leq \frac{1}{2}$ *and*

$$\{1 - \surd(1 - 2\gamma_0)\}\eta_0/\gamma_0 < \rho_0 \ .$$

Then the sequence $f_k(x)$ *is defined and there is a solution of the equation* $f(x)$ *such that* $\|f(x) - f_0(x)\|_\infty \leq \rho_0$ *and* $\|f(x) - \tilde{f}_k(x)\|_\infty \to 0$ *as* $k \to \infty$.

Proof. See the chapter by Antosiewicz and Rheinboldt in Todd (1962, p.513). The result is usually attributed to Kantorovitch (see also Moore, in Anselone (1964)).

Newton's method for the integral equation (1.51) requires the repeated evaluation of $K_k(x,y)(k = 0, 1, 2, \ldots)$. Since it may be a tedious process to determine $(\partial/\partial v)\, F(x,y;v)$ and set $v = f_k(y)$, we may replace $K_q(x,y)$ by $K_0(x,y)$ in eqn (1.54) . The resulting method is known as the modified Newton method.

2

NUMERICAL ANALYSIS

In Chapter 2, the primary purpose is to survey the basic numerical analysis which we require in order to set up equations for an approximate solution of an integral equation. Until the material of section 2.20, little space is devoted to numerical methods for solving the equations for the approximate solution, for we are principally concerned with the problems of approximation and approximate integration which form the foundations of many methods for integral equations.

The material of section 2.2 is fairly elementary, but has been included as a prerequisite for the material of section 2.17, and we have also used it to indicate the basis for quadrature rules given in section 2.11. The other material on approximation of functions has been included because of its relevance (either practical or theoretical) to collocation and Rayleigh-Ritz-Galerkin methods for solving integral equations. The underlying techniques of sections 2.13.1. and 2.19, though applied here to numerical integration, can be used more extensively in constructing approximate solutions of integral equations.

In order to obtain error bounds or error estimates for our approximate solutions we often need an error analysis for numerical integration or the approximation of functions. I have been careful to give elementary results of this type, but have also endeavoured, for the sake of the mathematician, to show the nature of errors when the usual conditions do not hold. I have attempted to fulfil the latter task in a way which avoids too heavy an analysis, and for this reason I have employed Jackson's theorem in many places where Peano theory might be more appropriate. However, Peano theory for quadrature is quite well documented (though omitted from many texts); it is included in section 2.16. Sections which will principally be of use in developing the *theory* of numerical methods for integral equations are sections 2.3-2.10, 2.12.1, and 2.14-2.16.

Basic reading for a non-mathematical scientist is provided by the following sections: 2.1,2.2,2.4,2.5,2.10-2.13.2, and 2.17 onwards.

2.1. Introduction

As Fox and Mayers (1969) remark, the solution of a problem numerically involves (i) formulation of the problem in mathematical terms, (ii) choice of a method, (iii) programming (for a computer), and (iv) testing and implementing the method. Of course, these items are sometimes interrelated.

We suppose that the problems which we are to discuss have been formulated as integral equations (we shall not, for the most part, discuss whether this is the most convenient formulation). However, there are essentially two types of problem: those which arise mathematically and those which arise from physical problems. In the latter case it is quite likely that the known functions in an equation will have been obtained from physical experiments. To fix our ideas, let us suppose that the kernel of a linear integral equation is given as a table of function values obtained experimentally. From the point of view of the purist, the kernel is not completely defined. Such a pedantic argument cannot obscure the fact that a problem has been posed and one now seeks a 'solution'. The theory of the subject assumes some importance here. We must ask whether the problem is well posed: is the 'solution' sensitive to changes in the kernel which are consistent with the tabulated values; and do slight changes in the tabulated values affect the 'solution' disproportionately? Our error analysis for numerical methods is often closely related to such questions.

Numerical methods seem entirely natural for the solution of a physical problem, because we can choose a numerical method which relies on the discrete data. In the case of an integral equation which is posed as a mathematical problem there is, again, a plausible argument for considering a numerical method of solution. For an integral equation is defined in terms of a mathematical notion of limiting processes. The theoretical methods of approach, indicated in Chapter 1, all depend at some point upon an infinite process such as the determination of an integral or the summation of an infinite series, and in practice we can only implement finite processes.

There are two approaches to the solution of a problem. We may adopt a numerical method from the beginning, or we may pursue the theoretical solution to a point where difficulties arise and then resort to a numerical method to complete the solution. In many cases, the first approach is preferable. Nevertheless, a difficult problem requires a considerable amount of analysis before we know which numerical methods will be suitable (or have been successful). For this reason, I do not apologize for including, where appropriate, some theoretical numerical analysis which enables us to assess the reliability of our numerical methods.

This much has been preamble. The main purpose of this chapter is to introduce, and in some cases to analyse, some of the basic numerical techniques which will be of use in the numerical solution of integral equations. The discussion will be presented with this particular application in mind.

First, we consider various types of approximations to functions, and then we consider the problem of numerical integration. Finally, in section 2.20, we mention some basic numerical techniques for solving, for example, systems of linear or non-linear equations or eigenvalue problems.

2.2. *Interpolation*

When a function has been obtained as a set of tabulated values, we may seek the function at a value of the argument which does not correspond to an entry in the table. This is a problem of *interpolation* (in which we include '*extrapolation*').

Suppose that $\{\phi_0(x), \phi_1(x), \ldots, \phi_n(x)\}$ is a system of $(n + 1)$ prescribed '*basic coordinate functions*' or '*basic functions*' defined for $a \le x \le b$. We may seek an approximation $\phi(x)$ of the form

$$\phi(x) \simeq (A\phi)(x) = \sum_{i=0}^{n} a_i \, \phi_i(x) . \tag{2.1}$$

The values of $a_0, a_1, \ldots, a_n$ may be computed using the tabulated values of $\phi(x)$ in some prescribed manner, and (2.1) may then be used to supply approximate values of $\phi(x)$.

We suppose that the function $\phi(x)$ is tabulated at the distinct *sample points* $x_0, x_1, \ldots, x_n$ in the interval $[a,b]$. We assume that $\det(\phi_i(x_j)) \neq 0$; if this is so for any possible choice of $\{x_j\}$, the set $\{\phi_i(x)\}$ is a *Haar system* on $[a,b]$. When $\det(\phi_i(x_j))$ is non-zero, there are unique values of $a_0, a_1, \ldots, a_n$ in eqn (2.1) such that

$$\phi(x_j) = \sum_{i=0}^{n} a_i\, \phi_i(x_j) \qquad (j = 0, 1, 2, \ldots, n),$$

and the function $(A\phi)(x)$ in (2.1) is then an *interpolant*.

Perhaps the commonest choice of basic functions $\{\phi_i(x)\}$ arises in *polynomial interpolation*, where $\phi_r(x)$ is a polynomial of degree r in x, and the interpolant is a polynomial of degree at most n. Most introductory books on numerical analysis discuss this case.

An explicit form for the polynomial of degree n interpolating $\phi(x)$ at the distinct points $\{x_i\}$ is given by the *Lagrangian form*. We denote this interpolant by $P_{0,1,2,\ldots,n}(x)$, and we have

$$P_{0,1,\ldots,n}(x) = \sum_{j=0}^{n} l_j(x)\, \phi(x_j), \tag{2.2a}$$

where

$$l_j(x) = \prod_{\substack{i=0 \\ i \neq j}}^{n} (x - x_i)/(x_j - x_i). \tag{2.2b}$$

In order to determine the difference between $\phi(x)$ and the approximation $(A\phi)(x) = P_{0,1,2,\ldots,n}(x)$ we require further information about $\phi(x)$, and introductory texts usually suppose that the $(n+1)$th derivative of $\phi(x)$ exists.

Example 2.1 (*polynomial interpolation*)

(a) *Constant interpolation.* $n = 0$, $x_0 = a$, $P_0(x) = \phi(a)$

$$(a \leq x \leq b),$$

$$\phi(x) - P_0(x) = (x - a)\phi'(\xi_x), \quad a < \xi_x < x,$$

if $\phi'(x)$ is continuous. Suffixes as in ξ_x will be supressed in (b)-(d)

(b) *Constant interpolation.* $n = 0$, $x_0 = \frac{1}{2}(a+b)$, $P_0(x) = \phi(\frac{1}{2}(a+b))$

$$(a \le x \le b),$$

$$\phi(x) - P_0(x) = (x - \tfrac{1}{2}(a+b))\phi'(\zeta), \quad a < \zeta < b,$$

if $\phi'(x)$ is continuous. ζ depends upon x.

(c) *Linear interpolation.* $n = 1$, $x_0 = a$, $x_1 = b$,

$$P_{0,1}(x) = \{(x-a)\phi(b) - (x-b)\phi(a)\}/(b-a) \quad (a \le x \le b),$$

$$\phi(x) - P_{0,1}(x) = \tfrac{1}{2}(x-a)(x-b)\,\phi''(\eta) \qquad (a < \eta < b)$$

if $\phi''(x)$ is continuous. η depends upon x.

(d) *Quadratic interpolation.* $n = 2$, $x_0 = a$, $x_1 = c = \frac{1}{2}(a+b)$, $x_2 = b$,

$$P_{0,1,2}(x) = \frac{(x-b)(x-c)}{(a-b)(a-c)}\phi(a) + \frac{(x-c)(x-a)}{(b-c)(b-a)}\phi(b) + \frac{(x-a)(x-b)}{(c-a)(c-b)}\phi(c) \qquad (a \le x \le b)$$

$$\phi(x) - P_{0,1,2}(x) = \frac{1}{6}(x-a)(x-b)(x-c)\,\phi'''(\nu) \quad (a < \nu < b)$$

if $\phi'''(x)$ is continuous. ν depends upon x. *

In the case of polynomial interpolation, the sample points x_i may be allowed to coincide if the function $\phi(x)$ is sufficiently differentiable and if we re-interpret the meaning of interpolation. If any point x_i is repeated, and occurs r_i times in the list of sample points, we will require that the sth derivative of the interpolant assumes the value $\phi^{(s)}(x_i)$ at $x = x_i$ for $s = 0,1,2,\ldots,(r_i - 1)$.

The expression (2.2) is indeterminate if some x_i coincide and in this case a new form must be found for $P_{0,1,2,\ldots,n}(x)$. When some of the sample points x_i coincide under the conditions above, $P_{0,1,2,\ldots,n}(x)$ will be called an *osculating* polynomial.

Example 2.2 (*osculating polynomials*)

(a) *Truncated Taylor series.* $x_0 = x_1 = x_2 = \ldots = x_n = a$.

$$P_{0,\ldots,0}(x) = \phi(a) + (x-a)\phi'(a) + \ldots + \frac{(x-a)^n}{n!}\phi^{(n)}(a);$$

$$\phi(x) - P_{0,\ldots,0}(x) = \frac{1}{n!}\int_a^x (x-t)^n \phi^{(n+1)}(t)\,dt = \frac{(x-a)^{n+1}}{(n+1)!}\phi^{(n+1)}(\xi)$$

for $a < \xi < x$, if $\phi(x) \in C^{n+1}[a,b]$ and $x \in [a,b]$.(ξ depends on x.)

(b) *Two-point Taylor interpolation.*

$n = 2m - 1$; $x_0 = x_1 = \ldots = x_{m-1} = a$, $x_m = x_{m+1} = \ldots = x_{2m-1} = b$.

Then

$$\phi(x) - P_{0,\ldots,0,n,\ldots,n}(x) = \frac{(x-a)^m(x-b)^m}{(2m)!}\phi^{(2m)}(\eta) \qquad (a < \eta < b)$$

if $\phi^{(2m)}(x)$ is continuous (see Davis 1965, p.37); η depends on x. *

Rather than approximate $\phi(x)$ by a single interpolating polynomial over an interval $[a,b]$, we sometimes prefer to subdivide the interval with a *partition*, say,

$$a = a_0 < a_1 < a_2 < \ldots < a_N = b ,$$

and interpolate $\phi(x)$ separately in each subinterval $[a_i, a_{i+1}]$ with an appropriate polynomial. We call this process *piecewise polynomial interpolation*. The situation is simplest when $a_i = a + ih$, where $h = (b - a)/N$. We shall again denote the resulting approximation by $(A\phi)(x)$.

Example 2.3 (*piecewise-polynomial interpolation*)

We suppose that $a_i = a_0 + ih$.

(a) *Piecewise-constant interpolation* (see Example 2.1).

$$\phi(x) \simeq (A\phi)(x) = \phi(a_i), \quad \text{for} \quad a_i \le x < a_{i+1} \qquad (\text{if } i \ne N-1)$$
$$= \phi(a_{N-1}), \text{for} \quad a_{N-1} \le x \le a_N .$$

Here $|\phi(x) - (A\phi)(x)| \le h \max_{a\le\xi\le b} |\phi'(\xi)|$ if $\phi'(x)$ is continous.

(b) *Piecewise-linear interpolation* (see Example 2.1).

$$\phi(x) \simeq (A\phi)(x) = \frac{x - a_i}{h}\phi(a_{i+1}) - \frac{x - a_{i+1}}{h}\phi(a_i), (a_i \le x \le a_{i+1}).$$

Here $|\phi(x) - (A\phi)(x)| \le \frac{h^2}{2} \max_{a\le\eta\le b} |\phi''(\eta)|$ if $\phi''(x)$ is continuous.

(c) If we take $(A\phi)(x)$ to be a piecewise two-point Taylor interpolating polynomial (see Example 2.2(b)) of degree $n = 2m - 1$, our approximation $(A\phi)(x)$ has a continuous $(m-1)$th derivative. If $m = 2$, we obtain a function which is piecewise-cubic and which has a continuous first derivative. *

In Example 2.3(c) we indicated a method of constructing a 'fairly smooth' approximation which was a piecewise polynomial. This description also applies to a *polynomial spline*, which is defined as follows. A (polynomial) spline of degree n is a function which has continuous $(n-1)$th derivatives and which is a piecewise polynomial of degree n. Thus a spline is a polynomial of degree n on certain subintervals, and the end-points of these subintervals are *knots* where the nth derivative may have a jump discontinuity. We have encountered particular examples of splines of degree n, where $n = 0$, $n = 1$, in Example 2.3(a) and (b) respectively. A method of interpolating using *cubic splines* is discussed by Rivlin (1969).

Example 2.4

Any spline of degree n with $(N+1)$ knots x_k $(k = 0,1,2,\ldots,N)$ may be written in the form

$$\sum_{k=0}^{n} \alpha_k x^k + \sum_{k=0}^{N} \beta_k (x - x_k)_+^n \quad , \tag{2.3}$$

where $z_+ = \max(z, 0)$, that is $z_+ = z$ if $z \geq 0$, 0 if $z < 0$. (Note that $z_+^0 = 1$ only if $z \geq 0$.) The form (2.3) of the spline is not always suitable for *computational* purposes (see Rice 1969, §10-7). Cox (1972) and de Boor (1972) discuss the use of alternative representations. *

A convenient representation of the spline $s(x)$ of degree n with knots $x_0 < x_1 < \ldots < x_N$ is in terms of *B-splines* or *fundamental splines* $m_{ni}(x)$ (Cox, 1972, uses a related notation) such that

$$s(x) = \sum_{i=0}^{N+n+1} c_i m_{ni}(x) \; .$$

Each B-spline $m_{ni}(x)$ is zero if $x \notin (x_{i-n-1}, x_i)$, and $m_{ni}(x)$ can be expressed as a divided difference in y of $(y - x)_+^n$ (as a function of y) or explicitly as

$$m_{ni}(x) = \sum_{r=i-n-1}^{i} \frac{(x_r - x)_+^n}{\pi'_{ni}(x_r)} \qquad (x_{-1} < x < x_{N+1}) ,$$

where

$$\pi_{ni}(x) = \prod_{r=i-n-1}^{i} (x - x_r) ,$$

and where the knots are supplemented by points

$$x_{-n-1} < x_{-n} < \ldots < x_{-1} < x_o \ ; \ x_{N+n+1} > x_{N+n} \ \ldots > x_N .$$

A stable method of computing $m_{ni}(x)$ is given by Cox (1972) and de Boor (1972), but for low values of n the B-splines can be written down in simple terms.

The B-splines have been used in collocation methods for integral equations by Phillips (1969), and for expansion methods generally they have the advantage that they are non-zero only in intervals of short length.

2.3. *Error in polynomial interpolation*

It is convenient, at this stage, to discuss the errors in polynomial interpolation and piecewise polynomial interpolation.

Suppose that $\phi(x)$ has a continuous $(n+1)$th derivative and $P_{0,1,2,\ldots,n}(x)$ is the polynomial of degree at most n which agrees with $\phi(x)$ at $x_0, x_1, \ldots, x_n$ in the interval $[a,b]$. Then, if $a \le x \le b$,

$$\phi(x) - P_{0,1,\ldots,n}(x) = \prod_{i=0}^{n} (x - x_i) \frac{\phi^{(n+1)}(\xi)}{(n+1)!}$$

for some value of ξ which depends on x and lies in (a,b), and where each x_i is counted according to its multiplicity. This result is a consequence of Rolle's theorem and its proof may be found in most introductory texts on numerical analysis. A bound on the error, for example,†

† We recall that $||\psi(x)||_\infty = \sup_{a\le x\le b} |\psi(x)|$.

$$\|f(x) - P_{0,\ldots,n}(x)\|_\infty \le \|\prod_{i=0}^{n}(x - x_i)\|_\infty \frac{\|\phi^{(n+1)}(x)\|_\infty}{(n+1)!} ,$$

is obtained when $\|\phi^{(n+1)}(x)\|_\infty$ is known. However, it is convenient to have a bound involving a lower-order derivative for the case where $\phi^{(n+1)}(x)$ does not exist or is not easily estimated. A good tool for such an analysis is provided by Peano's theorem (Davis 1965; Baker 1970), but when we use it the analysis becomes a little complicated. A simpler theory is provided in the case of distinct sample points $\{x_i\}$ by using *Jackson's theorem* (Cheney 1966; Rivlin 1969), which is stated below.

2.3.1.

We suppose $\phi(x) \in C[a,b]$ and we write

$$E_n(\phi) = \inf_{p_n} \{\sup_{a\le x\le b} |\phi(x) - p_n(x)|\} = \inf_{p_n} \|\phi(x) - p_n(x)\|_\infty ,$$

where $p_n(x)$ denotes some polynomial of degree n, and the infimum is taken over all polynomials of this degree. Associated with $\phi(x)$ is its *modulus of continuity* $\omega(\phi;\delta) = \sup\{ |\phi(x) - \phi(y)| : a\le x, y\le b, |x-y|\le\delta \}$.

THEOREM 2.1. *Jackson's theorem for polynomial approximation states that:*

(i) $E_n(\phi) \le \frac{1}{4}\pi(b-a)L/(n+1)$ *if* $|\phi(x_1) - \phi(x_2)| \le L|x_1 - x_2|$ *for all* x_1, x_2 *in* $[a,b]$. *(We say that* $\phi(x)$ *is 'Lipschitz-continuous' when it satisfies the preceding condition.) If* $\omega(\phi;\delta)$ *is the modulus of continuity of* $\phi(x)$, *then* $E_n(\phi) \le \omega(\phi; \frac{1}{2}(b-a)\pi/(n+1))$.

Furthermore (Cheney 1966, p.148):

(ii) $E_n(\phi) \le \{\frac{1}{4}\pi(b-a)\}^k \dfrac{\|\phi^{(k)}(x)\|_\infty}{(n+1)n\ldots(n-k+2)}$ *if* $\phi^{(k)}(x)$ *is bounded on* $[a,b]$ *and* $n \ge k$.

We also state, for its interest, the following version of Jackson's theorem. We denote by E_ρ the ellipse with foci (-1, 0) and (1, 0), with major semi-axis a and minor semi-axis $b = (a^2 - 1)$ and $\rho = a + b$. If $\phi(z)$ is analytic, with $|\mathrm{Re}\,(\phi(z))| \le M_0$ in the interior of E_ρ, and is real along the major axis (on the real line), then

$$E_n(\phi) \le \frac{8}{\pi\rho^{n+1}} M_0 .$$

If we employ a polynomial $p_n^*(x)$ of degree n for which

$$E_n(\phi) = ||\phi(x) - p_n{}^*(x)||_\infty$$

we may write, in the notation of (2.2),

$$\begin{aligned}\|\phi(x) - P_{0,1,2,\ldots,n}(x)||_\infty &= \|\phi(x) - \sum_{j=0}^{n} l_j(x)\, \phi(x_j)||_\infty \\ &\le \|\phi(x) - p_n{}^*(x)||_\infty + \|\sum_{j=0}^{n} l_j(x)\{p_n{}^*(x_j) - \\ &\qquad -\phi(x_j)\}||_\infty \\ &\le \{1 + ||\sum_{j=0}^{n}\left|l_j(x)\right|||_\infty\}\, E_n(\phi) \quad ,\end{aligned}$$

since

$$p_n{}^*(x) = \sum_{j=0}^{n} l_j(x)\, p_n{}^*(x_j)$$

when $x_0, x_1, \ldots, x_n$ are distinct. The number

$$L_n = ||\sum_{j=0}^{n}\left|l_j(x)\right|||_\infty$$

is known as the *Lebesgue constant for polynomial interpolation.* The size of L_n depends on n and the distribution of the points $\{x_i\}$, and on a and b since $||\phi(x)||_\infty = \sup_{a\le x\le b}|\phi(x)|$.

<u>Remark</u>. It may be shown (Rivlin 1969; Powell 1967) that when $a = -1$, $b = 1$ and $x_i = \cos\{(2i + 1)\pi/(2n + 2)\}$ $(i = 0, 1, \ldots, n)$, $L_n \le 2\pi \ln n + 4$. (Actually, this result is pessimistic: for $n \le 1000$, $L_n < 5.4$.) Now, $n^{-1} \ln n \to 0$ as $n \to \infty$. Thus with the preceding choice of $\{x_i\}$ and $[a,b]$, $\lim||\phi(x) - P_{0,1,2,\ldots,n}(x)||_\infty = 0$ if $\phi(x)$ is Lipschitz-continuous (with a Lipschitz constant L, see (i) above). These sample points $\{x_i\}$ are the zeros of $T_{n+1}(x)$, the *Chebyshev polynomial* of the first kind of degree $n + 1$, which we may define by the relation $T_{n+1}(x) = \cos\{(n + 1)\cos^{-1}x\}$ if $|x| \le 1$ (see also eqn (2.6) below). When $a = 0$, $b = 1$ we obtain a similar result

for interpolation at the zeros of the *shifted* Chebyshev polynomial $T_{n+1}^*(x) = T_{n+1}(2x - 1)$, since the values of L_n in each case are the same.

For a general distribution of sample points $\{x_i\}$ and an arbitrary continuous function $\phi(x)$ it is *not* true that

$$\lim_{n\to\infty} ||\phi(x) - P_{0,1,2,\ldots,n}(x)||_\infty = 0$$

(Natanson 1965; Cheney 1966). On the other hand, there are positive results about the convergence in the mean of $P_{0,1,2,\ldots,n}(x)$ to $\phi(x)$; see Natanson (1965, p.56). In particular we note in passing that if $a = -1$, $b = 1$, and $x_0, \ldots, x_n$ are the zeros of the *Legendre polynomial* $P_{n+1}(x)$ (see Example 2.6), then

$$\lim_{n\to\infty} ||\phi(x) - P_{0,1,2,\ldots,n}(x)||_2 = 0 .$$

The results we have stated may be used in a study of piecewise-polynomial interpolation. If we employ piecewise-constant or piecewise-linear interpolation using subintervals of $[a,b]$ of equal length, then the approximation converges uniformly to $\phi(x)$ on $[a,b]$ as the width of the subintervals decreases to zero (provided that $\phi(x) \in C[a,b]$). The same results holds for piecewise-polynomial interpolation of a given degree n, using unequal subintervals, as the maximum width of the subintervals tends to zero.

2.4. *Least-squares fitting*

Given the basic functions $\phi_0(x), \ldots, \phi_n(x)$ we may seek an approximation to $\phi(x)$ of the form

$$\sum_{i=0}^{n} a_i \phi_i(x) .$$

In our discussion of interpolation in section 2.2, we attempt to choose coefficients $\{a_i\}$ such that the approximation agrees with $\phi(x)$ at $(N + 1)$ distinct sample points $x_0, \ldots, x_N$, where $N = n$. If $n < N$ the equations

$$\sum_{j=0}^{n} a_j \phi_j(x_i) = \phi(x_i) \qquad (i = 0,1,2, \ldots, N)$$

can be inconsistent, and we may (therefore) seek a least-squares fit by choosing the coefficients such that

$$\sum_{j=0}^{n} a_j \phi_j(x_i) - \phi(x_i) = \varepsilon_i (i = 0,1,2, \ldots, N)$$

where

$$\sum_{j=0}^{N} |\varepsilon_j|^2 \text{ is minimized.} \tag{2.4}$$

If we write $\underset{\sim}{a} = [a_0, \ldots, a_n]^T$, $\underset{\sim}{\phi} = [\phi(x_0), \ldots, \phi(x_n)]^T$, and $\underset{\sim}{A} = [\phi_j(x_i)]$, (2.4) assumes the standard form for a linear least-squares problem, namely,

$$\text{minimise} \quad \|\underset{\sim}{A}\,\underset{\sim}{a} - \underset{\sim}{\phi}\|_2 , \tag{2.5}$$

where $\|\underset{\sim}{v}\|_2 = (\underset{\sim}{v}^*\underset{\sim}{v})^{\frac{1}{2}} = \{\sum_{i=0}^{n} |v_i|^2\}^{\frac{1}{2}}$. This problem may not have a unique solution. † Some care has to be exercised in solving the equations arising from any least-squares problem, and the reader is advised to refer (for example) to Golub and Reinsch (1970) (see section 2.20). If the functions $\phi_r(x)$ in the above discussion are polynomials of degree r , the resulting approximation is a polynomial of degree n . There is then a special choice of the sample points which considerably simplifies the construction of a least-squares fit. We discuss this case below.

Let us first note that any polynomial of degree n may be written in the form $\sum_{r=0}^{n} a_r T_r(x)$, where $T_r(x)$ is the *Chebyshev polynomial* (of the first kind) of degree r . A convenient algorithm for evaluating

---

† The least-squares problem (2.4) does have a unique solution if the set $\{\phi_i(x)\}$ is a Haar system on an interval containing $x_0, x_1, \ldots, x_N$.

$\sum_{r=0}^{n} a_r T_r(x)$ can be constructed from the relations

$$T_0(x) = 1 \quad ,$$

$$T_1(x) = x \quad , \tag{2.6}$$

$$T_{r+1}(x) = 2x\, T_r(x) - T_{r-1}(x) \quad (r \geq 1) \quad ,$$

which serve to define $T_r(x)$. The shifted polynomials $T_r^*(x)$ satisfy a different, but similar, set of relations.

Now, suppose $x_j = \cos\{(2j + 1)\pi/(2N + 2)\}$ $(j = 0,1,2, \ldots, N)$. Then, relative to these points, the least-squares polynomial fit of degree n may be written in the form

$$\sum_{r=0}^{n} a_r T_r(x) \quad ,$$

where

$$a_r = \{2v_r/(N + 1)\} \sum_{j=0}^{N} \phi(x_j)\, T_r(x_j) \quad ,$$

with $v_0 = \frac{1}{2}$, $v_r = 1$ $(r \geq 1)$. We refer the reader to Baker and Radcliffe (1970) and to Fox and Parker (1968, p.30) for further details.

In this section and the preceding one we have given formulae for approximating $\phi(x)$ when its values, and possibly those of its derivatives, are known at certain discrete points. We give another approximation of this type in section 2.9, but in sections 2.5 and 2.7 we discuss approximations which are suitable for functions which are defined mathematically and which depend on the behaviour of $\phi(x)$ over the whole range $[a,b]$.

2.5. *Fourier series*

The least-squares polynomial fit discussed at the end of section 2.4 is related to a generalization (section 2.7 below) of the Fourier series.

The nth partial sum of the Fourier series of a function $\phi(x)$ defined for $-\pi \leq x \leq \pi$ has the form

$$(A\phi)(x) = \tfrac{1}{2}\, a_0 + \sum_{k=1}^{n} (a_k \cos kx + b_k \sin kx) \,, \tag{2.7}$$

where

$$a_k = \{1/\pi\}\int_{-\pi}^{\pi} \phi(t) \cos kt \; dt \qquad (k \geq 0) \tag{2.8a}$$

and

$$b_k = \{1/\pi\}\int_{-\pi}^{\pi} \phi(t) \sin kt \; dt \qquad (k \geq 1) \tag{2.8b}$$

are the *Fourier coefficients*. The Fourier series has the form (2.7) with n replaced by ∞.

2.5.1.
If we define $\phi_0(x) = 1$, $\phi_1(x) = \sin x$, $\phi_2(x) = \cos x$, and in general $\phi_r(x) = \sin\{\tfrac{1}{2}(r + 1)x\}$ if r is odd, and $\phi_r(x) = \cos(\tfrac{1}{2}rx)$ if r is even, we may write the Fourier sum (2.7) in the form

$$(A\phi)(x) = \sum_{r=0}^{n} \frac{(\phi,\phi_r)}{(\phi_r,\phi_r)}\, \phi_r(x) \tag{2.9}$$

where, relative to $[-\pi,\pi]$,

$$(\psi,\phi_r) = \int_{-\pi}^{\pi} \psi(t)\, \overline{\phi_r(t)} \; dt$$

is the usual innerproduct. The sines and cosines are orthogonal, so that $(\phi_r,\phi_s) = 0$ if $r \neq s$. The Fourier coefficients are precisely those coefficients $\{a_k\},\{b_k\}$ which minimize

$$\|\phi(x) - \{\frac{a_0}{2} + \sum_{k=1}^{n} (a_k \cos kx + b_k \sin kx)\}\|_2 \,,$$

where $\|\psi(x)\|_2 = (\psi,\psi)^{\frac{1}{2}} = \{\int_{-\pi}^{\pi} |\psi(x)|^2 \, dx\}^{\frac{1}{2}}$.

2.5.2
The substitution $x = \cos\theta$ (for $\theta \in [0,\pi]$) in a function $\psi(x)$ defined for $-1 \leq x \leq 1$ yields a function $\phi(\theta) = \psi(\cos\theta)$, $\theta \in [0,\pi]$. We

define $\phi(-\theta) = \phi(\theta)$ and obtain a continuous even function for $\theta \in [-\pi, \pi]$. Since $\phi(\theta)$ is even, the coefficients (2.8b) are zero, and the sum (2.7) approximating $\phi(\theta)$ is

$$\tfrac{1}{2}a_0 + \sum_{k=1}^{n} a_k \cos k\theta ,$$

where a_k is a Fourier coefficient for $\phi(\theta)$. If we now write $\theta =$ arc cos (x), the approximation of $\phi(\theta)$ by its partial Fourier sum assumes the form

$$\psi(x) \simeq \tfrac{1}{2}a_0 + \sum_{k=1}^{n} a_k T_k(x) , \tag{2.10}$$

where $T_k(x)$ is $\cos(k \cos^{-1} x)$. If we rewrite a_k (defined by eqn (2.8a) in terms of $\phi(\theta)$) in terms of the function $\psi(x)$, we obtain

$$a_k = \frac{2}{\pi}\int_{-1}^{1} \frac{1}{\sqrt{(1-t^2)}}\, \psi(t)\, T_k(t)\, dt . \tag{2.11}$$

The sum (2.10) is known as the nth Fourier-Chebyshev sum (Davis 1965; Fox and Parker 1968).

Remark. Whereas we have indicated the connection of (2.10) with (2.7), it is possible to obtain (2.10) directly by the techniques of section 2.7.

2.6. *Error bounds*

If $\phi(x) = \phi(x + 2\pi)$ for all x and $\phi(x)$ is r-times continuously differentiable we can bound the error in the Fourier sum (2.7). If $n > r$

$$\|\phi(x) - (A\phi)(x)\|_\infty \leq \frac{16}{\pi}\left(1 + \frac{1}{\pi^2}\ln n\right) \|\phi^{(r)}(x)\|_\infty / n^r .$$

The analysis for the corresponding result for eqn (2.10) is given by Rivlin (1969, p.61), who obtains the following result: if $(A_n\phi)(x)$ is the nth Fourier-Chebyshev sum (2.10) and $\phi^{(r)}(x)$ is bounded, then

$$\|\phi(x) - (A_n\phi)(x)\|_\infty \leq (1 + \Lambda_n)E_n(\phi)$$

where $\Lambda_n \leq 3 + (\frac{4}{\pi^2}) \ln n$ and, by Jackson's theorem (if $n \geq k$)

$$E_n(\phi) \leq \{\tfrac{1}{4}\pi(b - a)\}^k \frac{(n-k+1)!}{(n + 1)!} \|\phi^{(k)}(x)\|_\infty .$$

(We gave this bound for $E_n(\phi)$ in section 2.3.1.). Λ_n is known as the *Lebesgue constant for Fourier-Chebyshev approximation*. For $n \leq 1000$, $\Lambda_n \leq 4.1$. We can also apply Peano theory to the analysis of the error in the Fourier-Chebyshev sum (Baker and Radcliffe 1970).

2.7. *Generalized Fourier series*

The Fourier series and the Fourier-Chebyshev series may be regarded as particular cases of a generalization which we will now describe.

Suppose that we have an inner product of the form

$$\langle\phi,\psi\rangle = \int_a^b \omega(t)\phi(t)\overline{\psi(t)}dt ,$$

where $\omega(t) \geq 0$ (eqn (1.18)), and suppose that we have a set of basic functions $\{\phi_i(x)\}$ such that $\langle\phi_i,\phi_j\rangle = 0$ if $i \neq j$. Then for all possible choices of $\alpha_0, \ldots, \alpha_n$,

$$\|\phi(x) - \sum_{i=0}^{n} \alpha_i\phi_i(x)\|_\omega \geq \|\phi(x) - \sum_{i=0}^{n} \frac{\langle\phi,\phi_i\rangle}{\langle\phi_i,\phi_i\rangle} \phi_i(x)\|_\omega$$

where $\|\psi(x)\|_\omega = \{\langle\psi,\psi\rangle\}^{\frac{1}{2}}$. In other words, the nth partial sum of the *generalized Fourier series*

$$(A_n\phi)(x) = \sum_{r=0}^{n} \frac{\langle\phi,\phi_r\rangle}{\langle\phi_r,\phi_r\rangle} \phi_r(x) \qquad (2.12)$$

minimizes the norm of the error, $\|\phi(x) - (A_n\phi)(x)\|_\omega$, where the norm is the one described above.

Example 2.5 *(the Fourier-Chebyshev series)*
With $\phi_r(x) = T_r(x)$ in (2.6), $[a,b] = [-1,1]$, and $\omega(x) = (1 - x^2)^{-\frac{1}{2}}$,

the series (2.12) becomes the Fourier-Chebyshev series (2.10), in which the coefficients are defined by (2.11). *

Example 2.6 *(the Fourier-Legendre series)*
The *Legendre polynomial* of degree n, denoted by $P_n(x)$, may be defined mathematically by the expression $(-1)^n(d/dx)^n\{(1-x^2)^n\}/2^n n!$ For

$$\langle\psi, \phi_r\rangle = \int_{-1}^{1} \psi(t)\, \overline{\phi_r(t)}\, dt$$

(the usual inner product), and $\phi_r(x) = P_r(x)$, the series (2.12) becomes the *Fourier-Legendre* series. Note that

$$\int_{-1}^{1} |P_r(t)|^2\, dt = 2/(2r+1)\ .$$

There are a number of mathematically equivalent definitions of $P_n(x)$ (Davis 1965, p.365). *

2.8. *Closure and completeness*

In sections 2.5.1 and 2.7 we have used different types of inner product on sets of functions. Our general situation is covered if we consider a space X of functions defined on $[a,b]$ with an inner product $\langle\phi,\psi\rangle$, which, for definiteness, the reader may suppose to be of the form

$$\int_{a}^{b} \omega(t)\, \phi(t)\, \overline{\psi(t)}\, dt\ .$$

The space X to be considered will be (the completion of) the set of functions $\phi(x)$ for which $\langle\phi,\phi\rangle$ is finite, and the norm is

$$\|\phi(x)\|_\omega = \langle\phi,\phi\rangle^{\frac{1}{2}}\ .$$

A sequence $\{\phi_i(x)\}$ of linearly independent functions in X is, by an extension of the definition of section 1.9, complete in X if

$$<\phi,\phi_i> = 0 \quad (i = 0, 1, 2, \ldots \quad)$$

implies that $\phi(x) = 0$. By a general theorem (Davis 1965), such a sequence is closed in X , which suggests that it is appropriate to approximate a given function $\phi(x) \in X$ by a linear combination

$$\sum_{i=0}^{n} a_i \, \phi_i(x) \, .$$

The approximation of this form which minimizes

$$\|\phi(x) - \sum_{i=0}^{n} a_i \, \phi_i(x)\|_\omega$$

is unique. In the case where $<\phi_i, \phi_j> = 0$ if $i \neq j$, this best approximation is given by the generalized Fourier series with $a_i = <\phi, \phi_i>/<\phi_i,\phi_i>$.

Suppose that $<\phi_i,\phi_j> = 0$ when $i \neq j$. If we accept the minimizing property

$$\|\phi(x) - \sum_{i=0}^{n} \frac{<\phi,\phi_i>}{<\phi_i,\phi_i>} \phi_i(x)\|_\omega \leq \|\phi(x) - \sum_{i=0}^{n} a_i \, \phi_i(x)\|_\omega \, ,$$

it is clear that when $\{\phi_i(x)\}$ is closed in X , then the generalized Fourier series of a function $\phi(x)$ in X converges to $\phi(x)$ in the norm as $n \to \infty$. We know that this need not imply uniform convergence of the generalized Fourier series to $\phi(x)$ (for example, least-square convergence does not imply uniform convergence). For this reason we have used Jackson's theorem to discuss the uniform convergence of the Fourier-Chebyshev series. There are, however, some inner products $<\phi,\psi>$ for which we obtain guaranteed uniform convergence. For example, if all our functions are sufficiently differentiable and, with $m \geq 1$,

$$<\phi,\psi>_m = \int_a^b \{\sum_{j=0}^{m} \phi^{(j)}(t) \, \psi^{(j)}(t)\} dt$$

and

$$|||\psi(x)|||_m^2 = <\psi,\psi>_m \, ,$$

there is a constant γ such that $\|\psi(x)\|_\infty \leq \gamma \|\psi(x)\|_m$ (see Varga (1970); though this will not be pursued here, a similar result will be employed in Chapter 5). Applying the latter result where $\psi(x)$ is the error in the nth partial sum can be used to establish uniform convergence.

Example 2.7

The powers of x, $\{1, x, x^2, \ldots\}$, are closed in $C[a,b]$ in $\| \ \|_2$, the integral mean-square norm. (They are also closed in the uniform norm, which is not an 'inner-product norm'.) *

Example 2.8

The functions 1, $\cos \pi x$, $\sin \pi x$, $\cos 2\pi x$, $\sin 2\pi x$, ..., $\cos r\pi x$, $\sin \pi r x$, ... are closed in $C[0,2]$ in the integral mean-square norm. *

2.9. *Discrete weighted least-squares approximations*

We now discuss a further generalization of the Fourier series. Let us define

$$\langle\phi,\psi\rangle = \sum_{i=0}^{N} \omega_i \, \phi(x_i) \, \overline{\psi(x_i)} ,$$

where $\omega_i > 0$ and $x_i (i = 0, 1, \ldots, N)$ are given. Since $\langle\phi,\phi\rangle^{\frac{1}{2}}$ is not a norm on $C[a,b]$ we shall require $\langle\phi_r,\phi_r\rangle \neq 0$ $(r = 0,1,2,\ldots,n)$. The replacement of $<\phi,\phi_r>$ and $<\phi_r,\phi_r>$ by $\langle\phi,\phi_r\rangle$ and $\langle\phi_r,\phi_r\rangle$ respectively in (2.12) then leads to a linear combination of $\phi_0(x), \ldots, \phi_n(x)$ which minimizes $\langle e,e\rangle$ where $e(x)$ is the error $\phi(x) - (A_n\phi)(x)$. We thus minimize the weighted sum of squares of the error

$$\sum_{i=0}^{N} \omega_i [e(x_i)]^2 .$$

Example 2.9

With $x_j = \cos\{(2j + 1)\pi/(2N + 2)\} (j = 0,1,2,\ldots,N), \omega_0 = \omega_1 = \ldots = \omega_N = 1$, and $\phi_r(x) = T_r(x)$ $(r = 0,1,2,\ldots,n)$, we obtain the Chebyshev series given in section 2.4. When $n = N$ we obtain an interpolating polynomial. *

Example 2.10

With $x_j = \cos(j\pi/N)$ $(j = 0,1,2,\ldots,N)$ and $\omega_0 = \omega_N = \frac{1}{2}$, $\omega_1 = \omega_2 = \ldots = \omega_{N-1} = 1$ we obtain the approximation

$$(A\phi)(x) = \sum_{r=0}^{n} a_r T_r(x) ,$$

where

$$a_r = \{2v_r/N\} \sum_{j=0}^{N} v_j f(x_j) T_r(x_j) ,$$

$$v_0 = v_N = \tfrac{1}{2},\ v_1 = v_2 = \ldots = v_{N-1} = 1 .$$

When $n = N$ we obtain an interpolating polynomial. *

The error analysis of section 2.6 for the Fourier-Chebyshev series extends to the approximations in Examples 2.9 and 2.10 if Λ_n is replaced by a slightly different value Λ_n^* which is of the same order of magnitude as the Lebesgue constant for the Fourier-Chebyshev sum (see also Baker and Radcliffe 1970).

Remark. The approximations given in Examples 2.9 and 2.10 may be computed more easily than the Fourier-Chebyshev series. For the latter series, the calculation of a_r requires the determination of certain integrals, whereas in the Examples in this section we calculate the coefficients a_r as sums with $(N + 1)$ terms. The only difficulty is that the $(N + 1)$ sample points x_j may be inconveniently placed in each case.

The values of the coefficients a_r given in Examples 2.9 and 2.10 may be regarded as approximations to the Fourier-Chebyshev coefficients. Other approximations are obtainable from any quadrature rules using available values of $\phi(x)$.

2.10. Approximation in more than one dimension

The concepts of the preceding sections may be applied to the problem of approximating a function of more than one variable. For simplicity, we suppose that we have a function of two variables $\Phi(x,y)$, and we suppose that we know some method of approximating a function $\phi(x)$ by an expression $(A\phi)(x)$. For definiteness only, we suppose that $(A\phi)(x)$ is an interpolating polynomial agreeing with $\phi(x)$ at $x_0, \ldots, x_n$. We may now derive an approximation to $\Phi(x,y)$ by constructing $(A_x\Phi)(x,y)$, where A_x denotes that the operator A is applied only in the x-variable. In the case of polynomial interpolation at $x_0, \ldots, x_n$

$$A_x \Phi(x,y) = \sum_{i=0}^{n} l_i(x)\, \Phi(x_i,y) ,$$

where $l_i(x)$ is defined in eqn (2.2b).

A more symmetric form of approximation is obtained if we also approximate in the y-variable, and form $A_x A_y \overline{\overline{\Phi(x,y)}}$ which we write as $A_x \overline{A}_y \Phi(x,y)$. (In the case of real-valued functions this is just $(A_x A_y \Phi)(x,y)$.) For example, corresponding to the polynomial interpolation above we obtain

$$\sum_{i=0}^{n} l_i(x) \sum_{j=0}^{n} \overline{l_j(y)}\, \Phi(x_i,x_j) .$$

If $\Phi(x,y)$ is Hermitian (that is, $\Phi(x,y) = \overline{\Phi(y,x)}$) then this approximation is also Hermitian. For some further reading, we refer to Garabedian (1965).

2.11. *Numerical integration*

We turn now to the problem of approximating an integral of the form

$$I(\phi) = \int_a^b \phi(y)\,dy . \tag{2.13}$$

We have already seen that we can construct an approximation $(A\phi)(x)$ to $\phi(x)$, and a natural way of obtaining a formula for approximating the integral (2.13) consists of integrating $(A\phi)(x)$. In particular, a class of formulae is obtained by integrating a polynomial which interpolates $\phi(x)$.

In general, we may seek to replace the calculation of $I(\phi)$ by the calculation of a sum, of the form

$$J(\phi) = \sum_{j=0}^{n} w_j \phi(y_j) , \tag{2.14}$$

which approximates the integral. Sometimes we may approximate the integral with a sum of the form

$$J(\phi) = \sum_{s=0}^{r} \sum_{j=0}^{n_s} \omega_{j,s} \phi^{(s)}(\xi_{j,s}) . \tag{2.15}$$

Clearly the *weights* w_j and the *abscissae* y_j in the sum (2.14) should depend on the interval of integration $[a,b]$. We will refer to (2.14) (or (2.15)) as a *quadrature rule* for approximating the integral $I(\phi)$. To be rather more precise, a quadrature rule is a prescription for choosing the weights and abscissae in (2.14) (or (2.15)) when the interval $[a,b]$ is known. We measure the error with the functional

$$E(\phi) = I(\phi) - J(\phi) . \tag{2.16}$$

We will consider one method of obtaining a quadrature rule. We first choose distinct points $y_0, y_1, \ldots, y_n$ in $[a,b]$. The polynomial of degree n interpolating $\phi(x)$ at these points has the form

$$A\phi(x) = \sum_{j=0}^{n} l_j(x) \phi(y_j)$$

given in (2.2). If we integrate this polynomial we obtain a rule (2.14), in which

$$w_j = \int_a^b l_j(y)\,dy .$$

(If we were to allow some of the y_j to coincide we would obtain a rule of the form (2.15).) Quadrature rules obtained in this way, by integrating interpolating polynomials, are called *interpolatory quadrature rules*. When y_j has the form $a + j(b-a)/n$ for $0 \leq j \leq n$ or

$y_j = a + (j+1)(b-a)/(n+2)$ for $0 \le j \le n$ the interpolatory rules are known as *Newton-Cotes* rules. Expressions for the error in an interpolatory rule are obtained, in many texts on numerical analysis, by integrating the error in the corresponding interpolating polynomial.

Example 2.11 *(examples of interpolatory quadrature rules)*
We write $E(\phi) = I(\phi) - J(\phi)$, where $I(\phi)$ and $J(\phi)$ are defined by (2.13) and (2.14).

(a) *Rectangle rule.* $n = 0$, $y_0 = a$, $w_0 = (b - a)$, $E(\phi) = \frac{1}{2}(b - a)^2 \times \phi'(\nu_0)$, where $a < \nu_0 < b$, if $\phi'(x)$ is continuous.

(b) *Mid-point rule.* $n = 0$, $y_0 = \frac{1}{2}(a + b)$, $w_0 = (b - a)$, $E(\phi) = \frac{1}{24}(b - a)^3\phi''(\nu_1)$, where $a < \nu_1 < b$, if $\phi''(x)$ is continuous.

(c) *Trapezium rule.* $n = 1$, $y_0 = a$, $y_1 = b$, $w_0 = w_1 = \frac{1}{2}(b - a)$. $E(\phi) = \frac{-1}{12}(b - a)^3\phi''(\nu_2)$, where $a < \nu_2 < b$, if $\phi''(x)$ is continuous.

(d) *Simpson's rule.* $n = 2$, $y_0 = a$, $y_1 = \frac{1}{2}(a + b)$, and $y_2 = b$, $w_0 = w_2 = \frac{1}{6}(b - a)$ $w_1 = \frac{2}{3}(b - a)$. $E(\phi) = -\frac{1}{90}\{\frac{1}{2}(b - a)\}^5\phi^{iv}(\nu_3)$, where $a < \nu_3 < b$, if $\phi^{iv}(x)$ is continous.

(e) *(Simpson's) 3/8 - rule.* $n = 3$, $y_0 = a$, $y_1 = \frac{1}{3}(2a + b)$, $y_2 = \frac{1}{3}(a + 2b)$, $y_3 = b$, $w_0 = w_3 = \frac{1}{8}(b - a)$, $w_1 = w_2 = \frac{3}{8}(b - a)$. $E(\phi) = -\frac{3}{80}\{\frac{1}{3}(b - a)\}^5\phi^{iv}(\nu_4)$, where $a < \nu_4 < b$, if $\phi^{iv}(x)$ is continuous. *

By their method of construction, the interpolatory rules discussed above must be exact for any polynomial of degree n, since $\phi(x) = (A\phi)(x)$ when $\phi(x)$ is a polynomial of degree n. In Example 2.11(d), however, we see an example of a rule which is exact for polynomials of a higher degree (we have $n = 2$ in that Example, but the rule is exact for any cubic). Another example of the same phenomenon occurs with the mid-point rule.

A quadrature rule is said to have a *degree of precision* equal to r if the rule is exact for every polynomial of degree r, but there is some polynomial of degree $(r + 1)$ for which it is not exact. By a suitable choice of the sample points $y_0, \ldots, y_n$, the degree of precision of an interpolatory rule may be made as large as $(2n + 1)$.

The rules which employ $(n + 1)$ function values and which have this degree of precision are the Gauss-Legendre rules, so named because the abscissae $y_0, \ldots, y_n$ are chosen as the zeros of the Legendre polynomial $P_{n+1}(x)$ if $a = -1$ and $b = 1$.

<u>Example 2.12</u> (*examples of Gauss-Legendre rules*) We use the notation of Example 2.11. The range of integration is $[a,b]$.
We write $\alpha = \frac{1}{2}(b - a)$ and $\beta = \frac{1}{2}(a + b)$; suffixes do not indicate the natural ordering of the abscissae given below.
(a) *One-point rule* : $n = 0$. We have the mid-point rule (Example 2.9(b)) whose degree of precision is 1.
(b) *Two-point rule* : $n = 1$. $w_0 = w_1 = \alpha$; $y_0 = \beta + \alpha/\surd 3$, $y_1 = \beta - \alpha/\surd 3$. The degree of precision is 3.
(c) *Three-point rule* : $n = 2$. $w_0 = w_2 = \frac{5}{9}\alpha$, $w_1 = \frac{8}{9}\alpha$;
$y_0 = \beta + \surd\frac{3}{5}\alpha$, $y_1 = \beta$, $y_2 = \beta - \surd\frac{3}{5}\alpha$. The degree of precision is 5.
(d) *Four-point rule* : $n = 3$. $w_0 = w_3 = \{18 + \surd 30\}\alpha/36$, $w_1 = w_2 = \{18 - \surd 30\}\frac{\alpha}{36}$, $y_0 = \beta - \nu\alpha$, $y_3 = \beta + \nu\alpha$, $y_1 = \beta - \omega\alpha$, $y_2 = \beta + \omega\alpha$,
where

$$\nu = \surd\left\{\frac{(15 - 2\surd 30)}{35}\right\} \quad \text{and} \quad \omega = \surd\left\{\frac{(15 + 2\surd 30)}{35}\right\} .$$

The degree of precision is 7.
In the case of a Gauss-Legendre rule employing $(n + 1)$ abscissae, the error $E(\phi)$ may be written as

$$(b - a)^{2n+3} \frac{\{(n + 1)!\}^4}{(2n + 3)\{(2n + 2)!\}^3} \phi^{(2n+2)}(\xi_n) ,$$

where $a < \xi_n < b$, if $\phi^{(2n+2)}(x)$ is continuous on $[a,b]$. *

Some examples of interpolatory rules which have a degree of precision greater than n but less than the maximum value of $2n + 1$ occur in the Radau integration formulae and the Lobatto integration formulae (Krylov 1962, pp. 166-74; Davis and Rabinowitz 1967, pp. 39-42; Stroud and Secrest 1966, Tables 11 and 12).

<u>Example 2.13</u>
Radau formulae for integrating over $[a,b]$ have the form (2.14) with

the restriction that $y_0 = a$, and that the degree of precision is $2n$. The two-point Radau rule is

$$J(\phi) = \frac{b - a}{4} \{f(a) + 3f(a + \tfrac{2}{3}(b - a))\} .$$

Lobatto formulae for integrating over $[a,b]$ have the form (2.14) with the restriction $y_0 = a, y_n = b$, and the degree of precision is $2n - 1$. The three-point Lobatto rule is Simpson's rule. *

Remark. The weights w_j and abscissae y_j of a quadrature rule (2.14) depend on the interval of integration $[a,b]$. In order to tabulate the weights and abscissae of some particular rule, it is often convenient to normalize the range of integration, and we may then set $a = 0$ and $b = 1$. (However, the Gauss-Legendre rules are usually tabulated for $a = -1$ and $b = 1$.)
The data in the examples in section 2.11 is given for a general interval $[a,b]$, and we shall see that this is appropriate for our discussion in section 2.12.

2.12. *Repeated quadrature rules*

We suppose that some quadrature rule $J(\phi)$ giving an approximation

$$\int_a^b \phi(y)dy \simeq \sum_{j=0}^{n} w_j \phi(y_j)$$

is known, and we shall refer to this approximation as a *basic* quadrature rule. Instead of applying the basic rule in a straightforward way, we may adapt it to yield a new rule, in a way which we now describe.

Suppose that we partition the interval $[a,b]$ with the points $a = a_0 < a_1 < \ldots < a_m = b$. Then

$$\int_a^b \phi(y)dy = \sum_{i=0}^{m-1} \int_{a_i}^{a_{i+1}} \phi(y)dy ,$$

and each term in this sum may be approximated by using the basic rule.

For simplicity, let us suppose that $a_i = a + ih (i = 0,1,2,\ldots,m)$, where $h = (b - a)/m$. Then we obtain, given $\{w_j;\ y_j\}$ as above,

$$\int_a^b \phi(y)\,dy \simeq \frac{1}{m}\sum_{i=0}^{m-1}\sum_{j=0}^{n} w_j\,\phi(a_i + \{y_j - a\}/m)\ . \qquad (2.17)$$

The sum on the right-hand side is a *repeated* or *composite* version of $J(\phi)$ and is denoted by $(m \times J)(\phi)$, where m denotes the number of equal subintervals $[a_i, a_{i+1}]$ each of length h.

<u>Example 2.14</u>

(a) *Repeated trapezium rule (the trapezoidal rule).*

$$\int_a^b \phi(y)\,dy = h \sum_{i=0}^{m}{}'' \phi(a + ih) + E_h^T\ ,$$

where $h = (b - a)/m$, and the sum gives the repeated trapezium rule. (The notation Σ' and Σ'' is commonly used in numerical analysis, where $\sum_{i=0}^{m}{}'' v_i = \tfrac{1}{2}(v_0 + v_m) + \sum_{i=1}^{m-1} v_i = \sum_{i=0}^{m-1}{}' v_i + \tfrac{1}{2}v_m$.)

The rule, obtained on neglecting E_h^T, is sometimes called the trapezoidal rule 'with step h'; $E_h^T = -h^2(b - a)\phi''(\zeta)/12$, if $\phi(x) \in C^2[a,b]$, for some $\zeta \in (a,b)$.

(b) *Repeated Simpson's rule (the parabolic rule).* With $h = (b - a)/m$,

$$\int_a^b \phi(y)\,dy = \frac{1}{3}\, h \sum_{i=0}^{m}{}'' \phi(a + ih) + \frac{2}{3}\, h \sum_{i=0}^{m-1} \phi(a + \{i + \tfrac{1}{2}\}h) + E_h^S\ .$$

The rule obtained on neglecting E_h^S is sometimes called the parabolic rule 'with step $\tfrac{1}{2}h$' (not h); $E_h^S = -\frac{1}{180}(\tfrac{1}{2}h)^4(b - a)\phi^{iv}(\xi)$, for some $\xi \in (a,b)$, if $\phi(x) \in C^4[a,b]$.

(c) *Repeated mid-point rule (with step h).* The rule satisfies

$$\int_a^b \phi(y)\,dy = h \sum_{i=0}^{m-1} \phi(a + \{i + \tfrac{1}{2}\}h) + E_h^M$$

where $h = (b - a)/m$, and $E_h^M = \frac{1}{24} h^2(b - a)\phi''(\eta)$ for some $\eta \in (a,b)$, if $\phi(x) \in C^2[a,b]$. *

Our purpose in using a repeated rule is evidently to obtain a better approximation than can be obtained with the basic rule. Let us examine the form of the error

$$(m \times E)(\phi) = I(\phi) - (m \times J)(\phi) , \tag{2.18}$$

for the rules given as Examples in section 2.9. In each of our Examples we gave an expression for the error in $J(\phi)$ which could be written in the form

$$E(\phi) = \gamma_r (b - a)^{r+2} \phi^{(r+1)}(\xi) , \tag{2.19}$$

where $a < \xi < b$. Here γ_r is an absolute constant which depends only on the quadrature rule, r is the degree of precision, and we suppose that $\phi^{(r+1)}(x)$ is continuous. (For the trapezium rule, $r = 1$ and $\gamma_1 = -\frac{1}{12}$, for example.) It is not true that the error in a rule may *always* be written in the form (2.19), but we shall proceed on this hypothesis because it is true for the Examples we have given.

Now, we may use (2.19) to obtain an expression for the error $(m \times E)(\phi)$, which may be written

$$\sum_{i=0}^{m-1} E_i(\phi) ,$$

where $E_i(\phi)$ denotes the errors in applying the basic rule to approximate

$$\int_{a+ih}^{a+(i+1)h} \phi(y)\,dy$$

with $h = (b - a)/m$. Thus if $\phi^{(r+1)}(x)$ is continuous where r is the degree of precision of $J(\phi)$, we have

$$(m \times E)(\phi) = \sum_{i=0}^{m-1} \gamma_r h^{r+2} \phi^{(r+1)}(\xi_i) ,$$

where $ih < \xi_i < (i + 1)h$. Because of the assumed continuity of $\phi^{(r+1)}(x)$ we may now write

$$(m \times E)(\phi) = m\gamma_r h^{r+2} \phi^{(r+1)}(\eta) ,$$

where $a < \eta < b$. Since $h = (b - a)/m$, we obtain

$$(m \times E)(\phi) = \frac{\gamma_r (b - a)^{r+2}}{m^{r+1}} \phi^{(r+1)}(\eta) \tag{2.20}$$

or

$$(m \times E)(\phi) = (b - a)\, \gamma_r\, h^{r+1} \phi^{(r+1)}(\eta) . \tag{2.21}$$

2.12.1 .

The behaviour of the error $(m \times E)(\phi)$ may be expressed in terms of the *order notation*. We write $\varepsilon_1 = O(h^s)$ (as $h \to 0$) if there is a constant M_s such that $|\varepsilon_1| \leq M_s h^s$ for $0 < h < h_0$, where h_0 is some fixed value. We also write $\varepsilon_2 = o(h^s)$ (as $h \to 0$) if $|\varepsilon_2/h^s| \to 0$ as $h \to 0$. (The terms o(1) which occur below, are terms which are negligible for small h.)

In view of (2.20) and (2.21) we may write

$$(m \times E)(\phi) = O(^1/m^{r+1}) \quad (\text{as } m \to \infty) \tag{2.22a}$$

and

$$(m \times E)(\phi) = O(h^{r+1}) \quad (\text{as } h \to 0) . \tag{2.22b}$$

Actually, when (2.19) is valid (and $\phi^{(r+1)}(x)$ exists) we may establish that $(m \times E)(\phi) = \gamma_r h^{r+1}\{(\phi^{(r)}(b) - \phi^{(r)}(a)) + o(1)\}$.

Eqns (2.22a) and (2.22b) remain valid, though (2.19) may not be true, provided r is less than or equal to the degree of precision of $J(\phi)$, and $\phi^{(r+1)}(x)$ is bounded and piecewise-continuous.

On the other hand, if $\phi^{(N+1)}(x)$ is continuous and $N \geq 0$ we can show (Baker and Hodgson 1971) that if $h = (b - a)/m$,

$$(m \times E)(\phi) = \sum_{k=r+1}^{N} \tau_k h^k + O(h^{N+1}) , \tag{2.23}$$

where τ_k vanishes (a) if $\phi^{(k)}(b) = \phi^{(k)}(a)$ or (b) if k is odd and the quadrature rule $J(\phi)$ is symmetric (for $i=0,1,\ldots,n, w_i = w_{n-i}$ and

$\frac{1}{2}(a+b) - x_i = x_{n-i} - \frac{1}{2}(a+b)$.) A particular case of this result is given by the Euler-Maclaurin formula, which we give in eqn (2.24) below.

2.12.2.

We would like to interpret (2.20) in practical terms. We see that if m is sufficiently large and $\phi(x)$ is sufficiently smooth the error in $(m \times J_1)(\phi)$ can be less than the error in $(m \times J_2)(\phi)$ if the rule $J_1(\phi)$ has a higher degree of precision than the rule $J_2(\phi)$. This appears to be one of the principal advantages for choosing to repeat a rule which has a high degree of precision.

The usual significance of the order notation introduced in section 2.12.1 to describe the behaviour of errors, is revealed in practice in the tables given in Examples 2.15 and 2.16 below.

We cannot say which of the rules $(m \times J_1)(\phi)$ or $(m \times J_2)(\phi)$ will be best for some fixed value of m, without knowing more above $\phi(x)$. In general, it is possible to compare only error bounds, and not the actual error. It is sometimes thought (often wrongly) that a quadrature rule with a high degree of precision should not be used when the integrand has a badly behaved low-order derivative. This view should be constrasted with the fact that $\lim_{m\to\infty} (m \times J)(\phi) = I(\phi)$ if $\phi(x)$ is bounded and Riemann-integrable, provided only that $\sum_{j=0}^{n} w_j = b - a$ and $a \leq y_j \leq b$ $(j = 0, 1, 2, \ldots, n)$ in the basic rule. All the rules which we have given in Examples satisfy these conditions. More convincingly, the Peano theory of section 2.16 can be used to establish that a high degree of precision is not, *of itself*, a disadvantage when the integrand has a low degree of continuity.

2.13. *Modifications of the trapezium rule*

The error in the repeated trapezium rule may be expressed in terms of the derivatives of the integrand $\phi(x)$, if $\phi(x)$ is sufficiently smooth. The Euler-Maclaurin formula (Davis and Rabinowitz 1967) gives, with $h = (b - a)/M$, the relation

$$\int_a^b \phi(y)\,dy = h\{\tfrac{1}{2}\phi(a) + \phi(a+h) + \phi(a+2h) + \dots + \tag{2.24}$$

$$+ \phi(a+(M-1)h) + \tfrac{1}{2}\phi(b)\} - \sum_{k=1}^{N} h^{2k}\frac{B_{2k}}{(2k)!}\{\phi^{(2k-1)}(b) - \phi^{(2k-1)}(a)\} + E,$$

where the constants B_{2k} are the Bernoulli numbers and where $E = O(h^{2N+1})$ if $\phi^{(2N+1)}(x)$ is continuous, and $E = O(h^{2N+2})$ if $\phi^{(2N+2)}(x)$ is continuous. The first term on the right-hand side of eqn (2.24) is clearly the repeated trapezium rule.

Eqn (2.24) provides the theoretical basis for the construction of some modifications to the trapezium rule. We shall describe, briefly, the classical Romberg scheme (Bauer, Rutishauser, and Stiefel 1963; Davis and Rabinowitz 1967) and Gregory's formula (Froberg 1970) which are associated with (2.24). The principle underlying the Romberg scheme will be applied in other contexts later.

2.13.1.

In order to motivate the Romberg scheme let us use eqn (2.24) with $M = mn$ to obtain (with $\sum_{r=0}^{M}{}'' v_r = \frac{1}{2}(v_0 + v_M) + \sum_{r=1}^{M-1} v_r$) the relation

$$\int_a^b \phi(y)\,dy = h\sum_{i=0}^{mn}{}'' \phi(a+ih) + \tau_1 h^2 + O(h^4), \tag{2.25}$$

where $h = (b-a)/mn$, and $\tau_1 = -\frac{1}{2}B_2\{\phi'(b)-\phi'(a)\}$, assuming that $\phi(x) \in C^4[a,b]$. Then

$$\int_a^b \phi(y)\,dy = \tfrac{1}{2}h\sum_{i=0}^{(2mn)}{}'' \phi(a+\tfrac{1}{2}ih) + \tfrac{1}{4}\tau_1 h^2 + O(h^4). \tag{2.26}$$

If we multiply (2.26) by 4 and subtract eqn (2.25) we eliminate τ_1 and obtain, on dividing by 3,

$$\int_a^b \phi(y)\,dy = \frac{2h}{3}\sum_{i=0}^{2mn}{}'' \phi(a+\tfrac{1}{2}ih) - \frac{h}{3}\sum_{i=0}^{mn}{}'' \phi(a+ih) + O(h^4). \tag{2.27}$$

If we put the first two terms on the right-hand side of (2.27) into a single sum, we have a quadrature rule approximating the integral with an error which is $O(h^4)$. The new sum is actually a (repeated) Simpson's rule, and if $\phi(x)$ is sufficiently differentiable we see, using (2.24), that its error may be written in the form (2.23).

In the classical Romberg scheme we extend the basic idea above and obtain a triangular array of entries $T_k^{(j)}$, each of which is an approximation to the integral $I(\phi)$. The array is generated from a column of values obtained from the trapezium rule with fixed n:

$m = 1$	$T_0^{(0)}$			
$m = 2$	$T_0^{(1)}$	$T_1^{(0)}$		
$m = 4$	$T_0^{(2)}$	$T_1^{(1)}$	$T_2^{(0)}$	
$m = 8$	$T_0^{(3)}$	$T_1^{(2)}$	$T_2^{(1)}$	$T_3^{(0)}$
	*	*	*	*

The entry $T_k^{(j)}$ is placed in the $(j + 1)$th position of the $(k + 1)$th column $(j, k = 0, 1, 2, \ldots)$. The entry $T_0^{(j)}$ in the first column is obtained from the trapezium rule repeated mn times, where $m = 2^j$ and n is fixed. In general, the entries in columns other than the first are formed using the recurrence relation

$$T_k^{(j)} = (4^k T_{k-1}^{(j+1)} - T_{k-1}^{(j)})/(4^k - 1) .$$

If $n = 1$, the entry $T_k^{(0)}$ will be called the kth *basic rule* of the classical Romberg scheme. If we write $J_k(\phi) = T_k^{(0)}$ (with fixed $n \geq 1$), $T_k^{(j)} = (m \times J_k)(\phi)$, where $m = 2^j$. The degree of precision of $J_k(\phi)$ is $2k + 1$, and the form of the error in $(m \times J_k)(\phi)$ can be obtained from eqn (2.23), assuming $\phi(x)$ is sufficiently differentiable. We see that the error in $T_k^{(j)}$ is then $O(h^{2k+2})$, where $h = (b-a)/2^j n$, given sufficient differentiability.

<u>Example 2.15</u>

Part of a Romberg table for the function $\phi(x) = e^x$ with $a = 0$, $b = 1$ is displayed below. The true value is $e-1 \simeq 1.718281828459$. The example is instructive because similar ideas are embodied in general schemes for the *deferred approach to the limit* (p. 151) and in Richardson's *repeated* h^2 *- extrapolation.*

1·8591409142				
1·7539310924	1·7188611518			
1·7272219046	1·7183188420	1·7182826879		
1·7205185921	1·7182841546	1·7182818421	1·7182818287	
1·7188411286	1·7182819742	1·7182818286	1·7182818284	1·7182818284

An extended table of errors is given in Table 2.1.

TABLE 2.1

Extended table of errors in a Romberg Scheme

$-1{\cdot}41\times10^{-1}$				
$-3{\cdot}56\times10^{-2}$	$-5{\cdot}79\times10^{-4}$			
$-8{\cdot}94\times10^{-3}$	$-3{\cdot}70\times10^{-5}$	$-8{\cdot}60\times10^{-7}$		
$-2{\cdot}24\times10^{-3}$	$-2{\cdot}33\times10^{-6}$	$-1{\cdot}38\times10^{-8}$	$-3{\cdot}93\times10^{-10}$	
$-5{\cdot}59\times10^{-4}$	$-1{\cdot}46\times10^{-7}$	$-3{\cdot}64\times10^{-10}$	$-1{\cdot}31\times10^{-10}$	$-1{\cdot}31\times10^{-10}$
$-1{\cdot}40\times10^{-4}$	$-9{\cdot}15\times10^{-9}$	$-4{\cdot}37\times10^{-11}$	$-4{\cdot}37\times10^{-11}$	$-4{\cdot}37\times10^{-11}$
$-3{\cdot}50\times10^{-5}$	$-6{\cdot}26\times10^{-10}$	$-4{\cdot}37\times10^{-11}$	$-4{\cdot}37\times10^{-11}$	$-4{\cdot}37\times10^{-11}$
$-8{\cdot}74\times10^{-6}$	$-1{\cdot}46\times10^{-11}$	$1{\cdot}46\times10^{-11}$	$1{\cdot}46\times10^{-11}$	$1{\cdot}46\times10^{-11}$
$-2{\cdot}18\times10^{-6}$	$-1{\cdot}02\times10^{-10}$	$-1{\cdot}02\times10^{-10}$	$-1{\cdot}02\times10^{-10}$	$-1{\cdot}02\times10^{-10}$

These values were obtained on a computer using between 10 and 11 significant decimal figures (39 binary digits) and the expected behaviour of the error is partially obscured by the effects of rounding error. *

In an automatic procedure for numerical integration using the Romberg scheme, the T-table is usually built up automatically (Gram 1964). However, for certain applications to the numerical solution of integral equations we explicitly need the weights and abscissae of the quadrature rules. The weights and abscissae of the kth Romberg rule may be obtained using relations given by Baker (1968) (see also Ralston

1965, p.124) or using an algorithm constructed by Welsch (1966*b*).

2.13.2.

We turn now to Gregory's rule, which can be derived from eqn (2.24), as we show below. The formula for Gregory's rule is usually given in terms of *finite differences*. Corresponding to a fixed step h the rth *forward difference* $\Delta^r\phi(x_0)$ is defined (for $r \geq 1$) by the equation

$$\Delta^r\phi(x_0) = \Delta^{r-1}\phi(x_0 + h) - \Delta^{r-1}\phi(x_0),$$

where $\Delta^0\phi(x) \equiv \phi(x)$. The rth *backward difference* $\nabla^r\phi(x_0)$ is equal to $\Delta^r\phi(x_0 - rh)$. Clearly

$$\Delta^r\phi(x_0) = \nabla^r\phi(x_0 + rh) = \sum_{s=0}^{r} (-1)^{r+s} {}_rC_s\, \phi(x_0 + sh),$$

where ${}_rC_s = r!/\{(r - s)!\, s!\}$.

Using the finite-difference notation we can express Gregory's rule in the form

$$J(\phi) = h \sum_{i=0}^{m}{}'' \phi(a + ih) + h\sum_{s\geq 1} c^*_{s+1}(\nabla^s\phi(b) + (-1)^s\Delta^s\phi(a)), \qquad (2.28)$$

where $h = (b - a)/m$ and $c^*_2 = -\frac{1}{12}$, $c^*_3 = -\frac{1}{24}$, $c^*_4 = -\frac{19}{720}$, $c^*_5 = -\frac{3}{160}$, etc. The values of c^*_s satisfy a well-defined relation (Welsch 1966; Henrici 1964, section 13.4). The second term on the right of eqn (2.28) represents a correction term for the trapezium rule. In practice we take at most pth-order differences so that the summation runs over $1 \leq s \leq p$ in the correction term (the infinite series may diverge). We shall then call the right-hand side of (2.28) the pth *Gregory rule*. If $p > m$ we require values of $\phi(x)$ for x lying outside $[a,b]$.

For some purposes it is useful to rearrange the correction terms

$$h\sum_{s=1}^{p} c^*_{s+1}(\nabla^s\phi(b) + (-1)^s\Delta^s\phi(a))$$

into the form

$$h \sum_{j=0}^{p} \Omega_j^{[p]} \{\phi(a + jh) + \phi(b - jh)\} \quad .$$

We tabulate in Table 2.2 a few of the values of $\Omega_j^{[p]}$.

TABLE 2.2

$p \backslash j$	0	1	2	3
1	$-\frac{1}{12}$	$\frac{1}{12}$		
2	$-\frac{1}{8}$	$\frac{1}{6}$	$-\frac{1}{24}$	
3	$-\frac{109}{720}$	$\frac{177}{720}$	$-\frac{87}{720}$	$\frac{19}{720}$

It is instructive to obtain the Gregory rules from eqn (2.24), since with careful tracking of error terms we can then show that the error in the pth Gregory rule is $O(h^{p+2})$ if $\phi(x)$ has a continuous derivative of order $p + 2$. Examining eqn (2.24), we set $N = \frac{1}{2}$ if p is even, and $N = \frac{1}{2}(p + 1)$ if p is odd, so that in each case the term E is $O(h^{p+2})$, with the assumed differentiability of $\phi(x)$. Each of the odd-order derivatives, $\phi'(a), \phi'(b)\ldots,$ up to $\phi^{(p-1)}(a)$, $\phi^{(p-1)}(b)$, or $\phi^{(p)}(a)$, $\phi^{(p)}(b)$, is now approximated by a finite-difference formula using differences of order p at most. We have the following formulae for the first and pth derivatives,

$$h\phi'(a) = \Delta\phi(a) - \tfrac{1}{2}\Delta^2\phi(a) + \tfrac{1}{3}\Delta^3\phi(a) \;\ldots\; - \frac{(-1)^p}{p}\,\Delta^p\phi(a) + O(h^{p+1}) \;,$$

$$h\phi'(b) = \nabla\phi(b) + \tfrac{1}{2}\nabla^2\phi(b) + \tfrac{1}{3}\nabla^3\phi(b) \;\ldots\; + \frac{\nabla^p\phi(b)}{p} + O(h^{p+1}) \;,$$

$$h^p\phi^{(p)}(a) = \Delta^p\phi(a) + O(h^{p+1}) \;,$$

$$h^p\phi^{(p)}(b) = \nabla^p\phi(b) + O(h^{p+1}) \;,$$

and corresponding formulae for the intermediate derivatives. Substituting the finite-difference terms in eqn (2.24) yields, on rearranging, a formula of the type

$$h \sum_{i=0}^{m}{}'' \phi(a + ih) + h \sum_{s=1}^{p} c_s^*(\nabla^s\phi(b) + (-\Delta)^s\phi(a)) + E_*,$$

where E_* is $O(h^{p+2})$, and if we discard E_* we obtain the pth Gregory rule.

Example 2.16

Table 2.3 shows absolute values of the error in the pth Gregory rule applied with step h to integrate $\phi(x) = e^x$ over $[0,1]$ on a computer using between 10 and 11 significant decimal figures.

TABLE 2.3

Number of differences

h \ p	0	1	2	3	4
$1/4$	$8{\cdot}9 \times 10^{-3}$	$2{\cdot}3 \times 10^{-3}$	$1{\cdot}0 \times 10^{-4}$	$6{\cdot}2 \times 10^{-5}$	$8{\cdot}6 \times 10^{-7}$
$1/8$	$2{\cdot}2 \times 10^{-3}$	$3{\cdot}0 \times 10^{-4}$	$8{\cdot}8 \times 10^{-6}$	$2{\cdot}0 \times 10^{-6}$	$5{\cdot}2 \times 10^{-8}$
$1/16$	$5{\cdot}6 \times 10^{-4}$	$3{\cdot}7 \times 10^{-5}$	$6{\cdot}2 \times 10^{-7}$	$6{\cdot}4 \times 10^{-8}$	$1{\cdot}2 \times 10^{-9}$
$1/32$	$1{\cdot}4 \times 10^{-4}$	$4{\cdot}7 \times 10^{-6}$	$4{\cdot}1 \times 10^{-8}$	$2{\cdot}1 \times 10^{-9}$	$1{\cdot}3 \times 10^{-10}$
$1/64$	$3{\cdot}5 \times 10^{-5}$	$5{\cdot}9 \times 10^{-7}$	$2{\cdot}8 \times 10^{-9}$	$1{\cdot}6 \times 10^{-10}$	$1{\cdot}0 \times 10^{-10}$
$1/128$	$8{\cdot}7 \times 10^{-6}$	$7{\cdot}4 \times 10^{-8}$	$1{\cdot}9 \times 10^{-10}$	$4{\cdot}4 \times 10^{-11}$	
$1/256$	$2{\cdot}2 \times 10^{-6}$	$9{\cdot}3 \times 10^{-9}$	$7{\cdot}3 \times 10^{-11}$	$4{\cdot}4 \times 10^{-11}$	
$1/512$	$5{\cdot}5 \times 10^{-7}$	$1{\cdot}2 \times 10^{-9}$	$7{\cdot}3 \times 10^{-11}$		
$1/1024$	$1{\cdot}4 \times 10^{-7}$	$2{\cdot}5 \times 10^{-10}$			
As $h \to 0$:	Error is $O(h^2)$	Error is $O(h^3)$	Error is $O(h^4)$	Error is $O(h^5)$	Error is $O(h^6)$

Entries of about 10^{-10}, and those values not given in the table (which are of the order of 10^{-10}, are affected by rounding error and may not be significant. *

The method of analysis used in the following Example is due to D.F. Mayers.

Example 2.17

Let us examine the form of the error in the approximation of an integral over $[0,rh]$ provided by the Gregory rule incorporating pth order differences (with $p \leq r$). It is sufficient to consider the approximation to an integral of the form

$\int_0^{rh} \phi(y)dy$, where $\phi(x) \in C^{p+3}[0,X]$ and $0 \leq rh \leq X$.

Our analysis proceeds as follows. We denote by $J_p[0,sh]$ the sum provided by the Gregory rule (using pth-order differences) as an approximation for $\int_0^{sh} \phi(y)dy$. Then $J_p[0,ph]$ is the approximation provided by the $(p+1)$-point Newton-Cotes formula for $\int_0^{ph} \phi(y)dy$ and

$$J_p^*[ph,rh] = J_p[0,rh] - J_p[0,ph]$$

is an approximation for $\int_{ph}^{rh} \phi(y)dy$. We investigate the error in approximating the latter integral by $J_p^*[ph,rh]$, and then use the known error in the Newton-Cotes formulae to obtain an expression for the error in $J_p[0,rh]$. The details of the analysis are as follows.

The theory of polynomial interpolation provides us with knowledge of the error term in the Newton backward difference formula. We have

$$\phi(rh+\theta h) = \{\phi_r + \theta\nabla\phi_r + \ldots + \frac{\theta(\theta+1)(\theta+2)\ldots(\theta+p-1)}{p!} \nabla^p \phi_r\} + \\ + \frac{\theta(\theta+1)\ldots(\theta+p)}{(p+1)!} h^{p+1} \phi^{(p+1)}(\xi),$$

where the term in braces denotes the interpolation formula, with $\phi_r = \phi(rh)$. In this equation, ξ depends upon θ and if (as we suppose) $-p \leq \theta \leq 0$ then $\xi \in ((r-p)h, rh)$. Moreover, $\phi^{(p+1)}(\xi)$ varies continuously as θ varies. Integrating with respect to θ over $[-1,0]$ we obtain

$$\int_{(r-1)h}^{rh} \phi(y)\,dy = h\int_{-1}^{0} \phi(rh+\theta h)\,d\theta =$$

$$= h\{\phi_r - \tfrac{1}{2}\nabla\phi_r - \frac{1}{12}\nabla^2\phi_r + \ldots + c_p^*\nabla^p\phi_r\} + E_{p,r} \quad ,$$

where

$$c_t^* = \frac{1}{t!}\int_{-1}^{0} \theta(\theta+1)\ldots(\theta+t-1)\,d\theta < 0 \quad (t=1,2,3,\ldots),$$

and

$$E_{p,r} = h\int_{-1}^{0} \frac{\theta(\theta+1)\ldots(\theta+p)}{(p+1)!}\, h^{p+1}\phi^{(p+1)}(\xi)\,d\theta \ .$$

By the integral mean-value theorem,

$$E_{p,r} = c_{p+1}^* h^{p+2}\, \phi^{(p+1)}(\eta_{p,r}) \ , \ \eta_{p,r} \in ((r-p)h,\ rh) \ .$$

With the above notation, clearly,

$$\int_{(s-1)h}^{sh} \phi(y)\,dy = h\{\tfrac{1}{2}(\phi_s + \phi_{s-1}) - \frac{1}{12}(\nabla\phi_s - \nabla\phi_{s-1}) +$$

$$+ \ldots + c_{p+1}^*(\nabla^p\phi_s - \nabla^p\phi_{s-1})\} + E_{p+1,s} \ ,$$

and hence

$$\int_{ph}^{rh} \phi(y)\,dy = \sum_{s=p+1}^{r} \int_{(s-1)h}^{sh} \phi(y)\,dy$$

$$= h\{\tfrac{1}{2}\phi_p + \phi_{p+1} + \ldots + \phi_{r-1} + \tfrac{1}{2}\phi_r -$$

$$- \frac{1}{12}(\nabla\phi_r - \nabla\phi_p) + \ldots + c_{p+1}^*(\nabla^p\phi_r - \nabla^p\phi_p)\} +$$

$$+ \sum_{s=p+1}^{r} E_{p+1,s} \ . \tag{2.29}$$

On the other hand, $J_p[0,ph]$ is the closed Newton-Cotes approximation to $\int_0^{ph} \phi(y)dy$, and we may denote its error term by R_p , where

$$\int_0^{ph} \phi(y)dy = h\{\tfrac{1}{2}\phi_0 + \phi_1 + \ldots + \phi_{p-1} + \tfrac{1}{2}\phi_p - \tfrac{1}{12}(\nabla\phi_p - \Delta\phi_0) + \ldots +$$
$$+ c^*_{p+1}(\nabla^p\phi_p + (-1)^{p+1}\Delta^p\phi_0)\} + R_p \; . \qquad (2.30)$$

Expressions for R_p are well known, and we will return to these later. For the present we add eqn (2.29) and eqn (2.30), and we find that

$$\int_0^{rh} \phi(y)dy = J_p[0,rh] + \{R_p + \sum_{s=p+1}^{r} E_{p+1,s}\} \; .$$

Thus the term in braces in the last equation is the required error term. We shall now simplify the expression for this error term, and consider, separately, the case where p is even and the case where p is odd.

If p is even,

$$R_p = d_p \, h^{p+3}\phi^{(p+2)}(\lambda) \; ,$$

where $d_p = \int_0^p (\theta - \tfrac{1}{2}p)\theta(\theta-1)\ldots(\theta-p)d\theta/(p+2)! < 0$, and $\lambda \in (0,rh)$.

Then

$$R_p + \sum_{s=p+1}^{r} E_{p+1,s} = h^{p+3}\{d_p \, \phi^{(p+2)}(\lambda) + \sum_{s=p+1}^{r} c^*_{p+2} \, \phi^{(p+2)}(\eta_{p,j})\}$$
$$= h^{p+3}\phi^{(p+2)}(\zeta_0)\{d_p + (r-p)c^*_{p+2}\} \qquad (p \le r)$$

for some $\zeta_0 \in (0,rh)$, by a mean-value theorem. On the other hand, if p is odd

$$R_p = e_p h^{p+2}\phi^{(p+1)}(\mu) \; ,$$

where

$$e_p = \int_0^p \theta(\theta-1)\ldots(\theta-p)d\theta/(p+1)! \; ,$$

and we can write

$$R_p + \sum_{s=p+1}^{r} E_{p+1,s} = e_p h^{p+2}\phi^{(p+1)}(\mu) + (r-p)c^*_{p+2}h^{p+3}\phi^{(p+2)}(\zeta_1) ,$$

where $\zeta_1 \in (0, rh)$.

If p is fixed and $h \to 0$ with rh fixed, the error terms are $O(h^{p+2})$. *

2.14 . *Classification of quadrature rules*

In the preceding sections we have described a number of quadrature formulae. We shall now abstract certain features of quadrature rules so that we can describe classes of rules in a compact notation. We are primarily interested in sums of the form given in (2.14), that is

$$J(\phi) = \sum_{j=0}^{n} w_j \phi(y_j) , \qquad (2.31)$$

which supply approximations to $I(\phi) = \int_a^b \phi(y)dy$, and which do not involve values of a derivative of $\phi(x)$.

We shall denote by J^+ the class of rules which have positive weights w_j , and abscissae y_j lying in $[a,b]$. The Gauss-Legendre rules, Romberg rules, and the examples given in sections 2.11-2.12 are all rules in J^+ . (However, certain of the Gregory rules have negative weights, or abscissae lying outside $[a,b]$. Certain Newton-Cotes rules have negative weights.) The class of quadrature rules $J(\phi)$ which are symmetric ($w_j = w_{n-j}$ and $y_j - a = b - y_{n-j}$ $(j = 0,1,2,\ldots, n)$) will be denoted, where needed, by writing J^S.

Certain quadrature rules are *Riemann sums* (Baker 1968). A Riemann sum for the integral $\int_a^b \phi(y)dy$ is a sum of the form

$$S(P,Q;\phi) = \sum_{i=0}^{n} (v_{i+1} - v_i)\phi(y_i) \qquad (2.32)$$

related to two partitions, $P = \{a = v_0 \le v_1 \le \cdots \le v_n \le v_{n+1} = b\}$ and $Q = \{y_0 \le y_1 \le \cdots \le y_{n-1} \le y_n\}$, which are meshed so that

$$a = v_0 \le y_0 \le v_1 \le y_1 \le v_2 \le \cdots \le v_n \le y_n \le v_{n+1} = b .$$

$J(\phi)$ is a Riemann sum if the weights w_i and abscissae y_i in (2.31) correspond to such partitions with $w_i = v_{i+1} - v_i$. Thus we require

$$\sum_{i=0}^{k-1} w_i + a \le y_k \le \sum_{i=0}^{k} w_i + a \quad (k = 0,1,2, \ldots,n)$$

and

$$\sum_{i=0}^{n} w_i = (b - a) .$$

It is clear that a rule which is a Riemann sum must be in J^+. We will denote the class of rules which are Riemann sums by J^R. The Gauss-Legendre rules, Romberg rules, and the specific examples given in sections 2.11-2.12 are rules in J^R.

Let us re-state, in our terms, a result of classical analysis.

If we have a family of rules $\{J_m(\phi)\} \subset J^R$ such that $\lim_{m\to\infty} \{ \max_{0\le i\le n} w_i \} = 0$, then $\lim_{m\to\infty} J_m(\phi) = I(\phi)$ provided that $\phi(x)$ is bounded and Riemann-integrable.

As a corollary we see that where $J(\phi) \in J^R$, $\lim_{m\to\infty} (m \times J)(\phi) = I(\phi)$ if $\phi(x)$ is bounded and Riemann-integrable. (For, if a rule is in J^R so is its composite version, and the behaviour of the weights of $(m \times J)(\phi)$ is known.) We stated a stronger result at the end of section 2.12.2.

2.15. *Convergence*

In Baker (1968) we see a simple bound on the error in approximating an integral by a Riemann sum. In the notation above, suppose $R = P \cup Q$ and the width of the partition R is $\Delta_R = \max_r \max \{v_{r+1} - y_r, y_r - v_r\}$.

Now for $\phi(x) \in C[a,b]$,

$$\int_a^b \phi(y)dy - S(P,Q;\phi) = \sum_{r=0}^{n} \int_{v_r}^{v_{r+1}} \phi(y)dy - w_r\phi(y_r)$$

$$= \sum_{r=0}^{n} w_r\{\phi(z_r) - \phi(y_r)\} ,$$

where $w_r = v_{r+1} - v_r, z_r \in [v_r, v_{r+1}]$, using the continuity of $\phi(x)$ and a mean-value theorem. We obtain

$$\left|\int_a^b \phi(y)dy - S(P,Q;\phi)\right| \leq \{\sum_{r=0}^{n} w_r \max_r |\phi(z_r) - \phi(y_r)|\}$$

$$\leq (b-a)\,\omega(\phi;\,\Delta_R)\ ,$$

where $\omega(\phi;\,\delta)$ is the modulus of continuity of $\phi(x)$,

$$\omega(\phi;\,\delta) = \sup_{\substack{x_1,x_2 \in [a,b] \\ |x_1-x_2| \leq \delta}} |\phi(x_1) - \phi(x_2)|\ .$$

Since $\lim_{\delta\to 0} \omega(\phi;\,\delta) = 0$ it follows that if we have a sequence of rules in J^R whose '*measure*' Δ_R tend to zero, the approximations which are provided tend to the value of the integral when the integrand is continuous. (It also follows that this convergence is *uniform* for a set of integrands which are equicontinuous.)

In this section we state a general condition for the convergence of a family $\{J_m(\phi)\}$ of quadrature rules to the value of the integral $I(\phi)$. The result which we give is a consequence of a theorem of Banach (Krylov 1962, p.61).

We suppose that

$$J_m(\phi) = \sum_{j=0}^{n} w_j\,\phi(y_j)$$

where $n = n(m)$, $w_j = w_j(m)$, and $y_j = y_j(m)$, where $a \leq y_j \leq b$ $(j = 0,1,2,\ldots,n)$. Then $\lim_{m\to\infty} J_m(\phi) = I(\phi)$ for every function $\phi(x)$ in $C[a,b]$ if and only if

(i) $\lim_{m\to\infty} J_m(\psi) = I(\psi)$

for every function $\psi(x)$ in some set S which is dense in the uniform norm in $C[a,b]$, and

(ii) there exists some constant W such that $\sum_{j=0}^{n} |w_j| \leq W$ for $m = 0,1,2,\ldots$.

In the rôle of S we may take, for example, the set of polynomials or the set of piecewise-linear functions. Note that a family of quadrature rules in J^+ with degree of precision $r \geq 0$ automatically satisfies (ii) since $\sum_{j=0}^{n} |w_j| = \sum_{j=0}^{n} w_j = (b - a)$ for all m. We refer to Krylov (1962, 12.3) for further details.

Example 2.18

Suppose $J_n(\phi)$ denotes the n-point Gauss-Legendre rule applied to approximate

$$I(\phi) = \int_a^b \phi(y)dy .$$

Then

$$\lim_{n\to\infty} J_n(\phi) = I(\phi)$$

if $\phi(x)$ is continuous. It is sufficient for $\phi(x)$ to be bounded and Riemann-integrable. *

Example 2.19

Let $T_k^{(j)}$ denote an element of the Romberg tableau (see section 2.13.1). Then

(i) $\lim_{j\to\infty} T_k^{(j)} = I(\phi)$ for fixed k, and

(ii) $\lim_{k\to\infty} T_k^{(j)} = I(\phi)$ for fixed j, if $\phi(x)$ is continuous, or even bounded and Riemann-integrable. *

Example 2.20

Let $J_{h,p}(\phi)$ denote the pth Gregory rule using step h. Then $\lim_{h\to 0} J_{h,p}(\phi) = I(\phi)$ for fixed p if $\phi(x)$ is continuous, or even bounded and Riemann-integrable, on $[a,b]$. *

In Chapter 3 we shall refer to the following result, which is due to Bückner (1950). (A simpler result, which will suffice for a first reading, is stated and proved in the following Example.)

We suppose that we have a family of rules

$$J_m(\phi) = \sum_{j=0}^{n} w_j \phi(y_j) ,$$

where $w_j > 0$ and $a \le y_j \le b$ $(j = 0,1,2,\ldots,n)$. We define

$$v_0 = a \ , \ v_j = v_{j-1} + (b - a)\, w_{j-1} \Big/ \sum_{j=0}^{n} w_j \ ,$$

and we assume that

$$\lim_{m\to\infty} J_m(\phi) = \int_a^b \phi(y)\,dy$$

for every function $\phi(x)$ which is continuous on $[a,b]$. Then (following Bückner) we may show that

(a) $$\lim_{m\to\infty} \sum_{j=0}^{n} w_j = b - a \ ;$$

(b) $$\lim_{m\to\infty} \max_{i=0,1,2,\ldots,n} w_i = 0 \ ;$$

(c) $$\lim_{m\to\infty} \max_{i=0,1,2,\ldots,n} |y_i - v_i| = 0 \ .$$

<u>Example 2.21</u>

Hodgson (1969, Theorem 1.18) established a particular case of the above result, which we state here.

If $J_m(\phi) \in J^R$ $(m = 0,1,2,\ldots)$ and

$$\lim_{m\to\infty} J_m(\phi) = \int_a^b \phi(y)\,dy \quad (n{=}n(m) \to \infty \text{ as } m \to \infty)$$

for every function $\phi(x) \in C[a,b]$, then the widths of the partitions $P \cup Q$ (section 2.14) tend to zero as $n \to \infty$. The width of the partition $P \cup Q$ (which is the common refinement of P and Q) is here referred to as the measure of $J_m(\phi)$.

Let us indicate a proof of this result. For an $\varepsilon > 0$ and an α fixed in (a,b) with $\alpha - \frac{1}{2}\varepsilon \ge a$ and $\alpha + \frac{1}{2}\varepsilon \le b$, define $\phi_\varepsilon(x)$ by setting $\phi_\varepsilon(x) = (2/\varepsilon)(x - \alpha + \frac{1}{2}\varepsilon)$ for $\alpha - \frac{1}{2}\varepsilon \le x \le \alpha$,

$\phi_\varepsilon(x) = (2/\varepsilon)(\alpha + \frac{1}{2}\varepsilon - x)$ for $\alpha \le x \le \alpha + \frac{1}{2}\varepsilon$ and $\phi_\varepsilon(x) = 0$ elsewhere in $[a,b]$. Then $\phi_\varepsilon(x) \in C[a,b]$, $\int_a^b \phi_\varepsilon(x)dx = \frac{1}{2}\varepsilon \neq 0$.

Thus $\lim_{m\to\infty} J_m(\phi_\varepsilon) = \frac{1}{2}\varepsilon \neq 0$, and hence there is an integer M such that for $m \ge M$ there is some abscissa x_k of $J_m(\phi)$ in $[\alpha - \frac{1}{2}\varepsilon, \alpha + \frac{1}{2}\varepsilon]$ where $k = k(m)$ (otherwise $J_m(\phi_\varepsilon)$ does not converge to a non-zero limit).

For some integer q , we apply the preceding result with $\varepsilon = (b - a) / q$ and $\alpha = \alpha_r = a + r\varepsilon$ $(r = 1,2,3,\ldots,(q - 1))$ in turn. We see that there exist $M_1, M_2, \ldots, M_{q-1}$ such that for $m \ge \max_r M_r$ there is some abscissa of $J_m(\phi)$ in each interval

$$[a + (r - \tfrac{1}{2})\varepsilon, a + (r + \tfrac{1}{2})\varepsilon] \quad (r = 1,2,3,\ldots,(q - 1)).$$

It follows (since all the abscissae of $J_m(\phi)$ are in $[a,b]$ at a separation of not more than a distance 2ε , and $v_0 = a$, $v_{n+1} = b$) that the measure of $J_m(\phi)$ is less than or equal to 2ε for all m larger than an integer M depending on ε . There follows the required result, since $\varepsilon > 0$ can be made arbitrarily small.

We obtain a proof of Bückner's result for rules $J_m(\phi) \in J^R$ as a simple corollary of the result established here (for $w_i > 0$, $|w_i| \le |y_i - v_i| + |v_{i+1} - y_i|$ is bounded by twice the measure of $J_m(\phi)$, and (c) follows immediately). *

2.16. *Precise error bounds*

Suppose that the quadrature rule $J(\phi)$ has a degree of precision $r \ge 0$ (which it has if $\sum_{j=0}^{n} w_j = b - a$ in eqn (2.14)). We can then use Taylor's theorem to obtain precise bounds for the error in our quadrature formula, in terms of bounds on derivatives of the integrand $\phi(x)$. The theory is an example of Peano theory (Davis 1965).

We suppose that r is the degree of precision of $J(\phi)$ and that s is a non-negative integer less than or equal to r . Then if $\phi^{(s+1)}(x)$ is bounded and piecewise-continuous we may write

$$\phi(x) = \phi(a) + (x - a)\,\phi'(a) + \ldots + \frac{(x - a)^s}{s!}\phi^{(s)}(a) + $$
$$+ \frac{1}{s!}\int_a^b (x - t)_+^s\, \phi^{(s+1)}(t)\mathrm{d}t\ .$$

This is Taylor's theorem with the integral remainder expressed in terms of the spline notation used in Example 2.4 (see, also, Example 2.2(a)).

If we write $E(\phi) = I(\phi) - J(\phi)$ then $E((x - a)^p) = 0$ for $p = 0,1,2,\ldots, s \leq r$ and since $E(\phi_1 + \phi_2) = E(\phi_1) + E(\phi_2)$ we obtain

$$E(\phi) = \frac{1}{s!}\int_a^b E_x((x - t)_+^s)\ \phi^{(s+1)}(t)\mathrm{d}t\ , \tag{2.33}$$

where $E_x(\ \ )$ signifies that we operate on the argument of E as a function of x only. Thus

$$\frac{1}{s!}E_x\ ((x - t)_+^s) = \frac{1}{s!}\ \{\int_a^b (y - t)_+^s\ \mathrm{d}y - \sum_{j=0}^{n} w_j(y_j - t)_+^s\}\ . \tag{2.34}$$

The left-hand side of eqn (2.34) is a function of t, which we shall denote by $M_{s+1}(t)$, so that we may rewrite eqn (2.33) in the form

$$E(\phi) = \int_a^b M_{s+1}(t)\ \phi^{(s+1)}(t)\mathrm{d}t \qquad (s=0,1,2,\ldots,r)\ . \tag{2.35}$$

The function $M_{s+1}(t)$ is usually known as a Peano kernel for the functional E. We shall refer to it as the sth *Peano kernel* of the quadrature rule $J(\phi)$, where s may take the values $0,1,2,\ldots$ up to the degree of precision r.

Eqn (2.35) is an exact expression for the error. We may use this expression to derive bounds for $|E(\phi)|$ in terms of norms of $\phi^{(s+1)}(x)$. Applying various forms of Hölder's inequality for integrals we obtain

$$|E(\phi)| \leq ||\phi^{(s+1)}(x)||_\infty\ \ ||M_{s+1}(x)||_1 \tag{2.36a}$$

$$|E(\phi)| \leq ||\phi^{(s+1)}(x)||_2\ \ ||M_{s+1}(x)||_2 \tag{2.36b}$$

and

$$|E(\phi)| \leq ||\phi^{(s+1)}(x)||_1\ \ ||M_{s+1}(x)||_\infty, \tag{2.36c}$$

where

$$\|\psi(x)\|_p = \{\int_a^b |\psi(x)|^p dx\}^{1/p} \quad \text{for } 1 \leq p < \infty \quad , \text{ and}$$

$$\|\psi(x)\|_\infty = \sup_{a \leq x \leq b} |\psi(x)| \ .$$

The constants

$$c_{s+1} \equiv c_{s+1}(J;\ a,b) = \|M_{s+1}(x)\|_1 \ , \tag{2.37a}$$

$$d_{s+1} \equiv d_{s+1}(J;\ a,b) = \|M_{s+1}(x)\|_2 \ , \tag{2.37b}$$

$$e_{s+1} \equiv e_{s+1}(J;\ a,b) = \|M_{s+1}(x)\|_\infty \ , \tag{2.37c}$$

which occur in eqns (2.36 a,b,c) are often known as *Peano constants*, and some of their values have been tabulated for various rules and standardized values of a and b (Stroud and Secrest 1966; Stroud 1966; Nikolskii 1966). It is not difficult to show that, for $s = 0,1,2,\ldots, r$,

$$c_{s+1}(J;\ a,b) = \{(b - a)/(d - c)\}^{s+2}\ c_{s+1}(J;\ c,d), \tag{2.38}$$

with *similar relations* for d_{s+1} and e_{s+1} . Consequently,

$$c_{s+1}(m \times J;\ a,b) = \frac{1}{m^{s+1}}\ c_{s+1}(J;\ a,b) \quad (s=0,1,2,\ldots,r) \tag{2.39}$$

and from this relation we deduce that the error in $(m \times J)(\phi)$, which we denote $(m \times E)(\phi)$, is $O(1/m^{s+1})$ when $\phi^{(s+1)}(x)$ is (say) piecewise-continuous and $s \leq r$.† If the quadrature rule $J(\phi)$ is in J^S

† Baker (1968) notes that it is sufficient that $\phi^{(s)}(x)$ is Lipschitz-continuous; if we relate $e_{s+1}(m \times J;\ a,b)$ to $e_{s+1}(J;\ a,b)$ we can show that it is sufficient that $\phi^{(s+1)}(x)$ exists everywhere in $[a,b]$ and is absolutely integrable.

(that is, if it is symmetric) then we may replace the norm of $\phi^{(s+1)}(x)$ in eqns (2.36a), (2.36b), (2.36c) by the norm of

$$(d/dx)^{s+1}\tfrac{1}{2}\{\phi(x) + \phi(-x)\}$$

(see Baker 1968).

Example 2.22 (*values of some Peano constants* for rules with $a=0$, $b=1$)
(a) *Mid-point rule.* $c_1 = 1/4$, $c_2 = 1/24$; $e_1 = 1/2\sqrt{3}$.
(b) *Trapezium rule.* $c_1 = 1/4$, $c_2 = 1/12$; $e_1 = 1/2\sqrt{3}$, $e_2 = 1/2\sqrt{30}$.
(c) *Simpson's rule.* $c_1 = 5/36$, $c_2 = 1/81$, $c_3 = 1/576$, $c_4 = 1/2880$; $e_1 = 1/6$, $e_2 = 1/12\sqrt{30}$, $e_3 = 1/48\sqrt{105}$, $e_4 = 1/576\sqrt{14}$. *

Example 2.23 (*values of some Peano constants* for Gauss-Legendre rules, with $a = -1$, $b = 1$)
(a) *Mid-point rule.* $c_1 = 1$, $c_2 = 1/3$. Compare with the values in Example 2.22, for $a = 0$ and $b = 1$.
(b) *Two-point rule.* $c_1 = \frac{1}{3}(5 - 2\sqrt{3}) \simeq 0.512$, $c_2 = \frac{4}{9\sqrt{3}}\{2\sqrt{3} - 3\}^{3/2} \simeq 0.081$, $c_3 = 1/12 - \sqrt{3}/27 \simeq 0.019$, $c_4 = 1/135$.
(c) *Three-point rule.* $c_1 \simeq 0.357$, $c_2 \simeq 0.038$, $c_3 \simeq 0.002$, $c_4 \simeq 0.001$, $c_5 \simeq 0.0002$, $c_6 \simeq 0.00007$. *

Remark. The values of the Peano constants provide one means of judging the quality of a quadrature rule. Stroud (1965) has compared the Romberg and Gauss-Legendre rules. Stroud and Secrest (1966) tabulate a number of Peano constants.

2.17. *Integration of products*

We turn now to the construction of approximation of the form

$$\int_a^b \omega(y)\,\phi(y)dy \simeq \sum_{j=0}^{n} v_j\,\phi(z_j) . \tag{2.40}$$

The right-hand side of (2.40) has the form of a quadrature rule ((2.14), (2.31)); the abscissae z_j depend on a,b , and the weights v_j now depend on a,b and also on the *weight function* $\omega(y)$ which occurs on the left in (2.40). If $\omega(x) \equiv 1$ the problem reduces to the classical problem of quadrature discussed in earlier sections.

We give two reasons for our interest in the more general problem. In the first instance we may require the value of

$$\int_a^b \phi_0(y)\,dy$$

where $\phi_0(x)$ is 'badly behaved'. If we suppose, for definiteness, that $\phi_0(x)$ behaves like $(x - a)^{-\frac{1}{2}}$ near $x = a$, but is elsewhere continuous, we may be able to find a continuous function $\phi(x)$ such that $\phi_0(x) = \phi(x)/(x - a)^{\frac{1}{2}}$. If we approximate $\phi(x)$ using a polynomial interpolant, say $P_{0,1,2,\ldots,n}(x)$, agreeing with $\phi(x)$ at $z_0,\ldots,z_n$, we may set

$$\begin{aligned}\int_a^b \phi_0(y)\,dy &= \int_a^b \omega(y)\ \phi(y)\,dy \\ &\simeq \int_a^b \omega(y)\ P_{0,1,2,\ldots,n}(y)\,dy \\ &= \sum_{j=0}^{n} v_j \phi(z_j),\end{aligned}$$

where $v_j = \int_a^b \omega(y)\ l_j(y)\,dy$ and $l_j(x) = \prod_{\substack{i=0\\ i\neq j}}^{n} (x - z_i)/(z_j - z_i)$.

Because $\phi(y)$ is continuous, we may hope to select $z_0,\ldots,z_n$ so that $P_{0,1,2,\ldots,n}(x)$ is a reasonable approximation to $\phi(x)$, whilst the corresponding task for $\phi_0(x)$ would be more difficult. In general, therefore, we may be interested in approximations like (2.40) where $\omega(x)$ is badly behaved and $\phi(x)$ is well behaved. We shall require that we can evaluate

$$\int_a^b \omega(y)y^s\,dy \quad (s = 0,1,2,\ldots,r),$$

where $r \geq 0$.

Example 2.24

Formulae of the form (2.40) which are exact when $\phi(x)$ is a polynomial of degree $(2n + 1)$ are called Gauss-type rules. They have been tabulated for a number of cases including the following ones:

(i) $a = 0$, $b = \infty$, $\omega(x) = \exp(-x)$ (Gauss-Laguerre formulae);
(ii) $a = -\infty$, $b = \infty$, $\omega(x) = \exp(-x^2)$ (Gauss-Hermite formulae);
(iii) $a = -1$, $b = 1$, $\omega(x) = (1 - x^2)^{-\frac{1}{2}}$ (Gauss-Chebyshev formulae). *

On the other hand, we have a special interest in formulae for approximating the integrals of products, because they occur in integral equations. Every linear integral equation (and also any non-linear integral equation of Hammerstein type) involves an integral of the form $\int_a^b K(x,y)\ \phi(y)dy$. If we fix x and write $\omega(y) = K(x,y)$ we obtain the form of the integral under discussion. In this case $K(x,y)$ may be well behaved, or it may be badly behaved (for example, weakly singular, so that $\omega(y)$ may have the form $|x-y|^{-\alpha}H(x,y)$ with $1 > \alpha > 0$).

As stated above, we may obtain approximations of the form (2.40) by constructing an approximation $(A\phi)(x)$, interpolating $\phi(x)$ at $z_0,\ldots, z_n$, then setting $\sum_{j=0}^{n} v_j\phi(z_j) = \int_a^b \omega(y)(A\phi)(y)dy$. Corresponding to the construction of the interpolatory quadrature rules discussed earlier, and as indicated earlier in this section, we may take $(A\phi)(x)$ to be an interpolating polynomial, say $P_{0,1,2,\ldots,n}(x)$, that agrees with $\phi(x)$ at points $z_0,\ldots, z_n$, which may be chosen at our discretion.

If $\phi(x) \in C^{n+1}[a,b]$, and $a \le z_i \le b$ $(i = 0,1,2,\ldots,n)$, then (see section 2.3) there is a value ξ depending on x, such that

$$\phi(x) - P_{0,1,\ldots,n}(x) = \{\prod_{i=0}^{n} (x - z_i)\} \frac{\phi^{(n+1)}(\xi)}{(n+1)!} . \tag{2.41}$$

Consequently, assuming that a,b are finite,

$$\int_a^b \omega(y)\ \phi(y)dy - \int_a^b \omega(y)\ P_{0,1,2,\ldots,n}(y)dy = \int_a^b \omega(y) \prod_{i=0}^{n} (y-z_i)\frac{\phi^{(n+1)}(\xi)}{(n+1)!}\ dy ,$$

where $\xi = \xi(y)$; thus when $v_j = \int_a^b \omega(y) \prod_{i\neq j} \{(y-z_i)/(z_j-z_i)\}dy$ in

(2.40), the error in the approximation is bounded by

$$||\phi^{(n+1)}(x)||_\infty \int_a^b |\omega(y) \prod_{i=0}^{n} (y-z_i)|dy/(n+1)! \quad , \tag{2.42}$$

provided $\int_a^b |\omega(y)|dy < \infty$.

Suppose that $z_0 = a$ and $z_1 = b$ and $\phi''(x)$ is continuous on $[a,b]$. Then

$$P_{0,1}(x) = \{(b - x)\ \phi(a) + (x - a)\ \phi(b)\}/(b - a)$$

and

$$\int_a^b \omega(y)\ \phi(y)dy \simeq v_0\ \phi(a) + v_1\ \phi(b) \ ,$$

where

$$v_0 = \int_a^b (b - y)\ \omega(y)dy/(b - a)$$

and

$$v_1 = \int_a^b (y - a)\ \omega(y)dy/(b - a) \ .$$

In view of eqn (2.42) the modulus of the error is bounded by the expression

$$\tfrac{1}{2}\ ||\phi''(x)||_\infty \int_a^b |\omega(y)\ (y - a)(y - b)|dy \quad . \tag{2.43}$$

An improved approximation may be obtained if we partition $[a,b]$ with the points $\{z_i\}$,

$$a = z_0 < z_1 < \ldots < z_n = b \ ,$$

and write

$$\int_a^b \omega(y)\ \phi(y)dy = \sum_{i=0}^{n-1} \int_{z_i}^{z_{i+1}} \omega(y)\ \phi(y)dy$$

$$= \sum_{i=0}^{n-1} \{v_0^{(i)}\ \phi(z_i) + v_1^{(i)}\ \phi(z_{i+1})\}\ , \qquad (2.44)$$

where $v_0^{(i)} = \int_{z_i}^{z_{i+1}} (z_{i+1} - y)\ \omega(y)dy/(z_{i+1} - z_i)$ and

$$v_1^{(i)} = \int_{z_i}^{z_{i+1}} (y - z_i)\ \omega(y)dy/(z_{i+1} - z_i)\ .$$

Collecting terms together, the approximation assumes the form of (2.40), and in view of (2.43) the corresponding error may be bounded by

$$\tfrac{1}{2}\ \|\phi''(x)\|_\infty \sum_{i=0}^{n} \int_{z_i}^{z_{i+1}} |\omega(y)(y - z_i)(y - z_{i+1})|dy\ . \qquad (2.45)$$

If $z_i = a + ih$, where $h = (b - a)/N$, the bound (2.45) on the error is $O(h^2)$, assuming $\int_a^b |\omega(y)|dy$ is finite. The integration formula using equally spaced z_i with step h will be known as the *generalized trapezium* rule since it is an extension of the (repeated) trapezium rule. An analogous procedure gives the *generalized mid-point* rule, which arises (with even n) on setting $z_i = a+ih$, $v_0 = v_2 = v_4 = \ldots = 0$ and (for $r = 0,1,2,\ldots$) $v_{2r+1} = \int_{z_{2r}}^{z_{2r+2}} \omega(y)dy$, in (2.40). This rule is exact if $\phi(x)$ is constant in each interval $[z_{2r}, z_{2r+2}]$.

There is a corresponding extension of Simpson's rule. We write $z_i = a + ih$, where $h = (b - a)/n$ with $n = 2m$, where m is an integer. In the interval $[z_{2k}, z_{2k+2}]$ we approximate $\phi(x)$ by the interpolating quadratic, say

$$P_{2k,2k+1,2k+2}(x) = \{\tfrac{1}{2}(x - z_{2k+1})(x - z_{2k+2})\ \phi(z_{2k}) -$$

$$- (x - z_{2k})(x - z_{2k+2})\ \phi(z_{2k+1}) +$$

$$+\tfrac{1}{2}(x - z_{2k})(x - z_{2k+1})\ \phi(z_{2k+2})\}/h^2\ ,$$

and we write

$$\int_a^b \omega(y)\ \phi(y)dy \simeq \sum_{k=0}^{m-1} \int_{z_{2k}}^{z_{2k+2}} \omega(y)\ P_{2k,2k+1,2k+2}(y)dy\ . \tag{2.46}$$

The approximation formula takes the form of a *generalized Simpson's rule*

$$\sum_{j=0}^{n} v_j\ \phi(z_j)\ , \tag{2.47}$$

where

$$v_0 = \int_a^{a+2h} (y - z_1)(y - z_2)\ \omega(y)dy/2h^2\ ,$$

$$v_n = \int_{b-2h}^{b} (y - z_{2m-2})(y - z_{2m-1})\ \omega(y)dy/2h^2\ ,$$

$$v_{2j+1} = -\int_{z_{2j}}^{z_{2j+2}} (y - z_{2j})(y - z_{2j+2})\ \omega(y)dy\ /h^2\ ,$$

and

$$v_{2j} = \{\int_{z_{2j-2}}^{z_{2j}} (y - z_{2j-2})(y - z_{2j-1})\ \omega(y)dy\ +$$

$$+ \int_{z_{2j}}^{z_{2j+2}} (y - z_{2j+1})(y - z_{2j+2})\ \omega(y)dy\}/2h^2$$

if $j \neq 0$ or m. This could be called the *generalized parabolic* rule. If $\phi'''(x)$ is continuous then the error in the formula is $O(h^3)$. (It is not $O(h^4)$ if $\phi^{iv}(x)$ is continuous, except in special cases, for example when $\omega(x) \equiv 1$. See Wang (1976) and the references there.)

We mentioned that, in applications, the function $\omega(y)$ of the foregoing discussion may have the form $(x-y)^{-\alpha}H(x,y)$ for selected values of x. We may illustrate the construction of the generalized trapezium rule by taking $\alpha = \frac{1}{2}$, $z_i = a+ih$, $h=(b-a)/n$, $H(x,y)=1$, and $x=z_k, k\epsilon(0,1,\ldots,n)$. We find an approximation of the form (2.44) where

$v_j^{(i)}$ depends on k , namely

$$\int_a^b (z_k-y)^{-\frac{1}{2}} \phi(y)dy \simeq \sum_{i=0}^{n-1} \{v_{0,k}^{(i)} \phi(z_i) + v_{1,k}^{(i)} \phi(z_{i+1})\} ,$$

where

$$v_{0,k}^{(i)} = \frac{1}{h} \int_{z_i}^{z_{i+1}} (z_{i+1} - y)(z_k - y)^{-\frac{1}{2}} dy$$

and

$$v_{1,k}^{(i)} = \frac{1}{h} \int_{z_i}^{z_{i+1}} (y - z_i)(z_k - y)^{-\frac{1}{2}} dy .$$

On setting $\psi_0(r) = \int_0^1 (1-s)\ |r-s|^{-\frac{1}{2}} ds$ and $\psi_1(r) = \int_0^1 s\ |r-s|^{-\frac{1}{2}} ds$,

$$v_{0,k}^{(i)} = h^{\frac{1}{2}}\ \psi_0(k-i),\ v_{1,k}^{(i)} = h^{\frac{1}{2}}\ \psi_1(k-i) ,$$

where $\psi_0(r) = \psi_2(r) - \psi_1(r)$ and

$$\psi_1(r) = \frac{2}{3} \left(\frac{(r-1)^2}{|r-1|^{\frac{1}{2}}} - \frac{r^2}{|r|^{\frac{1}{2}}}\right) + r\psi_2(r)$$

with

$$\psi_2(r) = 2\left\{ \frac{r}{|r|^{\frac{1}{2}}} - \frac{(r-1)}{|r-1|^{\frac{1}{2}}}\right\}.$$

The corresponding formula with $\omega(y)$ in the form $\ln|z_k-y|$ has coefficients obtained on setting

$$\int_a^b \ln |z_k-y|\ \phi(y)dy \simeq \sum_{i=0}^{n-1} \{ \hat{v}_{0,k}^{(i)} \phi(z_i) + \hat{v}_{1,k}^{(i)} \phi(z_{i+1})\}$$

where we now define $\psi_0^*(r) = \int_0^1 (1-s)\ \ln|r-s|\ ds$, $\psi_1^*(r) = \int_0^1 s \ln |r-s|\ ds$,

and obtain

$$\hat{v}_{0,j}^{(i)} = \tfrac{1}{2} h \ln h + h\ \psi_0^*(k-j)$$

$$\hat{v}^{(i)}_{1,k} = \tfrac{1}{2}h \ln h + h\psi^*_1(k - j) .$$

We now have

$$\psi^*_0(r) = \psi^*_2(r) - \psi^*_1(r)$$

with

$$\psi^*_1(r) = \tfrac{1}{2}\{(r-1)^2 \ln|r-1| - r^2\ln|r|\} +$$

$$+ \tfrac{1}{4} \{r^2 - (r-1)^2\} + r\psi^*_2(r)$$

and

$$\psi^*_2(r) = r \ln |r| - (r-1) \ln |r-1| - 1 .$$

In calculating the weights $v^{(i)}_{j,k}$ and $\hat{v}^{(i)}_{j,k}$ from the formulae above, care must be taken to avoid cancellation error. (In the functions employed above, limiting values are taken when $r = 0$ or when $r = 1$.)

Observe, in passing, that the approximations

$$\int_a^b |z_k-y|^{-\frac{1}{2}} \phi(y)dy \simeq \sum_{j=0}^{\frac{1}{2}n-1} w^{(2j+1)}_k \phi(a+(2j+1)h) \quad (n \text{ even}),$$

and

$$\int_a^b \ln |z_k-y| \phi(y)dy \simeq \sum_{j=0}^{\frac{1}{2}n-1} \hat{w}^{(2j+1)}_k \phi(a+(2j+1)h) \quad (n \text{ even}),$$

with $w^{(2j+1)}_k = v^{(2j)}_{0,k} + v^{(2j)}_{1,k} + v^{(2j+1)}_{0,k} + v^{(2j+1)}_{1,k}$, and a corresponding expression for $\hat{w}^{(2j+1)}_k$, provide generalizations of the repeated mid-point rule. Thus, as we see easily from first principles,

$$w^{(2j+1)}_k = 2\, h^{\frac{1}{2}} \left[\frac{k - 2j}{|k - 2j|^{\frac{1}{2}}} - \frac{k - 2j - 2}{|k-2j-2|^{\frac{1}{2}}} \right],$$

where the limiting values are taken to give $w_{2j+2}^{(2j+1)} = w_{2j}^{(2j+1)} = 2(2h)^{\frac{1}{2}}$.

2.17.1 .

The extension of the generalized trapezium rule and generalized parabolic rule so as to obtain a rule which is exact for polynomials of degree m, or piecewise polynomials of degree m, will now be apparent. We here, for example, set $h = (b-a)/n$, where $n = Nm$ and N and m are integers, and write $z_i = a + ih$ $(i = 0,1,2,\ldots,n)$. For $z_{km} \le x \le z_{km+m}$ we approximate $\phi(x)$ by the interpolating polynomial of degree m, say $P_{km,(km+1),\ldots,(k+1)m}(x)$, which agrees with $\phi(x)$ at $z_{km}, z_{km+1},\ldots,z_{(k+1)m}$. Then we set

$$\int_a^b \omega(y)\,\phi(y)\,dy = \sum_{k=0}^{N-1} \int_{z_{km}}^{z_{\overline{k+1}m}} \omega(y)\, P_{km,(km+1),\ldots,(k+1)m}(y)\,dy \quad ; \tag{2.48}$$

if we express the interpolating polynomials in Lagrangian form, we then obtain an approximation which can be expressed in the form (2.40). The right-hand side of eqn (2.48) becomes

$$\sum_{j=0}^{n} v_j\, \phi(z_j) ,$$

where, if $j = km + i$ where $i \neq 0$,

$$v_j = v_{km+i} = \int_{z_{km}}^{z_{(k+1)m}} \prod_{\substack{j=km \\ j\neq km+i}}^{(k+1)m} \{(y-z_j)/(z_{km+i}-z_j)\}\, \omega(y)\,dy , \tag{2.49a}$$

and if j is a multiple of m, say $j = km$,

$$v_j = v_{km} = \int_{z_{km}}^{z_{(k+1)m}} \prod_{j=km+1}^{(k+1)m} \{(y-z_j)/(z_{km}-z_j)\}\, \omega(y)\,dy \; +$$

$$+ \int_{z_{(k-1)m}}^{z_{km}} \prod_{j=(k-1)m}^{km-1} \{(y-z_j)/(z_{km}-z_j)\}\, \omega(y)\,dy . \tag{2.49b}$$

There is no need to restrict the points z_i so that they lie in the range of integration. If some points are chosen to lie outside $[a,b]$ in a larger interval $[A,B]$, we can imagine that we wish to approximate $\int_A^B \Omega(y)\phi(y)\,dy$, where $\Omega(x) = \omega(x)$ if $a \le x \le b$,

$\Omega(x) = 0$ otherwise.

2.17.2 .

It may be gathered, from the derivation of eqns (2.42) and (2.45), that the error in polynomial interpolation may be used to derive error bounds for the generalized interpolatory rules discussed above. (Peano theory is also directly applicable to the error analysis. If we now write $E(\phi)$ for $\int_a^b \omega(y)\,\phi(y)\,dy - \sum_{j=0}^{n} v_j \phi(z_j)$, eqn (2.33) remains valid provided $E(x^t) = 0$ for $t = 0,1,2,\ldots,r$, and $s \le r$.)

To simplify matters, we shall suppose that $z_i = a + ih$ and here we choose to apply Jackson's theorem to obtain error bounds for the approximation (2.48).

Suppose that $p^*_{m,k}(x)$ is a polynomial of degree m such that

$$\sup_{z_{km} \le x \le z_{(k+1)m}} |\phi(x) - p^*_{m,k}(x)| = \inf_{p_m(x)} \sup_{z_{km} \le x \le z_{(k+1)m}} |\phi(x) - p_m(x)| , \qquad (2.50)$$

where the infimum is taken over all possible polynomials of degree m . Then Jackson's theorem may be used to bound

$$e_{m,k}(\phi) = \max_{z_{km} \le x \le z_{(k+1)m}} |\phi(x) - p^*_{m,k}(x)| .$$

We have, for $k = 0,1,2,\ldots,(N-1)$,

$$e_{m,k}(\phi) \le \frac{(\pi m h/4)^{s+1} \|\phi^{(s+1)}\|_\infty}{(m+1)m \ldots (m-s+2)} , \qquad (2.51)$$

if $s \le m$ and $\phi^{(s+1)}(x)$ is continuous (or merely bounded). This is easily checked. Furthermore, it can be seen that

$$\left|\int_a^b \omega(y)\,\phi(y)\,dy - \sum_{j=0}^{n} v_j\, \phi(z_j)\right|$$

$$\leq \sum_{k=0}^{N-1} \left|\int_{z_{km}}^{z_{(k+1)m}} \omega(y)\{\phi(y) - p^*_{m,k}(y)\}dy\right| +$$

$$+ \left|\int_{z_{km}}^{z_{(k+1)m}} \omega(y)\{p^*_{m,k}(y) - P_{km,\ldots,(k+1)m}(y)\}\, dy\right|$$

$$\leq \{\sum_{j=0}^{n} |v^*_j| + \int_a^b |\omega(y)|dy\} \max_k |e_{m,k}(\phi)| \quad , \tag{2.52}$$

where $v^*_j = v_j$ if $j = km + i$, where $i \neq 0$ and where v^*_j is the sum of the moduli of the two terms in the right-hand side of (2.49b) when $j = km$ $(k = 0,1,\ldots, N-1)$. We can show that

$$\sum_{j=0}^{n} |v^*_j| \leq L_m \int_a^b |\omega(y)|dy ,$$

where

$$L_m = \sup_{0\leq t\leq m} \sum_{r=0}^{m} |l_r(t)| \quad \text{and} \quad l_r(t) = \prod_{\substack{s=0\\ s\neq r}}^{m} \{(t-s)/(r-s)\} .$$

Thus, we have

$$\sum_{j=0}^{n} |v^*_j| \leq \sum_{k=0}^{N-1} \sum_{r=0}^{m} \int_{z_{km}}^{z_{(k+1)m}} |\omega(y) l_{k,r}(y)|dy ,$$

where

$$l_{k,r}(x) = \prod_{\substack{j=km\\ j\neq km+r}}^{(k+1)m} \{(x-z_j)/(z_{km+r}-z_j)\} .$$

Hence,

$$\sum_{j=0}^{n} |v^*_j| \leq \sum_{k=0}^{N-1} \int_{z_{km}}^{z_{(k+1)m}} |\omega(y)| \sum_{r=0}^{m} |l_{k,r}(y)|dy$$

and by a change of variables

$$\sup_{z_{km} \le x \le z_{(k+1)m}} \sum_{r=0}^{m} |l_{k,r}(x)| = L_m$$

(independently of k) so that the required result follows.

We now obtain the error bound

$$\left|\int_a^b w(y)\ \phi(y)\,dy - \sum_{j=0}^{n} v_j\ \phi(z_j)\right|$$
$$\le (1+L_m) \int_a^b |w(y)|\,dy\ \sup_k |e_{m,k}(\phi)| \quad , \qquad (2.53)$$

where we can substitute (2.51) as a bound for $|e_{m,k}(\phi)|$. (Precisely the same result (2.53) can be obtained if we write

$$\sum_{j=0}^{n} v_j\ \phi(z_j) = \int_a^b w(y)\ A\phi(y)\,dy \ ,$$

where $A\phi(x) = P_{km,\ km+1,\ \ldots\ (k+1)m}(x)$ when $km \le x \le (k+1)m$ $(k = 0,1,\ldots,N-1)$; the error in approximating the integral is $\int_a^b w(y)\ \{\phi(y) - A\phi(y)\}dy$, and we can bound $||\phi(x) - A\phi(x)||_\infty$ by $\{1 + L_m\} \sup|e_{m,k}(\phi)|$, using the material of section 2.3.1.)

From the bounds (2.53) and (2.51) we see clearly that the error in the approximate integration is $O(h^{s+1})$ if $\phi^{(s+1)}(x)$ is continuous (or even bounded) and $s \le m$, as $h \to 0$, $N \to \infty$, and m is fixed. Under certain conditions the error bounds which we have given will be pessimistic, and the order term may be improved upon (de Hoog and Weiss 1973) but to achieve such results requires more complicated analysis than we propose to use here.

Example 2.25 (*calculation of Fourier coefficients*)
Filon's method (Davis and Rabinowitz 1967) uses the technique described above to obtain formulae for approximating

$$\int_0^1 \phi(y)\ \cos k\pi y\ dy \text{ and } \int_0^1 \phi(y)\ \sin k\pi y\ dy \ .$$

Fosdick (1968) gives a Peano kernel for Filon quadrature formulae. An algorithm has been given by Chase and Fosdick (1969). *

2.17.3

Eqns (2.49a) and (2.49b) are not always convenient for the computation of the weights v_j . Integration formulae can instead be derived using the moments or generalized moments of a function. Suppose that we seek a formula of the form

$$\int_a^b \omega(y)\ \phi(y)\,dy \simeq \sum_{j=0}^{n} v_j\ \phi(z_j) \quad ,$$

which is exact for polynomials of degree n . This is so if we make a choice of $z_0, \ldots, z_n$ and determine $v_0, \ldots, v_n$ so that

$$\int_a^b \omega(y)\ y^r\ dy = \sum_{j=0}^{n} v_j\ z_j^{\ r} \quad (r=0,1,2,\ldots,n) . \tag{2.54}$$

More generally we may take $(n + 1)$ linearly independent polynomials $\{p_r(x)\}$ each of degree n , for $r = 0,1,2,\ldots, n$, and require that

$$\int_a^b \omega(y)\ p_r(y)\,dy = \sum_{j=0}^{n} v_j\ p_r(z_j)\ (r=0,1,2,\ldots, n). \tag{2.55}$$

The quantities on the left-hand sides of (2.54) and (2.55) are called the moments and generalized moments, respectively, of the function $\omega(y)$. When $z_0, \ldots, z_n$ are fixed, eqn (2.54) (or (2.55)) becomes a linear system for determining $v_0, \ldots, v_n$.

<u>Example 2.26</u>

If we set

$$p_r(x) = \prod_{\substack{j=0\\ j\neq r}}^{n} (x-z_r)(z_j-z_r)^{-1}\ (r=0,1,2,\ldots,n)$$

in (2.55) we find $v_j = \int_a^b \omega(y)\ p_j(y)\,dy\ (j=0,1,2,\ldots,n)$. *

The general rule with weights (2.49a) and (2.49b) is obtained if we take $n = m$ and subdivide the range of integration with points

$z_j = a + j(b-a)/Nm$, then apply the above procedure over each interval $[z_{km}, z_{(k+1)m}]$. The illustration preceding section 2.17.1 provides a demonstration of such a procedure, with $\psi_1(r)$ and $\psi_2(r)$ providing generalized moments on appropriate intervals.

2.18. *Integration in more than one dimension*

We now consider the problem of integration over a hypercube, or equivalently, over a right-angled parallelepiped. We begin by discussing approximations for an integral of the form $\int_c^d \int_a^b \Phi(x,y)\,dy\,dx$. The problem is similar to that of quadrature, in so far as we seek '*cubature*' *rules* of the form

$$\int_a^b \int_c^d \Phi(x,y)\,dy \quad dx \simeq \sum_{j=0}^{n} \Omega_j \; \Phi(x_j,y_j) \quad . \tag{2.56}$$

It is sometimes considered convenient to require that $a \le x_j \le b$ and $c \le y_j \le d \; (j=0,1,2,\ldots, n)$. We may construct an approximation (2.56) in a natural way if we have one or two quadrature rules giving approximations of the form

$$\int_c^d \phi(y)\,dy \simeq \sum_{j=0}^{p} A_j \; \phi(y_j) \tag{2.57}$$

and

$$\int_a^b \psi(x)\,dx \simeq \sum_{j=0}^{q} B_j \; \psi(x_j) \quad . \tag{2.58}$$

We may then think it natural to write

$$\begin{aligned}\int_a^b \{\int_c^d \Phi(x,y)\,dy\}\,dx &\simeq \int_a^b \{\sum_{j=0}^{p} A_j \; \Phi(x,y_j)\}\,dx \\ &\simeq \sum_{k=0}^{q} \sum_{j=0}^{p} B_k \, A_j \; \Phi(x_k,y_j) \ ,\end{aligned} \tag{2.59}$$

and we then obtain an approximation which can be put in the form (2.56). If we denote the quadrature rule in (2.57) by $J(\phi)$ and the quadrature rule in (2.58) by $Q(\phi)$, it is convenient to describe the double sum in (2.59) with the notation $Q_x J_y(\phi)$. We say that the cubature rule $Q_x J_y(\phi)$ is the (*cartesian*) *product* (in the order specified) of the quadrature rules (2.57) and (2.58). A different cartesian product is obtained, in general, from $J_x Q_y(\phi)$.

Example 2.27 (*products of rules with themselves*)

(a) *The product mid-point rule.*

$$\int_a^b\{\int_c^d \Phi(x,y)\,dy\}dx \simeq (b-a)(d-c)\,\Phi(\tfrac{1}{2}(a+b),\,\tfrac{1}{2}(c+d))\ .$$

(b) *The product trapezium rule.*

$$\int_a^b\{\int_c^d \Phi(x,y)\,dy\}\,dx \simeq \tfrac{1}{4}(b-a)(d-c)\{\Phi(a,c) + \Phi(a,d) + \Phi(b,c) + \Phi(b,d)\}\ .$$

*

Not every cubature rule is the product of two quadrature rules. However, if we set $J^{[i]}(\phi) = (d-c)\ \phi(y_i)$ and $Q^{[i]}(\psi) = (b-a)\ \psi(x_i)$ ($i = 0,1,2,\ldots,\ n$) then

$$Q_x^{[i]}\ J_y^{[i]}\ (\Phi) = (b-a)(d-c)\ \Phi(x_i,\ y_i)\ ,$$

and the sum in eqn (2.56) may be written

$$\sum_{r=0}^{n} \Omega_r\ \Phi(x_r,y_r) = \sum_{r=0}^{n} \Omega_r Q_x^{[r]}\ J_y^{[r]}(\Phi)/\{(b-a)(d-c)\}\ . \tag{2.60}$$

Thus any cubature rule (2.56) is the weighted sum of certain cartesian products. This result indicates that there is a large variety of cubature rules to be found, and we seek a method of comparing the usefulness of each of them. As a prelude to this we define the (polynomial) *degree of precision* of a cubature rule. If the approximation (2.56) is exact when $\Phi(x,y) = x^r y^s$ and $0 \le r + s \le \rho$, but is not exact for $\Phi(x,y) = x^r y^s$ where, $r + s = \rho + 1$, then ρ is known as the degree of precision of the cubature rule.

We find a number of special rules of the form (2.56) in the

literature, and often they have been constructed so that they have a high degree of precision whilst requiring few function evaluations. Of itself, a high degree of precision does not guarantee high accuracy. However, a cubature rule may be used to supply a composite version, in which the basic rule is applied to each of m^2 equal rectangles which form the area of integration. If we use the representation (2.60) and denote a composite version of a quadrature rule $J(\phi)$ by $(m \times J)(\phi)$ then a composite version of the rule (2.56) may be defined as

$$\sum_{r=0}^{n} \Omega_r (m \times K_x^{[r]})\ (m \times J_y^{[r]})(\Phi)/\{(b-a)(d-c)\} .$$

The degree of precision (ρ, say) affects the behaviour of the error in this composite version as $m \to \infty$, when the integrand is smooth. For, if $\Phi(x,y)$ is sufficiently smooth, the error is $O(1/m^{\rho+1})$. It is then an advantage to make ρ as large as possible. Furthermore, the error can be expanded as an asymptotic series

$$\sum_{k=\rho+1}^{N} \tau_k (1/m)^k + O(1/m^{N+1})$$

when $\Phi(x,y)$ possesses bounded partial derivatives of total order $N+1$, for $N \geq 0$ (Baker and Hodgson 1971). It may be shown that τ_k vanishes for some values of k when the cubature rule has certain properties or the integrand is of a certain type.

Remark. If a cubature rule has a degree of precision greater than or equal to zero, and it is the product of two quadrature rules, we can obtain error bounds in terms of the derivatives of the integrand $\Phi(x,y)$ when we know the Peano constants of the quadrature rules (Nikolskii 1966; Stroud and Secrest 1966). We denote the rules in (2.57) and (2.58) by $J(\phi)$, $K(\psi)$, and suppose that the degree of precision of $J(\phi)$ is r_J and the degree of precision of $K(\psi)$ is r_K. We also suppose that $(\partial/\partial x)^{s+1}\ \Phi(x,y)$ and $(\partial/\partial y)^{t+1}\ \Phi(x,y)$ are continuous for $a \leq y \leq b$, $c \leq x \leq d$ and bounded by $N_{s+1}^{(x)}$ and $N_{t+1}^{(y)}$ respectively, where $0 \leq s \leq r_K$ and $0 \leq t \leq r_J$. We assume that the abscissae of $J(\phi)$ and $K(\psi)$ are in the ranges of integration, and the weights of $J(\phi)$ satisfy $\sum_{j=0}^{q} |A_j| \leq A$.

Then

$$\left|\int_c^d \left\{\int_a^b \Phi(x,y)\,dy\right\}dx - K_x J_y(\Phi)\right| \le (b-a)\; c_{s+1}(K;c,d)\; N_{s+1}^{(x)} +$$

$$+ Ac_{t+1}(J;a,b)\; N_{t+1}^{(y)} \quad . \qquad (2.61)$$

The form of the result (2.61) permits us to bound the error in composite rules. We could also use (2.60) with $r = s = 0$, or similar relationships for special rules, to extend (2.61) to rules which are not themselves product rules.

Remark. The book of Stroud (1972) covers much that is known about approximate multiple integration.

The discussion given above for the problem of cubature should serve as an indication of how we may deal with integration in more than two dimensions. In general we seek an approximation of the form

$$\int_V \Phi(\underset{\sim}{x})\; dS \simeq \sum_{j=0}^{n} \Omega_j\; \Phi(\underset{\sim}{x}_j) \;, \qquad (2.62)$$

where $\underset{\sim}{x} = [x_0,\ldots,x_d]^T$, say, and V is the volume $a_i \le x_i \le b_i$ $(i = 0,1,2,\ldots, d)$. One method of constructing such an approximation consists of taking the cartesian product of $d+1$ quadrature rules $J^{(i)}(i = 0,1,2,\ldots, d)$. Unfortunately the number of function evaluations required by such a rule rises rapidly with d and from the practical viewpoint some other method eventually becomes essential. A discussion of methods for constructing integration rules over hypercubes has been given by Lyness in a series of papers (Lyness 1965*a*,*b*, and *c*) , and the basis of Monte Carlo methods is discussed by Davis and Rabinowitz (1967). Thacher (1964) has given a rule with degree of precision 2 for approximate integration over a hypercube in d dimensions using $(2d+1)$ function evaluations. He made some comparison of his rule with Monte Carlo methods and product rules.

2.19. Asymptotic upper and lower bounds

One of the more difficult aspects of employing a numerical method lies in the problem of determining the accuracy of a result and of modifying the method if the accuracy is insufficient. In the problem of quadrature we are able to bound the error in terms of derivatives of the integrand, but the bounds may be pessimistic or difficult to evaluate. In practice we may seek agreement between different numerical results obtained in different ways.

Before proceeding further we define the term *asymptotic bound.* Suppose that we have real quantities T, and $T(\eta)$ depending on a real $\eta > 0$. If there is a value $\eta_0 > 0$ such that either (a) $\eta < \eta_0$ implies that $T(\eta) \geq T$ or (b) $\eta < \eta_0$ implies that $T(\eta) \leq T$, then we say that the values of $T(\eta)$ give asymptotic (upper or lower) bounds for T, as $\eta \to 0$. The same terminology is applied when $T, T(\eta)$ are complex-valued if both the real and imaginary parts of $T(\eta)$ give asymptotic bounds for the real and imaginary parts of T, respectively. We shall outline a scheme (see Bulirsch and Stoer 1966) for obtaining asymptotic upper and lower bounds for a computed quantity, not necessarily the value of an integral.† The scheme may be regarded as a systematic application of the criterion of 'consistency' between certain numerical results. If rounding error does not interfere, the scheme works when the error in our approximations has a certain type of asymptotic behaviour. Unfortunately, the error in an approximation does not always behave in the manner we require.

Let us suppose that for different positive values of a parameter η we obtain real-valued approximations $T(\eta)$ to a real T. We suppose that the *discretization parameter* η and the numerical method are such that $\lim_{\eta \to 0} T(\eta) = T$. To give an example, we could take $T = \int_a^b \phi(y)\,dy$ and $T(\eta) = (m \times J)(\phi)$ with $\eta = 1/m$, where $J(\phi)$ is any quadrature rule which is a member of J^+. We observe that η is not required to take arbitrary real values. In addition to the condition of convergence as $\eta \to 0$, we shall require that, as $\eta \to 0$,

$$T(\eta) = T + \tau_k \eta^k + o(\eta^k) \quad (\tau_k \neq 0) , \tag{2.63}$$

† The principle involved has wider applications later.

where $k \geq 1$. This condition is certainly satisfied if

$$T(\eta) = T + \tau_k\eta^k + O(\eta^{k+1}) \quad (\tau_k \neq 0) \ , \tag{2.64}$$

as $\eta \to 0$. In view of (2.62), we see that $T(\eta)$ tends to T from above if $\tau_k > 0$, and from below if $\tau_k < 0$, as $\eta \to 0$. Thus, for all sufficiently small η , $T(\eta)$ provides either an upper bound or a lower bound for the quantity T . No such statement can be made, without further investigation, if $\tau_k = 0$. (Furthermore, this is a theoretical statement which supposes that rounding error does not obscure the asymptotic behaviour.) Now if $\tau_k \neq 0$ we shall obtain asymptotic bounds for T from the opposite side if we can find a quantity, $U(\eta)$ say, such that

$$U(\eta) = T - \alpha\tau_k\,\eta^k + o(\eta^k) \quad , \tag{2.65}$$

where $\alpha > 0$, or such that

$$U(\eta) = T - \alpha\tau_k\,\eta^k + O(\eta^{k+1}) \ . \tag{2.66}$$

To this end let us suppose that we can construct $T(h)$ and $T(\gamma h)$ for some value of γ to be fixed later. Then

$$T(\gamma\eta) = T + \tau_k(\gamma\eta)^k + o(\eta^k) \ .$$

If we define $U(\eta) = \{T(\eta) - \beta T(\gamma\eta)\}/(1 - \beta)$, where $\beta \neq 1$ then $\lim_{\eta\to 0} U(\eta) = T$, and $U(\eta) = T - \alpha\tau_k\eta^k + o(\eta^k)$, where $\alpha = (1-\beta\gamma^k)/(\beta-1)$.
We now seek β and $\gamma > 0$ such that $\infty > \alpha > 0$. If $0 < \beta < 1$, we require that $\beta\gamma^k > 1$. If $\beta > 1$ we require $\beta\gamma^k < 1$. Possible values of β and γ are: (a) $\gamma = 2$, $\beta = \frac{3}{4}$, for $k = 1,2,3,\ldots$, (b) $\gamma = \frac{1}{2}$, $\beta = 2$, for $k = 2,3,4,\ldots$. Certain choices of β and γ have advantages over others, either through the effect on the amount of computation required and the effect on rounding error, or for 'analytical' reasons. In this respect it would seem unreasonable to have an upper bound which is much better than a lower bound, or vice

versa. We may therefore seek to make α as close as possible to 1, or as close as possible to γ^k. (In the latter case we would take $T(\gamma\eta)$ and $U(\eta)$ as our pair of asymptotic upper and lower bounds, instead of $T(\eta)$ and $U(\eta)$,)

Our analysis† has been based on eqn (2.63), and it is equally valid for (2.64). In practice we may suppose (hopefully) that η is sufficiently small for the asymptotic theory to apply, when we observe that $T(\eta)$ is monotonic as η is decreased further.

We now have the basis for a systematic examination of the consistency of successive values of $T(\eta)$. Suppose that we compute the T-sequence $T(\eta_0)$, $T(\gamma\eta_0)$, $T(\gamma^2\eta_0)$, $T(\gamma^3\eta_0)$. It is not difficult to obtain a U-sequence $U(\eta_0)$, $U(\gamma\eta_0)$, $U(\gamma^2\eta_0)$, ... from these values. When the U-sequence and T-sequence are monotonic we use the latest members of the sequences to estimate an interval containing the true value T, supposing that (2.63) is valid with $\tau_k \neq 0$.

Example 2.28

In Table 2.4 an example of a T-sequence and a U-sequence arising in quadrature is given. We have

$$T = \int_0^1 \frac{\exp\{1/(y+0\cdot1)\}dy}{(y+0\cdot1)^2} \simeq 2\cdot202398 \times 10^4 .$$

We set $\eta = h$ and let $T(h)$ denote the repeated trapezium rule applied with step h to approximating T. With $h = 1, \frac{1}{2}, \frac{1}{4}, \ldots,$ we obtained the following values. Here, $U(h) = 2T(\frac{1}{2}h) - T(h)$.

† Our discussion has been based on the case where T and $T(\eta)$ are real. In the case where T and $T(\eta)$ are complex, the analysis applies separately to the real and imaginary parts; thus, we might find asymptotic *upper* bounds for the real part of T and asymptotic *lower* bounds for the imaginary part, for example.

TABLE 2.4

h	T-sequence	U-sequence
1	$1{\cdot}10132 \times 10^6$	$1{\cdot}47069 \times 10^1$
1/2	$5{\cdot}50670 \times 10^5$	$7{\cdot}33124 \times 10^1$
1/4	$2{\cdot}75371 \times 10^5$	$4{\cdot}32225 \times 10^2$
1/8	$1{\cdot}37902 \times 10^5$	$2{\cdot}29017 \times 10^3$
1/16	$7{\cdot}00960 \times 10^4$	$7{\cdot}75538 \times 10^3$
1/32	$3{\cdot}89257 \times 10^4$	$1{\cdot}50765 \times 10^4$
1/64	$2{\cdot}70011 \times 10^4$	$1{\cdot}96793 \times 10^4$
1/128	$2{\cdot}33402 \times 10^4$	$2{\cdot}13763 \times 10^4$
1/256	$2{\cdot}23583 \times 10^4$	$2{\cdot}18575 \times 10^4$
1/512	$2{\cdot}21079 \times 10^4$	$2{\cdot}19821 \times 10^4$
1/1024	$2{\cdot}20450 \times 10^4$	

Both sequences are monotonic here, and the T-sequence provides upper bounds and the U-sequence provides lower bounds. *

In some cases we have a stronger property than (2.63), of the form

$$T(\eta) = T + \sum_{j=k}^{N} \tau_j \eta^j + o(\eta^N) \qquad (\eta \text{ real}), \qquad (2.67)$$

or, in particular,

$$T(\eta) = T + \sum_{j=k}^{N} \tau_j \eta^j + O(\eta^{N+1}), \qquad (2.68)$$

with $N > k$. It is then possible to perform the *deferred approach* to the limit and simultaneously obtain a number of asymptotic upper and lower bounds. The values of T and $T(\eta)$ are permitted to be complex.

We have encountered, in Romberg's scheme, an example of the

deferred approach to the limit. (Romberg's scheme corresponds to setting $\eta = h^2$, $T = I(\phi)$, and denoting the trapezium rule with step h by $T(\eta)$.) In general, we see from (2.67) that the error in $T(\eta)$ as an approximation to T is $O(\eta^k)$. We seek approximations derived from values of $T(\eta)$, with errors which are successively (Bulirsch and Stoer 1966)

$$O(\eta^{k+1}),\ O(\eta^{k+2}),\ \ldots,\ O(\eta^N),$$

and $o(\eta^N)$ or $O(\eta^{N+1})$ depending on which of (2.67) or (2.68) is valid. Let us suppose that (2.67) holds, and that $\eta_0, \eta_1, \ldots, \eta_p$ are distinct values, with $p \leq N - k$. It is clear that we can then choose $\alpha_0, \alpha_1, \ldots, \alpha_p$ so that

$$\sum_{i=0}^{p} \alpha_i T(\eta_i) = T + \sum_{j=k+p}^{N} \nu_j \eta^j + o(\eta^N), \tag{2.69}$$

where ν_j ($j = k+p, k+p+1, \ldots N$) depends on $\alpha_0, \alpha_1, \ldots, \alpha_p$. The values of $\alpha_0, \alpha_1, \ldots, \alpha_p$ can be determined from the equations

$$\sum_{j=0}^{p} \alpha_j \tau_r \eta_j^r = 0 \quad (r = k, k+1, \ldots, k+p-1),$$

$$\sum_{j=0}^{r} \alpha_j = 1. \tag{2.70}$$

When $\eta_j = \gamma_j \eta$ for $j = 0, 1, 2, \ldots, p$, the left-hand side of (2.69) is a function of η which we may denote by $T_p(\eta)$. (In particular we may take $\gamma_j = \frac{1}{2}^j$.) Asymptotic bounds for $T_p(\eta)$ can be constructed in the same way as for $T(\eta)$, if $\nu_{k+p} \neq 0$. In computing $T_p(\eta)$ we are extrapolating on $T(\eta_i)$ ($i = 0,\ldots,p$) to approximate $T = T(0)$, thus the process of the classical Romberg scheme is (repeated) h^2-*extrapolation*.

2.20. *Numerical methods*

In the preceding sections we have been concerned principally with the background theory of certain aspects of numerical analysis.

If the reader wishes to solve integral equations numerically, he will require a knowledge of certain numerical techniques. The theoretical discussion of these techniques is beyond the scope of this book. In this section I propose to draw attention to certain established numerical techniques and to introduce the reader to sources of algorithms for dealing with problems which may arise in the numerical solution of integral equations. The book by Fox and Mayers (1969) entitled *Computing methods for scientists and engineers* and the book *Modern computing methods* published by Her Majesty's Stationery Office (HMSO 1961) provide useful surveys of a number of numerical techniques. On the other hand, a number of coded algorithms are published in the literature. I must mention especially the journals *Numerische Mathematik*, with its Handbook Series of algorithms, some now collected in the *Handbook for automatic computation* (vol. 2; eds Wilkinson and Reinsch, 1971) , *B.I.T. (Nordisk Tidskrift för Informationsbehandling)*, the *Communications of the Association for Computing Machinery*, and *The Computer Journal*. Algorithms have also been published by the Centre National de la Recherche Scientifique (Paris), and the Numerical Algorithms Group (NAG 1973).

2.20.1.

Much of our previous discussion of formulae for numerical integration is related to the discretization of an integral equation. In a number of situations we shall require evaluation of integrals by the application of an effective algorithm. The books by Davis and Rabinowitz (1967 and 1975) provide a selection of Fortran programs, and a bibliography of Algol procedures. Davis and Rabinowitz (1967, p.103; 1975, p.362) give a short discussion of the effect of round-off error in numerical integration. A number of reasonably effective algorithms have been based on Simpson's rule, and also on Romberg's scheme, and on a scheme of Clenshaw and Curtis (NAG 1973).

The Romberg scheme which we have described provides one method (probably not the 'best' method) of calculating an integral of the form $\int_a^b \phi(y)dy$, where a,b are finite and $\phi(x)$ is smooth. If the function $\phi(x)$ is smooth for $c \leq x \leq b$ and has large derivatives for $a \leq x \leq c$ it may be better to use the scheme to calculate

$\int_a^c \phi(y)dy$ and $\int_c^b \phi(y)dy$ separately. A very efficient algorithm for numerical integration should take account of local variations in the behaviour of $\phi(x)$. The problem of determining integrals where one or both of the limits of integration are infinite is a difficult one. We feel that the Gauss-Hermite and Gauss-Laguerre formulae (see Example 2.24) are often unreliable, except possibly as an indication that an integral may be negligible. For an integral of the form $\int_a^\infty \phi(y)dy$ we suggest seeking a value c such that $\int_c^\infty \phi(y)dy = \int_0^\infty e^{-y}\{e^y\phi(y+c)\}dy$ is negligible when estimated by Gauss-Laguerre quadrature. Then

$$\int_a^\infty \phi(y)dy \simeq \int_a^c \phi(y)dy .$$

A similar procedure may be employed when evaluating $\int_{-\infty}^{\infty} \phi(y)dy$.

2.20.2.

We now turn to problems of linear algebra which arise frequently in numerical work. Useful surveys are provided by Forsythe and Moler (1967), the review by Forsythe (1967), and the chapters by Martin and Wilkinson in Walsh (1966). Further reading is provided by the books of Fox (1964), Wilkinson (1965), Varga (1962), and Wachspress (1966). Details of the methods which we mention below can be found in the above references, and also in Wilkinson and Reinsch (1971).

The solution of the system of equations

$$\sum_{j=0}^{n} A_{ij}\, x_j = b_i \qquad (i = 0,1,2,\ldots,n)$$

(where $\det(A_{ij}) \neq 0$) may be obtained using *Gaussian elimination*, the equations being suitably ordered by the use of interchanges (*partial pivoting* or *pivotal condensation*) as the calculation proceeds. In practice we work with the augmented matrix $[\underset{\sim}{A}|\underset{\sim}{b}]$, and interchange the rows. If we also interchange the columns of $\underset{\sim}{A}$ and re-order the components $\{x_i\}$ the process of interchanging is called *complete pivoting*. A pivoting strategy is used in order to limit the effects of rounding error, but even though $|\det(A_{ij})|$ may be large, the solution of the equations can be sensitive to rounding error. The system is then said to be ill-conditioned, and the degree of ill-conditioning can be measured by various *condition numbers*.

The condition number commonly employed has the form $\mu(\underset{\sim}{A}) = \|\underset{\sim}{A}\| \, \|\underset{\sim}{A}^{-1}\|$ where $\underset{\sim}{A}$ is a (square) non-singular matrix and $\| \ \|$ is a matrix norm (which is usually subordinate to a vector norm); if $\underset{\sim}{A}$ is singular $\mu(\underset{\sim}{A})$ is undefined. When $\mu(\underset{\sim}{A})$ is 'large' (its exact size depends on the choice of norm) the system is said to be ill conditioned. This diagnosis follows from an error analysis in which it is shown that if $\underset{\sim}{A}\underset{\sim}{x} = \underset{\sim}{b}$, $(\underset{\sim}{A} + \delta\underset{\sim}{A})(\underset{\sim}{x} + \delta\underset{\sim}{x}) = \underset{\sim}{b} + \delta\underset{\sim}{b}$, $\underset{\sim}{A}$ is non-singular and $\delta\underset{\sim}{A}$ is relatively small, then $\|\delta\underset{\sim}{x}\| / \|\underset{\sim}{x}\|$ can be bounded in terms of $\|\delta\underset{\sim}{A}\| / \|\underset{\sim}{A}\|$ and $\|\delta\underset{\sim}{b}\| / \|\underset{\sim}{b}\|$ by a quantity which is large when $\mu(\underset{\sim}{A})$ is large. For details see, for example, Wilkinson (1963, p.91 *et seq.*). The reader interested in the solution of Fredholm equations of the second kind should familiarize himself with such results by referring, say, to Wilkinson (1963), or Forsythe and Moler (1967, pp. 20-6).

The numerical solution can be affected by *scaling* of the equations or the unknowns $\{x_i\}$ (this is sometimes called *equilibration*), but no optimum strategy for scaling appears to be known at present. The process of *iterative refinement* or *iterative improvement* of a solution appears to be quite useful, since it leads to an estimate of a condition number and, in general, an accurate solution (Forsythe and Moler 1967, p.49 *et seq.*). Let us write $\underset{\sim}{A} = [A_{ij}]$. The basis of the method of iterative refinement is the construction of a sequence of approximations $\underset{\sim}{x}^{(k)}$, $(k = 0,1,2,\ldots,)$ to the solution $\underset{\sim}{x} = [x_0, \ldots, x_n]^T$. We determine $\underset{\sim}{x}^{(0)}$ by Gaussian elimination, and solve the equation

$$\underset{\sim}{A}\{\underset{\sim}{x}^{(k+1)} - \underset{\sim}{x}^{(k)}\} = \underset{\sim}{r}^{(k)} = \underset{\sim}{b} - \underset{\sim}{A}\,\underset{\sim}{x}^{(k)} ,$$

where $\underset{\sim}{r}^{(k)}$ is computed to a high accuracy. These equations can be solved efficiently, using the information obtained in computing $\underset{\sim}{x}^{(0)}$.

When the number of linear equations is large it is sometimes necessary to use *iterative methods* to obtain the solution, because of the problem of storing the elements $\{A_{ij}\}$, $\{b_i\}$, in a way which is convenient for Gaussian elimination.

We mention in passing the least-squares problem, which consists of finding a vector $\underset{\sim}{x}$ to minimize $\|\underset{\sim}{A}\underset{\sim}{x} - \underset{\sim}{b}\|_2$, where $\underset{\sim}{A}$ is square or rectangular. (If $\underset{\sim}{A}$ is not of full rank the solution $\underset{\sim}{x}$ is not unique and we choose the vector of minimimum norm $\|\underset{\sim}{x}\|_2$.) If $\underset{\sim}{A}$ is rectangular and of full rank, the solution can in theory be obtained from the *normal equations* represented by $\underset{\sim}{A}^*\underset{\sim}{A}\underset{\sim}{x} = \underset{\sim}{A}^*\underset{\sim}{b}$. As a practical technique this is not recommended. We prefer to use either a Householder technique (Björck and Golub 1967), a *Gram-Schmidt* technique (Björck 1967), or the *singular-value decomposition* of $\underset{\sim}{A}$ (Golub and Reinsch 1970).

We shall return to a discussion of the use of the singular-value decomposition in our discussion of equations of the first kind in Chapter 5. Let us observe here that the *singular vectors* $\{\underset{\sim}{u}, \underset{\sim}{v}\}$ are non-zero vectors satisfying

$$\underset{\sim}{A}\,\underset{\sim}{v} = \sigma\,\underset{\sim}{u}$$

$$\underset{\sim}{A}^*\underset{\sim}{u} = \sigma\,\underset{\sim}{v}\ ,$$

where $\sigma \geq 0$. Such a value σ will be called a *singular value* of $\underset{\sim}{A}$. Observe that if $\underset{\sim}{A}$ has m rows and n columns and $\underset{\sim}{O}$ is interpreted to be either of order m or of order n according to position, then

$$\left[\begin{array}{c|c} \underset{\sim}{O} & \underset{\sim}{A} \\ \hline \underset{\sim}{A}^* & \underset{\sim}{O} \end{array}\right] \begin{bmatrix} \underset{\sim}{u} \\ \\ \underset{\sim}{v} \end{bmatrix} = \sigma \begin{bmatrix} \underset{\sim}{u} \\ \\ \underset{\sim}{v} \end{bmatrix}$$

so that σ is an eigenvalue of a symmetric matrix. Also, $\underset{\sim}{A}^*\underset{\sim}{A}\underset{\sim}{v} = \sigma^2\underset{\sim}{v}$ and $\underset{\sim}{A}\underset{\sim}{A}^*\underset{\sim}{u} = \sigma^2\underset{\sim}{u}$. Let us now suppose, for definiteness, that $m \geq n$. Suppose that $\underset{\sim}{v}_1,\ldots,\underset{\sim}{v}_n$ are orthogonal eigenvectors of $\underset{\sim}{A}^*\underset{\sim}{A}$, and $\underset{\sim}{v}_1,\ldots,\underset{\sim}{v}_k$ correspond to the non-zero eigenvalues $\sigma_1^2,\ldots,\sigma_k^2$, where k is the rank of $\underset{\sim}{A}$. If we set $\underset{\sim}{u}_i = \sigma_i^{-1}\underset{\sim}{A}\underset{\sim}{v}_i$ then (Noble 1969) $\underset{\sim}{u}_i$, $\underset{\sim}{v}_i$ are singular vectors of $\underset{\sim}{A}$, and $\underset{\sim}{u}_1,\ldots,\underset{\sim}{u}_k$ are orthogonal eigenvectors of $\underset{\sim}{A}\underset{\sim}{A}^*$ corresponding to non-zero eigenvalues $\sigma_1^2,\ldots,\sigma_k^2$

Further eigenvectors $u_{k+1}, \ldots, u_m$ (corresponding to zero eigenvalues) can be obtained, if $k < m$, so that $u_1, \ldots, u_m$ is a complete orthogonal set of eigenvectors of AA^*.

The existence of a singular-value decomposition $U^*A\,V = \Sigma$ or (equivalently) $A = U\,\Sigma\,V^*$, can be established theoretically by using eigenvalue theory for A^*A and $A\,A^*$. See, for example, Forsythe and Moler (1967); we shall not pursue this point here.

If we set $U = [u_1|u_2|\ldots|u_m]$ and $V = [v_1|v_2|\ldots|v_n]$ then $U^*AV = \Sigma$ where Σ is a rectangular matrix with non-zero elements only in the 'diagonal' (i, i)th position. (The 'diagonal' elements of Σ are the singular values σ_i.) This decomposition $U^*AV = \Sigma$ is called the singular-value decomposition of A, see also Noble (1969).

The vector x of minimum norm $\|x\|_2$ which minimizes $\|Ax - b\|_2$ is given by setting

$$x = \sum_{\sigma_i \neq 0} \{u_i^* b/\sigma_i\} v_i .$$

The analogy with Fredholm equations of the first kind (see the concluding part of section 1.8) will be clear.

We may also be concerned with the solution of the eigenvalue problem

$$\sum_{j=0}^{n} A_{ij}\, x_j^{(r)} = \lambda^{(r)}\, x_i^{(r)} \qquad (i = 0,1,2,\ldots, n)$$

in which we seek the eigenvalues $\lambda^{(0)}, \lambda^{(1)}, \ldots, \lambda^{(n)}$ and, possibly corresponding eigenvectors $x^{(0)}, x^{(1)}, \ldots, x^{(n)}$, where $x^{(r)} = [x_0^{(r)}, \ldots, x_n^{(r)}]^T$. When the matrix A is real and symmetric (that is, $A_{ij} = A_{ji}$ for $i, j = 0,1,2,\ldots, n$) we may use *Householder's method* to reduce the problem to a situation where the matrix is real, symmetric, tridiagonal, and similar to the original matrix. (A matrix $T = |T_{ij}|$ is tridiagonal when $T_{ij} = 0$ if $|i - j| > 1$.) The eigenvalues of a real, symmetric, tridiagonal matrix may be located using a *Sturm sequence* property, and eigenvectors can then be found by an iterative technique. We refer to algorithms given by Wilkinson and others in the *Handbook Series (Linear Algebra)*, published in *Numerische Mathematik*. The *QR method* and variants of *Jacobi's* method may be useful for the eigenvalue problem when the matrix is real and symmetric. In the case of a general matrix $[A_{ij}]$ the eigenvalue problem may be treated using some variant of the *QR algorithm* (Francis 1960) after reducing the problem to one in which the matrix is upper Hessenberg, that is of the form $[A_{ij}]$, where $A_{ij} = 0$ if $i > j + 1$. Peters and Wilkinson (1970*b*) supply an algorithm for real and complex matrices.

The *power method* is an iterative method, and versions of this method can be applied to find an eigenvalue and a corresponding eigenvector when an approximation to the eigenvalue is known. The method of *inverse iteration* is a valuable variant of the power method, and we refer to Walsh (1966, p.54) for a brief description.

Remark. We shall sometimes encounter a generalized eigenvalue problem of the form $\underset{\sim}{A}\underset{\sim}{x} = \lambda \underset{\sim}{B}\underset{\sim}{x}$. Peters and Wilkinson (1970*a*) give a discussion of this problem. The case where $\underset{\sim}{A} = \underset{\sim}{A}^*$ and $\underset{\sim}{B} = \underset{\sim}{B}^*$ is positive definite is easier to treat than the more general case. The *QZ* algorithm (see Moler and Stewart 1973) is a generalization of the *QR* algorithm. It may be recommended (see Moler and Stewart for a discussion) for the generalized eigenvalue problem. The method is based on the fact that there are unitary (complex) matrices $\underset{\sim}{Q}, \underset{\sim}{Z}$ such that $\underset{\sim}{Q}\underset{\sim}{A}\underset{\sim}{Z} = \underset{\sim}{R}$ and $\underset{\sim}{Q}\underset{\sim}{B}\underset{\sim}{Z} = \underset{\sim}{L}$ where $\underset{\sim}{R}$ and $\underset{\sim}{L}$ are both upper-triangular. The generalized eigenvalues $\{\lambda_i\}$ are then, given by $\lambda_i = R_{ii}/L_{ii}$ (and λ_k is infinite if $L_{kk} = 0$, $R_{kk} \neq 0$). In the *QZ* algorithm an attempt is made to obtain $\underset{\sim}{R}$ and $\underset{\sim}{L}$, with modifications to avoid complex arithmetic.

Many aspects of the numerical solution of the eigenvalue problem have been investigated in depth in the work of Wilkinson (1966). We shall investigate the problem of assessing the sensitivity of eigenvalues to perturbations (say, in the elements of the matrix) since the accuracy obtainable is inherently governed by this feature.

Before discussing this problem we recall certain aspects (not all well known) of the mathematical theory of the eigenvalue problem.

An eigenvalue $\lambda^{(r)}$ of the matrix $\underset{\sim}{A}$ is said to have *algebraic multiplicity* p_r if it is a zero of multiplicity p_r of the polynomial $\det(\underset{\sim}{A} - \lambda\underset{\sim}{I})$. (We say the eigenvalue is simple if $p_r = 1$.) The *geometric multiplicity* m_r of the eigenvalue $\lambda^{(r)}$ is the maximum number of linearly independent eigenvectors $\underset{\sim}{x}_j^{(r)}$ such that

$(\underset{\sim}{A} - \lambda^{(r)}\underset{\sim}{I})\underset{\sim}{x}_j^{(r)} = \underset{\sim}{0}$ - it can be shown that $1 \le m_r \le p_r$. If $m_r = p_r$ (in particular, if $p_r = 1$) for every eigenvalue $\lambda^{(r)}$ ($r \in \{0,1,2,\ldots, n\}$) we can find a basis for the space of column vectors which consists of eigenvectors of $\underset{\sim}{A}$, but in the case $\Sigma m^{(r)} < \Sigma p^{(r)} = n + 1$ this is not possible. It is the latter case which causes complications in the mathematical theory, but we can form a basis if we augment the eigenvectors by vectors $\underset{\sim}{x} \neq \underset{\sim}{0}$ which for some eigenvalue $\lambda^{(r)}$, and some integer $k > 1$, satisfy $(\underset{\sim}{A}-\lambda^{(r)}\underset{\sim}{I})^k \underset{\sim}{x} = \underset{\sim}{0}$. Such vectors are called generalized eigenvectors, or principal vectors.

A set of generalized eigenvectors can be generated by considering, for each eigenvalue $\lambda^{(r)}$, various 'chains' of vectors satisfying recurrence relations of the form

$$(\underset{\sim}{A} - \lambda^{(r)}\underset{\sim}{I})\underset{\sim}{z}_0 = \underset{\sim}{0} , \quad (\underset{\sim}{A} - \lambda^{(r)}\underset{\sim}{I})\underset{\sim}{z}_p = \underset{\sim}{z}_{p-1} \quad (p = 1,2,3,\ldots) ,$$

where $\underset{\sim}{z}_0$ is some eigenvector of $\underset{\sim}{A}$. A generalised eigenvector $\underset{\sim}{x}$ is said to be of *grade* (or rank) k if $(\underset{\sim}{A} - \lambda^{(r)}\underset{\sim}{I})^{k-1}\underset{\sim}{x} \neq \underset{\sim}{0}$ and $(\underset{\sim}{A} - \lambda^{(r)}\underset{\sim}{I})^k\underset{\sim}{x} = \underset{\sim}{0}$. A discussion is given by Pease (1965, pp.76-82).

The reader may recall that, given a matrix $\underset{\sim}{A}$, there exists a matrix $\underset{\sim}{Z}$ such that $\underset{\sim}{Z}^{-1} \underset{\sim}{A}\underset{\sim}{Z} = \underset{\sim}{J}$ is the (upper) Jordan canonical form of $\underset{\sim}{A}$:

$$\underset{\sim}{J} = \begin{bmatrix} \underset{\sim}{C}_1(\lambda^{(1)}) & & & \underset{\sim}{0} \\ & \ddots & & \\ & & \ddots & \\ \underset{\sim}{0} & & & \underset{\sim}{C}_q(\lambda^{(p)}) \end{bmatrix}$$

where $\underset{\sim}{C}_t(\lambda^{(r)})$ is a matrix of order p_r of the form

$$\underset{\sim}{C}_t(\lambda^{(r)}) = \begin{bmatrix} \lambda^{(r)} & \alpha_1^{(r)} & & 0 \\ & \lambda^{(r)} & \ddots & \\ & & \ddots & \alpha_{p_r-1}^{(r)} \\ 0 & & & \lambda^{(r)} \end{bmatrix}$$

with $\alpha_s^{(r)} = 1$ or 0, $\Sigma\alpha_s^{(r)} = p_r - m_r$. The columns of $\underset{\sim}{Z}$ are eigenvectors or generalized eigenvectors of $\underset{\sim}{A}$ and the diagonal elements of $\underset{\sim}{J}$ are the eigenvalues of $\underset{\sim}{A}$. (The matrix $\underset{\sim}{Z}$ is not uniquely defined by the conditions imposed here.)

To bring the Jordan canonical form into a proper correlation with the recurrence relations for the generalized eigenvectors, write $\underset{\sim}{A}\underset{\sim}{Z}\underset{\sim}{e}_j = \underset{\sim}{Z}\underset{\sim}{J}\underset{\sim}{e}_j$ where $\underset{\sim}{e}_j$ is the jth column of the identity matrix. (It is helpful to consider the case $\lambda^{(r)} = \lambda^{(0)}$, $p_0 = n + 1$, $\alpha_s^{(0)} = 1$ for $s \geq 1$.) The required recurrence relations then appear, and the columns of $\underset{\sim}{Z}$ are shown to be eigenvectors or generalized eigenvectors of $\underset{\sim}{A}$.

Further, if we write $\underset{\sim}{W} = \underset{\sim}{Z}^{-1}$ then $\underset{\sim}{A}^T\underset{\sim}{W}^T = \underset{\sim}{W}^T\underset{\sim}{J}^T$ and it can be seen that the columns of $\underset{\sim}{W}^T$ are eigenvectors or generalized eigenvectors of $\underset{\sim}{A}^T$ (but, since $\underset{\sim}{J}^T$ is a lower canonical form, the grades of the generalized eigenvectors are complementary to those of the corresponding columns of $\underset{\sim}{Z}$). What does emerge fairly readily is that if $\lambda^{(r)}$ is an eigenvalue of $\underset{\sim}{A}$ with multiplicities p_r, m_r respectively, it is an eigenvalue of $\underset{\sim}{A}^T$ with the same multiplicities. Further suppose that $\underset{\sim}{x}^{(r)}$ satisfies $\underset{\sim}{A}\underset{\sim}{x}^{(r)} = \lambda^{(r)}\underset{\sim}{x}^{(r)}$, where $\lambda^{(r)}$ is simple, and suppose that the columns of $\underset{\sim}{Z}$ are $\underset{\sim}{z}_0, \underset{\sim}{z}_1, \ldots, \underset{\sim}{z}_n$ with $\underset{\sim}{z}_t = \underset{\sim}{x}^{(r)}$. If $\underset{\sim}{A}^T\underset{\sim}{w}^{(r)} = \lambda^{(r)}\underset{\sim}{w}^{(r)}$, then $\underset{\sim}{w}^{(r)T}\underset{\sim}{z}_s = 0$ for $s \neq t$ (in particular, $\underset{\sim}{w}^{(r)}$ may be chosen as the tth column of $\underset{\sim}{W}^T$).

Let us now return to the question of the sensitivity of eigenvalues of a matrix $\underset{\sim}{A}$. We suppose that $\underset{\sim}{A}$ has a complete set of eigenvectors, so that the Jordan form $\underset{\sim}{J}$ is diagonal and the matrix $\underset{\sim}{Z}$ is a 'modal matrix' whose columns are eigenvectors of $\underset{\sim}{A}$. If the eigenvalues of $\underset{\sim}{A} + \varepsilon\underset{\sim}{B}$ are $\lambda^{(r)}(\varepsilon)$, these are also the eigenvalues of $\underset{\sim}{Z}^{-1}(\underset{\sim}{A} + \varepsilon\underset{\sim}{B})\underset{\sim}{Z} = \underset{\sim}{J} + \varepsilon\underset{\sim}{Z}^{-1}\underset{\sim}{B}\underset{\sim}{Z}$ and since $\underset{\sim}{J}$ is diagonal, we can use Gerschgorin's theorem to show that when $|\varepsilon|$ is sufficiently small the eigenvalues $\lambda^{(r)}(\varepsilon)$ satisfy relations $|\lambda(\varepsilon) - \lambda| \leq |\varepsilon| \; \|\underset{\sim}{B}\|_\infty \; \mu(\underset{\sim}{Z})$ where $\mu(\underset{\sim}{Z}) = \|\underset{\sim}{Z}\|_\infty \; \|\underset{\sim}{Z}^{-1}\|_\infty$ (see, for example, Isaacson and Keller (1966, p.138) and also Wilkinson (1965, p.87), who do not require $|\varepsilon|$ to be small).

The preceding analysis is valid if it is possible to diagonalize $\underset{\sim}{A}$, and it gives the result $|\lambda(\varepsilon) - \lambda| = O(\varepsilon)$ which is not true for general matrices. This may be seen if we consider the 2×2 matrix with diagonal entries λ,λ and off-diagonal entries $1, \varepsilon$ respectively; then $\lambda(\varepsilon) = \lambda \pm (\varepsilon)^{\frac{1}{2}}$, and small values of ε give rise to larger perturbations of λ. Moreover the preceding analysis takes no account of possible variations in the sensitivity of different eigenvalues.

Now let us suppose that λ is a simple eigenvalue of the matrix $\underset{\sim}{A}$ with a corresponding eigenvector $\underset{\sim}{x}$. For convenience we suppose $\underset{\sim}{x}$ is the first column $\underset{\sim}{z}_0$ of $\underset{\sim}{Z}$, where $\underset{\sim}{Z}^{-1}\underset{\sim}{A}\underset{\sim}{Z} = \underset{\sim}{J}$. We shall consider the asymptotic behaviour of the eigenvalue $\lambda(\varepsilon)$ of $\underset{\sim}{A} + \varepsilon\underset{\sim}{B}$ which tends to λ as $\varepsilon \to 0$. Such limiting behaviour is easily established but we shall also assume that $\lambda(\varepsilon) = \lambda + \varepsilon\mu_1 + \varepsilon^2\mu_2 + \dots$ and that a corresponding eigenvector $\underset{\sim}{x}(\varepsilon)$ satisfies a relation $\underset{\sim}{x}(\varepsilon) = \underset{\sim}{x} + \varepsilon\underset{\sim}{v}_1 + \varepsilon^2\underset{\sim}{v}_2 + \dots$, where $\underset{\sim}{v}_1, \underset{\sim}{v}_2, \dots$ are certain vectors. For a justification of these assumptions see Wilkinson (1965, pp.66-7). The vectors $\underset{\sim}{v}_i$ can be expressed in terms of the columns $\underset{\sim}{z}_j$ of $\underset{\sim}{Z}$ as $\underset{\sim}{v}_i = \Sigma t_{ij}\underset{\sim}{z}_j$ and after re-normalizing we have an eigenvector $\underset{\sim}{x}(\varepsilon)$ which can be written

$$\underset{\sim}{x}(\varepsilon) = \underset{\sim}{z}_0 + \sum_{r=1}^{n} (\varepsilon t_{r1} + \varepsilon^2 t_{r2} + \dots)\underset{\sim}{z}_r ,$$

with $\underset{\sim}{z}_0 = \underset{\sim}{x}$. Now $(\underset{\sim}{A} + \varepsilon\underset{\sim}{B})\underset{\sim}{x}(\varepsilon) = \lambda(\varepsilon)\underset{\sim}{x}(\varepsilon)$; expanding in powers of ε and equating coefficients of ε we find

$$\underset{\sim}{A}\Big(\sum_{r=1}^{n} t_{r1}\underset{\sim}{z}_r\Big) + \underset{\sim}{B}\underset{\sim}{z}_0 = \lambda\Big(\sum_{r=1}^{n} t_{r1}\underset{\sim}{z}_r\Big) + \mu_1\underset{\sim}{z}_0 .$$

Pre-multiplying by $\underset{\sim}{w}_0^T$, which is the first row of $\underset{\sim}{X}^{-1}$ we have, since $\underset{\sim}{w}_0^T\underset{\sim}{A} = \lambda\underset{\sim}{w}_0^T$ and $\underset{\sim}{w}_0^T\underset{\sim}{z}_r = 0$ for $r = 1,2,3,\dots, n$,

$$\mu_1 \underset{\sim}{w}_0^T\underset{\sim}{z}_0 = \underset{\sim}{w}_0^T\underset{\sim}{B}\underset{\sim}{z}_0 .$$

(In this analysis we have modified slightly the discussion of Wilkinson (1965, pp.68-9). We now have $|\mu_1| \le \|\underset{\sim}{B}\|_2\|\underset{\sim}{w}_0\|_2\|\underset{\sim}{z}_0\|_2/|\underset{\sim}{w}_0^T\underset{\sim}{z}_0|$, and this inequality places a bound on the dominant term of $|\lambda(\varepsilon) - \lambda|$. The result provides motivation for the definition of a condition number $\mu(\lambda)$ of a simple eigenvalue λ of $\underset{\sim}{A}$ as the value

$$\mu(\lambda) = |\mathbf{w}_0^T \mathbf{z}_0| / \|\mathbf{w}_0\|_2 \; \|\mathbf{z}_0\|_2 ,$$

where $A\mathbf{z}_0 = \lambda \mathbf{z}_0$ and $\mathbf{w}_0^T A = \lambda \mathbf{w}_0^T$. (The value of $\mu(\lambda)$ is not defined if λ is not simple.) It is readily shown that $\mu(\lambda) \in (0, 1]$, and when $\mu(\lambda)$ is close to 1 we say that λ is 'well conditioned'. If $\mu(\lambda)$ is small, λ is sensitive to perturbation and 'ill conditioned'.

2.20.3 .

Another problem which we shall encounter is the *linear programming problem* of determining $\mathbf{x} = [x_0, \ldots, x_n]^T$ so that $\mathbf{c}^T\mathbf{x} = \sum_{i=0}^{n} c_i x_i$ has a maximum value subject to the constraints

$$\sum_{j=0}^{n} A_{ij}\, x_j \le b_i \qquad (i = 0,1,2,\ldots, m_1) ,$$

$$\sum_{j=0}^{n} A_{ij}\, x_j \ge b_i \qquad (i = m_1+1, \ldots, m_2) ,$$

$$\sum_{j=0}^{n} A_{ij}\, x_j = b_i \qquad (i = m_2+1, \ldots, m) ,$$

$x_j \ge 0$ $(j = 0,1,2,\ldots, n_1)$, $x_j \le 0$ $(j = n_1+1, \ldots, n_2)$, where $n_2 \le n$. This is a general formulation of the linear programming problem, and it may be recast in one of special 'standard forms'. For example, we may have $m_1 = m_2$ and $n_1 = n_2$, or alternatively the constraints may take the form

$$\sum_{j=0}^{n} A_{ij}\, x_j = b_i \qquad (i = 1,2,3,\ldots, m) ,$$

$$x_j \ge 0 \qquad (j = 0,1,2,\ldots, n) .$$

The theory of linear programming is discussed by Dantzig (1963) and Hadley (1962). Rabinowitz (1968) has written a survey of applications of linear programming to numerical analysis.

2.20.4.

The problem of solving a non-linear equation or a system of non-linear equations in several variables arises in the numerical solution of non-linear integral equations.

We begin by considering the solution of a single non-linear equation of the form $\phi(x) = 0$. Such an equation may sometimes be solved by an iterative method, of which the secant methods, methods of false position, and Newton's method are simple examples.

For Newton's method we suppose that $\phi(x)$ has a derivative $\phi'(x)$ which we may evaluate. We shall also suppose that $\phi''(x)$ exists in a neighbourhood of a zero α . We may then construct the sequence $\{x_n\}$ defined by the iteration

$$x_{n+1} = x_n - \frac{\phi(x_n)}{\phi'(x_n)} ,$$

where x_0 is a suitable first approximation to the required zero (α , say) of $\phi(x)$. If x_0 is sufficiently close to α , the iteration converges to this zero, and if α is a simple zero then

$$|x_{n+1} - \alpha| = 0(|x_n - \alpha|^2) .$$

If α is a multiple zero then $|x_{n+1} - \alpha| = 0(|x_n - \alpha|)$. If x_0 is real and $\phi(x)$ is real, we are unable to locate any complex-valued zeros, since all members of the sequence $\{x_n\}$ are also real. It is necessary to assign x_0 a complex value if α is complex.

To avoid the repeated evaluation of $\phi'(x)$ we may use the iteration

$$x_{n+1} = x_n - \frac{\phi(x_n)}{\phi'(x_0)} ,$$

in which case $|x_{n+1} - \alpha| = 0(|x_n - \alpha|)$ when x_0 is sufficiently close to α . It is clear that both methods can fail if $\phi'(x)$ is allowed to vanish at a point which is not a zero.

As a further alternative to evaluating $\phi'(x_n)$ we may use the approximation

$$\phi'(x_n) \simeq \frac{\phi(x_n) - \phi(x_{n-1})}{x_n - x_{n-1}} .$$

If we use the iteration

$$x_{n+1} = x_n - \frac{\phi(x_n)\ \{x_n - x_{n-1}\}}{\phi(x_n) - \phi(x_{n-1})}$$

we have a *secant method*. There are a number of variations of the secant or chord methods. More involved schemes are discussed by Traub (1964).

The methods described above for the solution of a single equation can be extended, in a number of ways, to the problem of solving a system of non-linear equations. We shall indicate how the Newton iteration can be extended.

We suppose that we seek a vector $\underset{\sim}{z} = [z_0, z_1, \ldots, z_n]^T$ such that

$$\phi_i(\underset{\sim}{z}) \equiv \phi_i(z_0, z_1, \ldots, z_n) = 0 \quad (i = 0,1,2,\ldots,n\ ) .$$

We suppose that the matrix $\underset{\sim}{J}(\underset{\sim}{\xi})$ with (i,j)th element

$$\frac{\partial\phi_i}{\partial z_j}(\xi_0, \xi_1, \ldots, \xi_n) \quad (i,\ j = 0,1,2,\ldots,n\ )$$

can be evaluated. $\underset{\sim}{J}(\underset{\sim}{\xi})$ is known as the *Jacobian* matrix.

If we choose an initial vector $\underset{\sim}{z}^{(0)}$ we may define a sequence $\{\underset{\sim}{z}^{(n)}\}$ using the relations

$$\underset{\sim}{\delta}^{(n)} = \underset{\sim}{z}^{(n+1)} - \underset{\sim}{z}^{(n)} = \omega[\underset{\sim}{J}(\underset{\sim}{z}^{(n)})]^{-1} \begin{bmatrix} \phi_0(\underset{\sim}{z}^{(n)}) \\ \phi_1(\underset{\sim}{z}^{(n)}) \\ \cdot \\ \cdot \\ \cdot \\ \phi_n(\underset{\sim}{z}^{(n)}) \end{bmatrix} \quad (n = 0,1,2,\ldots)$$

provided that $\det[\underset{\sim}{J}(\underset{\sim}{z}^{(n)})] \neq 0$ for $n = 0,1,2,\ \ldots$. We take $\omega = -1$

in *Newton's method* but more generally the parameter ω may take values which vary with n . The construction of $J(\underset{\sim}{z}^{(n)})$ and the solution of the equations

$$[J(\underset{\sim}{z}^{(n)})]\ \underset{\sim}{\delta}^{(n)} = \omega[\phi_0(\underset{\sim}{z}^{(n)})\ ,\ \phi_1(\underset{\sim}{z}^{(n)}),\ \ldots,\ \phi_n(\underset{\sim}{z}^{(n)})]^T$$

can be laborious, and we may decide to replace $J(\underset{\sim}{z}^{(n)})$ by $J(\underset{\sim}{z}^{(0)})$ for $n = 0,1,2,\ldots, (n_1 - 1)$ before 'updating' by calculating $J(\underset{\sim}{z}^{(n_1)})$. When $n = 0, 1, \ldots, (n_1-1)$ in this process, we repeatedly solve equations of the form

$$J(\underset{\sim}{z}^{(0)})\ \underset{\sim}{\delta}^{(n)} = \omega\underset{\sim}{\phi}^{(n)}$$

to find the change $\underset{\sim}{\delta}^{(n)}$ in the current estimate of the root. Thus it will be seen that we solve, in succession, a sequence of systems of linear equations in which the vector $\underset{\sim}{\phi}^{(n)}$ changes but the matrix $J(\underset{\sim}{z}^{(0)})$ remains fixed for a number of steps. (This can be done efficiently if we use Gaussian elimination and store upper and lower triangular matrices which are produced when determining $\underset{\sim}{\delta}^{(0)}$.)

Forsythe and Moler (1967) devote a chapter to the solution of systems of non-linear equations, and Ortega and Rheinboldt (1970) give an excellent comprehensive account in their book, which is devoted to this topic. Ortega and Rheinboldt also give an extensive bibliography.

2.20.5 .

In certain parts our discussion of the numerical solution of linear integral equations will bear a strong resemblance to numerical linear algebra. On the other hand, in our discussion of Volterra integral equations, the methods are reminiscent of techniques used for initial value problems in ordinary differential equations. Such techniques are discussed by Fox (1962), Henrici (1962, 1963), Lapidus and Seinfeld (1971), and Lambert (1973), for example. Of interest for their extensions to integral equations are *Runge-Kutta* formulae, *predictor-corrector* formulae, and the theory of *stability*. The reader may also recognize a similarity between aspects of the numerical solution of equations of Fredholm type and corresponding features in the numerical treatment of boundary-value problems of ordinary differential equations. The latter are discussed by Keller (1968), who devotes a chapter to integral equation methods, and by Fox (1957).

3

EIGENVALUE PROBLEMS

Chapter 3 is devoted to the numerical solution of the eigenvalue problem

$$\int_a^b K(x,y)f(y)dy = \kappa f(x) ,$$

which was introduced in sections 1.4.1 and 1.6. Emphasis is almost entirely on the case where $K(x,y)$ is well behaved. The available methods are discussed in sections 3.1 - 3.12, and underlying theory is discussed in the ensuing sections.

The methods most favoured in practice fall into two classes; those based on integration formulae (sections 3.2-3.4) and those which are expansion methods (sections 3.6-3.11) in particular the Rayleigh-Ritz, Galerkin, and collocation methods. The first class of methods are generally simpler to implement than the second class. In the expansion methods, an approximate eigenfunction is obtained as a linear combination of chosen functions $\phi_0(x)$, $\phi_1(x)$, ..., $\phi_n(x)$ and it is necessary to compute $\int_a^b K(x,y)\phi_i(y)dy$ $(i = 0,1,2,\ldots, n)$ for certain values of x . This is also true of the product integration method of section 3.4.

In suitable cases, the methods of sections 3.3, 3.4, 3.6-3.11 can be applied even when the $K(x,y)$ is weakly singular.

In the theory of sections 3.13-3.33 we consider convergence behaviour, and error bounds and error estimates. We establish that the quadrature method of section 3.2. can be used to provide asymptotic upper and lower bounds (see section 2.19), and that the techniques of section 2.13.1 can be applied to increase the accuracy of our approximations and to estimate the error. In the case of the Rayleigh-Ritz method (applied where the kernel is Hermitian) one-sided estimates of the eigenvalues are obtainable directly.

Approximations to small or multiple eigenvalues are generally inferior to approximations to large, well-separated eigenvalues. It is indicated that a condition number governs the accuracy obtainable in approximating a simple eigenvalue. Problems in which $K(x,y)$ is Hermitian are generally quite tractable.

3.1. *Introduction*

In this chapter we shall discuss the numerical solution of the eigenvalue problem

$$\int_a^b K(x,y)\, f(y)\mathrm{d}y = \kappa\, f(x) \qquad (a \le x \le b)\ , \tag{3.1}$$

where $K(x,y)$ is a given kernel and a and b are finite. We suppose that $K(x,y)$ is, at worst, piecewise-continuous with possible jump discontinuities along a finite number of continuous curves $y = \phi_i(x)$ (cf. section 1.3). Since a and b are assumed finite, we can employ a change of variables to rewrite eqn (3.1) in the form

$$\int_0^1 K_0(x,y)\, f_0(y)\mathrm{d}y = \kappa\, f_0(x) \qquad (0 \le x \le 1)\ ,$$

where $K_0(x,y) = (b - a)\, K((b - a)x + a,\ (b - a)\, y + a)$ and $f_0(x) = f((b - a)x + a)$.

The numerical methods which we shall describe for solving (3.1) will yield a finite number of approximate eigenvalues $\tilde{\kappa}_0, \tilde{\kappa}_1, \ldots, \tilde{\kappa}_n$ and corresponding approximate eigenfunctions. Since there may be a countably infinite number of solutions of eqn (3.1), we cannot claim to solve (3.1) completely. We can usually obtain approximations to the eigenvalues of largest modulus, and their corresponding eigenfunctions. (The eigenfunctions corresponding to eigenvalues of large modulus are often smoother than those corresponding to small eigenvalues, and it is usually easier to approximate the smoother functions.) However, if two or more eigenvalues are close to one another, or if an eigenvalue is a multiple eigenvalue, we may find difficulty in approximating the corresponding eigenfunctions. The latter difficulty need not arise if the kernel $K(x,y)$ is Hermitian.

If κ is a simple eigenvalue in (3.1) then there is a non-null function $\phi(x)$ (the left eigenfunction of section 1.6) such that $\int_a^b \phi(y)K(y,x)\mathrm{d}y = \kappa\, \phi(x) (a \le x \le b)$. It can be shown that the condition number

$$\sigma \equiv \sigma(\kappa) = \left|\int_a^b f(x)\phi(x)\mathrm{d}x\right| / \{||f(x)||_2 \; ||\phi(x)||_2\} \tag{3.2}$$

gives an indication of the sensitivity of κ to perturbation of the problem (3.1), and hence the difficulty of computing an approximation to κ. Indeed, $\sigma \in (0,1]$, and if σ is small the eigenvalue κ is ill conditioned. The similarities with the condition number for the algebraic eigenproblem are clear.

In our description of numerical methods for the solution of (3.1) we shall begin with the '*quadrature methods*'. These are not always the best methods, but they are, perhaps, the simplest to implement in a straightforward way and without much preliminary analysis. We then proceed to discuss various other methods in which the approximate eigenfunction is obtained as a linear combination of prescribed functions $\phi_0(x)$, $\phi_1(x), \ldots, \phi_n(x)$. We shall call these methods '*expansion methods*'.

Another method, which is at first sight completely different, consists of replacing the kernel $K(x,y)$ by an approximating degenerate kernel $K_N(x,y)$, and solving the eigenproblem for $K_N(x,y)$. This method is sometimes known as *Courant's method*. The approximation methods indicated in section 2.10 can be used to implement Courant's method. However, our analysis of the numerical methods which we present in section 3.13 onwards, shows that many numerical methods can be considered equivalent to some form of Courant's method. For this reason, our classification of methods is rather imprecise.

3.2. *The quadrature method*

We suppose that we have a quadrature rule (section 2.11),

$$J(\phi) = \sum_{j=0}^{n} w_j \, \phi(y_j) \, ,$$

which yields an approximation

$$\int_a^b \phi(y)\mathrm{d}y \simeq \sum_{j=0}^{n} w_j \, \phi(y_j) \, , \tag{3.3}$$

where $a \leq y_i \leq b$ for $i = 0,1,2,\ldots, n$. This quadrature rule may be employed to approximate the integral occurring in eqn (3.1). We then have

$$\sum_{j=0}^{n} w_j\, K(x,y_j)\, f(y_j) \simeq \kappa\, f(x) \qquad (a \leq x \leq b)\ . \tag{3.4}$$

This suggests that we may obtain a useful approximate eigenvalue $\tilde{\kappa}$ and a corresponding approximate eigenfunction $\tilde{f}(x)$ if we solve the equation

$$\sum_{j=0}^{n} w_j\, K(x,y_j)\, \tilde{f}(y_j) = \tilde{\kappa}\, \tilde{f}(x) \qquad (a \leq x \leq b)\ . \tag{3.5}$$

In order to solve eqn (3.5) we set $x = y_i$ $(i = 0,1,2,\ldots, n)$ and obtain

$$\sum_{j=0}^{n} w_j\, K(y_i,y_j)\, \tilde{f}(y_j) = \tilde{\kappa}\, \tilde{f}(y_i) \quad (i = 0,1,2,\ldots, n)\ . \tag{3.6}$$

If we write $\underset{\sim}{K} = [K(y_i,y_j)]$, $\underset{\sim}{\tilde{f}} = [\tilde{f}(y_0), \tilde{f}(y_1), \ldots, \tilde{f}(y_n)]^T$, and $\underset{\sim}{D} = \text{diag}\,(w_0,w_1,\ldots,w_n)$, then the equations in (3.6) yield the matrix eigenvalue problem

$$\underset{\sim}{K}\underset{\sim}{D}\underset{\sim}{\tilde{f}} = \tilde{\kappa}\underset{\sim}{\tilde{f}}\ . \tag{3.7}$$

There are $(n + 1)$ eigenvalues $\tilde{\kappa}_r$ $(r = 0,1,2,\ldots, n)$ and corresponding eigenvectors $\underset{\sim}{\tilde{f}}_r$ satisfying (3.7). If $\tilde{\kappa}_r$ is non-zero, the eigenvector $\underset{\sim}{\tilde{f}}_r$ may be extended to a function $\tilde{f}_r(x)$ by substituting in (3.5) and writing

$$\tilde{f}_r(x) = (1/\tilde{\kappa}_r) \sum_{j=0}^{n} w_j K(x,y_j)\, \tilde{f}_r(y_j)\ . \tag{3.8}$$

(Alternatively, we may perform polynomial, piecewise-polynomial, or spline interpolation on the values $\tilde{f}_r(y_0),\ldots,\tilde{f}_r(y_n)$. In this case, the resulting function need not satisfy (3.5) exactly, though it does satisfy (3.6).) The construction of approximations satisfying (3.8) is sometimes known as *Nyström's method*, and $\tilde{f}_r(x)$ is the *Nyström extension* of the eigenvector $\tilde{f}_r$ (suitably scaled).

Example 3.1

Consider
$$\int_0^1 K(x,y)\, f(y)dy = \kappa f(x) \quad (0 \le x \le 1)\ ,$$

with $K(x,y) = \frac{1}{2}x(2 - y)$ $(0 \le x \le y \le 1)$, and $K(x,y) = K(y,x)$ $(0 \le y \le x \le 1)$. The true eigenvalues are the roots of the equation $1/\sqrt{\kappa} + \tan(1/\sqrt{\kappa}) = 0$, and corresponding eigenfunctions are $\sin(x/\sqrt{\kappa})$. Solving the transcendental equation gives the three largest eigenvalues: $\kappa_1 = 2{\cdot}4296 \times 10^{-1}$, $\kappa_2 = 4{\cdot}1426 \times 10^{-2}$, $\kappa_3 = 1{\cdot}5709 \times 10^{-2}$. If we use the three-point trapezoidal rule on $[0,1]$, in the quadrature method, so that

$$J(\phi) = \tfrac{1}{4}\,\phi(0) + \tfrac{1}{2}\,\phi(\tfrac{1}{2}) + \tfrac{1}{4}\,\phi(1)\ ,$$

we obtain the matrix eigenvalue problem

$$\begin{bmatrix} 0 & 0 & 0 \\ 0 & 3/16 & 1/16 \\ 0 & 1/8 & 1/8 \end{bmatrix} \begin{bmatrix} \tilde{f}(0) \\ \tilde{f}(\frac{1}{2}) \\ \tilde{f}(1) \end{bmatrix} = \tilde{\kappa} \begin{bmatrix} \tilde{f}(0) \\ \tilde{f}(\frac{1}{2}) \\ \tilde{f}(1) \end{bmatrix}$$

and the three eigenvalues are

$$\tilde{\kappa}_1 = \tfrac{1}{4}\ ,\ \tilde{\kappa}_2 = 1/16,\ \tilde{\kappa}_3 = 0\ .$$

The eigenvector corresponding to $\tilde{\kappa}_3$ is $\tilde{f}_3 = [1,0,0]^T$; because $\tilde{\kappa}_3 = 0$ we cannot use (3.8) to extend the vector $\tilde{f}_3$ to a function $\tilde{f}_3(x)$. (The presence of the eigenvalue $\tilde{\kappa}_3$, and the corresponding eigenvector, is misleading. See Example 3.7 for a similar case. Here, zero is not a true eigenvalue but is consistently obtained as an approximate eigenvalue.)

If the eigenvectors $\tilde{f}_r$ are extended to functions using (3.8) (in the case $\tilde{\kappa}_r \neq 0$), then the resulting approximations $\tilde{f}_r(x)$ are not continuously differentiable. For this reason, other methods of extending $\tilde{f}_r$ to a function seem more attractive here. *

Example 3.2

We consider the problem

$$\int_0^{2\pi} K_\rho(x,y)\, f(y)dy = \kappa f(x) \quad (0 \le x \le 2\pi) ,$$

where $0 \le \rho < 1$ and $K_\rho(x,y) = (\frac{1-\rho^2}{2\pi})\, \{1 - 2\rho \cos(x-y) + \rho^2\}^{-1}$.
We may write

$$K_\rho(x,y) = \frac{1}{2\pi} + \frac{1}{\pi} \sum_{j=1}^{\infty} \rho^j \cos j(x-y)$$

and deduce (if we compare with Example 1.5) that the true eigenvalues are

$$\kappa_0 = 1,\ \kappa_1 = \kappa_2 = \rho\ ,\ \kappa_3 = \kappa_4 = \rho^2\ ,\ \ldots,\ \kappa_{2r-1} = \kappa_{2r} = \rho^r\ \ldots$$

and corresponding eigenfunctions have the form, say,

$$f_0(x) = 1,\quad f_{2r}(x) = \alpha \cos rx + \gamma \sin rx$$

$$f_{2r-1}(x) = \delta \cos rx + \beta \sin rx \quad (\alpha\beta \ne \gamma\delta),\quad r \ge 1 .$$

We can apply the primitive quadrature rule

$$\int_0^{2\pi} \phi(y)dy \simeq h \sum_{i=0}^{n-1} \phi(ih) \qquad (h = 2\pi/n)$$

to obtain approximate eigenvalues and eigenfunctions. In the notation above, $\underset{\sim}{D} = h\underset{\sim}{I}$ (where $\underset{\sim}{I}$ is the identity matrix) and $\underset{\sim}{K}\underset{\sim}{D} = h\underset{\sim}{K}$ is a symmetric *circulant* (Aitken 1964). Thus $h\underset{\sim}{K}$ has the form, with $c_r = c_{n-r}$,

$$\begin{bmatrix} c_0 & c_1 & \cdots & c_{n-1} \\ c_{n-1} & c_0 & \cdots & c_{n-2} \\ \cdot & \cdot & & \cdot \\ \cdot & \cdot & & \cdot \\ \cdot & \cdot & & \cdot \\ c_1 & c_2 & \cdots & c_0 \end{bmatrix} .$$

If ω_r is an nth root of unity ($\omega_r = \cos 2\pi r/n + i \sin 2\pi r/n$, $r = 0,1,2,\ldots, (n-1)$), then $\tilde{f}_r = [1, \omega_r, \ldots, \omega_r^{n-1}]^T$ is an eigenvector of this matrix, with the corresponding eigenvalue

$$\tilde{\kappa}_r = \sum_{j=0}^{n-1} c_j \omega_r^j .$$

Thus, with this indexing, the approximate eigenvalues are

$$\tilde{\kappa}_0 = 1 + \frac{2\rho^n}{1-\rho^n}, \quad \tilde{\kappa}_r = \tilde{\kappa}_{n-r} = \frac{\rho^r + \rho^{n-r}}{1-\rho^n} \quad (r = 1,2,3,\ldots, (n-1)).$$

Since $\rho < 1$, $\lim_{n\to\infty} \rho^n = 0$ and $\lim_{n\to\infty} \tilde{\kappa}_0 = \kappa_0$, $\lim_{n\to\infty} \tilde{\kappa}_r = \rho^r \in \{\kappa_s\}$, and the eigenvectors $\tilde{f}_r$ give exact values of suitable eigenfunctions. (Some details are left to the reader.) *

The basic procedure of the quadrature method has been outlined, and we saw that it resulted in the matrix eigenvalue problem

$$K D \tilde{f} = \tilde{\kappa}\tilde{f} .$$

The success of the approximation hinges to a great extent on the choice of quadrature rule. There is some advantage in choosing a rule with positive weights w_j for $j = 0,1,2,\ldots, n$, so that (see section 2.14), $J(\phi) \in J^+$, and $D > 0$. For, suppose that $K(x,y)$ is real and symmetric, then the matrix K is real and symmetric but KD is not. However, if $D > 0$, we can consider the matrix $D^{\frac{1}{2}} K D^{\frac{1}{2}}$ which is real and symmetric and similar to KD. Thus

$$(D^{\frac{1}{2}} K D^{\frac{1}{2}}) D^{\frac{1}{2}} \tilde{f} = \tilde{\kappa} D^{\frac{1}{2}} \tilde{f}$$

and we can consider the eigenvalue problem $D^{\frac{1}{2}} K D^{\frac{1}{2}} \phi = \tilde{\kappa} \phi$; if we obtain an eigenvector ϕ we may recover $\tilde{f}$ in the form of $D^{-\frac{1}{2}}\phi$. Similarly, if $K(x,y)$ is Hermitian then $D^{\frac{1}{2}} K D^{\frac{1}{2}}$ is Hermitian. Symmetry is exploited in a number of algorithms for finding eigenvalues of matrices.

3.2.1.

When no symmetry is involved, there might seem to be little or no advantage in considering the eigenvalue problem for $\underset{\sim}{D}^{\frac{1}{2}}\underset{\sim}{K}\underset{\sim}{D}^{\frac{1}{2}}$ in place of $\underset{\sim}{K}\,\underset{\sim}{D}$.

However, the *condition number* $\mu(\tilde{\kappa})$ of a simple eigenvalue $\tilde{\kappa}$ of $\underset{\sim}{D}^{\frac{1}{2}}\underset{\sim}{K}\underset{\sim}{D}^{\frac{1}{2}}$ is an approximation to a similar condition number $\sigma(\kappa)$ which can be defined for the simple eigenvalue κ of the kernel $K(x,y)$.

Indeed, for such a value κ we have

$$\int_a^b K(x,y)f(y)\mathrm{d}y = \kappa f(x) ,$$

and for some normalized function $\phi(x)$, a left-eigenfunction of $K(x,y)$,

$$\int_a^b K(y,x)\ \phi(y)\mathrm{d}y = \kappa\phi(x)$$

(or

$$\int_a^b K^*(x,y)\ \overline{\phi(y)}\mathrm{d}y = \bar{\kappa}\overline{\phi(x)}\ ),$$

and we define (see eqn (3.2))

$$\sigma(\kappa) = \left|\int_a^b f(x)\ \phi(x)\mathrm{d}x \Big/ \|f(x)\|_2\ \|\phi(x)\|_2\right| .$$

Thus $\sigma(\kappa) = |(f,\ \psi)|/\{\|f(x)\|_2 \times \|\psi(x)\|_2\}$, where $\psi(x) = \overline{\phi(x)}$. The significance of $\sigma(\kappa)$ as a condition number is indicated by the analysis of section 3.20.

Now the condition number $\mu(\tilde{\kappa})$ of the eigenvalue $\tilde{\kappa}$ of $\underset{\sim}{A} = \underset{\sim}{D}^{\frac{1}{2}}\underset{\sim}{K}\underset{\sim}{D}^{\frac{1}{2}}$ is obtained as follows. If $\underset{\sim}{K}\underset{\sim}{D}\tilde{\underset{\sim}{f}} = \tilde{\kappa}\tilde{\underset{\sim}{f}}$, and $\underset{\sim}{K}^T\underset{\sim}{D}\,\tilde{\underset{\sim}{\phi}} = \tilde{\kappa}\tilde{\underset{\sim}{\phi}}$ we find $(\underset{\sim}{D}^{\frac{1}{2}}\underset{\sim}{K}\underset{\sim}{D}^{\frac{1}{2}})\underset{\sim}{D}^{\frac{1}{2}}\,\tilde{\underset{\sim}{f}} = \tilde{\kappa}\,\underset{\sim}{D}^{\frac{1}{2}}\tilde{\underset{\sim}{f}}$ and $(\underset{\sim}{D}^{\frac{1}{2}}\underset{\sim}{K}^T\underset{\sim}{D}^{\frac{1}{2}})\underset{\sim}{D}^{\frac{1}{2}}\,\tilde{\underset{\sim}{\phi}} = \tilde{\kappa}\,\underset{\sim}{D}^{\frac{1}{2}}\,\tilde{\underset{\sim}{\phi}}$, so that $\mu(\tilde{\kappa})$ = $= |\tilde{\underset{\sim}{\phi}}^T\underset{\sim}{D}\tilde{\underset{\sim}{f}}|/\|\underset{\sim}{D}^{\frac{1}{2}}\,\tilde{\underset{\sim}{f}}\|_2\,\|\underset{\sim}{D}^{\frac{1}{2}}\,\tilde{\underset{\sim}{\phi}}\|_2$.

If we accept the validity of the approximations

$$\tilde{\underset{\sim}{\phi}}^T\ \underset{\sim}{D}\tilde{\underset{\sim}{f}} \simeq \underset{\sim}{\phi}^T\underset{\sim}{D}\underset{\sim}{f} \simeq \int_a^b \phi(x)\ f(x)\mathrm{d}x ,$$

and similarly $\|\underset{\sim}{D}^{\frac{1}{2}}\underset{\sim}{f}\|_2 \simeq \|f(x)\|_2$, etc., we see that $\mu(\tilde{\kappa}) \simeq \sigma(\kappa)$.

Observe that the value $\sigma(\kappa)$ is defined only if κ is simple and then $\sigma(\kappa) \in (0,1]$. If $\sigma(\kappa)$ is small, then the eigenvalue κ is ill conditioned. If κ is not a simple eigenvalue, the situation is more complicated; in general, a multiple eigenvalue of a non-Hermitian kernel is likely to be ill conditioned.

Example 3.3
If $K(x,y)$ is Hermitian and κ is a simple eigenvalue, $\sigma(\kappa) = 1$. If $\underset{\sim}{A}$ is Hermitian and $\tilde{\kappa}$ is simple, then $\mu(\tilde{\kappa}) = 1$. *

3.2.2.
We have noted that it is an advantage, in the quadrature method, to choose a quadrature rule $J(\phi)$ which is in J^+ . Our choice of rule is further influenced by the *discretization errors*

$$d(\phi) = \sup_{i=0,1,2,\ldots,n} \left| \int_a^b K(y_i,y)\ \phi(y)dy - \sum_{j=0}^{n} w_j K(y_i,y_j)\ \phi(y_j) \right|$$

and

$$D(\phi) = \sup_{a\leq x\leq b} \left| \int_a^b K(x,y)\ \phi(y)dy - \sum_{j=0}^{n} w_j K(x,y_j)\ \phi(y_j) \right| ,$$

where $\phi(x)$ is any sufficiently smooth function. A comparison of the discretization errors of different rules may be made in terms of the values of their Peano constants, but for composite rules it is simpler to argue in terms of the order notation. Thus, when the product $K(x,y)\ \phi(y)$ is sufficiently smooth, the repeated trapezium rule with step h gives discretization errors which are $O(h^2)$ and the parabolic rule with step h gives discretization errors which are $O(h^4)$, so one may be inclined to favour the latter rule when h is small.

The discretization errors are expressed in terms of an arbitrarily smooth function $\phi(x)$, but if a realistic assessment of the merits of various quadrature rules is to be made, we should consider the *local truncation errors*:

$$t(f_r) = \sup_{i=0,1,2,\ldots,n} \left| \int_a^b K(y_i,y)\ f_r(y)dy - \sum_{j=0}^{n} w_j K(y_i,y_j) f_r(y_j) \right|$$

and

$$T(f_r) = \sup_{a \le x \le b} \left| \int_a^b K(x,y)\, f_r(y)dy - \sum_{j=0}^{n} w_j K(x,y_j)\, f_r(y_j) \right| ,$$

which correspond to some particular eigenfunction $f_r(x)$. Since the eigenfunctions $f_r(x)$ are not necessarily scaled, we must suppose $||f_r(x)||_2 = 1$, say. Even though $f_r(x)$ is unknown, we may be able to determine its smoothness using the arguments indicated in section 1.4.1. Bounds on the derivatives of the normalized function $f_r(x)$ can be estimated when an estimate of κ_r is known.

If the differentiability properties of $f_r(x)$ are known, we can assess the size of $t(f_r)$ or $T(f_r)$ resulting from a given choice of rule. In broad terms, we need be concerned with $T(f_r)$, rather than $t(f_r)$, only if the Nyström extension (3.8) is to be employed.

When all partial derivatives of $K(x,y)$ are continuous, the eigenfunctions corresponding to non-zero eigenvalues are also continuously differentiable. In such a case we may consider using a high-order Gauss-Legendre rule for our quadrature formulae. Since the weights and abscissae of such rules are not very convenient for use, we may instead favour composite Gauss-Legendre rules. Such a choice seems satisfactory, other things being equal, if no estimates of accuracy are required, and the kernel $K(x,y)$ is easy to evaluate, or if *a priori* error bounds are available. Usually, however, we seek consistency checks for results obtained by different means, and in this case the quadrature rules using equally spaced abscissae appear attractively economical. One reason for this is that there are a number of rules using the same abscissae but having different weights. Thus the repeated trapezium rule employing a step $h = (b-a)/n$, where $n = 2m$ produces the same matrix $\underset{\sim}{K}$ as the m-times repeated Simpson's rule. It is only necessary to change the matrix $\underset{\sim}{D}$ of weights to obtain one matrix problem (eqn (3.7)) from the other. Other members of the Romberg family can also be used. On the other hand, if we now halve h and repeat our calculation, we have already computed a quarter of the elements required in the new matrix $\underset{\sim}{K}$.

When $K(x,y)$ suffers a discontinuity in some derivative, say, the determination of the order of $t(f_r)$ or $T(f_r)$ may present a problem. To obtain an informed assessment of the relative merits of various quadrature rules, familiarity with the material of section 2.16 is necessary. By way of illustration, it can be shown that though the kernel in Example 3.1 has a derivative $K_y(x,y)$ with a jump discontinuity, both the repeated trapezium rule and the repeated Simpson's rule (repeated with step h) yield $O(h^2)$ local truncation errors. In this example, we would favour the use of the repeated trapezium rule since the techniques of section 3.2.3 (below) can then be applied to 'process' the computed approximations to obtain higher-order accuracy or asymptotic upper and lower bounds. (The use of the repeated Simpson's rule does not appear to permit the successful application of these techniques with this example, though some modification of the process might be available.)

Example 3.4

We consider the problem

$$\int_0^1 e^{xy} f(y)dy = \kappa f(x) \quad .$$

We obtained the following values of the *largest* approximate eigenvalue by using the trapezium rule with step h. For $h = 0{\cdot}1$, $\tilde{\kappa} = 1{\cdot}35482$; for $h = 0{\cdot}05$, $\tilde{\kappa} = 1{\cdot}35348$; for $h = 0{\cdot}025$, $\tilde{\kappa} = 1{\cdot}35314$; for $h = 0{\cdot}0125$, $\tilde{\kappa} = 1{\cdot}35306$. Values obtained using the n-point Gauss-Legendre rule were: for $n = 2$, $\kappa = 1{\cdot}35208$, for $n \geq 4$, $\tilde{\kappa} = 1{\cdot}35303$ and apparently all these figures are correct. It is known (Brakhage 1961) that $1{\cdot}3527 < \kappa < 1{\cdot}3534$. *

Example 3.5

Consider the equation

$$\int_0^1 K(x,y)\, f(y)dy = \kappa f(x) \quad (0 \leq x \leq 1) ,$$

where $K(x,y) = (x + y)$. If we employ the repeated trapezium rule with step h we obtain two non-zero eigenvalues

$$\tilde{\kappa}_1, \tilde{\kappa}_2 = \frac{1}{2} \pm \{\frac{1}{3}(1 + \tfrac{1}{2}h^2)\}^{\frac{1}{2}}$$

and corresponding eigenvectors are such that

$$\tilde{f}_1(rh),\ \tilde{f}_2(rh) = rh \pm \{\tfrac{1}{3}(1 + \tfrac{1}{2}h^2)\}^{\frac{1}{2}} .$$

(Baker, Fox, Wright, and Mayers 1964). The true values are $\kappa_1, \kappa_2 = \frac{1}{2} \pm 1/\surd 3$ and $f_1(rh), f_2(rh) = rh \pm 1/\surd 3$ (suitably normalized). *

3.2.3.

The previous discussion has indicated that if the approximations obtained by the quadrature method are to be 'processed', the size of the local truncation errors may not provide overriding criteria in the choice of a quadrature rule.

Under certain conditions we can use the deferred approach to the limit, or obtain asymptotic upper and lower bounds, for the eigenvalues and eigenfunctions of a kernel. For simplicity, we shall first suppose that $K(x,y)$ is Hermitian and that we compute approximate eigenvalues using the repeated trapezium rule (with step $h = (b-a)/n$) in the quadrature method. Let us suppose that the real eigenvalues are indexed $\{\kappa_r^{\pm}\}$, $\{\tilde{\kappa}_r^{\pm}\}$, where

$$\kappa_1^+ \geq \kappa_2^+ \geq \ldots \geq 0 ,$$

$$\kappa_1^- \leq \kappa_2^- \leq \ldots < 0 , \tag{3.9}$$

$$\tilde{\kappa}_1^+ \geq \tilde{\kappa}_2^+ \geq \ldots \geq 0 ,$$

$$\tilde{\kappa}_1^- \leq \tilde{\kappa}_2^- \leq \ldots < 0 . \tag{3.10}$$

Though there may be a countable number of true non-zero eigenvalues $\kappa_r^{\pm}$, there are only $(n + 1)$ approximate eigenvalues $\tilde{\kappa}_r^{\pm} = \tilde{\kappa}_r^{\pm}(h)$. For the sake of illustration we suppose that we obtain *non-zero* approximate eigenvalues as follows. With $h = (b-a) = h_0$, we find $\tilde{\kappa}_1^+(h_0)$, say. With $h = \frac{1}{2} h_0$, we find two approximate eigenvalues, say $\tilde{\kappa}_1^-(\frac{1}{2} h_0)$ and $\tilde{\kappa}_1^+(\frac{1}{2} h_0)$. With $h = \frac{1}{4} h_0$, we find four approximate eigenvalues, say $\tilde{\kappa}_1^-(\frac{1}{4} h_0)$, $\tilde{\kappa}_3^+(\frac{1}{4} h_0)$, $\tilde{\kappa}_2^+(\frac{1}{4} h_0)$, and $\tilde{\kappa}_1^+(\frac{1}{4} h_0)$. Then, if $K(x,y)$

is sufficiently smooth, we can show that (see section 3.21), if $\tilde{\kappa}_1^+ \neq 0$, then $\lim_{h\to 0} \tilde{\kappa}_1^+(h) = \kappa_1^+$ and (with corresponding results for $\kappa_r^{\pm}$)

$$\tilde{\kappa}_1^+(h) = \kappa_1^+ + a_1^+ h^2 + b_1^+ h^4 + o(h^6) .$$

More generally, we have a relation of the form

$$\tilde{\kappa}_1^+(h) = \kappa_1^+ + \sum_{s=1}^{N} \mu_s h^{2s} + o(h^{2N+1}) \qquad (N \geq 1) . \tag{3.11}$$

The values $\tilde{\kappa}_1^+(h_0)$, $\tilde{\kappa}_1^+(\frac{1}{2}h_0)$, and $\tilde{\kappa}_1^+(\frac{1}{4}h_0)$ can be combined, using the process of deferred approach to the limit, to give an approximation to κ_1^+ which has an error which is $O(h_0^6)$, in the case we are describing. Similarly, if $\tilde{\kappa}_1^- \neq 0$, we may use $\tilde{\kappa}_1^-(\frac{1}{2}h_0)$ and $\tilde{\kappa}_1^-(\frac{1}{4}h_0)$ to compute $\frac{1}{3}\{4\tilde{\kappa}_1^-(\frac{1}{4}h_0) - \tilde{\kappa}_1^-(\frac{1}{2}h_0)\}$, which has an error which is $O(h_0^4)$. Taking $h=\frac{1}{8}h_0$ will provide an opportunity to extend the scheme.

Example 3.6

The positive approximate eigenvalue in Example 3.5 is

$$\tilde{\kappa}_1^+ \equiv \tilde{\kappa}_1^+(h) = \tfrac{1}{2} + \{\tfrac{1}{3}(1 + \tfrac{1}{2}h^2)\}^{\frac{1}{2}} \quad ,$$

so that (by the binomial expansion theorem),

$$\begin{aligned}\tilde{\kappa}_1^+ &= \tfrac{1}{2} + \frac{1}{\sqrt{3}}\{1 + \tfrac{1}{4}h^2 - \frac{1}{32}h^4 + \frac{1}{128}h^6 - \ldots\} \\ &= \kappa_1^+ + \frac{1}{\sqrt{3}}\{\tfrac{1}{4}h^2 - \frac{1}{32}h^4 + \frac{1}{128}h^6 - \ldots\} . \quad *\end{aligned}$$

Example 3.7

Consider

$$\int_0^1 K(x,y)\, f(y)\,dy = \kappa f(x) ,$$

where $K(x,y) = x(1-y)$ $(0 \leq x \leq y \leq 1)$, $K(x,y) = K(y,x)$ $(0 \leq y \leq x \leq 1)$. The true eigenvalues and eigenfunctions are given by

$$\kappa_r^+ = (r\pi)^{-2}, \quad f_r^+(x) = \sin r\pi x \qquad (r = 1, 2, 3, \ldots) .$$

If we use the trapezium rule with step $h = 1/(m+1)$ in the quadrature method, we obtain the eigenvalue problem for the matrix $\underset{\sim}{A} = \underset{\sim}{K}\,\underset{\sim}{D}$. Now, $\underset{\sim}{A}$ has elements

$$A_{ij} = \begin{cases} ih^2(1 - jh) & (i \leq j) \\ jh^2(1 - ih) & (j \leq i) \end{cases}$$

for $i,j = 0,1,2,\ldots,(m + 1)$, where $h = 1/(m + 1)$. The matrix $\underset{\sim}{A}$ has two zero eigenvalues which correspond (for all h) to eigenvectors of the form $[1,0,\ldots,0]^T$ and $[0,0,\ldots,1]^T$, say. (Such eigenvectors cannot be thought to approximate smooth eigenfunctions, and zero is not a true eigenvalue.) The other eigenvalues of $\underset{\sim}{A}$ are given by

$$\tilde{\kappa}_r^+ = h^2/\{4\sin^2(\tfrac{1}{2}r\pi h)\} \quad (r = 1,2,3,\ldots,m),$$

with corresponding eigenvectors such that $\tilde{f}_r^+(sh) = \sin \pi rsh$. In this instance, therefore, these approximate eigenvectors give an exact representation of the true eigenfunctions. Using the Taylor series for $\sin x$ we may show that

$$\tilde{\kappa}_r^+(h) = (1/\pi r)^2\{1 - \frac{1}{3!}(\frac{\pi rh}{2})^2 + \frac{1}{5!}(\frac{\pi rh}{2})^4 + \ldots\}^{-2}$$

$$= \kappa_r^+\{1 + \frac{1}{3}(\frac{\pi rh}{2})^2 + \frac{1}{15}(\frac{\pi rh}{2})^4 + \ldots\},$$

and this provides a basis for repeated h^2-extrapolation and the construction of asymptotic upper and lower bounds for the eigenvalues.

In the table below, we give some values of $\tilde{\kappa}_1^+(h)$, for $h = 1/10$, 1/20, 1/40, ..., 1/160, and extrapolated values obtained by the deferred approach to the limit using repeated h^2-extrapolation. In analogy with the notation for a Romberg table we write $T_0^{(j)} = \tilde{\kappa}_1^+(0{\cdot}1 \times 2^{-j})$ and set $T_k^{(j)} = \{4^k T_{k-1}^{(j+1)} - T_{k-1}^{(j)}\}/\{4^k - 1\}$. The analogue of the first table in Example 2.15 is then a 'T-table' of the form

0·1021586455			
0·1015297742	0·1013201505		
0·1013732830	0·1013211193	0·1013211839	
0·1013342055	0·1013211796	0·1013211836	0·1013211836
0·1013244389	0·1013211834	0·1013211836	0·1013211836

Here, the first and third columns give upper bounds, and the second column gives lower bounds. Using the ideas given in section 2.19, we can construct a 'U-table', setting $U_k^{(j)} = 2T_{k+1}^{(j)} - T_k^{(j)}$. (The columns of the T-table are 'T-sequences' and the columns of the U-table are corresponding 'U-sequences'.) In this way we obtain the following table:

0·1009009030			
0·1012167918	0·1013220881		
0·1012951279	0·1013212399	0·1013211834	
0·1013146723	0·1013211872	0·1013211836	0·1013211836

From the first columns we estimate $0{\cdot}1013146723 \le \kappa_1^+ \le 0{\cdot}1013244389$. From the last columns we estimate $\kappa_1^+ = 0{\cdot}1013211836$, and all these figures are correct. The reader may care to compute results for κ_2^+, κ_3^+. *

In the case of a non-Hermitian kernel $K(x,y)$, we may still apply both the deferred approach to the limit and methods of obtaining asymptotic upper and lower bounds to the eigenvalue problem, provided the true eigenvalue being approximated is simple and non-zero. However, there is rather more difficulty in the general case in ordering the eigenvalues and matching up approximations obtained with different values of h. Thus if we obtain, in succession, non-zero values

$$\tilde{\kappa}_1(h_0),\ \tilde{\kappa}_2(h_0);$$
$$\tilde{\kappa}_1(\tfrac{1}{2}h_0),\ \tilde{\kappa}_2(\tfrac{1}{2}h_0),\ \tilde{\kappa}_3(\tfrac{1}{2}h_0),\ \tilde{\kappa}_4(\tfrac{1}{2}h_0);$$
$$\tilde{\kappa}_1(\tfrac{1}{4}h_0),\ \tilde{\kappa}_2(\tfrac{1}{4}h_0),\ \tilde{\kappa}_3(\tfrac{1}{4}h_0),\ \tilde{\kappa}_4(\tfrac{1}{4}h_0),\ \ldots,\ \tilde{\kappa}_8(\tfrac{1}{4}h_0);$$

we wish to know that the indexing matches up the eigenvalues correctly. The indexing is correct in this way if, say,

$$\tilde{\kappa}_r(h) = \kappa_r + \sum_{s=1}^{N} \mu_{r,s} h^{2s} + O(h^{2N+1}) \tag{3.12}$$

for $h \leq h_0$ and for some $N \geq 1$. If the eigenvalues and approximate eigenvalues are all real, it is natural to index them according to their position on the real line, as in the case of a Hermitian kernel. More generally, for want of a system, we may consider indexing the eigenvalues so that $|\kappa_r| < |\kappa_s|$ implies $r > s$, and $|\kappa_r| = |\kappa_s|$ with $\arg(\kappa_r) > \arg(\kappa_s)$ implies that $r > s$. Usually, we must infer, from our numerical results, whether

$$\text{(a)} \quad \lim_{h\to 0} \tilde{\kappa}_r(h) = 0 \tag{3.13}$$

or

$$\text{(b)} \quad \lim_{h\to 0} \tilde{\kappa}_r(h) = \lim_{h\to 0} \tilde{\kappa}_s(h) \neq 0 \text{ for } r \neq s . \tag{3.14}$$

In the latter case, the non-zero value κ_r is a multiple eigenvalue. Either of these circumstances are rather special. In case (a) it is not always correct to assume that zero is a true eigenvalue. In case (a) , or in case (b) (where κ_r appears to be a multiple eigenvalue),we cannot, in general, justify the deferred approach to the limit for the eigenvalues $\tilde{\kappa}_r(h)$, even if the kernel $K(x,y)$ is smooth. In case (b), however, the process of the deferred approach can be modified, and the modified process is valid if $K(x,y)$ is sufficiently smooth. For we may be able to determine the multiplicity m, of the true eigenvalue κ_r , by noting that (counting algebraic multiplicities to find m)

$$\lim_{h\to 0} \tilde{\kappa}_{r_i}(h) = \kappa_r \qquad (i = 1,2,3,\ldots, m) .$$

In this case we define

$$\hat{\kappa}_r(h) = \frac{1}{m} \sum_{i=1}^{m} \tilde{\kappa}_{r_i}(h) \ . \tag{3.15}$$

Then $\lim_{h \to 0} \hat{\kappa}_r(h) = \kappa_r$, and we can apply the deferred approach to the limit to the values of $\hat{\kappa}_r(h)$ if $K(x,y)$ is sufficiently smooth (Baker 1971).

We have been rather imprecise about the conditions of differentiability which should be imposed upon the kernel $K(x,y)$ to guarantee the success of the deferred approach to the limit, and up to the present we have only considered the application of the process to the determination of eigenvalues. It is interesting to note that the kernel in Example 3.7 has a discontinuous derivative $(\partial/\partial y)K(x,y)$ at $x = y$, but this does not affect the validity of the deferred approach to the limit for the eigenvalues. On the other hand, it is easy to construct examples where the analytical behaviour of the kernel does mean that we do not obtain the expected advantages from the process. For suppose that the kernel is smooth and $\tilde{\kappa}_r(h) = \kappa_r + a_r h^2 + O(h^4)$. If $a_r \neq 0$, we have approximations to κ_r with errors which are $O(h^2)$. The extrapolated values $\frac{1}{3}\{4\ \tilde{\kappa}_r(\frac{1}{2}h) - \tilde{\kappa}_r(h)\}$ have errors which are $O(h^4)$ and this appears to be an advantage. But, if $\tilde{\kappa}_r(h) = \kappa_r + O(h^2)$, then the extrapolated values have errors which are, at worst, $O(h^2)$. Of course, if $\lim_{h\to 0} \tilde{\kappa}_r(h) = \kappa_r$ then $\lim_{h\to 0} \frac{1}{3}\{4\tilde{\kappa}_r(\frac{1}{2}h) - \tilde{\kappa}_r(h)\} = \kappa_r$, so that there is a sense in which h^2-extrapolation cannot 'fail' even if it does not 'succeed'. Sufficient conditions for 'success' are stated later, in section 3.21.

<u>Example 3.8</u>

The continuous kernel $K(x,y) = (xy)^{\frac{1}{4}}$ (on $0 \leq x,y \leq 1$) is a degenerate kernel with one non-zero eigenvalue,

$$\kappa = \int_0^1 y^{\frac{1}{2}}\, dy \ .$$

The quadrature method using the trapezium rule with step h , produces one non-zero eigenvalue,

$$\tilde{\kappa} = \tilde{\kappa}(h) = h \sum_{i=0}^{n}{}'' (ih)^{\frac{1}{2}} .$$

The relation between $\tilde{\kappa}$ and κ is therefore expressed by the relation between

$$\int_0^1 y^{\frac{1}{2}}\,dy \quad \text{and} \quad h \sum_{i=0}^{n}{}'' \,(ih)^{\frac{1}{2}} \quad .$$

Since $f(x) = x^{\frac{1}{2}}$ has a discontinuous and unbounded first derivative, the Euler-Maclaurin expansion (see section 2.13) does not apply, and we *cannot* write $\tilde{\kappa}(h) = \kappa + ah^2 + O(h^4)$. Indeed, Fox (1967) established that

$$\int_0^1 y^{\frac{1}{2}}dy - h \sum_{i=0}^{n}{}'' \,(ih)^{\frac{1}{2}} = ah^{\frac{3}{2}} + bh^2 + O(h^4) \ ,$$

so that $\kappa - \tilde{\kappa}(h) = ah^{3/2} + bh^2 + O(h^4)$. *

For further comments see section 3.2.4.

We now proceed to discuss the possibility of applying our extrapolation techniques to the determination of the eigenfunctions. Again we use the trapezium rule with step h . We would like to be able to write, for a suitable extension $\tilde{f}_r(x)$ of $\tilde{\mathbf{f}}_r$,

$$\tilde{f}_r(x) = f_r(x) + a(x)h^2 + b(x)h^4 + O(h^6) \ , \tag{3.16}$$

for example, or more generally (see section 3.21),

$$\tilde{f}_r(x) = f_r(x) + \sum_{s=1}^{N} \phi_{r,s}(x)h^{2s} + O(h^{2N+1}) \tag{3.17}$$

say. Unfortunately, $\tilde{f}_r(x)$ and $f_r(x)$ are not really well defined, since they may be multiplied by arbitrary scalars. In practice, we can only expect essential convergence of $\tilde{f}_r(x)$ to $f_r(x)$ - at best . (See section 1.5 .) In order to avoid this problem we may normalize $\tilde{f}_r(x)$ so that $\tilde{f}_r(\bar{x}) = 1$, where $\bar{x}$ is some fixed value. Naturally, we choose a value of $\bar{x}$ which is such that $f_r(\bar{x})$ appears, from our numerical results, to be non-zero. It is simpler if we take $\bar{x}$ to be one of the points of the form $ih+a$ (where i is an integer) which occurs as an abscissa when the trapezium rule is applied. Then $\tilde{f}_r(\bar{x})$ is always a component of $\tilde{\mathbf{f}}_r$, for $h = h_0, \frac{1}{2}h_0, \ldots$. The indexing of the eigenfunctions is induced by the system of indexing the eigenvalues, which was described above. We require that κ_r and $\tilde{\kappa}_r(h)$ are simple eigenvalues, and that $\lim_{h\to 0} \tilde{\kappa}_r(h) \neq 0$.

Then, if (3.17) can be established for certain values of x it forms a basis for the application of the deferred approach to the limit.

Example 3.9

We consider the largest eigenvalue κ_1^+ and a corresponding eigenfunction $f_1^+(x)$ for the kernel

$$K(x,y) = x(1-y)/y \qquad (0 \le x < y \le 1) ,$$

$$K(x,y) = (1-x) \qquad (0 \le y \le x \le 1) .$$

The problem is slightly artificial. This kernel is not Hermitian, but the eigenvalue problem may be written in the form

$$\int_0^1 x^{-\frac{1}{2}} K(x,y) y^{\frac{1}{2}} \phi_r(y)\,dy = \kappa_r \phi_r(x) ,$$

which gives the true eigenvalues $\{\kappa_r\}$ and eigenfunctions of the form $\phi_r(x) = f_r(x)/x^{\frac{1}{2}}$. The kernel $K_s(x,y) = x^{-\frac{1}{2}} K(x,y) y^{\frac{1}{2}}$ is symmetric. From this we deduce that the eigenvalues of $K(x,y)$ are real.

The quadrature method with step h gives the values shown in Table 3.1 for $\tilde{\kappa}_1^+$ and the vector $\tilde{f}_1^+$ normalized so that $\tilde{f}_1^+(\frac{1}{2}) = 1$.

TABLE 3.1

	$\tilde{f}_1^+(x)$				Extrapolated	
$x \backslash h$	1/4	1/8	1/32	1/64	(a)	(b)
0	0·000000	0·000000	0·000000	0·000000	0·000000	0·000000
0·125		0·608787	0·608144	0·607984	0·607929	0·607931
0·250	0·939693	0·936909	0·936236	0·936069	0·936012	0·936014
0·375		1·049062	1·048678	1·048582	1·048550	1·048551
0·500	1·000000	1·000000	1·000000	1·000000	1·000000	1·000000
0·625		0·835682	0·836001	0·836081	0·836108	0·836107
0·750	0·592396	0·594311	0·594775	0·594890	0·594930	0·594929
0·875		0·307274	0·307643	0·307734	0·307766	0·307765
1·000	0·000000	0·000000	0·000000	0·000000	0·000000	0·000000
$\tilde{\kappa}_1(h)$	0·267145	0·271135	0·272117	0·272362		

The extrapolated vector (a) was obtained by performing h^2-extrapolation on the results obtained with $h = 1/8$ and $h = 1/16$; (b) was obtained similarly using the results for $h = 1/16$ and $h = 1/32$. The values (b) are correct values of a true eigenfunction, to the number of figures given.

Whenever we can perform repeated h^2-extrapolation on the values $\tilde{\kappa}_1(h)$ (here assumed non-zero), we can perform it on the values $\tilde{\lambda}_1(h) = \{\tilde{\kappa}_1(h)\}^{-1}$, and take the reciprocal of the extrapolated value.

Repeated h^2-extrapolation on λ_1-values here gives the values:

3·743289			
3·688194	3·669829		
3·674888	3·670453	3·670494	
3·671590	3·670490	3·670493	3·670493

The approximation $\lambda_1 \simeq 3{\cdot}670493$ gives $\kappa_1 \simeq 0{\cdot}272443$. *

We note that in Example 3.9, as in Example 3.7, the kernel has a discontinuity in its derivative $(\partial/\partial y)K(x,y)$ when $x = y$. In such a case, we should not use (3.8) to generate a function $\tilde{f}_r(x)$ from the components of the computed eigenvector $\tilde{\underset{\sim}{f}}_r = [\tilde{f}_r(0), \tilde{f}_r(h), \ldots, \tilde{f}_r(1)]^T$. If we do generate a function in this way, we cannot expect to find (for example) that

$$\tilde{f}_r(x) = f(x) + a(x)h^2 + b(x)h^4 + 0\,(h^6)$$

except when x is a multiple of h. If x is a multiple of h, we may deal with the computed vectors, not with functions.

On the other hand, if the kernel $K(x,y)$ is sufficiently smooth, we can use (3.8) and perform the deferred approach to the limit to estimate values of $f_r(x)$ where x is allowed to range over *all* values in the range of integration.

<u>Example 3.10</u>

The approximate eigenvectors which correspond to non-zero κ, in Example 3.7, provide exact representations of the true eigenfunctions, in the sense that $\tilde{\underset{\sim}{f}}_r = \underset{\sim}{f}_r$ where $\underset{\sim}{f}_r = [f_r(0), f_r(h), \ldots, f_r(1)]^T$ and

$f_r(x)$ is a true eigenfunction. If we extend $\tilde{f}_r$ to a function $\tilde{f}_r(x)$ using eqn (3.8), we find that $\tilde{f}_r(x) - f_r(x) = O(h)$, that is $\|\tilde{f}_r(x) - f_r(x)\|_\infty = O(h)$, and there is no development of the error $\tilde{f}_r(x) - f_r(x)$ as a series in powers of h^2. *

3.2.4.

The key to the techniques described above lies in the validity of relationships of the form

$$\tilde{\kappa}_r(h) = \kappa_r + \sum_{s=1}^{N} \mu_{r,s} h^{2s} + O(h^{2N+1}) \tag{3.18}$$

and, for some values of x,

$$\tilde{f}_r(x) = f_r(x) + \sum_{s=1}^{N} \phi_{r,s}(x) h^{2s} + O(h^{2N+1}), \tag{3.19}$$

where $\tilde{\kappa}_r(h)$ and $\tilde{f}_r(x)$ are obtained using the trapezium rule with step h in the quadrature method. If we employ other repeated rules we obtain similar relations, under suitable conditions. Thus the repeated Simpson's rule using a step h yields relations of this sort with $\mu_{r,1} = 0$ and $\phi_{r,1}(x) \equiv 0$. If we employ the rule $(m \times J)(\phi)$ in the quadrature method, where the degree of precision of $J(\phi)$ is ρ, and $J(\phi)$ has positive weights, we expect to obtain eigenvalues $\tilde{\kappa}_r(m^{-1})$ (say) such that

$$\tilde{\kappa}_r(m^{-1}) = \kappa_r + \sum_{s=\rho+1}^{N} \nu_{r,s} (1/m)^s + O((1/m)^{N+1}), \tag{3.20}$$

with a similar relation for the eigenfunctions. If $J(\phi)$ is a symmetric rule, $\nu_{r,s} = 0$ if s is odd. Such relations are valid as $m \to \infty$ if $\kappa_r = \lim_{m\to\infty} \tilde{\kappa}_r(m^{-1})$ is simple and non-zero, and *if* the kernel is sufficiently differentiable.

The previous discussion has indicated that if the approximations obtained by the quadrature method are to be 'processed', the size of the local truncation errors may not provide overriding criteria in the choice of a quadrature rule. The choice of rule may be influenced by what can be claimed for the asymptotic form of the errors in the initial approximations. Thus, the mechanical process of applying repeated h^2-extrapolation to approximate eigenvalues obtained using Simpson's rule (or a composite form of another high-order rule) will not generally yield high-order accuracy if $(\partial/\partial y)K(x,y)$ has a discontinuity at $x=y$. It might be possible to apply some similar process, of course. (Likewise, the construction of asymptotic upper and lower bounds must be approached with caution.) The success of the technique in Example 3.7 is due to the use of the repeated trapezium rule (although the rectangle rules could also be used successfully).

3.2.5.

The methods described in section 3.2.3, arise out of the correction terms for quadrature rules (given by the Euler-Maclaurin formula or, more generally, by eqn (2.23)). In the case of a simple eigenvalue we can employ difference correction terms, such as those provided for the trapezium rule by the Gregory rules, to revise our estimates of the eigenvalues. We consider only the use of the trapezium rule and leave the extension of the method to the reader.

Let us first suppose that $K(x,y) = \overline{K(y,x)}$, and

$$\int_a^b K(x,y)f_r(y)\,dy = \kappa_r\, f_r(x)$$

and that we have computed the solution of the equations

$$h \sum_{j=0}^{n}{}'' K(a+ih,\ a+jh)\ \tilde{f}_r(a+jh) = \tilde{\kappa}_r \tilde{f}_r(a+ih) \quad (i = 0,1,2,\ldots,\ n)\ ,$$

where $h = (b-a)/n$. (We suppose that $\tilde{\underset{\sim}{f}}_r$ is a real vector.) If there are no discontinuities in the derivatives $(\partial/\partial y)^q K(a+ih,y)$ and $(\partial/\partial x)^q f_r(x)$ we can estimate correction terms for the approximation

$$\int_a^b K(a+ih,y)\ \tilde{f}_r(y)\,dy \simeq h \sum_{j=0}^{n}{}'' K(a+ih,\ a+jh)\ \tilde{f}_r(a+jh)$$

in terms of the first, second, ..., up to qth forward and backward differences (in the y-direction) of

$$\phi_i(y) = K(a+ih,y)\,\tilde{f}_r(y)$$

at $y = a$ and $y = b$ respectively. These differences involve the values $\tilde{f}_r(a+ih)$ which have been computed. The correction terms can be expressed in the form (see eqn (2.28))

$$\delta_i = h \sum_{s=1}^{p} c_s^*(\nabla^s \phi_i(b) + (-1)^s \Delta^s \phi_i(a)) \quad ,$$

where $p \leq q$. The correction to the computed eigenvalue $\tilde{\kappa}_r$ is given by computing the new approximation

$$\tilde{\kappa}_r + \{ \sum_{i=0}^{n}{}'' \delta_i \tilde{f}_r(a+ih) / \sum_{i=0}^{n}{}'' |\tilde{f}_r(a+ih)|^2 \} ,$$

the term in braces being a correction for $\tilde{\kappa}_r$. The number of differences occurring in δ_i can be progressively increased, if desired, so that only significant differences are taken into account, and the number of differences (p) could be made to depend on i, r and n.

Consider now the more general case; we no longer assume that $K(x,y) = \overline{K(y,x)}$. If $\kappa_r \neq 0$ is a simple eigenvalue of $K(x,y)$ then there is a left eigenfunction $\phi_r(x)$ with

$$\int_a^b K(y,x)\phi_r(y)\,dy = \kappa_r \phi_r(x) \quad .$$

If h is sufficiently small, there is an approximation $\tilde{\kappa}_r$ to κ_r which is simple, with a corresponding function $\tilde{\phi}_r(x)$ such that

$$h \sum_{j=0}^{n}{}'' K(a+jh,\, a+ih)\tilde{\phi}_r(a+jh) = \tilde{\kappa}_r\, \tilde{\phi}_r(a+ih) \quad (i = 0,1,2,\ldots,\, n) \quad .$$

To obtain a correction to $\tilde{\kappa}_r$ we compute the new approximation

$$\tilde{\kappa}_r + \left[\sum_{i=0}^{n}{}'' \delta_i \tilde{\phi}_r(a+ih) / \sum_{i=0}^{n}{}'' \{\tilde{f}_r(a+ih)\tilde{\phi}_r(a+ih)\} \right]$$

with δ_i defined as in the case above, where $K(x,y)$ was Hermitian. (If $K(x,y) = \overline{K(y,x)}$, the process reduces to that above. For a justification of the procedure we may refer to eqn (3.164), for example.)

Fox and Goodwin (1953) describe a method of relaxation based on this procedure, which permits correction of the eigenfunctions also.

Under certain conditions, the method we are describing becomes rather cumbersome. For the kernel $K(x,y)$ arising in Example 3.9 there is a discontinuity in the first and higher y-derivatives of $K(x,y)$ at $x = y$. In this case the expression for δ_i has to be amended by treating the interval $[a, a + ih]$, and the interval $[a + ih, b]$ separately, and incorporating forward and backward differences at $x = a + ih$. This would mean that if $p \geq 2$ we require values of $\tilde{f}_r(y)$ and $K(x,y)$ for $y = a - h, a - 2h, \ldots, a - (p-1)h$ and $y = b + h, b + 2h, \ldots, b + (p-1)h$. The values of the function $\tilde{f}_r(x)$ at points outside the range $[a,b]$ would have to be obtained by extrapolation.

Example 3.11

We perform the deferred correction process, using the Gregory rule correction terms with differences of order up to 3, to compute approximations to the first two eigenvalues of the kernel

$$K(x,y) = e^{xy} \quad (0 \leq x, y \leq 1), \text{ given in Table 3.2.}$$

TABLE 3.2

h	Results from the trapezium rule		Corrected values
$\frac{1}{10}$	$\kappa_1^+(h)$	1·354819	1·3530242
	$\kappa_2^+(h)$	0·108334	0·1059948
$\frac{1}{20}$	$\kappa_1^+(h)$	1·353476	1·3530297
	$\kappa_2^+(h)$	0·106572	0·1059837

(continued)

(Table 3.2 continued)

h	Results from the trapezium rule		Corrected values
$\frac{1}{40}$	$\kappa_1^+(h)$	1·353142	1·3530301
	$\kappa_2^+(h)$	0·106131	0·1059832
$\frac{1}{80}$	$\kappa_1^+(h)$	1·353058	1·3530302
	$\kappa_2^+(h)$	0·106020	0·1059832

(Coefficients used to calculate the correction terms have been given in section 2.13.2.) *

3.2.6.

We remark, in passing, that in the quadrature method it is sometimes possible to use quadrature rules involving derivatives of the integrand. Such rules have rarely been employed in practice.

As an example, we may use the first derivative correction terms to the trapezoidal rule in the Euler-Maclaurin series. From the eigenvalue problem

$$\int_a^b K(x,y)\, f_r(y)\,dy = \kappa_r f_r(x) \quad , \tag{3.21}$$

we obtain, if $(\partial/\partial x)\, K(x,y) = K_x(x,y)$ is continuous,

$$\int_a^b K_x(a,y)\, f_r(y)\,dy = \kappa_r f_r{}'(a) \quad , \tag{3.22}$$

$$\int_a^b K_x(b,y)\, f_r(y)\,dy = \kappa_r f_r{}'(b) \quad . \tag{3.23}$$

We then set up approximate equations for (3.21)-(3.23) involving $\tilde{f}_r(a + ih)$ $(i = 0,1,2,\ldots, n,\ h = (b - a)/n)$, $\tilde{f}_r{}'(a)$, and $\tilde{f}_r{}'(b)$. We require, of course, certain values of derivatives of $K(x,y)$. If

$K(x,y)$ has a jump at $x = y$, the truncated Euler-Maclaurin series should be applied to $[a,x]$ and $[x,b]$ in (3.21), and we will require values of $\tilde{f}_r'(x)$. We must then set up an equation for

$$\tilde{\kappa}_r \tilde{f}_r'(x) \quad (x = a + ih, \quad i = 0,1,2,\ldots, n) .$$

Whereas this method may seem a little complicated for general use, it seems perfectly feasible to adapt the method described in section 3.2.5 in order to use correction terms, involving derivatives, provided by the Euler-Maclaurin series. Again, however, it is a disadvantage if $K(x,y)$ has a discontinuity in one of its derivatives at points inside the range of integration.

3.2.7.

Before continuing, the considerations which enter an assessment of the relative merits of various quadrature rules for use in the methods described so far will be summarized.

In general, and especially if $K(x,y)$ is real and symmetric or Hermitian, we prefer a rule with positive weights. Usually we prefer rules with abscissae lying in the range of integration and which do not involve values of derivatives. When the kernel is given as a tabulated function, the tabulation points may affect the choice of rule. (If the tabulation points do not 'match' the abscissae of the rule it will be necessary to perform interpolation on the table.)

Let us suppose that the kernel is defined analytically and is Hermitian. In this case we can determine (as described below) *a posteriori* and *a priori* error bounds for the eigenvalues obtained using a given quadrature rule. In general, we seek a rule which provides a low truncation error $t(f_r)$ (see section 3.2.2) and we usually assess this by heuristic arguments. Alternatively, it is possible to argue (see section 3.20) that we seek a rule which gives a small error in the approximations

$$\int_a^b K(y,z)\, K(z,y_j)\,dz \simeq \sum_{k=0}^{n} w_k K(y,y_k)\, K(y_k,y_j) \quad (j = 0,1,2,\ldots,n, a\le y\le b) ,$$

and the rule may be chosen appropriately using the results given in Chapter 2.†

We will return to a rigorous discussion of error bounds and error estimates later. Since it may be difficult to construct rigorous precise error bounds, we may prefer methods for assessing the reliability of computed results by some type of consistency check. In other words, we compute approximate solutions, and then either repeat the calculation or amend the computed results in some way. Such methods of checking may provide their own limitations on the quadrature rules which we use. It will be obvious, for example, that if the kernel is difficult to evaluate and we wish to use asymptotic bounds or the deferred approach to the limit, then a rule using equally spaced abscissae provides some economy of computing compared with the repeated Gauss-Legendre rules. Similarly, finite-difference correction terms used in section 3.2.5 can easily be calculated if the abscissae in the rule are equally spaced. Derivative correction terms (used in place of differences, in the same way) can be employed with any rule if the analytic manipulations can be performed successfully.

Finally, let us remark again that it is often possible, without using asymptotic bounds, to obtain *a posteriori* error estimates for our approximate eigenvalues. These error estimates sometimes reflect a belief (often correct) that eigenvalues of large modulus are usually obtained to a higher precision than those of smaller modulus. Whilst I shall give, below, theorems which state that in principle any of the eigenvalues may be located by a sufficiently refined technique, it may in practice be extremely difficult to locate some particular small eigenvalue with any degree of accuracy.

3.3. *A modification of the quadrature method*

In this section I describe a slight variation of the quadrature method described in section 3.2. The method of this section is designed to reduce the effect of any discontinuities in the kernel $K(x,y)$, or one of its y-derivatives, at $x = y$. The method requires the

† It is sometimes sufficient to consider $y \in \{y_i\}$ rather than $a \leq y \leq b$.

evaluation of

$$\int_a^b K(x,y)\,dy \quad \text{for} \quad x = y_0, y_1, \ldots, y_n ,$$

where the y_i are the abscissae of our quadrature rule.

Suppose that $K(x,y)$ is discontinuous on $x = y$, but continuous elsewhere. More generally, suppose that $(\partial/\partial y)^s K(x,y)$ is discontinuous only at $x = y$ for some s. Before discretization, we may write the eigenvalue problem (3.1) for $K(x,y)$ in the form

$$\int_a^b K(x,y)\,\{f_r(y) - f_r(x)\}\,dy + f_r(x)\int_a^b K(x,y)\,dy = \kappa_r f_r(x) \quad . \tag{3.24}$$

In this equation, if x is fixed, we may evaluate

$$A(x) = \int_a^b K(x,y)\,dy \tag{3.25}$$

analytically, or numerically (taking into account any discontinuity in the kernel or its derivative). We shall approximate the first term in (3.24) in the hope that its integrand is smoother than the integrand $K(x,y)f(y)$ in (3.1).

If we now choose a quadrature rule of the form used in section 3.2, we may replace (3.24) by an equation for approximate eigenvalues $\overset{\circ}{\kappa}_r$ and corresponding approximate eigenfunctions $\overset{\circ}{f}_r(x)$:

$$\sum_{j=0}^{n} w_j\, K(x,y_j)\{\overset{\circ}{f}_r(y_j) - \overset{\circ}{f}_r(x)\} + \overset{\circ}{f}_r(x)\,A(x) = \overset{\circ}{\kappa}_r\overset{\circ}{f}_r(x) \ . \tag{3.26}$$

If we set $x = y_i$ $(i = 0,1,2,\ldots, n)$ in (3.26) we obtain an eigenvalue problem which may be written in matrix notation as

$$(\underset{\sim}{K}\underset{\sim}{D}-\underset{\sim}{\Delta})\,\overset{\circ}{\underset{\sim}{f}}_r = \overset{\circ}{\kappa}_r\,\overset{\circ}{\underset{\sim}{f}}_r , \tag{3.27}$$

where $\overset{\circ}{\underset{\sim}{f}}_r = [\overset{\circ}{f}_r(y_0),\ldots, \overset{\circ}{f}_r(y_n)]^T$, $\underset{\sim}{K} = [K(y_i,y_j)]$, $\underset{\sim}{D} = \text{diag}(w_0,\ldots,w_n)$,

and $\underset{\sim}{\Delta}$ is the diagonal matrix with entries

$$\Delta_{ii} = \sum_{j=0}^{n} w_j \, K(y_i,y_j) - A(y_i) \; . \tag{3.28}$$

To compute the entries (3.28) we evaluate the function $A(x)$ in (3.25) for $x = y_0, y_1, \ldots, y_n$. These values should be obtained with some accuracy, either exactly or using an adaptive method of numerical integration (section 2.20.1). If we simply use the quadrature with weights w_i, and abscissae y_i $(i = 0,1,2,\ldots, n)$ we merely recover the method of section 3.2.

In the case where the weights w_i are positive, we can replace (3.27) by

$$(\underset{\sim}{D}^{\frac{1}{2}}\underset{\sim}{K}\underset{\sim}{D}^{\frac{1}{2}} - \underset{\sim}{\Delta})\, \underset{\sim}{\phi}_r = \overset{\circ}{\kappa}_r \, \underset{\sim}{\phi}_r$$

$$\underset{\sim}{\phi}_r = \underset{\sim}{D}^{\frac{1}{2}} \, \overset{\circ}{\underset{\sim}{f}}_r \; . \tag{3.29}$$

This is obtained from (3.27) since diagonal matrices commute, and so $\underset{\sim}{D}^{\frac{1}{2}}\underset{\sim}{\Delta}\underset{\sim}{D}^{-\frac{1}{2}} = \underset{\sim}{\Delta}$. The matrix in (3.29) is Hermitian when $K(x,y)$ is Hermitian and $\underset{\sim}{D} \geq 0$. We refer to the solution of (3.27) or (3.29) as a *modified quadrature method.*

Example 3.12

The kernel appearing in Example 3.7 is rather remarkable since use of the trapezium rule in either the quadrature method or the modified quadrature method leads to the same approximate eigenvalues and eigenvectors, since $A(x)$ can be evaluated exactly using the trapezium rule, at the points used. Employing the parabolic rule in the modified quadrature method, and noting that $A(x) = 0{\cdot}5 \; K(x,x)$ in this example, we obtained the approximations 0·101553, 0·101376, and 0·101335 for the largest eigenvalue, using matrices of order 20, 40, and 80 respectively.*

Example 3.13

We consider the case where

$$K(x,y) = -\ln x \qquad (0 \leq y \leq x \leq 1) \; ,$$
$$K(x,y) = -\ln y \qquad (0 \leq x \leq y \leq 1) \; .$$

The eigenvalue problem for this kernel is equivalent to solving

$$xf''(x) + f'(x) + \lambda f(x) = 0$$

with $|f(0)| < \infty$, $f(1) = 0$, and $\kappa = 1/\lambda$. The true values for the eigenvalues are given by $\kappa_r = 4/\gamma_r^2$, where γ_r is the rth zero of the Bessel function $J_0(x)$. We obtained the values shown in Table 3.3 for the three largest eigenvalues, using the modified quadrature method and employing the trapezium rule with step h.

TABLE 3.3

h	$\tilde{\kappa}_1^+$	$\tilde{\kappa}_2^+$	$\tilde{\kappa}_3^+$
$\frac{1}{10}$	0·69559	0·13922	0·06223
$\frac{1}{20}$	0·69264	0·13330	0·05642
$\frac{1}{40}$	0·69191	0·13178	0·05419
$\frac{1}{80}$	0·69172	0·13140	0·05361
	κ_1^+	κ_2^+	κ_3^+
True.	0·69166028	0·13127123	0·05341381

It will be seen that here we obtained four correct figures in κ_1^+ after solving the eigenvalue problem of a full matrix of order about 80.

A mechanical application† of the process employed in repeated h^2-extrapolation to the full-precision values of $\tilde{\kappa}_1^+(h)$ produced the

† We believe the error in $\tilde{\kappa}_1^+(h)$ may be $O(-h^2 \ln h)$.

following table:

0·69559029			
0·69264228	0·61165961		
0·69190567	0·69166014	0·69166017	
0·69172161	0·69166026	0·69166027	0·69166027 . *

We note that the kernel $K(x,y)$ need not possess a discontinuity for the success of the modified method. It may be advantageous to use (3.26) when $K(x,y)$ has a 'hump' on $x = y$ so that a large contribution to the integral

$$\int_a^b K(x,y)f_r(y)\,dy$$

comes from the vicinity of $y = x$.

To compare the use of the modified method with the quadrature method, let us define

$$t^*(f_r) = \max_{i=0,1,2,\ldots,n} \left| \int_a^b K(y_i,y)f_r(y)\,dy - \sum_{j=0}^{n} K(y_i,y_j)\{f_r(y_j)-f_r(y_i)\} - A(y_i)f_r(y_i) \right| , \tag{3.30}$$

where $f_r(x)$ is some fixed (normalized) eigenfunction corresponding to an eigenvalue κ_r which we wish to approximate. Some basis for comparing the method of section 3.2 with the method of this section is provided by a comparison of $t^*(f_r)$ and the analogous local truncation error $t(f_r)$ defined in section 3.2.2 (the quadrature rules in each case need not be the same). The size of these quantities depends upon the unknown eigenfunction $f_r(x)$. It would be difficult to *assert* that one method will be better than the other in general, but the modified method would have some claim to superiority if a low-order y-derivative of the kernel becomes large when $x = y$, and the eigenfunctions are nevertheless 'smooth'.

3.3.1.

Since the iterates of a kernel are frequently smoother than the original kernel, it may sometimes be profitable to examine the eigenvalue problem for $K^p(x,y)$ $(p \geq 2)$, and use the method of section 3.2. It is then necessary to determine which of the pth roots of the eigenvalues of the iterated kernel are actually eigenvalues of $K(x,y)$; this can sometimes be judged from a crude approximation to the eigenvalues of $K(x,y)$. In a practical problem, the determination of an iterated kernel may be a difficult task.

3.4. *Product-integration methods*

In this section I discuss the application of methods of approximate product integration to the eigenvalue problem. The method requires, in general, that the kernel $K(x,y)$ should be defined analytically. When $K(x,y)$ is Hermitian, we do not always obtain an eigenvalue problem of a Hermitian matrix. This is a disadvantage of the method.

If we fix x , we may obtain formulae

$$\int_a^b K(x,y)\ \phi(y)\,dy \simeq \sum_{j=0}^{n} v_j(x)\ \phi(z_j) \tag{3.31}$$

which are exact when $\phi(x)$ is a polynomial of degree n , or alternatively when $\phi(x)$ is, say, piecewise-constant, or piecewise-linear. Methods for obtaining such formulae are indicated in section 2.17. A natural discretization for the eigenvalue problem (3.1) is then posed by the equations

$$\sum_{j=0}^{n} v_j(x)\ \tilde{f}_r(z_j) = \tilde{\kappa}_r\tilde{f}_r(x) \quad , \tag{3.32}$$

from which we obtain

$$\sum_{j=0}^{n} v_j(z_i)\ \tilde{f}_r(z_j) = \tilde{\kappa}_r\tilde{f}_r(z_i) \quad (i = 0,1,2,\ldots, n). \tag{3.33}$$

This is the eigenvalue problem for the matrix $\underset{\sim}{V}$ with elements $v_j(z_i)$.

The method is equivalent, in the form we have described, to a *collocation method* (see below). It appears to be most valuable if the kernel $K(x,y)$ or one of its y-derivatives is badly behaved (or, perhaps, oscillatory). The method, like that of section 3.3, can be applied when the kernel $K(x,y)$ is weakly singular. (In Chapter 5 we employ similar methods to equations of the second kind in which the kernel is weakly singular; such methods have immediate analogues for the eigenvalue problem.) It should be emphasized that the lack of symmetry in the matrix $v_j(z_i)$ when $K(x,y)$ is symmetric is a slight disadvantage.

Example 3.14
Consider the problem $\int_0^1 K(x,y)\, f(y)\mathrm{d}y = \kappa f(x)$, where

$$K(x,y) = \begin{cases} x(1-y) & (0 \le x < y \le 1) \\ y(1-x) & (0 \le y \le x \le 1) \end{cases} .$$

We shall use a product-integration formula and obtain formulae (3.31) which are exact (a) if $\phi(x)$ is piecewise-constant, and (b) if $\phi(x)$ is piecewise-linear. (For the kernel in this example, some product-integration rules assume a simple form.)

(a) We set $z_t = th$ $(h = 1/n)$ and suppose that $\phi(x)$ is piecewise-constant, and $\phi(x) = \phi_i$ for $z_i \le x < z_{i+1}$ $(i = 0,1,2,\ldots,(n-1))$. Then

$$\int_0^1 K(z_i,y)\, \phi(y)\mathrm{d}y = \sum_{j=0}^{n-1} \phi_j \int_{z_j}^{z_{j+1}} K(z_i,y)\mathrm{d}y$$

$$= \sum_{j=0}^{n-1} \phi_j\, hK(z_i,z_{j+\frac{1}{2}})$$

(using the mid-point rule, since $K(z_i,y)$ is linear in y for $z_j \le y \le z_{j+1}$), and we may set $v_j(z_i) = hK(z_i,z_{j+\frac{1}{2}})$ $(i,j = 0,1,2,\ldots,(n-1))$ in eqn (3.33), for the case under consideration. We solve the eigenvalue problem for the matrix

$$\begin{bmatrix} v_0(z_0), & \ldots, & v_{n-1}(z_0) \\ \cdot & & \\ \cdot & & \\ \cdot & & \\ v_0(z_{n-1}), & \ldots, & v_{n-1}(z_{n-1}) \end{bmatrix} .$$

The components ϕ_j of an eigenvector approximate $f(z_j)$, where $f(x)$ is an eigenfunction. Table 3.4 shows some numerical results (obtained with the QR method). The order of the matrix is denoted by n.

TABLE 3.4

n	$\tilde{\kappa}_1^+$	$\tilde{\kappa}_2^+$	$\tilde{\kappa}_3^+$
5	0·09265	0·01812	0·00610
10	0·09906	0·02316	0·00925
20	0·10075	0·02477	0·01070
40	0·10118	0·02519	0·01112

We give the correct results at the end of this Example.

(b) Let us now construct, for the given kernel, a product-integration rule which is exact if $\phi(x)$ is continuous and linear in each interval (z_i, z_{i+1}) $(i = 0,1,2,\ldots,(n-1))$. We write $\phi(th)$ as ϕ_t. Using Simpson's rule,

$$\int_{z_j}^{z_{j+1}} K(z_i,y)\ \phi(y)\mathrm{d}y =$$

$$\tfrac{1}{6}h\ \{K(z_i,z_j)\ \phi_j + 4K(z_i,z_{j+\frac{1}{2}})\phi_{j+\frac{1}{2}} + K(z_i,z_{j+1})\ \phi_{j+1}\} ,$$

because the integrand is quadratic in y in $[z_j,z_{j+1}]$. Since $\phi_{j+\frac{1}{2}} = \frac{1}{2}\{\phi_{j+1}+\phi_j\}$ and $K(z_i,z_{j+\frac{1}{2}}) = \frac{1}{2}\{K(z_i,z_j) + K(z_i,z_{j+1})\}$ we find

$$v_0(z_i) = \frac{1}{6}h\{2K(z_i,0) + K(z_i,h)\} ,$$

$$v_n(z_i) = \frac{1}{6}h\{2K(z_i,1) + K(z_i,1-h)\} ,$$

and, for $r = 1,2,3,\ldots,(n-1)$,

$$v_r(z_i) = \frac{1}{6}h\{K(z_i,z_{r-1}) + 4\,K(z_i,z_r) + K(z_i,z_{r+1})\} .$$

Some values of the approximation to the largest eigenvalue obtained for decreasing values of h are given in Table 3.5.

TABLE 3.5

h	$\tilde{\kappa}_1^+$
1	0·00000
$\frac{1}{2}$	0·08333
$\frac{1}{4}$	0·09628
$\frac{1}{8}$	0·10003
$\frac{1}{16}$	0·10100
$\frac{1}{32}$	0·10124
$\frac{1}{64}$	0·10130

From a matrix of order 40 we obtain the approximate eigenvalues 0·10127, 0·02528, and 0·01120 for the three largest eigenvalues, the true values being 0·10132, 0·02533, and 0·01126 to this number of figures. *

3.5. *Hämmerlin's approximate kernel method*

Hämmerlin (1960) has described an explicit application of 'Courant's method.' In Courant's method, we solve the eigenvalue problem of a kernel $\tilde{K}(x,y)$ which approximates the given kernel $K(x,y)$. In the case of Hämmerlin's version, the problem is reduced to a matrix eigenvalue problem in which the matrix elements depend in a fairly simple way on the values of $K(x,y)$. The matrix is not necessarily Hermitian when $K(x,y)$ is Hermitian.

Suppose $x_i = a + ih$, where $h = 1/n$. The simplest version of the method can be described as follows. We may approximate the kernel $K(x,y)$ by a piecewise-constant kernel $\tilde{K}(x,y)$ defined by

$$\tilde{K}(x,y) = K(x_i,x_j) \quad \begin{cases} x_i \leq x < x_{i+1} \\ x_j \leq y < x_{j+1} \, . \end{cases} \tag{3.34}$$

The eigenvalue problem for $\tilde{K}(x,y)$ is solved exactly by obtaining the solution to the problem

$$h \sum_{j=0}^{n-1} \tilde{K}(x_i,x_j)\, \tilde{f}(x_j) = \tilde{\kappa}\tilde{f}(x_i)$$

with

$$\tilde{f}(x) = \tilde{f}(x_i) \qquad (x_i \leq x < x_{i+1}, \quad i = 0,1,2,\ldots,n-1)$$

and

$$\tilde{f}(b) = \tilde{f}(x_{n-1}) \, .$$

This result may be established by noting that the eigenfunctions of $\tilde{K}(x,y)$ which correspond to non-zero $\tilde{\kappa}$ are piecewise-constant. In this case the method is equivalent to the quadrature method using the rectangle rule.

If we re-define $\tilde{K}(x,y)$ by setting (with $x_{i+\frac{1}{2}} = a + (i+\frac{1}{2})h$)

$$\tilde{K}(x,y) = K(x_{i+\frac{1}{2}}, x_{j+\frac{1}{2}}) \quad \begin{cases} x_i \le x < x_{i+1} \\ x_j \le y < x_{j+1} \end{cases} , \tag{3.35}$$

we obtain equations which are equivalent to the use of the mid-point rule in the quadrature method of section 3.2. In the preceding cases the matrix of the eigenvalue problem is Hermitian if $K(x,y)$ is Hermitian.

A more complicated version of Hämmerlin's method will now be described. Instead of forming a kernel $\tilde{K}(x,y)$ which is piecewise-constant, we form an approximate kernel which is continuous and bilinear, that is, linear in x and y for $x_i \le x \le x_{i+1}$, $x_j \le y \le x_{j+1}$ (where $x_i = a+ih$).
Thus,

$$\begin{aligned}\tilde{K}(x,y) = (1/h^2)\ \{&(x_{i+1}-x)(x_{j+1}-y)\ K(x_i,x_j) + \\ &+ (x-x_i)(y-x_j)\ K(x_{i+1},x_{j+1}) + \\ &+ (x_{i+1}-x)(y-x_j)\ K(x_i,x_{j+1}) + \\ &+ (x-x_i)(x_{j+1}-y)\ K(x_{i+1},x_j)\}\end{aligned} \tag{3.36}$$

for $x_i \le x \le x_{i+1}$, $x_j \le y \le x_{j+1}$.

The eigenfunctions of this kernel $\tilde{K}(x,y)$ are piecewise-linear if they correspond to non-zero eigenvalues. Consequently, we can reduce the eigenvalue problem for $\tilde{K}(x,y)$ to a matrix eigenvalue problem. If the eigenfunction $\tilde{f}_r(x)$ takes the values $\tilde{f}_i = \tilde{f}_r(x_i)$, we find that

$$\begin{aligned}&\tfrac{1}{6}h\Big[\{2K(x_i,x_0) + K(x_i,x_1)\}\ \tilde{f}_0 + \\ &\quad + \sum_{j=1}^{n-1} \{K(x_i,x_{j-1}) + 4\ K(x_i,x_j) + K(x_i,x_{j+1})\}\ \tilde{f}_j + \\ &\quad + \{K(x_i,x_{n-1}) + 2\ K(x_i,x_n)\}\ \tilde{f}_n\Big] \\ &= \tilde{\kappa}\tilde{f}_i \qquad\qquad (i = 0,1,2,\ldots, n).\end{aligned} \tag{3.37}$$

The matrix of coefficients in this matrix problem is clearly not Hermitian when $K(x,y)$ is Hermitian. But in such circumstances the kernel $\tilde{K}(x,y)$ *is* Hermitian, so that the matrix problem could be reformulated in a symmetric way.

Example 3.15

For the kernel appearing in Example 3.14, the bilinear version of the method described above gives the same results as those given in Example 3.14(b). *

Example 3.16

We apply the bilinear method to the kernel $K(x,y) = e^{xy}$ $(0 \le x,y \le 1)$ and obtain the following approximate values of the eigenvalue of largest size. The order of the matrix is denoted by n, and the following values correspond to halving h at each stage.

n=2	3	5	9	17	33	65	Estimated true value
1·47573	1·37932	1·35927	1·35457	1·35341	1·35313	1·35305	1·35303

These values seem to be comparable with those obtained from the quadrature method, using repeated trapezium rule. *

In his paper, Hämmerlin considers the possibility of unequally spaced $\{x_i\}$.

Remark. It is possible to recover symmetry in the bilinear version of Hämmerlin's method applied where $K(x,y)$ is symmetric. One such method will be noted here, though it is somewhat artless for practical use. Since $\tilde{K}(x,y)$ is bilinear for $x_i \le x \le x_{i+1}$, $x_j \le y \le x_{j+1}$ $(i,j = 0,1,2,\ldots, n-1)$, and $x_i = a+ih$, with $h = (b-a)/n$, the quadrature method using the n-times repeated Gauss-Legendre two-point rule gives the non-zero eigenvalues of $\tilde{K}(x,y)$ and corresponding eigenfunctions exactly. There is an obvious modification where the x_i are unequally spaced. In any case, the kernel $\tilde{K}(x,y)$ is Hermitian because $K(x,y)$ is Hermitian, so the quadrature method with positive weights can be applied to give a Hermitian matrix $\underset{\sim}{D}^{\frac{1}{2}}\underset{\sim}{K}\underset{\sim}{D}^{\frac{1}{2}}$. The order of this matrix is, however, $2n$ and this is an obvious disadvantage in practice. However, we see that Hämmerlin's method is 'equivalent'† to a particular application of the method of section 3.2 for a kernel $\tilde{K}(x,y)$ which approximates $K(x,y)$.

3.6. *Expansion methods*

In the quadrature methods discussed in sections 3.2-3.4, the approximate eigenfunctions are represented by vectors which supply values of the functions at the abscissae of quadrature rules. We now turn our attention to methods in which the approximate eigenfunctions are represented in the form

$$\tilde{f}(x) = \sum_{i=0}^{n} \tilde{a}_i \, \phi_i(x) \quad , \tag{3.38}$$

where $\phi_0(x), \phi_1(x), \ldots, \phi_n(x)$ are pre-selected functions. The solution of the matrix eigenvalue problem, which we set up, gives the approximate eigenvalues $\tilde{\kappa}$ and a corresponding eigenvector

$$\tilde{\underset{\sim}{a}} = [\tilde{a}_0, \tilde{a}_1, \ldots, \tilde{a}_n]^T \quad ,$$

from which values of $\tilde{f}(x)$ may be computed.

The difference between the quadrature method and those methods which we call 'expansion methods' is, however, a marginal one. If we take $h = 1/(n+1)$ and $\phi_i(x) = \delta_{ij}$ for $jh \leq x < (j+1)h$ ($\delta_{ij} = 1$ if $i=j$, zero otherwise; $i,j = 0,1,2,\ldots, n$, with equality at the right end-point if $j = n$) then $\tilde{a}_i = \tilde{f}(ih)$ for $i = 0,1,2,\ldots, n$. Similarly, in Nyström's method we may set $\phi_i(x) = K(x,y_i)w_i$ $(i=0,1,2,\ldots, n)$ and the method appears to be similar to an 'expansion method'. (Here, though the scaling of $\tilde{f}(x)$ by a factor $1/\tilde{\kappa}$ is involved, eigenfunctions are always subject to non-zero scaling factors.)

Expansion methods are best suited for kernels which are defined analytically, because there are a number of integrations involved. In general, there are two decisions to be made about expansion methods. The first consists of a choice of 'basic functions' $\phi_i(x)$ $(i=0,1,2,\ldots,n)$.

† The equivalence is theoretical rather than practical.

Secondly, we require some criteria for $\tilde{\kappa}$ and $\tilde{f}(x)$ which will lead us to a matrix eigenvalue problem. Usually, the matrix problem can be simplified by choosing special types of basic functions $\{\phi_i(x)\}$.

Example 3.17

Let us choose $\phi_i(x)$ $(i = 0,1,2,\ldots, n)$ and approximate the kernel $K(x,y)$ $(a \leq x, y \leq b)$ by the degenerate kernel $\tilde{K}(x,y) =$ $= \sum_{i=0}^{n} \phi_i(x)\, \psi_i(y)$, for $a \leq x, y \leq b$. The eigenfunctions of $\tilde{K}(x,y)$ have the form $\tilde{f}(x) = \sum_{i=0}^{n} \tilde{a}_i \phi_i(x)$. The eigenvalues and eigenfunctions of $\tilde{K}(x,y)$ can be taken as approximations to the eigenvalues and eigenfunctions of $K(x,y)$. The eigenvalue problem of $\tilde{K}(x,y)$ is solved by considering the matrix eigenvalue problem $\tilde{A}\tilde{a} = \tilde{\kappa}\tilde{a}$, where

$$\tilde{A}_{ij} = \int_a^b \phi_j(y)\, \psi_i(y)\,dy \quad .$$

The central problem here consists of choosing the functions $\psi_i(x)$ $(i = 0,1,2,\ldots, n)$, once the functions $\phi_i(x)$ have been prescribed, in order to obtain good approximations. The method we described here is sometimes called 'Courant's method'. (Compare this with the particular case in section 3.5.) *

The technique described in Example 3.17 is somewhat involved, because it is necessary to deal directly with the problem of approximation of a function of two variables before the matrix eigenvalue problem can be written down.

We shall derive, using a heuristic approach, methods of setting up various matrix eigenvalue problems for the approximate solutions of (3.1). The theory of the subject, which will be broached later, establishes that these methods are frequently equivalent to the technique indicated in Example 3.17.

3.7. *The collocation method*

We first consider the 'collocation method'. This method is simple to motivate, but it suffers from one or two disadvantages. Some of the methods described below are of greater use, particularly if $K = K^*$.

We suppose that $\phi_0(x), \phi_1(x), \ldots, \phi_n(x)$ are linearly independent. Now we consider the equation

$$\int_a^b K(x,y)\, f(y)\,dy = \kappa f(x) \qquad (a \le x \le b) \ , \tag{3.39}$$

and we approximate an eigenfunction $f(x)$ by $\sum_{i=0}^{n} \tilde{a}_i \phi_i(x)$

and the eigenvalue κ by $\tilde{\kappa}$, so that

$$\sum_{j=0}^{n} \tilde{a}_j \int_a^b K(x,y)\, \phi_j(y)\,dy \simeq \tilde{\kappa} \sum_{j=0}^{n} \tilde{a}_j\, \phi_j(x) \ . \tag{3.40}$$

In the collocation method we choose $x_0, x_1, \ldots, x_n$ and require that the approximation above is an equality if $x = x_i$ $(i = 0,1,2,\ldots, n)$. Thus we require

$$\sum_{j=0}^{n} \psi_j(x_i)\, \tilde{a}_j = \tilde{\kappa} \sum_{j=0}^{n} \phi_j(x_i)\tilde{a}_j \ , \tag{3.41}$$

where

$$\psi_j(x) = \int_a^b K(x,y)\, \phi_j(y)\,dy \ . \tag{3.42}$$

In this way we obtain a matrix eigenvalue problem of the form

$$\underset{\sim}{A}\, \underset{\sim}{\tilde{a}} = \tilde{\kappa}\, \underset{\sim}{B}\, \underset{\sim}{\tilde{a}} \ , \tag{3.43}$$

where $A_{i,j} = \psi_j(x_i)$ and $B_{i,j} = \phi_j(x_i)$. This matrix eigenvalue problem is in a 'non-standard' form and is sometimes referred to as a generalized eigenvalue problem. The solution of the problem (3.43) may be obtained by using the QZ algorithm (Moler and Stewart 1973), which is a generalization of the QR algorithm. Other methods for dealing with (3.43) are discussed by Peters and Wilkinson (1973).

Suppose that $\underset{\sim}{B}$ is non-singular. Since $\underset{\sim}{B}^{-1}$ exists, we can consider the standard eigenvalue problem

$$(\underset{\sim}{B}^{-1}\underset{\sim}{A})\tilde{\underset{\sim}{a}} = \tilde{\kappa}\,\tilde{\underset{\sim}{a}} \; . \tag{3.44}$$

If the functions $\phi_i(x)$ form a Haar system (see Chapter 2), the matrix $\underset{\sim}{B}$ is non-singular for all choices of $x_0, x_1, \ldots, x_n$. If they do not form a Haar system, it will be necessary to restrict the choice of the x_i . Alternatively we may be able to consider the problem $\tilde{\lambda}\tilde{\underset{\sim}{a}} = (\underset{\sim}{A}^{-1}\underset{\sim}{B})\tilde{\underset{\sim}{a}}$, where $\tilde{\lambda} = \tilde{\kappa}^{-1}$. Even though $\underset{\sim}{A}^{-1}$ or $\underset{\sim}{B}^{-1}$ may exist, the problem of determining $\underset{\sim}{B}^{-1}\underset{\sim}{A}$ or $\underset{\sim}{A}^{-1}\underset{\sim}{B}$ can be ill conditioned, and consequently a method for solving (3.43) directly is generally more appropriate. The *QZ* method does not fail when $\underset{\sim}{B}$ is singular, and though the problem (3.43) can display various forms of ill-conditioning, these are usually revealed when employing the *QZ* algorithm in a suitable way.

Note that symmetry of the kernel does not usually give rise to symmetry of $\underset{\sim}{A}$ in (3.43), and $\underset{\sim}{B}$ is generally unsymmetric.

Example 3.18

With $h = 1/(n+1)$, set $x_i = a+ih$, $\phi_i(x) = \delta_{i,j}$ if $x \in [x_j, x_{j+1})$ and $\phi_i(1) = \delta_{n,i}$, where $\delta_{r,s}$ is the Kronecker delta. Thus the functions $\phi_i(x)$ are step functions. The collocation method is then equivalent to a product-integration method in which integrals of the form $\int_a^b K(x,y)\,\phi(y)\,dy$ are represented exactly if

$$\phi(x) = \sum_{i=0}^{n} \phi(x_i)\,\phi_i(x) \quad .$$

In this case the matrix $\underset{\sim}{B}$ is the identity matrix, which is particularly convenient. *

Example 3.19

We consider the approximate solution of (3.39) in the case a=0, b=1, and $K(x,y) = e^{xy}$, using (3.41) with $\phi_r(x) = x^r$ $(r = 0,1,2,\ldots,n)$ and $x_i = i/n$ $(i = 0,1,2,\ldots, n)$. The functions $\psi_j(x)$ of (3.42) satisfy a recurrence relation

$$\psi_{j+1}(x) = \{ e^{x}-(j+1)\psi_j(x)\} /x \qquad (x \neq 0)$$
$$(j = 0,1,2,\ldots,(n-1)) ,$$

which, in theory, can be used to generate the elements of the matrix $\underset{\sim}{A}$, given

$$\psi_0(x) = \int_0^1 e^{xy}\,dy \ .$$

In practice, the forward use of this recurrence relation is unstable. The backward use of the recurrence, in which two values are assigned to $\psi_n(x)$ and the value of $\psi_n(x)$ which gives rise to the correct value of $\psi_0(x)$ is deduced, is susceptible to rounding error through cancellation. The use of the recurrence cannot therefore be recommended and it seems advisable here to calculate approximate values of $\psi_j(x_i)$ by adaptive quadrature.

The construction of the elements of $\underset{\sim}{B}$ is also susceptible to rounding, and we found it advisable to construct $\underset{\sim}{B}$ in double precision before rounding to single precision.

Inaccuracies in the formation of $\underset{\sim}{A}$ using the recurrence have an unusual effect. The eigenvalues κ of largest modulus were approximated in a recognizable fashion, but spurious large eigenvalues of very large modulus (of order 10^4, say) also appear. The accuracy of smaller eigenvalues also appeared to be affected. The use of double-precision delays, but does not eliminate, this effect.

The use of the *QZ* algorithm and the solution of the eigenproblem for $\underset{\sim}{B}^{-1}\underset{\sim}{A}$ yielded comparable results for the choice of n considered. With fairly careful control of rounding we obtained results using the *QZ* method:

$(n = 4)$	$1{\cdot}35303 \times 10^{0}$	$(n = 9)$	$1{\cdot}35303 \times 10^{0}$	$1{\cdot}52975 \times 10^{-8}$
	$1{\cdot}05986 \times 10^{-1}$		$1{\cdot}05983 \times 10^{-1}$	$1{\cdot}92885 \times 10^{-9}$
	$3{\cdot}56074 \times 10^{-3}$		$3{\cdot}56075 \times 10^{-3}$	$1{\cdot}60601 \times 10^{-10}$
	$7{\cdot}67495 \times 10^{-5}$		$7{\cdot}63797 \times 10^{-5}$	$1{\cdot}24756 \times 10^{-10}$.
	$1{\cdot}17789 \times 10^{-6}$		$1{\cdot}21276 \times 10^{-6}$	
			$1{\cdot}86506 \times 10^{-8}$	

The true eigenvalues are unknown (see, however, Example 3.4). *

From the preceding Example it will be seen that the practical details of the solution of (3.43) play an important rôle. In particular, the choice of functions $\phi_r(x)$ and of sample points x_i plays some part in determining the difficulty or ease of obtaining an approximate solution. It appears to be convenient if $\underset{\sim}{B}$ can be chosen to be orthogonal, or 'almost orthogonal'.

A basis equivalent in theory to $\phi_r(x) = x^r$ is obtained if we instead set $\phi_r(x) = T_r(x)$. A convenient choice of sample points $\{x_i\}$ to match this selection is discussed in section 3.7.1, below.

<u>Remark</u>. The method described in this section gives values of the coefficients $\{\tilde{a}_r\}$ in an expansion for $\tilde{f}(x)$. With little change in the method we may obtain values of $\tilde{f}(x)$ if $\underset{\sim}{B}^{-1}$ exists. For

$$\tilde{f}(x_i) = \sum_{r=0}^{n} \tilde{a}_r \, \phi_r(x_i) \qquad (i = 0,1,2,\ldots, n) ,$$

so that

$$\underset{\sim}{\tilde{f}} = \underset{\sim}{B} \, \underset{\sim}{\tilde{a}} , \tag{3.45}$$

where $\underset{\sim}{\tilde{f}} = [\tilde{f}(x_0), \tilde{f}(x_1), \ldots, \tilde{f}(x_n)]^T$. The eigenvalue problem $\underset{\sim}{A}\underset{\sim}{\tilde{a}} = \tilde{\kappa}\underset{\sim}{B}\underset{\sim}{\tilde{a}}$ can therefore be rewritten as

$$\underset{\sim}{A}\underset{\sim}{B}^{-1}\underset{\sim}{\tilde{f}} = \tilde{\kappa}\underset{\sim}{\tilde{f}} . \tag{3.46}$$

(It is obvious, in any case, that $\underset{\sim}{B}^{-1}\underset{\sim}{A}$ and $\underset{\sim}{A}\underset{\sim}{B}^{-1}$ are similar matrices.)

3.7.1.

If the eigenfunctions of the kernel $K(x,y)$ are smooth we may achieve some success by taking basic functions $\phi_r(x)$ $(r = 0,1,2,\ldots, n)$ which are related to the Chebyshev polynomials $T_r(x)$. We also take a corresponding special choice of $x_0, x_1, \ldots, x_n$.

For convenience, we suppose $a = -1$ and $b = 1$. We shall consider two possibilities. In the first case we take

$$\phi_0(x) = \tfrac{1}{2} , \; \phi_r(x) = T_r(x) \quad (r = 1,2,3,\ldots, n)$$

and

$$x_k = \cos \left(\frac{2k+1}{n+1} \cdot \frac{\pi}{2}\right) \quad (k = 0,1,2,\ldots, n) .$$

Then the matrix $\underset{\sim}{B}$ introduced above has an inverse whose elements can be written down explicitly. We have

$$(\underset{\sim}{B}^{-1})_{i,j} = \frac{2}{n+1} T_i(x_j) \quad (i, \quad = 0,1,2,\ldots, n) .$$

Secondly, we may take

$$\phi_0(x) = \tfrac{1}{2}, \; \phi_r(x) = T_r(x) \quad (r = 1,2,3,\ldots,(n-1)),$$

$$\phi_n(x) = \tfrac{1}{2}T_n(x) ,$$

$$x_k = \cos \frac{k\pi}{n} \quad (k = 0,1,2,\ldots, n) ,$$

and the corresponding inverse of $\underset{\sim}{B}$ is given, with $d_0 = d_n = \frac{1}{2}$ and $d_1 = d_2 = \ldots = d_{n-1} = 1$ by $(\underset{\sim}{B}^{-1})_{i,j} = \frac{2}{n}T_i(x_j)d_i$ $(i,j = 0,1,2,\ldots, n)$. For an eigenvalue problem in which the range of integration $[a,b]$ is not $[-1,1]$, a change of variables should first be employed to normalize the interval. (For the case where $[a,b] = [0,1]$, this is equivalent to using shifted Chebyshev polynomials $T_r^*(x)$ and shifted abscissae $x_i^* = \frac{1}{2}(x_i+1)$ directly in the way described above. It is, of course, possible to generalize these shifted polynomials instead of reformulating the integral equation.)

Example 3.20

We consider the eigenvalue problem for the kernel

$$K(x,y) = x(1-y) \; (0 \leq x \leq y \leq 1), \; K(x,y) = y(1-x) \; (0 \leq y \leq x \leq 1)$$

studied in Example 3.7.

We seek an eigenfunction of the form

$$\tilde{f}(x) = \sum_{r=0}^{n} \tilde{a}_r \, T_r^* \, (x) \quad (0 \leq x \leq 1) .$$

To determine

$$\int_0^1 K(x,y)\ T_r^*(y)\mathrm{d}y \quad (r = 0,1,2,\ldots, n)$$

we may employ the following relations,

$$y = \tfrac{1}{2}\{1+T_1^*(y)\},\quad 1 - y = \tfrac{1}{2}\{1-T_1^*(y)\} \quad ,$$

$$T_r^*(y)T_s^*(y) = \tfrac{1}{2}\{T_{r+s}{}^*(y) + T^*_{|s-r|}(y)\} \quad ,$$

$$\int_\alpha^\beta T_r^*(y)\mathrm{d}y = \tfrac{1}{4}\left[\frac{T_{r+1}{}^*(y)}{r+1} - \frac{T_{|r-1|}{}^*(y)}{r-1}\right]_{y=\alpha}^{y=\beta} \qquad (r \neq 1)\ .$$

We choose the zeros of $T_{n+1}{}^*(x)$ as the collocation points, these we denote by x_0^* , x_1^* , ..., x_n^* . Thus

$$x_r^* = \tfrac{1}{2}\ \{\cos\ (\frac{2r+1}{n+1}\ .\ \frac{\pi}{2}) + 1\},\ (r = 0,1,2,\ldots, n)\ ,$$

and we have $\phi_0(x) = \frac{1}{2}\ T_0^*(x),\ \phi_r(x) = T_r^*(x)$, $(r = 1,2,3,\ldots, n)$.
The matrix $\underset{\sim}{B}$ with (ij)th element $\phi_j(x_i^*)$ then has an inverse whose (ij)th element is

$$(\underset{\sim}{B}^{-1})_{i,j} = 2\ T_i^*(x_j^*)/(n+1) \qquad (i,j = 0,1,2,\ldots, n)\ .$$

(We note in passing that the functions $\psi_r(x) = \int_0^1 K(x,y)\ \phi_r(y)\mathrm{d}y$ are polynomials of degree $r + 2$.) Solving the matrix eigenvalue problem for $\underset{\sim}{B}^{-1}\underset{\sim}{A}$, where $A_{ij} = \psi_j(x_i^*)$, we obtain the approximations to the largest six eigenvalues shown in Table 3.6

TABLE 3.6

n	$\tilde{\kappa}_1^+$	$\tilde{\kappa}_2^+$	$\tilde{\kappa}_3^+$	$\tilde{\kappa}_4^+$	$\tilde{\kappa}_5^+$	$\tilde{\kappa}_6^+$
5	0·101321419	0·025374687	0·010543776	0·006802108	0·000634805	0·000561300
10	0·101321184	0·025330296	0·011257912	0·006331931	0·004062947	0·002728165
20	0·101321184	0·025330296	0·011257911	0·006332574	0·004052847	0·002814476
30	0·101321184	0·025330295	0·011257911	0·006332574	0·004052847	0·002814477
40	0·101321184	0·025330296	0·011257911	0·006332573	0·004052849	0·002814475

It will be noted that there is little point in taking a large-order matrix when approximations to the largest eigenvalues are sought.

This observation is generally valid for cases where $K(x,y)$ and, hence, the eigenfunction corresponding to the largest eigenvalue, are smooth. *

3.8. *The Galerkin method*

In the Galerkin method, as in the collocation method, we can begin by choosing linearly independent 'basic functions' $\phi_0(x)$, $\phi_1(x), \ldots, \phi_n(x)$ (such as trigonometric or algebraic polynomials) and we obtain the approximate eigenfunction as a linear combination of the functions $\phi_i(x)$. In the collocation method we required values of the functions $\psi_j(x)$ $(j = 0,1,2,\ldots, n)$ for $x = x_0, x_1, \ldots, x_n$. Now, in the Galerkin method, we will have to calculate integrals in which the integrands involve the functions $\psi_j(x)$ $(j = 0,1,2,\ldots, n)$.

As in section 3.7, any approximate eigenfunction

$$\tilde{f}(x) = \sum_{j=0}^{n} \tilde{a}_j \phi_j(x)$$

and corresponding value $\tilde{\kappa}$ give us the approximation

$$\sum_{j=0}^{n} \tilde{a}_j \int_a^b K(x,y)\, \phi_j(y)\,dy \simeq \tilde{\kappa} \sum_{j=0}^{n} \tilde{a}_j\, \phi_j(x) . \tag{3.47}$$

If we multiply this approximation by $\overline{\phi_i(x)}$ and integrate $(i = 0,1,2,\ldots,n)$ we obtain

$$\sum_{j=0}^{n} \tilde{a}_j(\psi_j, \phi_i) \simeq \tilde{\kappa} \sum_{j=0}^{n} \tilde{a}_j(\phi_j, \phi_i) \quad (i = 0,1,2,\ldots, n) , \tag{3.48}$$

where, as before,

$$\psi_j(x) = \int_a^b K(x,y)\, \phi_j(y)\,dy \tag{3.49}$$

and

$$(\psi,\phi) = \int_a^b \psi(x)\, \overline{\phi(x)}\,dx .$$

In the simplest form of the Galerkin method we choose $\tilde{a}_0, \tilde{a}_1, \ldots, \tilde{a}_n$ and $\tilde{\kappa}$ so that the approximation (3.48) above becomes an equality. Thus we require

$$\underset{\sim}{A}\underset{\sim}{\tilde{a}} = \tilde{\kappa}\underset{\sim}{B}\underset{\sim}{\tilde{a}} \quad , \tag{3.50}$$

where

$$A_{i,j} = (\psi_j, \phi_i) \ , \ B_{i,j} = (\phi_j, \phi_i) \ . \tag{3.51}$$

As in the collocation method, in general we obtain a non-standard eigenvalue problem. However, since $\phi_0(x)$, $\phi_1(x)$, ..., $\phi_n(x)$ are linearly independent here, the matrix $\underset{\sim}{B}$ is always non-singular. (The matrix $\underset{\sim}{B}$ is a Gram matrix; see Davis (1965, p.177).) Some simplification occurs if we select $\phi_0(x)$, $\phi_1(x)$, ..., $\phi_n(x)$ to be a sequence of orthonormal functions, so that $\underset{\sim}{B}$ is the identity matrix.

It is worth observing that the matrix $\underset{\sim}{A}$ is Hermitian when $K(x,y)$ is Hermitian. For then

$$A_{i,j} = \int_a^b \int_a^b K(x,y) \ \phi_j(y) \ \overline{\phi_i(x)dx} \ dy \ ,$$

and it is easily shown that $A_{i,j} = \overline{A}_{j,i}$. Similarly, $\underset{\sim}{B}$ is always Hermitian. In the case where $K(x,y)$ is Hermitian, the Galerkin equations above are precisely the equations obtained from the Rayleigh-Ritz method described below (section 3.11). The Rayleigh-Ritz method gives one-sided bounds for the true eigenvalues of $K(x,y)$.

<u>Example 3.21</u>

Consider

$$\int_0^1 K(x,y) \ f(y)dy = \kappa \ f(x) \quad ,$$

where $K(x,y) = y \ (x \geq y)$, $K(x,y) = x \ (y > x)$. We take $n = 1$ with $\phi_0(x) = 1$, $\phi_1(x) = \sqrt{2}\cos\pi x$ and find

$$\underset{\sim}{A} = \begin{bmatrix} 1/3 & -\sqrt{2}/\pi^2 \\ -\sqrt{2}/\pi^2 & 1/\pi^2 \end{bmatrix}, \quad \underset{\sim}{B} = \begin{bmatrix} 1 & 0 \\ 0 & 1 \end{bmatrix}$$

We find $\tilde{\kappa}_1^+ = 0{\cdot}4017$ and $\tilde{f}_1^+(x) = 0{\cdot}903 - 0{\cdot}609 \times \cos \pi x$. True values are: $\kappa_1^+ = 4/\pi^2 \simeq 0{\cdot}40528$ and $f_1^+(x) = \alpha \sin \frac{1}{2}\pi x$ (where α is arbitrary). Observe that $\kappa_1^+ \geq \tilde{\kappa}_1^+$. Similarly, $\tilde{\kappa}_2^+ = 0{\cdot}0330 \leq \kappa_2^+ = 4/9\pi^2 \simeq 0{\cdot}04503$. *

To permit more flexibility, the simple Galerkin method described above may be extended. We may replace the inner product (ψ,ϕ) in (3.48) by the inner product (see p. 21)

$$\langle\psi,\phi\rangle = \int_a^b \omega(x)\ \psi(x)\ \overline{\phi(x)}dx ,$$

where $\omega(x) \geq 0$ for $a \leq x \leq b$. The resulting Galerkin method yields the equation

$$\underset{\sim}{A}\underset{\sim}{\tilde{a}} = \tilde{\kappa}\underset{\sim}{B}\underset{\sim}{\tilde{a}} ,$$

where $B_{i,j} = \langle\phi_j, \phi_i\rangle$ and

$$A_{i,j} = \langle\psi_j, \phi_i\rangle = \int_a^b\int_a^b \omega(x)\ K(x,y)\ \phi_j(y)\overline{\phi_i(x)}dxdy \quad .$$

Again, if $\phi_0(x), \phi_1(x), \ldots, \phi_n(x)$ are linearly independent we are assured that $\underset{\sim}{B}$ is non-singular.

The general form of the Galerkin method permits us to use Chebyshev polynomials as the 'basic functions' $\{\phi_i(x)\}$, the Chebyshev polynomials being orthogonal with respect to the weight function $(1 - x^2)^{-\frac{1}{2}}$ on $[-1, 1]$. (The shifted Chebyshev polynomials appropriate to the interval $[a,b]$ are orthogonal with respect to the weight function $\{(b-x)(x-a)\}^{-\frac{1}{2}}$.)

Example 3.22.

We treat the problem considered in Example 3.20, seeking an approximate eigenfunction of the form $\tilde{f}(x) = \sum_{r=0}^{n}{}' \tilde{a}_r\ T_r^*(x)$, and setting

$$\langle\psi,\phi\rangle = \int_0^1 \omega(x)\ \psi(x)\ \overline{\phi(x)}dx ,$$

where $\omega(x) = 1/(x - x^2)^{\frac{1}{2}}$.

In order to obtain a sequence $\{\phi_r(x)\}$ which is orthonormal with respect to the inner product $<\phi,\psi>$, we set

$$\phi_0(x) = \sqrt{(1/\pi)}\ T_0^*(x)\ ,\ \phi_r(x) = \sqrt{(2/\pi)}\ T_r^*(x)\ ,\ (r = 1,2,3,\ldots,n).$$

As in Example 3.20, the functions

$$\psi_r(x) = \int_0^1 K(x,y)\ \phi_r(y)dy \quad (r=0,1,2,\ldots,n)$$

are polynomials of degree $r + 2$. Consequently, the inner products $<\psi_r,\phi_s>$ have the form $\int_0^1 \omega(x)\ P_{r+s+2}(x)dx$, where $P_n(x)$ denotes some polynomial of degree n. It follows that, for this example, we can evaluate each element $<\psi_r,\phi_s>$ using a shifted Gauss-Chebyshev quadrature formula with $n + 2$ points, since the degree of precision of this rule is $2n + 3$. The required quadrature rule is a simple modification of the Gauss-Chebyshev rule given in section 2.17, and it takes the form

$$\int_0^1 \frac{1}{(x - x^2)^{\frac{1}{2}}} f(x)dx \simeq \frac{\pi}{n+2} \sum_{j=0}^{n+1} f(x_j^*)\ ,$$

where x_0^*, x_1^*, ..., x_{n+1}^* are the zeros of $T_{n+2}^*(x)$.

It follows that, for the particular kernel we are considering, we may replace $<\psi,\phi>$ by the pseudo-inner product

$$\frac{\pi}{n+2} \sum_{j=0}^{n+1} \psi(x_j^*)\phi(x_j^*)\ .$$

With $n = 10$ we obtain the following approximate eigenvalues: 0·10132118363, 0·02533029609, 0·01125791015, 0·00633193041, 0·00405082389, 0·00272816456, 0·00195572392, 0·00096998501, 0·00069278777, 0·00005629060, 0·00003975246.

Corresponding to the largest eigenvalue we obtain an eigenfunction

$\tilde{f}(x) = \sum_{r=0}^{n} \tilde{a}_r \phi_r(x)$ with

$$\tilde{a}_0 = 1\cdot000000 , \quad \tilde{a}_2 = -0\cdot748158$$
$$\tilde{a}_4 = 0\cdot041935 , \quad \tilde{a}_6 = -0\cdot000894$$
$$\tilde{a}_8 = 0\cdot000010 , \quad \tilde{a}_{10} = -0\cdot000000 ,$$

and $\tilde{a}_{2r+1} \simeq 0$ $(r = 0,1,2,3,4)$. Setting $\phi_0(x) = \sqrt{(1/\pi)}T^*_r(x)$ and $\phi_r(x) = \sqrt{(2/\pi)}\, T^*_r(x)$ we evaluate this eigenfunction at $x = 0$, $0\cdot 1$, $0\cdot 2$, ..., 1 and scale the function so that the maximum of these values is 1. The resulting table gives an approximate eigenfunction which is correct to 8 decimal places. *

In practice, the elements of the matrices involved in Galerkin's method may have to be calculated by a numerical integration procedure. Each element might then be calculated by an adaptive use of integration rules, but there is evidence (Delves and Walsh 1974,pp,117) that a fixed formula should be used to evaluate every double integral A_{ij} ,B_{rs}.

If it is convenient, there is some point in calculating the double integrals needed for the elements of $\underset{\sim}{A}$ in such a way as to store certain values of

$\psi_i(x) = \int_a^b K(x,y)\, \phi_i(y)dy$. For if $\tilde{f}(x) = \sum_{i=0}^{n} \tilde{a}_i\phi_i(x)$ is one approximation to an eigenfunction, a second approximation is provided by

$$\int_a^b K(x,y)\, \tilde{f}(y)dy = \sum_{i=0}^{n} \tilde{a}_i\psi_i(x) .$$

3.8.1.

We pause to remark that our motivation of the Galerkin method has been heuristic. Without devoting much space to the topic at present, we shall briefly place it in perspective. The theoretical aspects will be pursued more fully later.

In the collocation method we have approximations $\tilde{f}(x)$ and $\tilde{\kappa}$ such that the residual $r(x)$ is given by

$$r(x) = \int_a^b K(x,y)\ \tilde{f}(y)\mathrm{d}y - \tilde{\kappa}\ \tilde{f}(x) \ ,$$

and we require that $r(x_i) = 0 \quad (i = 0,1,2,\ldots, n)$. In Galerkin's method, we impose a different condition on $r(x)$, and this leads to different approximations. In the simple Galerkin method, the approximations are chosen such that

$$(r,\phi_i) \equiv \int_a^b r(x)\ \overline{\phi_i(x)}\mathrm{d}x = 0 \quad (i = 0,1,2,\ldots,n\ ) \ .$$

In the extension of this method, we require that

$$\langle r,\phi_i\rangle = \int_a^b \omega(x)r(x)\overline{\phi_i(x)}\mathrm{d}x$$

should vanish for $i = 0,1,2,\ldots, n$.

Other conditions could be imposed on $r(x)$, and these would lead to different methods. For example, a least-squares criterion results in the method described in the next section.

3.9. *A least-squares method*

We now suppose that we have computed an approximate eigenfunction $\tilde{f}(x)$, and we wish to use the approximate eigenfunction to obtain an approximate eigenvalue. If $\tilde{\kappa}$ is the approximation which we seek, we may set ourselves the problem of choosing $\hat{\kappa}$ so as to minimize

$$\int_a^b |r(x)|^2\ \mathrm{d}x \ ,$$

where

$$r(x) = \int_a^b K(x,y)\ \tilde{f}(y)\mathrm{d}y - \hat{\kappa}\ \tilde{f}(x) \ . \tag{3.52}$$

The required choice of $\hat{\kappa}$ is given by

$$\hat{\kappa} \equiv \hat{\kappa}(\tilde{f}) = \frac{\int_a^b \int_a^b \overline{\tilde{f}(x)}\ K(x,y)\ \tilde{f}(y)\ dx dy}{\int_a^b |\tilde{f}(x)|^2 dx} \quad . \tag{3.53}$$

More generally, we may wish to minimize $\int_a^b \omega(x)\ |r(x)|^2\ dx$, where $\omega(x) \geq 0$ for $a \leq x \leq b$ (and $\omega(x) > 0$ almost everywhere). We then have

$$\hat{\kappa} \equiv \hat{\kappa}(\tilde{f}) \quad \frac{\int_a^b \int_a^b \omega(x)\ \overline{\tilde{f}(x)}\ K(x,y)\ \tilde{f}(y)\ dx\ dy}{\int_a^b \omega(x)\ |\tilde{f}(x)|^2 dx} . \tag{3.54}$$

The proof that the values of $\hat{\kappa}(\tilde{f})$ given above have the required property will be deferred until later.

It is possible to consider the problem of choosing a normalized function $\dot{f}(x)$ in order to minimize

$$\int_a^b |r(x)|^2 dx \ ,$$

where $r(x)$ is given by (3.52) with $\hat{\kappa} = \hat{\kappa}(\dot{f})$. The resulting equations are in general somewhat complicated and we shall not pursue this question in the present section, where we are concentrating on practical aspects of various methods.

The expression (3.53) for $\hat{\kappa}(\tilde{f})$, where $\tilde{f}(x)$ is non-null, is known as a (generalized) Rayleigh quotient. In compact notation, $\hat{\kappa}(\tilde{f}) = (K\tilde{f},\tilde{f})/(\tilde{f},\tilde{f})$. The approximation $\tilde{f}(x)$ may be produced by any method, for instance the quadrature method or the collocation method or Galerkin's method. Since these methods usually produce their own versions of an approximation $\tilde{\kappa}$, $\hat{\kappa}(\tilde{f})$ can be used to provide a check on $\tilde{\kappa}$. But for the Rayleigh-Ritz method $\hat{\kappa}(\tilde{f})$ *is* $\tilde{\kappa}$ and therefore provides no check.

3.10. The Rayleigh quotient

We now suppose that $K(x,y)$ is Hermitian, that is, $K(x,y) = \overline{K(y,x)}$.

Then we know (section 1.6) that

$$\kappa_1^+ = \sup \{(K\phi,\phi) \mid ||\phi(x)||_2 = 1\}$$

or equivalently

$$\kappa_1^+ = \sup_{\phi(x)\not\equiv 0} \left\{ \frac{\int_a^b \int_a^b K(x,y)\ \phi(y)\ \overline{\phi(x)}\ dxdy}{\int_a^b |\phi(x)|^2 dx\ .} \right\} . \tag{3.55}$$

If we have an approximate non-zero eigenfunction $f(x)$ and consider the choice $\phi(x) = \tilde{f}(x)$ in the expression in braces in eqn (3.55), we obtain

$$\kappa_1^+ \geq \frac{\int_a^b \int_a^b K(x,y)\ \tilde{f}(y)\ \overline{\tilde{f}(x)}\ \ dxdy}{\int_a^b |\tilde{f}(x)|^2 dx} . \tag{3.56}$$

When $K(x,y)$ is Hermitian, and $\tilde{f}(x)$ is non-null, the expression on the right-hand side of (3.56) is known as a Rayleigh quotient,

$$\kappa_R(\tilde{f}) = (K\tilde{f},\tilde{f})/(\tilde{f},\tilde{f}) \quad ; \tag{3.57}$$

since $K(x,y)$ is Hermitian, $\kappa_R(\tilde{f}) = (\tilde{f},K^*\tilde{f})/(\tilde{f},\tilde{f}) = (\tilde{f},K\tilde{f})/(\tilde{f},\tilde{f})$. Thus $\kappa_R(\tilde{f})$ is real and $\kappa_R(\tilde{f})$ is the same as the value $\hat{\kappa}(\tilde{f})$ given in eqn (3.53).

Example 3.23

If we have

$$\int_a^b K(x,y)\ f(y)dy = \kappa f(x) \quad ,$$

where $K(x,y) = \overline{K(y,x)}$, then setting $\tilde{f}(x) \equiv 1$ gives us

$$\kappa_1^+ \geq \int_a^b \int_a^b K(x,y)\ dx\ dy/(b-a) .$$

For $a = 0$, $b = 1$, and the kernel $K(x,y) = x(1-y)$ $(0 \le x \le y \le 1)$, $K(x,y) = y(1-x)$ $(0 \le y \le x \le 1)$, we obtain $\kappa_1^+ \ge 0{\cdot}0825$. (The true value of κ_1^+ is $1/\pi^2 \simeq 0{\cdot}10132$.) *

<u>Example 3.24</u>

For the kernel in Example 3.21 we obtain, easily, $\kappa_1^+ \ge \frac{1}{3}$. (We observe in passing that we also have, see Chapter 1, the upper bound $\kappa_1^+ \le \rho(K) \le \|K\|$, so that $\kappa_1^+ \le \{\int_a^b \int_a^b |K(x,y)|^2 \,dx\,dy\}^{\frac{1}{2}}$, and $\kappa_1^+ \le \sup_{a \le x \le b} \int_a^b |K(x,y)|\,dy$, etc.) *

At this point we have associated the Rayleigh quotient only with the largest positive eigenvalue κ_1^+. We may also show that

$$\kappa_1^- \le \kappa_R(\tilde{f}) . \tag{3.58}$$

Furthermore, from our discussion of section 3.9 and the remark that $\kappa_R(\tilde{f}) = \hat{\kappa}(\tilde{f})$, we may expect $\kappa_R(\tilde{f})$ to give us a reasonable approximation to $\kappa_r^{\pm}$ if $\tilde{f}(x)$ is a reasonable approximation to $f_r^{\pm}(x)$.

3.11. *The Rayleigh-Ritz equations*

If $K(x,y)$ is Hermitian and $\phi_0(x), \phi_1(x), \ldots, \phi_n(x)$ are orthonormal, we may apply the simple Galerkin method and obtain the eigenvalue problem

$$\underset{\sim}{A}\underset{\sim}{\tilde{a}} = \tilde{\kappa}\underset{\sim}{\tilde{a}} , \tag{3.59}$$

where $A_{ij} = \overline{A}_{ji} = (K\phi_j, \phi_i)$ (see eqn (3.50), with $\underset{\sim}{B}$ the identity matrix). Eqn (3.59) may also be derived by the Rayleigh-Ritz method for the eigenvalue problem. (The Rayleigh-Ritz method applies only to Hermitian kernels.) We shall derive these equations from a theoretical viewpoint later; at present we are concerned only with the practical details. I have already described the method as a particular case of Galerkin's method, so that it only remains to state the particular features which arise in this case.

The most important feature is that the approximate eigenvalues $\{\tilde{\kappa}\}$ give one-sided bounds for the true eigenvalues. If we suppose that the approximate eigenvalues are

$$\tilde{\kappa}_1^+ \geq \tilde{\kappa}_2^+ \geq \ldots \geq 0 \quad ,$$

$$\tilde{\kappa}_1^- \leq \tilde{\kappa}_2^- \leq \ldots < 0 \quad ,$$

and that the corresponding true eigenvalues are

$$\kappa_1^+ \geq \kappa_2^+ \geq \ldots \geq 0 \quad ,$$

$$\kappa_1^- \leq \kappa_2^- \leq \ldots < 0 \quad .$$

then $|\tilde{\kappa}_r^{\pm}| \leq |\kappa_r^{\pm}|$. Secondly, we remark that the computed eigenfunctions are such that $\tilde{\kappa}_r^{\pm} = \kappa_R(\tilde{f}_r^{\pm})$. Finally, we should remark that the approximations will (in general) improve as we increase n .
In our study of convergence behaviour, which appears later, we discover that $\lim_{n\to\infty} \tilde{\kappa}_r^{\pm} = \kappa_r^{\pm}$ if $\phi_0(x), \phi_1(x) , \ldots, \phi_n(x)$ are part of a complete orthogonal sequence of functions. Under the same conditions, if $\kappa_r^{\pm}$ is a simple eigenvalue then the approximating eigenfunction $\tilde{f}_r^{\pm}(x)$ converges essentially in $\| \quad \|_2$ to the true eigenfunction.

The previous conclusions are all valid if $\phi_0(x), \phi_1(x), \ldots, \phi_n(x)$ are part of a dense system of linearly independent functions and we consider the more general eigenvalue problem

$$\underset{\sim}{A}\underset{\sim}{\tilde{a}} = \tilde{\kappa}\underset{\sim}{B}\underset{\sim}{\tilde{a}}$$

(in which the Gram matrix $\underset{\sim}{B}$ is known to be positive definite). Observe that since $\underset{\sim}{A}$ and $\underset{\sim}{B}$ are both Hermitian ($K(x,y)$ being Hermitian), and $\underset{\sim}{B}$ is positive definite, the preceding eigenvalue problem is rather more tractable than the similar problem which arises with the general collocation method.

Example 3.24

For the eigenvalue problem where $K(x,y) = \frac{1}{2}x(2-y)$ $(0 \le x \le y \le 1)$, $K(x,y) = \frac{1}{2}y(2-x)$ $(0 \le y \le x \le 1)$ we may take $\phi_{r-1}(x) = \sqrt{2}\sin \pi r x$ $(r = 1,2,3,\ldots)$. Then $\underset{\sim}{B}$ is the identity matrix, and we obtain eqn (3.59) with $A_{j-1,j-1} = 2/(j\pi)^2$, $A_{i-1,j-1} = (-1)^{i+j}/(ij\pi^2)$ if $i \ne j$. We obtain (as shown in Table 3.7) values of approximations to the three largest eigenvalues of $K(x,y)$, where N $(=n+1)$ is the order of the matrix.

TABLE 3.7

N	20	40	60	True
$\tilde{\kappa}_1^+$	$2\cdot3976 \times 10^{-1}$	$2\cdot4133 \times 10^{-1}$	$2\cdot4187 \times 10^{-1}$	$2\cdot4296 \times 10^{-1}$
$\tilde{\kappa}_2^+$	$4\cdot0679 \times 10^{-2}$	$4\cdot1057 \times 10^{-2}$	$4\cdot1171 \times 10^{-2}$	$4\cdot1426 \times 10^{-2}$
$\tilde{\kappa}_3^+$	$1\cdot5411 \times 10^{-2}$	$1\cdot5557 \times 10^{-2}$	$1\cdot5607 \times 10^{-2}$	$1\cdot5709 \times 10^{-2}$

Corresponding to these values, the values obtained using the trapezium rule in the quadrature method were, for a matrix of order 60, $2\cdot4297 \times 10^{-1}$, $4\cdot1447 \times 10^{-2}$, $1\cdot5731 \times 10^{-2}$. The true values, to the number of figures given, are obtained using the modified quadrature method with the parabolic rule with a matrix of order 60. The true eigenvalues are related to the zeros of $\gamma + \tan\gamma$; see Example 3.1. *

The matrices which occur when applying the Rayleigh-Ritz method to integral equations are generally full. In special cases we obtain sparse matrices (as for example, using Legendre polynomials, with $\phi_r(x) = P_r(2x-1)$ applied to the problem of Example 3.7, when $\underset{\sim}{A}$ is sparse and $\underset{\sim}{B}$ is diagonal); these special cases seem to be quite rare in practice.

3.12. *Further remarks*

A comparison of the merits of various expansion methods can be made only when a choice of basic functions $\phi_0(x)$, $\phi_1(x)$, ..., $\phi_n(x)$ has been made in the expansion method. The closer that these functions approximate the true eigenfunctions, the better the approximate eigenvalues.

It is sometimes suggested that after selecting $\phi_0(x)$, we should choose further basic functions by setting

$$\phi_{r+1}(x) = \int_a^b K(x,y)\ \phi_r(y)\mathrm{d}y \quad (r = 0,1,2,\ldots,(n-1)) \quad .$$

(The resulting functions may not be linearly independent.) This sequence $\{\phi_r(x)\}$ may remind the reader of the power method used to find the dominant eigenvalue of a matrix. Similarly, the eigenvalue κ of largest modulus may be estimated using the relation

$$|\kappa| \simeq \|\phi_{r+1}(x)\|_2 / \|\phi_r(x)\|_2$$

and choosing the sign of κ by inspection. The expression $\|\phi_{r+1}(x)\|_2/\|\phi_r(x)\|_2$ is simply $\{(K\phi_r,K\phi_r)/(\phi_r,\phi_r)\}^{\frac{1}{2}}$. If $K(x,y)$ is Hermitian then the expression in braces is a Rayleigh quotient for $K^*K(x,y) = K^2(x,y)$; the eigenvalues of $K^2(x,y)$ are the squares of the eigenvalues of $K(x,y)$.

We may know something about the eigenfunctions of $K(x,y)$ which permits us to make an inspired choice of the functions $\{\phi_i(x)\}$. In the absence of such knowledge we may choose from (a) trigonometric functions, (b) Legendre polynomials, (c) Chebyshev polynomials, (d) piecewise-constant or continuous piecewise-linear functions, or (e) more 'sophisticated' spline functions.

If the choice is one of (a), (b), (c), we may increase the accuracy of our results by increasing the order of the matrix, often using matrix elements obtained with n_1 basic functions to construct the matrix elements when $n_2 > n_1$ basic functions are employed. Thus the Galerkin method using normalized Chebyshev polynomials of degree n_1-1 produces a matrix

$\underset{\sim}{A}_1$, and if we increase n_1 to n_2 we obtain an eigenvalue problem for a matrix

$$\underset{\sim}{A}_2 = \begin{bmatrix} \underset{\sim}{A}_1 & \underset{\sim}{C}_1 \\ & \\ \underset{\sim}{B}_1 & \underset{\sim}{G}_1 \end{bmatrix}$$

of which $\underset{\sim}{A}_1$ is a submatrix. In order to perform a similar process using piecewise constant functions, we might take, on an interval $[0,1]$,

$$\phi_i(x) = 1 \quad \text{if} \quad ih \leq x < (i+1)h \text{ , and if } i = n_1 \text{ and } x = 1$$

$$= 0 \quad \text{otherwise}$$

for $i = 0,1,2,\ldots, n_1$ and

$$\phi_i(x) = 1 \quad \text{if} \quad jh \leq x < (j+\tfrac{1}{2})h$$

$$= 0 \quad \text{otherwise}$$

for $i = n_1+1, \ldots, n_2 \leq 2n_1+1$, $j = i-n_1-1$, etc.

where $h = 1/(n_1+1)$. (This arrangement does not seem very elegant.)

The choice of 'basic functions' is determined in part by whether we seek eigenvalues and eigenfunctions, or merely eigenvalues. In the general Galerkin method, some choices may give some approximate eigenfunctions which are 'good' approximations to true eigenfunctions if we measure the error in the norm $||\cdot||_2$, but are not 'good' approximations if the error is measured in the norm $||\cdot||_\infty$. Some of the eigenvalues are likely to be good approximations to true eigenvalues in this case.

If we wish to employ collocation methods, it is not *absolutely* necessary to restrict the choice of basic functions and collocation points in order that $\underset{\sim}{B}^{-1}$ exists and can be determined accurately, since the QZ method can be applied to the non-standard eigenvalue problem

$\tilde{A}\tilde{a} = \tilde{\kappa}\tilde{B}\tilde{a}$. However, we would be suspicious of the *theoretical* foundations of any scheme in which $\tilde{B}$ is permitted to be singular.

When a linear space spanned by $\phi_0(x), \ldots, \phi_n(x)$ has been chosen for the computation of approximate solutions, some attention should be given to the actual choice of basic functions. Thus if spline approximations are sought, it may be considered that the B-splines (see Chapter 2, p. 91) give a suitable choice of functions $\phi_i(x)$ in the discussion above, since in computing the functions

$$\psi_i(x) = \int_a^b K(x,y)\ \phi_i(y)dy \quad (i = 0,1,2,\ldots, n)$$

the interval of integration $[a,b]$ can be replaced by a much smaller interval. (In the Galerkin method, the computation of $(K\psi_i,\phi_j)$ reduces to integration over a small rectangle.)

3.12.1.

I trust that the reader will have obtained the flavour of my opinion of the methods discussed, by considering the Examples given in this chapter. I have tried to emphasize consistency checks, and for this purpose it may be reasonable to employ both (say) Galerkin methods and quadrature methods, where possible, in order to check one set of results against another. There is little doubt that quadrature methods are easier to employ directly, but on occasion Galerkin methods (or even collocation methods, using suitable basic functions) may give higher accuracy.

In order to say more, it is necessary to undertake a mathematical analysis of the properties of the methods, and of the algorithms used to solve the eigenvalue problems. The latter problem is one of numerical linear algebra, which we shall not pursue further. Instead, we shall now turn to the numerical analysis of the approximations obtained for the eigenvalue problem of an integral equation.

3.13. *Theoretical results*

In the remaining sections of this chapter we establish certain theoretical results concerning the numerical methods introduced above. I have attempted, when describing the numerical methods, to present these methods as techniques to be implemented in suitable cases. The theoretical justification for the methods and the discussion of error bounds or error estimates have, thus far, received little emphasis, and I shall now rectify this situation.

The analysis is simplified if we suppose that the kernel $K(x,y)$ is well behaved, and at various points we shall impose restrictions which are sufficient to keep the discussion fairly simple. Our general assumption is that the kernel is continuous.†

We seek to establish various convergence theorems and asymptotic results for the approximate eigenvalues and eigenfunctions obtained by the methods which we have described (such convergence theorems provide a fairly basic kind of justification for the methods themselves). Where upper or lower bounds can be guaranteed - as in the Rayleigh-Ritz method- we shall demonstrate that this is so. Of practical interest is the construction of good error bounds for the computed approximations. This is usually possible only if the kernel $K(x,y)$ is Hermitian or, more generally, normal. In other cases we have to rely on asymptotic error bounds or error estimates. In practice such estimates will often suffice, since exact error bounds can be pessimistic, and we regard the theory as useful when it provides motivation for various consistency checks which can be used to estimate the accuracy of computed approximations.

3.13.1.

In the following sections we deal first with the quadrature method, and then with expansion methods (the methods of collocation, Galerkin, and Rayleigh-Ritz). However, it is appropriate at this stage to make some general comments about convergence proofs and error bounds for approximate solutions of the eigenproblem.

It must be appreciated that a (square-integrable) kernel $K(x,y)$ with finite domain $a \leq x,y \leq b$ has, in general, an indeterminate

---

† To deal with weakly singular kernels (see section 1.11) we require a rather more advanced theory, in general. For an indication of the theory required, see Chapter 5 [section 5.8].

number of non-zero eigenvalues; it[¶] may have countably many such eigenvalues or it may have none. Each of our numerical methods yields a finite number of approximate eigenvalues (some of which may be zero). Consequently, it is somewhat difficult to make general statements comparing the set of true eigenvalues with the set of approximate eigenvalues.

In the case of the eigenfunctions, comparison is again a little difficult (because the true and approximate eigenfunctions can both be scaled in an arbitrary fashion). For this reason, we appeal to a notion of 'essential convergence', introduced in section 1.5 (later, we shall generalize our concepts further).

I believe that some insight into the manner in which we shall proceed can be obtained by considering the matrix eigenvalue problem.

Suppose that the Hermitian matrix $\underset{\sim}{A}$ of order n has eigenvalues

$$\lambda_1^+ \geq \lambda_2^+ \cdots \geq \lambda_m^+ \geq 0 ,$$

$$\lambda_1^- \leq \lambda_2^- \cdots \leq \lambda_r^- < 0 ,$$

and the matrix $\underset{\sim}{A}+\underset{\sim}{E}$ is Hermitian and has non-negative eigenvalues μ_i^+ and negative eigenvalues μ_i^-. The minimax characterization of the eigenvalues of a Hermitian matrix gives the result

$$\max_i | \lambda_i^+ - \mu_i^+ | \leq ||\underset{\sim}{E}||_2 ,$$

with a similar result for $\lambda_i^- - \mu_i^-$ (assuming that $\lambda_i^{\pm}$ and $\mu_i^{\pm}$ both exist for a given i). Again, suppose that $\underset{\sim}{A}\underset{\sim}{x}_i = \lambda_i \underset{\sim}{x}_i$ and we have a value $\tilde{\lambda}$ and vector $\tilde{\underset{\sim}{x}} \neq \underset{\sim}{0}$ such that $\underset{\sim}{A}\tilde{\underset{\sim}{x}} = \tilde{\lambda}\tilde{\underset{\sim}{x}} + \underset{\sim}{\eta}$. If $\underset{\sim}{A}$ is Hermitian (or even normal) then

$$\min_i |\lambda_i - \tilde{\lambda}| \leq ||\underset{\sim}{\eta}||_2 / ||\tilde{\underset{\sim}{x}}||_2 .$$

¶ It is more precise, though a little pedantic, to speak of the eigenvalues of the corresponding integral operator on a given function space, say $L^2[a,b]$, rather than the eigenvalues "of the kernel".

More generally, if λ is a simple eigenvalue of an arbitrary matrix $\underset{\sim}{A}$, there are vectors $\underset{\sim}{x},\underset{\sim}{y}$ with $\underset{\sim}{A}^*\underset{\sim}{y} = \bar{\lambda}\underset{\sim}{y}$, and $\underset{\sim}{A}\underset{\sim}{x} = \lambda\underset{\sim}{x}$, and

$$|\underset{\sim}{y}^*\tilde{\underset{\sim}{x}}| \; |\tilde{\lambda} - \lambda| \le \|\underset{\sim}{y}\|_2 \; \|\underset{\sim}{\eta}\|_2$$

All these results indicate that a given perturbation to the eigenproblem results, *in the instances cited*, in a change of the same order in a given eigenvalue. However, this is not true in all cases as will be seen by considering the matrices

$$\begin{bmatrix} \lambda & 1 \\ & \\ 0 & \lambda \end{bmatrix} \text{ and } \begin{bmatrix} \lambda & 1 \\ & \\ \varepsilon & \lambda \end{bmatrix}$$

(the eigenvalues of the latter matrix are $\lambda \pm \varepsilon^{\frac{1}{2}}$!) .

Our analysis for integral equations will build upon our knowledge of the matrix eigenvalue problem. In the first instance, the eigenvalue problem for a degenerate kernel

$$\sum_{i=0}^{N} X_i(x)Y_i(y)$$

reduces (see Chapter 1,p. 32) to the eigenvalue problem for a matrix with entries

$$\int_a^b X_j(x)Y_i(x)\,dx \; .$$

The observations concerning the matrix case therefore give immediate insight. Secondly, we seek to extend the results for perturbations to the matrix eigenvalue problem, obtaining relationships between the eigenvalues of the kernel $K(x,y)$ and those of $K(x,y)+E(x,y)$, or relating the eigenvalues to approximations with

$$\int_a^b K(x,y)\tilde{f}(y)\,dy = \tilde{\kappa}\tilde{f}(x)+\eta(x)$$

(here the analogue of a condition number for the matrix case plays a rôle when $K(x,y)$ is non-Hermitian). As in the matrix case, symmetry is an advantage, so that the case where $K(x,y) = K^*(x,y)$ is special.† The results for perturbations of the integral equations are applied by showing that our numerical methods are equivalent to solving exactly the eigenvalue problem for a perturbed kernel (which is usually degenerate) in place of $K(x,y)$, or by considering the residual $\eta(x)$ arising from a given approximation.

In the case of a Hermitian kernel, I show how to obtain rigorous (though possibly pessimistic) error bounds. More generally, we establish convergence results (section 3.15) and *hence* the asymptotic error behaviour (section 3.20 *et seq.*) of the approximations obtained by methods discussed earlier in this chapter.

3.14. *The quadrature method for degenerate kernels*

Some insight into the quadrature method (described in section 3.2) is obtained by considering the case where $K(x,y)$ is a piecewise-continuous degenerate kernel, of the form

$$K(x,y) = \sum_{k=0}^{N} X_k(x)\, Y_k(y) . \tag{3.60}$$

We assume (routinely, and without loss of generality) that each of the sets $\{X_0(x), X_1(x), \ldots X_N(x)\}$ and $\{Y_0(x), Y_1(x), \ldots, Y_N(x)\}$ consists of linearly independent piecewise-continuous functions defined for $a \leq x \leq b$. As we observed in section 1.6 (see eqn (1.26)), the non-zero eigenvalues of the kernel (3.60) on the domain $a \leq x,y \leq b$ are the eigenvalues of a matrix $\underset{\sim}{A}$ whose (i,j)th element is

$$\int_a^b X_j(x)\, Y_i(x)\, dx .$$

† The treatment of sections 3.15.4-3.15.7 and 3.16, 3.17, 3.18, and 3.19 covers Hermitian kernels, in particular. In the course of this treatment I give a number of results which are applicable in the more general case, and are used later in a more general context. Such results are indicated in the text; their application to the Hermitian case provides immediate motivation.

If we apply the quadrature method to the kernel (3.60), eqn (3.7) assumes a special form. Since, from (3.60),

$$K(y_i, y_j) = \sum_{k=0}^{N} X_k(y_i)\, Y_k(y_j) \tag{3.61}$$

the matrix $\underset{\sim}{K}$ is a product, namely $\underset{\sim}{X}^T \underset{\sim}{Y}$, where $X_{i,j} = X_i(y_j)$ and $Y_{i,j} = Y_i(y_j)$. (The matrices $\underset{\sim}{X}$ and $\underset{\sim}{Y}$ are in general rectangular matrices.) Consequently the non-zero eigenvalues of $\underset{\sim}{K}\underset{\sim}{D} = \underset{\sim}{X}^T\underset{\sim}{Y}\underset{\sim}{D}$ are also the non-zero eigenvalues of the matrix $\tilde{\underset{\sim}{A}} = \underset{\sim}{Y}\underset{\sim}{D}\underset{\sim}{X}^T$ (see the Remark below). The (i,j)th element of $\tilde{\underset{\sim}{A}}$ is simply

$$\sum_{k=0}^{n} w_k Y_i(y_k) X_j(y_k)$$

and this quantity is evidently an approximation to

$$A_{i,j} = \int_a^b Y_i(x) X_j(x)\,dx .$$

<u>Remark</u>. We can show that the non-zero eigenvalues of $\underset{\sim}{K}\underset{\sim}{D}$ are those of $\tilde{\underset{\sim}{A}}$ by considering the following equations (where $\underset{\sim}{K} = \underset{\sim}{X}^T\underset{\sim}{Y}$). We have $\tilde{\kappa}\tilde{\underset{\sim}{f}} = \underset{\sim}{K}\underset{\sim}{D}\tilde{\underset{\sim}{f}} = \underset{\sim}{X}^T\underset{\sim}{Y}\underset{\sim}{D}\tilde{\underset{\sim}{f}}$, so that if $\tilde{\kappa} \neq 0$, $\tilde{\underset{\sim}{f}} = \underset{\sim}{X}^T\underset{\sim}{\alpha}$, where $\underset{\sim}{\alpha} = \underset{\sim}{Y}\underset{\sim}{D}\tilde{\underset{\sim}{f}}/\tilde{\kappa}$. But $\tilde{\kappa}\underset{\sim}{Y}\underset{\sim}{D}\tilde{\underset{\sim}{f}} = \underset{\sim}{Y}\underset{\sim}{D}\underset{\sim}{X}^T\,(\underset{\sim}{Y}\underset{\sim}{D}\tilde{\underset{\sim}{f}})$ so that $\tilde{\kappa}\underset{\sim}{\alpha} = \underset{\sim}{Y}\underset{\sim}{D}\underset{\sim}{X}^T\underset{\sim}{\alpha}$. Thus it follows that if $\tilde{\kappa}$ is a non-zero eigenvalue of $\underset{\sim}{K}\underset{\sim}{D}$, it is an eigenvalue of $\underset{\sim}{Y}\underset{\sim}{D}\underset{\sim}{X}^T = \tilde{\underset{\sim}{A}}$. (Similarly, if $\tilde{\kappa}$ is a non-zero eigenvalue of $\tilde{\underset{\sim}{A}}$, the corresponding eigenvector $\underset{\sim}{\alpha}$ has a form $\underset{\sim}{Y}\underset{\sim}{D}\underset{\sim}{\beta}$, where $\underset{\sim}{\beta} = \underset{\sim}{X}^T\underset{\sim}{\alpha}/\tilde{\kappa}$, and we can show that $\tilde{\kappa}$ is an eigenvalue of $\underset{\sim}{K}\underset{\sim}{D}$.) *

Since $\underset{\sim}{A}$ and $\tilde{\underset{\sim}{A}}$ are both matrices of order $N+1$, they can be compared directly. From such a comparison we can establish a relationship between true eigenvalues κ and approximate eigenvalues $\tilde{\kappa}$.

To effect this comparison, we first note that $\underset{\sim}{A} = \tilde{\underset{\sim}{A}}+\underset{\sim}{E}$, where $E_{ij} = \int_a^b X_j(x) Y_i(x)\,dx - \sum_{k=0}^{n} w_k X_j(y_k) Y_i(y_k)$. Let us suppose that $\max |E_{ij}| \to 0$ as $n \to \infty$. Then if we take n large enough, every true non-zero eigenvalue κ can be approximated arbitrarily closely by some approximate eigenvalue $\tilde{\kappa}$. Furthermore, every non-zero 'cluster point' or 'point of accumulation' of approximate eigenvalues, as $n \to \infty$, is a true eigenvalue κ . These results may be established by appealing to matrix perturbation theory (Wilkinson 1965; Kato 1966).

We may expect to be able, for some finite value of n , to bound the distance between some true eigenvalue κ and the closest approximate eigenvalue $\tilde{\kappa}$ in terms of the size of the elements E_{ij} . To do so, however, we require a knowledge of the structure of the matrix $\underset{\sim}{A}$. It is particularly instructive to *choose* a degenerate kernel $K(x,y)$ in such a way that

$$\underset{\sim}{A} = \begin{bmatrix} \kappa & 1 & & & \\ & \kappa & 1 & & \\ & & \kappa & 1 & \\ & & & \kappa & 1 \\ & & & & \kappa \end{bmatrix} \qquad (3.62)$$

and each function $X_i(x)$ or $Y_j(x)$ $(i,j = 0,1,2,\ldots,N)$ is a suitable smooth function. By matching the kernel to the quadrature rule (so that the rule integrates the product $X_j(x)Y_i(x)$ exactly unless $i = N$ and $j = 0$), we can arrange that

$$\underset{\sim}{E} = \begin{bmatrix} 0 & \text{---} & 0 \\ & & \\ 0 & \text{---} & 0 \\ \varepsilon\ 0 & \text{---} & 0 \end{bmatrix} \qquad (\varepsilon \neq 0) . \qquad (3.63)$$

Under such conditions, the true and approximate eigenvalues are related by the equation $\tilde{\kappa} = \kappa + \omega\varepsilon^{1/(N+1)}$, where ω is an $(N+1)$th root of unity. Thus, in general, the size of $\kappa-\tilde{\kappa}$ is not of the order of the error ε due to replacing an integral by a quadrature rule. However, in the case we have illustrated, a multiple eigenvalue κ has split into $(N+1)$ distinct approximate eigenvalues $\tilde{\kappa}_i$ $(i = 0,1,2,\ldots,N)$. If $\kappa \neq 0$ is a *simple* eigenvalue, there *is* an approximation $\tilde{\kappa}$ such that $\kappa-\tilde{\kappa}$ is of the order of the discretization error in replacing an integral by a quadrature rule. In the case of a degenerate kernel, $|\tilde{\kappa}-\kappa| \leq Le$, where $e = \|\underset{\sim}{E}\|_2$, say, and L is a constant; if κ is close to some other true eigenvalue κ', L may be large.

It is interesting to consider further the case where $K(x,y)$ has the form of eqn (3.60). We suppose that the functions $\{X_k(x)\}$, $\{Y_k(x)\}$ are polynomials of degree at most s, say. We now permit the matrix $\underset{\sim}{A}$ to have an arbitrary canonical form. If we apply the quadrature method using the repeated trapezium rule with step h, we find that

$$E_{ij} = \sum_{r=1}^{s} E_{ij}^{(r)} h^{2r} . \tag{3.64}$$

This result follows from the Euler-Maclaurin formula applied to the integrand $X_j(x)Y_i(x)$. In view of the expression for E_{ij},

$$\underset{\sim}{A} = \underset{\sim}{\tilde{A}}+h^2\underset{\sim}{E}^{(1)}+h^4\underset{\sim}{E}^{(2)} + \ldots + h^{2s}\underset{\sim}{E}^{(s)} . \tag{3.65}$$

For matrices $\underset{\sim}{\tilde{A}},\underset{\sim}{A}$ related in this fashion, one can relate pairs of true and approximate eigenvalues $\kappa,\tilde{\kappa}$ by means of an expression of the form

$$\tilde{\kappa} = \kappa + \mu_1 h^{2/r}+ \mu_2 h^{4/r} + \mu_3 h^{6/r} + \ldots , \tag{3.66}$$

where r is the (algebraic) multiplicity of κ. Some of the coefficients $\{\mu_k\}$ may vanish. Wilkinson (1965) discusses this type of perturbation theory.

The important feature of eqn (3.66) is that μ_1 may not vanish. For this reason, repeated h^2-extrapolation, and the related techniques, may not be a success unless $r=1$, *even though* $K(x,y)$ *is smooth.* The case where $K(x,y)$ is badly behaved has already been illustrated using a degenerate kernel, in Example 3.8.

3.15. *The quadrature method for general kernels*

The quadrature method for a general kernel can be analysed by introducing certain degenerate kernels. If we impose mild restrictions on the choice of quadrature rule, we can establish various results on the convergence of the approximate solutions when the kernel $K(x,y)$ is continuous. If we merely assume continuity of $K(x,y)$ the convergence results do not reflect the true rate of convergence when $K(x,y)$ has high-order derivatives and the degree of precision of the quadrature rule is of a moderate size. In general, suitable results on the rate of convergence must be obtained by a different strategy.

We can obtain fairly strong results when $K(x,y)$ is Hermitian, and we deal with this case first. Some of the analysis for this case is useful when $K(x,y)$ is non-Hermitian. (We shall state whether or not we require $K(x,y)$ to be Hermitian as the analysis proceeds.)

As a preliminary to our analysis we introduce a degenerate kernel $K_{[n]}(x,y)$ whose non-zero eigenvalues are precisely the non-zero eigenvalues of the matrix $\tilde{K}\tilde{D}$. With care in the definition of a slightly different kernel $K_n(x,y)$ we can also ensure that the eigenfunctions $\{f_{nr}(x)\}$ of $K_n(x,y)$ are such that $f_{nr}(y_i) = \tilde{f}_r(y_i)$. Thus, pointwise errors $|f_r(y_i) - \tilde{f}_r(y_i)|$ in the approximate eigenfunctions (suitably scaled) can be measured in terms of $|f_r(y_i) - f_{nr}(y_i)|$.

Throughout the analysis we suppose either that $J(\phi)$ is a Riemann sum (section 2.14), which we write $J(\phi) \in J^R$, or that $J(\phi)$ has positive weights and abscissae in the range of integration $(J(\phi) \in J^+)$. Our policy will be to treat the former case, which is slightly easier, and indicate the extensions for $J(\phi) \in J^+$; the discussion for $J(\phi) \in J^+$ can be omitted on a first reading. †

---

† In Chapters 4 and 5 we consider similar theories for $J(\phi) \in J^+$ without treating J^R as a special case; in this chapter we take the opportunity to emphasize the motivation, by dealing with J^R. It is instructive to work through the results for J^R by considering a special case such as the repeated trapezium rule.

Suppose that $J(\phi) \in J^R$. As indicated, we begin by defining a corresponding kernel $K_{[n]}(x,y)$, associated with a given kernel $K(x,y)$, which will permit us to investigate the behaviour of the approximate eigenvalues. (We subsequently define a new kernel $K_n(x,y)$, similar to $K_{[n]}(x,y)$, in order to make the analysis more satisfactory.) Symmetry of the given kernel $K(x,y)$ is not an assumption.

For $J(\phi) = \sum_{j=0}^{n} w_j \phi(y_j) \in J^R$ we may define (see section 2.14)

$$v_0 = a,\ v_k = v_{k-1} + w_{k-1} \quad (k = 1,2,3,\ldots,n+1)\ , \tag{3.67}$$

and we find that

$$a = v_0 \le y_0 \le v_1 \le \ldots \le v_n \le y_n \le v_{n+1} = b\ .$$

In terms of the values v_i, we may define the kernel $K_{[n]}(x,y)$ by setting

$$K_{[n]}(x,y) = K(y_i,y_j) \quad \text{if} \quad \begin{cases} v_i \le x < v_{i+1} \\ \quad \text{or} \quad x = b \quad \text{and} \quad i = n \\ \\ v_j \le y < v_{j+1} \\ \quad \text{or} \quad y = b \quad \text{and} \quad j = n\ . \end{cases}$$

The kernel $K_{[n]}(x,y)$ is degenerate; this may be seen if we write

$$K_{[n]}(x,y) = \sum_{j=0}^{n} \phi_j(x)\ K_{[n]}(y_j,y)\ ,$$

where $\phi_i(x) = 1$ if $v_i \le x < v_{i+1}$ or $x = b$ and $i = n$, and $\phi_i(x) = 0$ otherwise. An alternative expression for $K_{[n]}(x,y)$ is obtained on writing

$$K_{[n]}(x,y) = \sum_{i=0}^{n} \sum_{j=0}^{n} K(y_i,y_j)\ \phi_i(x)\ \phi_j(y)\ . \tag{3.68}$$

Now suppose that

$$\int_a^b K_{[n]}(x,y)\, f_n(y)\mathrm{d}y = \kappa_n\, f_n(x) \quad ,$$

where $f_n(x) \equiv f_{n,r}(x)$, $\kappa_n \equiv \kappa_{n,r}$ for some $r \in \{0,1,2,\ldots,n\}$, and where $\kappa_n \neq 0$. Then, clearly, $f_n(x)$ is a linear combination of $\phi_0(x),\phi_1(x),\ldots,\phi_n(x)$. (We refer to the discussion of degenerate kernels in Chapter 1.) If we set

$$f_n(x) = \sum_{i=0}^{n} \tilde{a}_i\, \phi_i(x)$$

then the eigenvalue equation for $K_{[n]}(x,y)$, which may be written

$$\sum_{j=0}^{n} \int_{v_j}^{v_{j+1}} K_{[n]}(x,y)\, f_n(y)\mathrm{d}y = \kappa_n f_n(x) \ ,$$

becomes

$$\sum_{j=0}^{n} (v_{j+1} - v_j) \sum_{i=0}^{n} K(y_i,y_j)\, \phi_i(x)\, \tilde{a}_j = \kappa_n \sum_{i=0}^{n} \tilde{a}_i \phi_i(x) \ .$$

Since $\phi_0(x),\phi_1(x),\ldots,\phi_n(x)$ are linearly independent and $v_{j+1} - v_j = w_j$, we find

$$\sum_{j=0}^{n} w_j\, K(y_i,y_j)\, \tilde{a}_j = \kappa_n\, \tilde{a}_i \qquad (i = 0,1,2,\ldots,n) \quad .$$

Now

$$\sum_{j=0}^{n} w_j\, K(y_i,y_j)\, \tilde{f}(y_j) = \tilde{\kappa}\tilde{f}(y_i) \ ,$$

so the non-zero eigenvalues $\{\kappa_n\}$ are eigenvalues $\{\tilde{\kappa}\}$ of the matrix $\underset{\sim}{K}\underset{\sim}{D}$, and the vectors $\underset{\sim}{\tilde{a}} = [\tilde{a}_0,\tilde{a}_1,\ldots,\tilde{a}_n]^T$ are eigenvectors of $\underset{\sim}{K}\underset{\sim}{D}$. Conversely, an approximate eigenvalue $\tilde{\kappa} \neq 0$ may be shown to be an eigenvalue of $K_{[n]}(x,y)$ and the corresponding eigenfunction is

$$f_n(x) = \sum_{i=0}^{n} \tilde{f}(y_i)\, \phi_i(x) \ .$$

Thus there is a correspondence between the eigenproblem for $K_{[n]}(x,y)$

and the eigenproblem for the matrix $\underset{\sim\sim}{KD}$. This correspondence actually preserves the multiplicities of the eigenvalues.

In order to compare the eigenvalues $\{\kappa\}$ of $K(x,y)$ with the approximate values $\{\tilde{\kappa}\}$ we can compare (as we show below) the kernels $K(x,y)$ and $K_{[n]}(x,y)$. This situation is, I suggest, better than having to compare the *kernel* $K(x,y)$ with a *matrix* $\underset{\sim\sim}{KD}$.

To compare $\tilde{f}(y_i)$ with the corresponding value $f(y_i)$ of a true eigenfunction, we would like to be able to relate $f_n(y_i)$ and $\tilde{f}(y_i)$. But at this point we may choose to pay some attention to details, since if

$$f_n(x) = \sum_{i=0}^{n} \tilde{f}(y_i)\ \phi_i(x) ,$$

and $y_k = v_{k+1}$ for some k, we find that

$$f_n(y_k) = \tilde{f}(y_{k+1}) .$$

This situation is a little unsatisfactory, but one way to circumvent it is to ensure that $\phi_i(x)$ is defined in such a way that $\phi_i(y_j) = \delta_{ij}$, where δ_{ij} is the Kronecker delta.

We can re-define the functions $\phi_i(x)$ $(i = 0,1,2,\ldots,n)$ and obtain the kernel $K_n(x,y)$ as follows. We set

$$\phi_i(y_j) = \delta_{ij} \qquad (i,j = 0,1,2,\ldots,n) , \tag{3.69a}$$

and if $x \notin \{y_k\}$ we set

$$\phi_i(x) = \begin{cases} 1 \text{ if } v_i \le x < v_{i+1} \\ 1 \text{ if } x = b \text{ and } i = n \\ 0 \text{ otherwise.} \end{cases} \tag{3.69b}$$

We then set

$$K_n(x,y) = \sum_{i=0}^{n} \sum_{j=0}^{n} K(y_i,y_j)\ \phi_i(x)\ \phi_j(y) . \tag{3.70}$$

The non-zero eigenvalues of $K_n(x,y)$ are precisely the non-zero eigenvalues $\tilde{\kappa}$ (as before with $K_{[n]}(x,y)$) and the corresponding eigenfunctions may be normalized so that $f_n(y_i) = \tilde{f}(y_i)$ $(i = 0,1,2,\ldots,n)$, where

$$\underset{\sim}{\mathrm{K}}\underset{\sim}{\mathrm{D}}\underset{\sim}{\tilde{\mathrm{f}}} = \tilde{\kappa}\underset{\sim}{\tilde{\mathrm{f}}}$$

and

$$\underset{\sim}{\tilde{\mathrm{f}}} = [\tilde{f}(y_0),\tilde{f}(y_1),\ldots,\tilde{f}(y_n)]^T .$$

It may be noted that $\underset{\sim}{\mathrm{K}}_n = [K_n(y_i,y_j)] = \underset{\sim}{\mathrm{K}} = [K(y_i,y_j)]$ when the definition (3.69) is employed in (3.70).

3.15.1. *Rules in* J^+

For completeness, we now consider the case where $J(\phi) \in \mathrm{J}^+$, that is $J(\phi) = \sum_{j=0}^{n} w_j\, \phi(y_j)$, where $w_j \geq 0$ and $a \leq y_j \leq b$ $(j = 0,1,2,\ldots,n)$.

We give an amended definition of $K_n(x,y)$ which can be used to analyse the quadrature method in this case. We define

$$w = \sum_{j=0}^{n} w_j/(b-a) ,$$

$$v_0=a, \quad v_k=v_{k-1}+(w_{k-1}/w) \quad (k=1,2,3,\ldots,(n+1)) , \tag{3.71}$$

and with these values of $v_0,v_1,\ldots,v_{n+1}$ we employ (3.69a) and (3.69b) to define functions $\phi_i(x)$ $(i=0,1,\ldots,n)$. We now use these functions to set

$$K_n(x,y) = w \sum_{i=0}^{n} \sum_{j=0}^{n} K(y_i,y_j)\, \phi_i(x)\, \phi_j(y) . \tag{3.72}$$

In the case $J(\phi) \in \mathrm{J}^R$, (3.72) reduces to (3.70). Analogous results to those established above in section 3.15 can be established for $J(\phi) \in \mathrm{J}^+$ using (3.72).

3.15.2. *Approximation of* $K(x,y)$ *by* $K_n(x,y)$

We require some information about the behaviour of $K(x,y) - K_n(x,y)$. Without assuming symmetry, we have

$$\sup_{a\le x,y\le b} |K(x,y) - K_n(x,y)|$$

$$= \max_{i,j} \sup_{\substack{v_i\le x<v_{i+1} \\ v_j\le y<y_{j+1}}} |K(x,y) - K(y_i,y_j)| \quad .$$

If $J(\phi)$ is a Riemann sum, and

$$\Delta = \max_i \{\max(v_{i+1}-y_i,\ y_i-v_i)\} \quad ,$$

then

$$\sup_{a\le x,y\le b} |K(x,y) - K_n(x,y)| \tag{3.73}$$

$$\le \sup_{a\le x,y\le b} \sup_{\substack{|\xi-x|\le\Delta \\ |\eta-y|\le\Delta}} |K(x,y) - K(\xi,\eta)| \quad .$$

Thus, if $K(x,y)$ satisfies a Lipschitz condition, or is differentiable,

$$\sup_{a\le x,y\le b} |K(x,y) - K_n(x,y)| = O(\Delta) \quad .$$

In particular, if $K_x(x,y)$ and $K_y(x,y)$ are bounded,

$$\sup_{a\le x,y\le b} |K(x,y) - K_n(x,y)| \le \Delta\{M_1 + M_2\} \quad , \tag{3.74}$$

where $M_1 = \sup_{a\le x,y\le b} |K_x(x,y)|$ and $M_2 = \sup_{a\le x,y\le b} |K_y(x,y)|$. Here, $J(\phi)$ is a Riemann sum and we can bound Δ using the relation $\Delta \le \max_i w_i$. When $J(\phi)$ is a rule which is not a Riemann sum, but has positive weights and abscissae lying in the interval $[a,b]$, a similar analysis applies. We must then involve the quantity w defined in (3.71) and

$$\Delta = \max_i \{\max(|v_{i+1} - y_i|,|y_i - v_i|)\} \quad , \tag{3.75}$$

which is to be defined using the quantities v_k in (3.71). We call Δ the *measure* of a quadrature rule $J(\phi)$ in J^R or in J^+.

There are a number of results which we can obtain in analogy with (3.73) for rules in J^+, in particular

$$\sup_{a\le x,y\le b} |K(x,y) - K_n(x,y)| \le (1+|w|) \sup_{\substack{|y-\eta|\le\Delta \\ |x-\xi|\le\Delta}} |K(x,y)-K(\xi,\eta)|, \quad (3.76)$$

where $K_n(x,y)$ is defined by (3.72). What is useful about this result is that for a convergent family of rules, $w\to 1$ and $\Delta\to 0$, so that if $K(x,y)$ is continuous $\|K(x,y) - K_n(x,y)\|_\infty \to 0$. This result will be applied later.

3.15.3.

We take the opportunity at this point to introduce the following definition, employed by Wielandt (1956).

DEFINITION 3.1. *A degenerate kernel*

$$K_N(x,y) = \sum_{k=0}^{N} X_k(x)\, Y_k(y)$$

is said to 'allow' a quadrature rule

$$J(\phi) = \sum_{j=0}^{n} w_j\, \phi(y_j)$$

if an only if the non-zero eigenvalues of $K_N(x,y)$ (on $a\le x,y\le b$) *are exactly the non-zero eigenvalues of the matrix* $\underset{\sim}{K}_N\underset{\sim}{D}$ *with elements* $w_j\, K_N(y_i,y_j)$.

<u>Example 3.25</u>

The kernels $K_n(x,y)$ defined in eqns (3.70) and (3.72) allow their respective quadrature rules $J(\phi)$. *

The following theorem permits us to establish the result in Example 3.25, and more general results which are of importance later.

THEOREM 3.1. *The degenerate kernel* $K_N(x,y) = \sum_{k=0}^{N} X_k(x)Y_k(y)$ *allows the quadrature rule* $J(\phi)$ *if*

$$\int_a^b X_j(x)Y_i(x)dx = \sum_{k=0}^{n} w_k X_j(y_k)Y_i(y_k) \quad (i,j = 0,1,2,\ldots,N) .$$

(The proof may be based on the fact that $\underset{\sim}{K} = \underset{\sim}{X}^T\underset{\sim}{Y}$ where $Y_{i,j} = Y_i(y_j)$ and $X_{i,j} = X_i(y_j)$. See p. 232.)

It should be noted that when $\kappa_r \equiv \kappa_r(K_N)$ is a non-zero eigenvalue of a degenerate kernel $K_N(x,y)$ allowing $J(\phi)$ then its algebraic multiplicity is the same as the algebraic multiplicity of $\kappa_r \equiv \kappa_r(\underset{\sim}{K}_N\underset{\sim}{D})$ as an eigenvalue of $\underset{\sim}{K}\underset{\sim}{D}$. For, the eigenvalues of $K_N(x,y)$ and $\underset{\sim}{K}_N\underset{\sim}{D}$ are the (complex) zeros of polynomials of degree $N+1$ and

$$\sum_r \{\kappa_r(K_N)\}^P = \sum_r \{\kappa_r(\underset{\sim}{K}_N\underset{\sim}{D})\}^P$$

for $P = 0,1,2,\ldots$. To establish this we note that

$$\sum_r \{\kappa_r(K_N)\}^P \equiv \mathrm{tr}(K_N^P) = \int_a^b K_N^P(t,t)dt = \sum_i [(\underset{\sim}{K}_N\underset{\sim}{D})^P]_{ii}$$

$$= \mathrm{tr}([\underset{\sim}{K}_N\underset{\sim}{D}]^P) \equiv \sum_r \{\kappa_r(\underset{\sim}{K}_N\underset{\sim}{D})\}^P$$

(see Smithies 1962, pp.79 *et seq.*).

3.15.4. *Eigenvalues of Hermitian kernels*

We now suppose that $K(x,y)$ is Hermitian, and we shall use the concepts introduced earlier to analyse the convergence of the approximate eigenvalues obtained using the quadrature method. We use a theorem which permits comparison of the eigenvalues of two Hermitian kernels.

We suppose that the eigenvalues of $K(x,y)$ are indexed

$$\kappa_1^+ \geq \kappa_2^+ \geq \ldots \geq 0 , \quad \kappa_1^- \leq \kappa_2^- \leq \ldots < 0 .$$

For clarity we shall denote the eigenvalue $\kappa_r^{\pm}$ of $K(x,y)$ by $\kappa_r^{\pm}(K)$. If, for some r , there is no rth positive (negative) eigenvalue $\kappa_r^+(K)$ $(\kappa_r^-(K))$ then the value should be set to zero in what follows.

According to our earlier analysis (sections 3.15 and 3.15.1) the non-zero approximate eigenvalues $\{\tilde{\kappa}\}$ obtained from the quadrature method (using any of the quadrature rules, with positive weights, given in Chapter 2) are eigenvalues of some degenerate kernel $K_n(x,y)$. Accordingly we may set

$$\tilde{\kappa}_r^+ = \kappa_r^+(K_n) \ , \ \tilde{\kappa}_r^- = \kappa_r^-(K_n) \quad ,$$

where $\tilde{\kappa}_r^{\pm}$ are the suitably indexed approximations, and $K_n(x,y)$ is Hermitian when $K(x,y)$ is Hermitian. As before, if there is no rth positive (negative) eigenvalue $\tilde{\kappa}_r^+$ $(\tilde{\kappa}_r^-)$ we set the value $\kappa_r^+(K_n)(\kappa_r^-(K_n))$ to zero. To compare $\kappa_r^{\pm}(K)$ with $\kappa_r^{\pm}(K_n)$ we appeal to the following theorem.†

THEOREM 3.2. *If* $K(x,y)$ *and* $K_n(x,y)$ *are Hermitian square-integrable kernels, with eigenvalues* $\kappa_r^{\pm}(K)$ *and* $\kappa_r^{\pm}(K_n)$ *, then*

$$|\kappa_r^{\pm}(K) - \kappa_r^{\pm}(K_n)| \leq \rho(K-K_n) \ ,$$

where $\rho(L)$ *denotes the spectral radius of* $L(x,y)$ *,*
$\rho(L) = \max\ (\kappa_1^+(L), -\kappa_1^-(L))$.

To establish this result, we require the following theorem.

THEOREM 3.3. *If* $K_1(x,y)$ *and* $K_2(x,y)$ *are Hermitian and square-integrable kernels and* $K_0(x,y) = K_1(x,y) + K_2(x,y)$ *, then*

$$\kappa_{p+q-1}^+ (K_0) \leq \kappa_p^+ (K_1) + \kappa_q^+ (K_2) \tag{3.77a}$$

and

$$\kappa_{p+q-1}^- (K_0) \geq \kappa_p^- (K_1) + \kappa_q^- (K_2) \ . \tag{3.77b}$$

---

† In this and the consequent results, it is sufficient for K, K_n , to be self-adjoint compact operators on $L^2[a,b]$.

Proof. The proof of Theorem 3.3 is established using a minimax characterization theorem (Smithies 1962, p.133) of the eigenvalues of a Hermitian kernel.

The required characterization may be stated as follows. Suppose that M_p is the space of linear combinations of chosen functions $\phi_1(x), \phi_2(x), \ldots, \phi_{p-1}(x)$. Thus

$$M_p = \{ \sum_{i=1}^{p-1} \alpha_i \phi_i(x) \}$$

is defined by a choice of $\phi_i(x)$ $(i = 1,2,3,\ldots,(p-1))$ of functions (which are not necessarily linearly independent). We denote by N_p the set of functions $\phi(x)$ such that $(\phi, \phi) = 1$ and $(\phi, \phi_i) = 0$ $(i = 1,2,3,\ldots,(p-1))$, and we define

$$\mu^+ (K,M_p) = \sup \{(K\phi, \phi) \mid \phi(x) \in N_p\} ,$$

where $(\nu, \phi) = \int_a^b \nu(x) \overline{\phi(x)} dx$.

Then

$$\kappa_p^+ (K) = \inf_{\phi_1(x),\ldots,\phi_{p-1}(x)} \mu^+ (K,M_p) \quad (p \geq 1) .$$

Note that if we define another space $M_p^1 \subseteq M_p$, the space N_p^1 corresponding to M_p^1 is such that $N_p^1 \supseteq N_p$. If $\phi_i(x) = f_i^+(x)$, the eigenfunction corresponding to $\kappa_i^+(K)$, then $\mu^+ (K,M_p) = \kappa_p^+(K)$, as noted in section 1.6.

Now suppose the 'first' $p-1$ eigenfunctions of $K_1(x,y)$ (those corresponding to $\kappa_1^+(K_1)$, $\kappa_2^+(K_1),\ldots,\kappa_{p-1}^+(K_1)$) form the basis for a linear space $M_p^{(1)}$, and, similarly, the 'first' $q-1$ eigenfunctions of $K_2(x,y)$ define a linear space $M_q^{(2)}$. The direct sum $M_p^{(1)} \oplus M_q^{(2)}$ denotes the linear space spanned by the 'first' $p-1$ eigenfunctions of $K_1(x,y)$ together with the 'first' $q-1$ eigenfunctions of $K_2(x,y)$, and its dimension is less than or equal to $p+q-2$.

Now from the characterization above

$$\kappa^+_{p+q-1}(K_0) = \inf_{M_{p+q-1}} \mu^+(K_0, M_{p+q-1})$$

$$\leq \mu^+(K_0, M_p^{(1)} \oplus M_q^{(2)})$$

$$\leq \mu^+(K_1, M_p^{(1)} \oplus M_q^{(2)}) + \mu^+(K_2, M_p^{(1)} \oplus M_q^{(2)})$$

$$\leq \mu^+(K_1, M_p^{(1)}) + \mu^+(K_2, M_q^{(2)})$$

$$= \kappa^+_p(K_1) + \kappa^+_q(K_2) \quad ,$$

since $\mu^+(K_1, M_p^{(1)}) = \kappa^+_p(K_1)$ and $\mu^+(K_2, M_q^{(2)}) = \kappa^+_q(K_2)$ (as remarked above).

We have established that

$$\kappa^+_{p+q-1}(K_0) \leq \kappa^+_p(K_1) + \kappa^+_q(K_2) \quad , \tag{3.78}$$

and the remaining part of the lemma is established on replacing $K_1(x,y)$ by $-K_1(x,y)$ and $K_2(x,y)$ by $-K_2(x,y)$. This is because $K_0(x,y)$ must then be replaced by $-K_0(x,y)$ and from (3.78) we obtain

$$\kappa^+_{p+q-1}(-K_0) \leq \kappa^+_p(-K_1) + \kappa^+_q(-K_2) \quad ,$$

so that the required result follows, on setting $\kappa^+_r(-K_s) = -\kappa^-_r(K_s)$.

We may now use Theorem 3.3 to establish Theorem 3.2, stated above. We set $p = r$, $q = 1$, $K_0(x,y) = K(x,y)$ and $K_1(x,y) = K_n(x,y)$ in eqns (3.77a), (3.77b), and obtain the inequalities

$$\begin{aligned} \kappa^+_r(K) - \kappa^+_r(K_n) &\leq \kappa^+_1(K - K_n) \quad , \\ \kappa^-_r(K) - \kappa^-_r(K_n) &\geq \kappa^-_1(K - K_n) \quad . \end{aligned} \tag{3.79}$$

Again, setting $p = r$, $q = 1$, $K_0(x,y) = K_n(x,y)$, and $K_1(x,y) = K(x,y)$

we obtain

$$\begin{aligned} \kappa_r^+(K_n) - \kappa_r^+(K) &\le \kappa_1^+(K_n-K) \ , \\ \kappa_r^-(K_n) - \kappa_r^-(K) &\ge \kappa_1^-(K_n-K) \ . \end{aligned} \tag{3.80}$$

Thus

$$-\kappa_1^+(K_n-K) \le \kappa_r^+(K) - \kappa_r^+(K_n) \le \kappa_1^+(K-K_n) \ ,$$

and

$$\kappa_r^+(K_n-K) = -\kappa_r^-(K-K_n) \ ,$$

so that

$$|\kappa_r^+(K) - \kappa_r^+(K_n)| \le \rho(K-K_n) \ .$$

We obtain a similar result for $|\kappa_r^-(K) - \kappa_r^-(K_n)|$ using the other inequalities, and the theorem is thus established.

Since we have identified $\tilde{\kappa}_r^{\pm}$ with $\kappa_r^{\pm}(K_n)$ (when $\tilde{\kappa}_r^{\pm}$ is non-zero), we can use Theorem 3.2 to bound $|\kappa_r^{\pm} - \tilde{\kappa}_r^{\pm}|$. However, $\rho(K-K_n)$ can be found only by solving an eigenvalue problem; we may therefore employ the following bounds.

LEMMA 3.1. *Suppose that* $L(x,y)$ *is piecewise-continuous, and* L *is the corresponding integral operator on the space of Riemann-integrable functions with norm* $\| \ \|_q$. *Then*

$$\rho(L) \le \|L\|_q \ , \tag{3.81}$$

for $q = 1,2,$ *or* ∞ .

This result is merely a particular case of the result stated in the second 'Remark' in section 1.5, and by setting $q = 1$ or ∞ , and $q = 2$, we may deduce inequalities (1.14a) and (1.14b). A condition of piecewise-continuity is more than sufficient for (3.81). For our present purposes, the bound (1.14a) will suffice.

3.15.5.

Now, the kernel $K_n(x,y)$ constructed to correspond to the quadrature method with a rule $J(\phi) \in J^R$ (using eqn (3.70)) satisfies

$$\max_{a\leq x,y\leq b} |K(x,y) - K_n(x,y)| \leq \sup_{\substack{|x-\eta|\leq\Delta \\ |y-\xi|\leq\Delta}} |K(x,y) - K(\xi,\eta)|$$

$$= \omega(K;\Delta) \text{ , say,}$$

where Δ is the quantity defined above, by (3.75). Now, $||K - K_n||_q$ $(q = 1,2,\infty)$ can be bounded in terms of $||K(x,y) - K_n(x,y)||_\infty$, in particular $||K - K_n||_\infty \leq (b - a) \; ||K(x,y) - K_n(x,y)||_\infty \leq (b - a)\omega(K;\Delta)$, from the preceeding bound. (The same inequality holds for $q = 1,2$.) We can therefore use Lemma 3.1 to bound $\rho(K - K_n)$ in Theorem 3.2, and we obtain

$$|\tilde{\kappa}_r^{\pm} - \kappa_r^{\pm}| \leq (b - a) \; \omega(K;\Delta) \; .$$

(There is an analogous result, if $J(\phi) \in J^+$, based on (3.76).)

Now if $K(x,y)$ is continuous, $\lim_{\Delta\to 0} \omega(K;\Delta) = 0$ so $|\tilde{\kappa}_r^{\pm} - \kappa_r^{\pm}| \to 0$.

But if $\{J_m(\phi)\}$ is a family of rules in J^R and

$$\lim_{m\to\infty} J_m(\phi) = \int_a^b \phi(y)\,dy \tag{3.82}$$

for every $\phi(x) \in C[a,b]$ we know (Example 2.21) that $\Delta_m \to 0$, where Δ_m is the measure (3.75) of $J_m(\phi)$. It follows that if we employ a family of such quadrature rules,

$$\lim_{m\to\infty} |\tilde{\kappa}_r^{\pm} - \kappa_r^{\pm}| = 0 \; .$$

This result is included in the theorem below. If $J_m(\phi) \in J^+$ and (3.82) holds, we know from Bückner's result (section 2.15) that $w \to 1$

and $\Delta \to 0$, in the definitions of section 3.15.1, and the right-hand side of (3.76) tends to zero as $m \to \infty$. The convergence theorem for rules in J^R can therefore be extended to rules in J^+, and we have the following result.

THEOREM 3.4. *Suppose that* $\{J_m(\phi)\}$ *is a family of quadrature rules such that* $J_m(\phi) \in J^+$ $(m=1,2,3,\ldots)$ *and*

$$\lim_{m\to\infty} J_m(\phi) = \int_a^b \phi(y)dy$$

for every function $\phi(x) \in C[a,b]$. *Suppose that non-zero eigenvalues* $\tilde{\kappa}_r^{\pm} \equiv \tilde{\kappa}_r^{\pm}(1/m)$ *are obtained from the quadrature method of section 3.12 using the rule* $J_m(\phi)$. *Then if* $K(x,y) = \overline{K(y,x)}$ *is continuous for* $a \le x, y \le b$,

$$\lim_{m\to\infty} \tilde{\kappa}_r^{\pm} = \kappa_r^{\pm} \tag{3.83}$$

for $r = 1,2,3,\ldots$, *if* $\kappa_r^{\pm}$ *exists.*

Remark. If $\kappa_r^{\pm}$ does not exist, it should be replaced by zero in eqn (3.83).

Quadrature rules which satisfy the conditions of Theorem 3.4 are obtained by setting $J_m(\phi)$ equal to, for example,

(a) an m-times repeated version $(m\times J)(\phi)$ where $J(\phi) \in J^+$ (for example, the trapezium or parabolic or mid-point rule repeated with step $h \propto 1/m$);

(b) the m-point Gauss-Legendre rule;

(c) the mth basic Romberg rule of the classical Romberg scheme.

3.15.6.

The technique used to establish Theorem 3.4 can be applied to yield error bounds for $|\kappa_r^{\pm} - \tilde{\kappa}_r^{\pm}|$. In this section, however, we consider the convergence of the approximate eigenfunctions. We continue to suppose that $K(x,y)$ is Hermitian, and that the quadrature rule is a Riemann sum, but the analysis extends to cover rules in J^+ when natural modifications are made.

Now, we solve the equations

$$\sum_{j=0}^{n} w_j\, K(y_i,y_j)\, \tilde{f}_r^{\pm}(y_j) = \tilde{\kappa}_r^{\pm}\, \tilde{f}_r^{\pm}(y_i) \qquad (i = 0,1,2,\ldots,n)$$

and obtain the eigenpair

$$\tilde{\kappa}_r^{\pm},\ \tilde{\mathbf{f}}_r^{\pm} = [\tilde{f}_r^{\pm}(y_0),\ \tilde{f}_r^{\pm}(y_1),\ \ldots,\ \tilde{f}_r^{\pm}(y_n)]^T .$$

We can extend the vector $\tilde{\mathbf{f}}_r^{\pm}$ to a function by setting

$$\tilde{\mathrm{f}}_r^{\pm}(x) = \sum_{j=0}^{n} \tilde{f}_r^{\pm}(y_j)\, \phi_j(x) , \tag{3.84}$$

where the functions $\{\phi_j(x)\}$ are those occurring in (3.69).† With this definition of $\tilde{\mathrm{f}}_r^{\pm}(x)$,

$$\int_a^b K_n(x,y)\, \tilde{\mathrm{f}}_r^{\pm}(y)\,dy = \tilde{\kappa}_r^{\pm}\, \tilde{\mathrm{f}}_r^{\pm}(x) \qquad (a\le x\le b) , \tag{3.85}$$

where $K_n(x,y)$ is the degenerate kernel defined by (3.70). Further, we have $\|f_r^{\pm}(x)\|_2 = 1$ and

$$\int_a^b K(x,y)\, f_r^{\pm}(y)\,dy = \kappa_r^{\pm}\, f_r^{\pm}(x) \qquad (a\le x\le b) , \tag{3.86}$$

and we suppose that $\kappa_r^{\pm} \neq 0$. Now suppose that $\{J_m(\phi)\}$ is a family of quadrature rules in J^R satisfying the conditions of Theorem 3.4. Then

$$\lim_{m\to\infty} \tilde{\kappa}_r^{\pm} \equiv \lim_{m\to\infty} \tilde{\kappa}_r^{\pm}(1/m) = \kappa_r^{\pm} ,$$

if the limit is non-zero. However, from (3.85) and (3.86),

---

† For general $J_m(\phi) \in J^+$, the values v_i in (3.69) are interpreted as those occurring in (3.71). At present, $J_m(\phi) \in J^R$.

$$\kappa_r^{\pm}\{f_r^{\pm}(x) - \tilde{f}_r^{\pm}(x)\} - \int_a^b K(x,y)\{f_r^{\pm}(y) - \tilde{f}_r^{\pm}(y)\}dy$$

$$= (\tilde{\kappa}_r^{\pm} - \kappa_r^{\pm})\,\tilde{f}_r^{\pm}(x) - \int_a^b \{K_n(x,y) - K(x,y)\}\,\tilde{f}_r^{\pm}(y)dy. \quad (3.87)$$

Without loss of generality we may assume that

$$||\tilde{f}_r^{\pm}(x)||_2^2 = \int_a^b |\tilde{f}_r^{\pm}(y)|^2 dy = 1 .$$

Then writing

$$\varepsilon_r^{\pm}(x) = f_r^{\pm}(x) - \tilde{f}_r^{\pm}(x) , \quad (3.88)$$

we have the equation

$$\kappa_r^{\pm}\,\varepsilon_r^{\pm}(x) - \int_a^b K(x,y)\,\varepsilon_r^{\pm}(y)dy = \delta_r^{\pm}(x) , \quad (3.89)$$

where $\delta_r^{\pm}(x)$ is the right-hand side in eqn (3.87).

It is clear that neither (3.88) nor (3.89) define $\varepsilon_r^{\pm}(x)$, since indeed $f_r^{\pm}(x)$ and $\tilde{f}_r^{\pm}(x)$ may be multiplied by arbitrary constants of modulus unity, whilst preserving unit norms.

However, eqn (3.89) does possess a (non-unique) solution even though $\kappa_r^{\pm}$ is an eigenvalue of $K(x,y)$, and this solution has the form

$$\varepsilon_r^{\pm}(x) = \sum_{i=1}^{p} \alpha_i\, f_{r_i}^{\pm}(x) + \gamma_r^{\pm}(x) , \quad (3.90)$$

where $\gamma_r^{\pm}(x)$ is defined below. Here we have assumed that $\kappa_r^{\pm}$ is a non-zero eigenvalue of multiplicity p, and that $f_{r_i}^{\pm}(x)$ $(i=1,2,3,\ldots,p)$ are orthonormal eigenfunctions of $K(x,y)$ corresponding to the eigenvalue $\kappa_r^{\pm}$. The coefficients α_i $(i=1,2,3,\ldots,p)$ are 'almost arbitrary' but subject to a condition imposed by the requirement

$$\int_a^b |\tilde{f}_r^{\pm}(y)|^2\, dy = 1 .$$

The function $\gamma_r^{\pm}(x)$ satisfies an integral equation of the second kind, namely,

$$\gamma_r^{\pm}(x) - \lambda_r^{\pm} \int_a^b K_r^{\pi}(x,y)\ \gamma_r^{\pm}(y)\,dy = \lambda_r^{\pm}\delta_r^{\pm}(x) , \qquad (3.91)$$

where $\delta_r^{\pm}(x)$ is the right-hand side of (3.87), $\lambda_r^{\pm} = 1/\kappa_r^{\pm}$ and

$$K_r^{\pi}(x,y) = K(x,y) - \kappa_r^{\pm} \sum_{i=1}^{p} f_{r_i}^{\pm}(x)\ \overline{f_{r_i}^{\pm}(y)} . \qquad (3.92)$$

From the construction of $K_r^{\pi}(x,y)$, $\kappa_r^{\pm}$ is not an eigenvalue of $K_r^{\pi}(x,y)$ since we ensure that $||f_{r_i}^{\pm}(x)||_2 = 1$.

Thus $\gamma_r^{\pm}(x)$ is the unique solution of eqn (3.91). Now, in (3.91),

$$\| \delta_r^{\pm}(x) \|_\infty \le |\tilde{\kappa}_r^{\pm} - \kappa_r^{\pm}|\ \ \|\tilde{f}_r^{\pm}(x)\|_\infty + \|K_n - K\|_\infty\ \|\tilde{f}_r^{\pm}(x)\|_\infty . \qquad (3.93)$$

To bound $|| \delta_r^{\pm}(x) ||_\infty$ we bound $\|\tilde{f}_r^{\pm}(x)\|_\infty$. We see that, from (3.85),

$$|\tilde{\kappa}_r^{\pm}|\ \ \|\tilde{f}_r^{\pm}(x)\|_\infty \le \sup_{a \le x \le b} \left\{ \int_a^b |K_n(x,y)|^2\, dy \right\}^{\frac{1}{2}} \|\tilde{f}_r^{\pm}(x)\|_2$$

so that, with $K_n(x,y)$ being defined by (3.70) and $J_m(\phi) \in J^R$,†

$$\|\tilde{f}_r^{\pm}(x)\|_\infty \le (b-a)^{\frac{1}{2}}\ \|K(x,y)\|_\infty / |\tilde{\kappa}_r^{\pm}| . \qquad (3.94)$$

Under the assumptions of Theorem 3.4, $\lim_{m\to\infty} \tilde{\kappa}_r^{\pm} = \kappa_r^{\pm}$, so that for m sufficiently large, $|\tilde{\kappa}_r^{\pm}| > \frac{1}{2}|\kappa_r^{\pm}|$. Furthermore

$$\|K_n - K\|_\infty \le (b-a) \sup_{a \le x,y \le b} |K_n(x,y) - K(x,y)|$$

$$= (b-a)\ \ \|K_n(x,y) - K(x,y)\|_\infty , \qquad (3.95)$$

† The modification in (3.94), required more generally if $J(\phi) \in J^+$, involves the insertion of a factor w(which tends to unity) in (3.94).

and also

$$|\kappa_r^{\pm} - \tilde{\kappa}_r^{\pm}| \le \|K_n - K\|_\infty$$

from Theorem 3.2 and Lemma 3.1.

Combining these inequalities, it is seen from (3.87) that

$$\lim_{m\to\infty} \|\delta_r^{\pm}(x)\|_\infty = 0 \ . \tag{3.96}$$

Since $\gamma_r^{\pm}(x)$ is related to $\delta_r^{\pm}(x)$ by means of eqn (3.91), in which $K_r^{\sqcap}(x,y)$ is independent of m,

$$\|\gamma_r^{\pm}(x)\|_\infty \le \{1 + |\lambda_r^{\pm}| \; \|R^{\sqcap}_{r,\lambda_r^{\pm}}\|_\infty\} \; \|\delta_r^{\pm}(x)\|_\infty / |\lambda_r^{\pm}| \ ,$$

where $R_r^{\sqcap}(x,y;\lambda)$ is the resolvent kernel of $K_r^{\sqcap}(x,y)$ (see eqn (1.34)) and $R^{\sqcap}_{r,\lambda}$ is the integral operator corresponding to $R_r^{\sqcap}(x,y;\lambda)$. In view of the result (3.96) for $\|\delta_r^{\pm}(x)\|_\infty$,

$$\lim_{m\to\infty} \|\gamma_r^{\pm}(x)\|_\infty = 0 \ . \tag{3.97}$$

For the case $J(\phi)\in J^R$ we establish the following result on taking $p=1$.

THEOREM 3.5. *Suppose that* $\{J_m(\phi)\}$ *and the Hermitian kernel* $K(x,y)$ *satisfy the condition in Theorem 3.4. Then the normalized function* $\tilde{f}_r^{\pm}(x)$, *defined by eqn (3.84), converges e.. entially to* $f_r^{\pm}(x)$, *in the uniform norm, as* $m \to \infty$, *provided* $\kappa_r^{\pm}$ *is a simple non-zero eigenvalue.*

On occasions, we may wish to extend an eigenvector $\tilde{\mathrm{f}}_r^{\pm}$ to an eigenfunctio.. using the Nyström extension defined by

$$\tilde{f}_r^{\pm}(x) = (1/\tilde{\kappa}_r^{\pm}) \sum_{j=0}^{n} w_j \, K(x,y_j) \, \tilde{f}_r^{\pm}(y_j) \ . \tag{3.98}$$

Theorem 3.5 remains valid if $\tilde{\mathrm{f}}_r^{\pm}(x)$ is replaced by $\tilde{f}_r^{\pm}(x)$, defined by eqn (3.98). (The function $\tilde{\mathrm{f}}_r^{\pm}(x)$ of eqn (3.84) is piecewise-constant.)

3.15.7.
When $\kappa_r^{\pm}$ is a multiple eigenvalue, it is more difficult to summarize the result of section 3.15.6 succinctly. In order to do so, let us define the distance of a function $\phi(x)$ from a linear space V of functions. We suppose that $\phi(x)$ and the elements of V all lie in a normed linear space. Then we may define

$$\text{dist}_p(\phi,V) = \inf_{\psi(x)} \{||\phi(x) - \psi(x)||_p \mid \psi(x) \in V\} \quad . \tag{3.99}$$

The result of section 3.15.6 may now be expressed in the following manner.

THEOREM 3.6. *With the conditions of Theorem 3.4,*

$$\text{dist}_\infty(\tilde{f}_r^{\pm}, V_r^{\pm}) \to 0 \quad as \quad m \to \infty ,$$

where $V_r^{\pm}$ *is the space of eigenfunctions of the Hermitian kernel* $K(x,y)$, *corresponding to the eigenvalue* $\kappa_r^{\pm}$ *(and the distance is measured in the uniform norm).*

It should be noted that for m sufficiently large, and any fixed r , $\tilde{\kappa}_r^{\pm}$ has the same multiplicity p as the true eigenvalue $\kappa_r^{\pm}$, and there is then a space of approximate eigenfunctions, with a basis $\tilde{f}_{r_i}^{\pm}(x)$ $(i = 1,2,3,\ldots,p)$ approximating $V_r^{\pm}$. The method of selecting an eigenfunction $\tilde{f}_r^{\pm}(x)$ from the space of approximate eigenfunctions does not affect the validity of Theorem 3.6.

Remark. To place our analysis in a more general context, we note that to establish the required convergence results for the eigenvalues it is sufficient to consider K,K_n as operators on $L^2[a,b]$ and show that $\lim ||K - K_n||_2 = 0$. Under such conditions the convergence in Theorem 3.6 would be replaced by the result that $\text{dist}_2(\tilde{f}_r^{\pm},V_r^{\pm}) \to 0$. Similar results can be established for the non-Hermitian case, and to deal with the convergence of the eigenvalues we only use the property that $||K - K_n||_2 \to 0$. (In general, if $K(x,y)$ and $K_n(x,y)$ are piecewise-continuous and $\|K - K_n\|_\infty \to 0$ then $\|K - K_n\|_2 \to 0$ also.)

3.15.8. *Convergence of eigenvalues in the general case*

We have seen above that when $J(\phi)$ is a Riemann sum, or at least has positive weights and abscissae in the range of integration, the quadrature method is essentially equivalent to the solution of the eigenvalue problem for a degenerate kernel. Now $K(x,y)$ and $K_n(x,y)$ are bounded and piecewise-continuous. Without assuming symmetry, we know that

$$\|K - K_n\|_q \to 0 \ (q = 2 \text{ or } \infty)$$

as we progress through certain sequences of quadrature rules $J_m(\phi)(m = 1,2,3,\ldots)$, since then

$$\lim_{m\to\infty} \|K(x,y) - K_n(x,y)\|_\infty = 0 .$$

Under these conditions we can show that the only non-zero limit point of non-zero approximate eigenvalues is a true eigenvalue, κ, and every true non-zero eigenvalue of $K(x,y)$ is a limit of a sequence of approximate eigenvalues.

The convergence proof which we now engage upon is closely related to the analysis of convergence for equations of the second kind. We characterize the eigenvalues of $K(x,y)$ as those values of κ for which the resolvent kernel $R(x,y;\lambda)$ does not exist when $\lambda = 1/\kappa$. In operator notation, κ is an eigenvalue of $K(x,y)$ if and only if the resolvent operator $(\kappa I - K)^{-1}$ does not exist.† In the statement

† Here the effect of the choice of underlying space is somewhat apparent. The resolvent operator $(\kappa I - K)^{-1}$ is defined, for our purposes, on the space of square-integrable functions, so that when $\kappa \neq 0$ the equation

$$\kappa f(x) - \int_a^b K(x,y)\, f(y)\mathrm{d}y = g(x)$$

has a unique square-integrable solution for every square-integrable function $g(x)$ if and only if κ is not an eigenvalue. The scope of our treatment is then quite general.

and proof of Theorem 3.7 we rely on operator notation, since the mathematics would otherwise appear unnecessarily cumbersome. Our key results are Theorems 3.8 and 3.9.

THEOREM 3.7. (a) *Suppose that* $\mu \neq 0$ *and* $(\mu I - K)^{-1}$ *exists. If* $K(x,y)$ *and* $K_n(x,y)$ *are square-integrable and*

$$\lim_{m\to\infty} \|K - K_n\|_2 = 0 \ ,$$

where $n = n(m)$, *then there exists an integer* m_0 *such that, for* $m \geq m_0$, $(\mu I - K_n)^{-1}$ *exists and*

$$\|(\mu I - K_n)^{-1}\|_2 \leq 2\,\|(\mu I - K)^{-1}\|_2 \ . \tag{3.100}$$

(b) *Conversely, if* $\mu \neq 0$ *and* $(\mu I - K_n)^{-1}$ *exists for all* $m \geq m_1$, $(\mu I - K)^{-1}$ *exists and there is an integer* m_2 *such that*

$$\|(\mu I - K)^{-1}\|_2 \leq 2\,\|(\mu I - K_n)^{-1}\|_2 \tag{3.101}$$

for $m \geq m_2$, *where* $n = n(m)$, *provided* $\lim_{m\to\infty} \|K - K_n\|_2 = 0$.

Proof. (a) Let us write $E_n = K - K_n$; E_n is an integral operator. Now $\mu I - K_n = \mu I - K + E_n = (\mu I - K)(I + F_n)$, where $F_n = (\mu I - K)^{-1} E_n$. F_n is an integral operator, and it exists since $(\mu I - K)^{-1}$ exists, by hypothesis.

Now

$$\mu I - K_n = (\mu I - K)(I + F_n) \tag{3.102}$$

and $(I + F_n)^{-1}$ exists if $\|F_n\|_2 < 1.$§ But $\|F_n\|_2$ may be made arbitrarily small by taking m large enough, since

$$\|F_n\|_2 \leq \|(\mu I - K)^{-1}\|_2 \ \|E_n\|_2 \tag{3.103}$$

§ See the discussion at the end of this section.

and $\lim_{m\to\infty} ||E_n||_2 = 0$. Hence, for m sufficiently large, $(\mu I - K_n)^{-1}$ exists, and

$$(\mu I - K_n)^{-1} = (I + F_n)^{-1} (\mu I - K)^{-1} . \tag{3.104}$$

Suppose that for $m \geq m_0$, $||F_n||_2 < \frac{1}{2}$. Then $||(I + F_n)^{-1}||_2 \leq 1/\{1 - ||F_n||_2\} < 2$ and hence, for $m \geq m_0$,

$$||(\mu I - K_n)^{-1}||_2 \leq 2\,||(\mu I - K)^{-1}||_2 . \tag{3.105}$$

Part (b) follows in a similar fashion by interchanging the rôles of K and K_n . Observe that the constant 2 may be replaced by $(1 + \alpha)$ for any $\alpha > 0$, for all μ in any closed bounded region which does not include eigenvalues of $K(x,y)$.

Our convergence theorems now follow from Theorem 3.7. In Theorem 3.8 we may suppose that $K(x,y)$, $K_n(x,y)$ are square-integrable kernels.

THEOREM 3.8. *Suppose that* $\lim_{m\to\infty} ||K - K_n||_2 = 0$. *Then* (a) *the only non-zero points of accumulation of eigenvalues of* $K_n(x,y)$ *are eigenvalues of* $K(x,y)$. *Also* (b) *every non-zero eigenvalue of* $K(x,y)$ *is the limit point of eigenvalues of* $K_n(x,y)$ *as* $m \to \infty$, *Indeed, given a non-zero eigenvalue* κ' *of* $K(x,y)$ *and any* $\varepsilon > 0$, *there exists an integer* M_0 *such that for* $m > M_0$ *and* $n = n(m)$, $K_n(x,y)$ *has some eigenvalue* $\tilde{\kappa}$ *with* $|\tilde{\kappa}-\kappa'| < \varepsilon$.

Proof. (a) Suppose that the eigenvalues of $K_n(x,y)$ (on $[a,b]$) are $\tilde{\kappa}_r \equiv \tilde{\kappa}_{n,r}(r=0,1,2,\ldots,n)$, $n=n(m)$, and $n_j=n(m_j)$. If $\kappa' \neq 0$ is an accumulation point of approximate eigenvalues, there is a subsequence $\tilde{\kappa}_{n_j r_j}$ such that

$$\lim_{j\to\infty} \tilde{\kappa}_{n_j r_j} = \kappa' \neq 0 . \tag{3.106}$$

Suppose that κ' is not an eigenvalue of $K(x,y)$. Then there is an integer M_0 such that for all $m \geq M_0$ either κ' is an eigenvalue of

$K_n(x,y)$ or $(\kappa' I - K_n)^{-1}$ exists and

$$\|(\kappa' I - K_{n_j})^{-1}\|_2 > 2\|(\kappa' I - K)^{-1}\|_2 \quad . \tag{3.107}$$

(For $\|(\mu I - K_{n_j})^{-1}\|_2$ can be made arbitrarily large† by making μ close enough to an eigenvalue of $K_{n_j}(x,y)$, and, here, $\lim \tilde{\kappa}_{n_j,r_j} = \kappa'$.) Now we have supposed that κ' is not an eigenvalue of $K(x,y)$. Hence $(\kappa' I - K)^{-1}$ exists. Further, $\|K - K_n\|_2 \to 0$. Then by Theorem 3.7(a), with $\mu = \kappa'$, there exists m_0 such that for all $m \geq m_0$ $(\kappa' I - K_{n_j})^{-1}$ exists and

$$\|(\kappa' I - K_{n_j})^{-1}\|_2 \leq 2\|(\kappa' I - K)^{-1}\|_2 \quad . \tag{3.108}$$

This is a contradiction of (3.107) so κ' must be an eigenvalue of $K(x,y)$.

(b) Now suppose that $\kappa \neq 0$ is an eigenvalue of $K(x,y)$ which is not a limit point of eigenvalues of $K_n(x,y)$. Then there is a countable sequence $\{K_{n_r}(x,y)\}$ $(r = 1,2,3,\ldots)$ for which $(\kappa I - K_{n_r})^{-1}$ exists. Thus, setting $K_r^0(x,y) = K_{n_r}(x,y)$ we have

$$\lim_{m\to\infty} \|K_m^0 - K\|_2 = 0$$

and $(\kappa I - K_m^0)^{-1}$ exists for all $m \geq 1$. Hence by Theorem 3.7(b), setting $\mu = \kappa$ and $K_{n(m)} = K_m^0$, $(\kappa I - K)^{-1}$ must exist. Then κ cannot be an eigenvalue of $K(x,y)$. Thus we obtain a contradiction and our result is established.

We can now apply our convergence theorem to the quadrature method. Observe that Theorem 3.8 assumes a rôle as a weak substitute for Theorem 3.2.

THEOREM 3.9. *If* $\{J_m(\phi)\}$ *is a family of quadrature rules satisfying*

---

† If $(\mu I - K_n)\phi = \tilde{f}$, where $K_n\tilde{f} = \tilde{\kappa}\tilde{f}$, then $\phi = (\mu - \tilde{\kappa})^{-1}\tilde{f}$ and hence $\|(\mu I - K_n)^{-1}\| \geq (\mu - \tilde{\kappa})^{-1}$.

the conditions of Theorem 3.4, and $K(x,y)$ is continuous, then every non-zero eigenvalue κ of $K(x,y)$ is the limit point of approximate eigenvalues $\tilde{\kappa}$ as $m \to \infty$, and the only non-zero limit points of approximate eigenvalues are true eigenvalues.

Proof. The non-zero eigenvalues of the kernel $K_n(x,y)$ in eqn (3.72) (or eqn (3.70)) are the non-zero values $\tilde{\kappa}$ found by the quadrature method for $K(x,y)$. Under the assumptions on $\{J_m(\phi)\}$, $\lim_{m\to\infty} \|K(x,y) - K_n(x,y)\|_\infty = 0$. Thus $\lim_{m\to\infty} \|K - K_n\|_2 = 0$ and Theorem 3.8 applies.

We note that it is possible for the matrix $\underset{\sim}{K}\underset{\sim}{D}$ of the quadrature method to have a zero eigenvalue for all n even though $K(x,y)$ has itself no zero eigenvalue. We see such a case in Example 3.6.

Though we shall not establish the result at this juncture, we note, now, that when κ is an eigenvalue of $K(x,y)$ with algebraic multiplicity p, then there are precisely p approximate eigenvalues $\tilde{\kappa}_i \equiv \tilde{\kappa}_i(1/m)$ (counting according to their algebraic multiplicity) which satisfy the condition $\lim_{m\to\infty} \tilde{\kappa}_i = \kappa$ in Theorem 3.8(a).

Remark. We take the opportunity, at this point, to divert from our general theme, and consider briefly a point raised by the proof of the Theorem 3.7. In obtaining (3.103) from (3.102) we make use of the fact that $(I + F_n)^{-1}$ exists if $\|F_n\|_2 < 1$. Effectively this statement is equivalent to the fact that the integral equation

$$\phi(x) + \int_a^b F_n(x,y)\,\phi(y)\mathrm{d}y = \psi(x)$$

possesses a solution for all (square-integrable) $\psi(x)$ if $\|F_n\|_2 < 1$. It was remarked in Chapter 1 (see section 1.9) that the solution can indeed be found by using the iteration

$$\phi_0(x) = \psi(x)\ ,$$

$$\phi_{k+1}(x) = \psi(x) - \int_a^b F_n(x,y)\,\phi_k(y)\mathrm{d}y\ .$$

Then $\lim_{k\to\infty} \phi_k(x) = \phi(x)$, where $\phi(x)$ is square-integrable. (If $F_n(x,y)$ and $\psi(x)$ are continuous, $\phi(x)$ is also continuous.) In general, if F_n is a linear operator acting on a normed linear function space V , the equation $(I + F_n)\phi(x) = \psi(x)$ has a solution $\phi(x) \in V$ when $\|F_n\| < 1$ provided either (see Theorem 1.1)

(a) F_n is a compact operator, or

(b) the space V is complete (see section 1.9).

This general statement permits us to extend Theorem 3.7 so that K_n is not required to be an integral operator.

3.15.9. *Convergence of eigenfunctions in the general case*

As in section 3.15.3, we suppose that the eigenvectors $\tilde{\mathrm{f}}_r$ obtained from the quadrature method are extended to piecewise-constant functions $\tilde{\mathrm{f}}_r(x)$ by using eqn (3.84). Thus the kernel $K_n(x,y)$ associated with a quadrature rule in J^R (or J^+) by the construction (3.70) (or (3.72)) satisfies

$$\int_a^b K_n(x,y)\ \tilde{\mathrm{f}}_r(y)\mathrm{d}y = \tilde{\kappa}_r\tilde{\mathrm{f}}_r(x) \quad .$$

If we suppose that $\tilde{\kappa}_r \to \kappa \neq 0$ as $m \to \infty$, then from the analysis in section 3.15.8 we know that κ is an eigenvalue of $K(x,y)$. Let us suppose that

$$\int_a^b K(x,y)\ f(y)\mathrm{d}y = \kappa f(x) \ ,$$

so that $f(x)$ is some eigenfunction of $K(x,y)$ corresponding to the eigenvalue κ . Then

$$\kappa\varepsilon(x) - \int_a^b K(x,y)\varepsilon(y)\mathrm{d}y = \delta(x) \ , \tag{3.109}$$

where

$$\varepsilon(x) = f(x) - \tilde{\mathrm{f}}_r(x)$$

and

$$\delta(x) = (\tilde{\kappa}_r - \kappa)\tilde{f}_r(x) - \int_a^b \{K_n(x,y) - K(x,y)\}\tilde{f}_r(y)dy \quad .$$

Remark. An equation similar to (3.109) was encountered in (3.89) (see section 3.15.6). Here, we no longer assume symmetry of $K(x,y)$.

Because of the form of $\delta(x)$, eqn (3.109) has a (non-unique) solution even though κ is an eigenvalue of $K(x,y)$. (It is easily shown that $\delta(x)$ is orthogonal to every left eigenfunction of $K(x,y)$ corresponding to the eigenvalue κ .)

Now the general solution of (3.109) may be written down in terms of a representative set of eigenfunctions $\{f_i(x)\}$ of $K(x,y)$. We suppose that

$$\int_a^b K(x,y)f_i(y)dy = \kappa f_i(x) \quad (i = 1,2,3,\ldots,s) \quad ,$$

where the geometric multiplicity of κ is s , and $f_1(x),\ldots,f_s(x)$ are linearly independent.

Then we may write

$$\kappa\varepsilon(x) = \sum_{i=1}^{s} \sigma_i f_i(x) + \delta(x) + \int_a^b R^{\#}(x,y)\delta(y)dy \; , \tag{3.110}$$

where $\sigma_1,\ldots,\sigma_s$ are unknown parameters and $R^{\#}(x,y)$ is a kernel which is dependent only on $K(x,y)$ and κ . (We refer for this result, which is not immediately evident, to Lovitt (1924, p.66). An alternative discussion may be found in Goursat (1964).)

Using this result we are in a position to establish the next theorem. Let us place the situation in perspective.

Theorems 3.2 and 3.8 certainly apply with square-integrable kernels $K(x,y)$, $K_n(x,y)$, and give convergence results for the eigenvalues if $||K - K_n||_2 \to 0$. Under this condition we would expect convergence in $||\;||_2$ of the eigenfunctions of $K_n(x,y)$ to the appropriate space of eigenfunctions of $K(x,y)$. However, in analysing the quadrature method for a continuous kernel $K(x,y)$ we introduce a piecewise-continuous kernel $K_n(x,y)$ such that $||K(x,y) - K_n(x,y)||_\infty \to 0$ and *a fortiori* $||K - K_n||_\infty \to 0$. We would then expect the mean-square convergence of eigenfunctions to be strengthened to uniform convergence (see the result already established for Hermitian kernels in Theorem 3.5). We have the following result.

THEOREM 3.10. (a) *Suppose that* $K(x,y)$, $\{K_n(x,y)\}$ *are Hermitian square-integrable kernels, where* $n = n(m)(m = 1,2,3,\ldots)$ *and* $||K - K_n||_2 \to 0$ *as* $m \to \infty$. *Then, if* F *is the linear space of eigenfunctions of* $K(x,y)$ *corresponding to an eigenvalue* $\kappa \neq 0$, *it is known that there is an eigenvalue* $\tilde{\kappa} \equiv \tilde{\kappa}_r$ *of each kernel* $K_n(x,y)$ *(where* r , *like* n , *depends on* m*) such that* $\lim_{m\to\infty} \tilde{\kappa} = \kappa$. *If* $\tilde{f}(x) \equiv \tilde{f}_r(x)$ *is a corresponding eigenfunction of* $K_n(x,y)$ *then* $\lim_{m\to\infty} \text{dist}_2 (\tilde{f},F) = 0$. (b) *Suppose, additionally, that* $K(x,y)$ *and* $K_n(x,y)$ *are piecewise-continuous and that* $||K - K_n||_\infty \to 0$, *that is,*

$$\lim_{m\to\infty} \sup_{a\leq x\leq b} \int_a^b |K(x,y) - K_n(x,y)|dy = 0.$$

Then

$$\lim_{m\to\infty} \text{dist}_\infty(\tilde{f},F) = 0$$

<u>Proof</u>. Observe that

$$||K - K_n||_2 \leq \{\int_a^b \int_a^b |K(x,y) - K_n(x,y)|^2 dx dy\}^{\frac{1}{2}} ,$$

and if the kernels are piecewise-continuous

$$\int_a^b |K(x,y) - K_n(x,y)|^2 dy \leq \{\int_a^b |K(x,y) - K_n(x,y)|dy\}^2$$

so that $||K - K_n||_2 \leq (b-a)^{\frac{1}{2}} \; ||K - K_n||_\infty$. The conditions of Theorem 3.10 therefore imply that $||K - K_n||_2 \to 0$ and the convergence theorem for the eigenvalues (Theorem 3.8) does hold in both (a) and (b).

In case (a), we shall establish that for the function $\delta(x)$ in (3.110), $||\delta(x)||_2 \to 0$ as $m \to \infty$. In case (b) we show that $||\delta(x)||_\infty \to 0$ as $m \to \infty$. The required results follow from (3.110).

To cover both cases, suppose $q = 2$ or ∞ , in case (a) and case (b) respectively. We suppose, without any loss of generality, that $\|\tilde{f}_r(x)\|_q \equiv 1$. (In case (b), $\tilde{f}_r(x)$ is piecewise-continuous.) From the expression for $\delta(x)$,

$$\|\delta(x)\|_q \leq \{|\tilde{\kappa}_r - \kappa| + \|K_n - K\|_q\} \|\tilde{f}_r(x)\|_q .$$

Thus

$$\|\delta(x)\|_q \to 0 , \quad \text{as} \quad m \to \infty .$$

Now, since $R^{\#}(x,y)$ is independent of m , the result follows from eqn (3.110).

<u>Remark</u>. We remark that in the case of a Hermitian kernel with $\kappa = \kappa_r^{\pm}$, $R^{\#}(x,y)$ is related to the resolvent kernel of $K_r^{\natural}(x,y)$ (see eqn (3.92)). In the non-symmetric case the analogue of $K_r^{\natural}(x,y)$ may involve not simply the eigenfunctions of $K(x,y)$ but also the generalized eigenfunctions (section 1.6). Thus $\tilde{f}_r(x)$ may contain a component of a generalized eigenfunction of $K(x,y)$, but the size of this component tends to zero in the limit as $m \to \infty$. We also note that at this point the true eigenfunctions of $K(x,y)$ may be the limit of a sequence of generalized eigenfunctions of $K_n(x,y)$ and need not appear as the limit of the eigenfunctions $\tilde{f}(x)$, unless this be established.

The following result is now a corollary of Theorem 3.10.

THEOREM 3.11. *If $\{J_m(\phi)\}$ is a family of quadrature rules satisfying the conditions imposed in Theorem 3.4, the conclusions of Theorem 3.10 apply and*

$$\lim_{m\to\infty} \text{dist}_q(\tilde{f}_r, F) = 0 \quad (q = 2 \text{ or } q = \infty) ,$$

where $\tilde{f}_r(x)$ is an approximate eigenfunction obtained from the eigenvector $\tilde{\mathbf{f}}_r$ by use of eqn (3.84).

To conclude this section we consider the problem of relating the multiplicity of a true eigenvalue κ to the number of approximate eigenvalues $\tilde{\kappa}$ which converge to κ as $m \to \infty$.

Suppose that κ has algebraic multiplicity p and geometric multiplicity s . Then $s \leq p$. Then for m sufficiently large there are precisely p approximate eigenvalues $\tilde{\kappa}_i \equiv \tilde{\kappa}_i(1/m)$ such that $\lim_{m\to\infty} \tilde{\kappa}_i = \kappa$. In counting these p eigenvalues we take account of the algebraic multiplicity of each $\tilde{\kappa}$. If each $\tilde{\kappa}_i$ is simple for all $m \geq m_0$, say, then there will be for each eigenvalue $\tilde{\kappa}_i$ a one-parameter family of eigenfunctions $\alpha\tilde{f}_i(x)$ $(\alpha \neq 0)$. In the general case, the total set of eigenfunctions corresponding to each $\tilde{\kappa}_i$ $(i=1,2,3,\ldots,p)$ span a linear space which we may call the total eigenspace of $\tilde{\kappa}_1,\tilde{\kappa}_2,\ldots,\tilde{\kappa}_p$. It may happen that this total eigenspace has dimension p even though the space of eigenfunctions corresponding to κ has dimension $s < p$. On the other hand, the sum of the geometric multiplicities of the approximations $\tilde{\kappa}_1,\tilde{\kappa}_2,\ldots,\tilde{\kappa}_p$ (which may coincide with one another) can be less than p even when the geometric multiplicity of κ equals its algebraic multiplicity p : for general values of m we cannot relate the sum of the geometric multiplicities of $\tilde{\kappa}_i$ to the geometric multiplicity of κ.

The situation with the algebraic multiplicities is less confused. For use later we note that if κ is simple and $\lim_{m\to\infty} \tilde{\kappa} = \kappa \neq 0$, where $\tilde{\kappa} \equiv \tilde{\kappa}(1/m)$, then for m sufficiently large $\tilde{\kappa}(1/m)$ is uniquely defined and is simple. This is a particular case of the following result.

THEOREM 3.12. *Let* κ *be a non-zero eigenvalue of the continuous kernel* $K(x,y)$ $(a \leq x,y \leq b)$, *and suppose that the algebraic multiplicity of* κ *is* p . *Suppose that the approximate eigenvalues* $\tilde{\kappa} \equiv \tilde{\kappa}(1/m)$ *are found by the quadrature method of section 3.2, using a family of quadrature rules* $\{J_m(\phi)\} \in J^+$ *such that*

$$\lim_{m\to\infty} J_m(\phi) = \int_a^b \phi(y)\,dy$$

for every $\phi(x)$ *in* $C[a,b]$. *Then for every sufficiently small value*

$\varepsilon > 0$ *there exists an integer* $m_0 \equiv m_0(\varepsilon)$ *such that for all* $m > m_0$ *there are precisely* p *approximate eigenvalues* $\tilde{\kappa}$ *(counting them according to their algebraic multiplicity) satisfying the inequality*

$$|\kappa - \tilde{\kappa}| \leq \varepsilon \ .$$

Proof. We know of no simple proof of this theorem. The following 'proof' relies on a formal operator which is discussed by Kato (1966). It is sufficient that $\{\tilde{\kappa}, \tilde{f}(x)\}$ be the eigensystem of any operator K_n where $||K-K_n||_2 \to 0$. We require a preliminary lemma.

LEMMA 3.2. *If* P_1 *and* P_2 *are projections from* $L^2[a,b]$ *onto the finite-dimensional subspaces* X_1 *and* X_2 *respectively, and* $||P_1 - P_2||_2 < 1$, *then the dimension of* X_1 *is equal to the dimension of* X_2.

For proofs of this result, consult Riesz and Nagy (1955, p.268) and Kato (1966, pp.33-4).

Now suppose that κ is an eigenvalue of $K(x,y)$ with multiplicity p, and *fix* $\varepsilon > 0$ small enough to ensure that the closed disc whose boundary is the circle

$$\Sigma = \{\mu \mid |\mu - \kappa| = \varepsilon\}$$

contains no other eigenvalue of $K(x,y)$ and does not contain zero. The (formal) operator

$$P_1 = \frac{1}{2\pi i} \int_{\Sigma} (\xi I - K)^{-1} \, d\xi \qquad (3.111)$$

(Dunford and Schwarz 1957, vol. 1, pp.566-80) is a projection, acting on $L^2[a,b]$, whose range is the space of dimension p spanned by the eigenfunctions and generalized eigenfunctions of $K(x,y)$ which correspond to κ (see Dunford and Schwarz 1957, p.579). For m sufficiently large, there corresponds a kernel $K_n(x,y)$ (defined by (3.72), say) such that

$$P_2^{[n]} = \frac{1}{2\pi i} \int_\Sigma (\xi I - K_n)^{-1} \, d\xi$$

exists and is a projection whose range has dimension equal to the sum of the algebraic multiplicities of eigenvalues $\tilde{\kappa}$ of $K_n(x,y)$ which lie in Σ. (In the present context, this is the sum of their algebraic multiplicities as eigenvalues of the matrix $\underset{\sim}{K}\underset{\sim}{D}$.) Now, we have

$$\|P_1 - P_2^{[n]}\|_2 \leq \int_\Sigma \|(\xi I - K)^{-1} - (\xi I - K_n)^{-1}\|_2 \, |d\xi|/2\pi$$

$$\leq \varepsilon \sup_{\mu\varepsilon\Sigma} \|(\mu I - K)^{-1} - (\mu I - K_n)^{-1}\|_2 \; .$$

In the notation used in the proof of Theorem 3.7, we write for any $\mu\varepsilon\Sigma$

$$\mu I - K_n = (\mu I - K)(I + F_n)$$

(where F_n depends upon μ). From (3.103), $\sup_{\mu\varepsilon\Sigma} \|F_n\|_2 < 1$ if m is sufficiently large. Then $(I + F_n)^{-1}$ exists, and

$$(\mu I - K)^{-1} - (\mu I - K_n)^{-1} = \{I - (I + F_n)^{-1}\} (\mu I - K)^{-1} \; .$$

Now

$$\|I - (I + F_n)^{-1}\|_2 = \|(I + F_n)^{-1} F_n\|_2 \leq \|(I + F_n)^{-1}\|_2 \|F_n\|_2 .$$

Now $\sup_{\mu\varepsilon\Sigma} \|F_n\|_2$ can be made arbitrarily small by taking m sufficiently large, and if $\|F_n\|_2 \leq \delta < 1$, $\|I-(I + F_n)^{-1}\|_2 < \delta(1 - \delta)^{-1}$. Thus for all $\mu\varepsilon\Sigma$ and m sufficiently large, say $m > m_0$,

$$\|I - (I + F_n)^{-1}\|_2 \|(\mu I - K)^{-1}\|_2 \leq 1/2\varepsilon \; .$$

and hence $\|P_1 - P_2^{[n]}\|_2 \leq \frac{1}{2}$. Here, $\varepsilon > 0$ is fixed and $\delta/(1-\delta)$ can be taken arbitrarily small by taking m large enough. The theorem follows on appealing to the Lemma quoted.

The theory above has been extended by Baker (1971) and by Atkinson (1967). The latter shows that the result of the theorem is true if the kernel $K(x,y)$ is weakly singular and the eigenvalues $\tilde{\kappa}$ are obtained by a method of product integration. (Atkinson's theory also applies to eigenvalues $\tilde{\kappa}$ obtained by other numerical methods.) For a discussion of Atkinson's theory, see also Anselone (1971, Chapter 4) who gives the background theory used above.

3.15.10.

I shall now give a fairly elementary proof of the following result, which is in part a summary of earlier results and will be used in section 3.20.1. The proof may be omitted, if desired.

THEOREM 3.13. *Suppose that* $\kappa \neq 0$ *is a simple eigenvalue of the continuous kernel* $K(x,y)$, *with a corresponding fixed and normalized eigenfunction* $f(x)$. *Suppose that* $J_m(\phi) = \sum_{j=0}^{n} w_j\phi(y_j) \in J^+$ $(m=1,2,3,\ldots)$ *where*

$$\lim_{m\to\infty} J_m(\phi) = \int_a^b \phi(y)\,dy$$

for every $\phi(x) \in C[a,b]$. *Then, for* m *sufficiently large, there exist* $\tilde{\kappa}$ *and* $\tilde{f}(x)$ *such that*

$$\tilde{\kappa}\tilde{f}(x) = \sum_{j=0}^{n} w_j K(x,y_j)\,\tilde{f}(y_j) ,$$

$$(\tilde{f},f) = 1, \lim_{m\to\infty} \tilde{\kappa} = \kappa , \quad \textit{and} \quad \lim_{m\to\infty} \|f(x) - \tilde{f}(x)\|_\infty = 0 .$$

Proof. We shall establish the result for $J_m(\phi) \in J^R$, the modifications for $J_m(\phi) \in J^+$ being left as an exercise. The convergence result for $\tilde{f}(x)$ will be deduced from the corresponding result for $\tilde{\mathrm{f}}(x)$. The normalization $(\tilde{f},f) = 1$ is one of convenience.

The convergence result $\tilde{\kappa} \to \kappa$ has been established in Theorem 3.9. We deduce that for m sufficiently large, $\tilde{\kappa} \neq 0$ (and this we assume). In our earlier analysis we employed (for $J_m(\phi) \in J^R$) the definition (3.84) of a function $\tilde{\mathrm{f}}(x) = \sum_{j=0}^{n} \tilde{f}(y_j)\phi_j(x)$ where, in the proof of Theorem 3.10(b),

$$\|\tilde{f}(x)\|_\infty = \max_j |\tilde{f}(y_j)| = 1 \quad ,$$

and

$$\tilde{\kappa}\tilde{f}(y_i) = \sum_{j=0}^{n} w_j\, K(y_i,y_j)\, \tilde{f}(y_j) \ .$$

According to Theorem 3.11, there exist $\alpha_1,\alpha_2,\alpha_3,\ldots$ such that

$$\lim_{m\to\infty} \|\alpha_m f(x) - \tilde{f}(x)\|_\infty = 0 \ ,$$

where $f(x)$ is a fixed eigenfunction of $K(x,y)$ corresponding to κ, normalized with $\int_a^b |f(x)|^2 dx = 1$.

The function

$$\tilde{f}(x) = \sum_{j=0}^{n} w_j K(x,y_j)\tilde{f}(y_j)/\tilde{\kappa}$$

generated by the 'Nyström extension' of the values $\tilde{f}(y_0),\tilde{f}(y_1),\ldots,\tilde{f}(y_n)$ does not necessarily satisfy the condition $(\tilde{f},f) = 1$. We shall therefore, in due course, generate a function $\tilde{\phi}(x)$ which is a scaled version of $\tilde{f}(x)$ and satisfies the conditions of the theorem when $\tilde{f}(x)$ is replaced by $\tilde{\phi}(x)$ in the statement of these conditions.

Now we know that $\|\alpha_m f(x) - \tilde{f}(x)\|_\infty \to 0$. But, from the triangle inequality $\|\alpha_m f(x) - \tilde{f}(x)\|_\infty + \|\tilde{f}(x)\|_\infty \geq |\alpha_m|\ \|f(x)\|_\infty$ and similarly $\|\alpha_m f(x) - \tilde{f}(x)\|_\infty + |\alpha_m|\ \|f(x)\|_\infty \geq \|\tilde{f}(x)\|_\infty$, whilst $\alpha = \|f(x)\|_\infty^{-1} < \infty$ and $\|\tilde{f}(x)\|_\infty = 1$ for all m . Hence, $\lim_{m\to\infty} |\alpha_m| = \alpha$.

For the Nyström extension $\tilde{f}(x)$,

$$\begin{aligned}\tilde{\kappa}\tilde{f}(x) &= \sum_{j=0}^{n} w_j K(x,y_j)\tilde{f}(y_j)\\ &= \alpha_m \sum_{j=0}^{n} w_j K(x,y_j) f(y_j) + \sum_{j=0}^{n} w_j K(x,y_j)\{\tilde{f}(y_j) - \alpha_m f(y_j)\}\end{aligned}$$

$$= \alpha_m \{\int_a^b K(x,y)f(y)dy + \tau_m(x)\} + \varepsilon_m(x) \quad ,$$

where

$$\tau_m(x) = \sum_{j=0}^{n} w_j K(x,y_j)f(y_j) - \int_a^b K(x,y)f(y)dy$$

and $\varepsilon_m(x)$ has an obvious meaning. We have

$$\|\varepsilon_m(x)\|_\infty = \|\sum_{j=0}^{n} w_j K(x,y_j)\ [\tilde{f}(y_j) - \alpha_m f(y_j)]\|_\infty$$

$$\leq (b-a)\ \|K(x,y)\|_\infty\ \|\tilde{f}(x) - \alpha_m f(x)\|_\infty \quad ,$$

so that $\lim_{m\to\infty} \|\varepsilon_m(x)\|_\infty = 0$. Further, the integrands $K(x,y)f(y)$ are equicontinuous for $a \leq x \leq b$ and since the quadrature rules $J_m(\phi)$ are convergent, $\|\tau_m(x)\|_\infty \to 0$. This is easily seen for rules in J^R , since if Δ is the measure of $J_m(\phi)$, $\Delta \to 0$ as $m \to \infty$ and (see section 2.15) $|\tau_m(x)| \leq (b-a) \sup_{|y'-y''|<\Delta} |K(x,y')f(y')-K(x,y'')f(y'')|$. (For rules in J^+ we may appeal to a general theorem that pointwise convergence is uniform on compact sets.)

We now have, after a minor rearrangement,

$$\tilde{\kappa}\tilde{f}(x) = \alpha_m \kappa f(x) + \alpha_m \tau_m(x) + \varepsilon_m(x) \quad ,$$

where $|\alpha_m| \to \alpha$, $\|\tau_m(x)\|_\infty \to 0$ and $\|\varepsilon_m(x)\|_\infty \to 0$. Immediately, we obtain $\|\tilde{\kappa}\tilde{f}(x) - \alpha_m \kappa f(x)\|_\infty \to 0$. Now let us define $\gamma_m = \tilde{\kappa}/(\alpha_m \kappa)$. Then $|\alpha_m \kappa| \to |\kappa\alpha|$, $\lim_{m\to\infty} \|\gamma_m \tilde{f}(x) - f(x)\|_\infty = 0$, and consequently $\|\gamma_m \tilde{f}(x) - f(x)\|_2 \to 0$ and $(\gamma_m \tilde{f},f) = (f,f) + (\gamma_m \tilde{f}-f,f) = 1 + \eta_m$, say, where $|\eta_m| \leq \|\gamma_m \tilde{f} - f\|_2$ and hence $|\eta_m| \to 0$. If we set $\beta_m = \gamma_m/(1+\eta_m)$ then $(\beta_m \tilde{f},f) = 1$, where $|\beta_m| \to 1/\alpha$.

Now we define $\tilde{\phi}(x) = \beta_m \tilde{f}(x)$. We have

$$\|\tilde{\phi}(x) - f(x)\|_\infty \leq |\beta_m - \gamma_m|\ \ \|\tilde{f}(x)\|_\infty + \|\gamma_m \tilde{f}(x) - f(x)\|_\infty \quad .$$

Here $\lim_{m\to\infty}|\beta_m - \gamma_m| = 0$, whilst for $a \le x \le b$ we have

$$|\tilde{f}(x)| = |\sum_j w_j K(x,y_j)\tilde{f}(y_j)|/|\tilde{\kappa}|$$

$$\le \sum_j w_j |K(x,y_j)|/|\tilde{\kappa}|$$

$$\le \|K(x,y)\|_\infty |b-a|/|\tilde{\kappa}|$$

which yields a uniform bound for $\|\tilde{f}(x)\|_\infty$ for all m sufficiently large, since $\tilde{\kappa}\to\kappa\neq 0$. Thus $(\tilde{\phi},f) = 1$ and $\|\tilde{\phi}(x) - f(x)\|_\infty \to 0$ and the theorem is established.

3.16. *Error bounds for computed eigenvalues of Hermitian kernels*

Having established various convergence theorems, we now consider bounds for the error in approximate eigenvalues computed by the quadrature method of section 3.2. Such bounds are in general available only for *Hermitian* kernels, though some apply also to normal kernels. We shall later construct error estimates for non-Hermitian kernels.

In the case of Hermitian kernels,† there are three types of statement possible: (a) a statement that $|\kappa_r^{\pm} - \tilde{\kappa}_r^{\pm}| < e$ $(r = 0,1,2,\ldots)$; (b) a statement that for some $\tilde{\kappa}_r^{\pm}$ there is an eigenvalue $\kappa_s^{\pm}$ such that $|\tilde{\kappa}_r^{\pm} - \kappa_s^{\pm}| \le \varepsilon_r^{\pm}$; (c) a statement that for some $\kappa_r^{\pm}$, there is an approximate eigenvalue $\tilde{\kappa}_s^{\pm}$ such that $|\kappa_r^{\pm} - \tilde{\kappa}_s^{\pm}| \le \delta_r^{\pm}$. We shall deal with each of these types of error bound in turn, assuming throughout that $K(x,y) = \overline{K(y,x)}$.

3.16.1.

The technique used to establish Theorem 3.4 can be adapted to yield error bounds for $|\kappa_r^{\pm} - \tilde{\kappa}_r^{\pm}|$. The proof, we note, relies heavily on Theorem 3.2.

Let us suppose that $K_n(x,y)$ is the kernel, allowing a quadrature

---

† The eigenvalues of a Hermitian kernel, and the approximations obtained using the quadrature method of section 3.2 with $J(\phi) \in J^+$, are real and can be ordered.

rule $J(\phi) \in J^R$, which was constructed in (3.70). (The reader may make appropriate modifications for the more general case $J(\phi) \in J^+$.) Then

$$\|K-K_n\|_\infty \le |b-a| \, \|K(x,y) - K_n(x,y)\|_\infty \; .$$

Now suppose that $K_x(x,y)$ exists and $\|K_x(x,y)\|_\infty \le M_1$. Then (since $K(x,y)$ is Hermitian), $K_y(x,y)$ exists and $\|K_y(x,y)\| \le M_1$. The bound (3.74) then reduces to

$$\|K(x,y) - K_n(x,y)\|_\infty \le 2M_1\Delta \; , \qquad (3.112)$$

where Δ is the 'measure' of $J(\phi)$, so that $\|K-K_n\|_\infty \le 2|b-a|M_1\Delta$. Using Theorem 3.2 and Lemma 3.1 we then obtain the following result.

THEOREM 3.14. *If* $K(x,y)$ *is Hermitian and* $\|K_x(x,y)\| \le M_1$ *, then*

$$|\kappa_r^{\pm} - \tilde{\kappa}_r^{\pm}| \le 2[b-a] \, M_1\Delta \quad (r = 1,2,3,\ldots) \; .$$

Here $\{\kappa_r^{\pm}\}$ *and* $\{\tilde{\kappa}_r^{\pm}\}$ *are, respectively, the true eigenvalues of* $K(x,y)$ *and the approximate eigenvalues obtained by the quadrature method using a quadrature rule* $J(\phi) \in J^R$ *, and* Δ *is the 'measure' of* $J(\phi)$.

<u>Remark</u>. Observe that (a) $\Delta \le \max_{0\le i\le n} w_i$, (b) the theorem applies if $\sum_{j=0}^{n} w_j = b-a$ and $J(\phi) \in J^+$; and (c) with care we can sometimes bound $\rho(K-K_n)$ by a closer quantity than $(b-a) \times \|K(x,y) - K_n(x,y)\|_\infty$, and we can then improve on the constant 2 appearing in the bound on $|\kappa_r^{\pm} - \tilde{\kappa}_r^{\pm}|$.

If we replace $J(\phi)$ by $(m\times J)(\phi)$ in the statement of Theorem 3.14, then Δ is replaced by Δ/m . The bound on $|\tilde{\kappa}_r^{\pm} - \kappa_r^{\pm}|$ is thus $O(1/m)$. However, we would expect to obtain better results, if $K(x,y)$ is smooth, on using an m-times repeated Gauss rule in place of an m-times repeated two-point trapezium rule, say. The bound in Theorem

3.14 therefore seems somewhat unrealistic when $K(x,y)$ is smooth. Wielandt (1956) progressed a considerable distance towards rectifying this situation by revising the construction of $K_n(x,y)$. In his work, $K_n(x,y)$ is sometimes quite an elaborate kernel, allowing (section 3.15.3) the rule $J(\phi)$, and providing a better approximation to $K(x,y)$ than the 'simple' piecewise-constant kernel defined in (3.72). We observe however, that the kernel $K_n(x,y)$ defined in (3.72) has as its eigenvalues the approximate eigenvalues $\tilde{\kappa}_r^{\pm}$; we can obtain the better results referred to above if we permit $K_n(x,y)$ to have a different set of eigenvalues and modify the theory slightly. The key to this modification lies in the matrix version of Theorem 3.2, which we now state, as a preliminary result.

THEOREM 3.15. *Suppose that* $\underset{\sim}{A}$, $\underset{\sim}{B}$ *are Hermitian matrices of the same order with eigenvalues*

$$\mu_1^+ \geq \mu_2^+ \geq \ldots \geq 0 \,, \quad \mu_1^- \leq \mu_2^- \leq \ldots < 0$$

$$\nu_1^+ \geq \nu_2^+ \geq \ldots \geq 0 \,, \quad \nu_1^- \leq \nu_2^- \leq \ldots < 0$$

respectively. Then

$$|\nu_r^{\pm} - \mu_r^{\pm}| \leq \rho(\underset{\sim}{A}-\underset{\sim}{B}) \,,$$

where $\rho(\underset{\sim}{A}-\underset{\sim}{B})$ *denotes the spectral radius of* $\underset{\sim}{A}-\underset{\sim}{B}$. *(In this inequality, a quantity* $\nu_r^{\pm}$ *or* $\mu_r^{\pm}$ *is to be set to zero if it does not exist.)*

<u>Proof</u>. The proof is an imitation of the proof of Theorem 3.2.

Now suppose that $K_n(x,y)$ is a Hermitian kernel which allows a quadrature rule $J(\phi) = \sum_{j=0}^{n} w_j \phi(y_j) \in J^R$. Thus the non-zero eigenvalues $\kappa_r^{\pm}(K_n)$ are precisely the eigenvalues (which we can also denote by $\tilde{\kappa}_r^{\pm}(K_n)$) of the Hermitian matrix $\underset{\sim}{A} = \underset{\sim}{D}^{\frac{1}{2}} \underset{\sim}{K}_n \underset{\sim}{D}^{\frac{1}{2}} = \left[w_i^{\frac{1}{2}} K_n(y_i,y_j) w_j^{\frac{1}{2}}\right]$. †

---

† Recall that $K_n(x,y)$ allows $J(\phi)$.

The approximate eigenvalues $\tilde{\kappa}_r^{\pm}$, computed using the quadrature method, may be written $\tilde{\kappa}_r^{\pm} = \tilde{\kappa}_r^{\pm}(K)$, and these are the eigenvalues of $\underset{\sim}{B} = \underset{\sim}{D}^{\frac{1}{2}}\underset{\sim}{K}\underset{\sim}{D}^{\frac{1}{2}}$.

From Theorem 3.15,

$$|\tilde{\kappa}_r^{\pm}(K) - \tilde{\kappa}_r^{\pm}(K_n)| \leq \rho(\underset{\sim}{A}-\underset{\sim}{B}) . \tag{3.113}$$

Now, $\rho(\underset{\sim}{A}-\underset{\sim}{B}) \equiv \rho(\underset{\sim}{D}^{\frac{1}{2}}\underset{\sim}{K}_n\underset{\sim}{D}^{\frac{1}{2}} - \underset{\sim}{D}^{\frac{1}{2}}\underset{\sim}{K}\underset{\sim}{D}^{\frac{1}{2}}) = \rho([\underset{\sim}{K}_n-\underset{\sim}{K}]\underset{\sim}{D}) \leq \|[\underset{\sim}{K}_n-\underset{\sim}{K}]\underset{\sim}{D}\|_\infty =$ $= \max_i \sum_j\{|K_n(y_i,y_j)-K(y_i,y_j)|w_j\} \leq [b-a] \;\; \|K(x,y) - K_n(x,y)\|_\infty$, since $\sum_{j=0}^{n} w_j = b-a$. Consequently

$$|\tilde{\kappa}_r^{\pm}(K) - \tilde{\kappa}_r^{\pm}(K_n)| \leq |b-a| \;\; \|K(x,y) - K_n(x,y)\|_\infty . \tag{3.114}$$

On the other hand, from Theorem 3.2 and the bounds $\rho(K-K_n) \leq \|K-K_n\|_\infty \leq |b-a| \;\; \|K(x,y) - K_n(x,y)\|_\infty$, we find that

$$|\kappa_r^{\pm}(K) - \kappa_r^{\pm}(K_n)| \leq |b-a| \;\; \|K(x,y) - K_n(x,y)\|_\infty . \tag{3.115}$$

Now, since $K_n(x,y)$ allows $J(\phi)$,

$$\tilde{\kappa}_r^{\pm}(K_n) = \kappa_r^{\pm}(K_n) , \tag{3.116}$$

and hence

$$\begin{aligned} |\kappa_r^{\pm}(K) - \tilde{\kappa}_r^{\pm}(K)| &\leq |\kappa_r^{\pm}(K) - \kappa_r^{\pm}(K_n)| + |\tilde{\kappa}_r^{\pm}(K_n) - \tilde{\kappa}_r^{\pm}(K)| \\ &\leq 2|b-a| \;\; \|K(x,y) - K_n(x,y)\|_\infty . \end{aligned} \tag{3.117}$$

We have established the following result.

THEOREM 3.16. *Suppose that* $J(\phi) \in J^R$, *that* $K(x,y)$ *is Hermitian, and that* $K_n(x,y)$ *is any Hermitian kernel allowing* $J(\phi)$. *Then*

$$|\kappa_r^{\pm} - \tilde{\kappa}_r^{\pm}| \leq 2|b-a| \;\; \|K(x,y) - K_n(x,y)\|_\infty .$$

We remark that when $J(\phi)$ is any quadrature rule

$$\sum_{j=0}^{n} w_j \phi(y_j) \quad \text{in} \quad J^+ ,$$

the result remains valid if the constant 2 in the last inequality is replaced by $1 + w$ (see:(3.71)). It is therefore unchanged if

$$\sum_{j=0}^{n} w_j = b-a .$$

Now Theorem 3.16 gives bounds on the error $|\kappa_r^{\pm} - \tilde{\kappa}_r^{\pm}|$ in terms of a kernel $K_n(x,y)$ of a particular type. In order to deduce bounds on $|\kappa_r^{\pm} - \tilde{\kappa}_r^{\pm}|$ we must construct, for a particular rule $J(\phi)$, a suitable kernel $K_n(x,y)$. As mentioned earlier, Wielandt (1956) undertakes some ingenious constructions of suitable kernels $K_n(x,y)$. We shall not follow Wielandt's treatment here, but instead give a similar type of analysis containing some simplifications.

We suppose that

$$J(\phi) = \sum_{j=0}^{n} w_j \phi(y_j)$$

is a quadrature rule in J^R, whose degree of precision is ρ, where $\rho \geq 2q$. (Thus we may take $J(\phi)$ to be a Gauss-Legendre rule employing $(q+1)$ function values, or a basic Romberg rule employing 2^q+1 function values.) It is now easy to see that the kernel

$$K_q(x,y) = \sum_{i=0}^{q}{}' \sum_{j=0}^{q}{}' \alpha_{ij} p_i(x) p_j(y) \tag{3.118}$$

will allow $J(\phi)$ if each $p_i(x)$ $(i = 0,1,2,\ldots,n)$ is a polynomial of degree at most q. Such a kernel may be constructed from $K(x,y)$ by use of Chebyshev polynomials, or shifted Chebyshev polynomials if $a \neq -1$ or $b \neq 1$. (See p.110 for the notation Σ'.)

3.16.2. *A kernel allowing* $J(\phi)$.

To simplify the analysis we suppose that $a = -1$ and $b = 1$. Associated with a function $\psi(x)$ of one variable is its Fourier-Chebyshev series (section 2.5) given by eqns (2.10) and (2.11), namely,

$$\tfrac{1}{2}a_0 + \sum_{k=1}^{q} a_k T_k(x) , \tag{3.119}$$

where $a_k = \frac{2}{\pi} \int_{-1}^{1} \frac{1}{\sqrt{(1-t^2)}} \psi(t)T_k(t)dt$. It is easily seen that the approximation (3.119), which we denote $(A_q\psi)(x)$, may be written as

$$(A_q\psi)(x) = \int_{-1}^{1} A_q(x,y)\psi(y)dy , \tag{3.120}$$

where

$$A_q(x,y) = 2 \sum_{k=0}^{q}{}' T_k(x)T_k(y)/\{\pi\sqrt{(1-y)^2}\} .$$

Now, we note that if a kernel $K(x,y)$ is continuous, say, then

$$K_{\{q\}}(x,y) = \int_{-1}^{1}\int_{-1}^{1} A_q(x,u)K(u,v)A_q(y,v)dudv \tag{3.121}$$

is a kernel allowing $J(\phi)$. We can rearrange (3.121) into the form (3.118) with $p_r(x) = T_r(x)$ and

$$\alpha_{i,j} = (\tfrac{2}{\pi})^2\int_{-1}^{1}\int_{-1}^{1} \frac{1}{\{(1-v^2)(1-u^2)\}^{\frac{1}{2}}} T_i(u)T_j(v)K(u,v)dudv .$$

If $K(u,v) = \overline{K(v,u)}$, $\alpha_{i,j} = \overline{\alpha_{j,i}}$ and $K_{\{q\}}(x,y)$ is Hermitian. (Indeed, since $A_q(y,v) = \overline{A_q(y,v)}$ we may write the integral operator $K_{\{q\}}$ as $A_qKA_q^*$, so that $K_{\{q\}}$ is obviously self-adjoint when $K = K^*$.)

We shall require a bound for $\|K(x,y) - K_{\{q\}}(x,y)\|_\infty$, and this is easily obtained, as follows, in terms of the Lebesque constant, L_q , for Fourier-Chebyshev approximation (section 2.6). We have

$$\begin{aligned}
&\|K(x,y) - (A_qKA_q^*)(x,y)\|_\infty \\
&\quad\le \|K(x,y) - (KA_q^*)(x,y)\|_\infty + \|(KA_q^*)(x,y) - (A_q(KA_q^*))(x,y)\|_\infty \\
&\quad\le \|K(x,y) - (KA_q^*)(x,y)\|_\infty + L_q\|K(x,y) - (A_qK)(x,y)\|_\infty
\end{aligned}$$

where

$$L_q = \sup_{-1\leq x\leq 1} \int_{-1}^{1} |A_q(x,y)|dy$$

(L_q is, indeed, the Lebesgue constant referred to above). Furthermore (with a corresponding bound for $|K(x,y) - (A_q K)(x,y)|$) we have for all $x\in [-1,1]$

$$|K(x,y) - (KA_q^*)(x,y)|$$

$$\leq (1+L_q)(\tfrac{\pi}{2})^t \frac{(q-t+1)!}{(q+1)!} \times \sup_y \left| \frac{\partial^t}{\partial y^t} K(x,y) \right| ,$$

when $t \leq q$ and we assume that the derivative exists and is bounded. (To obtain this result we use Jackson's theorem; see section 2.6.) Combining our inequalities we obtain, if $K(x,y) = K^*(x,y)$, the result

$$||K(x,y) - K_{\{q\}}(x,y)||_\infty \leq (1+L_q)^2(\tfrac{\pi}{2})^t \frac{(q-t+1)!}{(q+1)!} \left\| \frac{\partial^t}{\partial y^t} K(x,y) \right\|_\infty \qquad (3.122)$$

For numerical purposes we require an estimate for L_q ; as noted in section 2.6, $L_q \leq 4\cdot 1$ for $q \leq 1000$, and in general

$$L_q \leq 3 + (\tfrac{4}{\pi^2}) \ln q \quad .$$

Powell (1966) has given an analytic expression for L_q in the form

$$L_q = (\frac{2q+2}{2q+1}) + \frac{2}{\pi} \sum_{\tau=1}^{q} \frac{1}{\tau} \tan (\frac{\pi\tau}{2q+1}) \quad .$$

The bound (3.122) applies to the kernel $K_q(x,y)$ constructed using the Fourier-Chebyshev expansions on $[-1,1]$. An extension of our result to include the non-Hermitian case is easily obtained:

THEOREM 3.17. *If* $K(x,y)$ *is t-times differentiable, we can construct a kernel* $K_q(x,y)$, *which admits a rule* $J(\phi)$ *with degree of precision* $\geq 2q$, *such that, for* $t\leq q$,

$$||K(x,y) - K_q(x,y)||_\infty \leq$$

$$\tfrac{1}{2}(1+L_q)^2\{\tfrac{\pi}{4}(b-a)\}^t \frac{(q-b+1)!}{(q+1)!} \{\left\| \frac{\partial^t}{\partial y^t} K(x,y) \right\|_\infty + \left\| \frac{\partial^t}{\partial x^t} K(x,y) \right\|_\infty \}. \qquad (3.123)$$

Remark. Observe that $K(x,y)$ is not required to be Hermitian in (3.123), but $K_q(x,y)$ may be taken to be Hermitian when $K(x,y)=K^*(x,y)$.

3.16.3. *Application to Hermitian kernels*

Using Theorems 3.16 and 3.17 we obtain the following result.

THEOREM 3.18. *Suppose that* $J(\phi) \in J^R$ *has a degree of precision* $\rho \geq 2q$, *and that* $K(x,y)$ *is Hermitian. Then*

$$|\kappa_r^{\pm} - \tilde{\kappa}_r^{\pm}|$$

$$\leq 2^{1-2t}(1+L_q)^2 \pi^t [b-a]^{t+1} \frac{(q-t+1)!}{(q+1)!} \left\| \frac{\partial^t}{\partial y^t} K(x,y) \right\|_\infty \tag{3.124}$$

when $t \leq q$ *and* $(\frac{\partial^t}{\partial y^t}) K(x,y)$ *exists and is bounded. Here,* L_q *is given above.*

Remark. The above theorem is valid if $q \geq 0$ and then applies for $J(\phi) \in J^+$. It is not difficult to modify the preceding result to cover the case where the quadrature rule is of the form $(m \times J)(\phi)$, where $J(\phi)$ has the required degree of precision greater than or equal to $2q$. The right-hand side of the bound (3.124) should then be divided by m^t (where, we note, $t \leq q$) .

3.16.4.

Under certain conditions we can use Jackson's theorem for polynomial approximation of analytic functions to obtain results similar to (3.124). We shall give an example in which there is no need to introduce a Lebesgue constant.

Suppose that $K(x,y)$ is a real symmetric kernel defined on $[a,b] = [-1,1]$. Suppose further that $K(x,y)$ has the form $k(x-y)$, where $k(z)$ is analytic on the ellipse E_σ with foci at $(-1,0)$, $(1,0)$ and with minor semi-axis $\frac{1}{2}(\sigma-\sigma^{-1})$ and major semi-axis $\frac{1}{2}(\sigma+\sigma^{-1})$. From a form of Jackson's theorem it follows that, on $[-1,1]$, $k(x-y)$ can be approximated by a polynomial $k_q(x-y)$, where $k_q(z)$ is a polynomial of degree q in z , such that

$$\max_{-1 \leq z \leq 1} |k(z) - k_q(z)| \leq 8 M_0^\sigma(k)/\{\pi\sigma^{q+1}\}.$$

Here, we require

$$|\mathrm{Re}\ k(z)| \leq M_0^{\sigma}(k)$$

for all z in the interior of E_{σ} . Appealing to Theorem 3.16, it follows that if $J(\phi)$ has degree of precision $\geq 2q$ then

$$|\kappa_r^{\pm} - \tilde{\kappa}_r^{\pm}| \leq \frac{32\ M_0^{\sigma}(k)}{\pi\ \sigma^{q+1}} \quad . \tag{3.125}$$

3.16.5.

The trapezium rule with step h is a quadrature rule which integrates trigonometric polynomials of a certain degree exactly (Davis and Rabinowitz 1967, p.57). Consequently, when considering values $\tilde{\kappa}_r^{\pm}$ obtained by the use of this rule, we can apply Theorem 3.16 after constructing a kernel $K_n(x,y)$ which is a trigonometric polynomial in x and in y . In special cases, the difference between $K(x,y)$ and $K_n(x,y)$ can be bounded using Jackson's theorem for trigonometric approximation. Wielandt (1956, Theorem 10) uses such an argument to analyse the error when using the composite rectangle rule (see Example 2.11). Wielandt also gives special treatment to the use of the repeated mid-point rule, with step h , for which one-sided bounds of the form $\tilde{\kappa}_r^{+} - \kappa_r^{+} \leq ch^2\ \|K_{xx}(x,y)\|_{\infty}$ and $\kappa_r^{-} - \tilde{\kappa}_r^{-} \leq ch^2\ \|K_{xx}(x,y)\|_{\infty}$ can be obtained. It does not appear to be possible to extend such one-sided bounds to arbitrary quadrature rules in J^R .

3.16.6.

We have not followed the analysis of Wielandt very closely, above, and the reader will benefit from a careful study of the original paper. Our analysis does reflect some of the spirit of Wielandt's work, and the quality of the bounds is open to some criticism (as noted by Wielandt).

Example 3.26.

Consider the kernel $K(x,y) = e^{xy}$ with $a = 0$, $b = 1$. Then

$$\left\| \left(\frac{\partial^t}{\partial y^t}\right) K(x,y) \right\|_{\infty} \leq \mathrm{e}$$

for all t , and the bound (3.124) becomes

$$|\kappa_r^{+} - \tilde{\kappa}_r^{+}| \leq 2 \left(\frac{\pi}{4}\right)^t (1+L_q)^2 \frac{(q-t+1)!}{(q+1)!} e ,$$

where $t \leq q$. *

As we noted in section 3.16.4, the analysis developed above gives rise to error bounds which are $O(1/m^p)$ $(p \leq q)$ when a composite rule $(m\times J)(\phi)$ is used in the quadrature method and the degree of precision of $J(\phi)$ is at least $2q$. One might expect that $\tilde{\kappa}_r^{+} = \kappa_r^{+} + O(1/m^{2q})$ under such conditions (particularly in view of a statement like (3.11)) when $K(x,y)$ possesses high-order derivatives. We are therefore led to believe that the error analysis of section 3.16.1 leads to results which are pessimistic when $K(x,y)$ is smooth and when we are interested in the error in a particular eigenvalue. The merits (and, simultaneously, the failings) of the analysis of section 3.16 appear to be associated with the fact that we obtain a uniform error bound for $|\kappa_r^{+} - \tilde{\kappa}_r^{+}|$ *for all* r . The bound cannot be better than the actual error in the worst approximations. Since $\tilde{\kappa}_r^{+}$ is zero for r large enough, the rate of decrease of κ_r^{+} as r increases plays a role here; Zahar-Itkin (1964, 1966) has established some results in this area but there is no complete analysis relating the behaviour of κ_r^{+} to the degree of differentiability of $K(x,y)$. The behaviour of κ_r^{+} as $r \to \infty$ can be established for kernels of the type appearing in Example 1.5, however. For further references see Noble (1971 *a,b*).

3.17. A posteriori error bounds dependent on the local error

In an attempt to avoid some of the shortcomings of the analysis in section 3.16, I shall now try to relate the error in an eigenvalue to the corresponding local truncation error. Such an attempt is wholly successful only if $K(x,y)$ is normal. Since (in practice) normality is difficult to detect unless $K(x,y)$ is Hermitian, only the most useful error bounds for Hermitian kernels will be stated.

The analysis depends on the size of the function

$$\eta(x) = \int_a^b K(x,y)\tilde{f}(y)dy - \tilde{\kappa}\tilde{f}(x) \quad (a \le x \le b) \quad ,$$

where $\tilde{f}(x)$ is an approximate eigenfunction corresponding to an approximate eigenvalue $\tilde{\kappa}$. In the following theorem, $\tilde{\kappa}$ and $\tilde{f}(x)$ may have been produced by any means whatever (even by guesswork).

THEOREM 3.19. *Suppose that $K(x,y)$ is Hermitian, and square-integrable, for $a \le x,y \le b$. If there is a non-zero number $\tilde{\kappa}$ and a non-null square-integrable function $\tilde{f}(x)$ such that*

$$\int_a^b K(x,y)\tilde{f}(y)dy - \tilde{\kappa}\tilde{f}(x) = \eta(x)$$

then there is an eigenvalue κ of $K(x,y)$ such that

$$|\tilde{\kappa} - \kappa| \le \|\eta(x)\|_2 / \|\tilde{f}(x)\|_2 \quad .$$

(In the statement of this theorem, κ may be a zero eigenvalue. The theorem is true if we suppose $K(x,y)$ to be normal.)

Proof. We know (Smithies 1962, pp.152 et seq.) that $K(x,y)$ possesses an orthonormal system of eigenfunctions $\{f_r(x)\}$ corresponding to eigenvalues $\{\kappa_r\}$. Every square-integrable function $\psi(x)$ can be expressed as a linear combination

$$\psi(x) = \sum_{r=1}^{\infty} (\psi,f_r)f_r(x)$$

provided we include normalized eigenfunctions corresponding to the eigenvalue zero, should zero be an eigenvalue of $K(x,y)$.

Thus we may write

$$\tilde{f}(x) = \sum_{r=1}^{\infty} \alpha_r f_r(x) \quad (\text{where } \alpha_r = (\tilde{f},f_r)) \tag{3.126}$$

with

$$\int_a^b K(x,y)\, f_r(y)dy = \kappa_r f_r(x) \quad ,$$

and the only possible limit point of $\{\kappa_r\}$ is zero. Then

$$\eta(x) = (K\tilde{f})(x) - \tilde{\kappa}\tilde{f}(x) = \sum_{r=1}^{\infty} \alpha_r(\kappa_r - \tilde{\kappa})f_r(x) \tag{3.127}$$

Since $\{f_r(x)\}$ is an orthonormal system,

$$\|\eta(x)\|_2^2 = \sum_{r=1}^{\infty} |\alpha_r(\kappa_r - \tilde{\kappa})|^2$$

$$\geq \sum_{r=1}^{\infty} |\alpha_r|^2 \inf_r |\kappa_r - \tilde{\kappa}|^2 ,$$

whence (with $\tilde{\kappa} \neq 0$) $\quad \|\eta(x)\|_2^2 \geq \|\tilde{f}(x)\|_2^2 \inf_r |\kappa_r - \tilde{\kappa}|^2$.

Since $\|\tilde{f}(x)\|_2 \neq 0$ the theorem may then be established.

3.17.1.

We wish to apply Theorem 3.19 to an analysis of the quadrature method. Now in this method we compute an eigenvector $\tilde{\underset{\sim}{f}}$ which we extend to an approximate eigenfunction $\tilde{f}(x)$. Since $\eta(x)$ depends on the smoothness of $\tilde{f}(x)$, the method of extending $\tilde{\underset{\sim}{f}}$ to a function $\tilde{f}(x)$ is of some importance. Rather than use a piecewise-constant extension we shall use the Nyström extension given, if $\tilde{\kappa} \neq 0$, by the relation

$$\tilde{f}(x) = (1/\tilde{\kappa}) \sum_{j=0}^{n} w_j K(x,y_j)\tilde{f}(y_j) , \tag{3.128}$$

(this extension is not altogether satisfactory unless $K(x,y)$ is smooth). We suppose that $J(\phi) \in J^R$, or that $J(\phi) \in J^+$ and $\sum_{j=0}^{n} w_j = b-a$.

We may interpret (3.128) in operator notation. If we define the quadrature operator

$$(H\psi)(x) = \sum_{j=0}^{n} w_j K(x,y_j)\psi(y_j) \tag{3.129}$$

then, from (3.128),

$$(H\tilde{f})(x) = \tilde{\kappa}\tilde{f}(x) . \tag{3.130}$$

Had we chosen to extend $\tilde{f}$ to a piecewise-constant function in such a way that it was an eigenfunction of the kernel $K_n(x,y)$ in eqn (3.70), we would have had

$$(K_n\tilde{f})(x) = \tilde{\kappa}\tilde{f}(x) \tag{3.131}$$

in place of (3.130). In later applications of our theory we shall obtain $\tilde{\kappa}, \tilde{f}(x)$ as solutions of eqn (3.131) where $K_n(x,y)$ is some degenerate kernel. All our applications are covered if we suppose that

$$(L\tilde{f})(x) = \tilde{\kappa}\tilde{f}(x) \quad , \tag{3.132}$$

where L is a compact operator on the space of square-integrable functions, $\tilde{\kappa} \neq 0$, and $\tilde{f}(x)$ is non-null. We shall not assume that $K(x,y) = K^*(x,y)$ until we come to apply Theorem 3.19 in section 3.17.4.

3.17.2. *General bounds for a residual* $\eta(x)$

We have the following result.

THEOREM 3.20. *Suppose that eqn (3.132) is satisfied under the conditions stated above. Then if*

$$\eta(x) = (K\tilde{f})(x) - \tilde{\kappa}\tilde{f}(x) \quad ,$$

$$\|\eta(x)\|_2 \leq \|(K-L)L\,\tilde{f}(x)\|_2/|\tilde{\kappa}| \quad .$$

Proof. The proof is straightforward since

$$\eta(x) = (K\tilde{f})(x) - (L\tilde{f})(x) \text{ with } \tilde{\kappa}\tilde{f}(x) = (L\tilde{f})(x) \quad .$$

Thus

$$\tilde{\kappa}\eta(x) = \{(K-L)L\tilde{f}\}(x) \quad .$$

Remark. We note that L and K need not be self-adjoint in Theorem 3.20, or throughout sections 3.17.2 and 3.17.3.

COROLLARY 3.1. *Under the conditions of Theorem 3.20*

$$\frac{\|\eta(x)\|_2}{\|\tilde{f}(x)\|_2} \leq \|(K-L)L\|_2/|\tilde{\kappa}| .$$

Remark. Corollary 3.1 gives some insight into our suggestion that a quadrature rule $J(\phi)$ should be chosen for use in the quadrature method in such a way that

$$\sup_{x,y} \left| \int_a^b K(x,z)K(z,y)\mathrm{d}z - \sum_{j=0}^{n} w_j K(x,y_j)K(y_j,y) \right|$$

is small. For, setting $L = H$ in Corollary 3.1, the norm of $(K-L)L$ is small when the above quantity is small (see Theorem 3.21).

To proceed with our analysis of the quadrature method, we identify L with the quadrature operator H defined by (3.129). To utilize Theorem 3.20 we require a bound on $\|(M\tilde{f})(x)\|_2$, where $M = (K-H)H$, that is,

$$(M\tilde{f})(x) = \sum_{j=0}^{n} w_j \Delta(x,y_j)\tilde{f}(y_j) \tag{3.133}$$

with[§] $$\Delta(x,y) = \int_a^b K(x,z)K(z,y)\mathrm{d}z - \sum_{k=0}^{n} w_k K(x,y_k)K(y_k,y) .$$

Recall that we suppose that $J(\phi) \in J^R$ or that $J(\phi) \in J^+$ and the degree of precision of the rule is at least zero, so that in either case $w_j > 0$ for $j = 0,1,2,\ldots,n$ and $\sum_{j=0}^{n} w_j = b-a$.

Now,

$$\|(M\tilde{f})(x)\|_2^2 = \sum_{j=0}^{n} \sum_{k=0}^{n} w_j w_k \tilde{f}(y_j) \overline{\tilde{f}(y_k)} A_{jk} ,$$

where

$$A_{jk} = \int_a^b \Delta(x,y_j) \overline{\Delta(x,y_k)}\mathrm{d}x .$$

§ We distinguish $\Delta(x,y)$ from Δ of eqn (3.27) and from the measure of $J(\phi)$.

Let us suppose that

$$\sup_{\substack{a\le x\le b\\ 0\le j\le n}} |\Delta(x,y_j)| \le \delta_1 \quad . \tag{3.134}$$

Then

$$\|(M\tilde{f})(x)\|_2^2 \le \{(b-a)\delta_1^2\}\left\{\sum_{j=0}^{n} w_j\right\}\left\{\sum_{j=0}^{n} w_j\,|\tilde{f}(y_j)|^2\right\} \quad .$$

(To obtain this result, note that $|A_{jk}| \le (b-a)\delta_1^2$ and

$$|\sum_k w_k\tilde{f}(y_k)|^2 \le \{\sum_k w_k\}\{\sum_k w_k|\tilde{f}(y_k)|^2\}$$

with $w_j > 0$ for all j .)

Then $\Sigma\, w_j = b - a$, so that

$$\|(M\tilde{f})(x)\|_2 \le (b-a)\delta_1 \ \{\sum_{j=0}^{n} w_j|\tilde{f}(y_j)|^2\}^{\frac{1}{2}} \quad . \tag{3.135}$$

The bound (3.135) is valid whether or not $K(x,y)$ is Hermitian, and we shall use this result in section 3.19, as well as below.

3.17.3.

Now, in order to apply Theorem 3.20 we must bound

$$\|(M\tilde{f})(x)\|_2 / \{\|\tilde{f}(x)\|_2\ |\tilde{\kappa}|\} \quad .$$

We can employ (3.135) if we relate $\sum_{j=0}^{n} w_j|\tilde{f}(y_j)|^2$ to $\int_a^b |\tilde{f}(x)|^2 dx$,

and we now compare these two quantities. We write $B_{jk}(x) = K(x,y_j)\overline{K(x,y_k)}$.

Since $\tilde{\kappa}\tilde{f}(x) = (H\tilde{f})(x)$

$$|\tilde{\kappa}|^2|\tilde{f}(x)|^2 = \sum_{j=0}^{n}\sum_{k=0}^{n} w_j w_k \tilde{f}(y_j)\overline{\tilde{f}(y_k)} B_{j,k}(x) \quad ,$$

and hence $|\tilde{\kappa}|^2\|\tilde{f}(x)\|_2^2 = \sum_{j=0}^{n}\sum_{k=0}^{n} w_j w_k \tilde{f}(y_j)\overline{\tilde{f}(y_k)}\ K^*K(y_k,y_j)$,

since $\int_a^b B_{jk}(x)dx = \int_a^b K(x,y_j)\ \overline{K(x,y_k)}\ dx = K^*K(y_k,y_j)$.

In a similar fashion

$$|\tilde{\kappa}|^2 \sum_{q=0}^{n} w_q |\tilde{f}(y_q)|^2$$

$$= \sum_{j=0}^{n} \sum_{k=0}^{n} w_j w_k \tilde{f}(y_j)\overline{\tilde{f}(y_k)} \sum_{q=0}^{n} w_q B_{j,k}(y_q) \quad .$$

Evidently

$$|\tilde{\kappa}|^2 \quad \{\|\tilde{f}(x)\|_2^2 - \sum_{j=0}^{n} w_j |\tilde{f}(y_j)|^2\}$$

$$= \sum_{j=0}^{n} \sum_{k=0}^{n} w_j w_k \tilde{f}(y_j)\overline{\tilde{f}(y_k)} \; \Delta^{\#}(y_j,y_k) \quad , \tag{3.136}$$

where

$$\Delta^{\#}(y_j,y_k) = \int_a^b B_{jk}(x)\mathrm{d}x - \sum_{q=0}^{n} w_q B_{jk}(y_q) \quad .$$

When $K(x,y)$ is Hermitian $\Delta^{\#}(y_j,y_k) = \Delta(y_j,y_k)$, and $K^*K(x,y) = K^2(x,y)$.

We suppose that

$$\sup_{\substack{0\le j\le n \\ 0\le k\le n}} |\Delta^{\#}(y_j,y_k)| \le \delta_2 \quad ,$$

so that a bound for the modulus of the right-hand side of eqn (3.136) is

$$\delta_2\{\sum_{j=0}^{n} w_j|\tilde{f}(y_j)|\}^2 \le \delta_2\{\sum_{j=0}^{n} w_j\}\{\sum_{j=0}^{n} w_j|\tilde{f}(y_j)|^2\}$$

$$= \delta_2 \, (b-a) \sum_{j=0}^{n} w_j|\tilde{f}(y_j)|^2 \quad . \tag{3.137}$$

Since (3.137) is a bound for the modulus of the left-hand side of eqn (3.136) we obtain, if $|\tilde{\kappa}|^2 > \delta_2(b-a)$,

$$\|\tilde{f}(x)\|_2^2 \ge [1-\{\delta_2(b-a)\}/|\tilde{\kappa}|^2]\{\sum_{j=0}^{n} w_j|\tilde{f}(y_j)|^2\} \quad . \tag{3.138}$$

3.17.4. *Application to the Hermitian case*

We now combine eqns (3.135) and (3.138) and Theorem 3.20 to obtain the following result (which is due, essentially, to Brakhage (1961*a*)). At this point we assume that $K(x,y) = \overline{K(y,x)}$.

THEOREM 3.21. *Suppose* $\tilde{\kappa}$ *is a non-zero approximate eigenvalue of a Hermitian kernel* $K(x,y)$ *, obtained by the quadrature method using a rule* $J(\phi) \in J^+$ *such that* $\sum_{j=0}^{n} w_j = (b-a)$. *Define*

$$\Delta(x,y_j) = \int_a^b K(x,z)K(z,y_j)\mathrm{d}z - \sum_{k=0}^{n} w_k K(x,y_k)K(y_k,y_j)$$

and suppose that

$$\max_{\substack{a\le x\le b\\ 0\le j\le n}} |\Delta(x,y_j)| \le \delta_1 \quad ,$$

$$\max_{\substack{0\le k\le n\\ 0\le j\le n}} |\Delta(y_k,y_j)| \le \delta_2 \le \delta_1 \quad .$$

If $|\tilde{\kappa}|^2 \ge (b-a)\delta_2$ *then there is a true eigenvalue* κ *such that*

$$|\kappa - \tilde{\kappa}| \le \frac{(b-a)\delta_1}{\surd\{|\tilde{\kappa}|^2 - \delta_2(b-a)\}} \tag{3.139}$$

<u>Remark</u>. In (3.139), κ may be zero.

The inequality (3.139) provides a type of bound on a computed eigenvalue $\tilde{\kappa}$, provided that $|\tilde{\kappa}|$ is sufficiently large, and Theorem 3.21 may be regarded as a corollary of Theorem 3.20. We should not expect to deduce such a strong result using Corollary 3.1 unless we realize that the bound in Corollary 3.1 remains valid when the norm $||(L-K)L||_2$ is regarded as the norm of an operator acting on the normed linear space of functions of the form $(L\psi)(x)$, rather than on the space of functions $\psi(x)$.

3.17.5.

Let us suppose that in Theorem 3.21 we replace $J(\phi)$ by $(m\times J)(\phi)$, where $J(\phi)$ has a degree of precision $r\geq 0$. If the Hermitian kernel $K(x,y)$ has a $(r+1)$th partial derivative $(\partial/\partial x)^{r+1} K(x,y)$ which is continuous for $a\leq x,y\leq b$ then we can determine δ_1, δ_2 such that $\delta_1 = O(1/m^{r+1})$ and $\delta_2 = O(1/m^{r+1})$. Thus for m sufficiently large, Theorem 3.21 may be applied to any eigenvalue $\tilde{\kappa} = \tilde{\kappa}_s^{\pm}$, *where* s *is fixed*. For, we know that as $m \to \infty$, $\tilde{\kappa}_s^{\pm} \to \kappa_s^{\pm}$, and $\delta_2 \to 0$, and hence, for some sufficiently large m,

$$|\tilde{\kappa}_s^{\pm}|^2 > \delta_2(b-a) .$$

When this condition is satisfied, the bound (3.139) is $O(1/m^{r+1})$.

Now $\lim_{m\to\infty} \tilde{\kappa}_s^{\pm} = \kappa_s^{\pm}$, so that for m sufficiently large the true eigenvalue closest to $\tilde{\kappa}_s^{\pm}$ is $\kappa_s^{\pm}$. Thus for m sufficiently large, (3.139) gives the result

$$|\kappa_s^{\pm} - \tilde{\kappa}_s^{\pm}| = O(1/m^{r+1}) . \tag{3.140}$$

This result should be contrasted with the result stated in section 3.16.1.

We should note, however, that the bound (3.140) is valid for some fixed m only if $|\tilde{\kappa}|$ is sufficiently large. The bounds studied in section 3.16 have no such restriction. Unfortunately, if we apply Theorem 3.21 to a kernel of the form, say,

$$K(x,y) = x(1-y) \quad (0\leq x\leq y\leq 1) ,$$

$$K(x,y) = y(1-x) \quad (0\leq y\leq x\leq 1)$$

when the quadrature rule is the trapezium rule with step h, it requires a rather careful use of Peano theory to establish $O(h^2)$ error bounds. Some loss of sharpness of the bounds is also encountered, partly because of the stages involved in the proof of the theorem and partly because the operator H for this kernel has piecewise-linear eigenfunctions. For this example the true eigenfunctions $f(x)$ can be set in place of $\tilde{f}(x)$ in Theorem 3.19 to yield fairly good error bounds!

3.18. A dual approach: Hermitian kernels

The statement of Theorem 3.21 defines a circle centred on $\tilde{\kappa}$ and containing a true eigenvalue κ, under suitable conditions. We now show that a dual approach yields a circle centred on κ and containing an approximate eigenvalue $\tilde{\kappa}$, under certain conditions. The kernel $K(x,y)$ is assumed Hermitian.

We require the following theorem, which is an analogue of Theorem 3.19.

THEOREM 3.22. *If* $\underset{\sim}{A} = \underset{\sim}{A}^*$ *is a matrix of order* N , *and* μ *and* $\underset{\sim}{x}$ *are such that* $\underset{\sim}{A}\underset{\sim}{x} - \mu\underset{\sim}{x} = \underset{\sim}{v}$, *where* $\|\underset{\sim}{x}\|_2 \neq 0$, *then there is an eigenvalue* λ_i *of* $\underset{\sim}{A}$ *such that*

$$|\lambda_i - \mu| \leq \|\underset{\sim}{v}\|_2 / \|\underset{\sim}{x}\|_2 \quad . \tag{3.141}$$

<u>Proof</u>. The proof of Theorem 3.22 follows similar lines to the proof of Theorem 3.19. We refer the reader to Isaacson and Keller (1966, p.141).

We employ Theorem 3.22 in the following analysis. Suppose that

$$(Hf)(x) = \sum_{j=0}^{n} w_j K(x,y_j) f(y_j) = \kappa f(x) + \tau(x) \quad , \tag{3.142}$$

where

$$\int_a^b K(x,y) f(y)\,dy = \kappa f(x) \quad , \tag{3.143}$$

$J(\phi) \in J^+$, and where $K(x,y) = K^*(x,y)$. Then there is an approximate eigenvalue of H , $\tilde{\kappa}$, such that, if $\underset{\sim}{f} \neq \underset{\sim}{0}$,

$$|\kappa - \tilde{\kappa}| \le \left\{ \frac{\sum_{j=0}^{n} w_j |\tau(y_j)|^2}{\sum_{j=0}^{n} w_j |f(y_j)|^2} \right\}^{\frac{1}{2}}$$

$$= (\underset{\sim}{\tau}^* \underset{\sim}{D} \underset{\sim}{\tau})^{\frac{1}{2}} / (\underset{\sim}{f}^* \underset{\sim}{D} \underset{\sim}{f})^{\frac{1}{2}} \quad , \tag{3.144}$$

where $\underset{\sim}{D} = \underset{\sim}{\text{diag}}\,(w_0, w_1, \ldots, w_n) > 0$, and $\underset{\sim}{\tau} = \underset{\sim}{KDf} - \kappa \underset{\sim}{f}$ is the vector of 'local truncation errors'.

The proof of this is straightforward. For, $\underset{\sim}{D} > 0$ and $\underset{\sim}{KDf} - \kappa \underset{\sim}{f} = \underset{\sim}{\tau}$ so that $(\underset{\sim}{D}^{\frac{1}{2}} \underset{\sim}{K} \underset{\sim}{D}^{\frac{1}{2}})(\underset{\sim}{D}^{\frac{1}{2}} \underset{\sim}{f}) - \kappa(\underset{\sim}{D}^{\frac{1}{2}} \underset{\sim}{f}) = \underset{\sim}{D}^{\frac{1}{2}} \underset{\sim}{\tau}$. The matrix $\underset{\sim}{A} = \underset{\sim}{D}^{\frac{1}{2}} \underset{\sim}{K} \underset{\sim}{D}^{\frac{1}{2}}$ is Hermitian, and Theorem 3.22 applies.

We now imitate the proof of (3.139), and obtain from (3.142) and (3.143) the relations

$$|\kappa|^2 \sum_{j=0}^{n} w_j |\tau(y_j)|^2 \le (b-a)^2 \hat{\delta}_1^2 \, ||f(x)||_2^2$$

and

$$|\kappa|^2 \{ \, ||f(x)||_2^2 - \sum_{j=0}^{n} w_j |f(y_j)|^2 \} \le (b-a) \hat{\delta}_2 \, ||f(x)||_2^2 \quad ,$$

where

$$\Delta(x,y) = \sum_{j=0}^{n} w_j K(x,y_j) K(y_j,y) - \int_a^b K(x,z) K(z,y) dz \quad , \tag{3.145}$$

$$\max_{\substack{0 \le j \le n \\ a \le y \le b}} |\Delta(y_j,y)| \le \hat{\delta}_1 \quad , \tag{3.146}$$

and

$$\max_{a \le x, y \le b} |\Delta(x,y)| \le \hat{\delta}_2 \quad . \tag{3.147}$$

The preceding results apply when $J(\phi) \in J^+$ and $\sum_{j=0}^{n} w_j = b-a$.

We may now use the bound (3.144) to establish the following theorem.

THEOREM 3.23. *If* κ *is a true eigenvalue of the Hermitian kernel*

$K(x,y)$, satisfying (3.143), *and* $|\kappa|^2 > (b-a)\hat{\delta}_2$, *then there is an approximate eigenvalue* $\tilde{\kappa}$ *such that*

$$|\kappa - \tilde{\kappa}| \leq \frac{(b-a)\hat{\delta}_1}{\surd\{|\kappa|^2 - (b-a)\hat{\delta}_2\}}$$

where $\hat{\delta}_1$ *and* $\hat{\delta}_2$ *are defined by eqns* (3.146) *and* (3.147), *respectively.*

3.19. *The rôle of Peano theory*

In both Theorem 3.21 and Theorem 3.23 (and also later) we require bounds on errors of the form

$$\int_a^b K(x,z)K(z,y)dz - \sum_{j=0}^{n} w_j K(x,y_j)K(y_j,y) \tag{3.148}$$

for particular ranges of x and y.

Such quantities may be bounded in terms of the Peano constants of the quadrature rule $J(\phi) = \sum_{j=0}^{n} w_j\phi(y_j)$. If $J(\phi)$ has a degree of precision r then there are constants $c_1, c_2, \ldots c_{r+1}$ such that for all $x,y \in [a,b]$, (3.148) may be bounded in modulus by

$$c_s \sup_{a\leq x,y\leq b} L_s(x,y) \quad (s\leq r+1),$$

where

$$L_s(x,y) = \sup_{a\leq z\leq b} \left|\frac{\partial^s}{\partial z^s} K(x,z)K(z,y)\right| \tag{3.149}$$

(and $c_s = c_s(J;a,b)$), assuming that the derivatives occurring here exist and are bounded. (We could examine the bounds $c_s\|L_s(x,y)\|_\infty$ for $s = 1,2,3,\ldots,(r+1)$ and choose the least one. A direct estimate of the size of (3.148) might be practicable in some cases.)

3.19.1.

We can also apply Peano theory directly to establish a result which is similar to Theorem 3.23. We use Theorem 3.22 as the basis for the theory, and suppose $K(x,y)$ Hermitian.

To apply Theorem 3.22, given the relations (3.142) and (3.143), we must estimate $\tau(x)$. Using the Peano constants, and supposing that any necessary derivatives exist and are bounded,

$$\left|\int_a^b K(x,y)f(y)\mathrm{d}y - \sum_{j=0}^{n} w_j K(x,y_j)f(y_j)\right|$$

$$\le c_s \sup_y \left|\frac{\partial^s}{\partial y^s}\{K(x,y)f(y)\}\right| \tag{3.150}$$

for $a \le x \le b$, $s \le r+1$.

Now from (3.143)

$$\kappa K(x,y)f(y) = \int_a^b K(x,y)K(y,z)f(z)\mathrm{d}z \quad ,$$

so that on differentiating with respect to y and applying the Cauchy-Schwartz inequality, we can obtain the bound

$$|\kappa| \sup_{a\le y\le b} \left|\frac{\partial^s}{\partial y^s}\{K(x,y)f(y)\}\right|$$

$$\le \{\int_a^b |L_s(x,t)|^2 \,\mathrm{d}t \int_a^b |f(x)|^2 \mathrm{d}x\}^{\frac{1}{2}} \quad , \tag{3.151}$$

where $L_s(x,t) = \sup_{a\le z\le b}\left|\frac{\partial^s}{\partial z^s} K(x,z)K(z,t)\right|$, as in (3.149). Since the left-hand side of (3.150) is $|\tau(x)|$, (3.151) yields a bound

$$|\kappa|^2 \sum_{j=0}^{n} w_j |\tau(y_j)|^2$$

$$\le c_s^2 \{\int_a^b \sum_j w_j |L_s(y_j,y)|^2 \mathrm{d}y\} \int_a^b |f(x)|^2 \mathrm{d}x$$

$$\le c_s^2 (b-a)^2 \|L_s(x,y)\|_\infty^2 \|f(x)\|_2^2 \quad . \tag{3.152}$$

Now

$$\left|\int_a^b |f(x)|^2 \mathrm{d}x - \sum_{j=0}^{n} w_j |f(y_j)|^2 \right|$$

$$\le c_s \sup_x \frac{\mathrm{d}^s}{\mathrm{d}x^s}\{|f(x)|^2\} \quad , \tag{3.153}$$

and, since $\overline{K(z,y)} = K(y,z)$,

$$|\kappa|^2|f(z)|^2 = \int_a^b\int_a^b f(x)\overline{f(y)}K(y,z)K(z,x)\,dx\,dy \ ,$$

so that

$$|\kappa|^2 \left|\frac{d^s}{dz^s}|f(z)|^2\right| \le \int_a^b\int_a^b L_s(x,y)f(x)\overline{f(y)}\,dx\,dy$$

$$\le (b-a)\ \|L_s(x,y)\|_\infty\ \|f(x)\|_2^2 \ , \qquad (3.154)$$

using the notation of (3.149).

Combining (3.153) and (3.154) we obtain

$$\left|\|f(x)\|_2^2 - \sum_{j=0}^{n} w_j|f(y_j)|^2\right|$$

$$\le (c_s/|\kappa|^2)\ (b-a)\ \|L_s(x,y)\|_\infty\ \|f(x)\|_2^2 \ ,$$

so that, if κ is sufficiently large,

$$\sum_{j=0}^{n} w_j|f(y_j)|^2 \ge \{1-(c_s/|\kappa|^2)\ (b-a)\ \|L_s(x,y)\|_\infty\}\ \|f(x)\|_2^2 \ . \qquad (3.155)$$

We now have an upper bound (3.152) for $|\kappa|^2\ \underset{\sim}{\tau}^*\underset{\sim}{D}\underset{\sim}{\tau}$ and a lower bound (3.155) for $\underset{\sim}{f}^*\underset{\sim}{D}\underset{\sim}{f}$. If we substitute these bounds in (3.144) we obtain a bound on $|\kappa - \tilde{\kappa}|$. To display this bound economically, we write

$$\mathrm{L}_s = \|L_s(x,y)\|_\infty \ . \qquad (3.156)$$

We then have

$$|\kappa - \tilde{\kappa}| \le \frac{c_s\mathrm{L}_s\ (b-a)}{\sqrt{\{|\kappa|^2 - c_s(b-a)\ \mathrm{L}_s\}}} \qquad (3.157)$$

provided that the square-root term in the denominator is positive. The bound (3.157) is precisely the bound which we would obtain for Theorem 3.23 on writing

$$\hat{\delta}_1, \hat{\delta}_2 \le c_s\mathrm{L}_s \ .$$

We should note that in (3.157), c_s is a Peano constant associated with $J(\phi)$ on $[a,b]$, that is

$$c_s \equiv c_s(J;a,b) .$$

If we replace $J(\phi)$ by $(m\times J)(\phi)$, the bound (3.157) is $O(1/m^{s+1})$, where $s \leq r+1$.

Remark. The analysis which led to (3.157) employed the assumption of continuity in the first $(s-1)$ z-derivatives of $K(x,z)K(z,y)$ and, say, the boundedness of the sth derivative. However, we know that an eigenfunction may be analytic even though $K(x,y)$ has a discontinuous derivative at $x = y$. Eqn (3.154) may be amended to deal with such a case, but the analysis becomes a trifle involved. The bound (3.154), of course, is crucial in determining whether or not a bound on $|\kappa - \tilde{\kappa}|$ is applicable to some particular value of κ.

The bounds in Theorem 3.23 and eqn (3.157) require the unknown value κ. To apply the results it is necessary to obtain a lower bound for κ. If we set $\kappa = \kappa_r^{\pm}$, such a bound may be obtained after computing $\tilde{\kappa}_r^{\pm}$ from Wielandt's analysis, since this yields results of the form

$$|\tilde{\kappa}_r^{\pm} - \kappa_r^{\pm}| \leq e .$$

It follows that

$$|\kappa_r^{\pm}| \geq |\tilde{\kappa}_r^{\pm}| - e . \qquad (3.158)$$

If we compute $\tilde{\kappa}_r^{\pm}$ using a quadrature rule $(m\times J)(\phi)$, where $J(\phi) \in J^{+}$, then $\lim_{m\to\infty} \tilde{\kappa}_r^{\pm} = \kappa_r^{\pm}$. Thus for sufficiently large m, (3.157) provides a bound on $|\kappa^{\pm} - \tilde{\kappa}^{\pm}|$. This may be compared with (3.140).

The reader may wish to explore the application of Peano theory to bound δ_1 and δ_2 in (3.139).

3.20. *Asymptotic error bounds for simple eigenvalues*

The analysis of sections 3.16-3.19 is not entirely satisfactory. We obtained rigorous error bounds for eigenvalues of Hermitian kernels, but these error bounds often prove either pessimistic or of limited application. In this section we investigate the asymptotic behaviour of the error when the eigenvalue κ is simple; *the kernel* $K(x,y)$ *is not required to be Hermitian.* The asymptotic theory justifies various 'reasonable' error estimates, and consistency checks, like the deferred approach to the limit and deferred correction, as $h \to 0$.

3.20.1.

We suppose that κ is a non-zero simple eigenvalue of a general continuous kernel $K(x,y)$ on $[a,b]$, so that

$$\int_a^b K(x,y)f(y)dy = \kappa f(x) \quad (a \le x < b) \quad .$$

Then there is a function $\psi(x)$ such that

$$\int_a^b K^*(x,y)\psi(y)dy = \bar{\kappa}\psi(x) \; , \; (a \le x \le b) \tag{3.159}$$

and we suppose that $\|\psi(x)\|_2 = \|f(x)\|_2 = 1$.

We also suppose that $\{J_m(\Phi)\}$ is a family of quadrature rules in J^+ such that

$$\lim_{m\to\infty} J_m(\phi) = \int_a^b \phi(y)dy$$

whenever $\phi(x)$ is continuous on $[a,b]$. For some fixed m, we write

$$J_m(\phi) = \sum_{j=0}^{n} w_j \phi(y_j)$$

(where $n \equiv n(m)$, $w_j \equiv w_j(m)$, and $y_j \equiv y_j(m)$), and we suppose that

$$\sum_{j=0}^{n} w_j K(x,y_j)\tilde{f}(y_j) = \tilde{\kappa}\tilde{f}(x) \quad . \tag{3.160}$$

Clearly $\tilde{\kappa}$ and $\tilde{f}(x)$ depend on m. Suppose that κ is a simple non-zero eigenvalue satisfying (3.19). Then, according to Theorem 3.13, we can, for each m, choose one of the values $\tilde{\kappa}$ satisfying (3.160), so that

$$\lim_{m\to\infty} \tilde{\kappa} = \kappa \tag{3.161}$$

and

$$\lim_{m\to\infty} \|\tilde{f}(x) - f(x)\|_\infty = 0 , \tag{3.162}$$

where $\tilde{f}(x)$ is normalized, so that

$$\lim_{m\to\infty} (\tilde{f},f) = 1 . \tag{3.163}$$

(We note, in passing, that, if (3.161) is satisfied, then $\tilde{\kappa}$ is a simple eigenvalue for sufficiently large m, and $\tilde{f}(x)$ is then uniquely defined up to scalar multiplication.)

Now

$$\int_a^b K(x,y)\tilde{f}(y)\mathrm{d}y = \tilde{\kappa}\tilde{f}(x) + \eta(x) ,$$

where

$$\eta(x) = \int_a^b K(x,y)\tilde{f}(y)\mathrm{d}y - \sum_{j=0}^{n} w_j K(x,y_j)\tilde{f}(y_j) .$$

Thus, from (3.159),

$$\int_a^b \overline{\psi(x)} \int_a^b K(x,y)\tilde{f}(y)\mathrm{d}y\mathrm{d}x = \kappa\int_a^b \overline{\psi(y)}\, \tilde{f}(y)\mathrm{d}y .$$

Now, from the definition of $\eta(x)$,

$$\int_a^b \overline{\psi(x)} \int_a^b K(x,y)\tilde{f}(y)\mathrm{d}y = \tilde{\kappa}\int_a^b \overline{\psi(x)}\tilde{f}(x)\mathrm{d}x + \int_a^b \overline{\psi(x)}\, \eta(x)\mathrm{d}x .$$

In compact notation, the last two equations give

$$(\kappa - \tilde{\kappa})(\tilde{f},\psi) = (\eta,\psi) \ , \tag{3.164}$$

and consequently

$$|\kappa - \tilde{\kappa}| \ |(\tilde{f},\psi)| \leq \|\eta(x)\|_2 \|\psi(x)\|_2$$

or

$$|\kappa - \tilde{\kappa}| \ |(\tilde{f},\psi)| \leq \|\eta(x)\|_2 \ , \tag{3.165}$$

since $\|\psi(x)\|_2 = 1$.

To bound $\|\eta(x)\|_2$ we employ Theorem 3.20 and the bound (3.135). Thus

$$|\tilde{\kappa}| \ \|\eta(x)\|_2 \leq (b-a)\delta_1 \ \{\sum_{j=0}^{n} w_j|\tilde{f}(y_j)|^2\}^{\frac{1}{2}} \ , \tag{3.166}$$

where

$$\sup_{\substack{a\leq x\leq b \\ 0\leq j\leq n}} |\Delta(x,y_j)| \leq \delta_1 \tag{3.167}$$

with

$$\Delta(x,y) = \int_a^b K(x,z)K(z,y)\mathrm{d}z - \sum_{j=0}^{n} w_j K(x,y_j)K(y_j,y) \ , \tag{3.168}$$

and 'symmetry' of $K(x,y)$ is, of course, not assumed.

Now as $m \to \infty$, $\tilde{\kappa} \to \kappa$, so that

$$\|\eta(x)\|_2 \leq \frac{(b-a)\delta_1\{\sum_{j=0}^{n} w_j|\tilde{f}(y_j)|^2\}^{\frac{1}{2}}}{|\kappa|} \{1 + o(1)\} \tag{3.169}$$

as $m \to \infty$.

We can also show that $\sum_{j=0}^{n} w_j|\tilde{f}(y_j)|^2 = 1 + o(1)$ as $m \to \infty$. For, $\|\tilde{f}(x) - f(x)\|_\infty \to 0$, and so

$$\sum_{j=0}^{n} w_j|\tilde{f}(y_j)|^2 = \sum_{j=0}^{n} w_j|f(y_j)|^2 + o(1) \quad , \text{ as } m \to \infty,$$

$$= \int_a^b |f(y)|^2 \, dy + o(1)$$

$$= 1 + o(1) \quad \text{as } m \to \infty ,$$

in view of the properties of $J_m(\phi)$ and the continuity of $f(x)$.

Using this result we find that

$$||\eta(x)||_2 \le (b-a)\delta_1\{1 + o(1)\}/|\kappa| \quad . \tag{3.170}$$

To employ eqn (3.165) we must construct an asymptotic lower bound for $|(\tilde{f},\psi)|$. Now since $||\tilde{f}(x) - f(x)||_\infty \to 0$ as $m \to \infty$, it follows *a fortiori* that $||\tilde{f}(x) - f(x)||_2 \to 0$. But

$$(\tilde{f},\psi) = (f,\psi) + (\tilde{f}-f,\psi) ,$$

so that

$$|(f,\psi)| \le |(\tilde{f},\psi)| + |(\tilde{f}-f,\psi)|$$

$$\le |(\tilde{f},\psi)| + ||\tilde{f}(x) - f(x)||_2 \, ||\psi(x)||_2 ,$$

and hence

$$|(\tilde{f},\psi)| \ge |(f,\psi)| \, (1 - o(1)) \tag{3.171}$$

as $m \to \infty$.

Combining (3.171) and (3.170) in (3.165) we obtain the following result.

THEOREM 3.24. *Suppose that* $\{J_m(\phi)\}$ *is a family of quadrature rules in* J^+ *such that*

$$\lim_{m\to\infty} J_m(\phi) = \int_a^b \phi(y)dy$$

for every $\phi(x)$ *in* $C[a,b]$. *Suppose further that the eigenvalue problem for a continuous kernel* $K(x,y)$ $(a \le x,y \le b)$ *is solved approximately by the quadrature method using* $J_m(\phi)$, *for* $m = 1,2,3,\ldots$, *and that* κ *is a simple non-zero eigenvalue of* $K(x,y)$. *Then there is a sequence of approximate eigenvalues* $\tilde{\kappa} \equiv \tilde{\kappa}(m^{-1})$ *such that*

$$|\kappa - \tilde{\kappa}| \le \frac{\delta_1(b-a)}{|\kappa|\ |(f,\psi)|}\{1 + o(1)\} \text{ as } m \to \infty ,$$

where $\psi(x)$ *satisfies* (3.159) *and* δ_1 *is defined by* (3.167) *and* (3.168).

The analysis of this section may be regarded as a limited extension of the analysis of section 3.17. We may interpret Theorem 3.24, in part, by saying that the sensitivity of an eigenvalue κ to errors produced in the quadrature method (for example) is governed to some extent by the condition number $|(f,\psi)|$, when $||f(x)||_2$ and $||\psi(x)||_2$ are normalized to unity.

A condition number $\sigma(\kappa)$ was introduced in section 3.2.1 as 'the "cosine" of the angle between corresponding left and right eigenfunctions'. In mathematical notation this gives $\sigma(\kappa) = |(f,\psi)| / ||f(x)||_2\ ||\psi(x)||_2$, and since

$$\int_a^b \overline{\psi(x)}\ K(x,y)\mathrm{d}x = \kappa\overline{\psi(y)}$$

Theorem 3.24 provides a justification for our remarks in section 3.2.1 on the significance of $\sigma(\kappa)$.

In practice we cannot compute $\sigma(\kappa)$, though we can estimate this condition number if we calculate *approximate* left and right eigenfunctions and find the cosine of the angle between them. In particular, if $\tilde{\kappa} \simeq \kappa$ and $\tilde{f} \simeq \tilde{f}$, where $D^{\frac{1}{2}}KD^{\frac{1}{2}}(D^{\frac{1}{2}}f) = (D^{\frac{1}{2}}f)$, then $\sigma(\kappa)$ is approximated by the condition number of $\tilde{\kappa}$ as an eigenvalue of $D^{\frac{1}{2}}KD^{\frac{1}{2}}$.

3.21. *Asymptotic expansions for the error in a simple eigenvalue and the corresponding eigenfunction*

We again consider the eigenvalue problem for a general (not necessarily

Hermitian) kernel $K(x,y)$.

We are interested in obtaining asymptotic expansions for the error in an approximate eigenvalue $\tilde{\kappa}$ when κ (the limit point of $\tilde{\kappa}$) is a simple and non-zero eigenvalue. For simplicity, we suppose that $\tilde{\kappa}$ and the approximate eigenfunction $\tilde{f}(x)$ have been obtained using the trapezium rule with step h in the method of section 3.2. Thus $\tilde{\kappa} \equiv \tilde{\kappa}(h)$ and $\tilde{f}(x) \equiv \tilde{f}_h(x)$. (We may suppose that $\tilde{f}(x)$ is obtained from the computed vector $\tilde{\underset{\sim}{f}}$ using the 'Nyström extension'.)

Now suppose that, for h sufficiently small, we can find a value $\mu \equiv \mu(h)$ and a function $\phi(x)$ ($\equiv \phi_h(x)$) such that

$$\sum_{j=0}^{n} w_j K(x,y_j)\phi(y_j) - \mu\phi(x) = \varepsilon(x) \tag{3.172}$$

with

$$\left.\begin{aligned} &w_0 = w_n = \tfrac{1}{2}h \;,\; w_j = h \;(j = 1,2,3,\ldots,n-1) \;, \\ &y_j = a + jh \;(j = 0,1,2,\ldots,n) \;,\; h = (b-a)/n \;, \end{aligned}\right\} \tag{3.173}$$

where $||\varepsilon(x)||_\infty = O(h^{2N+1})$ and N is a non-negative integer. (In order to find such a pair μ, $\phi(x)$, we shall require, here, that $K(x,y)$ is sufficiently differentiable, and at least piecewise-continuous.)

We shall actually show how to construct μ and $\phi(x)$ in such a way that

$$\lim_{h\to 0} \mu(h) = \kappa \;, \tag{3.174}$$

where κ is a simple non-zero eigenvalue. Thus we also have, say,

$$\int_a^b K(x,y)f(y)\mathrm{d}y = \kappa f(x) \;, \tag{3.175}$$

and since κ is simple there is a corresponding function $\psi(x)$ such that, for $a\le x\le b$,

$$\int_a^b K^*(x,y)\psi(y)\mathrm{d}y = \overline{\kappa}\psi(x) \ , \tag{3.176}$$

where $(f,\psi) \neq 0$.

In view of the choice of weights and abscissae (3.173), there is a function $\tilde{\psi}(x) \equiv \tilde{\psi}_h(x)$ and an approximation $\tilde{\bar{\kappa}} \equiv \tilde{\bar{\kappa}}(h)$ to $\overline{\kappa}$ such that

$$\sum_{j=0}^{n} w_j \overline{K(y_j,x)}\ \tilde{\psi}(y_j) = \tilde{\bar{\kappa}}\tilde{\psi}(x) \quad , \tag{3.177}$$

where $\lim_{h\to 0} \tilde{\bar{\kappa}}(h) = \overline{\kappa}$ and $\lim_{h\to 0} \|\tilde{\psi}_h(x) - \psi(x)\|_\infty = 0$.

We may set $\tilde{\bar{\kappa}} = \overline{\tilde{\kappa}}$, where

$$\sum_{j=0}^{n} w_j K(x,y_j)\tilde{f}(y_j) = \tilde{\kappa}\tilde{f}(x) \tag{3.178}$$

and $\lim_{h\to 0} \tilde{\kappa} = \kappa$. (This is easily established on setting $x = y_i$ $(i = 0,1,2,\ldots,n)$ in (3.177) and (3.178), and using matrix theorems. From (3.177), $\underset{\sim}{K}^*\underset{\sim}{D}\underset{\sim}{\tilde{\psi}} = \tilde{\bar{\kappa}}\underset{\sim}{\tilde{\psi}}$, so $\underset{\sim}{D}^{\frac{1}{2}}\underset{\sim}{K}^*\underset{\sim}{D}^{\frac{1}{2}}(\underset{\sim}{D}^{\frac{1}{2}}\underset{\sim}{\tilde{\psi}}) = \tilde{\bar{\kappa}}(\underset{\sim}{D}^{\frac{1}{2}}\underset{\sim}{\tilde{\psi}})$, whilst $(\underset{\sim}{D}^{\frac{1}{2}}\underset{\sim}{K}^*\underset{\sim}{D}^{\frac{1}{2}})^* = \underset{\sim}{D}^{\frac{1}{2}}\underset{\sim}{K}\underset{\sim}{D}^{\frac{1}{2}}$. Also, from (3.178), $(\underset{\sim}{D}^{\frac{1}{2}}\underset{\sim}{K}\underset{\sim}{D}^{\frac{1}{2}})\underset{\sim}{D}^{\frac{1}{2}}\underset{\sim}{\tilde{f}} = \tilde{\kappa}\,\underset{\sim}{D}^{\frac{1}{2}}\underset{\sim}{\tilde{f}}$; the result follows.)

Now, in (3.172) we set $x = y_i$, multiply by $w_i\ \overline{\tilde{\psi}(y_i)}$ $(i = 0,1,2,\ldots,n)$, and sum over i to obtain

$$\sum_{j=0}^{n} w_j\{\sum_{i=0}^{n} w_i K(y_i,y_j)\overline{\tilde{\psi}(y_i)}\}\ \phi(y_j) - \mu\sum_{i=0}^{n} w_i\ \overline{\tilde{\psi}(y_i)}\ \phi(y_i)$$
$$= \sum_{i=0}^{n} w_i\ \overline{\tilde{\psi}(y_i)}\ \varepsilon(y_i) \ .$$

Taking conjugates throughout (3.177) and substituting in the first term in the above equation, we obtain

$$(\tilde{\kappa} - \mu)\sum_{j=0}^{n} w_j\ \overline{\tilde{\psi}(y_j)}\ \phi(y_j) = \sum_{j=0}^{n} w_j\ \overline{\tilde{\psi}(y_j)}\ \varepsilon(y_j) \tag{3.179}$$

(in obtaining (3.179), we set $\tilde{\kappa} = \overline{\tilde{\bar{\kappa}}}$) .

From (3.179) we obtain the inequality

$$|\tilde{\kappa} - \mu| \left| \sum_{j=0}^{n} w_j \overline{\tilde{\psi}(y_j)} \phi(y_j) \right|$$

$$\leq \{ \sum_{j=0}^{n} w_j |\tilde{\psi}(y_j)|^2 \sum_{j=0}^{n} w_j |\varepsilon(y_j)|^2 \}^{\frac{1}{2}} \quad . \qquad (3.180)$$

(At this point, on setting $\phi(x) \equiv f(x)$ and $\mu \equiv \kappa$, we may discern the duality between (3.180) and (3.164),)

3.21.1.

We shall use (3.180) later. We turn now to establishing a special choice of $\phi(x)$ and μ which, with (3.180), will permit us to establish the following result.

THEOREM 3.25. *If* $K(x,y)$ *is sufficiently differentiable and* κ *is a simple non-zero eigenvalue of the kernel* $K(x,y)$ *on* $a \leq x, y \leq b$, *then there exist values* $\mu_1, \mu_2, \ldots, \mu_N$ *such that*

$$\tilde{\kappa} \equiv \tilde{\kappa}(h) = \kappa + \mu_1 h^2 + \mu_2 h^4 + \ldots + \mu_N h^{2N} + O(h^{2N+1}) \quad ,$$

where $\tilde{\kappa}$ *is an eigenvalue obtained from the quadrature method using the trapezium rule with step* h .

Remark. Precise differentiability conditions sufficient for this theorem will emerge from the proof. We should remark, also, that the relation between $\tilde{\kappa}(h)$ and κ given in the theorem applies between *one* of the $(n+1)$ eigenvalues $\tilde{\kappa}$ and the true eigenvalue κ *when* h *is sufficiently small*. In practice we do not know where κ is and we must therefore select $\tilde{\kappa}$ by a systematic indexing of the $n+1$ eigenvalues $\tilde{\kappa}$ which occur with a fixed value of h . To this extent, Theorem 3.25 provides *motivation* for applying (with some convenient values of h) the deferred approach to the limit or the construction of asymptotic bounds, as described earlier, rather than a complete justification of a particular implementation of the process.

Proof. The proof of Theorem 3.25 hinges on our ability to demonstrate the existence of values $\mu_1, \mu_2, \ldots, \mu_N$ and functions

$$\phi_1(x),\ \phi_2(x),\ \ldots,\ \phi_N(x)$$

such than, on setting

$$\mu = \kappa + \sum_{r=1}^{N} \mu_r h^{2r} \tag{3.181}$$

and

$$\phi(x) = f(x) + \sum_{r=1}^{N} h^{2r}\phi_r(x)\ , \tag{3.182}$$

the assumptions associated with (3.172) are valid - principally, that

$$\|\varepsilon(x)\|_\infty = O(h^{2N+1})\ .$$

When this is accomplished we can introduce (3.180) and pursue the analysis further. If the reader is prepared to accept the existence of suitable μ_i and $\phi_i(x)$, he may proceed to section 3.21.3.

It will be convenient to write, at times, $\kappa = \mu_0$, $f(x) \equiv \phi_0(x)$. In order to establish the required result, we employ the Euler-Maclaurin formula (see eqn (2.24)). This formula leads to the following result.

THEOREM 3.26. *If* $\phi_i(x) \in C^{2M+1}[a,b]$ *and* $(\partial/\partial y)^{2M+1} K(x,y)$ *is continuous for* $a \le x, y \le b$ *then*

$$\int_a^b K(x,y)\phi_i(y)\,dy$$
$$= \sum_{j=0}^{n} w_j K(x,y_j)\phi_i(y_j) - \sum_{k=1}^{M} h^{2k}\alpha_{ik}(x) + O(h^{2M+1})\ , \tag{3.183}$$

where

$$\alpha_{ik}(x) = \frac{B_{2k}}{(2k)!}\left[(\partial/\partial y)^{2k-1}K(x,y)\phi_i(y)\right]_{y=a}^{y=b} \quad \begin{matrix} i = 0,1,\ldots,N; \\ k = 1,2,\ldots,M \end{matrix}\ , \tag{3.184}$$

and the order term is uniform for x *in* $[a,b]$. *Here the weights* w_j *and abscissae* y_j *are those in* (3.173).

Proof. The proof of this theorem may be deduced from our remarks in section 2.13.

Now suppose that $\phi(x)$ has the form (3.182), where $f(x)$ $(\equiv \phi_0(x))$ and $\phi_i(x)$ $(i = 1,2,\ldots,N)$ are sufficiently differentiable. If $\phi_i(x) \in C^{2(N-i)+1}[a,b]$ $(i = 0,1,\ldots,N)$ eqn (3.183) is valid with M replaced by $N-i$. Then (for the trapezium rule)

$$\sum_{j=0}^{n} w_j K(x,y_j)\phi(y_j)$$

$$= \sum_{i=0}^{N} \sum_{j=0}^{n} w_j K(x,y_j) h^{2i} \phi_i(y_j)$$

$$= \sum_{i=0}^{N} h^{2i}\left\{\int_a^b K(x,y)\phi_i(y)\,dy + \sum_{k=1}^{(N-i)} h^{2k}\alpha_{ik}(x) + O(h^{2(N-i)+1})\right\} .$$

If we write

$$\alpha_{i0}(x) = \int_a^b K(x,y)\phi_i(y)\,dy$$

then

$$\sum_{j=0}^{n} w_j K(x,y_j)\phi(y_j) = \left\{\sum_{i=0}^{N} h^{2i} \sum_{j=0}^{i} \alpha_{j,i-j}(x)\right\} + O(h^{2N+1}) . \tag{3.185}$$

Furthermore, from eqns (3.181) and (3.182) we find

$$\mu\phi(x) = \sum_{i=0}^{N} \mu_i h^{2i} \sum_{j=0}^{N} h^{2j}\phi_j(x)$$

$$= \sum_{i=0}^{N} h^{2i} \sum_{j=0}^{i} \mu_j \phi_{i-j}(x) + O(h^{2N+1}) . \tag{3.186}$$

The order terms in (3.185) and (3.186) are uniform in x.

Now we seek $\mu_1,\mu_2,\ldots,\mu_N$ and $\phi_1(x),\phi_2(x),\ldots,\phi_N(x)$ such that the leading terms of (3.185) and (3.186) agree. Equating powers of h, we require that

$$\sum_{j=0}^{k} \alpha_{j,k-j}(x) \equiv \sum_{j=0}^{k} \mu_j \phi_{k-j}(x) \tag{3.187}$$

for $k = 0,1,2,\ldots,N$. Thus for $k = 0$, $\mu_0 = \kappa$, $\phi_0(x) = f(x)$ we require

$$\alpha_{00}(x) \equiv \int_a^b K(x,y)f(y)\mathrm{d}y = \kappa f(x) \quad , \tag{3.188}$$

and this condition is automatically satisfied. With $k = 1$ we obtain

$$\int_a^b K(x,y)\phi_1(y)\mathrm{d}y + \alpha_{01}(x) = \mu_1 f(x) + \kappa\phi_1(x) \ ,$$

where (with $B_2 = \frac{1}{6}$),

$$\alpha_{01}(x) = \tfrac{1}{2}B_2\left[\partial/\partial y\,\{K(x,y)f(y)\}\right]_{y=a}^{y=b} \ .$$

We shall have to impose differentiability conditions on $K(x,y)$ which guarantee the existence of a smooth function $\alpha_{01}(x)$. We shall review sufficient conditions later (section 3.21.2) when the process for constructing $\phi(x)$ has become clear.

Now μ_1 and $\phi_1(x)$ must satisfy the equation

$$\int_a^b K(x,y)\phi_1(y)\mathrm{d}y - \kappa\phi_1(x) = \mu_1 f(x) - \alpha_{01}(x) \ . \tag{3.189}$$

But κ is an eigenvalue of $K(x,y)$, so that a solution $\phi_1(x)$ of (3.189) exists only if

$$\int_a^b \{\mu_1 f(x) - \alpha_{01}(x)\}\,\overline{\psi(x)}\mathrm{d}x = 0 \ , \tag{3.190}$$

where

$$\int_a^b K^*(x,y)\psi(y)\mathrm{d}y = \bar{\kappa}\psi(x)$$

(see the Fredholm alternative result, stated in section 1.4.2).

Eqn (3.190) determines μ_1. We obtain

$$\mu_1 = \frac{\tfrac{1}{2}B_2\displaystyle\int_a^b \overline{\psi(x)}\,\left[K(x,y)f'(y) + K_y(x,y)f(y)\right]_{y=a}^{y=b}\mathrm{d}x}{\displaystyle\int_a^b \overline{\psi(x)}\,f(x)\mathrm{d}x} \tag{3.191}$$

assuming sufficient differentiability.

With μ_1 determined by (3.191) there are many solutions $\phi_1(x)$ to (3.189). For definiteness we choose $\phi_1(x)$ to be orthogonal to $f(x)$. (This choice may be regarded, ultimately as defining a normalization of $\phi(x)$.)

Setting $k = 2$ in (3.187) we obtain the condition

$$\alpha_{02}(x) + \alpha_{11}(x) + \int_a^b K(x,y)\, \phi_2(y)dy$$
$$= \kappa\phi_2(x) + \mu_1\phi_1(x) + \mu_2 f(x);$$

μ_2 must be chosen so that

$$\int_a^b \overline{\psi(x)}\, \{\mu_1\phi_1(x) + \mu_2 f(x) - \alpha_{02}(x) - \alpha_{11}(x)\}dx = 0 .$$

Then $\phi_2(x)$ may be determined (in theory) if we require it to be orthogonal to $f(x)$.

The general conditions for determining $\mu_k, \phi_k(x)$ assume the form $(k = 1,2,3,\ldots,N)$

$$(\phi_k, f) = 0 ,$$
$$\int_a^b K(x,y)\phi_k(y)dy - \kappa\phi_k(x) = \chi_k(x) , \tag{3.192}$$
$$\int_a^b \overline{\psi(x)}\chi_k(x)dx = 0 , \tag{3.193}$$
$$\chi_k(x) = \sum_{j=1}^{k} \mu_j\phi_{k-j}(x) - \sum_{j=0}^{k-1} \alpha_{j,k-j}(x) . \tag{3.194}$$

Eqns (3.193) and (3.194) serve to define μ_k in terms of $\kappa,\mu_1,\mu_2,\ldots,\mu_{k-1}$ and $f(x),\phi_1(x),\ldots,\phi_{k-1}(x)$, and we may then determine $\phi_k(x)$ to satisfy (3.192).

3.21.2.

Now, in order to apply Theorem 3.26 with $M = N - k$ $(k = 0,1,2,\ldots,N)$, we require that $\phi_k(x) \in C^{2(N-k)+1}[a,b]$, and that $(\partial/\partial y)^{2(N-k)+1}K(x,y)$ should be continuous for $a \le x,y \le b$. For notational convenience we write

$$C_k[a,b] = C^{2(N-k)+1}[a,b] \; . \qquad (3.195)$$

From (3.192) we see that $\phi_k(x) \in C_k[a,b]$ if $(\partial/\partial x)^{2(N-k)+1}K(x,y)$ is continuous for $a \le x,y \le b$ and $\chi_k(x) \in C_k[a,b]$. This latter condition is satisfied if $f(x)$, $\phi_1(x)$, ..., $\phi_{k-1}(x) \in C_k[a,b]$ and

$$\alpha_{rs}(x) \in C_k[a,b] \; , \qquad (3.196)$$

where $r \ge 0$, $s \ge 1$ and $r + s = k$. In (3.196) we have (for $s = 1,2,3,\ldots,k$)

$$\alpha_{rs} = \frac{B_{2s}}{(2s)!} \left[\left(\frac{\partial}{\partial y}\right)^{2s-1} K(x,y)\phi_r(y)\right]_{y=a}^{y=b} \; .$$

This far we have gathered together the various differentiability requirements, and we now postulate the following result.

THEOREM 3.27. *For* $k = 0,1,2,\ldots,N,$ $\phi_k(x) \in C_k[a,b]$ *(see eqn* (3.195)*) if* $(\partial^{r+s}/\partial x^r \partial y^s)\, K(x,y)$ *is continuous*† *for* $r + s \le 2N + 1$ *and* $a \le x, y \le b$.

Proof. We shall refer to the remarks preceding the statement of this theorem. The proof will be by induction.

With $k = 0$, $\phi_0(x)$ is $f(x)$, and this eigenfunction is in $C_0[a,b] \equiv C^{2N+1}[a,b]$ when $(\partial/\partial x)^{2N+1}\, K(x,y)$ is continuous for $a \le x,y \le b$.

Now suppose that $k \le N$. We take our induction hypothesis to be that $\phi_{k-j}(x) \in C_{k-j}[a,b]$, for $j = 1,2,3,\ldots,k$. Thus $\alpha_{rs}(x)$ exists for $r = 0,1,2,\ldots,$ $s = 1,2,3,\ldots,$ and $r + s = k$. Furthermore $\alpha_{rs}(x) \in C_k[a,b]$ if

$$(\partial/\partial x)^{2(N-k)+1}(\partial/\partial y)^{2s-1}K(x,y)$$

is continuous for $a \le x,y \le b$ where $r + s = k$, and in particular if

---

† This condition is stronger than necessary.

$$(\partial^{2N}/\partial x^{2(N-k)+1}\partial y^{2k-1})K(x,y)$$

is continuous.

Thus by the remarks above, $\phi_k(x) \in C_k[a,b]$. This establishes the result by induction.

3.21.3.

The purpose of sections 3.21.1 and 3.21.2 has been to show that under the differentiability conditions of Theorem 3.27 we can find

$$\phi(x) = f(x) + \sum_{k=1}^{N} h^{2k}\phi_k(x)$$

and

$$\mu = \kappa + \sum_{k=1}^{N} h^{2k}\mu_k \quad ,$$

such that

$$\sum_{j=0}^{n} w_j K(x,y_j)\phi(y_j) - \mu\phi(x) = \varepsilon(x) \quad ,$$

where $\|\varepsilon(x)\|_\infty = O(h^{2N+1})$.

To complete the proof of Theorem 3.25 we now return to eqn (3.180) and use the function $\phi(x)$ and the value μ in that bound.

We have

$$|\tilde{\kappa} - \mu| \left| \sum_{j=0}^{n} w_j \overline{\tilde{\psi}(y_j)}\phi(y_j) \right| \leq \left\{ \sum_{j=0}^{n} w_j |\tilde{\psi}(y_j)|^2 \sum_{j=0}^{n} w_j |\varepsilon(y_j)|^2 \right\}^{\frac{1}{2}} \tag{3.197}$$

With our choice of $\phi(x)$, $\lim_{h\to 0} \|\phi(x) - f(x)\|_\infty = 0$ and by construction $\lim_{h\to 0} \|\tilde{\psi}(x) - \psi(x)\|_\infty = 0$, since $\tilde{\psi}(x)$ is suitably normalized. Thus

$$\sum_{j=0}^{n} w_j |\tilde{\psi}(y_j)|^2 = \sum_{j=0}^{n} w_j |\psi(y_j)|^2 + o(1) = \int_a^b |\psi(y)|^2 dy + o(1)$$

$$= 1 + o(1) \ .$$

Similarly

$$\sum_{j=0}^{n} w_j \overline{\tilde{\psi}(y_j)}\, \phi(y_j)$$

$$= \sum_{j=0}^{n} w_j\, \overline{\psi(y_j)} f(y_j) + o(1) = \int_a^b f(y)\, \overline{\psi(y)} dy + o(1) \quad .$$

By the construction of μ , and $\phi(x)$, $\|\varepsilon(x)\|_\infty = O(h^{2N+1})$ and so

$$\left\{ \sum_{j=0}^{n} w_j |\varepsilon(y_j)|^2 \right\}^{\frac{1}{2}} = O(h^{2N+1}) \; . \tag{3.198}$$

From (3.197) we then obtain

$$|\tilde{\kappa} - \mu| = O(h^{2N+1}) \tag{3.199}$$

as $h \to 0$.

Eqn (3.199) summarizes Theorem 3.25, for it states that

$$\tilde{\kappa} = \mu + O(h^{2N+1}) \; ,$$

that is,

$$\tilde{\kappa} = \kappa + \mu_1 h^2 + \ldots + \mu_N h^{2N} + O(h^{2N+1}) \; .$$

Our theorem is therefore established when $(\partial^{r+s}/\partial x^r \partial y^s)\, K(x,y)$ is assumed continuous for $a \le x,y \le b$, $r + s \le 2N + 1$.

Remarks.

(a) It is essential to note that $\mu_1, \mu_2, \ldots, \mu_N$ are independent of h .

(b) It is not in general true that the sum

$$\tilde{\kappa} - \sum_{i=0}^{\infty} \mu_i h^{2i}$$

exists and is equal to κ when μ_i exists for $i = 0,1,2,\ldots$.

3.21.4. *The corresponding eigenfunction*

Using the function $\phi(x)$ discussed above we can establish the existence of functions $\phi_i(x)$ such that, when $\overset{\circ}{f}(x)$ is a suitably scaled version of an approximate eigenfunction $\tilde{f}(x)$

$$\overset{\circ}{f}(x) = f(x) + \sum_{i=1}^{N} h^{2i}\phi_i(x) + O(h^{2N+1})$$

for certain values of x. (We suppose that $\kappa = \lim \tilde{\kappa}$ is a simple eigenvalue and non-zero.) We establish this result for values of x which are of the form $a + ih$ $(i = 0,1,2,\ldots,n\ ,\ h = (b-a)/n)$.

A satisfactory form of normalizing $\tilde{f}(x)$ may be described as follows. We suppose that $\bar{x}$ is a fixed value in $[a,b]$ such that $f(\bar{x}) \neq 0$, and we suppose that h tends to zero through values $h_0 > h_1 > h_2 > \ldots$ in such a way that $(\bar{x}-a)/h_k$ is an integer for each k. Then for each $h \in \{h_k\}$, $\overset{\circ}{\underset{\sim}{f}}$ may be normalized so that $\overset{\circ}{f}(\bar{x}) = 1$, provided h is sufficiently small. This is a satisfactory normalization, generally achievable in practice, with $\overset{\circ}{f}(x) = \tilde{f}(x)/\tilde{f}(\bar{x})$.

THEOREM 3.28. *Suppose that* $\overset{\circ}{f}(x)$ *is normalized as above and that the simple eigenvalue* κ *is non-zero. If* $(\partial^{r+s}/\partial x^r \partial y^s)\,K(x,y)$ *is continuous for* $a \leq x,y \leq b$ *and* $r + s \leq 2N + 1$ *then Theorem 3.25 applies and, further, there exist functions* $\hat{\phi}_i(x)$ $(i = 1,2,3,\ldots,N)$ *such that, with* $\hat{f}(x) = f(x)/f(\bar{x})$,

$$\overset{\circ}{f}(x) = \hat{f}(x) + \sum_{i=1}^{N} h^{2i}\hat{\phi}_i(x) + O(h^{2N+1})$$

whenever $x \in \{ih \mid i=0,1,\ldots,n\}$.

<u>Proof</u>. With the tools at present at our disposal, the proof is a little complicated. The part of the proof which is involved is placed in section 3.21.5 in order that the reader may perceive the outline of the proof.

Using μ and the function $\phi(x)$ constructed in the previous subsection,

$$\sum_{j=0}^{n} w_j K(x,y_j)\phi(y_j) - \mu\phi(x) = \varepsilon(x) \quad ,$$

where $\|\varepsilon(x)\|_\infty = O(h^{2N+1})$. Thus

$$\sum_{j=0}^{n} w_j K(x,y_j)\phi(y_j) - \tilde{\kappa}\phi(x) = \varepsilon_1(x) \quad ,$$

where $\varepsilon_1(x) = \varepsilon(x) - (\tilde{\kappa}-\mu)\phi(x)$. Given $h_0 > 0$ there is a constant m_0 such that $\|\phi(x)\|_\infty \le m_0$ for $h \le h_0$, and $\tilde{\kappa} = \mu + O(h^{2N+1})$, so $\|\varepsilon_1(x)\|_\infty = O(h^{2N+1})$; we can also suppose that h is sufficiently small that $\tilde{\kappa}$ is simple and non-zero. Now $\tilde{f}(x)$ satisfies the equation

$$\sum_{j=0}^{n} w_j K(x,y_j)\tilde{f}(y_j) = \tilde{\kappa}\tilde{f}(x) \quad ,$$

so that

$$\underset{\sim}{K}\underset{\sim}{D}\underset{\sim}{\delta} - \tilde{\kappa}\underset{\sim}{\delta} = \underset{\sim}{\varepsilon}_1 \quad , \tag{3.200}$$

where $\underset{\sim}{\varepsilon}_1 = [\varepsilon_1(y_0), \varepsilon_1(y_1), \ldots, \varepsilon_1(y_n)]^T$ and $\underset{\sim}{\delta} = \underset{\sim}{\phi} - \underset{\sim}{\tilde{f}}$. The vector $\underset{\sim}{\delta}$ is not completely defined by (3.200) since if $\underset{\sim}{\delta}_0$ is one solution, another is $\underset{\sim}{\delta}_0 + \alpha\underset{\sim}{\tilde{f}}$. We shall study a particular solution.

Suppose that

$$\underset{\sim}{K}^*\underset{\sim}{D}\underset{\sim}{\tilde{\psi}} = \bar{\tilde{\kappa}}\underset{\sim}{\tilde{\psi}} \tag{3.201a}$$

and

$$\underset{\sim}{K}\underset{\sim}{D}\underset{\sim}{\tilde{f}} = \tilde{\kappa}\underset{\sim}{\tilde{f}} \tag{3.201b}$$

(where $\bar{\kappa} = \bar{\tilde{\kappa}}$) . Then the solution $\underset{\sim}{\delta}_0$ of eqn (3.200) which satisfies $\underset{\sim}{\delta}_0^*\underset{\sim}{D}\underset{\sim}{\tilde{\psi}} = 0$ is obtained (see section 3.21.5) by solving

$$\{\underset{\sim}{K} - \tilde{\kappa}(\underset{\sim}{\tilde{f}}\underset{\sim}{\tilde{\psi}}^*/\underset{\sim}{\tilde{f}}^T\underset{\sim}{D}\overline{\underset{\sim}{\tilde{\psi}}})\}\,\underset{\sim}{D}\underset{\sim}{\delta}_0 - \tilde{\kappa}\underset{\sim}{\delta}_0 = \underset{\sim}{\varepsilon}_1 \ . \tag{3.202}$$

If we can show that

$$\| [\{\underset{\sim}{K} - \tilde{\kappa}(\underset{\sim}{\tilde{f}}\underset{\sim}{\tilde{\psi}}^*/\underset{\sim}{\tilde{f}}^T \underset{\sim}{D}\underset{\sim}{\bar{\tilde{\psi}}})\} \underset{\sim}{D} - \tilde{\kappa}\underset{\sim}{I}]^{-1} \|_\infty \tag{3.203}$$

is $O(1)$ then it follows that $||\delta_0||_\infty = O(h^{2N+1})$.

Now eqn (3.202) resembles an attempt to solve an equation

$$\int_a^b K^{\#}(x,y)\delta_0(y)dy - \kappa\delta_0(x) = \varepsilon_1(x) \quad ,$$

where $K^{\#}(x,y) = K(x,y) - \kappa\{f(x)\ \overline{\psi(y)}\ /\ \int_a^b f(x)\ \overline{\psi(x)}dx\}$.

This similarity can be used (see section 3.21.5) to relate $|| (K^{\#} - \kappa I)^{-1} ||_\infty$ and (3.203), and this relation permits us to show that the expression (3.203) is $O(1)$ as $h \to 0$. Since $\|\underset{\sim}{\varepsilon}_1||_\infty = O(h^{2N+1})$ it follows that $\|\underset{\sim}{\delta}_0||_\infty = O(h^{2N+1})$ and hence $\underset{\sim}{\delta} \equiv \underset{\sim}{\phi} - \underset{\sim}{\tilde{f}} = \alpha\ \underset{\sim}{\tilde{f}} + O(h^{2N+1})$ for some α . If $\underset{\sim}{\tilde{f}}$ is suitably normalized,

$$\underset{\sim}{\tilde{f}} = \underset{\sim}{\phi} + O(h^{2N+1})$$

$$= \underset{\sim}{f} + h^2\underset{\sim}{\phi}_1 + \ldots + h^{2N}\underset{\sim}{\phi}_N + O(h^{2N+1}) \ , \tag{3.204}$$

where $\underset{\sim}{\phi}_i = [\phi_i(y_0),\phi_i(y_1),\ \ldots,\ \phi_i(y_n)]^T$ and $\phi_i(x)$ is independent of h .

Now, from (3.204), if $\bar{x}\ \varepsilon\{a+jh\}$ we have the relation $\tilde{f}(\bar{x}) = f(\bar{x}) + h^2\phi_1(\bar{x}) + \ldots + h^{2N}\phi_N(\bar{x}) + O(h^{2N+1})$, and since $f(\bar{x}) \neq 0$, $\tilde{f}(\bar{x}) \neq 0$ if h is sufficiently small. The values of the function $\overset{\circ}{f}(x)$ $(x\ \varepsilon\ \{a+jh\})$, being normalized according to the conditions of the theorem, are obtained by forming

$$\overset{\circ}{\underset{\sim}{f}} = \underset{\sim}{\tilde{f}}/\tilde{f}(\bar{x}) \tag{3.205}$$

$$= \{\underset{\sim}{f} + \sum_{k=1}^{N} h^{2k}\underset{\sim}{\phi}_k + O(h^{2N+1})\}/\{f(\bar{x}) + \sum_{k=1}^{N} h^{2k}\phi_k(\bar{x}) + O(h^{2N+1})\} \ .$$

Thus, with an appropriately scaled eigenfunction $\hat{f}(x)$ satisfying $\hat{f}(\bar{x}) = 1$, we can derive functions $\{\hat{\phi}_k(x)\}$ from $\{\phi_k(x)\}$ so that

$$\overset{\circ}{\underset{\sim}{f}} = \hat{\underset{\sim}{f}} + \sum_{k=1}^{N} h^{2k}\hat{\underset{\sim}{\phi}}_k + O(h^{2N+1}) \quad ,$$

where $\hat{\underset{\sim}{\phi}}_k = [\hat{\phi}_k(y_0), \hat{\phi}_k(y_1), \ldots, \hat{\phi}_k(y_n)]^T$ and $\hat{\phi}_k(x)$ is independent of h $(k = 1,2,3,\ldots,N)$. This is the result of the theorem.

3.21.5. *Appendix*

We shall treat this section as an appendix to section 3.21.4 and study, further, a few details in the proof of Theorem 3.28.

(a) Let us first consider the vector $\underset{\sim}{\delta}_0$ defined by (3.202) and show that it is one solution of (3.200). From (3.202),

$$(\underset{\sim}{K}\underset{\sim}{D} - \tilde{\kappa}\underset{\sim}{I})\underset{\sim}{\delta}_0 = \underset{\sim}{\varepsilon}_1 + \beta(\tilde{\underset{\sim}{\psi}}^*\underset{\sim}{D}\underset{\sim}{\delta}_0)\tilde{\underset{\sim}{f}} \quad ,$$

where $\beta = \tilde{\kappa}/(\tilde{\underset{\sim}{f}}^T\underset{\sim}{D}\bar{\tilde{\underset{\sim}{\psi}}})$. (Since $\tilde{\kappa}$ is supposed non-zero and simple, $\tilde{\underset{\sim}{f}}^T\underset{\sim}{D}\bar{\tilde{\underset{\sim}{\psi}}} \neq 0$, and $0 < |\beta| < \infty$.) We shall show that $\tilde{\underset{\sim}{\psi}}^*\underset{\sim}{D}\underset{\sim}{\delta}_0 = 0$. For

$$\{(\tilde{\underset{\sim}{\psi}}^*\underset{\sim}{D}\underset{\sim}{K}) - \tilde{\kappa}\tilde{\underset{\sim}{\psi}}^*\}\underset{\sim}{D}\underset{\sim}{\delta}_0 = \tilde{\underset{\sim}{\psi}}^*\underset{\sim}{D}\underset{\sim}{\varepsilon}_1 + \beta(\tilde{\underset{\sim}{\psi}}^*\underset{\sim}{D}\tilde{\underset{\sim}{f}})(\tilde{\underset{\sim}{\psi}}^*\underset{\sim}{D}\underset{\sim}{\delta}_0) \; .$$

Since $\tilde{\underset{\sim}{\psi}}^*\underset{\sim}{D}\underset{\sim}{K} = \tilde{\kappa}\tilde{\underset{\sim}{\psi}}^*$,

$$\beta(\tilde{\underset{\sim}{\psi}}^*\underset{\sim}{D}\tilde{\underset{\sim}{f}})(\tilde{\underset{\sim}{\psi}}^*\underset{\sim}{D}\underset{\sim}{\delta}_0) = -\tilde{\underset{\sim}{\psi}}^*\underset{\sim}{D}\underset{\sim}{\varepsilon}_1 \; .$$

Now, in order that eqn (3.200) should possess a solution, it is necessary that $\underset{\sim}{\phi}^*\underset{\sim}{\varepsilon}_1 = 0$ for every vector $\underset{\sim}{\phi}$ such that $\underset{\sim}{\phi}^*(\underset{\sim}{K}\underset{\sim}{D} - \tilde{\kappa}\underset{\sim}{I}) = \underset{\sim}{0}^T$. Such a vector $\underset{\sim}{\phi}$ satisfies $\underset{\sim}{D}\underset{\sim}{K}^*\underset{\sim}{\phi} = \bar{\tilde{\kappa}}\underset{\sim}{\phi}$; but $\underset{\sim}{K}^*\underset{\sim}{D}\tilde{\underset{\sim}{\psi}} = \bar{\tilde{\kappa}}\tilde{\underset{\sim}{\psi}}$, so $\underset{\sim}{\phi}$ must have the form $\gamma\underset{\sim}{D}\tilde{\underset{\sim}{\psi}}$ and the condition reduces to

$$\tilde{\underset{\sim}{\psi}}^*\underset{\sim}{D}\underset{\sim}{\varepsilon}_1 = 0 \; .$$

Since $\beta(\tilde{\underset{\sim}{\psi}}^*\underset{\sim}{D}\tilde{\underset{\sim}{f}})(\tilde{\underset{\sim}{\psi}}^*\underset{\sim}{D}\underset{\sim}{\delta}_0) = -\tilde{\underset{\sim}{\psi}}^*\underset{\sim}{D}\underset{\sim}{\varepsilon}_1$ it follows that $\tilde{\underset{\sim}{\psi}}^*\underset{\sim}{D}\underset{\sim}{\delta}_0 = 0$ and hence that $\underset{\sim}{\delta}_0$ is a solution of (3.200). The general solution is thus $\alpha\tilde{\underset{\sim}{f}} + \underset{\sim}{\delta}_0$.

(b) We now look more closely at the expression (3.203) and indicate how we may prove that it is $O(1)$.

Now the vectors $\tilde{\underset{\sim}{\psi}}, \tilde{\underset{\sim}{f}}$ in (3.201) may be extended to piecewise-constant functions $\tilde{\psi}(x)$ and $\tilde{f}(x)$ such that

$$\int_a^b K_n(x,y)\,\tilde{f}(y)\,dy = \tilde{\kappa}\tilde{f}(x)$$

and

$$\int_a^b K_n^*(x,y)\tilde{\psi}(y)\,dy = \bar{\tilde{\kappa}}\tilde{\psi}(x) \quad ,$$

where $K_n(x,y)$ is defined in (3.70), and $K_n^*(x,y) = \overline{K_n(y,x)}$. If $\tilde{\psi}(x)$ and $\tilde{f}(x)$ are suitably scaled $\lim_{n\to\infty} \|\psi(x) - \tilde{\psi}(x)\|_\infty$ $= \lim_{n\to\infty} \|f(x) - \tilde{f}(x)\|_\infty = 0$, where

$$\int_a^b K(x,y)f(y)\,dy = \kappa f(x) \quad ,$$

$$\int_a^b K^*(x,y)\psi(y)\,dy = \bar{\kappa}\psi(x) \quad ,$$

and $\lim_{n\to\infty} \tilde{\kappa} = \kappa$. Consequently

$$\tilde{\mathbf{f}}^T \mathbf{D} \bar{\tilde{\boldsymbol{\psi}}} = \sum_{j=0}^{n} w_j \tilde{f}(y_j)\overline{\tilde{\psi}(y_j)} = \int_a^b f(x)\,\overline{\psi(x)}\,dx + o(1) \quad ,$$

and $\tilde{\kappa} = \kappa + o(1)$.

We define $K^{\#}(x,y)$ as above, and

$$\hat{K}_n^{\#}(x,y) = K_n(x,y) - \tilde{\kappa}\tilde{f}(x)\overline{\tilde{\psi}(y)}/\{\int_a^b f(x)\overline{\tilde{\psi}(x)}\,dx\} \quad ;$$

then

$$\lim_{n\to\infty} \|\hat{K}_n^{\#}(x,y) - K^{\#}(x,y)\|_\infty = 0 \quad .$$

Now we regard $K^{\#} - \kappa I$ and $\hat{K}_n^{\#} - \tilde{\kappa} I$ as operators on $R[a,b]$, the space of bounded Riemann-integrable functions with norm

$$\|\phi(x)\|_\infty = \sup_{a\le x\le b} |\phi(x)| \quad .$$

Because of the limiting behaviour of $\tilde{\mathbf{f}}^T \mathbf{D} \bar{\tilde{\boldsymbol{\psi}}}$ and of $\tilde{\kappa}$,

$$\lim_{n\to\infty} \|(K^{\#} - \kappa I) - (\hat{K}_n^{\#} - \tilde{\kappa} I)\|_\infty = 0 .$$

Using this relation we can show that, for all n sufficiently large,

$$\|(\hat{K}_n^{\#} - \tilde{\kappa} I)^{-1}\|_\infty \le 2\,\|(K^{\#} - \kappa I)^{-1}\|_\infty . \qquad (3.206)$$

The proof follows the lines of the proof of (3.101), the replacement of $\|\ \|_2$ in (3.101) by $\|\ \|_\infty$ causing no problem. It should be noted, however, that the difference

$$E_n = \hat{K}_n^{\#} - K^{\#} - (\tilde{\kappa}-\kappa)I$$

is not an integral operator, and is not compact, and the same is true of

$$F_n = (\kappa I - K^{\#})^{-1} E_n .$$

However, $R[a,b]$ is a complete space so the method of proof goes through (see remark (b) at the end of section 3.15.8, and Theorem 1.1).

Now when $g(x)$ is a piecewise-constant extension of a vector $[g(y_0), g(y_1), \ldots, g(y_n)]^T$ (of the same type as $\tilde{f}(x)$ and $\tilde{\psi}(x)$), the equation

$$(\hat{K}_n^{\#} - \tilde{\kappa} I)\gamma(x) = g(x)$$

can be solved by finding $\underset{\sim}{\gamma}$ such that

$$(\underset{\sim}{\hat{K}}{}_n^{\#} \underset{\sim}{D} - \tilde{\kappa} \underset{\sim}{I})\underset{\sim}{\gamma} = \underset{\sim}{g}$$

and extending $\underset{\sim}{\gamma}$ to a piecewise-constant function. Thus

$$\|(\underset{\sim}{\hat{K}}{}_n^{\#} \underset{\sim}{D} - \tilde{\kappa} \underset{\sim}{I})^{-1}\|_\infty = \sup_{\|g(x)\|_\infty = 1} \|\gamma(x)\|_\infty \le \|(\hat{K}_n^{\#} - \tilde{\kappa} I)^{-1}\|_\infty ,$$

where the last norm is an operator norm on $R[a,b]$ and is bounded by $2\,\|(K^{\#} - \kappa I)^{-1}\|_\infty$. Thus the norm (3.203) is $O(1)$.

3.22. *Extensions*

Suppose that the eigenvalues $\tilde{\kappa}$ $(\equiv \tilde{\kappa}(1/m))$ and eigenfunctions $\tilde{f}(x)$ are obtained using a quadrature rule $(m\times J)(\phi)$, where $J(\phi) \in J^{+}$ is a quadrature rule with a degree of precision $r \geq 0$. Then the analysis of section 3.21 may be modified to show that

$$\tilde{\kappa} = \kappa + \sum_{k=r+1}^{2N} v_k (1/m)^k + O(1/m^{2N+1}) \tag{3.207}$$

with a similar relation for the eigenfunctions. When $J(\phi)$ is a symmetric rule, v_k vanishes if k is odd, so that

$$\tilde{\kappa} = \kappa + \sum_{k=\frac{1}{2}(r+1)}^{N} \nu_k (1/m)^{2k} + O(1/m^{2N+1}) . \tag{3.208}$$

These relationships are valid when κ is simple, and when $(\partial^{r+s}/\partial x^r\, \partial y^s)\, K(x,y)$ is continuous for $a \leq x,y \leq b$.

In the case of the trapezium rule with step h the differentiability conditions which we imposed in section 3.21 are more stringent than necessary, since Theorem 3.25 and Theorem 3.28 remain valid when $K(x,y)$ is a certain type of Green's function. Indeed Theorem 3.28 is true if $K(x,y) = K_1(x,y)$ for $a \leq x \leq y \leq b$, $K_2(x,y)$ for $a \leq y < x \leq b$, where $K_1(x,y)$ and $K_2(x,y)$ have continuous mixed derivatives of order $2N+1$ on $a \leq x \leq y \leq b$, $a \leq y \leq x \leq b$ respectively. Such conditions, and a modification to Theorem 3.25 which is necessary when κ is a multiple eigenvalue, were discussed by Baker (1971), using rather more advanced tools than have been used in the proofs above.

3.22.1.

Perhaps I should now indicate, briefly, the relationship of the preceding theory with the method of deferred correction indicated in section 3.2.5, for use with the *trapezium rule*. We assume sufficient differentiability for the validity of the theory of section 3.21, to obtain an asymptotic analysis.

According to eqn (3.164) we have

$$\kappa = \tilde{\kappa} + (\eta,\psi)/(\tilde{f},\psi) ,$$

where $K^*\psi(x) = \bar{\kappa}\psi(x)$, and $\tilde{f}(x) = f(x) + O(h^2)$ whilst

$$\eta(x) = \int_a^b K(x,y)\tilde{f}(y)dy - \sum_{j=0}^{n} w_j K(x,y_j)\tilde{f}(y_j) \quad .$$

Thus $\|\underset{\sim}{\eta}\| = O(h^2)$ and (given sufficient differentiability) $\|\underset{\sim}{\eta}'\|$, $\|\eta'(x)\|$, $\|\eta''(x)\|$, and $\|\underset{\sim}{\eta}''\|$ are $O(h^2)$, and if $\tilde{\underset{\sim}{\psi}}$ is suitably normalized $\|\tilde{\underset{\sim}{\psi}} - \underset{\sim}{\psi}\| = O(h^2)$ and $\|\tilde{\underset{\sim}{\psi}}^{(r)} - \underset{\sim}{\psi}^{(r)}\| = O(h^2)$, etc.

We can now write

$$(\eta,\psi) = \tilde{\underset{\sim}{\psi}}^*\underset{\sim}{D}\underset{\sim}{\eta} + O(h^4)$$

and

$$(\tilde{f},\psi) = \tilde{\underset{\sim}{\psi}}^*\underset{\sim}{D}\tilde{\underset{\sim}{f}} + O(h^2) \quad .$$

We shall show how we obtain the first of these results, and the technique for proving the second is similar. We have

$$|(\eta,\psi) - (\eta,\tilde{\psi})| \leq \|\eta(x)\|_2 \, \|\psi(x)-\tilde{\psi}(x)\|_2 = O(h^4)$$

and

$$|(\eta,\tilde{\psi}) - \tilde{\underset{\sim}{\psi}}^*\underset{\sim}{D}\underset{\sim}{\eta}| \leq h^2 \, \| \, (d/dx)^2 \, \overline{\eta(x)\tilde{\psi}(x)}\|_\infty/12 \quad .$$

Since $\|\eta^{(r)}(x)\|_\infty = O(h^2)$ $(r = 1,2)$ and $\|\tilde{\psi}^{(r)}(x)\|_\infty = O(1)$ the result follows.

If we produce an approximation $\underset{\sim}{\delta}$ to $\underset{\sim}{\eta}$ using Gregory correction terms so that $\|\underset{\sim}{\delta} - \underset{\sim}{\eta}\|_\infty = O(h^4)$ then $(\eta,\psi) = \tilde{\underset{\sim}{\psi}}^*\underset{\sim}{D}\underset{\sim}{\delta} + O(h^4)$. If we collect the order terms we now find

$$\kappa = \tilde{\kappa} + (\tilde{\underset{\sim}{\psi}}^*\underset{\sim}{D}\underset{\sim}{\delta})/(\tilde{\underset{\sim}{\psi}}^*\underset{\sim}{D}\tilde{\underset{\sim}{f}}) + O(h^4) \quad ,$$

and the first two terms on the right give a 'corrected value' of $\tilde{\kappa}$.

3.23. *Expansion methods*

We now proceed to analyse the approximate eigenvalues and eigenfunctions obtained from expansion methods. Many of the techniques used to analyse convergence in the quadrature method will be applied to study the convergence in the expansion methods.

However, the Rayleigh-Ritz method for a Hermitian kernel is of special interest because we can guarantee one-sided bounds on the true eigenvalues. To establish this we have to employ a rather different approach to the ones adopted earlier.

The Rayleigh-Ritz method is a special case of Galerkin's method, although Galerkin's method can be applied to non-Hermitian kernels and the motivation for the one method differs from that of the other. Indeed, to an applied mathematician, the motivation for the Rayleigh-Ritz method may be the more appealing; it is of interest to us since it leads us to a proof that we obtain one-sided bounds for the true eigenvalues.

Both Galerkin's method and the method of collocation can be studied in the context of 'projection methods'. Though the resulting mathematics is both elegant and general, we shall not devote much space to this approach in the present chapter.

3.24. *The Rayleigh-Ritz method for Hermitian kernels*

In this section we suppose that $K(x,y)$ is Hermitian, $(K(x,y) = \overline{K(y,x)}\ ,\ a \le x,y \le b)$ and that its real eigenvalues are

$$\kappa_1^- \le \kappa_2^- \le \ldots < 0 \le \ldots \le \kappa_2^+ \le \kappa_1^+$$

(assuming the kernel has both positive and negative eigenvalues).

If κ_r^+ exists, it may be characterized by the relation

$$\kappa_r^+ \equiv \kappa_r^+(K) = \sup_{S_r} \inf_{\substack{\phi(x)\varepsilon S_r \\ \|\phi(x)\|_2=1}} \int_a^b\int_a^b K(x,y)\overline{\phi(x)}\,\phi(y)dxdy$$

$$= \sup_{S_r} \inf_{\substack{\phi(x)\varepsilon S_r \\ \|\phi(x)\|_2=1}} (K\phi,\phi) \quad , \tag{3.209}$$

where S_r denotes some linear space, of square-integrable functions, with dimension r . (This characterization differs from the one used in the proof of Theorem 3.3 in section 3.15. Cochran (1972) and Schlesinger (1957) show the connection between the two characterizations.)

Remark. Cochran (1971) has suggested that (3.209) should be linked with the name of Poincaré, who obtained an analogous result for certain differential operators, and with the name of Fisher, who obtained the result for self-adjoint matrices.

If we replace $K(x,y)$ by $-K(x,y)$ we obtain, from (3.209),

$$\kappa_r^- \equiv \kappa_r^-(K) = \inf_{S_r} \sup_{\substack{\phi(x)\varepsilon S_r \\ \|\phi(x)\|_2=1}} (K\phi,\phi) \quad . \tag{3.210}$$

Note that, since $K = K^*$, $(K\phi,\phi) = (\phi,K\phi)$.

As we indicated above, (3.209) and (3.210) provide a natural extension to a similar characterization of the eigenvalues of a matrix $\underset{\sim}{A} = \underset{\sim}{A}^*$ which is of order N , say. If we suppose that the eigenvalues of $\underset{\sim}{A}$ are $\mu_1 \geq \mu_2 \geq \ldots \geq \mu_N$ (repeated according to their multiplicity) then

$$\mu_r \equiv \mu_r(\underset{\sim}{A}) = \sup_{V_r \subseteq \mathbb{C}^N} \inf_{\substack{x\varepsilon V_r \\ \underset{\sim}{x}^*\underset{\sim}{x}=1}} \underset{\sim}{x}^*\underset{\sim}{A}\underset{\sim}{x} \tag{3.211}$$

where V_r denotes an r-dimensional subspace of the space $\mathbb{C}^N$ of complex N-vectors (Householder 1964, p.76; Halmos 1958, p.181).

A direct extension of (3.211) to integral operators is not satisfactory because there may be a countable set of eigenvalues $\{\kappa_r^+\}$ clustered at zero (and the indexing would not then enable us to reach the negative eigenvalues). However, (3.211) extends naturally to give (3.209) and the analogue of (3.210) is easily obtained. The matrix relations are best expressed, for our purposes, by supposing that the eigenvalues of $\underset{\sim}{A} = \underset{\sim}{A}^*$ are $\mu_1^+ \geq \mu_2^+ \geq \ldots \geq \mu_M^+ \geq 0$, $\mu_1^- \leq \mu_2^- \leq \ldots \leq \mu_m^- < 0$, where $M + m = N$. Then

$$\mu_r^+ = \mu_r^+(\underset{\sim}{A}) = \sup_{V_r} \inf_{\substack{\underset{\sim}{x} \in V_r \\ \underset{\sim}{x}^*\underset{\sim}{x}=1}} \underset{\sim}{x}^*\underset{\sim}{A}\underset{\sim}{x} , \tag{3.212}$$

$$\mu_r^- = \mu_r^-(\underset{\sim}{A}) = \inf_{V_r} \sup_{\substack{\underset{\sim}{x} \in V_r \\ \underset{\sim}{x}^*\underset{\sim}{x}=1}} \underset{\sim}{x}^*\underset{\sim}{A}\underset{\sim}{x} . \tag{3.213}$$

3.24.1.

Suppose that we are given (square-integrable) functions $\phi_0(x)$, $\phi_1(x)$, $\phi_2(x)$, ... such that

$$(\phi_i,\phi_j) \equiv \int_a^b \phi_i(x)\overline{\phi_j(x)}dx = \delta_{ij} \quad \text{(the Kronecker delta).}$$

The functions $\phi_0(x)$, ..., $\phi_n(x)$ form a basis for a linear space S_{n+1} comprising functions of the form

$$\phi(x) = \sum_{i=0}^{n} \alpha_i \phi_i(x) ,$$

where

$$||\phi(x)||_2 = \sum_{i=0}^{n} |\alpha_i|^2 = \underset{\sim}{\alpha}^*\underset{\sim}{\alpha}$$

with $\underset{\sim}{\alpha} = [\alpha_0,\alpha_1,\ldots,\alpha_n]^T$. The correspondence $\phi(x) \leftrightarrow \underset{\sim}{\alpha}$ establishes an isomorphism between S_{n+1} and $\mathbb{C}^{n+1}$, the space of complex $(n+1)$-vectors.

Suppose that S_r is an r-dimensional subspace of S_{n+1}. In view of (3.209),

$$\kappa_r^+(K) \geq \sup_{S_r \subseteq S_{n+1}} \inf_{\substack{\phi(x) \varepsilon S_r \\ \|\phi(x)\|_2 = 1}} (K\phi, \phi) . \tag{3.214}$$

In this expression, $(K\phi,\phi)$ has the form $(\sum_j \alpha_j K\phi_j, \sum_i \alpha_i\phi_i) =$

$= \sum_i \sum_j \bar{\alpha}_i \alpha_j (K\phi_j, \phi_i)$ (since $\phi(x) \varepsilon S_r \subseteq S_{n+1}$), which is of the form

$\underset{\sim}{\alpha}^* \underset{\sim}{A} \underset{\sim}{\alpha}$, where $\underset{\sim}{A} = \underset{\sim}{A}^* = [(K\phi_j, \phi_i)] = [(\phi_j, K\phi_i)]$.

The isomorphism $\phi(x) \leftrightarrow \underset{\sim}{\alpha} = [\alpha_0, \alpha_1, \ldots, \alpha_n]^T$, where $\alpha_i = (\phi, \phi_i)$ maps each subspace S_r into an r-dimensional subspace V_r of $\mathbb{C}^{n+1}$, and as noted above, $\|\phi(x)\|_2 = \underset{\sim}{\alpha}^* \underset{\sim}{\alpha}$. Consequently, (3.214) may be rewritten

$$\kappa_r^+(K) \geq \sup_{V_r \subseteq \mathbb{C}^{n+1}} \inf_{\substack{\underset{\sim}{\alpha} \varepsilon V_r \\ \underset{\sim}{\alpha}^* \underset{\sim}{\alpha} = 1}} \underset{\sim}{\alpha}^* \underset{\sim}{A} \underset{\sim}{\alpha} .$$

From (3.212) it follows that

$$\kappa_r^+(K) \geq \mu_r^+(\underset{\sim}{A}) . \tag{3.215}$$

Now, $\underset{\sim}{A} = [(K\phi_j, \phi_i)]$ is the matrix constructed in the Rayleigh-Ritz method, so that $\mu_r^+(\underset{\sim}{A}) = \tilde{\kappa}_r^+$, and (3.215) gives the result

$$\kappa_r^+ \geq \tilde{\kappa}_r^+ ,$$

provided κ_r^+, $\tilde{\kappa}_r^+$ both exist.

An exactly similar analysis can be performed for $\kappa_r^-(K)$, using (3.210), to obtain

$$\kappa_r^-(K) \leq \inf_{S_r \subseteq S_{n+1}} \sup_{\substack{\phi(x) \varepsilon S_r \\ \|\phi(x)\|_2 = 1}} (K\phi, \phi)$$

$$= \inf_{V_r \subseteq \mathbb{C}^{n+1}} \sup_{\substack{\alpha \in V_r \\ \alpha^*\alpha=1}} \alpha^* A \alpha ,$$

where V_r is isomorphic to S_r, whence

$$\kappa_r^- \equiv \kappa_r^-(K) \le \mu_r^-(A) = \tilde{\kappa}_r^- , \tag{3.216}$$

provided κ_r^- and $\tilde{\kappa}_r^-$ exist.

We have thus established the following result.

THEOREM 3.29. *Suppose that* $K(x,y)$ *is Hermitian and square-integrable. If* $(\phi_i, \phi_j) = \delta_{i,j}$ $(i=0,1,2,\ldots,n)$ *and* $A_{i,j} = [(K\phi_j, \phi_i)]$ *is the matrix used in the Rayleigh-Ritz method, then* $|\kappa_r^{\pm}| \ge |\tilde{\kappa}_r^{\pm}|$ *if* $\tilde{\kappa}_r^{\pm}, \kappa_r^{\pm}$ *both exist.*

3.24.2.

In the analysis of section 3.24.1 we assumed, for simplicity, that the functions $\phi_i(x)$ were orthonormal. In practice this may not be the case, and the Rayleigh-Ritz method then entails the solution of the eigenvalue problem

$$(A - \tilde{\kappa}B)\alpha = 0 ,$$

where $A_{i,j} = (K\phi_j, \phi_i)$, $B_{i,j} = (\phi_j, \phi_i)$, and $\phi_0(x), \phi_1(x), \ldots, \phi_n(x)$ are linearly independent. We shall show that the proof of Theorem 3.29 may be modified to cover this case. We shall consider only the positive eigenvalues since the modifications for negative eigenvalues are clear.

We require a slightly modified form of (3.209), namely,

$$\kappa_r^+(K) = \sup_{S_r} \inf_{\substack{\phi(x) \in S_r \\ \|\phi(x)\|_2 \ne 0}} \frac{(K\phi, \phi)}{(\phi, \phi)} . \tag{3.217}$$

The equivalence of (3.209) and (3.217) is easily established by setting $\psi(x) = \phi(x)/\|\phi(x)\|_2$ in (3.217).

Now suppose $\phi(x) = \sum_{i=0}^{n} \alpha_i \phi_i(x)$, where $\phi_0(x), \phi_1(x), \ldots, \phi_n(x)$ are linearly independent. Then

$$(\phi,\phi) = \sum_{i=0}^{n} \sum_{j=0}^{n} \alpha_i \bar{\alpha}_j (\phi_i, \phi_j) = \underset{\sim}{\alpha}^* \underset{\sim}{B} \underset{\sim}{\alpha} ,$$

where $B_{ij} = (\phi_j, \phi_i)$ and $\underset{\sim}{B} = \underset{\sim}{B}^*$ is positive definite (Davis 1965, pp. 176 *et seq*). When S_{n+1} is defined as the space spanned by $\phi_0(x), \phi_1(x), \ldots, \phi_n(x)$,

$$\kappa_r^+(K) \geq \sup_{S_r \subseteq S_{n+1}} \inf_{\substack{\phi(x) \in S_r \\ \|\phi(x)\|_2 \neq 0}} \frac{(K\phi,\phi)}{(\phi,\phi)}$$

$$= \sup_{V_r \subseteq \mathbb{C}^{n+1}} \inf_{\substack{\underset{\sim}{\alpha} \in V_r \\ \underset{\sim}{\alpha}^* \underset{\sim}{B} \underset{\sim}{\alpha} \neq 0}} \frac{\underset{\sim}{\alpha}^* \underset{\sim}{A} \underset{\sim}{\alpha}}{\underset{\sim}{\alpha}^* \underset{\sim}{B} \underset{\sim}{\alpha}}$$

$$= \sup_{V_r \subseteq \mathbb{C}^{n+1}} \inf_{\substack{\underset{\sim}{\alpha} \in V_r \\ \underset{\sim}{\alpha} \neq \underset{\sim}{0}}} \frac{\underset{\sim}{\alpha}^* \underset{\sim}{A} \underset{\sim}{\alpha}}{\underset{\sim}{\alpha}^* \underset{\sim}{B} \underset{\sim}{\alpha}} , \qquad (3.218)$$

where V_r is an r-dimensional subspace of $\mathbb{C}^{n+1}$. It is not surprising to find that when $\underset{\sim}{A} = \underset{\sim}{A}^*$, and $\underset{\sim}{B} = \underset{\sim}{B}^*$ is positive definite, (3.218) characterizes the rth positive eigenvalue $\tilde{\kappa}_r^+$ satisfying

$$(\underset{\sim}{A} - \tilde{\kappa}\underset{\sim}{B})\tilde{\underset{\sim}{a}} = \underset{\sim}{0} , \qquad (3.219)$$

for some $\tilde{\underset{\sim}{a}} \neq \underset{\sim}{0}$. We have thus obtained the necessary extension of Theorem 3.29.

THEOREM 3.30. *Suppose that* $K(x,y)$ $(a \leq x,y \leq b)$ *is square-integrable and* $\phi_0(x), \phi_1(x), \ldots, \phi_n(x)$ *are square-integrable on* $[a,b]$. *If* $\tilde{\kappa}_r^+$ $(\tilde{\kappa}_r^-)$ *is the* r*th positive (negative) eigenvalue satisfying* $(\underset{\sim}{A} - \tilde{\kappa}\underset{\sim}{B})\tilde{\underset{\sim}{a}} = 0$, *then* $\kappa_r^+ \geq \tilde{\kappa}_r^+$ $(\kappa_r^- \leq \tilde{\kappa}_r^-)$ *provided* $\kappa_r^+, \tilde{\kappa}_r^+$ $(\kappa_r^-, \tilde{\kappa}_r^-)$ *both exist.*

3.25. *Error bounds for the Rayleigh-Ritz method for Hermitian kernels*

The foundations for an error analysis and a convergence proof for the Rayleigh-Ritz method have already been laid, to a certain extent, in our discussion of the quadrature method.

We shall suppose that the functions $\phi_0(x), \phi_1(x), \ldots, \phi_n(x)$ employed in the Rayleigh-Ritz method are orthonormal, so that $(\phi_i, \phi_j) = \delta_{ij}$. Then the non-zero eigenvalues $\{\tilde{\kappa}_r^{\pm}\}$ of the matrix $\underset{\sim}{A} = [(K\phi_j, \phi_i)]$ are precisely the non-zero eigenvalues of the Hermitian degenerate kernel

$$\tilde{K}_{(n)}(x,y) = \sum_{i=0}^{n} \sum_{j=0}^{n} A_{i,j}\phi_i(x)\,\overline{\phi_j(y)} \quad . \tag{3.220}$$

(The kernel $\tilde{K}_{(n)}(x,y)$ may be written in the form

$$\sum_{j=0}^{n} \left(\sum_{i=0}^{n} A_{i,j}\phi_i(x) \right) \overline{\phi_j(y)} \; ,$$

so that its non-zero eigenvalues are those of the matrix with (i,j)th element $\displaystyle\int_a^b \overline{\phi_i(y)} \sum_{k=0}^{n} A_{k,j}\phi_k(y)\,dy = A_{i,j}$.)

The eigenvalues $\kappa_r^{\pm}, \tilde{\kappa}_r^{\pm}$ can now be compared using Theorem 3.1; we obtain $|\kappa_r^{\pm} - \tilde{\kappa}_r^{\pm}| \leq \rho(K - \tilde{K}_{(n)})$. Since $K(x,y)$ and $\tilde{K}_{(n)}(x,y)$ are both square-integrable, $\rho(K - \tilde{K}_{(n)}) \leq \|K - \tilde{K}_{(n)}\|_2$. Thus it is clear that

$$|\kappa_r^{\pm} - \tilde{\kappa}_r^{\pm}| \leq \|K - \tilde{K}_{(n)}\|_2 \tag{3.221}$$

when $\kappa_r^{\pm}$ and $\tilde{\kappa}_r^{\pm}$ both exist. Since the approximations are one-sided we obtain the following result.

THEOREM 3.31. *Under the conditions of Theorem 3.29,*

$$0 \leq \kappa_r^{+} - \tilde{\kappa}_r^{+} \leq \|K - \tilde{K}_{(n)}\|_2$$

and

$$0 \le \tilde{\kappa}_r^- - \kappa_r^- \le \|K - \tilde{K}_{(n)}\|_2 ,$$

where $K, \tilde{K}_{(n)}$ *are the integral operators with Hermitian kernels*

$K(x,y)$ *and* $\tilde{K}_{(n)}(x,y)$ *where* $\tilde{K}_{(n)}(x,y) = \sum_{i=0}^{n} \sum_{j=0}^{n} (K\phi_j, \phi_i)\phi_i(x)\, \overline{\phi_j(y)}$,

respectively, on $a \le x, y < b$.

Now because of the special form of $K_{(n)}(x,y)$ we can bound $\|K - \tilde{K}_{(n)}\|_2$ fairly readily. We have

$$\|K - \tilde{K}_{(n)}\|_2^2$$

$$\le \int_a^b\int_a^b |K(x,y) - \tilde{K}_{(n)}(x,y)|^2 dx\, dy \qquad (3.222)$$

$$= \int_a^b\int_a^b |K(x,y)|^2\, dx\, dy - \int_a^b\int_a^b \{K(x,y)\, \overline{\tilde{K}_{(n)}(x,y)} +$$

$$+ \overline{K(x,y)}\tilde{K}_{(n)}(x,y)\} dx\, dy + \int_a^b\int_a^b |\tilde{K}_{(n)}(x,y)|^2\, dx\, dy\, . \qquad (3.223)$$

But, as is easily checked,

$$\int_a^b \int_a^b K(x,y)\, \overline{\tilde{K}_{(n)}(x,y)}\, dx\, dy$$

$$= \int_a^b \int_a^b \overline{K(x,y)}\, \tilde{K}_{(n)}(x,y)\, dx\, dy$$

$$= \int_a^b \int_a^b |\tilde{K}_{(n)}(x,y)|^2\, dx\, dy$$

$$= \sum_{i=0}^{n} \sum_{j=0}^{n} |A_{i,j}|^2 , \qquad (3.224)$$

where $A_{i,j} = (K\phi_j, \phi_i)$, since $\phi_0(x), \phi_1(x), \ldots, \phi_n(x)$ are orthonormal. Thus,

$$\int_a^b \int_a^b |K(x,y) - \tilde{K}_{(n)}(x,y)|^2 \, dx \, dy$$

$$= \int_a^b \int_a^b |K(x,y)|^2 \, dx \, dy - \sum_{i=0}^{n} \sum_{j=0}^{n} |A_{i,j}|^2 \, . \qquad (3.225)$$

Combining (3.222) and (3.225) we obtain

$$||K - \tilde{K}_{(n)}||_2 \leq \left(\int_a^b \int_a^b |K(x,y)|^2 dx dy - \sum_{i=0}^{n} \sum_{j=0}^{n} |A_{i,j}|^2 \right)^{\frac{1}{2}} \qquad (3.226)$$

This inequality may be substituted in the bounds in Theorem 3.31.

Now the right-hand side of (3.226) may be written (see Smithies (1962), p.109, for example)

$$\{\sum_r |\kappa_r^{\pm}|^2 - \sum_r |\tilde{\kappa}_r^{\pm}|^2\}^{\frac{1}{2}} \quad ,$$

where the summations extend over all the true or approximate eigenvalues respectively. Thus the first set of inequalities in Theorem 3.31 assumes the form $0 \leq \kappa_r^+ - \tilde{\kappa}_r^+ \leq \{\sum_r [|\kappa_r^{\pm}|^2 - |\tilde{\kappa}_r^{\pm}|^2]\}^{\frac{1}{2}}$. In practice it is more convenient to employ the bound (3.226), as indicated in the following theorem.

THEOREM 3.32. *Under the conditions of Theorem 3.29,*

$$0 \leq \tilde{\kappa}_r^- - \kappa_r^- \leq \left(\int_a^b \int_a^b |K(x,y)|^2 dx \, dy - \sum_{i=0}^{n} \sum_{j=0}^{n} |A_{i,j}|^2 \right)^{\frac{1}{2}}$$

and

$$0 \leq \kappa_r^+ - \tilde{\kappa}_r^+ \leq \left(\int_a^b \int_a^b |K(x,y)|^2 dx \, dy - \sum_{i=0}^{n} \sum_{j=0}^{n} |A_{i,j}|^2 \right)^{\frac{1}{2}} .$$

3.25.1.

Now we noted that $\sum_i |\kappa_i^{\pm}|^2 - \sum_i |\tilde{\kappa}_i^{\pm}|^2$ is equal to the right-hand side of (3.226). Thus

$$|\kappa_r^{\pm}|^2 - |\tilde{\kappa}_r^{\pm}|^2 \leq \sum_i \{|\kappa_i^{\pm}|^2 - |\tilde{\kappa}_i^{\pm}|^2\} = \int_a^b \int_a^b |K(x,y)|^2 dx \, dy - \sum_{i=0}^{n} \sum_{j=0}^{n} |A_{i,j}|^2 \, .$$

Now, $|\kappa_r^{\pm}|^2 - |\tilde{\kappa}_r^{\pm}|^2 = |\kappa_r^{\pm} - \tilde{\kappa}_r^{\pm}|\,|\kappa_r^{\pm} + \tilde{\kappa}_r^{\pm}|$ and $|\kappa_r^{\pm} + \tilde{\kappa}_r^{\pm}| \geq 2\,|\tilde{\kappa}_r^{\pm}|$. Thus we obtain the following result, which is due to Wielandt.

THEOREM 3.33. *Under the conditions of Theorem 3.29*

$$|\kappa_r^{\pm} - \tilde{\kappa}_r^{\pm}| \leq \frac{1}{2|\tilde{\kappa}_r^{\pm}|}\left(\int_a^b\int_a^b |K(x,y)|^2 dx\, dy - \sum_{i=0}^{n}\sum_{j=0}^{n} |A_{i,j}|^2\right),$$

provided $\tilde{\kappa}_r^{\pm} \neq 0$.

The result in Theorem 3.33 is better than the result in Theorem 3.32 provided that

$$\frac{1}{2|\tilde{\kappa}_r^{\pm}|}\left(\int_a^b\int_a^b |K(x,y)|^2\, dx\, dy - \sum_{i=0}^{n}\sum_{j=0}^{n} |A_{i,j}|^2\right)^{\frac{1}{2}} \leq 1. \quad (3.227)$$

As a final note we observe that Cochran (1972, p.134) gives the result

$$|\tilde{\kappa}_r^{\pm}| < |\kappa_r^{\pm}| \leq \left\{|\tilde{\kappa}_r^{\pm}|^2 + \left(\int_a^b\int_a^b |K(x,y)|^2 dx\, dy - \sum_{i=0}^{n}\sum_{j=0}^{n} |A_{i,j}|^2\right)\right\}^{\frac{1}{2}}, \quad (3.228)$$

which he attributes to Trefftz (1933). See Cochran (1972) for further reading.

3.26. *Convergence of the eigenvalues and eigenfunctions*

Since the Rayleigh-Ritz method is a particular case of Galerkin's method, certain results can be gathered from sections which appear later.

When $\phi_0(x), \phi_1(x), \ldots, \phi_n(x)$ are part of a complete orthonormal set $\{\phi_i(x)\}$, the convergence of the eigenvalues is immediate. For, since the set of functions $\{\phi_i(x)\}$ is complete, every (continuous) kernel $K(x,y)$ can be approximated arbitrarily closely in the mean-square norm by its Fourier sum in terms of the functions $\phi_i(x)\,\overline{\phi_j(y)}$. To be precise, if $K(x,y)$ is square-integrable (and the functions $\{\phi_i(x)\}$ are complete),

$$\lim_{n\to\infty} \int_a^b \int_a^b |K(x,y) - \tilde{K}_{(n)}(x,y)|^2 \, dx \, dy = 0 ,$$

where

$$\tilde{K}_{(n)}(x,y) = \sum_{i=0}^{n} \sum_{j=0}^{n} A_{i,j}\phi_i(x) \overline{\phi_j(y)} ,$$

and $A_{i,j} = (K\phi_j,\phi_i)$. For a proof that this is so refer to Courant and Hilbert (1953, p.56) who discuss the case where $K(x,y)$ is continuous (see also section 3.26.1). Note that the choice $A_{i,j} = (K\phi_j,\phi_i)$ is the one which minimizes

$$||K(x,y) - \sum_{i=0}^{n} \sum_{j=0}^{n} A_{i,j}\phi_i(x)\phi_j(y)||_2 .$$

In view of Theorem 3.31 and eqn (3.222) we obtain the following result.

THEOREM 3.34. *If* $K(x,y) = K^*(x,y)$ *is square-integrable, and* $\{\phi_0(x),\phi_1(x),\ldots\}$ *is a complete orthonormal system in* $L^2[a,b]$ *, then* $\lim_{n\to\infty}|\kappa_r^{\pm} - \tilde{\kappa}_r^{\pm}| = 0$.

For the convergence of the eigenfunctions we modify the discussion of (3.85) by replacing $||\ ||_\infty$ by $||\ ||_2$ throughout the analysis, and substituting $\tilde{K}_{(n)}(x,y)$ for $K_n(x,y)$. We establish, in this way, the following result.

THEOREM 3.35. *Under the conditions of Theorem 3.34,* $\text{dist}_2(\tilde{f}_r^{\pm}, V_r^{\pm}) \to 0$ *as* $n \to \infty$ *, where* $V_r^{\pm}$ *is the space of eigenfunctions of* $K(x,y)$ *corresponding to* $\kappa_r^{\pm} \neq 0$ *, and the distance is measured in the mean-square norm. (Here*

$$\tilde{f}_r^{\pm}(x) = \sum_{i=0}^{n} \alpha_i\phi_i(x) ,$$

where $\underset{\sim}{A}\underset{\sim}{\alpha} = \tilde{\kappa}_r^{\pm}\underset{\sim}{\alpha}$ *.)*

Note that when $\kappa_r^{\pm}$ has multiplicity p we may find an orthogonal basis $f_{r_i}^{\pm}(x)$ $(i = 1,2,3,\ldots,p)$ for $V_r^{\pm}$, and the preceding theorem then states that

$$\lim_{n\to\infty} \left\| \tilde{f}_r^{\pm}(x) - \sum_{i=1}^{p} (\tilde{f}_{r_i}^{\pm}, f_{r_i}^{\pm}) f_{r_i}^{\pm}(x) \right\|_2 = 0 .$$

For a discussion of this result we refer also to Schlesinger (1957).

In rather special circumstances it may happen that $\|K - \tilde{K}_{(n)}\|_\infty \to 0$. (In particular this is the case when $\|K(x,y) - \tilde{K}_{(n)}(x,y)\|_\infty \to 0$.) Under such conditions, Theorem 3.35 is valid when the distance is measuredin the uniform norm. Even under the conditions of Theorem 3.35, the convergence of the eigenfunctions turns out to be 'relatively uniform' (see section 1.5), as noted by Cochran (1972).

It should be noted, however, that a new approximate eigenfunction $\overset{\circ}{f}_r^{\pm}(x)$ may be obtained from $\tilde{f}_r^{\pm}(x)$ if $\tilde{\kappa}_r^{\pm}$ is non-zero and $\kappa_r^{\pm}$ is simple and non-zero. We set

$$\overset{\circ}{f}_r^{\pm}(x) = (1/\tilde{\kappa}_r^{\pm}) \int_a^b K(x,y)\tilde{f}_r^{\pm}(y)\,dy . \tag{3.229}$$

Writing $\tilde{\lambda}_r^{\pm} = 1/\tilde{\kappa}_r^{\pm}$, we have

$$\begin{aligned} & f_r^{\pm}(x) - \overset{\circ}{f}_r^{\pm}(x) \\ &= \lambda_r^{\pm} \int_a^b K(x,y) f_r^{\pm}(y)\,dy - \tilde{\lambda}_r^{\pm} \int_a^b K(x,y)\tilde{f}_r^{\pm}(y)\,dy \\ &= (\lambda_r^{\pm} - \tilde{\lambda}_r^{\pm}) \int_a^b K(x,y) f_r^{\pm}(y)\,dy - \tilde{\lambda}_r^{\pm} \int_a^b K(x,y)(\tilde{f}_r^{\pm}(y) - f_r^{\pm}(y))\,dy . \end{aligned}$$

Thus if $K(x,y)$ is *continuous* (say),

$$\begin{aligned} & \|f_r^{\pm}(x) - \overset{\circ}{f}_r^{\pm}(x)\|_\infty \\ & \quad \le \sup_x \left(\int_a^b |K(x,y)^2|\,dy \right)^{\frac{1}{2}} \Big\{ \|f(x)\|_2 |\lambda_r^{\pm} - \tilde{\lambda}_r^{\pm}| + \\ & \quad\quad + |\tilde{\lambda}_r^{\pm}|\ \|\tilde{f}_r^{\pm}(x) - f_r^{\pm}(x)\|_2 \Big\} , \end{aligned}$$

and since $\lim_{n\to\infty} |\lambda_r^{\pm} - \tilde{\lambda}_r^{\pm}| = 0$, $\lim_{n\to\infty} \|f_r^{\pm}(x) - \overset{\circ}{f}_r^{\pm}(x)\|_\infty = 0$ when $\lim_{n\to\infty} \|f_r^{\pm}(x) - \tilde{f}_r^{\pm}(x)\|_2 = 0$. The scaling of $\tilde{f}_r^{\pm}(x)$ causes no real problem, and provided $\tilde{f}_r^{\pm}(x)$ converges essentially to $f_r^{\pm}(x)$ in the mean-square norm, $\overset{\circ}{f}{}_r^{+}(x)$ converges essentially in the uniform norm. (The reader will note that this result holds if $\sup_x \int_a^b |K(x,y)|^2 dy < \infty$, and in particular if $K(x,y)$ is continuous.)

In practice, we may wish to compute $\overset{\circ}{f}{}_r^{+}(x)/\|\overset{\circ}{f}{}_r^{+}(x)\|_2$.

3.26.1. *Mean-square convergence of* $\tilde{K}_{(n)}(x,y)$

As an appendix to section 3.26, let us indicate how we may establish that $\|K(x,y) - \tilde{K}_{(n)}(x,y)\|_2 \to 0$ as $n \to \infty$ (as asserted above). We suppose that $\phi_0(x), \phi_1(x), \phi_2(x), \ldots$ form a complete orthonormal system in $L^2[a,b]$, and $K(x,y)$ is square-integrable, and Hermitian.

Of all kernels of the form

$$K_n(x,y) = \sum_{i=0}^{n} \sum_{j=0}^{n} \alpha_{ij} \phi_i(x) \overline{\phi_j(y)} ,$$

that which minimizes $\|K(x,y) - K_n(x,y)\|_2^2$ is $\tilde{K}_{(n)}(x,y)$. (This is easily established by partial differentiation to locate the minimum.) Our task is, therefore, to show that we can construct a kernel $K_n(x,y)$ of the required form, such that $\|K(x,y) - K_n(x,y)\|_2 \to 0$.

Referring to Smithies (1962, p.115) we take the total eigensystem $\{\kappa_r, f_r(x)\}$ of $K(x,y)$, with $\|f_r(x)\|_2 = 1$ for $r \geq 0$, and choosing $\varepsilon > 0$ there exists an integer M such that

$$\|K(x,y) - \sum_{r=0}^{M} \kappa_r f_r(x) \overline{f_r(y)}\|_2 < \tfrac{1}{2}\varepsilon .$$

By completeness of the functions $\phi_i(x)$ we can, for each $r \in \{0,1,2,\ldots,M\}$, choose a value n_r such that for $n \geq n_r$ and $h_{n,r}(x) = \sum_{i=0}^{n} (f_r,\phi_i)\phi_i(x)$, $\|h_{n,r}(x) - f_r(x)\|_2 < \delta$, where δ is picked so that $2\rho(K)(2\delta+\delta^2) < \varepsilon$. Set $n = \max\{n_0,n_1,\ldots,n_M\}$. It can then be shown that with

$$K_n(x,y) = \sum_{j=0}^{n} \kappa_r h_{n,r}(x) \overline{h_{n,r}(y)} ,$$

we have a kernel of the required form such that

$$||K_n(x,y) - \sum_{r=0}^{M} \kappa_r f_r(x)\, \overline{f_r(y)}||_2 < \tfrac{1}{2}\varepsilon \quad .$$

Consequently

$$||K(x,y) - K_n(x,y)||_2 < \varepsilon \quad .$$

Thus, choosing $\varepsilon > 0$, there exists, for n sufficiently large, a kernel of the form $\sum_{i=0}^{n} \sum_{j=0}^{n} \alpha_{ij}\phi_i(x)\, \overline{\phi_j(y)}$ deviating in mean-square function norm from $K(x,y)$ by less than ε, as required.

3.27. *A further error bound*

There is some point in computing the function

$$\hat{f}_r^{\pm}(x) = \int_a^b K(x,y)\tilde{f}_r^{\pm}(y)\,dy \quad , \tag{3.230}$$

to obtain an error bound. (The function (3.230) is simply a scaled version of the function $\overset{\circ}{f}{}_r^{\pm}(x)$ in eqn (3.229), and it has the form $\sum_{i=0}^{n} \tilde{a}_i(K\phi_i)(x)$, where $\tilde{f}_r^{\pm}(x) = \sum_{i=0}^{n} \tilde{a}_i\phi_i(x)$.)

Suppose that $\tilde{f}_r^{\pm}(x)$ is any approximate eigenfunction corresponding to $\tilde{\kappa}_r^{\pm}$ such that

$$\int_a^b |\tilde{f}_r^{\pm}(x)|^2\, dx = 1 \quad .$$

Set

$$\begin{aligned}\eta(x) &= \int_a^b K(x,y)\tilde{f}_r^{\pm}(y)\,dy - \tilde{\kappa}_r^{\pm}\tilde{f}_r^{\pm}(x) \\ &= \hat{f}_r^{\pm}(x) - \tilde{\kappa}_r^{\pm}\tilde{f}_r^{\pm}(x) \quad ,\end{aligned}$$

and $||\eta(x)||_2$ may be computed or estimated using a method of numerical integration. According to Theorem 3.19 there is a true eigenvalue of $K(x,y)$ such that

$$|\kappa - \tilde{\kappa}_r^{+}| \leq ||\tilde{f}_r^{+}(x) - \tilde{\kappa}_r^{+}\tilde{f}_r^{+}(x)||_2 ,$$

since $||\tilde{f}_r^{+}(x)||_2 = 1$. (This bound can also be obtained from Theorem 3.20 on setting $L = \tilde{K}_{(n)}$.)

3.28. *The Rayleigh quotient*

When $K(x,y)$ is Hermitian and the set $\{\phi_i(x)\}$ used in the Rayleigh-Ritz method reduces to a single normalized function $\phi_0(x)$ (the case $n = 0$) the matrix $\underset{\sim}{A}$ reduces to the scalar $(K\phi_0,\phi_0) = \tilde{\kappa}$. If $\phi_0(x)$ is not normalized $\underset{\sim}{A}$ reduces to $(K\phi_0,\phi_0)$ and $\underset{\sim}{B}$ reduces to (ϕ_0,ϕ_0) so that the single eigenvalue $\tilde{\kappa}$ becomes

$$\kappa_R = \frac{(K\phi_0,\phi_0)}{(\phi_0,\phi_0)} = \frac{(\phi_0,K^*\phi_0)}{(\phi_0,\phi_0)} . \tag{3.231}$$

The value κ_R is known as the *Rayleigh quotient* and Theorem 3.31 and Theorems 3.32 and 3.33 apply with $n = 0$ to the value κ_R .

It is worth noting, in view of Theorem 3.19, that if $\phi_0(x)$ is fixed, μ is at our disposal, and

$$\eta(x) = \int_a^b K(x,y)\phi_0(y)dy - \mu\phi_0(x) ,$$

then $||\eta(x)||_2$ is minimized if we choose $\mu = \kappa_R$.

3.29. *Galerkin's method and the Rayleigh-Ritz method*

The practical implementation of quite general forms of the Galerkin method has been described in section 3.8, and in section 3.8.1 some theoretical insight into the construction of the method was given. Galerkin's method does not require any symmetry in $K(x,y)$.

To deal with the Galerkin method in some generality we set (as in section 3.8)

$$\langle\phi,\psi\rangle = \int_a^b w(x)\phi(x)\overline{\psi(x)}\,dx \quad , \tag{3.232}$$

where $w(x) > 0$. If $w(x) \equiv 1$ then $\langle\phi,\psi\rangle = (\phi,\psi)$.

As we suggested in Example 3.22, the Galerkin method may be further extended by replacing $\langle\phi,\psi\rangle$ by

$$\prec\phi,\psi\succ = \sum_{j=0}^{N} w_j\phi(x_j)\overline{\psi(x_j)} \quad , \tag{3.233}$$

where $w_j > 0$ and $a \le x_j \le b$ $(j = 0,1,2,\ldots,N)$, in the Galerkin equations $\underset{\sim}{A}\underset{\sim}{a} = \tilde{\kappa}\underset{\sim}{B}\underset{\sim}{a}$ with $A_{i,j} = \langle K\phi_j,\phi_i\rangle$ and $B_{i,j} = \langle\phi_j,\phi_i\rangle$. The values $\{w_j\},\{x_j\}$ are usually associated with a quadrature rule which permits us to approximate $\langle\psi,\phi\rangle$ by $\prec\psi,\phi\succ$.

The general statements which we make for the Galerkin method in terms of $\langle\psi,\phi\rangle$ extend to cover the method where $\prec\psi,\phi\succ$ is used in its stead.

The statements of section 3.8.1 are easily verified. If

$$\tilde{f}(x) = \sum_{i=0}^{n} \tilde{a}_i\phi_i(x)$$

and

$$r(x) = \tilde{\kappa}\tilde{f}(x) - \int_a^b K(x,y)\tilde{f}(y)dy$$

the conditions $\langle r,\phi_i\rangle = 0$ $(i = 0,1,2,\ldots,n)$ reduce to the set of equations

$$\tilde{\kappa}\sum_{j=0}^{n} \tilde{a}_j\langle\phi_j,\phi_i\rangle = \sum_{j=0}^{n} \tilde{a}_j\langle K\phi_j,\phi_i\rangle$$

$(i = 0,1,2,\ldots,n)$ and these are the Galerkin equations. Perhaps, therefore, it is a little surprising to note that Galerkin's method is equivalent to a form of Courant's method. The method is equivalent to constructing a degenerate kernel, $\tilde{K}_n(x,y)$ say, and solving

$$\int_a^b \tilde{K}_n(x,y)\tilde{f}(y)dy = \tilde{\kappa}\tilde{f}(x) \ .$$

This equivalence can be used to demonstrate convergence of the approximate values to true values, under certain conditions.

To construct a suitable kernel $\tilde{K}_n(x,y)$ let us first suppose that $\{\phi_i(x)\}$ is a sequence of functions such that $\langle\phi_i,\phi_j\rangle = \delta_{ij}$ (the Kronecker delta). We may then set

$$\tilde{K}_n(x,y) = \sum_{i=0}^{n} \phi_i(x) \int_a^b w(z)K(z,y)\overline{\phi_i(z)}\, dz . \qquad (3.234)$$

For, this kernel $\tilde{K}_n(x,y)$ has the form $\sum_{i=0}^{n} \phi_i(x)\psi_i(y)$ and its non-zero eigenvalues are those of the matrix $\tilde{A}$, where

$$\tilde{A}_{i,j} = \int_a^b \phi_j(x)\psi_i(x)dx .$$

Thus

$$\tilde{A}_{i,j} = \int_a^b \int_a^b \phi_j(x)w(z)K(z,x)\,\overline{\phi_i(z)}\, dz\, dx$$

$$= \langle K\phi_j,\phi_i\rangle$$

$$= A_{i,j} .$$

Since $B = I$, the non-zero eigenvalues of $\tilde{K}_n(x,y)$ on $a \le x,y \le b$ are precisely the non-zero eigenvalues $\tilde{\kappa}$ which are obtained by the Galerkin method.

Remark. It should be noted here that when $w(x) \equiv 1$ and $K(x,y)$ is Hermitian, the kernel $\tilde{K}_n(x,y)$ defined by eqn (3.234) is not the same as the kernel $\tilde{K}_{(n)}(x,y)$ defined in section 3.25 by eqn (3.220). Clearly there is more than one kernel $\tilde{K}_n(x,y)$ which satisfies our requirements.

We have gained some insight by constructing the kernel $\tilde{K}_n(x,y)$ in (3.234) and we can now consider the case where $\phi_0(x),\phi_1(x),\ldots,\phi_n(x)$ are linearly independent but not necessarily orthonormal. We shall again use $\tilde{K}_n(x,y)$ to denote the degenerate kernel which we require.

The general Galerkin method is closely associated with a method of approximating a function $\phi(x)$ by a linear combination of

$\phi_0(x),\phi_1(x),\ldots,\phi_n(x)$. We suppose that

$$A_n\phi(x) = \sum_{i=0}^{n} \alpha_i\phi_i(x) \quad ,$$

where the approximation operator A_n assigns the coefficients $\alpha_0,\alpha_1,\ldots,\alpha_n$ to $\phi(x)$ in such a way that $\|\phi(x) - (A_n\phi)(x)\| = \min$, where $\|\psi(x)\|^2 = \langle\psi,\psi\rangle$. Thus $\underset{\sim}{B}\underset{\sim}{\alpha} = \underset{\sim}{\nu}$, where $\nu_i = \langle\phi,\phi_i\rangle$, $B_{ij} = \langle\phi_j,\phi_i\rangle$. When $\langle\phi_i,\phi_j\rangle = \delta_{ij}, \alpha_i = \langle\phi,\phi_i\rangle$.

To construct the kernel $\tilde{K}_n(x,y)$ we form $A_{n,x}K(x,y)$, where the subscript x denotes that A_n is applied in the x-variable. Clearly, $A_{n,x}K(x,y)$ has the form

$$\sum_{i=0}^{n} \phi_i(x)\psi_i(y) \quad ,$$

and when $\langle\phi_i,\phi_j\rangle = \delta_{i,j}$,

$$\psi_i(y) = \int_a^b w(x)\overline{\phi_i(x)}K(x,y)\mathrm{d}x \quad ,$$

so that we obtain (3.234).

Let us set

$$\tilde{K}_n(x,y) = A_{n,x}K(x,y) \quad ,$$

that is

$$\tilde{K}_n(x,y) = \sum_{i=0}^{n} \phi_i(x)\psi_i(y) \ , \qquad (3.235)$$

where

$$\sum_{j=0}^{n} B_{i,j}\psi_j(y) = \gamma_i(y) \qquad (3.236)$$

and

$$\gamma_i(y) = \int_a^b w(x)\overline{\phi_i(x)}K(x,y)\mathrm{d}x \quad , \qquad (3.237a)$$

whilst

$$B_{i,j} = \int_a^b w(x)\overline{\phi_i(x)}\phi_j(x)\mathrm{d}x \ . \tag{3.237b}$$

Now the non-zero eigenvalues $\{\mu\}$ of $\tilde{K}_n(x,y)$, defined in this way, are precisely the non-zero eigenvalues of the matrix $\tilde{\underset{\sim}{A}}$, where

$$\tilde{A}_{i,j} = \int_a^b \phi_j(x)\psi_i(x)\mathrm{d}x \ .$$

The corresponding eigenfunctions of $\tilde{K}_n(x,y)$ have the form $\sum_{i=0}^{n} \alpha_i\phi_i(x)$, where $\underset{\sim}{\alpha}$ is an eigenvector of $\tilde{\underset{\sim}{A}}$. We have to show that the eigenvalue problems

$$\tilde{\underset{\sim}{A}}\underset{\sim}{\alpha} = \mu\underset{\sim}{\alpha} \tag{3.238}$$

and

$$\underset{\sim}{A}\tilde{\underset{\sim}{a}} = \tilde{\kappa}\underset{\sim}{B}\tilde{\underset{\sim}{a}} \tag{3.239}$$

are equivalent.

Now the matrix $\underset{\sim}{B}$ is non-singular, so that (3.238) is equivalent to

$$\underset{\sim}{B}\tilde{\underset{\sim}{A}}\underset{\sim}{\alpha} = \mu\underset{\sim}{B}\underset{\sim}{\alpha}$$

in the sense that there is a one-to-one correspondence between eigenvalues and corresponding eigenvectors. The (i,j)th element of $\underset{\sim}{B}\tilde{\underset{\sim}{A}}$ is

$$\sum_{k=0}^{n} B_{i,k}\tilde{A}_{k,j} = \sum_{k=0}^{n} B_{i,k}\int_a^b \psi_k(x)\phi_j(x)\mathrm{d}x$$
$$= \int_a^b \gamma_i(x)\phi_j(x)\mathrm{d}x \ ,$$

in view of eqn (3.236). Using (3.237a) to write out $\gamma_i(x)$ we discover that

$$\int_a^b \gamma_i(x)\phi_j(x)\mathrm{d}x = A_{i,j}$$

so that $\underset{\sim}{B}\tilde{\underset{\sim}{A}} = \underset{\sim}{A}$. In this way we have established the following result.

THEOREM 3.36. *Suppose that* $\underset{\sim}{A}\tilde{\underset{\sim}{a}} = \tilde{\kappa}\underset{\sim}{B}\tilde{\underset{\sim}{a}}$ *, where* $\underset{\sim}{A} = [\langle K\phi_j, \phi_i\rangle]$ *,* $\underset{\sim}{B} = [\langle\phi_j, \phi_i\rangle]$ *, and* $\phi_0(x), \phi_1(x), \ldots, \phi_n(x)$ *are linearly independent. If* $\tilde{\kappa} \neq 0$ *then*

$$\int_a^b \tilde{K}_n(x,y)\tilde{f}(y)\mathrm{d}y = \tilde{\kappa}\tilde{f}(x) , \tag{3.240}$$

where $\tilde{f}(x) = \sum_{i=0}^{n} \tilde{a}_i\phi_i(x)$ *and* $\tilde{K}_n(x,y)$ *is defined by eqns* (3.235) - (3.237). *Conversely, if* (3.240) *is satisfied with* $\tilde{\kappa} \neq 0$ *then* $\tilde{f}(x) = \sum_{i=0}^{n} \tilde{a}_i\phi_i(x)$ *, where* $\underset{\sim}{A}\tilde{\underset{\sim}{a}} = \tilde{\kappa}\underset{\sim}{B}\tilde{\underset{\sim}{a}}$. *Multiplicities are preserved.*

<u>Remark</u>. Let us suppose that $\langle\phi_i,\phi_j\rangle = \delta_{ij}$. Then the kernel $K(x,y)$ occurring in Theorem 3.36 has the form

$$\tilde{K}_n(x,y) = \sum_{i=0}^{n} \phi_i(x)\psi_i(y) ,$$

where

$$\psi_i(y) = \int_a^b w(x)\, \overline{\phi_i(x)}\, K(x,y)\mathrm{d}y .$$

In the statement of Theorem 3.36, the kernel $\tilde{K}_n(x,y)$ can be replaced by any kernel of the form

$$\hat{K}_n(x,y) = \sum_{i=0}^{n} \phi_i(x)\chi_i(y) ,$$

provided that

$$\int_a^b \phi_j(x)\chi_i(x)\mathrm{d}x = \int_a^b \phi_j(x)\psi_i(x)\mathrm{d}x \quad (i,j = 0,1,2,\ldots,n) .$$

Except when $w(x) \equiv 1$ and $K(x,y) = \overline{K(y,x)}$ there appears to be little real point, for our purposes, in replacing $\psi_i(x)$ by some other function $\chi_i(x)$. However, when $w(x) \equiv 1$ we can set

$$\chi_i(y) = \sum_{j=0}^{n} (\psi_i, \overline{\phi_j})\, \overline{\phi_j(y)}$$

and $\hat{K}_n(x,y)$ becomes

$$\sum_{i=0}^{n} \sum_{j=0}^{n} A_{i,j} \phi_i(x)\, \overline{\phi_j(y)} \quad ,$$

where $A_{i,j} = (K\phi_j, \phi_i)$. This was the kernel defined, when $K(x,y)$ was Hermitian, by (3.220).

3.30. *Convergence of the Galerkin method*

To analyse the convergence of the Galerkin methods we shall consider some particular cases.

Case (a). Suppose $w(x) \equiv 1$ and, hence, $\langle\phi,\psi\rangle = (\phi,\psi)$. Assume that $\phi_0(x), \phi_1(x), \ldots$ form a complete orthonormal set of functions in $L^2[a,b]$. (Each function $\phi_i(x)$ may be continuous, if required.) Then if $K(x,y)$ is a square-integrable kernel, $\lim_{n\to\infty} \|\tilde{K}_n(x,y) - K(x,y)\|_2 = 0$,

where $\tilde{K}_n(x,y)$ is defined above, in eqn (3.234) with $w(x) \equiv 1$.

Let us indicate a proof of the result stated here. Since $K(x,y)$ is a square-integrable kernel (section 1.3), the function

$$\psi_i(x) = \int_a^b K(z,x)\, \overline{\phi_i(z)} dz$$

is square-integrable, and $\|\psi_i(x)\|_2 \leq \|K(x,y)\|_2 \; \|\phi_i(x)\|_2$, so that $\phi_i(x)\psi_i(y)$ is square-integrable on $a \leq x,y \leq b$. We wish to establish that

$$\sum_{i=0}^{\infty} \phi_i(x)\psi_i(y)$$

is convergent, in mean-square norm, to $K(x,y)$.

Now, $\sum_{i=0}^{\infty} \|\psi_i(x)\|_2^2 < \infty$. Because $K(x,y) - \sum_{i=0}^{n} \phi_i(x)\psi_i(y)$ is, for $y \in [a,b]$, orthogonal to $\phi_j(x) (j=0,1,2\ldots,n)$, $\|\phi_k(x)\|_2 = 1$

for $k = 0,1,2,\ldots$ and so

$$\int_a^b |K(x,y)|^2 dx = \int_a^b |K(x,y) - \sum_{i=0}^{n} \phi_i(x)\psi_i(y)|^2 dx + \sum_{i=0}^{n} |\psi_i(y)|^2 .$$

(Apply 'Pythagoras' theorem', noting that

$$\{K(x,y) - \sum_{i=0}^{n} \phi_i(x)\psi_i(y)\} , \psi_0(y)\phi_0(x),\ldots,\psi_n(y)\phi_n(x)$$

are orthogonal functions in x.) Thus

$$\sum_{i=0}^{n} |\psi_i(y)|^2 \leq \int_a^b |K(x,y)|^2 dx$$

and so

$$\sum_{i=0}^{n} \|\psi_i(x)\|^2 \leq \|K(x,y)\|_2^2 < \infty ,$$

independently of n.

Now, $\sum_{i=0}^{\infty} \phi_i(x)\psi_i(y)$ is convergent in mean-square, since

$$\sum_{i=0}^{\infty} \|\phi_i(x)\psi_i(y)\|_2^2 = \sum_{i=0}^{\infty} \|\psi_i(x)\|_2^2 < \infty$$

(the functions $\phi_i(x)\psi_i(y)$ are orthogonal). But $K(x,y) - \sum_{i=0}^{\infty} \phi_i(x)\psi_i(y)$ is orthogonal to $\phi_j(x)$ for $j = 0,1,\ldots,n$ so vanishes (almost everywhere) for $y \in [a,b]$, that is,

$$K(x,y) \overset{\circ}{=} \sum_{i=0}^{\infty} \phi_i(x)\psi_i(y)$$

or

$$\int_a^b \int_a^b |K(x,y) - \sum_{i=0}^{\infty} \phi_i(x)\psi_i(y)|^2 dx\, dy = 0 .$$

The stated result expresses this relation : $\|K(x,y) - \tilde{K}_n(x,y)\|_2 \to 0$.

In view of the above result, $\|K - \tilde{K}_n\|_2 \to 0$ as $n \to \infty$ and we can employ Theorems 3.8 and 3.10 to establish the convergence of the approximations $\tilde{\kappa}, \tilde{f}(x)$. The convergence for the eigenfunctions holds in the mean-square norm. The analogue of Theorem 3.12 is also valid.

Now, our conclusion was established for the case where the functions $\{\phi_i(x)\}$ were assumed orthonormal. Provided this system is *closed*, and the functions *linearly independent*, the conclusions remain valid. For given $\phi_0(x), \phi_1(x), \ldots, \phi_n(x)$ there is an equivalent orthonormal system $\phi_0^*(x), \phi_1^*(x), \ldots, \phi_n^*(x)$ which is such that

$$A_n \phi(x) = \sum_{i=0}^{n} (\phi, \phi_i^*) \phi_i^*(x) \quad , \tag{3.241}$$

and if $\{\phi_i(x)\}$ is closed then $\{\phi_i^*(x)\}$ is closed and orthonormal. The kernel $\tilde{K}_n(x,y)$ could be rewritten

$$\sum_{i=0}^{n} \phi_i^*(x) \psi_i^*(y)$$

and the above treatment would be valid.

In view of the connection between the Galerkin equations using the inner product (ϕ, ψ) and $\tilde{K}_n(x,y)$, we have conditions which guarantee the convergence of this form of the Galerkin method. In particular, convergence takes place if $K(x,y)$ is continuous and

(i) $\phi_r(x) = P_r(x)$, the Legendre polynomial;

(ii) $\phi_r(x) = \sum_{j=0}^{r} a_j x^j, a_r \neq 0$;

(iii) $\phi_r(x) = e^{i\pi r x}$.

As remarked, the convergence results for the eigenfunctions hold in the norm $\| \ \|_2$.

<u>Case</u> (b). Suppose $w(x)$ is not identically unity. Feasible choices are $w(x) = 1/\sqrt{(1-x^2)}$, $\phi_r(x) = T_r(x)$ $(r = 0,1,2,\ldots)$, and $w(x) = \sqrt{(1-x^2)}$, $\phi_r(x) = U_r(x)$ (the Chebyshev polynomial of second kind, Davis (1965, p.164)) for $r = 0,1,2,\ldots$, supposing $a=-1$, $b=1$.

We shall consider the case where $a = -1$, $b = 1$,

$w(x) = 1/\sqrt{(1-x^2)}$, $\phi_0(x) = \sqrt{(1/\pi)}$, $\phi_r(x) = \sqrt{(2/\pi)}T_r(x)$, $(r = 1,2,3,\ldots)$.
Then $A_n\phi(x)$ is the Fourier-Chebyshev sum (section 2.5) of $\phi(x)$.
As in section 3.16

$$(A_n\phi)(x) = \int_{-1}^{1} A_n(x,y)\phi(y)dy ,$$

and we can write

$$\tilde{K}_n(x,y) = \int_{-1}^{1} A_n(x,z)K(z,y)dz ,$$

where

$$A_n(x,y) = \frac{2}{\pi} \sum_{k=0}^{n}{}' T_k(x)T_k(y)/\sqrt{(1-y^2)} .$$

For fixed y ,

$$|K(x,y) - \tilde{K}_n(x,y)| \leq \{1 + L_n\}E_n(K(\cdot,y)) , \qquad (3.242)$$

where

$$E_n(K(\cdot,y)) = \inf_{P_n \in P_n} \sup_{-1\leq x\leq 1} |P_n(x) - K(x,y)| ,$$

the supremum being taken over all polynomials $P_n(x)$ of degree at most n , and where $L_n = \|A_n\|_\infty$. If $\sup_{-1\leq y\leq 1} \{1 + L_n\}E_n(K(\cdot,y)) \to 0$ as $n \to \infty$ then

$$\|K(x,y) - \tilde{K}_n(x,y)\|_\infty \to 0 .$$

Suppose in particular that $K_x(x,y) = (\partial/\partial x)K(x,y)$ is continuous for $-1\leq x,y\leq 1$. Then

$$\|K(x,y) - \tilde{K}_n(x,y)\|_\infty \leq \{1 + L_n\} \frac{\pi}{2} \frac{\|K_x(x,y)\|_\infty}{n} \qquad (3.243)$$

(using Jackson's theorem, section 2.3.1). Since L_n behaves like

$\ln n$, $\lim_{n\to\infty} ||K(x,y) - \tilde{K}_n(x,y)||_\infty \to 0$. *A fortiori* $||K - \tilde{K}_n||_2 \to 0$, indeed $||K - \tilde{K}_n||_\infty \to 0$. Thus every eigenvalue of $K(x,y)$ appears as a cluster point of approximate eigenvalues $\tilde{\kappa}$; only such cluster points are true eigenvalues; and the approximate eigenfunctions converge to the appropriate space of eigenfunctions, in the uniform norm (see Theorems 3.8 and 3.10).

In general, if $\phi_r(x)$ is a polynomial of degree precisely r , then these results hold provided that

$$||A_n||_\infty E_n(K(\cdot,y)) \to 0 \tag{3.244}$$

uniformly for $-1 \leq y \leq 1$, where

$$||A_n||_\infty = \sup_{||\phi(x)||_\infty = 1} ||(A_n\phi)(x)||_\infty ,$$

and $a = -1$, $b = 1$.

3.31. *Convergence of the collocation method*

The collocation method can be analysed in much the same fashion as the Galerkin method. The general ideas apply to all 'projection methods', and both Galerkin's method and the method of collocation fit into this class. We shall give a slightly different flavour to the analysis here, to show how the material of section 3.29 may be extended.

Suppose that $\phi_0(x),\phi_1(x),\ldots,\phi_n(x)$ are piecewise-continuous functions (where $\phi_i(x) \equiv \phi_{i,n}(x)$) which give rise to a non-singular matrix $\underset{\sim}{B} = [\phi_j(x_i)]$ with some choice of collocation points $x_0,x_1,\ldots,x_n$. Then, given a function $\phi(x)$, there is a unique interpolant $(A_n\phi)(x) = \sum_{i=0}^{n} a_i\phi_i(x)$ such that $(A_n\phi)(x_i) = \phi(x_i)$ $(i = 0,1,2,\ldots,n)$. (The coefficients $a_0,a_1,\ldots,a_n$ are obtained as the solution of the equations

$$\sum_{j=0}^{n} B_{i,j}a_j = \phi(x_i) \quad (i = 0,1,2,\ldots,n).)$$

Using this interpolation operator in the same rôle as the approximation operator occurring in section 3.29 we construct the kernel

$$\tilde{K}_n(x,y) = A_x K(x,y) = \sum_{i=0}^{n} \phi_i(x)\psi_i(y) \quad , \tag{3.245}$$

where

$$\sum_{j=0}^{n} \phi_j(x_i)\psi_j(y) = K(x_i,y) \quad . \tag{3.246}$$

If $\|K - \tilde{K}_n\|_2 \to 0$ (and in particular if $\|K(x,y) - \tilde{K}_n(x,y)\|_2 \to 0$) then we can appeal to Theorems 3.8 and 3.10 to establish the convergence of the approximate eigenvalues and eigenfunctions. If we establish that $\|K - \tilde{K}_n\|_\infty \to 0$ then the convergence for the eigenfunctions is uniform.

The following are some particular cases which arise.

Case (a). Suppose $a = -1$, $b = 1$, that $\phi_r(x)$ is an algebraic polynomial of degree precisely r, and that $x_0, x_1, \ldots, x_n$ are the zeros of the Legendre polynomial $P_{n+1}(x)$. Then if $K(x,y)$ is continuous, $\|K - \tilde{K}_n\|_2 \to 0$ as $n \to \infty$.

Case (b). Suppose that $a = -1$, $b = 1$, $\phi_r(x) = T_r(x)$ (or $\phi_r(x) = \gamma_r T_r(x)$, $\gamma_r \neq 0$), and $x_0, x_1, \ldots, x_n$ are *either* the zeros of $T_{n+1}(x)$ *or* the extrema of $T_n(x)$. Then $\|K - \tilde{K}_n\|_\infty \to 0$ as $n \to \infty$ if $K(x,y)$ satisfies the condition

$$|K(x_1,y) - K(x_2,y)| \le L|x_1 - x_2|^\alpha \tag{3.247}$$

uniformly for $-1 \le y \le 1$, where $\alpha > 0$. In particular, this uniform Lipschitz condition is satisfied if the x-derivative $K_x(x,y)$ is continuous for $-1 \le x,y \le 1$. A corresponding result for an arbitrary finite interval $[a,b]$ is obtained if the Chebyshev polynomials $(T_r(x), T_n(x), T_{n+1}(x))$ are replaced by the appropriate shifted polynomials.

Case (c). Suppose, for the general interval $[a,b]$, $\phi_i(x) \equiv \phi_{i,n}(x)$ is defined by

$$\phi_i(x) = \begin{cases} 1 & a + ih \leq x < a + (i+1)h \\ 0 & \text{otherwise} \end{cases}, \qquad (3.248)$$

for $i = 0,1,2,\ldots(n-1)$, and

$$\phi_n(x) = \begin{cases} 1 & b - h \leq x \leq b \\ 0 & \text{otherwise} \end{cases} \qquad (3.249)$$

Then if $K(x,y)$ is continuous for $a \leq x,y \leq b$, $\|K - \tilde{K}_n\|_\infty \to 0$ as $n \to \infty$.

These results are established by considering the behaviour of $\|K(x,y) - \tilde{K}_n(x,y)\|$ as $n \to \infty$. Let us consider case (a) above. If $x_0, x_1, \ldots, x_n$ are the zeros of $P_{n+1}(x)$, $[a,b]$ is $[-1,1]$ and

$$\sum_{j=0}^{n} l_j(x)\phi(x_j)$$

is the polynomial interpolating $\phi(x)$ at these points x_j, then

$$\begin{aligned} &\Big\|\phi(x) - \sum_{j=0}^{n} l_j(x)\phi(x_j)\Big\|_2 \\ &\leq \|\phi(x) - p_n^*(x)\|_2 + \Big\|\sum_{j=0}^{n} l_j(x)\{p_n^*(x_j) - \phi(x_j)\}\Big\|_2 , \end{aligned} \qquad (3.250)$$

where $p_n^*(x)$ is the polynomial of degree n which best approximates $\phi(x)$ in the *uniform* norm. In (3.250) we may write immediately

$$\|\phi(x) - p_n^*(x)\|_2 \leq \sqrt{2}\,\|\phi(x) - p_n^*(x)\|_\infty .$$

Also, by Cauchy's inequality for sums,

$$\Big\|\sum_{j=0}^{n} l_j(x)\{p_n^*(x_j) - \phi(x_j)\}\Big\|_2^2$$

$$\leq \int_{-1}^{1} \sum_{j=0}^{n} (|p_n^*(x_j) - \phi(x_j)|^2\{l_j(x)\}^2)dx$$

$$\leq \|p_n^*(x) - \phi(x)\|_\infty^2 \sum_{j=0}^{n} \int_{-1}^{1} \{l_j(x)\}^2 dx$$

$$= 2\,\|p_n^*(x) - \phi(x)\|_\infty^2 \quad ,$$

since

$$\int_{-1}^{1} \sum_{j=0}^{n} \{l_j(x)\}^2 dx = \int_{-1}^{1} \Big| \sum_{j=0}^{n} l_j(x) \Big|^2 dx = 2 \; .$$

(We can use Gauss-Legendre quadrature to establish this, or note that

$$\int_{-1}^{1} l_i(x) l_j(x) dx = 0 \qquad (i \neq j)$$

since $x_0, x_1, \ldots, x_n$ are the zeros of $P_{n+1}(x)$.) Thus, from (3.250)

$$\Big\| \phi(x) - \sum_{j=0}^{n} l_j(x)\phi(x_j) \Big\|_2 \leq (2\sqrt{2}) \| p_n^*(x) - \phi(x) \|_\infty \quad .$$

Now we replace $\phi(x)$ by $K(x,y)$ and we find, for fixed y,

$$\int_{-1}^{1} |K(x,y) - \tilde{K}_n(x,y)|^2 dx \leq 8\{E_n(K(\cdot,y))\}^2, \qquad (3.251)$$

where $E_n(K(\cdot,y))$ was defined in section 3.29. A version of Jackson's theorem (Cheney 1966, p.147) yields a bound

$$E_n(K(\cdot,y)) \leq \omega(K(\cdot,y),\ \pi/(n+1)) \quad , \qquad (3.252)$$

where

$$\omega(K(\cdot,y),\ \delta) = \sup_{\substack{|x_1 - x_2| \leq \delta \\ |x_1|,|x_2| \leq 1}} |K(x_1,y) - K(x_2,y)| \; . \qquad (3.253)$$

Since $\lim_{\delta \to 0} \sup_{|y| \leq 1} \omega(K(\cdot,y),\ \delta) = 0$ when $K(x,y)$ is continuous, it follows that $\int_{-1}^{1}\int_{-1}^{1} |K(x,y) - \tilde{K}_n(x,y)|^2 \to 0$ as $n \to \infty$.

The proof of (b) imitates the discussion of the Galerkin method using Chebyshev polynomials and a weight function (see section 3.29) and the proof of (c) is straightforward using the (uniform) continuity of $K(x,y)$ for $a \leq x,y \leq b$.

3.31.1. *The dependence on the local truncation error*

We have chosen to establish the convergence of approximate eigenvalues and eigenfunctions (in particular, those obtained by collocation) by considering the eigensystem of a kernel $\tilde{K}_n(x,y)$. Let us now investigate the dependence of the approximate eigenvalues on the 'local truncation error' which, here, is the residual $\eta(x)$ obtained when κ , $f(x)$ are substituted in the eigenvalue equation for $\tilde{K}_n(x,y)$. Thus,

$$\eta(x) = \int_a^b \tilde{K}_n(x,y)f(y)dy - \kappa f(x) .$$

Let us suppose that we have established that as $n \to \infty$, $||K(x,y) - \tilde{K}_n(x,y)||_2 \to 0$ and hence by a suitable application of Theorem 3.8, $\tilde{\kappa} \to \kappa$, where we suppose that κ is a simple non-zero eigenvalue of $K(x,y)$ with a corresponding normalized eigenfunction $f(x)$. We would hope to be able to find an asymptotic theory for $\kappa - \tilde{\kappa}$ which parallels the analysis for the quadrature method. The essence of our argument should be apparent since it is similar to our discussion of the quadrature method.

We indicate the course of the analysis using operator notation. If $\tilde{K}_n\tilde{f}(x) = \tilde{\kappa}\tilde{f}(x)$, where $\tilde{\kappa} \to \kappa$ and κ is simple and non-zero, then $\tilde{\kappa}$ is a simple and non-zero eigenvalue of $\tilde{K}_n$ for n sufficiently large. Further, there exists a function $\tilde{\psi}(x)$ such that $||\tilde{\psi}(x)||_2 = 1$ and $\tilde{K}^*_n\tilde{\psi}(x) = \bar{\tilde{\kappa}}\tilde{\psi}(x)$. Since $\lim ||K(x,y) - \tilde{K}_n(x,y)||_2 = 0$, we obtain immediately the result $\lim ||K^*(x,y) - \tilde{K}^*_n(x,y)||_2 = 0$. Hence, applying Theorem 3.10 $||\tilde{\psi}(x) - \psi(x)||_2 \to 0$, where $K^*\psi(x) = \bar{\kappa}\psi(x)$. (We suppose that $\tilde{\psi}(x)$ is suitably scaled, and $||\psi(x)||_2 = 1$. Also $Kf(x) = \kappa f(x)$, with $||f(x)||_2 = 1$, and $\eta(x) = \tilde{K}_n f(x) - \kappa f(x) = (\tilde{K}_n - K)f(x)$.)

Now, $(\tilde{K}_n f,\tilde{\psi}) - \kappa(f,\tilde{\psi}) = (\eta,\tilde{\psi})$. But $(\tilde{K}_n f,\tilde{\psi}) = (f,\tilde{K}^*_n\tilde{\psi}) = (f,\bar{\tilde{\kappa}}\tilde{\psi}) = \tilde{\kappa}(f,\tilde{\psi})$ and hence $(\tilde{\kappa} - \kappa)(f,\tilde{\psi}) = (\eta,\tilde{\psi})$. Now

$|(f,\tilde{\psi}-\psi)| \leq \|\tilde{\psi}(x) - \psi(x)\|_2$ so that $(f,\tilde{\psi}) = (f,\psi) + (f,\tilde{\psi}-\psi) =$ $= (f,\psi) + o(1)$ and $|(\eta,\tilde{\psi})| \leq \|\eta(x)\|_2$, so that $|\kappa - \tilde{\kappa}| \leq$ $\leq \{\|\eta(x)\|_2/|(f,\psi)|\}(1 + o(1))$. Thus $|\kappa - \tilde{\kappa}| = O(\|\eta(x)\|_2)$ as $n \to \infty$.

3.31.2. *Order of convergence*

In section 3.31.1 we relate $\kappa - \tilde{\kappa}$ to the local truncation error (or 'residual') $\eta(x)$. We have shown that $|\kappa - \tilde{\kappa}| = O(\|\eta(x)\|_2)$, where

$$\eta(x) = \int_a^b \tilde{K}_n(x,y)f(y)dy - \kappa f(x),$$

that is (see above),

$$\eta(x) = \int_a^b \tilde{K}_n(x,y)f(y)dy - \int_a^b K(x,y)f(y)dy \quad ,$$

and κ is a simple non-zero eigenvalue of $K(x,y)$ coresponding to the normalized eigenfunction $f(x)$.

In order to obtain the rate of convergence of $\tilde{\kappa}$ to a simple eigenvalue κ we must examine $\|\eta(x)\|_2$. Clearly $\|\eta(x)\|_2 \leq \|K-\tilde{K}_n\|_2$ and, hence

$$\|\eta(x)\|_2 \leq \|K(x,y) - \tilde{K}_n(x,y)\|_2$$

where, in the collocation method, $\tilde{K}_n(x,y) = A_x K(x,y)$ and A_x is the appropriate interpolation operator applied in the x-variable. The latter bound can lead to a pessimistic result. Suppose, for example, that $K(x,y)$ is the Green's function

$$K(x,y) = x \quad (0 \leq x < y \leq 1), K(x,y) = y \ (0 \leq y \leq x \leq 1),$$

and that we employ the collocation method using a choice from (b) above. The corresponding operator A applied to a function $\phi(x)$ yields the polynomial of degree n which interpolates $\phi(x)$ at either the zeros of $T_n(x)$ or at the points of extrema of $T_{n+1}(x)$. Because of the discontinuity at $x = y$, we can readily establish only that

$$\|A_x K(x,y) - K(x,y)\|_\infty = O((\ln n)/n)$$

(since the Lebesgue constant L_n for interpolation at the Chebyshev zeros or extrema is $O(\ln n)$). We are led to believe that

$$\|\tilde{K}_n(x,y) - K(x,y)\|_\infty = O((\ln n)/n)$$

and to deduce that $\|\tilde{K}_n - K\|_2$ (and hence $\|\eta(x)\|_2$) is of the same order. We obtain much more realistic results if we rewrite $\eta(x) = A_x g(x) - g(x)$, where

$$g(x) = \int_a^b K(x,y)f(y)\mathrm{d}y = \kappa f(x) .$$

Here $f(x)$ is a fixed eigenfunction. Since, in the preceding example $f(x) \in C^\infty[0,1]$ (despite the discontinuity in $K_x(x,y)$) , we have

$$\|\eta(x)\|_2 \le \kappa M_r \|f^{(r)}(x)\|_\infty (1 + L_n)/n^r ,$$

where $\|f(x)\|_2 = 1$ $(r = 1,2,3,\ldots)$. (M_r is a constant in Jackson's theorem, and L_n behaves like $\ln n$.) Thus for any *fixed* finite r ,

$$\|\eta(x)\|_2 = O((\ln n)/n^r)$$

as $n \to \infty$, and $\kappa - \tilde{\kappa}$ is of the same order.

3.32. *A product-integration method*

The product-integration method described in section 3.4 may be thought of as a special case of a collocation method. Let us consider approximations satisfying eqns (3.33), namely

$$\sum_{j=0}^{n} v_j(z_i)\tilde{f}_r(z_j) = \tilde{\kappa}_r\tilde{f}_r(z_i) .$$

For definiteness, we suppose that $n = Nm$ (where N,m are integers), $h = (b-a)/n$, and $z_i = a + ih$ $(i = 0,1,2,\ldots,n)$, and we suppose that the coefficient functions $v_j(x)$ $(j = 0,1,2,\ldots,n)$ are obtained by constructing approximations of the form

$$\int_a^b K(x,y)\phi(y)dy \simeq \sum_{j=0}^{n} v_j(x)\phi(z_j) \quad ,$$

which are exact whenever $\phi(x) \in C[a,b]$ is a polynomial of degree m in $[z_{km}, z_{(k+1)m}]$ $(k = 0,1,2,\ldots,(N-1))$. (For details see sections 2.17 - 2.17.3.) Thus with $A\phi(x)$ defined, for $z_{km} \le x \le z_{(k+1)m}$, as the polynomial of degree m interpolating $\phi(x)$ at $z_{km}, z_{km+1}, \ldots, z_{(k+1)m}$ $(k = 0,1,2,\ldots,(N-1))$ we have

$$\sum_{j=0}^{n} v_j(x)\phi(z_j) = \sum_{k=0}^{N-1} \int_{z_{km}}^{z_{(k+1)m}} K(x,y)A\phi(y)dy \; .$$

In computing the approximate eigensystem we evaluate $\tilde{\kappa}_r, \tilde{f}_r(z_0)$, $\tilde{f}_r(z_1), \ldots, \tilde{f}_r(z_n)$. If $\tilde{\kappa}_r \neq 0$ we can extend the values $\tilde{f}_r(z_i)$ $(i = 0,1,\ldots,n)$ to a function $\tilde{f}_r(x)$ by setting

$$\tilde{f}_r(x) = (\tilde{\kappa}_r^{-1}) \sum_{j=0}^{n} v_j(x)\tilde{f}_r(z_j) \; .$$

To consider the method as a collocation method we consider instead the function $\hat{f}_r(x) = A\tilde{f}_r(x)$. The method is equivalent to finding such a function $\hat{f}_r(x)$ $(= A\tilde{f}_r(x))$ such that, for $i = 0,1,2,\ldots, n$

$$\int_a^b K(z_i,y)\hat{f}_r(y)dy - \tilde{\kappa}_r\hat{f}_r(z_i) = 0 \; ,$$

and hence

$$\int_a^b A_x K(x,y)\hat{f}_r(y)dy = \tilde{\kappa}_r A\hat{f}_r(x) \quad ,$$

where A_x denotes that A operates in the x-variable.

Let us now consider the question of convergence. We require some preliminary analysis.

For a function $\phi(x) \in C[a,b]$,

$$||\phi(x) - A\phi(x)||_\infty = \sup_k \sup_{z_{km} \le x \le z_{(k+1)m}} |\phi(x) - A\phi(x)|$$

and

$$\sup_{z_{km}\le x\le z_{(k+1)m}} |\phi(x)-A\phi(x)| \le \{1 + \sup_{z_{km}\le x\le z_{(k+1)m}} \sum_{j=km}^{(k+1)m} |l_{k,j}(x)|\}\times e_{m,k}(\phi),$$

where

$$e_{m,k}(\phi) = \inf_{p_m} \sup_{z_{km}\le x\le z_{(k+1)m}} |\phi(x) - p_m(x)|$$

(the infimum being taken over all polynomials of degree m) and

$$l_{k,j}(x) = \prod_{\substack{r\ne j\\ r=km}}^{(k+1)m} \{(x-z_r)/(z_j-z_r)\} .$$

We obtain

$$\|\phi(x) - A\phi(x)\| \le \{1 + L_m\}\sup_k|e_{m,k}(\phi)| ,$$

where

$$L_m = \sup_{0\le t\le m} \sum_{r=0}^{m} |l_r(t)|, l_r(t) = \prod_{\substack{j\ne r\\ r=0}}^{m} \{(t-r)/(j-r)\},$$

and we can obtain a bound for $e_{m,k}(\phi)$ from Jackson's theorem (section 2.3.1). Note that, in particular, $\sup_k |e_{m,k}(\phi)| \le \omega(\phi;\pi h/2)$,

where $\omega(\phi;\delta)$ is the modulus of continuity of $\phi(x)$, and that we can obtain bounds for $\sup_k |e_{m,k}(\phi)|$ which are $O(h^{s+1})$, if $\phi^{(s+1)}(x)$ is bounded and $s \le m$, as $h \to 0$ with m fixed.

Now from the previous remarks

$$\|K(x,y) - A_x K(x,y)\| \le \{1 + L_m\} \sup_{\substack{|x_1-x_2|\le\delta\\ a\le x_1,x_2,y\le b}} |K(x_1,y) - K(x_2,y)| ,$$

where $\delta = \pi h/2$. If m is fixed, $n, N \to \infty$, and $h \to 0$ then

$$||K(x,y) - A_x K(x,y)||_\infty \to 0 .$$

There immediately follows the convergence of the approximate eigenvalues $\tilde{\kappa}_r$ to true eigenvalues, in the sense of Theorem 3.8, and the corresponding uniform convergence of the normalized eigenfunctions $\hat{f}_r(x)$. Since

$$\tilde{f}_r(x) = (1/\tilde{\kappa}_r) \int_a^b K(x,y)\hat{f}_r(y)dy ,$$

the convergence of $\tilde{f}_r(x)$ (in the uniform norm) also follows from the type of argument employed previously in the discussion of eqn (3.229), in section 3.26.

In the case of a simple non-zero eigenvalue κ we know (section 3.31.2) that the rate of convergence of $\tilde{\kappa}$ to κ is governed by the corresponding local truncation error, $\eta(x)$. Thus in particular,

$$|\tilde{\kappa} - \kappa| = O(||Af(x) - f(x)||_2) ,$$

where $f(x)$ is a normalized eigenfunction corresponding to the simple non-zero eigenvalue κ of $K(x,y)$ – in particular, as $h \to 0$ with m fixed, $|\tilde{\kappa} - \kappa| = O(h^{s+1})$ if $f^{(s+1)}(x)$ is continuous $(s \leq m)$ and the order of convergence is the same (see section 2.17.2) as the order of error in replacing

$$\int_a^b K(x,y)f(y)dy$$

by $\sum_j v_j(x)f(z_j)$.

3.33. *A generalized Rayleigh quotient and a least-squares solution*

Suppose $\tilde{f}(x)$ is an approximate eigenfunction of $K(x,y)$ $(a \leq x,y \leq b)$. We may wish to choose a value $\tilde{\kappa}$ in such a way that, when we define

$$\eta(x) \equiv \eta_{\tilde{\kappa}}(x) = \int_a^b K(x,y)\tilde{f}(y)dy - \tilde{\kappa}\tilde{f}(x)$$

and

$$\langle\phi,\psi\rangle = \int_a^b w(x)\phi(x)\,\overline{\psi(x)}dx$$

(with $w(x) \geq 0$) , we minimize $\langle\eta,\eta\rangle/\langle\tilde{f},\tilde{f}\rangle$. Thus we seek the value $\tilde{\kappa}$ which minimizes the weighted least-squares norm of the residual $\eta(x)$ (when $\tilde{f}(x)$ is normalized).

The required value of $\tilde{\kappa}$ is

$$\kappa_R^* \equiv \kappa_R^*(\tilde{f}) = \langle K\tilde{f},\tilde{f}\rangle/\langle\tilde{f},\tilde{f}\rangle \ .$$

If $K(x,y) = \overline{K(y,x)}$ and $w(x) \equiv 1$, κ_R^* is the Rayleigh quotient κ_R . We therefore call κ_R^* a generalized Rayleigh quotient.

It is not difficult to show that κ_R^* is the required value. Since $\langle\tilde{f},\tilde{f}\rangle$ is fixed we regard ourselves as minimizing

$$\begin{aligned}\langle\eta,\eta\rangle &= \langle K\tilde{f} - \tilde{\kappa}\tilde{f},\ K\tilde{f} - \tilde{\kappa}\tilde{f}\rangle \\ &= \langle K\tilde{f},K\tilde{f}\rangle - \tilde{\kappa}\langle\tilde{f},K\tilde{f}\rangle - \overline{\tilde{\kappa}}\langle K\tilde{f},\tilde{f}\rangle + |\tilde{\kappa}|^2\langle\tilde{f},\tilde{f}\rangle \ .\end{aligned}$$

If we set $\tilde{\kappa} = \alpha + i\beta$ then we require

$$(\partial/\partial\alpha)\ \langle\eta,\eta\rangle = (\partial/\partial\beta)\ \langle\eta,\eta\rangle = 0$$

and obtain

$$2\alpha\langle\tilde{f},\tilde{f}\rangle = \langle\tilde{f},K\tilde{f}\rangle + \langle K\tilde{f},\tilde{f}\rangle \ ,$$

$$2\beta\langle\tilde{f},\tilde{f}\rangle = i\{\langle\tilde{f},K\tilde{f}\rangle - \langle K\tilde{f},\tilde{f}\rangle\} \ ,$$

so that $\alpha + i\beta = \langle K\tilde{f},\tilde{f}\rangle/\langle\tilde{f},\tilde{f}\rangle$. We leave it as an exercise to show that the stationary value is a minimum.

The formulation of a general least-squares solution for the

eigenvalue problem is now possible. We may formulate the following problem:

Given square-integrable functions $\phi_0(x), \phi_1(x), \ldots, \phi_n(x)$ choose $\tilde{a}_0, \tilde{a}_1, \ldots, \tilde{a}_n$ and $\tilde{\kappa}$ in order to minimize $\langle \eta, \eta \rangle$ where $\eta(x) \equiv \eta_{\tilde{\kappa}}(x)$ is defined above and $\tilde{f}(x) = \sum_{i=0}^{n} \tilde{a}_i \phi_i(x)$.

It is clear that this problem is solved if we choose $\tilde{a}_0, \tilde{a}_1, \ldots, \tilde{a}_n$ in such a way as to minimize $N = \langle K\tilde{f} - \kappa_R^*(\tilde{f})\tilde{f}, K\tilde{f} - \kappa_R^*(\tilde{f})\tilde{f} \rangle$, and we should then set up, and solve, the equations $\partial N / \partial \tilde{a}_i = 0$ $(i = 0, 1, 2, \ldots, n)$. We shall not pursue this further (except to note that when the function $\tilde{f}(x)$ has been chosen to minimize N , we set $\tilde{\kappa} = \kappa_R^*(\tilde{f})$).

4

LINEAR EQUATIONS OF THE SECOND KIND

In this chapter we consider the numerical solution of the Fredholm equation of the second kind, $f(x)-\lambda\int_a^b K(x,y)f(y)dy=g(x)$, where a,b are finite, $K(x,y)$ and $g(x)$ are continuous, and λ is a regular value. More difficult equations of a similar type will be treated in Chapter 5.

Numerical methods for the equation of the second kind are discussed in sections 4.1-4.14, the most favoured methods being those based on numerical integration formulae (sections 4.3-4.6) and the collocation, Ritz-Galerkin, and linear programming methods of sections 4.11-4.13.

The simple quadrature method of section 4.3 is perhaps the most readily applied in practice, and variations on this method (including the use of a deferred correction process, the deferred approach to the limit, and the construction of asymptotic upper and lower bounds) are described.

The second half of the chapter is concerned with a theoretical study of some of the principal methods. This study opens in section 4.15 with a comparison of the solution $f(x)$ with the solution $f_n(x)$ of a perturbed equation $f_n(x)-\lambda\int_a^b K_n(x,y)f_n(y)dy=g_n(x)$. Results relating $f(x)$ to $f_n(x)$ can be used to study (in section 4.16) the quadrature method of section 4.12. Such results may also (sections 4.24-4.28) be applied to the collocation and Ritz-Galerkin type methods

The theories of sections 4.17 and 4.29 provide an introduction to more general forms of analysis. Here, an approximate function $\tilde{f}(x)$ is regarded as the solution of an equation in which the operator is not necessarily an integral operator.

4.1. *Introductory remarks*

In discussing the numerical solution of linear integral equations of the second kind, the discussion will be limited to equations of Fredholm type, and I shall first describe the techniques and then deal with the numerical analysis of the methods. In this treatment, we can afford to be fairly brief, since the methods for equations of the second kind are, in general, fairly obvious extensions of the methods for eigenvalue problems. There are naturally some differences in detail in the treatment of Fredholm equations of the second kind, however, or cases where special techniques are applicable. On the other hand, the treatment of Volterra equations may be viewed quite differently from the treatment of Fredholm equations. We explore some sequential step-by-step or block-by-block methods for the solution of Volterra integral equations in Chapter 6.

4.2. *The Fredholm equation of the second kind*

In the following sections methods for the approximate solution of the equation

$$f(x) - \lambda \int_a^b K(x,y)f(y)\mathrm{d}y = g(x) \tag{4.1}$$

will be discussed. We shall require that λ is a regular value of the kernel and that $K(x,y)$ and $g(x)$ are at least piecewise-continuous. In this chapter we suppose (unless we state otherwise) that $g(x)$ is continuous for $a \le x \le b$, that $K(x,y)$ is continuous for $a \le x,y \le b$, and that we seek the solution $f(x)$ for $a \le x \le b$. Some of the methods introduced here can be extended to treat the case where $K(x,y)$ is weakly singular; this case is discussed in Chapter 5.

The accuracy attainable with any method for the approximate solution of eqn (4.1) may be limited by the equation itself. When small perturbations in $g(x)$ cause large changes in $f(x)$, eqn (4.1) is said to be ill conditioned, and any numerical method must be applied with caution if accurate answers are to be obtained. A desirable feature of a method is that it can be applied in a way which gives warning of ill-conditioning.

Example 4.1.
Suppose that

$$\mu f(x) - \int_a^b K(x,y)f(y)\mathrm{d}y = g(x) ,$$

where μ is close to an eigenvalue κ_r. If the corresponding eigenfunction is $f_r(x)$, then

$$\mu f_r(x) - \int_a^b K(x,y)f_r(y)\mathrm{d}y = (\mu-\kappa_r)f_r(x) .$$

If $g(x)$ is changed to $g(x) + \alpha f_r(x)$, then the solution $f(x)$ is changed to $f(x) + \alpha/(\mu-\kappa_r)f_r(x)$, and even though α is small, the change in $f(x)$ can be large. The equation is therefore ill conditioned when μ is close to κ_r. However, an equation can be ill conditioned even though the kernel has no non-zero eigenvalues. *

A particular integral equation may have special features which, if known, can be exploited in the numerical solution. Suppose, for example, that the behaviour of $g(x)$ and $K(x,y)$ is such that it is known that $f(x) = f_0(x) + f_1(x)$, where $f_1(x)$ is badly behaved (for example, $f_1'(a)$ is infinite) and the form of $f_1(x)$ is known, whilst $f_0(x)$ is smooth. Since the form of $f_1(x)$ is known, we can obtain an equation for $f_0(x)$. Thus

$$f_0(x) - \lambda \int_a^b K(x,y)f_0(y)\mathrm{d}y = g_0(x) , \tag{4.2}$$

where

$$g_0(x) = g(x) - f_1(x) + \lambda\int_a^b K(x,y)f_1(y)\mathrm{d}y$$

and $g_0(x)$ can be determined.

More generally, any knowledge of the behaviour of the solution may be exploited. Thus, there are special classes of integral equation which have received extensive study in the literature: examples follow.

Example 4.2.

If $K(x,y) = G(x,y)\Phi(y)$ in (4.1), where $\Phi(x) > 0$ for $a \le x \le b$ and $G(x,y)$ is Hermitian, then on setting $\psi(x) = \{\Phi(x)\}^{\frac{1}{2}}f(x)$ we may rewrite eqn (4.1) in the form

$$\psi(x) - \lambda\int_a^b \{\Phi(x)\}^{\frac{1}{2}}G(x,y)\{\Phi(y)\}^{\frac{1}{2}}\psi(y)dy = \{\Phi(x)\}^{\frac{1}{2}}g(x) .$$

The kernel of this equation is $\{\Phi(x)\}^{\frac{1}{2}}G(x,y)\{\Phi(y)\}^{\frac{1}{2}}$, which is Hermitian. *

Example 4.3.

A matrix $\underset{\sim}{A} = [A_{i,j}]$ $(i,j = 0,1,2,\ldots,n)$ is said to be centrosymmetric if $A_{i,j} = A_{n-i,n-j}$. In analogy with this, a kernel $K(x,y)$ is said to be centrosymmetric on $[a,b]$ if $K(a+s,a+t) = K(b-s,b-t)$. (Consequently, a kernel $K(x,y)$ which can be written in the form $k(|x - y|)$ is centrosymmetric.) If we write $\alpha = \frac{1}{2}(a+b)$, $\beta = \frac{1}{2}(b-a)$ then, if $K(x,y)$ is centrosymmetric, $K(\alpha+s,\alpha+t) = K(\alpha-s,\alpha-t)$ for $|s|$, $|t| \le \beta$, and eqn (4.1) can be written

$$f(\alpha+s) - \lambda\int_{-\beta}^{\beta} K(\alpha+s,\alpha+t)f(\alpha+t)dt = g(\alpha+s) ,$$

for $|s| \le \beta$. If we write

$$\sigma(s) - \lambda\int_0^{\beta} L(s,t)\sigma(t)dt = \{g(\alpha+s) + g(\alpha-s)\}$$

and

$$\delta(s) - \lambda\int_0^{\beta} M(s,t)\delta(t)dt = \{g(\alpha+s) - g(\alpha-s)\}$$

(where $L(s,t) = L(s,-t) = \{K(\alpha+s,\alpha+t) + K(\alpha-s,\alpha+t)\}$, and $M(s,t) = -M(s,-t) = \{K(\alpha+s,\alpha+t) - K(\alpha-s,\alpha+t)\}$) then $f(\alpha+s) = \frac{1}{2}\{\sigma(s) + \delta(s)\}$. If $g(s+\alpha) = g(\alpha-s)$ then $\delta(s) \equiv 0$ and some simplification occurs. *

Example 4.4.

A kernel $K(x,y)$ which is a function, $k(x-y)$ say, of the difference $x - y$ is a convolution kernel or a difference kernel. (Thus a real symmetric difference kernel is centrosymmetric.) Equations in which the kernel $K(x,y)$ is a difference kernel of various special types have been studied extensively. In particular Roark, Shampine, and Wing have studied the eigenvalue problem for kernels of the form $K(x,y) = k(\mu|x - y|)$ for some value of μ, where $k(z)$ has the form $k(z) = \int_0^A h(t) \cos zt \, dt$ (see Roark and Shampine 1965); Ganado (1971) has extended their ideas to equations of the second kind. *

4.3. *The quadrature method*

In this section I indicate how the techniques described for the eigenvalue problem in section 3.2 may be applied to eqn (4.1). In eqn (4.1), λ is a known value, assumed to be a regular value of the kernel.

Suppose that we have made a choice of quadrature rule, of the form $J(\phi) = \sum_{j=0}^{n} w_j \phi(y_j)$, for approximating an integral $\int_a^b \phi(y) dy$. Such an integration rule can be used to replace eqn (4.1) by the equation

$$\tilde{f}(x) - \lambda \sum_{j=0}^{n} w_j K(x,y_j)\tilde{f}(y_j) = g(x) \quad (a \le x \le b) , \tag{4.3}$$

in which the solution is a function $\tilde{f}(x)$, which may be regarded as an approximation to $f(x)$. Eqn (4.3) is a functional equation whose solution may be found (if $a \le y_i \le b$ for $i = 0,1,2,\ldots,n$) by setting $x = y_i$ $(i = 0,1,2,\ldots,n)$ in (4.3) to obtain the equations

$$\tilde{f}(y_i) - \lambda \sum_{j=0}^{n} w_j K(y_i,y_j)\tilde{f}(y_j) = g(y_i) \quad (i = 0,1,2,\ldots,n). \tag{4.3a}$$

If values $\tilde{f}(y_0), \tilde{f}(y_1),\ldots, \tilde{f}(y_n)$ satisfying these equations can be found, then we obtain the solution of eqn (4.3) on setting, for all $x \in [a,b]$,

$$\tilde{f}(x) = \lambda \sum_{j=0}^{n} w_j K(x,y_j)\tilde{f}(y_j) + g(x) . \tag{4.4}$$

However, we may be content to represent the approximate solution simply by the function values which define the vector $\tilde{\underset{\sim}{f}} = [\tilde{f}(y_0),\tilde{f}(y_1), \ldots, \tilde{f}(y_n)]^T$. The vector $\tilde{\underset{\sim}{f}}$ satisfies the equation

$$(\underset{\sim}{I} - \lambda\underset{\sim}{K}\underset{\sim}{D})\tilde{\underset{\sim}{f}} = \underset{\sim}{g} , \tag{4.5}$$

where $\underset{\sim}{g} = [g(y_0),g(y_1), \ldots, g(y_n)]^T$, $\underset{\sim}{K} = [K(y_i,y_j)]$, and $\underset{\sim}{D} = \text{diag}\,(w_0,w_1, \ldots, w_n)$. We shall refer to the function $\tilde{f}(x)$ defined by (4.4) as the 'Nyström extension' of $\tilde{\underset{\sim}{f}}$.

We have assumed that λ is a regular value in eqn (4.1), so that there is a unique solution $f(x)$. Nevertheless, for an arbitrary rule $J(\phi)$ it may happen that the matrix $(\underset{\sim}{I} - \lambda\underset{\sim}{K}\underset{\sim}{D})$, occurring in eqn (4.5), is singular. One can guarantee (under mild restrictions) that $(\underset{\sim}{I} - \lambda\underset{\sim}{K}\underset{\sim}{D})$ is non-singular if $J(\phi)$ is a 'sufficiently accurate' quadrature rule. The conditions will be made more rigorous later; they depend on λ and $K(x,y)$. However, our choice of quadrature rule should also depend upon the function $g(x)$, since the behaviour of this function has an influence on the behaviour of $f(x)$, and hence the accuracy obtainable.

Example 4.5.

Consider eqn (4.1) in which $a = 0$, $b = 1$ and $K(x,y) = x(1 - y)$ for $0 \le x \le y \le 1$, $K(x,y) = y(1 - x)$ for $0 \le y \le x \le 1$. We suppose that $g''(x)$ exists, and that $\lambda \notin \{(r\pi)^2 | r = 1,2,3,\ldots\}$ so that λ is a regular value and the solution $f(x)$ is uniquely defined. For this example, (4.1) is equivalent to the differential equation $f''(x) + \lambda f(x) = g''(x)$, where $f(0) = g(0)$, $f(1) = g(1)$.

For the quadrature rule $J(\phi)$, take the trapezium rule with step h ,

$$J(\phi) = \sum_{j=0}^{n}{}'' h\phi(jh) ,$$

where $h = 1/n$. Then (4.3a) becomes

$$\tilde{f}(ih) - \lambda h \sum_{j=0}^{n}{}'' K(ih,jh)\tilde{f}(jh) = g(ih)$$

$(i = 0,1, \ldots ,n)$. The matrix $(\underset{\sim}{I} - \lambda \underset{\sim}{K}\underset{\sim}{D})$ is non-singular if $|\lambda|<\infty$ and

$$\lambda \notin \{h^{-2}4 \sin^2 \tfrac{1}{2}(\pi rh) \,|\, r = 1,2,3,\ldots,(n-1)\}$$

(see Example 3.7, p.179) so that, if λ is a regular value, we obtain a unique solution $\underset{\sim}{\tilde{f}} = [\tilde{f}(0),\tilde{f}(h), \ldots, \tilde{f}(1)]^T$ on taking h sufficiently small.

Note that we can show that $\tilde{f}(0) = g(0)$, $\tilde{f}(1) = g(1)$ and $h^{-2}\,\delta^2\tilde{f}(ih) + \lambda\tilde{f}(ih) = h^{-2}\,\delta^2 g(ih)$ $(i = 1,2,3,\ldots,n)$, where δ is the central difference operator, defined by

$$\delta\psi(ih) = \psi(\overline{i+\tfrac{1}{2}}h) - \psi(\overline{i-\tfrac{1}{2}}h)$$

so $\delta^2\tilde{f}(ih) = \tilde{f}(\{i+1\}h) - 2\tilde{f}(ih) + \tilde{f}(\{i-1\}h)$. *

Example 4.6.

The equation

$$f(x) + \frac{1}{\pi}\int_{-1}^{1} \frac{d}{d^2 + (x-y)^2} f(y)\,dy = g(x) \quad (|x| \leq 1)$$

is known as Love's equation, and it arises in electrostatics. The non-zero value of d is given. We shall consider the numerical solution of the above equation with $d = -1$, $g(x) \equiv 1$, writing $K(x,y) = -1/[\pi\{1 + (x-y)^2\}]$, $\lambda = -1$ in (4.1). Using the trapezium rule with step h for the quadrature rule we obtain the equations

$$\tilde{f}(ih-1) - h \sum_{j=0}^{n}{}'' \frac{\tilde{f}(jh-1)}{\pi\{1+(i-j)^2h^2\}} = 1 ,$$

where $h = 2/n$, $(i = 0,1,2,\ldots,n)$.

In this example the matrix $\underset{\sim}{K} = [K(ih-1, jh-1)]$ of eqn (4.5) is centrosymmetric, and the quadrature rule is symmetric. It is easy to see that since $\underset{\sim}{g} = [1,1,\ldots,1]^T$ the solution vector $\tilde{\underset{\sim}{f}}$ is symmetric, that is, $\tilde{f}(1-ih) = \tilde{f}(ih-1)$ $(i = 0,1,2,\ldots,n)$. The system of linear equations above can therefore be reduced to a system of $[(\frac{1}{2}n+1)]$ equations in $[(\frac{1}{2}n+1)]$ unknowns.¶

In Table 4.1 the values computed for $\tilde{f}(x)$ at $x = 0, \pm\frac{1}{2}, \pm\frac{3}{4}, \pm 1$ with $n = 8$, 16, 32, and 64 are given.

TABLE 4.1

n	8	16	32	64
$x = \pm 1$	1·63639	1·63887	1·63949	1·63964
$x = \pm\frac{3}{4}$	1·74695	1·75070	1·75164	1·75187
$x = \pm\frac{1}{2}$	1·83641	1·84089	1·84201	1·84229
$x = \pm\frac{1}{4}$	1·89332	1·89804	1·89922	1·89952
$x = 0$	1·91268	1·91744	1·91863	1·91893

In place of the trapezium rule we can employ, for $n = 8$, 16, 32, 64, the basic $(n+1)$-point Romberg rule of the classical scheme. The corresponding results are displayed in Table 4.2. Note that to calculate these results, the weights of the Romberg rules are required explicitly in the matrix $\underset{\sim}{D}$ (see section 2.13.1).

¶ We denote the integer part of a real value x by $[x]$.

TABLE 4.2

	Basic Romberg rule with			
x	9 points	17 points	33 points	65 points
± 1	1·63973	1·63969	1·63970	1·63970
$\pm \frac{3}{4}$	1·75183	1·75196	1·75195	1·75195
$\pm \frac{1}{2}$	1·84215	1·84239	1·84238	1·84238
$\pm \frac{1}{4}$	1·89969	1·89961	1·89961	1·89962
0	1·91934	1·91903	1·91903	1·91903

These values were obtained using the symmetry and solving systems of order 5, 9, 17, 33 respectively. *

Example 4.7.

If we apply the quadrature method to eqn (4.1) when the kernel $K(x,y)$ is degenerate, say $K(x,y) = \sum_{k=0}^{N} X_k(x)Y_k(y)$, the true and approximate solutions can be compared. For the true solution is of the form

$$f(x) = g(x) + \lambda \sum_{k=0}^{N} a_k X_k(x) ,$$

where $(\underset{\sim}{I} - \lambda \underset{\sim}{A})\, \underset{\sim}{a} = \underset{\sim}{b}$, $A_{ij} = \int_a^b X_j(x)Y_i(x)dx$, and $b_i = \int_a^b g(x)Y_i(x)dx$

(see section 1.7). The solution to the equation

$$\tilde{f}(x) - \lambda \sum_{j=0}^{n} w_j K(x,y_j)\tilde{f}(y_j) = g(x)$$

is, similarly, of the form

$$\tilde{f}(x) = g(x) + \lambda \sum_{k=0}^{N} \tilde{a}_k X_k(x) ,$$

where $(\underset{\sim}{I} - \lambda\underset{\sim}{\tilde{A}})\underset{\sim}{\tilde{a}} = \underset{\sim}{\tilde{b}}$, $\tilde{A}_{ij} = \sum_{k=0}^{n} w_k X_j(y_k)Y_i(y_k)$, and

$\tilde{b}_i = \sum_{k=0}^{n} w_k g(y_k)Y_i(y_k)$.

Two features are apparent here. First, if $||\underset{\sim}{A} - \underset{\sim}{\tilde{A}}||$ and $||\underset{\sim}{b} - \underset{\sim}{\tilde{b}}||$ are to be small, the quadrature rule must be chosen with reference both to $K(x,y)$ and to $g(x)$. Secondly, it may be difficult to approximate $\underset{\sim}{a}$ accurately if the problem of solving $(\underset{\sim}{I} - \lambda\underset{\sim}{A})\underset{\sim}{a} = \underset{\sim}{b}$ is ill conditioned. This implies that, for certain choices of λ and $K(x,y)$, it will be difficult to approximate $f(x)$ accurately, due to ill-conditioning inherent in the problem. *

We may assume that the first priority in the choice of a quadrature rule is to produce a solution which is reasonably accurate. However, we may be prepared to 'process' our first approximations to a solution by using a method of deferred approach or deferred correction, and the accuracy of the initial approximation $\underset{\sim}{\tilde{f}}$ is not then of prime importance. Difficulty in the choice of rule often arises when $K(x,y)$, or $f(x)$, is known, or suspected, to be badly behaved. As a general rule, we should choose $J(\phi)$ so that the local truncation error

$$\tau(x) = f(x) - \lambda \sum_{j=0}^{n} w_j K(x,y_j)f(y_j) - g(x) \qquad (4.6)$$

is 'predictable'. In particular, if we seek an accurate solution from a single approximation $\underset{\sim}{\tilde{f}}$, we should choose a rule such that $||\underset{\sim}{\tau}||_\infty = \sup_{0\le i\le n} |\tau(y_i)|$ is small. In assessing whether this requirement is satisfied, a clear idea of the properties of quadrature rules is valuable, and we must explore the differentiability properties of $f(x)$. In this respect we should note that

$$f(x) = g(x) + \lambda\int_a^b R(x,y;\lambda)g(y)dy ,$$

where $R(x,y;\lambda)$ is the resolvent kernel; it is quite likely that if $g(x)$ is badly behaved then $f(x)$ will 'inherit' this bad behaviour.

Example 4.8.

The solution of the equation

$$f(x) + \int_0^1 \sqrt{(xy)}f(y)\,dy = \sqrt{x}$$

is $f(x) = (\frac{2}{3})\sqrt{x}$. Even though $f'(x)$ is unbounded at $x=0$, the trapezium rule in the quadrature method gives the correct answer (see Example 4.7) because the bad behaviour of $K(x,y) = \sqrt{(xy)}$ and $f(y) = \frac{2}{3}\sqrt{y}$ 'cancel' one another in the integrand. The equation $f(x) + \int_0^1 \sqrt{x}\, f(y)\,dy = \sqrt{x}$ also has a solution $f(x) = \alpha\sqrt{x}$ but the use of the trapezium rule with step h yields a solution whose error is $O(h^{3/2})$, and we obtain a more modest (but acceptable?) rate of convergence as $h \to 0$. *

4.3.1. *Deferred approach to the limit*

In our choice of quadrature rule we may favour a rule which permits us to apply the deferred approach to the limit in a convenient way. If the solution $f(x)$ and the kernel $K(x,y)$ are sufficiently smooth we can apply a composite rule $(m \times J)(\phi)$ for varying m and apply the deferred approach to the limit to the resulting approximations. In particular, we can apply the repeated trapezium rule with step h to obtain approximate values of $\tilde{f}(x) \equiv \tilde{f}_h(x)$. Suppose that we solve the equations

$$\tilde{f}_h(a+ih) - \lambda h \sum_{j=0}^{n}{}'' K(a+ih,a+jh)\tilde{f}_h(a+jh) = g(a+ih) \qquad (4.7)$$

$(i = 0,1,2,\ldots,n,\ h = (b-a)/n)$ for $h \in \{h_0, \tfrac{1}{2}h_0, \tfrac{1}{4}h_0\}$ say. In this way, we form a table of approximate values (Table 4.3).

TABLE 4.3

$x =$	a	$a+\frac{1}{4}h_0$	$a+\frac{1}{2}h_0$	$a+\frac{3}{4}h_0$	$a+h_0$	$a+\frac{5}{4}h_0$	$a+\frac{3}{2}h_0$	...	b
$h = h_0$	×				×				×
$h = \frac{1}{2}h_0$	×		×		×		×		×
$h = \frac{1}{4}h_0$	×	×	×	×	×	×	×	×	×

Table 4.3 is a 'ghost table', the symbol × indicating the entry of a numerical value of the function $\tilde{f}_h(x)$ in a real table.

Thus, if we calculate the vectors $\tilde{\underset{\sim}{f}}_h$ for $h = h_0, \frac{1}{2}h_0, \frac{1}{4}h_0$ we obtain three estimates of each value $f(a), f(a+h_0), f(a+2h_0), \ldots, f(b)$, two estimates of each value $f(a+\frac{1}{2}h_0), f(a+\frac{3}{2}h_0), \ldots, f(b-\frac{1}{2}h_0)$, and one estimate of each value $f(a+\frac{1}{4}h_0), f(a+\frac{3}{4}h_0), \ldots, f(b-\frac{1}{4}h_0)$. Suppose that (as can be shown under certain conditions)

$$\tilde{f}_h(a+ih) = f(a+ih) + \sum_{k=1}^{N} h^{2k}\phi_k(a+ih) + O(h^{2N+1}), \qquad (4.8)$$

uniformly in i, where $\phi_k(x)$ $(k = 0,1,2,\ldots,N)$ are independent of h. Then we can combine the estimates of the values $f(a+ih_0)$ using the deferred approach to the limit, to form

$$\hat{f}_{h_0}^{[1]}(a+ih_0) = \frac{4\tilde{f}_{\frac{1}{2}h_0}(a+ih_0) - \tilde{f}_{h_0}(a+ih_0)}{3},$$

$$\hat{f}_{h_0}^{[2]}(a+ih_0) = \frac{16\hat{f}_{\frac{1}{2}h_0}^{[1]}(a+ih_0) - \hat{f}_{h_0}^{[1]}(a+ih_0)}{15},$$

and so on, in analogy with the classical Romberg scheme (q.v.). Then $f(a+ih_0) - \tilde{f}(a+ih_0) = O(h_0^2)$, $f(a+ih_0) - \hat{f}^{[1]}(a+ih_0) = O(h_0^4)$, and $f(a+ih_0) - \hat{f}^{[2]}(a+ih_0) = O(h_0^6)$, where the order terms hold uniformly in i, as h_0 is decreased.

Under slightly stronger conditions than are required for the validity of (4.8) we can show that, when $\tilde{f}(x)$ is obtained from $\tilde{\underset{\sim}{f}}$ using the Nyström extension (4.4),

$$\tilde{f}(x) = f(x) + \sum_{k=1}^{N} h^{2k}\phi_k(x) + O(h^{2N+1}) \qquad (4.9)$$

uniformly for $a \leq x \leq b$, where $\phi_k(x)$ $(k = 0,1,2,\ldots, N)$ are, as before, independent of h. (We cannot in general establish the validity of (4.9) when $K(x,y)$ has a discontinuous derivative at $x = y$.)

We can now see that the values missing from the table above, namely $\tilde{f}_{h_0}(a+\frac{1}{4}h_0), \tilde{f}_{h_0}(a+\frac{1}{2}h_0), \tilde{f}_{h_0}(a+\frac{3}{4}h_0), \ldots,$ and $\tilde{f}_{\frac{1}{2}h_0}(a+\frac{1}{4}h_0)$, $\tilde{f}_{\frac{1}{2}h_0}(a+\frac{3}{4}h_0), \ldots,$ can be obtained by interpolation using the Nyström extension that is, by using eqn (4.4). If the relation (4.9) holds we can perform the deferred approach to the limit on each column of the 'filled-in' table and obtain $\hat{f}^{[2]}(x)$ for $x = a,\ a + \frac{1}{4}h_0,\ a + \frac{1}{2}h_0,$ $a + \frac{3}{4}h_0, \ldots, b$; indeed, for any $x \in [a,b]$.

If the trapezium rule with step h is replaced in (4.7) by the composite rule $(m \times J)(\phi)$ then, under suitably strong conditions, we obtain

$$\tilde{f}(x) \equiv \tilde{f}_{1/m}(x) = f(x) + \sum_{k=r+1}^{M} (1/m)^k \psi_k(x) + O(1/m^{M+1}), \qquad (4.10)$$

where the degree of precision of $J(\phi)$ is r, and we assume that $r \geq 0$. If $J(\phi)$ is a symmetric rule then $\psi_k(x)$ vanishes if k is odd. To employ (4.10) as the basis for a method of deferred approach we must in general compute values of the Nyström extension (4.4). Unless $J(\phi)$ employs equally spaced abscissae, it is unlikely that the components of the vectors $\tilde{\underset{\sim}{f}}$ (computed using two different values of m) will have a subset which approximate common values of $f(x)$. For example, the abscissae of the m-times repeated two-point Gauss-Legendre rule on $[-1, 1]$ have the form $\pm(\frac{1}{\sqrt{3}})(m = 1)$; $\frac{1}{2}\{1 \pm(\frac{1}{\sqrt{3}})\}$, $\frac{1}{2}\{-1 \pm(\frac{1}{\sqrt{3}})\}$ $(m = 2)$; etc., and there are no common points for different values of m. In such circumstances, we may use eqn (4.4) to form for different values of m, the approximations $\tilde{f}_{1/m}(z_0)$, $\tilde{f}_{1/m}(z_1), \ldots,$ $\tilde{f}_{1/m}(z_N)$, where $z_1, z_2, \ldots, z_N$ are fixed points at which approximate values of $f(x)$ are to be found. If we write $x = z_i$ $(i = 0,1,2,\ldots,N)$ in (4.10), we have the theoretical basis for performing a process of deferred approach to the limit on the values $\tilde{f}_{1/m}(z_i)$, with $m = m_0, m_1, \ldots$. This same technique can be applied if the *trapezium rule* is used with steps $h_0, h_1, h_2, \ldots$ which are not in a common ratio.

<u>Example 4.9.</u>
Consider Love's equation (see Example 4.6) with $d = -1$ and $g(x) = 1$. If we perform repeated h^2-extrapolation on the values obtained by applying the trapezium rule with steps $h = \frac{1}{8}, \frac{1}{16}, \frac{1}{32}, \frac{1}{64}$ we obtain the following approximate values for $\hat{f}_{\frac{1}{8}}^{[3]}(x)$:

$x = \pm 1$	$\pm\frac{3}{4}$	$\pm\frac{1}{2}$	$\pm\frac{1}{4}$	0	
1·63970	1·75195	1·84238	1·89962	1·91903	*

4.3.2. *Asymptotic upper and lower bounds*
Possibly of greater value than the use of the deferred approach is the opportunity to compute asymptotic upper and lower bounds on the solution (see section 2.18). The values of the approximations $\tilde{f}(x)$ which are needed for the construction of these asymptotic bounds are the same as those needed for the method of deferred approach.

When the relation (4.8) holds we can compute asymptotic upper and lower bounds for solution values of the form $f(a+ih)$ (in particular, for $f(a)$, $f(a+h_0)$, ..., and $f(b)$ in the situation illustrated above). When one of the more general relations (4.9) or (4.10) is valid, we can compute asymptotic upper and lower bounds for every value $f(x)$, where $a \le x \le b$.

<u>Example 4.10.</u>
The values used in Example 4.9 can be employed to estimate upper and lower bounds on $f(x)$ for $x = -1(\frac{1}{4})1$. The table of values of

$\check{f}_h(x) = 2\tilde{f}_{\frac{1}{2}h}(x) - \tilde{f}_h(x)$ is shown in Table 4.4.

TABLE 4.4

$x \backslash h =$	$\frac{1}{4}$	$\frac{1}{8}$	$\frac{1}{16}$	Estimated true value
$\pm$ 1·00	1·64135	1·64011	1·63979	1·63970
$\pm$ 0·75	1·75445	1·75258	1·75210	1·75195
$\pm$ 0·50	1·84537	1·84313	1·84257	1·84238
$\pm$ 0·25	1·90176	1·90040	1·89982	1·89962
0	1·92220	1·91982	1·91923	1·91903

This table should be compared with the first table of values of $\tilde{f}_h(x)$ in Example 4.6. Corresponding values of $\tilde{f}_h(x)$ are rather more accurate and lie the other side of the true values from the figures here. Both features are typical. *

4.3.3. *Deferred correction*

If $K(x,y)f(y)$ is sufficiently differentiable in y, we can employ a deferred correction method for the approximate solution of eqn (4.1). We shall indicate how Gregory's integration rule can be used in this context.

There are a number of slight variants of a deferred correction method. The two principal types are the 'deferred correction' methods (sometimes called 'difference correction' methods) and the 'iterated deferred correction' methods, in which a method of the first type is used iteratively.

To illustrate the first type, suppose that an initial approximation $\tilde{f}(x) = \tilde{f}^{(0)}(x)$ has been constructed using the quadrature method and employing the repeated trapezium rule with step h. It is sufficient merely to compute the vector $\tilde{\underset{\sim}{f}}^{(0)} = [\tilde{f}^{(0)}(a), \tilde{f}^{(0)}(a+h), \ldots, \tilde{f}^{(0)}(b)]^T$,

where

$$\tilde{f}^{(0)}(a+ih) - \lambda h \sum_{j=0}^{n}{}'' K(a+ih, a+jh)\tilde{f}^{(0)}(a+jh) = g(a+ih) \qquad (4.11)$$

$(i = 0,1,2,\ldots, n; \quad h = (b-a)/n)$.

Now, suppose that we use the Gregory rule in place of the repeated trapezium rule to set up equations for a new approximation $\tilde{f}(x)$. We may set

$$\tilde{f}(a+ih) - \lambda\{h \sum_{j=0}^{n}{}'' K(a+ih, a+jh)\tilde{f}(a+jh) + \delta_i(\tilde{f})\} = g(a+ih) \qquad (4.12)$$

$$(i = 0,1,2,\ldots, n) ,$$

where (section 2.13.2) we set

$$\delta_i(\tilde{f}) = h \sum_{s=1}^{p} c_s^*\{\nabla^s\Theta_i(\tilde{f};b) + (-1)^s\Delta^s\Theta_i(\tilde{f};a)\} \qquad (4.13)$$

with $p \leq n$, say, and $\nabla^s\Theta_i(\tilde{f};b) = \nabla^s\Theta_i(x)]_{x=b}$, etc., where

$$\Theta_i(x) \equiv \Theta_i(\tilde{f};x) = K(a+ih,x)\tilde{f}(x) . \qquad (4.14)$$

Thus, $\delta_i(\tilde{f})$ represents Gregory correction terms to the repeated trapezium rule. These correction terms are 'reasonable' if $K(x,y)f(y)$ has a continuous pth order y-derivative, and it is possible to modify $\delta_i(\tilde{f})$ in (4.13) by replacing p by a value p_i , which depends on i . (The process becomes more involved if $K(x,y)$ has discontinous derivatives, at $x = y$, say, since $\delta_i(\tilde{f})$ must take account of the discontinuity.)

Now, we may regard $\tilde{\underset{\sim}{f}}^{(0)}$, defined by eqn (4.11), as an approximation to $\tilde{\underset{\sim}{f}} = [\tilde{f}(a),\tilde{f}(a+ih), \ldots, \tilde{f}(b)]^T$ defined by (4.12). An improved approximation may be obtained, under certain conditions, if we write

$$\tilde{f}^{(1)}(a+ih) - \lambda h \sum_{j=0}^{n}{}'' K(a+ih, a+jh)\tilde{f}^{(1)}(a+jh)$$
$$= g(a+ih) + \lambda\delta_i(\tilde{f}^{(0)}) .$$

We actually compute $\underset{\sim}{\varepsilon}^{(1)} = \tilde{\underset{\sim}{f}}^{(1)} - \tilde{\underset{\sim}{f}}^{(0)}$, by solving the equations

$$\varepsilon_i^{(1)} - \lambda h \sum_{j=0}^{n}{}'' K(a+ih,\ a+jh)\varepsilon_j^{(1)} = \lambda\delta_i(\tilde{f}^{(0)}) .$$

in matrix notation

$$(\underset{\sim}{I} - \lambda\underset{\sim}{K}\underset{\sim}{D})\ \underset{\sim}{\varepsilon}^{(1)} = \lambda\underset{\sim}{\delta}(\tilde{f}^{(0)}) .$$

It may be noted that if $\tilde{\underset{\sim}{f}}^{(0)}$ is obtained using Gaussian elimination, information obtained during this calculation may be used to calculate $\underset{\sim}{\varepsilon}^{(1)}$ in an efficient manner.† Having computed $\underset{\sim}{\varepsilon}^{(1)}$, we recover $\tilde{\underset{\sim}{f}}^{(1)}$ by setting $\tilde{\underset{\sim}{f}}^{(1)} = \tilde{\underset{\sim}{f}}^{(0)} + \underset{\sim}{\varepsilon}^{(1)}$. The process of computing $\tilde{\underset{\sim}{f}}^{(1)}$ from $\tilde{\underset{\sim}{f}}^{(0)}$ is called a *deferred correction*.

We are now in a position to describe methods of iterated deferred correction. In this method, we implement the previous step, but now continue by constructing the vectors $\tilde{\underset{\sim}{f}}^{(r)}$ such that

$$(\underset{\sim}{I} - \lambda\underset{\sim}{K}\underset{\sim}{D})\tilde{\underset{\sim}{f}}^{(r)} = \underset{\sim}{g} + \lambda\underset{\sim}{\delta}(\tilde{f}^{(r-1)})$$

(where $\underset{\sim}{\delta}(\tilde{f}^{(r-1)}) = [\delta_0(\tilde{f}^{(r-1)}),\ \delta_1(\tilde{f}^{(r-1)}),\ \ldots,\ \delta_n(\tilde{f}^{(r-1)})]^T$ and $\delta_i(\tilde{f}^{(r-1)})$ is defined by substituting $\tilde{f}^{(r-1)}$ for $\tilde{f}$ in (4.13) and (4.14), and where $\underset{\sim}{K} = [K(a+ih,\ a+jh)]$, $\underset{\sim}{D} = \mathrm{diag}(\tfrac{1}{2}h,\ h,\ \ldots,\ h,\ \tfrac{1}{2}h)$, etc.). We actually compute, at each stage, $\underset{\sim}{\varepsilon}^{(r)} = \tilde{\underset{\sim}{f}}^{(r)} - \tilde{\underset{\sim}{f}}^{(r-1)}$ satisfying

$$(\underset{\sim}{I} - \lambda\underset{\sim}{K}\underset{\sim}{D})\ \underset{\sim}{\varepsilon}^{(r)} = \lambda\underset{\sim}{\delta}(\varepsilon^{(r-1)}) \quad (r > 1)$$

and then, in effect, form

$$\tilde{\underset{\sim}{f}}^{(r)} = \tilde{\underset{\sim}{f}}^{(r-1)} + \underset{\sim}{\varepsilon}^{(r)} = \underset{\sim}{\varepsilon}^{(r)} + \ldots + \underset{\sim}{\varepsilon}^{(1)} + \underset{\sim}{f}^{(0)},$$

and we also compute $\underset{\sim}{\delta}(\tilde{\varepsilon}^{(r)})$ for the next stage. This iteration might

† Compare this with algorithms for the iterative refinement of solutions of systems of linear equations.

be employed until convergence is obtained, but since convergence cannot be guaranteed we may accept, say, $\tilde{\underset{\sim}{f}}^{(p)}$ as the final approximation, when (4.13) is used to define $\underset{\sim}{\delta}(\tilde{f})$ and p is independent of i.

The scheme provides some scope for flexibility. If computing $\tilde{\underset{\sim}{f}}^{(r)}$ directly, we require the $\underset{\sim}{\delta}(\tilde{f}^{(r-1)})$ whose components are defined by (4.13) and (4.14) on replacing $\tilde{f}$ by $\tilde{f}^{(r-1)}$. As we noted above, it is reasonable to permit p to depend on i in the definition (4.13); it is also reasonable to permit p to depend on r when defining $\delta_i(\tilde{f}^{(r-1)})$. It is equally possible to set p equal to the value, depending on r and on i, which makes the contribution of the pth difference negligible in $\delta_i(\tilde{f}^{(r-1)})$, but a more systematic procedure, which has a sound theoretical basis when $K(x,y)f(y)$ is sufficiently differentiable, entails setting

$$\delta_i(\tilde{f}^{(r-1)}) = h\sum_{s=1}^{r} c_s^*\{\nabla^s\Theta_i(\tilde{f}^{(r-1)};b) + (-\Delta)^s\Theta_i(\tilde{f}^{(r-1)};a)\}\ . \qquad (4.15)$$

Thus, our last suggestion entails setting $p = r$, and we tend to *favour* this if p is not to remain fixed. (See Theorem 4.22, below.)

Whilst we have expressed the Gregory correction terms (4.13) and (4.15) using finite difference notation, we should note that $\delta_i(\tilde{f})$ can be expressed in Lagrangian form, in terms of the function $\Theta_i(x) = \Theta_i(\tilde{f};x)$ of (4.14), as

$$h\sum_{j=0}^{p}\Omega_j^{[p]}\ \{\Theta_i(a+jh) + \Theta_i(b-jh)\}\ , \qquad (4.16)$$

where $\Omega_j^{[p]}$ is given in section 2.13.2. Whichever formulation is used (and the Lagrangian expression appears computationally convenient) it appears to be necessary to reduce to a minimum cancellation errors due to rounding which occur when computing $\delta_i(\tilde{f}^{(r-1)})$. If the matrix $\underset{\sim}{I} - \lambda\underset{\sim}{K}\underset{\sim}{D}$ is ill conditioned there may be some justification for computing each vector $\underset{\sim}{\varepsilon}^{(r-1)}$ $(r = 1,2,3,\ldots)$ by Gaussian elimination, followed by iterative refinement using some double-length arithmetic (Forsythe and Moler 1967). In some circumstances, it is necessary to use double-length arithmetic for the Gaussian elimination, depending on the computer employed.

Remark. When solving systems of linear equations, the process of iterative refinement can often be used to indicate the presence of any ill-conditioning (Forsythe and Moler 1967).

Example 4.11.

Consider the equation

$$f(x) - \int_0^1 e^{xy} f(y)\,dy = 1 - \frac{(e^x - 1)}{x} .$$

Various versions of a deferred correction process were applied to this equation, using Gregory-rule correction terms to the trapezium rule throughout. The experiments were performed using different degrees of precision in floating-point arithmetic, and the maximum error $\|\underset{\sim}{f} - \underset{\sim}{\tilde{f}}^{(r-1)}\|_\infty$ tabulated. No attempt was made to avoid possible loss of precision in forming $(e^x-1)/x$ near $x = 0$.

Case (a). A single deferred correction, with 10 - 11 significant decimal figures throughout, and employing fixed numbers of differences, gave the following maximum errors (Table 4.5).

TABLE 4.5

h	Trapezium rule	1 difference	2 differences	3 differences
$\frac{1}{8}$	$2{\cdot}42 \times 10^{-3}$	$2{\cdot}97 \times 10^{-4}$	$4{\cdot}22 \times 10^{-6}$	$1{\cdot}33 \times 10^{-6}$
$\frac{1}{16}$	$6{\cdot}01 \times 10^{-4}$	$3{\cdot}81 \times 10^{-5}$	$3{\cdot}63 \times 10^{-7}$	$4{\cdot}54 \times 10^{-8}$
$\frac{1}{32}$	$1{\cdot}50 \times 10^{-4}$	$4{\cdot}85 \times 10^{-6}$	$2{\cdot}81 \times 10^{-8}$	$1{\cdot}81 \times 10^{-9}$
$\frac{1}{64}$	$3{\cdot}75 \times 10^{-5}$	$6{\cdot}13 \times 10^{-7}$	$2{\cdot}42 \times 10^{-9}$	$6{\cdot}86 \times 10^{-10}$

Case (b). A process of iterative deferred correction was performed, using r differences when computing $\underset{\sim}{\delta}(\tilde{f}^{(r-1)})$ $(r = 0,1,2,3,4,5)$. The experiment was performed using 6-7 significant decimal figures

(Table 4.6) and repeated using 15-16 significant decimal figures (Table 4.7) and the maximum errors $\|\underset{\sim}{f} - \underset{\sim}{\tilde{f}}^{(r)}\|_\infty$ were computed.

TABLE 4.6

h	Trapezium rule	1 iteration	2 iterations
$\frac{1}{8}$	$1{\cdot}6 \times 10^{-3}$	$2{\cdot}2 \times 10^{-4}$	$1{\cdot}0 \times 10^{-5}$
$\frac{1}{16}$	$4{\cdot}0 \times 10^{-4}$	$3{\cdot}0 \times 10^{-5}$	$7{\cdot}6 \times 10^{-6}$
$\frac{1}{32}$	$1{\cdot}1 \times 10^{-4}$	$1{\cdot}2 \times 10^{-5}$	$1{\cdot}4 \times 10^{-5}$
$\frac{1}{64}$	$3{\cdot}7 \times 10^{-5}$	$2{\cdot}0 \times 10^{-5}$	$2{\cdot}0 \times 10^{-5}$
	3 iterations	4 iterations	5 iterations
	No improvement		

TABLE 4.7

h	Trapezium rule	1 iteration	2 iterations
$\frac{1}{8}$	$1{\cdot}6 \times 10^{-3}$	$2{\cdot}2 \times 10^{-4}$	$8{\cdot}6 \times 10^{-6}$
$\frac{1}{16}$	$3{\cdot}9 \times 10^{-4}$	$2{\cdot}7 \times 10^{-5}$	$5{\cdot}5 \times 10^{-7}$
$\frac{1}{32}$	$9{\cdot}9 \times 10^{-5}$	$3{\cdot}4 \times 10^{-6}$	$3{\cdot}4 \times 10^{-8}$
$\frac{1}{64}$	$2{\cdot}5 \times 10^{-5}$	$4{\cdot}2 \times 10^{-7}$	$2{\cdot}2 \times 10^{-9}$

h	3 iterations	4 iterations	5 iterations
$\frac{1}{8}$	$1{\cdot}7 \times 10^{-6}$	$6{\cdot}3 \times 10^{-8}$	$1{\cdot}6 \times 10^{-8}$
$\frac{1}{16}$	$5{\cdot}3 \times 10^{-8}$	$1{\cdot}1 \times 10^{-9}$	$1{\cdot}3 \times 10^{-10}$
$\frac{1}{32}$	$1{\cdot}7 \times 10^{-9}$	$1{\cdot}8 \times 10^{-11}$	$1{\cdot}0 \times 10^{-12}$
$\frac{1}{64}$	$5{\cdot}2 \times 10^{-11}$	$3{\cdot}0 \times 10^{-13}$	$1{\cdot}1 \times 10^{-14}$

Case (c). A process of iterative correction was performed using 5 differences ($p = 5$ in (4.13)) to compute $\delta(\tilde{f}^{(r-1)})$ ($r = 0,1,2,3,4,5$). The figures obtained using 15-16 significant decimal figures are displayed in Table 4.8.

TABLE 4.8

h	Trapezium	1 iteration	2 iterations
$\frac{1}{8}$	$1{\cdot}6 \times 10^{-3}$	$1{\cdot}3 \times 10^{-5}$	$7{\cdot}8 \times 10^{-8}$
$\frac{1}{16}$	$3{\cdot}9 \times 10^{-4}$	$8{\cdot}0 \times 10^{-7}$	$1{\cdot}1 \times 10^{-9}$
$\frac{1}{32}$	$9{\cdot}9 \times 10^{-5}$	$5{\cdot}0 \times 10^{-8}$	$1{\cdot}7 \times 10^{-11}$
$\frac{1}{64}$	$2{\cdot}5 \times 10^{-5}$	$3{\cdot}1 \times 10^{-9}$	$2{\cdot}6 \times 10^{-13}$
	3 iterations	4 iterations	5 iterations
	$1{\cdot}6 \times 10^{-8}$	No further improvement	
	$1{\cdot}3 \times 10^{-10}$		
	$1{\cdot}0 \times 10^{-12}$		
	$1{\cdot}1 \times 10^{-14}$		

It is quite clear from these figures that the word-length of the computer obscures effects which would be expected when using exact arithmetic. *

Example 4.12.

The methods of deferred correction were applied to the equation

$$f(x) - \frac{1}{\pi}\int_{-1}^{1} \frac{f(y)}{1 + (x-y)^2}\, dy = 1$$

using the trapezoidal rule with step h and the Gregory-rule correction terms. Using 15-16 significant decimal figures the approximate values shown in Table 4.9 were computed.

TABLE 4.9

x	Trapezium rule ($h = \frac{1}{32}$)	1 iteration	2 iterations
0	1·918932396	1·919033247	1·919032035
0·25	1·899516469	1·899616260	1·899615225
0·50	1·842291248	1·842385317	1·842384891
0·75	1·751876332	1·751954266	1·751954812
1·00	1·639643383	1·639694113	1·639695272
x	3 iterations	4 iterations	5 iterations
0	1·919031988	1·919031993	1·9190319931
0·25	1·899615162	1·899615168	1·8996151676
0·50	1·842384789	1·842384795	1·8423847951
0·75	1·751954702	1·751954700	1·7519547009
1·00	1·639695229	1·639695222	1·6396952217

In these results, rth differences were used to compute $\underset{\sim}{\delta}(\tilde{f}^{(r-1)})$. *

We could employ Gregory's rule directly and consider eqns (4.12), rewritten in the form

$$\tilde{f}(a+ih) - \lambda h \sum_{j=0}^{n} \omega_{j,p} K(a+ih,\ a+jh)\tilde{f}(a+jh) = g(a+ih) \ , \qquad (4.17)$$

where $\omega_{0,p} = \frac{1}{2} + \Omega_0^{[p]} = \omega_{n,p}$ and $\omega_{j,p} = 1 + \Omega_j^{[p]}$ $(j = 1,2,3,\ldots,(n-1))$ the weights $\Omega_j^{[p]}$ being those in (4.16). There are, perhaps, three principal reasons for favouring the iterated deferred correction method. In the first place, the successive vectors $\tilde{\underset{\sim}{f}}^{(r-1)}$ $(r = 1,2,3,\ldots)$, which are produced in the iterative method, provide various estimates of $\underset{\sim}{f}$, and they can be compared against one another to give an idea of their reliability. In the second place, if we rewrite eqn (4.12) in the form (4.17) we must have in mind a fixed value of p for each correction term $\delta_i(\tilde{f})$, whilst there is some flexibility in the choice of p in the iterative method. Finally, if the diagonal elements of $\underset{\sim}{D}$ are positive and $\underset{\sim}{K}$ is Hermitian, we can rewrite

$$(\underset{\sim}{I} - \lambda\underset{\sim}{K}\underset{\sim}{D})\tilde{\underset{\sim}{f}} = \underset{\sim}{g} \tag{4.18}$$

in the form

$$(\underset{\sim}{I} - \lambda\underset{\sim}{D}^{\frac{1}{2}}\underset{\sim}{K}\underset{\sim}{D}^{\frac{1}{2}})(\underset{\sim}{D}^{\frac{1}{2}}\underset{\sim}{f}) = \underset{\sim}{D}^{\frac{1}{2}}\underset{\sim}{g}\ , \tag{4.19}$$

in which the matrix $\underset{\sim}{I} - \lambda\underset{\sim}{D}^{\frac{1}{2}}\underset{\sim}{K}\underset{\sim}{D}^{\frac{1}{2}}$ is Hermitian. Thus, since the weights in the trapezium rule are positive,

$$(\underset{\sim}{I} - \lambda\underset{\sim}{D}^{\frac{1}{2}}\underset{\sim}{K}\underset{\sim}{D}^{\frac{1}{2}})\underset{\sim}{D}^{\frac{1}{2}}\underset{\sim}{\varepsilon}^{(r-1)} = \underset{\sim}{D}^{\frac{1}{2}}\underset{\sim}{\delta}(f^{(r-1)})\ .$$

Though the preservation of symmetry is possibly not as important as in the eigenvalue problem, it may be of consequence. Thus it is of some relevance that the weights $\omega_{j,p}$ in (4.17) are of mixed sign for various values of p, so that it is not *generally* possible to preserve symmetry in (4.17) when $K(x,y)$ is Hermitian.

We have described the deferred correction methods in terms of difference corrections to the repeated trapezium rule. If $K(x,y)f(y)$ is sufficiently smooth, a similar technique can, in principle, be applied with any rule which is exact for polynomials of a certain degree. It is fairly simple, for example, to obtain difference corrections to the parabolic rule. For a repeated Gauss rule it may be preferable to employ *derivatives* of $K(x,y)\tilde{f}^{(r-1)}(y)$ to obtain correction terms;[†] clearly this is practicable only if $K(x,y)$ and $g(x)$ are simple functions.

4.4. *A modification to the quadrature method*

If $K(x,y)$ is badly behaved at $x = y$, we can (as in section 3.3) employ a device to reduce the error in the quadrature method. We write eqn (4.1) in the form

$$f(x) - \lambda f(x)\int_a^b K(x,y)\,dy - \lambda\int_a^b K(x,y)\{f(y) - f(x)\}\,dy = g(x) \tag{4.20}$$

and use a quadrature rule $J(\phi)$ to obtain

$$\mathring{f}(x)\{1-\lambda A(x)\} - \lambda\sum_{j=0}^{n} w_j K(x,y_j)\{\mathring{f}(y_j) - \mathring{f}(x)\} = g(x)\ , \tag{4.21a}$$

where

$$A(x) = \int_a^b K(x,y)\,dy\ . \tag{4.21b}$$

Thus

$$\mathring{f}(x)\{1+\lambda\Delta(x)\} - \lambda\sum_{j=0}^{n} w_j K(x,y_j)\mathring{f}(y_j) = g(x)\ , \tag{4.22}$$

where

$$\Delta(x) = \sum_{j=0}^{n} w_j K(x,y_j) - A(x)\ . \tag{4.23}$$

Setting $x = y_i$ in eqn (4.22) we obtain

$$\mathring{f}(y_i)(1+\lambda\Delta(y_i)) - \lambda\sum_{j=0}^{n} w_j K(y_i,y_j)\mathring{f}(y_j) = g(y_i)\ (i = 0,1,2,\ldots,n)\ .$$

† Such correction terms are available from the generalized Euler-Maclaurin formula (see, for example, Baker and Hodgson 1971).

In matrix notation

$$\{\underset{\sim}{I} + \lambda(\underset{\sim}{\Delta} - \underset{\sim}{K}\underset{\sim}{D})\}\overset{\circ}{\underset{\sim}{f}} = \underset{\sim}{g} \,, \tag{4.24}$$

where $\underset{\sim}{\Delta} = \operatorname{diag}\{\Delta(y_0), \Delta(y_1), \ldots, \Delta(y_n)\}$. As we noted in section 3.3, $A(x)$ should be computed accurately if the algorithm is to show an improvement over the one in section 4.3. A method based on the use of (4.24) or (4.22) will be called a *modified quadrature method*.

The approximate solution of eqn (4.1) by computing $\overset{\circ}{\underset{\sim}{f}}$ has been proposed for the case where $K(x,y)$ has a discontinuous derivative at $x = y$ (and also where $K(x,y)$ is weakly singular: see Chapter 5). In practice $\overset{\circ}{f}(x)$ may yield a better solution than the approximation $\tilde{f}(x)$ obtained from the quadrature method, even if $K(x,y)$ is well behaved. In particular, if $K(x,y)$ has a 'hump' at $x = y$ we may prefer the method of this section to that of section 4.3.

Example 4.13.

We consider the example of Love's equation treated earlier in Example 4.6. In order to compute $\underset{\sim}{\Delta}$, we require

$$A(x) = \int_{-1}^{1} K(x,y)\,dy = \frac{-1}{\pi}\tan^{-1}(2/x^2) \,.$$

Employing the trapezium rule repeated with step $h = 2/n$ we obtain the values shown in Table 4.10 for $\overset{\circ}{f}(x)$ at $x = (-1)(0{\cdot}25)1$.

TABLE 4.10

x \ n	8	16	32	64
$\pm$ 1·00	1·63838	1·63937	1·63961	1·63967
$\pm$ 0·75	1·75071	1·75164	1·75188	1·75194
$\pm$ 0·50	1·84135	1·84213	1·84232	1·84237
$\pm$ 0·25	1·89874	1·89940	1·89956	1·89960
0	1·91821	1·91883	1·91898	1·91902

These values compare favourably with the first table of values in Example 4.6. For comparison with the second table of values in Example 4.6, we computed $\overset{\circ}{f}(x)$ using the various basic Romberg rules and we obtained the values shown in Table 4.11.

TABLE 4.11

x \ Rules	9-point	17-point	33-point	65-point
± 1·00	1·63967	1·63970	1·63970	1·63970
± 0·75	1·75189	1·75196	1·75195	1·75195
± 0·50	1·84235	1·84239	1·84238	1·84238
± 0·25	1·89964	1·89961	1·89961	1·89962
0	1·91907	1·91903	1·91903	1·91903

*

The success of the method may be judged to depend in part on the smoothness of $f(x)$, since the method is based on the idea that as $y \to x$ in the third term on the left-hand side of eqn (4.20), this term will tend to zero. If $f(x)$ is discontinuous at some abscissae y_i, this may cause difficulty. On the other hand, if $f(x)$ is Lipschitz-continuous with a small exponent α, we can expect $f(y) - f(x)$ to tend slowly to zero as $y \to x$. (These rather naïve ideas have to be applied rather cautiously or they give misleading conclusions.)

Example 4.14.

Consider the equation

$$f(x) - \int_0^1 K(x,y)f(y)\mathrm{d}y = g(x) ,$$

where $K(x,y) = \min(x,y)$ and $g(x)$ is arbitrary. If we apply the modified quadrature method with the trapezium rule we obtain a vector $\overset{o}{\underset{\sim}{f}}$ which agrees with $\underset{\sim}{\tilde{f}}$, since $\Delta(y_i) = 0$ in this case. *

Example 4.15.

The order of accuracy in the values $\tilde{f}(y_i)$ computed using the method of section 4.3 with a rule $\sum_{j=0}^{n} w_j \phi(y_j)$ depends on the order of the values $\max_{0 \le i \le n} |\tau(y_i)|$, where

$$\tau(x) = \lambda \int_a^b K(x,y)f(y)dy - \lambda \sum_{j=0}^{n} w_j K(x,y_i)f(y_i) .$$

The error in the Nyström extension $\tilde{f}(x)$ where x ranges in $[a,b]$ depends on $\|\tau(x)\|_\infty = \sup_{a \le x \le b} |\tau(x)|$.

In the case of the modified quadrature method discussed in this section, there is an analogue (4.22) of the Nyström extension, which can be used if $K(x,y)$ is well behaved. We shall not consider this extension here, but the errors of the approximation $\overset{o}{f}(x)$ at $y_0, y_1, \ldots, y_n$ depend (assuming $K(x,x)$ bounded for $a \le x \le b$) on

$$\lambda \max_{0 \le i \le n} \left| \int_a^b K(y_i,y)\{f(y)-f(y_i)\}dy - \lambda \sum_{\substack{j \ne i \\ j=0}}^{n} w_j K(y_i,y_j)\{f(y_j)-f(y_i)\} \right| . \qquad (4.25)$$

Suppose that $K(x,y)$ is continuous and satisfies a Lipschitz condition in y (with exponent 1), and that $f(x)$ is also Lipschitz-continuous. Then

$$\int_a^b K(x,y)f(y)dy - h \sum_{j=0}^{n}{}'' K(x,a+jh)f(a+jh) = O(h) ,$$

that is, $\|\tau(x)\|_\infty = O(h)$ if we use the trapezium rule, and in this case the error (4.25) is also $O(h)$.

By way of contrast, consider the case where $K(x,y)$ suffers a finite jump discontinuity at $x = y$ but $(\partial/\partial y)^2 K(x,y)$ is continuous for $a \le y \le x \le b$, and for $a \le x \le y \le b$, and $f(x) \in C^2[a,b]$.

Then $||\tau(x)||_\infty$ and $\max\limits_{0\le i\le n} |\tau(a+ih)|$ are $O(h)$, and $||f(x)-\tilde{f}(x)||_\infty = O(h)$.

The modified quadrature method gives values with

$$\max_{0\le i\le n} |f(a+ih) - \overset{\circ}{f}(a+ih)| = O(h^2)$$

since the quantity (4.25) is $O(h^2)$. On the other hand, the quadrature method gives values $\tilde{f}(a+ih)$ with $O(h^2)$ accuracy if we *replace* $K(x,y)$ by $K_\mu(x,y) = \frac{1}{2}\{K(x,y+)+K(x,y-)\}$, where $K(x,y+) = \lim\limits_{y\to x+} K(x,y)$, etc., in the equations of the form (4.3).

It is of interest to consider the corresponding results for Simpson's rule, or a repeated rule of higher degree of precision. *

Rigorous discussion of the order of the local truncation errors provides insight into the relative merits of the method of this section and section 4.3, since the rate of convergence of the approximations corresponds to the order of these errors. Such a discussion is not, however, conclusive, since such arguments are asymptotic as $h \to 0$, and the order notation is in any case not very precise when used for purposes of comparison. Furthermore, we may seek to apply a method of deferred approach to the limit, or construct asymptotic bounds on the solution, using the modified quadrature method. This might sometimes be justified in cases where the validity of such techniques cannot be established for the quadrature method of section 4.3.

4.5. *A further modification*

If $K(x,y) = \psi(y)G(x,y)$ in (4.1), where $\psi(y)$ is badly behaved (for example, $\psi(y) = y^{\frac{1}{2}}$ so that $\psi'(0)$ is unbounded) then it seems reasonable to employ a quadrature rule for approximating products, of the form

$$\int_a^b \phi(y)\psi(y)dy \simeq \sum_{j=0}^{n} v_j\phi(z_j) . \tag{4.26}$$

In place of using a quadrature rule to produce the equations

$$\tilde{f}(y_i) - \lambda \sum_{j=0}^{n} w_j K(y_i, y_j)\tilde{f}(y_j) = g(y_i) ,$$

use of the approximation (4.26) will lead to equations of the form

$$\tilde{f}(z_i) - \lambda \sum_{j=0}^{n} v_j G(z_i, z_j)\tilde{f}(z_j) = g(z_i) \quad (i = 0,1,2,\ldots,n) ,$$

where $K(x,y) = \psi(y)G(x,y)$. If $\psi(y) > 0$, then we might construct a Gaussian quadrature rule to give the approximation (4.26). In any case, we have a particular case of a type of product integration method, and its success is clearly dependent on the smoothness of the solution $f(x)$, and of $G(x,y)$. Thus, if $G(x,y)$ suffers a discontinuity at $x = y$, the technique of the next section (see Example 4.18) may be more appropriate.

4.6. *A product-integration method*

As remarked in section 3.4, we may devise approximations of the form

$$\int_a^b K(z_i, y)\phi(y)\mathrm{d}y \simeq \sum_{j=0}^{n} v_j(z_i)\phi(z_j) ,$$

which are exact when $\phi(x)$ is a polynomial of a certain degree, or when $\phi(x)$ is piecewise-polynomial. The use of such an approximation permits us to replace the integral equation

$$f(x) - \lambda\int_a^b K(x,y)f(y)\mathrm{d}y = g(x)$$

by the approximations

$$\tilde{f}(z_i) - \lambda \sum_{j=0}^{n} v_j(z_i)\tilde{f}(z_j) = g(z_i) \quad (i = 0,1,2,\ldots; n) . \qquad (4.27)$$

In order to compute the values $v_j(z_i)$ $(i,j = 0,1,2,\ldots,n)$ it is in general necessary that $K(x,y)$ should be defined analytically. The success of the method depends in part upon the smoothness of the function $f(x)$, and hence, of course, upon λ and on the behaviour of $g(x)$ and $K(x,y)$.

Example 4.16.

Young (1954*b*) applies a product integration method to the problem

$$f(x) + \frac{1}{\pi}\int_{-1}^{1} \frac{f(y)}{\{1 + (x - y)^2\}} dy = 1 ,$$

which is obtained by setting $d = 1$ in the general equation in Example 4.6. Although the formulation of Young's method is rather different from the one we described, the processes are equivalent. On solving a system of 11 equations in 11 unknowns, Young obtains the values

x :	0	± 0·2	± 0·4	± 0·6	± 0·8	± 1
$\tilde{f}(x)$:	0·65741	0·66151	0·67389	0·69448	0·72249	0·75572

for this example. The product-integration rule was devised to be exact for $\int_{-1}^{1} K(x,y)\phi(y)dy$ if $\phi(x)$ was continuous and quadratic in each of the intervals $[-1,-\frac{3}{5}]$, $[-\frac{3}{5},-\frac{1}{5}]$, $[-\frac{1}{5},\frac{1}{5}]$, $[\frac{1}{5},\frac{3}{5}]$, $[\frac{3}{5},1]$. We refer to the original paper for details.

Corresponding figures computed using the quadrature method of section 4.3 with the trapezium rule with step h are given in Table 4.12.

TABLE 4.12

$h \backslash x$	0	$\pm\frac{1}{5}$	$\pm\frac{2}{5}$	$\pm\frac{3}{5}$	$\pm\frac{4}{5}$	± 1
$\frac{1}{10}$	0·65787	0·66197	0·67432	0·69481	0·72261	0·75553
$\frac{1}{20}$	0·65752	0·66163	0·67340	0·69546	0·72252	0·75567
$\frac{1}{40}$	0·65744	0·66154	0·67392	0·69450	0·72249	0·75571
$\frac{1}{80}$	0·65742	0·66152	0·67389	0·69449	0·72249	0·75572

From these figures it would appear that the performance of the product-integration method is superior, for a given amount of work (even if advantage is taken of centro-symmetry). *

Example 4.17.

With $K(x,y) = \min(x,y)$ consider the equation

$$f(x) + \int_0^1 K(x,y)f(y)\,dy = xe + 1 .$$

If we divide $[0,1]$ into n equal intervals of length $h = 1/n$, with $z_i = ih$, the approximation

$$\int_0^1 K(z_i,y)\phi(y)\,dy \simeq \sum_{j=0}^{n} v_j(z_i)\phi(z_j)$$

can be constructed to be exact if $\phi(x)$ is continuous and linear in each interval $[ih,(i+1)h]$ $(i = 0,1,2,\ldots,(n-1))$. We find that $v_{j-1}(z_{i-1}) = (\min(i,j)-1)h^2$ if $i \neq j$ and $v_{j-1}(z_{j-1}) = (j-\frac{7}{6})h^2$, when $j \neq 1$ and $j \neq n+1$, and $v_0(z_{i-1}) = \frac{1}{6}h^2\min(1,i-1)$, $v_n(z_{i-1}) = \frac{1}{6}h^2\min(3i-3,\ 3n-1)$. Solution of eqns (4.27) with $\lambda = -1$, and $g(x) = xe+1$, then yields the differencein the approximation from the true solution $\exp(x)$ shown in Table 4.13.

TABLE 4.13

x \ h	0·1	0·05
0	0·0	0·0
0·2	$-1{\cdot}9 \times 10^{-4}$	$-4{\cdot}8 \times 10^{-5}$
0·4	$-3{\cdot}5 \times 10^{-4}$	$-8{\cdot}9 \times 10^{-5}$
0·6	$-4{\cdot}8 \times 10^{-4}$	$-1{\cdot}2 \times 10^{-4}$
0.8	$-5{\cdot}6 \times 10^{-4}$	$-1{\cdot}4 \times 10^{-4}$
1·0	$-5{\cdot}9 \times 10^{-4}$	$-1{\cdot}5 \times 10^{-4}$

The errors are $O(h^2)$, as would be expected in general when $f(x)$ has a continuous second derivative. *

The above method may generally be regarded as a collocation method (see section 4.11) if the product-integration rule

$$\int_a^b K(x,y)\phi(y)dy \simeq \sum_{j=0}^{n} v_j(x)\phi(z_j)$$

is constructed in a suitable fashion. We indicate how a particular such method can be viewed in this light. Suppose that $z_i = a+ih$, $(i=0,1,2,\ldots,n+1)$, where $h=(b-a)/(n+1)$ and suppose that

$$v_j(x) = \int_{z_j}^{z_{j+1}} K(x,y)dy \quad (j=0,1,2,\ldots,n) .$$

Then the corresponding product-integration rule is exact whenever $\phi(x)$ is constant in each interval $[z_i,z_{i+1}]$ $(i=0,1,2,\ldots,n)$. If the solution of eqn (4.27) exists, we write $\phi_i(x) = \delta_{ij}$ for $z_j \le x < z_{j+1}$ whilst $\phi_i(b) = \delta_{in}$ and set

$$\tilde{f}(x) = \sum_{j=0}^{n} \tilde{a}_j\phi_j(x) ,$$

with $\tilde{a}_j = \tilde{f}(z_j)$. It can be seen that these approximate values $\tilde{a}_j = \tilde{f}(z_j)$ are determined by (4.27) in such a way that

$$\sum_{j=0}^{n} \tilde{a}_j\phi_j(z_i) - \lambda\int_a^b K(z_i,y) \sum_{j=0}^{n} \tilde{a}_j\phi_j(y)dy = g(z_i) .$$

As we shall see subsequently, the construction of these equations for $\tilde{a}_0,\ldots\tilde{a}_n$ amounts to applying a collocation method.

4.6.1.

The method of section 4.5 differs from that of section 4.6, and it cannot be viewed as a collocation method. I shall now indicate how the idea of section 4.5 can be extended to give a general method, which is frequently of use when $K(x,y)$ is weakly singular (see section 5.4).†

---

† Like the technique of section 4.5, the method we now describe cannot generally be viewed as a collocation method.

We suppose that $K(x,y) = L(x,y)M(x,y)$, where $L(x,y)$ is well behaved and $M(x,y)$ is badly behaved (it may even be weakly singular). Then, in devising a product integration rule we may associate $L(x,y)$ with $f(x)$ and for each x we cast $L(x,y)f(y)$ in the rôle of a function $\Phi(x,y)$, assuming that $f(x)$ is well behaved. We now employ a product integration formula to construct an approximation of the form

$$\int_a^b M(x,y)\Phi(x,y)dy \simeq \sum_{j=0}^{n} v_j(x)\Phi(x,z_j) \quad .$$

We set $\Phi(x,y) = L(x,y)f(y)$ and $x = z_i$ to obtain

$$\int_a^b M(z_i,y)L(z_i,y)f(y)dy \simeq \sum_{j=0}^{n} v_j(z_i)L(z_i,z_j)f(z_j) \quad . \tag{4.28}$$

Using this approximation in the integral equation we are led to the equations

$$\tilde{f}(z_i) - \lambda \sum_{j=0}^{n} v_j(z_i)L(z_i,z_j)\tilde{f}(z_j) = g(z_i) \tag{4.29}$$

for approximate values for $f(z_i)$ $(i = 0,1,2,\ldots,n)$.

This technique has been explored (and discussed further) by Atkinson (1966) in the case where $K(x,y)$ is weakly singular. We return to the method in Chapter 5. It is important to note, however, that this method (like most others) depends for its success on the smoothness of the solution $f(x)$, and also (here) of $L(x,y)$.

Example 4.18.
Consider an equation of the form

$$f(x) - \lambda\int_0^1 G(x,y)\psi(y)f(y)dy = g(x) \quad ,$$

where $\psi(x) > 0$ for $0 \leq x \leq 1$ and $G_y(x,y)$ suffers a discontinuity at $x = y$. Such an example is provided by the case $G(x,y) = \min(x,y)$, $\psi(x) = e^x$. Because of the discontinuity in $G_y(x,y)$ we prefer not to use an approximation of the form (4.26) (as indicated in section 4.5).

Instead we use the approximations

$$\int_0^1 \min(x,y)\phi(y)dy \simeq \sum_{j=0}^{n} v_j(x)\phi(jh)$$

derived in Example 4.17, and solve the equations

$$\tilde{f}(ih) - \lambda \sum_{j=0}^{n} v_j(ih)\psi(jh)\tilde{f}(jh) = g(ih) .$$

In the case $\lambda = 1$, $g(x) = \frac{5}{4}e^{2x} - \frac{1}{2}xe^2 - \frac{1}{4}$, where $f(x) = e^x$, the errors shown in Table 4.14 were obtained using steps $h = \frac{1}{10}$, $h = \frac{1}{20}$, $h = \frac{1}{40}$.

TABLE 4.14

h \ x	0·2	0·4	0·6	0·8	1·0
$\frac{1}{10}$	$1{\cdot}2\times10^{-2}$	$2{\cdot}4\times10^{-2}$	$3{\cdot}4\times10^{-2}$	$4{\cdot}1\times10^{-2}$	$4{\cdot}3\times10^{-2}$
$\frac{1}{20}$	$3{\cdot}1\times10^{-3}$	$6{\cdot}0\times10^{-3}$	$8{\cdot}5\times10^{-3}$	$1{\cdot}0\times10^{-2}$	$1{\cdot}1\times10^{-2}$
$\frac{1}{40}$	$7{\cdot}8\times10^{-4}$	$1{\cdot}5\times10^{-3}$	$2{\cdot}1\times10^{-3}$	$2{\cdot}5\times10^{-3}$	$2{\cdot}7\times10^{-3}$

The $O(h^2)$ behaviour of the error is apparent in this Table. *

4.7. *Reduction to a degenerate kernel*

Suppose that

$$f(x) - \lambda\int_a^b K(x,y)f(y)dy = g(x) \quad (a \le x \le b) , \tag{4.30}$$

where

$$K(x,y) = \int_a^b L(x,z)M(z,y)dz . \tag{4.31}$$

We may employ a quadrature rule

$$\int_a^b \phi(y)\,dy \simeq \sum_{j=0}^{n} w_j \phi(y_j) \tag{4.32}$$

to compute approximate values of $K(x,y)$. Thus we form the approximation $K_n(x,y)$, where

$$K_n(x,y) = \sum_{j=0}^{n} w_j L(x,y_j) M(y_j,y) \quad . \tag{4.33}$$

Clearly, $K_n(x,y)$ is a degenerate kernel and if we replace eqn (4.30) by the equation

$$f_n(x) - \lambda \int_a^b K_n(x,y) f_n(y)\,dy = g(x) \quad , \tag{4.34}$$

we can compute $f_n(x)$ exactly on solving a system of $(n + 1)$ equations in $(n + 1)$ unknowns (see eqn (1.28)). The coefficients in this system of equations involve the quantities

$$\int_a^b L(x,y_j) M(y_i,x)\,dx \quad (i,j = 0,1,2,\ldots,n)$$

and, in practice, we may have to evaluate these integrals numerically, using an adaptive integration scheme.

Example 4.19.

Consider the *coupled* integral equations

$$f_1(x) - \int_a^b L(x,y) f_0(y)\,dy = g_1(x) \quad ,$$

$$f_0(x) - \int_a^b M(x,y) f_1(y)\,dy = g_0(x) \quad .$$

If we eliminate $f_0(x)$ from the first equation, we obtain the integral equation (4.30) for $f_1(x)$, in which (4.31) holds, $\lambda = 1$, and

$$g(x) = g_1(x) + \int_a^b L(x,y)g_0(y)\mathrm{d}y \ . \quad *$$

4.7.1.

Anselone and Gonzalez-Fernandes (1965) have noted that the technique described above can be applied to general integral equations if we proceed as follows. First, suppose that both $+\lambda$ and $-\lambda$ are regular values of $K(x,y)$. Then, the equation $(I-\lambda K)f(x) = g(x)$ is equivalent to the equation $(I-\lambda^2K^2)f(x) = (I+\lambda K)g(x)$ (see section 1.11) and the kernel $K^2(x,y)$ has the required form.

More generally, if we construct a kernel

$$N(x,y) = \int_a^b L(x,z)M(z,y)\mathrm{d}z \ ,$$

we may seek a value c such that the operator $(I+\lambda K+cN)$ is invertible. We then form, given $(I-\lambda K)f(x) = g(x)$, the equation

$$(I+\lambda K+cN)(I-\lambda K)f(x) = (I+\lambda K+cN)g(x) = \gamma(x) \ ,$$

say. Thus

$$f(x) - \int_a^b \{\lambda^2K^2(x,y) - cN(x,y) + \lambda c(NK)(x,y)\}f(y)\mathrm{d}y = \gamma(x) \ , \tag{4.35a}$$

where

$$\left.\begin{aligned} K^2(x,y) &= \int_a^b K(x,z)K(z,y)\mathrm{d}z \simeq \sum_{j=0}^n w_jK(x,y_j)K(y_j,y) \ , \\ N(x,y) &= \int_a^b L(x,z)M(z,y)\mathrm{d}z \simeq \sum_{j=0}^n w_jL(x,y_j)M(y_j,y) \ , \\ (NK)(x,y) &= \int_a^b N(x,z)K(z,y)\mathrm{d}z \simeq \sum_{j=0}^n w_jN(x,y_j)K(y_j,y) \ , \end{aligned}\right\} \tag{4.35b}$$

and

$$\gamma(x) = g(x) + \lambda\int_a^b K(x,y)g(y)\mathrm{d}y + c\int_a^b N(x,y)g(y)\mathrm{d}y \ .$$

If we replace each of the above kernels by the approximations obtained using quadrature rules, we have an integral equation which may be solved in closed form. To achieve this involves the solution of a system of linear equations, and, possibly, approximation of the term $\gamma(x)$.

Though the method of section 4.7.1 is ingenious it does not appear to us to have been used extensively. The original paper should be consulted for details.

4.8. *Bateman's method*

In section 4.7, I described how we could construct a degenerate kernel to replace a kernel $K(x,y)$ of the special type (4.31). Many numerical methods can be association with such an approach (in which the inhomogeneous term $g(x)$ may also be approximated).

Bateman's method depends on constructing an approximation $K_n(x,y)$ to the kernel $K(x,y)$ and replacing the equation

$$f(x) - \lambda\int_a^b K(x,y)f(y)dy = g(x) \tag{4.36}$$

by

$$f_n(x) - \lambda\int_a^b K_n(x,y)f_n(y)dy = g(x) . \tag{4.37}$$

The kernel $K_n(x,y)$ is constructed in such a way that it is degenerate, so that (4.37) may again be solved in a closed form (see eqn (1.28)). I have some reservations about Bateman's method in general, since it appears to fail in special cases and no satisfactory general convergence theory appears to be available. However, Thompson (1957) has given a justification for the method when $K(x,y)$ is a Green's function.

In Bateman's method, we choose $x_0,x_1,\ldots,x_n$, y_0,y_1, ..., y_n in $[a,b]$ and form

$$K_n(x,y) = \sum_{i=0}^{n}\sum_{j=0}^{n} A_{ij}K(x_j,y)K(x,y_i) , \tag{4.38}$$

where $\{A_{ij}\}$ are chosen so that

$$K_n(x_i,y) = K(x_i,y) \quad (i = 0,1,2,\ldots,n, \quad a \le y \le b)$$

and (4.39)

$$K_n(x,y_j) = K(x,y_j) \quad (j = 0,1,2,\ldots,n, \quad a \le x \le b) \ .$$

Thus, it is sufficient to set $\underset{\sim}{A} = \underset{\sim}{K}^{-1}$ where $\underset{\sim}{K} = [K(x_i,y_j)]$, but of course it is possible for the choice to fail due to the vanishing of $\det(\underset{\sim}{K})$. In general, we expect $\underset{\sim}{K}$ to become progressively more 'singular' as we increase n , if $K(x,y)$ is continuous.

If $K_n(x,y)$ can be constructed to satisfy our requirements, then

$$f_n(x) = g(x) + \sum_{j=0}^{n} \gamma_j K(x,y_j) \quad , \tag{4.40}$$

where $\gamma_0, \gamma_1, \ldots, \gamma_n$ are chosen so that

$$f_n(x_i) - \lambda\int_a^b K(x_i,y)f_n(y)\mathrm{d}y = g(x_i) \ . \tag{4.41}$$

Thus the method is a collocation method (see section 4.11), and we discuss it no further but refer to the bibliography given by Thompson, and the discussion of Kantorovitch and Krylov (1964, p.155) .

4.9. *Expansion methods and quadrature methods*

An expansion method, for the solution of an integral equation, is one in which an approximate solution $\tilde{f}(x)$ is obtained as a (linear) combination $\sum_{j=0}^{n} a_j\phi_j(x)$, say, of prescribed functions $\phi_0(x), \phi_1(x), \ldots, \phi_n(x)$.

Most of the methods for the numerical solution of integral equations can be regarded as 'expansion methods' in *some* sense. Thus, whilst the quadrature method for the Fredholm equation of the second kind yields a vector $\tilde{\underset{\sim}{f}}$ of function values $\tilde{f}(y_0), \tilde{f}(y_1), \ldots, \tilde{f}(y_n)$, these values are used in the Nyström extension to yield the approximation

$$\tilde{f}(x) = g(x) + \sum_{j=0}^{n} \alpha_j K(x, y_j) , \qquad (4.42)$$

where $\alpha_j = \lambda w_j \tilde{f}(y_j)$. It follows that $\tilde{f}(x)$ is a linear combination of prescribed functions $g(x)$, $K(x, y_j)$ $(j = 0,1,2,\ldots,n)$.

At this point, however, we note that with some quadrature rules, the quadrature method is particularly convenient if we would like to generate an approximation of the form

$$\tilde{f}(x) = \sum_{i=0}^{n} \tilde{\alpha}_i \phi_i(x)$$

from the vector $\tilde{\mathbf{f}} = [\tilde{f}(y_0), \tilde{f}(y_1), \ldots, \tilde{f}(y_n)]^T$, where the abscissae y_i and functions $\phi_i(x)$ are specially matched. Consider the problem of approximating a function (defined by a table of values) by means of a sum of Chebyshev polynomials of degree up to n . We have seen, in Chapter 2, that this may be performed quite simply if the function is tabulated either at the zeros of the Chebyshev polynomial of degree $n + 1$ or at the points of extrema of the polynomial of degree n . The preceding ideas were used by El-gendi (1969) in an application of a quadrature method. The Clenshaw-Curtis integration rule yields an approximation of the form

$$\int_{-1}^{1} \phi(y)\mathrm{d}y = \sum_{j=0}^{n} w_j \phi(\cos j\pi/n) , \qquad (4.43)$$

with

$$w_j \equiv w_j^{(n)} = \left(\frac{4}{n}\right) \sum_{k=0}^{[n/2]} \frac{\cos(2jk\pi/n)}{1-4k^2} \quad (j \neq 0, n) ,$$

and $w_0 = w_n = 1/(n^2 - 1)$. If we employ this rule in the quadrature method we obtain approximate values $\tilde{f}(\cos(j\pi/n))$, $j = 0,1,2,\ldots, n$. From these values, we can obtain an approximation

$$\tilde{f}(x) = \sum_{k=0}^{n}{}'' \tilde{a}_k T_k(x)$$

on setting (see Example 2.10, of section 2.9)

$$\tilde{a}_k = (2/n) \sum_{j=0}^{n}{}'' \tilde{f}(\cos(j\pi/n))T_k(\cos(j\pi/n)) .$$

Example 4.20.

El-gendi (1969) applied his method, described above, to the equation

$$f(x) + \frac{1}{\pi}\int_{-1}^{1} \frac{f(y)}{\{1 + (x-y)^2\}} dy = 1$$

(see Example 4.16). With $n = 8$ he obtained the coefficients $\tilde{a}_0 = 1{\cdot}4151850$, $\tilde{a}_2 = 0{\cdot}0493851$, $\tilde{a}_4 = -0{\cdot}0010481$, $\tilde{a}_6 = -0{\cdot}0002310$, $\tilde{a}_8 = 0{\cdot}0000195$, and $\tilde{a}_1 = \tilde{a}_3 = \tilde{a}_5 = \tilde{a}_7 = 0$.

Evaluation of the approximation $\tilde{f}(x) = \sum_{r=0}^{8}{}'' \tilde{a}_r T_r(x)$ at equal intervals in x gives the following function values, rounded to 5 decimal figures:

$x = 0$	$\pm\frac{1}{5}$	$\pm\frac{2}{5}$	$\pm\frac{3}{5}$	$\pm\frac{4}{5}$	± 1
0·65740	0·66151	0·67390	0·69448	0·72248	0·75570 .

Observe that if the known symmetry of the solution is employed to deduce that $\tilde{a}_{2r+1} = 0$, these results can be obtained on solving 5 equations in the 5 unknowns $\tilde{a}_{2r} (r = 0,1,2,3,4)$. The results compare well with those given in Example 4.16. *

In the following sections we consider some methods which can, with more justification, be regarded as 'expansion methods'. In particular, we shall discuss the methods of Rayleigh, Ritz, and Galerkin, and the collocation, least-squares, and linear programming methods in section 4.11 and subsequent sections.

4.10. *An infinite system of equations*

Suppose that

$$f(x) - \lambda\int_a^b K(x,y)f(y)dy = g(x) \tag{4.44}$$

and that there is a countable set of linearly independent functions $\phi_0(x),\phi_1(x),\phi_2(x),\ldots$ such that

$$g(x) = \sum_{s=0}^{\infty} \gamma_s\phi_s(x) \quad ,$$

$$f(x) = \sum_{r=0}^{\infty} \alpha_r\phi_r(x) \quad ,$$

and

$$\int_a^b K(x,y)\phi_r(y)dy = \psi_r(x) \quad ,$$

where

$$\psi_r(x) = \sum_{s=0}^{\infty} B_{s,r}\phi_s(x) \quad .$$

Then, formally, we obtain from eqn (4.44) the relation†

$$\sum_{s=0}^{\infty} \alpha_s\phi_s(x) - \lambda \sum_{s=0}^{\infty} \sum_{r=0}^{\infty} B_{s,r}\alpha_r\phi_s(x) = \sum_{s=0}^{\infty} \gamma_s\phi_s(x) \quad .$$

Since the functions $\{\phi_s(x)\}$ are linearly independent, we deduce that

$$\alpha_s - \lambda \sum_{r=0}^{\infty} B_{s,r}\alpha_r = \gamma_s \quad (s = 0,1,2,\ldots) \quad . \tag{4.45}$$

Eqns (4.45) form an infinite system of equations in an infinite number of unknowns $\alpha_0,\alpha_1,\alpha_2, \ldots,$ and, in general, (4.45) is no more tractable than (4.44). However, methods for *approximating* the solution of (4.45) are fairly obvious; and the most obvious of these is obtained by *truncation*, replacing (4.45) by

$$\tilde{\alpha}_s - \lambda \sum_{r=0}^{n} B_{s,r}\tilde{\alpha}_r = \gamma_s \quad (s = 0,1,2,\ldots,n) \quad . \tag{4.46}$$

† A change in the order of summation is involved at this stage.

The analysis of such a technique is given by Kantorovitch and Krylov (1964).

In a particular case of the above method, Scraton (1969) sets $\phi_r(x) = T_r(x)$ $(r = 1,2,3,\ldots)$ and $\phi_0(x) = \frac{1}{2}T_0(x)$ and discusses an iterative method for solving the system (4.45). Scraton's work deals with the calculation of the coefficients $B_{r,s}$, and he considers especially the case where $K(x,y)$ has a discontinuity or a discontinuous derivative at $x = y$. Observe that the values γ_r must also be obtained, and approximations to these values can be obtained as described in Chapter 2.

Ganado (1971) treats the problem (4.44) where $K(x,y)$ is a difference kernel and the functions $\phi_r(x)$ are trigonometric functions.

Example 4.21.

Fox and Parker (1968, p.158) consider the equation

$$f(x) - \tfrac{1}{2}\int_{-1}^{1}|x-y|f(y)\mathrm{d}y = g(x) \quad .$$

With the notation above, we set $\phi_r(x) = T_r(x)$ for $r \geq 0$. The integrals $\psi_r(x) = \int_{-1}^{1}|x-y|T_r(y)\mathrm{d}y$ can be expressed in terms of $T_0(x), T_2(x)$ for $r = 0$, in terms of $T_1(x), T_3(x)$ for $r = 1$, and in terms of $T_0(x), T_2(x)$, and $T_4(x)$ for $r = 2$. For all $r \geq 3$ $\psi_r(x)$ may be expressed in terms of $T_0(x)$, (if r is even), $T_1(x)$ (if r is odd), and $T_{r-2}(x)$, $T_r(x)$, and $T_{r+2}(x)$. Since we write

$$\psi_r(x) = \sum_{s=0}^{\infty} B_{sr}T_s(x) \quad ,$$

it follows that the matrix $\underset{\sim}{B}$ has the form indicated, with a symbol x denoting non-zero entries:

$$\underset{\sim}{B} = \begin{bmatrix} x & 0 & x & 0 & x & 0 & x & 0 & x & --- \\ 0 & x & 0 & x & 0 & x & 0 & x & 0 & --- \\ x & 0 & x & 0 & x & 0 & 0 & 0 & 0 & \\ 0 & x & 0 & x & 0 & x & 0 & 0 & 0 & \\ 0 & 0 & x & 0 & x & 0 & x & 0 & 0 & \\ 0 & 0 & 0 & x & 0 & x & 0 & x & 0 & \\ & & & & & & & & & \end{bmatrix}$$

An approximate solution to eqn (4.45) may be obtained by assuming that $\alpha_{n+r} = 0$ when $r \geq 1$. If α_n is now assigned an arbitrary value, we can find approximate values of $\alpha_{n-1}, \alpha_{n-2}, \ldots \alpha_0$ in terms of $\alpha_n, \gamma_1, \ldots, \gamma_n$, and use the first equation to find α_n; otherwise we can apply a systematic Gaussian elimination method. See Fox and Parker (1968) for further comments along such lines.

We here observe that the assumption that $\alpha_{n+r} = 0$ is equivalent to truncation of (4.45), that is, the formulation of (4.46).

We employed the above scheme to compute the approximate solution of the equation

$$f(x) - \tfrac{1}{2}\int_{-1}^{1} |x-y| f(y)\mathrm{d}y = \mathrm{e}^x \quad .$$

The Chebyshev coefficients γ_r in the expression

$$\mathrm{e}^x = \sum_{r=0}^{\infty} \gamma_r T_r(x)$$

have been tabulated accurately for $r \leq 17$, and can be written down. From this, it follows that our scheme can be related to a Galerkin method (q.v). We set $n = 3,7,11$ and solved the resulting equations (4.46) for $\tilde{\alpha}_0, \tilde{\alpha}_1, \tilde{\alpha}_2, \ldots, \tilde{\alpha}_n$, then tabulated the solution $\tilde{f}(x) = \sum_{r=0}^{n} \tilde{\alpha}_r T_r(x)$ at equal intervals in x. The results shown in Table 4.15 were obtained.

TABLE 4.15

n	3	7	11
$\tilde{\alpha}_0$	$4{\cdot}21269$	$4{\cdot}20392$	(as n=7)
$\tilde{\alpha}_1$	$8{\cdot}35595 \times 10^{-1}$	$8{\cdot}35644 \times 10^{-1}$	(as n=7)
$\tilde{\alpha}_2$	$1{\cdot}13543$	$1{\cdot}13456$	(as n=7)
$\tilde{\alpha}_3$	$7{\cdot}44972 \times 10^{-2}$	$7{\cdot}45275 \times 10^{-2}$	(as n=7)
$\tilde{\alpha}_4$		$2{\cdot}81751 \times 10^{-1}$	(as n=7)
$\tilde{\alpha}_5$		$1{\cdot}44452 \times 10^{-3}$	(as n=7)
$\tilde{\alpha}_6$		$2{\cdot}75830 \times 10^{-4}$	$2{\cdot}75838 \times 10^{-4}$
$\tilde{\alpha}_7$		$1{\cdot}16752 \times 10^{-5}$	$1{\cdot}16754 \times 10^{-5}$
$\tilde{\alpha}_8$			$1{\cdot}41938 \times 10^{-6}$
$\tilde{\alpha}_9$			$5{\cdot}12564 \times 10^{-8}$
$\tilde{\alpha}_{10}$			$4{\cdot}47074 \times 10^{-9}$
$\tilde{\alpha}_{11}$			$1{\cdot}40884 \times 10^{-10}$

Tabulation of the errors at $x = -1, -0{\cdot}9, \ldots, 0{\cdot}9, 1$, gave the following maximum errors: $3{\cdot}8 \times 10^{-2}$ (in the case $n = 3$); $1{\cdot}4 \times 10^{-6}$ (in the case $n = 7$); and $4{\cdot}4 \times 10^{-11}$ ($n = 11$). The coefficients γ_r were supplied with 10 significant figures.

Observe that the analytical solution has the form $\frac{1}{2}xe^{x}+Ae^{x}+Be^{-x}$. To establish this, differentiate the integral equation to obtain a first-order integro-differential equation and then differentiate once more to find $f''(x) - f(x) = e^{x}$. The given form now follows. The values of A, B may be determined (after some manipulation) on substituting the form of $f(x)$ in the integral equation. Thus, writing $F(x) = \int^{x} f(t)dt$ and $G(x) = \int^{x} tf(t)dt$, we find that we require $F(1) + F(-1) = 0$ and $G(-1) + G(1) = 0$. We thus obtain $B = (e^4+6e^2+1)/\,8(e^2+1)$ and $A = B + (e^2+1)^{-1}$. *

4.11. *The collocation method*

The Rayleigh-Ritz, Galerkin, collocation, and least-squares methods are methods which avoid the problem of having to solve an infinite system of equations (4.45), by producing directly a finite system of equations in a finite number of unknowns. In some cases, there are similar methods for the approximate solution of (4.45), but we do not discuss these.

In the collocation method for eqn (4.44) we seek an approximate solution $\tilde{f}(x)$ in the form of a linear combination $\sum_{i=0}^{n} \tilde{a}_i\phi_i(x)$ of a *finite* number of linearly independent functions $\{\phi_i(x)\}$. The coefficients $\tilde{a}_i$ $(i = 0,1,2,\ldots,n)$ are chosen so that

$$\tilde{f}(z_i) - \lambda\int_a^b K(z_i,y)\tilde{f}(y)\,dy = g(z_i) \quad ,$$

where $z_0, z_1, \ldots, z_n$ are selected points in $[a,b]$. The equations for $\tilde{a}_0, \tilde{a}_1, \ldots, \tilde{a}_n$ are, accordingly,

$$\sum_{j=0}^{n} \{\phi_j(z_i) - \lambda\psi_j(z_i)\}\tilde{a}_j = g(z_i) \quad (i = 0,1,2,\ldots,n) \quad , \tag{4.47}$$

where

$$\psi_j(z_i) = \int_a^b K(z_i,y)\phi_j(y)\,dy \quad .$$

If we write $\tilde{\underset{\sim}{a}} = [\tilde{a}_0, \tilde{a}_1, \ldots, \tilde{a}_n]^T$, $\underset{\sim}{g} = [g(z_0), g(z_1), \ldots, g(z_n)]^T$ then eqns (4.47) may be written

$$[\underset{\sim}{B} - \lambda\underset{\sim}{A}]\tilde{\underset{\sim}{a}} = \underset{\sim}{g} \quad , \tag{4.48}$$

where $B_{ij} = \phi_j(z_i)$ and $A_{ij} = \psi_j(z_i)$.

It is clear that, for some particular choice of 'sample points' $z_0, z_1, \ldots, z_n$, the matrix $\underset{\sim}{B} - \lambda \underset{\sim}{A}$ can be singular even if λ is a regular value of the kernel $K(x,y)$.

If $\tilde{\underset{\sim}{f}} = [\tilde{f}(z_0), \tilde{f}(z_1), \ldots, \tilde{f}(z_n)]^T$, where

$\tilde{f}(x) = \sum_{i=0}^{n} \tilde{a}_i \phi_i(x)$ and $\tilde{\underset{\sim}{a}}$ satisfies (4.48), then

$$\underset{\sim}{B}\tilde{\underset{\sim}{a}} = \underset{\sim}{f} ,$$

and if $\underset{\sim}{B}$ is non-singular, then, from (4.48),

$$[\underset{\sim}{I} - \lambda \underset{\sim}{A}\underset{\sim}{B}^{-1}]\tilde{\underset{\sim}{f}} = g \quad . \tag{4.49}$$

Solution of (4.49) then gives the values $\tilde{f}(z_i)$ directly, but, depending on the circumstances, it may be more appropriate to calculate the coefficients $\tilde{a}_i$, from (4.48).

In general, the criteria for choosing the functions $\phi_i(x)$ are much the same for the equation of second kind as for the eigenvalue problem. If $g(x)$ is badly behaved, we may consider taking $\phi_0(x) = g(x)$. However, if $\phi_0(x)$ can be approximated accurately in terms of $\phi_1(x), \phi_2(x)$, ..., $\phi_n(x)$ the matrix $\underset{\sim}{B} - \lambda \underset{\sim}{A}$ is likely to be ill conditioned and the coefficients $\tilde{a}_i$ may be poorly determined as a consequence.

Both the choice of functions $\phi_r(x)$ and the choice of collocation points z_i have an effect on the conditioning of the problem (4.48) and the choice of one should be matched with the choice of the other. (It appears that a choice which gives a well-conditioned matrix $\underset{\sim}{B}$ is advantageous.) Other considerations enter this choice, besides the conditioning of $\underset{\sim}{B} - \lambda \underset{\sim}{A}$.

If the functions $\phi_i^{(1)}(x)$ $(i = 0,1,2,\ldots,n)$ and the functions $\phi_i^{(2)}(x)$ $(i = 0,1,2,\ldots,n)$ each span the same space, there is (given exact arithmetic) nothing to choose between them. In practice, however, one seeks an expansion $\Sigma \tilde{a}_i \phi_i(x)$ for $\tilde{f}(x)$ in order to be able to calculate $\tilde{f}(x)$ readily with varying values of x . It is well known that for this purpose some choices of $\phi_i(x)$ are superior to others in practice. Thus, the Chebyshev polynomials $(\phi_r^{(1)}(x) = T_r(x))$ are in theory equivalent to the powers of x $(\phi_r^{(2)}(x) = x^r)$ but in practice we generally prefer the Chebyshev polynomials. (In section 5.21 is an example where both choices produce difficulties.)

The choice of collocation points can be critical in determining the accuracy obtainable. We know that polynomial interpolation to a continuous function at equally spaced sample points may give poor approximations as the degree of polynomial is increased irrespective of the use of exact arithmetic. This suggests that when using polynomials for the functions $\phi_r(x)$, the points z_i in the collocation method should *not* be taken to be equally spaced. It seems to be generally accepted that the zeros of $P_{n+1}(x)$, or of $T_{n+1}(x)$, or the extrema of $T_n(x)$, shifted to the range $[a,b]$, may be chosen as collocation points. Kadner (1967) discusses the choice of collocation points which are, in a certain theoretical sense, optimal.

As we noted in Chapter 3, when setting $\phi_r(x) = T_r(x)$, there are some particularly convenient choices of $z_0, z_1, \ldots, z_n$ for which $\underset{\sim}{B}^{-1}$ is known explicitly. Eqn (4.49) may then be set up very readily, if we desire $\tilde{\underset{\sim}{f}}$ (compare this with El-gendi's method described in section 4.9).

Polynomial approximations will not in general yield good results if $f(x)$ is poorly behaved. One may then favour piecewise-polynomial or spline approximations. In the latter case, equally spaced collocation points appear to be acceptable when the knots of the spline occur at the collocation points. (Trivially, we may take $z_i = a + ih$ $(h=(b-a)/n)$ and set $\phi_i(x) = 1$ if $z_i \le x < z_{i+1}$, $\phi_i(x) = 0$ otherwise, if $i \neq n$, and, if $i = n$, $\phi_n(b) = 1$.)

Remark. As we noted earlier, the functions $\phi_i(x)$ and points z_i can be selected so that the collocation method is equivalent to a product integration method. There is, of course, some difference in practice.

Finally, we note that, if $f(x)$ is believed to be periodic, then it is sensible to take periodic functions $\phi_i(x)$. (However, knowledge of the interval of periodicity is desirable!)

Example 4.22.

Kasphitskaya and Luchka (1968) discuss the use of trigonometric sums taking certain equally spaced collocation points. In a numerical example, they consider the integral equation

$$f(x) - \int_0^{2\pi} K(x,y)f(y)\,dy = g(x) \quad ,$$

where $K(x,y) = x+y$ if $x \le y$, $1+x+y$ if $x > y$, and $g(x) = \sin(\frac{1}{2}x) + 2\cos(\frac{1}{2}x) - 4x - 4\pi - 2$. The true solution is $f(x) = \sin(\frac{1}{2}x)$, and they seek an approximate solution $\tilde{f}(x) = \tilde{a}_0 + \tilde{a}_1\cos x + \tilde{a}_2\sin x$. Taking 0, $\frac{2\pi}{3}$, and $\frac{4\pi}{3}$ as the collocation points they find $\tilde{a}_0 = 0{\cdot}6321$, $\tilde{a}_1 = -0{\cdot}4662$, $\tilde{a}_2 = -0{\cdot}0405$. The approximation is not very accurate, giving a maximum error of about 0·17, which occurs at the end-points. (Note that the period of the solution is different from that of the approximation.) *

Phillips (1969) has made an analysis of the collocation method for equations of the second kind. He shows, in particular, that the collocation method can be applied successfully to kernels which are weakly singular, provided that the solution $f(x)$ is continuous. He also examines the problem of evaluating each value $\psi_j(x_i)$ numerically.

We note in passing that the approximation $\tilde{f}(x) = \sum_{i=0}^{n} \tilde{a}_i\,\phi_i(x)$ may be used to generate a further approximation $\mathring{f}(x)$, say. If we substitute $\tilde{f}(x)$ in the integrand of the integral equation we obtain

$$\mathring{f}(x) = g(x) + \lambda\int_a^b K(x,y)\tilde{f}(y)\,dy$$

$$= g(x) + \lambda\sum_{j=0}^{n}\tilde{a}_j\,\psi_j(x) \quad . \qquad (4.50)$$

In general, $\mathring{f}(x) \neq \tilde{f}(x)$, but $\mathring{f}(z_i) = \tilde{f}(z_i)$ $(i = 0,1,2,\ldots,n)$. (For $\mathring{\underset{\sim}{f}} = \underset{\sim}{g}+\lambda\underset{\sim}{A}\underset{\sim}{\tilde{a}} = \underset{\sim}{B}\underset{\sim}{\tilde{a}} = \underset{\sim}{\tilde{f}}$.)

Example 4.23.

We may take $\phi_r(x) = T_r(x)$ $(r = 0,1,2,\ldots,n)$, and the points

$\cos\{(2i+1)\pi/2(n+1)\}$ $(i = 0,1,2,\ldots,n)$ as collocation points, to apply the collocation method to the equation

$$f(x) - \tfrac{1}{2}\int_{-1}^{1} |x-y| f(y)\mathrm{d}y = \mathrm{e}^{x}$$

treated in Example 4.21. The functions

$$\psi_r(x) = \int_{-1}^{1} |x-y| T_r(y)\mathrm{d}y$$

may be obtained analytically in this Example, although in general we may need to evaluate values $\psi_r(z_i)$ approximately. Here, we have

$$\psi_0(x) = \tfrac{3}{2} + \tfrac{1}{2}T_2(x) ,$$

$$\psi_1(x) = \tfrac{1}{12}\, T_3(x) - \tfrac{3}{4}T_1(x) ,$$

$$\psi_2(x) = \tfrac{1}{24}\, T_4(x) - \tfrac{1}{3}\, T_2(x) - \tfrac{3}{8}$$

and, for $r \geq 3$,

$$\psi_r(x) = \frac{T_{r+2}(x)}{2(r+1)(r+2)} - \frac{T_r(x)}{r^2 - 1} + \frac{T_{r-2}(x)}{2(r-1)(r-2)} +$$

$$+ \frac{1 - (-1)^r}{r^2 - 1}\, T_1(x) - \frac{1 + (-1)^r}{r^2 - 4}$$

Setting up and solving the collocation equations, using the collocation points indicated, we obtained the following coefficients $\tilde{a}_0,\tilde{a}_1,\ldots,\tilde{a}_n$ in the approximation $\sum_{r=0}^{n} \tilde{a}_r T_r(x)$: $(n = 3)$ 4·212715, 8·353411 × 10^{-1}, 1·135396, 7·311610 × 10^{-2}; $(n = 7)$ 4·203920, 8·356440 × 10^{-1}, 1·134557; 7·452748 × 10^{-2}, 2·817511 × 10^{-2}, 1·444522 × 10^{-3}, 2·758290 × 10^{-4}, 1·162430 × 10^{-5}, etc. On tabulating the errors in successive approximations at $-1(0{\cdot}1)1$, for varying n , the following maximum errors were observed: in the case $n = 3$, $3\cdot5 \times 10^{-2}$; for $n = 7$, $1\cdot5 \times 10^{-6}$; for $n = 15$, $1\cdot4 \times 10^{-13}$.

It is interesting to compare the method with the use of the quadrature method. We tabulate results computed using the trapezium rule with step h, and a further column of results computed using h^2-extrapolation on the values obtained with $h = 1/50$ and $h = 1/100$. The extrapolated results given in Table 4.16 are correct to the figures given.

TABLE 4.16

x \ h	1/25	1/50	1/100	Extrapolated
-1·0	4·45897	4·45622	4·45553	4·45530
-0·8	3·86002	3·85760	3·85699	3·85679
-0·6	3·43406	3·43189	3·43135	3·43116
-0·4	3·16799	3·16600	3·16551	3·16534
-0·2	3·05604	3·05417	3·05370	3·05354
0	3·09969	3·09787	3·09742	3·09726
0·2	3·30798	3·30616	3·30571	3·30557
0·4	3·69820	3·69632	3·69585	3·69569
0·6	4·29687	4·29489	4·29439	4·29423
0·8	5·14134	5·13919	5·13865	5·13847
1·0	6·28172	6·27935	6·27875	6·27856

(It is of interest to observe that the h^2-extrapolation on the tabulated values can be justified even though the y-derivative of the kernel $|x-y|$ is discontinuous.)

The quadrature method is generally simpler to apply than the collocation method, particularly when the functions $\psi_r(x)$ must be evaluated numerically, but in this case the collocation method is highly competitive. *

Example 4.24.

Phillips (1969) applies the collocation method to Love's equation (see Example 4.6) and assigns varying values to the parameter d in the general problem. For $d = \pm 1$, Phillips applies a collocation method using Chebyshev polynomials ($\phi_r(x) = T_r(x)$, $r \leq n$) and takes the points $x_i = \cos(i\pi/n)$ (the points of extrema of $T_n(x)$) for the collocation points. Phillips reports reasonable accuracy for $d = \pm 1$, but, for small values of d, the solution $f(x)$ is poorly behaved and Phillips then uses cubic spline approximations. The reader is referred to Phillips' work for a quantitative comparison of the results. See also Example 4.27 for the behaviour of $f(x)$ for a small value of d. *

Example 4.25.

If we apply the collocation method to the equation

$$f(x) - \int_0^1 K(x,y)f(y)\,dy = x\sin(1),$$

where $K(x,y) = x(1-y)$ $(0 \leq x \leq y \leq 1)$, $K(x,y) = y(1-x)$ $(0 \leq y < x \leq 1)$, we may take $\phi_0(x) = \frac{1}{2}$, $\phi_r(x) = T^*_r(x)$ $(r = 1,2,3,\ldots,n)$ with $x_i = \frac{1}{2}\left[\cos\{(2i+1)\pi/(2\overline{n+1})\}+1\right]$. For $n = 4$, the maximum of the errors $|f(x) - \tilde{f}(x)|$ for $x \in \{z_i\}$ is $1{\cdot}1 \times 10^{-8}$ and, for $n = 9$, the maximum is $7{\cdot}3 \times 10^{-12}$. (See also Example 4.28.) *

4.11.1. *Elliott's method*

Elliott (1963) combines a collocation method (see section 4.11, preceding) with a replacement of the kernel by a degenerate kernel.

When $a = -1$, $b = 1$ in eqn (4.36), we may set

$$K_m(x,y) = \sum_{r=0}^{m}{}'' \, b_r(x)T_r(y),$$

where $b_r(x) = (2/m)\sum_{j=0}^{m}{}'' \, K(x, \cos(\pi j/m))\cos(\pi rj/m)$. Elliott then proposes that we determine $\tilde{a}_0, \tilde{a}_1, \ldots, \tilde{a}_n$ such that

$$\tilde{f}(\cos(i\pi/n)) - \lambda\int_{-1}^{1} K_m(\cos(i\pi/n), y)\tilde{f}(y)\mathrm{d}y = g(\cos(i\pi/n))$$

$$(i=0,1,2,\ldots,n)$$

where

$$\tilde{f}(x) = \tfrac{1}{2}\tilde{a}_0 + \sum_{r=1}^{n} \tilde{a}_r T_r(x) \quad .$$

We shall thus obtain a linear system of equations for $\tilde{a}_0, \tilde{a}_1, \ldots, \tilde{a}_n$ in terms of the values $K(\cos(i\pi/n), \cos(j\pi/m))$ and $g(\cos(i\pi/n))$ $(i,j = 0,1,2, \ldots, n, m$ respectively).

I tend to favour El-gendi's method or Scraton's method in preference to Elliott's (1963) method. In particular, Scraton appears able to deal with kernels which are badly behaved across $x = y$. This criticism of Elliott's method has been recognised by him, since Elliott and Warne (1967) discuss and program a variation of the method, in which the option is provided of treating a discontinuity at $x = y$ in the kernel or its derivative. The treatment of a discontinuity is achieved by obtaining an approximation

$$K_{m,1}(x,y) = \sum_{r=0}^{m}{}'' b_{r,1}(x)T_r(y)$$

to $K(x,y)$ for $x \leq y$, and for $x>y$ an approximation

$$K_{m,2}(x,y) = \sum_{r=0}^{m}{}'' b_{r,2}(x)T_r(y) \quad .$$

To this end, we set $K(x,y) = K_1(x,y)$ for $x \leq y$ and $K(x,y) = K_2(x,y)$ for $y \leq x$, where $K_1(x,y)$ and $K_2(x,y)$ are both defined *on the whole square* $-1 \leq x, y \leq 1$. It is necessary for the success of the method that $K_1(x,y)$ and $K_2(x,y)$ are *smooth* on the whole square (see the Example below). Then the procedure by which $K_m(x,y)$ is obtained from $K(x,y)$ is applied to give $K_{m,1}(x,y)$ from $K_1(x,y)$ and $K_{m,2}(x,y)$

from $K_2(x,y)$.

Example 4.26.

In the case $K(x,y) = |x-y|$ $0 \leq x,y \leq 1$, the function $K_y(x,y)$ is discontinuous at $x = y$. We therefore set $K_1(x,y) = y-x$ and $K_2(x,y) = x-y$ for $0 \leq x, y \leq 1$. Deciding on appropriate definitions for $K_1(x,y)$ and $K_2(x,y)$ is not always as simple. Thus if $K(x,y) = 1/\{|x-y|+\frac{1}{2}\}$ for $0 \leq x,y \leq 1$, then setting $K_1(x,y) = 1/(y-x+\frac{1}{2})$ $(0 \leq x,y \leq 1)$ and $K_2(x,y) = 1/(x-y+\frac{1}{2})$ gives kernels which are infinite at points in the square on $[0,1]$.

In the case of the equation

$$f(x) - \frac{1}{2}\int_{-1}^{1} |x-y|f(y)\mathrm{d}y = \mathrm{e}^{x}$$

which we considered in Example 4.23, the method of Elliott and Warne reduces to collocation if we set $K_1(x,y) = -(x-y)$ and $K_2(x,y) = (x-y)$, for any $m \geq 1$. The collocation points used in their method are the points $\cos(i\pi/n)$ $(i = 0,1,2,\ldots,n)$. With $n = 3$, the computed Chebyshev coefficients for $\tilde{f}(x)$ are: $4{\cdot}19646$, $8{\cdot}36677 \times 10^{-1}$, $1{\cdot}15731$, $7{\cdot}45397 \times 10^{-2}$. The errors at the points $-1(0{\cdot}1)1$ were computed for varying n , as in Example 4.23, and the following maximum errors were observed: $(n=3)$ $5{\cdot}8 \times 10^{-2}$, $(n=7)$ $2{\cdot}5 \times 10^{-6}$, $(n=15)$ $1{\cdot}1 \times 10^{-13}$.

We note that *for this example*, selecting a value of m larger than 1 involves extra computational effort with no gain in accuracy. This is not however a universal feature of the method. *

Example 4.27.

The procedure of Elliott and Warne was applied to Love's equation with $d = -1$, namely,

$$f(x) - \frac{1}{\pi}\int_{-1}^{1} \frac{1}{1+(x-y)^2} f(y)\mathrm{d}y = 1 .$$

The degree of the approximation $K_m(x,y)$ as a polynomial in y and the degree of $\tilde{f}(x) = \sum_{r=0}^{n} \tilde{a}_r T_r(x)$ were taken to be equal, with $m = n$. For $n = 4$, the computed coefficients $\tilde{a}_{2r}$ were $3{\cdot}550319$, $-1{\cdot}419680 \times 10^{-1}$, $7{\cdot}045497 \times 10^{-3}$; for $n = 8$, $3{\cdot}548891$, $-1{\cdot}400434 \times 10^{-1}$, $4{\cdot}963829 \times 10^{-3}$, $3{\cdot}720487 \times 10^{-4}$, $-4{\cdot}212621 \times 10^{-5}$; and for $n = 16$, $3{\cdot}548890$, $-1{\cdot}400431 \times 10^{-1}$, $4{\cdot}961994 \times 10^{-3}$, $3{\cdot}762979 \times 10^{-4}$, $-4{\cdot}368691 \times 10^{-5}$, $-1{\cdot}623246 \times 10^{-6}$, $4{\cdot}966225 \times 10^{-7}$, $-6{\cdot}190306 \times 10^{-9}$, and $-5{\cdot}041943 \times 10^{-9}$. (The procedure permits economy of effort when the solution is known to be odd or even.) Corresponding tabulated values of the function are shown in Table 4.17.

TABLE 4.17

$n \backslash x$	0	$\pm\frac{1}{4}$	$\pm\frac{1}{2}$	$\pm\frac{3}{4}$	± 1
4	1·924173	1·903125	1·842621	1·750588	1·640237
8	1·919038	1·899618	1·842378	1·751958	1·639696
16	1·919032	1·899615	1·842385	1·751955	1·639695

The satisfactory situation observed here was not reproduced when we considered Love's equation with $d = 0{\cdot}1$ (see Example 4.24). For this example, with $n = 32$, the Chebyshev coefficients $\tilde{a}_{2r}$ decreased very slowly, being $1{\cdot}13$, $7{\cdot}95 \times 10^{-2}$, $4{\cdot}24 \times 10^{-2}$, $1{\cdot}93 \times 10^{-2}$, $6{\cdot}54 \times 10^{-3}$, $9{\cdot}65 \times 10^{-4}$, $-6{\cdot}69 \times 10^{-4}$,..., etc., with $\tilde{a}_{32} = -1{\cdot}89 \times 10^{-6}$. This behaviour indicates a badly behaved solution, which indeed varies quite rapidly close to ± 1. The *computed* function values with $n = 32$ are

$x = 0$	$\pm\frac{1}{4}$	$\pm\frac{1}{2}$	$\pm\frac{3}{4}$	± 1
$5{\cdot}144 \times 10^{-1}$	$5{\cdot}155 \times 10^{-1}$	$5{\cdot}199 \times 10^{-1}$	$5{\cdot}343 \times 10^{-1}$	$7{\cdot}125 \times 10^{-1}$. *

4.12. *Ritz-Galerkin type methods and minimization methods*

We consider together a number of methods which are basically rather similar. At this point we restrict ourselves to a basic description of the methods, and their motivation, deferring a theory of the quality of approximation to be expected.

We suppose that we seek an approximate solution

$$\tilde{f}(x) = \sum_{i=0}^{n} \tilde{a}_i \phi_i(x)$$

in the form of a linear combination of prescribed, linearly independent, functions $\{\phi_i(x)\}$ defined on $[a,b]$. If we substitute such a function $\tilde{f}(x)$ in the integral equation

$$f(x) - \lambda \int_a^b K(x,y) f(y) \mathrm{d}y = g(x) \quad ,$$

we find that

$$\tilde{f}(x) - \lambda \int_a^b K(x,y) \tilde{f}(y) \mathrm{d}y - g(x) = r(x) \quad , \qquad (4.51)$$

and in general it is not possible to choose $\tilde{a}_0, \tilde{a}_1, \ldots, \tilde{a}_n$ to ensure that $r(x) \equiv 0$. In the collocation method, we required that $r(z_i) = 0$ $(i = 0,1,\ldots,n)$. Alternative conditions on $r(x)$ lead to different equations for $\tilde{a}_0, \tilde{a}_1, \ldots, \tilde{a}_n$ and new approximations $\tilde{f}(x)$. Because the methods to be described here (which include collocation as a special case) can be derived from this viewpoint, they are discussed by Collatz (1966) under the heading 'error distribution principles', the term 'error' referring to $r(x)$. The aim is to make $r(x)$ small, in some sense.

In each of the generalized Galerkin methods described below, we require the functions

$$\psi_i(x) = \int_a^b K(x,y) \phi_i(y) \mathrm{d}y$$

and the coefficients $\tilde{a}_i (i = 0,1,\ldots,n)$ are obtained by solving a system

$$[\underset{\sim}{B} - \lambda\underset{\sim}{A}]\ \tilde{\underset{\sim}{a}} = \underset{\sim}{\gamma} \tag{4.52}$$

for $\tilde{\underset{\sim}{a}} = [\tilde{a}_0,\tilde{a}_1,\ldots,\tilde{a}_n]^T$. The definition of $\underset{\sim}{A},\underset{\sim}{B}$, and $\underset{\sim}{\gamma}$ is different in each of the cases described, but the matrices $\underset{\sim}{A},\underset{\sim}{B}$ depend on the functions $\{\phi_i(x)\}$ and $\{\psi_i(x)\}$.

To permit a compact presentation of the various Galerkin-type methods, we use the notation

$$(\psi,\phi) = \int_a^b \psi(x)\ \overline{\phi(x)}dx\ , \tag{4.53a}$$

$$\langle\psi,\phi\rangle = \int_a^b \omega(x)\psi(x)\ \overline{\phi(x)}dx\ , \tag{4.53b}$$

where $\omega(x) > 0$ for $a < x < b$ is a prescribed integrable function, and

$$\prec\psi,\phi\succ = \sum_{j=0}^{N} \Omega_j\psi(\xi_j)\ \overline{\phi(\xi_j)}\ , \tag{4.53c}$$

where $\Omega_j > 0$ and $a \le \xi_j \le b$ $(j = 0,1,2,\ldots,N,\ N \ge n)$. (We often choose Ω_j and ξ_j to correspond to a quadrature rule for product integration with weight $w(x)$, so that $\langle\psi,\phi\rangle \simeq \prec\psi,\phi\succ$ for some $\omega(x)$ chosen above.)

The Galerkin-type methods can now be discussed. We begin with the generalizations of the classical Galerkin method.

1. In the classical Galerkin method, we determine $\tilde{a}_0,\tilde{a}_1,\ldots,\tilde{a}_n$ such that $(r,\phi_i) = 0$ $(i = 0,1,2,\ldots,n)$. To ensure this, we set $B_{i,j} = (\phi_j,\phi_i)$, $A_{i,j} = (\psi_j,\phi_i)$, and $\gamma_i = (g,\phi_i)$ $(i,j = 0,1,2,\ldots,n)$ as the elements of $\underset{\sim}{A},\underset{\sim}{B}$, and $\underset{\sim}{\gamma}$ in (4.52). This is easily established by writing down the condition $(r,\phi_i) = 0$. It can be shown that in the case where $\lambda K(x,y)$ is Hermitian, this method is also a Rayleigh-Ritz method, which can be obtained from a variational formulation.

2. As a generalization of (1), we may instead require that $\langle r,\phi_i\rangle = 0$ $(i = 0,1,2,\ldots,n)$. Then, we set $B_{i,j} = \langle\phi_j,\phi_i\rangle$, $A_{i,j} = \langle\psi_j,\phi_i\rangle$ and $\gamma_i = \langle g,\phi_i\rangle$ in (4.52).

3. If we require $\langle r,\phi_i\rangle = 0 \quad (i = 0,1,\ldots,n)$, the corresponding equations are obtained by changing $<\,,\,>$ to $\langle\,,\rangle$ in the definitions in case 2.
4. For the classical method of moments, we pick $(n+1)$ linearly independent integrable functions $\chi_i(x) \quad (i = 0,1,2,\ldots,n)$ and choose $\tilde{a}_0,\tilde{a}_1,\ldots,\tilde{a}_n$ so that $(r,\chi_i) = 0 \quad (i = 0,1,2,\ldots,n)$. This is achieved by setting $B_{i,j} = (\phi_j,\chi_i)$, $A_{i,j} = (\phi_j,\chi_i)$, and $\gamma_i = (g,\chi_i) \quad (i,j = 0,1,2,\ldots,n)$. Thus 1 and 2 are special cases, obtained on setting $\chi_i(x) = \phi_i(x)$ and $\chi_i(x) = \omega(x)\phi_i(x)$ respectively.
5. Analogous to case 3, there is a discretized version of case 4, obtained by replacing $(\,,\,)$ by $\langle\,,\rangle$ throughout 4.

In all the preceding cases, we obtain equations of the form (4.52). The existence of a unique solution to such equations is not guaranteed, *a priori*.

The approximation $\tilde{f}(x)$ obtained with a Galerkin method may not necessarily be a good *uniform* approximation to $f(x)$. At the end of this section we note that an approximation

$$g(x) + \lambda \sum_{j=0}^{n} \tilde{a}_j\psi_j(x)$$

is readily obtained from the Galerkin approximation, and is generally better as a uniform approximation. Careful calculation of the elements of $\underset{\sim}{A}$ permits easy computation of this new approximation.

If $\lambda K(x,y) = \overline{\lambda K(y,x)}$ in the equation

$$f(x) - \lambda\int_a^b K(x,y)f(y)\mathrm{d}y = g(x) ,$$

we may employ a Rayleigh-Ritz technique. In this instance, we may show that the required solution $f(x)$ minimizes the functional

$$J\{\phi\} = (\phi-\lambda K\phi,\phi) - 2\,Re(\phi,g) ,$$

where

$$K\phi(x) = \int_a^b K(x,y)\phi(y)dy .$$

The Rayleigh-Ritz technique involves the choice of

$$\tilde{f}(x) = \sum_{r=0}^{n} \tilde{a}_r \phi_r(x)$$

which minimizes $J\{\tilde{f}\}$ over all possible choices of $\tilde{a}_0, \tilde{a}_1, \ldots, \tilde{a}_n$. If we write down (in the real case, say) the equations

$$(\partial/\partial\tilde{a}_s)J\{\tilde{f}\} = 0, \; (s=0,1,2,\ldots,n)$$

in order to determine $\tilde{a}_0, \tilde{a}_1, \ldots, \tilde{a}_n$, we find that we obtain (see section 4.23) the classical Galerkin equations of the form (4.52) where $B_{i,j} = (\phi_j, \phi_i)$ and $A_{i,j} = (\psi_j, \phi_i)$. The Rayleigh-Ritz method is known as a variational method because it relies on characterizing the solution of the integral as that function which gives a certain functional $J\{\phi\}$ an extreme value as ϕ varies in some class of functions. (In the Russian usage, the term variational method appears to encompass all the collocation and Galerkin-type methods.) In the Rayleigh-Ritz method, the theoretical motivation can be used to obtain bounds in $\|f(x)-\tilde{f}(x)\|_2$ (see Delves, in Delves and Walsh 1974, p.113). We regard the approach, when generalized, as providing useful motivation for the construction of practical schemes such as the Galerkin method (see Chapter 5) and we note that Delves (1974, p.112) writes down a variational formulation when symmetry of $K(x,y)$ is not assumed.

Remark. The implementation of the Galerkin-type methods is rather more complicated than the implementation of a collocation method. Both methods involve the solution of a system of the form $(\underset{\sim}{B}-\lambda\underset{\sim}{A})\underset{\sim}{a} = \gamma$, but in the Galerkin method the coefficient B_{ij} is an integral and A_{ij} is the integral of a product of functions one of which, $\psi_j(x)$, is itself obtained as an integral. It is computationally (and theoretically) attractive to choose functions $\phi_i(x)$ which ensure that $\underset{\sim}{B}$ is diagonal. In computing the elements of $\underset{\sim}{A}$ in the Galerkin-type methods we might (assuming the analytic expressions unknown) consider it economical to employ an adaptive multiple integration routine to calculate each element as efficiently as possible. However, Delves (Delves and Walsh 1974, p.117) suggests that the same integration rule should be used for each of the elements of $\underset{\sim}{A}$, and of $\underset{\sim}{B}$.

We propose now to extend our ideas by considering the minimization of various 'norms' of $r(x)$ in (4.51). We begin with least-squares methods. In the classical least-squares method, we choose $\tilde{a}_0, \tilde{a}_1, \ldots, \tilde{a}_n$ to minimize $(r,r) = \|r(x)\|_2^2$, but, more generally, we may determine the coefficients which minimize $\langle r,r \rangle$ or $\langle\!\langle r,r \rangle\!\rangle$; in the latter case we obtain the standard least-squares problem of finite-dimensional linear algebra, once that the functions $\psi_i(x)$ are known. Note that $\langle\!\langle r,r \rangle\!\rangle$ is a seminorm on the space $C[a,b]$.

In theory, the coefficients $\tilde{a}_i$ $(i = 0,1,2,\ldots,n)$ required for the solution of the generalized least-squares problems can be obtained by considering the equations

$$(\partial/\partial\tilde{a}_k)N(r) = 0 \quad (k = 0,1,2,\ldots,n)$$

where $N(r) = (r,r)$, $\langle r,r \rangle$ or $\langle\!\langle r,r \rangle\!\rangle$ as appropriate. A solution of these linear equations provides coefficients $\tilde{a}_i$ of a function $\tilde{f}(x)$. In the case where $N(r) = \langle\!\langle r,r \rangle\!\rangle$, we obtain equations known as normal equations by this process. It is now well-known (see the work of Golub (1965) and of Peters and Wilkinson (1970*a*)) that different approaches to the problem are numerically more stable and the normal equations should be avoided. However, the process of least-squares solution inevitably involves *some* manifestation of ill-conditioning of the original least-squares problem. We tend, therefore, to suspect the least-squares methods, and would not suggest them for *general* use.

In the cases discussed, the least-squares methods can all be regarded as equivalent to a particular application of one of the methods of moments 4 or 5 . In the case $N(r) = (r,r)$, we take $\chi_i(x) = (I-\lambda K)\phi_i(x)$ and apply method 4, or we can formulate the method as a Rayleigh-Ritz method for an associated equation. Indeed, to obtain the coefficients $\{\tilde{a}_i\}$ which minimize (r,r) , we may proceed as follows. When eqn (4.51) is assumed to have a unique solution, it follows that, equivalently,

$$f(x) - \int_a^b \{\lambda K(x,y) + \bar{\lambda}K^*(x,y) - |\lambda|^2 K^*K(x,y)\} f(y)\,dy$$

$$= g(x) - \bar{\lambda}\int_a^b K^*(x,y)g(y)\,dy \quad , \qquad (4.54)$$

where $K^*(x,y) = \overline{K(y,x)}$ and $K^*K(x,y) = \int_a^b K^*(x,z)K(z,y)\,dz$. If we now apply the Galerkin method 1 to the above equation we obtain equations for $\tilde{a}_0,\tilde{a}_1,\ldots,\tilde{a}_n$ which minimize $\|r(x)\|_2$, as required.†

Remark. Before proceeding with some examples, we observe that a further class of methods is produced if we endeavour to choose $\tilde{f}(x)$ to minimize $\|r(x)\|_\infty$ or $\|r(x)\|_1$. These approaches lead to rather difficult problems, and the situation is simplified if we choose $z_0,z_1,\ldots,z_N$ in $[a,b]$ and attempt to minimize $\max|r(z_i)|$ or

$$\sum_{j=0}^{N} \Omega_j\, r(z_j) \quad ,$$

where $\Omega_j > 0$. In particular, the latter problems can be posed as problems in linear programming (LP) , which are discussed in the next section.

Example 4.28.

A major part of the computation in setting up the Ritz-Galerkin equation is involved in the formation of the matrices $\underset{\sim}{A}$ and $\underset{\sim}{B}$. In Examples 3.20, 3.21, and 3.22 I give some examples in which these matrices can be set up fairly readily, in the context of the eigenvalue problem. It is left to the reader to perform numerical experiments with the numerical solution of various equations of the second kind corresponding to different functions $g(x)$. We content ourselves with a simple example.

Consider the equation

$$f(x) - \int_0^1 K(x,y)f(y)\,dy = g(x) \quad ,$$

† Because of the 'symmetry' of $\lambda K(x,y) + \bar{\lambda}K^*(x,y) - |\lambda|^2K^*K(x,y)$, the Galerkin method applied to (4.54) is also a Rayleigh-Ritz method.

where $K(x,y) = x(1-y)$ if $0 \leq x \leq y \leq 1$ and $K(x,y) = y(1-x)$ if $0 \leq y \leq x \leq 1$. If $g(x) = x \sin 1$, the true solution is $\sin x$.

We seek an approximate solution

$$\tilde{f}(x) = \sum_{r=0}^{n} \tilde{a}_r T_r^*(x) \quad .$$

We determine the coefficients $\tilde{a}_r$ $(r = 0,1,2,\ldots,n)$ using Galerkin's method with the inner product

$$\langle\phi,\psi\rangle = \int_0^1 (x-x^2)^{-\frac{1}{2}}\phi(x)\psi(x)\mathrm{d}x \quad .$$

The maximum error $|\tilde{f}(x) - f(x)|$ at points of the form $i/100$ $(i = 0,1, \ldots, 100)$ was computed, and we found for $n = 3$ an error of $1{\cdot}7 \times 10^{-4}$ and for $n = 7$ an error of $3{\cdot}8 \times 10^{-10}$. To find an *estimate* of the error in $f(x)$ when calculating with $n = 3$, we can re-apply the method with some value $n \geq 3$ (so that, in general, $\tilde{a}_4 \neq 0$) ; then the size of $\tilde{a}_4$ gives an estimate of the error. Thus, for $n = 9$, $\tilde{a}_4 = 1{\cdot}5 \times 10^{-4}$. (If $\tilde{a}_4 = 0$ we take the estimate from the next non-vanishing coefficient.) The estimate is based on the assumption that the coefficients $\tilde{a}_r$ are of the 'correct' order of magnitude. For if

$$f(x) = \sum_{r=0}^{\infty} \alpha_r T_r^*(x) \quad ,$$

then

$$\left|f(x) - \sum_{r=0}^{n} \alpha_r T_r^*(x)\right|$$

can be estimated by considering the size of the next contribution $\alpha_{n+1}T_{n+1}^*(x)$ in the Chebyshev series, and $|T_{r+1}^*(x)| \leq 1$ for $0 \leq x \leq 1$; in general, we estimate α_{n+1} by $\tilde{a}_{n+1}$. *

Example 4.29.

Collatz (1966*a*) applies the Rayleigh-Ritz method to the equation

$$f(x) - \int_{-1}^{1} K(x,y)f(y)\mathrm{d}y = 1/(1+x^2) \quad ,$$

where $K(x,y) = (x-y)^2-1$ if $|x-y| \leq 1$, $K(x,y) = 0$ if $|x-y| > 1$. Because of the symmetry of the solution when the function $g(x)$ is symmetric, the basic functions $\phi_i(x)$ should be taken so that $\phi_i(x) = \phi_i(-x)$. Only the elements on and above the diagonal of the matrices $\underset{\sim}{A}$ and $\underset{\sim}{B}$ need to be computed because the kernel is Hermitian, and $\underset{\sim}{A}^* = \underset{\sim}{A}$, $\underset{\sim}{B}^* = \underset{\sim}{B}$.

Taking $n = 1$, $\phi_0(x) = 1$, $\phi_1(x) = x^2$, the Rayleigh-Ritz method yields $\tilde{f}(x) = 0{\cdot}4428 - 0{\cdot}2164\, x^2$. We shall indicate how a mechanical application of the general theory (with $\lambda = 1$) results in this approximation.

If $\phi_r(x) = \phi_r(-x)$, then

$$\psi_r(x) = \int_{-1}^{1} K(x,y)\phi_r(y)\mathrm{d}y = \psi_r(-x)$$

is, for $x \geq 0$,

$$\int_{x-1}^{1} \{(y-x)^2-1\}\phi_r(y)\mathrm{d}y \ .$$

with $\phi_0(x) = 1$, $\phi_1(x) = x^2$, we find

$$\psi_0(x) = -\frac{4}{3} + x^2 - \frac{1}{3}|x|^3$$

and

$$\psi_1(x) = -\frac{4}{15} - \frac{1}{3}x^2 + \frac{1}{3}|x|^3 - \frac{1}{30}|x|^5 \ .$$

It is easily verified that

$$(\phi_0,\phi_0) = 2 \ , \ (\psi_0,\phi_0) = -2\int_0^1 (\tfrac{4}{3} - x^2 + \tfrac{1}{3}x^3)\mathrm{d}x = \frac{-13}{6} \ ,$$

$$(\psi_0,\phi_1) = -2\int_0^1 x^2(\tfrac{4}{3} - x^2 + \tfrac{1}{3}x^3)\mathrm{d}x = \frac{-9}{15} \quad ,$$

and $(\phi_0,\phi_1) = \frac{2}{3}$. Further, $(\phi_1,\phi_1) = \frac{2}{5}$ and $(\psi_1,\phi_1) = -\frac{5}{24}$.

We also require

$$\gamma_0 = \int_{-1}^{1} 1/(1+x^2)dx = 2 \tan^{-1} 1 = \tfrac{1}{2}\pi$$

and

$$\gamma_1 = \int_{-1}^{1} x^2/(1+x^2)dx = 2(1-\tan^{-1}1) = (\frac{4-\pi}{2}) .$$

Thus the system of equations for $\tilde{a}_0,\tilde{a}_1$, is

$$\frac{25}{6}\tilde{a}_0 + \frac{19}{15}\tilde{a}_1 = \frac{\pi}{2} .$$

$$\frac{19}{15}\tilde{a}_0 + \frac{73}{120}\tilde{a}_1 = \frac{4-\pi}{2}$$

and the solution for $\tilde{a}_0,\tilde{a}_1$ is straightforward.

Note that, if we take $\phi_r(x) = x^{2r}$ $(r = 0,1,2,\ldots,n)$ in the Rayleigh-Ritz or Galerkin methods applied to this example, we obtain matrices $\underset{\sim}{B} - \lambda\underset{\sim}{A}$ which are full matrices. If we apply a quadrature method to the problem, the resulting matrix is a band matrix (because $K(x,y)$ vanishes if $|x-y| > 1$) . The 'sparseness' can be restored to some extent if we apply the Ritz-Galerkin methods using spline functions.

Note that the kernel in this example is a centro-symmetric one, so that the original integral equation can be written in a different form, as indicated in Example 4.3. *

The Ritz-Galerkin-type methods supply an approximate solution in the form

$$\tilde{f}(x) = \sum_{i=0}^{n} \tilde{a}_i\phi_i(x) \quad ,$$

and in the formation of the matrix $\underset{\sim}{A}$ we find certain values of the functions

$$\psi_i(x) = \int_a^b K(x,y)\phi_i(y)\mathrm{d}y \quad (i = 0,1,\ldots,n).$$

(These function values are required to evaluate (ψ_r,ϕ_s), $\langle\psi_r,\phi_s\rangle$, or $\prec\psi_r,\phi_s\succ$.) If the values of $\psi_i(x)$ can be stored conveniently, we can readily compute the further approximation

$$g(x) + \lambda\int_a^b K(x,y)\tilde{f}(y)\mathrm{d}y = g(x) + \lambda\sum_{i=0}^{n} \tilde{a}_i\psi_i(x)$$

and use it as a check on the approximation

$$\tilde{f}(x) = \sum_{i=0}^{n} \tilde{a}_i\phi_i(x) \quad .$$

The new approximation is generally a better uniform approximation than $\tilde{f}(x)$ (which need not converge to $f(x)$ in the uniform norm), though obviously there will be cases where $\tilde{f}(x)$ is best.

Example 4.30.
Consider the equation

$$f(x) - \int_0^1 K(x,y)f(y)\mathrm{d}y = g(x) \;,$$

with the kernel of Example 4.28, and $g(1)g(0)$ non-vanishing. The functions $\sin \pi rx$ $(r = 0,1,2,\ldots)$ are complete in $L^2[0,1]$ and we may set $\phi_r(x) = \sin \pi rx$ and apply the Rayleigh-Ritz (classical Galerkin) method. For the resulting choice $\{\tilde{a}_r\}$ the function

$$\tilde{f}(x) = \sum_{r=0}^{n} \tilde{a}_r \sin \pi rx$$

vanishes at $x = 0$ and $x = 1$, although in this Example $f(0) = g(0) \neq 0$, $f(1) = g(1) \neq 0$. The approximation

$$g(x) + \sum_{r=0}^{n} \tilde{a}_r \sin \pi rx$$

assumes the correct values at $x = 0$ and $x = 1$ and it can be shown that as $n \to \infty$ the uniform error in this approximation tends to zero whilst we only have $\|f(x) - \tilde{f}(x)\|_2 \to 0$ for the original approximation. *

4.13. *Linear programming methods*

Young (1970) has considered the use of linear programming (LP) methods for the solution of integral equations of the second kind. Rabinowitz (1968) also gives general background material related to the approximate solution of functional equations using LP methods.

The LP methods may be used to solve the following problems.

(a) By an appropriate choice of $\tilde{a}_0, \tilde{a}_1, \ldots, \tilde{a}_n$, minimize $\|\underset{\sim}{r}\|_\infty$, where

$$\|\underset{\sim}{r}\|_\infty = \max_{i=0,1,\ldots,N} |r(z_i)|$$

with $z_0, z_1, \ldots, z_N$ fixed points in $[a,b]$,

$$r(x) = \tilde{f}(x) - \lambda\int_a^b K(x,y)\tilde{f}(y)\,dy - g(x) \tag{4.55}$$

and

$$\tilde{f}(x) = \sum_{i=0}^{n} \tilde{a}_i\phi_i(x) \quad (n \le N)\,. \tag{4.56}$$

(b) Minimize $\|\underset{\sim}{r}\|_1$, where (with a modification of the usual notation)

$$\|\underset{\sim}{r}\|_1 = \sum_{j=0}^{N} \Omega_j |r(z_j)|$$

with given $\Omega_j > 0$ $(j = 0,1,2,\ldots,N)$ and with $r(x)$ defined as above.

We shall indicate briefly how these problems are converted into a form where linear programming can be applied. (We shall refer to the problems (a) and (b) above as the linear l_∞-approximation problem and the linear l_1-approximation problem respectively.)

We write, for $i = 0,1,2,\ldots,N$, $j = 0,1,2,\ldots,n$

$$v_{i,j} = \phi_j(z_i) - \lambda(K\phi_j)(z_i) \; ,$$

where

$$(K\phi_j)(x) = \int_a^b K(x,y)\phi_j(y)\,\mathrm{d}y$$

and $z_0, z_1, \ldots, z_N$ are fixed points in $[a,b]$. The discrete linear Chebyshev problem (a) then assumes the following form: 'Choose $\tilde{\underset{\sim}{a}} = [\tilde{a}_0, \tilde{a}_1, \ldots, \tilde{a}_n]^T$ to minimize $\|\underset{\sim}{r}\|_\infty = \|\underset{\sim}{V}\tilde{\underset{\sim}{a}}-\underset{\sim}{g}\|_\infty$, where $\underset{\sim}{V} = [v_{ij}]$ and $\underset{\sim}{g} = [g(z_0), g(z_1), \ldots, g(z_N)]^T$.' This problem always has at least one solution, and, if the rank of $\underset{\sim}{V}$ is $n + 1$, this solution is unique. To reformulate the problem as a linear programming problem we pose it in the following manner: 'Determine $\tilde{a}_0, \tilde{a}_1, \ldots, \tilde{a}_n$ and $\tilde{a}_{n+1}$ which minimize $\rho = \tilde{a}_{n+1}$ subject to the constraints

$$\tilde{a}_{n+1} \geq 0$$

$$\left.\begin{array}{l} \displaystyle\sum_{j=0}^{n} v_{ij}\tilde{a}_j - \tilde{a}_{n+1} \leq g(z_i) \\[3mm] \displaystyle\sum_{j=0}^{n} v_{ij}\tilde{a}_j + \tilde{a}_{n+1} \geq g(z_i) \end{array}\right\} \qquad (i = 0,1,2,\ldots,N)' \; .$$

(It is clear that, in this formulation, $\rho = \tilde{a}_{n+1}$ plays the role of $\|\underset{\sim}{r}\|_\infty$.) In general we shall take $N \geq n$, and if $N \gg n$ it is usual to recommend, on grounds of efficiency, that the solution to the linear programming problem is obtained by solving the dual problem

$$\text{maximize} \quad \sum_{i=0}^{N} g(z_i)\{\sigma_i - \tau_i\}$$

subject to the constraints

$$\sigma_i \geq 0 \; , \quad \tau_i \geq 0 \quad (i = 0,1,2,\ldots,N)$$

$$\sum_{i=0}^{N} (\sigma_i+\tau_i) \leq 1 ,$$

$$\sum_{i=0}^{N} (\sigma_i-\tau_i)v_{i,j} = 0 \quad (j = 0,1,2,\ldots,n) .$$

The re-formulation of the original linear programming problem as its dual problem is often employed because the computational effort involved in the solution depends principally on the number of constraints appearing when the LP problem is put into a standard form. Further, the coefficients $\tilde{a}_0,\tilde{a}_1,\ldots,\tilde{a}_n$ can be obtained from the final simplex tableau computed for the dual problem (Barrodale and Young 1966) .

The linear l_1-approximation problem (b) may be formulated as the LP problem

$$\text{minimize } \sum_{i=0}^{N} (s_i+t_i)\,\Omega_i$$

subject to the constraints

$$\sum_{j=0}^{n} v_{i,j}\tilde{a}_j+s_i-t_i = g(z_i) \quad (i = 0,1,2,\ldots,N) ,$$

$$s_i \geq 0 \quad t_i \geq 0 \qquad (i = 0,1,2,\ldots,N) .$$

Rabinowitz (1968) remarks that the solution of this problem by the technique of Barrodale and Young (1966) becomes time-consuming if $N \gg n$, and he writes down the dual formulation. However, Barrodale and Roberts (1973) have more recently shown that if attention is paid to the specific nature of the primal LP problem, an algorithm can be obtained which is more efficient than usual techniques applied to the dual.

Example 4.31.
Young has applied method (a) to the equation

$$f(x) + \int_0^1 K(x,y)f(y)\mathrm{d}y = x^2 ,$$

where $K(x,y) = x(1-y)$ $(0 \le x \le y \le 1)$, $K(x,y) = y(1-x)$ $(0 \le y < x \le 1)$. Using 41 equally spaced points $\{z_j\}$ in $[0,1]$ and taking $\phi_i(x) = x^i$, Young reports an accuracy of $1{\cdot}5 \times 10^{-6}$ with a polynomial of degree 5.

Young also considers Love's equation with $d = \pm 1$, taking $\phi_i(x) = x^{2i}$, and he obtains good accuracy with polynomials of degree 10. *

The LP method effectively produces a 'pseudo-solution' to the collocation equations obtained when the number of collocation points is permitted to exceed the number of unknown parameters, and the resulting equations are possibly over-determined. We would expect that the co-efficients $\{\tilde{a}_i\}$ would be poorly determined for certain choices of functions $\{\phi_i(x)\}_0^n$ or points $\{z_i\}_0^N$ occurring in the LP method. There has been little investigation of this aspect.

Example 4.32.
A number of numerical examples are discussed by Barrodale (in Delves and Walsh 1974) and in most of these examples it is possible to obtain an estimate of the accuracy of the solution. We observe that the computed solution $\tilde{f}(x)$ satisfies an equation

$$\tilde{f}(x) - \lambda\int_a^b K(x,y)\tilde{f}(y)dy = g(x) + r(x) ,$$

where in the l_∞-method we have computed $\max_i |r(z_i)|$ giving an estimate for $\|r(x)\|_\infty$; in the l_1-method,

$$\sum_{i=0}^{N} \Omega_i \, |r(z_i)|$$

can be used (with a suitable choice of $\{\Omega_i\},\{z_i\}$) to estimate $\int_a^b |r(y)|dy$. We consider the former case and see that if

$$e(x) = \tilde{f}(x) - f(x)$$

then

$$e(x) - \lambda\int_a^b K(x,y)e(y)dy = r(x) .$$

With usual operator notation, $(I-\lambda K)e(x) = r(x)$ so that $e(x) = (I-\lambda K)^{-1}r(x)$ (or, using the resolvent kernel, $e(x) = r(x) + \lambda R_\lambda r(x)$). Thus

$$\|e(x)\|_\infty \le \|(I-\lambda K)^{-1}\|_\infty \|r(x)\|_\infty ,$$

and we require a bound on $\|(I-\lambda K)^{-1}\|_\infty$ in order to estimate $\|e(x)\|_\infty$, given that $\|r(x)\|_\infty \simeq \max_i |r(z_i)|$. Since $(I-\lambda K)(I-\lambda K)^{-1} = I$ we see that $(I-\lambda K)^{-1} = I + \lambda K(I-\lambda K)^{-1}$ and hence

$$\|(I-\lambda K)^{-1}\| \le 1 + |\lambda| \; \|K\| \; \|(I-\lambda K)^{-1}\| .$$

If $|\lambda| \; \|K\|_\infty < 1$ then we obtain (see also Theorem 1.1),

$$\|(I-\lambda K)^{-1}\|_\infty \le (1-|\lambda| \; \|K\|_\infty)^{-1} .$$

(This bound is not very precise but is easily obtained.) Here

$$\|K\|_\infty = \sup_x \int_a^b |K(x,y)|\,dy$$

and the criterion $|\lambda| \; \|K\|_\infty < 1$ (which is fairly restrictive) may be checked.

For the examples due to Barrodale, referred to above, such a theory applies. The kernels treated in these examples are $K(x,y) = x(1-y)$ $(0 \le x \le y \le 1)$, with $K(y,x) = K(x,y)$ and $\|K\|_\infty = \frac{1}{8}$ and the kernel $d/[\pi\{d^2 + (x-y)^2\}]$ $(-1 \le x,\ y \le 1)$ of Love's equation, with $\|K\|_\infty = (2/\pi)\tan^{-1}(1/d) < 1$. Numerical results are tabulated by Barrodale.

To see the restrictiveness of the theory in the form given here, consider the equation

$$f(x) - \tfrac{1}{2}\int_{-1}^{1} |x-y|f(y)\,dy = e^x$$

discussed earlier. We have $K(x,y) = |x-y|$ and $\|K\|_\infty = \|K\|_1 = 2$ so that $|\lambda| \; \|K\|_\infty = 1$ and the theory does not apply.

Although we do not deduce rigorous error bounds for the LP method applied to the preceding equation, the technique is reasonably successful. Seeking a solution of the form

$$\tilde{f}(x) = \sum_{r=0}^{n} \tilde{a}_r T_r(x)$$

we took $z_i = \cos(i\pi/2n)$ $(i = 0,1,2,\ldots,2n)$. With $n = 3$ we find coefficients $\tilde{a}_r$ equal to $4\cdot23499$, $8\cdot36043 \times 10^{-1}$, $1\cdot14025$, and $7\cdot55958 \times 10^{-2}$ and the residuals r_i of the form

$$r_i = \Sigma\tilde{a}_r\{T_r(z_i) - \lambda KT_r(z_i)\} - g(z_i)$$

had values $-2\cdot3 \times 10^{-2}$, $2\cdot3 \times 10^{-2}$, $1\cdot9 \times 10^{-2}$, $-2\cdot3 \times 10^{-2}$, ..., $-2\cdot3 \times 10^{-2}$. Thus, $\|\underset{\sim}{r}\|_\infty \simeq 2\cdot3 \times 10^{-2}$. When the computed solution was compared with the true solution at intervals of $0\cdot1$, the maximum error observed was $5\cdot9 \times 10^{-2}$. In the case $n = 7$ we found $\|\underset{\sim}{r}\|_\infty = 1\cdot2 \times 10^{-6}$ and the maximum tabulated error in $\tilde{f}(x)$ was $1\cdot8 \times 10^{-6}$. *

4.14. *Further remarks; conditioning*

The methods so far described for the numerical solution of the Fredholm equation of the second kind, with the exception of the LP methods and discrete least-squares methods, result in the numerical solution of a finite system of linear equations. In practice it is fairly important that the system of linear equations should not be more sensitive to errors (that is, ill conditioned) than the original integral equation. This can quite easily be achieved with a quadrature method, but some care has to be exercised in the choice of functions $\phi_i(x)$ if this is to be accomplished using the collocation and Galerkin-type methods. If the functions $\phi_i(x)$ are orthogonal, or if the vectors with components $\phi_i(z_j)$ $(j = 0,1,2,\ldots,n)$ are orthogonal where $z_0,\ldots,z_n$ are the collocation points, then the situation appears to be satisfactory (see Mikhlin (1971), the contribution of Baker on expansion methods in Delves and Walsh (1974), and Delves (1974)). The underlying purpose of such a choice is to ensure that the matrix $\underset{\sim}{B}$ in eqns (4.52) is well conditioned with respect to inversion.

In all the numerical methods proposed it is possible to estimate, computationally, the residual $r(x)$ obtained on substituting $\tilde{f}(x)$ in the integral equation

$$r(x) = \tilde{f}(x) - \lambda\int_a^b K(x,y)\tilde{f}(y)dy - g(x) \quad .$$

Then

$$\tilde{f}(x) - f(x) = r(x) + \lambda\int_a^b R(x,y;\lambda)r(y)dy \quad ,$$

where $R(x,y;\lambda)$ is the resolvent kernel; and it is generally possible to obtain *a posteriori* estimates of $\tilde{f}(x) - f(x)$ once $r(x)$ is known, using this relation (see Example 4.32 for such a technique).

Some authors propose the direct implementation of functional iterations for the solution of equations of the second kind. Thus we might propose the use of the Neumann iteration

$$f_{n+1}(x) = \lambda\int_a^b K(x,y)f_n(y)dy + g(x) \quad .$$

Since this iteration does not always converge, other iterative schemes have been devised. (Refer to PICC (1960, p.655) and Luchka (1965) for details of such methods.)

We feel that any sustained application of an analytical iterative technique can be cumbersome and ill suited to a numerical method. However, iterative methods may be useful if they are applied for a few steps to 'refine' a solution computed by some numerical method. The method of deferred correction is an example of such an iteration. Several iterative methods for the numerical solution of equations of the second kind have been discussed by Atkinson (1973*b*). Our comments are not meant to exclude the practicability of iterative methods for the solution of systems of algebraic equations which yield an approximate solution.

4.15. *The theory of methods for Fredholm equations of the second kind*

In the following sections, we shall analyse some of the numerical methods so far proposed for the approximate solution of the equation

$$f(x) - \lambda\int_a^b K(x,y)f(y)\mathrm{d}y = g(x) . \tag{4.57}$$

In this analysis, we shall assume, in general, that $K(x,y)$ and $g(x)$ are continuous for $a \leq x,\ y \leq b$ and $a \leq x \leq b$ respectively, and that λ is a regular value.

Following our philosophy in the analysis in Chapter 3 we shall demonstrate that (in general) our numerical methods are closely connected with the process of replacing $K(x,y)$ by a degenerate kernel $K_n(x,y)$, replacing $g(x)$ by an approximation $g_n(x)$, and then solving the equation

$$f_n(x) - \lambda\int_a^b K_n(x,y)f_n(y)\mathrm{d}y = g_n(x) . \tag{4.58}$$

This type of analysis must be modified slightly (for example, to deal with the Nyström extension of $\tilde{\underset{\sim}{f}}$ in the quadrature method), but it provides the basis for our theory. Other approaches to the theory are possible, and some will be examined in Chapter 5, and sections 4.17 and 4.29.

To establish the convergence of our approximate solution $f_n(x)$ to $f(x)$ we must analyse the behaviour of $||K-K_n||$ and $||g-g_n||$. Once we have established convergence, it is fairly simple to establish the rate at which convergence takes place, and an asymptotic theory may be constructed.

Anselone and others have developed a general theory which can be applied to the analysis of quadrature methods and various product-integration methods. This theory provides a rather more flexible tool than ours for the analysis of quadrature methods, and it can be applied where our analysis appears to fail. However, the Anselone theory is fairly abstract (see Anselone 1971) and we shall confine ourself to a brief indication of this theory at a later stage, in particular in section 4.17. What follows in this section is an analysis of the effect of replacing a given integral equation by one nearby. In succeeding sections we shall apply this analysis to investigate the behaviour of approximate solutions obtained by methods which we have outlined earlier.

In what follows, we shall employ norms of operators acting on spaces of functions. Some ambiguity arises unless the space is specified, and we assume this to be $L^2[a,b]$ if the norm $\| \ \|_2$ is used, and $R[a,b]$ if the norm $\| \ \|_\infty$ is used, unless otherwise indicated.

4.15.1.

Let us consider the relationship between the equations

$$f(x) - \lambda\int_a^b K(x,y)f(y)\mathrm{d}y = g(x) \tag{4.59}$$

and

$$f_n(x) - \lambda\int_a^b K_n(x,y)f_n(y)\mathrm{d}y = g_n(x)\ , \tag{4.60}$$

where $K_n(x,y)$ approximates $K(x,y)$ and $g_n(x)$ approximates $g(x)$.

We suppose that λ is a regular value of $K(x,y)$ so that eqn (4.59) possesses a unique solution (which is continuous since $K(x,y)$ and $g(x)$ are assumed to be continuous). We shall permit $K_n(x,y)$ and $g_n(x)$ to be piecewise-continuous functions, with finite jump discontinuities along lines parallel to the x- and y-axes. Under what conditions does eqn (4.60) possess a unique piecewise-continuous solution $f_n(x)$? (This question may be answered by reference to Theorem 3.7, and its proof.) We have the following result.

THEOREM 4.1. *If* $\|K-K_n\|_p = \delta$ *and*

$$|\lambda|\delta\ \|(I-\lambda K)^{-1}\|_p < 1$$

with $p = 2$, *then eqn (4.60) has a unique square-integrable solution* $f_n(x)$. *This solution is piecewise-continuous because of the conditions on* $K_n(x,y)$ *and* $g_n(x)$.

<u>Proof</u>. As indicated by the statement of the theorem, we may work in the

space of square-integrable functions. As in eqn (3.102), we have, in operator notation,

$$(I-\lambda K_n) = (I-\lambda K)(I+D_n) \quad ,$$

where

$$D_n = \lambda\,(I-\lambda K)^{-1}(K-K_n)$$

and $(I-\lambda K_n)^{-1}$ exists on the space if $\|D_n\|_p < 1$. Thus there is a solution $f_n(x)$ which is square integrable. Hence

$$\lambda\int_a^b K_n(x,y)f_n(y)\mathrm{d}y + g_n(x)$$

is piecewise-continuous; $f_n(x)$ is therefore piecewise-continuous.

If $K_n(x,y)$ and $g_n(x)$ are continuous, then $f_n(x)$ is continuous.

From Theorem 4.1, we know that the resolvent operator $(I-\lambda K_n)^{-1}$ exists if $\|K-K_n\|_p$ is sufficiently small. (If we have a bound on $\|g(x) - g_n(x)\|_p$, we can also determine the behaviour of $\|f(x) - f_n(x)\|_p$.) In most applications of Theorem 4.1, we find that

$$\|K(x,y) - K_n(x,y)\|_2$$

can be made small by taking n sufficiently large; an example of such a situation occurs in the analysis of certain Galerkin methods. Setting

$$E_n(x,y) = K(x,y) - K_n(x,y),$$

we have, of course,

$$\|E_n\|_2 \le \left\{\int_a^b\int_a^b |E_n(x,y)|^2\,\mathrm{d}x\mathrm{d}y\right\}^{\frac{1}{2}} \equiv \|E_n(x,y)\|_2 .$$

The requirement that $p = 2$ in Theorem 4.1 is, as might be inferred, unnecessarily restrictive, in the sense that other choices of the value p give alternative results. (It is not clear which choice gives the strongest result, however.)

The following Theorem is readily established by parodying the proof of Theorem 4.1, replacing the case $p=2$ with the case $p=\infty$ and considering Riemann-integrable functions in place of square-integrable functions.

THEOREM 4.2. *Theorem 4.1 is true with* $p = \infty$, *the operators acting on the space* $R[a,b]$.

Remark. Observe that if $K_n(x,y)$ is continuous and

$$|\lambda| \; \|K-K_n\|_\infty \; \|(I-\lambda K)^{-1}\|_\infty < 1$$

with subordinate norms on $R[a,b]$, then it follows immediately that the same condition holds for subordinate norms on $C[a,b]$, and (after some thought) vice versa.

If we have a bound on $\|g(x) - g_n(x)\|_\infty$, we can provide a bound on $\|f(x) - f_n(x)\|_\infty$ under the condition of Theorem 4.2. This we shall now show. We suppose that λ is a regular value of $K(x,y)$.

THEOREM 4.3. *Suppose that, for* $p = 2$ *or* $p = \infty$,

$$|\lambda| \; \|K-K_n\|_p \; \|(I-\lambda K)^{-1}\|_p < 1 .$$

Then, given (4.59) and (4.60),

$$\|f(x) - f_n(x)\|_p \le \frac{\mu_p \|f(x)\|_p}{1-|\lambda|\mu_p \|K-K_n\|_p \|T\|_p^{-1}} \times$$

$$\times \left\{\frac{\|g(x)-g_n(x)\|_p}{\|g(x)\|_p} + \frac{|\lambda| \; \|K-K_n\|_p}{\|T\|_p}\right\}$$

where $T \equiv (I-\lambda K)$, *and* μ_p *is the condition number*

$$\mu_p \equiv \mu_p(I-\lambda K) = \|(I-\lambda K)\|_p \; \|(I-\lambda K)^{-1}\|_p . \tag{4.61}$$

Proof. The proof of Theorem 4.3 can be reconstructed by reference to the usual error analysis for finite systems of linear algebraic equations (see Isaacson and Keller 1966, p.37). We shall not go through the details here.

It is usual to interpret a result like Theorem 4.3 by stating that the sensitivity of the eqn (4.59) is governed in part by the condition number $\mu_p(I-\lambda K)$. Large condition numbers imply possible ill-conditioning. In practice it is not possible to compute $\mu_p(I-\lambda K)$. However, it is fairly clear that $\mu_p(I-\lambda K_n)$ provides an estimate of $\mu_p(I-\lambda K)$ (see Theorem 4.6).

Most of our theorems have dual results in which the roles of K and K_n are interchanged. The following theorem can be expressed in terms of $\mu_p(I-\lambda K_n)$, and provides the basis of the dual of Theorem 4.3. The proof of Theorem 4.4 gives insight into the proof of Theorem 4.3.

THEOREM 4.4. *Suppose that*

$$|\lambda|\ \|(I-\lambda K_n)^{-1}\|_p\ \|K_n-K\|_p < 1$$

for $p = 2$ *or* $p = \infty$. *Then*

$$\|f(x) - f_n(x)\|_p \leq \left\{\frac{\|(I-\lambda K_n)^{-1}\|_p}{1-|\lambda|\ \|(I-\lambda K_n)^{-1}\|_p\|K_n-K\|_p}\right\} \times$$

$$\times \left\{\|g(x)-g_n(x)\|_p + |\lambda|\ \|K-K_n\|_p\|f_n(x)\|_p\right\}.$$

Proof. The proof of this result is less involved than the proof of Theorem 4.3, because the rôle of the condition number is not given explicitly.

We note that the conditions imply the boundedness of

$$\|(I-\lambda K_n)^{-1}\|_p$$

(with the existence of the inverse operator) and hence the existence of

$f_n(x)$. Our first step must be to establish the existence of $(I-\lambda K)^{-1}$, with, incidentally, a bound on its norm, and hence the existence of a unique function $f(x)$ with $(I-\lambda K)f(x) = g(x)$ (we use the geometric series theorem of section 1.9).

We have

$$(I-\lambda K) = (I-\lambda K_n) + \lambda(K_n-K)$$

$$= (I-\lambda K_n)\{I+\lambda(I-\lambda K_n)^{-1}(K_n-K)\} \quad , \tag{4.62}$$

and with the given conditions, the operator in braces has an inverse, by the geometric series theorem. Thus

$$(I-\lambda K)^{-1} = \{I+\lambda(I-\lambda K_n)^{-1}(K_n-K)\}^{-1}(I-\lambda K_n)^{-1} \tag{4.63}$$

and, furthermore,

$$\|(I-\lambda K)^{-1}\|_p \leq \|(I-\lambda K_n)^{-1}\|_p \times$$

$$\times \ \|\{I+\lambda(I-\lambda K_n)^{-1}(K_n-K)\}^{-1}\|_p \quad . \tag{4.64}$$

The second norm on the right-hand side of (4.64) may be bounded by

$$1/\{1-|\lambda| \ \|(I-\lambda K_n)^{-1}\|_p \ \|K_n-K\|_p\} \tag{4.65}$$

since this quantity is positive (recall $\|(I+G_n)^{-1}\|_p \leq 1/\{1-\|G_n\|_p\}$ when $\|G_n\|_p < 1$).

If we difference eqns (4.59) and (4.60), we obtain an equation for

$$\varepsilon(x) = f(x) - f_n(x) \ . \tag{4.66}$$

We have

$$\varepsilon(x) - \lambda\int_a^b K(x,y)\varepsilon(y)dy = \delta(x) \ ,$$

where

$$\delta(x) = \{g(x) - g_n(x)\} + \lambda\int_a^b (K(x,y)-K_n(x,y))f_n(y)dy .$$

Thus

$$\varepsilon(x) = (I-\lambda K)^{-1}\delta(x)$$

and

$$\|\varepsilon(x)\|_p \le \|(I-\lambda K)^{-1}\|_p \|\delta(x)\|_p .$$

Since

$$\|\delta(x)\|_p \le \|g(x) - g_n(x)\|_p + |\lambda| \|K-K_n\|_p \|f_n(x)\|_p ,$$

we obtain

$$\|\varepsilon(x)\|_p \le \|(I-\lambda K)^{-1}\|_p\{ \|g(x) - g_n(x)\|_p + $$

$$+ |\lambda| \|K-K_n\|_p \|f_n(x)\|_p\} . \qquad (4.67)$$

We obtain a bound for $\|(I-\lambda K)^{-1}\|_p$ in terms of $\|(I-\lambda K_n)^{-1}\|_p$ by combining (4.64) and (4.65); substituting the result in (4.67) gives the result required.

There are a large number of similar bounds which can be obtained for $\|f(x) - f_n(x)\|_p$. For example, in Theorem 4.4 we can interchange $K(x,y)$ and $K_n(x,y)$, $g(x)$ and $g_n(x)$, and $f(x)$ and $f_n(x)$ to obtain a bound subject to the conditions of Theorem 4.3. We use a result of this type in Theorem 4.7 below. Not all the bounds available are of interest to us. Theorem 4.4 is of some consequence, since it yields a bound which is, in principle, computable if $\|K-K_n\|_p$ is sufficiently small, if $K_n(x,y)$ is degenerate, and we solve (4.60) in place of (4.59).

At this stage, we are principally concerned with the question of convergence. We have the following result.

THEOREM 4.5. *If* $\|K-K_n\|_p \to 0$ *and* $\|g(x) - g_n(x)\|_p \to 0$ *as* $n \to \infty$ *then* $\|f(x) - f_n(x)\|_p \to 0$, *for* $p = 2$ *or* $p = \infty$.

Proof. We may establish this result either as a corollary of Theorem 4.3, or from eqn (4.67).

We also have the result below.

THEOREM 4.6. *If* $\lim_{n\to\infty} \|K-K_n\|_p = 0$, *then* $\lim_{n\to\infty} \mu_p(I-\lambda K_n) = \mu_p(I-\lambda K)$.

Proof. This result follows from eqn (4.61), and from the relationships of a type similar to (4.62) and (4.63). The result $\mu_p(I-\lambda K_n) \to \mu_p(I-\lambda K)$ follows from the results $\|(I-\lambda K_n)\|_p \to \|(I-\lambda K)\|_p$ and $\|(I-\lambda K_n)^{-1}\|_p \to \|(I-\lambda K)^{-1}\|_p$. Indeed, to establish these identities, we have the *basic result:* $I-\lambda K = I-\lambda K_n + \lambda(K_n-K)$. It follows that $\|I-\lambda K\|_p - |\lambda|\,\|K_n-K\|_p \le \|I-\lambda K_n\|_p \le \|I-\lambda K\|_p + |\lambda|\,\|K_n-K\|_p$. Thus $\|(I-\lambda K_n)\|_p \to \|(I-\lambda K)\|_p$.

As a preliminary for the next step we see that if we interchange the rôles of K_n and K in (4.64) we have

$$\begin{aligned}\|(I-\lambda K_n)^{-1}\|_p &\le \|(I-\lambda K)^{-1}\|_p \\ &\quad \times \|\{I+\lambda(I-\lambda K)^{-1}(K-K_n)\}^{-1}\|_p \\ &\le \|(I-\lambda K)^{-1}\|_p / \{1-|\lambda|\ \|(I-\lambda K)^{-1}\|_p\ \|K-K_n\|_p\}\end{aligned}$$

provided that $\|K-K_n\|_p$ is sufficiently small. Thus

$$\|(I-\lambda K_n)^{-1}\|_p \le 2\|(I-\lambda K)^{-1}\|_p$$

for n sufficiently large, and in passing we have established that $\|(I-\lambda K_n)^{-1}\|_p = O(1)$. Now our 'basic' identity gives the relation

$$(I-\lambda K_n)^{-1} = (I-\lambda K)^{-1} + \lambda(I-\lambda K_n)^{-1}(K_n-K)(I-\lambda K)^{-1}$$

when $(I-\lambda K_n)^{-1}$ exists (that is for n sufficiently large). Thus, with $\|(I-\lambda K_n)^{-1}\|_p \leq 2\|(I-\lambda K)^{-1}\|_p$ for n sufficiently large, we obtain the bounds

$$\|(I-\lambda K)^{-1}\|_p - 2\|(I-\lambda K)^{-1}\|_p^2 \, \|K_n-K\|_p|\lambda|$$

$$\leq \|(I-\lambda K_n)^{-1}\|_p \leq \|(I-\lambda K)^{-1}\|_p$$

$$+ 2\|(I-\lambda K)^{-1}\|_p^2 \|K_n-K\|_p|\lambda| ,$$

and hence $\|(I-\lambda K_n)^{-1}\|_p \to \|(I-\lambda K)^{-1}\|_p$. The theorem now follows.

In some of the analysis below we are concerned with the *rate* of convergence of $\|f(x) - f_n(x)\|_p$ to zero. Let us suppose that the function $\rho_n(x)$ is defined by the relation

$$f(x) - \lambda\int_a^b K_n(x,y)f(y)\mathrm{d}y = g_n(x) + \rho_n(x) . \tag{4.68}$$

From eqns (4.60) and (4.68), on differencing, we obtain

$$\varepsilon(x) - \lambda\int_a^b K_n(x,y)\varepsilon(y)\mathrm{d}y = \rho_n(x) ,$$

where $\varepsilon(x) = f(x) - f_n(x)$. If $\|K-K_n\|_p \to 0$, then, for n sufficiently large, $f_n(x)$ exists and

$$\varepsilon(x) = (I-\lambda K_n)^{-1}\rho_n(x) .$$

Thus

$$\|\varepsilon(x)\|_p \leq \|(I-\lambda K_n)^{-1}\|_p \, \|\rho_n(x)\|_p .$$

Now

$$\|(I-\lambda K_n)^{-1}\|_p = O(1) \quad \text{as} \quad n \to \infty ,$$

so that

$$\|\varepsilon(x)\|_p = O\,(\|\rho_n(x)\|_p)$$

as $n \to \infty$. Indeed, if we interchange the rôles of $K(x,y)$ and $K_n(x,y)$ in eqns (4.64) and (4.65), we obtain the result

$$\|(I-\lambda K_n)^{-1}\|_p \le \|(I-\lambda K)^{-1}\|_p \;\|\{I+\lambda(I-\lambda K)^{-1}(K-K_n)\}^{-1}\|_p$$

$$\le \frac{\|(I-\lambda K)^{-1}\|_p}{1-|\lambda|\;\|(I-\lambda K)^{-1}\|_p\;\|K-K_n\|_p}$$

for n sufficiently large.

We can combine our inequalities to obtain a result of the following type.

THEOREM 4.7. (a) *Subject to the conditions of Theorem 4.3,*

$$\|f(x) - f_n(x)\|_p \le \frac{\|(I-\lambda K)^{-1}\|_p\,\|\rho(x)\|_p}{1-|\lambda|\;\|(I-\lambda K)^{-1}\|_p\,\|K-K_n\|_p} \;.$$

(b) *If* $\|K-K_n\|_p \to 0$ *as* $n \to \infty$ *then*

$$\|f(x) - f_n(x)\|_p = O(\|\rho(x)\|_p)\;.$$

Remark. We shall not use this theorem in our discussion of the quadrature method, but will return to it later.

4.16. *Analysis of the quadrature method*

We suppose that $J(\phi)$ is a quadrature rule

$$J(\phi) = \sum_{j=0}^{n} w_j\phi(y_j) \tag{4.69}$$

for the approximation of the integral $\int_a^b \phi(y)\,dy$. *In the first place, we shall suppose that* $J(\phi)$ *is a Riemann sum* $(J(\phi) \in J^R)$. Later, we suppose that $J(\phi) \in J^+$ (but this restriction can also be relaxed if we use a theory due to Anselone (1971)).

Our purpose in this section is to establish a number of results concerning the behaviour of the approximation $\tilde{f}(x)$ obtained using a quadrature rule (4.69), as the choice of this rule runs through a convergent family of such formulae. The functions $K(x,y)$ and $g(x)$ are assumed continuous. In brief, we employ the theory of section 4.15 to establish, under such conditions, that $\sup_i |f(y_i)-\tilde{f}(y_i)| \to 0$. We deduce that $\|\tilde{f}(x)-f(x)\|_\infty \to 0$, where $\tilde{f}(x)$ is the Nyström extension of the values $\tilde{f}(y_0),\tilde{f}(y_1),\ldots,\tilde{f}(y_n)$. Implied by this is the result that the matrix $(\underset{\sim}{I}-\lambda\underset{\sim}{K}\underset{\sim}{D})$ is invertible if (4.69) is a sufficiently good rule, and we show, in the circumstances described above, that $\|(\underset{\sim}{I}-\lambda\underset{\sim}{K}\underset{\sim}{D})^{-1}\|_\infty = O(1)$. From such a result we can deduce rates of convergence of the errors $\sup_i|\tilde{f}(y_i)-f(y_i)|$ and $\|\tilde{f}(x)-f(x)\|_\infty$ in terms of the local truncation error

$$\tau(x) = f(x) - \lambda \sum_{j=0}^{n} w_j K(x,y_j)f(y_j)-g(x) .$$

The asymptotic behaviour of $\tilde{f}(x)-f(x)$ also follows, as we see in section 4.18.

4.16.1.

Recall that since $J(\phi)$ is a Riemann sum, we have

$$v_i \le y_i \le v_{i+1} \quad (i = 0,1,2,\ldots,n) ,$$

where

$$v_0 = a$$

$$v_j = a + \sum_{i=0}^{j-1} w_i \quad (j = 1,2,3,\ldots,(n+1)) , \qquad (4.70)$$

and $\Delta = \max\{\max(v_{i+1}-y_i , \; y_i-v_i)\}$ is the 'measure' of $J(\phi)$. Furthermore, as in the eigenvalue problem, we may define step functions $\phi_i(x)$ $(i = 0,1,2,\ldots,n)$ such that

$$\phi_i(y_j) = \delta_{ij} \quad (i,j = 0,1,2,\ldots,n) \quad , \tag{4.71}$$

$$\phi_i(x) = 1 \quad (v_i < x < v_{i+1}, \; i = 0,1,2,\ldots,n) \; , \tag{4.72}$$

and

$$\phi_i(x) = 0, \; x \notin [v_i, v_{i+1}] \quad . \tag{4.73}$$

This may leave some of the values $\phi_i(v_i)$ or $\phi_i(v_{i+1})(i = 0,1,2,\ldots,n)$ undefined, and it is sufficient to define these remaining values in any way such that

$$\sum_{i=0}^{n} \phi_i(x) = 1$$

for all $x \in [a,b]$. For definiteness we set $\phi_i(v_i) = 1$, if $v_i \notin \{y_k\}$; $\phi_i(v_j) = 0$, if $i \neq j$ or $i \neq n$, and $v_j \notin \{y_k\}$; and $\phi_n(b) = 1$.

The following properties of the functions $\phi_i(x)$ will be required.

(a) For $i,j = 0,1,2,\ldots,n,$

$$\int_a^b \phi_i(x)\phi_j(x)\mathrm{d}x = 0 \quad \text{if} \quad i \neq j \; . \tag{4.74}$$

(b) For $i = 0,1,2,\ldots,n,$

$$\int_a^b \{\phi_i(x)\}^2 \mathrm{d}x = w_i \quad . \tag{4.75}$$

(c) For $i,j = 0,1,2,\ldots,n$

$$\phi_i(y_j) = \delta_{ij} \quad , \tag{4.76}$$

and consequently, if $f_n(x) = \sum_{i=0}^{n} a_i\phi_i(x)$ then

$$f_n(y_i) = a_i \quad . \tag{4.77}$$

Now we define the kernel

$$K_n(x,y) = \sum_{i=0}^{n} \sum_{j=0}^{n} K(y_i,y_j)\phi_i(x)\phi_j(y) \tag{4.78}$$

and the function

$$g_n(x) = \sum_{i=0}^{n} g(y_i)\phi_i(x) , \tag{4.79}$$

and we consider the equation

$$f_n(x) - \lambda\int_a^b K_n(x,y)f_n(y)\mathrm{d}y = g_n(x) \tag{4.80}$$

(for $a \le x \le b$) . Clearly, if a solution $f_n(x)$ exists it has the form

$$f_n(x) = \sum_{i=0}^{n} \tilde{a}_i\phi_i(x) , \tag{4.81}$$

since $g_n(x)$ and $(K_nf_n)(x)$ are both linear combinations of the functions $\phi_i(x)$. To obtain equations for the coefficients $\tilde{a}_0,\tilde{a}_1,\ldots,\tilde{a}_n$, we substitute (4.81) in (4.80), and we find

$$\sum_{i=0}^{n} \tilde{a}_i\phi_i(x) - \lambda \sum_{k=0}^{n} \tilde{a}_k(K_n\phi_k)(x) = \sum_{i=0}^{n} g(y_i)\phi_i(x) ,$$

where

$$\begin{aligned}(K_n\phi_k)(x) &= \sum_{i=0}^{n} \phi_i(x) \int_a^b \sum_{j=0}^{n} K(y_i,y_j)\phi_j(y)\phi_k(y)\mathrm{d}y \\ &= \sum_{i=0}^{n} \phi_i(x)w_kK(y_i,y_k)\end{aligned}$$

using eqns (4.74) and (4.75). Thus

$$\begin{aligned}\sum_{i=0}^{n} \tilde{a}_i\phi_i(x) - \lambda \sum_{i=0}^{n} \sum_{j=0}^{n} \tilde{a}_jw_jK(y_i,y_j)\phi_i(x) \\ = \sum_{i=0}^{n} g(y_i)\phi_i(x) .\end{aligned}$$

Now the functions $\phi_i(x)$ $(i = 0,1,2,\ldots,n)$ are linearly independent, so we deduce that

$$\tilde{a}_i - \lambda \sum_{j=0}^{n} w_j K(y_i,y_j)\tilde{a}_j = g(y_i) \quad (i=0,1,2,\ldots,n). \tag{4.82}$$

In matrix notation,

$$(\underset{\sim}{I}-\lambda \underset{\sim}{K}\underset{\sim}{D})\underset{\sim}{\tilde{a}} = \underset{\sim}{g} \ . \tag{4.83}$$

Thus, if the integral equation (4.80) has a solution, it may be obtained by solving (4.83) and setting

$$f_n(x) = \sum_{i=0}^{n} \tilde{a}_i\phi_i(x) \ .$$

Conversely (since the steps are reversible), if (4.83) has a solution, then (4.80) has the solution

$$f_n(x) = \sum_{i=0}^{n} \tilde{a}_i\phi_i(x) \ .$$

In the quadrature method we solve the equation

$$(\underset{\sim}{I}-\lambda \underset{\sim}{K}\underset{\sim}{D})\underset{\sim}{\tilde{f}} = \underset{\sim}{g} \ . \tag{4.84}$$

Such a vector $\underset{\sim}{\tilde{f}}$ *may* be extended to a function

$$\tilde{f}(x) = \sum_{i=0}^{n} \tilde{f}(y_i)\phi_i(x) \ ,$$

where

$$\underset{\sim}{\tilde{f}} = [\tilde{f}(y_0),\tilde{f}(y_1),\ldots,\tilde{f}(y_n)]^T \ ,$$

and $\tilde{f}(x)$ is then a solution of the integral equation

$$f_n(x) - \lambda\int_a^b K_n(x,y)f_n(y)\,dy = g_n(x) \ . \tag{4.85}$$

Moreover, if the solution of (4.85) is unique, then (4.84) has a unique solution and

$$\tilde{\mathrm{f}}(x) \equiv f_n(x) \ . \tag{4.86}$$

If we now appeal to Theorem 4.1, or Theorem 4.2, we can establish the following result.

THEOREM 4.8. *Suppose* $K(x,y)$ *and* $g(x)$ *are continuous, and* λ *is a regular value of* $K(x,y)$, *then the equation*

$$(\underset{\sim}{I}-\lambda \underset{\sim}{K}\underset{\sim}{D})\tilde{\underset{\sim}{f}} = \underset{\sim}{g}$$

arising from the quadrature method with $J(\phi) \in J^R$ *has a unique solution if the measure* Δ *of* $J(\phi)$ *is sufficiently small.*

<u>Proof</u>. Since $||K-K_n||_\infty \leq |b-a| \ ||K(x,y)-K_n(x,y)||_\infty$ and, clearly, $||K(x,y)-K_n(x,y)||_\infty$ can be made sufficiently small by taking Δ sufficiently small (see (4.89) below), the result follows from Theorem 4.2.

We now consider the question of convergence. We suppose $J(\phi) \in J^R$ $(m = 0,1,2,\ldots)$ and we consider the function

$$\tilde{\mathrm{f}}(x) = \sum_{i=0}^{n} \tilde{f}(y_i)\phi_i(x) \ , \tag{4.87}$$

where $n \equiv n(m)$, $w_i \equiv w_i(m)$, $y_i \equiv y_i(m)$, and $J(\phi) \equiv J_m(\phi)$. If $\lim_{m\to\infty} \Delta_m = 0$ then $\lim_{m\to\infty} ||f(x)-\tilde{\mathrm{f}}(x)||_\infty = 0$. We state the result formally in the following theorem.

THEOREM 4.9. *If* $K(x,y)$ *and* $g(x)$ *are continuous and* $\{J_m(\phi)\}$ *is a family of rules in* J^R *such that* $\lim_{m\to\infty} \Delta_m = 0$, *then*

$$\lim_{m\to\infty} ||f(x)-\tilde{\mathrm{f}}(x)||_\infty = 0, \quad \textit{where} \quad \tilde{\mathrm{f}}(x) = \sum_{i=0}^{n} \tilde{f}(y_i)\phi_i(x) \ .$$

Proof. Since

$$\|g(x)-\tilde{g}(x)\|_\infty \le \sup_i \sup_{|x-y_i|<\Delta} |g(x)-g(y_i)| \ , \tag{4.88}$$

$$\|K(x,y)-K_n(x,y)\|_\infty \le \sup_{i,j} \sup_{\substack{|x-y_i|<\Delta \\ |y-y_j|<\Delta}} |K(x,y)-K(y_i,y_j)| \ , \tag{4.89}$$

and $\Delta \to 0$, we see that $\|g(x)-\tilde{g}(x)\|_\infty \to 0$ and $\|K(x,y)-K_n(x,y)\|_\infty \to 0$ as $m \to \infty$.

The result now follows from Theorem 4.5 since $\tilde{\mathrm{f}}(x) \equiv f_n(x)$ (for m sufficiently large to ensure that they exist).

We know that Theorem 4.9 applies if we set $J_m(\phi)$ equal to

(a) an m-times repeated rule $(m \times J_0)(\phi)$, where $J_0(\phi) \in \mathrm{J}^R$, or,

(b) an m-point Gauss-Legendre or m-point basic Romberg rule.

Since we may not be interested in the function $f(x)$ but merely the computed values $\tilde{f}(y_i)$ $(i = 0,1,2,\ldots,n)$ we state the following result.

COROLLARY 4.1. *Subject to the conditions of Theorem 4.9,*

$$\lim_{m\to\infty} \max_{i=0,1,2,\ldots,n(m)} |f(y_i)-\tilde{f}(y_i)| = 0 \ .$$

To compute an approximate solution for values of x which are not abscissae for some particular m , we may evaluate the Nyström extension $\tilde{f}(x)$ (which we distinguish from the function $\tilde{\mathrm{f}}(x)$), where

$$\tilde{f}(x) = g(x) + \lambda \sum_{j=0}^{n} w_j K(x,y_j)\tilde{f}(y_j) \ . \tag{4.90}$$

Under the conditions of Theorem 4.9, we may show that

$$\lim_{m\to\infty} \|\tilde{f}(x)-f(x)\|_{\infty} = 0 \quad .$$

To establish this result, we introduce the kernel

$$\hat{K}_n(x,y) = \sum_{i=0}^{n} K(x,y_i)\phi_i(y) \quad . \tag{4.91}$$

Then the functions of (4.87) and (4.90) are related by

$$\tilde{f}(x) = g(x) + \lambda \int_a^b \hat{K}_n(x,y)\tilde{f}(y)dy \ , \tag{4.92}$$

as is easily verified. Furthermore,

$$|\hat{K}_n(x,y)-K(x,y)| \leq \max_i \sup_{|y-y_i|<\Delta} |K(x,y_i)-K(x,y)|$$

so that $\lim\limits_{m\to\infty} \|\hat{K}_n(x,y)-K(x,y)\|_{\infty} = 0$.

Now

$$f(x) = g(x) + \lambda \int_a^b K(x,y)f(y)dy \tag{4.93}$$

so that we obtain, from eqns (4.92) and (4.93), the relation

$$e(x) = \lambda \left\{\int_a^b \hat{K}_n(x,y)\tilde{f}(y)dy - \int_a^b K(x,y)f(y)dy\right\} \ , \tag{4.94}$$

where

$$e(x) = \tilde{f}(x)-f(x) \ .$$

From eqn (4.94), we obtain

$$e(x) = \lambda\left\{\int_a^b K(x,y)(\tilde{f}(y)-f(y))dy + \int_a^b (\hat{K}_n(x,y)-K(x,y))\tilde{f}(y)dy\right\} \ , \tag{4.95}$$

so that

$$||e(x)||_\infty \le |\lambda| \quad \{||K||_\infty \, ||\tilde{f}(x)-f(x)||_\infty + ||\hat{K}_n-K||_\infty \, ||\tilde{f}(x)||_\infty\} \quad .$$

Now $\lim_{m\to\infty} ||\tilde{f}(x)-f(x)||_\infty = 0$, and consequently $\lim_{m\to\infty} ||\tilde{f}(x)||_\infty = ||f(x)||_\infty$. Further, $\lim_{m\to\infty} ||\hat{K}_n-K||_\infty = 0$ since $||\hat{K}_n(x,y)-K(x,y)||_\infty \to 0$. Thus $||e(x)||_\infty \to 0$, as $m \to \infty$. We have established the following result.

THEOREM 4.10. *Suppose that the conditions of Theorem 4.9 are valid. If* $\tilde{f}(x)$ *is the Nyström extension (4.90) of the vector* $\tilde{\mathbf{f}} = [\tilde{f}(y_0),\tilde{f}(y_1),\ldots,\tilde{f}(y_n)]^T$ *, then*

$$\lim_{m\to\infty} ||\tilde{f}(x)-f(x)||_\infty = 0 \ .$$

4.16.2.

Let us now consider the rate of convergence of $\tilde{f}(x)$ to $f(x)$ when $J_m(\phi) \in J^R$ $(m = 0,1,2,\ldots)$ and $\lim_{m\to\infty} \Delta_m = 0$.

The local truncation errors $\tau(y_i)$ $(i = 0,1,2,\ldots,n)$ may be defined by setting

$$\tau(y_i) = f(y_i)-\lambda \sum_{j=0}^{n} w_j K(y_i,y_j)f(y_j)-g(y_i) \ , \tag{4.96}$$

so that $\tau(y_i) = \lambda \{\int_a^b K(y_i,y)f(y)\,dy - \sum_{j=0}^{n} w_j K(y_i,y_j)f(y_j)\}$. Now

$$\boldsymbol{\tau} = (\mathbf{I}-\lambda\mathbf{K}\mathbf{D})\mathbf{f} - \mathbf{g} \ , \tag{4.97}$$

where $\mathbf{f} = [f(y_0),f(y_1),\ldots,f(y_n)]^T$. But

$$(\mathbf{I}-\lambda\mathbf{K}\mathbf{D})\tilde{\mathbf{f}} = \mathbf{g} \ . \tag{4.98}$$

Thus, from eqns (4.97) and (4.98)

$$(\mathbf{I}-\lambda\mathbf{K}\mathbf{D})(\mathbf{f}-\tilde{\mathbf{f}}) = \boldsymbol{\tau} \tag{4.99}$$

and, when $(\mathbf{I}-\lambda\mathbf{K}\mathbf{D})^{-1}$ exists,

$$||\underset{\sim}{f}-\underset{\sim}{\tilde{f}}||_\infty \leq \|(\underset{\sim}{I}-\lambda\underset{\sim}{K}\underset{\sim}{D})^{-1}\|_\infty \, ||\underset{\sim}{\tau}\|_\infty \; . \tag{4.100}$$

Now we show that $\|(\underset{\sim}{I}-\lambda\underset{\sim}{K}\underset{\sim}{D})^{-1}\|_\infty = O(1)$. Suppose that $\gamma_n(x)$ is a linear combination of $\phi_0(x),\phi_1(x),\ldots,\phi_n(x)$ and

$$\psi_n(x) - \lambda \int_a^b K_n(x,y)\psi_n(y)\mathrm{d}y = \gamma_n(x) \quad . \tag{4.101}$$

Then, as shown above,

$$(\underset{\sim}{I}-\lambda\underset{\sim}{K}\underset{\sim}{D})\underset{\sim}{\psi}_n = \underset{\sim}{\gamma}_n \; . \tag{4.102}$$

We suppose m is large enough and that (4.102) and (4.101) have unique solutions for all $\gamma_n(x)$. We can choose $\underset{\sim}{\gamma}_n$ so that

$$||\underset{\sim}{\psi}_n||_\infty = ||(\underset{\sim}{I}-\lambda\underset{\sim}{K}\underset{\sim}{D})^{-1}||_\infty \; ||\underset{\sim}{\gamma}_n||_\infty . \tag{4.103}$$

But

$$||\underset{\sim}{\psi}_n||_\infty = ||\psi_n(x)||_\infty \quad \text{and} \quad ||\underset{\sim}{\gamma}_n||_\infty = ||\gamma_n(x)||_\infty \quad ,$$

so that

$$||\psi_n(x)||_\infty = ||(\underset{\sim}{I}-\lambda\underset{\sim}{K}\underset{\sim}{D})^{-1}||_\infty \; ||\gamma_n(x)||_\infty \; . \tag{4.104}$$

On the other hand, from the definition of the norms and from eqn (4.101), we have

$$||\psi_n(x)||_\infty \leq ||(I-\lambda K_n)^{-1}||_\infty \; ||\gamma_n(x)||_\infty \; . \tag{4.105}$$

Comparing eqn (4.104) with (4.105) we obtain the bounds

$$||(\underset{\sim}{I}-\lambda\underset{\sim}{K}\underset{\sim}{D})^{-1}||_\infty \leq ||(I-\lambda K_n)^{-1}||_\infty \tag{4.106}$$

$$\leq \frac{||(I-\lambda K)^{-1}||_\infty}{1-|\lambda| \; ||(I-\lambda K)^{-1}||_\infty \; ||K-K_n||_\infty} \tag{4.107}$$

(as shown in the proof of Theorem 4.7), if m is sufficiently large for the denominator of (4.107) to be positive. Thus

$$\|(\underset{\sim}{I}-\lambda \underset{\sim}{K}\underset{\sim}{D})^{-1}\|_\infty = O(1) ,$$

and returning to the bound in (4.100) we have the result

$$\|\underset{\sim}{f}-\tilde{\underset{\sim}{f}}\|_\infty = O(\|\underset{\sim}{\tau}\|_\infty) . \tag{4.108}$$

The relations (4.107) and (4.108) are of importance. We state the following results.

THEOREM 4.11. (a) *If* $K(x,y)$ *is continous and* $J_m(\phi) \in J^R$ *has a width* Δ_m *such that*

$$|\lambda| \, \|(I-\lambda K)^{-1}\|_\infty \, \|K-K_n\|_\infty < 1$$

then

$$\|(\underset{\sim}{I}-\lambda \underset{\sim}{K}\underset{\sim}{D})^{-1}\|_\infty \le \frac{\|(I-\lambda K)^{-1}\|_\infty}{1-|\lambda| \, \|(I-\lambda K)^{-1}\|_\infty \, \|K-K_n\|_\infty} .$$

(b) *Suppose* $K(x,y)$ *is continuous and for* $m = 0,1,2,\ldots,$ $J_m(\phi) \in J^R$, $\lim_{m\to\infty} \Delta_m = 0$. *Then*

$$\sup_{i=0,1,2,\ldots,n(m)} |\tilde{f}(y_i)-f(y_i)| = O(\|\underset{\sim}{\tau}\|_\infty) ,$$

where

$$\underset{\sim}{\tau} = [\tau(y_0),\tau(y_1),\ldots,\tau(y_n)]^T$$

and

$$\tau(y_i) = f(y_i) - \lambda \sum_{j=0}^{n} w_j K(y_i,y_j) f(y_j) - g(y_i).$$

Remark. In Theorem 4.11(b), we presume $f(x)$ to be the true solution of the integral equation (4.93), but the result remains true if any function $\phi(x)$ is set in place of $f(x)$ throughout the statement of this result (see Theorem 4.14).

The similarity of Theorem 4.11 with Theorem 4.7 may be recognized.

Theorem 4.11(b) gives some practical insight. In particular, we should note that the behaviour of the errors $\tilde{f}(y_i)-f(y_i)$ depends only on the values $\tau(y_j)$ $(j = 0,1,2,\ldots,n)$ and not on other values of the function

$$\tau(x) = f(x) - \lambda \sum_{j=0}^{n} w_j K(x,y_j)f(y_j)-g(x) . \qquad (4.109)$$

Our next result gives the behaviour of the error in $\tilde{f}(x)$ when we use a repeated rule.

THEOREM 4.12. *Suppose that* $J_m(\phi) = (m \times J_0)(\phi)$, *where* $J_0(\phi) \in \mathrm{J}^R$ *has a degree of precision* ρ . *We denote the weights and abscissae of* $J_m(\phi)$ *by* w_j, y_j $(j = 0,1,\ldots,n)$, *where* $n = n(m)$, *etc. If* $K(x,y)$ *and* $g(x)$ *have differentiability properties which ensure that*

$$\left|\int_a^b K(y_i,y)f(y)dy - \sum_{j=0}^{n} w_j K(y_i,y_j)f(y_j)\right| \le M/m^r$$

for $i = 0,1,2,\ldots,n(m)$ *and* $m = 0,1,2,\ldots$ *(where* M *is a constant and* $r \le \rho + 1$*) then*

$$\max_{i=0,1,2,\ldots,n(m)} |f(y_i)-\tilde{f}(y_i)| = O(1/m^r) .$$

Example 4.33.

Suppose that for each h considered of the form $h = (b-a)/2m$ $(m = 0,1,\ldots)$ the functions $K(a+ih,y)f(y)$ are 4 times differentiable for $y \in [a,b]$ and $i = 0,1,2,\ldots,$ $(n = 2m)$. Then the use of the repeated Simpson's rule with step h gives approximations $\tilde{f}(x)$ with $\sup_i|\tilde{f}(a+ih)-f(a+ih)| = O(h^4)$. *

At a later point, we shall consider asymptotic expressions for the error, and the behaviour of the error in the Nyström extension. Since all the analysis of section 4.16 has been restricted to rules in J^R we pause to consider what changes are necessary if we consider rules in J^+.

4.16.3.

If we have a rule $J(\phi) \in J^+$ which is not a Riemann sum, then some changes must be made in the preceding analysis.

We again consider an equation of the form (4.80), but with revised definitions of $K_n(x,y)$ and $g_n(x)$. Furthermore, the solution $f_n(x)$ is no longer (in the proof which we follow †) one which takes the values $\tilde{f}(y_i)$ at $x = y_i$ $(i = 0,1,2,\ldots,n)$.

The notation introduced in this subsection is largely independent of that used earlier. We continue to suppose that $\tilde{f}$ is the vector $[\tilde{f}(y_0),\tilde{f}(y_1),\ldots,\tilde{f}(y_n)]^T$ of approximate solution values obtained from the equations $(\underset{\sim}{I}-\lambda\underset{\sim}{K}\underset{\sim}{D})\tilde{\underset{\sim}{f}} = \underset{\sim}{g}$. To define $K_n(x,y)$ and $g_n(x)$ we now set, in place of (4.70),

$$v_0 = a\,,$$

$$v_j = a+(b-a)\sum_{i=0}^{j-1} w_i \Big/ \sum_{i=0}^{n} w_i \quad (j = 0,1,2,\ldots,n) \tag{4.110}$$

and $\Delta = \max_i \max\{|v_{i+1}-y_i|\,,\ |y_i-v_i|\}$ is the measure of $J(\phi) \in J^+$. In terms of the values v_j $(j = 0,1,2,\ldots,(n+1))$, we define

$$\phi_j(x) = \begin{cases} 1 & \text{if } v_j \le x < v_{j+1} \text{ or } x = b \text{ and } j = n \\ \\ 0 & \text{otherwise.} \end{cases} \tag{4.111}$$

Thus

† Other proofs are possible. Compare with the similar analysis in Chapters 3 and 5.

$$\int_a^b \phi_i(x)\phi_j(x)\mathrm{d}x = (v_{j+1}-v_j)\delta_{ij}$$
$$=\{(b-a)w_j/\sum_{k=0}^{n} w_k\}\,\delta_{ij}. \tag{4.112}$$

For convenience, we set

$$w \equiv w^{(m)} = \{\sum_{j=0}^{n} w_j/(b-a)\}\ , \tag{4.113}$$

so that $w_j \leq 2w\Delta$ for $j=0, 1, \ldots, n$, and

$$\int_a^b \phi_i(x)\phi_j(x)\mathrm{d}x = w_j\delta_{ij}/w\ .$$

We now define

$$K_n(x,y) = w\sum_{i=0}^{n}\sum_{j=0}^{n} K(y_i,y_j)\phi_i(x)\phi_j(y) \tag{4.114}$$

and

$$g_n(x) = \sum_{i=0}^{n} g(y_i)\phi_i(x)\ . \tag{4.115}$$

We do not assume that $K_n(y_i,y_j) = K(y_i,y_j)$, or that $g_n(y_i) = g(y_i)$, since this is not true in general. However, $\|K(x,y)-K_n(x,y)\|_\infty$ and $\|g_n(x)-g(x)\|_\infty$ can be bounded in terms of w and Δ if the modulus of continuity of $K(x,y)$ and $g(x)$ are known.

If we have a family of quadrature rules $\{J_m(\phi)\}$ in J^+ and

$$\lim_{m\to\infty} J_m(\phi) = \int_a^b \phi(y)\mathrm{d}y$$

for every continuous function $\phi(x)$ then (using the properties in section 2.15) we may show that, assuming continuity of $K(x,y)$,

$$\lim_{m\to\infty} \|K(x,y)-K_n(x,y)\|_\infty = 0, \tag{4.116}$$

and, assuming continuity of $g(x)$,

$$||g(x)-g_n(x)||_\infty \to 0, \tag{4.117}$$

as $m \to \infty$.

As a consequence of the relations (4.116) and (4.117), we see that $f_n(x)$ exists for m sufficiently large and

$$\lim_{m\to\infty} ||f(x)-f_n(x)||_\infty = 0 . \tag{4.118}$$

However, since $f_n(y_i) \neq \tilde{f}(y_i)$ $(i = 0,1,2,\ldots,n)$ this does not establish the convergence of the quadrature method.

Here we may remark that, if λ is a regular value of $K(x,y)$, the equation

$$f_n(x) - \lambda\int_a^b K_n(x,y)f_n(y)dy = g_n(x) \tag{4.119}$$

has a unique solution $f_n(x) = \sum_{i=0}^{n} \tilde{a}_i\phi_i(x)$, if Δ is sufficiently small, and w is sufficiently close to 1, where

$$(\underset{\sim}{I}-\lambda\underset{\sim}{K}\underset{\sim}{D})\underset{\sim}{\tilde{a}} = \underset{\sim}{g} , \tag{4.120}$$

and $\underset{\sim}{g} = [g(y_0),g(y_1),\ldots,g(y_n)]^T$; the reader may verify this. The solution of eqn (4.120) is then unique, so that

$$\underset{\sim}{\tilde{a}} = \underset{\sim}{\tilde{f}} \tag{4.121}$$

and

$$f_n(x) = \sum_{i=0}^{n} \tilde{f}(y_i)\phi_i(x) , \tag{4.122}$$

where $\tilde{f}(y_0),\tilde{f}(y_1),\ldots,\tilde{f}(y_n)$ are the approximate values obtained using the quadrature method.

We would like to establish that $\lim_{m\to\infty} \sup_i |f(y_i)-\tilde{f}(y_i)| = 0$.

We may define $\xi_j^{(n)} \equiv \xi_j = \frac{1}{2}(v_j+v_{j+1})$. Then, $f_n(\xi_j) = \tilde{f}(y_j)$, and since $\lim_{m\to\infty}||f(x)-f_n(x)||_\infty = 0$ we see that $\lim_{m\to\infty} \sup_i |f(\xi_i)-f_n(\xi_i)| = 0$. But $|\tilde{f}(y_j)-f(y_j)| = |f_n(\xi_j)-f(y_j)| \le |f_n(\xi_j)-f(\xi_j)| + |f(\xi_j)-f(y_j)|$. Since $\Delta \to 0$ and $f(x)$ is continuous, $\lim_{m\to\infty} \sup_i |f(\xi_i)-f(y_i)| = 0$. It follows that $\lim_{m\to\infty} \sup|\tilde{f}(y_j)-f(y_j)| = 0$, as asserted.

We can actually establish the stronger result, namely, that $\lim_{m\to\infty} ||f(x)-\tilde{f}(x)||_\infty = 0$ (when $\lim_{m\to\infty} \Delta_m = 0$ and $\lim_{m\to\infty} w = 1$), where $\tilde{f}(x)$ is the Nyström extension. We have

$$\tilde{f}(x) = g(x) + \lambda \sum_{j=0}^{n} w_j K(x,y_j)\tilde{f}(y_j) \tag{4.123}$$

$$= g(x) + \lambda \int_a^b \hat{K}_n(x,y) f_n(y)\,dy , \tag{4.124}$$

where

$$\hat{K}_n(x,y) = w \sum_{j=0}^{n} K(x,y_j)\phi_j(y) . \tag{4.125}$$

Again, we can show that, if $J_m(\phi) \to \int_a^b \phi(y)\,dy$ for every continuous $\phi(x)$ as $m \to \infty$, then

$$\lim_{m\to\infty} ||K(x,y)-\hat{K}_n(x,y)||_\infty = 0 . \tag{4.126}$$

We may now follow the proof of Theorem 4.10, replacing $\tilde{f}(x)$ by $f_n(x)$, as follows. If

$$e(x) = f(x) - \tilde{f}(x)$$

then

$$e(x) = \lambda\left\{\int_a^b \hat{K}_n(x,y) f_n(y)\,dy - \int_a^b K(x,y) f(y)\,dy\right\} \tag{4.127}$$

and

$$\|e(x)\|_\infty \leq |\lambda| \{\|K\|_\infty \|f_n(x)-f(x)\|_\infty + \|\hat{K}_n-K\|_\infty \|f_n(x)\|_\infty\}, \quad (4.128)$$

so that, since $\|f_n(x)-f(x)\|_\infty \to 0$ and $\|\hat{K}_n-K\|_\infty \to 0$, it follows that $\|e(x)\|_\infty \to 0$.

THEOREM 4.13. *Suppose that* $\{J_m(\phi)\}$ *is a family of quadrature rules in* J^+ *such that*

$$\lim_{m\to\infty} J_m(\phi) = \int_a^b \phi(y)\,dy$$

for every continuous function $\phi(x)$. *Then if* $K(x,y)$ *and* $g(x)$ *are continuous*

$$\lim_{m\to\infty} \|\tilde{f}(x)-f(x)\|_\infty = 0,$$

where $\tilde{f}(x)$ *is the Nyström extension of the values* $\tilde{f}(y_i)$ $(i = 0,1,2,\ldots,n)$ *obtained from the quadrature method using the rule* $J_m(\phi)$.

We have an analogue of Theorem 4.11 for rules in J^+. The following version provides an extension of Theorem 4.11(b), as foreshadowed in an earlier remark. At present, our interest in part (b) of the next theorem arises only in the case $\phi(x) = f(x)$.

THEOREM 4.14. *Suppose the conditions of Theorem 4.13 are satisfied.*

(a) *We can choose* m_0 *so that for all* $m \geq m_0$

$$|\lambda|\, \|(I-\lambda K)^{-1}\|_\infty \|K-K_n\| < 1$$

(where $n = n(m)$ *and* $K_n(x,y)$ *is defined by eqn (4.114)). Then*

$$\|(\underset{\sim}{I}-\lambda\underset{\sim}{K}\underset{\sim}{D})^{-1}\|_\infty \leq \frac{\|(I-\lambda K)^{-1}\|_\infty}{1-|\lambda|\, \|(I-\lambda K)^{-1}\|_\infty \|K-K_n\|_\infty}$$

for $m \geq m_0$

(b) *Suppose* $\phi(x)$ *depends on* m, *for* $m = 0,1,2,\ldots$ *and* $\rho(y_i) = \phi(y_i) - \lambda \sum_{j=0}^{n} w_j K(y_i,y_j)\phi(y_j) - g(y_i)$ *with* $n = n(m)$, *etc.* *and* $\underset{\sim}{\rho} = [\rho(y_0),\rho(y_1),\ldots,\rho(y_n)]^T$. *Then, as* $m \to \infty$,

$$\sup_{i=0,1,2,\ldots,n} |\tilde{f}(y_i) - \phi(y_i)| = O(||\underset{\sim}{\rho}||_\infty) .$$

(c) *The condition number* $\mu_\infty(\underset{\sim}{I}-\lambda\underset{\sim}{K}\underset{\sim}{D})$ *converges to* $\mu_\infty(I-\lambda K)$ *given in (4.61), as* $m \to \infty$.

Proof. The proof of part (a) follows the lines of the proof of (4.107), with the revised definition of $K_n(x,y)$ when $J_m(\phi) \in J^+$. For the proof of part (b) we have the following argument.

Since $(\underset{\sim}{I}-\lambda\underset{\sim}{K}\underset{\sim}{D})\tilde{\underset{\sim}{f}} = \underset{\sim}{g}$ and $(\underset{\sim}{I}-\lambda\underset{\sim}{K}\underset{\sim}{D})\underset{\sim}{\phi} = \underset{\sim}{g} + \underset{\sim}{\rho}$ we see that $(\underset{\sim}{\phi}-\tilde{\underset{\sim}{f}}) = (\underset{\sim}{I}-\lambda\underset{\sim}{K}\underset{\sim}{D})^{-1}\underset{\sim}{\rho}$, provided m is large enough to ensure that $(\underset{\sim}{I}-\lambda\underset{\sim}{K}\underset{\sim}{D})^{-1}$ exists. Now, from part (a), $||(\underset{\sim}{I}-\lambda\underset{\sim}{K}\underset{\sim}{D})^{-1}||_\infty = O(1)$; indeed we have $||(\underset{\sim}{I}-\lambda\underset{\sim}{K}\underset{\sim}{D})^{-1}||_\infty \leq ||(I-\lambda K_n)^{-1}||_\infty$, where K_n is the integral operator with the appropriate kernel $K_n(x,y)$, and the last norm is $O(1)$ as $m \to \infty$. Thus $||\underset{\sim}{\phi}-\tilde{\underset{\sim}{f}}||_\infty = O(||\underset{\sim}{\rho}||_\infty)$, and this is the result required.

The proof of part (c) depends on a number of preliminary observations, and involves the kernels

$$K_n(x,y) = w \sum_{i=0}^{n} \sum_{j=0}^{n} K(y_i,y_j)\phi_i(x)\phi_j(y)$$

associated with the quadrature rules. We discuss the proof of these observations later.

(i) We observe that, with the subordinate operator norm defined on $R[a,b]$, $||I-\lambda K_n||_\infty = ||\underset{\sim}{I}-\lambda\underset{\sim}{K}\underset{\sim}{D}||_\infty + \nu_n$ where, as $m \to \infty$, $|\nu_n| \to 0$.

(ii) We observe that if $R_n(x,y;\lambda)$ is the resolvent kernel of $K_n(x,y)$, then

$$R_n(x,y;\lambda) - K_n(x,y) = \lambda\int_a^b R_n(x,z;\lambda)K_n(z,y)dz = \lambda\int_a^b K_n(x,z)R_n(z,y;\lambda)dz$$

and hence there exist certain values $M_n(y_i,y_j;\lambda)$ such that

$$R_n(x,y;\lambda) = \omega \sum_{i=0}^{n} \sum_{j=0}^{n} M_n(y_i,y_j;\lambda)\phi_i(x)\phi_j(y) .$$

Further, if $\underset{\sim}{M}$ denotes the matrix of elements $M_n(y_i,y_j;\lambda)$, then $(\underset{\sim}{I}-\lambda\underset{\sim}{K}\underset{\sim}{D})^{-1}$ exists and is $(\underset{\sim}{I}+\lambda\underset{\sim}{M}\underset{\sim}{D})$, if λ is a regular value of $K_n(x,y)$.

(iii) The result of (i) can be applied using the result of (ii) to give

$$\|(I-\lambda K_n)^{-1}\|_\infty \equiv \|I+\lambda R_n\|_\infty = \|\underset{\sim}{I}+\lambda\underset{\sim}{M}\underset{\sim}{D}\|_\infty + \eta_n ,$$

where $|\eta_n| \to 0$ as $m \to \infty$.

Combining these observations, we deduce that

$$\|\underset{\sim}{I}-\lambda\underset{\sim}{K}\underset{\sim}{D}\|_\infty \|(\underset{\sim}{I}-\lambda\underset{\sim}{K}\underset{\sim}{D})^{-1}\|_\infty = \mu_\infty(I-\lambda K_n) + \zeta_n ,$$

where

$$|\zeta_n| \le \|I-\lambda K_n\|_\infty |\eta_n| + \|(I-\lambda K_n)^{-1}\|_\infty |\nu_n| + |\eta_n\nu_n|$$

and since $\|I-\lambda K_n\|_\infty = O(1)$ and $\|(I-\lambda K_n)^{-1}\| = O(1)$, $|\zeta_n| \to 0$. By Theorem 4.6, $\lim_{m\to\infty} \mu_\infty(I-\lambda K_n) = \mu_\infty(I-\lambda K)$, and we see that

$\|\underset{\sim}{I}-\lambda\underset{\sim}{K}\underset{\sim}{D}\|_\infty \times \|(\underset{\sim}{I}-\lambda\underset{\sim}{K}\underset{\sim}{D})^{-1}\|_\infty$ tends to $\mu_\infty(I-\lambda K) = \|I-\lambda K\|_\infty \|(I-\lambda K)^{-1}\|_\infty$ as required.

The observations employed in the preceding proof require some substantiation, but the reader may wish to proceed to section 4.16.4 directly. For (i) we may show that

$$\|I-\lambda K_n\|_\infty = 1+|\lambda| \ \|K_n\|_\infty = 1 + |\lambda| \max_i \sum_{j=0}^{n} w_j|K(y_i,y_j)| \,\omega$$

To see that this is so we note that

$$\|I-\lambda K_n\|_\infty \le \|I\|_\infty + |\lambda| \ \|K_n\|_\infty = 1 + |\lambda| \ \|K_n\|_\infty ,$$

and

$$||K_n||_\infty = \sup_x \int_a^b |K_n(x,y)|dy ,$$

where $K_n(x,y)$ has the special form given. Thus

$$||K_n||_\infty = \sum_{j=0}^{n} w_j |K(y_k,y_j)|$$

for some $k \in \{0,1,2,\ldots,n\}$ which maximizes the sum, and if we set $\phi(x) = \text{sign}\ \lambda K(y_k,x)$ for $x \neq y_k$ and $\phi(y_k) = 1$ then $||\phi(x)||_\infty = 1$, $\phi(x) \in R[a,b]$, and $||(I-\lambda K_n)\phi(x)||_\infty = 1 + |\lambda|\ ||K_n||_\infty$, establishing $||I-\lambda K_n||$.

It is readily seen that, if we write $K_{ij} = K(y_i,y_j)$, we can set

$$||\underset{\sim}{I}-\lambda\underset{\sim}{K}\underset{\sim}{D}||_\infty = \max_i \sum_{j=0}^{n} |\delta_{ij}-\lambda w_j K_{ij}|$$

$$= 1 + |\lambda| \max_i \sum_{j=0}^{n} w_j |K_{ij}| + \nu_n ,$$

with $|\nu_n| \leq 4|\lambda| w\ ||K(x,y)||_\infty \Delta_m$, where Δ_m is the measure of $J_m(\phi)$ and $\max_j (w_j) \leq 2\Delta_m w$, when m is large enough (and hence Δ_m is small enough) to ensure $|\nu_n| < 1$. Clearly, $\lim_{m\to\infty}|\nu_n| = 0$.

To establish (iii), we shall wish to pursue an analysis corresponding to that above, and with the kernel $K_n(x,y)$ replaced by its resolvent kernel. We first show that the resolvent kernel is a degenerate kernel of the same form as $K_n(x,y)$ with K_{ij} replaced by suitable values M_{ij}.

Consider (ii). The equations for the resolvent kernel are

$$R_n(x,y;\lambda) - K_n(x,y) = \lambda\int_a^b K_n(x,z)\, R_n(z,y;\lambda)dz = \lambda\int_a^b R_n(x,z;\lambda)K_n(z,y)\ dz$$

and the form of $K_n(x,y)$ reveals immediately that for appropriate values $M_{i,j} = M_n(y_i,y_j;\lambda)$ we have

$$R_n(x,y;\lambda) = \hat{w} \sum_{i=0}^{n} \sum_{j=0}^{n} M_n(y_i,y_j;\lambda)\ \phi_i(x)\phi_j(y) .$$

On substituting in the equations for the resolvent we find that $\underset{\sim}{M} - \underset{\sim}{K} = \lambda\underset{\sim}{M}\,\underset{\sim}{D}\,\underset{\sim}{K} = \lambda\underset{\sim}{K}\,\underset{\sim}{D}\,\underset{\sim}{M}$, so that $(\underset{\sim}{I} - \lambda\underset{\sim}{K}\underset{\sim}{D})^{-1}$ is $(\underset{\sim}{I} + \lambda\underset{\sim}{M}\underset{\sim}{D})$.

Finally we establish (iii), using (ii). We parallel the argument employed to establish (i), and find the required result with $|\eta_m| \leq 4|\lambda| \sup |M_{i,j}| w \Delta_m$. Since $(\underset{\sim}{I} - \lambda \underset{\sim}{K}\underset{\sim}{D})\underset{\sim}{M} = \underset{\sim}{K}$, and $||(\underset{\sim}{I} - \lambda\underset{\sim}{K}\underset{\sim}{D})^{-1}||_\infty = O(1)$ it follows that $\sup |M_{i,j}| = O(1)$ and $\eta_m \to 0$ as $m \to \infty$

4.16.4. *The resolvent of the quadrature operator*

In section 4.17 I shall discuss the approach of Anselone, Atkinson, Mysovskih, and Brakhage to the analysis of the quadrature method. This approach avoids the use of the degenerate kernels $K_n(x,y)$ introduced above, and the remainder of this section serves as an introduction to the type of *quadrature operator* on which their discussion is based.

In establishing the convergence properties discussed above we have dealt in the main with the vectors $\tilde{\underset{\sim}{f}}$ and $\underset{\sim}{f}$. Convergence rates for $\|\tilde{\underset{\sim}{f}}-\underset{\sim}{f}\|_\infty$ arise from the fact that $\|(\underset{\sim}{I}-\lambda\underset{\sim}{K}\underset{\sim}{D})^{-1}\|_\infty$ has been shown to be $O(1)$. (We then say that the quadrature method is (asymptotically) a 'stable' scheme.) In Theorem 4.13 we obtained a convergence result for $\|\tilde{f}(x)-f(x)\|_\infty$ via the convergence of $\|\tilde{\underset{\sim}{f}}-\underset{\sim}{f}\|_\infty$.

Underlying the result of Theorem 4.13 is the following property, which we first discuss informally. Suppose that

$$\tilde{f}(x) - \lambda \sum_{j=0}^{n} w_j K(x,y_j)\tilde{f}(y_j) = g(x) \quad , \tag{4.129}$$

where the weights and abscissae employed are those of a quadrature rule $J_m(\phi) \in J^+$. Then, if

$$\lim_{m\to\infty} J_m(\phi) = \int_a^b \phi(y)\,dy$$

for every $\phi(x) \in C[a,b]$, $\tilde{f}(x)$ converges to $f(x) = (I-\lambda K)^{-1}g(x)$, so $\|\tilde{f}(x)\|_\infty \to \|f(x)\|_\infty$, where $\|f(x)\|_\infty \leq \|(I-\lambda K)^{-1}\|_\infty \|g(x)\|_\infty$. It follows that

$$\sup_{\|g(x)\|_\infty \neq 0} \frac{\|\tilde{f}(x)\|_\infty}{\|g(x)\|_\infty} = O(1)$$

as $m \to \infty$. In other terms, $\|\tilde{f}(x)\|_\infty = O(\|g(x)\|_\infty)$ as $m \to \infty$, and if we denote by H the 'quadrature operator'† such that

$$(H\phi)(x) = \sum_{j=0}^{n} w_j K(x,y_j)\phi(y_j) \quad , \tag{4.130}$$

then $\|(I-\lambda H)^{-1}\|_\infty = O(1)$, where the operator acts on the space $C[a,b]$ with uniform norm. The operator $(I-\lambda H)^{-1}$ may be called the resolvent operator for $I-\lambda H$, and it has the form $I+\lambda M$, where M is a quadrature operator.

THEOREM 4.15. *Subject to the conditions of Theorem 4.13, there exists an integer* m_0 *such that, for* $m \geq m_0$, $\|(\underset{\sim}{I}-\lambda \underset{\sim}{K}\underset{\sim}{D})^{-1}\|_\infty$ *exists, and*

$$\|(I-\lambda H)^{-1}\|_\infty \leq 1 + |\lambda| \sum_{j=0}^{n} |w_j| \, \|K(x,y)\|_\infty \, \|(\underset{\sim}{I}-\lambda \underset{\sim}{K}\underset{\sim}{D})^{-1}\|_\infty \quad .$$

Proof. The existence of $(\underset{\sim}{I}-\lambda \underset{\sim}{K}\underset{\sim}{D})^{-1}$ for all $m \geq m_0$ has been stated above, in Theorem 4.14(a). For any non-trivial continuous function $g(x)$,

$$\tilde{f}(x) = g(x) + \lambda \sum_{j=0}^{n} w_j K(x,y_j)\tilde{f}(y_j) \quad ,$$

so that

$$\|\tilde{f}(x)\|_\infty \leq \|g(x)\|_\infty + |\lambda| \sum_{j=0}^{n} |w_j| \, \|K(x,y)\|_\infty \|\tilde{\underset{\sim}{f}}\|_\infty \, . \tag{4.131}$$

But

$$\|\tilde{\underset{\sim}{f}}\|_\infty \leq \|(\underset{\sim}{I}-\lambda \underset{\sim}{K}\underset{\sim}{D})^{-1}\|_\infty \|\underset{\sim}{g}\|_\infty \leq \|(\underset{\sim}{I}-\lambda \underset{\sim}{K}\underset{\sim}{D})^{-1}\|_\infty \, \|g(x)\|_\infty \quad .$$

We now obtain a bound on

$$\sup_{\|g(x)\|_\infty \neq 0} \|\tilde{f}(x)\|_\infty / \|g(x)\|_\infty$$

which leads to the result of the theorem.

---

† The term numerical integral operator is sometimes used.

As a corollary, we obtain immediately the first part of the following theorem.

THEOREM 4.16. (a) *Subject to the conditions of Theorem 4.13,* $\|(I-\lambda H)^{-1}\|_\infty = O(1)$ *as* $m \to \infty$, *where* $(I-\lambda H)^{-1}$ *acts on* $C[a,b]$.

(b) *If*

$$\phi(x) - \lambda \sum_{j=0}^{n} w_j K(x,y_j)\phi(y_j) = g(x) + \rho(x)$$

where $\phi(x)$ *depends on* m, $n = n(m)$ *etc., then*

$$\|\tilde{f}(x)-\phi(x)\|_\infty = O(\|\rho(x)\|_\infty) \quad .$$

Proof. The proof of part (b) follows from the relations $e(x) = \phi(x)-\tilde{f}(x)$, $e(x)-\lambda(He)(x) = \rho(x)$, together with the results in part (a).

There is a basic similarity between Theorem 4.16, Theorem 4.11, and Theorem 4.7. These results could all be placed in a general framework of a theory of 'stable' approximations to the integral equation. The three theorems prove useful for a comparison between the rates of convergence of different methods for the solution of the Fredholm equation of the second kind, but it should be noted that the difference between Theorem 4.16 and Theorem 4.14 seems to be important. (It could happen that the behaviour of $K(x,y)$ and $g(x)$ is such that $\sup_{0\le i\le n} |\tilde{f}(y_i)-f(y_i)| = O(h^2)$ as $h \to 0$ whereas $\|\tilde{f}(x)-f(x)\|_\infty = O(h)$, say, where h represents the step length used in the quadrature rule $J_m(\phi)$ and $m \to \infty$ as $h \to 0$.)

4.16.5. *Error estimates and bounds*

The preceding results can be employed to construct error bounds in terms of the local truncation error, which can be written

$$\tau(x) = \lambda\int_a^b K(x,y)f(y)dy - \lambda \sum_{j=0}^{n} w_j K(x,y_j)f(y_j) \quad . \tag{4.132}$$

We have $(\underset{\sim}{I}-\lambda\underset{\sim}{K}\underset{\sim}{D})\tilde{\underset{\sim}{f}} = \underset{\sim}{g}$ and $(\underset{\sim}{I}-\lambda\underset{\sim}{K}\underset{\sim}{D})\underset{\sim}{f} = \underset{\sim}{g} + \underset{\sim}{\tau}$, where $\underset{\sim}{\tau} = [\tau(y_0),\tau(y_1),\ldots,\tau(y_n)]^T$, and hence $\underset{\sim}{f} - \tilde{\underset{\sim}{f}} = (\underset{\sim}{I}-\lambda\underset{\sim}{K}\underset{\sim}{D})^{-1}\underset{\sim}{\tau}$ when the inverse exists. Similarly, $f(x)-\tilde{f}(x) = (I-\lambda H)^{-1}\tau(x)$. Theorem 4.14 can be used to bound $\|(\underset{\sim}{I}-\lambda\underset{\sim}{K}\underset{\sim}{D})^{-1}\|_\infty$ and hence (Theorem 4.15) $\|(I-\lambda H)^{-1}\|_\infty$, but in a practical situation we may wish to estimate $\|(\underset{\sim}{I}-\lambda\underset{\sim}{K}\underset{\sim}{D})^{-1}\|_\infty$ directly or bound it rigorously (Wilkinson 1965, p.263). Then

$$\|\underset{\sim}{f}-\tilde{\underset{\sim}{f}}\|_\infty \le \|(\underset{\sim}{I}-\lambda\underset{\sim}{K}\underset{\sim}{D})^{-1}\|_\infty \|\underset{\sim}{\tau}\|_\infty ,$$

and

$$\|f(x)-\tilde{f}(x)\|_\infty \le \|(I-\lambda H)^{-1}\|_\infty \|\tau(x)\|_\infty .$$

To bound $\|\tau(x)\|_\infty$ or $\|\underset{\sim}{\tau}\|_\infty$, we may employ the Peano constants for the rule $J(\phi)$ and corresponding bounds on the y-derivatives of $K(x,y)f(y)$. (Alternatively, difference correction terms may be used to estimate $\tau(x)$.) The derivatives referred to involve the unknown function $f(x)$, and we have, for example,

$$f'(x) = g'(x) + \lambda\frac{d}{dx}\int_a^b K(x,y)f(y)dy ,$$

and if we assume, say, that $K_x(x,y)$ is continuous then

$$\|f'(x)\|_\infty \le \|g'(x)\|_\infty + |\lambda|\ |b-a|\ \|K_x(x,y)\|_\infty \|f(x)\|_\infty .$$

(To evaluate the bound on the right we seek a bound for $\|f(x)\|_\infty$. In some cases this is available. For example, if $|\lambda|\ \|K\|_\infty < 1$

$$\|f(x)\|_\infty \le \|g(x)\|_\infty/\{1-|\lambda|\ \|K\|_\infty\} , \tag{4.133}$$

but in general the entire process becomes unpleasantly complicated.)

We shall now show (see also section 4.17) that the local truncation error can be bounded without engaging in a search for estimates of the derivatives of $f(x)$. A bound on $\|f(x)\|_\infty$ is still required, however.

We have, in view of the equation

$$f(x) = g(x) + \lambda\int_a^b K(x,y)f(y)\,dy ,$$

that

$$\int_a^b K(x,y)f(y)\,dy - \sum_{j=0}^{n} w_j K(x,y_j)f(y_j)$$

$$= \int_a^b K(x,y)g(y)\,dy - \sum_{j=0}^{n} w_j K(x,y_j)g(y_j) +$$

$$+ \lambda\{\int_a^b\int_a^b K(x,z)K(z,y)\,dz\; f(y)\,dy - \int_a^b \sum_{j=0}^{n} w_j K(x,y_j)K(y_j,y)f(y)\,dy\} .$$

Hence, in (4.132),

$$|\tau(x)| \le |\lambda| \left| \int_a^b K(x,y)g(y)\,dy - \sum_{j=0}^{n} w_j K(x,y_j)g(y_j)\right| +$$

$$+ |\lambda|^2 \|f(x)\|_\infty \int_a^b \left| \int_a^b K(x,z)K(z,y)\,dz - \sum_{j=0}^{n} w_j K(x,y_j)K(y_j,y)\right| dy .$$

The last term can be bounded by

$$|b-a|\; |\lambda|^2 \;\|f(x)\|_\infty \sup_{\substack{x \\ a\le y\le b}} |\Delta(x,y)| ,$$

where the quantity

$$\Delta(x,y) = \int_a^b K(x,z)K(z,y)\,dz - \sum_{j=0}^{n} w_j K(x,y_j)K(y_j,y)$$

can be estimated using Peano theory.

Bounds on the terms in the preceding inequality can readily be obtained, with the exception of a bound for $\|f(x)\|_\infty$. For this quantity we may take the estimate $\|\tilde{f}(x)\|_\infty$, or use (4.133), when $|\lambda|\;\|K\|_\infty < 1$; see, also, Theorem 4.18.

In earlier sections, I have already described certain methods of deferred approach to the limit and deferred correction. These contain some built-in techniques for estimating the errors, a process which seems more attractive than endeavouring to obtain rigorous bounds in difficult situations.

However, the choice of numerical method may be based on the rate of convergence of $\tilde{f}(x)$ to $f(x)$ either over $[a,b]$ or on the set of abscissae y_i. To determine these rates, we need only to know (a) that certain y-derivatives of $K(x,y)f(y)$ *do* exist and are bounded, and (b) the rate of decrease of certain Peano constants of the rule $J_m(\phi)$ as $m \to \infty$. The material in Chapter 2 (section 2.16) is adequate to settle the latter issue, for many rules, if used in conjunction with tabulated values of Peano constants (Stroud and Secrest 1966).

4.17. *The Anselone-Brakhage-Mysovskih approach*

We deferred until section 4.16.4 the introduction of the quadrature operator H of (4.130), because it possesses some peculiar properties unlike those of the integral operators K_n which we associated with our general analysis of section 4.16. In this section, we divert briefly from our general theme to consider the quadrature operator H and some of its properties.

In particular, suppose $K(x,y)$ is non-null, and is continuous so that

$$\|K\|_\infty = \sup_{a \le x \le b} \int_a^b |K(x,y)|\,dy = \int_a^b |K(\hat{x},y)|\,dy$$

for some $\hat{x} \in [a,b]$, and suppose that $J_m(\phi) = \sum_{j=0}^{n} w_j \phi(y_j)$ (with $n = n(m)$ etc.) defines, for $m = 1,2,3,\ldots$, a quadrature rule in J^+. We may suppose that

$$\lim_{m\to\infty} J_m(\phi) = \int_a^b \phi(y)\,dy$$

for each $\phi(x) \in C[a,b]$, if we wish. As we show in Example 4.34, given the function $K(\hat{x},y)$ it is possible to construct, for each m, a continuous function $f_m(x)$ and a corresponding value ε_m such that:

$$\|f_m(x)\|_\infty = 1 ,$$

$$\lim_{m\to\infty} \varepsilon_m = 0 ,$$

and

$$\|(Hf_m)(\hat{x})-(Kf_m)(\hat{x})\|_\infty \geq \|K\|_\infty - \varepsilon_m .$$

From the last inequality there follows the interesting result that

$$\lim_{m\to\infty} \|H-K\|_\infty \geq \|K\|_\infty > 0 , \tag{4.134}$$

where the subordinate norms are taken for operators on $C[a,b]$. The same result (4.134) holds if the operator norms are taken as subordinate norms of operators on $R[a,b]$, as we readily deduce.

The fact that $\lim_{m\to\infty} \|H-K\|_\infty \neq 0$ contrasts with the result that $\lim_{m\to\infty} \|K_n-K\|_\infty = 0$ (with norms subordinate to $R[a,b]$, and when the family of rules $J_m(\phi)$ are convergent). Here, $K_n(x,y)$ is the kernel associated with the rule $J_m(\phi)$, as indicated above.

<u>Example 4.34</u>.
Suppose that $\gamma = \max_{a\leq y\leq b} |K(\hat{x},y)|$ where

$$\|K\|_\infty = \int_a^b |K(\hat{x},y)|\,dy ,$$

and for convenience suppose (with some loss of generality) that the function $K(\hat{x},y)$ vanishes for $a \leq y \leq b$ only at $y = z_0 , z_1, z_2,\ldots,z_N$, where N is finite. We define the discontinuous function $f_m^*(x) = \text{sign}\,\{K(\hat{x},x)\}$ (with $f_m^*(z_i) = 0,\ i = 0,1,2,\ldots,N)$. Given m and $\varepsilon_m > 0$, we set $\delta_m = \varepsilon_m/\{2(N+n)\gamma\}$, where $n = n(m)$, and we define

$$f_m(x) = f_m^*(x) \ \text{ if } \ |x-z_k| > \delta_m (k = 0,1,2,\ldots,N)$$
$$\text{and } |x-y_i| > \delta_m (i = 0,1,2,\ldots,n)$$

(see Fig. 4.1). We suppose, for convenience, that each of the intervals $[z_k-\delta_m, z_k+\delta_m]$ $(k = 0,1,2,\ldots,N)$ $[y_j-\delta_m, y_j+\delta_m]$ $(j = 0,1,2,\ldots,n)$ lies wholly in $[a,b]$ and that the intervals are disjoint. In each interval $[z_k-\delta_m, z_k+\delta_m]$ we define $f_m(x)$ to be the linear function agreeing with $f_m{}^*(x)$ at the end-points $z_k-\delta_m$ and $z_k+\delta_m$. In each interval $[y_j-\delta_m, y_j]$ or $[y_j, y_j+\delta_m]$ we define $f_m(x)$ to be the linear function agreeing with $f_m{}^*(x)$ at $x = y_j-\delta_m$ or $x = y_j+\delta_m$ respectively and zero at $x = y_j$.

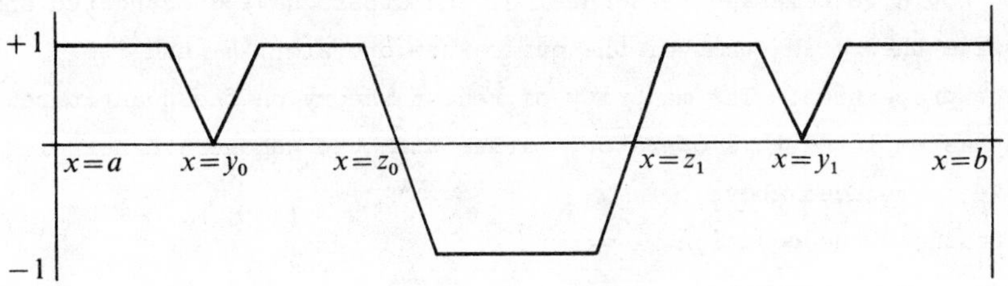

FIGURE 4.1

Graph of $f_m(x)$

Then

$$\left|\int_a^b K(\hat{x},y)f_m(y)\,dy - \int_a^b |K(\hat{x},y)|\,dy\right|$$

$$\leq \quad 2\gamma(N+n)\delta_m \leq \varepsilon_m ,$$

so that

$$\left|\int_a^b K(\hat{x},y)f_m(y)\,dy\right| \geq \|K\|_\infty - \varepsilon_m .$$

Further, $\sum_{j=0}^{n} w_j K(x,y_j) f_m(y_j) = 0$ and $\|f_m(x)\|_\infty = 1$. Thus we have verified the general statements above by constructing a suitable function $f_m(x)$ under convenient assumptions. *

4.17.1. *A rigorous bound avoiding derivatives of* $f(x)$

We suggested in section 4.16.5 that it may be more fruitful to concentrate on the theoretical basis for error *estimates* rather than the construction of rigorous error *bounds*. However, it is convenient at this time to produce a rigorous error bound in terms of the quadrature operator H defined in eqn (4.130).

We treat the material in this section (originally due, in essence, to Mysovskih (1961) and Brakhage (1960)) as something of a diversion, but it should be remarked that Anselone and others have constructed an impressive theory in terms of the quadrature operator H and its abstract properties. The majority of recent theory on the quadrature method has employed this operator, rather than the degenerate kernels which we introduced above.

Consider the equation

$$f(x) - \lambda\int_a^b K(x,y)f(y)\mathrm{d}y = g(x) ,$$

that is,

$$(I-\lambda K)f(x) = g(x) .$$

Suppose that $J(\phi) = \sum_{j=0}^{n} w_j\phi(y_j)$ is *any* quadrature rule for $\int_a^b \phi(y)\mathrm{d}y$ such that $y_i \in [a,b]$ $(i = 0,1,2,\ldots,n)$ and

$$\sup_{a\le x\le b} \sum_{j=0}^{n} \left|w_j\Delta(x,y_j)\right| < \frac{1}{|\lambda^2|\ \|(I-\lambda K)^{-1}\|_\infty} ,$$

where

$$\Delta(x,y) = \int_a^b K(x,z)K(z,y)\mathrm{d}z - \sum_{k=0}^{n} w_k K(x,y_k)K(y_k,y) ,$$

and λ is a regular value of the continuous kernel $K(x,y)$. (Here

$(I-\lambda K)^{-1}$ is the resolvent operator.) Then the equation

$$(I-\lambda H)\tilde{f}(x) = g(x) \quad ,$$

that is,

$$\tilde{f}(x) - \lambda \sum_{j=0}^{n} w_j K(x,y_j)\tilde{f}(y_j) = g(x) \quad ,$$

has an unique solution for every continuous function $g(x)$. Effectively this is equivalent to the statement that $(I-\lambda H)^{-1}$ exists on $C[a,b]$, and the result may be deduced from the following theorem.

THEOREM 4.17. *Suppose that*

$$\|\lambda^2(H-K)H\|_\infty < 1/\|(I-\lambda K)^{-1}\|_\infty \; ,$$

where H,K *are the operators introduced above and act on the space* $C[a,b]$ *with uniform norm. Then*

(a) $(I-\lambda H)^{-1}$ *exists; also*

(b) $$\|(I-\lambda H)^{-1}\|_\infty \le \frac{\|I+\lambda(I-\lambda K)^{-1}H\|_\infty}{1-\|(I-\lambda K)^{-1}\|_\infty \|\lambda^2(H-K)H\|_\infty} \; ;$$

(c) $$\|f(x)-\tilde{f}(x)\|_\infty \le$$

$$\|(I-\lambda K)^{-1}\|_\infty \frac{|\lambda| \; \|(Hg)(x)-(Kg)(x)\|_\infty + |\lambda|^2\|(H-K)H\|_\infty \|f(x)\|_\infty}{1-|\lambda|^2 \|(I-\lambda K)^{-1}\|_\infty \|(H-K)H\|_\infty}$$

<u>Proof</u>. To avoid the use of a general theorem of analysis (see section 1.9 or Theorem 4.34) we shall first establish a preliminary result. Suppose that

$$E\phi(x) = \sum_{j=0}^{n} w_j E(x,y_j)\phi(y_j) \; .$$

(The continuous functions $E(x,y_j)$ are permitted to depend upon n .)

Then E is a bounded operator on the space $C[a,b]$ and

$$\|E\|_\infty = \sup_x \sum_{j=0}^{n} \{|w_j|\ |E(x,y_j)|\} .$$

We prove that $(I-E)^{-1}$ exists if $\|E\|_\infty < 1$. If $\|\underset{\sim}{E}\underset{\sim}{D}\|_\infty$ denotes the subordinate l_∞-matrix norm, where $\underset{\sim}{E} = [E(y_i,y_j)]$ and $\underset{\sim}{D} = \underset{\sim}{\mathrm{diag}}\,(w_0,\ldots,w_n)$, then

$$\|\underset{\sim}{E}\underset{\sim}{D}\|_\infty = \max_i \sum_{j=0}^{n} |w_j E(y_i,y_j)| \le \|E\|_\infty .$$

It follows that if $\|E\|_\infty < 1$ then $\|\underset{\sim}{E}\underset{\sim}{D}\|_\infty < 1$ and hence $(\underset{\sim}{I}-\underset{\sim}{E}\underset{\sim}{D})^{-1}$ exists (and is given by $\sum_{r=0}^{\infty} \{\underset{\sim}{E}\underset{\sim}{D}\}^r$).

Consider, then, the equation $\psi(x)-E\psi(x) = \phi(x)$, where $\phi(x) \in C[a,b]$. We set $\underset{\sim}{\psi} = [\psi(y_0),\psi(y_1),\ldots,\psi(y_n)]^T$ and $\underset{\sim}{\psi} = (\underset{\sim}{I}-\underset{\sim}{E}\underset{\sim}{D})^{-1}\underset{\sim}{\phi}$, since $(\underset{\sim}{I}-\underset{\sim}{E}\underset{\sim}{D})^{-1}$ exists, where

$$\underset{\sim}{\phi} = [\phi(y_0),\phi(y_1),\ldots,\phi(y_n)]^T .$$

Then

$$\psi(x) = \phi(x) + \sum_{j=0}^{n} w_j E(x,y_j)\psi(y_j) ,$$

and $\psi(x)$ is uniquely defined for any $\phi(x)$. Thus $(I-E)^{-1}$ exists.

Let us now establish the result of the theorem.

We may show that $(I-\lambda K)^{-1} = I+\lambda(I-\lambda K)^{-1}K$. We deduce, on expanding, that

$$\{I+\lambda(I-\lambda K)^{-1}H\}\ (I-\lambda H) = I-\lambda^2(I-\lambda K)^{-1}\ (H-K)H = I-E ,$$

say, where $E = \lambda^2(I-\lambda K)^{-1}\ (H-K)H$ and

$$\|E\|_\infty \le |\lambda|^2 \|(I-\lambda K)^{-1}\|_\infty \|(H-K)H\|_\infty . \qquad (4.135)$$

Thus $\|E\|_\infty < 1$ by our assumption. Then $(I-E)^{-1}$ exists and

$$||(I-E)^{-1}||_{\infty} \leq \{1-||E||_{\infty}\}^{-1} \quad ;$$

we see that

$$||(I-E)^{-1}||_{\infty} = ||\{I-\lambda^2(I-\lambda K)^{-1}(H-K)H\}^{-1}||_{\infty}$$

$$\leq 1/\{1-|\lambda^2| \, ||(I-\lambda K)^{-1}||_{\infty}||(H-K)H)||_{\infty}\} \; .$$

It follows that

$$\left[(I-E)^{-1}\{I+\lambda(I-\lambda K)^{-1}H\}\right] (I-\lambda H) = I \; ,$$

so $(I-\lambda H)^{-1}$ exists and is the operator $\left[(I-E)^{-1}\{I+\lambda(I-\lambda K)^{-1}H\}\right]$, and

$$||(I-\lambda H)^{-1}||_{\infty} \leq ||(I-E)^{-1}||_{\infty} \, ||\{I+\lambda(I-\lambda K)^{-1}H\}||_{\infty} \tag{4.136}$$

and part (b) follows.

To obtain the result (c), we note that

$$\tilde{f}(x)-f(x) = \lambda\{(H\tilde{f})(x)-(Kf)(x)\}$$

$$= \lambda(H-K)\tilde{f}(x)+ \lambda K(\tilde{f}(x)-f(x)) \quad ,$$

whence, with $e(x) = \tilde{f}(x)-f(x)$,

$$(I-\lambda K)e(x) = \lambda(H-K)\tilde{f}(x)$$

$$= \lambda(H-K)\{g(x)+\lambda(H\tilde{f})(x)\} \quad .$$

Then

$$e(x) = \lambda(I-\lambda K)^{-1}(H-K)\{g(x)+\lambda(Hf)(x)+\lambda(He)(x)\} \; , \tag{4.137}$$

so that

$$\{I-\lambda^2(I-\lambda K)^{-1}(H-K)H\}e(x)$$

$$= (I-E)e(x) = \lambda(I-\lambda K)^{-1}(H-K)\{g(x)+\lambda(Hf)(x)\} \quad . \qquad (4.138)$$

We have the required result on solving for $e(x)$ and taking norms, using the inequality from part (b).

<u>Remark</u>. To apply Theorem 4.17(c), we require (in the notation of eqn (3.133))

$$\|(H-K)H\|_\infty = \sup_{a\le x\le b} \sum_{j=0}^{n} |w_j\Delta(x,y_j)| \quad .$$

If z-derivatives of $K(x,z)K(z,y)$ are available, $|\Delta(x,y_j)|$ can be bounded in terms of Peano constants of $J(\phi)$ and these bounds. Derivatives of $f(x)$ are thus not required, though we do require a bound on $\|f(x)\|_\infty$, and bounds on the y-derivaties of $K(x,y)g(y)$ are required in order to compute $\|(Hg)(x)-(Kg)(x)\|_\infty$.

In order to compute a bound on $\|f(x)\|_\infty$ we may write (cf. p. 455)

$$\|f(x)\|_\infty \le \|(I-\lambda K)^{-1}\|_\infty \ \|g(x)\|_\infty \, .$$

If $|\lambda| \ \|K\|_\infty < 1$, we have $\|(I-\lambda K)^{-1}\|_\infty \le \{1-|\lambda| \ \|K\|_\infty\}^{-1}$; otherwise we can, (in principle) bound $\|(I-\lambda K)^{-1}\|_\infty$ in terms of $\|(I-\lambda H)^{-1}\|_\infty$ when the following result holds.

THEOREM 4.18. *Suppose that*

$$\|\lambda^2(K-H)K\|_\infty < 1/\|(I-\lambda H)^{-1}\|_\infty \, .$$

Then

$$\|(I-\lambda K)^{-1}\|_\infty \le \frac{\|I+\lambda(I-\lambda H)^{-1}K\|_\infty}{1-\|(I-\lambda H)^{-1}\|_\infty \ \|\lambda^2(K-H)K\|_\infty} \quad .$$

The complexity of the bounds present a daunting aspect to the problem of computing rigorous error bounds in general. A bound on

$\|(I-\lambda H)^{-1}\|_\infty$ is computable from $\|(\underset{\sim}{I}-\lambda\underset{\sim}{K}\underset{\sim}{D})^{-1}\|_\infty$.

It is interesting to note that Theorem 4.17(c) provides, directly, a convergence theorem, and a statement of the rate of convergence of $\|f(x)-\tilde{f}(x)\|_\infty$ to zero, under assumptions on the continuity or differentiability of $K(x,y)$ and $g(x)$. In particular, we have Theorem 4.19, which is established using the following result.

LEMMA 4.1. *If* $K(x,y)$ *is continuous and*

$$\{J_m(\phi) = \sum_{j=0}^{n} w_j\phi(y_j)\ ;\ m = 0,1,2,..;\ n = n(m)\}$$

is a family of quadrature rules such that

$$\lim_{m\to\infty} J_m(\phi) = \int_a^b \phi(y)dy$$

for every $\phi(x) \in C[a,b]$, *then* $\lim_{m\to\infty} \sup_{a\le x,y\le b} |\Delta(x,y)| = 0$ *where*

$$\Delta(x,y) = K^2(x,y) - HK(x,y)$$

Proof. This lemma is of importance in the proof of the following theorem, and it merits some attention. If $J_m(\phi) \in J^R$ $(m = 0,1,2,\ldots)$ then the lemma is established fairly immediately using the width Δ_m of $J_m(\phi)$. For, if $\psi_{1,2}(x) = K(x_1,x)K(x,x_2)$, where x_1 and x_2 are held fixed,

$$\left|\int_a^b \psi(z)dz - \sum_{j=0}^{n} w_j\psi(y_j)\right| \le 2|b-a|\omega(\psi;\Delta_m),$$

where $\omega(\psi;\delta)$ is the modulus of continuity of $\psi(z)$. Under the conditions of the lemma, $\lim_{m\to\infty}\Delta_m = 0$ (see section 2.15.1). As $\Delta_m \to 0$, $\omega(\psi;\Delta_m) \to 0$, but to establish the lemma we require that

$$\lim_{m\to\infty} \sup_{a\le x_1,x_2\le b} \omega(\psi_{1,2};\Delta_m) = 0,$$

and this follows because the continuity of $K(x_1,x)K(x,x_2)$ is *uniform* for $a \le x_1, x_2 \le b$.

In other words, the last statement may be re-expressed by saying that the (two-parameter) system of functions $\{\psi_{1,2}(x)\}$ are equicontinuous. Having noted this fact, the lemma can then be established in full generality by appealing to the result that pointwise-convergence of a bounded linear functional is uniform on equicontinuous bounded sets.

THEOREM 4.19. *Under the conditions of the preceding lemma, and if* $g(x)$ *is continuous for* $a \le x \le b$,

$$\lim_{m\to\infty} \|f(x)-\tilde{f}(x)\|_\infty = 0$$

Proof. Theorem 4.17(c), and the preceding lemma, yield this result immediately. For $Kg(x)$ is continuous, whilst $\|(H-K)H\|_\infty \le \sum_{j=0}^{n}|w_j|\,\|\Delta(x,y)\|_\infty$. Thus $\|Kg(x)-Hg(x)\|_\infty \to 0$ whilst $\|(H-K)H\|_\infty \to 0$ by the result of Lemma 4.1 and the uniform boundedness of $\Sigma|w_j|$.

4.18. *Asymptotic expansions*

Theorem 4.11, Theorem 4.14, or Theorem 4.16 may be used to generate a theory of the asymptotic behaviour of the error in the approximate solution obtained by the quadrature method.

Suppose that $K(x,y)$ and $g(x)$ are continuous and λ is a regular value, and that $J(\phi) \in J^+$. From Theorem 4.14(b), we know that $\|\underset{\sim}{f}-\underset{\sim}{\tilde{f}}\|_\infty = O(\|\underset{\sim}{\tau}\|_\infty)$, where $\underset{\sim}{\tau} = (\underset{\sim}{I}-\lambda\underset{\sim}{K}\underset{\sim}{D})\underset{\sim}{f}-\underset{\sim}{g}$. (This result follows on setting $\phi(x) = f(x)$ in Theorem 4.14(b).) It follows that, if $K(x,y)$ and $f(x)$ are sufficiently differentiable and $J(\phi)$ is the trapezium rule with step h, then $\|\underset{\sim}{\tau}\|_\infty = O(h^2)$, and hence $\|\underset{\sim}{f}-\underset{\sim}{\tilde{f}}\|_\infty = O(h^2)$. That is, there is a constant C independent of h and i such that (for $i = 0,1,2,\ldots,n$)

$$|\tilde{f}(a+ih)-f(a+ih)| \le Ch^2 . \qquad (4.139)$$

In this section, I propose to strengthen this result by showing that, when certain differentiability conditions are satisfied, there exist functions $\phi_k(x)$ such that

$$\tilde{f}(x)-f(x) - \sum_{k=1}^{N} \phi_k(x)h^{2k} = O(h^{2N+1}) \qquad (4.140)$$

uniformly for $x \in \{a+ih,\ i = 0,1,2,\ldots,n\}$, where $h = (b-a)/n$.

Under stronger conditions (4.140) holds uniformly for $a \leq x \leq b$. Whilst (4.140) may be established for $x \in \{a+ih\}$ by employing Theorem 4.14(b), we require Theorem 4.16(b) (or a similar result) to establish (4.140) for $a \leq x \leq b$.

Since the principal purpose here is to demonstrate the manner of proof, we shall confine ourselves to establishing the following rather weak theorem and indicating various modifications which are possible.

THEOREM 4.20. *Suppose that* λ *is a regular value of the continuous kernel* $K(x,y)$, *and the solution of the equation*

$$f(x) - \lambda\int_a^b K(x,y)f(y)dy = g(x)$$

is in $C^{2N+1}[a,b]$. *Suppose also that the derivatives*

$$\left(\frac{\partial}{\partial x}\right)^r\left(\frac{\partial}{\partial y}\right)^s K(x,y) \quad (0 \leq r+s \leq 2N+1) \qquad (4.141)$$

are continuous for $a \leq x,\ y \leq b$.

If

$$\tilde{f}_h(x)-\lambda h \sum_{j=0}^{n}{}'' K(x,a+jh)\tilde{f}_h(a+jh) = g(x) \quad ,$$

there exist functions $\phi_k(x)$ $(k = 1,2,3,\ldots,N)$ *such that*

$$\|\tilde{f}_h(x)-\{f(x) + \sum_{k=1}^{N} h^{2k}\phi_k(x)\}\|_\infty = O(h^{2N+1}) \ .$$

<u>Proof</u>. We shall write

$$\phi(x) = \sum_{k=0}^{N} h^{2k}\phi_k(x) \qquad (4.142)$$

and seek a choice of functions $\phi_k(x)$ such that $\phi_0(x) = f(x)$ and

$$||\tilde{f}_h(x)-\phi(x)||_\infty = O(h^{2N+1}) .$$

Let us anticipate the choice of such functions and suppose that $\phi_k(x) \in C^{2(N-k)+1}[a,b]$. Then, from the Euler-Maclaurin expansion (2.24) and the conditions on $K(x,y)$,

$$||\lambda\int_a^b K(x,y)\phi_k(y)dy - \lambda h \sum_{j=0}^{n}{}'' K(x,a+jh)\phi_k(a+jh) - \sum_{j=1}^{N-k} h^{2j}\psi_j(\phi_k;x)||_\infty = O(h^{2(N-k)+1}), \quad (4.143)$$

where

$$\psi_j(\phi_k;x) = -\lambda\frac{B_{2j}}{(2j)!}\left[\frac{\partial^{2j-1}}{\partial y^{2j-1}}\{K(x,y)\phi_k(y)\}\right]_{y=a}^{y=b} \quad (j=1,2,3,\ldots,(N-k)).(4.144)$$

The notation is simplified if we write

$$(H\phi)(x) = h \sum_{j=0}^{n}{}'' K(x,a+jh)\phi(a+jh) .$$

Then eqn (4.143) may be written as

$$(\lambda K)\phi_k(x)-(\lambda H)\phi_k(x)-\sum_{j=1}^{N-k} h^{2j}\psi_j(\phi_k;x) = O(h^{2(N-k)+1}) .$$

Now, $(I-\lambda H)\tilde{f}_h(x) = g(x)$, and we may imagine that we can determine $\phi(x)$ in (4.142) so that $(I-\lambda H)\phi(x) = g(x)+\rho(x)$, where $||\rho(x)||_\infty = O(h^{2N+1})$. According to Theorem 4.16(b), we may then assert that $||\tilde{f}_h(x)-\phi(x)||_\infty = O(h^{2N+1})$. We therefore propose to construct a suitable sequence of functions $\phi_1(x),\phi_2(x),\ldots,\phi_N(x)$, with $\phi_0(x) = f(x)$. We explain further what constitutes a suitable sequence of functions.

From eqn (4.142), we have the relation

$$(I-\lambda H)\phi(x) = (I-\lambda H)\{f(x) + \sum_{k=1}^{N} h^{2k}\phi_k(x)\} \quad .$$

In this expression, we have

$$(I-\lambda H)f(x) = (I-\lambda K)f(x) + \sum_{k=1}^{N} h^{2k}\psi_k(f;x) + O(h^{2N+1}) = g(x) + O(h^2)$$

so that

$$(I-\lambda H)\phi(x) = g(x)+\rho(x) \quad ,$$

where $\|\rho(x)\|_\infty = O(h^2)$. We are to choose $\phi_1(x),\phi_2(x), \ldots$ to ensure that this $O(h^2)$ term is 'reduced' to $O(h^{2N+1})$, and to do this we examine the coefficients of successive powers of h^2 in $\rho(x)$ and equate them to zero.

The notation is simplified if we write $f(x) = \phi_0(x)$ and we then have

$$\begin{aligned}(I-\lambda H)\phi(x) &= (I-\lambda H)\{\sum_{k=0}^{N} h^{2k}\phi_k(x)\} + O(h^{2N+1}) \\ &= \sum_{k=0}^{N} h^{2k}(I-\lambda H)\phi_k(x) + O(h^{2N+1}) \\ &= \sum_{k=0}^{N} h^{2k}\{(I-\lambda K)\phi_k(x) + \sum_{j=1}^{N-k} h^{2j}\psi_j(\phi_k;x)\}+O(h^{2N+1}) \\ &= \sum_{q=0}^{N} h^{2q}\{(I-\lambda K)\phi_q(x) + \sum_{p=1}^{q} \psi_p(\phi_{q-p};x)\}+O(h^{2N+1})\end{aligned} \tag{4.145}$$

(on collecting like powers of h)

$$= (I-\lambda K)f(x) + h^2\{(I-\lambda K)\phi_1(x) + \psi_1(f;x)\} +$$

$$+ h^4\{(I-\lambda K)\phi_2(x)+\psi_1(\phi_1;x)+\psi_2(f;x)\} + \ldots + O(h^{2N+1}) \, , \tag{4.146}$$

on again writing $\phi_0(x) = f(x)$. In this last expression, the first term is $g(x)$ and the remainder is $\rho(x)$. Equating to zero the coefficients of $h^2, h^4, \ldots$ up to h^{2N} in $\rho(x)$, we obtain the equations

$$(I-\lambda K)\phi_1(x) = -\psi_1(f;x) \tag{4.147}$$

$$(I-\lambda K)\phi_2(x) = -\{\psi_1(\phi_1;x) + \psi_2(f;x)\} , \tag{4.148}$$

etc. In general, we have the recursion

$$(I-\lambda K)\phi_q(x) = -\sum_{p=1}^{q} \psi_p(\phi_{q-p};x) \quad (q = 1,2,3,\ldots,N) \tag{4.149}$$

with $\phi_0(x) = f(x)$.

It will be seen that $\phi_q(x)$ is defined by an integral equation of the second kind (uniquely solvable since λ is a regular value) in which the inhomogeneous term is expressed in terms of $\phi_0(x), \phi_1(x), \ldots, \phi_{q-1}(x)$.

In deriving these conditions, we assumed that

$$\phi_q(x) \in C^{2(N-q)+1}[a,b]$$

and we must check that these conditions are satisfied by the solutions of eqns (4.149).

Clearly, $\phi_q(x)$ is r-times continuously differentiable if $(\partial/\partial x)^r K(x,y)$ is continuous for $a \le x,\ y \le b$ and $(d/dx)^r \psi_p(\phi_{q-p};x)$ is continuous, for $p = 1,2,3,\ldots,q$, on $a \le x \le b$. It is simple to prove by induction that the condition (4.141) is sufficient to guarantee this.

In this way, we have established that there exist functions $\phi_1(x), \phi_2(x), \ldots, \phi_N(x)$ defined by (4.149) which satisfy the statement of the theorem.

4.18.1.

There are a number of ways in which Theorem 4.20 can be extended. We may, for example, consider the behaviour of approximations obtained from the quadrature method using a composite rule $(m \times J_0)(\phi)$. If $\tilde{f}_{1/m}(x)$ is the Nyström extension of the vector $\tilde{\underset{\sim}{f}}$ then, under the conditions of Theorem 4.20, we know that there exist functions $\phi_k(x)$ such that

$$||\tilde{f}_{1/m}(x)-\{f(x) + \sum_{k=r+1}^{2N} (\tfrac{1}{m})^k \phi_k(x)\}||_\infty = O({}^1/_m{}^{2N+1}),$$

where r is the degree of precision of $J_0(\phi)$. Here $\phi_k(x) = 0$ when k is odd provided that the rule $J_0(\phi)$ is symmetric.

Of more interest, for results obtained from the trapezium rule, is the determination of weak conditions under which there exist functions $\phi_k(x)$ such that

$$\max_{i=0,1,2,\ldots,n} |\tilde{f}_h(a+ih)-\{f(a+ih) + \sum_{k=1}^{N} h^{2k}\phi_k(a+ih)\}| = O(h^{2N+1}), \quad (4.150)$$

where $h = (b-a)/n$ tends to zero. To establish this result, it is necessary to find a function $\phi(x)$ in (4.142) such that

$$\max_{i=0,1,2,\ldots,n} |\phi(a+ih)-h \sum_{j=0}^{n}{}'' K(a+ih,a+jh)\phi(a+jh)-g(a+ih)| = O(h^{2N+1}), (4.151)$$

and (4.150) then follows using Theorem 4.14(b).

We would like to establish (4.150) for equations in which the solution $f(x)$ is well behaved *but* $K(x,y)$ is a Green's functions, say, so that derivatives of $K(x,y)$ are discontinuous across the line $x = y$.

The integral equation

$$f(x) - \lambda\int_a^b K(x,y)f(y)dy = g(x)$$

may then be written in the form

$$f(x)-\lambda\int_a^x K_1(x,y)f(y)\,dy - \lambda\int_x^b K_2(x,y)f(y)\,dy = g(x) , \qquad (4.152)$$

where

$$K(x,y) = \begin{cases} K_1(x,y) & (a \le y \le x \le b) \\ K_2(x,y) & (a \le x \le y \le b) . \end{cases} \qquad (4.153)$$

Clearly the conditions of Theorem 4.20 are not then satisfied (and the conclusion is also false). Actually, in (4.152), high-order accuracy cannot be obtained by using the quadrature method in its basic form but it can be obtained by extrapolation based on the validity of (4.151).

For a kernel of the type described, it is necessary to re-define the function $\psi_j(\phi_k;x)$ used above by setting, in place of (4.144),

$$\psi_j(\phi_k;x) = -\lambda\frac{B_{2j}}{(2j)!}\left\{\left[\frac{\partial^{2j-1}}{\partial y^{2j-1}}\{K_1(x,y)\phi_k(y)\}\right]_{y=a}^{y=x} + \right.$$
$$\left. + \left[\frac{\partial^{2j-1}}{\partial y^{2j-1}}\{K_2(x,y)\phi_k(y)\}\right]_{y=x}^{y=b}\right\} .$$

Functions $\phi_k(x)$ can then be determined to satisfy (4.151) (see Coldrick 1972). We have the following result.

THEOREM 4.21. *Suppose that in the conditions of Theorem 4.20 applied to eqn (4.152), the differentiability condition (4.141) is replaced by the following conditions:*

(a) $(\partial/\partial x)^{2N+1}K_1(x,y)$ *is continuous for* $a \le y \le x \le b$

$(\partial/\partial x)^{2N+1}K_2(x,y)$ *is continuous for* $a \le x \le y \le b$,

(b) *for* $r+s = 2N$,

$$(\partial/\partial x)^r(\partial/\partial y)^s K_1(x,y) \quad \textit{is continuous for} \quad a \le y \le x \le b$$

$$(\partial/\partial x)^r(\partial/\partial y)^s K_2(x,y) \quad \textit{is continuous for} \quad a \le x \le y \le b \;.$$

Then, eqn (4.150) is satisfied for some choice of functions $\{\phi_k(x)\}$.

Proof. We refer to Coldrick (1972).

4.19. *Deferred correction*

The theory of section 4.18 may be used to justify the construction of asymptotic upper and lower bounds or the application of the deferred approach to the limit. The process of deferred correction has strong connections with the latter process since we may consider it as an attempt to approximate numerically the correction terms

$$\sum_{k=1}^{N} h^{2k}\phi_k(x)$$

for $\tilde{f}(x)$.

We consider the theoretical basis for methods of deferred correction (of section 4.3.3) under the assumption that $K(x,y)$ and $g(x)$, and hence $f(x)$, are smooth functions possessing high-order derivatives.

We suppose that the approximation $\tilde{\mathbf{f}}$ is initially obtained using the trapezium rule with step h . Thus, with $\mathbf{K} = [K(a+ih,a+jh)]$, $\mathbf{D} = \mathbf{diag}\,(\tfrac{1}{2}h,h,\ldots,h,\tfrac{1}{2}h)$,

$$(\mathbf{I}-\lambda\mathbf{KD})\tilde{\mathbf{f}} = \mathbf{g} \;.$$

Now $(\mathbf{I}-\lambda\mathbf{KD})\mathbf{f} = \mathbf{g}+\boldsymbol{\tau}$, where

$$\tau_i = \lambda\{\int_a^b K(a+ih,y)f(y)\,dy - h\sum_{j=0}^{n}{}'' K(a+ih,a+jh)f(a+jh)\} \quad (i = 0,1,2,\ldots,n)$$

are the local truncation errors.

Consequently

$$(\underset{\sim}{I}-\lambda\underset{\sim}{K}\underset{\sim}{D})\ (\underset{\sim}{f}-\tilde{\underset{\sim}{f}}) = \underset{\sim}{\tau}\ , \tag{4.154}$$

and we could compute $(\underset{\sim}{f}-\tilde{\underset{\sim}{f}})$ if we knew $\underset{\sim}{\tau}$!

We write

$$\Psi_i(x) = K(a+ih,x)f(x)$$

and suppose that $(d/dx)^{p+2}\Psi_i(x)$ is continuous for $a \leq x \leq b$. We can use Gregory's formula to write (see eqn (2.28))

$$\tau_i = h\lambda\{\frac{1}{12}(\nabla\Psi_i(b) - \Delta\Psi_i(a)) + \frac{1}{24}(\nabla^2\Psi_i(b) + \Delta^2\Psi_i(a)) + \ldots +$$

$$+ c_p^*(\nabla^p\Psi_i(b) + (-1)^p\Delta^p\Psi_i(a))\} + O(h^{p+2})\ . \tag{4.155}$$

In this expression, the first term depends on $f(x)$ as a factor in $\Psi_i(x)$, and we may write

$$\tau_i = \lambda\delta_i^{(p)}(f) + O(h^{p+2})\ .$$

Consistent with this notation, we define

$$\delta_i^{(p)}(\phi) = h\{\frac{1}{12}(\nabla\theta_i(b)-\Delta\theta_i(a)) + \ldots +$$

$$+ c_p^*(\nabla^p\theta_i(b) + (-1)^p\Delta^p\theta_i(a))\}\ ,$$

where $\theta_i(\phi;x) \equiv \theta_i(x) = K(a+ih,x)\phi(x)$.

Now since $f(x)$ is unknown, we cannot compute $\delta_i^{(p)}(f)$. But, in the processes of deferred correction, we do compute $\delta_i^{(p)}(\tilde{f})$, having first obtained $\tilde{\underset{\sim}{f}}$ using the repeated trapezium rule. We know that $\tau_i = \lambda\delta_i^{(p)}(f) + O(h^{p+2})$; to what accuracy is $\lambda\delta_i^{(p)}(\tilde{f})$ an approximation to $\lambda\delta_i^{(p)}(f)$?

We observe that $\delta_i^{(p)}(\phi)$ may be written (see section 2.13.2) in the form

$$h \sum_{j=0}^{p} \Omega_j^{[p]} \{K(a+ih,a+jh)\phi(a+jh) + K(a+ih,b-jh)\phi(b-jh)\} \ .$$

Consequently, $|\delta_i^{(p)}(f) - \delta_i^{(p)}(\phi)| = O(h^{q+1})$ for fixed p, if

$$||\underset{\sim}{\phi}-\underset{\sim}{f}||_\infty = O(h^q),$$

and in particular,

$$|\delta_i^{(p)}(f)-\delta_i^{(p)}(\tilde{f})| = O(h^3) \ . \qquad (4.156)$$

We are now in a position to consider a single deferred correction to $\tilde{\underset{\sim}{f}}$, in which we compute (indirectly) a vector $\tilde{\underset{\sim}{f}}^{(1)}$ satisfying

$$(\underset{\sim}{I}-\lambda\underset{\sim}{K}\underset{\sim}{D})\underset{\sim}{f}^{(1)} = \underset{\sim}{g} + \lambda\underset{\sim}{\delta}^{(p)}(\tilde{f})$$

$$= \underset{\sim}{g} + (\underset{\sim}{\tau}+\underset{\sim}{\sigma}^{(1)}) \ ,$$

where we have seen, from (4.156), that $||\underset{\sim}{\sigma}^{(1)}||_\infty = O(h^3)$.
But

$$(\underset{\sim}{I}-\lambda\underset{\sim}{K}\underset{\sim}{D})\underset{\sim}{f} = \underset{\sim}{g}+\underset{\sim}{\tau}$$

so

$$(\underset{\sim}{I}-\lambda\underset{\sim}{K}\underset{\sim}{D})(\tilde{\underset{\sim}{f}}^{(1)}-\underset{\sim}{f}) = \underset{\sim}{\sigma}^{(1)} \ ,$$

and since $||(\underset{\sim}{I}-\lambda\underset{\sim}{K}\underset{\sim}{D})^{-1}||_\infty = O(1)$ as $h \to 0$, it follows that

$$||\tilde{\underset{\sim}{f}}^{(1)}-\underset{\sim}{f}||_\infty = O(\ ||\underset{\sim}{\sigma}^{(1)}||_\infty) = O(h^3) \ .$$

The above analysis reveals that, if $p \geq 1$, a single deferred correction yields an approximation $\tilde{\underset{\sim}{f}}^{(1)}$ from $\tilde{\underset{\sim}{f}}$ where the vector $\tilde{\underset{\sim}{f}}^{(1)}$ converges to $\underset{\sim}{f}$ as $h \to 0$ with an error which is $O(h^3)$, and it appears that the order of convergence in h is independent of $p \geq 1$.

Let us now consider an iterative process, in which we compute $\tilde{\underset{\sim}{f}}^{(r)}$, such that, for $r \geq 1$,

$$(\underset{\sim}{I}-\lambda\underset{\sim}{K}\underset{\sim}{D})\tilde{\underset{\sim}{f}}^{(r)} = \underset{\sim}{g}+\lambda\underset{\sim}{\delta}^{(q)}(\tilde{f}^{(r-1)})$$

and

$$\tilde{\underset{\sim}{f}}^{(0)} = \tilde{\underset{\sim}{f}} .$$

In this scheme (see the discussion of section 4.13.3), we (a) permit q to be fixed and suppose $r \leq q \leq p$, or (b) if q varies we permit the choice $q = r$, where $r = 1,2,3,\ldots,p$.

We supposed that the $(p+1)$th derivatives of $K(a+ih,x)f(x)$ $(i = 0,1,2,\ldots,n;\ h = (b-a)/n)$ are continuous. As a basis for a proof by induction, let us suppose to be true what we have shown for $r = 1,2$, namely that

$$\|\underset{\sim}{f}-\tilde{\underset{\sim}{f}}^{(r-1)}\|_{\infty} = O(h^{1+r})$$

as $h \to 0$. Then, from the foregoing remarks,

$$\|\underset{\sim}{\delta}^{(q)}(\tilde{f}^{(r-1)})-\underset{\sim}{\delta}^{(q)}(f)\|_{\infty} = O(h^{2+r}) , \tag{4.157}$$

and hence, if $r \leq p$,

$$(\underset{\sim}{I}-\lambda\underset{\sim}{K}\underset{\sim}{D})\tilde{\underset{\sim}{f}}^{(r)} = \underset{\sim}{g}+(\underset{\sim}{\tau}+\underset{\sim}{\sigma}^{(r)}) ,$$

where $\|\underset{\sim}{\sigma}^{(r)}\|_{\infty} = O(h^{2+r})$. It follows as before that $\|\underset{\sim}{f}-\tilde{\underset{\sim}{f}}^{(r)}\|_{\infty} = O(h^{2+r})$. We have the following result.

THEOREM 4.22. *Suppose the vectors* $\tilde{\underset{\sim}{f}}^{(r)}$ $(r = 0,1,2,\ldots,p)$ *with*

$$(\underset{\sim}{I}-\lambda\underset{\sim}{K}\underset{\sim}{D})\tilde{\underset{\sim}{f}}^{(r)} = \underset{\sim}{g}+\lambda\underset{\sim}{\delta}^{(q)}(\tilde{f}^{(r-1)})$$

are the successive approximations obtained by the process of deferred correction using differences of order q *in the Gregory correction terms to the trapezium rule with step* h , *where* $q = r$ *or* $q = p$. *If the* $(p+2)$th-*derivatives of the functions* $K(a+ih,y)f(y)$ *are continuous for* $a \le y \le b$ $(i = 0,1,2,\ldots,n)$, $h = (b-a)/n$ *and* $n = n_1 < n_2, < \ldots$ *then, as* $n \to \infty$ *through these values,*

$$\|\tilde{f}^{(r)}-f\|_\infty = O(h^{2+r}) \quad (r \le p) .$$

4.20. The modified quadrature method

We described in section 4.4 a procedure using a quadrature rule

$$J_m(\phi) = \sum_{j=0}^{n} w_j \phi(y_j)$$

for setting up the equations

$$\{I+\lambda(\Delta-KD)\}\mathring{f} = g \tag{4.158}$$

where Δ is a diagonal matrix whose ith diagonal entry is

$$\Delta(y_i) = \sum_{k=0}^{n} w_k K(y_i,y_k) - \int_a^b K(y_i,y)\,dy \quad .$$

If the quadrature rule is one of a convergence family of rules $\{J_m(\phi)\}$ in J^+ , we know that if $K(x,y)$ is continuous, $\|(I-\lambda KD)^{-1}\|_\infty = O(1)$. If we can show that

$$\|(I+\lambda(\Delta-KD))^{-1}\|_\infty = O(1) , \tag{4.159}$$

and then that

$$\|(I+\lambda(\Delta-KD))f-g\|_\infty \to 0 \tag{4.160}$$

as $m \to \infty$, then we shall be able to establish that $\|f-\mathring{f}\|_\infty \to 0$. The rate of convergence to zero is the same as the rate of convergence in (4.160).

THEOREM 4.23. *Suppose the conditions of Theorem 4.13 are satisfied. If* $\underset{\sim}{f} = [f(y_0), f(y_1), \ldots, f(y_n)]^T$ *and* $\overset{\circ}{\underset{\sim}{f}}$ *is defined by (4.158), where* $n = n(m)$ *etc., then*

$$\lim_{m\to\infty} \|\underset{\sim}{f} - \overset{\circ}{\underset{\sim}{f}}\|_\infty = 0 .$$

Proof. From Theorem 4.14, we know that there is an integer m such that for all $m \geq m_0$, $\underset{\sim}{F} = (\underset{\sim}{I} - \lambda \underset{\sim}{K}\underset{\sim}{D})^{-1}$ exists, and its norm is $O(1)$ as $m \to \infty$. Now for $m \geq m_0$

$$\{\underset{\sim}{I} + \lambda(\underset{\sim}{\Delta} - \underset{\sim}{K}\underset{\sim}{D})\} = (\underset{\sim}{I} - \lambda\underset{\sim}{K}\underset{\sim}{D})(\underset{\sim}{I} + \lambda\underset{\sim}{F}\underset{\sim}{\Delta}) , \tag{4.161}$$

so that the matrix on the left has an inverse if $(\underset{\sim}{I} + \lambda\underset{\sim}{F}\underset{\sim}{\Delta})^{-1}$ exists. According to Theorem 4.14(a), $\|\underset{\sim}{F}\|_\infty$ is uniformly bounded for $m \geq m_0$, so that if m is such that

$$|\lambda| \sup_m \|\underset{\sim}{F}\|_\infty \ \|\underset{\sim}{\Delta}\|_\infty < 1 \tag{4.162}$$

then $(\underset{\sim}{I} + \lambda\underset{\sim}{F}\underset{\sim}{\Delta})^{-1}$ exists. But

$$\|\underset{\sim}{\Delta}\|_\infty \leq \sup_{a\leq x\leq b} \left| \int_a^b K(x,y)\,dy - \sum_{j=0}^{n} w_j K(x,y_j) \right| \tag{4.163}$$

and by the continuity (and hence uniform continuity) of $K(x,y)$ for $a \leq x, y \leq b$, the right-hand side of (4.163) can be made arbitrarily small by taking m sufficiently large. The existence of an inverse for the matrix in (4.161) is thus established, and

$$\begin{aligned} &\|\{\underset{\sim}{I} + \lambda(\underset{\sim}{\Delta} - \underset{\sim}{K}\underset{\sim}{D})\}^{-1}\|_\infty \\ &\leq \|(\underset{\sim}{I} - \lambda\underset{\sim}{K}\underset{\sim}{D})^{-1}\|_\infty \ \|(\underset{\sim}{I} + \lambda\underset{\sim}{F}\underset{\sim}{\Delta})^{-1}\|_\infty \\ &= O(1) \quad \text{as} \quad m \to \infty . \end{aligned}$$

The scheme (4.158) is thus 'asymptotically stable', and the theorem now results from the consistency result (4.160). We have, with

$$(\underset{\sim}{I}+\lambda(\underset{\sim}{\Delta}-\underset{\sim}{K}\underset{\sim}{D}))\underset{\sim}{f}-\underset{\sim}{g} = \underset{\sim}{\tau} ,$$

$$\|\underset{\sim}{\tau}\|_\infty \le \sup_{a\le x\le b}\left|\int_a^b K(x,y)\{f(y)-f(x)\}dy - \sum_{j=0}^{n} w_j K(x,y_j)\{f(y_j)-f(x)\}\right|$$

which tends to zero, from the conditions on $J_m(\phi)$, because the continuity in y of the function $K(x,y)[f(y)-f(x)]$ is uniform for $a \le x \le b$.

As we asserted, the theorem finally follows from the relation

$$\{\underset{\sim}{I}+\lambda(\underset{\sim}{\Delta}-\underset{\sim}{K}\underset{\sim}{D})\}\ (\underset{\sim}{f}-\overset{\circ}{\underset{\sim}{f}}) = \{\underset{\sim}{I}+\lambda(\underset{\sim}{\Delta}-\underset{\sim}{K}\underset{\sim}{D})\}\ \underset{\sim}{f}-\underset{\sim}{g} = \underset{\sim}{\tau} ,$$

whence

$$\|\underset{\sim}{f}-\overset{\circ}{\underset{\sim}{f}}\|_\infty = O(\|\underset{\sim}{\tau}\|_\infty) .$$

4.21. A product-integration method

In section 4.6, a product-integration method was described which may be viewed as a collocation method, but this technique was extended (see eqn (4.29)) to a method outside this framework.

We suppose that $K(x,y) = M(x,y)L(x,y)$, and that the integral equation $(I-\lambda K)f(x) = g(x)$ is replaced by

$$(I-\lambda H)\tilde{f}(x) = g(x) , \qquad (4.164)$$

where

$$(H\phi)(x) = \sum_{j=0}^{n} v_j(x)L(x,z_j)\phi(z_j) . \qquad (4.165)$$

The reader should compare (4.164) with eqn (4.29). In order to solve (4.164), we set $x = z_i$ $(i = 0,1,2,\ldots,n)$ and solve the resulting equations for $\underset{\sim}{\tilde{f}} = [\tilde{f}(z_0),\tilde{f}(z_1),\ldots,\tilde{f}(z_n)]^T$; then

$$\tilde{f}(x) = g(x) + \lambda \sum_{j=0}^{n} v_j(x)L(x,z_j)(z_j) \quad .$$

We shall suppose, for definiteness, that the functions $v_i(x)$ and the points z_i $(i = 0,1,2,\ldots,n)$ are chosen with $z_i = a+ih$, $h = (b-a)/n$, and so that $(\mathrm{H}\phi)(x) \equiv (K\phi)(x)$ whenever $L(x,y)\phi(y)$ is continuous and linear in y for $z_i \leq y \leq z_{i+1}$.

THEOREM 4.24. *With* $z_i, v_i(x)$ $(i = 0,1,2,\ldots,n)$ *chosen as above,*

$$\lim_{n\to\infty} \|\tilde{f}(x)-f(x)\|_\infty = 0$$

when $L(x,y), M(x,y)$, *and* $f(x)$ *are continuous and* λ *is a regular value of* $K(x,y)$.

Proof. We sketch the proof which may be carried out using the type of argument employed in Theorem 4.17. We replace H by the operator H defined in (4.165) and show that, as $n \to \infty$,

$$\|(\mathrm{H}-K)\mathrm{H}\|_\infty \to 0 \tag{4.166}$$

and

$$\|(\mathrm{H}-K)g(x)\|_\infty \to 0 \,. \tag{4.167}$$

Eqn (4.167) is the more readily established, and follows from eqn (2.53) and Jackson's theorem (p.93), which gives $2\,\|M\|_\infty \sup\,\omega(\psi;\alpha h)$ as a bound for the left-hand side with $\psi(y) = L(x,y)g(y)$ and the supremum taken over x, whilst $\alpha = \frac{1}{4}\pi(b-a)$. By uniform continuity, the functions $\psi(x)$ are equicontinuous and so the required convergence takes place as $h \to 0$. A correspondingly direct ε tack on (4.166) leads to some manipulative details, but is possible. However, we note that estimation of $\|(\mathrm{H}-K)\mathrm{H}\|_\infty$ by the following argument is not convenient. We have

$$(\mathrm{H}-K)\mathrm{H}\phi(x) = \sum_{k=0}^{n} E(x,z_k)\phi(z_k)$$

and $$||(\mathrm{H} - K)\mathrm{H}||_\infty = \sup_x \sum_{k=0}^{n} |E(x,z_k)| \le (n+1)\sup_{x,j}|E(x,z_j)|,$$

where

$$E(x,z_k) = \sum_{j=0}^{n} v_j(x)L(x,z_j)\{L(z_j,z_k)v_k(z_j)\} -$$

$$- \int_a^b K(x,y)\{L(y,z_k)v_k(y)\}\mathrm{d}y .$$

Unfortunately, the indicated line of argument now becomes difficult (without additional assumptions) because of the factor $(n+1)$ in our last bound.

We shall pursue a more abstract argument, instead. We observe that the operator H has finite-dimensional range and is compact on $C[a,b]$. Thus the set of functions D of the form $\mathrm{H}\,\phi(x)$ where $||\phi(x)||_\infty \le 1$ is uniformly bounded and equicontinuous, and therefore $||(\mathrm{H} - K)\mathrm{H}\,\psi(x)||_\infty \to 0$ *uniformly* for $\psi(x) \in D$ (the convergence for any particular $\psi(x) \in C[a,b]$ being established by the argument which gives (4.167)). Thus, $||(\mathrm{H} - K)\mathrm{H}\,||_\infty = \sup\{\,||(\mathrm{H} - K)\mathrm{H}\phi(x)||_\infty : ||\phi(x)||_\infty = 1\} \to 0$, as required.

4.22. *Expansion methods*

In previous sections we have considered the theoretical basis for methods for solving

$$f(x) - \lambda\int_a^b K(x,y)f(y)\mathrm{d}y = g(x) \qquad (4.168)$$

using methods of numerical integration. In the following sections we consider the theoretical basis for some of the expansion methods introduced in sections 4.9-4.12. We shall be concerned, primarily, with establishing convergence and rates of convergence of the methods of collocation and general Galerkin-Ritz methods, but we begin in section 4.24 by establishing the variational basis of the Rayleigh-Ritz method. We shall generally suppose that

$$\int_a^b\int_a^b |K(x,y)|^2 dx dy < \infty .$$

4.23. *Variational formulation of the Rayleigh-Ritz method*

Suppose that $K(x,y)$ is Hermitian and λ is real in eqn (4.168). If λ is a regular value, it is possible to characterize the unique solution of (4.168) as the function which gives a stationary value to the expression

$$J\{\phi\} = \lambda\int_a^b\int_a^b K(x,y)\overline{\phi(x)}\phi(y) dx\, dy - \int_a^b |\phi(x)|^2 dx +$$

$$+ \int_a^b (\phi(x)\overline{g(x)} + \overline{\phi(x)}g(x)) dx \tag{4.169}$$

(the functions $\phi(x)$ and $g(x)$ are assumed square-integrable).†

THEOREM 4.25. *If* $K(x,y) = \overline{K(y,x)}$, *and* λ *is a regular value, then*

$$\left[\frac{\partial}{\partial\varepsilon} J\{\phi+\varepsilon\psi\}\right]_{\varepsilon=0} = 0 \quad (\varepsilon \text{ real})$$

for all *functions* $\psi(x)$ *with* $\int_a^b |\psi(x)|^2 dx < \infty$, *if and only if* $\phi(x) = f(x)$.

<u>Proof</u>. With $(\phi,\psi) = \int_a^b \phi(x)\overline{\psi(x)} dx$, we may write

$$J\{\phi\} = \lambda(K\phi,\phi)-(\phi,\phi)+(\phi,g)+(g,\phi) ,$$

when $\phi(x)$, $g(x)$, and $K\phi(x)$ are square-integrable. Then, if $\psi(x)$ is square-integrable, $K\psi(x)$ is square-integrable from our assumption

---

† The reader may verify that this characterization holds if $K(x,y)$ is weakly-singular and Hermitian.

about $K(x,y)$, and

$$J\{\phi+\varepsilon\psi\} = \lambda(K(\phi+\varepsilon\psi),\phi+\varepsilon\psi)-(\phi+\varepsilon\psi,\phi+\varepsilon\psi)+(\phi+\varepsilon\psi,g)+(g,\phi+\varepsilon\psi) \quad .$$

for all ε , and

$$J\{\phi+\varepsilon\psi\} = J\{\phi\}+\varepsilon\{(\lambda K\phi-\phi+g,\psi)+(\psi,\bar{\lambda}K^*\phi-\phi+g)\}+|\varepsilon|^2\, Q\{\psi\} ,$$

where $Q\{\psi\} = \lambda(K\psi,\psi)-(\psi,\psi)$, since ε is assumed real. Now, $K = K^*$ and so

$$\left[\frac{\partial}{\partial\varepsilon}J\{\phi+\varepsilon\psi\}\right]_{\varepsilon=0} = 2\mathrm{Re}(\lambda K\phi-\phi+g,\psi) = 2\mathrm{Re}(r,\psi) \quad ,$$

where

$$r(x) = g(x)-\phi(x) + \lambda\int_a^b K(x,y)\phi(y)\mathrm{d}y \quad . \tag{4.170}$$

The condition

$$\left[(\partial/\partial\varepsilon)J\{\phi+\varepsilon\psi\}\right]_{\varepsilon=0} = 0$$

is therefore expressible as the condition $\mathrm{Re}(r,\psi) = 0$.

If $\phi(x) = f(x)$, then $r(x) = 0$ and $(r,\psi) = 0$ for all $\psi(x)$. On the other hand, suppose

$$2\mathrm{Re}(r,\psi) = 0 \quad \forall \quad \psi(x) .$$

For any $\phi(x)$ set there is a corresponding function $r(x)$ defined by (4.170); we set $\psi(x) = r(x)$ and we find that

$$\mathrm{Re}(r,r) \equiv ||r(x)||_2^2 = 0 ,$$

so that $r(x) = 0$ and hence $\phi(x)$ must be the solution $f(x)$. The theorem is thus established.

Theorem 4.25 provides a variational characterization of the solution $f(x)$ of eqn (4.168). The restriction that ε is real is imposed for simplicity, and can be relaxed. We may say that the functional $J\{\phi\}$ has a stationary value when $\phi(x) = f(x)$. This characterization is used to motivate the Rayleigh-Ritz method, as we shall now see.

We shall suppose that we seek an approximate solution

$$\tilde{f}(x) = \sum_{k=0}^{n} \tilde{a}_k \phi_k(x) \quad ,$$

where it is convenient to assume that the functions $\phi_i(x)$ are orthonormal: $(\phi_i,\phi_j) = \delta_{i,j}$ $(i,j = 0,1,2,\ldots,n)$. In view of Theorem 4.25, we are led to choose the coefficients $\tilde{a}_i$ $(i = 0,1,2,\ldots,n)$ to make $J\{\tilde{f}\}$ a stationary value of $J\{\phi\}$ when $\phi(x)$ is required to be a linear combination of the functions $\phi_k(x)$. Thus, if $\tilde{a}_k = \alpha_k + i\beta_k$ $(k = 0,1,2,\ldots,n)$, where $i = \sqrt{-1}$, and α_k, β_k are real, we choose these coefficients to satisfy the requirement

$$\frac{\partial}{\partial \alpha_k} J\{\phi\} = \frac{\partial}{\partial \beta_k} J\{\phi\} = 0 \quad , \tag{4.171}$$

where

$$\phi(x) = \sum_{k=0}^{n} (\alpha_k + i\beta_k)\phi_k(x) \quad .$$

The presentation is simplified if we suppose λ and the functions $K(x,y), g(x), f(x), \phi_k(x)$, and the values $\tilde{a}_k$ are real, so that we then require that

$$\left[\frac{\partial}{\partial a_k} J\{\phi\}\right]_{\underset{\sim}{a}=\underset{\sim}{\tilde{a}}} = 0 \;, \quad \phi(x) = \sum_{k=0}^{n} a_k \phi_k(x) \quad .$$

Now

$$J\{\phi\} = \lambda \sum_{i=0}^{n} \sum_{j=0}^{n} a_i a_j (K\phi_i,\phi_j) - \sum_{i=0}^{n} a_i^2 + 2\sum_{i=0}^{n} a_i(\phi_i,g) \;, \tag{4.172}$$

so that the requirement

$$\left.\frac{\partial}{\partial a_k} J\{\phi\}\right]_{\underset{\sim}{a}=\underset{\sim}{\tilde{a}}} = 0 \quad (k = 0,1,2,\ldots,n)$$

gives us the conditions

$$\lambda \sum_{j=0}^{n} \tilde{a}_j (K\phi_k,\phi_j) - \tilde{a}_k + (g,\phi_k) = 0 \quad ,$$

where $k = 0,1,2,\ldots,n$. Thus, the conditions are satisfied if and only if

$$\underset{\sim}{\tilde{a}} - \lambda \underset{\sim}{A}\underset{\sim}{\tilde{a}} = \underset{\sim}{\gamma} \quad ,$$

where $\gamma_k = (g,\phi_k)$, $A_{i,j} = (K\phi_i,\phi_j) = A_{j,i}$. If $(\phi_i,\phi_j) = \delta_{i,j}$, eqn (4.53) reduces to this form.

In the case where the functions involved are complex-valued, the conditions (4.171) reduce to

$$\underset{\sim}{\alpha} - \lambda \underset{\sim}{A}\underset{\sim}{\alpha} = \underset{\sim}{\mu} \;,$$

$$\underset{\sim}{\beta} - \lambda \underset{\sim}{A}\underset{\sim}{\beta} = \underset{\sim}{\nu} \;,$$

where $\underset{\sim}{\mu} = [\mu_0,\mu_1,\ldots,\mu_n]^T$, and $\underset{\sim}{\nu} = [\nu_0,\nu_1,\ldots,\nu_n]^T$, and

$$\mu_i = \mathrm{Re}(g,\phi_i) \;,$$

$$\nu_i = \mathrm{Im}(g,\phi_i) \;.$$

Thus

$$(\underset{\sim}{I} - \lambda \underset{\sim}{A})(\underset{\sim}{\alpha} + i\underset{\sim}{\beta}) = \underset{\sim}{\gamma} \;.$$

The simplification obtained in assuming that $(\phi_i,\phi_j) = \delta_{i,j}$

is not essential. In general, the functions $\{\phi_i(x)\}$ should merely be linearly independent. We then find

$$J\{\phi\} = \lambda \underset{\sim}{a}^* \underset{\sim}{A} \underset{\sim}{a} - \underset{\sim}{a}^* \underset{\sim}{B} \underset{\sim}{a} + \underset{\sim}{a}^* \underset{\sim}{\gamma} + \underset{\sim}{\gamma}^* \underset{\sim}{a}, \tag{4.173}$$

where

$$A_{i,j} = (K\phi_j, \phi_i), B_{i,j} = (\phi_j, \phi_i), \ \gamma_i = (g, \phi_i) \ ,$$

and

$$\phi(x) = \sum_{k=0}^{n} a_k \phi_k(x) \ .$$

Conditions (4.171) then give us

$$(\underset{\sim}{B} - \lambda \underset{\sim}{A}) \underset{\sim}{a} = \underset{\sim}{\gamma} \ ,$$

as in eqn (4.52), for $\underset{\sim}{a} = \tilde{\underset{\sim}{a}}$. Here, $\underset{\sim}{A} = \underset{\sim}{A}^*$ and $\underset{\sim}{B} = \underset{\sim}{B}^*$.

The derivation of the Rayleigh-Ritz equations (in the way outlined above) is of some interest since the principle can be extended, and yields a method of constructing approximate equations to be solved in place of the original equation. In what follows, however, we shall merely consider the Rayleigh-Ritz method as a particular case of Galerkin's method.

The Rayleigh-Ritz method is certainly defined whenever

$$\int_a^b \int_a^b |K(x,y)|^2 \, dx dy < \infty \ ,$$

$$\int_a^b |\phi_k(x)|^2 \, dx < \infty \ (k = 0,1,2,\ldots,n) \ ,$$

and

$$\int_a^b |g(x)|^2 \, dx < \infty.$$

In practice, even the first condition need not be satisfied (this is the case for some kernels with weak singularities). However, we shall now continue with the assumptions made earlier in this chapter and in general require that the functions involved be continuous.

The Rayleigh-Ritz equations turn out to be a particular case of the Galerkin equations. Like the collocation method, the method of moments, and the method of least squares, the Galerkin equations are derived by requiring that the approximation $\tilde{f}(x)$ is chosen so that the residual

$$r(x) = \tilde{f}(x)-\lambda(K\tilde{f})(x)-g(x)$$

satisfies certain conditions. The variational formulation of the Rayleigh-Ritz method plays little part in our general analysis of expansion methods, providing only a somewhat convenient stepping stone in the construction of approximate methods for non-linear integral equations or integro-differential equations.

4.24. *Collocation*

The collocation method and the methods of Ritz-Galerkin type (sections 4.11 and 4.12) can all be viewed as 'projection methods' (see section 4.29 and Phillips (1969)) and we can use a similar style of analysis to discuss all these methods. The basic principles involved appear in the analysis of any one of the methods, and we have chosen to begin with a study of the collocation method.

Recall that, in the collocation method, we choose functions $\phi_k(x)$ $(k = 0,1,2,\ldots,n)$, here assumed continuous or piecewise-continuous and bounded, and collocation points z_k $(k = 0,1,2,\ldots,n)$, and we obtain the approximation

$$\tilde{f}(x) = \sum_{k=0}^{n} \tilde{a}_k\phi_k(x) \quad , \tag{4.174}$$

where (see eqn (4.46))

$$\sum_{j=0}^{n} (\phi_j(x_i)-\lambda\psi_j(z_i))\tilde{a}_j = g(z_i) \quad (i = 0,1,2,\ldots,n) \tag{4.175}$$

with

$$\psi_j(x) = \int_a^b K(x,y)\phi_j(y)dy \ . \qquad (4.176)$$

With the notation used in section 4.11 (in particular, $B_{i,j} = \phi_j(z_i)$, $A_{i,j} = \psi_j(z_i)$) we can write (4.175) in the form $\underset{\sim}{B}\tilde{\underset{\sim}{a}}-\lambda\underset{\sim}{A}\tilde{\underset{\sim}{a}} = \underset{\sim}{g}$. We assume that $\underset{\sim}{B}^{-1}$ exists.

Now, the analysis of the error $\|f(x)-\tilde{f}(x)\|_\infty$ can be based on the following theorem, which associates $\tilde{f}(x)$ with a certain kernel $\tilde{K}(x,y)$ and a function $\tilde{g}(x)$. For the method of collocation we set

$$\tilde{K}(x,y) = \sum_{j=0}^{n} \phi_j(x)\chi_j(y) \ , \qquad (4.177)$$

where

$$\sum_{j=0}^{n} \phi_j(z_k)\chi_j(y) = K(z_k,y) \quad (k = 0,1,2,\ldots,n) \qquad (4.178)$$

and

$$\tilde{g}(x) = \sum_{j=0}^{n} \gamma_j\phi_j(x) \ , \qquad (4.179)$$

where

$$\sum_{j=0}^{n} \gamma_j\phi_j(z_k) = g(z_k) \quad (k = 0,1,2,\ldots,n) \qquad (4.180)$$

(that is, $\tilde{g}(z_k) = g(z_k)$ for $k = 0,1,2,\ldots,n$) . In order to guarantee that $\tilde{K}(x,y)$ and $\tilde{g}(x)$ are uniquely defined we have assumed that $\underset{\sim}{B}^{-1}$ exists, that is,

$$\det[\phi_i(z_j)] \neq 0 \ .$$

The kernel $\tilde{K}(x,y)$ was encountered previously in eqn (3.245). In terms of the above notation, we have the following result.

THEOREM 4.26. (a) *Suppose that the system of eqns (4.175) has a unique solution. Then*

$$\tilde{f}(x) - \lambda\int_a^b \tilde{K}(x,y)\tilde{f}(y)\mathrm{d}y = \tilde{g}(x) . \tag{4.181}$$

(b) *Conversely, if this integral equation has a unique solution, it is given by (4.174) where the coefficients* $\tilde{a}_j$ $(j = 0,1,2,\ldots,n)$ *satisfy eqns (4.175).*

<u>Proof</u>. (a) Suppose that

$$\tilde{f}(x) = \sum_{k=0}^{n} \tilde{a}_k\phi_k(x) .$$

Then if we set

$$r(x) = \tilde{f}(x)-\lambda(\tilde{K}\tilde{f})(x)-\tilde{g}(x) ,$$

$r(x)$ is clearly a linear combination of $\phi_0(x), \phi_1(x),\ldots,\phi_n(x)$, say

$$r(x) = \sum_{k=0}^{n} \rho_k\phi_k(x) .$$

Consequently

$$r(z_i) = \sum_{k=0}^{n} \rho_k\phi_k(z_i)$$

or

$$\underset{\sim}{r} = \mathrm{B}\underset{\sim}{\rho} , \tag{4.182}$$

where $\underset{\sim}{r} = [r(z_0),r(z_1),\ldots,r(z_n)]^T$, $\underset{\sim}{\rho} = [\rho_0,\rho_1,\ldots,\rho_n]^T$.

But

$$r(z_i) = \tilde{f}(z_i)-\lambda\int_a^b \tilde{K}(z_i,y)\tilde{f}(y)\mathrm{d}y - \tilde{g}(z_i)$$

$$= \sum_{k=0}^{n} \tilde{a}_k\phi_k(z_i)-\lambda \sum_{j=0}^{n} \phi_j(z_i)\int_a^b \chi_j(y) \sum_{k=0}^{n} \tilde{a}_k\phi_k(y)dy-g(z_i) .$$

Thus

$$\underset{\sim}{r} = \underset{\sim}{B}\tilde{\underset{\sim}{a}}-\lambda\underset{\sim}{D}\tilde{\underset{\sim}{a}}-\underset{\sim}{g} ,$$

where

$$D_{i,k} = \sum_{j=0}^{n} \phi_j(z_i)\int_a^b \chi_j(y)\phi_k(y)dy = \int_a^b K(z_i,y)\phi_k(y)dy ,$$

using (4.176), $D_{i,k} = \psi_k(z_i)$, or $\underset{\sim}{D} = \underset{\sim}{A}$. Consequently, $\underset{\sim}{r} = \underset{\sim}{0}$ and since $\underset{\sim}{B}$ is non-singular in (4.182) we see that $\underset{\sim}{\rho} = \underset{\sim}{0}$ so that $r(x) \equiv 0$, and the integral equation is satisfied.

(b) If the integral equation is satisfied, its solution is clearly a linear combination (4.174) of $\phi_0(x),\phi_1(x),\ldots,\phi_n(x)$. Setting $x = z_i$ $(i = 0,1,2,\ldots,n)$ in (4.181) we readily obtain the collocation equations (4.175).

From Theorem 4.26(b) we can deduce conditions under which the collocation equations have a unique solution, using Theorem 4.2. Thus there is a unique approximation $\tilde{f}(x)$ if $\underset{\sim}{B}^{-1}$ exists and

$$\sup_{a\le x,y\le b} |K(x,y)-\tilde{K}(x,y)| = ||K(x,y)-\tilde{K}(x,y)||_\infty$$

is sufficiently small. For, as the reader will recall,

$$||K-\tilde{K}||_\infty = \sup_x\int_a^b |K(x,y)-\tilde{K}(x,y)|dy$$

$$\le |b-a| \; ||K(x,y)-\tilde{K}(x,y)||_\infty .$$

Moreover, when such a condition is satisfied, Theorem 4.4 can be used to supply a theoretical error bound for $||f(x)-\tilde{f}(x)||_\infty$, and the analysis of section 4.15 applies, with $\tilde{K}(x,y)$ replacing $K_n(x,y)$.

Let us analyse in some detail a primitive form of collocation using the functions

$$\phi_i(x) = \begin{cases} 1 \text{ if } a+ih \le x < a+(i+1)h \text{ , or } x = b \text{ and } i = n \\ 0 \text{ otherwise} \end{cases}$$

where $h = 1/(n+1)$, and $z_i = a+ih$ $(i = 0,1,2,\ldots,n+1)$.

We suppose that $K(x,y)$ and $g(x)$ are continuous, and that λ is a regular value of $K(x,y)$. For the corresponding kernel $\tilde{K}(x,y)$ and the function $\tilde{g}(x)$ introduced in section 4.23 we can write

$$\tilde{K}(x,y) = K(z_i,y)(z_i \le x < z_{i+1}) \tag{4.183}$$

$$\tilde{g}(x) = g(z_i) \quad (z_i \le x < z_{i+1}) \tag{4.184}$$

where $i = 0,1,2,\ldots,n$, with equality at the right end-point if $i = n$.

Clearly, $\lim_{h\to 0} \|\tilde{g}(x)-g(x)\|_\infty = 0$ and $\lim_{h\to 0}\|\tilde{K}(x,y)-K(x,y)\|_\infty = 0$ using the continuity of $g(x)$ and the continuity of $K(x,y)$ on $[a,b]$. If we apply Theorem 4.14 with $K_n(x,y)$ replaced by $\tilde{K}(x,y)$, $g_n(x)$ by $\tilde{g}(x)$, and $f_n(x)$ by $\tilde{f}(x)$ we can establish the following result.

THEOREM 4.27. *In this case of collocation, described above*

$$\lim_{h\to 0} \|f(x)-\tilde{f}(x)\|_\infty = 0 ,$$

when $K(x,y)$ *and* $g(x)$ *are continuous.*

We lose something by considering $\sup_{x,y} |K(x,y)-\tilde{K}(x,y)|$, since it is sufficient that $\|K-\tilde{K}\|_\infty \to 0$. We do not require continuity of $K(x,y)$ in order to establish Theorem 4.27. For

$$\|K-\tilde{K}\|_\infty = \sup_{a\le x\le b} \int_a^b |K(x,y)-\tilde{K}(x,y)|\,dy$$

$$= \max_i \sup_{ih\le x<(i+1)h} \int_a^b |K(x,y)-K(ih,y)|\,dy \quad .$$

If we impose the condition (used by Atkinson (1967), and others) that

$$\int_a^b |K(x_1,y)-K(x_2,y)|\,dy \to 0 \qquad (4.185)$$

uniformly for $a \leq x_1, x_2 \leq b$ as $|x_1-x_2| \to 0$, then clearly $\|K-\tilde{K}\|_\infty \to 0$, and when $g(x)$ is continuous we can again establish Theorem 4.27. We shall refer to such results later. (If $K(x,y)$ is continuous then clearly (4.185) is valid.)

On the other hand, suppose that $(\partial/\partial x)\,K(x,y)$ is continuous (or merely bounded) for $a \leq x, y \leq b$, and that $g'(x)$ is also bounded. Then

$$||K(x,y)-\tilde{K}(x,y)||_\infty \leq h\ ||(\partial/\partial x)K(x,y)||_\infty$$

and

$$||g(x)-\tilde{g}(x)||_\infty \leq h\,||g'(x)||_\infty \;.$$

Consequently

$$||f(x)-\tilde{f}(x)||_\infty = O(h) \;.$$

To obtain a faster rate of convergence the choice of functions $\phi_k(x)$ must be made more 'sophisticated', and additional conditions must be imposed on $K(x,y)$ and $g(x)$. For example, (see Phillips 1969) it is possible to use cubic splines to obtain $O(h^4)$ convergence when $(\partial/\partial x)^4 K(x,y)$ and $g^{iv}(x)$ are continuous. We also have the following result.

THEOREM 4.28. *Suppose, for* $n = 0,1,2,\ldots,$ *that the functions* $\phi_k(x)$ $(k = 0,1,2,\ldots,n)$ *form a linearly independent set of algebraic polynomials of degree* $\leq n$. *Suppose that* $z_k = \frac{1}{2}(b-a)x_k + \frac{1}{2}(a+b)$ *where either*

(a) $x_k = \cos\left(\frac{k\pi}{n}\right)$ $(k = 0,1,2,\ldots,n)$

or

$$\text{(b)} \quad x_k = \cos\left[(2k+1)\pi/\{2(n+1)\}\right] \quad (k = 0,1,2,\ldots,n) .$$

Then if (for $r \geq 1$) $(\partial/\partial x)^r K(x,y)$ is bounded for $a \leq x, y \leq b$ and $g^{(r)}(x)$ is bounded for $a \leq x \leq b$, it follows that

$$\|\tilde{f}(x)-f(x)\|_\infty = O((\ln n)/n^r)$$

as $n \to \infty$.

<u>Proof</u>. Suppose that $\tilde{g}(x)$ is the polynomial of degree n such that $\tilde{g}(z_i) = g(z_i)$, $i = 0,1, \ldots ,n$. Then

$$\|g(x)-\tilde{g}(x)\|_\infty \leq (1+L_n)e_n(g) ,$$

where L_n is the appropriate Lebesgue constant (section 2.2) and $e_n(g) = \inf_{p_n \in P_n} \|g(x)-p_n(x)\|_\infty$ can be bounded by Jackson's theorem in terms of $\|g^{(r)}(x)\|_\infty$, where $r \leq n$. For the choice of z_k given x_k in (a) or (b), $L_n = O(\ln n)$ whilst, also, $e_n(g) = O(1/n^r)$ and

$$\|g(x)-\tilde{g}(x)\|_\infty = O(\ln n /n^r) .$$

In a similar fashion,

$$\|K(x,y)-\tilde{K}(x,y)\|_\infty \leq (1+L_n)\{\tfrac{1}{4}\pi(b-a)\}^r \frac{\|(\partial/\partial x)^r K(x,y)\|_\infty}{\{(n+1)!/(n-r+1)!\}}$$

$$= O((\ln n)/n^r) \text{ as } n \to \infty .$$

The required result now follows from Theorem 4.3.

4.24.1.

Theorem 4.28 raises some points of interest. In the first place, it does provide *some* basis for comparing the rates of convergence of a version of the collocation method with the rates of convergence obtainable in the quadrature method. However, Theorem 4.28 is clearly not a very strong result because it does not predict the quite good rates of convergence obtainable when $K(x,y)$ is a Green's function with discontinuities in low order derivatives across $x = y$.

Secondly, it is apparent from Theorem 4.28 that the choice of functions $\phi_k(x)$ is of no *theoretical* consequence provided these functions are linearly independent polynomials of degree at most n . In practice there are advantages in taking $\phi_0(x) = \frac{1}{2}$, $\phi_r(x) = T_r(x)$ on $[-1, 1]$, and the conditioning of the numerical method *does* depend on the choice of basic functions.

If we regard the size of the condition number

$$\|(\underset{\sim}{B}-\lambda\underset{\sim}{A})\|_\infty \, \|(\underset{\sim}{B}-\lambda\underset{\sim}{A})^{-1}\|_\infty$$

as a measure of the susceptibility of eqns (4.175) to ill-conditioning, the following theorem can be regarded as relevant. As a prelude to this theorem, consider the operator Φ which transforms the vector $\underset{\sim}{a} = [a_0, a_1, \ldots, a_n]^T$ into the function

$$\sum_{k=0}^{n} a_k \phi_k(x) \quad .$$

On the space of such functions Φ^{-1} exists; $\underset{\sim}{a} = \underset{\sim}{B}^{-1}\underset{\sim}{\phi}$, since $\underset{\sim}{B}$ is non-singular by assumption. We define

$$\|\Phi\|_\infty = \sup_{\underset{\sim}{a}\neq\underset{\sim}{0}} \frac{\|\sum_{k=0}^{n} a_k\phi_k(x)\|_\infty}{\|\underset{\sim}{a}\|_\infty}$$

$$= \|\sum_{k=0}^{n} |\phi_k(x)| \, \|_\infty \tag{4.186}$$

and

$$\|\Phi^{-1}\|_\infty = \sup_{\substack{\phi(x)=\Sigma a_k\phi_k(x) \\ \phi(x)\neq 0}} \frac{\|\underset{\sim}{B}^{-1}\underset{\sim}{\phi}\|_\infty}{\|\phi(x)\|_\infty}$$

$$\leq \sup_{\underset{\sim}{\phi}\neq\underset{\sim}{0}} \frac{\|\underset{\sim}{B}^{-1}\underset{\sim}{\phi}\|_\infty}{\|\underset{\sim}{\phi}\|_\infty}$$

or

$$\|\Phi^{-1}\|_\infty \le \|\underset{\sim}{B}^{-1}\|_\infty . \tag{4.187}$$

Using the operators Φ and Φ^{-1} we may establish the following theorem.

THEOREM 4.29. *Suppose that* $\underset{\sim}{B}^{-1}$ *and* $(\underset{\sim}{B}-\lambda\underset{\sim}{A})^{-1}$ *exist. Then*

$$\|\underset{\sim}{B}-\lambda\underset{\sim}{A}\|_\infty \le \|\underset{\sim}{B}\|_\infty \|\Phi\|_\infty \|\Phi^{-1}\|_\infty \|(I-\lambda\tilde{K})\|_\infty$$

and

$$\|(\underset{\sim}{B}-\lambda\underset{\sim}{A})^{-1}\|_\infty \le \|\underset{\sim}{B}^{-1}\|_\infty \|\Phi\|_\infty \|\Phi^{-1}\|_\infty \|(I-\lambda\tilde{K})^{-1}\|_\infty .$$

<u>Proof</u>. If $\phi(x) = \sum_{k=0}^{n} a_k\phi_k(x)$ then $(I-\lambda\tilde{K})\phi(x) = s(x) = \sum_{k=0}^{n} \sigma_k\phi_k(x)$, where $\underset{\sim}{B}\underset{\sim}{\sigma} = (\underset{\sim}{B}-\lambda\underset{\sim}{A})\underset{\sim}{a}$. Thus

$$\begin{aligned}(\underset{\sim}{B}-\lambda\underset{\sim}{A})\underset{\sim}{a} &= \underset{\sim}{B}\Phi^{-1}s(x)\\ &= \underset{\sim}{B}\Phi^{-1}(I-\lambda\tilde{K})\phi(x)\\ &= \underset{\sim}{B}\Phi^{-1}(I-\lambda\tilde{K})\Phi\underset{\sim}{a}\end{aligned}$$

for every vector $\underset{\sim}{a}$. On taking norms,

$$\|(\underset{\sim}{B}-\lambda\underset{\sim}{A})\|_\infty \le \|\underset{\sim}{B}\|_\infty \|\Phi^{-1}\|_\infty \|\Phi\|_\infty \|(I-\lambda\tilde{K})\|_\infty$$

(since we may take $\|\underset{\sim}{a}\|_\infty \ne 0$ and cancel this term). Furthermore,

$$(\underset{\sim}{B}-\lambda\underset{\sim}{A})^{-1} = \Phi^{-1}(I-\lambda\tilde{K})^{-1}\Phi\,\underset{\sim}{B}^{-1} ,$$

and the second result follows.

The following result is a corollary of Theorem 4.29.

COROLLARY 4.2. *Suppose that for any non-singular matrix* $\underset{\sim}{C}$,

$\mu(\underset{\sim}{C}) = \|\underset{\sim}{C}\|_\infty \|\underset{\sim}{C}^{-1}\|_\infty$, *that* $\nu_\infty(\Phi) = \|\Phi\|_\infty \|\Phi^{-1}\|_\infty$ *and that* $\mu(I-\lambda\tilde{K}) = \|(I-\lambda\tilde{K})\|_\infty \|(I-\lambda\tilde{K})^{-1}\|_\infty$. *Then*

$$\mu(\underset{\sim}{B}-\lambda\underset{\sim}{A}) \leq \{\nu_\infty(\Phi)\}^2\mu(\underset{\sim}{B})\mu(I-\lambda\tilde{K}) \quad .$$

We note also that if $\|K-\tilde{K}\|_\infty \to 0$, $\mu(I-\lambda\tilde{K}) \to \mu(I-\lambda K)$.

This corollary to Theorem 4.29 provides a bound on the condition number of $(\underset{\sim}{B}-\lambda\underset{\sim}{A})$ but does not give precise information on the behaviour of this quantity. If we are wise, however, we will avoid a choice of functions $\phi_k(x)$ for which $\nu_\infty(\Phi)$ or $\mu(\underset{\sim}{B})$ is large. The problem of bounding $\mu(\underset{\sim}{I}-\lambda\underset{\sim}{A}\underset{\sim}{B}^{-1})$, which is required if we solve (4.49) in place of (4.48), is left as an exercise, for the reader.

4.25. *Ritz-Galerkin-type methods*

The general theory of the Ritz-Galerkin-type methods introduced in section 4.12 follows similar lines to the general theory of collocation methods, as presented in section 4.23. In order to give a fairly unified theory I introduce the notation $[\phi,\psi]$ which can be taken to represent, consistently, any one of the quantities

$$(\phi,\psi) = \int_a^b \phi(x)\overline{\psi(x)}dx \ ,$$

or

$$\langle\phi,\psi\rangle = \int_a^b w(x)\phi(x)\overline{\psi(x)}dx \ ,$$

where $w(x) \geq 0$ for $a \leq x \leq b$, with $\omega(x) \not\equiv 0$ almost everywhere, or

$$\langle\phi,\psi\rangle = \sum_{k=0}^{N} \Omega_k\phi(x_k)\overline{\psi(x_k)}$$

where $\Omega_k > 0$, $a \leq x_k \leq b$ $(k = 0,1,2,\ldots,n)$. In general, therefore, we might write

$$[\phi,\psi] = \int_a^b w(x)\phi(x)\overline{\psi(x)}dx + \sum_{k=0}^{N} \Omega_k\phi(x_k)\overline{\psi(x_k)} \ . \quad (4.188)$$

We suppose that (4.188) is an inner-product , or, more generally,

$$\int_a^b w(y)\,dy + \prod_{k=0}^{N} \Omega_k > 0$$

in (4.188), with $w(x) \geq 0$, $\Omega_k \geq 0$.

Now we suppose that $\phi_0(x),\phi_1(x),\ldots,\phi_n(x)$ are linearly independent, and we seek an approximation

$$\tilde{f}(x) = \sum_{k=0}^{n} \tilde{a}_k\phi_k(x) \ .$$

The coefficients $\tilde{a}_k$ define a function

$$r(x) = \tilde{f}(x)-\lambda\int_a^b K(x,y)\tilde{f}(y)\,dy - g(x) \ .$$

In the classical Galerkin method we require $\tilde{a}_0,\tilde{a}_1,\ldots,\tilde{a}_n$ such that $(r,\phi_k) = 0$ $(k = 0,1,2,\ldots,n)$. A general method of Ritz-Galerkin type corresponds to a choice of $[\phi,\psi]$ of the form (4.188) and the determination of $\tilde{a}_0,\tilde{a}_1,\ldots,\tilde{a}_n$ so that

$$[r,\phi_k] = 0 \quad (k = 0,1,2,\ldots,n) \ . \tag{4.189}$$

The coefficients $\tilde{a}_0,\tilde{a}_1,\ldots,\tilde{a}_n$ are therefore required to satisfy equations

$$(\underset{\sim}{B}-\lambda\underset{\sim}{A})\tilde{\underset{\sim}{a}} = \underset{\sim}{\gamma} \ , \tag{4.190}$$

where $\tilde{\underset{\sim}{a}} = [\tilde{a}_0,\tilde{a}_1,\ldots,\tilde{a}_n]^T$, $\underset{\sim}{\gamma} = [\gamma_0,\gamma_1,\ldots,\gamma_n]^T$, $B_{i,j} = [\phi_j,\phi_i]$,

$$\text{and } A_{i,j} = [\psi_j,\phi_i] \ , \tag{4.191}$$

with

$$\psi_j(x) = K\phi_j(x) \ , \quad \text{and} \quad \gamma_i = [g,\phi_i] \ . \tag{4.192}$$

The matrix $\underset{\sim}{B}$ is automatically positive definite when the functions

$\{\phi_k(x)\}$ are linearly independent and $w(x) > 0$, or when $\phi_0(x), \phi_1(x), \ldots, \phi_n(x)$ form a Haar set and $w(x) \equiv 0$.

We define

$$\tilde{g}(x) = \sum_{j=0}^{n} \mu_j \phi_j(x) , \tag{4.193}$$

where $\underset{\sim}{B}\underset{\sim}{\mu} = \underset{\sim}{\gamma}$ and $\gamma_i = [g, \phi_i]$. We also set

$$\tilde{K}_n(x,y) = \sum_{j=0}^{n} \phi_j(x)\chi_j(y) , \tag{4.194}$$

where

$$\sum_{j=0}^{n} B_{ij}\chi_j(y) = \Gamma_i(y) \quad (i = 0,1,2,\ldots,n) \tag{4.195}$$

with

$$\Gamma_i(y) = [K(.,y),\ \phi_i],$$

that is,

$$\Gamma_i(y) = \int_a^b w(x)K(x,y)\overline{\phi_i(x)}dx + \sum_{k=0}^{N} \Omega_k K(x_k,y)\overline{\phi_i(x_k)} ,$$

using the general definition (4.188).

With these definitions, the approximation can be associated with an integral equation, analogous to the result of Theorem 4.26. The theorem is stated somewhat differently to provide some variety.

THEOREM 4.30. (a) *If the resolvent* $(I-\lambda\tilde{K}_n)^{-1}$ *exists, the solution of the equation*

$$\tilde{f}(x) - \lambda\int_a^b \tilde{K}_n(x,y)\tilde{f}(y)dy = \tilde{g}(x)$$

is given by $\tilde{f}(x) = \sum_{k=0}^{n} \tilde{a}_k\phi_k(x)$, *where*

$$(\underset{\sim}{B}-\lambda\underset{\sim}{A})\underset{\sim}{a} = \underset{\sim}{\gamma} .$$

(b) *Conversely, if* $(\underset{\sim}{B}-\lambda\underset{\sim}{A})^{-1}$ *exists and* $\underset{\sim}{\tilde{a}}$ *satisfies the above equation, the associated function* $\tilde{f}(x)$ *satisfies the given integral equation.*

Proof. (a) Suppose that $\tilde{f}(x)$ is the solution of the integral equation, then, from the form of $\tilde{K}_n(x,y)$ and $\tilde{g}(x)$,

$$\tilde{f}(x) = \sum_{j=0}^{n} (\mu_j+\lambda\nu_j)\phi_j(x) = \sum_{j=0}^{n} \tilde{\alpha}_j\phi_j(x) , \tag{4.196}$$

where

$$\nu_i = \int_a^b \chi_i(y)\tilde{f}(y)dy .$$

If we multiply (4.196) by $\chi_i(x)$ and integrate, we obtain an equation for ν_i $(i = 0,1,2,...,n)$ and we find

$$\underset{\sim}{\nu} = \underset{\sim}{C}\underset{\sim}{\mu} + \lambda\underset{\sim}{C}\underset{\sim}{\nu} ,$$

where

$$C_{i,j} = \int_a^b \chi_i(x)\phi_j(x)dx .$$

We have now seen that

$$\tilde{f}(x) = \sum_{j=0}^{n} \tilde{\alpha}_j\phi_j(x) ,$$

where $\underset{\sim}{\tilde{\alpha}} = \underset{\sim}{\mu}+\lambda\underset{\sim}{\nu}$ and $(\underset{\sim}{I}-\lambda\underset{\sim}{C})\underset{\sim}{\nu} = \underset{\sim}{C}\underset{\sim}{\mu}$. Hence, $(\underset{\sim}{I}-\lambda\underset{\sim}{C})\underset{\sim}{\tilde{\alpha}} = (\underset{\sim}{I}-\lambda\underset{\sim}{C})\underset{\sim}{\mu}+\lambda(\underset{\sim}{I}-\lambda\underset{\sim}{C})\underset{\sim}{\nu} = (\underset{\sim}{I}-\lambda\underset{\sim}{C})\underset{\sim}{\mu} + \lambda\underset{\sim}{C}\underset{\sim}{\mu}$, or

$$(\underset{\sim}{I}-\lambda\underset{\sim}{C})\underset{\sim}{\tilde{\alpha}} = \underset{\sim}{\mu} : \tag{4.197}$$

Now, the matrix $\underset{\sim}{B}$ with $B_{i,j} = [\phi_i,\phi_j]$ is non-singular, so that (4.197) is equivalent to

$$(\underset{\sim}{B}-\lambda\underset{\sim}{B}\underset{\sim}{C})\underset{\sim}{\tilde{\alpha}} = \underset{\sim}{B}\underset{\sim}{\mu} = \underset{\sim}{\gamma} \; .$$

On examining (4.195), we find that $\underset{\sim}{B}\underset{\sim}{C} = \underset{\sim}{A}$. Indeed,

$$\sum_{k=0}^{n} B_{i,k}C_{k,j} = \int_a^b \Gamma_i(y)\phi_j(y)dy = [K\phi_j,\phi_i] \; .$$

Thus $\underset{\sim}{\tilde{\alpha}} = \underset{\sim}{\tilde{a}}$, where

$$(\underset{\sim}{B}-\lambda\underset{\sim}{A})\underset{\sim}{\tilde{a}} = \underset{\sim}{\gamma} \; ,$$

and part (a) follows. Part (b) can be establish by reversing the steps.

The above analysis successfully provides a basis for a general treatment of Ritz-Galerkin-type methods (though we shall see, in section 4.29, that another approach is more powerful).

The question of convergence of $\tilde{f}(x)$ to $f(x)$ in some norm is determined by whether or not $||K-\tilde{K}_n|| \to 0$ and $||\tilde{g}(x)-g(x)|| \to 0$ as $n \to \infty$.† To make much headway, we must select a particular choice of functions $\{\phi_k(x)\}$ and/or a particular form for $[\phi,\psi]$ in (4.188).

4.26. *The classical Galerkin method*

In this section, we discuss the classical Galerkin method which corresponds to the choice

$$[\phi,\psi] = (\phi,\psi) \equiv \int_a^b \phi(x)\overline{\psi(x)}dx$$

in (4.188). (We set $w(x) \equiv 1$, and $\Omega_k = 0$ for $k = 0,1,2,\ldots,N$) .

We are then concerned with the kernel

$$\tilde{K}_n(x,y) = \sum_{i=0}^{n} \phi_i(x)\chi_i(y) \quad ,$$

where

$$\sum_{j=0}^{n} (\phi_j,\phi_i)\chi_j(y) = \int_a^b K(x,y)\overline{\phi_i(x)}dx \; , \tag{4.198}$$

† N may be a function of n in (4.185).

and with the function

$$\tilde{g}(x) = \sum_{i=0}^{n} \gamma_i \phi_i(x) ,$$

where

$$\sum_{j=0}^{n} (\phi_j,\phi_i)\gamma_j = (g,\phi_i) \quad (i = 0,1,2,\ldots,n) . \qquad (4.199)$$

We shall suppose that the functions $\phi_0(x),\phi_1(x),\phi_2(x), \ldots$ are *complete* in $L^2[a,b]$ (they may all be continuous functions) and that the first $(n+1)$ functions are employed in the Galerkin method. For example, with $a = 0$ and $b = 1$ we may take $\phi_r(x) = T_r(2x-1)$, or $\phi_r(x) = P_r(2x-1)$, where $T_r(x)$ is the Chebyshev polynomial of degree r and $P_r(x)$ is the Legendre polynomial.

THEOREM 4.31. *If* $\phi_k(x)$ $(k = 0,1,2,\ldots)$ *is a complete set of linearly independent functions then*

$$\lim_{n\to\infty} \| g(x)-\tilde{g}(x) \|_2 = 0 .$$

and

$$\lim_{n\to\infty} \| K(x,y)-\tilde{K}_n(x,y) \|_2 = 0 .$$

Proof. It is sufficient to assume that

$$\int_a^b |g(x)|^2 dx < \infty$$

and

$$\int_a^b\int_a^b |K(x,y)|^2 dx\, dy < \infty .$$

Now $\tilde{g}(x) = \sum_{i=0}^{n} \gamma_i\phi_i(x)$, where $\gamma_0,\gamma_1,\ldots,\gamma_n$ are chosen so that

$(g-\tilde{g},\phi_k) = 0$ $(k = 0,1,2,\ldots,n)$. This choice of coefficients γ_i is precisely that for which

$$\|g(x)-\tilde{g}(x)\|_2 \leq \|g(x) - \sum_{i=0}^{n} \alpha_i\phi_i(x)\|_2$$

for any choice $\alpha_0,\alpha_1,\ldots,\alpha_n$. By *completeness* of the functions $\phi_i(x)$, we know that, given $\varepsilon > 0$, there is an integer m_0 and a choice $\alpha_0,\alpha_1,\ldots,\alpha_{m_0}$ for which the right-hand side in the preceding inequality is $< \varepsilon$; hence $\|g(x)-\tilde{g}(x)\|_2 < \varepsilon$ for all $n \geq m_0$, and $\|g(x)-\tilde{g}(x)\|_2 \to 0$ as $n \to \infty$.

For the behaviour of $\|K(x,y)-\tilde{K}_n(x,y)\|_2$, we may see section 3.30 or consider the following argument, which establishes that $\|K(x,y)-\tilde{K}_n(x,y)\|_2 \to 0$.

Fixing $\varepsilon > 0$, there is a degenerate kernel

$$K^{\#}(x,y) = \sum_{i=0}^{M} X_i(x)Y_i(y)$$

with

$$\int_a^b\int_a^b |K(x,y)-K^{\#}(x,y)|^2 dx\, dy < \tfrac{1}{2}\varepsilon .$$

(Here $M = M(\varepsilon)$; the functions $X_i(x),Y_i(x)$ are square-integrable and may be taken to be continuous.) We take a specific such kernel $K^{\#}(x,y)$, with *fixed* functions $X_i(x),Y_i(x)$ $(i = 0,1,\ldots,M)$, and we define

$$\tilde{X}_i(x) = \sum_{j=0}^{n} \mu_{ij}\phi_j(x) ,$$

where the coefficients μ_{ij} are chosen to minimize $\|X_i(x)-\tilde{X}_i(x)\|_2$. Then, given $\delta > 0$ we can ensure by taking n sufficiently large (say, $n \geq n_{0,i}$) that $\|X_i(x)-\tilde{X}_i(x)\|_2 < \delta$ (this follows from the argument used for the approximation of $g(x)$, above).

If we set $n_0 \equiv n_0(\delta) = \max_{i=0,1,2,\ldots,M} n_{0,i}$, then, for $n \geq n_0$,

$$\|X_i(x)-\tilde{X}_i(x)\|_2 \leq \delta \qquad (i = 0,1,\ldots,M) .$$

With

$$K_n^{\#}(x,y) = \sum_{i=0}^{M} \tilde{X}_i(x)Y_i(y) ,$$

we see that

$$K_n^{\#}(x,y) = \sum_{i=0}^{n} \phi_i(x)Z_i(y)$$

for some functions $Z_i(y)$. We have

$$||K^{\#}(x,y)-K_n^{\#}(x,y)||_2 \leq \sum_{i=0}^{M} ||\{X_i(x)-\tilde{X}_i(x)\}Y_i(y)||_2$$

$$\leq \sum_{i=0}^{M} ||X_i(x)-\tilde{X}_i(x)||_2 \, ||Y_i(x)||_2$$

$$< \delta \sum_{i=0}^{M} ||Y_i(x)||_2 .$$

In this argument, we may choose any $\delta > 0$. Given $\varepsilon > 0$, we set

$$\delta = \varepsilon/\{2 \sum_{i=0}^{M} ||Y_i(x)||_2\} ,$$

and n_0 is then a function of ε , say $n_0 = n_0[\varepsilon]$. We obtain, for $n \geq n_0[\varepsilon]$,

$$\|K(x,y)-K_n^{\#}(x,y)||_2$$

$$\leq \|K(x,y)-K^{\#}(x,y)||_2 + ||K^{\#}(x,y)-K_n^{\#}(x,y)||_2 < \varepsilon .$$

But, for (almost all) $y \in [a,b]$ and $n \geq n_0$,

$$\int_a^b |K(x,y)-\tilde{K}_n(x,y)|^2 dx \leq \int_a^b |K(x,y)-K^{\#}(x,y)|^2 dx \quad .$$

So on integrating with respect to y,

$$\|K(x,y)-\tilde{K}_n(x,y)\|_2 < \varepsilon \quad (n \geq n_0[\varepsilon]) \; .$$

The result is thus established.

Since $\|g(x)-\tilde{g}(x)\|_2 \to 0$ and $\|K(x,y)-\tilde{K}_n(x,y)\|_2 \to 0$ as $n \to \infty$, we can set $g_n(x) = \tilde{g}(x)$ and $K_n(x,y) = \tilde{K}_n(x,y)$ in Theorem 4.3. We have the following result.

THEOREM 4.32. *If the functions* $\{\phi_k(x)\,,\ k = 0,1,2,\ldots\}$ *form a complete linearly independent set, and* λ *is a regular value of* $K(x,y)$, *then the approximation* $\tilde{f}(x)$ *obtained by the classical Galerkin method is such that* $\lim_{n\to\infty} \|f(x)-\tilde{f}(x)\|_2 = 0$. *Moreover, if* $K(x,y)$ *and* $g(x)$ *are continuous and*

$$\overset{\circ}{f}(x) = g(x) + \lambda\int_a^b K(x,y)\tilde{f}(y)dy \; ,$$

then $\lim_{n\to\infty} \|f(x)-\overset{\circ}{f}(x)\|_\infty = 0$.

Proof. The first part of the theorem is an immediate consequence of Theorem 4.31 together with Theorem 4.3 (allowing n to tend to infinity). Now

$$f(x)-\overset{\circ}{f}(x) = \lambda\int_a^b K(x,y)\{f(y)-\tilde{f}(y)\}dy \; ,$$

so that

$$|f(x)-\overset{\circ}{f}(x)| \quad \leq$$

$$\leq |\lambda|(\int_a^b |K(x,y)|^2 dy)^{\frac{1}{2}} (\int_a^b |f(y)-\tilde{f}(y)|^2 dy)^{\frac{1}{2}} ,$$

whence

$$\|f(x)-\mathring{f}(x)\|_\infty \leq |\lambda| \sup_{a\leq x\leq b} (\int_a^b |K(x,y)|^2 dy)^{\frac{1}{2}} \|f(x)-\tilde{f}(x)\|_2$$

and the result of the theorem clearly follows.

In certain special cases, it may happen that $\|f(x)-\tilde{f}(x)\|_\infty \to 0$. To establish such a result we could endeavour to show, in the preceding analysis, that $\|g(x)-\tilde{g}(x)\|_\infty \to 0$ and that $\|K(x,y)-\tilde{K}_n(x,y)\|_\infty \to 0$ (or, more generally, that $\|K-K_n\|_\infty \to 0$) as $n \to \infty$.

4.27. *The Galerkin methods using Chebyshev polynomials*

As we have seen, the Chebyshev polynomials satisfy three orthogonality relations which make them convenient for use in the Galerkin methods when $a = -1$, $b = 1$. For example,

$$\int_{-1}^1 (1-x^2)^{-\frac{1}{2}} T_r(x) T_s(x) dx = 0$$

if $r \neq s$, $= \frac{1}{2}\pi$ if $r = s \neq 0$, $= \pi$ if $r = s = 0$, and there are two corresponding *discrete* orthogonality relations based on the points which give the extrema of $T_N(x)$ or are the zeros of $T_{N+1}(x)$.

Thus, with

$$[\phi,\psi] = \int_{-1}^1 (1-x^2)^{-\frac{1}{2}} \phi(x)\overline{\psi(x)} dx ,$$

and the choice $\phi_0(x) = \sqrt{\frac{1}{2}}$, $\phi_r(x) = T_r(x)$, the Galerkin equations for the equation

$$f(x) - \lambda \int_{-1}^1 K(x,y) f(y) dy = g(x)$$

assume a special form with $\underset{\sim}{B} = \frac{1}{2}\pi \underset{\sim}{I}$ in (4.190).

The matrix $\underset{\sim}{B}$ is diagonal if $\phi_0(x) = 1$, $\phi_r(x) = T_r(x)$ $(r \geq 1)$

and

$$[\phi,\psi] = \int_{-1}^{1} (1-x^2)^{-\frac{1}{2}}\phi(x)\overline{\psi(x)}dx \ ,$$

or (4.200)

$$[\phi,\psi] = \sum_{j=0}^{N} \Omega_j\phi(x_j)\overline{\psi(x_j)}$$

with either (a) $\Omega_0 = \frac{1}{2}$, $\Omega_j = 1$ $(j \geq 0)$, $x_j = \cos((2j+1)\pi/(2N+2))$, $n \leq N$, or (b) $\Omega_0 = \Omega_N = \frac{1}{2}$, $\Omega_j = 1$ $(1 \leq j \leq N-1)$, $x_j = \cos(j\pi/N)$, $n \leq N$, because of the discrete orthogonality relations referred to above.

As we remarked for the collocation method, the choice $\phi_r(x) = T_r(x)$ could in theory, but not in practice, be replaced by any (equivalent) basis for the polynomials of degree at most n , and the theoretical behaviour of the methods is then determined by the definition of $[\phi,\psi]$. In analogy with Theorem 4.28, we have the following result.

THEOREM 4.33. *Suppose* $a = -1$, $b = 1$, λ *is a regular value of* $K(x,y)$, *and*

$$[\phi,\psi] = \int_{-1}^{1} (1-x^2)^{-\frac{1}{2}}\phi(x)\overline{\psi(x)}dx \ .$$

With $\phi_0(x),\phi_1(x),\ldots,\phi_n(x)$ *a basis for the polynomials of degree* n ,

$$||\tilde{f}(x)-f(x)||_\infty = O((\ln n)/n^r) \ ,$$

when $(\partial/\partial x)^r K(x,y)$ *is bounded for* $a \leq x, y \leq b$, $g^{(r)}(x)$ *is bounded for* $a \leq x \leq b$, *and*

$$\tilde{f}(x) = \sum_{r=0}^{n} \tilde{a}_r\phi_r(x)$$

is determined from eqn (4.190).

Proof. The proof follows the proof of Theorem 4.28, and depends only on the behaviour of the appropriate Lebesgue constant Λ_n (Baker and Radcliffe 1970; Rivlin 1969; see also section 2.6).

Remark. We *speculate* that the Lebesgue constant associated with the discrete inner products (a) and (b) above is $O(\ln n)$ when N/n is held fixed and $n \to \infty$, so that Theorem 4.33 can be extended to cover this case. *

We also have the analogue of Theorem 4.29, where Φ is replaced by the operator which assigns to $\phi(x)$ its Chebyshev approximation

$$\sum_{k=0}^{n}{}' a_k T_k(x) \quad (\text{with } a_k = [\phi,T_k]/[T_k,T_k]) \quad ,$$

using any of the definitions of $[\phi,\psi]$ employed in (4.200).

4.28. The method of least squares and the method of moments

In the classical method of moments, we generalize the classical Galerkin method, by choosing two linearly independent sets of functions $\{\phi_i(x),\ i = 0,1,2,\ldots,n\}$ and $\{\psi_i(x)\}$. We then seek an approximation

$$\sum_{i=0}^{n} \tilde{a}_i \phi_i(x)$$

such that the residual

$$r(x) = \tilde{f}(x) - \lambda(K\tilde{f})(x) - g(x) \tag{4.201}$$

is orthogonal, in the usual sense, to $\psi_j(x)$ $(j = 0,1,2,\ldots,n)$. The resulting equations for $\tilde{a}_0,\tilde{a}_1,\ldots,\tilde{a}_n$ have the form

$$\sum_{j=0}^{n} \{(\phi_j,\psi_i) - \lambda(K\phi_j,\psi_i)\}\tilde{a}_j = (g,\psi_i) \quad (i = 0,1,2,\ldots,n) \ , \tag{4.202}$$

and Theorem 4.30 is applicable if the definitions of $\tilde{K}_n(x,y)$ and $\tilde{g}(x)$ and $\underset{\sim}{A}$, $\underset{\sim}{B}$, $\underset{\sim}{\gamma}$ are suitably revised. It is necessary to set

$$\gamma_i = (g,\psi_i),\ B_{ij} = (\phi_j,\psi_i)\ ,\ A_{ij} = (K\phi_j,\psi_i)$$

and, in (4.195)

$$\Gamma_i(y) = (K(.,y),\psi_i) \equiv \int_a^b K(x,y)\psi_i(x)\mathrm{d}x \quad .$$

The method of least squares can be regarded either as an application of the method of moments with

$$\psi_i(x) = \phi_i(x)-\lambda\int_a^b K(x,y)\phi_i(y)\mathrm{d}y \tag{4.203}$$

or as an application of the classical Galerkin method to eqn (4.54). In compact notation, we can write (4.54) as

$$\{I-(\lambda K+\bar{\lambda}K^*-|\lambda|^2K^*K)\}f(x) = (I-\bar{\lambda}K^*)g(x)\ . \tag{4.204}$$

The Galerkin method applied to this equation gives the equations

$$\sum_{j=0}^{n} ((I-\lambda K-\bar{\lambda}K^*+|\lambda|^2K^*K)\phi_j,\phi_i)\tilde{a}_j = ((I-\bar{\lambda}K^*)g,\phi_i)$$

or

$$\sum_{j=0}^{n} ((I-\lambda K)\phi_j,(I-\lambda K)\phi_i)\tilde{a}_j = (g,(I-\lambda K)\phi_i)$$

or

$$\sum_{j=0}^{n} ((I-\lambda K)\phi_j,\psi_i)\tilde{a}_j = (g,\psi_i)\ , \tag{4.205}$$

where $\psi_i(x)$ is defined above. The method is thus seen as a method of moments.

We should also show that the Galerkin method for (4.204) is indeed the least-squares method for the original equation. Suppose $r(x)$ to be defined by eqn (4.201); then

$$\|r(x)\|_2^2$$

$$= (\tilde{f}-\lambda K\tilde{f}-g,\tilde{f}-\lambda K\tilde{f}-g) = ((I-\lambda K)\tilde{f},(I-\lambda K)\tilde{f}) +$$

$$- (g,(I-\lambda K)\tilde{f})-((I-\lambda K)\tilde{f},g) + (g,g) \qquad (4.206)$$

and the coefficients $\tilde{a}_k$ are to be chosen to minimize this quantity.

Now, consider the Rayleigh-Ritz function $J\{\phi\}$ (section 4.22) associated with eqn (4.204). It assumes the form $J^*[\phi]$ say, where

$$J^*[\phi]$$

$$= (\phi,(I-\bar{\lambda}K^*)g)+((I-\bar{\lambda}K^*)g,\phi)-((I-\lambda K-\bar{\lambda}K^* + |\lambda|^2K^*K)\phi,\phi)$$

$$= ((I-\lambda K)\phi,g)+(g,(I-\lambda K)\phi) - ((I-\lambda K)\phi,(I-\lambda K)\phi) \quad .$$

Consequently

$$J^*[\tilde{f}] = -\|r(x)\|_2^2 + (g,g) \; , \qquad (4.207)$$

and, so, $J^*[\tilde{f}]$ has a stationary value if and only if $\|r(x)\|_2^2$ does too. There is only one stationary value of $\|r(x)\|_2^2$ which is, it transpires, necessarily a minimum. The equivalence of the method of least-squares with the Rayleigh-Ritz method applied to eqn (4.204) is thus established.

4.29. Projection methods

All the expansion methods discussed in sections 4.24-4.28 may be regarded as 'projection methods'. We now explain the meaning of this term.

We suppose X to be a complete normed linear space of functions, for example, $X = C[a,b]$ with the uniform norm. We suppose that a linear projection operator A is defined on X, mapping it into some finite dimensional subspace X_n; thus, the operator A is assumed linear and bounded: $A : X \to X_n$, and then $A^2 = A$. (The given operator A is a projection if $A^2 = A$.) By way of example, suppose that X_n is the space of polynomials of degree at most n, and the approximation $A = A^I$ transforms a function $f(x) \in X(=C[a,b])$ into the interpolating polynomial of degree n which agrees with $f(x)$ at prescribed points $z_0, z_1, \ldots, z_n$ in $[a,b]$.

Now, we suppose that the solution of the equation

$$f(x)-\lambda(Kf)(x)-g(x) = 0$$

is in the space X (= $C[a,b]$, say), and we seek an approximation

$$\tilde{f}(x) = \sum_{i=0}^{n} \tilde{a}_i \phi_i(x) ,$$

where the functions $\phi_i(x)$ $(i = 0,1,\ldots,n)$ are a basis for a linear space Y_n. Frequently, we choose the functions so that $Y_n = X_n$; in any case, $Y_n \subseteq X$, $K(Y_n) \subseteq X$ and $g(x) \in X$.

The general projection method is now formulated by requiring that the coefficients $\tilde{a}_i$ $(i = 0,1,\ldots,n)$ are chosen so that

$$A(\tilde{f}(x)-\lambda(K\tilde{f})(x)-g(x)) = 0 . \tag{4.208}$$

With the choice $X = C[a,b]$, $X_n = Y_n = P_n$ = the space of polynomials of degree $\le n$, the projection method obtained on setting $A = A^I$ in (4.208) is precisely the collocation method with collocation points $z_0, z_1, \ldots, z_n$, where $\phi_0(x), \phi_1(x) \ldots, \phi_n(x)$ form a basis for the polynomials of degree n (for example, $\phi_0(x) = \frac{1}{2}$, $\phi_r(x) = T_r(x)$ for $r \ge 1$).

By an appropriate choice of X, X_n, Y_n, and A, we can obtain a formulation (4.208) of all the Ritz-Galerkin-type methods. Only in the case of the method of moments does the choice of X_n differ from the choice of Y_n.

Let us suppose that $X_n = Y_n$ and thus exclude the method of moments. Then $\tilde{f}(x) = A\tilde{f}(x)$, and since

$$A\tilde{f}(x)-\lambda AK\tilde{f}(x)-Ag(x) = 0 , \tag{4.209}$$

we see that

$$\tilde{f}(x)-\lambda AK\tilde{f}(x)-Ag(x) = 0 . \tag{4.210}$$

Of course, $Ar(x) = 0$, where

$$\tilde{f}(x)-\lambda K\tilde{f}(x)-g(x) = r(x) ,$$

and

$$f(x)-\lambda(Kf)(x)-g(x) = 0 . \tag{4.211}$$

If $A = A^{I}$ is the polynomial interpolation operator above, then the condition $A^{I}r(x) = 0$ is simply the condition that

$$r(z_0) = r(z_1) = \ldots = r(z_n) = 0$$

and so the method defined by (4.208) is obviously a collocation method. The analysis extends to the case where $\phi_1(x),\phi_2(x),\ldots,\phi_n(x)$ form a Haar system. On the other hand, suppose that $A\phi(x)$ is a Fourier-type approximation to $\phi(x)$, and that

$$A\phi(x) = \sum_{i=0}^{n} \{(\phi,\phi_i)/(\phi_i,\phi_i)\}\phi_i(x) ,$$

where

$$(\phi,\psi) = \int_a^b \phi(x)\overline{\psi(x)}\mathrm{d}x$$

and $(\phi_i,\phi_j) = 0$, $i \neq j$. Then $Ar(x) = 0$ if and only if $(r,\phi_i) = 0$ $(i = 1,2,3,\ldots,n)$. Thus (4.208) is equivalent to choosing

$$\tilde{f}(x) = \sum_{i=0}^{n} \tilde{a}_i\phi_i(x)$$

so that $(r,\phi_i) = 0$, giving us a classical Galerkin method. The general Ritz-Galerkin-type method is an extension of this example (see below).

In our earlier work, we have recast eqn (4.210) as an integral equation

$$\tilde{f}(x)-\lambda(\tilde{K}\tilde{f})(x)-\tilde{g}(x) = 0 , \qquad (4.212)$$

in which $\tilde{g}(x) = \mathrm{A}g(x)$ and $\tilde{K}$ is an integral operator with kernel†

$$\tilde{K}(x,y) = \mathrm{A}_x K(x,y) .$$

The comparison of (4.212) with (4.211) is then accomplished by means of Theorem 4.3. However, as we pointed out earlier, we lose something by considering $\tilde{K}(x,y)-K(x,y)$. At the cost of some involvement in operator theory, we gain something by applying a result similar to Theorem 4.3 directly, with $K_n = \mathrm{A}K$ and $g_n(x) = \mathrm{A}g(x)$. In appropriate cases $\|K-\mathrm{A}K\|_p \to 0$. We have the following result, which we may wish to apply with $K_n = \mathrm{A}K$, where A depends upon n.

THEOREM 4.34. *Suppose that* X *is a normed linear space with norm* $\|\ \|$, *and operator norms are subordinate to this norm. If* K, K_n *are bounded linear operators mapping* X *into* X *with*

$$|\lambda|\ \|(I-\lambda K)^{-1}\|\ \|K_n-K\| < 1$$

and either (i) X *is a Banach space or* (ii) K, K_n *are compact (completely continuous), then* $(I-\lambda K_n)^{-1}$ *exists. Further, if* $(I-\lambda K)f = g$ *and* $(I-\lambda K_n)f_n = g_n$, *then*

$$\|f_n-f\| \le \frac{\|(I-\lambda K)^{-1}\|}{1-|\lambda|\ \|(I-\lambda K)^{-1}\|\ \|K-K_n\|}\left(\|g-g_n\| + |\lambda|\ \|K-K_n\|\ \|f\|\right) .$$

---

† The suffix in A_x denotes that we fix y and treat $K(x,y)$ as a function of x to obtain $\mathrm{A}_x K(x,y)$, for each $y \in [a,b]$.

Proof. For completeness only, we give a proof in some detail. The proof may be omitted if desired. We shall use some of the definitions of section 1.9.

Suppose, to begin, that K is compact, and $(I-\lambda K)^{-1}$ exists as implied by the assumptions. Then, $(I-\lambda K)^{-1}$ can be written $I+\lambda R_\lambda$ for some operator R_λ, and it follows that $R_\lambda = K + \lambda K R_\lambda$. We first establish that R_λ is compact.

Since R_λ is bounded, it maps a bounded set into a bounded set and K transforms this set into a (relatively sequentially) compact set, so that KR_λ is a compact operator. Since the sum of K and λKR_λ is the sum of compact operators, it too is compact; that is, R_λ is compact.

Now, consider the general case, and write

$$I-\lambda K_n = I-\lambda K + \lambda(K-K_n) = (I-\lambda K)\{I+\lambda(I-\lambda K)^{-1}(K-K_n)\} .$$

If $T_\lambda = \{I+\lambda(I-\lambda K)^{-1}(K-K_n)\}$ has an inverse then $(I-\lambda K_n)^{-1} = T_\lambda^{-1}(I-\lambda K)^{-1}$. We see that $T_\lambda = I+\lambda(K-K_n)+\lambda^2 R_\lambda(K-K_n)$, where $(I-\lambda K)^{-1} = I+\lambda R_\lambda$. (Thus, $S_\lambda \equiv I-T_\lambda$ is compact if K and K_n are compact, since the compactness of R_λ follows from that of K.) In any case we see that

$$||T_\lambda - I|| \equiv ||S_\lambda|| \leq |\lambda| \; ||(I-\lambda K)^{-1}|| \times ||K-K_n|| = \rho < 1$$

by assumption.

Now, $T_\lambda = I-S_\lambda$. The existence of T_λ^{-1} follows from a general theorem which states that $(I-S_\lambda)^{-1}$ exists if S_λ is linear and $||S_\lambda|| < 1$ provided either (i) the underlying linear space X is complete (X is a Banach space) or (ii) S_λ is compact.

The general theorem required to establish the preceding result is the 'Geometric Series theorem' which we stated in Theorem 1.1. We refer to p. 52 for a proof. Observe that, under the assumed conditions (in particular, $||S_\lambda|| = \rho < 1$) we have

$$||(I-S_\lambda)^{-1}|| \leq 1/(1-\rho) .$$

Having established the existence of T_λ^{-1}, the existence of $(I-\lambda K_n)^{-1}$ follows as indicated.

Now $(I-\lambda K_n)(f_n-f) = g_n-g+\lambda(K-K_n)f$, and on taking norms,

$$||f_n-f|| \leq ||(I-\lambda K_n)^{-1}|| \, \{||g-g_n|| + |\lambda| \, ||K_n-K|| \, ||f||\} .$$

But

$$||(I-\lambda K_n)^{-1}|| \leq ||(I-\lambda K)^{-1}|| \, ||T_\lambda^{-1}|| .$$

The result stated in the theorem now follows immediately upon using our result

$$||T_\lambda^{-1}|| \leq \{1-||S_\lambda||\}^{-1} \leq \{1-|\lambda| \, ||(I-\lambda K)^{-1}|| \, ||K-K_n||\}^{-1} .$$

The general nature of the conditions of Theorem 4.34 imply that we need not worry unduly about conditions on K_n and K in applications. In this section K is generally compact on a suitable space X such as $C[a,b]$ with norm $|| \; ||_\infty$, or $L^2[a,b]$ with norm $|| \; ||_2$, and since A_nK has finite-dimensional range it is automatically compact. In the instances quoted, X is also complete.

THEOREM 4.35. *Suppose that* K *is a compact integral operator on the normed linear space* X *of functions with norm* $|| \; ||$. *If* $\{A_n\}$ $(n = 0,1,2,\ldots)$ *is a sequence of projection operators on* X *such that*

$$\lim_{n\to\infty} ||A_n\phi(x)-\phi(x)|| = 0 \quad \textit{for each} \quad \phi(x) \in X ,$$

then $\lim_{n\to\infty} ||A_nK-K|| = 0$ *(where the operator norm is subordinate to the norm of* X*).*

Proof.

$$\|A_nK-K\| = \sup_{\|\phi(x)\| = 1} \|(A_nK)\phi(x)-(K\phi)(x)\|$$

$$= \sup_{\psi(x)\varepsilon D} \|A_n\psi(x)-\psi(x)\| ,$$

where $D = \{(K\phi)(x) \mid \|\phi(x)\| = 1$ and $\phi(x) \varepsilon X\}$. Now D is the image, under K , of a bounded subset of X and K is compact. Hence D is a set which is compact in the space X. Now A_n is a bounded linear operator and hence the elementwise convergence $\|A_n\psi(x)-\psi(x)\| \to 0$ is uniform on D , that is, by Theorem 1.4,

$$\sup_{\psi(x)\varepsilon D} \|A_n\psi(x)-\psi(x)\| \to 0 .$$

The theorem thus follows.

THEOREM 4.36. *Under the conditions of Theorem 4.35, if* $\tilde{f}(x)$ *is computed from (4.210) with* $A \equiv A_n$ *, the range of* A_n *being finite-dimensional (where* A_n *is a projection) then*

$$\lim_{n\to\infty} \|f(x)-\tilde{f}(x)\| = 0 .$$

Proof. We use the result of Theorem 4.35 in Theorem 4.34.

The reader may detect some similarity in the above material and the earlier treatment of the quadrature operator H (cf. section 4.17). Whilst the operator H is not an integral operator, the operator AK has in the analysis of earlier sections been treated as an integral operator with kernel $A_xK(x,y)$. However, if $K(x,y)$ is weakly singular then $A_xK(x,y)$ may not exist; consider the case $A = A^I$ above. We choose, in the analysis of this section, to treat AK as an abstract operator (not always an integral operator with a kernel). Now $\|H-K\| \not\to 0$, but $\|HK-K\| \to 0$ and $\|KH-H\| \to 0$. Here, because A_n is a projection operator, $\|A_n-I\| \not\to 0$; but $\|A_nK-K\| \to 0$.

The analysis is of use because it provides a unified treatment of many expansion methods. Thus, the general Galerkin-type method is arrived at if we define

$$A_n\phi(x) = \sum_{k=0}^{n} a_k\phi_k(x) ,$$

where

$$\sum_{k=0}^{n} [\phi_k,\phi_j]a_k = [\phi,\phi_j] \quad (j = 0,1,2,\ldots,n)$$

and take, say, $X = R[a,b]$. (For the classical Galerkin method with $[\phi,\psi] = (\phi,\psi)$, we may take $X = L^2[a,b]$ with the norm $\|\ \|_2$.) Furthermore, some proofs of convergence are actually simplified.

Consider, for example, the convergence of $\|A_nK-K\|_2$ to zero in the classical Galerkin method. If

$$\int_a^b\int_a^b |K(x,y)|^2 dx\, dy < \infty ,$$

K is certainly compact on $X = L^2[a,b]$ with norm $\|\ \|_2$, and the completeness of $\phi_0(x),\phi_1(x),\ldots$ implies elementwise (pointwise) convergence; that is, $\|A_n\ \phi(x)-\phi(x)\|_2 \to 0$. Theorem 4.35 may now be applied to show that $\|A_nK-K\|_2 \to 0$.

Again, it is possible to relax the differentiability requirements on $K(x,y)$ which have been required by our earlier treatment. Consider for example the application of collocation using

$$\phi_0(x) = \tfrac{1}{2} , \ \phi_r(x) = T_r(x) \quad (r = 1,2,\ldots,n) \quad (\text{with } a = -1,\ b = 1) ,$$

taking the collocation points to be the zeros of $T_{n+1}(x)$ or the points of extrema of $T_n(x)$. Then $\|A_n\|_\infty$ is the Lebesgue constant, which is $O(\ln n)$; we suppose X is the space $C[a,b]$ with uniform norm, and we also suppose that $K(x,y)$ has properties which ensure that D consists of functions which are sufficiently smooth to ensure that

$$\sup_{\phi(x)\varepsilon D} ||A_n\psi(x)-\psi(x)||_\infty \to 0 .$$

This will be the case if every function $\psi(x)$ in D is in the class $\mathrm{Lip}_L\alpha$ (where L is uniform for $\psi(x) \varepsilon D$) and $\alpha > 0$. For then $||A\psi(x)-\psi(x)||_\infty = O((\ln n)/n^\alpha) \to 0$ as $n \to \infty$, uniformly for $\psi(x) \varepsilon D$. In our previous analysis, we had to consider the smoothness of $K(x,y)$ for every y , but we now consider functions in D .

Remark. Let us consider the case $K(x,y) = x$ $(-1 \le x \le y \le 1)$, $K(x,y) = 0$ $(-1 \le y \le x \le 1)$. (For each $y \ne 1$, $K(x,y)$ is discontinuous, so we do not *expect* that $||K(x,y)-\tilde{K}(x,y)||_\infty \to 0$.) Then for $\psi(x)\varepsilon\ D$ there is some function $\phi(x)$ with $||\phi(x)||_\infty = 1$ and

$$\psi(x) = \int_{-1}^{1} K(x,y)\phi(y)dy = \int_x^1 x\phi(y)dy .$$

Thus $\psi'(x) = \int_x^1 \phi(y)dy - x\phi(x)$, so that $\|\psi'(y)\|_\infty \le 3\ \|\phi(x)\|_\infty \le 3$ for all $\psi(x)\ \varepsilon\ D$ and $D \subset \mathrm{Lip}_3\ 1$.

A general result is expressed in the following theorem.

THEOREM 4.37. *Suppose that*

$$D = \{(K\phi)(x) \mid \ ||\phi(x)|| = 1 , \phi(x)\ \varepsilon\ X\} .$$

Then $||A_nK-K|| \to 0$ *if for* $n=0,1,2,\ldots$ A_n *is a projection from* X *to* X_n *and*

$$\lim_{n\to\infty} \{1+\|A_n\|\}\ \mathrm{dist}\ (D,X_n) = 0 ,$$

where

$$\mathrm{dist}\ (D,X_n) = \sup_{\psi(x)\varepsilon D}\ \inf_{\sigma(x)\varepsilon X_n} ||\psi(x) - \sigma(x)||.$$

Proof. We have seen that

$$\|A_nK-K\| = \sup_{\psi(x)\varepsilon D} \|A_n\psi(x)-\psi(x)\| .$$

Suppose $\sigma(x) \in X_n$ is the best approximation in $\|\ \|$ to $\psi(x)$; then

$$\|A_n\psi(x)-\psi(x)\| = \|A_n\psi(x)-A_n\sigma(x)+\sigma(x)-\psi(x)\|$$

$$\leq \{1+\|A_n\|\}\ \|\psi(x)-\sigma(x)\| ,$$

and the result follows.

FURTHER DISCUSSION OF THE TREATMENT OF FREDHOLM EQUATIONS

In this chapter we consider various developments of the material presented in Chapter 4. These developments proceed in a number of directions. Thus the early sections (sections 5.1-5.17) are concerned with the treatment of linear equations. The practical treatment of equations of the second kind with some complication not previously discussed (such as non-uniqueness of the solution, or the presence of a weak or intrinsic singularity) is discussed first. In particular, one section (5.10) introduces the reader to the literature[§] on intrinsically singular equations. The application of abstract theories to a study of numerical methods for linear equations with weak singularities is introduced in sections 5.12-5.14. The flavour of both a Russian and an American school of analysis is given.

The treatment of linear equations concludes with the discussion of equations of the first kind in sections 5.15-5.17. The difficulty associated with this equation is emphasized, and the way various naive methods can be adapted is indicated. The use of regularization techniques and their relationship to various adaptations is also discussed, and further references are offered in the text.[¶]

The remainder of the chapter (sections 5.18-5.25) is concerned with non-linear equations. Since it is difficult to cover all types of non-linear equations, the scope is limited to an indication of how general techniques for linear equations can be adapted to the non-linear case. This chapter ends in section 5.25 with a theoretical study based on a Russian school of analysis.

---

§ The recent translation from the Russian original of the work of Ivanov (1976) provides a welcome addition to this literature.

¶ Since the preparation of this manuscript, a number of interesting papers on the treatment of equations of the first kind have appeared. I have taken the opportunity to mention some of these in the References, and would refer in particular to the work of Hilgers, and of Nashed and Wahba.

5.1. *Introduction*

In Chapter 4, we discussed methods for the numerical solution of linear Fredholm equations of the second kind, of the form

$$f(x) - \lambda\int_a^b K(x,y)f(y)dy = g(x) , \tag{5.1}$$

assuming that the kernel $K(x,y)$ and the inhomogeneous term $g(x)$ in (5.1) were continuous and that λ was a regular value of the kernel $K(x,y)$. In this chapter, we extend the ideas developed in Chapter 4 to enable us to treat a variety of more complicated problems. The techniques developed in Chapters 3 and 4 are not always successful for the types of problem discussed in this chapter. Our attention will be restricted, here, to integral equations in which the limits of integration are constants; such equations we call Fredholm equations.† In the next chapter we consider various types of Volterra equations (see Chapter 1), including non-linear equations and equations with weak singularities in the discussion.

The case where λ in (5.1) is a characteristic value of $K(x,y)$, so that any solution of (5.1) is not unique, has received little attention in the literature, and our treatment of this problem is limited. The practical difficulties which arise when λ is, or is close to, a characteristic value, are in some ways similar to those found with equations of the first kind. We discuss equations of the latter type in section 5.15.

In sections 5.3-5.8 we study eqn (5.1), where $K(x,y)$ and/or $g(x)$ are badly behaved, for example, when $K(x,y)$ is weakly singular. The theory of the quadrature method where $K(x,y)$ is badly behaved can be presented in terms of a general theory due to Anselone. I shall indicate the basis of the Anselone theory in section 5.11, and give

---

† Although this terminology is fairly common, it is sometimes reserved for linear integral equations in which the Fredholm alternative (see Chapter 1) applies. I reserve the terminology for the latter use if the integral in the equation is taken along a contour.

a somewhat different theory in section 5.12.

Also of interest is the non-linear equation

$$f(x) - \int_a^b F(x,y;f(y))\,dy = g(x) , \qquad (5.2)$$

and we shall investigate some numerical methods for the solution of this equation and more general non-linear equations. Some types of integro-differential equation are also discussed, in passing, when dealing with the treatment of Fredholm equations of the first kind.

5.2. *Linear equations*

Linear integral equations of the second kind are not always presented in the form (5.1). It sometimes happens that the integral in the equation occurs as a contour integral, the free variable lying on the contour. If the contour C is parametrized by the length of arc measured from a fixed point on the curve, the equation can be transformed to the form (5.1). If C is closed the integrand $K(x,y)f(y)$ is then periodic with a period equal to the length of C. In these circumstances, use of the trapezium rule in the quadrature method yields high-order local truncation errors if $K(x,y)$ and $f(x)$ are smooth.

Example 5.1.
To illustrate the formulation of integral equations using contour integrals, consider the equation (of the first kind, and with a weakly singular kernel)

$$\int_C \ln|z-\zeta|\; f(\zeta)\; |d\zeta| = g(z), \quad z \in C$$

where $z = x + iy$ and ζ are complex, and C is a circle with centre at $z = 0$ and radius r. For an example of such an equation, see Symm (1966). A point P on C may be parametrized by the angle θ which the line from the origin to P makes with the positive x-axis in the plane. Thus $\zeta = re^{i\theta}$ and $|d\zeta| = r|d\theta|$, so that

$$r\int_0^{2\pi} \ln|z - re^{i\theta}|\, f(re^{i\theta})d\theta = g(z)\ .$$

We now write $z = re^{i\phi}$, where r is fixed and ϕ is the free variable, and obtain

$$r\int_0^{2\pi} \ln|r(e^{i\phi} - e^{i\theta})|\, f(re^{i\theta})\, d\theta = g(re^{i\phi}) \qquad (0\le\phi\le 2\pi)\ . \quad *$$

5.2.1.

We sometimes encounter, in place of (5.1), a *system* of integral equations

$$f_i(x) - \lambda \sum_{j=0}^{N} \int_a^b K_{i,j}(x,y)f_j(y)dy = g_i(x) \quad (i = 0,1,2,\ldots,N) \qquad (5.3)$$

(which can clearly be written in matrix and vector notation where the components are functions). Such a system of equations can be expressed as a single integral equation in which the functions involved may be discontinuous. A general theory for (5.1) is then obviously applicable, though the numerical treatment of (5.3) is usually more direct.

Example 5.2.

The system of equations

$$f_1(x) - \lambda\int_a^b K_{1,1}(x,y)f_1(y)dy - \lambda\int_a^b K_{1,2}(x,y)f_2(y)dy = g_1(x), \quad (a\le x\le b)$$

$$f_2(x) - \lambda\int_a^b K_{2,1}(x,y)f_1(y)dy - \lambda\int_a^b K_{2,2}(x,y)f_2(y)dy = g_2(x), \quad (a\le x\le b)$$

may be reformulated as

$$f(x) - \lambda\int_a^{2b-a} K(x,y)f(y)dy = g(x) \quad (a\le x\le 2b-a)\ ,$$

where $f(x) = f_1(x)$ for $a\le x\le b$, $f(x) = f_2(x-(b-a))$ for $b<x\le 2b-a$, and

$$K(x,y) = \begin{cases} K_{1,1}(x,y) & a \leq x,\ y \leq b \\ K_{1,2}(x,y-(b-a)) & a \leq x \leq b,\ b < y \leq 2b - a \\ K_{2,1}(x-(b-a),y) & b < x \leq 2b - a,\ a \leq y \leq b \\ K_{2,2}(x-(b-a),y-(b-a)) & b < x,y \leq 2b - a\ , \end{cases}$$

$$g(x) = \begin{cases} g_1(x) & a \leq x \leq b \\ g_2(x-(b-a)) & b < x \leq 2b - a\ . \end{cases}$$

In general, the kernel $K(x,y)$ is discontinuous when $x = b$ or $y = b$, and $g(x)$ is discontinuous at $x = b$. *

It is convenient to rewrite (5.3) in the form (5.1) in order to establish *theoretical* properties of the solution of the system of integral equations. In the numerical treatment of (5.3) however, we generally decide to treat the system of equations directly. Thus, for the quadrature methods, we may write

$$\tilde{f}_i(x) - \lambda \sum_{j=0}^{N} \sum_{k=0}^{n} w_k K_{i,j}(x,y_k)\tilde{f}_j(y_k) = g_i(x) \tag{5.4}$$

and compute the values $\tilde{f}_j(y_k)$ $(j = 0,1,2,\ldots,N,\ k = 0,1,2,\ldots,n)$. This is equivalent to applying a suitable composite rule to the reformulated single integral equation but certain problems are automatically avoided when the system is treated directly. In practice, the weights w_k and abscissae y_k could depend on i,j.

In the same spirit, expansion methods may be applied where we seek approximations of the form

$$\tilde{f}_i(x) = \sum_{j=0}^{n_i} \tilde{a}_{i,j}\phi_{i,j}(x)\ , \tag{5.5}$$

where $\phi_{i,j}(x)$ $(i = 0,1,2,\ldots,N,\ j = 0,1,2,\ldots,n_i)$ are chosen functions. Thus, the classical Galerkin method is extended to (5.3) when we determine the coefficients $\tilde{a}_{i,j}$ $(i = 0,1,2,\ldots,N,\ j = 0,1,2,\ldots,n_i)$ such that (for example)

$$\int_a^b \{\tilde{f}_i(x) - \lambda \sum_{j=0}^{N} \int_a^b K_{i,j}(x,y)\tilde{f}_j(y)\mathrm{d}y - g_i(x)\}\overline{\phi_{i,k}(x)}\mathrm{d}x = 0 \tag{5.6}$$

$$(i = 0,1,2,\ldots,N,\ k = 0,1,2,\ldots,n_i) .$$

The extension of methods applicable to a single equation to the treatment of a system of equations should appear fairly natural, and we do not pursue the problem further.

5.2.2.

In theoretical terms, there is little difference between eqn (5.1) and the equation

$$f(\underset{\sim}{x}) - \lambda \int_{a_0}^{b_0}\int_{a_1}^{b_1} \ldots \int_{a_n}^{b_n} K(\underset{\sim}{x},\underset{\sim}{z})f(\underset{\sim}{z})\mathrm{d}z_0\mathrm{d}z_1 \ldots \mathrm{d}z_n = g(\underset{\sim}{x}) , \tag{5.7}$$

where $\underset{\sim}{x} = [x_0,x_1,\ldots,x_n]^T$ and $\underset{\sim}{z} = [z_0,z_1,\ldots,z_n]^T$.

Numerical methods for this multidimensional problem can usually be derived from the methods for (5.1) without much difficulty. Thus, the quadrature method for eqn (5.7) involves finding an integration rule

$$\int_{a_0}^{b_0}\int_{a_1}^{b_1} \ldots \int_{a_n}^{b_n} \phi(\underset{\sim}{z})\mathrm{d}z_0\mathrm{d}z_1\ldots\mathrm{d}z_n \simeq \sum_{j=0}^{N} \Omega_j\phi(\underset{\sim}{\sigma}_j) , \tag{5.8}$$

and writing

$$\tilde{f}(\underset{\sim}{\sigma}_i) - \lambda \sum_{j=0}^{N} \Omega_j K(\underset{\sim}{\sigma}_i,\underset{\sim}{\sigma}_j)\tilde{f}(\underset{\sim}{\sigma}_j) = g(\underset{\sim}{\sigma}_i) . \tag{5.9}$$

If the integration rule (5.8) is a composite rule such as the product version of the trapezium rule with step h, and $K(\underset{\sim}{x},\underset{\sim}{z})$ and $g(\underset{\sim}{z})$ are smooth, then we may show, say, that

$$\tilde{f}(\underset{\sim}{x}) = f(\underset{\sim}{x}) + h^2\phi_1(\underset{\sim}{x}) + h^4\phi_2(\underset{\sim}{x}) + \ldots .$$

Such an expansion justifies the use of the deferred approach to the limit, just as in section 4.3 but with scalar variables replaced by vectors.

Difference correction terms to an integration rule (5.8) may be complicated if the dimension, n, is large, and in these circumstances a deferred correction method for (5.9) will be cumbersome. In general, the solution of (5.9) presents practical problems when n is large. Thus (5.6) represents a system of equations in $(N + 1)$ unknowns, and, if a product rule is used in (5.8), this number varies exponentially with n. Efficient rules for hypercubature have to be sought, and a useful reference for these is Stroud (1971).

The expansion methods discussed for eqn (5.1) have obvious analogues for (5.7). If the choice of functions $\phi_0(\underset{\sim}{x}),\phi_1(\underset{\sim}{x}),\ldots,\phi_N(\underset{\sim}{x})$ is a fortunate one it may be possible to determine an approximation $\sum_{i=0}^{N} \tilde{a}_i\phi_i(\underset{\sim}{x})$ which gives a good representation of the true solution $f(\underset{\sim}{x})$. In the methods of collocation and Ritz-Galerkin type we derive systems of equations in the $(N + 1)$ unknowns $\tilde{a}_j(j = 0,1,2,\ldots,N)$, and although the dimension number n does not arise directly in the order of the matrix, it is usually necessary to choose N large when n is large, in order to obtain good accuracy.

All the expansion methods require a choice of functions $\phi_0(\underset{\sim}{x}),\phi_1(\underset{\sim}{x}),\ldots,\phi_N(\underset{\sim}{x})$. If we choose a sequence of functions $\Phi_0(x),\Phi_1(x)$, ... suitable for the case $n = 0$ we can generate the functions $\phi_i(\underset{\sim}{x})$ by considering all products of the form $\Phi_{i_0}(x_0)\Phi_{i_1}(x_1)\ \ldots\ \Phi_{i_n}(x_n)$, but other choices of functions $\phi_i(\underset{\sim}{x})$ may be better.

Example 5.3.

Consider the functions

$$1;x_0,x_1,\ldots,x_n;x_0^2,x_0x_1,x_0x_2,\ldots,x_0x_n,x_1^2\ ,$$

$$x_1x_2,\ldots,x_1x_n,\ldots,x_{n-1}x_n,x_n^2;\ x_0^3,x_0^2x_1,\ldots,x_0^2x_n,\ldots,x_n^3;\ldots,$$

where $x_0,x_1,\ldots,x_n$ are real variables. These functions are complete in the space of continuous functions with maximum norm, and they can be generated from the monomials $1,x,x^2,\ldots$ in one variable, in an obvious way. Just as the monomials in one variable may not be suitable for numerical work, it can be better, in n variables, to replace x_i^r by $T_r(x_i)$ in this array of products. (The range is presumed to be normalized so that $|x_i| \leq 1$, $\underset{\sim}{x} = [x_0,x_1,\ldots,x_n]^T$.) *

When the number of dimensions of the range of integration is large, it may be impracticable to extend the methods discussed for eqn (5.1), since the labour involved becomes prohibitive. Monte Carlo methods may be effective in such cases; we do not discuss such methods here.

5.3. *'Mild' and weak singularities in linear equations*

In our earlier discussions of the eqn (5.1), the emphasis has been on those equations in which $K(x,y)$ and $g(x)$ are continuous. Consider the case where $K(x,y)$ and $g(x)$ are piecewise-continuous, with finite jump discontinuities only on lines parallel to the coordinate axes; these 'singularities' will be called 'mild'. In this case, the equation can generally be re-formulated as a system of equations and the remarks of section 5.2.1 then apply. Such an approach (using eqn (5.4), say) may also be profitable when $K(x,y)$ and $g(x)$ are continuous but have low-order derivatives which have 'mild' discontinuities. There are a number of other possibilities which can occur and we shall examine various alternative approaches.

We must separate, in our discussion, techniques which can be used to re-formulate the integral equation so that it appears in a more convenient form, and any special precautions which should be taken (or obstacles which occur) if conventional numerical methods are applied. We shall also separate the effect of bad behaviour in $g(x)$ and bad behaviour in $K(x,y)$.

Any bad behaviour in $g(x)$ is of importance if it leads to correspondingly bad behaviour in the solution $f(x)$. In general, $f(x)$

may be expected to inherit the form and nature of 'singularities' in $g(x)$, since

$$f(x) = g(x) + \lambda\int_a^b R(x,y;\lambda)g(y)\mathrm{d}y \ ,$$

and only if $R(x,y;\lambda)$ is badly behaved can the 'singularities' of $g(x)$ be eliminated in this equation. Since the accuracy of the quadrature method, for example, and also that of the methods described further in sections 5.5-5.7, is determined in part by the smoothness of $f(x)$, it is an advantage to obtain an equation for a function which is smoother than $f(x)$.

One technique, already introduced in Chapter 4, is to write $f(x) = \phi(x) + g(x)$ and solve the equation

$$\phi(x) - \lambda\int_a^b K(x,y)\phi(y)\mathrm{d}y = \lambda\int_a^b K(x,y)g(y)\mathrm{d}y \ .$$

Now, the right-hand term of this equation may be difficult to obtain. It may be easier to employ a simple function $\gamma(x)$ which exhibits the bad characteristics of $g(x)$ inherited by $f(x)$. We set $f(x) = \psi(x) + \gamma(x)$, where $g(x) - \gamma(x)$ is well behaved and

$$\psi(x) - \lambda\int_a^b K(x,y)\psi(y)\mathrm{d}y = g(x) - \gamma(x) + \lambda\int_a^b K(x,y)\gamma(y)\mathrm{d}y \ . \qquad (5.10)$$

We then employ a numerical method to approximate $\psi(x)$ and hence $f(x)$. Such techniques are successful only if $\int_a^b K(x,y)g(y)\mathrm{d}y$ or $\int_a^b K(x,y)\gamma(y)\mathrm{d}y$ are smooth functions, and in particular when $K(x,y)$ is smooth. In this case (where $K(x,y)$ is at least continuous) treatment of an eqn (5.10) provides a general remedy for bad behaviour in $g(x)$. With a suitable choice of $\gamma(x)$, the function $\psi(x)$ is well behaved and can be approximated accurately by one of our methods.

Treatment of the original equation (5.1) is of course possible if the bad behaviour of $g(x)$ stems from a discontinuity in a derivative rather than in the function itself. We shall not then *expect* high-order accuracy from a quadrature method, in general, since $f(x)$ may inherit the behaviour of $g(x)$. If we use an expansion method for (5.1) and $\phi_1(x),\ldots,\phi_n(x)$ are smooth functions, we may take $\phi_0(x) = g(x)$, but the conditioning of the resulting algebraic problems must be monitored. In general it seems advisable to choose well behaved independent functions $\phi_i(x)$ $(i = 0,1,2,\ldots,n)$ to obtain

$$\tilde{f}(x) = \sum_{i=0}^{n} \tilde{a}_i\phi_i(x) ,$$

from which we compute

$$g(x) + \lambda\int_a^b K(x,y)\tilde{f}(y)\mathrm{d}y$$

as the accepted approximation.

In general, our numerical methods require a certain smoothness of $g(x)$. The collocation and quadrature methods use values of $g(x)$ which are guaranteed bounded if $g(x)$ is continuous, and the Ritz-Galerkin methods generally require that

$$\int_a^b |g(x)|^2\mathrm{d}x < \infty .$$

When such conditions are violated it may be *necessary* to treat an equation of the form (5.10), instead of the original equation.

5.3.1.

Now, suppose that there are no discontinuities in $g(x)$, and consider the effect of the behaviour of $K(x,y)$.

Discontinuities in the kernel $K(x,y)$, which lie on lines parallel to the y-axis, can be inherited by $f(x)$. An example is given by setting

$$K(x,y) = H(x,y)/|x - x_1|^{\alpha}$$

with $x_1 \in [a,b]$, $0 < \alpha < \frac{1}{2}$, and $H(x,y)$ continuous, in (5.1). In this case, the quadrature method (section 4.3) and collocation method are not generally appropriate. If we take $\phi_0(x)$, $\phi_1(x)$, ..., $\phi_n(x) \in C[a,b]$, the Ritz-Galerkin-type methods are applicable, and (as earlier) an approximation

$$\tilde{f}(x) = \sum_{i=0}^{n} \tilde{a}_i \phi_i(x)$$

can probably be improved by obtaining

$$g(x) + \lambda \int_a^b K(x,y)\tilde{f}(y)\,dy ,$$

which has the required properties at $x = x_1$. Since

$$\psi_i(x) = \int_a^b K(x,y)\phi_i(y)\,dy$$

is unbounded at $x = x_1$, there is a rôle for product integration in computing the quantities (ψ_i, ϕ_j) required in the method. For the corresponding case of a kernel of the form $H(x,y)/|y - y_1|^{\alpha}$, (with $0 \leq \alpha < 1$) we can employ product-integration methods and, with suitable basic functions $\phi_i(x)$, the expansion methods.

The case where $K(x,y)$ is weakly singular and of the form

$$K(x,y) = H(x,y)/|x - y|^{\alpha} , \tag{5.11}$$

where $H(x,y)$ is continuous and $0 < \alpha < 1$, is not necessarily difficult. This case is an important one in practice and has received a certain amount of attention in the literature (see, for example, Atkinson 1972*a*; Phillips 1969; Kussmaul and Werner 1967). I shall devote the remainder of the section to this case, and in sections 5.5-5.7 we explore the methods further (giving some numerical examples). A theory of such methods is covered in sections 5.11-5.14.

For a weakly singular kernel (5.11) it is sometimes possible to re-formulate the integral equation (5.1) so that the kernel is well behaved (at least continuous). Thus, if $K(x,y)$ is of the form (5.11), where $0 < \alpha < \frac{1}{2}$. and $H(x,y)$ is continuous, the iterated kernel $K^2(x,y)$ is bounded and continuous (Goursat 1964; Mikhlin 1964). Suppose

$(I-\lambda K)^{-1}$ and $(I+\lambda K)^{-1}$ *both* exist (that is, λ and $-\lambda$ are both regular values of $K(x,y)$). Then we can obtain (see sections 1.11 and 4.7), from the equation

$$f(x) - \lambda\int_a^b K(x,y)f(y)\mathrm{d}y = g(x) \quad , \tag{5.12}$$

the equation

$$f(x) - \lambda^2\int_a^b K^2(x,y)f(y)\mathrm{d}y = g(x) + \lambda\int_a^b K(x,y)g(y)\mathrm{d}y \ . \tag{5.13}$$

In the case under discussion, $K^2(x,y)$ is well behaved, but we immediately note the problems which may arise in evaluating both $K^2(x,y)$ and $\int_a^b K(x,y)g(y)\mathrm{d}y$. Moreover, the analysis breaks down if $(I+\lambda K)^{-1}$ does not exist. However, it may then be possible to determine a value μ for which the operator $(I+\lambda K+\mu K^2)^{-1}$ exists, so that the original equation is equivalent to

$$f(x) - \int_a^b \{(\lambda^2-\mu)K^2(x,y) + \lambda\mu\, K^3(x,y)\}f(y)\mathrm{d}y$$

$$= g(x) + \lambda\int_a^b K(x,y)g(y)\mathrm{d}y + \mu\int_a^b K^2(x,y)g(y)\mathrm{d}y \ , \tag{5.14}$$

where $K^3(x,y) = \int_a^b K(x,z)K^2(z,y)\mathrm{d}z$. If $0<\alpha<\frac{1}{2}$ the kernel for this equation is bounded; if $\frac{1}{2}\leq\alpha<1$, $K^2(x,y)$ is still unbounded but the process can be repeated. A slightly different technique was indicated in section 1.11.

Example 5.4.

Consider the case where $K(x,y) = |x-y|^{-\frac{1}{2}}$, and λ is real, and a regular value, in eqn (5.1). Then in view of the results of Example 1.41, and section 1.11,

$$f(x) - \lambda^3 \int_0^1 K^3(x,y)f(y)\mathrm{d}y = d_\lambda(x) \ ,$$

where $d_\lambda(x) = (I-\lambda\omega K)(I-\lambda\omega^2 K)g(x)$ and $\omega = \exp 2\pi i/3$. The kernel $K^3(x,y)$ is here continuous. To obtain $d_\lambda(x)$ directly, write $\mu = \lambda^2$ in (5.14). *

The techniques suggested here clearly depend on an ability to manipulate the functions involved, analytically. If $K(x,y)$ is a complicated function, the techniques can only be applied with difficulty (if at all), because $K^p(x,y)$ $(p = 2,3,4,\ldots)$ will then be difficult to obtain. A theoretical investigation of the equation may be helpful in deciding on the continuity properties of a solution $f(x)$, however.

Example 5.5.
Suppose that $K(x,y) = H(x,y)/|x-y|^\alpha$, where $\alpha \in [0,1)$, $H(x,y)$ is continuous for $a\le x,y\le b$ and $H(x,y)$ is Lipschitz-continuous in x with exponent 1, or $H_x(x,y)$ exists, and $g(x)$ is Lipschitz-continuous (with exponent 1), or $g'(x)$ exists, for $a\le x\le b$.

It can then be shown that $f(x)$, defined by (5.12), is continuous. Hence,

$$\int_a^b K(x,y)f(y)\,dy$$

satisfies a Lipschitz condition with exponent $(1-\alpha)$. It follows that $f(x)$ satisfies this same Lipschitz condition. (Similar arguments can be applied to transformed equations such as eqn (5.14).) In the case where $K(x,y) = H(x,y)\ln|x-y|$, the modulus of continuity $\omega(f;\delta)$ of $f(x)$ is $O(-\delta \ln \delta)$. These results are rather weak in the case where $H(x,y)$ and $g(x)$ are particularly smooth. On the other hand, bad behaviour of $g(x)$ is generally inherited by $f(x)$ (see Example 5.20 for such a case). *

For a weakly singular kernel the basic quadrature method is clearly inapplicable because $K(x,x)$ is infinite, assuming that $H(x,x) \neq 0$ in (5.11). The product-integration method of section 4.21 can be adapted and extended to this case however, and we devote sections 5.5-5.6 to this important topic.

The modified quadrature method of section 4.4 can also be extended to weakly singular kernels as we now indicate (see also Kantorovitch and Krylov (1964, p.102) for a slightly different extension). When $K(x,y)$ is weakly singular we are faced with unbounded values in eqn (4.22), and we propose the following modification. We now set

$$\{1+\lambda\Delta^*(y_i)\}\tilde{f}(y_i) - \lambda \sum_{\substack{j\neq i\\ j=0}}^{n} w_j K(y_i,y_j)\tilde{f}(y_j) = g(y_i) , \tag{5.15}$$

where

$$\Delta^*(y_i) = \sum_{\substack{j\neq i\\ j=0}}^{n} w_j\, K(y_i,y_j) - \int_a^b K(y_i,y)\,dy \quad . \tag{5.16}$$

This is equivalent to setting $K(x,y)\,\{f(y) - f(x)\}$ equal to zero at $x = y$, in (4.20), before approximating with a quadrature rule, and it requires some justification in theory (see section 5.11.2) and in practice (section 5.4).

Finally, consider the expansion methods. The Ritz-Galerkin-type methods and the collocation methods can sometimes be applied if $K(x,y) = H(x,y)/|x-y|^{\alpha}$, where $\alpha \in (0,1)$, and the inhomogeneous term $g(x)$ is (say) continuous. To see whether any complication can arise, recall that in these methods we require the functions

$$\psi_i(x) = \int_a^b K(x,y)\phi_i(y)\,dy ,$$

and in the Galerkin method (for example) we require (ψ_i,ϕ_j) . If the functions $\phi_i(x)$ $(i = 0,1,2,\ldots,n)$ are continuous then the functions $\psi_i(x)$ are continuous, and no problem arises; the methods can, in theory, be applied without difficulty (the continuity of $\psi_i(x)$ follows from Lemma 5.3, which appears in section 5.13 below). In the case where $0 < \alpha < 1$ and the functions $\phi_i(x)$ are square-integrable (hence, possibly discontinuous) the functions $\psi_i(x)$ are square-integrable, but possibly discontinuous.

<u>Example 5.6</u>.

Consider the following situation.

If $\alpha = 1-\rho$ where $\frac{1}{2}>\rho>0$, and $K(x,y) = [x-y]^{-\alpha}$, $\phi_0(x) = x^{-\rho}$, $a = 0$, and $b = 1$, then $\psi_0(x)$ is unbounded at $x = 0$ and $\psi_0(x)$ is therefore discontinuous, whilst $\phi_0(x) \in L^2[0,1]$. *

Example 5.7.

Suppose that $\phi_i(x) \in L^2[0,1]$, and $K(x,y) = H(x,y)/|x-y|^{\alpha}$ with $0<\alpha<1$, $a = 0$, $b = 1$, and $H(x,y)$ continuous. Then

$$\int_0^1 |\psi_i(x)|^2 dx = \int_0^1 \left|\int_0^1 K(x,y)\phi_i(y)dy\right|^2 dx$$

$$\leq \int_0^1 \left(\int_0^1 |K(x,y)\phi_i(y)|dy\right)^2 dx$$

$$= \int_0^1 \left(\int_0^1 |K(x,y)|^{\frac{1}{2}}|K(x,y)|^{\frac{1}{2}} |\phi_i(y)|dy\right)^2 dx$$

$$\leq \int_0^1 \left(\int_0^1 |K(x,y)| \; |\phi_i(y)|^2 dy\right) \int_0^1 |K(x,y)| \; dy)dx$$

by the Cauchy-Schwarz inequality. We set

$$M = \sup_{0\leq x\leq 1} \int_0^1 |K(x,y)|dy = ||K||_{\infty} .$$

Then

$$\int_0^1 |\psi_i(x)|^2 dx \leq M\int_0^1 \int_0^1 |K(x,y)|dx \; |\phi_i(y)|^2 dy$$

(by Fubini's theorem (see Smithies 1962, p.12) for the change in order of integration)

$$\leq MN \int_0^1 |\phi_i(y)|^2 dy ,$$

where $N = \sup_{0\leq y\leq 1} \int_0^1 |K(x,y)|dx$. It is a simple exercise to show that $M < \infty$ and $N < \infty$ for $0<\alpha<1$ and hence $\psi_i(x) \in L^2[0,1]$. Moreover,

$||K||_2 \leq (MN)^{\frac{1}{2}}$, although $\|K(x,y)\|_2$ may be unbounded if $\frac{1}{2}<\alpha<1$. *

For practical purposes, we conclude, the Ritz-Galerkin (and collocation) methods are applicable to weakly singular kernels provided the functions $\phi_i(x)$ are continuous. If we propose to compute

$$(\psi_i,\phi_j) = \int_a^b \psi_i(x)\,\overline{\phi_j(x)}\,dx$$

using numerical integration formulae we may, however, experience some difficulty in obtaining good approximations, since $\psi_i(x)$ is not in general very smooth.

5.4. *The modified quadrature method applied to equations with weakly singular kernels*

A given quadrature rule

$$\int_a^b \phi(y)\,dy \simeq \sum_{j=0}^{n} w_j\phi(y_j) \quad (a\leq y_j\leq b,\ j=0,1,2,\ldots,n)$$

may be applied to give an approximate solution of the equation

$$f(x) - \lambda\int_a^b \frac{H(x,y)}{|x-y|^\alpha} f(y)\,dy = g(x) \quad , \tag{5.17}$$

where $H(x,y)$ is continuous and $0<\alpha<1$.
We first write

$$\{1-\lambda\int_a^b \frac{H(x,y)}{|x-y|^\alpha}\,dy\}\, f(x) - \lambda\int_a^b \frac{H(x,y)}{|x-y|^\alpha}\{f(y)-f(x)\}\,dy = g(x) \ . \tag{5.18}$$

If $f(x) \in \text{Lip}\ \beta$, where $\beta > \alpha$, the integrand in the second term on the left tends continuously to zero as x tends to y . The quadrature rule may then be applied to give approximate values $\tilde{f}(y_i) \simeq f(y_i)$ $(i = 0,1,2,\ldots,n)$, where

$$\{1-\lambda A(y_i)\}\tilde{f}(y_i) - \lambda \sum_{\substack{j\neq i\\ j=0}}^{n} \frac{H(y_i,y_j)w_j}{|y_i-y_j|^{\alpha}}\{\tilde{f}(y_j) - \tilde{f}(y_i)\} = g(y_i) ,$$

with $A(x) = \int_a^b H(x,y)|x-y|^{-\alpha}dy$.

The values $\tilde{f}(y_i)$ $(i = 0,1,2,\ldots, n)$ can be used to define a function $\tilde{f}(x)$ by piecewise-constant or piecewise-linear interpolation (or a mixture of these if $a < y_0$ or $b > y_n$) , or by polynomial or spline interpolation.

Example 5.8.
If $K(x,y) = \ln |x+y|$, we can define $H(x,y) = |x-y|^{\alpha} \ln |x+y|$ for any $\alpha \in (0,1)$, and $H(x,y)$ is continuous. The kernel $K(x,y)$ is then seen to be weakly singular, and also square-integrable.

Consider the equation

$$f(x) - \int_0^1 \ln |x+y| f(y)dy = g(x) .$$

The modified quadrature method may be applied if we determine, for $x \in [0,1]$,

$$A(x) = \int_0^1 \ln |x+y|dy = (x+1) \ln (x+1) - x \ln x-1 .$$

The function $g(x)$ was chosen to correspond to the solution $f(x) = x$, and for this (rather simple) example we applied the modified quadrature method using (a) the $(n+1)$-point repeated mid-point rule, (b) the n-times repeated trapezium rule, and (c) the repeated Simpson's rule with $n+1$ points. The maximum errors, $||\tilde{\mathrm{f}}-\mathrm{f}||_{\infty}$, are shown in Table 5.1.

TABLE 5.1

n	(a) Midpoint ($h=1/(n+1)$)	(b) Trapezium ($h=1/n$)	(c) Simpson ($h=1/2n$)
2	$1{\cdot}16 \times 10^{-2}$	$5{\cdot}52 \times 10^{-2}$	$1{\cdot}16 \times 10^{-2}$
4	$4{\cdot}27 \times 10^{-3}$	$1{\cdot}37 \times 10^{-2}$	$3{\cdot}45 \times 10^{-3}$
8	$1{\cdot}34 \times 10^{-3}$	$3{\cdot}43 \times 10^{-3}$	$9{\cdot}74 \times 10^{-4}$
16	$4{\cdot}31 \times 10^{-4}$	$1{\cdot}05 \times 10^{-3}$	$2{\cdot}64 \times 10^{-4}$
32	$1{\cdot}33 \times 10^{-4}$	$3{\cdot}14 \times 10^{-4}$	$6{\cdot}95 \times 10^{-5}$
64	$3{\cdot}98 \times 10^{-5}$	$9{\cdot}18 \times 10^{-5}$	$1{\cdot}80 \times 10^{-5}$
128	$1{\cdot}16 \times 10^{-5}$	$2{\cdot}64 \times 10^{-5}$	$4{\cdot}58 \times 10^{-6}$

The order of the error can be judged to be perhaps $O(h^2 \ln h)$ in each case; it is not $O(h^4)$ in the case (c) because the solution is $f(x) = x$ and the error in Simpson's rule applied to the integrand $(x-y) \ln |x+y|$ is not $O(h^4)$. *

Example 5.9.

The equation

$$f(x) - \int_0^1 \ln|x-y| f(y) dy = g(x) ,$$

in which $g(x)$ is chosen so that the solution is $f(x) = x$, is similar to the equation considered above. We take

$$g(x) = x - 0{\cdot}5\{x^2 \ln (x) + (1-x^2) \ln (1-x) - (x+0{\cdot}5)\} ,$$

and use the modified quadrature method and the repeated Simpson's rule with step h to compute an approximate solution. The errors shown in

Table 5.2 are then observed, using a computer working to 10 decimal digits:

TABLE 5.2

$h \backslash x$	0	0·25	0·5	0·75	1
$\frac{1}{2}$	$-1{\cdot}3 \times 10^{-2}$		$3{\cdot}6 \times 10^{-12}$		$1{\cdot}3 \times 10^{-2}$
$\frac{1}{4}$	$-3{\cdot}6 \times 10^{-3}$	$7{\cdot}5 \times 10^{-5}$	$3{\cdot}6 \times 10^{-12}$	$-7{\cdot}5 \times 10^{-5}$	$3{\cdot}6 \times 10^{-3}$
$\frac{1}{8}$	$-1{\cdot}0 \times 10^{-3}$	$1{\cdot}9 \times 10^{-5}$	$7{\cdot}3 \times 10^{-12}$	$-1{\cdot}9 \times 10^{-5}$	$1{\cdot}0 \times 10^{-3}$
$\frac{1}{16}$	$-2{\cdot}7 \times 10^{-4}$	$3{\cdot}1 \times 10^{-6}$	$-1{\cdot}5 \times 10^{-11}$	-3.1×10^{-6}	$2{\cdot}7 \times 10^{-4}$
$\frac{1}{32}$	$-7{\cdot}1 \times 10^{-5}$	$4{\cdot}4 \times 10^{-7}$	$7{\cdot}3 \times 10^{-12}$	$-4{\cdot}4 \times 10^{-7}$	$7{\cdot}1 \times 10^{-5}$
$\frac{1}{64}$	$-1{\cdot}8 \times 10^{-5}$	$5{\cdot}8 \times 10^{-8}$	$-2{\cdot}2 \times 10^{-11}$	$-5{\cdot}8 \times 10^{-8}$	$1{\cdot}8 \times 10^{-5}$
	(maximum)		(zero)		(maximum)

(The values at $x = 0{\cdot}5$ are correct almost to within the level of rounding error.) The extreme error is attained for $x = 0$ and $x = 1$ and its rate of decrease is approximately that of $-h^2 \ln h$, as h tends to zero. However, the errors at $x = 0{\cdot}25,\ 0{\cdot}75$ decrease faster than this.

It is plausible to suggest that the rate of convergence of the maximum error in the computed values is determined by the order of the maximum of the local truncation errors. These local truncation errors are given, for $i = 0,1,2,\ldots,\ n = 2m$, with $h = 1/n$, by

$$\tau_i = \int_0^1 \ln|ih-y|\ (ih-y)dy -$$

$$- \left[\frac{2h}{3} \sum_{j=0}^{m}{}'' \ln|ih-2jh|\ (ih-2jh) + \frac{4h}{3} \sum_{j=0}^{m-1} \ln|ih-(2j+1)h|\{ih-(2j+1)h\}\right] .$$

the term in square brackets being an expression for the repeated Simpson's rule applied to the integral term. In the case $i = 0$ we know (see Fox 1967) that $\tau_0 = O(h^2)$ but it appears, from a heuristic argument similar to that used by Fox, that $\tau_1 = \tau_3 = \ldots = O(-h^2 \ln h)$, and that $\max_i |\tau_i| = O(-h^2 \ln h)$, since $\tau_0 = \tau_2 = \tau_4 = \ldots = O(h^2)$.

Indeed, to examine the size of τ_{2i+1} , the contributions from approximating

$$\int_{2ih}^{2(i+1)h} (y-\overline{2i+1}h) \ln|y-\overline{2i+1}h| \, dy$$

and the corresponding integrals over $[0,2ih]$ and over $[2(i+1)h,1]$ can be assessed separately. The first contribution is straightforward, and the generalized Euler-Maclaurin formula for the remainders in Simpson's rule applied to the integrand $\phi_i(y) = (y-\overline{2i+1}h)\ln|y-\overline{2i+1}h|$ on $[0,2ih]$ and $[2(i+1)h,1]$ gives, *formally*, with $k = 2i+1$ for convenience,

$$\tau_{2i+1} = \int_{2ih}^{(2i+2)h} (y-kh) \ln|y-kh| \, dy +$$

$$+ \frac{h^4}{180}[\{\phi_i'''(0) - \phi_i'''((k-1)h)\} + \{\phi_i'''((k+1)h) - \phi_i'''(1)\}] +$$

$$+ \frac{h^6}{1512}[\{\phi_i^v(0) - \phi_i^v((k-1)h)\} + \{\phi_i^v((k+1)h) - \phi_i^v(1)\}] + \ldots$$

$$= \int_{-h}^{h} z \ln|z| \, dz + \frac{h^4}{180}\{\phi_i'''(0) - \phi_i'''(1)\} +$$

$$+ \frac{h^6}{1512}\{\phi_i^v(0) - \phi_i^v(1)\} + \ldots \quad (i = 0,1,2,\ldots,(m-1);\ k = 2i+1) ,$$

which indicates a result of the required form. (Observe that the Euler-Maclaurin expressions do not always converge, so the argument is not rigorous.) Similar arguments apply in the case of the error obtained when using the repeated trapezium rule or repeated mid-point rule, both in this Example and in the preceding Example, and we conclude that the rate of convergence is not improved if we use the repeated Simpson's rule in place of either of these rules. *

5.5. *Simple product-integration formulae for weakly singular kernels*

An approximate solution to the equation

$$f(x) - \lambda\int_a^b K(x,y)f(y)\,dy = g(x) \quad , \tag{5.19}$$

where $K(x,y) = H(x,y)/|x-y|^{\alpha}$ is weakly singular, can be obtained using the simple form of product integration described in section 4.6. The method which is described there is mathematically equivalent to a collocation method.

In section 5.5.1 I show how our method can be modified to deal with practical problems, and this is illustrated by equations in which the integral is a contour integral. The application of the method of section 4.6.1 to weakly singular equations is discussed in section 5.6; this technique seems to be particularly useful.

Initially, suppose that the interval $[a,b]$ is partitioned by means of the points $\{a = z_0 < z_1 < \ldots < z_n = b\}$. It is convenient to suppose that $z_i = a+ih$, $h = (b-a)/n$ (though in practice the points z_i may be unequally spaced). We construct integration rules

$$\int_a^b K(z_i,y)\phi(y)\,dy \simeq \sum_{j=0}^{n} v_j(z_i)\phi(z_j) \quad , \tag{5.20}$$

which are designed to be exact if $\phi(x)$ is a certain type of function: either (a) piecewise-constant, being constant in each interval $[z_0,z_1]$, $[z_1,z_2]$, ..., $[z_{n-1}, z_n]$, or (b) continuous and piecewise-linear, being linear in each interval $[z_i,z_{i+1}]$ $(i = 0,1,2,\ldots, n-1)$ or (c) some smooth function, such as a polynomial of degree n on $[a,b]$. Comments on such a choice of the quantities $v_j(z_i)$ have been made in Chapter 2 (see section 2.17) where we showed, for example, how to generalize the repeated trapezium and repeated Simpson rules. When the weights $v_j(z_i)$ are known, we solve the equations

$$\tilde{f}(z_i) - \lambda \sum_{j=0}^{n} v_j(z_i)\tilde{f}(z_j) = g(z_i) \qquad (i=0,1,\ldots,n) \tag{5.21}$$

first encountered in eqn (4.27)).

Example 5.10.
Consider the case

$$K(x,y) = |x-y|^{-\frac{1}{2}} \quad .$$

We seek an approximation of the form

$$\int_a^b \frac{1}{|z_i-y|^{\frac{1}{2}}} \phi(y)\,dy \simeq \sum_{j=0}^{n} v_j(z_i)\phi(z_j)$$

($z_i = a+ih$, $h = (b-a)/n$) , which is an extension of the repeated trapezium rule. Thus the approximation is required to be exact whenever $\phi(x)$ is continuous and is linear in each subinterval $[a,a+h]$, $[a+h, a+2h]$, ... $[b-h,b]$. We substitute $w(y) = K(z_i,y)$ in eqn (2.44) and obtain the required approximation where

$$v_0(z_k) = \int_{z_0}^{z_1} (z_1-y)K(z_k,y)\,dy/h \ ,$$

$$v_n(z_k) = \int_{z_{n-1}}^{z_n} (y-z_{n-1})K(z_k,y)\,dy/h \ ,$$

and

$$v_j(z_k) = \frac{1}{h}\{\int_{z_{j-1}}^{z_j} (y-z_{j-1})K(z_k,y)\,dy \ +$$

$$+ \int_{z_j}^{z_{j+1}} (z_{j+1}-y)K(z_k,y)\,dy\} \ , \quad \text{for} \quad j\neq 0, \ j\neq n \quad .$$

Accordingly, we require, for the kernel discussed in this example, the quantities

$$\frac{1}{h}\int_{z_{j-1}}^{z_j} \frac{(z_j-y)}{|z_k-y|^{\frac{1}{2}}}\,dy \ , \text{ and } \ \frac{1}{h}\int_{z_{j-1}}^{z_j} \frac{(y-z_{j-1})}{|z_k-y|^{\frac{1}{2}}}\,dy$$

in order to determine $v_j(z_i)$, where $z_k = a+kh$. These quantities may be calculated conveniently (as was noted by Atkinson (1966)). With $y = z_{j-1} + h\xi$ the quantities displayed above can be written

$$h^{\frac{1}{2}} \int_0^1 \frac{(1-\xi)}{|q-\xi|^{\frac{1}{2}}} d\xi \, , \quad h^{\frac{1}{2}} \int \frac{\xi}{|q-\xi|^{\frac{1}{2}}} d\xi \, ,$$

where $q = k - j + 1$, $-n + 1 \leq q \leq n$. The $(n+1)^2$ quantities $v_j(z_k)$ can thus be expressed, in this example, in terms of a much smaller number of definite integrals which can be calculated explicitly. Similar results can be established for the kernel $\ln|x-y|$ (see Atkinson 1966), or any difference kernel $K(x,y) = k(x-y)$. *

Example 5.11.
Consider the equation

$$f(x) - \lambda \int_{-1}^{1} |x-y|^{-\frac{1}{2}} f(y) dy = g(x) \, , \quad (|x| \leq 1) \, ,$$

where $\lambda = \frac{2}{3}$ and $g(x) = (1-x^2)^{\frac{3}{4}} - \pi\sqrt{2}(2-x^2)/4$. The exact solution is $f(x) = (1-x^2)^{\frac{3}{4}}$.

An approximate solution was obtained using the generalized trapezium rule as the product-integration rule for

$$\int_{-1}^{1} |z_i-y|^{-\frac{1}{2}} \phi(y) dy \quad ;$$

this is exact when $\phi(x)$ is continuous and linear in each interval $[z_i, z_{i+1}]$ ($z_k = kh - 1$, $h = 2/n$; $i = 0,1,2,\ldots,(n-1)$). (See Example 5.10.) The values $\tilde{f}(z_i)$ $(i = 0,1,2,\ldots,n)$ were computed, and the errors are tabulated below

$n = 5$	$x = \pm 1$	$\pm 0{\cdot}6$	$\pm 0{\cdot}2$
error:	$-2{\cdot}55 \times 10^{-2}$	$2{\cdot}25 \times 10^{-2}$	$1{\cdot}27 \times 10^{-1}$

The maximum errors, $\|\underset{\sim}{f}-\underset{\sim}{\tilde{f}}\|_\infty$, for varying values of n were $n = 10$, $6\cdot18 \times 10^{-2}$; $n = 20$, $2\cdot30 \times 10^{-2}$; $n = 30$, $1\cdot22 \times 10^{-2}$; $n = 40$, $7\cdot66 \times 10^{-3}$; $n = 50$, $5\cdot48 \times 10^{-3}$; $n = 60$, $4\cdot25 \times 10^{-3}$; $n = 70$, $3\cdot42 \times 10^{-3}$.

Observe that the rate of convergence is slow, but $f(x)$ is not differentiable at $x = \pm 1$.

Phillips (1969) has obtained a numerical solution for this problem using polynomial collocation, and reports on the accuracy expected theoretically and obtained in practice. An error of about 3×10^{-2} can be obtained using a polynomial of degree 7 and solving a system of 8 linear equations, and an error of order 10^{-3} is obtained on solving a system of 30 equations. *

5.5.1.

When we consider the integral equation

$$f(z) - \lambda\int_C K(z,\sigma)f(\sigma)d\sigma = g(z), \quad z \in C, \tag{5.22}$$

where C is a contour in the complex plane, certain complications can occur in the application of the technique of section 5.5. We permit $K(z,\sigma)$ to be unbounded at $z = \sigma$, but integrable.

When considering eqn (5.22) we commence by choosing points $s_0, s_1, \ldots, s_n$ on C ($s_n = s_0$ if C is closed), and the arc of C joining the points s_r, s_{r+1} is denoted by C_r, with length h_r. Then

$$f(z) - \lambda \sum_{r=0}^{n-1}\int_{C_r} K(z,\sigma)f(\sigma)d\sigma = g(z) \quad .$$

Suppose that z_r is a point on C_r. We may determine a piecewise-constant function $\tilde{f}(t)$, $t \in C$, such that $\tilde{f}(z) = \tilde{f}(z_r)$ on the interior of the segment C_r. Ideally, we would like to determine the values $\tilde{f}(z_r)$ so that (say)

$$\tilde{f}(z_p) - \lambda \sum_{r=0}^{n-1} \tilde{f}(z_r)\int_{C_r} K(z_p,\sigma)d\sigma = g(z_p) \; , \quad (p = 0,1,2,\ldots,n)$$

but in general the nature of C may make the determination of $\int_{C_r} K(z_p,\sigma)d\sigma$ difficult. (Similar difficulties can, of course, arise when C is a segment $[a,b]$ of the real line.) We may therefore write(if $p \neq r$)

$$\int_{C_r} K(z_p,\sigma)d\sigma \simeq \sum_{j=0}^{n_r} w_{j,r} K(z_p,\sigma_{j,r}) \quad , \tag{5.23}$$

and apply a quadrature rule to determine the integral over portions of arc C_r which do not contain z_p (the integrand is then bounded). For $p = r$ some prior manipulation is necessary, in the weakly singular case, to avoid infinite quantities.

Example 5.12.

A simple approximation for $\int_{C_r} \ln|z_p-\sigma|d\sigma$ $(r \neq p)$, where σ is arc-length, is $h_r \ln|z_p-z_r|$; Symm (1966) uses the approximation

$$\int_{C_r} \ln|z_p-\sigma| \, |d\sigma| \simeq \frac{h_r}{6}(\ln|z_p-s_r| + 4\ln|z_p-z_r| + \ln|z_p-s_{r+1}|) \; ,$$

where z_r is the mid-point of C_r (this is an application of Simpson's rule) where h_r is the length of C_r , provided that $r \neq p$. For $r = p$, Symm (1966) writes

$$\int_{C_r} \ln|z_r-\sigma| \, |d\sigma| \simeq |z_r-s_r|(\ln|z_r-s_r| - 1) + |z_r-s_{r+1}|(\ln|z_r-s_{r+1}| - 1) \; ,$$

which is obtained by replacing C_r by the union of the straight lines from s_r to z_r and z_r to s_{r+1} and integrating exactly. *

In general, the calculation of $\int_{C_r} K(z_r,\sigma)d\sigma$ may cause some difficulty and the effect of making various approximations to this quantity has not been fully analysed.

The situation is seen in a slightly different light if the closed contour C in eqn (5.22) is parametrized in terms, say, of the arc-length τ from a fixed point σ_0 of C. Thus, for example, the equation

$$f(z) - \lambda \int_C \ln|z-\sigma|f(\sigma)|d\sigma| = g(z) \quad (z \in C) \quad , \tag{5.24}$$

where $d\sigma$ is the arc differential at σ, is recast as

$$f(z) - \lambda\int_0^L \ln|z-\xi(\tau)|f(\xi(\tau))d\tau = g(z) \quad (z \in C) \quad , \tag{5.25}$$

where τ is the length of arc from σ_0 to $\sigma = \xi(\tau)$.† In eqns (5.24) and (5.25), $z = x+iy$ and $\sigma = \zeta+i\eta$ are points in the complex plane lying on C. Parameterization by τ gives $\xi = \upsilon(\tau) + i\nu(\tau) \equiv \xi(\tau)$. If $z = \xi(t)$, then the integrand in (5.25) becomes infinite when $\tau = t$. It follows that the equation is weakly singular; indeed, we recognize that $|t-\tau|^{\frac{1}{2}} \ln|z-\xi(\tau)|$ is continuous. For, with $z = \xi(t)$, $\ln|z-\xi(\tau)| = \frac{1}{2}\ln[\{\upsilon(t)-\upsilon(\tau)\}^2 + \{\nu(t)-\nu(\tau)\}^2]$, in the notation established above.

Suppose that $s_r = \xi(t_r)$ $(r = 0,1,2,\ldots,n)$ and $z_p = \xi(t_p^*)$, where $t_p^* = \frac{1}{2}(t_p+t_{p+1})$ $(p = 0,1,2,\ldots,n)$. Then the integrals

$$\int_{C_r} \ln|z_p-\sigma|\ |d\sigma|$$

may be written as

$$\int_{t_r}^{t_{r+1}} \ln|z_p-\xi(\tau)|d\tau \quad .$$

---

† Because of the definition of τ, $|\xi'(\tau)| = 1$ if the contour is smooth.

If $\xi(t)$ is a simple function, it is possible that the integrand can be evaluated exactly, but more generally it must be approximated, along similar lines to those indicated in Example 5.12. Thus, we may write,

$$\int_{t_r}^{t_{r+1}} \ln|z_p-\xi(\tau)|d\tau = \int_{t_r}^{t_{r+1}} \ln|\xi(t_p^*)-\xi(\tau)|d\tau \quad .$$

If $p \neq r$ we can, in principle, use any quadrature rule to approximate this integral. Suppose, instead, that $p = r$ so that $z_p \in C_r$. We assume that $\xi(t)$ is twice differentiable, and $|\xi'(t)| = 1$. Then

$$\xi(t) - \xi(\tau) = (t-\tau)\xi'(t)\{1+O(t-\tau)\}$$

and

$$\ln|\xi(t) - \xi(\tau)| = \ln|t-\tau| + \ln|\xi'(t)| + \log|1+O(\tau-t)| \quad .$$

We set $t = t_r^*$, and we obtain

$$\int_{t_r}^{t_{r+1}} \ln|\xi(t_r^*) - \xi(\tau)|d\tau = \int_{t_r}^{t_{r+1}} \ln|t_r^*-\tau|d\tau +$$

$$+ \ln|\xi'(t_r^*)|(t_{r+1}-t_r) + O(t_{r+1}-t_r)^2 \quad . \qquad (5.26)$$

The first term on the right in (5.26) can be evaluated explicitly, whilst the second term vanishes, for a smooth arc, with $|\xi'(t_r^*)| = 1$.

<u>Remark</u>. In the foregoing exposition, we have sought an approximation $f(x)$ in the form of a step function. We can generalize the method to obtain a piecewise-polynomial approximation (say piecewise-linear and -continuous, or, following Hayes, Kahaner, and Kellner (1972), piecewise-quadratic) if we find

$$\int_{t_r}^{t_{r+1}} \tau^q \ln|\xi(t) - \xi(\tau)|d\tau \quad (q = 0,1,2,\ldots;\ t \in \{t_r^*\}) \quad .$$

5.6. *Practical product-integration methods applied to weakly singular kernels*

The technique described in section 5.5 requires formulae of the type (5.20) for the approximation of the integrals

$$\int_a^b K(z_i,y)\phi(y)\mathrm{d}y \quad .$$

If the function $K(x,y)$ is a complicated function it may be unrealistic to ask for such an approximation. However, the method of section 4.6.1 (q.v.) can be applied to treat the case of a weakly singular kernel, and it provides a useful technique, using approximations to such integrals as

$$\int_a^b |x-y|^{-\alpha}\phi(y)\mathrm{d}y \quad , \quad \alpha \in [0,1) \quad ,$$

and

$$\int_a^b \ln|x-y|\phi(y)\mathrm{d}y \ ,$$

which can be obtained fairly readily.

In its simplest form, the technique depends on representing $K(x,y)$ in some way as the product $L(x,y)M(x,y)$, where $L(x,y)$ is *at least* continuous and $M(x,y)$ is weakly singular. In terms of implementing the method, it is an advantage if $M(x,y)$ is a relatively simple function like $|x-y|^{-\alpha}$ or $\ln|x-y|$.

The methods under discussion are treated by Atkinson (1966) and the quality of results is somewhat variable. An improved version also discussed by Atkinson will be discussed below in section 5.6.1; Kershaw (1963) introduced a similar method. Another modification is introduced in section 5.6.2. Whilst the theory of such methods is now understood, the methods should be applied skilfully to obtain good results. Our first Example illustrates one of the basic problems. In this Example we have the kernel $K(x,y) = \ln|x-y|$ and could readily set $M(x,y) = K(x,y)$ and $L(x,y) \equiv 1$ (effectively applying the method of section 5.5, it transpires). Instead, we choose to write, somewhat artificially, $M(x,y) = |x-y|^{-\frac{1}{2}}$ and $L(x,y) = |x-y|^{\frac{1}{2}}K(x,y)$, in order to investigate the numerical behaviour of the method under possibly adverse conditions. Observe that $L(x,y)$ is here continuous, *but not smooth*. Even with a skilful choice of $L(x,y)$ and $M(x,y)$, the full benefits of the method can only be realized if the solution $f(x)$ is smooth.

The following Example serves as an introduction to the basic method, in which we write $K(x,y) = L(x,y)M(x,y)$ (as described above) and determine formulae

$$\int_a^b M(z_i,y)\phi(y)\,dy \simeq \sum_{j=0}^{n} v_j(z_i)\phi(z_j) \quad , \tag{5.27}$$

where $a \le z_0 < z_1 < \ldots < z_n \le b$. We employ (5.27) to replace the equation

$$f(x) - \lambda\int_a^b K(x,y)f(y)\,dy = g(x) \quad (a \le x \le b)$$

by the system of equations

$$\tilde{f}(z_i) - \lambda \sum_{j=0}^{n} v_j(z_i)L(z_i,z_j)\tilde{f}(z_j) = g(z_i) \quad .$$

The method under discussion, and its derivatives described below, is very easy to implement once one has a standard set of formulae of the form (5.27), corresponding to a selection of weakly singular functions $M(x,y)$.

Example 5.13.

Consider the equation

$$f(x) - \lambda\int_0^1 \ln|x-y|f(y)\,dy = g(x) \quad .$$

The kernel $K(x,y) = \ln|x-y|$ can be written as a product $L(x,y)M(x,y)$, where $M(x,y) = |x-y|^{-\frac{1}{2}}$ becomes unbounded at $x = y$ and

$L(x,y) = (x-y)^{\frac{1}{2}} \ln|x-y|$ is continuous for $0 \le x, y \le 1$. Now in Example 5.10 we saw how to construct approximations of the form

$$\int_a^b \frac{1}{|z_i - y|^{\frac{1}{2}}} \phi(y)dy \simeq \sum_{j=0}^{n} v_j(z_i)\phi(z_j) ,$$

where $z_i = a+ih$, $h = (b-a)/n$. These weights $v_j(z_i)$ can be employed to yield the equations

$$\tilde{f}(z_i) - \lambda \sum_{j=0}^{n} v_j(z_i) \{L(z_i,z_j)\tilde{f}(z_j)\} = g(z_i) \quad (r = 0,1,\ldots,n) ,$$

which arise naturally on setting $\phi(x) = L(z_i,x)f(x)$ in the integral equation. (The general process may be justified by conditions on $L(x,y)$ and the assumption that $f(x)$ is at least continuous.) In the example considered here, we obtained the equations

$$\tilde{f}(z_i) - \lambda \sum_{j=0}^{n} v_j(z_i)|z_i-z_j|^{\frac{1}{2}} \ln|z_i-z_j|\tilde{f}(z_j) = g(z_i) \quad (i = 0,1,\ldots,n) .$$

For the choice of $g(x)$ which corresponds to $\lambda = 1$, $f(x) = x$ (see Example 5.9) we computed approximations with the errors, $\tilde{f}(x) - f(x)$ shown in Table 5.3.

TABLE 5.3.

x \ n	10	20	30	40
0·0	$-3{\cdot}6 \times 10^{-2}$	$-1{\cdot}6 \times 10^{-2}$	$-1{\cdot}1 \times 10^{-2}$	$-8{\cdot}2 \times 10^{-3}$
0·2	$1{\cdot}6 \times 10^{-2}$	$9{\cdot}0 \times 10^{-3}$	$6{\cdot}4 \times 10^{-3}$	$5{\cdot}0 \times 10^{-3}$
0·4	$5{\cdot}4 \times 10^{-2}$	$2{\cdot}9 \times 10^{-2}$	$2{\cdot}0 \times 10^{-2}$	$1{\cdot}6 \times 10^{-2}$
0·6	$9{\cdot}5 \times 10^{-2}$	$5{\cdot}0 \times 10^{-2}$	$3{\cdot}5 \times 10^{-2}$	$2{\cdot}7 \times 10^{-2}$
0·8	$1{\cdot}6 \times 10^{-1}$	$8{\cdot}2 \times 10^{-2}$	$5{\cdot}6 \times 10^{-2}$	$4{\cdot}3 \times 10^{-2}$
1·0	$6{\cdot}3 \times 10^{-2}$	$3{\cdot}3 \times 10^{-2}$	$2{\cdot}3 \times 10^{-2}$	$1{\cdot}7 \times 10^{-2}$
maximum error	$1{\cdot}6 \times 10^{-1}$	$1{\cdot}3 \times 10^{-1}$	$9{\cdot}2 \times 10^{-2}$	$7{\cdot}2 \times 10^{-2}$

From these figures, we suppose that the error is $o(1)$ as $h \to 0$, and the accuracy obtained compares unfavourably with that obtained in Example 5.9 for comparable work. (However, weights based on $M(x,y) = \ln|x-y|$ could give the exact answer.) *

The implementation of our basic method requires the 'splitting' of a given kernel $K(x,y)$ into the product $L(x,y)M(x,y)$ and the determination of a product-integration rule. The following examples illustrate how the split might be achieved, and serve as an introduction to a modified method.

Example 5.14.

For the kernel $\ln|x-y|$ we may take

$$M(x,y) = |x-y|^{-\alpha}, \; L(x,y) = |x-y|^{\alpha}\ln|x-y|$$

for any $\alpha \in (0,1)$. Also, we may take $\alpha = 0$ and interchange the roles of $L(x,y), M(x,y)$, so $M(x,y) = K(x,y)$ and the calculation of the weights $v_j(z_i)$ requires

$$\int_{z_k}^{z_{k+1}} (y-z_k)^r \ln|z_i-y|\,dy \;, \quad (r = 0,1,2,\ldots) \;,$$

depending on the degree of precision required. (Kershaw (1963) notes that such quantities can be calculated by using a recurrence relation but that this recurrence is *unstable*.) *

Example 5.15.

The kernel $K(x,y) = Y_0(|x-y|)$ (where $Y_0(x)$ is the Bessel function of the second kind and of zero order) may be written

$$L_0(|x-y|)\ln|x-y| + L_1(|x-y|) \quad ,$$

where $L_0(x)$ and $L_1(x)$ are analytic. Consequently,

$$L(x,y) = |x-y|^{\alpha} Y_0(|x-y|)$$

is continuous for $0 < \alpha < 1$ and we may set $M(x,y) = |x-y|^{-\alpha}$. *

5.6.1.

In Example 5.15 we have an instance where the kernel is weakly singular but can be written as the sum of a weakly singular kernel and a continuous kernel (that is, $L_1(|x-y|)$). For such an example, we are tempted to modify the method described to date, and we shall discuss a method for treating kernels of the form

$$K(x,y) = \sum_{k=0}^{N} L_k(x,y)M_k(x,y) ,$$

where $M_k(x,y)$ is (at worst) weakly singular $(k = 0,1,2,\ldots,N)$ and $L_k(x,y)$ is continuous (or at least bounded, but preferably differentiable).

Example 5.16.

Kershaw (1963) considers an integral equation of the form

$$f(x) - \lambda\int_0^L K(x,y)f(y)\,dy = g(x) \quad (0 \le x \le L) ,$$

where $K(x,y)$ is a function which is expressible in terms of complete elliptic integrals of the first and second kinds. The kernel $K(x,y)$ can be written in the form $K(x,y) = P(x,y)\ln|x-y| + Q(x,y)$, where $P(x,y)$ is continuous, and $Q(x,y)$ is continuous except when $x = y$, where finite-jump discontinuities can occur. The function $g(x)$ is such that the solution $f(x)$ is continuous.

Kershaw (attributing the idea to G.F. Miller) devises integration formulae of the type

$$\int_0^L \ln|z_i-y|\phi(y)\,dy \simeq \sum_{j=0}^{n} \nu_{1,j}(z_i)\phi(z_j)$$

$$(z_i=ih,\ h=L/n)$$

and

$$\int_0^L Q(z_i,y)\,\phi(y)\,dy \simeq \sum_{j=0}^{n} \nu_{2,j}(z_i)\phi(z_j) \qquad (i = 0, 1, \ldots,n),$$

each approximation being exact when $\phi(x)$ is a type of piecewise-polynomial. (Alternatively, such formulae can be constructed by the methods we have discussed above.) The integral equation is then replaced by the set of equations

$$\tilde{f}(z_i) - \lambda \sum_{j=0}^{n} \{\nu_{1,j}(z_i)P(z_i,z_j) + \nu_{2,j}(z_i)\}\tilde{f}(z_j) = g(z_i) .$$

Kershaw considers a number of cases and tabulates the approximate solutions. *

In general, suppose that $K(x,y)$ can be written in the form $\sum_{k=0}^{N} K_k(x,y)$, where $K_k(x,y) = L_k(x,y)M_k(x,y)$. We suppose that $L_k(x,y)$ is continuous and $M_k(x,y)$ is either continuous or, more generally, weakly singular. For a choice $z_0,z_1,\ldots,z_n$ in $[a,b]$ we construct the product integration rules

$$\int_a^b M_k(z_i,y)\phi(y)\mathrm{d}y \simeq \sum_{j=0}^{n} \nu_{k,j}(z_i)\phi(z_j) , \tag{5.28}$$

which are used to formulate the approximating equations

$$\tilde{f}(z_i) - \lambda \sum_{k=0}^{N} \sum_{j=0}^{n} \nu_{k,j}(z_i)L_k(z_i,z_j)\tilde{f}(z_j) = g(z_i) . \tag{5.29}$$

In particular we may take $z_i = a+ih$, $h = (b-a)/n$, and each of the product integration rules may be devized to be exact for functions $\phi(x)$ which are, say, (a) constant in $[z_i,z_{i+1})$ $(i = 0,1,2,\ldots,(n-1))$; or (b) continuous, and linear in $[z_i,z_{i+1}]$, etc.

Different approximate values will in general result from different choices of kernels $K_k(x,y)$, since the decomposition

$$K(x,y) = \sum_{k=0}^{N} K_k(x,y)$$

is clearly not uniquely defined. Similarly, each function $K_k(x,y)$ may be written in a number of ways as a product $L_k(x,y)M_k(x,y)$. Our aim will be to construct a 'splitting' of $K(x,y)$ in which each function $L_k(x,y)$ is particularly *smooth*.

Example 5.17.
We observe (see Atkinson (1966)) that the kernel

$$K(x,y) = \ln|\cos x - \cos y| \;, \quad 0 \le x,\ y \le \pi \;,$$

can be written as

$$K(x,y) = \sum_{k=0}^{4} M_k(x,y) L_k(x,y) \quad ,$$

where

$$L_0(x,y) = \ln[2 \sin\{\tfrac{1}{2}(x-y)\}/(x-y)] \;,$$

$$L_1(x,y) = \ln[\sin(\tfrac{1}{2}(x+y))/\{(x+y)(2\pi-x-y)\}] \;,$$

$$L_2(x,y) = L_3(x,y) = L_4(x,y) = 1 \quad ,$$

$$M_0(x,y) = M_1(x,y) = 1 \;,$$

$$M_2(x,y) = \ln|x-y| \;,$$

$$M_3(x,y) = \ln(2\pi-x-y) \;,$$

$$M_4(x,y) = \ln(x+y) \quad .$$

On the other hand, $K(x,y) = |x-y|^{-\frac{1}{2}} L(x,y)$, where $L(x,y) = |x-y|^{\frac{1}{2}} \times \ln|\cos x - \cos y|$ is continuous for $0 \le x,\ y \le \pi$. (If $\cos x = \cos y$ then $x=y+2k\pi$ $(k = 0,1,2,\ldots)$ and when $0 \le x,\ y \le \pi$ this implies that $x = y$.) Using the first split

$$\sum_{k=0}^{4} M_k(x,y) L_k(x,y) \quad ,$$

and generalizations of the trapezium rule for

$$\int_0^{\pi} M_k(x,y)\phi(y)dy \quad (k = 0,1,2,3,4) \quad ,$$

Atkinson has reported substantially better results than those obtained with the second split $|x-y|^{-\frac{1}{2}}L(x,y)$. (He considered the case where $g(x) = 1$, $\lambda = 1$, and $f(x) = (1+\pi \ln 2)^{-1}$.) *

In general we shall expect that the order of accuracy obtained with the product integration methods is the same as the order of the local truncation errors. If $z_0, z_1, \ldots, z_n$ are equally spaced,

$$f(x) - \lambda \sum_{k=0}^{N} \sum_{j=0}^{n} v_{k,j}(x)L_k(x,z_j)f(z_j) = g(x) + \tau(x) \quad ,$$

and

$$\tilde{f}(x) - \lambda \sum_{k=0}^{N} \sum_{j=0}^{n} v_{k,j}(x)L_k(x,z_j)\tilde{f}(z_j) = g(x) \quad ,$$

then we shall expect that

$$\|f(x) - \tilde{f}(x)\|_\infty = O(\|\tau(x)\|_\infty) \text{ , as } n \to \infty \quad .$$

We shall also expect that

$$\max|f(z_i) - \tilde{f}(z_i)| = O(\|\underset{\sim}{\tau}\|_\infty) \quad ,$$

where $\underset{\sim}{\tau}$ is the vector with components $\tau(z_i)$ $(i = 0,1,2,\ldots,n)$, as $n \to \infty$. It sometimes happens that $\|\underset{\sim}{\tau}\|_\infty$ is smaller than might be expected at first sight from the elementary analysis of Chapter 2 (section 2.17).

Example 5.18.

Consider the use of a split of the form

$$K(x,y) = \sum_{k=0}^{N} L_k(x,y)M_k(x,y) \quad .$$

From our analysis in Chapter 2 of the error in approximate product integration, we might expect $||\tau(x)||_\infty$, and hence the error obtainable, to be $O(h^3)$ at best, if we use the generalized parabolic rule with step h . We would expect lower-order accuracy if $f(x)$ is badly behaved or if one of the functions $L_k(x,y)$ has a badly behaved derivative, say. The results given by Atkinson (1966), using the first split of the kernel $\ln|\cos x - \cos y|$ reported in Example 5.17, correspond to roughly $O(h^4)$ accuracy (which is the order to be expected using the ordinary repeated Simpson's rule on a smooth integrand). Atkinson explains this result somewhat briefly (in part in terms of the simple nature of the solution).

de Hoog and Weiss (1973*b*) have given an anlysis of the error in approximate product integration, and they discuss the application of this analysis to the approximate solution of integral equations. They mention in particular the split used by Atkinson, which we have just referred to, and show that their analysis establishes that $||\underset{\sim}{\tau}||_\infty$, the maximum of the local truncation errors at quadrature points, is $O(h^4 \ln h)$. This leads to the expectation of $O(h^4 \ln h)$ convergence.

Consider, also, the solution, using the basic technique of section 5.6, of an equation of the form

$$f(x) - \lambda\int_a^b L(x,y)|x-y|^{-\alpha}f(y)\,dy = g(x) ,$$

where $f(x)$ and $L(x,y)$ have continuous fourth derivatives. Using the generalized parabolic rule as the product integration formula for

$$\int_a^b |x-y|^{-\alpha}\phi(y)\,dy ,$$

we would expect $O(h^{4-\alpha})$ convergence, according to the general results of de Hoog and Weiss. *

5.6.2.

In practice, determining product-integration rules required in sections 5.6 and 5.6.1 may cause some difficulty. Atkinson (1972*a*) proposed a hybrid method (linked to the methods of sections 5.4-5.6) to alleviate this difficulty. The proposal of Atkinson (like that of section 5.4) is prompted by the observation that with a kernel $K(x,y) = H(x,y)/|x-y|^{\alpha}$, the integrand causes *severe* difficulty only when y is close to x. If a region $|y-x| \leq \delta$ $(\delta>0)$ is removed from the range of integration, then a conventional quadrature rule can be used to approximate the integral over $[a,y-\delta]$, and $[y+\delta,b]$. (Clearly the quality of approximation depends on the size of δ.) For the integral over $[y-\delta,y+\delta]$ we revert to the use of product-integration formulae, in which account is taken of the nature of the singularity.

Suppose, by way of example, that $z_0 = a$, $z_n = b$, and $z_r = z_{r-1} + h$, where $h = (b-a)/n$. We may set $\delta = h$ and use the repeated trapezium rule or its generalization. Thus we may write

$$\tilde{f}(z_i) - \frac{\lambda h}{2} \sum_{\substack{j=0\\ j\neq i\\ j\neq i-1}}^{n-1} \{K(z_i,z_j)\tilde{f}(z_j) + K(z_i,z_{j+1})\tilde{f}(z_{j+1})\} -$$

$$- \lambda(1-\delta_{i,0})\int_{z_{i-1}}^{z_i} \frac{1}{|z_i-y|^{\alpha}} \{\frac{y-z_{i-1}}{h} H(z_i,z_i)\tilde{f}(z_i) +$$

$$+ \frac{z_i-y}{h} \cdot H(z_i,z_{i-1})\tilde{f}(z_{i-1})\}dy -$$

$$- \lambda(1-\delta_{i,n})\int_{z_i}^{z_{i+1}} \frac{1}{|z_i-y|^{\alpha}} \{\frac{y-z_i}{h} H(z_i,z_{i+1})\tilde{f}(z_{i+1}) +$$

$$+ \frac{z_{i+1}-y}{h} H(z_i,z_i)\tilde{f}(z_i)\}dy$$

$$= g(z_i) \quad (i = 0,1,2,\ldots,n)$$

where δ_{ij} is the Kronecker delta, and $K(x,y) = H(x,y)/|x-y|^{\alpha}$. There are extensions of this example which will suggest themselves to the reader. Observe that the method of Example 5.12 is related to the technique described here.

5.7. *Expansion methods for weakly singular kernels*

We recall (see Chapter 4, section 4.11) that in a collocation method we obtain a function

$$\tilde{f}(x) = \sum_{i=0}^{n} \tilde{a}_i \phi_i(x)$$

such that the function

$$r(x) = \tilde{f}(x) - \lambda \int_a^b K(x,y)\tilde{f}(y)\mathrm{d}y - g(x) \quad (a \le x \le b)$$

vanishes at $n+1$ distinct collocation points $\xi_0, \xi_1, \ldots, \xi_n$ (say) in $[a,b]$. This we may do by solving the equations

$$\sum_{j=0}^{n} \tilde{a}_j \phi_j(\xi_i) - \lambda \sum_{j=0}^{n} \tilde{a}_j \psi_j(\xi_i) = g(\xi_i) \quad (i = 0,1,2,\ldots,n) \quad ,$$

where

$$\psi_i(x) = \int_a^b K(x,y)\phi_i(y)\mathrm{d}y \quad (i = 0,1,2,\ldots,n) \quad .$$

The simple product-integration method defined by eqn (5.21) is equivalent to such a collocation method if we employ a rule

$$\int_a^b K(x,y)\phi(y)\mathrm{d}y \simeq \sum_{j=0}^{n} v_j(x)\phi(z_j)$$

such as the generalized mid-point rule, generalized trapezium rule, or generalized parabolic rule, or indeed any rule which is chosen to be exact for functions of the form

$$\sum_{i=0}^{n} \tilde{a}_i \phi_i(x)$$

with some choice $\{\phi_i(x)\}$. The collocation points are then $z_0, z_1, \ldots, z_n$. We shall suppose that the rule is of the type described.

Whilst the methods of section 5.6 extend the technique associated with (5.21) in one direction, viewing the method (under the conditions stated) as a collocation method provides a different extension. For we see immediately that we have a choice of collocation points, not necessarily $z_0, z_1, \ldots, z_n$.

Let us suppose indeed that we have product-integration formulae of the type described, and let us suppose that the associated functions $\phi_0(x), \phi_1(x), \ldots, \phi_n(x)$ form a Haar system (so that $\det[\phi_i(x_j)]$ is non-zero for any distinct points $x_0, x_1, \ldots, x_n$ in $[a,b]$). Then we write $\tilde{f}(x) = \sum_{i=0}^{n} \tilde{a}_i \phi_i(x)$, and the collocation equations can be written

$$\tilde{f}(\xi_i) - \lambda \sum_{j=0}^{n} v_j(\xi_i)\tilde{f}(z_j) = g(\xi_i) \quad (i = 0,1,2,\ldots,n) \quad .$$

Suppose that $\underset{\sim}{A}^0$ is the matrix with elements $\{\phi_j(z_i)\}$, $\underset{\sim}{A}'$ is the matrix with elements $\{\phi_j(\xi_i)\}$, and $\underset{\sim}{V}$ is the matrix with elements $\{v_j(\xi_i)\}$, and suppose $\tilde{\underset{\sim}{f}}^0$ is the vector with components $\tilde{f}(z_i), \tilde{\underset{\sim}{f}}^1$ is the vector with components $\tilde{f}(\xi_i)$, and $\tilde{\underset{\sim}{a}} = [\tilde{a}_0, \tilde{a}_1, \ldots, \tilde{a}_n]^T$. We can write down equivalent equations for $\tilde{\underset{\sim}{a}}, \tilde{\underset{\sim}{f}}^0$, and $\tilde{\underset{\sim}{f}}^1$, from the equations above. For we have

$$\tilde{\underset{\sim}{f}}^1 - \lambda \underset{\sim}{V}\tilde{\underset{\sim}{f}}^0 = \underset{\sim}{g} ,$$

$$\underset{\sim}{A}^0 \tilde{\underset{\sim}{a}} = \tilde{\underset{\sim}{f}}^0 ,$$

and

$$\underset{\sim}{A}' \tilde{\underset{\sim}{a}} = \tilde{\underset{\sim}{f}}^1 ,$$

where $\underset{\sim}{g} = [g(\xi_0), g(\xi_1), \ldots, g(\xi_n)]^T$ and we can write, for example, $\tilde{\underset{\sim}{f}}^0 = \underset{\sim}{A}^0 \underset{\sim}{A}'^{-1} \tilde{\underset{\sim}{f}}^1$ and $\tilde{\underset{\sim}{f}}^1 - \lambda \underset{\sim}{V} \underset{\sim}{A}^0 \underset{\sim}{A}'^{-1} \tilde{\underset{\sim}{f}}^1 = \underset{\sim}{g}$.

In particular cases this technique is quite simple. Thus the use of the generalized trapezium rule is equivalent to finding a piecewise-linear function $\tilde{f}(x)$ and this piecewise-linearity can be used to write down the components of $\tilde{\underset{\sim}{f}}^0$ in terms of those of $\tilde{\underset{\sim}{f}}^1$ without forming any matrices $\tilde{A}^0\tilde{A}^{-1}$. Such a technique has been employed by Atkinson for intrinsically singular problems (see section 5.10).

Collocation methods are generally applicable in the form described in section 4.11. The presence of weak singularities in $K(x,y)$ has little effect on the practical implementation of the method, except possibly to cause difficulties in the computation of the functions

$$\psi_i(x) = \int_a^b K(x,y)\phi_i(y)\,dy ,$$

provided that $\phi_0(x),\phi_1(x),\ldots,\phi_n(x)$ are continuous. (Observe that Phillips (1969, 1972) uses Gaussian quadrature with weight $x^{-\frac{1}{2}}$ to compute $\psi_i(x)$ in the case $K(x,y) = |x-y|^{-\frac{1}{2}}$.) The Galerkin methods can generally be used if the functions $\phi_i(x)$ $(i = 0,1,2,\ldots,n)$ are square-integrable, but it is more convenient if they are continuous. In practice, the choice of $\{\phi_i(x)\}$ used for a continuous kernel will usually suffice, say a basis for spline (or general piecewise-polynomial) functions, or a set of classical polynomials such as the Legendre polynomials, or the Chebyshev polynomials. In the latter cases, the functions $\phi_i(x)$ satisfy an orthogonality relation which can sometimes be used to advantage in the calculation of the functions $\psi_i(x)$.

My purpose in the remainder of this section is to provide one or two examples of the expansion methods for equations with weakly singular kernels.

<u>Example 5.19</u>.

Suppose that $a = 0$, $b = \pi$, with $K(x,y) = \ln|\cos x - \cos y|$ and suppose (see Example 5.22) that λ is a regular value of the kernel. We shall assume that $g(x) \in C[0,\pi]$. If we take $\phi_r(x) = \cos rx$, $\phi_r(x)$ is an eigenfunction of $K(x,y)$, and the Rayleigh-Ritz (classical Galerkin) technique produces an approximate solution

$$\tilde{f}(x) = \sum_{r=0}^{n} \tilde{a}_r \cos rx .$$

(The coefficients $\tilde{a}_r$ are given directly in terms of

$$\int_0^\pi g(y)\cos ry\,dy$$

and the rth eigenvalue.) As $n \to \infty$, $||f(x) - \tilde{f}(x)||_2 \to 0$ as may easily be verified. However,

$$\tilde{f}(0) = \sum_{r=0}^{n} \tilde{a}_r = -\tilde{f}(\pi) \quad ,$$

whereas it is only for certain inhomogeneous terms $g(x)$ that $f(0) = -f(\pi)$. We deduce that the mean-square convergence is not necessarily uniform.

To restore uniform convergence we construct the new approximation

$$\hat{f}(x) = g(x) + \lambda\int_0^\pi \ln|\cos x - \cos y|\tilde{f}(y)dy \quad ,$$

which in this case is $g(x) + \lambda \sum_{r=0}^{n} \tilde{a}_r\kappa_r \cos rx$. *

In general, the Galerkin methods supply approximations of the form

$$\tilde{f}(x) = \sum_{r=0}^{n} \tilde{a}_r\phi_r(x) \quad ,$$

which converge in mean to $f(x)$ if $\phi_0(x), \phi_1(x), \ldots$ are complete in $L^2[a,b]$. If we form

$$\hat{f}(x) = g(x) + \lambda\int_a^b K(x,y)f(y)dy$$

then

$$|f(x) - \hat{f}(x)| = |\lambda\int_a^b K(x,y)\{f(y)-\tilde{f}(y)\}dy| \quad ,$$

and if $K(x,y) = H(x,y)|x-y|^{-\alpha}$, where $\alpha \in [0,\frac{1}{2})$, we have

$$||f(x) - \hat{f}(x)||_\infty \leq |\lambda| \sup_{a\leq x\leq b} \sqrt{\{\int_a^b |K(x,y)|^2 dy\}} ||f(x) - \tilde{f}(x)||_2 ,$$

which establishes uniform convergence of $\hat{f}(x)$ to $f(x)$, given mean-square convergence. This analysis does not necessarily apply if $\alpha \geq \frac{1}{2}$, since $\int_a^b |K(x,y)|^2 dy$ is then unbounded, in general.

Example 5.20.

Let $a = -1$, $b = 1$, $K(x,y) = \ln|x-y|$ in the equation of second kind. We may choose $\phi_r(x) = P_r(x)$, the Legendre polynomial of degree r. Now

$$(r+1)P_{r+1}(x) = (2r+1)x\, P_r(x) - r\, P_{r-1}(x), \quad (r = 1,2,3,\ldots)$$

and $P_r(1) = 1$, $P_r(-1) = (-1)^r$, so that (Garwick 1952) when

$$\psi_r(x) = \int_{-1}^{1} \ln|x-y| P_r(y)\,dy$$

we find that

$$(r+2)\psi_{r+1}(x) = -(r-1)\psi_{r-1}(x) + (2r+1)x\psi_r(x) \quad (r \geq 2) ,$$

whilst

$$\psi_0(x) = (1-x)\ln|1-x| + (1+x)\ln(1+x) - 2 ,$$

$$\psi_1(x) = \tfrac{1}{2}(1-x^2)\ln\left|\frac{1-x}{1+x}\right| - x .$$

Similar results are available for the kernel $K(x,y) = |x-y|^{-\alpha}$. Thus (see Garwick (1952) for details) if $a = -1$, $b = 1$, and $\phi_r(x) = P_r(x)$, we find for the functions $\psi_r(x)$ corresponding to this kernel, the relation

$$(r-\alpha)\psi_{r+1}(x) + (r-1+\alpha)\psi_{r-1}(x) = (2r+1)x\psi_r(x) \quad (r \geq 2) ,$$

and we find

$$\psi_0(x) = \{\text{sign}(1-x)|1-x|^{1-\alpha}+\text{sign}(1+x)|1+x|^{1-\alpha}\}/(1-\alpha) \quad ,$$

$$\psi_1(x) = x\psi_0(x) + \{|1-x|^{2-\alpha}-|1+x|^{2-\alpha}\}/(2-\alpha) \quad . \quad *$$

The Chebyshev polynomials (shifted for the interval $[a,b]$ if necessary) are possibly to be preferred, in comparison with the Legendre polynomials. When $\phi_r(x) = T_r(x)$ $(-1 \le x \le 1)$ and

$$\psi_r(x) = \int_{-1}^{1} K(x,y)\phi_r(y)dy$$

we may use the recurrence relation

$$T_{r+1}(x) = 2xT_r(x) - T_{r-1}(x)$$

to generate the functions $\psi_r(x)$. We write

$$I_{r,s}(x) = \int_{-1}^{1} K(x,y)y^s T_r(y)dy$$

and obtain

$$I_{r+1,s}(x) = 2I_{r,s+1}(x) - I_{r-1,s}(x) \; , \quad (r \ge 2, \; s \ge 0) \quad ,$$

where $\psi_r(x) \equiv I_{r,0}(x)$. If we evaluate the moments $I_{0,s}(x)$ directly, the recurrence relation can be used (Phillips 1972) to generate the values of $I_{r,0}(x)$. (The process is *theoretically* equivalent to expressing $T_r(x)$ explicitly as a polynomial

$$\sum_{s=0}^{r} c_{r,s}\, x^s \; ;$$

obviously

$$\psi_r(x) = \sum_{s=0}^{r} c_{r,s} I_{0,s}(x).)$$

Example 5.21.

Phillips (1972) considers the equation

$$f(x) - \frac{2}{3}\int_{-1}^{1} |x-y|^{-\frac{1}{2}} f(y)\,dy = g(x) \quad ,$$

where $g(x)$ is chosen so that $f(x) = (1-x^2)^{\frac{3}{4}}$, and uses $\phi_r(x) = T_r(x)$ $(r = 0,1,2,\ldots,n)$ in a collocation method using the zeros of $T_{n+1}(x)$ as collocation points. With $n = 31$, an accuracy $\|f(x) - \tilde{f}(x)\|_\infty \le 3{\cdot}8 \times 10^{-3}$ was estimated. From the approximation

$$\tilde{f}(x) = \sum_{r=0}^{n} \tilde{a}_r T_r(x)$$

we may obtain the new approximation

$$\{g(x) + \frac{2}{3}\int_{-1}^{1} |x-y|^{-\frac{1}{2}} \tilde{f}(y)\,dy\} \quad ,$$

which, for $n = 31$, has a maximum error of 1×10^{-4} . In this problem, the solution $f(x)$ has a derivative which is unbounded at $x = \pm 1$, this behaviour being inherited from $g(x)$. The initial approximation in terms of Chebyshev polynomials does not display such behaviour, but the improved approximation does so because of the contribution from $g(x)$. Because of the bad behaviour of $f(x)$, a spline approximation might be thought more suitable than a Chebyshev sum, and according to the work of Phillips (1969, 1972) this is indeed the case. *

For most choices of functions $\phi_i(x)$ the success obtained with the collocation method or classical Galerkin method appears to depend on the smoothness of $f(x)$ and of

$$\int_a^b K(x,y) f(y)\,dy \quad .$$

If $f(x)$ is known to be badly behaved, re-formulation of the problem may be attempted before applying the method (see sections 4.2 and 5.3).

5.8. *The eigenvalue problem for a weakly singular kernel*

Numerical methods for the approximate solution of the eigenvalue problem

$$\int_a^b K(x,y)f(y)\,dy = \kappa f(x) \quad (a \le x \le b) \tag{5.30}$$

(or, $\lambda\int_a^b K(x,y)f(y)\,dy = f(x)$) , where $K(x,y)$ is weakly singular, are usually direct analogues of the methods which we have discussed above for the inhomogeneous problem.

For completeness, I shall give a brief review of methods for treating (5.30). The material has some bearing on our discussion in section 5.9.

From the modified quadrature method applied to (5.30) we obtain the eigenvalue problem

$$\tilde{\kappa}\tilde{f}(y_i) = \sum_{\substack{j\neq i\\ j=0}}^{n} w_j K(y_i,y_j)\tilde{f}(y_j) - \Delta^*(y_i)\tilde{f}(y_i) \quad (i = 0,1,2,\ldots,n), \tag{5.31}$$

using the notation employed in eqn (5.15). Similarly, the product integration methods of sections 5.5, 5.6, 5.6.1, and 5.6.2 have their analogues here. Corresponding to eqn (5.21) we obtain

$$\sum_{j=0}^{n} v_j(z_i)\tilde{f}(z_j) = \tilde{\kappa}\tilde{f}(z_i) \quad (i = 0,1,2,\ldots,n) \quad ,$$

where $z_0,z_1,\ldots,z_n$ are distinct points in $[a,b]$, and

$$\int_a^b K(x,y)\phi(y)\,dy \simeq \sum_{j=0}^{n} v_j(x)\phi(z_j)$$

is a product-integration rule. The analogue of eqn (5.29) is, similarly, the eigenvalue problem

$$\sum_{k=0}^{N}\sum_{j=0}^{n} v_{k,j}(z_i)L_k(z_i,z_j)\tilde{f}(z_j) = \tilde{\kappa}\tilde{f}(z_i) \quad (i = 0,1,2,\ldots,n) \; , \tag{5.32}$$

where we employ product-integration rules of the form (5.28) and the 'splitting'

$$K(x,y) = \sum_{k=0}^{N} L_k(x,y)M_k(x,y)$$

Expansion methods such as the Galerkin and collocation methods will be discussed below, but we now pause, briefly, to examine the methods mentioned above.

In the case where $\tilde{\kappa}$ is an approximation to a *simple* non-zero eigenvalue κ, corresponding to a normalized eigenfunction $f(x)$ in (5.30), we shall probably expect $\kappa-\tilde{\kappa}$ to be of the same order of magnitude as the local truncation error corresponding to our scheme. In particular, using (5.32), we have a function $\tau(x)$ associated with the particular eigenpair, such that

$$\sum_{k=0}^{N} \sum_{j=0}^{n} \nu_{k,j}(x)L_k(x,z_j)f(z_j) = \kappa f(x) + \tau(x) \quad .$$

We might expect† that $\kappa-\tilde{\kappa} = O(||\tau(x)||_2)$ unless $\kappa = 0$ or κ is a multiple eigenvalue, as in the case of a continuous kernel and the quadrature method. In order to judge the size of $\tau(x)$ we require knowledge of the differentiability properties of each function $L_k(x,y)$ and of the eigenfunction $f(x)$. (Observe that the heuristic justification of the modified quadrature method also requires some smoothness of the eigenfunction.)

At this point we note a distinction between the eigenvalue problem (5.30) and the corresponding inhomogeneous problem

$$f(x) - \lambda\int_a^b K(x,y)f(y)\,dy = g(x) \quad .$$

The differentiability properties of the solution of the inhomogeneous problem depend in part on the function $g(x)$; it is possible for the solution of this problem to be arbitrarily differentiable for some function $g(x)$, or for it to be discontinuous for some other function

---

† We have not seen this conjecture established, and the result which we shall give later is somewhat different.

$g(x)$. On the other hand, the differentiability properties of the eigenfunctions in (5.30) (and hence the local truncation errors) are determined completely by the kernel $K(x,y)$. Moreover, if $K(x,y)$ is weakly singular then we would not, at least at first without some thought, expect high-order differentiability of the eigenfunctions.

A complete investigation of the smoothness properties of eigenfunctions of weakly singular kernels does not seem to exist, but the situation is not as bad as might at first appear. For, if (5.30) is satisfied by an eigenfunction $f(x)$ then it follows that

$$\int_a^b K^p(x,y)f(y)\,dy = \kappa^p f(x) \quad (a \le x \le b) \quad ,$$

for $p = 1,2,3,\ldots,$ where $K^p(x,y)$ is the pth iterated kernel. Now, $K^p(x,y)$ is smoother than $K(x,y)$ for $p > 1$, and $f(x)$ is an eigenfunction of this smoother kernel. Indeed, if $\kappa \neq 0$ and

$$K(x,y) = H(x,y)/|x-y|^\alpha \quad ,$$

where $H(x,y)$ is continuous for $a \le x,\ y \le b$, then $K^p(x,y)$ is continuous for sufficiently large p and it follows immediately that the eigenfunction $f(x)$ in (5.30) is (at least) continuous, since it is an eigenfunction of $K^p(x,y)$.

<u>Remark</u>. Given the continuity of $f(x)$ we can, indeed, write

$$|\kappa|\ |f(x_1) - f(x_2)| \le \|f(x)\|_\infty \int_a^b |K(x_1,y) - K(x_2,y)|\,dy$$

and

$$\int_a^b |K(x_1,y) - K(x_2,y)|\,dy \le \|H(x,y)\|_\infty \int_a^b \left||x_1-y|^{-\alpha} - |x_2-y|^{-\alpha}\right| dy \ +$$

$$+ \sup_{a\le y\le b} |H(x_1,y) - H(x_2,y)| \left\| \int_a^b |x-y|^{-\alpha}\,dy \right\|_\infty \quad ,$$

and we can deduce that if $H(x,y)$ satisfies, uniformly in y, a Lipschitz condition in x, with exponent β, then $f(x)$ satisfies a Lipschitz condition with exponent $\gamma = \min(1-\alpha,\beta)$ if $\kappa \neq 0$. The same argument can be applied with $K^p(x,y)$ in place of $K(x,y)$.

Example 5.22.

Collatz (1966*a*, p.511) gives the eigenfunctions for the kernel

$$K(x,y) = \ln|\cos x - \cos y| \quad (0 \leq x,\ y \leq \pi)$$

in the form $f_0(x) = \pi^{-\frac{1}{2}}$, $f_r(x) = \sqrt{(2/\pi)}\cos rx$ $(r = 1,2,\ldots,)$ with $\kappa_0 = -\pi \ln 2$, $\kappa_r = -\pi/r$ $(r = 1,2,3,\ldots)$. Observe that the eigenfunctions are infinitely differentiable. *

The accuracy of approximations to the eigenvalues $\{\kappa\}$ is governed partly by the smoothness of the eigenfunction but also by the multiplicity of the eigenvalue. Thus $|\kappa-\tilde{\kappa}| = O(\|\tau(x)\|_2)$ is suggested above, if the eigenvalue is *simple*, or if the kernel is *Hermitian*. (In this context, we note that Kershaw (private communication) has found certain eigenvalues of the kernel $\ln|y+1| + \ln|x-y|$ on $[-1,1]$ by considering instead the kernel $\ln|x-y|$, which is Hermitian.)

The integral operator K defined by a weakly singular kernel $K(x,y)$ is compact on $L^2[a,b]$, and there is no theoretical problem about applying the Rayleigh-Ritz method (in the case of a Hermitian kernel), or the classical Galerkin technique, to obtain an approximate eigenfunction

$$\tilde{f}(x) = \sum_{i=0}^{n} \tilde{a}_i\phi_i(x) .$$

In theory we may take $\phi_0(x),\phi_1(x),\ldots$ from any set of linearly independent square-integrable functions, and with

$$\psi_j(x) = \int_a^b K(x,y)\phi_j(y)\,dy$$

solve the equations

$$\sum_{j=0}^{n} \tilde{a}_j(\psi_j,\phi_i) = \tilde{\kappa} \sum_{j=0}^{n} \tilde{a}_j(\phi_j,\phi_i) \quad (i = 0,1,2,\ldots,n) \quad , \tag{5.33}$$

but we shall encounter computational simplicity if the functions $\phi_i(x)$ are continuous. In the case of the collocation equations

$$\sum_{j=0}^{n} \tilde{a}_j\psi_j(z_i) = \tilde{\kappa} \sum_{j=0}^{n} \tilde{a}_j\phi_j(z_i) \quad , \tag{5.34}$$

the functions $\psi_j(x)$ are continuous (and hence $\psi_j(z_i)$ is bounded, as required in (5.34)) if the functions $\phi_j(x)$ are continuous.

The Galerkin and collocation techniques are projection methods and existing theories appear to give error bounds in terms of the smoothness of the eigenfunctions.

Observe that in the case of a Hermitian weakly singular kernel $K(x,y)$ the Rayleigh-Ritz technique gives one-sided bounds for the eigenvalues just as in the continuous case. This follows immediately from the standard literature in the case $K(x,y) = H(x,y)/|x-y|^{\alpha}$, where $\alpha \in [0,\frac{1}{2})$, since such kernels are square-integrable. More generally, we can use a functional analytic approach (Collatz 1966*b*). For all $\alpha \in [0,1)$, the integral operator corresponding to a kernel $K(x,y) = H(x,y)/|x-y|^{\alpha}$ is compact on the space $L^2[a,b]$ provided that $H(x,y)$ is square-integrable. If, further, $K(x,y) = K^*(x,y)$ (almost everywhere) then Courant's maximum-minimum principle characterizes the eigenvalues and the required result follows (Collatz 1966*b*, p.161). In particular, the properties of the Rayleigh quotient for a Hermitian kernel (see Chapter 3) apply in the general case described.

Example 5.23.
Consider the eigenvalue problem

$$\kappa f(x) = \int_{-1}^{1} \ln|x-y| f(y)\,dy \ .$$

(This eigenvalue problem arises in a problem in water-waves in a dock.) We report here on results obtained using the generalized trapezium rule

$$\int_{-1}^{1} \ln|x-y|\phi(y)dy \simeq \sum_{j=0}^{n-1} \{\alpha_j(x)\phi(z_j) + \beta_j(x)\phi(z_{j+1})\}$$

$$= \sum_{j=0}^{n} \nu_j(x)\phi(z_j) \quad , \quad \text{say,}$$

where $z_j = -1 + jh,\ h = 2/n$. We find that $\nu_0(x) = \alpha_0(x)$, $\nu_n(x) = \beta_{n-1}(x)$, and, for $j = 1,2,3,\ldots,(n-1)$, $\nu_j(x) = \alpha_j(x) + \beta_{j-1}(x)$, where

$$\alpha_j(z_i) = \tfrac{1}{2}h \ln h + h\int_0^1 (1-t)\ln|i-j-t|dt \quad ,$$

$$\beta_j(z_i) = \tfrac{1}{2}h \ln h + h\int_0^1 t \ln|i-j-t|dt \quad ,$$

for $j = 0,1,2,\ldots,(n-1)$. We can write $\alpha_j(z_i) = \frac{1}{2}h \ln h + h\psi_0(i-j) + h\ \psi_1(i-j)$ and $\beta_j(z_i) = \frac{1}{2}h \ln h + h\psi_1(i-j)$, where

$$\psi_0(k) = k \ln|k| - (k-1)\ln|k-1| - 1$$

and

$$\psi_1(k) = \tfrac{1}{2}\{(k-1)^2\ln|k-1| - k^2\ln|k|\} + \tfrac{1}{4}\{k^2-(k-1)^2\} + k\psi_0(k) \ .$$

(Thus, $k\psi_1(k) = \psi_0(1-k) - \psi_1(1-k)$.)

The eigenvalues κ are approximated by the eigenvalues of the matrix $[\nu_j(z_i)]$. Whilst this matrix has n^2 elements, a large number of these are repetitions and the matrix can be constructed economically using at most the values $\psi_0(k)$ and $\psi_1(k)$ for $|k| \le N$, say, where k is integer. Thus the matrix of order 4 has entries

$-6\cdot3516 \times 10^{-1}$	$-3\cdot4611 \times 10^{-1}$	$1\cdot7713 \times 10^{-1}$	$1\cdot9043 \times 10^{-3}$
$-3\cdot0182 \times 10^{-1}$	$-1\cdot2703 \times 10^{0}$	$-3\cdot4611 \times 10^{-1}$	$3\cdot1512 \times 10^{-2}$
$3\cdot1512 \times 10^{-2}$	$-3\cdot4611 \times 10^{-1}$	$-1\cdot2703 \times 10^{0}$	$-3\cdot0182 \times 10^{-1}$
$1\cdot9043 \times 10^{-1}$	$1\cdot7713 \times 10^{-1}$	$-3\cdot4611 \times 10^{-1}$	$-6\cdot3516 \times 10^{-1}$

(and its two eigenvalues of largest modulus are $-1\cdot6542$ and $-1\cdot2954$).

Table 5.4 shows the five eigenvalues of largest modulus obtained with varying orders of the matrix.

TABLE 5.4

n 20	40	80
$-1\cdot7609$	$-1\cdot7634$	$-1\cdot7640$
$-1\cdot5567$	$-1\cdot5636$	$-1\cdot5654$
$-7\cdot7582 \times 10^{-1}$	$-7\cdot8518 \times 10^{-1}$	$-7\cdot8753 \times 10^{-1}$
$-5\cdot9638 \times 10^{-1}$	$-6\cdot0873 \times 10^{-1}$	$-6\cdot1187 \times 10^{-1}$
$-4\cdot3595 \times 10^{-1}$	$-4\cdot5091 \times 10^{-1}$	$-4\cdot5477 \times 10^{-1}$

The corresponding eigenfunctions are respectively even and odd alternatively, and the order of the errors in the eigenvalues appear to be $O(h^2)$ or $O(-h^2 \ln h)$ as $h \to 0$.

Approximate eigenvalues corresponding to odd eigenfunctions are tabulated by A. Davis (1970). These eigenvalues are obtained by an expansion method, approximating *odd* eigenfunctions in the form

$$\tilde{f}(x) = \sum_{r=1}^{n} \tilde{a}_r \phi_r(x) \quad ,$$

where $\phi_r(x) = (4r-1)^{\frac{1}{2}} P_{2r-1}(x)$, where $P_k(x)$ is the Legendre polynomial of degree k . The classical Galerkin method is then applied (using the fact that, here, $(\phi_r,\phi_s) = \frac{1}{2}\delta_{rs})$ and the required approximate eigenvalues $\tilde{\kappa}$ are obtained on solving

$$\sum_{p=1}^{n} \frac{(4p-1)^{\frac{1}{2}}(4s-1)^{\frac{1}{2}}\tilde{\alpha}_p}{(p+s-1)(p+s)\{1-4(p-s)^2\}} = \tilde{\kappa}\tilde{\alpha}_s \quad (s = 1,2,3,\ldots,n) .$$

(To establish this form for the equations required some manipulation.) We computed the following approximations for the first three of these eigenvalues, and found: $-1{\cdot}5660$, $-6{\cdot}1296 \times 10^{-1}$, $-3{\cdot}8034 \times 10^{-1}$. In computing these figures we found agreement to 7-figure accuracy of the results obtained with $n = 20, 40, 80$. (Davis tabulates the first 20 of these eigenvalues with $n = 70$.)

The Galerkin technique yields embarassingly high accuracy when compared with the use of the generalized trapezium rule, but it must be remembered that derivation of the form of the matrix elements $(K\phi_i,\phi_j)$ requires some manipulative effort, whereas the construction of the generalized trapezium rule is fairly automatic.

5.9. *Equations of the second kind which are not uniquely solvable*

Suppose that $K(x,y)$ is weakly singular (of the form $H(x,y)/|x-y|^{\alpha}$, $0 < \alpha < 1$, with $H(x,y)$ continuous), or continuous, and λ *is a characteristic value* of $K(x,y)$. The equation

$$f(x) - \lambda\int_a^b K(x,y)f(y)\,dy = g(x) , \tag{5.35}$$

where $K(x,y)$ is continuous or weakly singular and where $g(x)$ is continuous, is then solvable only for functions $g(x)$ which satisfy certain conditions. Any solution is not then unique.

To be precise, suppose the geometric multiplicity of λ is r . Then there are r linearly independent eigenfunctions $\phi_i(x)$ $(i = 0,1,2,\ldots,(r-1))$ such that

$$\lambda\int_a^b K(x,y)\phi_i(y)\,dy = \phi_i(x)$$

and r linearly independent functions $\psi_i(x)$ such that

$$\bar{\lambda}\int_a^b \overline{K(y,x)}\psi_i(y)\,dy = \psi_i(x) \quad .$$

A solution exists only if

$$\int_a^b g(x)\overline{\psi_i(x)}\,dx = 0 \quad (i = 0,1,2,\ldots,(r-1)) \ ,$$

and, if $f(x)$ is one solution, any choice of $\alpha_0,\alpha_1,\ldots,\alpha_{r-1}$ provides another solution

$$f(x) + \sum_{i=0}^{r-1} \alpha_i\phi_i(x) \ .$$

In brief, the Fredholm alternative is valid for weakly singular kernels even though they may not be square-integrable. The intimate connection with the eigenvalue problem (section 5.8) is apparent. We shall suppose, throughout, that $K(x,y)$ and $g(x)$ satisfy the conditions outlined above.

In the case where (5.35) has a solution we see that the solution is uniquely specified by r independent conditions, such as

$$\int_a^b f(x)\overline{\phi_i(x)}\,dx = 0 \quad (i = 0,1,2,\ldots,(r-1)) \ .$$

Example 5.24.
The following problem arises in conformal mapping (Todd and Warschawski 1955). Suppose the equation $z = \zeta(s)$ represents a closed Jordan curve C where the arc-length measured from a fixed point is s , $0 \le s \le L$. Suppose the transformation $w = f(z)$ maps the interior R of C conformally onto the open disc $|w| < 1$ and that $f(z_1) = 0, f(z_2) = 1$. We suppose $f(z)$ continuous on $C \cup R$. The function $\theta(s) = \arg f(\zeta(s))$ then satisfies *Gerschgorin's integral equation*

$$\theta(s) - \int_0^L K(s,t)\theta(t)\,dt = \beta(s) \quad ,$$

where

$$\beta(s) = 2 \arg\left[\{z_2 - \zeta(s)\}/\{z_1 - \zeta(s)\}\right]$$

and

$$K(s,t) = \frac{1}{\pi}\frac{\sin(\tau-\phi)}{|\zeta(t) - \zeta(s)|}, \quad \tau \equiv \tau(t), \quad \phi \equiv \phi(t,s) .$$

Here, $\tau = \tau(t)$ is the angle of the tangent to the curve at $\zeta(t)$ and $\phi = \arg\{\zeta(t) - \zeta(s)\}$.

It can be shown that

$$\int_0^L K(s,t)\,\mathrm{d}t = 1$$

and that the value 1 is a simple eigenvalue of this kernel $K(s,t)$ $(0 \le s, t \le L)$ and solutions exist only if a condition of the form

$$\int_0^L \beta(t)\mu(t)\,\mathrm{d}t = 0$$

is satisfied, where $\mu(t)$ is a certain function. The solution is then determined only to within an arbitrary additive constant (which can be fixed by specifying, say, $\theta(p_0)$ for some suitable value p_0) . *

Let us suppose that in the general situation described above, $r = 1$ so that λ is a simple characteristic value. The integral equation (5.35) then has a solution if $g(x)$ satisfies a single orthogonality condition. Suppose that we seek a numerical solution by a standard method described above. For convenience, and to fix our ideas, we shall suppose $K(x,y)$ is continuous, so that we can use the quadrature method. Thus we consider the equations

$$\tilde{f}(y_i) - \lambda \sum_{j=0}^{n} w_j K(y_i,y_j)\tilde{f}(y_j) = g(y_i) . \qquad (5.36)$$

The associated homogeneous equations in the eigenvalue problem are of the form

$$\tilde{f}_r(y_i) = \tilde{\lambda}_r \sum_{j=0}^{n} w_j K(y_i, y_j) \tilde{f}_r(y_j) \quad , \tag{5.37}$$

and (assuming, as we may, that the appropriate methods of Chapter 3 or section 5.8 are to be recommended) we would expect that there is an approximate characteristic value $\tilde{\lambda}_r$ close to λ , for some r and for large n . Consequently the matrix with (i,j)th element equal to

$$\delta_{i,j} - \lambda w_j K(y_i, y_j)$$

is either singular, or 'almost singular' and hence ill conditioned, at least for large n . In the former case there will be no solution unless $g(x)$ satisfies a discrete orothgonality relation

$$\sum_{j=0}^{n} w_j g(y_j) \overline{\tilde{\psi}(y_j)} = 0 \quad , \tag{5.38}$$

where $\tilde{\psi}(x)$ is a certain approximate eigenfunction of $K^*(x,y)$.

Now suppose that nevertheless we proceed to solve the system of equations (5.36) by Gaussian elimination *with row interchanges*, in the usual fashion (Fox 1964). We shall find, if the quadrature rule is sufficiently accurate, that the $(n+1)$ equations for $\tilde{f}(y_0), \tilde{f}(y_1), \ldots, \tilde{f}(y_n)$ are reduced to n equations in which the matrix is upper triangular, and a final equation for $\tilde{f}(y_n)$ in which the coefficient of $\tilde{f}(y_n)$ is zero or is small (moreover, this equation may be inconsistent, in the former case). If we assign the value γ to $\tilde{f}(y_n)$ and express the remaining function values in terms of γ we have a one-parameter family of approximate solutions, each of which can be extended to a function $\tilde{f}(x)$. Any normalizing condition used in the original problem can be employed to determine γ . An entirely analogous procedure can be employed if an expansion method is used, or if the multiplicity of λ is greater than 1. (If the multiplicity is r we would expect that some minor of order $(n-r+1)$ in the matrix determining

$\tilde{f}$ is singular or nearly so.) It should be remarked that determining the rank of the matrix of coefficients is a delicate matter, and careful monitoring of Gaussian elimination may be less reliable than the use of the singular-value decomposition (as in least-squares problems).

We have seen that a knowledge of the eigensystem of $K(x,y)$ and $K^*(x,y)$ provides some assistance in the solution of the equation when λ is a characteristic value. If the linearly independent set of functions $\psi_i(x)$ $(i = 0,1,2,\ldots,(r-1))$ are known, they can be employed in a numerical method to determine a solution. Such a method was proposed by Atkinson (1966), and we now outline the underlying ideas, using the preceding notation and assumptions.

Consider the equations

$$f(x) - \lambda\{\int_a^b K(x,y)f(y)dy + \sum_{i=0}^{r-1} \alpha_i\psi_i(x)\} = g(x) \quad , \tag{5.39}$$

where λ is a characteristic value of $K(x,y)$, $K^*\psi_i(x) = \bar{\lambda}\psi_i(x)$ $(i = 0,1,2,\ldots,(n-1))$, as above, where $\psi_0(x),\psi_1(x),\ldots,\psi_{r-1}(x)$ are linearly independent, and

$$\alpha_i = L_i(f) \quad (i = 0,1,2,\ldots,(r-1)) \tag{5.40}$$

where $L_i(f)$ is a bounded linear functional on $C[a,b]$.† In particular, we may set

$$L_i(f) = f(p_i) \tag{5.41}$$

for some distinct points $p_i \in [a,b]$, or (say)

$$L_i(f) = (f,\psi_i) \equiv \int_a^b f(x)\overline{\psi_i(x)}dx \quad . \tag{5.42}$$

---

† We are assuming that $g(x) \in C[a,b]$, but if $g(x) \in L^2[a,b]$ then $f(x) \in L^2[a,b]$ and the choice (5.42) gives a bounded linear functional on $L^2[a,b]$.

Eqns (5.39) and (5.40) have a *unique* continuous solution if $g(x) \in C[a,b]$, $(g,\psi_i) = 0$ $(i = 0,1,2,\ldots,(r-1))$ and $K(x,y)$ is continuous or weakly singular, provided that the matrix with (i,j)th element equal to $L_i(\phi_j)$ $(i,j = 0,1,2,\ldots,(r-1))$ is non-singular. Moreover, the solution of eqn (5.39) under such conditions is a solution of the equation

$$f(x) - \lambda\int_a^b K(x,y)f(y)\,dy = g(x) \quad . \tag{5.43}$$

Here, $g(x)$ is assumed to satisfy the requirement that

$$(g,\psi_i) \equiv \int_a^b g(x)\overline{\psi_i(x)}\,dx = 0 \quad (i = 0,1,2,\ldots,(r-1)) \ .$$

Remark. To prove the statements in the preceding paragraph, we write eqn (5.39) as

$$f(x) - \lambda\int_a^b K(x,y)f(y)\,dy = g(x) + \lambda\sum_{i=0}^{r-1} L_i(f)\psi_i(x) \quad ,$$

and we see from the Fredholm alternative that this equation has one or more solutions if and only if, for $k = 0,1,2,\ldots,(r-1)$,

$$\int_a^b \{g(x) + \lambda\sum_{i=0}^{r-1} L_i(f)\psi_i(x)\}\overline{\psi_k(x)}\,dx = 0 \quad , \tag{5.44}$$

In view of the condition on $g(x)$, (5.44) reduces to the requirement that

$$\int_a^b \sum_{i=0}^{r-1} L_i(f)\psi_i(x)\overline{\psi_k(x)}\,dx = 0 \ ,$$

that is, since $\det\,[(\psi_i,\psi_j)] \neq 0$,

$$L_k(f) = 0 \quad (k = 0,1,2,\ldots,(r-1)) \quad .$$

Now there is a family of solutions

$$f(x) = f_0(x) + \sum_{j=0}^{r-1} a_j\phi_j(x)$$

of eqn (5.43) (where $f_0(x)$ is a particular solution) and $L_k(f) = 0$ $(k = 0,1,2,\ldots,(r-1))$, if

$$\sum_{j=0}^{r-1} a_j L_k(\phi_j) = -L_k(f_0) \quad (k = 0,1,2,\ldots,(r-1)) \quad .$$

Since det $[L_i(\phi_j)] \neq 0$ we can find a set of values $a_0,a_1,\ldots,a_{r-1}$ to satisfy this condition and the corresponding function $f(x)$ is a solution of eqn (5.43) satisfying (5.40).

Moreover, the solution of eqns (5.39) and (5.40) is unique if there is *no* non-trivial solution of the homogeneous equation

$$\phi(x) - \lambda\int_a^b K(x,y)\phi(y)\mathrm{d}y - \lambda\sum_{i=0}^{r-1} L_i(\phi)\psi_i(x) = 0 \quad .$$

This is the case where $g(x) = 0$ above, so we see that for any solution,

$$L_k(\phi) = 0 \quad (k = 0,1,2,\ldots,(r-1)) \tag{5.45}$$

by the argument employed to establish that $L_k(f) = 0$, above. It follows that

$$\phi(x) - \lambda\int_a^b K(x,y)\phi(y)\mathrm{d}y = 0$$

or

$$\phi(x) = \sum_{i=0}^{r-1} \beta_i\phi_i(x) \; , \tag{5.46}$$

for some choice $\beta_0,\beta_1,\ldots,\beta_{r-1}$. Combining (5.45) and (5.46) we see that any solution of the homogeneous equation is such that

$$\sum_{j=0}^{r-1} \beta_j L_i(\phi_j) = 0$$

and since $\det [L_i(\phi_j)] \neq 0$ this implies $\beta_0 = \beta_1 = \ldots = \beta_{r-1} = 0$. Thus $\phi(x) = 0$ is the only solution.

We have a basis for a numerical method for finding a solution of (5.43) by considering the uniquely solvable equations (5.39) and (5.40), and applying a numerical method to the latter equations. Thus a product-integration method (section 5.5) leads us to an equation of the form

$$\tilde{f}(x) - \lambda \sum_{j=0}^{n} v_j(x)\tilde{f}(z_j) - \lambda \sum_{i=0}^{r-1} L_i(\tilde{f})\psi_i(x) = g(x) \quad ,$$

which yields equations for $\tilde{f}(z_0), \tilde{f}(z_1), \ldots, \tilde{f}(z_n)$ on setting $x = z_i$ $(i = 0,1,2,\ldots,n)$. To apply the method, however, it is essential to know the functions $\phi_i(x)$ and $\psi_i(x)$ and to verify that $\det [L_i(\phi_j)] \neq 0$. If we set $L_i(\phi_j) = \phi_j(p_i)$ for $p_0, p_1, \ldots, p_n \in [a,b]$, there is *some* choice of these distinct points p_i which ensures that $\det [L_i(\phi_j)] \neq 0$. In practice some experimentation may be needed to find a suitable selection of points.

For further details of Atkinson's method we refer to Atkinson (1966), who gives examples and discusses convergence properties.

5.10. *Singular integral equations*

We discuss here two types of linear integral equation which seem likely to cause difficulty. These are equations of the form

$$f(x) - \lambda \int_A^B K(x,y)f(y)\,dy = g(x) \quad , \tag{5.47}$$

where one or both of the limits A and B are infinite, and equations of the form

$$f(x) - \lambda \int_a^b K(x,y)f(y)\,dy = g(x) \quad , \tag{5.48}$$

where $K(x,y)$ is intrinsically singular and the integral exists only as a Cauchy principal value. (In some cases the integrals occurring are expressed as line integrals which can be re-expressed in the form

(5.47) or (5.48).)

For equations of the type (5.47) and (5.48), the integral operators involved are not necessarily compact, and the Fredholm alternative governing the existence of a unique solution may not apply. This seems to be the source of some of the problems arising in the solution of these equations, but in practical numerical terms we may also have difficulties in approximating integrals over infinite ranges of integration, or approximating Cauchy principal values.

There are not many English-language papers to be found dealing with eqns (5.47) or (5.48). A part of the Russian literature has, however, been translated into English, and Noble (1971*a*,*b*) lists a number of papers. I shall limit myself here to indicating some of the more accessible discussions to be found in the literature. It seems important to view critically the reliability of any method which has not been rigorously justified, and any theoretical insight into the behaviour of the solution is valuable. (For a recent reference see Ivanov, 1976.)

The impression which I intend to convey is that the treatment of singular integral equations is not completely understood at the present. Thus, in 1972, Stenger was able to write of the Wiener-Hopf integral equation,

$$f(x) - \int_0^\infty k(x-y)f(y)\,dy = g(x) \quad (x > 0) \quad , \tag{5.49}$$

'Currently there is considerable interest in Wiener-Hopf integral equations. However, there appear to exist very few *effective* methods for solving such equations ...'. (The italics are mine.)

We shall begin by considering equations of the form (5.47), and in particular integral equations (5.49) of Wiener-Hopf type, where the kernel is a difference kernel. The theoretical background for such equations is quite reasonably developed, and this equation and the equation

$$f(x) - \int_{-\infty}^\infty k(x-y)f(y)\,dy = g(x) \tag{5.50}$$

are mentioned briefly by Cochran (1972), who gives references.

The questions of solvability for (5.49), where $k(x)$ and $g(x)$ are sufficiently well behaved, are determined by the index

$$\nu = -\ \mathrm{ind}(1-K) = -\frac{1}{2\pi}\int_{-\infty}^{\infty} \arg(1-K(\lambda))d\lambda$$

given with

$$K(\lambda) = \int_{-\infty}^{\infty} \exp(i\lambda y)k(y)dy \ .$$

If $1-K(\lambda) \neq 0$ for $|\lambda| < \infty$, and $\nu = 0$, then (Krein 1962) eqn (5.49) has a unique (well behaved) solution. The situation can be compared with that in the theory of intrinsically singular equations (see Chapter 1).

Stenger (1972, 1973) has presented a numerical method for solving the equation

$$f(x) - \int_{0}^{\infty} k(x-y)f(y)dy = g(x) \quad (x > 0)$$

which is based on the classical use of Fourier transform theory. (Uniqueness of the solution is assumed and certain mild conditions are imposed on the functions $k(x)$, $g(x)$ to ensure the same conditions on the solution.) The method discussed by Stenger (1973) (who gives additional references) is suitable whenever it is possible to obtain the transforms

$$K(x) = \int_{-\infty}^{\infty} \exp(ixy)k(y)dy$$

and

$$G_{+}(x) = \int_{0}^{\infty} \exp(ixy)g(y)dy \ ,$$

and is motivated by the theoretical treatment of (5.49). The determination of an approximate solution $\tilde{f}(x)$ proceeds by approximating the transform $F_{+}(x)$ and inverting the transform. It would appear that some use might be found for the fast Fourier transform (Singleton 1967) in an efficient algorithm using the method described by Stenger.

The theoretical background to Wiener-Hopf equations is given by Krein (1962), to whom reference may be made for a rigorous treatment. (Extensions to corresponding equations in more than one variable are mentioned in Goldenstein and Gohberg (1960).) Whilst the method of Stenger (1973) follows the theoretical treatment using Fourier transforms, the approach of Shinbrot (1969) is somewhat different. Stenger (1972, p.723) suggests that such a technique might form the basis for a numerical method. It is also known (see Cochran (1972, p.301) for references) that certain Wiener-Hopf integral equations can be solved by considering non-linear, but otherwise well behaved, integral equations. Cochran (1972) supplies other references of interest.

The theory for the solvability of the general equation (5.47) is not complete. However, for the numerical treatment of a general equation of the form (5.47), a natural technique involves replacing the infinite limit(s) of integration by finite quantities. The resulting equation is a well behaved equation, and its approximate solution can be achieved by any of the techniques discussed earlier. Some practical problems may be encountered if it is necessary to take the interval of integration large, and heuristically it seems possible that we shall find difficulty if the solution $f(x)$ of (5.30) is unbounded as x tends to (say) infinity. A useful technique appears to be one mentioned by Atkinson (1969) in which we choose a non-vanishing function $w(x)$ and write $f(x) = w(x)\phi(x)$ to obtain a new equation. Thus

$$f(x) - \int_0^\infty K(x,y)f(y)\,dy = g(x) \tag{5.51}$$

becomes

$$\phi(x) - \int_0^\infty \frac{w(y)}{w(x)}K(x,y)\phi(y)\,dy = \frac{g(x)}{w(x)} \quad . \tag{5.52}$$

Krein (1963) employs such a technique to treat an equation in which $f(x)$ grows exponentially, but it is clear that some prior analysis is necessary to determine the behaviour of the solution $f(x)$.

The analysis of Atkinson (1969) is concerned with the error

incurred in replacing (5.49) by an equation with finite limits of integration, under the assumptions that

$$\sup_{y\geq 0}|K(x,y)| < \infty \quad (0 \leq x < \infty) \quad ,$$

$$\sup_{x\geq 0}\int_0^\infty |K(x,y)|\,dy < \infty \quad ,$$

$$\lim_{h\to 0}\sup_{\substack{|x_1-x_2|\leq h\\ x_1,x_2\geq 0}}\int_0^\infty |K(x_1,y) - K(x_2,y)|\,dy = 0 \quad ,$$

and, for all $b > 0$,

$$\lim_{x\to\infty}\int_0^b |K(x,y)|\,dy = 0 \quad .$$

In circumstances where these conditions are not satisfied it may be possible to study an equation of the form (5.52) instead. The treatment of Atkinson (1969) is fairly general; others, in particular Gohberg and Feldman (1965), have considered similar questions for Wiener-Hopf equations.

Remark. In the preceding discussion a numerical method is based on modifying the original integral equation to one with finite limits. There remains, however, the possibility of a direct numerical treatment using Gauss-Hermite or Gauss-Laguerre quadrature formulae to modify the conventional quadrature method, or (say) a collocation method to give an approximate solution of the form

$$\sum_{r=0}^{n} \tilde{a}_r\phi_r(x) \quad .$$

(Clearly the functions $\phi_r(x)$ should be carefully chosen in view of the behaviour of the solution. If $f(x)$ is exponentially increasing then one of the functions $\phi_r(x)$ should have the same behaviour, for example.) I am not aware of any systematic study of such numerical methods.

In section 1.11 brief mention was made of the theory of certain classical integral equations in which the integral is interpreted as a Cauchy principal value, because of an intrinsic singularity. We refer to Muskhelishvili (1953) for a systematic treatment, and for a succinct introduction reference may be made to Mikhlin (1957) and Pogorzelski (1966). For a study of a system of singular integral equations (and hence, in particular, a single such equation) see Vekua (1967).

We noted, in section 1.11, that an equation with an intrinsically singular kernel may sometimes be converted into an equivalent equation which satisfies the Fredholm alternative. Such is the case, for example, with the equation involving Hilbert's kernel,

$$\alpha f(x) + \frac{\beta}{2\pi}\int_0^{2\pi} f(y)\ \cot\tfrac{1}{2}(y-x)\,dy = g(x) \quad ,$$

where α and β are constants.

If (Mikhlin 1957, p.122) we use operator notation to write

$$M\phi(x) = \alpha\phi(x) - \frac{\beta}{2\pi}\int_0^{2\pi} \phi(y)\cot\tfrac{1}{2}(y-x)\,dy$$

and apply M to both sides of the given integral equation, then we obtain the equivalent equation

$$(\alpha^2+\beta^2)f(x) - \frac{\beta^2}{2\pi}\int_0^{2\pi} f(y)\,dy = Mg(x) \quad ,$$

which is notable for its simplicity, when $\alpha^2+\beta^2 \neq 0$.

Example 5.25.

If we consider the more general equation

$$\alpha(x)f(x) + \frac{\beta(x)}{2\pi}\int_0^{2\pi} f(y)\cot\tfrac{1}{2}(y-x)\,dy + \int_0^{2\pi} K(x,y)f(y)\,dy = g(x) \quad ,$$

(where $K(x,y)$ is weakly singular or continuous) and apply to both sides of the equation the operator M defined by

$$M\phi(x) = \alpha(x)\phi(x) - \frac{\beta(x)}{2\pi}\int_0^{2\pi}\phi(y)\cot\tfrac{1}{2}(y-x)\,dy \quad ,$$

we obtain a Fredholm equation with a weakly singular kernel. Any solution of the singular equation is also a solution of this Fredholm equation, and the converse is true provided that there are no non-trivial solutions of the equation $M\phi(x) = 0$.

The 'Hilbert kernel' $\cot\frac{1}{2}(y-x)$ is related (Pogorzelski 1966; Mikhlin 1964*a*, p.119; Muskhelishvili 1953, p.116) to the Cauchy kernel $(t-\tau)^{-1}$. *

To illustrate further the conversion of an intrinsically singular equation to (at worst) a weakly singular equation, we consider the equation

$$a(t)f(t) - \frac{1}{\pi i}\int_\Gamma \frac{K_*(t,\tau)f(\tau)}{t-\tau}d\tau = g(t) \quad ,$$

where Γ is, as in section 1.11, a smooth closed curve, and we can suppose $K_*(t,\tau)$ is Lipschitz (Hölder)-continuous in t and τ with exponents in $[0,1)$.

This equation can be written in the form

$$Af(t) \equiv a(t)f(t) + \frac{b(t)}{\pi i}\int_\Gamma \frac{f(\tau)}{\tau-t}d\tau + \lambda\int_\Gamma K(t,\tau)f(\tau)d\tau = g(t) \quad . \tag{5.53}$$

The parameter λ is included for convenience in the subsequent discussion. In the derived equation λ has the value $1/\pi i$ in eqn (5.53), whilst $b(t) = K_*(t,t)$, and $K(t,\tau) = (K_*(t,\tau) - K_*(t,t))/(\tau-t)$. We can write $K(t,\tau) = H(t,\tau)/|t-\tau|$ with $\alpha \in (0,1)$. The index of eqn (5.53), namely,

$$m = \frac{1}{2\pi i}\left[\ln\frac{a(t)-b(t)}{a(t)+b(t)}\right]_\Gamma , \tag{5.54}$$

is defined by the *dominant part* (see section 1.11) of the equation. In the case where $b(t)$ vanishes for all $t \in \Gamma$, the index m is zero. When this is the case, the intrinsic singularity is not present, the kernel is at most weakly singular and the equation is called a Fredholm equation. More generally, if $m = 0$ then eqn (5.53) is called a quasi-Fredholm equation, or 'pseudo-regular'.

We gain some insight by considering the case where $b(t)$ vanishes, since Fredholm's theorems then apply and if λ is a characteristic value the equation has a (non-unique) solution only if $g(t)$ satisfies certain conditions. On the other hand, consider the case where $\lambda = 0$, so that the contribution from $K(t,\tau)$ is absent in (5.53). If $m \geq 0$ then the equation has a solution for any inhomogeneous term $g(t)$ which is Lipschitz-continuous,† however, only in the case $m = 0$ is this solution unique, since for $m > 0$ there are m linearly independent solutions of the homogeneous problem. For $m < 0$ and $\lambda = 0$, the equation is uniquely solvable if $g(t)$ satisfies $-m$ orthogonality conditions (Muskhelishvili 1953). In the more general case where $\lambda \neq 0$ the situation is somewhat similar, but if $m = 0$ then (depending on λ) the solution of (5.53) may not exist, or may not be unique. Indeed, for the quasi-Fredholm equation the Fredholm alternative applies (Muskhelishvili 1953, p.152). Necessary and sufficient conditions for the solvability of the general eqn (5.53) are given by Pogorzelski (1966, p.513) and Muskhelishvili (1953, p.143), and it will be found that if $m > 0$ then any solution is not unique. To obtain these theoretical results we may convert the given equation (5.53) into a weakly singular equation. Thus we may analyse

$$\{a^2(t)-b^2(t)\}f(t) + \frac{1}{\pi^2}\int_\Gamma\int_\Gamma \frac{b(\theta)K_*(\theta,\tau)d\theta}{(\theta-t)(\tau-\theta)} f(\tau)d\tau$$

$$= a(t)g(t) - \frac{1}{\pi i}\int_\Gamma \frac{b(\tau)g(\tau)}{(\tau-t)} d\tau$$

if $m \geq 0$, whilst if $m < 0$ we obtain

$$\{a^2(t)-b^2(t)\}\psi(t) + \frac{1}{\pi^2}\int_\Gamma\int_\Gamma \frac{K_*(t,\theta)b(\tau)d\theta}{(\theta-t)(\tau-\theta)} \psi(\tau)d\tau = g(t) \quad ,$$

† Solutions are sought from the class of Lipschitz (Hölder)-continuous functions.

and $f(t)$ can be obtained from any given solution $\psi(t)$ (see Pogorzelski 1966, pp. 512-13).

The problem of converting a singular integral equation into a Fredholm equation (that is, one with a kernel which is at most weakly singular) may be solved by determining an operator M, in particular (say)

$$M\phi(t) = \alpha\phi(t) + \frac{\beta(t)}{\pi i}\int_{\Gamma}\frac{\phi(\tau)}{\tau - t}d\tau ,$$

which *reduces* (Muskhelishvili 1953, p.120; Pogorzelski, 1966) the original singular equation $Af(t) = g(t)$ to such a Fredholm integral equation. In the case $m \geq 0$ we consider the equation $MAf(t) = Mg(t)$, which, if written out, assumes the form of an integral equation. Since the index of MA is the sum of the indices of M and A, and a Fredholm equation has zero index, we shall require that

$$\frac{1}{2\pi i}\left[\ln\frac{\alpha(t)-\beta(t)}{\alpha(t)+\beta(t)}\right]_{\Gamma} = -m ,$$

where $\alpha(t)$, $\beta(t) \neq 0$ for $t \in \Gamma$. The original equation and the new equation are then completely equivalent in the usual sense if $m \geq 0$. Indeed, if $m \geq 0$ then all solutions of $Af(t) = g(t)$ are solutions of $MAf(t) = Mg(t)$ and vice versa.

However, if $m < 0$ there are non-trivial solutions of the equation $M\phi(t) = 0$ and any such function may be added to a particular solution of $MAf(t) = Mg(t)$ to give a new solution. The solution of $Af(t) = g(t)$ may be unique, however, (assuming its existence). In this case $(m < 0)$ we now choose M, with index $-m$, so that $AM\psi(t) = g(t)$ is a (well behaved) Fredholm equation, and set $f(t) = M\psi(t)$; to every function $\psi(t)$ there corresponds a solution $f(t)$ and to every solution $f(t)$ there corresponds (at least) one such function $\psi(t)$. (The equation $AM\psi(t) = g(t)$ is not necessarily uniquely solvable.) For further remarks, see Muskhelishvili (1953) and Vekua (1967).

Remark. In the case considered above, the contour Γ is closed. If Γ is not closed then we can define a new contour Γ' including Γ as as a section and set functions to zero for argument values on Γ' but not on Γ, provided our theories are adapted to the case of discontinuous coefficients.

We shall not, here, consider non-linear singular equations, which are mentioned by Pogorzelski (1966).

5.10.1.

In the numerical treatment of intrinsically singular equations we may convert the equation to a weakly singular equation (as indicated in the theoretical treatment) or we may prefer to adopt a direct numerical approach. *Formally*, all those methods which can be used for the approximate solution of systems of Fredholm integral equations can be applied to the approximate solution of singular equations (with some modifications in the case of quadrature methods). However, not all of these methods have been justified at the present time, and some are unreliable.

Ivanov (1965) describes and justifies a number of direct methods in a paper written in Russian. The paper of Ivanov (1963) has been translated into English and deals with methods for systems of equations. Other Russian papers by Ivanov, (for example, see Noble (1972*a*,*b*) for references) are of interest. Hildebrand (1941) considers a class of singular integral equations, Collatz (1966*a*) deals briefly with our topic, and papers by Gabdulhaev (1968, 1970) and the paper on Cauchy transforms by Atkinson (1972*b*) contain relevant and interesting results.

Gabdulhaev (1968) considers a quadrature method applied to the equation

$$a(t)f(t) + \frac{b(t)}{\pi i}\int_\Gamma \frac{f(\tau)}{\tau - t}d\tau + \frac{\lambda}{2\pi i}\int_\Gamma k(t,\tau)f(\tau)d\tau = g(t)$$

in the case where Γ is the unit circle centred on the origin. Then the approximate solution values are interpolated, if the index of the equation is zero, by means of trigonometric polynomials, but otherwise by constructing an approximation

$$\sum_{k=0}^{n} \alpha_k t^k - \sum_{k=-n}^{-1} \alpha_k t^{k-m} \quad , \qquad (5.55)$$

where m is the index (5.54).

The quasi-Fredholm equation with zero index seems relatively easy to treat. It is clear of course that if A° is then the dominant part of the operator A in eqn (5.53), the equation

$$Af(t) = A^{\circ}f(t) + \int_{\Gamma} K(t,\tau)f(\tau)d\tau = g(t)$$

can be replaced using product-integration formulae by

$$A^{\circ}\tilde{f}(t_i) + \sum_{j=0}^{n} \nu_j(t_i)\tilde{f}(t_j) = g(t_i) \quad , \qquad (5.56)$$

using the inverse of A° (assumed known) to write

$$\tilde{f}(t_i) + \sum_{j=0}^{n} u_j(t_i)\tilde{f}(t_j) = \gamma(t_i) \quad , \qquad (5.57)$$

where $u_j(t) = (A^{\circ})^{-1}\nu_j(t)$ and $\gamma(t) = (A^{\circ})^{-1}g(t)$. Such a procedure seems to be equivalent to converting the original singular equation to a weakly singular equation and applying a product-integration technique.

A product-integration technique can be applied directly to the singular equation, using product-integration formulae for

$$\int_{\Gamma} (\tau-t)^{-1}\phi(\tau)dt$$

and for

$$\int_{\Gamma} K(t,\tau)\phi(\tau)d\tau \quad .$$

Thus use of the generalized trapezium rule amounts to the determination of a piecewise-linear function $\tilde{f}(t)$ such that $A\tilde{f}(t_i) = g(t_i)$, where $\tilde{f}(t)$ is linear for $t \in \Gamma$ between the points t_i, t_{i+1} . Formulated in this manner, our method is seen as a collocation technique (see section 5.7). For added generality we may require instead that

$$A\tilde{f}(s_i) = g(s_i) \quad (s_i \in \Gamma) \; , \qquad (5.58)$$

where $\{s_i\}_0^n$ are points on Γ. If $s_i \neq t_i$ we may suppose that each point s_i separates two adjacent points t_i on Γ. A theoretical study of such methods, which includes the above, has been given by Gabdulhaev (1970) in the case of an equation formulated as

$$a(x)f(x) + \frac{1}{2\pi}\int_0^{2\pi} H(x,y) \cot \tfrac{1}{2}(y-x)f(y)\mathrm{d}y = g(x) \text{ .}^{\dagger}$$

This analysis establishes error bounds and convergence results on the hypothesis (which cannot always be justified) that the inverse of a certain operator exists. The theory for such a technique is not, therefore, complete, and the theoretical difficulties appear to be reflected by computational difficulties in practice. Thus, Atkinson (1972*b*) has reported on poor results associated with ill-conditioning in the case $s_i = t_i$, but success when s_i is taken midway between t_i and t_{i+1}.

The situation described above casts doubt on the validity of general collocation methods as well as on the product-integration technique in the presence of an intrinsic singularity, as described. A method of moments has, however, been justified by Ivanov (1956), in the case where Γ is a circle, with $m \geq 0$ or $m < 0$.

Take A° to be the dominant part of the operator A, which we suppose has zero index. We form $\chi_j(t) = A^{\circ}\phi_j(t)$, where $\phi_0(t)$, $\phi_1(t)$, $\phi_2(t)$, ... form a complete set of functions. An approximate solution for the equation $Af(t) = g(t)$, of the form $\tilde{f}(t) = \sum_0^n \tilde{a}_k\phi_k(t)$, is obtained on solving the equations

$$\sum_{k=0}^{n} (A\phi_k,\chi_j)\tilde{a}_k = (g,\chi_j) \quad (j = 0,1,2,\ldots,n) \text{ ,} \tag{5.59}$$

where $(\phi,\psi) = \int_{\Gamma} \phi(\tau)\overline{\psi(\tau)}\mathrm{d}\,\tau$.

A detailed investigation of methods for intrinsically singular integral equations is beyond our present scope and our coverage is incomplete, but before concluding we mention the technique discussed by

† Gabdulhaev assumes the unique solvability of this equation.

Fang-Wang Hap (1965) for eqn (5.53). This involves replacing the function $K(t,\tau)$ (assumed square-integrable) by a nearby degenerate kernel

$$\sum_{i=0}^{n} u_i(t)v_i(\tau) \quad .$$

A particular case of such a technique involving the use of Jacobi polynomials is discussed by Kappenko (1967).

5.11. *Theory of methods for continuous and weakly singular kernels*

I now give some results which can be established for various of the methods discussed earlier. In Chapter 4 we discussed the case where the kernel $K(x,y)$ is continuous, so our attention is now directed particularly at the case of *weakly singular* kernels. (Clearly, continuous kernels are subsumed in the discussion.)

The following subsections are devoted to statements, without proof, of various results. In subsequent sections extensions of the theories already encountered are presented, for use in establishing our results. We then proceed to deduce these results.

Unless otherwise stated, we consider the equation of the second kind

$$f(x) - \lambda\int_a^b K(x,y)f(y)\mathrm{d}y = g(x) \qquad (5.60)$$

in which a,b are finite, $g(x)$ is continuous for $a \leq x \leq b$, and λ is a regular value of the kernel $K(x,y)$.

5.11.1.

We first consider the modified quadrature method.

THEOREM 5.1. *Suppose that* $K(x,y) = H(x,y)/|x-y|^{\alpha}$, *where* $H(x,y)$ *is continuous and* $0 < \alpha < 1$, *and suppose that the modified quadrature method of section 5.4 is applied, using the quadrature rule* $J_m(\phi) = (m \times J)(\phi)$, *where*

$$\int_0^1 \phi(y)dy \simeq J(\phi) \equiv \sum_{j=0}^{N} w_j\phi(y_j) ,$$

$J(\phi)$ *has degree of precision at least* N , *and where*

$0 = y_0 \leq y_1 \leq \cdots \leq y_{N-1} \leq y_N = 1$, $w_j \geq 0$ for $j = 0,1,2,\ldots,N$.

Then

$$\lim_{m\to\infty} \max_{\substack{0\leq k\leq m-1 \\ 0\leq j\leq N}} |f(a+kh+hy_j) - \tilde{f}(a+kh+hy_j)| = 0 ,$$

where $h = (b-a)/m$.

In practice, we imagine that $J(\phi)$ will be a simple rule such as the trapezium rule or Simpson's rule, or even the rectangular rule. The latter rule is not covered by the preceding theorem (since $y_N \neq 1$, for example), but a corresponding result is stated and proved by Kussmaul and Warner (1967) and our theorem is inspired by their general method of proof. (Kussmaul and Warner show that, in the presence of a certain type of smooth periodicity, the normally low-accuracy rectangular rule provides $O(h^3)$ accuracy in the computed values of $\tilde{f}(x)$. Such periodicity can occur when the equation is obtained from a contour integral equation, with a closed contour.)

5.11.2.

We now consider the product integration method. The following general result is simplified in the ensuing Remarks.

THEOREM 5.2. *Suppose that the product-integration method of section 5.6 is applied to give an approximate solution* $\tilde{f}(x)$ *satisfying the equations*

$$\tilde{f}(x) - \lambda \sum_{k=0}^{N} \sum_{j=0}^{n} v_{kj}(x)L_k(x,z_j)\tilde{f}(z_j) = g(x) ,$$

where $z_i \in [a,b]$ $(i = 0,1,2,\ldots,n)$, $z_i \neq z_j$ if $i \neq j$, *and*

$$K(x,y) = \sum_{k=0}^{N} L_k(x,y)M_k(x,y) . \tag{5.61}$$

We suppose that, for $k = 0,1,2,\ldots,N$

(i) $\sup_{a\le x\le b} \int_a^b |M_k(x,y)|\,dy < \infty$;

(ii) $\int_a^b |M_k(x_1,y) - M_k(x_2,y)|\,dy \to 0$ *uniformly as* $|x_1 - x_2| \to 0$ *with* $a \le x_1,\ x_2 \le b$; *and*

(iii) $L_k(x,y)$ *is continuous for* $a \le x,\ y \le b$.

We further suppose that the weights $v_{k,j}(x)$ *are chosen such that, for every function* $\phi(x) \in C[a,b]$,

$$\sum_{j=0}^{n} v_{kj}(x)\phi(z_j) = \int_a^b M_k(x,y)\,\{(A_m^k\phi)(y)\}dy\ , \tag{5.62}$$

where A_m^k *is a linear approximation operator* $(k = 0,1,2,\ldots,N\ ;\ m = 0,1,2,\ldots)$ *with the property that, for each* $\phi(x) \in C[a,b]$

$$\lim_{m\to\infty} \|\phi(x) - (A_m^k\phi)(x)\|_\infty = 0\ .$$

Here, $n = n(m) \to \infty$ *as* $m \to \infty$.

Under the preceding conditions, if λ *is a regular value of* $K(x,y)$, *then* $\tilde{f}(x)$ *is defined uniquely for* m *sufficiently large and*

$$\lim_{m\to\infty} \|f(x) - \tilde{f}(x)\|_\infty = 0\ .$$

Remark. Suppose that λ is a regular value of the kernel

$$K(x,y) = \sum_{k=0}^{N} L_k(x,y)M_k(x,y)\quad ,$$

where, for $k = 0,1,2,\ldots,N$, $M_k(x,y) = G_k(x,y)/|x-y|^{\alpha_k}$ $(0 \le \alpha_k < 1)$ and $G_k(x,y)$ and $L_k(x,y)$ are continuous†. Then conditions (i) and (ii) of the preceding theorem are satisfied. Suppose further that the

† We may sometimes place $G_k(x,y)$ as a factor in $L_k(x,y)$ rather than $M_k(x,y)$.

values $\{z_i, v_{kj}(z_i)\}$ are defined, with $z_i = a+i(b-a)/n$, by the generalized trapezium rule or by the generalized parabolic rule. Then the preceding theory is applicable, and

$$\lim_{n\to\infty} \|f(x) - \tilde{f}(x)\|_{\infty} = 0 .$$

Remark. If $N = 0$, Theorem 5.2 and Theorem 5.3 below apply to the simple product-integration method described in section 5.6. The method described in section 5.5 corresponds to the case where $N = 0$ and $L_0(x,y) = 1$, and the quadrature method of section 4.3 is covered by setting $N = 0$ and $M_0(x,y) = 1$.

The rate of convergence in the product-integration method is determined by the local truncation error for the method (see section 2.17), and thus by the choice of product-integration rules and the smoothness of the functions $L_k(x,y)f(y)$ $(k = 0,1,2,\ldots,N)$. To make the situation precise, suppose that in the statement of Theorem 5.2, $A_m^k\phi(x)$ is, in each interval $[z_0,z_r]$, $[z_r,z_{2r}]$, ..., $[z_{n-r},z_n]$ a polynomial of degree r, such that $A_m^k\phi(z_s) = \phi(z_s)$ $(s = 0,1,2,\ldots,n)$, where $z_s = a+sh$, $h = (b-a)/n$, and $n = mr$ where m is an integer. The case $r = 1$ (or, $r = 2$) corresponds to piecewise-linear (or,-quadratic) interpolation, which generates the generalized trapezium rule (or, parabolic rule). If, for $0 \le s \le r$, $(\partial/\partial y)^{s+1}\{L_k(x,y)f(y)\}$ is bounded for $a \le x$, $y \le b$ and $k = 0,1,2,\ldots,N$, then $\|f(x) - \tilde{f}(x)\|_{\infty} = O(h^{s+1})$. This result can sometimes be strengthened.

We shall also establish, in section 5.14.3, a somewhat stronger version of the following result.

THEOREM 5.3. *Suppose that* $K(x,y)$ *is given by* (5.61), *where conditions* (i) *to* (iii) *of Theorem* 5.2 *are satisfied, and suppose that*

$$\int_a^b K(x,y)f_r(y)\,dy = \kappa_r f_r(x) \quad (a \le x \le b) .$$

Suppose, for $r = 0,1,2,\ldots,$ *that the conditions stated for* (5.62) *are satisfied and*

$$\sum_{k=0}^{N} \sum_{j=0}^{n} \nu_{k,j}(x) L_k(x,z_j)\tilde{f}_r(z_j) = \tilde{\kappa}_r \tilde{f}_r(x) \quad (a \le x \le b; \quad r = 0,1,2,\ldots,n) .$$

Then, as $m \to \infty$, *every eigenvalue* κ *is a limit point of approximate eigenvalues* $\tilde{\kappa}_r$, *and every non-zero cluster point of approximate eigenvalues is an eigenvalue* κ . *Moreover, if as* $m \to \infty$, $\tilde{\kappa}_r \to \kappa$ *(with* $r = r(m)$*)* , *where* κ *is simple and non-zero, then the corresponding function* $\tilde{f}_r(x)$ *converges essentially in the uniform norm, to an eigenfunction* $f(x)$ *corresponding to* κ .

5.11.3.

Certain aspects of the theory of those expansion methods which may be viewed as projection methods are covered by the theory in section 4.29. These results can be applied even in the presence of weak singularities in $K(x,y)$, though we do require $g(x)$ to be fairly smooth. (We suppose $g(x)$ to be at least continuous, in order to be able to establish uniform convergence results.)

Recall that, in the projection methods, we have a bounded linear projection A_n , associated with an approximation operator, such that

$$A_n \phi(x) = \sum_{k=0}^{n} a_k \phi_k(x) \tag{5.63}$$

for some choice $a_0, a_1, \ldots, a_n$ corresponding to $\phi(x)$. We may suppose A_n is defined for $\phi(x) \in C[a,b]$. The projection method produces an approximation

$$\tilde{f}(x) = \sum_{k=0}^{n} \tilde{a}_k \phi_k(x)$$

such that

$$A_n \{\tilde{f}(x) - \lambda \int_a^b K(x,y)\tilde{f}(y)dy - g(x)\} = 0 . \tag{5.64}$$

From this we deduce (see equation (4.210)) that

$$\tilde{f}(x) - \lambda(A_n K)\tilde{f}(x) = A_n g(x)$$

and hence (Theorem 4.36) that

$$\lim_{n\to\infty} \|f(x) - \tilde{f}(x)\|_\infty = 0$$

provided that

$$\lim_{n\to\infty} \|g(x) - (A_n g)(x)\|_\infty = 0$$

and

$$\lim_{n\to\infty} \|\psi(x) - (A_n \psi)(x)\|_\infty = 0 \tag{5.65}$$

for every function $\psi(x)$ of the form $(K\Phi)(x)$ with $\Phi(x) \in C[a,b]$ and $\|\phi(x)\|_\infty = 1$.

To establish the results involving A_n we must refer to the appropriate properties of interpolation (for the collocation method) or of Fourier-type series (for the Ritz-Galerkin-type methods). If

$$\lim_{n\to\infty} \|\phi(x) - (A_n\phi)(x)\|_\infty = 0$$

for *every* $\phi(x) \in C[a,b]$ then (5.65) follows from the compactness of the integral operator K on $C[a,b]$. In practice, A_n does not always have this favourable property, and we are more likely to require that the modulus of continuity of $\phi(x)$ behave in a certain way (so that, for example, $\phi(x)$ satisfies a Dini-Lipschitz condition). Then we shall require that $g(x)$ satisfies such a criterion, and to establish (5.65) we resort to the following result (which can be employed in Theorem 4.37).

THEOREM 5.4. *Suppose that*

$$F = \{(K\phi)(x) \,\big|\, \|\phi(x)\|_\infty = 1\,,\ \phi(x) \in C[a,b]\}\ .$$

Define, for any function $H(x,y)$ *continuous for* $a \le x,\ y \le b$, *the quantity*

$$\Omega_\infty(H;\delta) = \sup_{\substack{|x_1-x_2|\le\delta \\ a\le x_1,x_2\le b}} \max_{a\le y\le b} |H(x_1,y) - H(x_2,y)| .$$

If $\psi(x) \in \mathrm{F}$ *and* $\omega(\psi;\delta)$ *is the modulus of continuity of* $\psi(x)$, *then*

(i) *if* $K(x,y)$ *is continuous,* $\omega(\psi;\delta) = O(\Omega_\infty(K;\delta))$, *as* $\delta \to 0$;

(ii) *if* $K(x,y) = H(x,y)/|x-y|^\alpha$, *where* $0 < \alpha < 1$ *and* $H(x,y)$ *is continuous, then* $\omega(\psi;\delta) = O(\delta^{1-\alpha}) + O(\Omega_\infty(H;\delta))$;

(iii) *if* $K(x,y) = H(x,y) \ln|x-y| + G(x,y)$, *where* $H(x,y)$ *and* $G(x,y)$ *are continuous,* $\omega(\psi;\delta) = O(\Omega_\infty(G;\delta)) + O(-\delta \ln \delta)$.

(For an indication of the proof of these results we refer to Lemma 5.3 (b) in section 5.13.)

As a corollary, and as a consequence of Jackson's theorem, we see that if X_n is the space of polynomials of degree $n-1$ then

$$\mathrm{dist}_\infty(\mathrm{F},X_n) = O(\Omega_\infty(K;1/n)) .$$

If X_n is a space of piecewise polynomials of degree $r-1$ (each function in X_n being a polynomial of degree $r-1$ on m subintervals whose maximum width is h) then

$$\mathrm{dist}_\infty(\mathrm{F},X_n) = O(\Omega_\infty(K;h)) ,$$

as $h \to 0$ with r fixed.

Remark. We saw that many projection methods associated with Chebyshev polynomials (of degree up to n) correspond to projections A_n with $||A_n||_\infty = O(\ln n)$. Cubic spline interpolation at equally spaced knots and at two additional regularly placed points gives a projection with $||A_n||_\infty = O(1)$† and corresponds to spline collocation using the knots and

---

† The corresponding result for piecewise-linear interpolation is fairly obvious. Phillips (1969) refers to the doctoral thesis of de Boor for the result quoted here for cubic splines. Analogous results for splines of other degrees can also be found in the literature, usually in the discussion of spline collocation for differential equations.

the two additional points as collocation points. Thus, from (iii) above, we can establish that $||K - A_n K||_\infty \to 0$ in each of these cases if $H_x(x,y)$ and $G_x(x,y)$ are bounded for all x,y or, more generally, if $H(x,y)$ and $G(x,y)$ are Lipschitz-continuous in x, uniformly for $a \leq y \leq b$. If, further, $g(x)$ is Lipschitz-continuous then $\|g(x) - (Ag)(x)\|_\infty \to 0$ and hence $||f(x) - \tilde{f}(x)||_\infty \to 0$.

In Theorem 5.5 we re-assess a theory which yields the above results. The analysis of projection methods can also be based on the Anselone theory of compact operators. We confine ourselves to convergence results, but others (Phillips 1972; Ikebe 1970) have considered projection methods in greater depth.

5.12. *An abstract theory*

In this section I present an abstract theory which may be used to establish the convergence of a number of numerical methods for integral equations. I shall illustrate its application by employing the general theory to prove Theorem 5.1. The theory is of Russian origin.

Suppose that X and X_m are normal linear subspaces with $X \supseteq X_m$ and that P_m is a projection (linear, bounded, with $P_m^2 = P_m$) mapping X onto X_m. For $g \in X$ we consider the equation

$$Tf \equiv f - \lambda Kf = g \,, \tag{5.66}$$

where K is a linear operator from X into X, and the associated equation

$$T_m f_m \equiv f_m - \lambda K_m f_m = g_m \,, \tag{5.67}$$

where $g_m = P_m g$ and K_m is a *compact* linear operator mapping X_m into itself. K and K_m are required to be close in the sense that

$$\|P_m K\tilde{\phi} - K_m\tilde{\phi}|| \leq \varepsilon_m || \tilde{\phi} || \tag{5.68}$$

for each $\tilde{\phi} \in X_m$.

We also require that elements of the form $K\phi \in X$ can be approximated closely by suitable elements in X_m. Thus we require that for every $\phi \in X$ there is an element $\tilde{\psi} \in X_m$ such that

$$\|K\phi-\tilde{\psi}\| \leq \gamma_m \|\phi\| . \tag{5.69}$$

Finally, we shall require that the distance between g and g_m is small. Let us require that for $g \in X$ there exists a $\tilde{g} \in X_m$ such that

$$\|g-\tilde{g}\| \leq \eta_m^{(1)} \|g\| , \tag{5.70a}$$

or, more directly, suppose that $g_m = P_m g$ satisfies a relation

$$\|g-g_m\| \leq \eta_m^{(2)} \|g\| , \tag{5.70b}$$

in what follows. (Of course, we can always set $\tilde{g} = 0$, $\eta_m^{(1)} = 1$, but we shall need something better than this.) The values $\eta_m^{(2)}$ and $\eta_m^{(1)}$ can be related, since $\tilde{g} = P_m\tilde{g} \in X_m$ and $\|g-g_m\| = \|g-P_m g\|$ $\leq \|g-\tilde{g}\| + \|P_m\tilde{g}-P_m g\| \leq (1+\|P_m\|)\|g-\tilde{g}\| \leq (1+\|P_m\|)\eta_m^{(1)}\|g\|$.

Thus, given (5.70a), we may obtain the result (5.70b), where

$$\eta_m^{(2)} \leq (1+\|P_m\|)\eta_m^{(1)} . \tag{5.71}$$

The values of $\eta_m^{(1)}$ *and* $\eta_m^{(2)}$ *are allowed to depend on the particular element* $g \in X$, whereas γ_m and ε_m must be independent of the elements $\tilde{\phi}, \phi \in X$ in (5.68) and (5.69).

5.12.1.

To permit the simple statement of our initial results we define for any operator L, whose domain in X includes X_m † (and with range in X),

$$\|L\|_{X_m} = \sup_{\substack{\|\tilde{\phi}\|=1 \\ \tilde{\phi}\in X_m}} \|L\tilde{\phi}\| , \tag{5.72}$$

---

† In this section, all operators are linear.

where the norms on the right-hand side are norms of elements in X. (The norm of $\tilde{\phi} \in X_m$ is the same as that in X since $X_m \subseteq X$.) Condition (5.68) above implies that $||P_m K - K_m||_{X_m} \leq \varepsilon_m$. The basic assumptions of the preceding subsection continue to hold.

THEOREM 5.5. (a) *Suppose that* $T = I - \lambda K$ *acts on the normed linear space* X *and has a bounded inverse* T^{-1}. *Suppose also that* $T_m = I - \lambda K_m$ *maps a normed linear space* $X_m \subseteq X$ *into itself, that* K_m *is compact on* X_m, *and*

$$q = ||T^{-1}|| \; ||T - T_m||_{X_m} < 1 .$$

Then T_m^{-1} *exists on* X_m, *and*

$$||T_m^{-1}||_{X_m} \leq \frac{||T^{-1}||}{(1-q)} ,$$

whilst if $T_m \tilde{f}_m = \tilde{g}_m \in X_m$ *then* $\tilde{f}_m$ *is unique in* X_m *and*

$$||\tilde{f}_m|| \leq \frac{||T^{-1}||}{(1-q)} ||\tilde{g}_m|| .$$

(b) *Subject to the conditions in part* (a), *given* $Tf = g$ *and* $T_m \tilde{f}_m = \tilde{g}_m$ *then*

$$||f - \tilde{f}_m|| \leq \frac{q}{1-q} ||f|| + \frac{||T^{-1}|| \; ||g - \tilde{g}_m||}{1-q} .$$

Proof. (a) We shall establish, below, an inequality of the form

$$\inf_{\substack{\tilde{\phi} \in X_m \\ ||\tilde{\phi}|| = 1}} ||T_m \tilde{\phi}||_{X_m} \geq \rho > 0 .$$

It then follows, since $T_m = I - \lambda K_m$, that there is no function $\tilde{\phi} \in X_m$ such that $(I - \lambda K_m)\tilde{\phi} = 0$ and $||\tilde{\phi}|| = 1$. Thus λ is not a characteristic value of K_m, and by 'the Fredholm alternative' it follows,† since K_m is compact that T_m^{-1} exists. Further

$$\|T_m^{-1}\|_{X_m} = \sup_{\|\phi\|_{X_m}=1} \{1/\|T_m\phi\|\} \leq \rho^{-1} .$$

(For, if $T_m\tilde{\phi} = \psi$ then $\tilde{\phi} = T_m^{-1}\psi$ and

$$\|T_m^{-1}\|_{X_m} = \sup_{\substack{\|\psi\|\neq 0 \\ \psi\in X_m}} \|\tilde{\phi}\|/\|\psi\|$$

$$= \sup_{\substack{\|\tilde{\phi}\|\neq 0 \\ \tilde{\phi}\in X_m}} \|\tilde{\phi}\|/\|T_m\tilde{\phi}\| = \sup_{\substack{\|\tilde{\phi}\|=1 \\ \tilde{\phi}\in X_m}} 1/\|T_m\tilde{\phi}\| ,$$

norms being in X_m.)

We shall now establish a value ρ which gives the result required in the above argument.

Suppose that $\|\tilde{\phi}\| = 1$, $\tilde{\phi} \in X_m$. Then

$$\|T_m\tilde{\phi}\| \geq \|T\tilde{\phi}\| - \|(T-T_m)\tilde{\phi}\|$$

$$\geq \inf_{\substack{\phi\in X \\ \|\phi\|=1}} \|T\phi\| - \sup_{\substack{\tilde{\phi}\in X_m \\ \|\tilde{\phi}\|=1}} \|(T-T_m)\tilde{\phi}\|$$

$$= 1/\|T^{-1}\| - \|(T-T_m)\|_{X_m}$$

$$= (1-q)/\|T^{-1}\| = \rho , \text{ say, where } \rho \geq 0$$

since $q < 1$. On taking the infimum for $\tilde{\phi} \in X_m$, we see that $\|T_m^{-1}\|_{X_m} \leq \|T^{-1}\|/(1-q)$. The first part of the theorem then follows.

---

† There is no requirement for completeness of X_m; see, for example, Taylor (1961, p.281, Theorem 5.5-F).

We now establish part (b), namely that, if $Tf = g$ and $T_m\tilde{f}_m = \tilde{g}_m$, then

$$\|f-\tilde{f}_m\| \le \frac{q}{1-q}\|f\| + \frac{\|T^{-1}\| \ \|g-\tilde{g}_m\|}{1-q} .$$

We have

$$T(f-\tilde{f}_m) = (T_m-T)\tilde{f}_m + g-\tilde{g}_m$$

so that

$$f-\tilde{f}_m = T^{-1}\{(T_m-T)\tilde{f}_m + g-\tilde{g}_m\} .$$

Then

$$\|f-\tilde{f}_m\| \le \|T^{-1}\| \ \{\|T_m-T\|_{X_m}\|\tilde{f}_m\| + \|g-\tilde{g}_m\|\}$$

since $\tilde{f}_m \in X_m$. Consequently

$$\|f-\tilde{f}_m\| \le q\,\|\tilde{f}_m\| + \|T^{-1}\| \ \|g-\tilde{g}_m\| . \tag{5.73}$$

Now

$$\|\tilde{f}_m\| \le \|f-\tilde{f}_m\| + \|f\| \le q\,\|\tilde{f}_m\| + \|T^{-1}\| \ \|g-\tilde{g}_m\| + \|f\|$$

whence (since $q < 1$),

$$\|\tilde{f}_m\| \le \{\|T^{-1}\| \ \|g-\tilde{g}_m\| + \|f\|\}/(1-q) .$$

Substituting this result in (5.73) we obtain the result of the theorem.

Theorem 5.5(b) provides the basis for a result on the convergence of $\tilde{f}_m$ to f as $m \to \infty$. This result we express in the following theorem. (In applying the conditions (i)-(iii) of Theorem 5.6 we note that we can write, using (5.71),

$$\eta_m^{(2)} \le (1+\|P_m\|)\eta_m^{(1)}$$

and can deduce alternative conditions in terms of $\|P_m\|$ and $\eta_m^{(1)}$.)

THEOREM 5.6. *Suppose for* $m = 1,2,3,\ldots$ $T_m f_m = g_m$, *where* $g_m = P_m g$, $Tf = g$, $T = I-\lambda K$, $T_m = I - \lambda K_m$, *and* K_m *is compact;* ε_m, γ_m, $\eta_m^{(1)}$ *and* $\eta_m^{(2)}$ *are defined by eqns* (5.68), (5.69), (5.70a), *and* (5.70b); *and* $T^{-1} = (I-\lambda K)^{-1}$ *exists. If*

(i) $\lim_{m\to\infty} \varepsilon_m = 0$,

(ii) $\lim_{m\to\infty} \gamma_m = 0$,

(iii) $\lim_{m\to\infty} \|P_m\|\,\gamma_m = 0$,

and

(iv) $\lim_{m\to\infty} \eta_m^{(2)} = 0$,

then

$$\lim_{m\to\infty} \|f-f_m\| = 0 .$$

<u>Proof</u>. For $\tilde{\phi} \in X_m \subseteq X$,

$$\|K\tilde{\phi}-K_m\tilde{\phi}\| \le \|P_m K\tilde{\phi}-K\tilde{\phi}\| + \|P_m K\tilde{\phi}-K_m\tilde{\phi}\| \le$$

$$\le \|P_m K\tilde{\phi}-K\tilde{\phi}\| + \varepsilon_m\|\tilde{\phi}\|$$

from (5.68). Now $\tilde{\phi} \in X$; let $\tilde{\psi}$ be the element of X_m such that $\|K\tilde{\phi}-\tilde{\psi}\| \le \gamma_m\|\tilde{\phi}\|$ by (5.69). Since P_m is a projection *onto* X_m , $\tilde{\psi} = P_m\tilde{\psi}$. We write $\chi = K\tilde{\phi}-\tilde{\psi}$. Then

$$||P_m K\tilde{\phi}-K\tilde{\phi}|| = ||P_m K\tilde{\phi}-P_m\tilde{\psi} + \tilde{\psi}-K\tilde{\phi}|| = ||P_m\chi-\chi|| \leq$$

$$\leq (1+ ||P_m||)\,||\chi|| \leq (1+ ||P_m||)\gamma_m\,||\tilde{\phi}|| .$$

Consequently we may write

$$||K\tilde{\phi}-K_m\tilde{\phi}|| \leq \{\varepsilon_m+\gamma_m(1+ ||P_m||)\}||\tilde{\phi}||$$

and hence

$$||K-K_m||_{X_m} \leq \{\varepsilon_m+\gamma_m(1+ ||P_m||)\} .$$

Observe that $\|T-T_m\|_{X_m} \leq |\lambda|\ ||K-K_m||_{X_m}$.

The conditions (i), (ii), (iii) ensure that for every $\rho \in (0,1)$ we may determine a value N such that $\|T^{-1}\|\ \|T-T_m\|_{X_m} \leq \rho$ for $m \geq N$. Thus for $m \geq N$, Theorem 5.5 is applicable, with $g_m = P_m g$ in the rôle of $\tilde{g}_m$. Now

$$||g-P_m g|| = ||g-g_m|| \leq \eta_m^{(2)}\ ||g||.$$

If $\eta_m^{(2)} \to 0$ then $||g-g_m|| \to 0$, where $g_m = P_m g$.

The convergence result $\|f-f_m\| \to 0$ now follows from applying Theorem 5.5(b) and considering the limit as $m \to \infty$. We also have the following result.

THEOREM 5.7. $\lim_{m\to\infty} ||f-f_m|| = 0$ *if* T , T_m , g_m , f , *and* f_m *are defined as in Theorem 5.6, and*

(i) $\lim_{m\to\infty} \varepsilon_m = 0$,

(ii) $\lim_{m\to\infty} \gamma_m = 0$, (iii) $\lim_{m\to\infty} \|P_m\|\,\gamma_m = 0$,

(iv) $\lim_{m\to\infty} (1+ \|P_m\|)\,\eta_m^{(1)} = 0$.

Condition (iv) *in this statement may not be satisfied when condition* (iv) *of the preceding Theorem is satisfied.*

5.12.2.

From the approach above we can establish the following result (Kantorovitch and Akilov 1964, p.555).

THEOREM 5.8. *Suppose that* X *is a complete normed linear space, that* P_m *is a projection of* X *onto* X_m *where* $\lim_{m\to\infty} P_m\phi = \phi$ *for each* $\phi \in X$, *that* K *is completely continuous,* $g_m = P_m g$ *and* $K_m = P_m K$. *If* λ *is a regular value of* K *then* $\lim_{m\to\infty}||f_m - f|| = 0$, *where* $(I-\lambda K)f = g$ *and* $(I-\lambda K_m)f_m = g_m$.

Since Theorem 5.8 is essentially a re-statement of Theorem 4.36 we must ask whether the theory of section 5.12.1 is any improvement on the theory of Chapter 4. What has happened is that, by paying attention to details, we have seen that it is sufficient to consider whether $\lim_{m\to\infty} ||K-K_m||_{X_m} = 0$ instead of whether $\lim_{m\to\infty} ||K-K_m||_X = 0$. In the next section, we shall see that in practice this distinction is of some assistance to us in a particular application. (Equally, it may be suggested that there is probably no need to establish Theorem 5.5 *ab initio*.) The treatment above gives some indication of the lines of development followed in much of the Russian analysis (see, for example, Kantorovitch and Akilov 1964).

5.13. *Convergence of the modified quadrature method*

We denote by $J(\phi)$ the quadrature rule which gives the approximation

$$I(\phi) \equiv \int_0^1 \phi(y)\,dy \simeq \sum_{j=0}^{N} w_j \phi(y_j) \equiv J(\phi) \quad .$$

We suppose that $0 \leq y_j \leq 1$ and $w_j > 0$ $(j = 0,1,2,\ldots,N)$ and that the degree of precision of the rule is at least N. (Thus $\sum_{j=0}^{N} w_j = 1$, in particular.) We also suppose that $y_0 = 0$, $y_N = 1$, as a matter of convenience.

Associated with $J(\phi)$ is the m-times repeated version $(m \times J)(\phi)$ which provides for an integral over $[a,b]$ the approximation

$$\int_a^b \phi(y)\mathrm{d}y \simeq h \sum_{k=0}^{m-1} \sum_{j=0}^{N} w_j \phi(t_k+hy_j) , \tag{5.74}$$

where $t_k = a+kh$, $h = (b-a)/m$.

For each value of m, we define the linear space X_m of functions which are continuous on $[a,b]$ and are polynomials of degree N in each interval $[t_k,t_{k+1}]$ $(k = 0,1,2,\ldots,(m-1))$. An associated projection P_m of the space $X = C[a,b]$ onto X_m is defined by performing piece-wise-polynomial interpolation, that is, by setting†

$$(P_m\phi)(x) = \sum_{j=0}^{N} l_j^{(k)}(x) f(t_k+hy_j) \tag{5.75}$$

$$(t_k \leq x \leq t_{k+1};\ k = 0,1,2,\ldots,(m-1)) ,$$

where

$$l_j^{(k)}(x) = \prod_{\substack{i\neq j\\ i=0}}^{N} [(x-t_k-hy_i)/\{h(y_j-y_i)\}] .$$

It is convenient to write

$$\zeta_{kj} = t_k+hy_j , \tag{5.76}$$

and $l_j(x) = l_j^{(0)}((x-a)/h)$, where $h > 0$.

† The superscripts here do not, of course, imply differentiation.

The function $(P_m\phi)(x)$ is continuous, because $y_0 = 0$, $y_N = 1$ in (5.74). The norms on the space X and X_m are taken to be the uniform norm (the L_∞-norm). Thus,

$$\|\phi\| = \|\phi(x)\|_\infty = \max_{a \le x \le b} |\phi(x)|$$

(in X and hence in X_m).

The subordinate norm $\|P_m\|$ may now be obtained. Suppose that $\|\phi(x)\|_\infty = 1$. Then

$$\|P_m\phi(x)\|_\infty = \max_{k=0,1,2,\ldots,(m-1)} \max_{x\varepsilon[t_k,t_{k+1}]} |P_m\phi(x)|$$

$$= \max_{k=0,1,2,\ldots,(m-1)} \max_{x\varepsilon[t_k,t_{k+1}]} \left|\sum_{j=0}^{N} l_j^{(k)}(x)\phi(t_k+hy_j)\right|$$

$$\le \max_{k} \max_{x\varepsilon[t_k,t_{k+1}]} \sum_{j=0}^{N} \left|l_j^{(k)}(x)\right|$$

since $\|\phi(x)\|_\infty = 1$. But

$$\max_{x\varepsilon[t_k,t_{k+1}]} \sum_{j=0}^{N} \left|l_j^{(k)}(x)\right| = \max_{x\varepsilon[t_k,t_{k+1}]} \sum_{j=0}^{N} \left|\prod_{\substack{i\ne j\\ i=0}}^{N} \frac{x-\zeta_{ki}}{h(y_j-y_i)}\right|$$

$$= \max_{s\varepsilon[0,1]} \sum_{j=0}^{N} \left|\prod_{\substack{i\ne j\\ i=0}}^{N} \frac{s-y_i}{y_j-y_i}\right|$$

(on using the transformation $x-t_k = sh$)

$$= \max_{s\varepsilon[0,1]} \sum_{j=0}^{N} |l_j(s)| = L_N \text{, say.}$$

Thus $\|P_m\|_\infty = \max_{\|\phi(x)\| = 1} \|P_m\phi(x)\|_\infty = L_N$, which is the Lebesgue constant associated with polynomial interpolation of degree N at

$y_0, y_1, \ldots, y_N$, over $[0,1]$. (Accordingly, $\|P_m\|$ is independent of h and m, and depends only on N, which is fixed in this discussion.)

We now define, for $\phi(x) \in X$

$$(T\phi)(x) = \phi(x) - \lambda \int_a^b K(x,y)\phi(y)\,dy \,, \tag{5.77}$$

where $K(x,y)$ is weakly singular; thus, $K(x,y) = H(x,y)/|x-y|^\alpha$, where $0 < \alpha < 1$ and $H(x,y)$ is continuous. (The integral operator K is compact on X and T^{-1} is defined on X if λ is a regular value.) We also define $(T_m\tilde{\phi})(x)$ for $\tilde{\phi}(x) \in X_m$, as follows. We have $\zeta_{k,j} = t_k + hy_j$, and define for $k = 0,1,2,\ldots,(m-1)$, $j = 0,1,2,\ldots,N$,

$$(T_m\tilde{\phi})(\zeta_{kj}) = (1+\lambda\Delta(\zeta_{kj}))\tilde{\phi}(\zeta_{kj}) - \lambda h \sum_{p=0}^{m-1} \Big\{ \sum_{\substack{q=0 \\ \zeta_{pq}\neq\zeta_{kj}}}^{N} w_q K(\zeta_{kj}, \zeta_{pq})\tilde{\phi}(\zeta_{pq}) \Big\}, \tag{5.78}$$

where

$$\Delta(\zeta_{kj}) = h \sum_{p=0}^{m-1} \Big\{ \sum_{\substack{q=0 \\ \zeta_{pq}\neq\zeta_{kj}}}^{N} w_q K(\zeta_{kj}, \zeta_{pq}) \Big\} - \int_a^b K(\zeta_{kj}, y)\,dy \,.$$

The condition that $(T_m\tilde{\phi})(x) \in X_m$ now completely defines $(T_m\tilde{\phi})(x)$ by piecewise-polynomial interpolation. We can then write

$$(K_m\tilde{\phi})(x) = -\{(T_m\tilde{\phi})(x) - \tilde{\phi}(x)\}/\lambda \,, \tag{5.79}$$

for every $\tilde{\phi}(x) \in X_m$, so that $T_m = I - \lambda K_m$.

Remark. Suppose that we apply the modified quadrature method, using the rule $(m \times J)(\phi)$, to the integral equation $(Tf)(x) = g(x)$. We then obtain the approximation

$$\tilde{\mathbf{f}}_m = [\tilde{f}(\zeta_{00}), \tilde{f}(\zeta_{01}), \ldots, \tilde{f}(\zeta_{m-1,N})]^T .$$

If the function values are extended to a piecewise-polynomial function so $\tilde{f}_m(x) = P_m\tilde{f}_m(x)$, $g_m = P_m g(x)$, then $(T_m\tilde{f}_m)(x) = g_m(x)$ for $x \in [a,b]$. A comparison of the equations $(T_m\tilde{f}_m)(x) = \tilde{g}_m(x)$ $(a \le x \le b)$ and $(Tf)(x) = g(x)$ will provide a convergence result for

$$\|f(x) - \tilde{f}_m(x)\|_\infty$$

and *a fortiori* for $\|\mathbf{f}-\tilde{\mathbf{f}}_m\|_\infty$.

Our convergence result can be established by appealing to Theorem 5.8 or Theorem 5.6 (or Theorem 5.7). We can construct bounds of the form (5.68) (5.69), and (5.70) and show that ε_m, γ_m, and $\eta_m^{(2)}$ tend to zero as $m \to \infty$. Since $\|P_m\| = L_N = O(1)$ our convergence result then follows from Theorem 5.6. The verification of each of the assumptions of this theorem proceeds by a sequence of lemmas.†

We first consider the bound (5.68). We have, for $\tilde{\phi}(x) \varepsilon X_m$, and since $(K_m\tilde{\phi})(x) = P_m K_m \tilde{\phi}(x)$,

$$\|P_m K\tilde{\phi} - K_m\tilde{\phi}\|_\infty \le \|P_m\| \times \max_{\zeta_{kj}} |(K\tilde{\phi})(\zeta_{kj}) - (K_m\tilde{\phi})(\zeta_{kj})| . \tag{5.80}$$

Now $\|P_m\| = L_N$, and we have the following result, which will enable us to verify an inequality (5.68) satisfying condition (i) of Theorem 5.7.

LEMMA 5.1. *Given* $\varepsilon > 0$, *we can find a value* $M \equiv M(\varepsilon)$ *such that when* $m \ge M(\varepsilon)$

$$\max_{\substack{k=0,1,2,\ldots,(m-1)\\ j=0,1,2,\ldots,N}} |(K\tilde{\phi})(\zeta_{kj}) - (K_m\tilde{\phi})(\zeta_{kj})| \le \varepsilon\|\tilde{\phi}(x)\|_\infty / L_N$$

† The proofs of these lemmas are sometimes involved, and the casual reader will probably prefer to accept them on trust.

for every $\tilde{\phi}(x) \in X_m$.

Proof. The proof is somewhat involved. In essence, we choose $\varepsilon > 0$, construct a bound of the form (5.68), and show that for all m sufficiently large, $\varepsilon_m \le \varepsilon/L_N$.

Recall that $K(x,y) = H(x,y)/|x-y|^\alpha$, where $H(x,y)$ is continuous. Consequently $K(x,y)$ is continuous on the regions $D_1 = \{a+\frac{1}{2}\delta \le y \le b,\ a \le x \le y-\frac{1}{2}\delta\}$ and $D_2 = \{a+\frac{1}{2}\delta \le x \le b,\ a \le y \le x-\frac{1}{2}\delta\}$, for any $\delta > 0$. $K(x,y)$ is therefore continuous in y (uniformly for x) in these regions, and we fix $\delta = \delta(\varepsilon)$ so that

$$0 < 16L_N \, ||H(x,y)||_\infty (2\delta)^{1-\alpha} \le \varepsilon(1-\alpha) \quad .$$

With δ fixed, we can now choose $\eta > 0$ so that

$$|K(x,y_1) - K(x,y_2)| \le \varepsilon/\{8(b-a)L_N\}$$

whenever $(x,y_1), (x,y_2) \in D_1$ or $(x,y_1), (x,y_2) \in D_2$, and $|y_1-y_2| \le \eta$. The choice of η depends on ε (and on $\delta = \delta(\varepsilon)$).

Now choose m_0 so that, with $h_0 = (b-a)/m_0, h_0 \le \min(\frac{\delta}{2}, \eta)$. For any $m \ge m_0$ and any fixed $x \in [a,b]$, we divide the set $S = \{0,1,2,\ldots,(m-1)\}$ into two disjoint sets S_1 and S_2 such that for all $k \in S_1$, $[t_k,t_{k+1}] \cap [x-\delta,x+\delta]$ is empty and $S_1 \cup S_2 = S$. Clearly S_1 and S_2 depend on x ; for the sets corresponding to $x = \zeta_{kj}$ we define

$$\delta_1 = \sum_{s\in S_1} \Big[\int_{t_s}^{t_{s+1}} K(\zeta_{kj},y)\{\tilde{\phi}(y) - \tilde{\phi}(\zeta_{kj})\}dy -$$

$$- h \sum_{q=0}^{N} w_q K(\zeta_{kj},\zeta_{sq})\{\tilde{\phi}(\zeta_{sq}) - \tilde{\phi}(\zeta_{kj})\}\Big] \quad ,$$

$$\delta_2 = \sum_{s\varepsilon S_2} \Big[\int_{t_s}^{t_{s+1}} K(\zeta_{kj},y)\{\tilde{\phi}(y) - \tilde{\phi}(\zeta_{kj})\}dy -$$

$$- h \sum_{\substack{\zeta_{sq}\neq\zeta_{kj} \\ q=0}}^{N} w_q K(\zeta_{kj},\zeta_{sq})\{\tilde{\phi}(\zeta_{sq}) - \tilde{\phi}(\zeta_{kj})\}\Big] .$$

Here, δ_1 and δ_2 depend on the point ζ_{kj}. The result of the lemma will be established provided we show that $|\delta_1+\delta_2| \leq \varepsilon \,\|\tilde{\phi}(x)\|_\infty/L_N$ for m sufficiently large. The manipulation required is, unfortunately, somewhat tedious. We set $h = (b-a)/m$. For $m \geq m_0$, $h \leq h_0$.

Now

$$\delta_1 = \sum_{q=0}^{N} w_q \sigma_q ,$$

where

$$\sum_{q=0}^{N} w_q = 1, \; w_r > 0 \quad (r = 0,1,2,\ldots,N)$$

and

$$\sigma_q = \sum_{s\varepsilon S_1} \int_{t_s}^{t_{s+1}} K(\zeta_{kj},y)\{\tilde{\phi}(y) - \tilde{\phi}(\zeta_{kj})\} -$$

$$- hK(\zeta_{kj},\zeta_{sq})\{\tilde{\phi}(\zeta_{sq}) - \tilde{\phi}(\zeta_{kj})\} ,$$

Thus $\sigma_q = \sigma_q^{(1)} + \sigma_q^{(2)}$, where

$$\sigma_q^{(1)} = \sum_{s\varepsilon S_1} \int_{t_s}^{t_{s+1}} \{K(\zeta_{kj},y) - K(\zeta_{kj},\zeta_{sq})\}\{\tilde{\phi}(y) - \tilde{\phi}(\zeta_{kj})\}dy \qquad (5.81)$$

and

$$\sigma_q^{(2)} = \sum_{s\varepsilon S_1} K(\zeta_{kj},\zeta_{sq})\Big[\int_{t_s}^{t_{s+1}} \{\tilde{\phi}(y) - \tilde{\phi}(\zeta_{kj})\}dy - h\tilde{\phi}(\zeta_{sq}) + h\tilde{\phi}(\zeta_{kj})\Big]. \qquad (5.82)$$

In the expression for $\sigma_q^{(1)}$, $\zeta_{sq} \in [t_s, t_{s+1}]$ so that, for $y \in [t_s, t_{s+1}]$,

$$|K(\zeta_{kj}, y) - K(\zeta_{kj}, \zeta_{sq})| \leq \varepsilon/\{8(b-a)L_N\}$$

from the choice of h. Thus

$$|\sigma_q^{(1)}| \leq 2 \, \|\tilde{\phi}(x)\|_\infty \int_a^b \varepsilon/\{8(b-a)L_N\} dy = \varepsilon \, \|\tilde{\phi}(x)\|_\infty/\{4L_N\} .$$

Now $\delta_1 = \sum_{q=0}^{N} w_q(\sigma_q^{(1)} + \sigma_q^{(2)})$, so that

$$|\delta_1| \leq \frac{\varepsilon \, \|\tilde{\phi}(x)\|_\infty}{4L_N} + \left| \sum_{q=0}^{N} w_q \sigma_q^{(2)} \right| . \tag{5.83}$$

We consider the problem of bounding $\sum_{q=0}^{N} w_q \sigma_q^{(2)}$. Since $\tilde{\phi}(x) \in X_m$ and the degree of precision of $J(\phi)$ is N,

$$\int_{t_s}^{t_{s+1}} \{\tilde{\phi}(y) - \tilde{\phi}(\zeta_{kj})\} dy = h \sum_{p=0}^{N} w_p \{\tilde{\phi}(\zeta_{sp}) - \tilde{\phi}(\zeta_{kj})\} ,$$

so that

$$\begin{aligned}
&\sum_{q=0}^{N} w_q \sigma_q^{(2)} \\
&= \sum_{s \in S_1} \sum_{q=0}^{N} w_q K(\zeta_{kj}, \zeta_{sq}) \times [h \sum_{p=0}^{N} w_p \{\tilde{\phi}(\zeta_{sp}) - \tilde{\phi}(\zeta_{kj})\} - h\tilde{\phi}(\zeta_{sq}) + h\tilde{\phi}(\zeta_{kj})] \\
&= \sum_{s \in S_1} \sum_{p=0}^{N} \sum_{q=0}^{N} w_p w_q h\{K(\zeta_{kj}, \zeta_{sq}) - K(\zeta_{kj}, \zeta_{sp})\}\tilde{\phi}(\zeta_{sp}) + \\
&+ \sum_{s \in S_1} \sum_{q=0}^{N} \sum_{p=0}^{N} w_q w_p h\{K(\zeta_{kj}, \zeta_{sp})\tilde{\phi}(\zeta_{sp}) - K(\zeta_{kj}, \zeta_{sq})\tilde{\phi}(\zeta_{sq})\}
\end{aligned} \tag{5.84}$$

The modulus of the first term is bounded, since $|\zeta_{sq}-\zeta_{sp}| \leq h$, and $h \leq \min(\frac{1}{2}\delta,\eta)$, by:

$$mh\,\|\tilde{\phi}(x)\|_\infty\,\varepsilon/\{8(b-a)L_N\} < \varepsilon\,\|\tilde{\phi}(x)\|_\infty/\{4L_N\},$$

since $mh = (b-a)$. The second term can be written

$$\sum_{s\varepsilon S_1}\{(\sum_{q=0}^{N} w_q)\sum_{p=0}^{N} w_p hK(\zeta_{kj},\zeta_{sp})\tilde{\phi}(\zeta_{sp}) -$$

$$-(\sum_{p=0}^{N} w_p)\sum_{q=0}^{N} w_q hK(\zeta_{kj},\zeta_{sq})\tilde{\phi}(\zeta_{sq})\} = 0\quad. \tag{5.85}$$

Combining these results we establish that

$$|\delta_1| \leq \varepsilon\,\|\tilde{\phi}(x)\|_\infty/\{2L_N\}. \tag{5.86}$$

We now obtain a bound for δ_2. We have

$$|\delta_2| \leq \sum_{s\varepsilon S_2}\int_{t_s}^{t_{s+1}}|K(\zeta_{kj},y)|\;|\tilde{\phi}(y)-\tilde{\phi}(\zeta_{kj})|dy\;+$$

$$+\sum_{p=0}^{N}\sum_{\substack{\zeta_{sp}\neq\zeta_{kj}\\ s\varepsilon S_2}} hw_p|K(\zeta_{kj},\zeta_{sp})|\;|\tilde{\phi}(\zeta_{sp})-\tilde{\phi}(\zeta_{kj})|$$

$$\leq 2\;\|\tilde{\phi}(x)\|_\infty\,\|H(x,y)\|_\infty\Big(\int_{\zeta_{kj}-(\delta+h)}^{\zeta_{kj}+(\delta+h)}\frac{1}{|y-\zeta_{kj}|^\alpha}dy\;+$$

$$+\sum_{p=0}^{N}\sum_{s\varepsilon S_2}{}^{*} hw_p\frac{1}{|\zeta_{sp}-\zeta_{kj}|^\alpha}\Big), \tag{5.87}$$

where Σ^* denotes that terms with $\zeta_{sp} = \zeta_{kj}$ are omitted. The intervals $[t_s,t_{s+1}]$ with $s \in S_2$ are included in an interval of width $2(\delta+h)$. Thus

$$|\delta_2| \le 2\,\|\tilde\phi(x)\|_\infty\;\|H(x,y)\|_\infty\Big(2\int_0^{\delta+h} y^{-\alpha}dy + \sum_{p=0}^{N} w_p \sum_{s\in S_2}^{*} h|\zeta_{kj} - \zeta_{sp}|^{-\alpha}\Big) \tag{5.88}$$

Let $\min\limits_{j\neq p}(\min(y_j-y_p, 1+y_p-y_j)) = \mu_*$; then

$$\sum_{s\in S_2}^{*} h|\zeta_{kj} - \zeta_{sp}|^{-\alpha} = \sum_{s\in S_2}^{*} h|(\zeta_{kj}-hy_p) - t_s|^{-\alpha}$$

$$\le \sum_{s\in S_2}^{*}\int_{t_s}^{t_{s+1}}|(\zeta_{kj}-hy_p) - y|^{-\alpha}dy\,/\mu_*$$

(using $t_s = a+sh$, and sectional monotonicity of the integrand)

$$\le \int_{\zeta_{kj}-(\delta+h)}^{\zeta_{kj}+(\delta+h)}|(\zeta_{kj} - hy_p) - y|^{-\alpha}dy\,/\mu_*$$

$$\le 2\int_0^{\delta+2h} t^{-\alpha}dt\,/\mu_* \,. \tag{5.89}$$

Since $\sum_{p=0}^{N} w_p = 1$, we then obtain, with $h \le \tfrac{1}{2}\delta$,

$$|\delta_2| \le 8\,\|\tilde\phi(x)\|_\infty\,\|H(x,y)\|_\infty\int_0^{2\delta} t^{-\alpha}dt/\mu_*$$

$$= 8\,\|\tilde\phi(x)\|_\infty\,\|H(x,y)\|_\infty(2\delta)^{1-\alpha}/(1-\alpha)\,\mu_*$$

$$\le \varepsilon\;\|\tilde\phi(x)\|_\infty/2L_N\,\mu_* . \tag{5.90}$$

Now combining (5.90) and (5.86) $|\delta_1+\delta_2| \le |\delta_1| + |\delta_2| \le \varepsilon/L_N\,\mu_*$ so that Lemma 5.1 is established, on assigning ε the value $\varepsilon\mu_*$.

From Lemma 5.1 and eqn (5.80) we deduce that, given $\varepsilon > 0$ there exists an integer $M(\varepsilon)$ such that

$$\|P_m K\tilde\phi(x) - K_m\tilde\phi(x)\|_\infty \le \varepsilon\;\|\tilde\phi(x)\|_\infty \tag{5.91}$$

for all $m \geq M(\varepsilon)$ and all $\tilde{\phi}(x) \in X_m$. Thus we can find ε_m satisfying (5.68) such that $\lim_{m\to\infty} \varepsilon_m = 0$. The next lemma is a preliminary to our verification of condition (iv) of Theorem 5.6.

LEMMA 5.2. *For every* $\phi(x) \in C[a,b]$ *with modulus of continuity* $\omega(\phi;\delta)$ *there exists a corresponding function* $\tilde{\phi}(x) \in X_m$ *such that*

$$\|\phi(x) - \tilde{\phi}(x)\|_\infty \leq \omega(\phi;\tfrac{b-a}{m}) = \eta_m^{(1)} \|\phi(x)\|_\infty ,$$

where $\eta_m^{(1)} \to 0$ *as* $m \to \infty$. †

Proof. Since $N \geq 1$, the continuous piecewise-linear interpolant defined by

$$\tilde{\phi}(x) = \frac{x-t_k}{h}\phi(t_{k+1}) + \frac{t_{k+1}-x}{h}\phi(t_k) \quad (t_k \leq x \leq t_{k+1}) ,$$

where $h = (b-a)/m$, is in X_m. By the mean-value theorem, since $\phi(x)$ is continuous, $\tilde{\phi}(x) = \phi(\tau_x)$ for some $\tau_x \in [t_k, t_{k+1}]$, when $x \in [t_k, t_{k+1}]$. Thus

$$\|\phi(x) - \tilde{\phi}(x)\|_\infty \leq \sup_{|x-\tau_x|\leq h} |\phi(x) - \phi(\tau_x)| \leq \omega(\phi;h) .$$

In view of the first part we can set, for $\phi(x) \equiv 0$ $\eta_m^{(1)} = 0$, and, if $\phi(x) \not\equiv 0$, $\eta_m^{(1)} = \omega(\phi;(b-a)/m)/\|\phi(x)\|_\infty$. (The dependence of $\eta_m^{(1)}$ on $\phi(x)$ is obvious!)

From the last result we have

$$\eta_m^{(2)} \leq (1+\|P_m\|)\eta_m^{(1)} = (1+L_N)\eta_m^{(1)}$$

in condition (iv) of Theorem 5.6. Consequently, as $m \to \infty$, $\eta_m^{(2)} \to 0$. Now $\eta_m^{(2)}$ may depend on $\phi(x)$, but the value γ_m is to be independent of $\phi(x)$ in (5.69). In the next lemma the compactness of the integral

† $\eta_m^{(1)}$ depends on $\phi(x)$, which is permitted.

operator K effectively plays a role in establishing condition (ii) of Theorem 5.6. (Since $\|P_m\| = L_N = O(1)$ as $m \to \infty$ with N fixed, condition (iii) will follow from condition (ii).)

LEMMA 5.3. (a) *For every* $\phi(x) \in C[a,b]$ *there exists a corresponding function* $\tilde{\psi}(x) \in X_m$ *such that*

$$\|(K\phi)(x) - \tilde{\psi}(x)\|_\infty \leq \Omega_1(K;(b-a)/m)\, \|\phi(x)\|_\infty \quad ,$$

where

$$\Omega_1(K,\delta) = \sup_{|x_1-x_2|\leq\delta} \int_a^b |K(x_1,y)-K(x_2,y)|\,dy$$

(b) *If* $K(x,y)$ *is weakly singular then it is 'uniformly y-integrable' and*

$$\lim_{\delta\to 0} \Omega_1(K,\delta) = 0 \ .$$

Proof. (a) From Lemma 5.2 we know that with

$$\psi(x) = \int_a^b K(x,y)\phi(y)\,dy$$

we can find $\tilde{\psi}(x) \in X_m$ such that $\|\psi(x) - \tilde{\psi}(x)\|_\infty \leq \omega(\psi;(b-a)/m)$

Now

$$\omega(\psi;\delta) = \sup_{|x_1-x_2|\leq\delta} |\psi(x_1) - \psi(x_2)|$$

$$= \sup_{|x_1-x_2|\leq\delta} \left| \int_a^b \{K(x_1,y) - K(x_2,y)\}\phi(y)\,dy \right|$$

$$\leq \Omega_1(K,\delta)\, \|\phi(x)\|_\infty \tag{5.92}$$

(by a Hölder inequality).

(b) Suppose that $K(x,y) = H(x,y)/|x-y|^{\alpha}$, where $0 < \alpha < 1$, and $H(x,y)$ is continuous for $a \le x, y \le b$. Then

$$\Omega_1(K;\delta) = \sup_{|x_1-x_2|\le\delta} \int_a^b \left| \frac{H(x_1,y)}{|x_1-y|^{\alpha}} - \frac{H(x_2,y)}{|x_2-y|^{\alpha}} \right| dy$$

$$\le \sup_{|x_1-x_2|\le\delta} \left(\int_a^b |H(x_1,y)| \left| \frac{1}{|x_1-y|^{\alpha}} - \frac{1}{|x_2-y|^{\alpha}} \right| dy + \right.$$

$$\left. + \int_a^b \frac{|H(x_1,y)-H(x_2,y)|}{|x_2-y|^{\alpha}} dy \right)$$

$$\le \|H(x,y)\|_{\infty} \sup_{|x_1-x_2|\le\delta} \int_a^b \left| \frac{1}{|x_1-y|^{\alpha}} - \frac{1}{|x_2-y|^{\alpha}} \right| dy +$$

$$+ \Omega_{\infty}(H;\delta) \sup_{a\le x\le b} \int_a^b \frac{1}{|x-y|^{\alpha}} dy , \tag{5.93}$$

where $\Omega_{\infty}(H;\delta) = \sup_{|x_1-x_2|\le\delta} \sup_{a\le y\le b} |H(x_1,y) - H(x_2,y)|$.

Now suppose, without loss of generality, that $x_1 > x_2$. Then with $|x_1-x_2| \le \delta$,

$$\int_a^b \left| |x_1-y|^{-\alpha} - |x_2-y|^{-\alpha} \right| dy$$

$$= (1-\alpha)^{-1} \left[4\left(\frac{x_1-x_2}{2}\right)^{1-\alpha} - \{(x_1-a)^{1-\alpha} - (x_2-a)^{1-\alpha}\} \right.$$

$$\left. - \{(b-x_2)^{1-\alpha} - (b-x_1)^{1-\alpha}\} \right]$$

$$\le \frac{4(\delta/2)^{1-\alpha}}{1-\alpha}$$

which is $O(\delta^{1-\alpha})$ as $\delta > 0$ tends to zero. Also $H(x,y)$ is bounded and continuous and hence uniformly continuous, so that $\lim_{\delta\to 0} \Omega_{\infty}(H;\delta) = 0$.

Thus part (b) follows, from (5.93).

It now follows that we may set $\gamma_m = \Omega_1(K;(b-a)/m)$ and conditions (ii) and (iii) follow. Using the results established in the preceding lemmas we can verify that the conditions of Theorem 5.6 are satisfied. The compactness of the integral operator K on $C[a,b]$ follows from the nature of the kernel (see section 5.14). What we require is compactness of the operator K_m; this follows because the range of K_m is a finite-dimensional subspace of $C[a,b]$ for each value of m. Theorem 5.1 is therefore a routine corollary of Theorem 5.6, once the preceding lemmas have been established, and $f_m(x)$ converges uniformly to $f(x)$ as $m \to \infty$.

The preceding theory has been presented to give some insight into the application of an abstract theory developed by a Russian school. It will be noted that a general theory, though powerful, transforms the burden of proof to the (possibly difficult) verification of a number of subsidiary conditions. This appears to be true of any abstract theory. In the next section we consider the application of a theory developed by an American school of analysis.

5.14. *Convergence of the general product-integration method*

Theorem 5.7 can be used to establish the convergence of the simple product-integration method of section 5.5. However, we shall present a proof of Theorem 5.2, and in doing so we take the opportunity to extend the analysis of section 4.17.1. By this means, we present an insight into the quite powerful theory developed by Anselone, Moore, Atkinson§, and others. For further reading on this, consult Anselone (1971).

In section 4.17.1 we considered replacing the equation $f(x) - \lambda(Kf)(x) = g(x)$ by an equation $\tilde{f}(x) - \lambda(H\tilde{f})(x) = g(x)$ and showed that if $\|(H-K)H\|_\infty$ is sufficiently small, $\tilde{f}(x)$ is uniquely defined and a bound on $\|f(x) - \tilde{f}(x)\|_\infty$ is available (see Theorem 4.17). The quadrature rule $J_m(\phi)$ used to construct $\tilde{f}(x)$ defines a corresponding operator $H \equiv H_m$, and for various families of quadrature rules

§ For a late reference, see Atkinson (1976).

$$\lim_{m\to\infty} J_m(\phi) = \int_a^b \phi(y)dy \quad \text{and} \quad \lim_{m\to\infty} ||(H_m-K)H_m||_\infty = 0 ,$$

so that $\lim_{m\to\infty} ||f(x) - \tilde{f}(x)||_\infty = 0$. This observation provides some insight into the motivation of the following abstract result.

THEOREM 5.9. *Let* X *be a Banach space (with norm* $||.||$ *) and suppose that* K *, and* $H_0, H_1, H_2, \ldots$ *are bounded linear operators from* X *onto* X *. Suppose that*

(i) K *is compact,*

(ii) $\{H_m, m \geq 0\}$ *is collectively compact,*
that is, the set

$$F := \{ \bigcup_{m=0}^{\infty} H_m\phi \mid ||\phi|| \leq 1\} \tag{5.94}$$

is (sequentially) compact in X *,*

and, further, that

(iii) $||H_m\phi - K\phi|| \to 0$ *as* $m \to \infty$ *for each* $\phi \in X$ *, and*

(iv) $(I-\lambda K)^{-1}$ *exists and is bounded, from* X *into* X .

Then (a) $\lim_{m\to\infty} ||(H_m - K)H_m|| = 0$ *, and (b) for sufficiently large* m $(I - \lambda H_m)^{-1}$ *exists and*

$$||(I-\lambda H_m)^{-1}|| \leq \frac{1+ ||\lambda(I-\lambda K)^{-1}H_m||}{1- ||(I-\lambda K)^{-1}|| \; ||\lambda^2(H_m-K)H_m||} .$$

It is of interest (if, perhaps, academic) to note that (i) is not required as an assumption since the compactness of K follows from the other assumptions. We also note, for use below, that (ii) ensures, by Theorem 1.3 (the uniform boundedness principle) that the operators $H_m (m = 0,1,2,\ldots)$ are uniformly bounded. (Hence, if $\{H_m\}$ is not uniformly bounded, (ii) cannot be satisfied.) Each operator H_m is compact, in view of (ii).

From (b) and condition (iii) of the theorem we can obtain a convergence result relating f and $\tilde{f}_m$, where $(I-\lambda K)f = g$ and $(I-\lambda H_m)\tilde{f}_m = g$. We obtain, for all sufficiently large m, the relation $\tilde{f}_m - f = \lambda(I-\lambda H_m)^{-1}(H_m-K)f$. Now

$$\|\tilde{f}_m - f\| \leq |\lambda| \; \|(I-\lambda H_m)^{-1}\| \; \|H_m f - Kf\| .$$

The last term tends to zero by (iii) and part (b) of the theorem provides a bound on $\|(I-\lambda H_m)^{-1}\|$. We can show that the latter bound is a uniform bound (for, there is a uniform bound on $\|H_m\|$ and $\|(I-\lambda K)^{-1}H_m\| \leq \|H_m\| \; \|(I-\lambda K)^{-1}\|$). Thus $\|\tilde{f}_m - f\| \to 0$ as $m \to \infty$. We then have the following corollary.

COROLLARY 5.1. *Under the conditions of Theorem 5.9,* $\|\tilde{f}_m - f\| \to 0$ *as* $m \to \infty$, *where* $(I-\lambda K)f = g$ *and* $(I-\lambda H_m)\tilde{f}_m = g$, *for any* $g \in X$.

Proof of Theorem 5.9. The manipulation required to establish part (b) is, as indicated below, identical to that employed in Theorem 4.17(b). We shall concentrate mainly on a proof of (a) (see also Anselone 1971, p.8).

Now

$$\|(H_m-K)H_m\| = \sup_{\|\phi\|=1} \|(H_m-K)H_m\phi\|$$

$$\leq \sup_{\psi \in F \subset X} \|(H_m-K)\psi\| .$$

For each fixed $\psi \in X$, $\|(H_m-K)\psi\| \to 0$ from (iii), and the set F is, in X, (sequentially) compact so we can deduce that $\|(H_m-K)\psi\| \to 0$ *uniformly* for $\psi \in$ F since $\{H_m\}$ is uniformly bounded (by virtue of Theorem 1.4).

To indicate briefly how part (b) may be established we write

$$\{I+\lambda(I-\lambda K)^{-1}H_m\} \; (I-\lambda H_m)$$

$$= I+\lambda^2(I-\lambda K)^{-1}(K-H_m)H_m .$$

If $\|\lambda^2(I-\lambda K)^{-1} (K-H_m)H_m\| < 1$ (which is guaranteed by requiring that

$\|(K-H_m)H_m\|$ is small enough, say $|\lambda^2| \; \|(I-\lambda K)^{-1}\| \; \|(K-H_m)H_m\| < 1$) the right-hand side has an inverse. It follows that no non-trivial function is annihilated by the operator on the right, and hence the same is true of the operator on the left (of which $I-\lambda H_m$ is a factor). Thus λ is not an eigenvalue of H_m and, since H_m is compact, $(I-\lambda H_m)^{-1}$ therefore exists. Then

$$\left\{ [I+\lambda^2(I-\lambda K)^{-1} (K-H_m)H_m]^{-1} [I+\lambda(I-\lambda K)^{-1}H_m] \right\} (I-\lambda H_m) = I \;, \qquad (5.95)$$

from which expression $(I-\lambda H_m)^{-1}$ can be obtained, as the operator in braces in (5.95). The result follows, since (see Theorem 1.1)

$$\|\{I+\lambda^2(I-\lambda K)^{-1} (K-H_m)H_m\}^{-1}\|$$

$$< \{1-|\lambda^2| \; \|(I-\lambda K)^{-1}\| \; \|(K-H_m)H_m\|\}^{-1}$$

when the right-hand side of this inequality is positive.

In order to apply Theorem 5.9 to analyse our numerical methods, we must choose a suitable Banach space of functions, make a suitable definition of H_m, and verify the conditions of the theorem. The most important of these issues appears to be a choice of X and a definition of H_m to satisfy (ii). It is usually not difficult to verify (iii), in applications.

In the present context, the following result is of some interest (see also Anselone 1966, p.8).

THEOREM 5.10. *Let* $K, H_m (m = 0,1,2,\ldots)$ *be bounded linear operators from* X *into* X. *Then if* $H_m\phi - K\phi \to 0 \; \forall \phi \in X$, $\{H_m | m = 0,1,2,\ldots\}$ *is collectively compact if and only if* $\{(H_m-K) | m = 0,1,2,\ldots\}$ *is collectively compact and* K *is compact.*

Remark. Any 'difficulty' in the proof of Theorem 5.9 lies only with the underlying notions of functional analysis. These are more advanced than the notions required by the Russian school of analysis to the extent that the concept of sequential compactness is exploited.

The object of the conditions is to ensure that $||(K-H_m)H_m||$ can be made sufficiently small, a condition which is certainly guaranteed *if* $||K-H_m|| \to 0$ (since $\{H_m\}$ is then uniformly bounded). The theory which was given in Chapter 4 was constructed using integral operators K_m (with kernels $K_m(x,y)$), for which $||K-K_m||_\infty \to 0$ when $m \to \infty$, as $||K(x,y)-K_m(x,y)||_\infty \to 0$. However, the quadrature operators H_m are such that $||K-H_m|| \not\to 0$, with $X = C[a,b]$ using the uniform norm.

Indeed, we have the following result (given by Anselone 1971, p.82).

THEOREM 5.11. $||H_m - K|| \to 0$ *if and only if the set* $\{H_m - K | m=0,1,2,\ldots\}$ *is (sequentially) compact and* $||(K-H_m)\phi|| \to 0$ *for every* $\phi \in X$.

5.14.1. *Application of the abstract theory*

We wish to apply the preceding results to the numerical solution of the integral equation

$$f(x) - \lambda\int_a^b K(x,y)f(y)\,dy = g(x) .$$

For this purpose, we seek a Banach space X of functions, containing $g(x)$ and $f(x)$ and on which the integral operator K is a compact operator. We therefore recall that

(i) if $X = L^2[a,b]$ with the norm $||\ ||_2$, K is certainly compact on X if $\int_a^b\int_a^b |K(x,y)|^2\,dx\,dy < \infty$; and

(ii) if $X = C[a,b]$ with the norm $||\ ||_\infty$, K is compact on X if

(a) $K(x,y)$ is continuous for $a \le x, y \le b$,

or

(b) $K(x,y)$ is weakly singular, $K(x,y) = H(x,y)/|x-y|^\alpha$ $(0 < \alpha < 1,\ H(x,y)$ continuous).

We are unlikely to be interested in choosing $L^2[a,b]$ as the underlying space when we consider numerical integration methods,† because values of functions in $L^2[a,b]$ can be infinite! Instead we shall consider the space $C[a,b]$ with the uniform norm; this is, of course, a Banach space. Since we are interested in compact sets in $C[a,b]$, we recall that a set of functions is compact in $C[a,b]$ if it is uniformly bounded, and the set is equicontinuous. (This is, in effect, the Arzela-Ascoli theorem.)

The next theorems permit us to recognize when a given integral operator is compact on $C[a,b]$, as required in condition (i) of Theorem 5.9.

THEOREM 5.12. *The integral operator* K, *where*

$$(K\phi)(x) = \int_a^b K(x,y)\phi(y)\,dy\ ,$$

is a compact operator on the space $C[a,b]$ *with uniform norm when*

(i) $$||K||_\infty = \sup_{a\le x\le b} \int_a^b K(x,y)\ dy < \infty\ .$$

(ii) $$\lim_{\delta\to 0}\ \sup_{\substack{x_1,x_2\in[a,b]\\ |x_1-x_2|\le\delta}} \int_a^b |K(x_1,y)-K(x_2,y)|\,dy = 0\ .$$

<u>Proof</u>. K is compact if the set $S = \{(K\phi)(x) \mid ||\phi(x)|| \le 1\}$ is (sequentially) compact. By (i) the set S is uniformly bounded; if $\psi(x)\ \varepsilon\ S$, $\|\psi(x)\|_\infty \le \|K\|_\infty \|\phi(x)\|_\infty \le \|K\|_\infty$. From (ii) we see that the set S is an equicontinuous set of functions; for, if $\psi(x)$ is any function in S,

$$\psi(x_1)-\psi(x_2) = \int_a^b \{K(x_1,y)-K(x_2,y)\}\phi(y)\,dy$$

† We may prefer to work in $L^2[a,b]$ in the case of Galerkin methods, however.

for some $\phi(x)$ with $\|\phi(x)\|_\infty \leq 1$. Thus

$$|\psi(x_1)-\psi(x_2)| \leq \int_a^b |K(x_1,y)-K(x_2,y)|\,dy \ . \tag{5.96}$$

Given $\varepsilon > 0$ there exists a δ such that

$$\int_a^b |K(x_1,y)-K(x_2,y)|\,dy \leq \varepsilon$$

when $|x_1-x_2| < \delta$ and $x_1,x_2 \in [a,b]$. Thus $|\psi(x_1)-\psi(x_2)| \leq \varepsilon$ for every $\psi(x) \in S$ and the conclusion follows. Thus S is uniformly bounded and equicontinuous, and hence compact in $C[a,b]$. (Indeed by the Arzela-Ascoli theorem, every sequence of functions in S contains a uniformly convergent subsequence with a limit in $C[a,b]$. Thus S is (sequentially) compact in $C[a,b]$.) The operator K is therefore a *compact* – that is, completely continuous – operator.

THEOREM 5.13. (a) *If, for* $i = 0,1,2,\ldots,N$, $K_i(x,y)$ *satisfies the conditions of the preceding theorem, then so also does the kernel* $K(x,y) = \sum_{i=0}^{N} K_i(x,y)$.

(b) *If* $M_i(x,y)$ *satisfies the conditions of the preceding theorem, so that*

$$\sup_{a\leq x\leq b} \int_a^b |M_i(x,y)|\,dy < \infty \quad ,$$

and

$$\sup_{\substack{|x_1-x_2|\leq\delta \\ a\leq x_1,x_2\leq b}} \int_a^b |M_i(x_1,y)-M_i(x_2,y)|\,dy \to 0$$

as $\delta \to 0$ *then the conditions are satisfied by the kernel* $K_i(x,y) = L_i(x,y)M_i(x,y)$, *where* $L_i(x,y)$ *is continuous for* $a \leq x,\ y \leq b$.

Finally, to see the relevance of the above results, we state the following corollary.

COROLLARY 5.2. *If* $K(x,y)$ *satisfies the conditions given in Theorem 5.2 then the integral operator* K *is compact on the space* $C[a,b]$ *with uniform norm.*

To establish Theorem 5.2 we consider a definition of an operator H_m associated with a product-integration method.

First suppose that for every $\phi(x)$ in $C[a,b]$, $A_m^k\phi(x) \equiv A_{m,x}^k[\phi(x)]$ is a continuous approximation $(k = 0,1,2,\ldots, N$, $m = 0,1,2,\ldots)$, such that

$$\lim_{m\to\infty} ||\phi(x) - A_{m,x}^k\phi(x)||_\infty = 0 \tag{5.97}$$

(where $k = 0,1,2,\ldots,N)$, and suppose that $A_{m,x}^k$ is a linear operator.

Now associated with the kernel $K_k(x,y) = L_k(x,y)M_k(x,y)$ we define

$$H_m^k(\phi) = \int_a^b M_k(x,y)A_{m,y}^k[L_k(x,y)\phi(y)]\,dy\ . \tag{5.98}$$

In this expression, $A_{m,y}^k$ denotes A operating on its argument *as a function of* y, for each $x \in [a,b]$.

We define, as an operator on $C[a,b]$,

$$H_m(\phi) = \sum_{k=0}^{N} H_m^k(\phi)\quad , \tag{5.99}$$

and this is the required operator. We shall show that $\{H_m\}$ satisfies the conditions of Theorem 5.9, by appealing to the following results; these may be accepted without proof on a first reading.

THEOREM 5.14. *If, for* $k = 0,1,2,\ldots,N$, $\{H_m^k | m = 0,1,2,\ldots\}$ *is a uniformly bounded collectively compact set of linear operators on a Banach space* X, *and*

$$H_m = \sum_{k=0}^{N} H_m^k\ ,$$

then $\{H_m | m = 0,1,2,\ldots\}$ *is uniformly bounded and collectively compact.*

Proof. $\|H_m\| \leq \sum_{k=0}^{N} \|H_m^k\|$ is uniformly bounded and the first result follows.

Consider now a sequence $H_{m(n)}\phi_n$ in $F = \bigcup_{m=0}^{\infty} \{H_m\phi | \|\phi\| \leq 1\}$.

We have

$$H_{m(0)}\phi_0 = H^0_{m(0)}\phi_0 + H^1_{m(0)}\phi_0 + \ldots + H^N_{m(0)}\phi_0 ,$$

and in general

$$H_{m(i)}\phi_i = H^0_{m(i)}\phi_i + H^1_{m(i)}\phi_i + \ldots + H^N_{m(i)}\phi_i$$

for $i \geq 0$. Since the set

$$F_k = \bigcup_{m=0}^{\infty} \{H^k_m\phi | \|\phi\| \leq 1\}$$

is sequentially compact in X, for each $k \in \{0,1,2,\ldots, N\}$, and in particular for $k = 0$, the sequence $H^0_{m(i)}\phi_i$ has an infinite convergent subsequence $\{H^0_{m(i_j)}\phi_{i_j}\}$ with limit ψ_0 in X. Consider, now, the infinite sequence $\{H^1_{m(i_j)}\phi_{i_j}\}$ which is a subsequence of $\{H_{m(i)}\phi_i\}$. Since F_1 is sequentially compact, $\{H^1_{m(i_j)}\phi_{i_j}\}$ has itself a convergent subsequence associated with a limit ψ_1 in X. (At the same time, the corresponding subsequence of $H^0_{m(i_j)}\phi_{i_j}$ converges to ψ_0.) Proceeding in this way we find a subsequence $\{H^N_{n_r}\phi_r\}$ of $\{H^N_{m(i)}\phi_i\}$ converging to ψ_N whilst $H^k_{n_r}\phi_r \to \psi_k$ $(k = 0,1,\ldots,N-1)$. Then $H_{n_r}\phi_r$ converges to $\psi_0+\psi_1+\ldots+\psi_N = \psi \in X$, so that the sequence $H_{m(i)}\phi_i$ $(i \geq 0)$ has a convergent subsequence with limit ψ in X. This establishes the sequential compactness of F and hence the collective compactness of $\{H_m\}$.

THEOREM 5.15. *With the definition of* H^k_m *given in eqn (5.98), and with* $K_k(x,y)$ *satisfying the conditions of Theorem 5.13, then*

(a)

$$\lim_{m\to\infty} \|H_m^k\phi(x)-K_k\phi(x)\|_\infty = 0 \qquad \phi(x) \in C[a,b] ,$$

where K_k *is the integral operator with kernel* $K_k(x,y)$; *and*
(b) $\{H_m^k | m = 0,1,2,\ldots\}$ *is a collectively compact set of operators on* $C[a,b]$.

<u>Proof</u>. (a) The convergence of $A_{m,y}^k[L_k(x,y)\phi(y)]$ to $L_k(x,y)\phi(y)$ for each fixed $\phi(x)$ is uniform for $a \le x \le b$. (The operators $A_{m,y}^k$ are linear and uniformly bounded and the functions $\{L_k(x,y)\phi(y)\}$, in y , parametrized by $x \in [a,b]$, form an equicontinuous set which is uniformly bounded. Then the set $\{L_k(x,y)\phi(y) | x \in [a,b]\}$ is compact, and we can apply Theorem 1.4.)

Now, for each fixed $\phi(x) \in C[a,b]$,

$$\left\|\int_a^b K_k(x,y)\phi(y)dy - (H_m^k\phi)(x)\right\|_\infty$$

$$= \left\|\int_a^b M_k(x,y)\{L_k(x,y)\phi(y)-A_{m,y}^k[L_k(x,y)\phi(y)]\}dy\right\|_\infty$$

$$\le \sup_{a\le x\le b}\int_a^b [M_k(x,y)]\,dy\ \|L_k(x,y)\phi(y) - $$

$$-A_{m,y}^k[L_k(x,y)\phi(y)]\|_\infty ,$$

and the right-hand side tends to zero as $m \to \infty$ from the foregoing observation.

(b) To establish (b) we apply the Arzela-Ascoli theorem and consider the set

$$F_k = \{(H_m^k\phi)(x) | m = 0,1,2,\ldots;\ \|\phi(x)\|_\infty \le 1\}$$

We are required to establish that this set is bounded and equicontinuous. The convergence result in (a) above implies the uniform boundedness of $\{H_m^k\}$, so the set F_k is *bounded*. (Thus if $\|H_m^k\|_\infty \le B$ then $\|\psi(x)\|_\infty \le B$ when $\psi(x) \in F_k$.)

Now we consider the equicontinuity of F_k. For convenience write

$$\tilde{\psi}_{k,m} \equiv \tilde{\psi}_k(x,y) = A^k_{m,y}[L_k(x,y)\phi(y)] \quad , \tag{5.100}$$

so that

$$\begin{aligned}
&|(H^k_m\phi)(x_1) - (H^k_m\phi)(x_2)| \\
&= \left|\int_a^b \{M_k(x_1,y)\tilde{\psi}_k(x_1,y) - M_k(x_2,y)\tilde{\psi}_k(x_2,y)\}dy\right| \\
&\leq \left|\int_a^b \{M_k(x_1,y) - M_k(x_2,y)\}\tilde{\psi}_k(x_1,y)dy\right. + \\
&+ \left.\int_a^b M_k(x_2,y)\{\tilde{\psi}_k(x_1,y) - \tilde{\psi}_k(x_2,y)\}dy\right| \quad .
\end{aligned}$$

Thus, for every $\phi(x)$ with $\|\phi(x)\|_\infty \leq 1$,

$$\begin{aligned}
&|(H^k_m\phi)(x_1) - (H^k_m\phi)(x_2)| \\
&\leq \sup_{\substack{a\leq x,y\leq b \\ \|\tilde{\phi}(x)\|_\infty \leq 1}} |\tilde{\psi}_k(x,y)| \int_a^b |M_k(x_1,y) - M_k(x_2,y)|dy + \\
&+ \sup_{a\leq x\leq b}\int_a^b |M_k(x,y)|dy \times \sup_{\substack{a\leq y\leq b \\ \|\tilde{\phi}(x)\|_\infty \leq 1}} |\tilde{\psi}_k(x_1,y) - \tilde{\psi}_k(x_2,y)| \quad .
\end{aligned} \tag{5.101}$$

From the hypotheses on $M_k(x,y)$, the equicontinuity of $\{(H^k_m\phi)(x)\}$ now follows if we establish a uniform bound for $|\tilde{\psi}_k(x,y)|$ (with $m \geq m_0$, say, and $a \leq x$, $y \leq b$), and the equicontinuity of $\{\tilde{\psi}_{k,m}\}$ as a family of functions in y parametrized by $x \in [a,b]$ and by $\phi(x)$ with $\|\phi(x)\|_\infty \leq 1$. These properties are established in the next Lemma, and we return to the general theme in section 5.14.2.

LEMMA 5.4. The functions $\{\tilde{\psi}_{k,m}\}$ *are uniformly bounded and equicontinuous.*

Proof. Now

$$\sup_{x,y,\phi(x)} |\tilde{\psi}_k(x,y)| \leq \|A_m^k\|_\infty \|L_k(x,y)\|_\infty \|\phi(x)\|_\infty$$

$$\leq \|A_m^k\|_\infty \|L_k(x,y)\|_\infty ,$$

which gives the first result required, since there is a uniform bound for $\|A_m^k\|_\infty$ by the Banach-Steinhaus theorem (q.v.).

To establish the equicontinuity of $\tilde{\psi}_{k,m}$ we first note the equicontinuity of $\{L_k(x,y)\phi(y)\}$, where $\|\phi(x)\|_\infty \leq 1$. Given $\varepsilon > 0$ there exists δ such that for $|x_1-x_2| \leq \delta$,

$$|L_k(x_1,y)\phi(y)-L_k(x_2,y)\phi(y)| \leq \|\phi(x)\|_\infty \sup_{\substack{a\leq y\leq b \\ |x_1-x_2|\leq\delta}} |L_k(x_1,y)-L_k(x_2,y)| \leq \varepsilon ,$$

using the (uniform) continuity of $L_k(x,y)$ and the condition $\|\phi(x)\|_\infty \leq 1$.

Now for each $x \in [a,b]$ and $\phi(x) \in C[a,b]$

$$\sup_{a\leq y\leq b} |A_{m,y}^k L_k(x,y)\phi(y)-L_k(x,y)\phi(y)| \to 0 \qquad (5.102)$$

as $m \to \infty$. Since the functions $L_k(x,y)\phi(y)$, as functions in y parametrized by $x \in [a,b]$ and with $\|\phi(x)\|_\infty \leq 1$, are uniformly bounded and equicontinuous, the convergence in (5.102) is uniform when x and $\|\phi(x)\|_\infty$ are bounded in this way. Thus, given $\varepsilon > 0$, there exists m_0 such that for all $m \geq m_0$, for all $x \in [a,b]$, and for all $\phi(x) \in C[a,b]$ with $\|\phi(x)\|_\infty \leq 1$, $\|A_{m,y}^k L_k(x,y)\phi(y)-L_k(x,y)\phi(y)\|_\infty \leq \varepsilon$.

We now have

$$|\psi_k(x_1,y)-\psi_k(x_2,y)| \leq |A^k_{m,y}[L_k(x_1,y)\phi(y)]-L_k(x_1,y)\phi(y)| +$$

$$+ |L_k(x_1,y)\phi(y)-L_k(x_2,y)\phi(y)| + |L_k(x_2,y)\phi(y)-A^k_{m,y}[L_k(x_2,y)\phi(y)]| \leq 3\varepsilon ,$$

if $|x_1-x_2| < \delta$, $x_1,x_2 \in [a,b]$ and $m \geq m_0$. It follows that the set $\{A^k_{m,y}[L_k(x,y)\phi(y)] \mid m \geq m_0\}$ is equicontinuous (and the equicontinuity of $F_k = \{(H^k_m\phi)(x) \mid m = 0,1,2,\ldots; \ \|\phi(x)\|_\infty \leq 1\}$ then follows from eqn (5.101)).

Remark. The collective compactness of $\{H^k_m \mid m = 0,1,2,\ldots\}$ is established through the boundedness and equicontinuity of F_k.

5.14.2.

We are now in a position to apply Theorem 5.9. We set $X = C[a,b]$ and take the uniform norm, and define

$$K\phi(x) = \int_a^b K(x,y)\phi(y)\mathrm{d}y$$

and

$$H_m\phi(x) = \sum_{j=0}^{n} \sum_{k=0}^{N} \nu_{j,k}(x)L_k(x,z_j)\phi(z_j) ,$$

where the functions $\nu_{j,k}(x)$ are defined in Theorem 5.2.

We have established that Theorem 5.9 applies, so that we deduce that for m sufficiently large $(I-\lambda H_m)^{-1}$ exists and is uniformly bounded in norm, and hence $\tilde{f}_m(x)$ is defined uniquely by the equation $\tilde{f}_m(x)-\lambda(H_m\tilde{f}_m)(x) = g(x)$. Now $f(x)-\lambda(H_mf)(x) = g(x)-\lambda(H_m-K)f(x)$, so that, if $f(x)-\tilde{f}_m(x) = e_m(x)$, $e_m(x)-\lambda(H_me_m)(x) = \lambda(H_m-K)f(x)$. Consequently $e_m(x) = \lambda(I-\lambda H_m)^{-1}(H_m-K)f(x)$ and, from Theorem 5.9, $\|(I-\lambda H_m)^{-1}\|_\infty = O(1)$ so that

$$\|e_m(x)\|_\infty = O(\ \|(H_m-K)f(x)\|_\infty) \tag{5.103}$$

as $m \to \infty$. Since $\|(H_m-K)f(x)\|_\infty \to 0$, when $f(x) \in C[a,b]$,

$\|e_m(x)\|_\infty \to 0$ as $m \to \infty$. We also establish that the rate of convergence is determined by the rate of convergence of $(H_m-K)f(x)$ to zero, that is by the *local truncation error*. Thus the smoothness of the solution $f(x)$ is a factor in the rate of convergence. Assuming the conditions of Theorem 5.2, the behaviour of the local truncation error is known from section 2.17, and the result of Theorem 5.2 follows completely.

From the foregoing remarks (and in particular from (5.103)) we see that $\|f(x)-\tilde{f}(x)\|_\infty = O(\|\tau(x)\|_\infty)$ as $m \to \infty$, where $\tau(x) \equiv \tau_m(x)$ is the local truncation error: $\tau(x) = f(x)-\lambda H_m f(x) - g(x)$. We also suggested, earlier, that $\sup|f(z_i)-\tilde{f}(z_i)| = O(\|\underset{\sim}{\tau}\|_\infty)$ where $\underset{\sim}{\tau}$ is the vector with components $\tau(z_i)$. Let us see how this follows from the analysis already established. We shall, for convenience, write A_m^0 as A_m (other choices of A_m would suffice equally).

We know that $\|(I-\lambda H_m)^{-1}\| = O(1)$ by our application of the general theory. However, the values $\tilde{f}(z_0), \tilde{f}(z_1), \ldots, \tilde{f}(z_n)$ form the components of a vector $\underset{\sim}{\tilde{f}}_m$ which may be determined by solving a system of equations of the form

$$(\underset{\sim}{I}-\lambda \underset{\sim}{H}_m)\underset{\sim}{\tilde{f}}_m = \underset{\sim}{g}$$

where $\underset{\sim}{g}$ has components $g(z_i)$, $(i = 0,1,2,\ldots,n)$, and in the case $N = 0$, $\underset{\sim}{H}_m$ has entries $v_{0,j}(z_i)L(z_i,z_j))$. If we replace $\underset{\sim}{\tilde{f}}_m$ by $\underset{\sim}{f} = [f(z_0), f(z_1),\ldots, f(z_n)]^T$ then we must replace $\underset{\sim}{g}$ by $\underset{\sim}{g} + \underset{\sim}{\tau}$, in the preceding equation, where $\underset{\sim}{\tau} = [\tau(z_0), \tau(z_1),\ldots, \tau(z_n)]^T$. It follows immediately that $(\underset{\sim}{I}-\lambda\underset{\sim}{H}_m)(\underset{\sim}{f}-\underset{\sim}{\tilde{f}}_m) = \underset{\sim}{\tau}$ and if we can show that $\|(\underset{\sim}{I}-\lambda\underset{\sim}{H}_m)^{-1}\|_\infty = O(1)$ then the result required follows, namely $\|\underset{\sim}{f}-\underset{\sim}{\tilde{f}}_m\| = O(\|\underset{\sim}{\tau}\|_\infty)$.

Now suppose $\gamma_0, \gamma_1, \ldots, \gamma_n$ are arbitrary numbers. Define $\gamma(x)$ by requiring $\gamma(z_i) = \gamma_i$ $(i = 0,1,2,\ldots,n)$ and $\gamma(x) = A_m\gamma(x)$ (where $n = n(m)$). Then

$$\|\gamma(x)\|_\infty \le \|A_m\|_\infty \|\underset{\sim}{\gamma}\|_\infty ,$$

with equality being obtained for certain values of $\gamma_0, \ldots, \gamma_n$.† Now

$$\|\tilde{\phi}_m\|_\infty \le \|\tilde{\phi}_m(x)\|_\infty \le \|(I-\lambda H_m)^{-1}\|_\infty \ \|\gamma(x)\|_\infty ,$$

where $(\underset{\sim}{I}-\lambda \underset{\sim}{H}_m)\tilde{\phi}_m = \gamma$ and $(I-\lambda H_m)\tilde{\phi}_m(x) = \gamma(x)$. Thus

$$\|\tilde{\phi}_m\| \le \|(I-\lambda H_m)^{-1}\|_\infty \ \|A_m\|_\infty \ \|\underset{\sim}{\gamma}\|_\infty .$$

In this argument, γ is arbitrary so that

$$\|(\underset{\sim}{I}-\lambda \underset{\sim}{H}_m)^{-1}\|_\infty = \sup_{\underset{\sim}{\gamma} \ne \underset{\sim}{0}} \frac{\|\tilde{\phi}_m\|_\infty}{\|\underset{\sim}{\gamma}\|_\infty} \le \|(I-\lambda H_m)^{-1}\|_\infty \|A_m\|_\infty$$

On taking limits as $m \to \infty$ we observe that $\lim_{m\to\infty} \|A_m\|_\infty = O(1)$, in view of the Banach-Steinhaus theorem, with condition (5.97) for the case $k = 0$, and $A_m = A_m^0$. Since $\|(I-\lambda H_m)^{-1}\| = O(1)$ it follows that $\|(\underset{\sim}{I}-\lambda \underset{\sim}{H}_m)^{-1}\|_\infty = O(1)$, as we required.

5.14.3. *Application to the eigenvalue problem*

Let us briefly consider the convergence of the product-integration method for a weakly singular kernel, as heralded by Theorem 5.3. For this purpose we have certain abstract results of Atkinson (1967), which we shall indicate but not investigate in depth. Our aim is to establish a result similar to Theorem 3.12.

We suppose, with Atkinson, that K and H_m $(m = 1,2,3,\ldots)$ are linear operators acting on a Banach space X with norm $\| \ \|$, that $H_m\phi \to K\phi$ for each $\phi \in X$, as $m \to \infty$, and that $\{H_m\}$ is collectively compact. We deduce that K is a compact (that is, completely continuous) operator and hence its eigenvalues $\{\kappa\}$ form a countable set $\sigma(K)$ with zero as the only possible point of accumulation.

Suppose that C is some circle (actually, any positively oriented

---

† Suppose that $\psi(x)$ is any function with $\psi(z_i) = \gamma_i$. Then $A_m\psi(x) = \gamma(x)$ and $\|\gamma(x)\|_\infty \le \|A_m\|_\infty \|\psi(x)\|_\infty$. Clearly we can choose $\psi(x)$ to be a piecewise-linear interpolant to the values γ_i such that $\|\psi(x)\|_\infty = \|\gamma\|_\infty$, and the result follows.

closed contour would suffice), and define

$$E(C;K) = \frac{1}{2\pi i}\int_C (\zeta - K)^{-1} d\zeta \ .$$

If none of the eigenvalues $\kappa \in \sigma(K)$ lie on C, and C does not pass through the origin, then $E(C;K)$ is a projection (Anselone 1971, Chapter 4) and its range $E(C;K)X$ is the linear space spanned by all eigenvectors and generalized eigenvectors which correspond to any eigenvalue $\kappa \in \sigma(K)$ which lies *inside* C. Thus, if C separates a given eigenvalue κ from other points in the closure of $\sigma(K)$, we write $E(C;K) = E(\kappa;K)$ and the dimension of $E(\kappa;K)X$ is the algebraic multiplicity of κ as an eigenvalue of K. (If C contains more than one eigenvalue than the dimension of $E(C;K)X$ is the sum of the algebraic multiplicities of eigenvalues in C.)

The following theorem, in which the assumptions of this paragraph prevail, may now be stated. The proof of the theorem will not be given here but employs the property that $\|(H_m - K)H_m\|$ and $\|(H_m - K)K\|$ tend to zero as $m \to \infty$. We note in passing that Theorem 3.12 is a corollary of the following result.

THEOREM 5.16. *Let* ρ, ε *be arbitrary positive numbers. Then there exists an integer* M *such that for* $m \geq M$ (a) *if* $\tilde{\kappa} \in \sigma(H_m)$ *with* $|\tilde{\kappa}| > \rho$, *there exists some* $\kappa \in \sigma(K)$ *with* $|\kappa| \geq \rho$ *and* $|\tilde{\kappa} - \kappa| \leq \varepsilon$, (b) *if* $\kappa \in \sigma(K)$ *the dimension of* $E(\kappa;K)X$ *is the dimension of* $E(C;H_m)X$, *where* C *is the circle centred on* κ *with radius* ε, *and* (c) $E(C;H_m)x \to x$ *for every* $x \in E(\kappa;K)X$ *as* $m \to \infty$.

This theorem is a re-statement of results given by Atkinson (1967), and we refer to that paper for a proof. For additional discussion see Chapter 4 of Anselone (1971), and Atkinson (1973).

Let us interpret and apply the preceding result. We have seen that if we take X as $C[a,b]$ with the uniform norm, the conditions of the theorem are fulfilled if K is an integral operator with a weakly singular or continuous kernel, and H_m is an operator arising from the type of product-integration method discussed above.

We shall denote by H_m an operator on $C[a,b]$ with

$$H_m\phi(x) = \sum_{k=0}^{N} \sum_{j=0}^{n} \nu_{k,j}(x)L_k(x,z_j)\phi(z_j) \quad ,$$

where

$$K(x,y) = \sum_{k=0}^{N} L_k(x,y)M_k(x,y) \ .$$

The theorem then yields the result that every non-zero eigenvalue κ of the kernel $K(x,y)$, having algebraic multiplicity q is approximated by eigenvalues $\tilde{\kappa}$ of H_m, tending in the limit to κ. The sum of the algebraic multiplicities of the eigenvalues $\tilde{\kappa}$, which lie inside a circle C which isolates $\kappa \neq 0$ in $\sigma(K)$ from zero and from other eigenvalues of $K(x,y)$, is precisely q (the algebraic multiplicity of κ) when m is sufficiently large. The convergence property (c) in the general theorem also establishes, in the case $q = 1$, that $\|\alpha_m\tilde{f}(x)-f(x)\|_\infty \to 0$ as $m \to \infty$, where $|\alpha_m| \to 1$ and $\|\tilde{f}(x)\|_2 = \|f(x)\|_2 = 1$. For further insight into the above results we refer to Anselone (1971) and his references.† In particular, see Theorem 4.17 of Anselone (1971).

Let us now see whether we may bound the error $|\kappa-\tilde{\kappa}|$ when κ is a simple eigenvalue.

We suppose that, for $x \in [a,b]$,

$$\int_a^b K(x,y)f(y)\,dy = \kappa f(x) \ ,$$

and

$$\int_a^b \psi(y)\overline{K(y,x)}\,dy = \bar{\kappa}\psi(x) \ ,$$

with $\|f(x)\|_2 = \|\psi(x)\|_2 = 1$. Suppose also that the approximate eigenfunction $\tilde{f}(x)$ gives rise to a function

---

† The reader may also be interested in the paper by Vainikko and Dementeva (1968).

$$\eta(x) = \int_a^b K(x,y)\tilde{f}(y)\,dy - \tilde{\kappa}\tilde{f}(x) .$$

The function $\eta(x)$ may be computed (since $\tilde{\kappa}$ is known). Now,

$$\int_a^b \eta(x)\overline{\psi(x)}\,dx = \int_a^b\int_a^b K(x,y)\tilde{f}(y)\overline{\psi(x)}\,dx\,dy - \tilde{\kappa}\int_a^b \tilde{f}(x)\overline{\psi(x)}\,dx ,$$

that is,

$$\begin{aligned}(\eta,\psi) &= (K\tilde{f},\psi)-\tilde{\kappa}(\tilde{f},\psi)\\ &= (\tilde{f},K^*\psi)-\tilde{\kappa}(\tilde{f},\psi)\\ &= (\kappa-\tilde{\kappa})(\tilde{f},\psi) .\end{aligned}$$

Thus we have, as in Chapter 3, an expression relating $\kappa-\tilde{\kappa}$ to $\eta(x)$ and involving the unknown function $\psi(x)$.

Remark. To *estimate* $\kappa-\tilde{\kappa}$ we might compute an approximate eigenfunction $\tilde{\psi}(x)$, and $\eta(x)$, but it is certainly not clear that $\tilde{\psi}(x)$ can be obtained, in general, from the conjugate transpose of the matrix which yields $\tilde{\underset{\sim}{f}}$.

Now suppose $\tilde{f}(x)$ is scaled so that $\|f(x)-\tilde{f}(x)\|_\infty \to 0$, where $f(x)$ is the fixed normalized eigenfunction corresponding to κ . Then $\|\tilde{f}(x)\|_2 \to 1$, and $(\tilde{f},\psi) \to (f,\psi)$.

Thus, to show this, we have

$$\|\tilde{f}(x)\|_2^2 = (\tilde{f},\tilde{f}) = (f,f) + (\tilde{f}-f,f) + (f,\tilde{f}-f) + (\tilde{f}-f,\tilde{f}-f) ,$$

so that

$$\left|\,\|\tilde{f}(x)\|_2^2 - \|f(x)\|_2^2\,\right| \le 2\,\|f(x)\|_2\|\tilde{f}(x)-f(x)\|_2 + \|\tilde{f}(x)-f(x)\|_2^2.$$

Now since $\tilde{f}(x)$ and $f(x)$ are continuous

$$\|f(x)-\tilde{f}(x)\|_2 \le (b-a)^{\frac{1}{2}}\,\|f(x)-\tilde{f}(x)\|_\infty \to 0 .$$

Thus $\|\tilde{f}(x)\|_2 \to \|f(x)\|_2 = 1$. Again,

$$|(\tilde{f},\psi)-(f,\psi)| \le \|\tilde{f}(x)-f(x)\|_2 \|\psi(x)\|_2 = \|\tilde{f}(x)-f(x)\|_2 ,$$

which tends to zero. Our claims are then established.

Now $(f,\psi) \neq 0$ since κ is a simple eigenvalue. Hence, for m sufficiently large, $(\tilde{f},\psi) \neq 0$. Thus

$$|\kappa-\tilde{\kappa}| \le |(\eta,\psi)|/|(\tilde{f},\psi)| \le \|\eta(x)\|_2/|(\tilde{f},\psi)| ,$$

since $\|\psi(x)\|_2 = 1$. Hence,†

$$|\kappa-\tilde{\kappa}| \le \frac{\|\eta(x)\|_2}{|(f,\psi)|}\{1+o(1)\} ,$$

and we have an asymptotic bound for $|\kappa-\tilde{\kappa}|$ in terms of $\eta(x)$. Now $\|\eta(x)\|_2$ could be computed but we can also obtain an asymptotic bound for this quantity in terms of the norm of $(K-H_m)H_m$. Thus we have

$$\eta(x) = K\tilde{f}(x)-\tilde{\kappa}\tilde{f}(x) = K\tilde{f}(x)-H_m\tilde{f}(x) = (K-H_m)H_m\tilde{f}(x)/\tilde{\kappa}$$

if m is sufficiently large so that $\tilde{\kappa} \neq 0$. Then

$$\|\eta(x)\|_2 \le (b-a)^{\frac{1}{2}} \|\eta(x)\|_\infty ,$$

and we see that

$$\|\eta(x)\|_\infty \le \|(K-H_m)H_m\|_\infty \|\tilde{f}(x)\|_\infty/|\tilde{\kappa}|$$

$$= \|(K-H_m)H_m\|_\infty \|\tilde{f}(x)\|_\infty\{1+o(1)\}/|\kappa| .$$

(If desired, the norm $\|\tilde{f}(x)\|_\infty$ can be bounded in terms of $\|\tilde{f}(x)\|_2$,

---

† Since the result given is an asymptotic bound, it remains true when the function $\eta(x)$ corresponds to an approximate eigenfunction $\tilde{f}(x)$ scaled so that $\|\tilde{f}(x)\|_2 = 1$.

$K(x,y)$, and κ.)

Observe that we might prefer to obtain bounds in terms of $\|\tau(x)\|_2$, where $\tau(x) = H_m f(x) - \kappa f(x) = (H_m - K)f(x)$, since the order of $\tau(x)$ can be determined using Peano theory and the differentiability of $f(x)$. The difference between $\|\tau(x)\|_2$ and $\|\eta(x)\|_2$ can be bounded if required.

5.15. *Fredholm equations of the first kind*

The Fredholm equation of the first kind,

$$\int_a^b K(x,y) f(y)\,dy = g(x) \ , \quad x \in [a,b], \qquad (5.104)$$

frequently poses difficulties when a theoretical or numerical solution is sought. In 1957, Tricomi wrote: 'Some mathematicians still have a kind of fear whenever they encounter a Fredholm equation of the first kind ... Today this fear is no longer justified.' In practical terms we take the contrary view, since accurate results are difficult to obtain. Without care we may well find ourselves computing 'approximate solutions' for problems which have no true solution.

In part the difficulty arises with the precise interpretation of the problem associated with eqn (5.104). Do we, for example, seek a *continuous* solution $f(x)$, or a *square-integrable* solution, or an *absolutely integrable solution*? It is possible that the last will exist though the first does not; the standard theory is an L^2-theory.

Example 5.26.
With $a = 0$, $b = 1$, $K(x,y) = x(1-y)$ if $x \le y$, and $K(y,x) = K(x,y)$, there is a solution of eqn (5.104) only if $g(0) = g(1) = 0$ and $g''(x)$ exists. The solution $f(x) = -g''(x)$ is then continuous if $g''(x)$ is continuous. *

Recall† that when $g(x)$ is square-integrable, and

$$\int_a^b\int_a^b |K(x,y)|^2 dx\,dy < \infty ,$$

and we seek a square-integrable solution $f(x)$, then this solution exists provided that (i) $(g,\psi) = 0$ for every square-integrable function

† See Smithies (1962) and Chapter 1, section 1.8, above.

$\psi(x)$ such that $(K^*\psi)(x) = 0$, and (ii) $\Sigma\, \mu_n^2 |(g,u_n)|^2 < \infty$, where $\{u_n(x), v_n(x);\ \mu_n\}$ is the singular system (section 1.8) of $K(x,y)$. (Then, an L^2-solution is obtained as $f(x) = \sum_{n=0}^{\infty} \mu_n (g,u_n) v_n(x)$, and this solution is unique in $L^2[a,b]$ provided that there is no square-integrable function $f_0(x)$ such that $(Kf_0)(x) = 0$.) Instead of this approach, a more direct enquiry is often informative.

Example 5.27.

Consider the equation

$$\int_{-1}^{1} [f(y)/\{a^2+(x-y)^2\}]\mathrm{d}y = 1 ,$$

for $|x| \leq 1$, where $a \neq 0$. This equation has no (integrable) solution.

(Proof: Consider, for any integrable function $f(x)$, the transform

$$F(z) = \int_{-1}^{1} [f(t)/\{a^2+(z-t)^2\}]\mathrm{d}t ,$$

defined for z in the complex plane. $F(z)$ is regular in the whole plane except when $z = x \pm ia$, $-1 < x < 1$. The segment $|x| < 1$ of the real line is in the region of regularity, so if $F(x) = 1$ for $|x| < 1$, $F(z) \equiv 1$ if $|\mathrm{Im}(z)| < a$. But from the expression for $F(z)$, $\lim_{z\to\infty} F(z) = 0$. Thus there is no solution to the equation $F(x) = 1$, $|x| \leq 1$. This example and proof were communicated by F.J. Ursell.) *

Example 5.28.

The equation

$$\int_0^1 (x+y)f(y)\mathrm{d}y = g(x)$$

has a solution only if $g(x)$ is linear and it then has infinitely many solutions. *

The kernels in the preceding examples are continuous and the theory for square-integrable kernels (Smithies 1962, p.164; Tricomi 1957, p.143; Goursat 1964, III(2), p.157) applies. We gave some indication of this theory in section 1.8, and above. Observe that the weakly singular kernel $H(x,y)/|x-y|^{\alpha}$ is square-integrable if $\alpha \in [0,\frac{1}{2})$ and $H(x,y)$ is square-integrable.

It is frequently suggested that the smoother the kernel $K(x,y)$, the more ill conditioned is the equation of the first kind. (This is explained heuristically for square-integrable kernels by examining the rate of increase of the singular values of the kernel $K(x,y)$ since the faster this rate the more ill conditioned is the problem, and smooth kernels appear to correspond to a rapid increase in the sequence of singular values.[†] The general conclusion seems to provide a crude rule of thumb rather than a precise guide to the nature of the problem. We shall see that difficulties can arise with weakly singular kernels, and that the solution can depend on arbitrary constants when the integral is a Cauchy principal value.

Example 5.29.

Consider the equation (Smirnov 1964, Vol. IV, p.174-5)

$$\int_{-a}^{a} \{f(y)/(x-y)\}dy = g(x) \quad (-a < x < a) ,$$

where the integral is a *Cauchy principal value* and $g(x)$ is continuous. The general solution is

$$f(x) = \{C- \frac{1}{\pi} \int_{-a}^{a} \frac{g(y)\sqrt{(a^2-y^2)}}{x-y}dy\}/\{\pi(a^2-x^2)\}^{\frac{1}{2}} ,$$

where C is arbitrary $(C = -\int_{-a}^{a} f(y)dy)$.

Thus *an additional condition is required* to define $f(x)$. *

Example 5.30.

Consider the equation

$$\int_{0}^{1} \ln|x-y|f(y)dy = g(x) , \ 0 \le x \le 1, \qquad (5.105)$$

† For a Hermitian kernel, the singular values are the moduli of the characteristic values. We may compare the case $K(x,y) = \ln|x-y|$ (see section 5.8) and the kernel of Example 3.7.

where $g'(x)$ is continuous. Differentiating with respect to x, we obtain

$$\int_0^1 \{f(y)/(x-y)\}dy = g'(x) ,$$

and the transformation $u = 2x-1$, $v = 2y-1$ yields the equation

$$\int_{-1}^1 \{\psi(v)/(u-v)\}dv = g'(\tfrac{1}{2}(u+1)) ,$$

where $\psi(v) = f(\tfrac{1}{2}(v+1))$. Using the result of the previous Example and determining the constant C from eqn (5.105) we find

$$f(x) = \left[\frac{-1}{2\ln 2}\int_0^1 \frac{g(y)}{\surd\{y(1-y)\}}dy+\int_0^1 \frac{g'(y)\surd\{y(1-y)\}}{y-x}dy\right]\Big/\left[\pi^2\surd\{x(1-x)\}\right] .*$$

Example 5.31.

Consider the equation $\int_{-a}^{a} \ln|x-y|f(y)dy = \pi$. There is no solution if $a = 2$. If $a = 1$, the solution is $f(x) = -1/\{\ln 2\surd(1-x^2)\}$, and in general if $a \neq 2, f(x) = 1/\{\ln(\tfrac{1}{2}a)\surd(a^2-x^2)\}$.

We may establish these results by appealing to Example 5.30. Writing $x = 2au-a$, $y = 2av-a$, where $a > 0$, we obtain

$$\int_0^1 \ln|u-v|\psi(v)dv = \frac{\pi}{2a} - \ln 2a \int_0^1 \psi(v)dv , \tag{5.106}$$

where $\psi(u) = f(2au-a)$. We regard the right hand-side in (5.106) as a function $\gamma(u)$ (with $\gamma'(u) = 0$) and the preceding Example establishes that any solution has the form $\psi(u) = \{u(1-u)\}^{-\frac{1}{2}}\Gamma$. (Since $\int_0^1 \{u(1-u)\}^{-\frac{1}{2}}du = \pi$, $\Gamma = \int_0^1 \psi(u)du/\pi$.) Multiply (5.106) by $\{u(1-u)\}^{-\frac{1}{2}}$ and integrate with respect to u, noting that

$$\int_0^1 \{u(1-u)\}^{-\frac{1}{2}}\ln|u-v|du = -2\pi\ln 2,$$

and we find

$$-2\pi\ln 2\int_0^1 \psi(v)\,dv = \{\pi^2/2a\}-\pi\ln 2a\int_0^1 \psi(v)\,dv\ .$$

If $a = 2$ this implies that $\frac{1}{4}\pi^2 = 0$ which is a contradiction, and no Γ exists. If $a = 1$ then $\int_0^1 \psi(v)\,dv = -\frac{1}{2}\pi/\ln 2$, and the solution follows. *

Some of the previous Examples illustrate a completely different approach from the use of singular values and singular functions. It should be noted that in Example 5.31, the solution given (for $a \neq 2$) is not square-integrable, so that the theory of section 1.8 is not applicable. Other approaches are, therefore, fruitful. In particular if the kernel $K(x,y)$ is a difference kernel, say $k(x-y)$, and $k(x)$ satisfies a certain type of differential equation, *Latta's method* can be used; references and a brief exposition are provided by Cochran (1972).

Example 5.32.

For $0 < \alpha < 1$, if

$$\int_{-1}^{1} f(y)\,|x-y|^{-\alpha}\,dy = \pi x \quad (x \in (-1,1))$$

then $f(x) = \alpha^{-1}x(1-x^2)^{\frac{1}{2}(\alpha-1)}\cos\frac{1}{2}\pi\alpha$. This result (Cochran 1972) can be established by Latta's method.

Shinbrot (1958) extends Latta's method to deal with equations in which $K(x,y) = k(x-y)$ and $k(x) = p(x)j(x)+q(x)$, where $p(x)$, $q(x)$ are polynomials and $j(x)$ satisfies a certain type of differential equation. (When $p(x)$ and $q(x)$ are not polynomials but are continuous, we may be able to approximate them by polynomials; a rôle for numerical analysis is apparent here.) Kernels of this type arise frequently in practice. *

The preceding Examples provide varying degrees of insight into the nature and existence of solutions of equations of the first kind. Clearly, any numerical method would have to be very versatile if it were to deal with all types of equations, since the solution may not exist in some sense. (If the only solutions are unbounded then clearly there are no continuous solutions!)

Any theoretical insight into the expected behaviour of the solution should certainly be fully exploited. In particular, certain equations of the first kind are equivalent to equations of the second kind, for which there is generally more theory available (see Bückner's contribution in Todd (1962) for an example of this kind).

The discussion of Fredholm integral equations of the first kind is sometimes set in an abstract framework of 'improperly' or 'incorrectly' posed problems. (See Lavrentiev (1967). At one time it was thought that such problems could not arise from physically meaningful problems, but this view is no longer considered correct.) In an abstract setting, such problems arise when we consider a compact operator K mapping an infinite-dimensional Banach space X into itself, and we seek a solution $f \in X$ to the equation $Kf = g$, where $g \in X$. Such an equation does not always have a solution since the range $K(X) \subset X$ $(K(X) \neq X)$.

<u>Example 5.33</u>.
Choose $X = L^2[a,b]$ with the norm $\| \ \|_2$. Suppose $K(x,y)$ is continuous for $a \leq x, y \leq b$. If $f(x) \in L^2[a,b]$ then

$$Kf(x) = \int_a^b K(x,y)f(y)\,dy$$

is continuous. Thus $K(X) \subseteq C[a,b] \subset L^2[a,b]$. If $g(x) \in L^2[a,b]$ is not continuous, then the equation cannot have a square-integrable solution. *

Further, let us consider a set $S \subset X$ of elements with bounded norm; if $f \in S$, $\|f\| \leq B$, say. Then, since K is compact, $K(S)$ is sequentially compact in X. Consider, then, an infinite sequence of elements $\{f_n\}$ in S; since $K(S)$ has compact closure, there exists a subsequence $\{f_{n_r}\}$ such that Kf_{n_r} converges in norm to some $\gamma \in X$. The sequence $\{f_{n_r}\}$ *need have no limit point*. Thus we can satisfy the equation $K\phi = \gamma + \varepsilon_{n_r}$, with ε_{n_r} arbitrarily small, by setting $\phi = f_{n_r}$, and the functions f_{n_r} need have no limit point.

<u>Example 5.34</u>.
Choose $X = C[a,b]$ with the uniform norm, and let $K(x,y)$ be continuous.

Take for S the functions $\{f_n(x) = \cos nx \mid n = 0,1,2,\ldots\}$. Each function has a (uniform) norm less than or equal to 1. There is no uniform limit of the functions $f_n(x)$ as $n \to \infty$. On the other hand, $\|Kf_n(x)\|_\infty \to 0$ (appealing to the Riemann-Lebesgue theorem) so that with $\gamma(x) \equiv 0$, $\|Kf_n(x) - \gamma(x)\|_\infty \to 0$, and we may take $n_r = n$ in the general discussion.

Suppose, further, that the equation $\int_a^b K(x,y)f(y)\,dy = g(x) \not\equiv 0$ has a solution $f(x)$. Then $\phi_n(x) = f(x) + \cos nx$ is a function with $\int_a^b K(x,y)\phi_n(y)\,dy = g(x) + \varepsilon_n$, where ε_n can be made arbitratily small by taking n large enough, but the difference between $\phi_n(x)$ and $f(x)$ is large. *

The abstract framework for a discussion of ill-posed problems depends on a choice of Banach space with norm $\|\ \|$. Clearly, different norms result in different conclusions. However, it is worth noting that we still have difficulties when $K(x,y)$ is weakly singular (though perhaps not as severe as in the continuous case) since $K(x,y)$ gives rise to a compact operator on the space $C[a,b]$ with uniform norm.

If we examine carefully the difficulty which arises in Example 5.34, we shall see that the problem arises because the set of functions considered as possible solutions is only restricted by a bound on the uniform norm. If the set in which the solution is sought were a compact set, any infinite sequence $\{f_n(x)\}$ would have a subsequence converging to a limit point. (Proceeding this way we could eliminate contributions from the functions $\cos nx$, since these functions are not equicontinuous and hence do not form a compact set in $C[a,b]$.) This observation prompts us to restrict the class of functions in which we seek an approximate solution, and actually provides the underlying motivation for certain regularization methods, in which we consider an integro-differential equation in place of the integral equation of the first kind. To adopt such an approach, the existence of a solution in a certain class of functions (for example, a continuously differentiable solution) should be assumed.

5.15.1. *Introduction to numerical methods*

In the following two sections we shall investigate some numerical methods for the approximate solution of Fredholm equations of the first kind. The material presented will provide an introduction to the research literature; since the problem is inherently ill posed the practical application of a method is a delicate matter.

A naive approach to the construction of schemes for the approximate solution of the equation

$$\int_a^b K(x,y)f(y)\,dy = g(x)$$

might involve constructing analogues of methods for the equation

$$f(x) - \lambda\int_a^b K(x,y)f(y)\,dy = g(x)$$

(in short, setting $\lambda = -1$ and dropping the free term $f(x)$). We shall refer to any such scheme as *naive*. In practice, such naive methods cannot *generally* be relied upon; in particular we shall show (section 5.16) that the quadrature method cannot be recommended, although in isolated cases such methods may succeed.† However, solutions of related least-squares problems are sometimes useful, as we shall see later.

It was mentioned that in some cases the natural analogues of methods for equations of the second kind may succeed with equations of the first kind. There are some *heuristic* arguments which suggest conditions under which we may be successful. Consider as an example of a naive method, the use of a quadrature rule $J(\phi) = \sum_{j=0}^{n} w_j\phi(y_j)$ to produce a system of equations (which we discuss in more detail later) of the form

$$\sum_{j=0}^{n} w_j K(y_i,y_j)\tilde{f}(y_j) = g(y_i) .$$

The matrix $\underset{\sim}{K}$ with entries $K(y_i,y_j)$ may be singular, but in general we may expect that the condition number of the coefficient matrix bears some resemblance to the inverse ratio of the first and $(n+1)$th singular

---

† See, for example, the observation by Miller (Delves and Walsh 1974, p.181 (section 7)).

values of $K(x,y)$; if $\mu_n \to \infty$ rapidly as $n \to \infty$ then we expect rapid onset of ill-conditioning. If therefore we can achieve *high accuracy* in the local truncation error with a *low-order matrix* we shall hope that our approximation is a good one. (If $f(x)$ is known to be a unique solution of the problem and

$$\sum_{j=0}^{n} w_j K(y_i, y_j) f(y_j) = g(y_i) + \tau_i ,$$

then we require the local truncation errors τ_i to be small.)

The preceding discussion is by no means rigorous, and with other naive methods it is less easy to see a connection between the conditioning of the problem and the singular values of the kernel. I believe the general impression to be correct, however.

An alternative approach, rather than adapting methods for equations of the second kind, is to mimic the theoretical development. In the case where the L^2-theory applies, we may imitate the construction of a series $\sum_{n=0}^{\infty} \mu_n (g,u_n) v_n(x)$ which converges, in the mean, to the solution (see the comment before Example 5.27). To mimic this approach involves the construction of a finite sum $\sum_{n=0}^{m} \tilde{\mu}_n (g,\tilde{u}_n)\tilde{v}_n(x)$, say, where $\tilde{\mu}_n \simeq \mu_n$, etc. The system $\{u_n(x), v_n(x); \mu_n\}$ is here the singular system of the kernel $K(x,y)$. It transpires that approximations $\tilde{\mu}_n, \tilde{u}_n(x), \tilde{v}_n(x)$ are sometimes implicit in a direct application of a singular value decomposition when we solve, in a least-squares sense (see section 2.20), over-determined equations arising from a naive method.

There are two disadvantages in basing methods on the solution $\sum_{n=0}^{\infty} \mu_n (g,u_n) v_n(x)$. In the first case the series converges in the mean but not necessarily uniformly, and the same may be true of an approximation $\Sigma\tilde{\mu}_n (g,\tilde{u}_n)\tilde{v}_n(x)$. Secondly, $\mu_n \to \infty$ as $n \to \infty$, and it is generally the case that accurate approximations to $\mu_n, u_n(x)$, and $v_n(x)$ are increasingly difficult to obtain as $n \to \infty$. For this reason it is sometimes proposed that $\tilde{\mu}_n$ be replaced by a value $\tilde{\nu}_n$ such that

$\tilde{v}_n$ has a finite limit point. Inaccuracies in $\tilde{u}_n(x), \tilde{v}_n(x)$ for large n are then damped out. One such technique is equivalent to replacing the equation

$$\int_a^b K(x,y) f(y) \, dy = g(x)$$

by the equation

$$\alpha^2 f_\alpha(x) + \int_a^b K^* K(x,y) f_\alpha(y) \, dy = K^* g(x) ,$$

where

$$K^* K(x,y) = \int_a^b \overline{K(z,x)} \, K(z,y) \, dz$$

and

$$K^* g(x) = \int_a^b \overline{K(z,x)} \, g(z) \, dz ,$$

and applying a numerical method to this equation of the second kind. Replacement of the equation of the first kind by a well-posed equation of the second kind is an example of a type of 'regularization technique'. It may be established that $\|f_\alpha(x) - f(x)\|_2 \to 0$ as $\alpha \to 0$, but this convergence is not necessarily uniform (and a quadrature method applied to the regularized equation of the second kind can also give an approximate solution which does not converge uniformly).

<u>Remark</u>. The functional equations $(\alpha^2 I + K^* K) f_\alpha(x) = K^* g(x)$, as $\alpha \to 0$, are related to the generalized inverse of K. Insight will be obtained in terms of the generalized inverse of a matrix if K, I are replaced by matrices and $f_\alpha(x), g(x)$ are replaced by vectors. The generalized inverse of a matrix is related to the least-squares problem. The vector $\underset{\sim}{x}$ of minimum norm, $\|\underset{\sim}{x}\|_2$, which minimizes $\|\underset{\sim}{A}\underset{\sim}{x} - \underset{\sim}{b}\|_2$ (where $\underset{\sim}{A}$ may be rectangular) is $\underset{\sim}{x} = \underset{\sim}{A}^+ \underset{\sim}{b}$, where $\underset{\sim}{A}^+$ is the generalized inverse of $\underset{\sim}{A}$ and one definition of $\underset{\sim}{A}^+$ is $\underset{\sim}{A}^+ = \lim_{\alpha \to 0} (\alpha^2 \underset{\sim}{I} + \underset{\sim}{A}^* \underset{\sim}{A})^{-1} \underset{\sim}{A}^*$.

A second type of regularization technique involves replacing the integral equation by an integro-differential equation with boundary conditions. The integro-differential equation depends on a parameter γ and has a corresponding solution $f_\gamma(x)$. (When $\gamma = 0$, an equation of the first kind is obtained.) Since $f_\gamma(x)$ satisfies an integro-differential equation it is naturally differentiable; $f_\gamma(x)$ provides, for each $\gamma > 0$, an approximation to $f(x)$ and the approximate solution is chosen from a restricted class of functions. If $f(x)$ (assumed unique) is in the same class then this is reasonable. For some such methods we can show that $\|f_\gamma(x)-f(x)\|_\infty \to 0$ as $\gamma \to 0$. (The argument is related to our remarks on finding an approximate solution from a compact set of functions.) Historically such methods apparently arose from finite-difference analogues of this approach.

Such topics will be investigated further below. Throughout we shall consider the mathematically posed problem of solving an equation without experimental errors. For further reading we refer to the excellent contribution by Miller (Delves and Walsh 1974) and the references which he gives.

5.16. *Quadrature methods and expansion methods*

Let us investigate briefly how a quadrature method may be set up for an equation of the first kind. We shall then indicate the *potentially* unsatisfactory nature of the method. Modifications of the method, intended to alleviate some of its shortcomings, will be discussed later.

From eqn (5.104) we have, on replacing the integral by a rule with weights w_j and abscissae y_j $(j = 0,1,2,\ldots,n)$, the approximation

$$\sum_{j=0}^{n} w_j K(x,y_j)f(y_j) \simeq g(x) .$$

If we write

$$\sum_{j=0}^{n} w_j K(x,y_j)\tilde{f}(y_j) = g(x) \tag{5.107}$$

and seek values of $\tilde{f}(y_j)$ $(j = 0,1,2,\ldots,n)$ satisfying (5.107), we shall not in general find a solution. A solution exists only if $g(x)$ is a linear combination of the functions $K(x,y_j)$ $(j = 0,1,2,\ldots,n)$

and, even if the integral equation has a solution, this is not guaranteed. However, if we select $z_0, z_1, \ldots, z_m$ we may be able to satisfy the equations

$$\sum_{j=0}^{n} w_j K(z_i, y_j) \tilde{f}(y_j) = g(z_i) \quad (i = 0,1,2,\ldots,m) . \tag{5.108}$$

We shall suppose that $m = n$, until further notice, and these equations then define $\tilde{f}(y_0), \ldots, \tilde{f}(y_n)$ provided $\det[K(z_i, y_j)] \neq 0$ (in particular, if the functions $K(x, y_j)$ form a Haar set). There is no necessity in these equations to choose $z_i = y_i$, though it may be thought natural to do so. We shall demonstrate by examples that eqns (5.108) do *not* lead to a generally reliable method of approximating the values of a solution of the integral equation. We shall proceed under the general assumption that $K(x,y)$ and $g(x)$ are continuous.

Example 5.35.
Consider the equation in Example 5.26, (with $K(x,y) = x(1-y)$ if $x \leq y$, $K(y,x) = K(x,y)$) and suppose that we use the repeated trapezium rule to set up (5.108). We take $z_i = y_i = ih$, where $h = 1/n$ $(a=0, b=1)$. Now in this example, $K(0,y) = K(1,y) = K(x,0) = = K(x,1) = 0$ for $0 \leq x, y \leq 1$. Thus eqns (5.108) reduce to the form

$$\sum_{j=1}^{n-1} hK(ih, jh) \tilde{f}(jh) = g(ih) \quad (i = 0,1,2,\ldots,n) ,$$

where corresponding to $i = 0$ we have $g(0) = 0$ and corresponding to $i = n$ we have $g(1) = 0$. Thus eqns (5.108) have, in this case, no solution unless $g(0) = 0$ and $g(1) = 0$. The values $\tilde{f}(0)$ and $\tilde{f}(1)$ are not determined by the equations. Suppose $g(0) = g(1) = 0$ (otherwise¶ there is no solution of the integral equation). Then we have $(n-1)$ equations for $\tilde{f}(h), \tilde{f}(2h), \ldots, \tilde{f}(1-h)$. Because of the particular form of the matrix of coefficients, with $hK(ih, jh) = ih^2(1-jh)$ for $i \leq j$ and $K(jh, ih) = K(ih, jh)$, we can solve these equations explicitly. Thus

¶ If $g(0)$ and $g(1)$ do not vanish, it is possible that eqn (5.104) can be satisfied *almost everywhere*, $K(x,y)$ being the kernel under discussion.

$$\tilde{\mathbf{f}} \equiv \begin{bmatrix} \tilde{f}(h) \\ \tilde{f}(2h) \\ \tilde{f}(3h) \\ \cdot \\ \cdot \\ \cdot \\ \tilde{f}(1-h) \end{bmatrix} = \frac{-1}{h^2} \begin{bmatrix} -2 & 1 & & & \\ 1 & -2 & 1 & & \\ & 1 & -2 & 1 & \\ & & \cdot & \cdot \quad \cdot & \\ & & & \cdot \quad \cdot & 1 \\ & & & 1 & -2 \end{bmatrix} \begin{bmatrix} g(h) \\ g(2h) \\ g(3h) \\ \cdot \\ \cdot \\ \cdot \\ g(1-h) \end{bmatrix}$$

and $\tilde{f}(ih) = -\{g(\{i-1\}h)-2g(ih)+g(\{i+1\}h)\}/h^2$, where $g(0) = 0$ and $g(1) = 0$. Clearly $\tilde{f}(ih)$ is an approximation for $f(ih) = -g''(ih)$, where $g''(x)$ is assumed to exist. (If $g^{iv}(x)$ exists, $\tilde{f}(ih) = -g''(ih)+O(h^2)$.)

Suppose that $h_0 = 1/n_0$ and that $h \to 0$ through the sequence $h_0, \frac{1}{2}h_0, \frac{1}{4}h_0, \ldots$. If $i \in \{1,2,3,\ldots,n_0-1\}$ we obtain a sequence of approximate values $\tilde{f}(\hat{x})$, where $\hat{x} = ih_0$, and under these conditions

$$\lim_{h\to 0} \tilde{f}(\hat{x}) = -g''(\hat{x}) ,$$

provided $g''(x) \in C[0,1]$. Thus the use of the trapezium rule is satisfactory here, *provided* that special account is taken of zero elements in the border of the matrix of coefficients in our version of eqn (5.108).

Let us now suppose that we employ the repeated Simpson's rule in place of the trapezium rule.† It is now necessary to take n even. If we discard zero elements as before, we are led to equations of the form

---

† The choice of Simpson's rule is not the most natural, because the integrand has a discontinuous derivative, at $x = y$. However, its local truncation error is at least $O(h^2)$, as is that of the trapezium rule, when $g^{iv}(x)$ is continuous.

$$\begin{bmatrix} \frac{4}{3}\tilde{f}(h) \\ \frac{2}{3}\tilde{f}(2h) \\ \frac{4}{3}\tilde{f}(3h) \\ \cdot \\ \cdot \\ \cdot \\ \frac{4}{3}\tilde{f}(1-h) \end{bmatrix} = \frac{-1}{h^2} \begin{bmatrix} -2 & 1 & & & \\ 1 & -2 & 1 & & \\ & 1 & -2 & 1 & \\ & & \ddots & \ddots & \ddots \\ & & & & 1 \\ & & & 1 & -2 \end{bmatrix} \begin{bmatrix} g(h) \\ g(2h) \\ g(3h) \\ \\ \\ \\ g(1-h) \end{bmatrix}.$$

Here, $h = 1/n$, where $n = 2m$. Suppose that $h_0 = 1/2m_0$ and $m \to \infty$ through the sequence of values $m = m_0, 2m_0, 3m_0 \ldots$, and we again hold $\hat{x} = ih_0$ *fixed* (with $1 \le i \le 2m_0 - 1$). Then the values $\tilde{f}(\hat{x})$ have no limit, but oscillate. Indeed, for the subsequence corresponding to $m = m_0, 3m_0, 9m_0, \ldots$ we find, as $m \to \infty$, through powers of three,

$$\tilde{f}(\hat{x}) \to \frac{-3}{4} g''(\hat{x}) \tag{5.109}$$

if $\hat{x} = ih_0$, where i is *odd*, whereas for $m = m_0, 2m_0, 4m_0, \ldots,$

$$\tilde{f}(\hat{x}) \to \frac{-3}{2} g''(\hat{x}) .$$

The disturbing feature of such results is that examining a subset of the available answers, with the (natural) assumption that consistency implies reliability, will lead to an acceptance of wrong results. In this example, taking $m=1,2,4,8,16,\ldots$ gives, for every number $\hat{x} \in [0,1]$ which is expressible as a terminating binary fraction, the value $\lim \tilde{f}(\hat{x}) = \frac{3}{2}f(\hat{x})$, although *uniform* convergence does not occur. *

The general situation illustrated in the preceding Example is as follows. Suppose that we have two quadrature rules

$$\int_a^b \phi(y)\,dy \simeq \sum_{j=0}^{n} w_j^{(1)} \phi(y_j)$$

and

$$\int_a^b \phi(y)\,dy \simeq \sum_{j=0}^{n} w_j^{(2)} \phi(y_j)$$

which employ the *same* abscissae y_j . (Various Riemann sums and the family of Romberg rules provide numerous such examples.) Using the first rule, eqns (5.108) become, in matrix notation.

$$\underset{\sim}{K}\underset{\sim}{D}^{(1)}\tilde{\underset{\sim}{f}}^{(1)} = \underset{\sim}{g} \ ,$$

where $\underset{\sim}{K} = [K(z_i,y_j)]$, $\underset{\sim}{D}^{(1)} = \underset{\sim}{\text{diag}}(w_0^{(1)}, \ldots, w_n^{(1)})$, and $\underset{\sim}{g} = [g(z_0),g(z_1), \ldots, g(z_n)]^T$. If $\det(\underset{\sim}{K}) \neq 0$ we obtain an initial approximation $\tilde{\underset{\sim}{f}}^{(1)} = [\tilde{f}^{(1)}(y_0),\tilde{f}^{(1)}(y_2),\ldots,\tilde{f}^{(1)}(y_n)]^T$. The second rule gives a similar approximation $\tilde{\underset{\sim}{f}}^{(2)}$ where $\underset{\sim}{K}\underset{\sim}{D}^{(2)}\tilde{\underset{\sim}{f}}^{(2)} = \underset{\sim}{g}$ and $\underset{\sim}{D}^{(2)} = \underset{\sim}{\text{diag}}(w_0^{(2)},w_1^{(2)}, \ldots, w_n^{(2)})$. It is clear that $\underset{\sim}{D}^{(1)}\tilde{\underset{\sim}{f}}^{(1)} = \underset{\sim}{D}^{(2)}\tilde{\underset{\sim}{f}}^{(2)}$, or

$$\tilde{\underset{\sim}{f}}^{(1)} = [\underset{\sim}{D}^{(1)}]^{-1}\underset{\sim}{D}^{(2)}\tilde{\underset{\sim}{f}}^{(2)} \ ,$$

and the *smoothness* of the functions represented by $\tilde{\underset{\sim}{f}}^{(1)}$ and $\tilde{\underset{\sim}{f}}^{(2)}$ *depends upon the choice of weights*.

There is a further problem which is illustrated by our example. If, for some i , $K(z_i,y_j) = 0$ for all y_j or if $K(z_i,y_j) = 0$ for some j and all z_i , $\det(\underset{\sim}{K}) = 0$. Moreover, $\underset{\sim}{K}$ can be made as nearly singular as 'required' by an 'unsuitable' choice of $\{z_i\}$ or $\{y_j\}$, so that the system of equations for $\tilde{\underset{\sim}{f}}$ is frequently singular or ill conditioned and cannot be solved accurately in practical situations.

Example 5.36.

In the use of the repeated trapezium rule in Example 5.35, we produce a solution $\tilde{f}(ih) = -\{g(\{i-1\}h)-2g(ih)+g(\{i+1\}h)\}/h^2 \simeq -g''(ih)$. Now it is well known that the calculation of the derivatives is unstable with respect to rounding error in $g(x)$. On a practical computer there is an optimal size for h , at which the error in replacing $g''(x)$ by the 'divided difference' is balanced by the propagation of rounding error, and it is unsatisfactory to take h too small.

In a more general framework, we see that in Example 5.35 we first produced a system of the form $h\underset{\sim}{K}\underset{\sim}{\tilde{f}} = \underset{\sim}{g}$ with $\underset{\sim}{\tilde{f}} = [\tilde{f}(h),\tilde{f}(2h),\ldots,\tilde{f}(1-h)]^T$, $\underset{\sim}{g} = [g(h),g(2h),\ldots,g(1-h)]^T$ and $[h\underset{\sim}{K}]^{-1}$ a tridiagonal matrix $-h^{-2}\underset{\sim}{T}$. (The diagonal elements of $\underset{\sim}{T}$ are -2 and co-diagonal elements of $\underset{\sim}{T}$ are unity.) Thus a condition number of $h\underset{\sim}{K}$ is $\|h\underset{\sim}{K}\|_\infty \; \|h^{-2}\underset{\sim}{T}\|_\infty$. But $\|h\underset{\sim}{K}\|_\infty \to \|K\|_\infty$ as $h \to 0$, whereas $\|h^{-2}\underset{\sim}{T}\| = 4/h^2 \to \infty$, so this condition number increases like h^{-2} as $h \to 0$.

Consider the equation in Example 5.28. For any quadrature rule with $n + 1$ points $(n > 1)$, the corresponding matrix $\underset{\sim}{K}$ is singular: it is of rank 2. *

5.16.1.

Some of the details of the preceding discussion do not apply exactly to the modified quadrature method, but the general conclusions appear to be the same. For the modified quadrature method it is now appropriate to take $z_i = y_i$, and replace the integral equation by the equations

$$\tilde{f}(y_i)\int_a^b K(y_i,y)\,dy - \sum_{\substack{j\neq i\\ j=0}}^{n} w_j K(y_i,y_j)\{\tilde{f}(y_j)-\tilde{f}(y_i)\} = g(y_i) .$$

We might, pernaps, consider these equations satisfactory when $K(x,y)$ is badly behaved at $x = y$, and the true solution to the integral equation is continuous. In practice the method appears to depend in an unreliable fashion on the choice of quadrature rule (but is possibly better than the quadrature method, in general).

<u>Example 5.37</u>.

Consider the equation

$$\int_0^1 \ln|x-y|\,f(y)\,dy = g(x) \quad ,$$

where $g(x)$ is chosen so that $f(x) = x$. We applied the modified quadrature method using various m-times repeated rules. The corresponding results are displayed in Tables 5.5 and 5.6.

TABLE 5.5

Mid-point rule with m points (the true solution is $f(x) = x$).

$x \backslash m$	3	9	27
$\frac{1}{6}$	0·1579	0·16637	0·16664
$\frac{1}{2}$	0·5000	0·50000	0·50000
$\frac{5}{6}$	0·8421	0·83363	0·83336

TABLE 5.6

Maximum errors

Order of matrix	3	5	9	17	33	
Mid-point rule	9×10^{-3}	6×10^{-3}	3×10^{-3}	2×10^{-3}	9×10^{-4}	
Trapezium rule	2×10^{-1}	6×10^{-2}	3×10^{-2}	2×10^{-2}	8×10^{-3}	
Simpson rule	4×10^{-2}	2×10^{-2}	9×10^{-3}	5×10^{-3}	2×10^{-3}	*

5.16.2.

We are also able to apply the product-integration method.

Example 5.38.

For the problem considered in Example 5.37 we may use approximations of the form

$$\int_0^1 |x-y|^{-\frac{1}{2}}\phi(y)\,dy \simeq \sum_{j=0}^{n} v_j(x)\phi(y_j) .$$

(For the numerical results below we took $y_j = j/n$ and the approximation was exact when $\phi(x)$ was linear in $[y_j, y_{j+1}]$ $(j = 0,1,2,\ldots,n-1)$.)

Then

$$\int_0^1 \ln|x-y| f(y)\,dy \simeq \sum_{j=0}^{n} \nu_j(x)|x-y_j|^{\frac{1}{2}} \ln|x-y_j| f(y_j) \; .$$

Taking $m = n$, we seek a solution to the equations

$$\sum_{j=0}^{n} \nu_j(z_i)|z_i-y_j|^{\frac{1}{2}} \ln|z_i-y_j| \tilde{f}(y_j) = g(z_i) \; .$$

The errors tabulated below were obtained for this example on taking $z_i = y_i = i/n \quad (i = 0,1,2,\ldots,n)$, and with

$$g(x) = \tfrac{1}{2}\{x^2 \ln x + (1-x^2)\ln(1-x)\} - \tfrac{1}{2}x - \tfrac{1}{4} \; .$$

It may be observed that the similar method for an equation of the second kind (see Example 5.13) did not yield very good accuracy.

Order of matrix	Maximum error	Order of matrix	Maximum error
10	$9{\cdot}9 \times 10^{-1}$	40	$7{\cdot}9 \times 10^{-1}$
20	$1{\cdot}0 \times 10^{0}$	50	$1{\cdot}7 \times 10^{0}$
30	$9{\cdot}1 \times 10^{-1}$		

There seems little virtue in the method for this example. For the present kernel it seems more appropriate to use approximations of the form

$$\int_0^1 \ln|x-y| \phi(y)\,dy \simeq \sum_{j=0}^{n} \nu_j(x)\phi(y_j) \; ,$$

but we consider practical conditions are better reflected by our treatment in this example. *

Example 5.39.

The product-integration method used in the preceding Example can be applied directly to the equation considered in Example 5.32, for $x \in [-1,1]$ with $\alpha = \frac{1}{2}$. We then have a simple idealized test of the ability of the general method to deal with unbounded solutions. We consider the results to be surprisingly good (see Table 5.7).

TABLE 5.7

x \ n	10	20	40	80	True
-1·0	$-2{\cdot}5\times10^{0}$	$-2{\cdot}9\times10^{0}$	$-3{\cdot}5\times10^{0}$	$-4{\cdot}147\times10^{0}$	-Infinity
-0·8	$-5{\cdot}0\times10^{-1}$	$-7{\cdot}0\times10^{-1}$	$-7{\cdot}1\times10^{-1}$	$-7{\cdot}210\times10^{-1}$	$-7{\cdot}303\times10^{-1}$
-0·6	$-4{\cdot}5\times10^{-1}$	$-4{\cdot}6\times10^{-1}$	$-4{\cdot}7\times10^{-1}$	$-4{\cdot}711\times10^{-1}$	$-4{\cdot}743\times10^{-1}$
-0·4	$-2{\cdot}8\times10^{-1}$	$-2{\cdot}9\times10^{-1}$	$-2{\cdot}9\times10^{-1}$	$-2{\cdot}939\times10^{-1}$	$-2{\cdot}954\times10^{-1}$
-0·2	$-1{\cdot}4\times10^{-1}$	$-1{\cdot}4\times10^{-1}$	$-1{\cdot}4\times10^{-1}$	$-1{\cdot}422\times10^{-1}$	$-1{\cdot}429\times10^{-1}$
0·0	$-3{\cdot}6\times10^{-15}$	$-5{\cdot}3\times10^{-15}$	$3{\cdot}8\times10^{-15}$	$-3{\cdot}55\times10^{-15}$	Zero

The true solution and the computed values are antisymmetric about $x = 0$. *

5.16.3.

Finally, let us observe that expansion methods can also be applied to equations of the first kind. If we seek an approximate solution

$\tilde{f}(x) = \sum_{i=0}^{n} \tilde{a}_i \phi_i(x)$, and set

$$\psi_i(x) = \int_a^b K(x,y)\phi_i(y)\,dy ,$$

we can attempt to determine the coefficients $\tilde{a}_0, \tilde{a}_1, \ldots, \tilde{a}_n$ by imposing $n + 1$ conditions. Thus, for the collocation method we choose $z_0, z_1, \ldots, z_m$, with $m = n$, and seeks values $\{\tilde{a}_i\}$ such that

$$\sum_{j=0}^{n} \tilde{a}_j \psi_j(z_i) = g(z_i) \quad (i = 0,1,2,\ldots,m) \quad .$$

In general this is possible if $\det\left[\psi_j(z_i)\right] \neq 0$, and if $K(z_i,y) = 0$ for $a \le y \le b$ and some selected point z_i , this condition is violated.

<u>Example 5.40</u>.

Consider the problem treated in Example 5.26, with $g(x) = x\sin(1) - \sin x$. The solution is $-g''(x) = -\sin x$. We employed a collocation method to find an approximate solution $\tilde{f}(x) = \sum_{r=0}^{n} \tilde{a}_r T_r(x)$, and chose for the collocation points $\{z_i\}$ the zeros of $T_{n+1}(x)$. The method was remarkably successful for this example. We determined the errors $|\tilde{f}(z_i)-f(z_i)|$, and the maximum values are: $n=4$, $1{\cdot}6 \times 10^{-5}$; $n=9$, $1{\cdot}7 \times 10^{-9}$; $n=19$, $1{\cdot}3 \times 10^{-6}$. (For the case $n=19$ the errors ranged between $1{\cdot}3 \times 10^{-6}$ and $3{\cdot}5 \times 10^{-11}$.) *

For a more general method than collocation we may take $m \geq n$ and use linear programming techniques to obtain a *minimax solution*. Alternatively, we may obtain a *least-squares solution* to the equations

$$\sum_{j=0}^{n} \tilde{a}_j \psi_j(z_i) = g(z_i) \quad (i = 0,1,2,\ldots,m) . \tag{5.110}$$

We return to the latter approach below. (This method can be viewed as a generalized Galerkin method.) For the classical Galerkin method we seek values $\tilde{a}_i$ such that, for $i = 0,1,2,\ldots, n$,

$$\sum_{j=0}^{n} \tilde{a}_j \int_a^b \overline{\phi_i(x)}\psi_j(x)\,dx = \int_a^b \overline{\phi_i(x)}g(x)\,dx . \tag{5.111}$$

Example 5.41.

Suppose that the kernel $K(x,y)$ is symmetric and the equation $Kf(x) = g(x)$ is uniquely solvable. If the functions $\{\phi_i(x)\}$ are the eigenfunctions of $K(x,y)$ then the approximation to $f(x)$ obtained from (5.111) is the (truncated) Fourier-type expansion for $f(x)$ in terms of the functions $\phi_i(x)$ $(i = 0,1,2,\ldots)$. In general the limiting series converges only in the mean (and not uniformly) to the solution.

In particular for the kernel in Example 5.26, the eigenfunctions are $\phi_r(x) = \sin\{\pi(r+1)x\}$ $(r = 0,1,2,\ldots)$ and if $f(x) = x$ any approximation in terms of $\{\phi_r(x)\}$ has an error of at least 1, at $x = 1$. *

Example 5.42.

Barrodale (Delves and Walsh 1974, p.104) obtains an approximation

$\tilde{f}(x) = \sum_{j=1}^{n} \alpha_j x^{j-1}$ to the solution of the equation

$$\int_0^1 e^{xy} f(y)\,dy = (e^{x+1}-1)/(x+1) .$$

The true solution is e^x. Barrodale reports on the accuracy obtained with a linear programming technique using 41 equally spaced points z_i in constructing a *minimax* solution of the equations corresponding to (5.110), with $\phi_j(x) = x^j$, $j \geq 0$, and investigates the effect of introducing errors in the free term $(e^{x+1}-1)/(x+1)$. Evaluating this function accurately to 12 decimal figures, the maximum errors $||\tilde{f}(x)-f(x)||_\infty$ reported are of order $10^{-3}(n=2)$, $10^{-5}(n=3)$, $10^{-8}(n=4)$, $10^{-11}(n=5$ and $n=6)$. This example therefore shows a successful application of this technique with small values of n. *

When the solution $f(x)$ of the equation of first kind exists and is a smooth function we *may* be able to obtain an accurate solution to a mathematically posed problem using a product-integration method or an expansion method. In general these techniques may be preferable to a simple application of a quadrature method. However, all naive methods result in an ill-conditioned system if tested to the limit.

5.17. *A least-squares approach and regularization methods*

We have seen that some of the possible numerical methods for equations of the first kind lead to approximations of the form

$$\sum_{j=0}^{n} \psi_j(x)\alpha_j \simeq g(x) , \tag{5.112}$$

where the coefficients α_j $(j = 0,1,2,\ldots,n)$ represent the approximate solution $\tilde{f}(x)$. (Thus for the quadrature method $\psi_j(x) = w_j K(x,y_j)$ and $\alpha_j = \tilde{f}(y_j)$, whilst for the collocation method

$\psi_j(x) = \int_a^b K(x,y)\phi_j(y)\,dy$ and $\tilde{f}(x) = \sum_{j=0}^{n} \alpha_j\phi_j(x)$.) A class of methods then arose if we chose $z_0,z_1,\ldots,z_m$ with $m = n$ and required that

$$\sum_{j=0}^{n} \psi_j(z_i)\alpha_j = g(z_i) \quad (i = 0,1,2,\ldots,m) . \tag{5.113}$$

As we have indicated, the principal difficulty with all such methods is that the matrix of coefficients $[\psi_j(z_i)]$ may be singular or ill conditioned. Thus a solution vector $\underset{\sim}{\alpha} = [\alpha_0, \alpha_1, \ldots, \alpha_n]^T$ may not exist, or if it exists it may not be unique.

It is convenient in such instances to obtain a unique vector $\underset{\sim}{\alpha} = [\alpha_0, \alpha_1, \ldots, \alpha_n]^T$ which provides a *minimal least-squares solution* to (5.113). In such circumstances there is no objection to permitting a choice $m \geq n$. What we are proposing is that, with $m \geq n$, we seek $\underset{\sim}{\alpha} = [\alpha_0, \alpha_1, \ldots, \alpha_n]^T$ which, with corresponding values

$$r_i = \sum_{j=0}^{n} \psi_j(z_i)\alpha_j - g(z_i) ,$$ (5.114)

minimizes

$$\sum_{i=0}^{m} |r_i|^2 .$$ (5.115)

If such a vector is not unique, we choose that which minimizes

$$\sum_{i=0}^{n} |\alpha_i|^2 .$$ (5.116)

The vector $\underset{\sim}{\alpha}$ satisfying (5.114)-(5.116) is called the minimal least-squares solution of the system (5.113). If the rank of the matrix $[\psi_j(z_i)]$ is $n+1$, then $\underset{\sim}{\alpha}$ is defined uniquely without condition (5.116).

5.17.1.

From the preceding remarks it is clear that our proposal is applicable to the collocation method, but we now concentrate on the quadrature method, and elaborate on the technique.

In matrix notation, we set

$$\underset{\sim}{r} = \underset{\sim}{K}\underset{\sim}{D}\tilde{\underset{\sim}{f}} - \underset{\sim}{g}$$ (5.117)

where $\underset{\sim}{D} = \underset{\sim}{diag}(w_0, w_1, \ldots, w_n)$, $\underset{\sim}{K} = [K(z_i, y_j)]$ is an $(m+1)\times(n+1)$ *rectangular matrix* with $m \geq n$, and $\underset{\sim}{g} = [g(z_0), g(z_1), \ldots, g(z_m)]^T$. We seek the vector $\tilde{\underset{\sim}{f}}$ such that $\|\underset{\sim}{r}\|_2$ is minimized and $\|\tilde{\underset{\sim}{f}}\|_2$ is minimized.

It is well known that if the rank of the $(m+1)\times(n+1)$ matrix $\underset{\sim}{K}\underset{\sim}{D}$ is $(n+1)$, with $m \geq n$, a solution to the problem is given by solving

$$(\underset{\sim}{K}\underset{\sim}{D})^*\underset{\sim}{K}\,\underset{\sim}{D}\tilde{\underset{\sim}{f}} = (\underset{\sim}{K}\underset{\sim}{D})^*\underset{\sim}{g} .$$ (5.118)

(If eqn (5.118) is written as a system of equations, these equations are known as the *normal equations*.) It is also well known that whilst (5.118) then provides a theoretical solution to the least squares problem, the practical solution of the normal equations is in general ill advised, because of the effects of pronounced ill conditioning. Moreover, it is frequently the case, here, that the rank of $\underset{\sim}{K}\underset{\sim}{D}$ is less than $(n+1)$, so that $(\underset{\sim}{K}\underset{\sim}{D})^{*}\underset{\sim}{K}\underset{\sim}{D}$ is singular.

5.17.2.

Because of its general applicability and the adaptations we shall make, we now address ourselves in general terms to the problem of obtaining a minimal least-squares solution. Suppose that $\underset{\sim}{A}$ has M rows, N columns, that $\underset{\sim}{b}$ is an M-vector and we seek the N-vector $\underset{\sim}{x}$ to minimize $\|\underset{\sim}{\rho}\|_2$, where $\underset{\sim}{\rho} = \underset{\sim}{A}\underset{\sim}{x} - \underset{\sim}{b}$, with $\|\underset{\sim}{x}\|_2$ also a minimum. (In the context above,† $M = m+1$, $N = n+1$, $\underset{\sim}{A} = \underset{\sim}{K}\underset{\sim}{D}$ and $\underset{\sim}{b} = \underset{\sim}{g}$ whilst $\underset{\sim}{x} = \underset{\sim}{f}$.) Wilkinson and Peters (1970) propose a number of algorithms for obtaining $\underset{\sim}{x}$, which is (as we indicated in section 2.20) given in the form $\sum_{\sigma_i \neq 0} (\underset{\sim}{u}_i^{*}\underset{\sim}{b}/\sigma_i)\underset{\sim}{v}_i$, where $\sigma_i, \underset{\sim}{u}_i, \underset{\sim}{v}_i$ are the singular values and singular vectors of $\underset{\sim}{A}$. (Thus $\underset{\sim}{A}^{*}\underset{\sim}{A}\underset{\sim}{v}_i = \sigma_i^2\underset{\sim}{v}_i$, $\underset{\sim}{A}\underset{\sim}{A}^{*}\underset{\sim}{u}_i = \sigma_i^2\underset{\sim}{u}_i$.) The solution vector $\underset{\sim}{x}$ can be re-expressed in terms of the generalized inverse $\underset{\sim}{A}^{+}$ of the (possibly rectangular) matrix $\underset{\sim}{A}$. See Peters and Wilkinson (1970 *a*) for the properties of $\underset{\sim}{A}^{+}$.

We shall pursue, briefly, an interest in the construction of $\underset{\sim}{A}^{+}$ since a short investigation sheds some light on computational schemes for solving the minimal least-squares problem. (It may be remarked that determining the singular vectors $\underset{\sim}{u}_i, \underset{\sim}{v}_i$ and values σ_i by examining the eigenproblem for $\underset{\sim}{A}^{*}\underset{\sim}{A}$ and for $\underset{\sim}{A}\underset{\sim}{A}^{*}$ is not generally economical. Further, eigenvectors annihilated by the matrices will not be required.)

<u>Example 5.43</u>.

Suppose that $\underset{\sim}{A} = \underset{\sim}{A}^{*}$ and $\underset{\sim}{A} = \underset{\sim}{X}\,\underset{\sim}{\Lambda}\,\underset{\sim}{X}^{*}$, where the columns of $\underset{\sim}{X}$ are orthogonal eigenvectors $\underset{\sim}{x}_i$ of $\underset{\sim}{A}$ and $\underset{\sim}{\Lambda}$ is a diagonal matrix with λ_i in the ith diagonal position. Then the solution of the minimal least-squares problem 'minimize $\|\underset{\sim}{A}\underset{\sim}{x}-\underset{\sim}{b}\|_2$ with minimum $\|\underset{\sim}{x}\|_2$' is given by choosing $\underset{\sim}{x} = \Sigma\alpha_i\underset{\sim}{x}_i$ with $\alpha_i = 0$ if $\lambda_i = 0$, $\alpha_i = \underset{\sim}{x}_i^{*}\underset{\sim}{b}/\lambda_i$ if

† Later we shall suggest a revision of this choice.

$\lambda_i \neq 0$. It will be seen that $\underset{\sim}{x} = \underset{\sim}{A}^+ \underset{\sim}{b}$, where $\underset{\sim}{A}^+ = \underset{\sim}{X} \underset{\sim}{\Lambda}^{\ddagger} \underset{\sim}{X}^*$ and $\underset{\sim}{\Lambda}^{\ddagger}$ is the diagonal matrix with entry λ_i^{-1} in the ith position if $\lambda_i \neq 0$, and zero otherwise. (When does $\underset{\sim}{\Lambda}^+ = \underset{\sim}{\Lambda}^{\ddagger}$?)   *

In constructing the generalized inverse of an arbitrary rectangular matrix, or solving the minimal least squares problem with coefficient matrix $\underset{\sim}{A}$, we may employ the singular-value decomposition (section 2.20). We seek a decomposition $\underset{\sim}{A} = \underset{\sim}{U}\underset{\sim}{\Sigma}\underset{\sim}{V}^*$ where, if $M > N$,

$$\underset{\sim}{\Sigma} = \left[\frac{\underset{\sim}{\Sigma}_0}{\underset{\sim}{0}} \right] ,$$

and

$$\underset{\sim}{\Sigma} = \left[\underset{\sim}{\Sigma}_0 | \underset{\sim}{0} \right]$$

if $M < N$ (whilst $\underset{\sim}{\Sigma} = \underset{\sim}{\Sigma}_0$ if $M = N$) , where $\underset{\sim}{\Sigma}_0$ is a square positive semi-definite diagonal matrix, $\underset{\sim}{\Sigma}$ is $M \times N$, and $\underset{\sim}{U},\underset{\sim}{V}$ are orthogonal matrices of order M,N respectively.

Example 5.44.

A singular-value decomposition can be obtained in the case $\underset{\sim}{A} = \underset{\sim}{A}^*$, from the decomposition $\underset{\sim}{X}\underset{\sim}{\Lambda}\underset{\sim}{X}^*$, mentioned in Example 5.43, by introducing a diagonal matrix $\underset{\sim}{\Omega}$ with entries ± 1 and with $\underset{\sim}{\Omega}\underset{\sim}{\Lambda}$ positive semi-definite. Then we may set $\underset{\sim}{U} = \underset{\sim}{X}\underset{\sim}{\Omega}$, $\underset{\sim}{\Sigma} = \underset{\sim}{\Omega}\underset{\sim}{\Lambda}$, and $\underset{\sim}{V} = \underset{\sim}{X}$. It will be observed that a singular-value decomposition is not unique: we could set $\underset{\sim}{U} = \underset{\sim}{X}$, $\underset{\sim}{V} = \underset{\sim}{X}\underset{\sim}{\Omega}$.   *

We shall indicate how $\underset{\sim}{A}^+\underset{\sim}{b}$ can be obtained from the singular-value decomposition. The (non-zero) diagonal elements of $\underset{\sim}{\Sigma}_0$ are the singular values of $\underset{\sim}{A}$ and the columns $\underset{\sim}{u}_i$ of $\underset{\sim}{U}$ and $\underset{\sim}{v}_i$ of $\underset{\sim}{V}$ are singular vectors. To form $\underset{\sim}{A}^+$ we may compute $\underset{\sim}{V}\underset{\sim}{\Sigma}^+\underset{\sim}{U}^*$, where $\underset{\sim}{\Sigma}^+$ is easily written down from $\underset{\sim}{\Sigma}$. Indeed, $\underset{\sim}{\Sigma}$ is an $M \times N$ 'diagonal' matrix; let us suppose the ith diagonal element is $\Sigma_{ii} = \sigma_i$. Define

$$\sigma_i^+ = 1/\sigma_i \text{ if } \sigma_i \neq 0 , \sigma_i^+ = 0 \text{ otherwise.}$$

Then $\underset{\sim}{\Sigma}^+$ is the $N \times M$ matrix with σ_i^+ in the ith position of the diagonal and zero elsewhere. Since the (non-zero) diagonal elements of $\underset{\sim}{\Sigma}$ (that is of $\underset{\sim}{\Sigma}_0$) are singular values of $\underset{\sim}{A}$, it is clear that

$$\underset{\sim}{V}\underset{\sim}{\Sigma}^{+}\underset{\sim}{U}^{*}\underset{\sim}{b} = \sum_{\sigma_i \neq 0} \{(\underset{\sim}{u}_i^{*}\underset{\sim}{b})/\sigma_i\}\underset{\sim}{v}_i$$

is the required minimal least-squares solution, $\underset{\sim}{A}^{+}\underset{\sim}{b}$.

The preceding remarks may be regarded as parenthetical. What we require in the minimal least-squares solution is (generally) not $\underset{\sim}{A}^{+}$ but the vector $\underset{\sim}{A}^{+}\underset{\sim}{b} = \sum_{\sigma_i \neq 0} \{(\underset{\sim}{u}_i^{*}\underset{\sim}{b})/\sigma_i\}\underset{\sim}{v}_i$, and the point to be emphasized is that a singular-value decomposition $\underset{\sim}{A} = \underset{\sim}{U}\underset{\sim}{\Sigma}\underset{\sim}{V}^{*}$ provides information from which this vector may be computed. What we also recognize is the rôle of columns of $\underset{\sim}{U}$ and of $\underset{\sim}{V}$ as singular vectors, and the nature of the non-zero diagonal elements of $\underset{\sim}{\Sigma}$ as singular values. The analogy with the singular values of a kernel will be emphasized later.† Finally, there are other methods of computing a minimal least-squares solution, but the use of a singular-value decomposition enables us to monitor closely which of the computed diagonal elements of $\underset{\sim}{\Sigma}$ should be discarded as zero or included as non-zero. In the presence of rounding errors, this feature is important. Golub and Reinsch (1970) supply an algorithm which gives a singular-value decomposition of a matrix.

5.17.3.

Consider the equations

$$\sum_{j=0}^{n} w_j K(z_i, y_j)\tilde{f}(y_j) = g(z_i) \quad (i = 0,1,2,\ldots,m,\ m \geq n)$$

where $z_i \in [a,b]$. These equations form a system $\underset{\sim}{K}\underset{\sim}{D}\tilde{\underset{\sim}{f}} = \underset{\sim}{g}$, where $\underset{\sim}{K}\underset{\sim}{D}$ is square or rectangular with entries $K(z_i,y_j)w_j$ and

$$\tilde{\underset{\sim}{f}} = [\tilde{f}(y_0),\tilde{f}(y_1),\ldots,\tilde{f}(y_n)]^T ,$$

$$\underset{\sim}{g} = [g(z_0),g(z_1),\ldots,g(z_n)]^T ,$$

and $\underset{\sim}{D}$ is diag $(w_0,w_1,\ldots,w_n)$. The minimal least-squares solution provides a unique vector $\tilde{\underset{\sim}{f}}$ from what may be an over-determined or singular system of equations. Such a vector may be obtained on setting

---

† To perceive the analogy, replace σ_i by μ_i^{-1}, $\underset{\sim}{v}_i$ by $v_i(x)$, $\underset{\sim}{u}_i^{*}\underset{\sim}{b}$ by (g,u_i) and we obtain the series for $f(x)$. Note that μ_i is to be compared with σ_i^{+} rather than σ_i due to the vagaries of terminology.

$M = m+1$, $N = n+1$, $\underset{\sim}{A} = \underset{\sim}{K}\underset{\sim}{D}$, $\underset{\sim}{b} = \underset{\sim}{g}$ in the preceding discussion and obtaining $\underset{\sim}{f} = \underset{\sim}{V}\underset{\sim}{\Sigma}^{+}\underset{\sim}{U}^{*}\underset{\sim}{g}$, where $\underset{\sim}{V}$ is a matrix of eigenvectors of $(\underset{\sim}{K}\underset{\sim}{D})^{*}\underset{\sim}{K}\underset{\sim}{D} = \underset{\sim}{D}^{*}\underset{\sim}{K}^{*}\underset{\sim}{K}\underset{\sim}{D}$. We shall argue, however, that a weighted least-squares approximation, involving a simple adjustment to this choice of $\underset{\sim}{A}$, is more appealing.

Suppose that the square-integrable theory applies and that the equation of the first kind has a unique solution $f(x)$. Then $f(x)$ is a linear combination of the singular functions $v_i(x)$ of $K(x,y)$, that is, of the eigenfunctions of

$$K^{*}K(x,y) = \int_a^b \overline{K(z,x)}K(z,y)\,dz \, .$$

A natural method of approximating these functions $v_i(x)$ would be to replace $K^{*}K(x,y)$ by an approximation

$$L(x,y) = \sum_{j=0}^{m} \Omega_j \overline{K(z_j,x)}K(z_j,y)$$

(involving the use of a quadrature rule

$$\int_a^b \phi(y)\,dy \simeq \sum_{j=0}^{m} \Omega_j \phi(z_j))$$

and applying the quadrature method to this approximate kernel. If we use the rule $\sum_{j=0}^{n} w_j \phi(y_j)$ in the quadrature method, we obtain approximations $\tilde{v}_i(x)$, where $(\underset{\sim}{K}^{*}\underset{\sim}{\Omega}\underset{\sim}{K})\underset{\sim}{D}\tilde{\underset{\sim}{v}}_r = \tilde{\mu}_r^{-2}\tilde{\underset{\sim}{v}}_r$ and hence (if $\underset{\sim}{D} \geq \underset{\sim}{0}$)

$$(\underset{\sim}{D}^{\frac{1}{2}}\underset{\sim}{K}^{*}\underset{\sim}{\Omega}\underset{\sim}{K}\underset{\sim}{D}^{\frac{1}{2}})\underset{\sim}{D}^{\frac{1}{2}}\tilde{\underset{\sim}{v}}_r = \tilde{\mu}_r^{-2}\underset{\sim}{D}^{\frac{1}{2}}\tilde{\underset{\sim}{v}}_r \; .$$

This follows since $L(y_i,y_j)$ is the (i,j)th element of $\underset{\sim}{K}^{*}\underset{\sim}{\Omega}\underset{\sim}{K}$. Similarly (if $\underset{\sim}{\Omega} \geq \underset{\sim}{0}$) we can obtain approximations $\tilde{u}_r(x)$ to the functions $u_r(x)$, with

$$(\underset{\sim}{\Omega}^{\frac{1}{2}}\underset{\sim}{K}\underset{\sim}{D}\underset{\sim}{K}^{*}\underset{\sim}{\Omega}^{\frac{1}{2}})\underset{\sim}{\Omega}^{\frac{1}{2}}\tilde{\underset{\sim}{u}}_r = \tilde{\mu}_r^{-2}\underset{\sim}{\Omega}^{\frac{1}{2}}\tilde{\underset{\sim}{u}}_r \; .$$

Note that, if $\underset{\sim}{A} = \underset{\sim}{\Omega}^{\frac{1}{2}}\underset{\sim}{K}\underset{\sim}{D}^{\frac{1}{2}}$, then $\underset{\sim}{A}^{*}\underset{\sim}{A} = \underset{\sim}{D}^{\frac{1}{2}}\underset{\sim}{K}^{*}\underset{\sim}{\Omega}\underset{\sim}{K}\underset{\sim}{D}^{\frac{1}{2}}$, $\underset{\sim}{A}\underset{\sim}{A}^{*} = \underset{\sim}{\Omega}^{\frac{1}{2}}\underset{\sim}{K}\underset{\sim}{D}\underset{\sim}{K}^{*}\underset{\sim}{\Omega}^{\frac{1}{2}}$;

$\sigma_r = \tilde{\mu}_r^{-1}$ is a singular value of $\underset{\sim}{A}$.

To approximate $\sum_i \{(g,u_i)\mu_i\} v_i(x)$ we may replace (g,u_i) by $\tilde{u}_i^* \Omega g$, μ_i by $\tilde{\mu}_i$ and $v_i(y_j)$ by $\tilde{v}_i(y_j)$. We shall see below that the same effect is obtained by considering a weighted least-squares problem.

Let us suppose $\Omega_j > 0$, $w_j > 0$ and consider the minimal least-squares problem with $A = \Omega^{\frac{1}{2}} K D^{\frac{1}{2}}$, $b = \Omega^{\frac{1}{2}} g$. This is a weighted least-squares problem (with weights† Ω_j and coefficient matrix $KD^{\frac{1}{2}}$) which can now be re-formulated. In a general framework, with $\rho = Ax - b$ we find x of minimal norm $(x^* x)^{\frac{1}{2}}$ which minimizes $\rho^* \rho$. In terms of f, g we minimize $\|\rho\|_2 = \|\Omega^{\frac{1}{2}}(KD^{\frac{1}{2}} D^{\frac{1}{2}} f - g)\|_2$ choosing the solution which minimizes $\|D^{\frac{1}{2}} \tilde{f}\|_2$.

Remark. The comparable continuous problem involves choosing $f(x)$ to minimize $\|Kf(x) - g(x)\|_2$ and $\|f(x)\|_2$. The quantity $\|\rho\|_2$ approximates the first, whilst $\|D^{\frac{1}{2}} \tilde{f}\|_2$ approximates the second, of the quantities to be minimized.

The basis of our proposed algorithm is now to set $A = \Omega^{\frac{1}{2}} K D^{\frac{1}{2}}$, $b = \Omega^{\frac{1}{2}} g$ and solve for the vector x of minimum norm which minimizes $\|Ax - b\|_2$. We recover $\tilde{f}$ in the form $D^{-\frac{1}{2}} \tilde{x}$. If $K(x,y) = K^*(x,y)$ and $m = n$ then considerable simplification occurs, if we take $\Omega = D$, $z_i = y_i$. We can then consider the eigenvalue problem for $D^{\frac{1}{2}} K D^{\frac{1}{2}}$ since this leads directly to the singular values. In such circumstances the method is based on that of Baker, Fox, Wright, and Mayers (1964).

In practice, our proposal for an algorithm encounters some difficulties, which are generally associated with a technique which is based on the series $\sum_{r=0}^{\infty} \{(g,u_r)\mu_r\} v_r(x)$. In the first place, convergence of this series is not necessarily uniform, and this is then reflected in an approximate method. Secondly, large values of μ_r are generally represented inaccurately by approximate values and correspond to oscillatory functions $v_r(x)$. The use of a singular-value decomposition enables

---

† The weights chosen here are employed in order to mimic the integral equation. If $g(x)$ contains observational errors, with known distribution, this could be used to motivate an alternative choice. Alternative choices of $z_0, z_1, \ldots, z_m$ might also follow.

careful control to be exercised on the latter issue, and components of inaccurate oscillatory functions can be filtered out, or their effect reduced. Such components would remain were a more direct least-squares method (based on Householder transformations) employed instead.

An example of the above technique is discussed in Example 5.44 (Method 1). See also Baker *et al.* (1964). For related remarks see Hanson (1971).

5.17.4.

Phillips (1962) and Twomey (1963) pursue a somewhat different approach from that adopted in section 5.17.3.

From Example 5.35 it can be seen that approximate solutions $\tilde{\underset{\sim}{f}}$ often represent highly oscillatory functions. The use of quadrature rules with unequal weights can cause this oscillation but such oscillations can also arise from ill-defined components of highly oscillatory functions in the expression for $\tilde{\underset{\sim}{f}}$. (These components can arise with any numerical method.)

Phillips, and Twomey, considered the quadrature method and they sought a method of eliminating oscillations, in the components of $\tilde{\underset{\sim}{f}}$, which they regard as spurious. To this end they seek $\tilde{\underset{\sim}{f}} = [\tilde{f}(y_0),\tilde{f}(y_1),\ldots,\tilde{f}(y_n)]^T$ which, with $\underset{\sim}{r} = \underset{\sim}{K}\underset{\sim}{D}\tilde{\underset{\sim}{f}}-\underset{\sim}{g}$, minimizes an expression of the form†

$$\|\underset{\sim}{r}\|_2^2 + \alpha^2 \sum_{i=0}^{n} \{\tilde{f}(y_{i+1})-2\tilde{f}(y_i)+\tilde{f}(y_{i-1})\}^2 . \qquad (5.119)$$

(Some 'boundary conditions' for $\tilde{f}(y_{-1})$ and $\tilde{f}(y_{n+1})$ are required here, and for convenience we set each to zero at present. This choice is referred to again later.) Phillips supposed that $\underset{\sim}{K}$ was the square, non-singular matrix $[K(y_i,y_j)]$, whilst Twomey supposed that $\underset{\sim}{K} = [K(z_i,y_j)]$ was rectangular, and Twomey's analysis reduces (essentially) to that of Phillips in the special case.

Following Twomey, in spirit if not to the letter, we can show that

---

† There are slight differences between the work of Phillips and that of Twomey, which we are obscuring in order to present a unified approach. In this we risk some injustice, since Twomey's work provides a refinement of that of Phillips.

the vector which minimizes (5.119) can be written as $\tilde{f} = (C^*C+\alpha^2 H)^{-1} C^* g$, where $C = KD$ and where we take H to be the symmetric matrix

$$H = \begin{bmatrix} 5 & -4 & 1 & & & & \\ -4 & 6 & -4 & 1 & & & \\ 1 & -4 & 6 & -4 & 1 & & \\ & & \cdot & \cdot & \cdot & \cdot & 1 \\ & & & \cdot & \cdot & \cdot & -4 \\ & & & & 1 & -4 & 5 \end{bmatrix} .$$

The amount of smoothing is controlled by the choice of $\alpha^2 > 0$ (which is a practical detail of great relevance), but for small α the solution of the equations

$$(C^*C + \alpha^2 H)\,\tilde{f} = C^* g \qquad (5.120)$$

shows all the disadvantages associated with the normal equations. (The matrix $(C^*C+\alpha^2 H)$ is not singular with $\alpha^2 > 0$ because H is positive definite.) Increasing α makes the smoothing effected by the sum in (5.119) more dominant, generally at the expense of increasing $\|r\|_2$.

To obtain the solution $\tilde{f}$ of the problem (5.120), we may propose a standard least-square approach, which alleviates (somewhat) the ill-conditioning associated with (5.120). We consider the definitions

$$A = \begin{bmatrix} KD \\ \text{-----} \\ \alpha T \end{bmatrix} , \qquad b = \begin{bmatrix} g \\ \text{--} \\ 0 \end{bmatrix} ,$$

where T is the matrix of order $(n+1)$:

$$T = \begin{bmatrix} -2 & 1 & & & \\ 1 & -2 & 1 & & \\ & \cdot & \cdot & \cdot & \\ & & \cdot & \cdot & \\ & & & 1 & -2 \end{bmatrix}$$

and consider the problem of minimizing $\|\varrho\|_2$, where $\varrho = \underset{\sim}{A}\underset{\sim}{x}-\underset{\sim}{b}$. Since the rank of $\underset{\sim}{T}$ is $n+1$, the rank of $\underset{\sim}{A}$ is $n+1$, and the vector $\underset{\sim}{x}$ which minimizes $\|\varrho\|_2$ is unique. Thus we see that $\underset{\sim}{A}^{+}\underset{\sim}{g}$ provides the vector $\underset{\sim}{\tilde{f}}$ which minimises $\|\varrho\|_2^2 = \|\underset{\sim}{r}\|_2^2 + \alpha^2\|\underset{\sim}{T}\underset{\sim}{\tilde{f}}\|_2^2$. Here $\underset{\sim}{T}^{*}\underset{\sim}{T} = \underset{\sim}{H}$, $\underset{\sim}{\tilde{f}}$ is the solution of (5.120), and $\underset{\sim}{r} = \underset{\sim}{K}\underset{\sim}{D}\tilde{f}-\underset{\sim}{g}$.

Note that $\underset{\sim}{A}$ is a matrix with $(m+n+2)$ rows and $(n+1)$ columns and its lower submatrix depends on α . We frequently seek a value of α by experimenting with a number of values of this parameter, and both (5.120) and the later formulation seem inconvenient for repeated calculations with varying α . Phillips and Twomey consider how acceptable values of α may be chosen.

We observe in passing that Twomey (1963) minimizes

$$\|\underset{\sim}{r}\|_2^2 + \alpha^2 \sum_{i=1}^{n-1}\{\tilde{f}(y_{i+1})-2\tilde{f}(y_i) + \tilde{f}(y_{i-1})\}^2 ,$$

where $\underset{\sim}{r} = \underset{\sim}{K}\underset{\sim}{D}\tilde{f}-\underset{\sim}{g}$, and the exclusion of the terms $i = 0$ and $i = n$ in the summation avoids the need for explicit boundary conditions. With

$$\underset{\sim}{T} = \begin{bmatrix} 1 & -2 & 1 & & & \\ \cdot & \cdot & \cdot & & & \\ & \cdot & \cdot & \cdot & & \\ & & \cdot & \cdot & \cdot & \\ & & & 1 & -2 & 1 \end{bmatrix}, \quad \underset{\sim}{H} = \underset{\sim}{T}^{*}\underset{\sim}{T} = \begin{bmatrix} 1 & -2 & 1 & 0 & 0 & 0 \\ -2 & 5 & -4 & 1 & 0 & 0 \\ 1 & -4 & 6 & -4 & 1 & 0 \\ 0 & 1 & -4 & 6 & -4 & 1 \end{bmatrix} \cdots$$

Observe, however, that $\underset{\sim}{H} = \underset{\sim}{T}^{*}\underset{\sim}{T}$ is now singular and hence $\alpha^2\underset{\sim}{H} + \underset{\sim}{C}^{*}\underset{\sim}{C}$ (with $\underset{\sim}{C} = \underset{\sim}{K}\underset{\sim}{D}$) can be singular. (Consider the case where $\underset{\sim}{C}\underset{\sim}{e}=\underset{\sim}{H}\underset{\sim}{e}=\underset{\sim}{0}$, where $\underset{\sim}{e} = [1,1,\dots,1]^{T}$.) The required $\underset{\sim}{\tilde{f}}$ is not then unique, though we may select the vector of minimum norm.

Twomey (1965) extends the ideas presented in his paper of 1963. Eldén (1974) reports on similar techniques to those discussed here and includes some recent references. In particular, we should observe that some methods adopt a philosophy of fixing $\|\underset{\sim}{r}\|_2$ and minimizing $\tilde{\underset{\sim}{f}}^{*}\underset{\sim}{H}\tilde{\underset{\sim}{f}}$, which is rather different from what is discussed here.

Example 5.45.

As a numerical example we treat the equation

$$\int_0^1 (x^2+y^2)^{\frac{1}{2}} f(y)\,dy = \tfrac{1}{3}\{(1+x^2)^{\frac{3}{2}} - x^3\} ,$$

whose solution is $f(x) = x$. This equation was considered by Fox and Goodwin (1953) and Baker *et al.* (1964).

In Tables 5.8 and 5.9 two of the numerical methods discussed above are applied. In method 1 we implement the technique described in section 5.17.3 employing the $(m+1)$-point and $(n+1)$-point repeated versions of the trapezium rule. We therefore set (for $i = 0,1,2,\ldots,m$, $j = 0,1,2,\ldots,n$)

$$A_{i,j} = (\Omega_i w_j)^{\frac{1}{2}}\{(i/m)^2 + (j/n)^2\}^{\frac{1}{2}} ,$$

where $\Omega_i = 1/m$ $(i = 1,2,3,\ldots,(m-1))$, $\Omega_0 = \Omega_m = 1/2m$, $w_i = 1/n$ $(i = 1,2,3,\ldots,(n-1))$, $w_0 = w_n = 1/2n$. An algorithm of Golub and Reinsch can be employed to compute the singular system of $\underset{\sim}{A} = \underset{\sim}{\Omega}^{\frac{1}{2}}\underset{\sim}{K}\underset{\sim}{D}^{\frac{1}{2}}$; we find vectors $\underset{\sim}{u}_r, \underset{\sim}{v}_r$ and scalars σ_r $(r = 0,1,2,\ldots,n)$ such that

$$\underset{\sim}{A}\underset{\sim}{v}_r = \sigma_r \underset{\sim}{u}_r ,$$

$$\underset{\sim}{A}^* \underset{\sim}{u}_r = \sigma_r \underset{\sim}{v}_r .$$

For various subsets I_s of the integers $\{0,1,2,\ldots,n\}$ we compute (see section 5.17.3)

$$\underset{\sim}{D}^{-\frac{1}{2}} \sum_{r\in I_s} \{\sigma_r^+(\underset{\sim}{\Omega}^{\frac{1}{2}}\underset{\sim}{g},\underset{\sim}{u}_r)\}\underset{\sim}{v}_r ,$$

where $\underset{\sim}{D} = \underset{\sim}{\text{diag}}\,(w_0,w_1,\ldots,w_n)$, $\underset{\sim}{\Omega} = \underset{\sim}{\text{diag}}\,(\Omega_0,\Omega_1,\ldots,\Omega_m)$, and $\sigma_r^+ = \sigma_r^{-1}$ if $\sigma_r \neq 0$, zero otherwise. The results tabulated under method 1 were computed for varying values of m, n with $I_s = \{0\}$, $\{0,1\}$, $\{0,1,2\}$, etc., as indicated. Corresponding results were obtained in Baker *et al.* (1964) using $m = n$ and computing the values σ_r as the eigenvalues of $\underset{\sim}{A}$. ($\underset{\sim}{A}^* = \underset{\sim}{A}$ if $m = n$, $z_i = y_i$, $w_i = \Omega_i$ and $K(x,y) = \overline{K(y,x)}$.)

In method 2 the algorithm of Golub and Reinsch is employed to solve the least-squares problem: minimize $\|\underset{\sim}{A}\underset{\sim}{\tilde{f}}-\underset{\sim}{b}\|_2$, where $\underset{\sim}{b}^* = [\underset{\sim}{g}^*|\underset{\sim}{0}^*]$ and

$$\underset{\sim}{A} = \begin{bmatrix} \underset{\sim}{K}\underset{\sim}{D} \\ \text{--} \\ \alpha\underset{\sim}{T} \end{bmatrix}$$

where $\underset{\sim}{D} = \text{diag}\,(w_0,\ldots,w_n)$, $K_{ij} = \{(i/m)^2+(j/n)^2\}^{\frac{1}{2}}$ $(i = 0,1,2,\ldots,m;$ $j = 0,1,2,\ldots,n)$, and $\underset{\sim}{T}$ is the tridiagonal matrix of order $(n+1)$ displayed in section 5.17.4.

Computed results are shown in Tables 5.8 and 5.9 for various values of α and $m = n = 9$. The theory would indicate that to obtain good results as α decreases, m and n should be increased. (There is little advantage in choosing $m = 2n$, the effect being similar to changing α.)

Phillips and Twomey have given further examples, computed by a different strategy. The literature contains some advice on the 'best' choice of α - a key practical point.

If the results can be summarized it seems appropriate to note that for this example they are disappointing unless low accuracy is required. Both techniques can be used to smooth out the worst effects of ill-conditioning.

TABLE 5.8

Table of errors: Method 1 ($h = \frac{1}{8}$, trapezium rule)

	x	{1}	{1,2}		{1,2,3}	{1,2,3,4}
$m=n$	0	$-2{\cdot}0\times10^{-1}$	$-1{\cdot}3\times10^{-2}$		$-6{\cdot}0\times10^{-2}$	$1{\cdot}9\times10^{-3}$
	0·25	$-2{\cdot}1\times10^{-1}$	$2{\cdot}5\times10^{-2}$		$4{\cdot}1\times10^{-2}$	$1{\cdot}8\times10^{-2}$
	0·50	$-6{\cdot}0\times10^{-2}$	$-2{\cdot}1\times10^{-2}$		$4{\cdot}7\times10^{-3}$	$3{\cdot}2\times10^{-2}$
	0·75	$5{\cdot}8\times10^{-2}$	$-8{\cdot}3\times10^{-2}$		$-8{\cdot}6\times10^{-2}$	$-7{\cdot}9\times10^{-2}$
	1	$5{\cdot}8\times10^{-1}$	$4{\cdot}3\times10^{-1}$		$4{\cdot}1\times10^{-1}$	$3{\cdot}8\times10^{-1}$
		{1,2,3}	{1,2,3,4,5}		{1,2,3}	{1,2,3,4}
$m=2n$	0	$-6{\cdot}1\times10^{-2}$	$-1{\cdot}8\times10^{-2}$	$m = 4n$	$-6{\cdot}1\times10^{-2}$	$5{\cdot}1\times10^{-3}$
	0·25	$4{\cdot}2\times10^{-2}$	$-3{\cdot}6\times10^{-2}$		$4{\cdot}2\times10^{-2}$	$1{\cdot}3\times10^{-2}$
	0·50	$7{\cdot}0\times10^{-3}$	$4{\cdot}9\times10^{-2}$		$7{\cdot}7\times10^{-3}$	$3{\cdot}5\times10^{-2}$
	0·75	$-8{\cdot}6\times10^{-2}$	$-5{\cdot}5\times10^{-2}$		$-8{\cdot}6\times10^{-2}$	$-7{\cdot}7\times10^{-2}$
	1	$4{\cdot}0\times10^{-1}$	$3{\cdot}5\times10^{-1}$		$4{\cdot}0\times10^{-1}$	$3{\cdot}8\times10^{-1}$

TABLE 5.9.
Table of errors - Method 2.

(a) ($m = n = 9$; $h = \frac{1}{8}$, trapezium rule)

x \ α	0·01	0·0001	0·00001	0·0000001	0·000000001	0
0	$-2·2\times10^{-2}$	$-5·2\times10^{-2}$	$-1·7\times10^{-2}$	$-1·5\times10^{-2}$	$-1·5\times10^{-2}$	$-1·4\times10^{-2}$
0·25	$7·4\times10^{-2}$	$3·7\times10^{-2}$	$-3·8\times10^{-3}$	$4·9\times10^{-2}$	$6·6\times10^{-2}$	$9·3\times10^{-2}$
0·50	$-1·0\times10^{-1}$	$6·1\times10^{-3}$	$2·2\times10^{-2}$	$5·0\times10^{-2}$	$2·5\times10^{-1}$	$7·3\times10^{-1}$
0·75	$-1·4\times10^{-1}$	$-1·3\times10^{-1}$	$-7·0\times10^{-2}$	$-1·1\times10^{-1}$	$1·8\times10^{-1}$	$1·2\times10^{0}$
1	$3·4\times10^{-1}$	$2·4\times10^{-1}$	$1·4\times10^{-1}$	$2·2\times10^{-1}$	$2·8\times10^{-1}$	$5·9\times10^{-1}$

(b) ($m = n = 9$; $h = \frac{1}{8}$, Simpson's rule)

x \ α	0·01	0·001	0·00001	0·0000001	0·000000001	0
0	$-2·1\times10^{-2}$	$-7·5\times10^{-2}$	$-2·4\times10^{-2}$	$-2·2\times10^{-2}$	$-2·2\times10^{-2}$	$-2·2\times10^{-2}$
0·25	$3·0\times10^{-2}$	$-3·6\times10^{-3}$	$-7·8\times10^{-3}$	$-4·8\times10^{-2}$	$-2·3\times10^{-2}$	$1·3\times10^{-2}$
0·50	$-2·5\times10^{-1}$	$-1·1\times10^{-1}$	$-8·6\times10^{-2}$	$-8·1\times10^{-2}$	$1·8\times10^{-1}$	$8·4\times10^{-1}$
0·75	$-3·5\times10^{-1}$	$-3·5\times10^{-1}$	$-2·6\times10^{-1}$	$-2·6\times10^{-1}$	$3·1\times10^{-2}$	$1·5\times10^{0}$
1	$1·9\times10^{-1}$	$5·6\times10^{-2}$	$-9·4\times10^{-2}$	$-8·2\times10^{-2}$	$-4·6\times10^{-2}$	$3·9\times10^{-1}$

*

5.17.5.

Tikhonov† (1963, 1964, 1965) and his co-workers have considered a different approach to equations of the first kind, which has some connections with the method produced by Phillips and Twomey.

The methods form a class known as regularization techniques. To demonstrate the principle let us assume that the equation

$$\int_a^b K(x,y)f(y)\,dy = g(x)$$

has a unique solution, and consider the equation of the second kind $(\gamma^2+K^*K)f_\gamma(x) = (K^*g)(x)$ with $\gamma > 0$. Because $\gamma > 0$ this equation is always solvable, and if $(K^*K)(x,y)$ and $g(x)$ are square-integrable

† Frequently transliterated as Tihonov.

we can express $f_\gamma(x)$ in terms of the singular functions $v_n(x)$ of $K(x,y)$ and show that

$$\lim_{\gamma\to 0} \|f_\gamma(x) - f(x)\|_2 = 0.$$

It does not follow, however, that

$$\|f_\gamma(x) - f(x)\|_\infty \to 0.$$

Example 5.46.

A similar technique to that proposed above is sometimes mentioned as a possibility for the case $K(x,y) = K^*(x,y)$. It involves replacing the equation

$$\int_a^b K(x,y)f(y)\,dy = g(x)$$

by

$$uf_u(x) + \int_a^b K(x,y)f_u(y)\,dy = g(x) ,$$

and considering the behaviour of $f_u(x)$ as $u \to 0$. Care must be exercised here to ensure that u tends to zero through a sequence of regular values. Since the eigenvalues of the Hermitian kernel $K(x,y)$ are real, it is legitimate to choose u from a sequence of purely imaginary values tending to zero - or more generally from numbers with non-zero imaginary part, at the cost of some inconvenience if $K(x,y)$ is real. If $K(x,y)$ is positive definite (or negative definite) then u may tend to zero through positive (respectively negative) real values. The convergence of $\|f_u(x)-f(x)\|_2$ to zero as $u \to 0$, assuming $f(x)$ exists and is unique in $L^2[a,b]$, follows very readily on examining the series

$$\sum_{i=0}^{\infty} \frac{(g,f_i)}{\kappa_i+u} f_i(x) ,$$

which converges in the mean to $f_u(x)$, where $\{\kappa_i,f_i(x)\}$ is the total eigensystem of $K(x,y)$ and $u \notin \{-\kappa_i\}$. However, the convergence is not necessarily uniform. Thus, consider the equation discussed in Example 5.26, with solution $f(x) = -g''(x)$. For the solution to exist, $g(0) = g(1) = 0$, since $K(0,y) = K(1,y)$ vanishes for $0 \le y \le 1$.

Since

$$uf_u(x) = g(x) - \int_a^b K(x,y)f_u(y)\,dy$$

it follows in this case that $f_u(0) = f_u(1) = 0$ for all $u \neq 0$. It is clear however, that $f(0)$ and $f(1)$ need not vanish ($g''(0)$ and $g''(1)$ need not vanish), and $\|f_u(x)-f(x)\|_\infty \geq \frac{1}{2}\{|f_u(0)-f(0)|+|f_u(1)-f(1)|\}$.

The same criticism applies to the case of the equation $(\gamma^2+K^*K)f_\gamma(x) = K^*g(x)$. This can be derived, formally, by transforming the equation $Kf(x) = g(x)$ to $K^*Kf(x) = K^*g(x)$ and applying the technique of this Example. Since $K^*K(x,y)$ is positive definite (when $f(x)$ is unique) we may set $u = \gamma^2$ where γ is real. *

Example 5.47.

Suppose the equation of the first kind has a unique solution, and that $\{u_n(x),v_n(x);\mu_n\}$ is the singular system of $K(x,y)$. (Thus $\mu_n^{-1}u_n(x) = (Kv_n)(x)$ and $\mu_n^{-1}v_n(x) = (Ku_n)(x)$.) Here $\mu_n > 0$ and $(K^*K)v_n(x) = \mu_n^{-2}v_n(x)$. It is easily shown that $K^*K = (K^*K)^*$ and $(K^*K)^2v_n(x) = \mu_n^{-4}v_n(x)$ so that the singular system of $(K^*K)(x,y)$ is expressed in terms of its eigenfunctions as $\{v_n(x),v_n(x);\mu_n^2\}$. Now for $\gamma > 0$ the equation $K^*K\,f_\gamma(x) = d_\gamma(x)$, where $d_\gamma(x) = (K^*g)(x) - \gamma^2 f_\gamma(x)$, has a solution. From the theory of equations of the first kind,

$$f_\gamma(x) = \sum_{n=0}^{\infty} \mu_n^2(d_\gamma,v_n)v_n(x)$$

(where we use the eigenfunctions for the singular system of $(K^*K)(x,y)$). Now

$$(d_\gamma,v_n) = (g,Kv_n)-\gamma^2(f_\gamma,v_n) = \mu_n^{-1}(g,u_n)-\gamma^2(f_\gamma,v_n) .$$

Thus

$$f_\gamma(x) = \sum_{n=0}^{\infty} \alpha_n v_n(x) ,$$

where $\alpha_n = (f_\gamma, v_n)$ and $(1+\gamma^2\mu_n^2)\alpha_n = \mu_n(g,u_n)$, and $\lim_{\gamma\to 0} \alpha_n = \mu_n(g,u_n) = (f,v_n)$. In the present case, this indicates the required convergence result, for

$$f_\gamma(x)-f(x) = \sum_{n=0}^{\infty} \{\alpha_n - \mu_n(g,u_n)\} v_n(x) .$$

Now, given $\varepsilon > 0$, choose $N = N(\varepsilon)$ so that

$$\sum_{n=N+1}^{\infty} |\mu_n(g,u_n)|^2 < \frac{1}{9}\varepsilon^2 .$$

Since $|\alpha_n| \le |\mu_n(g,u_n)|$, $\sum_{n=N+1}^{\infty} |\alpha_n|^2 < \frac{1}{9}\varepsilon^2$, and

$$\|f_\gamma(x)-f(x)\|_2 \le \left\|\sum_{n=0}^{N} \{\alpha_n - \mu_n(g,u_n)\} v_n(x)\right\|_2 +$$

$$+ \left\|\sum_{n=N+1}^{\infty} \alpha_n v_n(x)\right\|_2 + \left\|\sum_{n=N+1}^{\infty} \mu_n(g,u_n) v_n(x)\right\|_2 < \gamma^2 \left\{\sum_{n=0}^{N} |\mu_n^3(g,u_n)|^2\right\}^{\frac{1}{2}} + \frac{2}{3}\varepsilon .$$

If we choose γ_0 (depending on N and hence on ε) so that the first term in the last expression is less than $\frac{1}{3}\varepsilon$, then $\|f_\gamma(x)-f(x)\|_2 < \varepsilon$ for all $\gamma < \gamma_0$.

In place of the equation $(\gamma^2 + K^*K)f_\gamma(x) = (K^*g)(x)$ we may consider various integro-differential equations.

Example 5.48.
Consider the equation

$$\gamma^2 f''(x) = \int_a^b (K^*K)(x,y) f_\gamma(y)\,dy - \int_a^b K^*(x,y) g(y)\,dy \tag{5.121}$$

with boundary conditions $f'_\gamma(a) = f'_\gamma(b) = 0$, and with

$$K^*K(x,y) = \int_a^b \overline{K(z,x)}\, K(z,y)\,dz .$$

We suppose, for simplicity, that $K(x,y)$ is continuous and that there is no non-trivial function $\phi(x)$ with $\int_a^b K(x,y)\phi(y)\,dy = 0$. (No loss of generality ensues if we suppose that $a = 0$, $b = 1$.)

The existence of a unique solution $f_\gamma(x)$ under the above conditions is demonstrated in Example 1.33. (Note however that if, with $\phi(x) \equiv 1$, $\int_a^b K(x,y)\phi(y)\,dy = 0$ then any solution of the boundary-value problem for the integro-differential equation contains an arbitrary additive constant, as does a solution of the equation $\int_a^b K(x,y)f(y)\,dy = g(x)$.)

It is not difficult (see below) to show that a solution $f_\gamma(x)$ of eqn (5.121), subject to the boundary conditions, provides a minimum for the functional

$$J_\gamma[\psi] = \gamma^2 \int_a^b |\psi'(x)|^2 dx + \int_a^b |(K\psi)(x)-g(x)|^2 dx$$

when $\psi'(x)$ is (say) continuous, but not constrained by the boundary conditions on $f_\gamma(x)$. If, on the other hand, we discretize the integro-differential equation we minimize a related function. Thus with $y_j = a+jh$, $h = (b-a)/n$, and a quadrature rule, such as the parabolic rule, with weights $w_j > 0$, we can set (for $i = 0,1,2,\ldots,n$)

$$\gamma^2 \left\{\frac{\tilde{f}_\gamma(y_{i+1})-2\tilde{f}_\gamma(y_i)+\tilde{f}_\gamma(y_{i-1})}{h^2}\right\}$$

$$= \sum_{j=0}^{n} w_j \sum_{k=0}^{n} w_k \overline{K(y_k,y_i)} K(y_k,y_j)\tilde{f}_\gamma(y_j) - \sum_{k=0}^{n} w_k \overline{K(y_k,y_i)} g(y_k) ,$$

and

$$(1/h)\{\tilde{f}_\gamma(y_{-1})-\tilde{f}_\gamma(y_0)\} = (1/h)\{\tilde{f}_\gamma(y_{n+1})-\tilde{f}_\gamma(y_n)\} = 0 .$$

The solution then minimizes

$$\tilde{J}_\gamma[\psi] = (\gamma^2/h^2) \sum_{i=0}^{n} \{\psi(y_{i+1})-\psi(y_i)\}^2 + \|\underset{\sim}{D}^{\frac{1}{2}}\{\underset{\sim}{K}\underset{\sim}{D}\underset{\sim}{\psi}-\underset{\sim}{g}\}\|_2^2$$

subject to the constraints $\psi(y_{-1}) = \psi(y_0)$, $\psi(y_{n+1}) = \psi(y_n)$. Here $\underset{\sim}{K} = [K(y_i,y_j)]$ is a square matrix, $\underset{\sim}{D} = \underset{\sim}{\mathrm{diag}}\,(w_0,w_1,\ldots,w_n)$ and $\underset{\sim}{\psi} = [\psi(y_0),\psi(y_1),\ldots,\psi(y_n)]^T$.

The procedure indicated above is followed by Tikhonov and Glasko (1964) who consider (as a numerical example) the equation

$$\frac{1}{\pi}\int_{-1}^{1}\{(x-y)^2+1\}^{-1}f(y)\,dy = g(x) \quad ,$$

where $g(x)$ is chosen so that $f(x) = (1-x^2)^2$. They tabulate values which they computed for $\alpha = 10^{-8}$, 5×10^{-9}, and 10^{-9}: in the latter case they obtain accuracy in about the first two decimal places. *

If we replace the equation of the first kind by an integro-differential equation it would appear that the boundary conditions which are imposed restrict the accuracy obtained. However, this difficulty can sometimes be avoided, as we shall hope to indicate below. If we treat an equivalent variational problem we shall find that the boundary conditions can be suppressed.

We shall endeavour to indicate the connection of the methods of Phillips and Twomey with the use of an integro-differential equation.

Consider the equation

$$\gamma^2 f_\gamma^{iv}(x) = \int_a^b K^*K(x,y)f_\gamma(y)\,dy - K^*g(x) \tag{5.122}$$

with the boundary conditions

$$f_\gamma'''(a) = f_\gamma'''(b) = f_\gamma''(a) = f_\gamma''(b) = 0 \;. \tag{5.123}$$

If (as we shall assume) there is no non-trivial function $\phi(x)$ such that

$$\int_a^b K(x,y)\phi(y)\,dy = 0$$

then the solution $f_\gamma(x)$ can be shown to exist and be unique. (We may follow a similar argument to the one employed for the preceding Example.) The integro-differential equation is associated with the minimization of a certain functional. We define

$$J_\gamma[\psi] = \gamma^2\int_a^b |\psi''(x)|^2 dx + \|K\psi(x)-g(x)\|_2^2 \tag{5.124}$$

for $\psi(x) \in C^2[a,b]$ and consider $J[\psi+\varepsilon\chi]$, where $\chi(x)$ is an arbitrary function in $C^2[a,b]$ and $\varepsilon = \zeta + i\eta$, where ζ, η are real. For a stationary value of $J_\gamma[\psi]$ when $\psi(x) = f_\gamma(x)$ we require that

$$(\partial/\partial\zeta)J_\gamma[\psi+\varepsilon\chi]\Big]_{\varepsilon=0} = (\partial/\partial\eta)J_\gamma[\psi+\varepsilon\chi]\Big]_{\varepsilon=0} = 0$$

when $\psi(x) = f_\gamma(x)$. For simplicity we shall suppose, for the present, that *all* functions involved are to be real, and $\eta = 0$ so that $\zeta = \varepsilon$. The analysis is readily extended to the complex case. If $\psi(x)\,\varepsilon\,C^4[a,b]$, we find that

$$(\partial/\partial\varepsilon)J_\gamma[\psi+\varepsilon\chi]\Big]_{\varepsilon=0}$$

$$= \gamma^2\int_a^b \chi''(x)\psi''(x)\,dx + \int_a^b \{K\psi(x)-g(x)\}K\chi(x)\,dx$$

$$= \int_a^b \{\gamma^2\psi^{iv}(x)+(K^*K)\psi(x)-(K^*g)(x)\}\chi(x)\,dx +$$

$$+ \gamma^2[\psi''(x)\chi'(x)]_a^b - \gamma^2[\psi'''(x)\chi(x)]_a^b$$

on applying integration by parts to the first term and using $(r,K\chi) = (K^*\mathbf{r},\chi)$ in the second term. Now $(\partial/\partial\varepsilon)J_\gamma[\psi+\varepsilon\chi]\Big]_{\varepsilon=0} = 0$ for *all* $\chi(x)\ \varepsilon\ C^2[a,b]$ if (and only if†) $\psi(x) = f_\gamma(x)$. For, the boundary conditions on $f_\gamma(x)$ ensure that the above terms involving $\chi(a), \chi(b), \chi'(a), \chi'(b)$ vanish. Moreover, there are no boundary conditions to be imposed on $\chi(x)$. If we define $f_\gamma(x)$ as the function such that

$$J_\gamma[f_\gamma] \leq J_\gamma[\phi] \quad \text{for all} \quad \phi(x)\ \varepsilon\ C^2[a,b]\,, \tag{5.125}$$

we see that the boundary conditions in the integro-differential equation

---

† Uniqueness has only been established for $\psi(x)\ \varepsilon\ C^4[a,b]$, but a standard variational argument can be used to show that (5.125) characterizes $f_\gamma(x)$ uniquely.

are, in a sense, *suppressible*: they have no explicit rôle in (5.125) even though $f_\gamma(x)$ satisfies the boundary conditions if it satisfies (5.125).

Now we seek a numerical approximation to $f_\gamma(x)$. If we discretize the integro-differential equation the boundary conditions will not *in general* disappear. However, if we approximate $J_\gamma[\psi]$ there is no necessity to introduce boundary conditions. In particular, if we approximate $\int_a^b |\psi''(x)|^2 dx$ in terms of $\psi(y_0), \psi(y_1), \ldots, \psi(y_n)$, and $y_0, y_1, \ldots, y_n \in [a,b]$, then boundary conditions are superfluous: by this means we obtain Twomey's type of method.

We may compare (5.124) with (5.119) to see the similarity of the technique indicated here with Twomey's method. A natural analogue of (5.124) would, however, involve the minimization of

$$\|D^{\frac{1}{2}}r\|_2^2 + \gamma^2 \sum_{j=1}^{n-1} w_j \{\tilde{f}_\gamma(y_{j+1}) - 2\tilde{f}_\gamma(y_j) + \tilde{f}_\gamma(y_{j-1})\}^2$$

(with $y_i = a+ih$, $y_0 = a$, $y_n = b$).

The use of finite differences and quadrature rules to approximate $J_\gamma[\psi]$ may be readily implemented, but the variational formulation lends itself more readily to a Rayleigh-Ritz approach. If we choose $\phi_0(x), \phi_1(x), \ldots, \phi_n(x) \in C^2[a,b]$, we can find $\tilde{f}_\gamma(x) = \sum_{r=0}^{n} \tilde{a}_r \phi_r(x)$, where $\tilde{a}_0, \tilde{a}_1, \ldots, \tilde{a}_n$ are chosen so that

$$\frac{\partial}{\partial \tilde{a}_s} J_\gamma\left[\sum_{r=0}^{n} \tilde{a}_r \phi_r\right] = 0 \quad (s = 0,1,2,\ldots,n), \qquad (5.126)$$

and, again, no boundary conditions appear in determining $\tilde{f}_\gamma(x)$.

5.17.6.

I have indicated a connection of the methods of Phillips and of Twomey with an integro-differential equation and an approach adopted by Tikhonov and his associates. (It would appear that the *specific* 'regularization methods' displayed by Tikhonov (1963) do not strictly include (5.124), though they are closely related.)

In a general context, Tikhonov considers the integro-differential equation

$$\gamma^2 \sum_{i=0}^{r+1} (-1)^{i+1} \frac{d^i}{dx^i}(p_i(x)f_\gamma^{(i)}(x)) - \int_a^b (K^*K)(x,y)f_\gamma(y)dy + (K^*g)(x) = 0, \quad (5.127)$$

where, as we assume throughout, $(d/dx)^i p_i(x) \in C[a,b]$, and where $p_i(x) \geq p_i > 0$ for $x \in [a,b]$ $(i = 0,1,2,\ldots,(r+1))$ subject to the boundary conditions

$$\pi^{(i)}(a) = \pi^{(i)}(b) = 0 \quad (i = 1,2,3,\ldots,(r+1)), \quad (5.128)$$

where

$$\pi^{(i)}(x) = \sum_{k=i}^{r+1} (-1)^{k-i} \left(\frac{d}{dx}\right)^{k-i} \{p_k(x) f^{(k)}(x)\}$$

and $r \geq -1$.

The case $r = 0$ is simple, when we have

$$\gamma^2\{p_1(x)f_\gamma''(x) + p_1'(x)f_\gamma'(x) - p_0(x)f_\gamma(x)\} - K^*Kf_\gamma(x) + K^*g(x) = 0 \quad (5.129)$$

with the 'natural' or 'suppressible' boundary conditions $f_\gamma'(a) = f_\gamma'(b) = 0$ (since $p_1(x) > 0$ for $x = a$, $x = b$) . We shall only consider simple cases.

Example 5.49.
In the case $r = 0$, set $p_1(x) = 1$, $p_0(x) = q > 0$ and we find $\gamma^2\{f_\gamma''(x) - qf_\gamma(x)\} - K^*Kf_\gamma(x) + (K^*g)(x) = 0$, with $f_\gamma'(a) = f_\gamma'(b) = 0$. The solution $f_\gamma(x)$ provides a minimum for the quadratic functional

$$\gamma^2\{ \|\psi'(x)\|_2^2 + q\|\psi(x)\|_2^2\} + \|K\psi(x) - g(x)\|_2^2 \quad .$$

If (unlike Tikhonov) we permit q to be zero, we obtain Example 5.48. *

In the analysis below we shall establish that $\lim_{\gamma\to 0} \|f_\gamma(x) - f(x)\|_\infty = 0$, where $f_\gamma(x)$ satisfies the integro-differential equation (5.127) in the case $r = 0$. (The method of proof appears to be general enough to extend to other values of r.) In the proof, we shall see that various variational aspects have an important role. We begin with a lemma concerning two different norms on the space $C^1[a,b]$, one of which appears in the variational functional. Note that $p_0(x) \geq p_0 > 0$, $p_1(x) \geq p_1 > 0$ for $a \leq x \leq b$.

LEMMA 5.5. *For every function* $\psi(x) \in C^1[a,b]$, $||\psi(x)||_\infty \leq M||\psi(x)||_D$, *where*

$$||\psi(x)||_D^2 = \int_a^b \{p_0(x)|\psi(x)|^2 + p_1(x)|\psi'(x)|^2\}dx \qquad (5.130)$$

with $p_i(x) \in C^i[a,b]$, $p_i(x) \geq p_i > 0$ *for* $a \leq x \leq b$, $i = 0,1$ *and* M *is a constant, independent of* $\psi(x)$.

Proof. Since $\psi'(x)$ is continuous,

$$\psi(x)-\psi(t) = \int_t^x \psi'(s)ds \quad \text{for} \quad a \leq x,\ t \leq b .$$

Integrating with respect to t,

$$(b-a)\psi(x) = \int_a^b \psi(t)dt + \int_a^b\int_t^x \psi'(s)ds\ dt .$$

Set $L(x,t) = t-a$ for $t < x$, $L(x,t) = t-b$ for $x \leq t$; then

$$(b-a)\psi(x) = \int_a^b \psi(t)dt + \int_a^b \psi'(t)L(t,x)dt .$$

Thus

$$|(b-a)\psi(x)| \leq \left|\int_a^b [p_0(t)]^{-\frac{1}{2}}\{[p_0(t)]^{\frac{1}{2}}\psi(t)\}dt\right| + $$
$$+ \left|\int_a^b \{[p_1(t)]^{-\frac{1}{2}}L(t,x)\}\{[p_1(t)]^{\frac{1}{2}}\psi'(t)\}dt\right| ,$$

and the result now follows if we use the Schwartz inequality (Smithies 1962, p.7) and set

$$M = \max_{a\leq x\leq b} \left|\int_a^b \{\frac{|L(t,x)|^2}{p_1(t)} + \frac{1}{p_0(t)}\}dt\right|^{\frac{1}{2}}/(b-a) . \qquad (5.131)$$

The conditions on $p_0(t)$ and $p_1(t)$ ensure that $M < \infty$.

THEOREM 5.17. *Suppose that* $K(x,y)$ *and* $g(x)$ *are continuous and that with the preceding conditions*

$$D\{\psi(x)\} = (d/dx)\{p_1(x)\psi'(x)\} - p_0(x)\psi(x) \quad .$$

Assume that the equation

$$\int_a^b K(x,y)f(y)\,dy = g(x)$$

has a unique square-integrable solution. Then the function $f_\gamma(x)$ *such that*

$$\gamma^2 D\{f_\gamma(x)\} = K^* K f_\gamma(x) - K^* g(x) \quad (a \le x \le b) \quad ,$$

$$f'_\gamma(a) = f'_\gamma(b) = 0 \quad ,$$

is unique and it provides a (unique) minimum for

$$J_\gamma[\psi] = \gamma^2 \,\|\psi(x)\|_D^2 + \|K\psi(x) - g(x)\|_2^2$$

with varying $\psi(x) \in C^1[a,b]$, *in the notation of* (5.130).

Proof. With the aid of Green's function† $G(x,y)$ for the problem $D\{\psi(x)\} = d(x)$, $\psi'(a) = \psi'(b) = 0$, we can obtain an integral equation for $f_\gamma(x)$ and establish its uniqueness. For $\gamma^2 f_\gamma(x) + GK^* K f_\gamma(x) = GK^* g(x)$, and by the Fredholm alternative this is uniquely solvable if γ^2 is not an eigenvalue of $-(GK^* K)(x,y)$. But any square-integrable eigenfunction $\phi(x)$ of this kernel is twice-differentiable and satisfies $\gamma^2 D\{\phi(x)\} = K^* K\phi(x)$, $\phi'(a) = \phi'(b) = 0$. If $\phi(x) \not\equiv 0$ is such a function then

---

† The existence of Green's function is assured by the conditions $p_i(x) \ge p_i > 0$ and the derivative boundary conditions. The continuity conditions on $p_i(x)$ are also the same as above.

$$\gamma^2(D\{\phi\},\phi) = (K^*K\phi,\phi) = ||K\phi(x)||_2^2 > 0 ,$$

where the inequality is strict. (For $f(x)$ is unique, and hence there are no non-trivial functions annihilated by K.) But $\gamma^2(D\{\phi\},\phi) < 0$ (since $\int_a^b p_0(x)|\phi(x)|^2 dx > 0$, and, using integration by parts and the boundary conditions, $\int_a^b \{p_1(x)\phi'(x)\}' \overline{\phi(x)} dx \leq 0$). Hence there is a contradiction and no non-trivial function $\phi(x)$ exists. Thus γ^2 is not an eigenvalue of $-GK^*K(x,y)$, and $f_\gamma(x)$ exists.

Consider the problem of minimizing $J_\gamma[\psi]$ for $\psi(x) \in C^1[a,b]$. We have, for real ε and for $\psi(x)$, $\nu(x) \in C^1[a,b]$,

$$J_\gamma[\psi+\varepsilon\nu] = J_\gamma[\psi] + 2\varepsilon\{\gamma^2\int_a^b p_0(x)\psi(x)\overline{\nu(x)}dx +$$

$$+ \gamma^2\int_a^b p_1(x)\psi'(x)\overline{\nu'(x)}dx +$$

$$+ \int_a^b \{K\psi(x)-g(x)\}\overline{K\nu(x)}dx\} + O(\varepsilon^2) .$$

Now let $\psi(x) \in C^2[a,b]$. Then (integrating by parts)

$$(\partial/\partial\varepsilon)J_\gamma[\psi+\varepsilon\nu]\Big]_{\varepsilon=0} = 2\{\int_a^b (-\gamma^2)D\{\psi(x)\}\overline{\nu(x)}dx +$$

$$+ \gamma^2\Big[p_1(x)\psi'(x)\overline{\nu(x)}\Big]_{x=a}^{x=b} + \int_a^b \{K\psi(x)-g(x)\}\overline{K\nu(x)}dx\} ,$$

and this vanishes for all $\nu(x) \in C^1[a,b]$ if and only if $\psi(x)$ satisfies the integro-differential equation for $f_\gamma(x)$ and $\psi'(a) = \psi'(b) = 0$, that is, if and only if $\psi(x) = f_\gamma(x)$. Finally, since $J_\gamma[\psi] \geq 0$ and $J_\gamma[\psi]$ has a unique extreme value over $C^2[a,b]$, $f_\gamma(x)$ provides a minimum value; that is, $J_\gamma[f_\gamma] \leq J_\gamma[\psi]$ for all $\psi(x) \in C^2[a,b]$. Indeed,

$$J_\gamma[f_\gamma] \leq J_\gamma[\psi] \quad \forall \ \psi(x) \in C^1[a,b] ,$$

as we shall now argue. For, suppose $\psi(x) \in C^1[a,b]$; then there exists a sequence $\psi_0(x),\psi_1(x),\ldots$ in $C^2[a,b]$ such that † $\|\psi(x)-\psi_n(x)\|_2 \to 0$ and $\|\psi'(x)-\psi_n'(x)\|_2 \to 0$, whence it follows that

$$J_\gamma[\psi] = \lim_{n\to\infty} J_\gamma[\psi_n] \ .$$

Now, $J_\gamma[\psi_n] \geq J_\gamma[f_\gamma]$ since $\psi_n(x) \in C^2[a,b]$. Thus $J_\gamma[\psi] \geq J_\gamma[f_\gamma]$ when $\psi(x) \in C^1[a,b]$.

THEOREM 5.18. *Suppose the equation*

$$\int_a^b K(x,y)f(y)\,dy = g(x) \quad (a \leq x \leq b)$$

has, for some particular $g(x) \in C[a,b]$ *, a unique square-integrable solution* $f(x)$ *where* $f(x) \in C^1[a,b]$. *Then* $\lim_{\gamma\to 0} \|f(x)-f_\gamma(x)\|_\infty = 0$.

Proof. We shall establish that the functions $\{f_\gamma(x)\}$ are equicontinuous and uniformly bounded (in $\|\ \|_\infty$) and that $(f_\gamma,v_i) \to (f,v_i)$, where $v_0(x),v_1(x),\ldots$ are the singular functions of $K(x,y)$. Since

---

† The argument here indicates a type of analysis, used by Mikhlin (1964*b*), enabling us to work in the completion of $C^2[a,b]$ induced by $\|\ \|_D$. (Let $\chi_n(x)$ be the polynomial of degree $(n-1)$ which minimizes $\|\psi'(x)-\chi_n(x)\|_2$, and set

$$\psi_n(x) = \psi(a) + \int_a^x \chi_n(t)\,dt \ ,$$

to obtain a sequence of the type required.)

It is more direct to observe that, for non-null $\nu(x) \in C^1[a,b]$,

$$J_\gamma[f_\gamma+\nu] = J_\gamma[f_\gamma] + \{\gamma^2\|\nu(x)\|_D^2 + \|K\nu(x)\|_2^2\} > J_\gamma[f_\gamma]$$

provided $p_0(x) \geq p_0 > 0$, or if $K\nu(x) = 0$ implies that $\nu(x) = 0$. Uniqueness of the minimum now follows immediately.

$f(x)$ is unique, these functions $v_i(x)$ provide a complete (orthonormal) basis. The required result then follows from Theorem 1.2.

Since, by assumption $f(x) \in C^1[a,b]$, $J_\gamma[f_\gamma] \leq J_\gamma[f]$. But

$$\gamma^2 \|f_\gamma(x)\|_D^2 \leq J_\gamma[f_\gamma]$$

and

$$J_\gamma[f] = \gamma^2 \|f(x)\|_D^2$$

since $Kf(x) = g(x)$. Thus $\|f_\gamma(x)\|_D \leq \|f(x)\|_D$ and hence, by Lemma 5.5,

$$\|f_\gamma(x)\|_\infty \leq M \|f(x)\|_D < \infty, \quad \text{for } \gamma \neq 0 .$$

To establish the equicontinuity, we have

$$\int_a^b |f'_\gamma(x)|^2 dx \leq \|\{p_1(x)\}^{-1}\|_\infty \ \|\sqrt{p_1(x)} f'_\gamma(x)\|_2^2$$

and hence

$$\|f'_\gamma(x)\|_2^2 \leq \gamma^{-2} \|\{p_1(x)\}^{-1}\|_\infty J_\gamma[f_\gamma] \leq \|\{p_1(x)\}^{-1}\|_\infty \|f(x)\|_D^2 . \tag{5.132}$$

However, for $x_1, x_2 \in [a,b]$ and $x_2 > x_1$,

$$|f_\gamma(x_1) - f_\gamma(x_2)| = \left|\int_{x_1}^{x_2} f'_\gamma(t)dt\right|$$

$$\leq \left\{\int_{x_1}^{x_2} dt \int_{x_1}^{x_2} |f'_\gamma(t)|^2 dt\right\}^{\frac{1}{2}}$$

$$\leq \sqrt{(x_2 - x_1)}\, \|f'_\gamma(x)\|_2 ,$$

and the uniform bound for $\|f'_\gamma(x)\|_2^2$ in (5.132) establishes the equicontinuity.

Let us now examine the behaviour of (f_γ, v_n) $(n = 0,1,2,\ldots)$, where $\{u_n(x), v_n(x); \mu_n\}$ is the singular system of $K(x,y)$ (and thus $\mu_n^2 K^* K v_n(x) = v_n(x)$) . If we set $d_\gamma(x) = \gamma^2 D\{f_\gamma(x)\} + K^* g(x)$, we have $K^* K f_\gamma(x) = d_\gamma(x)$ and we know that there is a unique (square-integrable) solution $f_\gamma(x)$ for $\gamma > 0$. The solution is then, according to the standard theory for equations of the first kind,

$$f_\gamma(x) = \sum_{n=0}^{\infty} \mu_n^2 (d_\gamma, v_n) v_n(x) ,$$

and

$$\begin{aligned}(f_\gamma, v_n) &= \mu_n^2 (d_\gamma, v_n) \\ &= \mu_n^2 \gamma^2 (D\{f_\gamma\}, v_n) + \mu_n^2 (K^* g, v_n) \\ &= \mu_n^2 \gamma^2 (D\{f_\gamma\}, v_n) + (f, v_n)\end{aligned}$$

since $K^* K f(x) = K^* g(x)$.

But

$$\begin{aligned}|\gamma^2 (D\{f_\gamma\}, v_n)|^2 &\le \gamma^2 \|D\{f_\gamma\}\|_2^2 = \|K^* K f_\gamma(x) - K^* g(x)\|_2^2 \\ &\le \|K\|_2^2 \|K f_\gamma(x) - g(x)\|_2^2 \\ &\le \|K\|_2^2 J_\gamma[f_\gamma] \le \|K\|_2^2 J_\gamma[f] \\ &= \gamma^2 \|K\|_2^2 \|f(x)\|_D^2 \to 0\end{aligned}$$

as $\gamma \to 0$. Thus $\lim_{\gamma \to 0} (f_\gamma, v_n) = (f, v_n)$. The result required now follows from Lemma 1.1.

The preceding convergence result gives some encouragement to the construction of variational schemes (based on the minimization of $J_\gamma[\psi]$) or, alternatively, the discretization of the corresponding integro-differential equation. Our analysis has been conducted under the assumption that $p_i(x) \ge p_i > 0$, however, the procedure of Twomey can be seen as the minimization of a discrete analogue of $J_\gamma[\psi]$ in the case where $p_0(x) = p_1(x) = 0$, $r = 1$.

Let us consider the effect of setting $p_0(x) \equiv 0$, whilst retaining the restriction $r = 0$, $p_1(x) \geq p_1 > 0$, in the preceding analysis. The proof of Theorem 5.17 involves the use of a Green's function for the differential operator, giving $D\{\psi(x)\}$, and derivative boundary conditions. In the case $p_0(x) \equiv 0$ there is only the *generalized* Green's function at our disposal. However, we have established in detail in Example 1.33 (for the case $p_1(x) \equiv 1$) that $f_\gamma(x)$ is defined uniquely by the integro-differential equation, and boundary conditions. We may readily convince ourselves, now, that Theorem 5.17 holds in its entirety for the case $p_0(x) \equiv 0$.

If we examine the proof of Lemma 5.5 we shall see that it does not apply in the case $p_0(x) \equiv 0$. However, inspection of the proof of Theorem 5.18 shows that use of Lemma 5.5 enters only when we establish a somewhat weaker result than given by the lemma, namely that $\|f_\gamma(x)\|_\infty \leq M \|f_\gamma(x)\|_D$. Thus a weaker result† than that given by the lemma might suffice, since $f_\gamma(x)$ satisfies the given integro-differential equation and boundary conditions. Thus, the strength of Lemma 5.5 may not be needed; it may be possible to establish that $\|f_\gamma(x)\|_\infty \leq M_* \|f_\gamma(x)\|_D$, where M_* is independent of γ . (The proof of the equicontinuity of $\{f_\gamma(x)\}$ is independent of the vanishing of $p_0(x)$, provided $p_1(x) \geq p_1 > 0$.)

Let us proceed without the aid of such an M_* . We can follow the lines of the proof of Theorem 5.18. The existence and uniqueness of $f_\gamma(x)$ has been discussed above. We are no longer able to establish a uniform bound for $\|f_\gamma(x)\|_\infty$. However, as before, $(f_\gamma,v_n) \to (f,v_n)$ and mean-square convergence of $f_\gamma(x)$ to $f(x)$ follows. (This convergence is uniform when the functions $f_\gamma(x)$ are uniformly bounded and equicontinuous, but we have not established the uniform boundedness.) I shall state our conclusion formally.

---

† Tikhonov and Glasko (1964) state such a result for the *special case* where $K(x,y) \geq k_* > 0$. For this case, $\|f_\gamma(x) - f(x)\|_\infty \to 0$ as $\gamma \to 0$.

THEOREM 5.19. *Suppose, in the conditions applying in Theorem 5.18, we now set* $p_0(x) = 0$. *Then* $\lim_{\gamma\to 0} \|f_\gamma(x)-f(x)\|_2 = 0$.

For further discussion, see Ribière (1967).

The convergence results above give some insight into variational methods which might be employed to obtain an approximation $\tilde{f}_\gamma(x)$. With a standard Rayleigh-Ritz approach, we compute an approximation

$$\tilde{f}_\gamma(x) = \sum_{i=0}^{n} \tilde{a}_i \phi_i(x)$$

by solving the equations $(\partial/\partial\tilde{a}_k)J_\gamma[\tilde{f}_\gamma] = 0$ $(k = 0,1,2,\ldots,n)$. We see that a number of such schemes arise on varying the choice of $p_i(x)$, and we see that the boundary conditions are suppressible, and need not be satisfied by $\tilde{f}_\gamma(x)$. (Though these boundary conditions are satisfied by $f_\gamma(x)$, they need not be satisfied by $f(x)$!) We see, also, from a familiarity with variational methods, that if $r = 1$ there is no requirement that $\tilde{f}_\gamma''(x) \in C[a,b]$; it is more than sufficient that $f_\gamma'(x) \in C[a,b]$. Thus $\tilde{f}_\gamma(x)$ could be a piecewise-linear function, in the second-order case. The approach outlined is in contrast with a 'finite-difference' treatment of the 'regularized problem' in which the approximation $\tilde{f}_\gamma(x)$ can, in principle, be computed by approximating the integro-differential equation or by minimizing a (discrete) functional which approximates a functional minimized by $f_\gamma(x)$.

Whereas $\|f_\gamma(x)-f(x)\|_\infty \to 0$ under certain conditions, we cannot be sure that $\|\tilde{f}_\gamma(x)-f(x)\|_\infty \to 0$ as $\gamma \to 0$. Indeed, it appears that the problem of determining $\tilde{f}_\gamma(x)$ becomes more and more ill conditioned as $\gamma \to 0$ so that $f_\gamma(x)$ and $\tilde{f}_\gamma(x)$ may be widely separated functions for small γ . On the other hand,

$$\Big| \|\tilde{f}_\gamma(x)-f_\gamma(x)\|_\infty - \|f_\gamma(x)-f(x)\|_\infty \Big|$$

$$\leq \|\tilde{f}_\gamma(x)-f(x)\|_\infty$$

$$\leq \|\tilde{f}_\gamma(x)-f_\gamma(x)\|_\infty + \|f_\gamma(x)-f(x)\|_\infty \quad ,$$

and, when $\|f(x)-f_\gamma(x)\|_\infty \to 0$, we see that a sufficient condition for $\|\tilde{f}_\gamma(x)-f(x)\|_\infty$ to tend to zero is that $\|f_\gamma(x)-\tilde{f}_\gamma(x)\|_\infty$ tend to zero.

In practice, this means that, as $\gamma \to 0$, the determination of $\tilde{f}_\gamma(x)$ must be set up and solved with *increasing accuracy*, which is determined by the rate of convergence of γ to zero. In a practical situation this is not possible because of rounding errors or errors in the given function $g(x)$.

Suppose, for example, that $r = 0$, $p_0(x) \geq p_0 > 0$, $p_1(x) \geq p_1 > 0$,

$$\gamma^2 D\{f_\gamma(x)\} = \{K^*Kf_\gamma(x)-K^*g(x)\} \quad ,$$

$$\gamma^2 D\{\tilde{f}_\gamma(x)\} = \{K^*K\tilde{f}_\gamma(x)-K^*g(x)\} - \rho(x) \quad ,$$

and $f_\gamma(x)$ and $\tilde{f}_\gamma(x)$ both satisfy the associated boundary conditions. If we write $\varepsilon_\gamma(x) = \tilde{f}_\gamma(x)-f_\gamma(x)$, and $G(x,y)$ is the appropriate Green's function, we find that

$$(\gamma^2 I + GK^*K)\varepsilon_\gamma(x) = G\rho(x) \quad ,$$

and $\|\varepsilon_\gamma(x)\| \to 0$ if and only if $\|(\gamma^2 I + GK^*K)^{-1}G\rho(x)\| \to 0$. Indeed, $\|\varepsilon_\gamma(x)\|_2 \leq \|(\gamma^2 I + GK^*K)^{-1}\|_2 \|G\rho(x)\|_2$ with a possibility of equality for some $\rho(x)$, whilst $\|(\gamma^2 I + GK^*K)^{-1}\|_2 = O(1/\gamma^2) \to \infty$. For safety, we should at least require that $\|\rho(x)\|_2 = O(\gamma^2)$ as $\gamma \to 0$. (For arguments along these lines see Bakushinskii (1965). See also Ribière (1967).)

The natural inclination in circumstances like those described may be to determine, for reasonably large values γ_i, approximations $\tilde{f}_{\gamma_i}(x)$, and perform rational interpolation in γ to determine $\lim_{\gamma\to 0} \tilde{f}_\gamma(x)$.

In closing this section it should be remarked that the method of regularization has some connections with various statistical approaches (Wahba 1969; Strand and Westwater 1968*a*) to equations of the first kind. Hilgers (1974) relates regularization methods to the theory of reproducing kernel Hilbert spaces, and gives further references.

5.18. *Non-linear integral equations*

Consider the non-linear Fredholm integral equation of the second kind

$$f(x) - \int_a^b F(x,y;f(y))dy = g(x) \quad (a \le x \le b) , \tag{5.133}$$

which is known as the Urysohn equation. In general we suppose that $F(x,y;v)$ is at least continuous and a,b are finite. Eqn (5.133) is a particular case of the more general equation

$$F_0(x;f(x)) - \int_a^b F_1(x,y;f(x),f(y))dy = 0 , \tag{5.134}$$

but there is the merit in (5.133) of the linear free term $f(x)$. We are able to consider the more general equation (5.134), but we in general require that $F_0(x;u)$ is non-vanishing unless $u = 0$, since otherwise certain features of equations of the first kind (see section 5.17) can appear.

Example 5.50.

Consider the equation

$$f(x) = 1 + xf(x)\int_0^1 \frac{\Psi(y)f(y)}{x + y}\, dy \quad (0 \le x \le 1) ,$$

which arises in radiative transfer (and is known as the H-equation). Here $\Psi(x)$ is known for $0 \le x \le 1$. The equation is readily presented in the form of (5.134) by writing $F_0(x;u) = u, F_1(x,y;u,v) =$ $= x\Psi(y)uv/(x+y) + 1$ and $a = 0$, $b = 1$. However, if we write $\phi(x) = 1/f(x)$ we obtain, for $0 \le x \le 1$

$$\phi(x) + \int_0^1 \frac{x\Psi(y)}{x + y}\,\frac{1}{\phi(y)}\, dy = 1 .$$

This equation is of the form (5.133) and its solution $\phi(x)$ is continuous if $f(x)$ is continuous and non-vanishing. *

We recognize from the preceding Example that a given non-linear integral equation can be presented in a number of ways and this is generally the case. To some extent (5.133) appears satisfactory because of the similarity with the linear equations of the second kind, but (5.134) can also be treated successfully under reasonable conditions.

5.18.1. *General comments on numerical methods*

The obvious numerical methods for treating non-linear equations are, in general, natural extensions of the methods for linear equations. The methods which suggest themselves fall into the usual classes, namely approximate integration methods and expansion methods, the latter including generalized Galerkin methods and variational (Rayleigh-Ritz) methods. In general, the formulation of such methods results in a set of non-linear algebraic equations which determine an approximate solution. The solution of a system of non-linear equations can itself provide a considerable problem, and an investigation in depth of the difficulties is beyond us here. (A comprehensive reference on the subject of systems of non-linear algebraic equations is Ortega and Rheinboldt (1970). See also Collatz (1966*b*) and Rall (1969).) We restrict ourselves, in the main, to some comments on Newton's method and a modified form of Newton's method, both of which were introduced in general terms in section 2.20.4.

<u>Remark</u>. When Newton's method is applied to systems of algebraic equations approximating (5.133), the resulting iteration is an analogue of the Newton iteration of section 1.12.1.

5.19. *The quadrature methods*

Given a quadrature rule

$$\int_a^b \phi(y)\,\mathrm{d}y \simeq \sum_{j=0}^{n} w_j \phi(y_j) ,$$

the quadrature method for linear equations is readily extended. We set, for eqn (5.133),

$$\tilde{f}(x) - \sum_{j=0}^{n} w_j F(x,y_j;\tilde{f}(y_j)) = g(x) , \tag{5.135}$$

and for (5.134)

$$F_0(x;\tilde{f}(x)) - \sum_{j=0}^{n} w_j F_1(x,y_j;\tilde{f}(x),\tilde{f}(y_j)) = 0 . \tag{5.136}$$

Eqns (5.135) and (5.136) are functional equations and to obtain a finite system of non-linear equations we set $x = y_i$ $(i=0,1,2,\ldots,n)$ and obtain, for each equation respectively,

$$\tilde{f}(y_i) - \sum_{j=0}^{n} w_j F(y_i, y_j; \tilde{f}(y_j)) = g(y_i) \tag{5.137}$$

and

$$F_0(y_i; \tilde{f}(y_i)) - \sum_{j=0}^{n} w_j F_1(y_i, y_j; \tilde{f}(y_i), \tilde{f}(y_j)) = 0 \,. \tag{5.138}$$

In the former case, eqn (5.135) provides a convenient interpolation formula for generating $\tilde{f}(x)$ when $x \notin \{y_i\}$.

Example 5.51.
Consider the equation

$$f(x) + \int_0^1 M(x,y) L(y, f(y))\, dy = g(x) \;, \quad x \in [0,1] ,$$

where $M(x,y) = x(1-y)$ for $0 \le x \le y \le 1$ and $M(y,x) = M(x,y)$. If we use the repeated trapezium rule with step $h = 1/n$, we obtain the equations

$$\tilde{f}(0) = g(0)$$

$$\tilde{f}(ih) + \sum_{j=0}^{i}{}'' jh^2(1-ih) L(jh, \tilde{f}(jh))$$

$$+ \sum_{j=i}^{n}{}'' ih^2(1-jh) L(jh, \tilde{f}(jh)) = g(ih) \quad (i = 1,2,3,\ldots,(n-1)) \,,$$

$$\tilde{f}(1) = g(1) \quad .$$

Here, the values $\tilde{f}(0)$ and $\tilde{f}(1)$ are given explicitly; they do not contribute to the remaining non-linear equations. These non-linear equations are in a form studied by Ortega and Rheinboldt (1970, p.20). *

In certain circumstances the modified quadrature method appears appropriate.

Example 5.52.
For the equation

$$f(x) + \int_a^b M(x,y)N(x,y,f(y))\,dy = g(x) ,$$

where $M(x,y)$ is integrable but unbounded at $x = y$, we may write

$$\int_a^b M(x,y)N(x,y,f(y))\,dy$$

$$= N(x,x,f(x))\int_a^b M(x,y)\,dy$$

$$+ \int_a^b M(x,y)\ \{N(x,y,f(y))-N(x,x,f(x))\}dy$$

and (assuming continuity of $N(x,y,f(y))$ at $x = y$) set

$$\tilde{f}(y_i) + N(y_i,y_i,\tilde{f}(y_i))\{\int_a^b M(y_i,y)\,dy - \sum_{j\neq i} w_j M(y_i,y_j)\} +$$

$$+ \sum_{j\neq i} w_j M(y_i,y_j)N(y_i,y_j,\tilde{f}(y_j)) = g(y_i) .$$ *

Product-integration methods also have a rôle where the integrand can be written as the sum of products of well-behaved functions and badly behaved functions, the latter being independent of both $f(x)$ and $f(y)$.

Example 5.53.
For the equation

$$f(x) - \int_a^b |x-y|^{-\alpha}\{f(y)\}^2 dy = g(x) \quad (0<\alpha<1)$$

we may use approximations of the form

$$\int_a^b |x-y|^{-\alpha}\phi(y)\,dy \simeq \sum_{j=0}^{n} \nu_j(x)\phi(z_j)$$

to write

$$\tilde{f}(z_i)-\sum_{j=0}^{n} \nu_j(z_i)\{\tilde{f}(z_j)\}^2 = g(z_i) \quad (i = 0,1,2,\ldots,n) . \quad *$$

5.19.1. *Iterative methods*

If we apply Newton's method to eqn (5.133) we employ the iteration defined by

$$r_k(x) = f_k(x)-\int_a^b F(x,y;f_k(y))\mathrm{d}y-g(x) ,$$

$$\varepsilon_k(x)-\int_a^b K_k(x,y)\varepsilon_k(y)\mathrm{d}y = r_k(x) ,$$

$$f_{k+1}(x) = f_k(x)-\varepsilon_k(x) ,$$

where $K_k(x,y) = (\partial/\partial v)F(x,y;v)\Big]_{v=f_k(y)}$. The analogue for eqn (5.137) is to employ the iteration defined by

$$\tilde{r}_k(y_i) = \tilde{f}_k(y_i)-\sum_{j=0}^{n} w_j F(y_i,y_j;\tilde{f}_k(y_j))-g(y_i) , \tag{5.139}$$

$$\tilde{\varepsilon}_k(y_i)-\sum_{j=0}^{n} w_j \tilde{K}_k(y_i,y_j)\tilde{\varepsilon}_k(y_j) = \tilde{r}_k(y_i) , \tag{5.140}$$

$$\tilde{f}_{k+1}(y_i) = \tilde{f}_k(y_i)-\tilde{\varepsilon}_k(y_i) \quad (i = 0,1,2,\ldots,n) \tag{5.141}$$

where, for $a \leq x,\ y \leq b$,

$$\tilde{K}_k(x,y) = (\partial/\partial v)F(x,y;v)\Big]_{v=\tilde{f}_k(y)} . \tag{5.142}$$

Conditions similar to those in Theorem 1.5 can be obtained for the Newton iteration which is defined by eqns (5.139)-(5.142).

In practice the repeated determination of $\tilde{K}_k(y_i,y_j)$ in (5.140) may be expensive, and in the modified Newton method we replace $\tilde{K}_k(y_i,y_j)$ in (5.140) by $\tilde{K}_0(y_i,y_j)$. (The iterates produced in this way differ from the original ones, of course!)

In general we may be uncertain about the existence of a solution to the non-linear equations (5.137). The following result provides some insight into this question.†

---

† See Moore's contribution in Anselone (1964), particularly p.70.

THEOREM 5.20. *Suppose that, for some initial vector* $\tilde{\mathbf{f}}_0 = [\tilde{f}_0(y_0), \tilde{f}_0(y_1), \ldots \tilde{f}_0(y_n)]^T$, *where* $\tilde{\mathbf{r}}_0 = [\tilde{r}_0(y_0), \tilde{r}_0(y_1), \ldots, \tilde{r}_0(y_n)]^T$ *is defined by* (5.139), $\tilde{\mathbf{K}}_0 = [\tilde{K}_0(y_i, y_j)]$ *and* $\mathbf{D} = \mathbf{diag}(w_0, w_1, \ldots, w_n)$, *the following conditions are satisfied*

(i) $\|(\mathbf{I}-\tilde{\mathbf{K}}_0\mathbf{D})^{-1}\|_\infty \le \beta_0$,

(ii) $\|(\mathbf{I}-\tilde{\mathbf{K}}_0\mathbf{D})^{-1}\tilde{\mathbf{r}}_0\|_\infty \le \eta_0$,

(iii) $$\max_{i=0,1,2,\ldots,n} \sum_{j=0}^{n} \left| w_j(\partial^2/\partial v^2)F(y_i,y_j;v)\Big]_{v=v_j} \right| \le c$$

whenever $\|\mathbf{v}-\tilde{\mathbf{f}}_0\|_\infty \le \{1+\surd(1-2\gamma_0)\}\eta_0/\gamma_0 \le 2\eta_0$, *with* $\mathbf{v} = [v_0, v_1, \ldots, v_n]^T$, *and with*

(iv) $\gamma_0 = \beta_0\eta_0 c$ *such that* $\gamma_0 < \frac{1}{2}$.

Then the system (5.137) *has a unique solution* $\tilde{\mathbf{f}} = [\tilde{f}(y_0), \tilde{f}(y_1), \ldots, \tilde{f}(y_n)]^T$ *satisfying the condition*

$$\|\tilde{\mathbf{f}}-\tilde{\mathbf{f}}_0\|_\infty < \{1+\surd(1-2\gamma_0)\}\eta_0/\gamma_0 .$$

If $\tilde{\mathbf{f}}_n$ *is the vector of values* $\tilde{f}_n(y_i)$ *obtained from* (5.141) *at the* nth *stage, then*

$$\|\tilde{\mathbf{f}}-\tilde{\mathbf{f}}_n\|_\infty \le \{1-\surd(1-2\gamma_0)\}^{n-1}\|\tilde{\mathbf{f}}-\tilde{\mathbf{f}}_0\|_\infty .$$

The iterative solution of the non-linear equations (5.138) can also be accomplished by the general Newton iteration or its modified form (section 2.20.1) provided the Jacobian of the system can be computed. Thus we are required to be able to compute the matrix whose elements are

$$(\partial/\partial v_j)\{F_0(y_i;v_i) - \sum_{p=0}^{n} w_p F_1(y_i,y_p;v_i,v_p)\}$$
$$(i,j = 0,1,2,\ldots,n)$$

in the (i,j)th position, with v_i set equal to $\tilde{f}_k(y_i)$. The partial derivatives $(\partial/\partial v)F_0(x;v)$, $(\partial/\partial u)F_1(x,y;u,v)$, and $(\partial/\partial v)F_1(x,y;u,v)$

must therefore be available.

If we resort to the modified Newton method to save evaluations of these derivatives, the usual quadratic convergence of Newton's method is lost. Moreover, other linearly or superlinearly convergent iterations may be available without the inconvenience of evaluating derivatives (Ortega and Rheinboldt 1970). A flexible approach is therefore required.

Example 5.54.
The analogue of an iteration of section 1.12.1 for eqn (5.137) is defined by setting $\tilde{f}_0(x) = g(x)$ for $x \in \{y_i\}$ and

$$\tilde{f}_{k+1}(y_i) = g(y_i) + \sum_{j=0}^{n} w_j F(y_i, y_j; \tilde{f}_k(y_j)) \quad (i = 0,1,2,\ldots,n) .$$

(There are some cases in which this iteration will not converge.) *

Example 5.55.
For the equation for $f(x)$ in Example 5.50 eqn (5.138) becomes

$$\tilde{f}(y_i) = 1 + y_i \tilde{f}(y_i) \sum_{j=0}^{n} w_j \frac{\Psi(y_j)\tilde{f}(y_j)}{y_i + y_j} ,$$

so that

$$\tilde{f}(y_i) = \left[1 - y_i \sum_{j=0}^{n} w_j \frac{\Psi(y_j)\tilde{f}(y_j)}{y_i + y_j} \right]^{-1} .$$

This form suggests the iteration defined by

$$\tilde{f}^{[k+1]}(y_i) = \left[1 - y_i \sum_{j=0}^{n} \frac{w_j \Psi(y_j)\tilde{f}^{[k]}(y_j)}{y_i + y_j} \right]^{-1} \quad (i=0,1,2,\ldots,n, k=0,1,2,\ldots)$$

with $\tilde{f}^{[0]}(x) \equiv 1$. *

5.19.2. *Choice of quadrature rule*

The choice of quadrature rule will in general depend upon the assumed smoothness of the integrand (which involves the unknown function $f(x)$). In the case of a non-linear integral equation, determining the smoothness of the solution may be too difficult a problem, and it may be necessary to obtain insight into the behaviour of $f(x)$ from the behaviour of some approximate solution.

If the method of deferred approach to the limit seems feasible, a composite rule (such as the repeated trapezium rule) is appropriate. Given a suitably smooth integrand in the integral equation, we can also expect to obtain asymptotic upper and lower bounds for the solution. (The general theory of Stetter (1965) applies to integral equations and provides a justification of these procedures under certain conditions. Such results generalize those we have established in Chapter 4.)

Example 5.56.
Consider the equation

$$f(x) + \int_0^1 \{x+f(y)\}^2 dy = x^2+2x+\frac{1}{3},$$

for which a solution is the simple function $f(x) = x$; a second solution is $f(x) = 5x-4$. Since Simpson's rule would yield the correct solutions for this equation, it provides a somewhat artificial example.

We may use the trapezium rule, with step $h = 1/n$, to form the approximating equations

$$\tilde{f}(ih) + h\sum_{j=0}^{n}{}''\{ih+\tilde{f}(jh)\}^2 = (ih)^2+2ih+\frac{1}{3} \quad (i=0,1,2,\ldots,n).$$

We solved these non-linear equations approximately by Newton's method, for $n = 8, 16, 32, 64$. For each value of n, the iteration was terminated at the fifth iterate. The initial estimate, corresponding to the choice $\tilde{f}_0(x) = x^2 + 2x + \frac{1}{3}$, determined which of the solutions was obtained. Values of $\tilde{f}_5(x)$ are shown in Table 5.10.

TABLE 5.10

$x \backslash n$	8	16	32	64
0	−0·0002	−0·00005	−0·000012	−0·000003
0·25	0·2485	0·24963	0·249908	0·249977
0·50	0·4990	0·49976	0·499939	0·499985
0·75	0·7495	0·74988	0·749969	0·749992
1·00	1·0000	1·00000	1·000000	1·000000

The errors appear to be $O(h^2)$ and if we take 4 times the values for $n = 32$, subtract the values obtained with $n = 16$, and divide by 3, we obtain to close accuracy the values for $n = 64$. This is the sort of result to be expected on iterating to convergence. *

5.19.3. *A particular equation*

Consider the equation

$$f(x) = 1 + \frac{\Lambda}{2} f(x) \int_0^1 \frac{x}{x+y} f(y)\,dy \tag{5.143}$$

which arises from the equation in Example 5.50 on setting $\Psi(x) \equiv \frac{1}{2}\Lambda$. The explicit solution is (Stibbs and Weir 1959)

$$f(x) = \exp\{\frac{-x}{\pi} \int_0^{\pi} \frac{\ln(1-\Lambda t \cot t)}{\cos^2 t + x^2 \sin^2 t}\,dt\}, \quad x \in [0,1].$$

We shall use this equation, with various values of Λ, to illustrate numerical results obtained by various methods.

We report here on some numerical experiments performed with the quadrature method applied to eqn (5.143). Some variations will be introduced.

If we apply a quadrature rule

$$\int_0^1 \phi(y)\,dy \simeq \sum_{j=0}^{n} w_j \phi(y_j)$$

$$(\text{with } 0 \le y_0 < y_1 < \ldots < y_n \le 1)$$

directly to the integral equation we find

$$\tilde{f}(x) = 1 + \frac{\Lambda}{2} \tilde{f}(x) \sum_{j=0}^{n} \frac{w_j x}{x + y_j} \tilde{f}(y_j) \quad (0 < x \le 1), \tag{5.144}$$

whilst $\tilde{f}(0) = 1$. Thus on setting $x = y_i$ we obtain a set of equations for the values $\tilde{f}(y_i)$, $(i = 0,1,2,\ldots,n)$. These equations can be written in the form

$$\tilde{f}(y_i)(1 - \frac{\Lambda}{2} \sum_{j=0}^{n} A_{i,j} \tilde{f}(y_j)) = 1, \tag{5.145}$$

where $A_{i,j} = w_j y_i/(y_i+y_j)$ (if $y_0 = 0$, then $A_{0,j} = 0$) for $i,j = 0,1,2,\ldots,n$.

The Newton iteration for eqns (5.145) is defined by setting up the Jacobian matrix with elements

$$J_{i,j}^{[k]} \equiv J_{i,j}(\tilde{\mathrm{f}}^{[k]}) = \delta_{i,j}\{1-\frac{\Lambda}{2}\sum_{j=0}^{n} A_{i,j}\tilde{f}^{[k]}(y_j)\}-\frac{\Lambda}{2}\tilde{f}^{[k]}(y_i)A_{i,j},$$

where

$$\tilde{\mathrm{f}}^{[k]} = [\tilde{f}^{[k]}(y_0),\tilde{f}^{[k]}(y_1),\ldots,\tilde{f}^{[k]}(y_n)]^T.$$

The iteration is then

$$\tilde{\mathrm{f}}^{[k+1]} = \tilde{\mathrm{f}}^{[k]} - [\mathrm{J}^{[k]}]^{-1}\mathrm{r}^{[k]},$$

where

$$r_i^{[k]} = \tilde{f}^{[k]}(y_i)(1-\frac{\Lambda}{2}\sum_{j=0}^{n} A_{i,j}\tilde{f}^{(k)}(y_j))-1,$$

and $\mathrm{J}^{[k]} = [J_{i,j}^{[k]}]$.

The iteration in Example 5.55 can also be applied to (5.145), and there are two variants. Thus we have a 'method of simultaneous displacements'

$$\tilde{f}^{[k+1]}(y_i) = (1-\frac{\Lambda}{2}\sum_{j=0}^{n} A_{i,j}\tilde{f}^{[k]}(y_j))^{-1} \quad (i = 0,1,2,\ldots,n)$$

(which is like the Jacobi iteration for linear equations). We also have a 'method of successive displacements'

$$\tilde{f}^{[k+1]}(y_i) = (1-\frac{\Lambda}{2}\sum_{j=0}^{i-1} A_{i,j}\tilde{f}^{[k+1]}(y_j)-\frac{\Lambda}{2}\sum_{j=i}^{n} A_{i,j}\tilde{f}^{[k]}(y_j))^{-1}$$

$$(i = 0,1,2,\ldots,n)$$

(which is like the Gauss-Seidel SOR iteration).

In analogy with the transformation of the integral equation in Example 5.50, we can also rewrite (5.145) in the form

$$\tilde{\phi}(y_i) = 1/\tilde{f}(y_i)$$

$$\tilde{\phi}(y_i) + \frac{\Lambda}{2}\sum_{j=0}^{n} A_{i,j}/\tilde{\phi}(y_j) = 1 . \tag{5.146}$$

The Newton iteration for the latter equations is defined by setting

$$\rho_i^{[k]} = \tilde{\phi}^{[k]}(y_i) + \frac{\Lambda}{2}\sum_{j=0}^{n} A_{i,j}/\tilde{\phi}^{[k]}(y_j) - 1$$

and

$$\tilde{\underset{\sim}{\phi}}^{[k+1]} = \tilde{\underset{\sim}{\phi}}^{[k]} - [\underset{\sim}{J}^{\{k\}}]^{-1}\underset{\sim}{\rho}^{[k]}$$

where, now,

$$J_{i,j}^{\{k\}} = \delta_{i,j} - \frac{\Lambda}{2} A_{i,j}/\{\tilde{\phi}^{[k]}(y_j)\}^2 .$$

Indeed, there are a number of ways of rewriting the integral equation (5.143) prior to application of the quadrature method. Thus, writing

$$(1-\Lambda)^{\frac{1}{2}} f(x) = 1 - \frac{\Lambda}{2} f(x)\int_0^1 \frac{yf(y)}{x + y}dy , \tag{5.147}$$

we have an integral equation in which the integrand is apparently better behaved, than that in (5.143), at $y = 0$. Noble (in Anselone 1964) shows how (5.147) may be derived from (5.143).

If we return to eqn (5.143) we can show that $f(x)$ behaves like $1 + Cx\ln(x)$ for x sufficiently small, and the integrand $xf(y)/(x+y)$ is smoothed by setting $x = u^2$ and $y = v^2$. The equation then becomes

$$f(u^2) = 1 + \frac{\Lambda}{2}f(u^2)\int_0^1 \frac{u^2}{u^2+v^2}f(v^2)2v\,dv .$$

Writing $\psi(u) = f(u^2)$ we obtain

$$\psi(u) = 1 + \Lambda\psi(u)\int_0^1 \frac{vu^2}{u^2+v^2}\psi(v)\,dv . \tag{5.148}$$

(More generally we may set $x = u^n$, $y = v^n$, $n > 2$.)

The quadrature method can be applied either to (5.147) or (5.148). In particular, for (5.148) we obtain

$$\tilde{\psi}(y_i) = 1 + \Lambda\tilde{\psi}(y_i) \sum_{j=0}^{n} B_{i,j}\tilde{\psi}(y_j) , \qquad (5.149)$$

where

$$B_{i,j} = w_j y_j y_i^2/(y_i^2+y_j^2) ,$$

so that the general form of the equations is the same as (5.145). A programme for solving (5.145) can easily be applied to solve (5.149).

In all the methods above, a choice of quadrature rule is required. In our experiments we chose (a) the m-times repeated trapezium rule - that is, the trapezoidal rule with step $h = 1/m$ - and (b) the m-times repeated version of the 12-point Gauss-Legendre rule, and solved the resulting equations by an iterative technique. The criterion for termination of the iteration was taken to be agreement of successive iterates $\tilde{f}^{[k]}$ to within approximately 10^{-10}, taking $\tilde{f}^{[0]} = [1,1,\ldots,1]^T$. In the case of the Gauss-Legendre rule we then used (5.144) as an extrapolation formula to tabulate values of $\tilde{f}(x)$ at equally spaced intervals of the argument.

Various aspects of our numerical computations are summarized below.

Example 5.57.

We considered the case $\Lambda = \frac{1}{2}$. Using the m-times repeated Gauss-Legendre rules in (5.147) and (5.149), we found (consistently to the number of figues shown) the results shown in Table 5.11 in the first row. The second and third rows give results with tolerances which have been computed by Rall (1965).

TABLE 5.11

x	0·25	0·50	0·75	1·00
Agreement for $m = 1,2,4$:	1·129653	1·187735	1·224933	1·251260
Rall:	1·12965 ± 0·00019	1·18774 ± 0·00020	1·22493 ± 0·00020	1·25122 ± 0·00024
Rall:	1·129654 ± 0·000008	1·187795 ± 0·000012	1·225058 ± 0·000009	1·251348 ± 0·000001

The reader will find it interesting to consult Rall (1965) for a theoretical discussion of the tolerance tabulated here, and computer timings for his calculations.

Since so many methods based on the Gauss-Legendre rules gave consistent agreement, we might be prepared to accept all the figures tabulated in the first row.† In obtaining these figures the Newton method required 4 iterations, the method of successive displacements required 9 iterations and the method of simultaneous displacements required 13 iterations. Of course, a single step of the Newton iteration involves more work than a single step of either of the latter two methods.

The accuracy obtained using the trapezium rule in (5.145) is indicated by the figures in Table 5.12.

TABLE 5.12

m-times repeated trapezium rule

$m \diagdown x$	0·25	0·50	0·75	1·00
8	1·130868	1·181849	1·225094	1·251269
16	1·129957	1·187845	1·224971	1·251260
32	1·129729	1·187762	1·224942	1·251260
64	1·129672	1·187742	1·224935	1·251260

Solving (5.146) by Newton's method, using the trapezium rule (repeated m-times with $m = 64$) to formulate the equations, we compute for $f(x) = 1/\phi(x)$ the figures below:

$x =$ 0·25	0·5	0·75	1·00
1·129671	1·187742	1·224935	1·251260

In computing these figures, the fourth and fifth iterates $\tilde{\underset{\sim}{\phi}}^{[4]}, \tilde{\underset{\sim}{\phi}}^{[5]}$ agree to 15 decimal figures. *

<u>Example 5.58</u>.

The case $\Lambda = 0{\cdot}95$ is considered by Moore (in Anselone 1964), who con-

† To do so would conflict with some of the tolerances quoted from Rall.

siders (5.146) to obtain $\tilde{\phi}(x)$ using a 6-point Gauss-Legendre rule, and compares his results with those of Stibbs and Weir (1959).

We solved (5.145) using an m-times repeated 12-point Gauss rule and computed the results shown in Table 5.13, with $\tilde{\phi}(x) = 1/\tilde{f}(x)$.

TABLE 5.13

m \ x	0·05	0·10	0·30	0·60	1·00
1	1·1115065	1·1952323	1·4604541	1·7647117	2·0771239
2	1·1115079	1·1952319	1·4604539	1·7647116	2·0771239
4	1·1115078	1·1952318	1·4604539	1·7647116	2·0771238
Stibbs and Weir (1959)	1·111508	1·195232	1·460454	1·764711	2·077123
Moore (1964)	1.111063	1·195159	1·460465	1·764719	2·077129

As for the case $\Lambda = \frac{1}{2}$, the accuracy appears better near to $x = 1$ than near to $x = 0$. In the iteration for $m = 4$ the Newton iteration took 6 steps, the simultaneous displacement method took 43, and the successive displacement method took 25 steps, to give comparable accuracy.

The accuracy of the figures tabulated above appears to be improved by using (5.149), applying a repeated 12-point Gauss rule. We then obtain the following figures.

$x =$	0.05	0·10	0·30	0·60	1·00	
$m = 2$	1·1115078	1·1952318	1·4604539	1·7647116	2·0771238	*

Example 5.59.

The case $\Lambda = 1$ in eqn (5.143) is discussed by Noble (in Anselone 1964), who shows that for a real solution $f(x)$ it is necessary (see eqn (5.147)) to have $\Lambda \leq 1$. He also shows that for $\Lambda = 1$

$$f(x) = \left(\tfrac{1}{2}\int_0^1 \frac{yf(y)}{x+y}dy\right)^{-1},$$

and this equation forms the basis of some of his numerical methods.

We solved eqns (5.145) using the trapezium rule repeated m times and obtained the results shown in Table 5.14, in the case $\Lambda = 1$.

TABLE 5.14

m \ x	0·25	0·50	0·75	1·00
8	1·5289326	1·9591330	2·3647683	2·7541935
16	1·5374709	1·9857031	2·4137208	2·8299838
32	1·5422413	1·9991947	2·4385159	2·8686700

To obtain these results we employed the Newton method, the other iterations discussed being too slowly convergent for practical use when $\Lambda = 1$. We suggest that this slow convergence is associated with the fact that $\Lambda = 1$ is a limiting value for a real solution.

The discussion of Noble, mentioned above, is too substantial to summarize here. However, it must be mentioned that he employs extrapolation on values computed with the repeated Simpson's rule applied to the equation as formulated above. For comparison, his extrapolated results are

x = 0·25	0·50	1·00
1·547325	2·012778	2·907810

Noble states that the error in these results is one unit in the last place. *

5.20. *Expansion methods*

The principles underlying the construction of expansion methods for non-linear equations are much the same as those for linear equations. However, it is sometimes convenient to manipulate the non-linear equation into a new form before applying these principles. We saw in our examples of the quadrature method (section 5.19) that an equation can often be re-formulated in different ways.

For eqns (5.133) and (5.134) we may seek an approximate solution $\tilde{f}(x) = \sum_{i=0}^{n} \tilde{a}_i \phi_i(x)$. We then find

$$\tilde{f}(x) - \int_a^b F(x,y;\tilde{f}(y))\,dy - g(x) = r(x) \qquad (5.150)$$

or

$$F_0(x;\tilde{f}(x)) - \int_a^b F_1(x,y;\tilde{f}(x),\tilde{f}(y))\,dy = r(x) \ , \qquad (5.151)$$

respectively, where $r(x)$ depends upon $\tilde{a}_0,\tilde{a}_1,\ldots,\tilde{a}_n$. The method of collocation results if we choose $z_0,z_1,\ldots,z_n \in [a,b]$ and determine $\tilde{a}_0,\tilde{a}_1,\ldots,\tilde{a}_n$ so that $r(z_i) = 0$. In general, the equations for $\tilde{a}_i$ $(i = 0,1,2,\ldots,n)$ are, of course, non-linear. The classical Galerkin method is produced if we choose $\tilde{a}_i$ $(i = 0,1,2,\ldots,n)$ so that

$$\int_a^b r(x)\overline{\phi_i(x)}\,dx = 0 \quad (i = 0,1,2,\ldots,n) \ .$$

More generally we can pick functions $\psi_i(x)$ and require

$$\int_a^b r(x)\overline{\psi_i(x)}\,dx = 0$$

or, with $\omega(x) > 0$, we can require that

$$\int_a^b \omega(x)r(x)\overline{\phi_i(x)}\,dx = 0 \ .$$

Similarly, with $\Omega_i > 0$ $(i = 0,1,2,\ldots,N)$ and $N > n$, we can require that

$$\sum_{j=0}^{N} \Omega_j \overline{\phi_k(z_j)} r(z_j) = 0 \quad (k = 0,1,2,\ldots,n) \ .$$

The Rayleigh-Ritz approach consists of finding a functional $J[\psi]$ which takes a stationary value when $\psi(x) = f(x)$, and then choosing $\tilde{a}_0,\tilde{a}_1,\ldots,\tilde{a}_n$, so that, for $k = 0,1,2,\ldots,n$, $(\partial/\partial\tilde{a}_k)J[\tilde{f}] = 0$. We discuss this method in a later section and shall show that it reduces to a classical Galerkin method for a suitable reformulation of the integral equation.

5.21. *Collocation*

We have described, briefly, the motivation of the collocation method, and our purpose is now served best by considering an example.

Example 5.60.
In this Example, we consider again the equation

$$f(x) = 1+\tfrac{1}{2}xf(x)\int_0^1 \frac{f(y)}{x+y}dy \quad (0 \le x \le 1) .$$

An approximate solution may be obtained by collocation, using shifted Chebyshev polynomials. Now, $z_0,z_1,\ldots,z_n$ are the collocation points in $[0,1]$ and we seek an approximation

$$\tilde{f}(x) = \sum_{r=0}^{n} \tilde{a}_r T_r^*(x) ,$$

with $T_r^*(x) = T_r(2x-1)$. The collocation equations for $\tilde{a}_0,\tilde{a}_1,\ldots,\tilde{a}_n$ are

$$\sum_{j=0}^{n} \tilde{a}_j T_j^*(z_i)-\tfrac{1}{2}z_i \sum_{j=0}^{n} \tilde{a}_j T_j^*(z_i)\{ \sum_{k=0}^{n} I_{i,k}^* \tilde{a}_k\} = 1 ,$$

where

$$I_{i,k}^* = \int_0^1 \{T_k^*(y)/(z_i+y)\}dy \quad (i = 0,1,2,\ldots,n) .$$

Thus the non-linear equations for $\tilde{a}_0,\tilde{a}_1,\ldots,\tilde{a}_n$ assume the form $\underset{\sim}{\phi}(\tilde{a}_0,\tilde{a}_1,\ldots,\tilde{a}_n) = \underset{\sim}{0}$, where the ith component of $\underset{\sim}{\phi}(\tilde{a}_0,\tilde{a}_1,\ldots,\tilde{a}_n)$ is

$$\phi_i(\tilde{a}_0,\tilde{a}_1,\ldots,\tilde{a}_n)$$
$$= \sum_{j=0}^{n} \tilde{a}_j\{T_j^*(z_i)- \sum_{k=0}^{n} \tfrac{1}{2}z_i T_j^*(z_i)I_{i,k}^* \tilde{a}_k\} -1 \quad (i= 0,1,2,\ldots,n) .$$

We propose a Newton iteration to solve these non-linear equations.

Denoting by $\underset{\sim}{\alpha}^{(r)}$ successive approximations $(r = 0,1,2,\ldots)$ to $\tilde{\underset{\sim}{a}} = [\tilde{a}_0,\tilde{a}_1,\ldots,\tilde{a}_n]^T$ we set $\underset{\sim}{\alpha}^{(0)} = [1,0,\ldots,0]^T$ and $\underset{\sim}{\alpha}^{(r+1)} = \underset{\sim}{\alpha}^{(r)} - \underset{\sim}{\delta}^{(r)}$, where

$$\underset{\sim}{A}^{(r)}\underset{\sim}{\delta}^{(r)} = \underset{\sim}{\phi}(\alpha_0^{(r)},\alpha_1^{(r)},\ldots,\alpha_n^{(r)})$$

and

$$A_{i,j}^{(r)} = \left.\frac{\partial\phi_i}{\partial\tilde{a}_j}\right]_{\tilde{\underset{\sim}{a}}=\underset{\sim}{\alpha}^{(r)}} .$$

To implement this technique we require

$$\frac{\partial\phi_i}{\partial\tilde{a}_j} = T_j^*(z_i) - \tfrac{1}{2}z_i \sum_{k=0}^{n} \tilde{a}_k \{I_{i,k}^* T_j^*(z_i) + I_{i,j}^* T_k^*(z_i)\} .$$

Thus the integrals $I_{k,j}^*$ $(k,\ j = 0,1,2,\ldots,n)$ are required throughout the calculation. Using the recurrence relation for the shifted Chebyshev polynomials we can show that

$$I_{k,j+1}^* + I_{k,j-1}^* = \frac{2}{j^2-1}\{(-1)^{j-1}-1\} - 2(2z_k+1)I_{k,j}^* \quad (j \geq 2) ,$$

where $I_{k,0}^* = \ln\{1+(1/z_k)\}$, $I_{k,1}^* = 2-(2z_k+1)\ln(1+(1/z_k))$, and $I_{k,2}^* = (1+8z+8z^2)I_{k,0}-8z-4$, if $z \equiv z_k \in [0,1]$. Unfortunately this recurrence relation is *unstable* and it cannot be used successfully for large values of j unless great care is exercised. Nevertheless, we experimented with its use in the collocation method for values of n up to about 25. (We consider the instability in the forward use of the recurrence relation takes effect at about $n \geq 8$, on the computer used for the experiments.)

For the collocation points z_i we took the zeros of $T_{n+1}^*(x)$ and computed the approximate solution for $n = 4,8$. The figures obtained are indicated below.

Accepted coefficients of $T_r^*(x)$, $r \geq 0$ are as follows: for $n = 4$: $1{\cdot}965932$, $0{\cdot}252573 \times 10^{-1}$, $-2{\cdot}989327 \times 10^{-2}$, $1{\cdot}103407 \times 10^{-2}$, $-3{\cdot}826714 \times 10^{-3}$; for $n = 8$: $1{\cdot}982014$, $0{\cdot}253314 \times 10^{-1}$, $-2{\cdot}796807 \times 10^{-2}$, $1{\cdot}52860 \times 10^{-2}$, $-5{\cdot}664463 \times 10^{-3}$, $3{\cdot}038697 \times 10^{-3}$,

$-1{\cdot}703630 \times 10^{-3}$, $9{\cdot}319601 \times 10^{-4}$, $-4{\cdot}154230 \times 10^{-4}$.

The slow rate of decrease in the coefficients indicates that $f(x)$ is not a very smooth function.

Values of $\tilde{f}(x)$ are shown in Table 5.15.

TABLE 5.15

$x \backslash n$	4	8
0	1·016614	1·005431
0·1	1·240222	1·247080
0·2	1·444053	1·448075
0·3	1·634205	1·638841
0·4	1·815604	1·824275
0·5	1·971998	2·005605
0·6	2·165960	2·184607
0·7	2·338888	2·362231
0·8	2·511004	2·538254
0·9	2·681354	2·713050
1·0	2·847810	2·887093

Since investigation of the solution of the integral equation shows that $f(x) = 1+Cx \ln x$ as $x \to 0$, we may seek an approximation of the form

$$\tilde{f}(x) = \sum_{r=0}^{n} \tilde{a}_r T_r^*(x) + \tilde{a}_{-1} x \ln x \ .$$

We computed the results shown in Table 5.16 for $n = 8,12$, using as collocation points the zeros $z_0, z_1, \ldots, z_n$ of $T_{n+1}^*(x)$ and, additionally, $z_{n+1} = 1$.

TABLE 5.16

$x \backslash n$	8	12
0	1·000451	1·000126
0·1	1·247184	1·247280
0·2	1·449926	1·450192
0·3	1·641803	1·642250
0·4	1·828212	1·828871
0·5	2·011314	2·012222
0·6	2·192217	2·193405
0·7	2·371558	2·373057
0·8	2·549735	2·551576
0·9	2·729230	2·729230
1·0	2·903584	2·906204

Because of the instability in the recurrence relation for $I^*_{k,j}$ we consider expressing an approximation $\tilde{f}(x) = \sum_{r=0}^{n} \tilde{a}_r T^*_r(x)$ as $\sum_{r=0}^{n} \tilde{b}_r x^r$. The integrals $\int_0^1 \{y^r/(z_i+y)\}dy$ can now be evaluated satisfacto: .ly. Unfortunately, for large values of n we find that the coefficients $\tilde{b}_r$ $(r = 0,1,2,\ldots,n)$ are such that small perturbations cause large changes in $\tilde{f}(x)$, and loss of significance occurs. The instability is thus transferred to a different part of the calculation.

We display some computed results below. Coefficients $\tilde{b}_r$, with $n = 8$, were computed as follows: 1·005431, 2·882288, −6·907930, $2{\cdot}902931 \times 10^1$, $-7{\cdot}653612 \times 10^1$, $1{\cdot}236338 \times 10^2$, $-1{\cdot}186921 \times 10^2$, $6{\cdot}208220 \times 10^1$, $-1{\cdot}361260 \times 10^1$. Values of $\tilde{f}(x)$ agreed with those tabulated above using $T^*_r(x)$ for $n = 8$. For $n = 16$ the coefficients $\tilde{b}_r$ varied in magnitude up to $\pm 10^4$ and a criterion for terminating the Newton iteration presented a problem. Including the $\ln x$ term and using a polynomial of degree 6 produced quite modest coefficients $\tilde{b}_r$, and values $\tilde{f}(x)$ with accuracy comparable with that for $n = 16$ in the results obtained using $T^*_r(x)$ $(r = 0,1,2,\ldots,n)$.

The possibility of rearranging the recurrence relation for the quantities $I^*_{k,j}$ (using a technique described in another context by Fox 1962, p.100) has been noted by Dr. Goodwin (private communication). We have

$$I^*_{k,j+1}+2(2z_k+1)I^*_{k,j}+I^*_{k,j-1} = \frac{2}{j^2-1}\{(-1)^{j-1} - 1\} ,$$

and for each k we set (suppressing k for convenience)

$$H_j \equiv H_{k,j} = I^*_{k,j+1}+ \lambda I^*_{k,j}$$

where $\lambda + \lambda^{-1} = 2(2z_k+1)$. Thus $\lambda > 1$ if $z_k \in (0,1]$, and

$$H_j + \lambda^{-1}H_{j-1} = \frac{2}{j^2-1}\{(-1)^{j-1} - 1\} .$$

It follows that

$$H_{2j} + \lambda^{-1}H_{2j-1} = \frac{-4}{4j^2-1}$$

and

$$H_{2j-1} = -\lambda^{-1}H_{2j-2} .$$

With $H_0 = I^*_{k,1} + \lambda I^*_{k,0}$ we can employ these relations to compute $H_1 = -\lambda^{-1}H_0$, $H_2 = \lambda^{-1}H_1 - \frac{4}{3}$, etc., in a stable manner, whilst (from the definition of H_j) we find immediately that

$$I^*_{k,j} = \sum_{r=0}^{\infty} \frac{(-1)^r}{\lambda^{r+1}} H_{j+r} .$$

Summation of the series to give $I^*_{k,N+1}$ permits $I^*_{k,N}, I^*_{k,N-1},\ldots,I^*_{k,1},I^*_{k,0}$ to be obtained using the relation $I^*_{k,j} = \lambda^{-1}(H_j - I^*_{k,j+1})$ and the values of $I^*_{k,1}$ and $I^*_{k,0}$ thus obtained provide a check on the calculation. In practice the method is slightly less satisfactory for small values of z_k than for values close to unity (with, incidentally, more terms being required in the summation to give $I^*_{k,j}$ when z_k is close to zero), but otherwise worked well in the cases attempted.

If we seek an approximate solution $\tilde{f}(x)$ which is a polynomial of degree n, ary basis for the space of polynomials of degree gives rise to a choice of funtions $\phi_r(x)$. The choice of basis and the choice of collocation points are not related, and we can continue to use the zeros of $T^*_{n+1}(x)$ with $\phi_r(x) = x^r$. Bases which seem potentially satisfactory are obtained on setting $\phi_r(x) = P_r(x)$ $(r = 0,1,2,\ldots,n)$ (the Legendre polynomials) or $\phi_r(x) = T_r(x)$ $(r = 0,1,2,\ldots,n)$.

We find, with

$$\hat{I}_{i,k} = \int_0^1 \frac{T_k(y)}{z_i+y}\, dy \ ,$$

that

$$\hat{I}_{i,n+1} + 2z_i\hat{I}_{i,n} + \hat{I}_{i,n-1} = 2\int_0^1 T_n(y)\,dy \ ,$$

whilst if

$$\ddot{I}_{i,k} = \int_0^1 \frac{P_k(y)}{z_i+y}\, dy$$

then the values of $\ddot{I}_{i,k}$ satisfy a homogeneous recurrence relation. Thus

$$\ddot{I}_{i,n+1} + \left(2-\frac{1}{n+1}\right)z_i\ddot{I}_{i,n} + \left(1-\frac{1}{n+1}\right)\ddot{I}_{i,n-1} = 0 \ .$$

Both recurrence relations are more stable, for $z_i \in (0,1)$, than the recurrence for $I^*_{i,k}$. *

Example 5.61.

For the equation treated in Example 5.60 we may find a function $\tilde{f}(x)$ which is continuous and piecewise-linear. We set

$$\tilde{f}(x) = \{(x-x_i)\tilde{f}(x_{i+1}) + (x_{i+1}-x)\tilde{f}(x_i)\}/h$$

for $x_i \le x \le x_{i+1}$, where $x_i = ih,\ h = 1/n$, and take the points x_i as collocation points to obtain $(n+1)$ non-linear equations for

$\tilde{f}(x_0),\tilde{f}(x_1),\ldots,\tilde{f}(x_n)$.

Thus

$$h\tilde{f}(x_i)$$

$$= h+\frac{x_i}{2}\tilde{f}(x_i)\sum_{j=0}^{n-1}\{\tilde{f}(x_{j+1})\int_{x_j}^{x_{j+1}}\frac{(t-x_j)}{x_i+t}dt+\tilde{f}(x_j)\int_{x_j}^{x_{j+1}}\frac{(x_{j+1}-t)}{x_i+t}dt\}\ ,$$

wherein

$$\int_{x_j}^{x_{j+1}}\frac{t-x_j}{x_i+t}dt = (x_{j+1}-x_j)+(x_j+x_i)\{\ln(x_i+x_j)-\ln(x_i+x_{j+1})\}\ ,\quad \text{etc.}$$

The proposed method is equivalent to a product-integration method using approximations

$$\int_0^1 (x+y)^{-1}\phi(y)\,dy \approx \sum_{j=0}^{n}\nu_j(x)\phi(x_j)$$

which are exact if $\phi(x)$ is continuous and linear on $[x_j,x_{j+1}]$ $(j = 0,1,2,\ldots,(n-1))$.

The resulting system of equations can be solved by Newton's method. With $x_i = ih,\ h = 1/n$ we computed the following results, employing a programme written by Miss J.A. Thomas.

n \ x	0·25	0·50	0·75	1·00
20	1·5347426	1·9806258	2·4051255	2·8169121
40	1·5405022	1·9952068	2·4314123	2·8576517

These results compare roughly with those obtained using the repeated trapezium rule in Example 5.59; the derivation of the equations is somewhat more complicated, however. *

5.22. *Galerkin methods*

The application of Galerkin methods is in general more laborious than the application of collocation methods because a double integration is involved.

For eqn (5.133) we obtain an approximation $\tilde{f}(x) = \sum_{r=0}^{n} \tilde{a}_r \phi_r(x)$,

where

$$\sum_{r=0}^{n} \{\int_a^b \phi_r(x)\overline{\phi_s(x)}\,dx\}\tilde{a}_r - \int_a^b\int_a^b \overline{\phi_s(x)} F(x,y;\ \sum_{r=0}^{n} \tilde{a}_r\phi_r(y))\,dy\,dx$$

$$= \int_a^b g(x)\overline{\phi_s(x)}\,dx\ . \qquad (5.152)$$

Some simplification occurs if

$$\int_a^b \phi_r(x)\overline{\phi_s(x)}\,dx = (\phi_r,\phi_s) = \delta_{r,s}\ .$$

but the double integral in (5.152) has to be expressed in terms of $\tilde{a}_0,\tilde{a}_1,\ldots,\tilde{a}_n$ in order to obtain a solution of the non-linear equations.

Some further simplification occurs if $F(x,y;f(y))=K(x,y)M(y;f(y))$. If we write

$$\psi_s(y) = \int_a^b \overline{\phi_s(x)} K(x,y)\,dx\ ,$$

then eqns (5.152) reduce to the form

$$\tilde{a}_s - \int_a^b \psi_s(y) M(y;\ \sum_{r=0}^{n} \tilde{a}_r\phi_r(y))\,dy = \int_a^b \overline{\phi_s(x)} g(x)\,dx\ ,$$

when $(\phi_r,\phi_s) = \delta_{r,s}$.

Example 5.62.

Consider the equation

$$f(x) - \int_0^1 \surd(x+y)\{f(y)\}^2\,dy = 0\ ,$$

which has a non-trivial solution in addition to the null solution. We obtain the Galerkin method if we set $g(x) = 0$, $K(x,y) = \surd(x+y)$, and $M(y,v) = v^2$, in the preceding equations.

To illustrate the process we choose $\phi_r(x) = x^r$. The Galerkin equations are

$$\sum_{r=0}^{n} \tilde{a}_r/(r+s+1) = \int_0^1 \gamma_s(y)\{\sum_{r=0}^{n} \tilde{a}_r y^r\}^2 \, dy \, , \tag{5.153}$$

where $\gamma_s(y) = \int_0^1 \sqrt{(x+y)}x^s \, dx$ $(s = 0,1,2,\ldots,n)$. With $n = 1$, $\gamma_0(y) = \frac{2}{3}\{(1+y)^{\frac{3}{2}} - y^{\frac{3}{2}}\}$ and $\gamma_1(y) = \frac{2}{5}\{(1+y)^{\frac{5}{2}} - y^{\frac{5}{2}}\} - y\gamma_0(y)$, and the right-hand side of (5.153) then becomes

$$\tilde{a}_0^2\int_0^1 \gamma_s(y)dy + 2\tilde{a}_0\tilde{a}_1\int_0^1 y\gamma_s(y)dy + \tilde{a}_1^2\int_0^1 y^2\gamma_s(y)dy \quad \text{for} \quad s = 0,1. \quad *$$

The difficulties encountered with expansion methods for non-linear equations vary with the complexity of the integral equation. Methods based on numerical integration are frequently simpler to apply.

5.23. *Variational formulation*

The general principle of variational methods is common to most, if not all, applications. The application to non-linear integral equations is of some interest, though practical difficulties may arise when specific equations are considered.

We consider the integral equation

$$F_0(x;f(x)) + \int_a^b F_1(x,y;f(x),f(y))dy = 0 \, . \tag{5.154}$$

This is related to eqn (5.134) since $g(x)$ can be incorporated in $F_0(x;f(x))$, or in $F_1(x,y;f(x),f(y))$. For convenience we suppose that $F_0(x;u), F_1(x,y;u,v)$ and the solution $f(x)$ are real-valued functions (of real variables). Eqn (5.154) can sometimes be associated with a functional of the form

$$J[\phi] = \int_a^b\int_a^b \Phi(x,y;\phi(x),\phi(y))dx \, dy + \int_a^b \Gamma(y;\phi(y))dy \tag{5.155}$$

in such a way that $J[f]$ is a stationary value of $J[\phi]$ if $f(x)$ is a solution of the integral equation. We suppose that such a solution exists in $C[a,b]$ and that the stationary value occurs as $\phi(x)$ varies within $C[a,b]$.

Example 5.63.

Suppose $F_0(x;f(x)) = f(x)-g(x)$ and $F_1(x,y;f(x),f(y)) = K(x,y) \times H(f(x)\times f(y))f(y)$, where $K(x,y)$ is symmetric and $H(x)$ is integrable. If

$$H^*(x) = \int_{x_0}^{x} H(t)\,dt$$

for any suitable x_0, then (as we shall see) we may write

$$\Gamma(y;\phi(y)) = \tfrac{1}{2}[\phi(y)]^2$$

and

$$\Phi(x,y;\phi(x),\phi(y)) = \tfrac{1}{2}K(x,y)H^*(\phi(x)\times\phi(y)).$$

This result can in theory be applied to the equation discussed in Example 5.62, according to Noble (in Anselone 1964). *

Example 5.64.

For the equation

$$f(x) = \int_0^1 \surd(x+y)\{f(y)\}^2 dy$$

we may write

$$\{f(x)\}^2 = \int_0^1 \surd(x+y)\{f(x)f(y)\}f(y)\,dy \ .$$

Associated with this equation we can define $J[\phi]$ by eqn (5.155), where $\Gamma(x;u) = \frac{1}{3}u^3$ and (see, with $H(t) = t$, Example 5.63) $\Phi(x,y;u,v) = -\frac{1}{4}\surd(x+y)u^2v^2$. This will now be demonstrated.

We set

$$J[\phi] = \int_0^1 \tfrac{1}{3}\{\phi(y)\}^3 dy - \int_0^1\int_0^1 \tfrac{1}{4}\surd(x+y)\{\phi(x)\phi(y)\}^2 dx\, dy \ .$$

We find, with $\phi(x), \eta(x) \in C[0,1]$, that

$$J[\phi+\varepsilon\eta] = J[\phi] + \varepsilon\{\int_0^1 [\phi(y)]^2 \eta(y)dy -$$

$$- \tfrac{1}{2}\int_0^1\int_0^1 \surd(x+y)\,\phi(x)\phi(y)(\eta(x)\phi(y)+\phi(x)\eta(y))dx\,dy\} + O(\varepsilon^2) .$$

The double integral in this expression can be written

$$\int_0^1\int_0^1 \surd(x+y)\,\phi(x)\phi(y)\eta(x)\phi(y)dy\,dx +$$

$$+ \int_0^1\int_0^1 \surd(x+y)\,\phi(x)\phi(y)\eta(y)\phi(x)dy\,dx ,$$

and if we interchange the variables of integration in the various integrals we find

$$J[\phi+\varepsilon\eta] = J[\phi]+\varepsilon\int_0^1 \eta(x)\{[\phi(x)]^2-\int_0^1 \surd(x+y)\,\phi(x)[\phi(y)]^2dy\}dx + O(\varepsilon^2) .$$

Thus $(\partial/\partial\varepsilon)J[\phi+\varepsilon\eta]$ vanishes at $\varepsilon = 0$ for all $\eta(x) \in C[a,b]$ if

$$[\phi(x)]^2 = \int_0^1 \surd(x+y)\{\phi(x)\phi(y)\}\phi(y)dy ,$$

that is, if $\phi(x)$ is a solution $f(x)$ of the given integral equation.

The analysis here is indicative of the analysis in the general case. *

In order to establish a relationship between (5.154) and (5.155) we shall consider the functional $J[\phi]$. We obtain a condition for $J[\phi]$ to have a stationary value, and we identify this condition as one which is satisfied if $\phi(x)$ is the solution of an integral equation. This equation must be identified with eqn (5.154) for $f(x)$.

Not all integral equations are presented in a form where they have a direct and obvious relationship with a functional $J[\phi]$, and it may be necessary to manipulate them (if possible) into an equivalent and more 'suitable' formulation. This explains why the analysis begins with $J[\phi]$ rather than with the integral equation (5.154).

We suppose that $\Phi(x,y;u,v)$ and $\Gamma(y;v)$ are functions with continuous derivatives $\Gamma_v(y;v) = (\partial/\partial v)\Gamma(y;v)$, $\Phi_u(x,y;u,v) = (\partial/\partial u)\Phi(x,y;u,v)$ and (similarly) $\Phi_v(x,y;u,v)$. We shall determine conditions on the choice of functions in (5.155) which ensure that $J[f]$ is a stationary value of $J[\phi]$ as $\phi(x)$ varies in, say $C[a,b]$. (We assume $f(x) \in C[a,b]$.)

We require that

$$(\partial/\partial\varepsilon)J[f+\varepsilon\eta]\Big]_{\varepsilon=0} = 0 \quad \forall \ \eta(x) \in C[a,b] .$$

Now to determine $J[f+\varepsilon\eta]$ we use the expressions

$$\Phi(x,y;f(x)+\varepsilon\eta(x),f(y)+\varepsilon\eta(y))$$

$$= \Phi(x,y;f(x),f(y))+\varepsilon\{\eta(x)\Phi_u(x,y;f(x),f(y)) +$$

$$+ \eta(y)\Phi_v(x,y;f(x),f(y))\} + o(\varepsilon) ,$$

as $\varepsilon \to 0$ and

$$\Gamma(y;f(y)+\varepsilon\eta(y)) = \Gamma(y;f(y))+\varepsilon\eta(y)\Gamma_v(y;f(y)) + o(\varepsilon) .$$

Thus from (5.155)

$$J[f+\varepsilon\eta] = J[f] + \varepsilon\{\int_a^b\int_a^b [\eta(x)\Phi_u(x,y;f(x),f(y)) +$$

$$+ \eta(y)\Phi_v(x,y;f(x),f(y))]dx\, dy + \int_a^b \eta(y)\Gamma_v(y;f(y))dy\} + o(\varepsilon)$$

and $(\partial/\partial\varepsilon)J[f+\varepsilon\eta]\Big]_{\varepsilon=0}$ is the coefficient of ε in the second term.

On changing some variables of integration we can write

$$(\partial/\partial\varepsilon)J[f+\varepsilon\eta]\Big]_{\varepsilon=0}$$

$$= \int_a^b \eta(x)\Big[\Gamma_v(x;f(x)) + \int_a^b\{\Phi_u(x,y;f(x),f(y)) +$$

$$+ \Phi_v(y,x;f(y),f(x))\}dy\Big]dx \qquad (5.156)$$

and this expression vanishes for all $\eta(x) \in C[a,b]$ if and only if

$$\Gamma_v(x;f(x)) + \int_a^b \{\Phi_u(x,y;f(x),f(y)) + \Phi_v(y,x;f(y),f(x))\}dy = 0 \quad (5.157)$$

This equation is the 'Euler equation' for $J[\phi]$, and it is an integral equation for $f(x)$ which is the same as (5.154) provided

$$\Gamma_v(x;v) = F_0(x;v)$$

and

$$\Phi_u(x,y;u,v) + \Phi_v(y,x;v,u) = F_1(x,y;u,v) \quad .$$

The first equation is always satisfied by setting

$$\Gamma(x;v) = \int_{v_0}^{v} F_0(x;t)dt \ ,$$

but it is generally more difficult to construct $\Phi(x,y;u,v)$ given $F_1(x,y;u,v)$. In order to find such a function $\Phi(x,y;u,v)$ it is often helpful to rearrange the given integral equation, possibly multiplying throughout by a factor as in Example 5.64.

Example 5.65.
Consider the linear equation

$$f(x) - \int_0^1 K(x,y)f(y)dy = g(x) \ ,$$

where $K(x,y) = K(y,x)$ is real, and $g(x)$ is real. The function $J[\phi]$ is given in section 4.23, and the theory can be applied in a straightforward manner. *

Example 5.66.
Suppose that $F_1(x,y;u,v) = K(x,y)H(u+v)$ where $K(x,y) = K(y,x)$ and $H(x)$ is integrable, with

$$H^*(x) = \int_{x_0}^{x} H(t)dt \ .$$

Then we may set

$$\Phi(x,y;u,v) = \tfrac{1}{2}K(x,y)H^{*}(u+v) .$$

This example is fairly simple because of the symmetry (see Collatz 1966*a* p.486). *

Example 5.67.
Consider the equation

$$f(x) - \int_a^b K(x,y)M(y;f(y))\,dy = g(x) ,$$

which can be written in the form of (5.154) if we write $F_0(x;u)=u-g(x)$ and $F_1(x,y;u,v) = -K(x,y)M(y,v)$. Note that $F_1(x,y;u,v)$ is independent of u. If we set $\Gamma(x;u) = \tfrac{1}{2}u^2-ug(x)$ we find $\Gamma_u(x;u)=u-g(x)$. Is it possible *in general* to find a function $\Phi(x,y;u,v)$ with $\Phi_u(x,y;f(x),f(y))+\Phi_v(y,x;f(y),f(x)) = -K(x,y)M(y;f(y))$? For the case $K(x,y) = K(y,x)$ see Example 5.68.

Note that the general linear equation of the second kind and the equation in Example 5.64 are of the type discussed here. *

5.23.1.

From what has gone before it is apparent that we may choose to re-express the integral equation before it appears conveniently as the Euler equation of a functional $J[\phi]$ of the form (5.155). In particular we saw in Example 5.64 that the equation

$$f(x) = \int_0^1 \surd(x+y)\{f(y)\}^2\,dy$$

was to be multiplied through by $f(x)$ to become

$$\{f(x)\}^2 = \int_0^1 \surd(x+y)f(x)\{f(y)\}^2\,dy .$$

This latter equation has the *repeated* solution $f(x) = 0$.

In general it seems possible that 'rearranging' the equation will introduce solutions which were not solutions of the original equation, and $J[\phi]$ may have stationary values at functions $\phi(x)$ which satisfy the Euler equation but do not satisfy the original integral equation, in addition to $\phi(x) = f(x)$.

Example 5.68.

Consider the equation

$$f(x) = \int_a^b K(x,y)M(y;f(y))\mathrm{d}y$$

in which we assume that $K(x,y) = K(y,x)$. We write $M_v(x;u) = (\partial/\partial v)M(x;v)$ and

$$M_v(x;f(x))f(x) - \int_a^b K(x,y)M_v(x;f(x))M(y;f(y))\mathrm{d}y = 0 . \qquad (5.158)$$

Associated with eqn (5.158) we define $J[\phi]$ in (5.155) with

$$\Gamma(x;u) = \int_{u_0}^{u} M_v(x;v)v \,\mathrm{d}v$$

and

$$\Phi(x,y;u,v) = -\tfrac{1}{2}K(x,y)M(x;u)M(y;v) .$$

It is easy to see that

$$\Phi_u(x,y;f(x),f(y)) = -\tfrac{1}{2}K(x,y)M_v(x;f(x))M(y;f(y))$$

$$= \Phi_v(y,x;f(y),f(x)) .$$

Using the symmetry of $K(x,y)$, the Euler equation of $J[\phi]$ is (5.158).

Note that when $M_v(x;\phi(x)) = 0$ for functions $\phi(x)$ not satisfying the original equation, these functions *also* provide stationary values of $J[\phi]$. *

It is also convenient, on occasion, to find a functional $J^*[\phi]$ which attains a stationary value not when $\phi(x) = f(x)$ but when $\phi(x)$ is related to $f(x)$ in a definite way, such that $f(x)$ is *recoverable* from this function.

Example 5.69.

Consider, again, the equation

$$f(x) = \int_a^b K(x,y)M(y;f(y))\,dy \quad ,$$

where $K(x,y) = K(y,x)$. If we write

$$K\phi(x) = \int_a^b K(x,y)\phi(y)\,dy$$

for any suitable $\phi(x)$, and define

$$J^*[\phi] = \int_a^b\int_a^b K(x,y)\phi(x)\phi(y)\,dx\,dy - 2\int_a^b\int_a^{(K\phi)(x)} M(x;\mu)\,d\mu\,dx$$

we can show that if $J^*[\psi]$ is a stationary value then

$$f(x) = \int_a^b K(x,y)\psi(y)\,dy$$

is a solution of the integral equation.

To substantiate this we note that

$$(\partial/\partial\varepsilon)J^*[\phi+\varepsilon\eta]\Big]_{\varepsilon=0}$$

$$= 2\int_a^b K\phi(x)\eta(x)\,dx - 2\int_a^b \{K\eta(y)\}M(y;K\phi(y))\,dy$$

$$= 2\int_a^b \eta(x)\{\int_a^b K(x,y)\phi(y)\,dy - \int_a^b K(x,y)M(y;K\phi(y))\,dy\}\,dx \ .$$

This vanishes for all $\eta(x)$ if $\phi(x) = \psi(x)$, where

$$K\psi(x) = \int_a^b K(x,y)M(y;K\psi(y))\,dy \ .$$

Since

$$f(x) = \int_a^b K(x,y)M(y;f(y))\,dy$$

we may set $f(x) = K\psi(x)$.

The process we have followed has an analogue in writing

$$F(x) = M(x;f(x)), \quad F(x) = M(x;\int_a^b K(x,y)F(y)\,dy)$$

and constructing a variational functional for this integral equation.

For further insight consult Green (1969) and Tricomi (1957). *

Example 5.70.
Consider the equation

$$f(x) = \int_0^1 \surd(x+y)\{f(y)\}^2 dy .$$

If we write $\psi(x) = \{f(x)\}^2$ we have

$$\surd\psi(x) = \int_0^1 \surd(x+y)\psi(y)dy .$$

It is not difficult to find a functional for this equation since the right-hand part is linear in $\psi(y)$, and $\surd(x+y)$ is symmetric in x and y. We write $\Gamma(x;u) = \frac{2}{3}u^{3/2}$ and $\Phi(x,y;u,v) = -\frac{1}{2}(x+y)^{\frac{1}{2}}uv$. Then $J[\phi]$ attains an extreme value if $\phi(x) = \psi(x) = \{f(x)\}^2$. See Collatz (1966*a*) for a treatment of this equation. *

5.24. *Variational methods*

The implementation of a variational method is fairly natural once we have found a functional $J[\phi]$ which can be associated with a solution $f(x)$ of our integral equation.

As usual we select linearly independent functions $\phi_0(x),\phi_1(x),\ldots,\phi_n(x)$ spanning a space S_n of (continuous) functions of the form

$$\phi(x) = \sum_{i=0}^{n} \alpha_i\phi_i(x) .$$

We seek a function

$$\tilde{\psi}(x) = \sum_{i=0}^{n} \tilde{a}_i\phi_i(x)$$

such that when $\psi(x) = \tilde{\psi}(x)$,

$$J[\psi] \text{ is stationary (over } S_n) .$$

Thus we require the values $\tilde{a}_0,\tilde{a}_1,\ldots,\tilde{a}_n$ such that

$$(\partial/\partial\alpha_k)J[\phi]\Big]_{\alpha_k=\tilde{\alpha}_k}=0 \quad (k = 0,1,2,\ldots,n)\ . \tag{5.159}$$

Depending on the construction of $J[\phi]$ we set $\tilde{f}(x) = \tilde{\psi}(x)$ (or, see section 5.23.1, *recover* $\tilde{f}(x)$ from $\tilde{\psi}(x)$) . The equations for $\tilde{a}_0,\tilde{a}_1,\ldots,\tilde{a}_n$ are, in general, non-linear.

From the theoretical viewpoint it is interesting to view the variational method in the following way. (For simplicity we suppose that the Euler equation of the functional $J[\phi]$ is an integral equation satisfied by $f(x)$ itself, rather than by a related function, as in section 5.23.1.) If $\tilde{a}_0,\tilde{a}_1,\ldots,\tilde{a}_n$ are chosen to satisfy (5.159) then

$$(\partial/\partial\varepsilon)J[\tilde{f}+\varepsilon\tilde{\eta}]\Big]_{\varepsilon=0} = 0 \tag{5.160}$$

for all $\tilde{\eta}(x)\ \varepsilon\ S_n$, and vice versa, where $\tilde{f}(x) = \sum_{k=0}^{n} \tilde{a}_k\phi_k(x)\ \varepsilon\ S_n$. The condition (5.160) is therefore equivalent to the requirement that (5.156) is satisfied when $f(x)$ is replaced by $\sum_{k=0}^{n} \tilde{a}_k\phi_k(x)$ and $\eta(x)$ is replaced by an arbitrary function $\tilde{\eta}(x)\ \varepsilon\ S_n$. It is therefore necessary and sufficient that

$$\int_a^b \tilde{\eta}(x)r(x)\mathrm{d}x = 0 \ \ \forall\ \ \tilde{\eta}(x)\ \varepsilon\ S_n$$

or, equivalently, that

$$\int_a^b \phi_k(x)r(x)\mathrm{d}x = 0 \quad (k = 0,1,2,\ldots,n)\ ,$$

where

$$r(x) = \Gamma_v(x;\tilde{f}(x))+\int_a^b \{\Phi_u(x,y;\tilde{f}(x),\tilde{f}(y))+\Phi_v(y,x;\tilde{f}(y),\tilde{f}(x))\}\mathrm{d}y\ .$$

In this manner we show that *the variational method is simply a Galerkin method applied to the Euler equation* for $J[\phi]$.

There appears to us to be a practical advantage in viewing the variational method in this way since the equations for $\tilde{a}_0,\tilde{a}_1,\ldots,\tilde{a}_n$ can be set up immediately, when the integral equation has been reformulated as an Euler equation for a suitable functional $J[\phi]$. It is not really necessary to construct $J[\phi]$, but merely to verify that the equation to be treated can be cast as an Euler equation. (On the other hand, it is not obvious that the variational approach will give a better approximation than can be obtained by applying the Galerkin method to a formulation of the integral equation which is *not* an Euler equation.)

Example 5.71.

For the equation $f(x) = \int_0^1 \surd(x+y)\{f(y)\}^2 dy$ in Example 5.64 we seek

$$\tilde{f}(x) = \sum_{k=0}^{n} \tilde{a}_k\phi_k(x)$$

by requiring that

$$\int_a^b \phi_i(x)\left[\{\sum_{k=0}^{n} \tilde{a}_k\phi_k(x)\}\rho_n(x)\right]dx = 0 \quad (i = 0,1,2,\ldots,n) ,$$

where

$$\rho_n(x) = \sum_{k=0}^{n} \tilde{a}_k\phi_k(x)-\int_0^1 \surd(x+y)\{\sum_{k=0}^{n} \tilde{a}_k\phi_k(y)\}^2 dy .$$

For the classical Galerkin method, we require

$$\int_a^b \phi_i(x)\rho_n(x)dx = 0 .$$

(Here $\phi_0(x),\phi_1(x),\ldots,\phi_n(x)$ are real valued functions.) *

5.25. *Theoretical study of methods for non-linear integral equations*

Our purpose in this section is to consider the theoretical basis for the numerical methods for non-linear integral equations. Anselone (1971) gives an introduction to a fairly abstract school of study which we shall not consider here. Instead we adopt the viewpoint of a Russian school (Vainikko 1967) because of my personal view that the basic functional analysis involved is simpler† conceptually (though somewhat involved manipulatively).

It is difficult to avoid some abstract analysis if we propose to study general non-linear integral equations. If we restrict our study to the Urysohn equation

$$f(x) - \int_a^b F(x,y;f(y))dy = g(x) \tag{5.161}$$

then certain simplifications occur which can be used to advantage in the development of the theory. We shall suppose that $F(x,y;v)$, $F_v(x,y;v)$ and $F_{vv}(x,y;v)$ are continuous for $a \le x,y \le b$ when v satisfies a condition of the type $|v - v_0| \le \omega$, where v_0 and ω are convenient numbers $(\omega > 0)$. We shall use (5.161) to introduce various concepts.

5.25.1. *Preliminaries*

We have already encountered the Newton iteration for the solution of (5.161). Recall (p. 81) that this is an iteration in which we solve, at the kth step, an equation of the form

$$(I - K_k)\varepsilon_k(x) = r_k(x) , \tag{5.162}$$

where K_k is the linear integral operator with the kernel $K_k(x,y) = F_v(x,y;f_k(y))$. This linear integral operator is known as the *Fréchet derivative* of the non-linear integral operator F ,

$$Ff(x) = \int_a^b F(x,y;f(y))dy , \tag{5.163}$$

(acting on $C[a,b]$ - or $R[a,b]$) and evaluated at $f_k(x)$. It is usual to write $K_k = F'(f_k)$, or $K_k = F'_{f_k}$, and $(K_k\varepsilon_k)(x) = F'(f_k)(\varepsilon_k(x))$.

† The Russian school of analysis, already encountered in the linear case, is essentially a development of an approach familiar to numerical analysts in the backward-error analysis for the approximate solution of systems of linear algebraic equations.

From the condition on $F(x,y;v)$, $K_k(x,y)$ exists and is continuous if $\|f_k(x)-v_0\|_\infty \leq \omega$.

The rôle of a Fréchet derivative is to provide an approximation of a non-linear operator by means of a linear one. Thus for example we see from the Taylor series with remainder, for $F(x,y;v)$, that

$$Ff_1(x)$$
$$= \int_a^b F(x,y;f_0(y))dy + \int_a^b K_0(x,y)\varepsilon_0(y)dy + O(\|\varepsilon_0(x)\|_\infty^2)$$

when $f_1(x) = f_0(x) + \varepsilon_0(x)$. In other words, $Ff_1(x) \simeq Ff_0(x)+K_0\varepsilon_0(x)$, where $K_0 = F'(f_0)$.

This leads us to the following definition. For refinements and further details see, for example, Collatz (1966*b*) or Rall (1969).

DEFINITION 5.1. *Suppose that* U,V *are normed linear spaces with norms* $\|\ \|_u$, $\|\ \|_v$, *and suppose that* $T:V \to U$ *is a linear or non-linear operator defined on* V *with range in* U. *If, for* $f_0 \in V$, *we can find a bounded linear operator* L *such that, uniformly for* $\gamma \in V$,

$$\|T(f_0+\gamma)-Tf_0-L\gamma\|_u = o(\|\gamma\|_v), \quad \text{as } \|\gamma\|_v \to 0,$$

then L *is the Fréchet derivative* $T'(\phi)$ *of* T *at* $\phi = f_0$.

Remark. In our applications we shall generally take U,V to be spaces of functions with the same norm $\|\ \|_\infty$. Thus we define

$$\nu(T;f_0,\gamma) = T(f_0(x)+\gamma(x))-Tf_0(x)-T'(f_0)\gamma(x)$$

and $\|\nu(T;f_0,\gamma)\|_\infty = o(\|\gamma(x)\|_\infty)$. (For simplicity, the reader may generally suppose that $U = V = R[a,b]$ with the uniform norm.)

Example 5.72.

(a) If the operator F is a *linear* operator, acting on a function space, then $F'_f \equiv F'(f) = F$. (b) If $F = F_0+F_1$ then $F'(f) = F'_0(f) + F'_1(f)$. (c) In particular, from (a) and (b), if $F_0 = I, F'_0(f) = I$

and hence if $T = I-F, T'(f) = I-F'(f)$. (d) Finally, if $F_0:U_0 \to V_1$ and F_0 is linear, and $F_1:U_1 \to U_0$ whilst $T = F_0F_1$ maps U_1 into V_1, then $T'(f) = F_0F_1'(f)$ for $f \in U_1$. *

Example 5.73.
Suppose that

$$Tf(x) = F_0(x;f(x))-\int_0^1 F_1(x,y;f(x),f(y))dy-g(x) ,$$

where $(\partial/\partial u)F_0(x;u)$, $(\partial/\partial u)F_1(x,y;u,v)$, and $(\partial/\partial v)F_1(x,y;u,v)$ are continuous. Then

$$T'(f)\varepsilon(x) = \{(\partial/\partial u)F_0(x;f(x))-\int_a^b(\partial/\partial u)F_1(x,y;f(x),f(y))dy\}\varepsilon(x)$$
$$-\int_a^b(\partial/\partial v)F_1(x,y;f(x),f(y))\,\varepsilon(y)dy.$$

The space of functions including $f(x)$, $\varepsilon(x)$ is taken to be the bounded Riemann-integrable functions with uniform norm.† *

DEFINITION 5.2. *The Newton iteration for the solution of* $Tf(x) = 0$, *where* T *is a linear or non-linear operator on a space* U, *is*

$$f_{k+1}(x) = f_k(x)-\{T'(f_k)\}^{-1}Tf_k(x) ,$$

and is defined if $T'(f_k)$ *exists and is invertible for* $k = 0,1,2,...,$ *given some* $f_0(x) \in U$.

DEFINITION 5.3. *The operator* T *is called continuously (Fréchet) differentiable at* $f_0(x)$ *if there is* $\varepsilon > 0$ *such that* $T'(\phi)$ *exists whenever* $||\phi(x)-f_0(x)|| < \varepsilon$ *and if, given* $\delta > 0$, *there is a value* ε', *with* $0 < \varepsilon' \leq \varepsilon$, *such that* $||T'(\psi)-T'(f_0)|| \leq \delta$ *whenever* $||\psi(x)-f_0(x)|| \leq \varepsilon'$. *(Thus, the continuity is in the argument* ϕ *near* $\phi = f_0$ *.) Operator norms are subordinate to the function space*

† The rigorous definition of the Fréchet derivative depends upon the space of functions and on the norm.

norms, as usual.

DEFINITION 5.4. *An operator* T *has a continuously invertible (Fréchet-)derivative at* $f_0(x)$ *if, for some* $\varepsilon > 0$, $T'(\phi)$ *exists when* $\|\phi(x)-f_0(x)\| < \varepsilon$ *and, given* $\delta > 0$ *there exists* ε' *with* $0 < \varepsilon' \leq \varepsilon$ *such that*

$$\| [T'(\psi)]^{-1} - [T'(f_0)]^{-1}\| \leq \delta$$

whenever $\|\psi(x)-f_0(x)\| < \varepsilon'$.

Remark. If T is continuously (Fréchet-)differentiable at $f(x)$, and $T'(f)$ has a bounded inverse, then T has a continuously invertible (Fréchet) derivative at $f(x)$. (The proof is an exercise in linear-operator theory.)

In what follows, we shall omit the qualifying word Fréchet since all derivatives will be Fréchet derivatives.

Example 5.74.
The operator F defined on $R[a,b]$ by (5.163) is continuously differentiable at every $f_0(x) \in R[a,b]$ when $F_v(x,y;v)$ is continuous for $a \leq x,\ y \leq b,\ |v| < \infty$. *

There are 'mean-value' theorems† which we can obtain from the first part of the following lemma.

LEMMA 5.6. *Suppose that* $T'(\phi)$ *exists whenever* $\phi(x) = f(x)+\theta\gamma(x)$, *for* $0 \leq \theta \leq 1$, *and suppose that* L *is any continuous linear operator, defined on the same space as* T. *Then*

$$\text{(a)} \quad \|T(f+\gamma)(x)-Tf(x)-L\gamma(x)\| \leq \|\gamma(x)\| \sup_{0\leq\theta\leq 1} \|T'(f+\theta\gamma)-L\| .$$

In particular, if $L = T'(f)$, L *is linear and continuous (since it is*

---

† Because of the differences between the abstract case and the study of functions of a real variable, this term is something of a misnomer; see, however, Collatz (1966*b*, p.282).

bounded) and $\nu(T;f,\gamma) = T(f(x)+\gamma(x))-Tf(x)-T'(f)\gamma(x)$. *Then*

$$\text{(b)} \qquad \|\nu(T;f,\gamma)\| \leq \|\gamma(x)\| \sup_{0\leq\theta\leq 1} \|T'(f+\theta\gamma)-T'(f)\| .$$

Part (b) is similar to a mean-value theorem but does not provide an equality. A further result arises in (a) if we set $L = 0$. The proof of Lemma 5.6 in the case $L = 0$ is classic (Collatz 1966*b* p.282) and the general result is obtained on replacing T by $T-L$ in this case.

Remark. If we suppose that $T:V \to U$ and we regard T' as an operator from $V \times V$ into U, we can define its second Fréchet derivative, and the 'mean-value theorem' can be expressed in terms of T'' , when this derivative exists (see, for example, Collatz 1966*b*, p.281).
Then we can write

$$\|\nu(T;f_0,\gamma)\| \leq \tfrac{1}{2} \|\gamma(x)\|^2 \sup_{0\leq\theta\leq 1} \|T''(f_0+\theta\gamma)\|$$

provided T'' exists.

We shall have occasion to employ the preceding lemma in the proof of Theorem 5.21 below. In anticipation of this purpose, it is convenient to introduce the notation

$$\nu_L(T;f,\gamma) = T(f+\gamma)(x)-Tf(x)-L\gamma(x). \qquad (5.164)$$

Thus Lemma 5.6 provides a bound on the norm of (5.164); if $L = T'(f)$ then $\nu_L(T;f,\gamma)$ is $\nu(T;f,\gamma)$.

5.25.2. *Numerical methods in a theoretical framework*

If we consider the equation

$$f(x) = \int_a^b F(x,y;f(y))\,dy + g(x) ,$$

we can write this in operator notation as

$$f(x) = (Tf)(x) , \qquad (5.165)$$

where $(Tf)(x) = (Ff)(x) + g(x)$ and F is defined by (5.163).

In an abstract setting we shall consider the equation $f(x)=(Tf)(x)$, where T is a (linear or) non-linear operator defined on a bounded region of X, where X is a Banach space of functions with norm $\| \ \|$. In applications we may take $X = R[a,b]$, with the uniform norm. For later use we define

$$W(e) = \{\phi(x) \mid \phi(x) \in X, \ \|\phi(x)-f(x)\| \le e\} \quad , \tag{5.166}$$

where $f(x)$ is some fixed solution of $f(x) = (Tf)(x)$.

<u>Example 5.75</u>.

(a) If the linear integral equation of the second kind is written in the form (5.165), the operator T is non-linear because of the free term $g(x)$.

(b) In order to write (5.134) in the form (5.165) it is necessary to produce a free term $f(x)$. For any value of α we have

$$f(x) = f(x)+\alpha\{F_0(x;f(x))-\int_a^b F_1(x,y;f(x),f(y))dy\}$$

and the right-hand side defines a non-linear operator T for any choice of α . Thus

$$T\phi(x) = \phi(x)+\alpha\{F_0(x;\phi(x))-\int_a^b F_1(x,y;\phi(x),\phi(y))dy\} . \quad *$$

In order to consider numerical methods for the solution of non-linear integral equations we introduce a sequence $\{X_n\}$ of closed sub-spaces of X . For a particular value of n, an approximate solution $f_n(x)$ is to lie in X_n . We also suppose that P_n is a linear projection operator from X *onto* X_n. It is convenient (Zaanen 1964, p171) to assume that, for each n, P_n is bounded; this we shall do. Since $f_n(x) \in X_n$, $P_nf_n(x) = f_n(x)$.

We will identify our numerical methods as being equivalent to the solution of

$$f_n(x) = P_nTf_n(x) + S_nf_n(x) \quad , \tag{5.167}$$

where S_n is in general a non-linear operator; S_n is to be defined from X_n into X_n and is required to be continuous on $W_n(\Omega) = W(\Omega) \cap X_n$ for some $\Omega > 0$. Eqn (5.167) is known to the Russian school as a Galerkin perturbation method (equation) for the operator equation (5.165). If $S_n \equiv 0$ then the method is a projection method, and the classical Galerkin method is a special case. By a suitable choice of P_n, S_n, X_n we can construct most of our numerical methods.

Example 5.76.
Suppose $X = R[a,b]$, and $\|\cdot\| = \|\cdot\|_\infty$. We may define X_n to be the space of polynomials of degree at most n. The norm for X_n is to be the same as that for X. We set $S_n \equiv 0$ and define $P_n\phi(x)$, for $\phi(x) \in R[a,b]$, (a) as the least-squares polynomial approximation of degree n to $\phi(x)$ on $[a,b]$, or (b) as the polynomial of degree n interpolating $\phi(x)$ at prescribed points $z_0, z_1, \ldots, z_n$ lying in $[a,b]$. In case (a), (5.167) gives a theoretical description of the classical Galerkin method and in case (b) we obtain the collocation method.

Consider the quadrature method for the equation

$$f(x) - \int_0^1 F(x,y;f(y))\,dy = g(x) .$$

Suppose that, with the definition of J^+ given in Chapter 2,

$$J_n(\phi) \equiv J(\phi) = \sum_{j=0}^{n} w_j \phi(y_j) \in J^+$$

and values of $\tilde{f}(x)$ at $x = y_i$ (with $y_i \in [0,1]$, $i = 0,1,2,\ldots,n$) are determined so that

$$\tilde{f}(y_i) - \sum_{j=0}^{n} w_j F(y_i, y_j; \tilde{f}(y_j)) = g(y_i) \quad (i = 0,1,2,\ldots,n) .$$

Using functions $\phi_i(x)$ $(i = 0,1,2,\ldots,n)$ similar to those defined in section 4.16.3, we set $f_n(x) = \sum_{i=0}^{n} \tilde{f}(y_i)\phi_i(x)$, to obtain an approximate solution.

Observe that with the definitions employed in section 4.16.3 the mapping P_n defined by $P_n\phi(x) = \sum_{i=0}^{n} \phi(y_i)\phi_i(x)$ is not in general a projection. For, the functions $\phi_i(x)$ are linearly independent and

$$P_n P_n \phi(x) = \sum_{i=0}^{n} (P_n\phi)(y_i)\phi_i(x) = P_n\phi(x)$$

only *if* $(P_n\phi)(y_i) = \phi(y_i)$, and this is only the case if $\phi_i(y_j) = \delta_{ij}$.

With the preceding remark in mind we shall define

$$\phi_i(x) = \begin{cases} 0 & \text{if } x \in \{y_k | k \neq i\} \\ 1 & \text{if } x = y_i \\ 1 & \text{if } v_i \leq x < v_{i+1},\ x \notin \{y_k | k \neq i\} \\ 0 & \text{otherwise,} \end{cases}$$

where (as in (4.110)) $v_0 = a$, $v_k = v_{k-1}+w_{k-1}(b-a)/\sum_{j=0}^{n} w_j$ and, for later use, $w = (b-a)^{-1}\sum_{j=0}^{n} w_j$. The functions $\phi_i(x)$ satisfy (4.112); they are integrable but not continuous.

The connection with an abstract Galerkin perturbation method has yet to be made. For the theoretical description we define $X = R[a,b]$, with the uniform norm, and X_n is the space of functions in X which are linear combinations of $\phi_0(x),\phi_1(x),\ldots,\phi_n(x)$. As noted, we define P_n by setting

$$(P_n\phi)(x) = \sum_{i=0}^{n} \phi(y_i)\phi_i(x) .$$

Then

$$f_n(x) = P_n F f_n(x) + P_n g(x) + S_n f_n(x) = P_n T f_n(x) + S_n f_n(x) ,$$

where T, F are defined by (5.165) and (5.163) respectively. S_n is defined by requiring, for $\Phi_n(x) \in X_n$, that

$$S_n\Phi_n(y_i) = \sum_{j=0}^{n} w_j F(y_i, y_j; \Phi_n(y_j)) - \int_a^b F(y_i, y; \Phi_n(y))\,dy$$

and $S_n\Phi_n(x) = P_n S_n \Phi_n(x) \in X_n$. Then $f_n(y_i) = \tilde{f}(y_i)$, and any bound on $||f(x)-f_n(x)||_\infty$ yields immediately a bound for $\max_i |f(y_i)-\tilde{f}(y_i)|$. *

Eqn (5.167) provides a basis for the Galerkin perturbation method, and Example 5.76 indicates specific realizations of the method.

One of the first questions to be answered is whether eqn (5.167) possesses a solution when there is a solution to the original eqn (5.165). We can write eqn (5.167) as $f_n(x) = T_n f_n(x)$, where $T_n = P_n T + S_n$; the following result can then be used to answer this question under certain conditions.

THEOREM 5.21. *Suppose that* $T, \hat{T}$ *are defined*† *on a Banach space* X *and that the equation* $f(x) = Tf(x)$ *has a solution* $f(x) \in X$ *and suppose, for some* $e > 0$, *that* (a) $\hat{T}'(\phi)$ *exists for* $\phi(x) \in W(e)$, *and* $I-\hat{T}'(f)$ *is invertible with*

$$||\{I-\hat{T}'(f)\}^{-1}|| \leq C .$$

Suppose, also, that (b)

$$\sup_{\phi(x) \in W(e)} ||\hat{T}'(\phi)-\hat{T}'(f)|| \leq p/C$$

and that (c)

$$||\hat{T}f(x)-Tf(x)|| \leq \Delta(1-p)/C ,$$

where $0 < p < 1$ *and* $0 < \Delta \leq e$. *Then the equation* $\hat{f}(x) = \hat{T}\hat{f}(x)$ *has a solution which is unique in* $W(\Delta)$

Remark. If $S_n = 0$ we can set $\hat{T} = T_n$ and the preceding result can be applied immediately. If S_n is non-trivial, then by our assumptions T_n is defined on X_n rather than on the whole of X and this difficulty must be circumvented. In a number of applications S_n can be defined equally well on X as on X_n. However certain conditions which we wish to impose on, say, $||S_n||$ may be difficult to verify unless we restrict the domain to X_n (compare with section 5.13).

† $T, \hat{T}$ may be defined on $W(\varepsilon)$ (rather than on the whole of X) for some $\varepsilon > e > 0$. $W(\varepsilon)$ is defined by (5.166).

Proof of Theorem 5.21. We have $f(x) = Tf(x)$ and, if $\hat{f}(x)$ exists, $\hat{f}(x) = \hat{T}\hat{f}(x)$. On this assumption, $\hat{f}(x)-f(x) = (\hat{T}\hat{f}(x)-\hat{T}f(x)) + (\hat{T}f(x)-Tf(x))$. Now subtract $\hat{T}'(f)(\hat{f}(x)-f(x))$ from each side and then apply the inverse of $\{I-\hat{T}'(f)\}$ to obtain

$$\hat{f}(x) = f(x)+\{I-\hat{T}'(f)\}^{-1}\{\hat{T}\hat{f}(x)-\hat{T}f(x)-\hat{T}'(f)(\hat{f}(x)-f(x))+\hat{T}f(x)-Tf(x)\} \tag{5.168}$$

The steps are retraceable, and the equation $\hat{f}(x) = \hat{T}\hat{f}(x)$ and the latter equation are equivalent. We shall establish that (5.168) has a unique solution in $W(\Delta)$; this provides the corresponding result for the equation $\hat{f}(x) = \hat{T}\hat{f}(x)$.

We may write (5.168) as $\hat{f}(x) = f(x)+A\hat{f}(x)$, where

$$A\phi(x) = \{I-\hat{T}'(f)\}^{-1}[\hat{T}\phi(x)-\hat{T}f(x)-\hat{T}'(f)(\phi(x)-f(x))+\{\hat{T}f(x)-Tf(x)\}]$$

for each $\phi(x) \in X$. To establish that the equation $\hat{f}(x) = f(x)+A\hat{f}(x)$ has a unique solution $\hat{f}(x) \in W(\Delta)$, consider the iteration $\hat{f}^{[n+1]}(x) = f(x) + A\hat{f}^{[n]}(x)$ with $\hat{f}^{[0]}(x) \in W(\Delta)$. It is sufficient to show that

(a) for $n = 1, 2, 3, \ldots,$ $\hat{f}^{[n]}(x) \in W(\Delta)$,

and

(b) $\|\hat{f}^{[n+1]}(x)-\hat{f}^{[n]}(x)\| \le p\,\|\hat{f}^{[n]}(x)-\hat{f}^{[n-1]}(x)\|$, where $p < 1$.

Consequently $\{\hat{f}^{[n]}(x)\}$ is a Cauchy sequence contained in $W(\Delta) \subset X$, and it has a limit $\hat{f}(x) \in W(\Delta)$, which is a solution of $\hat{f}(x) = \hat{T}\hat{f}(x)$. (The argument is a standard 'contraction mapping' argument; compare with section 1.2.) Moreover if $\hat{f}(x) = \hat{T}\hat{f}(x)$ and $\hat{\phi}(x) = \hat{T}\hat{\phi}(x)$, where $\hat{f}(x),\hat{\phi}(x) \in W(\Delta)$, we can show that

$$\|\hat{f}(x)-\hat{\phi}(x)\| \le p\,\|\hat{f}(x)-\hat{\phi}(x)\|$$

whence (since $p < 1$) $\hat{f}(x) = \hat{\phi}(x)$.

The detailed proof of (a) and (b) now follows. For (a), we have

$$||\hat{f}^{[n+1]}(x)-f(x)|| = ||A\hat{f}^{[n]}(x)|| .$$

Our induction hypothesis, satisfied for $n = 0$, is that $\hat{f}^{[n]}(x) \in W(\Delta)$. Then

$$||\hat{f}^{[n+1]}(x)-f(x)|| \leq C\,||\nu(\hat{T};f,\hat{f}^{[n]}-f)|| + C\,||\hat{T}f(x)-Tf(x)||$$

$$\leq C\Delta \sup_{0\leq\theta\leq 1} ||\hat{T}'(f+\theta(\hat{f}^{[n]}-f))-\hat{T}'(f)|| + \Delta(1-p) ,$$

using Lemma 5.6(b), the conditions of the theorem, and the property $\hat{f}^{[n]}(x) \in W(\Delta)$. Then

$$||\hat{f}^{[n+1]}(x)-f(x)|| \leq p\Delta+\Delta(1-p) = \Delta ,$$

so that $\hat{f}^{[n+1]}(x) \in W(\Delta)$. For (b), write $L = \hat{T}'(f)$. We observe that

$$||\hat{f}^{[n+1]}(x)-\hat{f}^{[n]}(x)|| = ||A\hat{f}^{[n]}(x)-A\hat{f}^{[n-1]}(x)||$$

$$= ||\,[I-\hat{T}'(f)]^{-1}\nu_L(\hat{T};\hat{f}^{[n-1]},\hat{f}^{[n]}-\hat{f}^{[n-1]})||$$

$$\leq C \sup_{0\leq\theta\leq 1} ||\hat{T}'(\hat{f}^{[n-1]}+\theta[\hat{f}^{[n]}-\hat{f}^{[n-1]}])-\hat{T}'(f)||\,||\hat{f}^{[n]}(x)-\hat{f}^{[n-1]}(x)||$$

using Lemma 5.6(a). Now for $0 \leq \theta \leq 1$, $\hat{f}^{[n-1]}(x)+\theta[\hat{f}^{[n]}(x)-\hat{f}^{[n-1]}(x)] \in W(\Delta) \subseteq W(e)$, so that (from condition (b) of Theorem 5.21)

$$||\hat{f}^{[n+1]}(x)-\hat{f}^{[n]}(x)|| \leq C\{p/C\}||\hat{f}^{[n]}(x)-\hat{f}^{[n-1]}(x)|| ,$$

where $p < 1$ and the result (b) of the preceding page then follows.

We may now deduce the following result.

THEOREM 5.22. *Suppose that the equation* $f(x) = Tf(x)$ *has a solution* $f(x) \in X$ *and* $I-T'(f)$ *is invertible with* $||\{I-T'(f)\}^{-1}|| \leq C$.

Suppose also that

$$\sup_{\|\phi(x)-f(x)\| \le e} \|T'(\phi)-T'(f)\| \le p/C ,$$

where $p < 1$. *Then* $f(x)$ *is the only solution in the set* $W(e) = \{\phi(x) \mid \|\phi(x)-f(x)\| \le e\}$.

Proof. In the preceding theorem set $T = \hat{T}$ and $\Delta = e$.

5.25.3. *Convergence*

We now state an abstract result, using the notation (and assumptions) already introduced. In particular, $f(x) = Tf(x)$ and we consider the Galerkin perturbation equation (5.167) in which P_n is a projection.

THEOREM 5.23. *Suppose that, for some* $e > 0$, T *(and hence* P_nT*) is continuous on* $W(e)$, *that* S_n *is continuous on* $W_n(e) = W(e) \cap X_n$, *where* $W(e)$ *is defined by* (5.166), *and that* $f(x)$ *in* $W(e)$ *is a solution of* $f(x) = (Tf)(x)$. *Suppose that the following conditions hold:*

(i) $\|f(x)-P_nf(x)\| \to 0$ *as* $n \to \infty$, *and hence an integer* n_0 *exists so that* $W_n(e)$ *is non-empty if* $n \ge n_0$;

(ii) $\|S_nP_nf(x)\| \to 0$ *as* $n \to \infty$;

(iii) T *is continuously differentiable at* $f(x)$ *and* $\{I-T'(f)\}$ *is continuously invertible;*

(iv) *given* $\delta > 0$, *we can find* N' , ε' , *such that*

$$\|(I-P_n)T'(\Phi)\| \le \delta$$

when $n \ge N'$ *and* $\Phi(x) \in W(\varepsilon')$ *where* $0 < \varepsilon' < e$;

(v) $S_n'(\Phi_n)$ *exists as an operator on* X_n , *when* $\Phi_n(x) \in W_n(\varepsilon')$; *and,*

(vi) *given* $\sigma > 0$, *we can find* N_1', ε_1' *such that* $0 < \varepsilon_1' \le \varepsilon'$ *and*

$$\|S_n'(\Phi_n)\|_{X_n} \le \sigma$$

when $n \geq N_1'$ *and* $\Phi_n(x) \in W_n(\varepsilon_1')$.

Under the preceding conditions the solution $f(x) = Tf(x)$ *is unique in* $W(\Delta)$ *for some* $\Delta > 0$ *and there is an integer* N *such that for each* $n \geq N$ *eqn* (5.167) *has a corresponding solution* $f_n(x) \in W_n(\Delta)$. *For this solution* $f_n(x) \in W_n(\Delta)$, $\lim_{n\to\infty} \|f(x)-f_n(x)\| = 0$ *and, moreover,*

$$\|f_n(x)-f(x)\| \leq \gamma\|(I-P_n)f(x)-S_nf_n(x)\|$$

for $n \geq N$ *and for some* γ *depending on* N .

We shall prove the preceding theorem, which is due to Vainikko, later.

The reader who is interested in applying the abstract theory, rather than verifying its correctness, may prefer to omit the proofs of the following results and progress immediately to section 5.25.5, perhaps pausing for the statement of the remaining theorems in the present sub-section.

For those readers who propose to pursue the details of the abstract theory, I outline the structure of the analysis below. Lemma 5.8 is a preliminary result, in which we establish various consequences of the assumptions of Theorem 5.23, which are employed in the later discussion. The proof of Theorem 5.23 then proceeds from an intermediate result (stated in Theorem 5.24). (From this intermediate result we obtain the result stated in Theorem 5.23 under the additional assumption that $S_n = 0$.) The more general theorem follows from this result as shown in section 5.25.4.

It will be noted that the conditions (iii)-(iv) of Theorem 5.23 involve Fréchet derivatives, which are linear operators. There is thus a strong connection with the hypotheses adopted in our study of linear equations in section 5.12. In particular, we note (without proof) the following result.

LEMMA 5.7. *Suppose that, for any* $\phi(x) \in X$, *we can find a function* $\psi_n(x)$ *in* X_n *such that*

$$\|T'(\Phi)\phi(x)-\psi_n(x)\| \leq \eta_n\|\phi(x)\|$$

whenever $\Phi(x) \in W(\varepsilon)$ *and that* $||P_n||\eta_n \to 0$ *as* $n \to \infty$. *Then condition* (iv) *of the preceding theorem can be satisfied.*

Some consequence of the conditions of Theorem 5.23 are noted in the following lemma.

LEMMA 5.8. *Under the conditions of Theorem* 5.23, *the following statements hold.*

(a) $\|(I-P_n)T'(f)\| \to 0$ *as* $n \to \infty$. (5.169)

(b) *We can find* N, ε, *and* C *such that when* $n \geq N$ *and* $\phi(x) \in W(\varepsilon)$, $T'(\phi)$, $\{I-T'(\phi)\}^{-1}$ *and* $\{I-P_nT'(\phi)\}^{-1}$ *all exist and are bounded on* X, *whilst* $\{I-P_nT'(\phi_n)-S_n'(\phi_n)\}^{-1}$ *exists and is bounded on* X_n, *for* $\phi_n(x) \in W_n(\varepsilon)$. *Further,*

(i) $\|\{I-T'(\phi)\}^{-1}\| \leq C$ (5.170)

for $\phi(x) \in W(\varepsilon)$,

(ii) $\|\{I-P_nT'(\phi)\}^{-1}\| \leq C$ (5.171)

for $\phi(x) \in W(\varepsilon)$ *and* $n \geq N$,

(iii) $\|\{I-P_nT'(\phi_n)-S_n'(\phi_n)\}^{-1}\|_{X_n} \leq C$ (5.172)

for $\phi_n(x) \in W_n(\varepsilon)$ *and* $n \geq N$, *and*

(iv) *if* $\|P_n\| \leq M < \infty$ *for all* $n \geq N$, *then*

$$\|[I-P_nT'(\phi)-S_n'(P_n\phi)P_n]^{-1}\| \leq C \quad (5.173)$$

for $\phi(x) \in W(\varepsilon)$ *and* $n \geq N$.

(c) *Given* $\delta > 0$ *there exist* ε, N *such that*

$$\|P_nT'(\phi)-P_nT'(f)\| \leq \delta \quad (5.174)$$

whenever $\phi(x) \in W(\varepsilon)$ *and* $n \geq N$.

Proof. Part (a) follows from condition (iv) of Theorem 5.23. Suppose throughout that $n \geq n_0$, for convenience referring to Theorem 5.23.

Part (b)(i) is an immediate consequence of condition (iii) of that theorem. Suppose that (5.170) is established with $C = C_0$, $\varepsilon = \varepsilon_0$. Now in condition (iv) of Theorem 5.23 suppose $\delta C_0 < 1$ and

$$\|(I-P_n)T'(\phi)\| \leq \delta$$

for $\phi(x) \in W(\varepsilon_0')$ with n sufficiently large, say $n \geq N_1$. Then

$$I-P_nT'(\phi) = I-T'(\phi)+\{(I-P_nT'(\phi))-(I-T'(\phi))\}$$

so that†

$$\|\{I-P_nT'(\phi)\}^{-1}\| \leq C_0/(1-\delta C_0) ,$$

when $\phi(x) \in W(\varepsilon_1)$, $\varepsilon_1 = \min(\varepsilon_0,\varepsilon_0')$,

$$C = C_1 = C_0/(1-\delta C_0) \geq C_0 , \quad n \geq N_1 .$$

The bound (5.171) holds *a fortiori* for the subordinate norm $\|\ \|_{X_n}$ when $C = C_1$. From condition (vi) of Theorem 5.23, choose δ_0' with $\delta_0'C_1 < 1$ such that $\|S_n'(\phi_n)\|_{X_n} \leq \delta_0'$ when $\phi_n(x) \in W_n(\varepsilon_1')$ for some ε_1' and $n \geq N_1' \geq N_1$. Then (5.172) follows with $\varepsilon = \varepsilon_2 = \min(\varepsilon_1,\varepsilon_1')$, $C = C_1/(1-\delta_0'C_1)$, and $N = N_1'$. Similarly if δ_0'' is chosen so that $M\delta_0''C_1 < 1$ and $\|S_n'(\phi_n)\|_{X_n} \leq \delta_0''$ when

---

† We have the usual type of argument:

$$I-P_nT'(\phi) = \{I-T'(\phi)\}\{I-H\} ,$$

where

$$H = \{I-T'(\phi)\}^{-1}\{P_nT'(\phi)-T'(\phi)\} .$$

Thus $\|H\| \leq \delta C_0 < 1$ so (section 1.9) $(I-H)^{-1}$ exists,

$$\|\{I-H\}^{-1}\| \leq \{1- \|H\| \}^{-1} .$$

The bound on $\|\{I-P_nT'(\phi)\}^{-1}\| = \|\{I-H\}^{-1}\{I-T'(\phi)\}^{-1}\|$ then follows.

$\phi_n(x) \in W_n(\varepsilon_1'')$ for some $N_1'' \geq N_1$ and ε_1'', and $n \geq N_1''$, we find (5.173) follows with $\varepsilon = \varepsilon_2^* = \min(\varepsilon_1, \varepsilon_1'')$, $C = C_1/(1-\delta_0''MC_1)$, $N = N_1$ and with $\phi = \phi_n$. To obtain the result stated, let N_1''' be such that for $n \geq N_1'''$, $||f-P_nf|| < \frac{1}{2}\varepsilon_2^*$, and take $\varepsilon = \varepsilon_2 = \varepsilon_2^*/(2M)$, $N = \max(N_1'', N_1''')$

To unify the theorem, we now take $\varepsilon = \min(\varepsilon_0, \varepsilon_1, \varepsilon_2)$, $C = \max(C_1/(1-\delta_0'C_1), C_1/(1-\delta_0''C_1M))$, and $N = \max(N_1', N_1'', N_1''')$.

To establish (5.174), we write

$$\|P_nT'(\phi)-P_nT'(f)\| \leq \|(I-P_n)T'(\phi)\| + \|T'(\phi)-T'(f)\| + \|(I-P_n)T'(f)\| ,$$

and the result follows using, in particular, condition (iv) of Theorem 5.23 to discuss the first and last terms, and condition (iii) (the continuity in Φ of $T'(\Phi)$) to discuss the middle term.

In section 5.25.5 we shall apply Theorem 5.23, and in the applications discussed we find either that $S_n = 0$ or that $\|P_n\| \leq M < \infty$ for all n. If we add these assumptions to the statement of Theorem 5.23, the proof is simplified. We shall therefore state and prove the following result.

THEOREM 5.24. *Suppose that either $\|P_n\| \leq M$ for all n, or $S_n = 0$ for all n. Then the statement of Theorem 5.23 is valid, and $f_n(x)$ is the only approximate solution in $W_n(\Delta)$ for n sufficiently large.*

<u>Proof</u>. We observe that if $f_n(x) \in X_n$ and

$$f_n(x) = P_nTf_n(x)+S_nf_n(x) ,$$

then $f_n(x)$ is a solution of the equation

$$f_n(x) = P_nTf_n(x)+S_nP_nf_n(x)$$

regarded as an equation to be solved in X. With $\hat{T} = P_nT+S_nP_n$, Theorem 5.21 can be applied.

First observe that by the result of Theorem 5.22 and the bound (5.170), if $f(x)$ is a solution of $f(x) = (Tf)(x)$ it is unique in a neighbourhood $W(\Omega)$ for some $\Omega > 0$. It is sufficient for T to be defined on $W(\Omega)$.

We seek to apply Theorem 5.21, when n is sufficiently large, with $\hat{T} = P_nT+S_nP_n$. We require $\hat{T}'(\phi)$ which is $P_nT'(\phi)+S_n'(P_n\phi)P_n$, for $\phi(x) \in X$. Now $\hat{T}'(\phi)$ exists for $\phi(x) \in W(\Omega)$, and according to (5.173) or (5.171) there exists an integer N_0 such that $\|\{I-\hat{T}'(f)\}^{-1}\| \leq C$ if $n \geq N_0$ when $\|P_n\| \leq M$ or $S_n = 0$. This is condition (a) of Theorem 5.21. We next look at condition (b).

We wish to establish that for some e, with $0 < e \leq \Omega$,

$$\sup_{\phi(x)\in W(e)} \|\hat{T}'(\phi)-\hat{T}'(f)\| \leq p/C ,$$

where $0 < p < 1$. We choose $p \in (0,1)$ and set $\delta = p/C$. We have

$$\|\hat{T}'(\phi)-\hat{T}'(f)\|$$

$$\leq \|P_nT'(\phi)-P_nT'(f)\| + \|S_n'(P_n\phi)-S_n'(P_nf)\|_{X_n} \|P_n\| . \qquad (5.175)$$

If $S_n = 0$ then the second term in (5.175) is missing. *Otherwise*, $\|P_n\| \leq M$, and from condition (vi) of Theorem 5.23, there exists an $\varepsilon_1' > 0$ and an integer N_1' such that for $n \geq N_1'$ and $\phi_n(x) \in W_n(\varepsilon_1')$, $\|S_n'(\phi_n)\|_{X_n} \leq \delta/4M$. With this value of ε_1', we know from condition (i) of Theorem 5.23 that there is an integer N_1'' such that

$$\|f(x)-P_nf(x)\| \leq \tfrac{1}{2}\varepsilon_1'$$

if $n \geq N_1''$. Thus $P_nf(x) \in W_n(\varepsilon_1')$ if $n \geq N_1''$. Further

$$\|f(x)-P_n\phi(x)\|$$

$$\leq \|f(x)-P_nf(x)\| + \|P_nf(x)-P_n\phi(x)\| \leq \tfrac{1}{2}\varepsilon_1'+M\|f(x)-\phi(x)\| ,$$

since, by hypothesis, $\|P_n\| \leq M$ (the case $S_n \neq 0$). Thus if $\phi(x) \in W(e_1)$, where $e_1 = \varepsilon_1'/2M$, $P_n\phi(x) \in W(\varepsilon_1')$.

Consequently,

$$\|S_n'(P_n\phi)-S_n'(P_nf)\|_{X_n} \|P_n\| \leq M(\|S_n'(P_n\phi)\|_{X_n} + \|S_n'(P_nf)\|_{X_n}) \leq \tfrac{1}{2}\delta$$

whenever $\phi(x) \in W(e_1)$ and $n \geq N_1$ with $N_1 = \max(N_1', N_1'')$.

Now, according to (5.174) there exists $e_2 > 0$ and an integer N_2 such that for all $n \geq N_2$, $\|P_n T'(\phi) - P_n T'(f)\| \leq \frac{1}{2}\delta$ whenever $\phi(x) \in W(e_2)$. If we set $N_3 = \max(N_1, N_2)$ and $e = \min(e_1, e_2, \Omega)$ we obtain, from (5.175)

$$\|\hat{T}'(\phi) - \hat{T}'(f)\| \leq \delta = p/C$$

whenever $n \geq N_3$ and $\phi(x) \in W(e)$. This is condition (b) of Theorem 5.21.

Finally we must show that we can find an integer N_4 and a value Δ such that $0 < \Delta \leq e$ and

$$\|\hat{T}f(x) - Tf(x)\| = \|(I-P_n)Tf(x) - S_n P_n f(x)\| \leq \Delta(1-p)/C$$

when $n \geq N_4$, where p was fixed above. Now,

$$\|\hat{T}f(x) - Tf(x)\| \leq \|(I-P_n)Tf(x)\| + \|S_n P_n f(x)\| = \|(I-P_n)f(x)\| +$$

$$+ \|S_n P_n f(x)\| .$$

From condition (i) of Theorem 5.23, given $\Delta \in (0,e]$ there is an integer N_5 such that

$$\|(I-P_n)f(x)\| \leq \Delta(1-p)/2C$$

whenever $n \geq N_5$. From condition (ii) of Theorem 5.23 there exists an integer N_6 such that

$$\|S_n P_n f(x)\| \leq \Delta(1-p)/2C$$

whenever $n \geq N_6$. Thus for $n \geq N_4 = \max(N_5, N_6)$,

$$\|\hat{T}f(x) - Tf(x)\| \leq \Delta(1-p)/C .$$

This is condition (c) of Theorem 5.21.

Now, when $N \equiv N(\Delta) = \max(N_4, N_3)$ and $0 < \Delta \leq e$, the conditions of Theorem 5.21 are satisfied with $\hat{T} = P_n T + S_n P_n$. Consequently there is a unique $f_n(x) \in W_n(\Delta)$ whenever $n \geq N = N(\Delta)$, with

$$f_n(x) = (P_n T + S_n P_n) f_n(x).$$

Since $\|f(x) - f_n(x)\| \leq \Delta$, and Δ can be taken arbitrarily small,

$$\lim_{n\to\infty} \|f_n(x) - f(x)\| = 0 .$$

Let us consider an asymptotic error bound. We can readily establish that

$$f_n(x) - f(x) = \{P_n T f_n(x) - P_n T f(x)\} + S_n P_n f_n(x) - (I - P_n) f(x)$$

$$= \left[P_n T'(f)\{f_n(x) - f(x)\} + P_n \nu(T; f, f_n - f)\right] + S_n P_n f_n(x) - (I - P_n) f(x) .$$

Since $S_n P_n f_n(x) = S_n f_n(x)$, we now find that, in the notation of Lemma 5.6,

$$\{I - P_n T'(f)\}\{f_n(x) - f(x)\}$$

$$= \nu(P_n T; f, f_n - f) + S_n f_n(x) - (I - P_n) f(x) . \tag{5.176}$$

Now we have established that $\|f_n(x) - f(x)\| \to 0$. For $\mu > 0$ and n sufficiently large

$$\|\nu(P_n T; f, f_n - f)\| \leq \mu \|f_n(x) - f(x)\|$$

by (5.174), and by Lemma 5.6(b) with $P_n T$ for T, where $\mu > 0$ can be chosen arbitrarily small. Using the constant C in (5.171), there follows from (5.176) the bound

$$(1 - \mu C) \|f_n(x) - f(x)\| \leq \|\{I - P_n T'(f)\}^{-1}\{S_n f_n(x) - (I - P_n) f(x)\}\| , \tag{5.177}$$

for n sufficiently large. A bound of the form stated in the theorem can thus be obtained, by choosing $\mu C < 1$,

$$\gamma = \sup \| \{I - P_n T'(f)\}^{-1} \| / (1-\mu C) \leq C/(1-\mu C) .$$

5.25.4.

For completeness we now proceed with the full proof of Theorem 5.23. Since the local uniqueness of $f(x)$ follows from the corollary to Theorem 5.22, we begin by establishing the existence of $f_n(x) \in W_n(\Delta)$ for some $\Delta > 0$ and $n \geq N$, where $f_n(x) = (P_n T + S_n) f_n(x)$.

We start by relating the equation $f(x) = Tf(x)$ and the equation $\hat{f}(x) = P_n T\hat{f}(x)$. We now write $\hat{T}_n = P_n T$. We regard both equations as equations having solutions in X, though $\hat{f}(x) \in X_n \subset X$. We apply Theorem 5.24, since this applies with $S_n = 0$, to establish the following result: for every sufficiently small $\Delta > 0$, there is a corresponding integer $N = N(\Delta)$ with $\|\hat{f}(x) - f(x)\| \leq \Delta$ for $n \geq N$, and

$$\|\hat{f}(x) - f(x)\| \to 0 \quad \text{as} \quad n \to \infty .$$

Observe that $\hat{f}(x)$ is the only solution of $\hat{f}(x) = P_n T\hat{f}(x)$ in $W_n(\Delta)$, for a fixed $\Delta > 0$ and all n sufficiently large.

We now compare the equations

$$\hat{f}(x) = P_n T\hat{f}(x)$$

and

$$f_n(x) = P_n T f_n(x) + S_n f_n(x) ;$$

both regarded as equations in X_n. We apply Theorem 5.21 with X replaced by X_n. The rôle of $W(e)$ is assumed by

$$\hat{W}_n(e) = \{\phi(x) \mid \phi(x) \in X_n , \|\phi(x) - \hat{f}(x)\| \leq e\} ,$$

and we place $\hat{T}_n = P_n T$ in the rôle of T, and $P_n T + S_n$ in the rôle of $\hat{T}$. Operator norms are therefore subordinate norms $\| \ \|_{X_n}$. We choose a sufficiently small value $\Delta > 0$,† $0 < p < 1$, and $n \geq N$ so that,

† What constitutes a sufficiently small Δ will emerge in the proof.

from the material above, there is a unique 'solution' $\hat{f}(x) \in W_n(\Delta)$. We will establish the existence of a unique $f_n(x) \in \hat{W}_n(\Delta)$, if n is sufficiently large. In this situation $\|f(x)-f_n(x)\| \leq 2\Delta$.

Now $P_nT'(\phi)+S_n'(\phi)$ exists for $\phi(x) \in W_n(2\Delta)$ and from (5.172) we can find a value C and integer $N_1(\Delta)$ such that for $n \geq N_1(\Delta) \geq N$

$$\|\{I-P_nT'(\hat{f})-S_n'(\hat{f})\}^{-1}\|_{X_n} \leq C . \tag{5.178}$$

(Note that (5.172) leads to such a bound whenever $\hat{f}(x) \in W_n(\Delta)$.)
The result (5.178) is the appropriate condition (a) of Theorem 5.21 with P_nT+S_n in the rôle of $\hat{T}$.

Now consider the appropriate form of condition (b) of Theorem 5.21. We can show that, when n is sufficiently large,

$$\sup_{\phi(x)\in\hat{W}_n(\Delta)} \|\{P_nT'(\phi)-S_n'(\phi)\}-\{P_nT'(\hat{f})-S_n'(\hat{f})\}\|_{X_n} \leq p/C . \tag{5.179}$$

For, in condition (vi) of Theorem 5.23, we set $\delta = p/3C$ and choose n sufficiently large so that

$$\sup_{\phi(x)\in\hat{W}_n(\Delta)} \|S_n'(\phi)\|_{X_n} \leq \delta . \tag{5.180}$$

(Δ will have been reduced if necessary to achieve this.) In (5.174) we set $\varepsilon = \frac{1}{2}\delta$ and ensure that $\sup_{\phi(x)\in W(2\Delta)} \|P_nT'(\phi)-P_nT'(f)\| \leq \frac{1}{2}\delta$ by taking Δ sufficiently small and $n \geq N$, where N is sufficiently large. Then we can set $\phi = \hat{f}$ in the last inequality, and *a fortiori*

$$\sup_{\phi(x)\in W_n(\Delta)} \|P_nT'(\phi)-P_nT'(\hat{f})\|_{X_n} \leq \delta , \tag{5.181}$$

by the triangle inequality. Now (5.179) follows from (5.181) and (5.180) by the choice of δ , using the triangle inequality.

Finally we establish that

$$\|S_n\hat{f}(x)\|_{X_n} \leq \Delta(1-p)/C \tag{5.182}$$

if n is sufficiently large. To see this we write

$$||S_n\hat{f}(x)|| \leq ||S_nP_nf(x)|| + ||S_n\hat{f}(x)-S_nP_nf(x)||$$

$$\leq ||S_nP_nf(x)|| + \sup_{0\leq\theta\leq 1} ||S_n'(\hat{f}+\theta[P_nf-\hat{f}])||_{X_n} ||\hat{f}(x)-P_nf(x)|| .$$

But, condition (ii) of Theorem 5.23 gives $||S_nP_nf(x)|| \to 0$ whilst

$$||\hat{f}(x)-P_nf(x)|| \leq ||\hat{f}(x)-f(x)|| + ||f(x)-P_nf(x)||$$

which tends to zero, and given $\delta > 0$ we can, by condition (vi) of Theorem 5.23, ensure that the supremum is less than δ if n is sufficiently large. It follows that $||S_n\hat{f}(x)|| \to 0$ as $n \to \infty$, and (5.180) is satisfied for any sufficiently small fixed $\Delta > 0$ provided n is sufficiently large.

Applying Theorem 5.21 with the aid of (5.178), (5.179), and (5.182) we establish the existence of a solution $f_n(x)$ satisfying

$$f_n(x) = P_nTf_n(x) + S_nf_n(x) ,$$

which is unique in $\hat{W}_n(\Delta)$, for n sufficiently large. Thus $||f(x)-f_n(x)|| \leq 2\Delta$ for sufficiently large n, depending on Δ, and hence, since Δ can be chosen arbitrarily small,

$$\lim_{n\to\infty} ||f(x)-f_n(x)|| = 0.$$

The asymptotic bound now follows as in the proof of Theorem 5.23.

Remark. In the above proof of the existence of $f_n(x)$, I show its uniqueness in $\hat{W}_n(\Delta)$, where $\hat{f}(x)$ is unique in $W_n(\Delta)$, for n sufficiently large. *This* does not appear, to me, to establish the uniqueness of $f_n(x)$ in $W_n(2\Delta)$. However, we can establish the required result as follows.

THEOREM 5.25. *Suppose that the conditions of Theorem 5.23 hold. If* $\Delta > 0$ *is sufficiently small, there is an integer* N *such that, for*

$n \geq N$, $f_n(x)$ is a unique solution of

$$f_n(x) = (P_nT+S_n)f_n(x)$$

in $W_n(\Delta)$.

Proof. Suppose (from Theorem 5.22) that n is sufficiently large that

$$\|f(x)-f_n(x)\| \leq \Delta ,$$

where Δ is sufficiently small. If we establish that there is no other solution of

$$f_n(x) = (P_nT+S_n)f_n(x)$$

within 2Δ of $f_n(x)$, then $f_n(x)$ is unique in $W_n(\Delta)$.

We therefore apply the result of Theorem 5.22, replacing $f(x)$ by $f_n(x)$, X by X_n, T by P_nT+S_n, and e by 2Δ . For n sufficiently large, (5.172) is valid. When n is sufficiently large and Δ sufficiently small, (5.174) and condition (vi) of Theorem 5.23 permit us to establish that

$$\sup_{\|\phi(x)-f_n(x)\| \leq 2\Delta} \|(P_nT'+S'_n)(\phi)-(P_nT'+S'_n)(f_n)\|_{X_n} \leq p/C$$

for some $p < 1$. Thus, Theorem 5.22 applies and $f_n(x)$ is unique in a sphere of radius 2Δ . Thus $f_n(x)$ is unique in $W_n(\Delta)$, as indicated above.

5.25.5. *Application of the abstract theory*

It was indicated in Example 5.76 how the abstract formulation of the Galerkin perturbation method could be applied to numerical methods for approximating the solution of non-linear integral equations.

For the Galerkin method, S_n is the zero operator, and Theorem 5.23 simplifies to the extent that conditions (ii), (v) and (vi), in the statement of the theorem, are satisfied trivially. Condition (i) is readily satisfied if P_n is a 'reasonable' approximation operator and the solution $f(x)$ is sufficiently smooth, but condition (iv) - and also sometimes (iii) - may require careful thought before we can see whether it is satisfied. (Condition (iii) depends upon the equation.)

We must also remark that the conditions of Theorem 5.23 involve the function $f(x)$ which is unknown. It may therefore be necessary to undertake some prior analysis (for example, to bound $||f(x)||_\infty$) before the conditions can be established.

Example 5.77.

Suppose that X is the space $C[0,1]$ with uniform norm and $T:X \to X$, where

$$T\phi(x) = \int_0^1 F(x,y;\phi(y))dy+g(x),\ g(x)\varepsilon\ C[0,1]\ ,$$

and $F(x,y;v),F_v(x,y;v)$ are continuous. Then

$$T'(\phi)\gamma(x) = \int_0^1 F_v(x,y;\phi(y))\gamma(y)dy\ .$$

Let us suppose that $P_n\phi(x)$ is a piecewise-linear interpolant to $\phi(x)$, linear in $[ih,(i+1)h]$ $(i = 0,1,2,\ldots, n-1)$, where $h = 1/n$, with $\phi(ih) = P_n\phi(ih)$ $(i = 0,1,2,\ldots, n)$. Further, $f(x)$ is continuous, so

$$||f(x)-P_nf(x)||_\infty \leq \omega(f;h) \to 0$$

as $n \to \infty$. Now we must also consider $||(I-P_n)T'(\Phi)||_\infty$, where $\Phi(x)$ varies such that $\Phi(x)\ \varepsilon\ W(\varepsilon)$, that is, such that

$$||\Phi(x)-f(x)||_\infty \leq \varepsilon\ .$$

(For each fixed $\Phi(x)$, $T'(\Phi)$ is compact, so that the functions $\{T'(\Phi)\gamma(x)\}$ are equicontinuous for $\gamma(x)\ \varepsilon\ C[a,b]$, $||\gamma(x)||_\infty < 1$, and hence $||(I-P_n)T'(\Phi)||_\infty \to 0$ for fixed $\Phi(x)$. We extend this argument to cover all $\Phi(x)\ \varepsilon\ W(\varepsilon)$.)

It is easily seen that if $\mu(x) = T'(\Phi)\gamma(x)$

$$|\mu(x_1)-\mu(x_2)| \leq \int_a^b |F_v(x_1,y;\Phi(y))-F_v(x_2,y;\Phi(y))|dy\ ||\gamma(x)||_\infty\ .$$

Now if $\Phi(x) \in W(\varepsilon)$ then for $a \leq y \leq b$

$$|\Phi(y)| \leq ||f(x)||_\infty + \varepsilon ;$$

and if $F_v(x,y;v)$ is continuous for $a \leq x,\ y \leq b,\ |v| \leq ||f(x)||_\infty + \varepsilon$ then it is uniformly continuous here. Thus for all $\Phi(x) \in W(\varepsilon)$ we can, given $\delta > 0$, find η such that for

$$|x_1 - x_2| \leq \eta, \quad |\mu(x_1) - \mu(x_2)| \leq \delta \, ||\gamma(x)||_\infty \leq \delta$$

when $||\gamma(x)||_\infty \leq 1$. Since

$$F^* = \{\mu(x) \mid \Phi(x) \in W(e),\ ||\gamma(x)||_\infty \leq 1\}$$

is an equicontinuous uniformly bounded family and $||(I-P_n)\mu(x)||_\infty \to 0$ for each $\mu(x) \in F^*$, the convergence is uniform and $||(I-P_n)T_n'(\Phi)||_\infty \to 0$ as $n \to \infty$ uniformly for all $\Phi(x) \in W(e)$. Thus condition (iv) of Theorem 5.23 can be established. *

Example 5.78.

Suppose that X is $C[0,1]$, with $||\phi(x)|| = ||\phi(x)||_\infty$ for $\phi(x) \in X$, and suppose T, defined as in the previous Example, maps X into X and that the solution $f(x)$ is in $C^1[0,1]$. We suppose that $F_{xv}(x,y;v)$ is continuous for $0 \leq x,\ y \leq 1,\ |v| < \infty$. If $P_n\phi(x)$ is the (shifted) Fourier-Chebyshev sum for $\phi(x)$, $||f(x) - P_n f(x)||_\infty \to 0$ since $f'(x) \in X$. The set

$$F = \{T'(\Phi)\gamma(x) \mid \Phi(x) \in W(\varepsilon),\ ||\gamma(x)||_\infty \leq 1\}$$

is equi-differentiable† and, since $||P_n||_\infty = O(\ln n)$, $||(I-P_n)T'(\Phi)||_\infty \to 0$

† The derivatives of functions in F are uniformly bounded. If $\psi(x) = T'(\Phi)\gamma(x)$,

$$\psi'(x) = \int_0^1 F_{xv}(x,y;\Phi(y))\gamma(y)\,dy ,$$

and

$$\sup_{||\gamma(x)||_\infty = 1} ||(I-P_n)T'(\Phi)\gamma(x)||_\infty \leq \{1 + O(\ln n)\} \sup_{||\gamma(x)||_\infty = 1} ||\psi'(x)||_\infty / n \to 0.$$

uniformly for $\Phi(x) \in W(\varepsilon)$. Thus, condition (iv) of Theorem 5.23 can be established. *

We now undertake to relate the conditions of Theorem 5.23 to the application of the quadrature method. Recall that, for the particular equation

$$f(x) = \int_a^b F(x,y;f(y))dy + g(x) , \tag{5.183}$$

we set

$$\tilde{f}(y_i) = \sum_{j=0}^{n} w_j F(y_i,y_j;\tilde{f}(y_j)) + g(y_i) .$$

In Example 5.76 we related these equations to 'suitable' definitions of T,X,X_n,S_n, and P_n and we employ those definitions in what follows. We suppose $J_n(\phi) \equiv J(\phi)$ where we suppress n and

$$J(\phi) = \sum_{j=0}^{n} w_j\phi(y_j) \in J^+ \quad (n = 0,1,2,\ldots)$$

and

$$\lim_{n\to\infty} J(\phi) = \int_a^b \phi(y)dy$$

for every $\phi(x) \in C[a,b]$. We shall suppose $g(x)$ and $F_v(x,y;v)$ are continuous; *then* $f(x)$ *is continuous*, but we take $X = R[a,b]$.

For any particular n we define, as in Example 5.76, $v_0,v_1,\ldots,v_n$,

$$w = (b-a)^{-1}\sum_{i=0}^{n} w_i,$$

and functions $\phi_0(x),\phi_1(x),\ldots,\phi_n(x)$. (We could write $w = w^{(n)}, \phi_i(x) = \phi_i^{(n)}(x)$ $(i = 0,1,2,\ldots,n)$, $y_j = y_j^{(n)}$, and $w_j = w_j^{(n)}$, etc., to denote the dependence on n , but for notational convenience the superscript is dropped.)

Recall that from the conditions on $J(\phi)$, $\lim_{n\to\infty} w = 1$, and

$\lim_{n\to\infty} \Delta = 0$, where

$$\Delta = \max_{0 \le i \le n} \{\max|v_{i+1}-y_i|, |y_i-v_i|\} .$$

Now, for $\phi(x) \in X$,

$$P_n\phi(x) = \sum_{i=0}^{n} \phi(y_i)\phi_i(x) \in R[a,b] .$$

(But $P_n\phi(x) \notin C[a,b]$; this explains the choice of X.) If $\phi(x)$ is continuous,

$$||\phi(x)-P_n\phi(x)||_\infty \le \omega(\phi;\Delta) , \quad (5.184)$$

where $\omega(\phi;\Delta)$ is the modulus of continuity of $\phi(x)$. In particular this bound holds with $\phi(x) = f(x)$ and we deduce immediately that $\lim_{n\to\infty} \|f(x)-P_nf(x)||_\infty = 0$, thus establishing condition (i) of Theorem 5.23. A minor modification of the discussion of Example 5.77 (with Δ in the rôle of h) also permits us to establish condition (iv). Conditions (ii), (v), and (vi) involve S_n and we now examine these conditions.

Now, for $\phi(x) \in X$,

$$S_nP_n\phi(x) = P_nS_nP_n\phi(x) \in X_n$$

where $P_n\phi(y) = \sum_{i=0}^{n} \phi(y_i)\phi_i(y)$, and (see p. 727)

$$S_nP_n\phi(x) = \sum_{i=0}^{n} \{S_nP_n\phi(y_i)\}\phi_i(x)$$

$$= \sum_{i=0}^{n} \phi_i(x)\{ \sum_{j=0}^{n} w_jF(y_i,y_j;\phi(y_j))-\int_a^b F(y_i,y;P_n\phi(y))dy \}$$

$$= \sum_{i=0}^{n} \phi_i(x) \sum_{j=0}^{n} (1-w^{-1})w_jF(y_i,y_j;\phi(y_j)) + \quad (5.185)$$

$$+ \sum_{i=0}^{n} \phi_i(x) \sum_{j=0}^{n} \int_{v_j}^{v_{j+1}} \{F(y_i,y_j;\phi(y_j))-F(y_i,y;\phi(y_j))\}dy ,$$

since

$$w_j F(y_i,y_j;\phi(y_j)) = w\int_{v_j}^{v_{j+1}} F(y_i,y_j;\phi(y_j))\,dy$$

and $P_n\phi(x) = \phi(y_j)$ for $x \notin \{y_k\}$, $v_j \leq x < v_{j+1}$. (By definition,

$$v_{j+1} = v_j + w_j\{(b-a)/\sum_{i=0}^{n} w_i\} = v_j + w_j/w \ .)$$

Now, as $n \to \infty$, $\Delta \to 0$ so that $\sup|y_j - v_j| \to 0$, and $w \to 1$ from the conditions on $J(\phi)$. Thus setting $\phi(x) = f(x) \in C[a,b]$, from the assumed continuity of $F(x,y;v)$ in $a \leq x$, $y \leq b$ and a neighbourhood of $v = f(y)$, we see that

$$\sup_{y\in[v_j,v_{j+1}]} |F(y_i,y_j;f(y_j)) - F(y_i,y;f(y_j))| \to 0 \qquad (5.186)$$

and, since $w \to 1$,

$$\{\sum_{k=0}^{n} w_k\}\sup|(1-w^{-1})F(y_i,y_j;f(y_j))| \to 0 \qquad (5.187)$$

as $n \to \infty$. Since $\|\phi_i(x)\|_\infty = 1$ $(i = 0,1,2,\ldots,n)$ we obtain the first condition required,

$$\lim_{n\to\infty} \|S_n P_n f(x)\|_\infty = 0 \ ,$$

using (5.186) and (5.187) to bound (5.185).

For $\gamma_n(x), \Phi_n(x) \in X_n$, $P_n\Phi_n(x) = \Phi_n(x)$, and we find $S_n'(\Phi_n)$ exists,

$$S_n'(\Phi_n)\gamma_n(x) = \sigma_1(x) + \sigma_2(x) \ ,$$

where, cf. (5.185),

$$\sigma_1(x) = \sum_{i=0}^{n} \phi_i(x)\{\sum_{j=0}^{n}\int_{v_j}^{v_{j+1}} \{F_v(y_i,y_j;\Phi_n(y_j)) - F_v(y_i,y;\Phi_n(y_j))\}dy\ \gamma_n(y_j)\}$$

and

$$\sigma_2(x) = \sum_{i=0}^{n} \phi_i(x)(1-w^{-1})\sum_{j=0}^{n} w_j F_v(y_i,y_j;\Phi_n(y_j))\gamma_n(y_j) \ .$$

Suppose $\|\gamma_n(x)\|_\infty = 1$. Given $\varepsilon > 0, W_n(\varepsilon)$ is non empty if n is sufficiently large, say $n \geq N_0$, which we suppose. Assuming *continuity* of $F_v(x,y;v)$ for $a \leq x, y \leq b$ and values of v in a neighbourhood containing values of $f(y)$, $a \leq y \leq b$, it follows that, with $\Phi_n(x) \in W_n(\varepsilon)$,

$$\sup_{n \geq N_0} |F_v(y_i,y_j;\Phi_n(y_j))| < \infty ,$$

and given $\sigma > 0$ we can find an integer $N_1 \geq N_0$ such that for $n \geq N_1$, Δ is sufficiently small that

$$\sup_{|y-y_j|<\Delta} \sup_{\Phi_n(x)\in W_n(\varepsilon)} |F_v(y_i,y_j;\Phi_n(y_j))-F_v(y_i,y;\Phi_n(y_j))| \leq \tfrac{1}{2}\sigma/(b-a).$$

Since $w \to 1$ we can also ensure that for all $n \geq N_2 \geq N_0$

$$|w^{-1}-1|\sup|F_v(y_i,y_j;\Phi_n(y_j))| \leq \tfrac{1}{2}\sigma/(b-a) ,$$

by choosing N_2 large enough. It follows, from the expressions for $\sigma_1(x)$ and $\sigma_2(x)$, that $|\sigma_1(x)| \leq \frac{1}{2}\sigma$ and $|\sigma_2(x)| \leq \frac{1}{2}\sigma$ for $n \geq N_1$ and for $n \geq N_2$ respectively.

Combining the bounds for $\|\sigma_1(x)\|_\infty$, $\|\sigma_2(x)\|_\infty$ we find for $n \geq \max(N_1,N_2)$, $\sup_{\Phi_n(x)\in W_n(\varepsilon)} \|S_n'(\Phi_n)\|_\infty \leq \sigma$, which is the required condition.

From the material above, and from Theorem 5.23, it is clear that if the quadrature method is applied to a *suitable* non-linear integral equation, the computed approximation is one for which

$$\|f(x)-\sum_{i=0}^{n} \tilde{f}(y_i)\phi_i(x)\|_\infty \to 0$$

under mild conditions on the quadrature rule. From this we can deduce immediately that

$$\sup_i|f(y_i)-\tilde{f}(y_i)| \to 0$$

as $n \to \infty$. To establish the preceding results, however, we must *check* that the operator equation $f(x) = Tf(x)$ itself satisfies the conditions imposed in Theorem 5.23. In particular, we have to establish that condition (iii) is satisfied. This can sometimes be established by showing that $I-T'(\phi)$ is invertible for suitable $\phi(x)$, or a direct approach may succeed.

Example 5.79.
Consider the equation

$$\phi(x) = 1 - \frac{\Lambda}{2}\int_0^1 \frac{x}{x+y}\,\frac{1}{\phi(y)}\,dy \quad (0 \le x \le 1) \tag{5.188}$$

(see the second equation in Example 5.50) which is a re-formulation of the equation

$$f(x) = 1 + \frac{\Lambda}{2}xf(x)\int_0^1 (x+y)^{-1}f(y)dy \quad . \tag{5.189}$$

We assume the existence of a solution $f(x) \in R[0,1]$ for $0 \le \Lambda \le 1$. We propose to determine bounds on $\phi(x)$ without solving the equation. Suppose in eqn (5.189) that $f(\hat{x}) = 0$ for $\hat{x} \in [0,1]$. The right-hand side of (5.189) is non-zero at $x = \hat{x}$, giving a contradiction. Thus $f(x)$ does not vanish, and if we write $\phi(x) = 1/f(x)$ we obtain a bounded function. We have $f(x) \in R[0,1]$ by assumption and

$$\phi(x) = 1 - \frac{\Lambda}{2}\left\{\int_0^1 \frac{1}{\phi(y)}\,dy - \int_0^1 \frac{y}{x+y}\,\frac{1}{\phi(y)}\,dy\right\} ,$$

so that, for $x_1, x_2 \in [0,1]$,

$$\phi(x_1)-\phi(x_2) = \frac{\Lambda}{2}\int_0^1 \frac{x_2-x_1}{(x_1+y)(x_2+y)}\,\frac{y}{\phi(y)}\,dy .$$

Thus $\phi(x)$ is continuous. Further, since $\phi(0) = 1$, $\phi(x)$ is positive and hence monotonic decreasing. We have

$$\phi(1) = 1 - \frac{\Lambda}{2}\int_0^1 \frac{1}{1+y}\,\frac{1}{\phi(y)}dy$$

$$= 1 - \frac{\Lambda}{2}\int_0^1 \frac{f(y)}{1+y}dy \ge 1 - \frac{\Lambda}{2}\|f(x)\|_1 ,$$

if $\Lambda \in [0,1]$. Since $f(x) > 0$ in $[0,1]$, $\int_0^1 f(x)\mathrm{d}x = \|f(x)\|_1$, and integrating the equation for $f(x)$ in the form

$$f(x) = 1 + \tfrac{\Lambda}{4}f(x)\int_0^1 f(y)\mathrm{d}y + \tfrac{\Lambda}{4}f(x)\int_0^1 \frac{x-y}{x+y}f(y)\mathrm{d}y$$

(integrating with respect to x) we find $\|f(x)\|_1 = 1 + \frac{1}{4}\Lambda\,\|f(x)\|_1^{\,2}$ and, selecting a square-root†,

$$\|f(x)\|_1 = 2\{1-(1-\Lambda)^{\frac{1}{2}}\}/\Lambda .$$

Thus

$$\phi(1) \geq 1-\{1-(1-\Lambda)^{\frac{1}{2}}\}$$

$$= (1-\Lambda)^{\frac{1}{2}}$$

(from the computed figures this bound is known to be pessimistic).

We now have the result¶

$$(1-\Lambda)^{\frac{1}{2}} \leq \phi(1) \leq \phi(x) \leq 1$$

for $0 \leq x \leq 1$, $0 \leq \Lambda \leq 1$. Let us consider the derivative $T'(\Phi)$, where

$$T\Phi(x) = 1 - \frac{\Lambda}{2}\int_0^1 \frac{x}{x+y}\,\frac{1}{\Phi(y)}\,\mathrm{d}y .$$

$T\Phi(x)$ is defined if $\Phi(x) > 0$, for $\Phi(x) \in R[0,1]$, and if $1-\Lambda = 2\varepsilon^2$ $(0 \leq \Lambda < 1)$, where $\varepsilon > 0$, $T'(\Phi)$ is defined on $W(\varepsilon) = \{\Phi(x) \in R[0,1] \,\big|\, \|\Phi(x)-\phi(x)\|_\infty < \varepsilon\}$. To see this we note that if $\Phi(x) \in W(\varepsilon)$ then

$$\Phi(x) \geq (1-\Lambda)^{\frac{1}{2}} - \varepsilon > 0 ,$$

and

$$T'(\Phi)\gamma(x) = +\frac{\Lambda}{2}\int_0^1 \frac{x}{x+y}\,\frac{\gamma(y)}{\{\Phi(y)\}^2}\mathrm{d}y$$

† We select that root which ensures that $\|f(x)\|_1 \to 1$ as $\Lambda \to 0$, as required by (5.189).

¶ Given $f(x) \geq 0$, the result $\phi(x) \geq (1-\Lambda)^{\frac{1}{2}}$ also follows from (5.147).

is defined for $\Phi(x) \in W(\varepsilon)$. Note that $T'(\Phi)$ is continuous for $\Phi(x) \in W(\varepsilon)$.

We have, for $I - T'(\phi)$,

$$\{I-T'(\phi)\}\gamma(x) = \gamma(x) - \frac{\Lambda}{2}\int_0^1 \frac{x}{x+y} \frac{\gamma(y)}{\{\phi(y)\}^2}dy ,$$

and this is invertible if 1 is not an eigenvalue of K , where

$$K(x,y) = \frac{\Lambda}{2} \frac{x}{x+y} \frac{1}{\{\phi(y)\}^2} .$$

But

$$\rho(K) \leq \|K\|_1 \leq \tfrac{1}{2}\Lambda \sup_{0\leq x\leq 1} \int_0^1 x[(x+y)\{\phi(y)\}^2]^{-1}dy \leq \tfrac{1}{2}\Lambda\int_0^1 |\phi(y)|^{-2}dy ,$$

since $\sup_{0\leq x,y\leq 1} |x/(x+y)| = 1$. Consequently $\rho(K) \leq \tfrac{1}{2}\Lambda\{\phi(1)\}^{-2} \leq \tfrac{1}{2}\Lambda(1-\Lambda)^{-1}$ and $\rho(K) < 1$ if $0 \leq \Lambda < \frac{2}{3}$. Thus Theorem 5.23 applies for this range of Λ; a more sophisticated analysis is required to investigate the case where $\frac{2}{3} \leq \Lambda \leq 1$. *

5.25.6. *Order of convergence*

In the case of a projection method applied under the conditions of Theorem 5.24, the asymptotic bound in the theorem often (though apparently not always) gives a realistic statement of the order of convergence. Since $S_n = 0$ we find

$$\|f_n(x)-f(x)\| = O(\|f(x)-P_nf(x)\|)$$

as $n \to \infty$, and if the differentiability properties of $f(x)$ can be determined we can generally discover, in advance, the expected rate of convergence of $f_n(x)$ to $f(x)$. Thus for collocation at the zeros of the (shifted) Chebyshev polynomial of degree $(n+1)$, taking $\phi_r(x)$ to be the shifted polynomial of degree r ,

$$\|f_n(x)-f(x)\|_\infty = O((\ln n)n^{-r})$$

when $f(x) \in C^r[a,b]$. The rates of convergence can generally be expected to be the same as in the case of a linear equation.

The theorem applied, as described in the preceding section, to the quadrature method does not provide realistic results on the rate of convergence. This seems due, principally, to the choice of P_n employed in the analysis.

Let us, for convenience, suppose that

$$Tf(x) = \int_a^b F(x,y;f(y))dy-g(x) \tag{5.190}$$

and

$$\tilde{f}(y_i) = \sum_{j=0}^{n} w_j F(y_i,y_j;\tilde{f}(y_j))-g(y_i) \quad . \tag{5.191}$$

Using the definitions employed in the analysis of the quadrature method we have

$$f_n(x) = P_n Tf_n(x) + S_n f_n(x) \; .$$

We suppose the conditions of Theorem 5.23 are satisfied, and

$$\|\{I-P_n T'(\Phi_n)-S_n'(\Phi_n)\}^{-1}\|_{X_n} \leq C$$

according to (5.172), for $\Phi_n(x) \in W_n(\varepsilon)$ and $n \geq N$.

Now suppose that

$$f(y_i) = \sum_{j=0}^{n} w_j F(y_i,y_j;f(y_j))-g(y_i)+\tau_i \quad (i = 0,1,2,\ldots,n) \; .$$

We would like to compare $\max_i |f(y_i)-\tilde{f}(y_i)|$ with $\max_i |\tau_i|$. We find, with $\varepsilon(y_i) = f(y_i)-\tilde{f}(y_i)$

$$\varepsilon(y_i) = \sum_{j=0}^{n} w_j \{F(y_i,y_j;f(y_j))-F(y_i,y_j;\tilde{f}(y_j))\} + \tau_i$$

$$= \sum_{j=0}^{n} w_j F_v(y_i,y_j;f(y_j))\,\varepsilon(y_j) + \tau_i + o(\|\underset{\sim}{\varepsilon}\|_\infty) \tag{5.192}$$

uniformly for $i = 0,1,2,\ldots,n$, where $\underset{\sim}{\varepsilon} = [\varepsilon(y_0),\varepsilon(y_1),\ldots,\varepsilon(y_n)]^T$, and $F_v(x,y;v)$ is supposed to be continuous in a region containing $(x,y,f(y))$ for $a \leq x$, $y \leq b$. We suppose $n \geq N_1$, where N_1 is taken sufficiently large† to ensure that this region contains the points $(y_i,y_j,\tilde{f}(y_j))$.

Now it follows from eqn (5.192) that with $\underset{\sim}{\rho} = [\rho_0,\rho_1,\ldots,\rho_n]^T$, $\underset{\sim}{\tau} = [\tau_0,\tau_1,\ldots,\tau_n]^T$

$$\varepsilon(y_i) - \sum_{j=0}^{n} w_j F_v(y_i,y_j;f(y_j))\varepsilon(y_j) = \rho_i , \tag{5.193}$$

with

$$\|\underset{\sim}{\rho}-\underset{\sim}{\tau}\|_\infty = o(\|\underset{\sim}{\varepsilon}\|_\infty) . \tag{5.194}$$

From these relations we wish to establish that $\|\underset{\sim}{\varepsilon}\|_\infty = O(\|\underset{\sim}{\tau}\|_\infty)$, but our previous theorems have been concerned with functions ($f_n(x)$, etc.), rather than with vectors ($\underset{\sim}{f}$, etc.). The connection is easily established with the aid of a simple lemma.

LEMMA 5.9. *Suppose that* $\gamma(x) = \sum_{i=0}^{n} \gamma(y_i)\phi_i(x)$, *where* $\phi_0(x),\phi_1(x),\ldots,\phi_n(x)$ *are defined in Example* 5.76. *Then* $\|\gamma(x)\|_\infty = \|\underset{\sim}{\gamma}\|_\infty$, *where* $\underset{\sim}{\gamma} = [\gamma(y_0),\gamma(y_1),\ldots,\gamma(y_n)]^T$.

Proof. For $\hat{x} \in [a,b]$, $\gamma(\hat{x}) = \gamma(y_i)$ if $\hat{x} = y_i$ otherwise $\gamma(\hat{x}) = \gamma(y_k)$ where $v_k \leq \hat{x} < v_{k+1}$ and k depends on $\hat{x}$ and on n. The result then follows.

Now, suppose that we multiply eqn (5.193) by $\phi_i(x)$ and sum over $i = 0,1,2,\ldots,n$. If we write

$$\varepsilon_n(x) = \sum_{i=0}^{n} \varepsilon(y_i)\phi_i(x), \quad \rho(x) = \sum_{i=0}^{n} \rho_i\phi_i(x)$$

we find

$$\varepsilon_n(x) - (P_n T'(P_n f) + S_n'(P_n f))\varepsilon_n(x) = \rho(x) \tag{5.195}$$

† Convergence of $\tilde{f}(x)$ to $f(x)$ is presumed to have been established, using Theorem 5.23. The existence of N_1 is then assured.

as is easily verified. Since $\rho(x) \in X_n$ we have

$$||\varepsilon_n(x)||_\infty \leq ||\{I-P_nT'(P_nf)-S_n'(P_nf)\}^{-1}||_{X_n} \; ||\rho(x)||_\infty$$

or according to (5.172) and the preceding lemma, $||\underset{\sim}{\varepsilon}||_\infty \leq C\,||\underset{\sim}{\rho}||_\infty$ provided n is sufficiently large. Now $||\underset{\sim}{\rho}-\underset{\sim}{\tau}||_\infty = o(\,||\underset{\sim}{\varepsilon}||_\infty)$ (that is, $||\underset{\sim}{\rho}-\underset{\sim}{\tau}||_\infty/\,||\underset{\sim}{\varepsilon}||_\infty \to 0$ as $n \to \infty$). Therefore, there is an integer N_* such that, for $n \geq N_*$, $||\underset{\sim}{\rho}-\underset{\sim}{\tau}||_\infty \leq q\,||\underset{\sim}{\varepsilon}||_\infty$, where $qC < \frac{1}{2}$. Thus $||\underset{\sim}{\varepsilon}||_\infty \leq C\,||\underset{\sim}{\tau}||_\infty + C\,||\underset{\sim}{\rho}-\underset{\sim}{\tau}||_\infty \leq C\,||\underset{\sim}{\tau}||_\infty + \frac{1}{2}\,||\underset{\sim}{\varepsilon}||_\infty$, or $||\underset{\sim}{\varepsilon}||_\infty \leq 2C\,||\underset{\sim}{\tau}||_\infty$ for n sufficiently large. Hence, since C is independent of $n \geq N_*$,

$$||\underset{\sim}{\varepsilon}||_\infty = O(\,||\underset{\sim}{\tau}||_\infty) \tag{5.196}$$

as $n \to \infty$. This is the same type of result as that obtained for linear equations.

The order of the local truncation errors τ_i can be estimated in terms of the differentiability of $F(x,y;v)$ and $f(y)$, and Peano theory for quadrature.

The preceding analysis can be extended, to provide asymptotic expansions for the errors $\varepsilon_i = \tilde{f}(y_i)-f(y_i)$ when suitable repeated quadrature rules are used. Stetter (1965) gives a general abstract theory to establish such results, and Coldrick (1972), in his doctoral thesis, gives a less abstract treatment for a general class of integral equations of Urysohn type.

6

VOLTERRA INTEGRAL EQUATIONS

This chapter is devoted to a study of step-by-step, or block-by-block, methods for the numerical solution of linear and non-linear Volterra equations of the second kind (sections 6.1-6.9 and 6.12-6.14) and linear Volterra equations of the first kind (sections 6.10-6.11 and 6.15-6.17).

The first half of this chapter - sections 6.1-6.11 - is largely devoted to a description of numerical methods, with Examples; a theoretical investigation of such methods is undertaken in the remainder of the chapter. This pattern is interrupted in places, in particular in section 6.3, where we study a stability problem of a type not discussed elsewhere in this book. Such a theory provides insight into the rate of growth of errors, due to rounding error or the discretization of the integral equation, when using a step-by-step or block-by-block method.

Stability problems are of considerable practical importance, and in places (not all fully explored) the stability problems encountered when using a given method may outweigh advantages which might otherwise be expected when using a particular method, given proofs of its convergence with high-order accuracy. The material presented here reflects a number of approaches to a study of such stability problems.

Most of the methods described here are based on natural applications of quadrature formulae (possibly with some special starting method of section 6.5 or 6.6), and apply to equations in which the functions are well-behaved. (Equations of the first kind present some problems; not all methods are equally suitable.) Cases where a weak singularity occurs can be solved approximately with the use of product-integration formulae (discussed in sections 6.9, 6.11, 6.14, and 6.17).

6.1. *Introduction*

We shall study Volterra integral equations of the form

$$f(x) - \lambda\int_0^x K(x,y)f(y)\,dy = g(x) \quad (0 \le x \le X) \tag{6.1}$$

and, more generally,

$$f(x) - \int_0^x F(x,y;f(y))\,dy = g(x) \quad (0 \leq x \leq X) \,, \tag{6.2}$$

which will be referred to as equations 'of the second kind'. We also study equations 'of the first kind', of the form

$$\int_0^x K(x,y)f(y)\,dy = g(x) \quad (0 \leq x \leq X).$$

The latter equation can cause some difficulty, and we defer consideration of this problem until section 6.10. Unless otherwise stated we suppose that $K(x,y)$, $g(x)$, and $F(x,y;f(y))$ are continuous for $0 \leq y \leq x \leq X$, where X is some constant.

Remark. An equation of the form (6.2) is readily obtained if we consider a differential equation

$$f'(x) = F(x;f(x)) \quad (0 \leq x \leq X) \tag{6.3}$$

with initial condition $f(0) = \gamma$, whence we obtain, for $x \in [0,X]$,

$$f(x) = \int_0^x F(y;f(y))\,dy + \gamma \quad .$$

Numerical methods for the solution of initial-value problems for ordinary differential equations of the form (6.3) generally proceed in a step-by-step or block-by-block fashion (Lambert 1973), and numerical methods for Volterra equations borrow much from the treatment of the initial-value problem for the differential equation (6.3).

Equations of the second kind are sometimes presented in forms which differ from (6.1) and (6.2). Thus the lower limit of integration is non-zero in the equation

$$f(x) - \int_a^x F(x,y;f(y))\,dy = g(x)$$

(which is valid for $a \leq x \leq X'$, say, where a is assumed finite), but on writing $\phi(x) = f(x+a)$ we obtain

$$\phi(x) - \int_0^x F(x+a,y+a;\phi(y))dy = g(x+a) \text{ for } 0 \le x \le X = X' - a \text{ .}$$

Also, the discussion in the literature is sometimes related to the '*canonical form*' of (6.2), namely,

$$\phi(x) = \int_0^x \Phi(x,y;\phi(y))dy \text{ ,} \tag{6.4}$$

to which (6.2) can be reduced on writing $\phi(x) = f(x)-g(x)$ and $\Phi(x,y;v) = F(x,y;v+g(y))$.

Whilst X may be infinite in eqns (6.1) and (6.2), these equations are generally valid for $0 \le x \le X < \infty$, and the solution $f(x)$ is sought for this range of values of x . When $X < \infty$; eqns (6.1) and (6.2) can be recast as Fredholm equations (in terms of X) . Thus, typically, (6.2) may be written in the form:

$$f(x) - \int_0^X \Psi(x,y;f(y))dy = g(x) \quad (0 \le x \le X) \text{ ,} \tag{6.5}$$

where $\Psi(x,y;v) = F(x,y;v)$ for $x \ge y$ and $\Psi(x,y;v) = 0$ for $x < y$. If we apply the methods of Chapter 4 or Chapter 5 to (6.5) we can obtain an approximate solution of (6.2), and if we particularly wish to use an expansion method to obtain an approximate solution

$$\tilde{f}(x) = \sum_{i=0}^{n} \tilde{a}_i \phi_i(x)$$

this is a feasible approach.

In the case of the linear equation (6.1) the equation corresponding to (6.5) is also linear, and if we apply the quadrature method to the Fredholm formulation of the equation (using the rule

$$\int_0^X \phi(y)dy \simeq \sum_{j=0}^{n} w_j \phi(y_j) \quad ,$$

where $0 \le y_0 < y_1 < \ldots < y_n \le X$) we then obtain the equations

$$\tilde{f}(y_i) - \lambda \sum_{j=0}^{i} w_j K(y_i,y_j)\tilde{f}(y_j) = g(y_i) \quad (i = 0,1,2,\ldots,n). \tag{6.6}$$

(If $y_0 = 0$, we suppose that the sum is zero when† $i = 0$.) This procedure appears to be reasonable, particularly if $K(x,x) = 0$, since the Fredholm formulation of (6.1) then has a kernel which is continuous for $0 \le x,\ y \le X$. However, the local truncation errors are of the form

$$\int_0^{y_i} K(y_i,y)f(y)\,dy - \sum_{j=0}^{i} w_j K(y_i,y_j)f(y_j) \quad (i = 0,1,2,\ldots,n),$$

and these quantities generally tend to be larger than necessary. We prefer to apply an integration formula which is specifically designed for integration over $[0,y_i]$, rather than using a formula for integrating over $[0,X]$ and setting the integrand to zero in $(y_i,X]$. In the next section we shall see how preferred schemes can be devised with the choice $y_r = rh$, $h > 0$. It transpires, eventually, that other considerations besides the size of the local truncation errors are important. (The reader may recall that for the initial-value problem for (6.3), certain multi-step schemes with sufficiently high-order local truncation errors are unstable and non-convergent - see Lambert (1973, p.38). Fortunately, the situation for integral equations is rather better, but a scheme must be chosen with care.)

Eqn (6.6) has the important feature that the coefficient matrix is triangular, and the values $\tilde{f}(y_0)$, $\tilde{f}(y_1)$, $\tilde{f}(y_2)$, ... can be computed in order. We then say that the solution can be computed in a 'step-by-step' fashion. Nearly all of the numerical methods for Volterra equations discussed in the literature have the same characteristic, or a similar one where blocks of values ($\{\tilde{f}(y_0), \tilde{f}(y_1), \tilde{f}(y_2)\}$, $\{\tilde{f}(y_3), \tilde{f}(y_4), \tilde{f}(y_5)\}$, ..., say) are computed sequentially.

There has been very little discussion of expansion methods (of the

---

† Sums of the form $\sum_{j=0}^{0} v_j$ are interpreted according to the context in this chapter, and are frequently taken to be zero.

type discussed in Chapter 4) being applied to Volterra equations. However, methods using product-integration formulae, which are often equivalent to a collocation method, do figure prominently in the literature; such methods are discussed in sections 6.8, 6.9, and 6.11.

6.2. *The linear equation of the second kind and Volterra type*

We shall now consider the linear integral equation (6.1), assuming that $K(x,y)$ is continuous for $0 \le y \le x \le X$, where the solution $f(x)$ is sought for $0 \le x \le X$. If $g(x) \in C[0,X]$, as we assume, a continuous solution $f(x)$ exists for all values of λ.

From (6.1) we obtain (on setting $x = 0$), $f(0) = g(0)$. Now, suppose that we choose a step $h > 0$ and positive integers k,N, with $Nh = X$, $k \le N$. With $x = kh$ in (6.1) we have

$$f(kh) - \lambda\int_0^{kh} K(kh,y)f(y)\,dy = g(kh) . \tag{6.7}$$

If we adapt the idea of the quadrature method for Fredholm equations, we now seek an approximation of the form, say,

$$\int_0^{kh} \phi(y)\,dy \simeq \sum_{j=0}^{k} w_{kj}\phi(jh) \quad (k \le N) \tag{6.8}$$

to approximate the integral in (6.7). Such a quadrature rule provides an equation

$$\tilde{f}(kh) - \lambda \sum_{j=0}^{k} w_{kj}K(kh,jh)\tilde{f}(jh) = g(kh) . \tag{6.9}$$

If we set $k = 1,2,3,\ldots,N$ in (6.9) and use the initial value $\tilde{f}(0) = f(0) = g(0)$ we obtain the equations

$$\begin{aligned} \tilde{f}(0) &= g(0) , \\ - \lambda w_{10}K(h,0)\tilde{f}(0) + (1-\lambda w_{11}K(h,h))\tilde{f}(h) &= g(h) , \end{aligned} \tag{6.10}$$

$$-\{\lambda w_{20}K(2h,0)\tilde{f}(0)+\lambda w_{21}K(2h,h)\tilde{f}(h)\}+(1-\lambda w_{22}K(2h,2h))\tilde{f}(2h) = g(2h) ,$$

$$\cdots \qquad (6.10)\ \text{(cntd.)}$$

$$-\sum_{j=0}^{N-1} \lambda w_{Nj}K(Nh,jh)\tilde{f}(jh)+(1-\lambda w_{NN}K(Nh,Nh))\tilde{f}(Nh) = g(Nh) ,$$

which can be solved by forward substitution, calculating $\tilde{f}(h)$, $\tilde{f}(2h)$, $\tilde{f}(3h)$, ..., in a step-by-step fashion.

The accuracy obtained by solving the system (6.10) depends upon the choice of h and of quadrature rule (6.8), and the smoothness of $K(x,y)$ and $g(x)$.

Example 6.1

Referring to section 6.1, we see that a trapezium rule

$$\int_0^X \phi(y)\,dy \simeq h \sum_{j=0}^{N}{}'' \phi(jh) \qquad (X = Nh)$$

could be used to obtain (6.6) by setting $\phi(x) = 0$ for $kh \leq x \leq X = Nh$. We obtain $w_{k0} = \frac{1}{2}h$ and $w_{kj} = h$ $(j = 1,2,3,\ldots,k)$. In practice this procedure is not to be greatly recommended unless $K(x,x) = 0$. If $K(x,x) \neq 0$, the local truncation errors are $O(h)$ as $h \to 0$. As we remarked in section 6.1, it is more appropriate to determine a quadrature rule (6.8) specifically adapted to integration over the range $[0,kh]$. Thus the choice $w_{k0} = w_{kk} = \frac{1}{2}h$, $w_{kj} = h$ $(j = 1,2,3,\ldots,(k-1))$ gives a local truncation error which is $O(h^3)$ as $h \to 0$ with k fixed, assuming sufficient differentiability. *

Remark. Let us consider (6.8) with $k = 1$. There are a few choices in the selection of a rule (6.8) such as

$$\int_0^h \phi(y)\,dy \simeq h\phi(0) ,$$

$$\int_0^h \phi(y)\,dy \simeq h\phi(h) ,$$

and

$$\int_0^h \phi(y)\,dy \simeq \frac{h}{2}\{\phi(0) + \phi(h)\} ,$$

and of these the trapezium rule has the best discretization error for small h. Thus

$$\int_0^h \phi(y)\,dy - \frac{h}{2}\{\phi(0) + \phi(h)\} = O(h^3) \qquad (6.11)$$

as $h \to 0$, if $\phi''(x)$ is continuous (or, say, if $\phi'(x)$ is Lipschitz-continuous) on $[0,h]$. Let us therefore choose $w_{1,0} = w_{1,1} = \frac{1}{2}h$.

If we now consider (6.8) with $k = 2$ we have (for example)

$$\int_0^{2h} \phi(y)\,dy = 2h\phi(h) + O(h^3) , \qquad (6.12)$$

or

$$\int_0^{2h} \phi(y)\,dy = \{\tfrac{1}{2}h\phi(0)+h\phi(h)+\tfrac{1}{2}h\phi(2h)\}+O(h^3) ,$$

or, again,

$$\int_0^{2h} \phi(y)\,dy = \frac{h}{3}\{\phi(0)+4\phi(h)+\phi(2h)\}+O(h^5) , \qquad (6.13)$$

if $\phi^{iv}(x)$ is continuous on $[0,2h]$. From the point of view of the local truncation error, Simpson's rule is to be preferred and leads to $w_{20} = w_{22} = \frac{1}{3}h$, $w_{21} = \frac{4}{3}h$. (Of course, the discretization error term in (6.13) drops to $O(h^4)$ if $\phi^{iv}(x)$ is very badly behaved, and so on. See section 2.16.)

6.2.1.

For the moment we shall discard the thought of using Simpson's rule, and consider the more simple alternative of using the repeated trapezium rule. Thus we may write

$$\int_0^{rh} \phi(y)\,dy = h \sum_{j=0}^{r}{}'' \phi(jh) + E_r \quad (r = 1,2,3,\ldots) , \tag{6.14a}$$

where E_r denotes the local discretization error. In describing the behaviour of E_r as $h \to 0$ we must exercise some care. Clearly *if* r *is fixed*, and $\phi''(x)$ is continuous, $E_r = O(h^3)$ as $h \to 0$. Frequently, however, we are interested in the behaviour of E_r as $h \to 0$ and rh is held fixed. We then have $\lim\limits_{\substack{r\to\infty \\ rh \text{ fixed}}} E_r = E$, $E = O(h^2)$, that is,

$$E_r = O(h^2) \quad \text{as} \quad h \to 0 \;, \quad \text{with } rh \text{ fixed.} \tag{6.14b}$$

The use of the repeated trapezium rule gives, as a simple example of (6.9),

$$\tilde{f}(kh) - \lambda h \sum_{j=0}^{k}{}'' K(kh,jh)\tilde{f}(jh) = g(kh) \quad (k = 1,2,3,\ldots), \tag{6.15}$$

where $\tilde{f}(0) = g(0)$.

Remark. Eqns (6.15) serve to define the values $\tilde{f}(kh)$ but there seems to be no obvious extension of these values to a function $\tilde{f}(x)$ as in the Nyström extension for Fredholm equations. It is not difficult to construct some form of extension, however, and in particular we can write

$$\tilde{f}(x) - \lambda h \sum_{j=0}^{k}{}'' K(x,jh)\tilde{f}(jh) - \frac{\lambda}{2}(x-kh)\{K(x,kh)\tilde{f}(kh) + K(x,x)\tilde{f}(x)\} = g(x) \tag{6.16}$$

for $kh \le x \le (k+1)h$ $(k = 0,1,2,\ldots)$ when h is sufficiently small. *

Whilst we concentrate for a while on the numerical solution of eqn (6.15) (which arise from an application of the repeated trapezium rule) we shall return later to the development of schemes based on the use of (6.13), and various other quadrature rules, in the construction of the weights $w_{k,j}$ in (6.9). For, as is clear, the computational effort in solving (6.9) is independent of the choice of weights $w_{k,j}$ (except that a reduction in effort accrues if some of these weights are zero). The 'best' choice of weights $w_{k,j}$ in (6.9) can therefore be considered to be the choice which produces the most accurate solution for a given step-size h. The choice in (6.15) can frequently be bettered, but by considering this case we gain some feel for the general process.

In the following Example we shall analyse the use of (6.15) in the particular case where $K(x,y) \equiv 1$ for $0 \leq y \leq x$. This simple version of eqn (6.1) will recur in subsequent Examples. (In particular we shall display numerical results obtained on applying various numerical methods to this case.) In view of the simplicity of the equation and the fact that its solution can be obtained analytically, the persistent appearance of this equation in our discussion requires some comment.

It is, of course, the very simplicity of the equation

$$f(x) - \lambda\int_0^x f(y)\,\mathrm{d}y = g(x)$$

which makes it ideal for mathematical analysis, and a study of this simple case *does give some genuine insight* into the more general equation (6.1). (The rôle of this simple example is similar to the rôle of the equation

$$f'(x) = \lambda f(x) + g'(x)$$

in the numerical treatment of initial value problems in ordinary differential equations.) What we must guard against is an unquestioning acceptance that a method which is suitable for this rather special integral equation is suitable for more complicated (and possibly non-linear) equations. In an endeavour to redress the balance we provide some numerical results obtained by applying the methods which we describe in this chapter to less specialized test equations.

Example 6.2.

Consider the equation

$$f(x) - \lambda\int_0^x f(y)\,dy = g(x) .$$

We can employ the Neumann series to show that the solution is

$$f(x) = g(x) + \lambda\int_0^x e^{\lambda(x-y)}g(y)\,dy .$$

If we employ (6.15) we have $\tilde{f}(0) = g(0)$ and

$$\tilde{f}(kh) - \lambda h \sum_{j=0}^{k}{}'' \tilde{f}(jh) = g(kh) \quad (k = 1,2,3,\ldots) .$$

Differencing the equations for $k = r$, $k = r + 1$ we obtain

$$(1-\tfrac{1}{2}\lambda h)\tilde{f}((r+1)h)-(1+\tfrac{1}{2}\lambda h)\tilde{f}(rh) = \Delta g(rh) \tag{6.17}$$

(where $\Delta g(rh) = g((r+1)h)-g(rh)$) $(r = 1,2,3,\ldots)$ and $\tilde{f}(0) = g(0)$. Thus, the values $\tilde{f}(kh)$ satisfy, in this instance, a two-term recurrence relation (6.17). The solution of (6.17) is given by

$$\tilde{f}(kh) = \gamma^k g(0) + (1-\tfrac{1}{2}\lambda h)^{-1}\sum_{s=0}^{k-1}\gamma^{k-s-1}\Delta g(sh) ,$$

where $\gamma = (1+\frac{1}{2}\lambda h)/(1-\frac{1}{2}\lambda h)$. (This result may be readily established by induction on k.)

Let us reconcile the expression for $\tilde{f}(kh)$ with the value of $f(kh)$, and establish convergence of the approximate value to the true value as $h \to 0$ with kh fixed. We assume that $g'''(x)$ is continuous.

We have

$$f(x) = g(x) + \lambda\int_0^x e^{\lambda(x-y)}g(y)\,dy$$

(and $f'''(x)$ is therefore continuous). Applying integration by parts to the integral, we find

$$f(x) = g(0)e^{\lambda x} + \int_0^x e^{\lambda(x-y)} g'(y)\,dy \ .$$

Now suppose that $x = kh$, where x is fixed (and k is an integer). We can write

$$f(kh) = g(0)e^{\lambda kh} + h \sum_{s=0}^{k-1} e^{\lambda(kh-sh-\frac{1}{2}h)} g'((s+\tfrac{1}{2})h) + O(h^2) \qquad (6.18)$$

(as $h \to 0$ with $kh = x$) on replacing the integral in the preceding expression for $f(x)$ by the repeated mid-point rule and its error term. (See Example 2.14(c).) Let us now write $g'((s+\frac{1}{2})h) = (1/h)\Delta g(sh) + O(h^2)$. Further,

$$(1-\tfrac{1}{2}\lambda h)^{-1} = e^{\frac{1}{2}\lambda h} + O(h^2)$$

and

$$\gamma = (1+\tfrac{1}{2}\lambda h)/(1-\tfrac{1}{2}\lambda h) = 1+\lambda h+\tfrac{1}{2}\lambda^2 h^2 + O(h^3) = e^{\lambda h} + O(h^3)$$

(using the binomial theorem for $(1-\frac{1}{2}\lambda h)^{-1}$). Hence, also,

$$\gamma^k = e^{\lambda kh} + O(h^2) \quad \text{as} \quad h \to 0$$

with $kh = x$ fixed (we again employ the binomial theorem to establish this).

We now have $(1-\frac{1}{2}\lambda h)^{-1}\gamma^{k-s-1} = e^{\lambda(k-s-\frac{1}{2})h} + O(h^2)$ for $s=0,1,2,\ldots,(k-1)$ as $h \to 0$ with $kh = x$. Thus from eqn (6.18)

$$f(kh) = \gamma^k g(0) + (1-\tfrac{1}{2}\lambda h)^{-1} \sum_{s=0}^{k-1} \gamma^{k-s-1} \Delta g(sh) + O(h^2) = \tilde{f}(kh) + O(h^2) \ .$$

In discussing the particular equations for $\tilde{f}(kh)$ which arose above, we obtained a two-term recurrence relation, expressing $\tilde{f}((k+1)h)$ in terms of $\tilde{f}(kh)$. However, a common feature of eqns (6.10) applied in the more general case is the need to record *all* the values $\tilde{f}(0), \tilde{f}(h), \ldots, \tilde{f}(kh)$ for use in computing $\tilde{f}((k+1)h)$. (We must emphasize that differencing the equations in this Example is an analytical device employed to obtain the analytical form of $\tilde{f}(kh)$, and it does not reflect general numerical practice.)

In the equation treated here, we have established that $\tilde{f}(kh)-f(kh) = O(h^2)$ as $h \to 0$ with $kh = x$ fixed. This is true for more general equations, and indeed we may show (by adapting the theory of section 4.18) that $\tilde{f}(kh) = f(kh)+h^2\phi_1(kh)+O(h^3)$, where $\phi_1(x)$ is a function independent of h . Thus, h^2-extrapolation can be employed in the deferred approach to the limit to improve the order of accuracy.

Statements of the order of accuracy are, of course, asymptotic statements concerning behaviour as $h \to 0$. In the next part of this Example we find it instructive to consider what happens when using a fixed non-zero step h .

We consider the equation $f(x) - \lambda\int_0^x f(y)\mathrm{d}y = g(x)$, and take $\lambda = 12$, $g(x) = 13e^{-x} - 12$. The solution is $f(x) = e^{-x}$ which is exponentially decaying. If we apply the trapezium rule with varying steps h in (6.15), we obtain the approximate values shown in Table 6.1.

TABLE 6.1

x \ h	$\frac{1}{5}$	$\frac{1}{10}$	$\frac{1}{20}$	true
0	1·0000	1·0000	1·0000	1·0000
0·2	0·7825	0·8304	0·8209	0·8187
0·4	1·0392	0·8665	0·6974	0·6703
0·6	-3·5330	3·6962	0·8723	0·5488
0·8	45·3292	50·8142	4·2989	0·4493
1·0	-493·3267	806·2106	46·1611	0·3679
1·2	543	12893	545	0·3012
1·4	-59736	206296	6480	0·2466
1·6	657106	3300733	77080	0·2019
1·8	-7228166	52811728	916910	0·1653
2·0	79509827	844987641	10907067	0·1353

These results are hardly satisfactory, though they do improve as h is decreased!

If we consider the equation with $\lambda = -12$, $g(x) = 13e^{x} - 12$, $f(x) = e^{x}$, the situation is somewhat improved. We now find 3 decimal places are correct up to $x = 2$ using $h = \frac{1}{20}$.

An explanation for the behaviour described and exhibited here lies in the equations for $\tilde{f}(kh)$. The values of $\tilde{f}(kh)$ satisfy (6.17), and we may see the effect of introducing an isolated perturbation in, say, $\tilde{f}(0)$. Suppose that this 'error' is ε_0. Then we will obtain for $\tilde{f}(h)$ the value $\gamma\varepsilon_0+\tilde{f}(h)$, where $\gamma = (1+\frac{1}{2}\lambda h)/(1-\frac{1}{2}\lambda h)$ and for $\tilde{f}(2h)$ the value $\gamma^2\varepsilon_0+\tilde{f}(2h)$. The effect of a perturbation ε_0 in $\tilde{f}(0)$ is to perturb $\tilde{f}(kh)$ by an amount $\gamma^k\varepsilon_0$, and the effect increases as k increases if $|\gamma| > 1$.

When $\lambda = 12$, $h = \frac{1}{5}$, we have $\gamma = -11$ and since the corresponding values of $\tilde{f}(x)$ are almost entirely in error for $x \geq 1$, they clearly display the oscillating behaviour of $(-11)^k$. As $h \to 0$, $|\gamma| \to 1$. (Indeed, $\gamma > 0$ if h is sufficiently small and so $\gamma \to 1$.) The effect of a single perturbation ε_0 is thus magnified over a step h by a factor which grows closer to unity as $h \to 0$.

If $|\gamma| > 1$ the scheme will be said to 'display instability'. It will readily be seen that if $\lambda > 0$, $\gamma > 1$ for all $h > 0$, but if $\lambda < 0$, $|\gamma| < 1$ for all $h > 0$ and the scheme no longer 'displays instability' and may be called 'stable'. However, since $\gamma \to 1$ as $h \to 0$ the instability or stability for very small h is not very pronounced and is sometimes referred to as 'weak'. The investigation of stability is a particularly important aspect of the numerical solution of Volterra equations which we shall discuss in section 6.3. *

6.2.2.

We limited our attention in section 6.2.1 to the use of the repeated trapezium rule in (6.9), but we indicated that with $k = 2$ in eqn (6.8) we may use Simpson's rule (and there are a number of other possibilities).

For $k = 3$ we now have quite a large number of choices of quadrature rule in (6.8) and (6.9). In particular we could employ (Simpson's) $\frac{3}{8}$th rule to write

$$\int_0^{3h} \phi(y)\,dy = \frac{3h}{8}\{\phi(0)+3\phi(h)+3\phi(2h)+\phi(3h)\}+O(h^5) \qquad (6.19)$$

or we may combine the (conventional) Simpson's rule with the trapezium rule to write either

$$\int_0^{3h} \phi(y)\,dy = \frac{h}{3}\{\phi(0)+4\phi(h)+\phi(2h)\}+\frac{h}{2}\{\phi(2h)+\phi(3h)\}+O(h^3) \quad (6.20)$$

or

$$\int_0^{3h} \phi(y)\,dy = \frac{h}{2}\{\phi(0)+\phi(h)\}+\frac{h}{3}\{\phi(h)+4\phi(2h)+\phi(3h)\}+O(h^3)\ . \quad (6.21)$$

For general values of k it is impractical to suggest that the integral $\int_0^{kh} \phi(y)\,dy$ be approximated by the closed Newton-Cotes rule using $k+1$ points. It can indeed be shown that such a proposal would lead to a rather bad method (although there is nothing amiss with the $\frac{3}{8}$th rule when $k = 3$).

We shall first suppose, for the sake of illustrating the general ideas, that we approximate the general integral $\int_0^{kh} \phi(y)\,dy$ by the repeated version of Simpson's rule if k is even, and by the repeated version of Simpson's rule together with the two-point trapezium rule when k is odd. This description is (see Tables 6.2 and 6.3) somewhat ambiguous and we shall later argue that certain ways of combining the trapezium rule with Simpson's rule are unsatisfactory.

We shall begin with a method which, it later transpires, is somewhat unsatisfactory (see section 6.4), even though it looks promising and has (as $h \to 0$) a fairly small local truncation error. Moreover, convergence (as $h \to 0$) of the approximate solution obtained by the scheme can be established, and it is only later (see Example 6.7 for example) that we shall be forced to appreciate its defects. We shall use the approximations

$$\int_0^{2kh} \phi(y)\,dy \simeq \frac{2h}{3}\sum_{j=0}^{k}{}''\phi(2jh)+\frac{4h}{3}\sum_{j=0}^{k-1}\phi(2jh+h) \qquad (6.22)$$

(the repeated Simpson's rule) for $k = 1,2,3,\ldots$, and †

$$\int_0^{(2k+1)h} \phi(y)\,dy \simeq \frac{1}{2}h\{\phi(0)+\phi(h)\} + \frac{2}{3}h \sum_{j=0}^{k}{}'' \phi(2jh+h) + \frac{4}{3}h \sum_{j=1}^{k} \phi(2jh) \qquad (6.23)$$

for $k = 0,1,2,\ldots$ (in which the trapezium rule is used to approximate $\int_0^h \phi(y)\,dy$, and the repeated Simpson's rule is used for the remainder of the integral). These approximations give local truncation errors which are respectively $O(h^4)$ and $O(h^3)$ as $h \to 0$ with kh fixed.

The corresponding array of weights $w_{k,j}$ in the system (6.10) is given in Table 6.2.

TABLE 6.2

$k \backslash j$	0	1	2	3	4	5 ...
1	$\frac{1}{2}h$	$\frac{1}{2}h$				
2	$\frac{1}{3}h$	$\frac{4}{3}h$	$\frac{1}{3}h$			
3	$\frac{1}{2}h$	$\frac{5}{6}h$	$\frac{4}{3}h$	$\frac{1}{3}h$		
4	$\frac{1}{3}h$	$\frac{4}{3}h$	$\frac{2}{3}h$	$\frac{4}{3}h$	$\frac{1}{3}h$	
5	$\frac{1}{2}h$	$\frac{5}{6}h$	$\frac{4}{3}h$	$\frac{2}{3}h$	$\frac{4}{3}h$	$\frac{1}{3}h$
.						
.						
.						

Example 6.3

Consider the equation

$$f(x) - \lambda\int_0^x f(y)\,dy = 1 ,$$

which has the solutic. $f(x) = e^{\lambda x}$. The system of weights (Table 6.2),

† The sums in (6.23) are taken to be zero if $k = 0$.

substituted in the corresponding equations (6.10) now give the equations

$$\tilde{f}(0) = 1 ,$$

$$(1-\tfrac{1}{2}\lambda h)\tilde{f}(h) - \tfrac{1}{2}\lambda h\tilde{f}(0) = 1 ,$$

$$(1-\tfrac{1}{3}\lambda h)\tilde{f}(2h) - \tfrac{4}{3}\lambda h\tilde{f}(h) - \tfrac{1}{3}\lambda h\tilde{f}(0) = 1 ,$$

etc. If we subtract the kth equation from the $(k+2)$th equation (for the purpose of analysis) we obtain

$$(1-\tfrac{1}{3}\lambda h)\tilde{f}((k+2)h)-\tfrac{4}{3}\lambda h\tilde{f}((k+1)h)-(1+\tfrac{1}{3}\lambda h)\tilde{f}(kh) = 0 \quad (k = 0,1,2,\ldots) .$$

In this instance the values $\tilde{f}(kh)$ satisfy a three-term recurrence equation of the form

$$\alpha\tilde{f}((k+2)h)+\beta\tilde{f}((k+1)h)+\gamma\tilde{f}(kh) = 0 .$$

The general solution of such a recurrence equation has the form $\tilde{f}(kh) = At_1^k+Bt_2^k$, where A,B are determined by the initial equations (which give $\tilde{f}(0)$, $\tilde{f}(h)$) and where t_1 and t_2 are the roots (assumed distinct) of the 'auxiliary equation' $\alpha t^2+\beta t+\gamma = 0$.

In the present case the roots of the auxiliary equation are

$$\{\tfrac{2}{3}\lambda h\pm\surd(1+\tfrac{1}{3}\lambda^2h^2)\}/(1-\tfrac{1}{3}\lambda h) .$$

Writing t_1,t_2 for these roots where $t_1 = 1+\lambda h+O(h^2)$ and $t_2 = -(1-\tfrac{1}{3}\lambda h) +O(h^2)$, the particular solution $\tilde{f}(kh) = At_1^k+Bt_2^k$ is determined by the conditions $\tilde{f}(0) = 1$ and $\tilde{f}(h) = (1+\tfrac{1}{2}\lambda h)/(1-\tfrac{1}{2}\lambda h) = \gamma$, say. Thus we find $A = (\gamma-t_2)/(t_1-t_2)$ and $B = 1-A$.

It is not difficult to see that $\lim_{h\to 0} A = 1$, and $\lim_{h\to 0} B = 0$, and $\lim_{h\to 0} t_1^k = e^{\lambda kh}$ when $h \to 0$ with kh fixed. Thus $|\tilde{f}(kh)-f(kh)| \to 0$ *as* $h \to 0$ *with* kh *fixed, and convergence is established.* If we pay attention to the details, the error may be shown to be $O(h^3)$.

The analysis in this example is identical, beyond a certain point, with the analysis of Milne's method for solving the equation $f'(x) = \lambda f(x)$ with $f(0) = 1$ and the 'starting value' $\tilde{f}(h) = \gamma$. The unsatisfactory nature of the method associated with the choice in Table 6.2, as a general purpose method, is due to its stability properties. When the method reduces to Milne's method (as above) these are well documented; we shall discuss these features later, in section 6.4. At this point let us merely remark that for small λh, the component Bt_2^k in the solution $\tilde{f}(kh)$ behaves like $(-1)^k Be^{-\frac{1}{3}\lambda kh}$, and for negative λ this contribution increases whereas the true solution $f(x)$ decays. However, the latter remark is true because $g(x) = 1$ in the example being considered. With other choices of the inhomogeneous term $g(x)$ the solution $f(x)$ may grow exponentially, and the presence of a component Bt_2^k in the approximate solution may not then cause any concern. *

The trapezium rule may be combined with Simpson's rule to give variations on (6.23). In particular we may combine with the choice (6.22) the approximation

$$\int_0^{(2k+1)h} \phi(y)\,dy \simeq \frac{2}{3}h \sum_{j=0}^{k}{}'' \phi(2jh) + \frac{4}{3}h \sum_{j=0}^{k-1} \phi(2jh+h) + \frac{1}{2}h\{\phi(2kh)+\phi(2kh+h)\},$$

(which has local truncation error $O(h^3)$ as $h \to 0$ with kh fixed) to give the array of weights $w_{k,j}$ shown in Table 6.3.

TABLE 6.3

$k \backslash j$	0	1	2	3	4	5
1	$\frac{1}{2}h$	$\frac{1}{2}h$				
2	$\frac{1}{3}h$	$\frac{4}{3}h$	$\frac{1}{3}h$			
3	$\frac{1}{3}h$	$\frac{4}{3}h$	$\frac{5}{6}h$	$\frac{1}{2}h$		
4	$\frac{1}{3}h$	$\frac{4}{3}h$	$\frac{2}{3}h$	$\frac{4}{3}h$	$\frac{1}{3}h$	
5	$\frac{1}{3}h$	$\frac{4}{3}h$	$\frac{2}{3}h$	$\frac{4}{3}h$	$\frac{5}{6}h$	$\frac{1}{2}h$,
etc.						

As we shall show subsequently, the choice in Table 6.3 has substantial advantages over the choice in Table 6.2.

Example 6.4.

We shall examine the use of the weights (Table 6.3) in the approximate solution of the equation discussed in Example 6.3. On setting up the equations corresponding to (6.10) for the equation

$$f(x) - \lambda\int_0^x f(y)\,dy = 1\ ,$$

and differencing these equations, we find

$$\tilde{f}((2k+1)h)-\tilde{f}(2kh)-\frac{\lambda h}{2}\{\tilde{f}(2kh)+\tilde{f}((2k+1)h)\} = 0 \quad (k = 0,1,2,\ldots)$$

and

$$\tilde{f}(2kh)-\tilde{f}((2k-1)h)-\frac{1}{6}\lambda h\{2\tilde{f}(2kh)+5\tilde{f}((2k-1)h)-\tilde{f}((2k-2)h)\} = 0 \quad (k=1,2,3,\ldots)$$

The first equation gives $\tilde{f}((2k+1)h) = \gamma\tilde{f}(2kh)$ $(k = 0,1,2,\ldots)$ where

$$\gamma = (1+\tfrac{1}{2}\lambda h)/(1-\tfrac{1}{2}\lambda h),$$

and replacing k by $k-1$ we may use this relation to eliminate $\tilde{f}((2k-1)h)$ from the second equation. Thus

$$(1-\tfrac{1}{3}\lambda h)(1-\tfrac{1}{2}\lambda h)\tilde{f}(2kh) = \{(1+\tfrac{1}{2}\lambda h)(1+\tfrac{5}{6}\lambda h)-\tfrac{1}{6}\lambda h(1-\tfrac{1}{2}\lambda h)\}\tilde{f}(2(k-1)h) \quad (k = 1,2,3,\ldots)\ ,$$

and if we multiply through by γ we deduce that the same relation holds between successive *odd*-indexed values $\tilde{f}(kh)$.

We have, then,

$$\tilde{f}((k+2)h) = \frac{1+(\frac{7}{6})\lambda h+(\frac{1}{2})\lambda^2h^2}{1-(\frac{5}{6})\lambda h+(\frac{1}{6})\lambda^2h^2}\,\tilde{f}(kh) \quad (k = 0,1,2,\ldots)\ .$$

The solution is therefore easily written down in terms of

$$\Gamma = \frac{1+(\frac{7}{6})\lambda h+(\frac{1}{2})\lambda^2 h^2}{1-(\frac{5}{6})\lambda h+(\frac{1}{6})\lambda^2 h^2}. \tag{6.24}$$

Now

$$\Gamma = e^{2\lambda h}+ \frac{1}{9}(\lambda h)^4+O(h^5)$$

as $h \to 0$ and

$$\tilde{f}(2rh) = \tilde{f}(0)\Gamma^r = \Gamma^r,$$

whilst

$$\tilde{f}((2r+1)h) = \tilde{f}(h)\Gamma^r = \gamma\tilde{f}(0)\Gamma^r$$

where $\gamma = (1+\frac{1}{2}\lambda h)/(1-\frac{1}{2}\lambda h)$. We have

$$\Gamma^r = e^{2r\lambda h}+\frac{1}{9}r(\lambda h)^4+O(h^4)$$

as $h \to 0$ with $2rh$ fixed and we find

$$\tilde{f}(2rh)-f(2rh) = \lambda^4 h^3(\tfrac{1}{9}rh)+O(h^4),$$

which, in particular, establishes $O(h^3)$ *convergence.* Similarly, if $2rh$ is fixed and $h \to 0$,

$$\tilde{f}((2r+1)h)-f((2r+1)h) = \gamma\tilde{f}(2rh)-e^{\lambda h}f(2rh)$$

and since $\gamma = e^{\lambda h}+O(h^3)$ this expression is $O(h^3)$.

From the above analysis we see that if, in this example, we evaluate $\tilde{f}_{h_0}(2rh_0)$ using a step $h = h_0$, and halve h to obtain $\tilde{f}_{\frac{1}{2}h_0}(4rh_0/2)$ using a step $\frac{1}{2}h_0$, the value $\{8\tilde{f}_{\frac{1}{2}h_0}(2rh)-\tilde{f}_{h_0}(2rh_0)\}/7$ has $O(h_0^4)$ accuracy. *

In the foregoing schemes, the use of the trapezium rule over a single end-interval has lowered the $O(h^4)$ accuracy of the parabolic rule to $O(h^3)$ (as $h \to 0$ with fixed upper limit of integration). However, the $\frac{3}{8}$th rule (6.19) gives

$$\int_a^{a+3h} \phi(y)\,dy = \tfrac{3}{8}h\{\phi(a)+3\{\phi(a+h)+\phi(a+2h)\}+\phi(a+3h)\}+O(h^5) \text{ (as } h \to 0), \quad (6.25)$$

and this rule can be combined with the repeated Simpson's rule to preserve high-order local truncation error. Thus Simpson's rule can be used to write down an equation giving $\tilde{f}(2h)$ and (6.25) can be used to give an equation for $\tilde{f}(3h)$. To write down an equation for $\tilde{f}(h)$ we shall continue, in the present discussion, to employ the trapezium rule for integrating over $[0,h]$. Other 'starting methods' for obtaining $\tilde{f}(h)$ will be discussed later. In obtaining an equation for $\tilde{f}((2r+1)h)$, where $r > 0$, we may use the repeated Simpson's rule and the $\frac{3}{8}$th rule.

In analogy with Table 6.2, we can now employ the $\frac{3}{8}$th rule over $[0,3h]$ when appropriate and obtain the weights $w_{k,j}$ displayed in Table 6.4 for use in eqns (6.10).

TABLE 6.4

$k \backslash j$	0	1	2	3	4	5	6
1	$\frac{1}{2}h$	$\frac{1}{2}h$					
2	$\frac{1}{3}h$	$\frac{4}{3}h$	$\frac{1}{3}h$				
3	$\frac{3}{8}h$	$\frac{9}{8}h$	$\frac{9}{8}h$	$\frac{3}{8}h$			
4	$\frac{1}{3}h$	$\frac{4}{3}h$	$\frac{2}{3}h$	$\frac{4}{3}h$	$\frac{1}{3}h$		
5	$\frac{3}{8}h$	$\frac{9}{8}h$	$\frac{9}{8}h$	$\frac{17}{24}h$	$\frac{4}{3}h$	$\frac{1}{3}h$	
6	$\frac{1}{3}h$	$\frac{4}{3}h$	$\frac{2}{3}h$	$\frac{4}{3}h$	$\frac{2}{3}h$	$\frac{4}{3}h$	$\frac{1}{3}h$,
etc.							

An alternative procedure is to place the $\frac{3}{8}$th rule at the end of appropriate intervals. Thus an analogue of Table 6.3 results if we choose the following weights $w_{k,j}$ given in Table 6.5.

TABLE 6.5

$k \backslash j$	0	1	2	3	4	5	6	7
1	$\frac{1}{2}h$	$\frac{1}{2}h$						
2	$\frac{1}{3}h$	$\frac{4}{3}h$	$\frac{1}{3}h$					
3	$\frac{3}{8}h$	$\frac{9}{8}h$	$\frac{9}{8}h$	$\frac{3}{8}h$				
4	$\frac{1}{3}h$	$\frac{4}{3}h$	$\frac{2}{3}h$	$\frac{4}{3}h$	$\frac{1}{3}h$			
* 5	$\frac{1}{3}h$	$\frac{4}{3}h$	$\frac{17}{24}h$	$\frac{9}{8}h$	$\frac{9}{8}h$	$\frac{3}{8}h$		
6	$\frac{1}{3}h$	$\frac{4}{3}h$	$\frac{2}{3}h$	$\frac{4}{3}h$	$\frac{2}{3}h$	$\frac{4}{3}h$	$\frac{1}{3}h$	
* 7	$\frac{1}{3}h$	$\frac{4}{3}h$	$\frac{2}{3}h$	$\frac{4}{3}h$	$\frac{17}{24}h$	$\frac{9}{8}h$	$\frac{9}{8}h$	$\frac{3}{8}h$.

Asterisked rows in Table 6.5 are those which differ from corresponding rows in Table 6.4. *The use of Table 6.5 has considerable advantages*† *over the use of Table 6.4* (which we would avoid in designing a practical scheme).

Example 6.5

We may again consider the equation

$$f(x) - \lambda \int_0^x f(y)\,dy = 1 \ .$$

If we apply the choice in Table 6.4 in setting up eqns (6.10), the analysis of the resulting equations is for all practical purposes

† These advantages are certainly not obvious until we discuss stability problems in later sections.

contained in that of Example 6.3, since differencing the equations leads to the same three-term recurrence

$$(1-\tfrac{1}{3}\lambda h)\tilde{f}((k+2)h) - \tfrac{4}{3}\lambda h\tilde{f}((k+1)h) - (1+\tfrac{1}{3}\lambda h)\tilde{f}(kh) = 0 . \qquad (6.26)$$

However, instead of holding for $k = 0,1,2,\ldots$ this recurrence is satisfied for $k = 2,3,4,\ldots,$ and the values of A and B in the solution $\tilde{f}(kh) = At_1^k+Bt_2^k$ (see Example 6.3) are determined by $\tilde{f}(2h)$ and $\tilde{f}(3h)$. If we work through the details we can establish that $\tilde{f}(2h) = f(2h)+O(h^4)$ and $\tilde{f}(3h) = f(3h)+O(h^4)$ but at a *fixed* evaluation point x, $\tilde{f}(x)-f(x) = O(h^4)$, as $h \to 0$. The use of the $\frac{3}{8}$th rule in place of the trapezium rule thus reduces the error from $O(h^3)$ to $O(h^4)$, but the 'stability' properties (section 6.3) of the method are the same as those of the method studied in Example 6.3, and the choice (in Table 6.4) is not to be generally recommended.

If we attempt to analyse the choice in Table 6.5, the analysis becomes complicated (even for the present simple example). We shall not pursue it here but leave it as an exercise for the reader to obtain the relations

$$(1-\tfrac{1}{3}\lambda h)\tilde{f}(2kh) - \tfrac{4}{3}\lambda h\tilde{f}((2k-1)h) - (1+\tfrac{1}{3}\lambda h)\tilde{f}(2(k-1)h) = 0$$

and

$$(1-\tfrac{3}{8}\lambda h)\{\tilde{f}((2k+1)h) - \tilde{f}((2k-1)h)\} - \tfrac{9}{8}\lambda h\{\tilde{f}((2k-1)h - \tilde{f}((2k-3)h)\}$$

$$= (1+\tfrac{3}{8}\lambda h)\{\tilde{f}(2(k-1)h) - \tilde{f}(2(k-2)h)\} + \tfrac{9}{8}\lambda h\{\tilde{f}(2kh) - \tilde{f}(2(k-1)h)\} \qquad (6.27)$$

for suitable k . The general theory can be used to establish $O(h^4)$ convergence. *

The general ideas underlying the use of quadrature rules in a repeated fashion to construct step-by-step methods for solving (6.1) will now be clear. Before proceeding we point out that there are many variations on this theme. For example, the trapezium-rule weights in Table 6.3 could be incorporated in one of many ways, and in Table 6.5 the rôles of Simpson's rule and the $\frac{3}{8}$th rule could have been partially interchanged (for $k \geq 5$) to give the variation on Table 6.5 displayed in Table 6.6.

TABLE 6.6

$k \backslash j$	0	1	2	3	4	5	6	7	8
5	$\frac{3}{8}h$	$\frac{9}{8}h$	$\frac{9}{8}h$	$\frac{17}{24}h$	$\frac{4}{3}h$	$\frac{1}{3}h$			
6	$\frac{3}{8}h$	$\frac{9}{8}h$	$\frac{9}{8}h$	$\frac{3}{4}h$	$\frac{9}{8}h$	$\frac{9}{8}h$	$\frac{3}{8}h$		
7	$\frac{3}{8}h$	$\frac{9}{8}h$	$\frac{9}{8}h$	$\frac{17}{24}h$	$\frac{4}{3}h$	$\frac{2}{3}h$	$\frac{4}{3}h$	$\frac{1}{3}h$	
8	$\frac{3}{8}h$	$\frac{9}{8}h$	$\frac{9}{8}h$	$\frac{3}{4}h$	$\frac{9}{8}h$	$\frac{9}{8}h$	$\frac{17}{24}h$	$\frac{4}{3}h$	$\frac{1}{3}h$
etc.									

Our concern so far has been with the local truncation error of formulae with equal-interval abscissae. Later (in section 6.6) we shall see that rules with unequally spaced abscissae sometimes have a rôle. However, all the examples we have presented above have employed various simple quadrature rules applied over subintervals. Before turning to numerical examples and new aspects of the quadrature methods, we note that Gregory's rule (see Chapter 1) can provide a choice of weights for use in eqn (6.10), and Gregory's rule differs in type from the rules previously employed. Thus we might use the weights $w_{k,j}$ of Table 6.5 for $k \leq 3$ and for $k \geq 4$ we might construct the weights $w_{k,j}$ from the Lagrangian form of a Gregory rule incorporating pth-order differences in the correction term for the repeated trapezium rule where, say, $p = 3$ or $p = 4$.

Example 6.6

Suppose that for every continuous function $\phi(x)$, and $p \leq r$, $(r = 1,2,3,\ldots)$

$$\sum_{j=0}^{r} \Omega_{rj}^{[p]} \phi(jh) = h \sum_{j=0}^{r}{}'' \phi(jh) - \{\frac{h}{12}\{\nabla\phi(rh) - \Delta\phi(0)\} + \frac{h}{24}\{\nabla^2\phi(rh) + \Delta^2\phi(0)\}$$

$$\ldots - hc^*_{p+1}\{\nabla^p\phi(rh) + (-\Delta)^p\phi(0)\}\} , \qquad (6.28)$$

where the right-hand terms in braces give the Gregory correction. (The expression for c^*_{p+1} is given by Henrici (1964, p.257) and an algorithm for calculating the weights $\Omega^{[p]}_{rj}$ is given by Welsch (1966).)

For any $p \leq k$ it is feasible to choose $w_{k,j} = \Omega^{[p]}_{k,j}$ in (6.10). In practice it appears unwise to vary p continually, as happens if we set $w_{k,j} = \Omega^{[k]}_{k,j}$ for all k. Instead let us choose a modest value of p and set $w_{k,j} = \Omega^{[k]}_{k,j}$ if $k < p$ and $w_{k,j} = \Omega^{[p]}_{k,j}$ for $k \geq p$.

With the latter choice of weights, let us apply our method to the equation

$$f(x) - \lambda\int_0^x f(y)\,dy = 1 .$$

To analyse the behaviour of the approximate values we difference the successive equations

$$\tilde{f}(kh) - \lambda \sum_{j=0}^{k} \Omega^{[p]}_{kj}\tilde{f}(jh) = 1$$

for $k = r$ and $k = r+1$, where $r \geq p$. It is convenient to revert to the finite difference form of the equations to achieve this differencing, and we find

$$(1-\tfrac{1}{2}\lambda h)\tilde{f}((r+1)h)-(1+\tfrac{1}{2}\lambda h)\tilde{f}(rh)+$$

$$+\ \lambda\nabla\{\tfrac{h}{12}\nabla\tilde{f}((r+1)h)+\tfrac{h}{24}\nabla^2\tilde{f}((r+1)h)\ldots -hc^*_{p+1}\nabla^p\tilde{f}((r+1)h)\} = 0$$

or, for $r \geq p$,

$$(1-\tfrac{1}{2}\lambda h)\tilde{f}((r+1)h)-(1+\tfrac{1}{2}\lambda h)\tilde{f}(rh)+$$

$$+\ \lambda h\{\tfrac{1}{12}\nabla^2+\tfrac{1}{24}\nabla^3\ldots -c^*_{p+1}\nabla^{p+1}\}\tilde{f}((r+1)h) = 0 .$$

This is the formula which would be obtained if the closed corrector of the Adams-Moulton-Bashforth formulae (incorporating $\nabla^{p+1}\tilde{f}((r+1)h)$) were applied to the equation $f'(x) = \lambda f(x)$. The convergence and stability properties of the latter method may be found in the literature (see, for example, Lambert 1973), and are quite good. *

What has been said above, concerning the use of Gregory's rule, has implied the use of these rules to construct weights $w_{k,j}$. However, the rules provide difference correction terms to the trapezium rule which can form the basis of a method of deferred corrections (somewhat similar to that described for Fredholm equations in Chapter 4). In our search for a completely adaptive method (matching the technique automatically to the problem) Gregory's formula may yet be established as a useful tool; there are certainly those who, from practical experience, would advocate its usefulness.

The deferred correction process has been described by Fox and Goodwin (1953) and by Mayers (in Fox 1962). We shall outline the process here, applied to the equation

$$f(x) - \lambda\int_0^x K(x,y)f(y)\mathrm{d}y = g(x) .$$

We may use the repeated trapezium rule with step h , and its Gregory correction terms, to write

$$\tilde{f}(0) = g(0) ,$$

$$-\tfrac{1}{2}\lambda hK(h,0)\tilde{f}(0)+\{1-\tfrac{1}{2}\lambda hK(h,h)\}\tilde{f}(h) = g(h)+C_1 ,$$

$$-\tfrac{1}{2}\lambda hK(2h,0)\tilde{f}(0)-\tfrac{1}{2}\lambda hK(2h,h)\tilde{f}(h) + \{1-\tfrac{1}{2}\lambda hK(2h,2h)\}\tilde{f}(2h) = g(2h)+C_2 ,$$

etc., where $C_1,C_2,\ldots$ represent correction terms to the repeated trapezium rule approximation to

$$\lambda\int_0^{rh} K(rh,y)f(y)\mathrm{d}y \quad (r = 1,2,3,\ldots) .$$

A typical stage of the technique involves the calculation of $\tilde{f}(kh)$ in terms of previously accepted approximations to $\tilde{f}(0),\tilde{f}(1),\ldots,\tilde{f}((k-1)h)$. (For small values of k a special starting procedure of section 6.5

may have to be used; we shall therefore assume k large enough for any of our purposes.) The kth step is then initiated by ignoring c_k and computing $\tilde{f}^{[0]}(kh)$ from the formula

$$\{1-\tfrac{1}{2}\lambda hK(kh,kh)\}\tilde{f}^{[0]}(kh) = g(kh)+\lambda h\sum_{j=1}^{k-1} K(kh,jh)\tilde{f}(jh)+\tfrac{1}{2}\lambda hK(kh,0)\tilde{f}(0) .$$

The finite differences (in the y-direction) of the values

$$\lambda K(kh,0)\tilde{f}(0),\lambda K(kh,h)\tilde{f}(h),\ldots,\lambda K(kh,(k-1)h)\tilde{f}((k-1)h),\lambda K(kh,kh)\tilde{f}^{[0]}(kh)$$

can be used in the Gregory correction to the trapezium rule to obtain a correction term $c_k^{[0]}$. In computing $c_k^{[0]}$ we can fix the number of differences or can adapt the number to the situation, incorporating non-negligible terms.

We can use the term $c_k^{[0]}$ to obtain a corrected value $\tilde{f}^{[1]}(kh)$, satisfying

$$\{1-\tfrac{1}{2}\lambda hK(kh,kh)\}\tilde{f}^{[1]}(kh) = c_k^{[0]}+g(kh)+\lambda h\sum_{j=1}^{k-1} K(kh,jh)\tilde{f}(jh)+\tfrac{1}{2}\lambda hK(kh,0)\tilde{f}(0) .$$

Of course, $\tilde{f}^{[1]}(kh) = \tilde{f}^{[0]}(kh) + c_k^{[0]}/\{1-\tfrac{1}{2}\lambda hK(kh,kh)\}$. The process can be continued, using $\tilde{f}^{[1]}(kh)$ to compute $c_k^{[1]}$ and computing $\tilde{f}^{[2]}(kh)$, continuing for a fixed number of iterations or until $\tilde{f}^{[r]}(kh)$ agrees to the required accuracy with $\tilde{f}^{[r-1]}(kh)$. When $\tilde{f}^{[r]}(kh)$ is accepted it occurs as $\tilde{f}(kh)$ in the formula used for calculating subsequent values.

There are some difficulties associated with the method if $K(x,y)f(y)$ is not a smooth function or if at any stage one seeks pth order differences with $p > k$. If we insist (perhaps unwisely) on determining the limiting value $\lim_{r\to\infty} \tilde{f}^{[r]}(kh)$, this may impose restrictions on the stepsize h . With favourable differentiability conditions it should be possible, however, to control the local accuracy of the formula. The only likely source of difficulty is then the effect of stability properties on the propagation of local errors, and in this context we would suggest it wise to fix some upper limit on p , the number of differences to be incorporated in the correction terms; taking $p = k$ or $p = [\frac{1}{2}k]$, for example, is likely to lead to instability which would make the method useless. It might also be unwise to vary p too frequently as k changes.

At the time of writing this book, further investigation is being conducted into such aspects of the use of Gregory formulae.

Example 6.7

We consider the numerical solution of the simple test equation

$$f(x) - \lambda\int_0^x f(y)\,dy = 1 .$$

The exact solution is $f(x) = e^{+\lambda x}$, and approximate solutions were computed using eqn (6.9) with varying choices of weights, and varying choices of step-size h.

We tabulate the numerical results which were obtained using the following weights:

Method (a) Repeated Simpson's rule or repeated Simpson's rule and single trapezium rule - using weights in Table 6.3,

Method (b) Repeated Simpson's rule or single trapezium rule and repeated Simpson's rule - using weights in Table 6.2,

Method (c) Repeated Simpson's rule or repeated Simpson's rule and a single $\frac{3}{8}$th rule - using weights given in Table 6.5,

Method (d) Repeated Simpson's rule or single $\frac{3}{8}$th rule and repeated Simpson's rule - using weights in Table 6.4,

Method (e) Repeated $\frac{3}{8}$th rule or repeated $\frac{3}{8}$th rule and single or double Simpson's rule,

Method (f) Repeated $\frac{3}{8}$th rule or single or double Simpson's rule and repeated $\frac{3}{8}$th rule,

Method (g) Gregory's rule employing pth differences in the correction of the repeated trapezium rule.

The errors at an interval which is a multiple of the step size h are shown in Table 6.7.

TABLE 6.7

$(h = \frac{1}{32}\,,\ \lambda = -12)$

x	Method (a)	Method (b)	Method (c)	Method (d)	Method (e)	Method (f)	Method (g) $p \leq 3$	Method (g) $p \leq 4$
0·25	$-5{\cdot}6\times10^{-4}$	$-4{\cdot}4\times10^{-3}$	$-1{\cdot}9\times10^{-4}$	$-5{\cdot}5\times10^{-4}$	$-2{\cdot}3\times10^{-5}$	$-2{\cdot}0\times10^{-4}$	$-1{\cdot}4\times10^{-4}$	$-1{\cdot}8\times10^{-4}$
0·5	$-5{\cdot}6\times10^{-5}$	$-1{\cdot}2\times10^{-2}$	$-1{\cdot}1\times10^{-5}$	$-1{\cdot}2\times10^{-3}$	$1{\cdot}6\times10^{-6}$	$-4{\cdot}3\times10^{-4}$	$-8{\cdot}4\times10^{-6}$	$-7{\cdot}4\times10^{-6}$
0·75	$-4{\cdot}2\times10^{-6}$	$-3{\cdot}3\times10^{-3}$	$-5{\cdot}9\times10^{-7}$	$-3{\cdot}2\times10^{-3}$	$2{\cdot}9\times10^{-7}$	$7{\cdot}8\times10^{-4}$	$-5{\cdot}0\times10^{-7}$	$-3{\cdot}3\times10^{-7}$
1·00	$-2{\cdot}8\times10^{-7}$	$-8{\cdot}8\times10^{-2}$	$-3{\cdot}2\times10^{-8}$	$-8{\cdot}6\times10^{-3}$	$2{\cdot}2\times10^{-8}$	$-1{\cdot}4\times10^{-4}$	$-2{\cdot}9\times10^{-8}$	$-1{\cdot}5\times10^{-8}$
1·25	$-1{\cdot}8\times10^{-8}$	$-2{\cdot}4\times10^{-1}$	$-1{\cdot}8\times10^{-9}$	$-2{\cdot}3\times10^{-2}$	$1{\cdot}4\times10^{-9}$	$-1{\cdot}4\times10^{-3}$	$-1{\cdot}7\times10^{-9}$	$-6{\cdot}8\times10^{-10}$
1·50	$-1{\cdot}1\times10^{-9}$	$-6{\cdot}4\times10^{-1}$	$-9{\cdot}4\times10^{-11}$	$-6{\cdot}3\times10^{-2}$	$9{\cdot}6\times10^{-11}$	$2{\cdot}2\times10^{-3}$	$-9{\cdot}3\times10^{-11}$	$-3{\cdot}0\times10^{-11}$
1·75	$-6{\cdot}2\times10^{-11}$	$-1{\cdot}7\times10^{0}$	$-5{\cdot}1\times10^{-12}$	$-1{\cdot}7\times10^{-1}$	$5{\cdot}7\times10^{-12}$	$5{\cdot}3\times10^{-5}$	$-5{\cdot}1\times10^{-12}$	$-1{\cdot}3\times10^{-12}$
2·00	$-3{\cdot}6\times10^{-12}$	$-4{\cdot}6\times10^{0}$	$-2{\cdot}7\times10^{-13}$	$-4{\cdot}5\times10^{-1}$	$3{\cdot}2\times10^{-13}$	$-4{\cdot}7\times10^{-3}$	$-2{\cdot}8\times10^{-13}$	$-5{\cdot}5\times10^{-14}$

TABLE 6.7 (continued)

($h = \frac{1}{64}$, $\lambda = -12$)

x	Method (b)	Method (e)	Method (g) $p = 0$	$p \leq 1$	$p \leq 2$	$p \leq 3$	$p \leq 4$
0·25	$-6\cdot4\times10^{-4}$	$-3\cdot1\times10^{-6}$	$4\cdot4\times10^{-4}$	$-1\cdot4\times10^{-5}$	$-1\cdot4\times10^{-6}$	$-7\cdot4\times10^{-6}$	$-7\cdot9\times10^{-6}$
0·50	$-1\cdot8\times10^{-3}$	$1\cdot4\times10^{-8}$	$4\cdot4\times10^{-5}$	$-2\cdot9\times10^{-6}$	$2\cdot1\times10^{-7}$	$-4\cdot1\times10^{-7}$	$-3\cdot9\times10^{-7}$
0·75	$-4\cdot8\times10^{-3}$	$8\cdot6\times10^{-9}$	$3\cdot2\times10^{-6}$	$-2\cdot5\times10^{-7}$	$2\cdot5\times10^{-8}$	$-2\cdot3\times10^{-8}$	$-1\cdot9\times10^{-8}$
1·00	$-1\cdot3\times10^{-2}$	$5\cdot8\times10^{-10}$	$2\cdot2\times10^{-7}$	$-1\cdot8\times10^{-8}$	$1\cdot9\times10^{-9}$	$-1\cdot2\times10^{-9}$	$-9\cdot3\times10^{-10}$
1·25	$-3\cdot5\times10^{-2}$	$5\cdot8\times10^{-11}$	$1\cdot3\times10^{-8}$	$-1\cdot2\times10^{-9}$	$1\cdot3\times10^{-10}$	$-6\cdot6\times10^{-11}$	$-4\cdot6\times10^{-11}$
1·50	$-9\cdot6\times10^{-2}$	$3\cdot8\times10^{-12}$	$7\cdot9\times10^{-10}$	$-7\cdot2\times10^{-11}$	$8\cdot2\times10^{-12}$	$-3\cdot5\times10^{-12}$	$-2\cdot2\times10^{-12}$
1·75	$-2\cdot6\times10^{-1}$	$2\cdot3\times10^{-13}$	$4\cdot6\times10^{-11}$	$-4\cdot3\times10^{-12}$	$4\cdot9\times10^{-13}$	$-1\cdot8\times10^{-13}$	$-1\cdot1\times10^{-13}$
2·00	$-7\cdot1\times10^{-1}$	$1\cdot4\times10^{-14}$	$2\cdot6\times10^{-12}$	$-2\cdot4\times10^{-13}$	$3\cdot1\times10^{-14}$	$-1\cdot2\times10^{-14}$	$-2\cdot5\times10^{-15}$

It should be noted that the accuracy of the results obtained depends upon the choice of quadrature rules, although the work involved in computing our approximations depends only upon h. In particular, the results obtained with Methods (b), (d), and (f) are obviously unsatisfactory, and we shall see later that there is a general theory which leads us to expect this, even though convergence can be established as $h \to 0$ in the absence of rounding errors.

We should also remark that the accuracy obtainable with the methods is affected by the use of low-order rules to calculate early members of the sequence $\tilde{f}(h)$, $\tilde{f}(2h)$, ... (Thus, in each case, the trapezium rule is used to calculate $\tilde{f}(h)$.) We shall describe, later, starting methods which can be used to calculate these values more accurately. To justify our interest in such starting methods, we observe that with $h = \frac{1}{32}$, Method (g), employing $p \leq 3$, but with $\tilde{f}(h)$ and $\tilde{f}(2h)$, computed exactly, gives the following errors:

$x = 0{\cdot}25$	$0{\cdot}50$	$0{\cdot}75$	$1{\cdot}00$
$-1{\cdot}0\times10^{-6}$,	$-1{\cdot}7\times10^{-6}$,	$-1{\cdot}7\times10^{-7}$,	$-1{\cdot}3\times10^{-8}$
$x = 1{\cdot}25$	$1{\cdot}50$	$1{\cdot}75$	$2{\cdot}00$
$-8{\cdot}3.10^{-10}$,	$-5{\cdot}1\ 10^{-11}$,	$-3{\cdot}1\ 10^{-12}$,	$-1{\cdot}8\ 10^{-13}$

The corresponding errors with $p \leq 4$ and $\tilde{f}(h)$, $\tilde{f}(2h)$, and $\tilde{f}(3h)$ obtained exactly are:

$6{\cdot}6\times10^{-6}$,	$9{\cdot}0\times10^{-7}$,	$7{\cdot}3\times10^{-8}$,	$5{\cdot}1\times10^{-9}$,
$3{\cdot}2\times10^{-10}$,	$2{\cdot}0\times10^{-11}$,	$1{\cdot}1\times10^{-12}$,	$6{\cdot}5\times10^{-14}$

(In these figures, the effect of the improved starting values decreases as the scheme progresses.) *

Example 6.8

I give here the numerical results computed for a selection of simple (linear) test equations. The equations considered are:

(a) $f(x) + \int_0^x (x-y)\cos(x-y)f(y)dy = \cos(x)$,

with solution $f(x) = \frac{1}{3}(2\cos\sqrt{3}x + 1)$,

(b) $f(x) - \int_0^x \sin(x-y)f(y)\,dy = x$, with solution $f(x) = x+\frac{1}{6}x^3$;

(c) $f(x) - \int_0^x (x-y)f(y)\,dy = \sin x$, with solution

$f(x) = \frac{1}{2}(\sin x + \sinh x)$;

(d) $f(x) - 2\int_0^x \cos(x-y)f(y)\,dy = e^x$, with solution $f(x) = e^x(1+x)^2$;

(e) $f(x) - \int_0^x f(y)\,dy = \cos x$, with solution

$f(x) = \frac{1}{2}(e^x+\cos x + \sin x)$;

and, finally,

(f) $f(x) + \int_0^x \cosh(x-y)f(y)\,dy = \sinh x$, with solution

$f(x) = (\frac{2}{\sqrt{5}}) \sinh(\frac{\sqrt{5}}{2} x)e^{-\frac{1}{2}x}$.

Results tabulated below were computed using the weights of repeated Simpson's rule with the $\frac{3}{8}$th rule (where appropriate) as indicated in Table 6.5, and a step h . Absolute errors are given in Table 6.8.

TABLE 6.8.

x	0·25	0·50	0·75	1·00	1·25	1·50	1·75	2·00
($h = \frac{1}{8}$)								
(a)	2×10^{-7},	5×10^{-7},	1×10^{-6},	2×10^{-6},	5×10^{-6},	8×10^{-6},	1×10^{-5},	2×10^{-5}
(b)	7×10^{-6},	2×10^{-5},	3×10^{-5},	5×10^{-5},	6×10^{-5},	7×10^{-5},	9×10^{-5},	1×10^{-4}
(c)	7×10^{-6},	2×10^{-5},	4×10^{-5},	5×10^{-5},	8×10^{-5},	1×10^{-4},	1×10^{-4},	2×10^{-4}
(d)	9×10^{-4},	1×10^{-3},	2×10^{-3},	3×10^{-3},	4×10^{-3},	6×10^{-3},	8×10^{-3},	1×10^{-2}
(e)	5×10^{-7},	1×10^{-6},	2×10^{-6},	4×10^{-6},	6×10^{-6},	9×10^{-6},	1×10^{-5},	2×10^{-5}
(f)	2×10^{-5},	2×10^{-5},	1×10^{-5},	1×10^{-5},	1×10^{-5},	1×10^{-5},	1×10^{-5},	1×10^{-5}
($h = \frac{1}{16}$)								
(a)	7×10^{-9},	2×10^{-8},	5×10^{-8},	1×10^{-7},	3×10^{-7},	4×10^{-7},	6×10^{-7},	8×10^{-7}
(d)	6×10^{-5},	9×10^{-5},	1×10^{-4},	2×10^{-4},	3×10^{-4},	4×10^{-4},	5×10^{-4},	7×10^{-4}
(f)	9×10^{-7},	1×10^{-6},	8×10^{-7},	7×10^{-7},	6×10^{-7},	6×10^{-7},	6×10^{-7},	7×10^{-7}

$(h = \frac{1}{32})$ TABLE 6.8 (continued)

(d)	4×10^{-6}, 6×10^{-6}, 9×10^{-6}, 1×10^{-5}, 2×10^{-5}, 3×10^{-5}, 3×10^{-5}, 5×10^{-5} *

Example 6.9

We consider here the numerical solution of certain simple test equations of the form

$$f(x) - \lambda\int_0^x f(y)\,dy = g(x)$$

for $x \geq 0$. We attempt the numerical solution of these equations using the weights in Table 6.3, obtained from Simpson's rule and the trapezium rule.

In the cases considered, the true solution satisfies exactly the equations for the approximate solution. It is not sufficient, however, that the local truncation errors be zero, since the numerical solution for the values $\tilde{f}(0), \tilde{f}(h), \tilde{f}(2h), \ldots$ (in sequence) breaks down if $\frac{1}{2}\lambda h = 1$ or if $\frac{1}{3}\lambda h = 1$. (The coefficient of $\tilde{f}(h)$ in the second equation or the coefficient of $\tilde{f}(2h)$ in the third equation vanishes in these situations.) Thus the choice of λh has some effect on the approximate solution.

In the next section we shall discuss various concepts of stability. In terms of a definition of Noble, which we discuss later, a scheme based on Table 6.3 is 'stable', whilst that based on Table 6.2 is 'unstable'. (See Example 6.7 for insight into this.) Our analysis of Example 6.13 will suggest, however, that to satisfy a different stability requirement, we require $|\lambda h| < 6$ when using Table 6.3 for our test equation with $\lambda < 0$. As becomes clear from discussions in Example 6.4 and Example 6.13, the growth of any error, introduced from rounding (for example) as we compute approximate values for $f(0), f(h), f(2h), \ldots$, in sequence, depends in part on the size of

$$\Gamma = \{1+\tfrac{7}{6}\lambda h+\tfrac{1}{2}\lambda^2h^2\}/\{1-\tfrac{5}{6}\lambda h+\tfrac{1}{6}\lambda^2h^2\} \quad .$$

Since Γ is a continuous function of λh when $\lambda h < 0$, no *dramatic* change in the rate of error propagation is likely as h is changed so that $|\lambda h|$ increases past 6 with $\lambda h < 0$. (Observe that Γ is an approximation to $\exp 2\lambda h$ which becomes increasingly accurate as $\lambda h \to 0$. The rôle of Γ as an approximation to $\exp 2\lambda h$ is amplified in the discussion of relative stability in section 6.3.)

Consider, then, the equation

$$f(x) - \lambda\int_0^x f(y)\,dy = x - \tfrac{1}{2}\lambda x^2$$

with solution $f(x) = x$. In the absence of rounding error, our choice of weights in Table 6.3 would yield an exact solution. The effect of setting $\tilde{f}(0) = f(0) + 10^{-5}$ (thus introducing a single perturbation) is illustrated by the following figures. (The influence of rounding errors in the computer is not significant in the tabulated results.)

(a) $\lambda = -12$, $h = 0{\cdot}25$, $\Gamma = 0{\cdot}4$, $\exp 2\lambda h = 2{\cdot}48\times10^{-3}$. For argument $x = 0{\cdot}25, 0{\cdot}50, 0{\cdot}75, 1{\cdot}00, \ldots, 4{\cdot}00, 4{\cdot}25, 4{\cdot}50, 4{\cdot}75, 5{\cdot}00$ we find errors: $-6{\cdot}0\times10^{-6}$, $7{\cdot}0\times10^{-6}$, $-1{\cdot}4\times10^{-6}$, $2{\cdot}8\times10^{-6}$, ..., $1{\cdot}1\times10^{-8}$, $-2{\cdot}3\times10^{-9}$, $4{\cdot}6\times10^{-9}$, $-9{\cdot}2\times10^{-10}$, $1{\cdot}8\times10^{-9}$ and we find corresponding relative errors: $-2{\cdot}4\times10^{-5}$, $1{\cdot}4\times10^{-6}$, $-1{\cdot}9\times10^{-6}$, $2{\cdot}8\times10^{-6}$, ..., $2{\cdot}9\times10^{-9}$, $-5{\cdot}4\times10^{-10}$, $1{\cdot}0\times10^{-9}$, $-1{\cdot}9\times10^{-10}$, $3{\cdot}7\times10^{-10}$.

(b) $\lambda = -12$, $h = 0{\cdot}5$, $\Gamma = 1$, $\exp 2\lambda h = 6{\cdot}14\times10^{-6}$. At argument values $x = 0{\cdot}50, 1{\cdot}00, 1{\cdot}50, \ldots, 4{\cdot}50, 5{\cdot}00$ we find the corresponding errors: $-7{\cdot}5\times10^{-6}$, $1{\cdot}3\times10^{-5}$, $-6{\cdot}7\times10^{-6}$, ..., $-6{\cdot}7\times10^{-6}$, $1{\cdot}3\times10^{-5}$ and relative errors: $-1{\cdot}5\times10^{-5}$, $1{\cdot}3\times10^{-6}$, $-4{\cdot}4\times10^{-6}$, ..., $-1{\cdot}5\times10^{-6}$, $2{\cdot}7\times10^{-6}$.

(c) $\lambda = -12$, $h = 0{\cdot}75$, $\Gamma = 1{\cdot}4$, $\exp 2\lambda h = 1{\cdot}52\times10^{-8}$. For, respectively, $x = 0{\cdot}75, 1{\cdot}50, 2{\cdot}25, 3{\cdot}00, \ldots, 11{\cdot}25, 12{\cdot}00, 12{\cdot}75, 13{\cdot}5$ we find errors: $-8{\cdot}2\times10^{-6}$, $1{\cdot}7\times10^{-6}$, $-1{\cdot}1\times10^{-5}$, $2{\cdot}4\times10^{-5}$, ..., $-8{\cdot}5\times10^{-5}$, $1{\cdot}9\times10^{-4}$, $-1{\cdot}2\times10^{-4}$, $2{\cdot}6\times10^{-4}$ and we find the corresponding relative errors: $-1{\cdot}1\times10^{-5}$, $1{\cdot}1\times10^{-6}$, $-4{\cdot}8\times10^{-6}$, $8{\cdot}0\times10^{-6}$, ..., $-7{\cdot}6\times10^{-6}$, $1{\cdot}6\times10^{-5}$, $-9{\cdot}4\times10^{-6}$, $2{\cdot}0\times10^{-5}$.

(d) $\lambda = -12$, $h = 1$, $\Gamma = 1{\cdot}69$, $\exp 2\lambda h = 3{\cdot}78\times10^{-11}$. Considering, in turn, $x = 1,2,3,\ldots,11,12,13$, we find the corresponding errors as shown; errors = $-8{\cdot}6\times10^{-6}$, $1{\cdot}9\times10^{-5}$, $-1{\cdot}4\times10^{-5}$, $1{\cdot}1\times10^{-4}$, $2{\cdot}6\times10^{-4}$, $-1{\cdot}9\times10^{-4}$. The propagation of errors depends upon the product λh rather than on λ or h separately. Thus we have:

(e) $\lambda = -1$, $h = 1$, $\Gamma = 0{\cdot}17$, $\exp 2\lambda h = 1{\cdot}35\times10^{-1}$. Considering $x = 1,2,3,\ldots,11,12,13$, yields the corresponding set of errors: $-3{\cdot}3\times10^{-6}$, $8{\cdot}3\times10^{-7}$, $2{\cdot}8\times10^{-7}$, ..., $2{\cdot}2\times10^{-10}$, $1{\cdot}1\times10^{-10}$, $3{\cdot}6\times10^{-11}$.

The preceding figures prompt the following observations. A perturbation in $\tilde{f}(0)$ is magnified if $|\Gamma| > 1$, though the rate of error growth is not catastrophic if $|\Gamma|$ is not very large. The effect of any growth depends upon the range $[0,X]$ over which the solution is obtained. (In some calculations, as here, we may require the use of an accurate quadrature rule with a large step size h to cover a large range in a few steps.) If the true solution is itself increasing, the relative error may not grow even if the actual error grows.

Observe that, in the tabulated figures, the error oscillates.

The reader may care to attempt similar tests on less simple equations. *

6.3. *Stability and ill-conditioning*

We have discussed, above, a number of formulae which permit the step-by-step computation of values $\tilde{f}(kh)$ of an approximate solution of (6.1). Now $\tilde{f}(0) = g(0)$ is (subject to rounding error) an exact value, but it quite frequently happens that in implementing such a method on a computer the errors in $\tilde{f}(kh)$ increase as k is increased. However, we may be fortunate and find that good accuracy is maintained.

It is our task in this section to identify the sources of errors in the computation of the values of $\tilde{f}(x)$. With such insight it is to be hoped that a catastrophic growth of errors in *computed* values for $\tilde{f}(kh)$ may be avoided, particularly if the growth of errors is due to the choice of the numerical method. Here we should note that in a practical computation we obtain values $\hat{f}(kh)$, say, rather than $\tilde{f}(kh)$, $(k = 0,1,2,\ldots)$ because of the presence of rounding errors.

In our discussion we might hope to gain much from the corresponding analysis of our step-by-step methods for the solution of initial-value problems in ordinary differential equations. Whilst this expectation is perhaps justified, the attempted extension of the ideas involved in ordinary differential equations, to the treatment of Volterra equations, appears to me to be sometimes contrived. My view is that much more work is necessary on this important aspect of Volterra equations, at the time of writing, to obtain results which assist in the design of efficient practical algorithms.

It is fairly easy, however, to list the *sources* of error in the approximate values $\hat{f}(kh)$: the introduction of rounding errors and local truncation errors results in the solution of a perturbed problem. But the integral equation may be 'ill conditioned', so that (in the linear case) a small perturbation in $g(x)$ or in $K(x,y)$ in (6.1) induces a very large change in the solution $f(x)$. To measure such changes it is necessary to fix our attention on the solution of the problem (6.1), say,

$$f(x) - \lambda\int_0^x K(x,y)f(y)\,dy = g(x) \tag{6.29}$$

on some definite interval $[0,X]$. Since the equation can be regarded as a linear Fredholm equation, the theory of Chapter 4 on condition numbers applies, and we can introduce a condition number $\mu(K) \equiv \mu(K,X)$, say, to indicate the degree of ill-conditioning.

<u>Example 6.10</u>
We may define

$$\mu(K;X) = \|I-\lambda K\|_\infty \ \|I+\lambda R_\lambda\|_\infty ,$$

where the norms are those of operators on $C[0,X]$ with the uniform norm and where K is here a Volterra operator and R_λ is the operator corresponding to its resolvent kernel. If $\mu(K;X)$ is large, relatively small perturbations $\delta(x)$ in $g(x)$, or $E(x,y)$ in $K(x,y)$, may lead to relatively 'large' perturbations $\delta f(x)$ measured by their uniform norm on $[0,X]$. See, for example, Theorem 4.3 and note that $(I-\lambda K)^{-1} = I+\lambda R_\lambda$. *

<u>Remark</u>. Because the condition number only provides a bound on the perturbation in $f(x)$, it provides a rather coarse judgement of the actual variation encountered in $f(x)$. In the case of a non-linear

equation it appears even more difficult to decide whether the equation is ill conditioned. It seems to be generally accepted, at least, that the equation

$$\phi(x) - \lambda\int_0^x F(x,y;\phi(y))\,dy = g(x) \tag{6.30}$$

for the function $\phi(x)$ is liable to ill-conditioning if the corresponding linear equation (6.29), with $K(x,y) = F_v(x,y;\phi(y))$ and $F_v(x,y;v) = (\partial/\partial v)F(x,y;v)$, is itself ill conditioned.

It is sometimes preferable to avoid the introduction of a condition number and analyse the particular integral equation on its merits, in detail.

Example 6.11

Consider the equation

$$f(x) - \lambda\int_0^x K(x,y)f(y)\,dy = g(x) ,$$

where $K(x,y) \geq M > 0$ for $0 \leq y \leq x$. When $g(x)$ is perturbed to a function $g(x) + \delta g(x)$, the corresponding solution changes from $f(x)$ to $f(x) + \delta f(x)$, where

$$\delta f(x) - \lambda\int_0^x K(x,y)\,\delta f(y)\,dy = \delta g(x) .$$

Thus

$$\delta f(x) = \delta g(x) + \lambda\int_0^x R_\lambda(x,y)\,\delta g(y)\,dy ,$$

where $R_\lambda(x,y)$, is given (section 1.7) by the Neumann series for the resolvent,

$$R_\lambda(x,y) = K(x,y) + \lambda K^2(x,y) + \lambda^2 K^3(x,y) + \ldots,$$

where $K^p(x,y)$ is the pth iterated kernel of $K(x,y)$. Since

$$K^p(x,y) = \int_y^x K(x,z)\, K^{p-1}(z,x)\,dz ,$$

it can be shown by induction that

$$K^p(x,y) \geq M^p(x-y)^{p-1}/(p-1)!$$

and hence (if $\lambda > 0$)

$$R_\lambda(x,y) \geq Me^{\lambda M(x-y)} .$$

Consequently

$$\delta f(x) \geq \delta g(x) + \lambda M\int_0^x e^{\lambda M(x-y)}\delta g(y)\,dy$$

in the case where $\delta g(x) > 0$ and $\lambda > 0$. In the latter case, $\delta f(x)$ has a possibly increasing component (behaving like

$$M\lambda e^{\lambda Mx}\int_0^x e^{-\lambda My}\delta g(y)\,dy) .$$

The practical conclusion which follows is that if $\lambda K(x,y)$ is large and positive, the equation is susceptible to ill-conditioning. The situation is worst when the particular choice of $g(x)$ corresponds to a true solution $f(x)$ which is decreasing, since both the relative error and the absolute error due to a perturbation in $g(x)$ are then likely to grow. *

Although the integral equation may be well conditioned, the errors in $\tilde{f}(kh)$ or in $\hat{f}(kh)$ may be large. We consider that the computational scheme displays practical 'instability' if the effect of the scheme is to magnify ('unreasonably') the effect of rounding errors and/or local truncation errors. If we wish, we can approach the question of practical 'stability', in the loose sense above, by experiment and heuristic argument. The practical effects of errors due to rounding and due to local truncation errors may be difficult to distinguish; indeed they cannot always be separated in a genuine practical problem. In Examples 6.2, 6.3, and 6.4, however, we give the analytical form of the approximate values $\tilde{f}(kh)$ without regard to inexact arithmetic, and a study of these examples reveals to what extent the local truncation errors cause $\tilde{f}(x)-f(x)$ to grow as x increases. (For some special equations the local truncation errors vanish and any error growth, which occurs when the approximate solution $\hat{f}(x)$ is calculated on a computer, is then due entirely to the effect of 'instability' combined with the rounding error of approximate arithmetic.) It is clear that the notion of 'stability', in the practical sense being employed here, is linked to computation with a practical (non-zero) step size h, using finite precision arithmetic to perform the calculations (Wilkinson 1963).

Another concept of stability ('zero-stability') is discussed by Lambert (1973) in the context of methods for eqn (6.3); it concerns the limiting behaviour of a method as $h \to 0$. For the present, we shall only consider calculations with a fixed step $h > 0$. Our reason for not discussing the analogue of zero-stability here is that *all* reasonable schemes based on (6.9) are zero-stable. Thus, any scheme which produces approximations $\tilde{f}(kh)$ which converge to $f(kh)$ as $h \to 0$ with kh fixed -but arbitrary -is zero-stable, when applied to solve eqn (6.3).

The concepts of 'stability' encountered in the literature (and there are many of these, though they are not always given explicitly in the case of integral equations) generally correspond to various mathematical models for assessing the practical stability effects of a computational scheme. We shall introduce some definitions, hoping to indicate the motivation for them and their usefulness (as well as any limitations they may possess). We shall deduce some of the 'stability criteria' which are found in the literature.

A general step-by-step method for computing a sequence of values $\tilde{\phi}_0, \tilde{\phi}_1, \tilde{\phi}_2, \ldots$ is defined in terms of functions $\Phi_k(\quad)$ if we set

$$\Phi_k(\tilde{\phi}_0, \tilde{\phi}_1, \tilde{\phi}_2, \ldots, \tilde{\phi}_k) = 0 \quad (k = 0,1,2,\ldots) .$$

We solve the kth equation for $\tilde{\phi}_k$ in terms of $\tilde{\phi}_0, \tilde{\phi}_1, \ldots, \tilde{\phi}_{k-1}$. In the case of a linear scheme we can readily re-formulate the equation to give $\tilde{\phi}_k$ explicitly, say as

$$\tilde{\phi}_k = \Psi_k(\tilde{\phi}_0, \tilde{\phi}_1, \ldots, \tilde{\phi}_{k-1}) \quad (k = 0,1,2,\ldots) .$$

(For an example of such a situation, see eqn (6.9), with $\lambda w_{kk}K(kh,kh) \neq 1$ $(k = 1,2,3,\ldots)$, and set $\tilde{\phi}_r = \tilde{f}(rh)$ $(r = 0,1,2,\ldots)$.)

Our first definition is concerned with the effect of an *isolated perturbation* in a value $\tilde{\phi}_k$. That is, we shall be concerned with the effect on $\tilde{\phi}_{k+r}$ $(r = 1,2,3,\ldots)$ of changing a particular value $\tilde{\phi}_k$ to $\tilde{\phi}_k + \xi_k$ (leaving $\tilde{\phi}_0, \tilde{\phi}_1, \ldots, \tilde{\phi}_{k-1}$ unchanged). Such a change can arise if at the kth stage we solve an equation of the form

$$\Phi_k(\tilde{\phi}_0, \tilde{\phi}_1, \ldots, \tilde{\phi}_k + \xi_k) = \rho_k \quad ,$$

where the *residual* ρ_k would be zero in the correct kth equation. We shall suppose that the subsequent change in $\tilde{\phi}_{k+r}$ $(r = 1,2,...)$ is η^k_{k+r}, so that

$$\Phi_{k+r}(\tilde{\phi}_0,\tilde{\phi}_1,...,\tilde{\phi}_k+\xi_k,\tilde{\phi}_{k+1}+\eta^k_{k+1},...,\tilde{\phi}_{k+r}+\eta^k_{k+r}) = 0 \quad (r = 1,2,3,...) \quad .$$

In certain non-linear schemes it is necessary to restrict the size of ξ_k to ensure that the sequence of perturbed values exists. In the remainder of the chapter we shall assume (unless stated to the contrary) that the scheme is linear, although extensions of our remarks to cover the more general case can be made. In the linear case it is easy to relate ξ_k to the residual ρ_k in the kth equation of the unperturbed scheme.

DEFINITION 6.1. *A computational scheme for computing a certain set of values* $\tilde{\phi}_0,\tilde{\phi}_1,\tilde{\phi}_2,...$ *in a step-by-step fashion will be said to be 'absolutely stable' (with respect to an isolated perturbation) if a perturbation* ξ_k *in* $\tilde{\phi}_k$ *(for any fixed* $k \geq 0$ $\tilde{\phi}_0,\tilde{\phi}_1,..,\tilde{\phi}_{k-1}$ *being unchanged) results in changes* $\eta^k_{k+1},\eta^k_{k+2},...$ *in* $\tilde{\phi}_{k+1},\tilde{\phi}_{k+2},...$ *such that*

$$|\eta^k_{k+r}| \leq \rho|\xi_k| \quad (r = 1,2,3,...)$$

for some $\rho \in [0,1]$. *If* $\rho \in [0,1)$ *the stability will be called 'strict'.*

DEFINITION 6.2. *In the case where* $\lim_{r\to\infty} |\eta^k_{k+r}| = 0$, *the scheme will be said to be 'damped'.*

Remark. We shall seek to apply the concepts of the preceding definitions to schemes associated with (6.9) for obtaining $\tilde{f}(h),\tilde{f}(2h),\tilde{f}(3h),...,$ (or, for obtaining computed approximations $\hat{f}(h),\hat{f}(2h),\hat{f}(3h),...)$. Thus the values $f(kh)$ of the solution $f(x)$ of (6.1) satisfy a perturbed form of (6.9), namely

$$f(kh) - \lambda \sum_{j=0}^{k} w_{kj}K(kh,jh)f(jh) = g(kh) + \tau_k \quad ,$$

where τ_k is the local truncation error, and we need to examine the effect of this perturbation on $f(kh)-\tilde{f}(kh)$. Since the preceding definitions are concerned with the behaviour of a scheme when an *isolated error* is introduced at the kth stage, they do not reflect the situation in practice (where, for example, truncation errors are introduced at *each stage* of the computation and cause associated perturbations). We shall show, later, that in some cases strict absolute stability with respect to an isolated perturbation does lead to some conclusions when we consider the more realistic situation.

A given numerical method applied to a particular integral equation may lead to an absolutely stable computational scheme for values $\tilde{f}(h)$, $\tilde{f}(2h)$, $\tilde{f}(3h)$, ... only if the integral equation satisfies certain conditions. It is sometimes desirable to identify classes of equations for which a given numerical method will lead to an absolutely stable computational scheme.

Example 6.12

An example of a scheme which is both damped and strictly absolutely stable (with respect to isolated perturbation) is provided by the use of (6.17) to solve the equation considered in Example 6.2 with $\lambda < 0$. An error ξ_k introduced in $\tilde{\phi}_k = \tilde{f}(kh)$ results in a change $\gamma\xi_k$ in $\tilde{\phi}_{k+1} = \tilde{f}((k+1)h)$ and, in general, a change $\gamma^r\xi_k$ in $\tilde{\phi}_{k+r} = \tilde{f}((k+r)h)$ for $r \geq 1$, where $\gamma = (1+\frac{1}{2}\lambda h)/(1-\frac{1}{2}\lambda h)$. We find that $|\gamma| < 1$ and the algorithm is damped and strictly absolutely stable provided $\lambda < 0$ (and $h > 0$) . However, if $\lambda > 0$, $h > 0$, then $\gamma > 1$ and the use of (6.17) is not stable in the preceding sense, and certainly not damped. *

The use of an algorithm which is absolutely stable (in the sense of the preceding definition) ensures that errors introduced through rounding (for example) do not grow 'catastrophically'. However, a linear growth of errors can occur with the use of an absolutely stable method since, setting $\tilde{\phi}_k = \tilde{f}(kh)$ in the above discussion, a perturbation ξ_k at the kth stage may be followed by the subsequent introduction of new errors ξ_{k+1}, ξ_{k+2}, ... (Thus the total perturbation in $\tilde{\phi}_{k+1}$ would be $\xi_{k+1} + \eta_{k+1}^k$.) The acceptability of any error growth may depend on the growth of the true solution $f(kh)$, since this affects the relative error growth. Our definition of stability is concerned with absolute errors.

In the case of the scheme discussed in Example 6.2, the effect of the total perturbations $\xi_{k+r}, \eta^{k}_{k+r}, \ldots, \eta^{k+r-1}_{k+r}$ can be analysed. We recall that the use of the repeated trapezium rule with step h for the equation

$$f(x) - \lambda\int_0^x f(y)\,dy = g(x)$$

gives rise to equations for $\tilde{\phi}_k = \tilde{f}(kh)$ which are analytically equivalent to $\tilde{\phi}_{k+1} - \gamma\tilde{\phi}_k = \delta_k$ (see eqn (6.17)), where

$$\delta_k = (1-\tfrac{1}{2}\lambda h)^{-1}\Delta g(kh)\ , \quad (k = 0,1,2,\ldots)\ .$$

Now suppose that we introduce a perturbation ξ_k in $\tilde{\phi}_k$. The consequent change η^{k}_{k+1} in $\tilde{\phi}_{k+1}$ is $\gamma\xi_k$, and an introduction of a new error ξ_{k+1} gives a total perturbation $\xi_{k+1}+\gamma\xi_k$. In general, we have, for the computed values $\hat{\phi}_k$ approximating values $\tilde{\phi}_k$,

$$\hat{\phi}_{k+1} = \gamma\hat{\phi}_k + \delta_k + \xi_{k+1}$$

and

$$\tilde{\phi}_{k+1} = \gamma\tilde{\phi}_k + \delta_k$$

so that for the total perturbations

$$\hat{\phi}_k - \tilde{\phi}_k = \varepsilon_k\ ,$$
$$\varepsilon_{k+1} = \gamma\varepsilon_k + \xi_{k+1}\ .$$

Thus ε_{k+1} behaves like the solution which would be obtained, with exact arithmetic, when we replace δ_k by ξ_{k+1} in the relation for $\tilde{\phi}_k$. In the situation in which $\tilde{\phi}_k$ is growing, as k increases, it is quite likely that replacing δ_k by ξ_{k+1} will also give an increasing solution ε_k, so we do not expect absolute stability under these conditions. Thus if $g(x) = 1$, $f(x) = e^{\lambda x}$ with $\lambda > 0$, $\gamma > 1$, $\tilde{f}(kh)$ increases (exponentially) with k, and the errors ε_k also grow whilst the relative errors $\{\varepsilon_k/\tilde{f}(kh)\}$ remain more modest.

Suppose, now, that $|\gamma| < 1$ and $|\xi_k| \le \xi$ $(k = 0,1,2,\ldots)$.

Then $|\varepsilon_{k+1}| \leq |\gamma| \, |\varepsilon_k| + \xi$ and $|\varepsilon_0| \leq \xi$. It is readily shown, by induction, that $|\varepsilon_k| \leq e_k$ $(k = 0,1,2,\ldots)$, where $e_0 = \xi$ and $e_{k+1} = |\gamma| e_k + \xi$. Clearly,

$$e_k = \xi(1-|\gamma|^{k+1})/(1-|\gamma|)$$

when $|\gamma| < 1$. We see now that, if $|\gamma| < 1$,

$$|\varepsilon_k| \leq \xi/(1-|\gamma|) ,$$

and the total effects of the perturbations, measured by ε_k, are bounded if the scheme is strictly absolutely stable. To express this situation we give the definition which follows. This definition relates to a step-by-step scheme for obtaining the value $\tilde{\phi}_k$ in terms of $\tilde{\phi}_0, \tilde{\phi}_1, \ldots, \tilde{\phi}_{k-1}$ $(k = 1,2,3,\ldots)$. The value $\hat{\phi}_k$ which is generated is obtained instead in terms of past approximate values $\hat{\phi}_0, \hat{\phi}_1, \ldots, \hat{\phi}_{k-1}$ subject to an *additional* error ξ_k. Since an additional error is introduced at each stage, the perturbation is 'persistent'. The formal definition now follows.

DEFINITION 6.3. *Suppose that, with a step-by-step scheme defined by equations* $\Phi_k(\tilde{\phi}_0, \tilde{\phi}_1, \ldots, \tilde{\phi}_k) = 0$ $(k = 1,2,3,\ldots)$, $\hat{\phi}_0 = \tilde{\phi}_0 + \xi_0$, *and* $\Phi_k(\hat{\phi}_0, \hat{\phi}_1, \ldots, \hat{\phi}_k - \xi_k) = 0$ *at the kth stage, where the only requirement of* ξ_k *is that* $|\xi_k| \leq \xi$ $(k = 0,1,2,\ldots)$. *If there is an absolute constant* $\Omega < \infty$ *such that* $|\hat{\phi}_k - \tilde{\phi}_k| \leq \Omega\xi$ $(k = 0,1,2,\ldots)$ *then the scheme will be said to display 'bounded error propagation'.*

Remark. Clearly every algorithm for a *terminating* sequence $\{\tilde{\phi}_k\}$ $(k = 0,1,2,\ldots,N)$ displays bounded error propagation when $|\hat{\phi}_k| < \infty$ $(k = 0,1,2,\ldots,N)$. The terminology is of primary interest if the sequence $\{\tilde{\phi}_k\}$ is infinite. On the other hand, as we saw in the discussion which preceded the definition, we are often interested in computational schemes which depend on a parameter h. The length of the sequence (if it is finite) then depends on h. We would like to be able to choose Ω, in the definition above, to be independent of the parameter h, and if this is possible the scheme may be said to display *uniformly* bounded error propagation.

It must also be emphasized, in passing, that our definitions apply to a particular scheme which arises on applying some method to a definite equation. We may wish to determine classes of equations for which a particular method yields, for *each* of the equations in the class considered, a scheme which satisfies one of our definitions.

The application of the preceding remarks to the numerical solution of the equation discussed in Example 6.2 is readily indicated and gives insight into its wider application. We have, in the case under consideration, $f(x) - \lambda \int_0^x f(y)dy = g(x)$, and consequently

$$f(kh) - \lambda h \sum_{j=0}^{k}{}'' f(jh) = g(kh) + \tau_k ,$$

where

$$\tau_k = \lambda\{\int_0^{kh} f(y)dy - h \sum_{j=0}^{k}{}'' f(jh)\} .$$

If we difference successive equations we obtain (compare with eqn (6.17) and the equations above):

$$f((k+1)h)-\gamma f(kh) = (1-\tfrac{1}{2}\lambda h)^{-1}\{\Delta g(kh)+(\tau_{k+1}-\tau_k)\} ,$$

whilst, in (6.17),

$$\tilde{f}((k+1)h)-\gamma\tilde{f}(kh) = (1-\tfrac{1}{2}\lambda h)^{-1}\Delta g(kh) .$$

Thus the analysis above can be applied when $|\gamma| < 1$ if we let $\tilde{\phi}_k$ denote $\tilde{f}(kh)$, and write $\hat{\phi}_k$ for $f(kh)$ and

$$\xi_{k+1} = (1-\tfrac{1}{2}\lambda h)^{-1}(\tau_{k+1}-\tau_k) .$$

The theory then establishes a bound on the growth of the errors $f(kh)-\tilde{f}(kh)$.

On the other hand, the use of finite precision arithmetic to compute $\tilde{f}(kh)$ will result in a set of approximate values, and we may now let $\hat{\phi}_k$ denote the value $\hat{f}(kh)$ computed for $\tilde{f}(kh)$ (which we again denote by $\tilde{\phi}_k$) . The computation of $\hat{\phi}_{k+1}$ can be expressed as

$$\hat{\phi}_{k+1} = \text{computed value of } \{(g(kh)+\lambda h \sum_{j=0}^{k-1}{}' \hat{f}(jh))/(1-\tfrac{1}{2}\lambda h)\} =$$

$$= \{g(kh)+\lambda h \sum_{j=0}^{k-1}{}' \hat{f}(jh)\}/(1-\tfrac{1}{2}\lambda h) + \nu_k ,$$

where a bound on ν_k can be obtained from an analysis of the finite precision arithmetic. (Such analysis usually gives a bound on the relative error in the computation; see Wilkinson (1963).) If $|\gamma| < 1$ and we write $\xi_{k+1} = \nu_{k+1} - \nu_k$ the preceding theory gives an analysis of the growth in $\tilde{\phi}_k - \hat{\phi}_k$, or $\tilde{f}(kh) - \hat{f}(kh)$.

Similarly, the true solution values $f(kh)$ and the computed values $\hat{f}(kh)$ can be compared directly if we write

$$\xi_{k+1} = (\nu_{k+1} - \nu_k) - (\tau_{k+1} - \tau_k)/(1 - \tfrac{1}{2}\lambda h) .$$

We have seen that for the simple method discussed in Example 6.2, strict absolute stability implies bounded error propagation. In order to extend our discussion to more complicated numerical methods, we propose to introduce a new type of stability. We begin by defining a block-by-block method, to provide some motivation for the stability definition.

DEFINITION 6.4. *A 'block-by-block method' is a step-by-step method for computing vectors (or blocks)* $\tilde{\underset{\sim}{\phi}}_0, \tilde{\underset{\sim}{\phi}}_1, \tilde{\underset{\sim}{\phi}}_2, \ldots,$ *in sequence.*

We shall subsequently discuss new methods which are block-by-block methods, but the simple step-by-step methods (like those already introduced in section 6.2, for computing values $\tilde{f}(0), \tilde{f}(h), \tilde{f}(2h), \ldots$ in sequence) provide examples of a block-by-block method. Suppose that in such a method we write $\tilde{\underset{\sim}{\phi}}_k = [\tilde{f}_{ks}, \tilde{f}_{ks+1}, \ldots, \tilde{f}_{(k+1)s-1}]^T$ for some $s \geq 1$, where $\tilde{f}_t = \tilde{f}(th)$. Then by grouping the successive values together, we can consider the step-by-step calculation of the individual values as a block-by-block process for computing $\tilde{\underset{\sim}{\phi}}_0, \tilde{\underset{\sim}{\phi}}_1, \ldots$ in order. (The computation of the block of values comprising the components of $\tilde{\underset{\sim}{\phi}}_k$ can *in this instance* be broken down into the successive calculation of the individual components.)

DEFINITION 6.5. *A block-by-block scheme for the calculation of a particular set of vectors* $\tilde{\underset{\sim}{\phi}}_0, \tilde{\underset{\sim}{\phi}}_1, \tilde{\underset{\sim}{\phi}}_2, \ldots$ *is 'block-stable' in the norm* $\|\ \|$ *(with respect to an isolated block perturbation) if for any* $k > 0$, *a sufficiently small change* $\underset{\sim}{\xi}_k$ *in* $\tilde{\underset{\sim}{\phi}}_k$, $\tilde{\underset{\sim}{\phi}}_0, \tilde{\underset{\sim}{\phi}}_1, \ldots, \tilde{\underset{\sim}{\phi}}_{k-1}$ *remaining unchanged, gives rise to changes* $\underset{\sim}{\eta}^k_{k+r}$ *in* $\tilde{\underset{\sim}{\phi}}_{k+r}$, *satisfying*

$\|\eta^{k}_{k+r}\| \le \rho\|\xi_{k}\|$ ($r = 1,2,3,\ldots$) *where* $0 \le \rho \le 1$; *this stability is called 'strict' if* $\rho < 1$. *In the case where* $\inf\{\rho \mid \|\eta^{k}_{k+r}\| < \rho\|\xi_k\|\} \le 1$, *the infemum being taken over all choices of norm and of* ρ, *the scheme may be called 'block-stable'; it is not, however, necessarily 'block-stable in norm' for any norm, unless the infemum is attained. Finally, a scheme is called 'damped' if* $\lim_{r\to\infty}\|\eta^{k}_{k+r}\| = 0$, *for all* k.

Remark. The above definition applies to any block-by-block method in which successive vectors $\tilde{\phi}_k$ whose components are blocks of s values (say, $\tilde{f}_{ks+1},\ldots,\tilde{f}_{ks+s-1}$) are obtained in sequence. Such a method may be called an s-component block method. Any s-component block method may also be regarded as an (ms)-component block method, where m is integer, so that a step-by-step method is a 1-component block method but it can also be regarded as an m-component block method for any m. A more 'genuine' example of a block-by-block method is provided by the product-integration methods for Volterra integral equations, discussed below.† Observe that an s-component method which is block-stable in the norm $\|\ \|_2$ (with $\|\eta^{k}_{k+r}\|_2 \le \rho\|\xi_k\|_2$ and $\rho \le 1$) is block-stable in the norm $\|\ \|_\infty$ if $\rho\sqrt{s} \le 1$, for

$$\|\eta^{k}_{k+r}\|_\infty \le \|\eta^{k}_{k+r}\|_2 \le \rho\|\xi_k\|_2 \le \rho s^{\frac{1}{2}}\|\xi_k\|_\infty .$$

If a step-by-step method is damped when it is regarded as a block-by-block method it is also damped as a step-by-step method. 'Damping' of a block-by-block method is independent of the choice of norm (assuming that the vectors $\tilde{\phi}_k$ have a finite number of components!).

In the following Examples we hope to demonstrate that (block-) stability and damping may be meaningful properties to require of an algorithm. The examples are linear and even then somewhat special; although this makes the discussion less general than might be desired, it has the advantage of resulting in tractable analysis.

---

† The discussion here will appear familiar to those who have studied the matrix method for analyzing the stability of numerical methods.

As a preliminary to the Examples, consider a linear recurrence relation of the form

$$\tilde{\phi}_{k+1} = \underset{\sim}{M}\tilde{\phi}_k + \underset{\sim}{\delta}_k \,, \tag{6.31}$$

where $\underset{\sim}{M}$ is a fixed square matrix. A perturbation ξ_k in $\tilde{\phi}_k$ results in a change η_{k+1}^{k} in $\tilde{\phi}_{k+1}$ where $\eta_{k+1}^{k} = \underset{\sim}{M}\xi_k$. Further, $\eta_{k+r}^{k} = \underset{\sim}{M}^r \xi_k$ is the resulting change in $\tilde{\phi}_{k+r}$ $(r = 2,3,4,\ldots)$ and the recurrence for the vectors $\tilde{\phi}_k$ is damped if $\lim\limits_{r\to\infty} \|\underset{\sim}{M}^r\| = 0$. Now for any subordinate matrix norm, $\lim\limits_{r\to\infty} \|\underset{\sim}{M}^r\|^{1/r} = \rho(\underset{\sim}{M})$, the spectral radius of $\underset{\sim}{M}$. Thus the method is damped if $\rho(\underset{\sim}{M}) < 1$. Moreover, this condition is necessary since if $\rho(\underset{\sim}{M}) = \lambda \geq 1$ and $\underset{\sim}{M}x = \lambda \underset{\sim}{x}$ we may take $\xi_k = \underset{\sim}{x}$ and find that $\eta_{k+r}^{k} = \lambda^r \xi_k$ $(r = 1,2,3,\ldots)$. Thus a necessary and sufficient condition for damping is that $\rho(\underset{\sim}{M}) < 1$.

Further, a necessary condition for block-stability is that $\rho(\underset{\sim}{M}) \leq 1$, since we require that $\lim\limits_{r\to\infty} \|\eta_{k+r}^{k}\| \leq \|\xi_k\|$. A sufficient condition for block-stability is, of course, the requirement that $\|\underset{\sim}{M}\| \leq 1$, since $\|\eta_{k+r}^{k}\| \leq \|\underset{\sim}{M}\| \|\xi_{k+r-1}\|$ and we have 'strict' block-stability if $\|\underset{\sim}{M}\| < 1$. Now we can show (see, for example, Ortega 1972) that for any matrix $\underset{\sim}{M}$ and any $\varepsilon > 0$, there is some subordinate norm such that $\|\underset{\sim}{M}\| \leq \rho(\underset{\sim}{M}) + \varepsilon$. Thus if $\rho(\underset{\sim}{M}) < 1$ we may take $\varepsilon = \frac{1}{2}\{1-\rho(\underset{\sim}{M})\}$ and for the corresponding subordinate norm, $\|\underset{\sim}{M}\| < 1-\varepsilon < 1$. On the other hand, we frequently encounter a situation where $\rho(\underset{\sim}{M}) = 1$, and this does not automatically ensure that there is a norm with $\|\underset{\sim}{M}\| = 1$. Suppose, however, that $\underset{\sim}{M}$ has distinct eigenvalues, or is similar to a diagonal matrix, or (more generally) is *of class M* (Ortega 1972, p.24).[†] Then the existence of a norm with $\|\underset{\sim}{M}\| = \rho(\underset{\sim}{M})$ is guaranteed, and if $\rho(\underset{\sim}{M}) = 1$ then the scheme based on (6.31) is block-stable in some norm.

Remark. When $\underset{\sim}{M}$ is similar to a diagonal matrix, the solution of (6.31) can be expressed in terms of the eigenvectors of $\underset{\sim}{M}$. (In the more general case where $\underset{\sim}{M}$ is not diagonalizable the solution can be

---

† A matrix $\underset{\sim}{M}$ is of class M if the Jordan blocks corresponding to eigenvalues of modulus $\rho(\underset{\sim}{M})$ are diagonal.

expressed in terms of eigenvectors and generalized eigenvectors, but this will not be pursued here.) We suppose that $\underset{\sim}{M}$ is of order m and $\underset{\sim}{M}\underset{\sim}{x}_i = \mu_i \underset{\sim}{x}_i$ $(i = 1,2,3,\ldots,m)$, where $\underset{\sim}{x}_1, \underset{\sim}{x}_2, \ldots \underset{\sim}{x}_m$ are linearly independent. Then $\alpha_1^{[k]}, \alpha_2^{[k]}, \ldots, \alpha_m^{[k]}$ may be found such that

$$\tilde{\underset{\sim}{\phi}}_k = \sum_{i=1}^{m} \alpha_i^{[k]} \underset{\sim}{x}_i \, , \tag{6.32}$$

and $\beta_1^{[k]}, \beta_2^{[k]}, \ldots, \beta_m^{[k]}$ such that $\underset{\sim}{\delta}_k = \sum_{i=1}^{m} \beta_i^{[k]} \underset{\sim}{x}_i$. Substitution in (6.31) gives

$$\sum_{i=1}^{m} \alpha_i^{[k+1]} \underset{\sim}{x}_i = \sum_{i=1}^{m} (\alpha_i^{[k]} \mu_i + \beta_i^{[k]}) \underset{\sim}{x}_i \, .$$

Since $\underset{\sim}{x}_1, \underset{\sim}{x}_2, \ldots, \underset{\sim}{x}_m$ are linearly independent we can write, for $i = 1,2,\ldots, m$,

$$\alpha_i^{[k+1]} = \mu_i \alpha_i^{[k]} + \beta_i^{[k]} \quad (k = 0,1,2,\ldots)$$

and we find that

$$\alpha_i^{[k]} = \mu_i^k \alpha_i^{[0]} + \sum_{r=0}^{k} \mu_i^r \beta_i^{[k-r]} \quad (k = 0,1,2,\ldots) \, . \tag{6.33}$$

It is clear that if $|\mu_i| > 1$ then a perturbation in $\alpha_i^{[0]}$, or in $\beta_i^{[s]}$ for some s, will be magnified in the subsequent values $\alpha_i^{[k]}$. A growth in the changes in the associated vectors $\tilde{\underset{\sim}{\phi}}_k$ can then occur.

Suppose, now, that $\|\underset{\sim}{M}\| < 1$ in (6.31), and suppose that the computation of the solution of (6.31) results in a sequence of vectors $\hat{\underset{\sim}{\phi}}_0 \simeq \tilde{\underset{\sim}{\phi}}_0, \hat{\underset{\sim}{\phi}}_1 \simeq \tilde{\underset{\sim}{\phi}}_1, \ldots$ satisfying the relations $\|\tilde{\underset{\sim}{\phi}}_0 - \hat{\underset{\sim}{\phi}}_0\| \leq \xi$ and

$$\hat{\underset{\sim}{\phi}}_{k+1} = \underset{\sim}{M}\hat{\underset{\sim}{\phi}}_k + \underset{\sim}{\delta}_k + \underset{\sim}{\xi}_{k+1} \, .$$

If we write $\underset{\sim}{\varepsilon}_k = \hat{\underset{\sim}{\phi}}_k - \tilde{\underset{\sim}{\phi}}_k$, then $\underset{\sim}{\varepsilon}_{k+1} = \underset{\sim}{M}\underset{\sim}{\varepsilon}_k + \underset{\sim}{\xi}_{k+1}$ and $\|\underset{\sim}{\varepsilon}_{k+1}\| \leq$ $\leq \|\underset{\sim}{M}\| \times \|\underset{\sim}{\varepsilon}_k\| + \|\underset{\sim}{\xi}_{k+1}\|$. Suppose that $\|\underset{\sim}{\xi}_{k+1}\| \leq \xi$ $(k = 0,1,2,\ldots,n)$. Then it can be established, quite readily, that with $\|\underset{\sim}{M}\| < 1$

$$\|\underset{\sim}{\varepsilon}_{k+1}\| \leq \left\{\frac{1 - \|\underset{\sim}{M}\|^{k+2}}{1 - \|\underset{\sim}{M}\|}\right\} \xi \quad (k = 0,1,2,\ldots,n) \, .$$

Setting $\Omega = (1 - \|\underset{\sim}{M}\|)^{-1}$ we then find, with $\|\underset{\sim}{\xi}_{k+1}\| \leq \xi$, that

$$\|\underset{\sim}{\varepsilon}_{k+1}\| < \Omega\xi \, .$$

The preceding analysis encourages us to make a formal definition, which is analogous to the definition of bounded error propagation, and relates to the practical situation in a realistic way.

DEFINITION 6.6. *A block-by-block method for obtaining* $\tilde{\Phi}_0, \tilde{\Phi}_1, \tilde{\Phi}_2, \ldots$ *by solving the equations* $\underset{\sim}{\Phi}_k(\tilde{\Phi}_0, \tilde{\Phi}_1, \ldots, \tilde{\Phi}_k) = \underset{\sim}{0}$ $(k = 0,1,2,\ldots)$ *displays error propagation which is 'bounded-in-norm' if, when* $\hat{\Phi}_0 = \tilde{\Phi}_0 + \underset{\sim}{\xi}_0$ *and* $\Phi_k(\hat{\Phi}_0, \hat{\Phi}_1, \ldots, \hat{\Phi}_{k-1}, \hat{\Phi}_k - \underset{\sim}{\xi}_k) = \underset{\sim}{0}$,

$$\|\tilde{\Phi}_k - \hat{\Phi}_k\| \leq \Omega\xi \quad (k = 0,1,2,\ldots)$$

for some absolute constant Ω , *whenever* $\|\underset{\sim}{\xi}_k\| \leq \xi$ $(k = 0,1,2,\ldots)$.

We observe that *the preceding definition is independent of the choice of norm,* since two norms on a finite dimensional space are equivalent. In particular, given a norm $\|\ \|$ on a given finite-dimensional vector-space, there are constants p,q such that $p\|\underset{\sim}{x}\| \leq \|\underset{\sim}{x}\|_\infty \leq q\|\underset{\sim}{x}\|$ for any vector $\underset{\sim}{x}$. Thus if $\|\underset{\sim}{\xi}_k\|_\infty \leq \xi^*$ $(k = 0,1,2,\ldots)$ then $\|\underset{\sim}{\xi}_k\| \leq \xi^*/p$ and with $\xi = \xi^*/p$ the terms of the preceding definition require that $\|\tilde{\Phi}_k - \hat{\Phi}_k\| \leq \Omega\xi^*/p$. Then $\|\tilde{\Phi}_k - \hat{\Phi}_k\|_\infty \leq q\,\|\tilde{\Phi}_k - \hat{\Phi}_k\| \leq$ $\leq \{q\Omega/p\}\xi^* = \Omega^*\xi^*$ for some absolute constant Ω^* related to Ω . The remarks which preceded the Definition 6.6 show that (6.31) displays error propagation which is bounded in the norm $\|\ \|$ if $\|\underset{\sim}{M}\| < 1$. The situation is not as satisfactory when $\|\underset{\sim}{M}\| = 1$, since we then obtain the inequalities

$$\|\underset{\sim}{\varepsilon}_{k+1}\| \leq \|\underset{\sim}{\varepsilon}_k\| + \|\underset{\sim}{\varepsilon}_{k+1}\| \leq (k+2)\xi ,$$

and there is clearly the possibility of a 'linear growth' in the norms $\|\underset{\sim}{\varepsilon}_k\|$ as k increases. However, our remarks show that it is sufficient to establish that *any* subordinate norm of $\underset{\sim}{M}$ is less than unity. Hence the condition $\rho(\underset{\sim}{M}) < 1$ is also sufficient, following our earlier remarks. Since $\|\underset{\sim}{M}\| \geq \rho(\underset{\sim}{M})$ for any subordinate norm, we see that *the condition* $\rho(\underset{\sim}{M}) < 1$ *is both necessary and sufficient* for error propagation which is bounded-in-norm.

On the other hand, block-stability is guaranteed if $\rho(\underset{\sim}{M}) \leq 1$, but for a number of schemes (see Example 6.15, however) it transpires that $\rho(\underset{\sim}{M}) = 1$. Suppose in this case that $\underset{\sim}{M}$ is of class M (see the reference given earlier) or, in particular, that the eigenvalues of $\underset{\sim}{M}$ are distinct. Then for some subordinate norm we have $\|\underset{\sim}{M}\| = 1$, and there is, as noted above, the possibility of linear error growth. This is not generally catastrophic. We recognize the feature by giving the following definition which is consistent with the preceding remarks, and in which the conditions prevailing are those of the last definition.

DEFINITION 6.7. *A block-by-block method is said to display error propagation with 'at most linear growth' if, in some norm,*

$$\|\hat{\underset{\sim}{\phi}}_k - \tilde{\underset{\sim}{\phi}}_k\| \leq \Omega(k+1)\xi \quad (k = 0,1,2,\ldots) ,$$

where Ω *is some absolute constant, and* $\|\underset{\sim}{\xi}_k\| \leq \xi \ (k = 0,1,2,\ldots)$.

The preceding remarks have been concerned with the stability properties of step-by-step or block-by-block schemes. (For further reading on this aspect see Chapter 4 of Ortega (1972), where a different terminology is employed.) Little emphasis has been placed on the relation of the schemes to the problem of solving an integral equation, except to note that the true solution of the integral equation gives a solution of a perturbed form of the step-by-step or block-by-block equations. In the following Examples, we see how the concepts above relate to particular applications of the method associated with eqns (6.9). In effect, this reduces to determining $\rho(\underset{\sim}{M})$ for a number of schemes of the form (6.31).

<u>Example 6.13</u>.

Let us analyse the method based on the weights in Table 6.3 for the numerical solution of

$$f(x) - \lambda\int_0^x f(y)\,dy = g(x) .$$

(We may, as in Example 6.4, consider the case where $g(x) \equiv 1$.) The method consists of step-by-step solution of eqn (6.10) using the weights of Table 6.3. Writing $\alpha = -\lambda h$ for convenience, the array of coefficients of $\tilde{f}(0)$, $\tilde{f}(h)$, $\tilde{f}(2h)$, ... in the system of equations has the form (6.34):

$$\begin{array}{cc|cc|cc}
1 & 0 & & & & \\
\frac{1}{2}\alpha & 1+\frac{1}{2}\alpha & & & & \\
\hline
\frac{1}{3}\alpha & \frac{4}{3}\alpha & 1+\frac{1}{3}\alpha & 0 & & \\
\frac{1}{3}\alpha & \frac{4}{3}\alpha & \frac{5}{6}\alpha & 1+\frac{1}{2}\alpha & & \\
\hline
\frac{1}{3}\alpha & \frac{4}{3}\alpha & \frac{2}{3}\alpha & \frac{4}{3}\alpha & 1+\frac{1}{3}\alpha & 0 \\
\frac{1}{3}\alpha & \frac{4}{3}\alpha & \frac{2}{3}\alpha & \frac{4}{3}\alpha & \frac{5}{6}\alpha & 1+\frac{1}{2}\alpha \\
\hline
\end{array} \; . \tag{6.34}$$

The array (6.34) may be partitioned as indicated. The values $\tilde{f}(0)$, $\tilde{f}(h)$, $\tilde{f}(2h)$, ... may be associated in blocks consisting of pairs of adjacent function values. Thus we may write $\underset{\sim}{\tilde{\phi}}_0 = [\tilde{f}(0), \tilde{f}(h)]^T$, $\underset{\sim}{\tilde{\phi}}_1 = [\tilde{f}(2h), \tilde{f}(3h)]^T$, etc., and if we write $\underset{\sim}{I}$ for the identity matrix of order 2, we can also write

$$\underset{\sim}{A} = \alpha \begin{bmatrix} \frac{2}{3} & \frac{4}{3} \\ \frac{2}{3} & \frac{4}{3} \end{bmatrix}, \quad \underset{\sim}{A}_0 = \alpha \begin{bmatrix} \frac{1}{3} & \frac{4}{3} \\ \frac{1}{3} & \frac{4}{3} \end{bmatrix}, \quad \underset{\sim}{B} = \alpha \begin{bmatrix} \frac{1}{3} & 0 \\ \frac{5}{6} & \frac{1}{2} \end{bmatrix},$$

and then, from the partitioning indicated above, we find that

$$(\underset{\sim}{I}+\underset{\sim}{B})\underset{\sim}{\tilde{\phi}}_k = -\underset{\sim}{A}(\underset{\sim}{\tilde{\phi}}_{k-1}+\ldots+\underset{\sim}{\tilde{\phi}}_1) - \underset{\sim}{A}_0\underset{\sim}{\tilde{\phi}}_0 + \underset{\sim}{g}_k \, ,$$

where $\underset{\sim}{g}_k = [g(2kh), g((2k+1)h)]^T$. Then for $k \geq 1$ we find

$$(\underset{\sim}{I}+\underset{\sim}{B})(\underset{\sim}{\tilde{\phi}}_{k+1}-\underset{\sim}{\tilde{\phi}}_k) = -\underset{\sim}{A}\underset{\sim}{\tilde{\phi}}_k + \Delta\underset{\sim}{g}_k \, ,$$

where $\Delta\underset{\sim}{g}_k = \underset{\sim}{g}_{k+1} - \underset{\sim}{g}_k$, and hence

$$(\underset{\sim}{I}+\underset{\sim}{B})\underset{\sim}{\tilde{\phi}}_{k+1} = \{\underset{\sim}{I}+(\underset{\sim}{B}-\underset{\sim}{A})\}\underset{\sim}{\tilde{\phi}}_k + \Delta\underset{\sim}{g}_k \, . \tag{6.35}$$

(Thus, a perturbation of $\underset{\sim}{\xi}_k$ in $\underset{\sim}{\tilde{\phi}}_k$ gives rise to a perturbation $\underset{\sim}{\eta}^k_{k+1}$ in $\underset{\sim}{\tilde{\phi}}_{k+1}$ satisfying

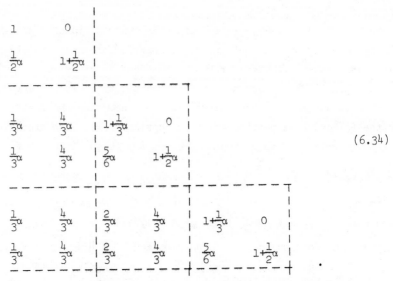

$$(\underset{\sim}{I}+\underset{\sim}{B})\underset{\sim}{\eta}_{k+1}^{k} = \{\underset{\sim}{I}+(\underset{\sim}{B}-\underset{\sim}{A})\}\underset{\sim}{\xi}_k \;.)$$

Since $\underset{\sim}{B}$ is the lower-triangular matrix above, we see that $\underset{\sim}{I}+\underset{\sim}{B}$ is non-singular if $\frac{1}{2}\lambda h \neq 1$ and $\frac{1}{3}\lambda h \neq 1$ – in particular if $\lambda < 0$ or if $0 \leq \lambda h < 2$ (which is satisfied when h is sufficiently small).

When $(\underset{\sim}{I}+\underset{\sim}{B})^{-1}$ exists, eqn (6.35) can be expressed in the form (6.31) with $\underset{\sim}{M} = (\underset{\sim}{I}+\underset{\sim}{B})^{-1}(\underset{\sim}{I}+\underset{\sim}{B}-\underset{\sim}{A})$ and $\underset{\sim}{\delta}_k = (\underset{\sim}{I}+\underset{\sim}{B})^{-1}\Delta\underset{\sim}{g}_k$, and our study of (6.31) applies. To determine which of our definitions apply, we compute $\rho(\underset{\sim}{M})$, and for this purpose we write the equation $\underset{\sim}{M}\underset{\sim}{x} = \mu\,\underset{\sim}{x}$ in the form

$$\mu(\underset{\sim}{I}+\underset{\sim}{B})\underset{\sim}{x} = (\underset{\sim}{I}+\underset{\sim}{B}-\underset{\sim}{A})\underset{\sim}{x}$$

and solve the determinantal equation $\det\{\underset{\sim}{I}+\underset{\sim}{B}-\underset{\sim}{A}-\mu(\underset{\sim}{I}+\underset{\sim}{B})\} = 0$. On solving this quadratic equation we determine the eigenvalues $\mu = 1$ and $\mu = (1-\frac{7}{6}\alpha+\frac{1}{2}\alpha^2)/(1+\frac{5}{6}\alpha+\frac{1}{6}\alpha^2) = \Gamma$, say, where $\alpha = -\lambda h$. Thus, for all λ, $\rho(\underset{\sim}{M}) \geq 1$ and the block-by-block algorithm does not display error propagation which is bounded-in-norm. For block-stability we now seek $\rho(\underset{\sim}{M}) = 1$ or, equivalently, $|\Gamma| \leq 1$. We thus require that

$$-2 \leq \left\{\frac{2\lambda h(1+\frac{1}{6}\lambda h)}{(1-\frac{1}{2}\lambda h)(1-\frac{1}{3}\lambda h)}\right\} \leq 0 \;.$$

Suppose that $\lambda < 0$, whence $\alpha = -\lambda h > 0$. The second inequality above gives the requirement $|\alpha| \leq 6$ whilst the first gives $2\alpha^2-\alpha+6 \geq 0$ (where $\alpha = -\lambda h$) which is always satisfied. If $\lambda h<0$, we have block stability *in a norm* in the algorithm for $\underset{\sim}{\Phi}_1, \underset{\sim}{\Phi}_2, \underset{\sim}{\Phi}_3, \ldots$ (in a norm associated with $\underset{\sim}{M}$) provided $|\lambda h| <6$, with 'block-stability' also for $\lambda h=-6$.

We may also consider the propagation of any error $\underset{\sim}{\xi}_0$ in $\tilde{\underset{\sim}{\Phi}}_0$, which governs the algorithm for $\tilde{\underset{\sim}{\Phi}}_0, \tilde{\underset{\sim}{\Phi}}_1, \tilde{\underset{\sim}{\Phi}}_2, \ldots$. (We suppose $\lambda < 0$ and $|\lambda h| < 6$ so that the above result holds.) A perturbation $\underset{\sim}{\xi}_0$ in $\tilde{\underset{\sim}{\Phi}}_0$ causes a change $\underset{\sim}{\eta}_1$ in $\tilde{\underset{\sim}{\Phi}}_1$, where $(\underset{\sim}{I}+\underset{\sim}{B})\underset{\sim}{\eta}_1 = -\underset{\sim}{A}_0\underset{\sim}{\xi}_0$. The eigenvalues of $-(\underset{\sim}{I}+\underset{\sim}{B})^{-1}\underset{\sim}{A}_0$ are easily found, since one is zero, but even if the spectral radius of this matrix is in $[-1,1]$ this does not establish the block-stability of the algorithm for $\tilde{\underset{\sim}{\Phi}}_0, \tilde{\underset{\sim}{\Phi}}_1, \ldots$ in the norm used earlier. The norm associated with the matrix $\underset{\sim}{M} = (\underset{\sim}{I}+\underset{\sim}{B})^{-1}(\underset{\sim}{I}+\underset{\sim}{B}-\underset{\sim}{A})$ is not known, and a bound for $\|\underset{\sim}{\eta}_1\| / \|\underset{\sim}{\xi}_0\|$ cannot, therefore, be found. It can however be shown that for h sufficiently small and $\lambda < 0$ the algorithm for $\tilde{\phi}_0, \tilde{\phi}_1, \tilde{\phi}_2, \ldots$ is block-stable.

For $\lambda > 0$ there are no values of λh, with $h > 0$, which ensure that $|\Gamma| \leq 1$, and block-stability is not possible. It is interesting to compare the above results with the insight which we obtain from Example 6.4. We showed there that $\tilde{f}((2k+1)h) = \gamma\tilde{f}(2kh)$ and $\tilde{f}((k+2)h) = \Gamma\tilde{f}(kh)$, where $\gamma = (1+\frac{1}{2}\lambda h)/(1-\frac{1}{2}\lambda h)$, and an obvious requirement for step-by-step stability is that $|\Gamma| \leq 1$ and $|\gamma| \leq 1$, the second condition being satisfied automatically if $\lambda < 0$. Since it is possible to choose λh so that $|\Gamma| \leq \rho < 1$ and $|\gamma| \leq \rho < 1$, damping of the step-by-step method can occur under certain conditions. Moreover, the propagation of errors in $\tilde{\phi}_0 = \tilde{f}(0)$ or in $\tilde{\phi}_1 = \tilde{f}(h)$ is readily examined.

A variation of the matrix analysis is given by Mayers (1962). *

Example 6.14.

An analysis of the block-stability of the method based on the weights of Table 6.2, for the approximate solution of the equation

$$f(x) - \lambda\int_0^x f(y)\,dy = 1,$$

is again fairly simple. In analogy with the preceding example we partition the matrix of coefficients in the form indicated, with $\alpha = -\lambda h$:

$$\begin{array}{cc|cc|cc|cc|}
1 & 0 & & & & & & \\
\frac{1}{2}\alpha & 1+\frac{1}{2}\alpha & & & & & & \\
\hline
\frac{1}{3}\alpha & \frac{4}{3}\alpha & 1+\frac{1}{3}\alpha & 0 & & & & \\
\frac{1}{2}\alpha & \frac{5}{6}\alpha & \frac{4}{3}\alpha & 1+\frac{1}{3}\alpha & & & & \\
\hline
\frac{1}{3}\alpha & \frac{4}{3}\alpha & \frac{2}{3}\alpha & \frac{4}{3}\alpha & 1+\frac{1}{3}\alpha & 0 & & \\
\frac{1}{2}\alpha & \frac{5}{6}\alpha & \frac{4}{3}\alpha & \frac{2}{3}\alpha & \frac{4}{3}\alpha & 1+\frac{1}{3}\alpha & & \\
\hline
\frac{1}{3}\alpha & \frac{4}{3}\alpha & \frac{2}{3}\alpha & \frac{4}{3}\alpha & \frac{2}{3}\alpha & \frac{4}{3}\alpha & 1+\frac{1}{3}\alpha & 0 \\
\frac{1}{2}\alpha & \frac{5}{6}\alpha & \frac{4}{3}\alpha & \frac{2}{3}\alpha & \frac{4}{3}\alpha & \frac{2}{3}\alpha & \frac{4}{3}\alpha & 1+\frac{1}{3}\alpha
\end{array} . \tag{6.36}$$

We write

$$\underset{\sim}{A} = \alpha\begin{bmatrix}\frac{2}{3} & \frac{4}{3}\\ \frac{4}{3} & \frac{2}{3}\end{bmatrix}, \quad \underset{\sim}{B} = \alpha\begin{bmatrix}\frac{1}{3} & 0\\ \frac{4}{3} & \frac{1}{3}\end{bmatrix}, \quad \underset{\sim}{A}_0 = \alpha\begin{bmatrix}\frac{1}{3} & \frac{4}{3}\\ \frac{1}{2} & \frac{5}{6}\end{bmatrix},$$

and $\underset{\sim}{\Phi}_0 = [\tilde{f}(0),\tilde{f}(h)]^T$, $\underset{\sim}{\Phi}_1 = [\tilde{f}(2h),\tilde{f}(3h)]^T$, etc., as before, to obtain

$$(\underset{\sim}{I}+\underset{\sim}{B})\underset{\sim}{\Phi}_k = -\underset{\sim}{A}(\underset{\sim}{\Phi}_{k-1}+\ldots+\underset{\sim}{\Phi}_1)-\underset{\sim}{A}_0\underset{\sim}{\Phi}_0+\underset{\sim}{g}_k$$

for $k \geq 1$. The block-stability of the algorithm for $\underset{\sim}{\Phi}_1,\underset{\sim}{\Phi}_2,\underset{\sim}{\Phi}_3,\ldots$ may be discussed by considering the spectral radius of the matrix $\underset{\sim}{M} = (\underset{\sim}{I}+\underset{\sim}{B})^{-1}(\underset{\sim}{I}+\underset{\sim}{B}-\underset{\sim}{A})$. We have $\underset{\sim}{M}\underset{\sim}{x} = \mu\underset{\sim}{x}$ for $\underset{\sim}{x} \neq \underset{\sim}{0}$ if and only if $\det\{(1-\mu)(\underset{\sim}{I}+\underset{\sim}{B})-\underset{\sim}{A}\} = 0$. We find that this condition simplifies to

$$\{(1-\tfrac{1}{3}\alpha)-\mu(1+\tfrac{1}{3}\alpha)\}^2 = (\tfrac{4}{3}\alpha)^2\mu ,$$

and as $h \to 0$ the limiting behaviour of the eigenvalues is like $e^{2\lambda h}$ and $e^{-\frac{2}{3}\lambda h}$. Thus for small values of h one of the eigenvalues is greater than unity and block-stability *is not then possible*. This is in contrast to the previous example, and although the local truncation errors associated with each method are of the same order, the stability properties of the method of Example 6.13 make it preferable. *

In the previous Example, successive vectors $\underset{\sim}{\Phi}_k$ had (in general) no common components. In the following Examples, the approach is more flexible.

Example 6.15.

Consider the situation studied in Example 6.12, where we have relations of the form

$$\tilde{f}((k+1)h) = \gamma\tilde{f}(kh) + \delta_k .$$

If we write

$$\underset{\sim}{\Phi}_k = [\tilde{f}(kh) \ , \tilde{f}((k+1)h)]^T$$

and

$$\underset{\sim}{\Phi}_{k+1} = [\tilde{f}((k+1)h) \ , \tilde{f}((k+2)h)]^T ,$$

then

$$\underset{\sim}{\Phi}_{k+1} = \underset{\sim}{M}\underset{\sim}{\Phi}_k + \underset{\sim}{\delta}_k ,$$

where

$$\underset{\sim}{M} = \begin{bmatrix} \gamma & 0 \\ 0 & \gamma \end{bmatrix} \text{ and } \underset{\sim}{\delta}_k = [\delta_k, \delta_{k+1}]^T .$$

Thus $\rho(\underset{\sim}{M}) = |\gamma|$, which is less than unity if $|\gamma| < 1$, that is, if $\lambda < 0$ and $h > 0$.

Observe that, in place of the previous formulation, we can write

$$\begin{bmatrix} \tilde{f}((k+1)h) \\ \tilde{f}((k+2)h) \end{bmatrix} = \begin{bmatrix} 0 & 1 \\ 0 & \gamma \end{bmatrix} \begin{bmatrix} \tilde{f}(kh) \\ \tilde{f}((k+1)h) \end{bmatrix} + \begin{bmatrix} 0 \\ \delta_{k+1} \end{bmatrix} \quad (k = 0,1,2,\ldots) ,$$

so that more than one way can be found of writing down a recurrence of the form (6.31). (It should be noted that a perturbation $\underset{\sim}{\xi}_k = [\zeta, \eta]^T$ with $\zeta \neq 0$ *could* be thought 'inadmissible' in the last formulation, since it would yield the equation $\tilde{f}((k+1)h) = \tilde{f}((k+1)h) + \zeta$.)

Consider, now, the situation discussed in Example 6.13. We have, as before, the equations

$$\tilde{f}(0) = g(0)$$

$$\tfrac{1}{2}\alpha\tilde{f}(0) + (1+\tfrac{1}{2}\alpha)\tilde{f}(h) = g(h)$$

$$\tfrac{1}{3}\alpha\tilde{f}(0) + \tfrac{4}{3}\alpha\tilde{f}(h) + (1+\tfrac{1}{3}\alpha)\tilde{f}(2h) = g(2h)$$

$$\tfrac{1}{3}\alpha\tilde{f}(0) + \tfrac{4}{3}\alpha\tilde{f}(h) + \tfrac{5}{6}\alpha\tilde{f}(2h) + (1+\tfrac{1}{2}\alpha)\tilde{f}(3h) = g(3h) , \quad \text{etc.}$$

If we difference successive equations, beginning with the third and fourth, we find

$$\tilde{f}(0) = g(0)$$

$$\tfrac{1}{2}\alpha\tilde{f}(0) + (1+\tfrac{1}{2}\alpha)\tilde{f}(h) = g(h)$$

$$\tfrac{1}{3}\alpha\tilde{f}(0) + \tfrac{4}{3}\tilde{f}(h) + (1+\tfrac{1}{3}\alpha)\tilde{f}(2h) = g(2h) ,$$

and, for $k = 0,1,2,\ldots,$

$$(1+\tfrac{1}{2}\alpha)\tilde{f}((2k+3)h)-(1-\tfrac{1}{2}\alpha)\tilde{f}((2k+2)h) = g((2k+3)h)-g((2k+2)h)$$

whilst, for $k = 1,2,3,\ldots,$

$$(1+\tfrac{1}{3}\alpha)\tilde{f}((2k+2)h)-(1-\tfrac{5}{6}\alpha)\tilde{f}((2k+1)h)-\tfrac{1}{6}\alpha\tilde{f}(2kh) = g((2k+2)h)-g((2k+1)h) .$$

Consequently,

$$(\underset{\sim}{I}+\hat{\underset{\sim}{B}})\begin{bmatrix}\tilde{f}((2k+3)h)\\ \tilde{f}((2k+2)h)\end{bmatrix} + \hat{\underset{\sim}{A}}\begin{bmatrix}\tilde{f}((2k+1)h)\\ \tilde{f}(2kh)\end{bmatrix} = \begin{bmatrix}\Delta g((2k+2)h)\\ \Delta g((2k+1)h)\end{bmatrix} \quad (k \geq 1) ,$$

where

$$\underset{\sim}{I}+\hat{\underset{\sim}{B}} = \begin{bmatrix}1+\frac{1}{2}\alpha & -(1-\frac{1}{2}\alpha)\\ 0 & (1+\frac{1}{3}\alpha)\end{bmatrix} \quad \text{and} \quad \hat{\underset{\sim}{A}} = \begin{bmatrix}0 & 0\\ -(1-\frac{5}{6}\alpha) & -\frac{1}{6}\alpha\end{bmatrix}.$$

For block-stability properties of the sequence of vectors

$$\tilde{\Phi}_k = [\tilde{f}((2k+1)h)\ ,\ \tilde{f}(2kh)]^T \quad (k \geq 1)$$

we examine the eigenvalues of $-(\underset{\sim}{I}+\hat{\underset{\sim}{B}})^{-1}\hat{\underset{\sim}{A}}$, and these are found to be $\mu = 0$ and $\mu = \Gamma = (1-\frac{7}{6}\alpha+\frac{1}{2}\alpha^2)/(1+\frac{5}{6}\alpha+\frac{1}{6}\alpha^2)$. Observe that the eigenvalue $\mu = 1$ obtained in the analysis of Example 6.13 has been *replaced* by an eigenvalue $\mu = 0$.

In discussing strict absolute stability or damping it is important to distinguish between the cases where, in (6.31), $\rho(\underset{\sim}{M}) = 1$ or $\rho(\underset{\sim}{M}) < 1$. This Example shows how differing conclusions can be reached depending on the prior manipulation of the system of equations. (Thus, in Example 6.13, $\rho(\underset{\sim}{M}) \geq 1$ whilst, here, $\rho(\underset{\sim}{M}) = |\Gamma|$ and $\rho(\underset{\sim}{M}) < 1$ if $|\Gamma| < 1$.) *

The results of the previous Example suggest a new definition.

DEFINITION 6.8. *A numerical technique for solving a class of Volterra integral equations will be said to be respectively 'stable' or 'strictly stable' for that class of integral equations if, using a given step-size* h *, the equations for the approximate values can be expressed (if necessary, by a suitable rearrangement) as block-by-block schemes which display, respectively, 'block-stability' or 'strict block-stability in a norm'.*

<u>Remark</u>. If a (step-by-step) technique for solving the equations

$$f(x) - \lambda\int_0^x f(y)\,dy = 1$$

is *strictly* stable for every positive step-size h, whenever $\mathrm{Re}(\lambda) < 0$, then it is sometimes called *A-stable*.

The quadrature method using the repeated trapezium rule is A-stable (see Lambert 1973, p.235).

The reader may agree with the author in the view that, whilst the concept of block-stability is somewhat artificial (considering, as it does, the effect of an isolated perturbation), the concept of error propagation which is bounded-in-norm is valuable in the context of a fairly realistic model. It is worthwhile recalling that, if $\rho(\underset{\sim}{M}) < 1$ in the scheme in (6.31), then the error propagation *is* bounded-in-norm. (This contrasts with the possibility of linear error growth in the case where $\rho(\underset{\sim}{M}) = 1$.)

6.3.1.

The treatment of the next Example is rather more complicated. We shall refer to the coefficients in (6.37):

$$\begin{array}{cc|cc|cc|cc}
1 & 0 & & & & & & \\
\frac{1}{2}\alpha & 1+\frac{1}{2}\alpha & & & & & & \\
\hline
\frac{1}{3}\alpha & \frac{4}{3}\alpha & 1+\frac{1}{3}\alpha & 0 & & & & \\
\frac{3}{8}\alpha & \frac{9}{8}\alpha & \frac{9}{8}\alpha & 1+\frac{3}{8}\alpha & & & & \\
\hline
\frac{1}{3}\alpha & \frac{4}{3}\alpha & \frac{2}{3}\alpha & \frac{4}{3}\alpha & 1+\frac{1}{3}\alpha & 0 & & \\
\frac{1}{3}\alpha & \frac{4}{3}\alpha & \frac{17}{24}\alpha & \frac{9}{8}\alpha & \frac{9}{8}\alpha & 1+\frac{3}{8}\alpha & & \\
\hline
\frac{1}{3}\alpha & \frac{4}{3}\alpha & \frac{2}{3}\alpha & \frac{4}{3}\alpha & \frac{2}{3}\alpha & \frac{4}{3}\alpha & 1+\frac{1}{3}\alpha & 0 \\
\frac{1}{3}\alpha & \frac{4}{3}\alpha & \frac{2}{3}\alpha & \frac{4}{3}\alpha & \frac{17}{24}\alpha & \frac{9}{8}\alpha & \frac{9}{8}\alpha & 1+\frac{3}{8}\alpha
\end{array} \qquad (6.37)$$

<u>Example 6.16</u>.

We consider the use of the weights in Table 6.5 arising from the repeated Simpson's rule and the $\frac{3}{8}$th rule for the numerical solution of our test equation

$$f(x) - \lambda\int_0^x f(y)\,dy = g(x) .$$

The triangular array of coefficients for the equations for $\tilde{f}(0), \tilde{f}(h), \ldots$ is now (6.37), where $\alpha = -\lambda h$. (Omitted superdiagonal elements are zero.) Writing

$$\tilde{\psi}_0 = [\tilde{f}(0), \tilde{f}(h)]^T, \ \tilde{\psi}_1 = [\tilde{f}(2h), \tilde{f}(3h)]^T, \ldots,$$

and

$$g_k = [g(2kh), g((2k+1)h)]^T$$

we have, for $k \geq 1$,

$$(I+B)\tilde{\psi}_{k+1} + C\tilde{\psi}_k + A(\tilde{\psi}_{k-1} + \ldots + \tilde{\psi}_1) + A_0\tilde{\psi}_0 = g_{k+1} ,$$

where

$$A = \alpha\begin{bmatrix} \frac{2}{3} & \frac{4}{3} \\ \frac{2}{3} & \frac{4}{3} \end{bmatrix}, \qquad B = \alpha\begin{bmatrix} \frac{1}{3} & 0 \\ \frac{9}{8} & \frac{3}{8} \end{bmatrix},$$

$$C = \alpha\begin{bmatrix} \frac{2}{3} & \frac{4}{3} \\ \frac{17}{24} & \frac{9}{8} \end{bmatrix}, \qquad A_0 = \alpha\begin{bmatrix} \frac{1}{3} & \frac{4}{3} \\ \frac{1}{3} & \frac{4}{3} \end{bmatrix}.$$

If we difference these equations we have, for $k \geq 2$

$$(I+B)(\tilde{\psi}_{k+1} - \tilde{\psi}_k) + C(\tilde{\psi}_k - \tilde{\psi}_{k-1}) + A\tilde{\psi}_{k-1} = \Delta g_k ,$$

or

$$(I+B)\tilde{\psi}_{k+1} - (I+B-C)\tilde{\psi}_k + (A-C)\tilde{\psi}_{k-1} = \Delta g_k$$

which is not of the form (6.31). We shall therefore write

$$\left[\begin{array}{c|c} I+B & 0 \\ \hline 0 & I \end{array}\right]\begin{bmatrix} \tilde{\psi}_{k+1} \\ \tilde{\psi}_k \end{bmatrix} = \left[\begin{array}{c|c} (I+B-C) & (C-A) \\ \hline I & 0 \end{array}\right]\begin{bmatrix} \tilde{\psi}_k \\ \tilde{\psi}_{k-1} \end{bmatrix} + \begin{bmatrix} \Delta g_k \\ 0 \end{bmatrix}$$

which has the required form with

$$\Phi_{k+1} = \begin{bmatrix} \tilde{\psi}_{k+1} \\ \tilde{\psi}_k \end{bmatrix} \qquad \Phi_k = \begin{bmatrix} \tilde{\psi}_k \\ \tilde{\psi}_{k-1} \end{bmatrix}$$

(We note that it will not be possible to establish strict block-stability in the uniform norm since ψ_k is a component of Φ_k and Φ_{k+1}.)

We set

$$\underset{\sim}{M} = \begin{bmatrix} \underset{\sim}{I} + \underset{\sim}{B} & \underset{\sim}{O} \\ \underset{\sim}{O} & \underset{\sim}{I} \end{bmatrix}^{-1} \begin{bmatrix} \underset{\sim}{I} + \underset{\sim}{B} - \underset{\sim}{C} & (\underset{\sim}{C} - \underset{\sim}{A}) \\ \underset{\sim}{I} & \underset{\sim}{O} \end{bmatrix}, \qquad (6.38)$$

and to apply the theory of (6.31) we shall show how $\rho(\underset{\sim}{M})$ depends upon α. ($\underset{\sim}{M}$ is defined if $(\underset{\sim}{I}+\underset{\sim}{B})^{-1}$ exists.)

Since the inverse in the expression (6.38) is an inverse of a triangular matrix it can be readily determined. Thus the elements of M can be written down explicitly, and $\|\underset{\sim}{M}\|$ and $\rho(\underset{\sim}{M})$ can be found. To find $\rho(\underset{\sim}{M})$ we may, alternatively, solve a generalized eigenvalue problem. The eigenvalues of $\underset{\sim}{M}$ satisfy the equation

$$\det \begin{bmatrix} (\underset{\sim}{I}+\underset{\sim}{B})(1-\mu)-\underset{\sim}{C} & (\underset{\sim}{C}-\underset{\sim}{A}) \\ \underset{\sim}{I} & -\mu\underset{\sim}{I} \end{bmatrix} = 0 ,$$

and it can be shown quite readily that $\mu = 1$ and $\mu = 0$ are eigenvalues. The other eigenvalues are zeros (μ_1 and μ_2, say) of the quadratic

$$(1+\frac{17\alpha}{24}+\frac{\alpha^2}{8})\mu^2-(1-\frac{13\alpha}{12}+\alpha^2)\mu+(\frac{5\alpha}{24}-\frac{\alpha^2}{8})$$

where $\alpha = -\lambda h$. Note that with appropriate naming of the roots, $\lim \mu_1 = 0$ and $\lim \mu_2 = 1$, as $h \to 0$ with λ fixed.

We find that for $\rho(\underset{\sim}{M}) = 1$ we require $-2 \leq \lambda h \leq 0$, which is more restrictive than the result of Example 6.13.

A different analysis will now be given, in which the eigenvalue $\mu = 1$ will be replaced by a zero eigenvalue.

Let us write down the triangular system of equations as before, but difference successive rows (beginning at the fourth and fifth). The triangular array of coefficients is shown in (6.39).

$$\begin{array}{cc|cc|cc|cc}
1 & 0 & & & & & & \\
\frac{1}{2}\alpha & 1+\frac{1}{2}\alpha & & & & & & \\
\hline
\frac{1}{3}\alpha & \frac{4}{3}\alpha & 1+\frac{1}{3}\alpha & 0 & & & & \\
\frac{1}{24}\alpha & -\frac{5}{24}\alpha & -(1-\frac{19}{24}\alpha) & 1+\frac{3}{8}\alpha & & & & \\
\hline
-\frac{1}{24}\alpha & \frac{5}{24}\alpha & -\frac{11}{24}\alpha & -(1-\frac{23}{24}\alpha) & 1+\frac{1}{3}\alpha & 0 & & \\
0 & 0 & \frac{1}{24}\alpha & -\frac{5}{24}\alpha & -(1-\frac{19}{24}\alpha) & 1+\frac{3}{8}\alpha & & \\
\hline
0 & 0 & -\frac{1}{24}\alpha & \frac{5}{24}\alpha & -\frac{11}{24}\alpha & -(1-\frac{23}{24}\alpha) & 1+\frac{1}{3}\alpha & 0 \\
0 & 0 & 0 & 0 & \frac{1}{24}\alpha & -\frac{5}{24}\alpha & -(1-\frac{19}{24}\alpha) & 1+\frac{3}{8}\alpha
\end{array} \tag{6.39}$$

We write

$$\hat{\underset{\sim}{A}} = \begin{bmatrix} -\frac{1}{24}\alpha & \frac{5}{24}\alpha \\ 0 & 0 \end{bmatrix}, \quad \hat{\underset{\sim}{B}} = \begin{bmatrix} \frac{1}{3}\alpha & 0 \\ -(1-\frac{19}{24}\alpha) & \frac{3}{8}\alpha \end{bmatrix}, \quad \hat{\underset{\sim}{C}} = \begin{bmatrix} -\frac{11}{24}\alpha & -(1-\frac{23}{24}\alpha) \\ \frac{1}{24}\alpha & -\frac{5}{24}\alpha \end{bmatrix}$$

and $\tilde{\psi}_k = [\tilde{f}(2kh), \tilde{f}((2k+1)h)]^T$ to obtain, for $k \geq 1$,

$$(\underset{\sim}{I}+\hat{\underset{\sim}{B}})\tilde{\psi}_{k+1} + \hat{\underset{\sim}{C}}\tilde{\psi}_k + \hat{\underset{\sim}{A}}\tilde{\psi}_{k-1} = \Upsilon_{k+1} ,$$

where Υ_{k+1} is a vector whose components are derived from the values $g(rh)$. We set $\underset{\sim}{\Phi}_{k+1} = [\tilde{\psi}_{k+1}^T, \tilde{\psi}_k^T]^T$, and we obtain

$$\left[\begin{array}{c|c} \underset{\sim}{I}+\hat{\underset{\sim}{B}} & \underset{\sim}{0} \\ \hline \underset{\sim}{0} & \underset{\sim}{I} \end{array}\right] \begin{bmatrix} \tilde{\psi}_{k+1} \\ \tilde{\psi}_k \end{bmatrix} = \left[\begin{array}{c|c} -\hat{\underset{\sim}{C}} & -\hat{\underset{\sim}{A}} \\ \hline \underset{\sim}{I} & \underset{\sim}{0} \end{array}\right] \begin{bmatrix} \tilde{\psi}_k \\ \tilde{\psi}_{k-1} \end{bmatrix} + \begin{bmatrix} \Upsilon_{k+1} \\ \underset{\sim}{0} \end{bmatrix}$$

so that the stability properties of the block-by-block scheme for the computation of the vectors $\underset{\sim}{\Phi}_k$ $(k \geq 1)$ are obtained by considering the eigenvalues of the matrix

$$\hat{\underset{\sim}{M}} = \left[\begin{array}{c|c} \underset{\sim}{I}+\hat{\underset{\sim}{B}} & \underset{\sim}{0} \\ \hline \underset{\sim}{0} & \underset{\sim}{I} \end{array}\right]^{-1} \left[\begin{array}{c|c} -\hat{\underset{\sim}{C}} & -\hat{\underset{\sim}{A}} \\ \hline \underset{\sim}{I} & \underset{\sim}{0} \end{array}\right] ,$$

where we suppose $\det(\underset{\sim}{I}+\hat{\underset{\sim}{B}}) \neq 0$ (the scheme fails otherwise). We find that these eigenvalues are μ_1,μ_2 and $0,0$, where μ_1,μ_2 are the zeros of the quadratic

$$\mu^2\left(1+\frac{17}{24}\alpha+\frac{\alpha^2}{8}\right)-\mu\left(1-\frac{13}{12}\alpha+\alpha^2\right)+\left(\frac{5}{24}\alpha-\frac{\alpha^2}{8}\right).$$

Thus (see above), we have strict block-stability, damping, and error propagation which is bounded-in-norm, if $\rho(\hat{\underset{\sim}{M}}) < 1$, that is if $-2 < \lambda h < 0$. *

Remark. In the preceding Example we studied recurrence relations of the form

$$\underset{\sim}{X}_2\tilde{\underset{\sim}{\psi}}_{k+1}+\underset{\sim}{X}_1\tilde{\underset{\sim}{\psi}}_k+\underset{\sim}{X}_0\tilde{\underset{\sim}{\psi}}_{k-1}=\underset{\sim}{\nu}_{k+1}\quad (k\geq 1).$$

More generally, the study of various quadrature methods produces recurrences of the form

$$\sum_{p=0}^{s}\underset{\sim}{X}_p\tilde{\underset{\sim}{\psi}}_{k+p-1}=\underset{\sim}{\nu}_{k+s-1}.$$

(In the Example above, we had $s = 2$.) We assume that $\det(\underset{\sim}{X}_{k+s-1}) \neq 0$. Introducing the vector $\tilde{\underset{\sim}{\Phi}}_{k+1} = [\tilde{\underset{\sim}{\psi}}^T_{k+s-1},\ldots,\tilde{\underset{\sim}{\psi}}^T_k]^T$ we can write down a recurrence relation of the form $\tilde{\underset{\sim}{\Phi}}_{k+1} = \underset{\sim}{M}_*\tilde{\underset{\sim}{\Phi}}_k + \underset{\sim}{\nu}_{k+1}$, where the eigenvalues of $\underset{\sim}{M}_*$ are the roots of the equation

$$\det\left(\sum_{p=0}^{s}\underset{\sim}{X}_p\mu^p\right)=0.$$

This observation facilitates the study of stability regions, and permits the derivation of a number of general observations.

Example 6.17.

In the preceding Examples we indicated how it is possible to investigate the block stability of an algorithm for a sequence of vectors $\underset{\sim}{\Phi}_k,\underset{\sim}{\Phi}_{k+1},\ldots$ in which the last components of $\underset{\sim}{\Phi}_{r+1}$ are the leading components of $\underset{\sim}{\Phi}_r$. The same technique can be applied to the problem of Example 6.13 as indicated by Mayers (1962) in Fox (1962, p.180). *

6.3.2. *Relative error*

The previous discussion has been concerned with the possible growth of absolute error in computing with an algorithm. To discuss the effect of local truncation errors on the relative error in the approximations, the behaviour of the solution of the integral equation must be known, at least qualitatively. It is, of course, the behaviour of the true solution which determines whether or not any growth of errors in $\tilde{f}(kh)-f(kh)$ as k increases (with fixed h) gives a large relative error.

If we compare our problem with the situation in the numerical solution of initial-value problems in ordinary differential equations, we are led to the following considerations. Consider the particular equation

$$f(x) - \lambda\int_0^x f(y)\mathrm{d}y = 1 .$$

This is an integral equation formulation of the problem $f'(x) = \lambda f(x)$ with initial condition $f(0) = 1$. We set $\tilde{f}(0) = 1$, and to obtain a 'starting value' $\tilde{f}(h)$ we write

$$\tilde{f}(h) - \tilde{f}(0) = \frac{\lambda h}{2}\{\tilde{f}(h)+\tilde{f}(0)\} \simeq \int_0^h f'(t)\mathrm{d}t .$$

This gives us $\tilde{f}(h) = \gamma\tilde{f}(0)$, where $\gamma = (1+\frac{1}{2}\lambda h)/(1-\frac{1}{2}\lambda h)$. To compute $\tilde{f}(2h),\tilde{f}(3h),\ldots$ we write

$$f((k+1)h)-f((k-1)h) = \int_{(k-1)h}^{(k+1)h} f'(t)\mathrm{d}t$$

and use Simpson's rule to obtain the equation

$$\tilde{f}((k+1)h)-\tilde{f}((k-1)h) = \frac{h}{3}\{\lambda\tilde{f}((k-1)h)+4\lambda\tilde{f}(kh)+\lambda\tilde{f}((k+1)h)\}\ (k=1,2,3,\ldots) .$$

The values of $\tilde{f}(kh)$ $(k = 1,2,3,\ldots)$ computed using this procedure, are exactly the same as those computed using the weights in Table 6.2, as indicated in Example 6.3.

Now the true solution of the functional equation is $f(x) = e^{+\lambda x}$, but the computed solution has the form $\tilde{f}(kh) = At_1^k + Bt_2^k$, where, as $h \to 0$ with $x = kh$ fixed, $t_1^k \to e^{\lambda x}$ and $|t_2^k| \to e^{-\frac{1}{3}\lambda x}$. Thus whereas $A \to 1$ and $B \to 0$ as $h \to 0$, the computed solution $\tilde{f}(x)$ has for small $h > 0$ a component behaving like the true solution $f(x)$ and another component behaving like $e^{-\frac{1}{3}\lambda x}$ in its modulus; the latter is commonly called a 'parasitic solution'. It should be observed that, when $\lambda > 0$, the solution of the equations

$$f'(x) = \lambda f(x), \quad f(0) = 1,$$

is exponentially increasing and the parasitic solution may be of no great consequence (if its modulus is exponentially decreasing). Difficulty occurs where $\lambda < 0$ since the true solution then decays, but the parasitic solution grows exponentially and the relative error will increase. The effect of this parasitic solution should decrease as $h \to 0$ (since then $B \to 0$) but in practice the rounding errors of approximate arithmetic do not permit this limiting behaviour. (Even if $\tilde{f}(0)$ and $\tilde{f}(h)$ were chosen to ensure that $B = 0$, rounding error would introduce a component of the parasitic solution.)

The risks inherent in the solution of a first-order differential equation by a recurrence relation of order greater than one (of order 2 in the previous example) are often attributed to the presence of more complementary solutions in the difference scheme (t_1^k and t_2^k in the previous Example) than are present in the solution of the differential equation. There does not seem to be a *simple* analogue of this argument in the treatment of integral equations. Since the homogeneous Volterra equation of the second kind has no non-trivial solution there seems to be no exact analogue of the complementary function and we should instead consider the family of solutions introduced by small perturbation in the integral equation. (Observe also that the scheme associated with the weights in Table 6.2 in Example 6.3 was reduced to a three-term recurrence relation only as the result of differencing the original equations.) It is appropriate to ask whether the insight obtained in the previous example has any further extension. We shall see that phenomena similar to those occurring in differential equations occur generally in the numerical solution of Volterra equations, but they are best studied by a direct approach, with consideration of specific formulae and specific examples of integral equations.

We shall first consider our simple test equation

$$f(x) - \lambda\int_0^x f(y)\,dy = g(x) \;.$$

In Examples 6.13 and 6.14 we have seen how certain numerical methods for the approximate solution of this integral equation can be associated with schemes of the form

$$\underset{\sim}{\Phi}_{k+1} = \underset{\sim}{M}\underset{\sim}{\Phi}_k + \underset{\sim}{\delta}_k \;, \tag{6.40}$$

where the components of $\underset{\sim}{\Phi}_k$ are approximate values of the solution. In these Examples, we have

$$\underset{\sim}{\Phi}_k = [\tilde{f}(kmh),\ \tilde{f}((km+1)h),\ldots\ \tilde{f}((k+1)mh-h)]^T$$

for some integer m, and in the case where $g(x) \equiv 1$, $\underset{\sim}{\delta}_k = \underset{\sim}{0}$ $(k = 0,1,2,\ldots)$.

We have seen that when $\underset{\sim}{M}$ is similar to a diagonal matrix and $\underset{\sim}{M}\underset{\sim}{x}_i = \mu_i\underset{\sim}{x}_i$ $(i = 1,2,\ldots,m)$, where $\underset{\sim}{x}_1,\underset{\sim}{x}_2,\ldots,\underset{\sim}{x}_m$ are linearly independent, the solution of (6.40) is given by $\tilde{\underset{\sim}{\Phi}}_k = \sum_{i=1}^{m} \alpha_i^{[k]}\underset{\sim}{x}_i$, where $\alpha_1^{[k]},\alpha_2^{[k]},\ldots,\alpha_m^{[k]}$ are given by (6.33). Thus, in the notation of (6.33),

$$\underset{\sim}{\Phi}_k = \sum_{i=1}^{m} \{\mu_i^k\alpha_i^{[0]} + \sum_{r=0}^{k} \mu_i^r\beta_i^{[k-r]}\}\underset{\sim}{x}_i \;.$$

The component $\tilde{f}((km+j)h)$ of $\underset{\sim}{\Phi}_k$ is $\underset{\sim}{e}_{j+1}^T\underset{\sim}{\tilde{\Phi}}_k$ where $\underset{\sim}{e}_j^T$ is the jth row of the identity matrix $(j=1,2,\ldots,m)$. Hence

$$\tilde{f}((km+j)h) = \sum_{i=1}^{m} \chi_{ij}\{\mu_i^k\alpha_i^{[0]} + \sum_{r=0}^{k} \mu_i^r\beta_i^{[k-r]}\} \;,$$

where $\chi_{ij} = \underset{\sim}{e}_{j+1}^T\underset{\sim}{x}_i$.

Now let us consider the case where $g(x) \equiv 1$, $\underset{\sim}{\delta}_k = \underset{\sim}{0}$ $(k=0,1,2,\ldots)$. Then $\beta_i^{[k]} = 0$ $(i = 1,2,3,\ldots,m\ ,\ k = 0,1,2,\ldots)$ and we obtain a relation

$$\tilde{f}((km+j)h) = \sum_{i=1}^{m} \gamma_{ij}\mu_i^k \;,$$

where $\gamma_{ij} = \chi_{ij}\alpha_i^{[0]}$. Now the corresponding solution of the integral equation gives us $f((km+j)h) = e^{\lambda(km+j)h}$, and hence

$$f(\{(k+1)m+j\}h)/f((km+j)h) = e^{\lambda mk} .$$

If $\tilde{f}((km+j)h)$ simulates the true solution, we shall expect a similar relation for $\tilde{f}(\{(k+1)m+j\}h)/\tilde{f}((km+j)h)$. We shall examine this ratio.

Now let us *assume* that $|\mu_1| = \rho(\underset{\sim}{M})$ and $|\mu_i| < |\mu_1|$ $(i = 2,3,4,\ldots,m)$. Then

$$\tilde{f}((km+j)h) = \mu_1^k \sum_{i=1}^{m} \gamma_{ij}(\mu_i/\mu_1)^k ,$$

so that, as $k \to \infty$, $\tilde{f}((km+j)h) \sim \gamma_{1j}\,\mu_1^k$ (assuming that $\gamma_{1j} \neq 0$) and $\tilde{f}(\{(k+1)m+j\}h)/\tilde{f}((km+j)h) \to \mu_1$. Thus we expect that $\mu_1 \simeq e^{\lambda mh}$, and if $\lambda > 0$ we expect that $\rho(\underset{\sim}{M}) = \mu_1 > 1$. (In Example 6.13, $\mu_1 = \Gamma \simeq e^{2\lambda h}$ if $\lambda > 0$, $\mu_1 = 1$ if $\lambda < 0$.) In cases where Φ_k and Φ_{k+1} have overlapping components the expected value of μ_1 is modified.)

Consider, now, the numerical solutions of the general equation

$$f(x) - \lambda\int_0^x f(y)\,dy = g(x) ,$$

where $\lambda > 0$ and suppose that $g(x)$ is chosen so that $f(x)$ is non-increasing with x . We suppose that the computed values are obtained using (6.40), where $\rho(\underset{\sim}{M}) = \mu_1 \simeq e^{\lambda mh}$. Then we do not expect block-stability (since $e^{\lambda mh} > 1$) ; we expect $|\tilde{f}(kh)-f(kh)|$ to increase with k, and we expect computed values $\hat{f}(kh)$ using rounded arithmetic to display unstable growth of $|\hat{f}(kh)-\tilde{f}(kh)|$. Indeed, the relative errors also grow. However, such features are inherent in the integral equation and the discretized version inherits its instability from the ill-conditioning of the integral equation because μ_1 is an approximation to $e^{\lambda mh}$. Similar discussions are found in the treatment of differential equations.

We supposed, previously, that $\mu_1 \simeq e^{\lambda mh}$. Now suppose as before that $|\mu_1| = \rho(\underset{\sim}{M})$ and $|\mu_1| > |\mu_i|$ $(i = 2,\ldots,m)$, but suppose that $|\mu_1| \gg e^{\lambda mh}$. As before the jth component of Φ_k is $\sum_{i=1}^{m} \gamma_{i,j}\mu_i^k$.

Even if $\gamma_{i,j} = 0$, perturbations $\underset{\sim}{\delta}_k$ in the computational scheme will introduce non-zero values $\beta_i^{[k]}$ into the values computed for $\underset{\sim}{e}_j^T \underset{\sim}{\Phi}_k$ and the associated error and *relative error* will grow as k increases. Since this phenomenon is not associated directly with the integral equation, we may consider μ_1 to be a 'parasitic eigenvalue of $\underset{\sim}{M}$'.

We have noted that even when $\mu_1 \simeq e^{\lambda mh}$ relative errors may grow as k increases, when the integral equation is inherently 'unstable' or ill conditioned. We cannot, in general, expect better stability properties of the computational scheme than are inherent in the integral equation. On the other hand, if $\rho(\underset{\sim}{M}) > e^{\lambda mh}$ the stability properties inherent in the integral equation are worsened. We therefore introduce the following definition.

DEFINITION 6.9. *The m-component block-by-block method* $\tilde{\underset{\sim}{\Phi}}_{k+1} = \underset{\sim}{M}\tilde{\underset{\sim}{\Phi}}_k + \underset{\sim}{\delta}_k$, *(where* $\tilde{\underset{\sim}{\Phi}}_k$ *and* $\tilde{\underset{\sim}{\Phi}}_{k+1}$ *are non-overlapping) is a relatively stable discretization of the equation*

$$f(x) - \lambda\int_0^x f(y)\,dy = 1$$

if $\rho(\underset{\sim}{M}) \leq e^{\lambda mh}$.

(<u>Remark</u>. If the last component of $\underset{\sim}{\Phi}_k$ is always the first component of $\underset{\sim}{\Phi}_{k+1}$ we replace $e^{\lambda mh}$ by $e^{\lambda(m-1)h}$ in the definition above, and so on.)

Kobayasi (1966), Linz (1967*a*), and Noble (1969) have endeavoured to show that certain methods based on numerical integration tend to lead to schemes which are in a related sense 'unstable'. In their discussions there is no need to restrict attention to the simple equation

$$f(x) - \lambda\int_0^x f(y)\,dy = g(x)$$

since similar notions apply more generally. We shall, however, consider particular methods (based on Simpson's rule and the trapezium rule) to simplify the presentation.

Let us consider the use of the weights (6.23) for the linear integral equation

$$f(x) - \lambda\int_0^x K(x,y)f(y)\mathrm{d}y = g(x) .$$

We have, for $k = 0,1,2,\ldots,$

$$\tilde{f}(2kh) - \frac{2\lambda}{3}\{h \sum_{j=0}^{k}{}'' K(2kh,2jh)\tilde{f}(2jh)\} - \frac{4\lambda}{3}\{h \sum_{j=0}^{k-1} K(2kh,(2j+1)h)\tilde{f}((2j+1)h)\} = g(2kh) \tag{6.41}$$

(the sums being zero for $k = 0$) and

$$\tilde{f}((2k+1)h) - \frac{1}{2}\lambda h[K((2k+1)h,0)\tilde{f}(0)+K((2k+1)h,h)\tilde{f}(h)] - \frac{2\lambda}{3}\{h \sum_{j=0}^{k}{}'' K((2k+1)h,(2j+1)h)\tilde{f}((2j+1)h) - \frac{4\lambda}{3}\{h \sum_{j=1}^{k} K((2k+1)h,2jh)\tilde{f}(2jh)\} = g((2k+1)h) . \tag{6.42}$$

We suppose this system of equations has a solution (which is guaranteed if $h\lambda \|K(x,y)\|_\infty \leq 1)$, for $k = 0,1,2,\ldots$.

Now we examine the terms in braces in eqns (6.41) and (6.42). First, however, we shall write $\tilde{f}(2jh) = \tilde{\phi}(2jh)$ and $\tilde{f}((2j+1)h) = \tilde{\psi}((2j+1)h)$ $(j = 0,1,2,\ldots)$. With this mental substitution, the first term in braces in (6.41) appears as a repeated trapezium rule approximation, *using a step* $2h$, for $\int_0^{2kh} \frac{1}{2}K(2kh,y)\tilde{\phi}(y)\mathrm{d}y$, and the second such term in (6.41) appears as a repeated mid-point rule, *with step* $2h$, applied to $\int_0^{2kh} \frac{1}{2}K(2kh,y)\tilde{\psi}(y)\mathrm{d}y$. The corresponding terms in (6.42) bear a similar relation to $\int_h^{(2k+1)h} \frac{1}{2}K(x,y)\tilde{\phi}(y)\mathrm{d}y$, and $\int_h^{(2k+1)h} \frac{1}{2} K(x,y)\tilde{\psi}(y)\mathrm{d}y$, where $x = (2k+1)h$.

We shall therefore suppose (though the assumption requires justification) that the values of $\tilde{\phi}(x)$ and $\tilde{\psi}(x)$ behave, for small h, like the solutions $\phi(x)$, $\psi(x)$ of equations (from eqns (6.41) and (6.42)):

$$\phi(x) - \frac{1}{3}\lambda\int_0^x K(x,y)\phi(y)\mathrm{d}y - \frac{2}{3}\lambda\int_0^x K(x,y)\psi(y)\mathrm{d}y = g(x) + \xi_1(x) , \tag{6.43}$$

$$\psi(x) - \frac{1}{3}\lambda\int_0^x K(x,y)\psi(y)dy - \frac{2}{3}\lambda\int_0^x K(x,y)\phi(y)dy = g(x) + \xi_2(x) \ , \qquad (6.44)$$

where $\xi_1(x)$ and $\xi_2(x)$ diminish for fixed x as $h \to 0$. Thus we suppose $\tilde{\phi}(2jh) \simeq \phi(2jh)$ and $\tilde{\psi}((2j+1)h) \simeq \psi((2j+1)h)$ $(j = 0,1,2,\ldots)$.

If eqns (6.43) and (6.44) are added together we find

$$\{\phi(x)+\psi(x)\}-\lambda\int_0^x K(x,y)\{\phi(y)+\psi(y)\}dy = 2g(x)+\xi_1(x)+\xi_2(x) \quad ,$$

and if $R(x,y;\lambda)$ is the resolvent kernel

$$\phi(x)+\psi(x) = 2g(x)+\xi_1(x)+\xi_2(x)+\lambda\int_0^x R(x,y;\lambda)\{2g(y)+\xi_1(y)+\xi_2(y)\}dy. \qquad (6.45)$$

Similarly, on subtracting (6.44) from (6.43), we find

$$\{\phi(x)-\psi(x)\} + \frac{1}{3}\lambda\int_0^x K(x,y)\{\phi(y)-\psi(y)\}dy = \xi_1(x)-\xi_2(x)$$

and

$$\phi(x)-\psi(x) = \xi_1(x)-\xi_2(x) - \frac{1}{3}\lambda\int_0^x R(x,y;-\tfrac{1}{3}\lambda)\{\xi_1(y)-\xi_2(y)\}dy. \qquad (6.46)$$

Thus, (6.45) and (6.46) can be added and subtracted to give $\phi(x)$ and $\psi(x)$, respectively, on halving. In particular we find

$$\tilde{f}(2jh) = \tilde{\phi}(2jh) \simeq \phi(2jh) \ ,$$

where

$$\begin{aligned}\phi(x) &= g(x)+\lambda\int_0^x R(x,y;\lambda)g(y)\,dy + \\ &+ \xi_1(x) + \frac{1}{2}\lambda\int_0^x R(x,y;\lambda)\{\xi_1(y)+\xi_2(y)\}dy - \qquad (6.47)\\ &- \frac{1}{6}\lambda\int_0^x R(x,y;-\tfrac{1}{3}\lambda)\{\xi_1(y)-\xi_2(y)\}dy \ .\end{aligned}$$

In (6.47), the first two terms represent the true solution $f(x)$, the second terms represent the form of perturbation inherently expected from the form of the integral equation and the last term is a 'parasitic term' involving the resolvent at a parameter value $-\frac{1}{3}\lambda$ which differs from the value λ in the original equation. It is a parasitic term which can cause practical 'instability', as can be seen in the case $K(x,y) = 1$ for $0 \le y \le x$, zero otherwise. In this case $R(x,y;\mu) = e^{\mu(x-y)}$ and for $\lambda < 0$ the resolvent corresponding to $\mu = -\frac{1}{3}\lambda$ gives rise to an increasing component. (The relative error in the computed solution may not grow even in these conditions however, since the behaviour of $f(x)$ is determined by $g(x)$ in addition to the resolvent. On the other hand, in the case where $\lambda > 0$ and the parasitic component is decaying, it may still dominate the true solution $f(x)$ and lead to a growth in the relative error.)

The previous argument has been partly heuristic. There is however, a rigorous foundation for the above ideas, and the practical conclusion generally found in the literature is that for stability reasons it is unwise to insert special rules over *leading* subintervals (as with the trapezium rule over $[0,h]$ in the scheme of Table 6.2). The question to be asked, therefore, is whether the scheme of Table 6.3 -which has the same order of local truncation error as Table 6.2 -has better stability characteristics, in the sense that we are investigating them here. The answer is affirmative.

For an analysis of the choice of weights of Table 6.3 eqns (6.41) and (6.43) are unchanged, but there is an obvious revision of (6.42), which leads to the equation

$$\psi(x) - \frac{1}{3}\lambda\int_0^x K(x,y)\phi(y)\,dy - \frac{2}{3}\lambda\int_0^x K(x,y)\psi(y)\,dy = g(x)+\xi_2(x) \quad (6.48)$$

in place of (6.44), and we now find

$$\phi(x)-\psi(x) = \xi_1(x)-\xi_2(x) \quad (6.49)$$

and

$$\{\phi(x)+2\psi(x)\}-\lambda\int_0^x K(x,y)\{\phi(y)+2\psi(y)\}dy = 3g(x)+\xi_1(x)+2\xi_2(x)\,, \quad (6.50)$$

from which $\phi(x)$ and $\psi(x)$ can be written down in terms of $g(x)$, $\xi_1(x)$, $\xi_2(x)$, and $R(x,y;\lambda)$. There is no parasitic term, in the sense that $R(x,y;\mu)$ is involved only for $\mu = \lambda$, and any instability is therefore inherent in the original equation.

Our discussion of this topic, and its extension to non-linear integral equations, will be resumed later. We conclude this section by introducing two definitions, and a theoretical result.

DEFINITION 6.10. *The 'repetition factor' associated with the approximations*

$$\int_0^{kh} \phi(y)\mathrm{d}y \simeq \sum_{j=0}^{k} w_{kj}\phi(jh) \quad (k = 1,2,3,\ldots)$$

is the smallest integer r *such that* $w_{s+r,j} = w_{s,j}$ $(j=p,(p+1),\ldots,(s-q))$ *where* p *and* q *are integers independent of* k .

Example 6.18.
The repetition factors of the weights in Table 6.3 and 6.5 is 1 and the repetition factor of Table 6.2 and of Table 6.4 is 2. *

In the spirit of Noble (1969*a*) we introduce the following definition.

DEFINITION 6.11. *A scheme for the evaluation of an approximate solution of a Volterra integral equation is called an unstable discretization of the integral equation if the error in the computed solution has dominant spurious components introduced by the numerical scheme (and not inherent in the integral equation).*

The above definition is a little vague because of the use of the terms 'spurious' and 'dominant'. However, we have seen in the discussion of the equation $f(x) - \lambda\int_0^x f(y)\mathrm{d}y = g(x)$ that when the approximate solution contains a component which behaves like the solution of an equation of the form

$$\phi(x) - \mu\int_0^x \phi(y)\mathrm{d}y = \gamma(x)$$

and $\mu > \lambda$, such a component is spurious and it can be dominant (that is, it can grow faster than the true solution $f(x)$) in the case where $g(x)$ and $\gamma(x)$ are suitably related.

Linz (1967*a*) conjectured that methods with repetition factor 1 are not unstable in the sense above. Kobayasi (1966) gives a rigorous treatment of such a theory whilst Noble (1969*a*) presents insight into the problem along lines indicated here. The theory lends backing to the view that quadrature methods *with a repetition factor greater than one should not generally*¶ *be employed.* See Baker and Keech (1975,1977).

However, it must be stated that methods with repetition factor 1 do not yield block-stable schemes for all values of h. Further, such schemes can indeed be relatively unstable in the sense of *our* definition above, with $\rho(\underset{\sim}{M}) > e^{\lambda mh}$. In the latter case it frequently happens that

$$\lim_{h \to 0} \{\rho(\underset{\sim}{M})/e^{\lambda mh}\} = 1 .$$

It may be noted, indeed, that the theory of Kobayasi, and its interpretations due to Linz and Noble, is asymptotic (as $h \to 0$) in nature, and may fail to provide a good model for practical purposes when the step-size h is large. The limitations of a discussion in terms of a matrix $\underset{\sim}{M}$ (with the consequent restriction to a linear equation) will also be apparent.

Example 6.19.
From the analysis of Example 6.4 and Example 6.13 we note that the stability properties of the scheme associated with the weights in Table 6.3 are determined by the size of

$$\Gamma = (1 + \tfrac{7}{6}\lambda h + \tfrac{1}{2}\lambda^2 h^2)/(1 - \tfrac{5}{6}\lambda h + \tfrac{1}{6}\lambda^2 h^2) .$$

Comparison of $|\Gamma|$ with 1 and $\exp 2\lambda h$ shows that block-stability occurs if $-6 \le \lambda h \le 0$ and a relatively stable discretization of

$$f(x) - \lambda\int_0^x f(y)\,dy = g(x)$$

occurs with the block-by-block method of Example 6.13 if $\lambda h \ge 3{\cdot}11$ (approximately). The choice of Table 6.3 is, according to the concepts of Kobayasi, Noble, and Linz, not unstable.

The choice of Table 6.2 gives a scheme (Example 6.14) which is nowhere block-stable but does display relative stability if (approximately) $3{\cdot}44 \le \lambda h$. The use of Table 6.2 is not recommended because its

¶ A scheme with repetition factor 2 and non-trivial stability region exists.

repetition factor is two, and it can be seen that a small step h is not consistent with either relative or absolute stability.

The scheme associated with Table 6.5 (see Example 6.16) is block-stable if $-2 \le \lambda h \le 0$ and gives a relatively stable discretization if $0{\cdot}23 \le \lambda h$ (approximately). This scheme has a repetition factor unity. *

6.4. *Quadrature methods for non-linear integral equations*

The discussion of section 6.2 was concerned with applying a certain class of quadrature methods to the numerical solution of linear Volterra integral equations of the second kind. In this section we shall consider what modifications are necessary to apply the same (or similar) ideas to the approximate solution of the non-linear Volterra equation

$$f(x) - \int_0^x F(x,y;f(y))\,dy = g(x) \quad (0 \le x \le X) . \tag{6.51}$$

We shall require the existence of a unique solution to this non-linear equation, and we suppose, for the present, that $F(x,y;v)$ and $g(x)$ are smooth functions. Indeed, if

(a) $g(x) \in C[0,X]$;

(b) $F(x,y;v)$ is uniformly continuous in x,y for all finite v and $0 \le x \le y \le X$;

(c) there exists a constant L such that

$$|F(x,y;v_1) - F(x,y;v_2)| \le L|v_1 - v_2| \tag{6.52}$$

uniformly for $0 \le x \le y \le X$;

then eqn (6.51) has a unique continuous solution on $[0,X]$. For proofs of this result see, for example, Davis (1962) or Tricomi (1957).

Now the analogue of eqn (6.9) is

$$\tilde{f}(kh) - \sum_{j=0}^{k} w_{k,j} F(kh,jh;\tilde{f}(jh)) = g(kh) \quad (k = 0,1,2,\ldots) , \tag{6.53}$$

where $w_{0,0}$ is zero, and $\tilde{f}(0) = g(0)$.† The possible choice of the

† In practice, the use of (6.53) for small values of k may be replaced by the use of a starting procedure (see section 6.5).

weights $w_{k,j}$ is the same as in the case of the linear equation (though we shall consider some additions to those given in section 6.2), and the solution of the successive equations (6.53) for $k = 1,2,3,\ldots$ can again proceed in a step-by-step fashion. Thus, with $\tilde{f}(0) = g(0)$ known we have, for $k = 1$,

$$\tilde{f}(h)-w_{1,1}F(h,h;\tilde{f}(h)) = g(h) + w_{1,0}F(h,0;\tilde{f}(0)) .$$

The latter equation is in general a *non-linear* equation for $\tilde{f}(h)$ (though the equation is linear in the case $w_{1,1} = 0$). If $\tilde{f}(h),\tilde{f}(2h),\ldots,$ $\tilde{f}((k-1)h)$ have been found (in sequence) we have the equation

$$\tilde{f}(kh)-w_{k,k}F(kh,kh;\tilde{f}(kh)) = g(kh)+ \sum_{j=0}^{k-1} w_{k,j}F(kh,jh;\tilde{f}(jh)) \tag{6.54}$$

for solution at the next step. Thus, in contrast with the case of a non-linear Fredholm equation (where, in general, we have to solve a *system* of simultaneous non-linear equations), the approximate solution of the non-linear Volterra equation proceeds by means of a sequence of steps in each of which we solve a *single* non-linear equation in a single unknown. It is this feature which makes the step-by-step methods computationally more attractive than the expansion methods which could be derived by regarding the Volterra equation as a Fredholm equation on a finite range $[0,X]$.

Eqn (6.54) may be written symbolically in the form†

$$\Phi_k(\tilde{f}(kh)) = 0$$

where $\Phi_k(z)$ is determined by $F(x,y;v)$ and the values $\tilde{f}(0),\tilde{f}(h),\ldots,$ $\tilde{f}((k-1)h)$. We may attempt to solve this equation by any of the standard techniques such as the secant method, regula falsi method, or by the Newton-Raphson method if $\Phi_k'(z)$ is readily available.

In practice an alternative technique appears to be popular, and this we now describe. For convenience we write $\tilde{f}_k$ for $\tilde{f}(kh)$ and γ_k for the right-hand term in (6.54). Then the equation $\Phi_k(\tilde{f}_k) = 0$

† Our notation now differs slightly from the use of Φ_k in section 6.3.

can be written

$$\tilde{f}_k = \gamma_k + w_{k,k}F(kh,kh;\tilde{f}_k) \tag{6.55}$$

A 'natural' iteration for the solution of (6.55) is obtained if we choose $\tilde{f}_k^{[0]}$ and write

$$\tilde{f}_k^{[r]} = \gamma_k + w_{k,k}F(kh,kh;\tilde{f}_k^{[r-1]}) \; (w_{k,k} \neq 0) \quad (r = 1,2,3,\ldots) \quad . \tag{6.56}$$

In this way we generate a sequence of approximations $\tilde{f}_k^{[r]}$ to $\tilde{f}_k$, and this sequence converges to $\tilde{f}_k$ if h is sufficiently small, if (as occurs with all reasonable methods) $w_{k,k} = O(h)$, and the condition (6.52) is satisfied. Observe that, in (6.56), γ_k is evaluated once for a particular value of k, and it does not vary with r.

The iteration (6.56) is similar to the use of a corrector formula in the predictor-corrector methods for the solution of initial-value problems in ordinary differential equations. Many features of such methods have their analogues here. Thus we need not iterate, in (6.56), until numerical convergence appears to have been achieved. We may instead choose m ($=m_k$, say, if the value depends upon k) and accept $\tilde{f}_k^{[m]}$ as the approximate value $\tilde{f}(kh)$.

Remark. At the next stage, when computing $\tilde{f}((k+1)h)$, we require the value of γ_{k+1} and this involves (see eqn (6.54)) a contribution $w_{k+1,k}F((k+1)h,kh,\tilde{f}(kh))$. In the special case where $F(x,y;v)$ is independent of x we may choose to substitute $w_{k+1,k}F((k+1)h,kh;\tilde{f}_k^{[m-1]})$ for this value (since this function value has already been obtained in (6.56) when evaluating $\tilde{f}_k^{[m]}$). More generally we shall have to evaluate the new value $F((k+1)h,kh;\tilde{f}_k^{[m]})$. Thus there are *special* cases where there is an analogue of the choice between (see Lambert 1973) $P(EC)^m$ and $P(EC)^mE$ methods for ordinary differential equations. The stability properties of such methods are not the same. *

In the use of the iteration (6.56) for the approximate value $\tilde{f}(kh)$, we require a value $\tilde{f}_k^{[0]}$ to start the iteration. Such a starting value may be obtained on setting $\tilde{f}_k^{[0]} = \tilde{f}((k-1)h)$, but the coarser the estimate used, the greater the number of iterations will be required in (6.56) (and such iterations tend to be computationally expensive if

$F(x,y;v)$ is a complicated function).

The starting approximation $\tilde{f}_k^{[0]} = \tilde{f}((k-1)h)$ can be employed; it results from the assumption that $\tilde{f}(x)$ is constant in $[(k-1)h,kh]$. Possibly more accurate initial estimates result from the hypothesis that $\tilde{f}(x)$ is linear in the interval $[(k-2)h,kh]$, or the hypothesis that $\tilde{f}(x)$ is quadratic in $[(k-3)h,kh]$. The former hypothesis leads to the starting value

$$\tilde{f}_k^{[0]} = 2\tilde{f}((k-1)h)-\tilde{f}((k-2)h) \quad (k \geq 2) , \tag{6.57}$$

whilst the latter leads to the starting value

$$\tilde{f}_k^{[0]} = 3\{\tilde{f}((k-1)h)-\tilde{f}((k-2)h)\} + \tilde{f}((k-3)h) \quad (k \geq 3) . \tag{6.58}$$

The formulae (6.57) and (6.58) appear to have much to recommend them in their computational simplicity, and they can be modified to give more accurate formulae. (It will be noticed that if $\tilde{f}(rh)$ is exact for $r = k-1,k-2,k-3$ then the error in $\tilde{f}_k^{[0]}$ in (6.58) is $O(h^3)$ provided we have sufficient differentiability. $O(h^r)$ accuracy can be attained if we use a formula employing $\tilde{f}((k-1)h),\ldots,\tilde{f}((k-r)h)$.)

An alternative type of method for computing a value $\tilde{f}_k^{[0]}$ is given by the use of a 'predictor formula' defined by the substitution in (6.54) of weights

$$w^p_{k,0}, w^p_{k,1},\ldots,w^p_{k,k-1} \quad (w^p_{k,k} = 0),$$

which give an approximation

$$\int_0^{kh} \phi(y)\,dy \simeq \sum_{j=0}^{k-1} w^p_{k,j}\phi(jh) . \tag{6.59}$$

We then set

$$\tilde{f}_k^{[0]} = g(kh)+ \sum_{j=0}^{k-1} w^p_{k,j}F(kh,jh;\tilde{f}(jh)) . \tag{6.60}$$

<u>Example 6.20.</u>

We suppose, by way of illustration, that $w_{k,j} = h$ $(j = 1,2,3,\ldots,(k-1))$ and $w_{k0} = w_{kk} = \frac{1}{2}h$. An appropriate set of weights w^p_{kj} for the predictor formula is then given by setting $w^p_{kj} = h$ $(j = 1,2,3,\ldots,(k-3))$, $w^p_{k,k-2} = w^p_{k,0} = \frac{1}{2}h$, and $w^p_{k,k-1} = 2h$. The associated approximation (6.59) results from the repeated trapezium rule applied over $[0,(k-2)h]$ and the mid-point rule with step $2h$ applied over $[(k-2)h,kh]$. The local truncation error in the formula is $O(h^2)$ for fixed kh as $h \to 0$. *

The quadrature rule (6.59) is open and, here, it is to be expected that the quadrature formula will be obtained by using a composite (repeated) rule over some interval $[0,sh]$ and some simple 'open' rule over $[sh,kh]$, where $s = s_k$. This would not appear to be an unduly 'unstable' combination and it has certain computational advantages.

By way of an example, the use of weights of Table 6.3 in (6.54) could perhaps be matched by a choice of weights $w^p_{k,j}$ of the form displayed in Table 6.9.

TABLE 6.9

k \ j	0	1	2	3	4	5	6	Local truncation error
1	h	0						$O(h^2)$
2	0	$2h$	0					$O(h^3)$
3	0	$\frac{3}{2}h$	$\frac{3}{2}h$	0				$O(h^3)$
4	0	$\frac{8}{3}h$	$-\frac{4}{3}h$	$\frac{8}{3}h$	0			$O(h^5)$
or 4	$\frac{1}{3}h$	$\frac{4}{3}h$	$\frac{1}{3}h$	$2h$	0			$O(h^3)$
5	$\frac{1}{3}h$	$\frac{4}{3}h$	$\frac{1}{3}h$	$\frac{3}{2}h$	$\frac{3}{2}h$	0		$O(h^3)$
6 (say)	$\frac{1}{3}h$	$\frac{4}{3}h$	$\frac{2}{3}h$	$\frac{4}{3}h$	$\frac{1}{3}h$	$2h$	0	$O(h^3)$

The local truncation errors are given as $h \to 0$ with k fixed, and assuming (as always, unless otherwise stated) sufficient differentiability.

Remark. The values of weights $w^p_{k,j}$ in the second, third, and fourth rows of Table 6.9 are those of the open Newton-Cotes formulae, using a rule with degree of precision $k-2$ (if k is odd) and $k-1$ (if k is even). The weights of these rules reappear for $k = 5,6,7;8,9,10, \ldots$ in combination with the (repeated) Simpson's rule used over the leading subintervals. For $k = 1$ there is obviously no non-trivial open Newton-Cotes rule, and we use a 'rectangular rule' employing the left end-point

$$\int_0^h \phi(y)\mathrm{d}y \simeq h\phi(0) .$$

This rule is 'half-open' being closed at the left end-point. *

The choice of weights in Table 6.9 employs the open Newton-Cotes rules where possible (that is when $k \neq 1$) but we could also choose to construct the half-open formulae of the type

$$\int_0^{ph} \phi(y)\mathrm{d}y \simeq \sum_{j=0}^{p-1} \omega_j \phi(jh) \tag{6.61}$$

for use in place of these rules. Thus the degree of precision of the approximation

$$\int_0^{3h} \phi(y)\mathrm{d}y \simeq \tfrac{3}{4}(h\phi(0)+3h\phi(2h)) , \tag{6.62}$$

is 2, and the local truncation error with (6.62) is $O(h^4)$. In general the weights in (6.61) would be chosen to give a degree of precision $(p-1)$. For even values of p the rules reduce to the corresponding open Newton-Cotes formulae, but the other rules of type (6.61) appear to offer some advantage in their increased order of accuracy.

We have suggested that the weights w_{kj}^p may be obtained by applying a composite rule over some interval $[0,s_k h]$ followed by an open (or half-open) rule over the interval $[s_k h, kh]$. This has been the case in the two examples we have given, and we have observed that a generally satisfactory method (section 6.3) for constructing the weights w_{kj} of the corrector follows along similar lines (the rule over the interval $[s_k h, kh]$ being closed, however).

It is of some computational advantage to arrange that

$$w_{k,j}^p = w_{k,j} \quad (j = 0,1,2,\ldots, \sigma_k \leq k-1)$$

(where it may be possible, in the situation above, to ensure that $\sigma_k = s_k$). The terms in the sum

$$\gamma_k^* = g(kh) + \sum_{j=0}^{\sigma_k} w_{k,j}^p F(kh,jh;\tilde{f}(jh)) \tag{6.63}$$

then occur in the expression for γ_k in the correction formula (6.56). We may record the value of γ_k^* when calculating the predicted value

$$\tilde{f}_k^{[0]} = \gamma_k^* + \sum_{j=\sigma_k+1}^{k-1} w_{k,j}^p F(kh,jh;\tilde{f}(jh)) \tag{6.64}$$

for the subsequent calculation of

$$\gamma_k = \gamma_k^* + \sum_{j=\sigma_k+1}^{k-1} w_{k,j} F(kh,jh;\tilde{f}(jh)) \tag{6.65}$$

for use in (6.56). If σ_k is close to $k-1$, this procedure becomes competitive with the use of an expression such as (6.58); the exact comparison depends upon the complexity of $F(x,y;v)$ and the number of corrections to be made in each case.

Example 6.21.
Garey (1972) suggests a scheme based on the following weights:

(a) For $k = 2r$, where r is integer,
$4w_{k,0} = 4w_{k,0}^p = w_{k,k-3} = w_{k,k-3}^p = w_{k,j} = w_{k,j}^p = 2w_{k,j+1}$
$= 2w_{k,j+1}^p = \frac{4}{3}h$, $(j = 1,3,5,\ldots,(k-5))$ and
$6w_{k,k-2}^p = w_{k,k-1}^p = 2h$, $w_{k,k}^p = 0$ whilst $2w_{k,k-2} = w_{k,k-1}$
$= 4w_{k,k} = \frac{4}{3}h$, if k is sufficiently large.

(b) For $k = 2r+1$, where r is integer:
$4w_{k,0} = 4w_{k,0}^p = w_{k,k-4} = w_{k,k-4}^p = w_{k,j} = w_{k,j}^p = 2w_{k,j+1}$
$= 2w_{k,j+1}^p = \frac{4}{3}h$, $(j = 1,3,5,\ldots,(k-6))$, whilst
$w_{k,k-3}^p = \frac{1}{3}$, $w_{k,k-2}^p = w_{k,k-1}^p = \frac{3}{2}h$ $(w_{k,k}^p = 0)$ and $w_{k,k-3} = \frac{17}{24}h$,
$w_{k,k-2} = w_{k,k-1} = 3w_{k,k} = \frac{9}{8}h$, if k is sufficiently large.

The predictor weights above are a variation on those in Table 6.9 and the corrector is a combination of Simpson's rule and the $\frac{3}{8}$ths rule. Garey (1972) gives another scheme, for comparison. Applied to the equation

$$f(x) = \exp(x^2) - x + x \exp \tfrac{1}{2}x^2 - \int_0^x xy\sqrt{f(y)}\,dy \ ,$$

which has the solution $f(x) = \exp x^2$, using a step $h = \frac{1}{10}$, Garey's results (using the weights displayed here) yield the relative errors which we display below. (It should be mentioned that the results were obtained by a prediction followed by a *single* correction, and more accurate results would probably be obtained on iterating to convergence.) With $h = 0{\cdot}1$, relative errors in the solution values at $x = 1,2,3,4$ are respectively $8{\cdot}5\times10^{-6}$, $1{\cdot}2\times10^{-5}$, $2{\cdot}8\times10^{-6}$, and $2{\cdot}6\times10^{-7}$ respectively. Whereas relative errors decrease with increasing x, the absolute errors increase. *

Example 6.22.

We employed the corrector defined by the weights in Table 6.5 (which corresponds to the use of the repeated Simpson's rule with the $\frac{3}{8}$th rule, where appropriate) to determine the approximate solution of the simple test equation

$$f(x) - \int_0^x e^{x-y}\{f(y)\}^2 dy = 2 - e^x \ .$$

The solution is $f(x) = 1$. The approximate solution was computed using a simple predictor formula and iterating to convergence. Using a step $h = \frac{1}{20}$, the following errors resulted:

	absolute errors		absolute errors
$x = 0{\cdot}25$	$2{\cdot}5\times10^{-6}$	$x = 0{\cdot}75$	$1{\cdot}2\times10^{-5}$
$x = 0{\cdot}50$	$7{\cdot}7\times10^{-6}$	$x = 1{\cdot}00$	$2{\cdot}4\times10^{-5}$

The use as a corrector of the repeated $\frac{3}{8}$th rule terminating with either (see Table 6.6) Simpson's rule or a double Simpson's rule, where appropriate, gave the following absolute errors:

$x = 0{\cdot}25$	$2{\cdot}3\times10^{-6}$	$x = 0{\cdot}75$	$1{\cdot}0\times10^{-5}$
$x = 0{\cdot}50$	$4{\cdot}9\times10^{-6}$	$x = 1{\cdot}00$	$2{\cdot}2\times10^{-5}$ *

In practice, if a predictor-corrector method is employed to iterate to convergence in the corrector, the requirement of convergence imposes a restriction on the maximum size of h which can be employed. Other techniques of root-finding (such as the secant method or Newton's method) may be employed for the solution of the closed formula associated with the corrector, but stability and accuracy requirements also impose restrictions on the permissible size of h. It is unwise to permit too large a step size, in general.†

It is, of course, quite proper to enquire whether closed (corrector) formulae are appropriate for the numerical solution of non-linear Volterra equations, since they must be solved iteratively. An alternative might be to use the weights w^{p}_{kj} of a predictor formula and solve the corresponding explicit equations for values $\tilde{f}(kh) \simeq f(kh)$. In general, however, closed quadrature formulae are more accurate than their counterparts of open type (though there are exceptions to this rule). Moreover, if we compare the situation with the numerical solution of the initial-value problem for the equation $\tilde{f}'(x) = F(x,f(x))$ we see that implicit (closed) formulae are frequently preferred over explicit (open) formulae. Whilst we consider that the case has never been conclusively made, we know of no suggestion yet being made in the literature to the effect that a predictor formula (defined by weights w^{p}_{kj}) should be employed alone for the step-by-step solution of a Volterra equation.

On the other hand, the use of a predictor-corrector technique in which the correction is made a finite number (m_k, say) of times gives rise in effect to an *explicit* (though complicated) method for determining the values $\tilde{f}(0),\tilde{f}(h),\ldots,\tilde{f}(kh)$. (We shall also see, later, that the methods of Runge-Kutta type are defined by formulae for the explicit calculation of the approximate values.)

---

† The choice of a large step-size h may result in the non-uniqueness of the approximate values $\{\tilde{f}(kh)\}$.

6.4.1.

Before concluding this section, let us examine the local truncation error associated with a corrector formula which is paired with a given predictor formula, and in which the iteration with the corrector is performed a fixed number of times. We shall consider the order of accuracy as $h \to 0$ with kh fixed. We *assume* that $\tilde{f}(h), \tilde{f}(2h), \ldots, \tilde{f}((k-1)h)$ are $O(h^p)$ approximations to the corresponding true function values, and $\tilde{f}(0) = f(0)$. Suppose that we set

$$\tilde{f}^{[0]}(kh) = \sum_{j=0}^{k-1} w_{k,j}^{p} F((kh, jh; \tilde{f}(jh)) + g(kh) \quad ,$$

where

$$\int_0^{kh} F(kh, y; f(y))\,dy = \sum_{j=0}^{k-1} w_{k,j}^{p} F(kh, jh; f(jh)) + O(h^q)$$

(as $h \to 0$ with kh fixed). We shall suppose that $q \le p$, and

$$\sum_{j=0}^{k-1} |w_{k,j}^{p}| \le kh\Omega \quad ,$$

where $\Omega < \infty$ and, for an explicit formula normally used as a predictor, $w_{k,k}^{p} = 0$.

We suppose, also, that $F(x,y;v)$ satisfies the Lipschitz condition

$$|F(x,y;v_1) - F(x,y;v_2)| \le L|v_1 - v_2|$$

which ensures that, for $j \le k-1$,

$$F(kh, jh; f(jh)) - F(kh, jh; \tilde{f}(jh)) = O(h^p) \,.$$

Thus we readily obtain

$$\tilde{f}^{[0]}(kh) = \sum_{j=0}^{k-1} w_{k,j}^{p} F(kh, jh; f(jh)) + g(kh) + \{\sum_{j=0}^{k} |w_{k,j}^{p}|\} O(h^p)$$

$$= f(kh) + O(h^q) + O(h^p) = f(kh) + O(h^q)$$

(as $h \to 0$ with kh fixed).

Now suppose that

$$\tilde{f}^{[r]}(kh) = \sum_{j=0}^{k} w_{k,j} F(kh,jh;\tilde{f}^{[r-1]}(kh))+g(kh) \quad (r = 1,2,3,\ldots,m) ,$$

where

$$\int_0^{kh} F(kh,y;f(y))\mathrm{d}y = \sum_{j=0}^{k} w_{k,j} F(kh,jh;f(jh))+O(h^p) ,$$

(as $h \to 0$ with kh fixed). Then we can establish by induction that

$$\tilde{f}^{[r]}(kh)-f(kh) = O(h^s) \quad ,$$

where $s = \min(p,q+r)$. In particular, when $r = p-q$, we have

$$\tilde{f}^{[p-q]}(kh)-f(kh) = O(h^p) .$$

The point of this discussion is to indicate that any predictor can be employed, provided it is corrected sufficiently often, without altering the order of the error as $h \to 0$. However, in a practical algorithm the choice of predictor may effect the efficiency, accuracy, and stability of a scheme.

6.5. *Starting values*

We have indicated (in sections 6.2 and 6.4) methods of choosing weights w_{kj} governing the construction of approximations satisfying the equations

$$\tilde{f}(kh)-\lambda \sum_{j=0}^{k} w_{k,j} K(kh,jh)\tilde{f}(jh) = g(kh) \tag{6.66}$$

in the case of the equation

$$f(x)-\lambda\int_0^x K(x,y)f(y)\mathrm{d}y = g(x) \quad ,$$

and

$$\tilde{f}(kh)- \sum_{j=0}^{k} w_{k,j} F(kh,jh;\tilde{f}(jh)) = g(kh) \tag{6.67}$$

in the case of the non-linear equation

$$f(x) - \int_0^x F(x,y;f(y))\,dy = g(x) \ . \qquad (6.68)$$

We are concerned, in this section, with a modification of the schemes of sections 6.2 and 6.4 necessary to develop a practical algorithm. As earlier, all functions involved are assumed highly smooth.

A glance at the formulae (6.66) and (6.67) will reveal that the accuracy obtainable in evaluating $\tilde{f}(kh)$ depends in part on the accuracy obtained when computing $\tilde{f}(h), \tilde{f}(2h), \ldots, \tilde{f}(rh)$ $(r < k-1)$. Further, a glance at the weights w_{kj} in Table 6.5, for example, will show that it is difficult to achieve high-order local truncation errors (that is, local truncation errors which are $O(h^p)$ for large p) with a single quadrature rule, until k is itself large. Consequently the errors in $\tilde{f}(kh)$ for small k may be expected to be somewhat large (and, we suggest, larger than needs to be tolerated). It is our purpose here to show how this situation can be remedied, and to indicate, more precisely, the reasons for doing so.

Perhaps it is appropriate at this stage to state a theorem which (*inter alia*) indicates the effect of 'starting errors', that is, the effect of errors in $\tilde{f}(0), \tilde{f}(h), \ldots, \tilde{f}(sh)$ for small values of s . We can then determine what size of starting errors is permissible. We shall give the proof of the theorem in a later section.

THEOREM 6.1. *Suppose that the values* $\tilde{f}(kh)$ *satisfy eqn (6.67) and, in (6.68), suppose that for some finite value* L , *independent of* x,y,

$$|F(x,y;v_1)-F(x,y;v_2)| \le L|v_1-v_2| \qquad (\ 0 \le y \le x \le X\ )$$

when v_1, v_2 *are finite. If , for some fixed* $q \le N$,

$$\tau(h) \equiv \tau_N(h) = \max_{r=q,q+1,\ldots,N} \left| \int_0^{rh} F(rh,y;f(y))\,dy - \sum_{j=0}^{r} w_{rj} F(rh,jh;f(jh)) \right| ,$$

where $Nh \le X, W \equiv W_N = \max_{0 \le j \le r \le N} \frac{|w_{rj}|}{h}$ *and* $\xi(h) = \sum_{j=0}^{q-1} |f(jh)-\tilde{f}(jh)|$

then

$$|f(rh)-\tilde{f}(rh)| \leq \frac{\tau_N(h)+hLW_N\xi(h)}{1-hLW_N} \exp\left(\frac{LW_N rh}{1-hLW_N}\right) \quad (r = q,(q+1),(q+2)\ldots,N) \tag{6.69}$$

provided that $hLW_N < 1$.

The preceding theorem can be used to establish the convergence of the values $\tilde{f}(rh)$ to $f(rh)$ as $h \to 0$, with fixed $rh \leq X$, if we employ quadrature rules generating weights w_{kj} such that $\lim_{h\to 0} W_N < \infty$ and $\lim_{h\to 0} \tau_N(h) = 0$ ($N \to \infty$ as $h \to 0$ when Nh is fixed).

Remark. Observe that if the weights w_{kj} are those of the kth closed Newton-Cotes rule, then $\lim_{h\to 0} W_N$ is not finite. For all the schemes that we have suggested are suitable, $\lim_{h\to 0} W_N < \infty$, and we can regard W as an absolute constant.

The theorem establishes the *rate* of convergence of the values $\tilde{f}(rh)$ to $f(rh)$ at a *fixed* point $x \in [0,X]$. For h sufficiently small, $r \geq q$, where $rh = x$. Setting $N = r$ in the theorem we find that

$$|f(x)-\tilde{f}(x)| \leq \frac{\tau_r(h)+hLW_r\xi(h)}{1-hLW_r} \exp\left(\frac{LW_r x}{1-hLW_r}\right).$$

Since $\lim_{h\to 0} W_r = \lim_{r\to\infty} W_r < \infty$, we may take limits as $h \to 0$ and observe that

$$f(x)-\tilde{f}(x) = O(\tau_r(h))+O(h\xi(h)) \tag{6.70}$$

as $h \to 0$ and $r \to \infty$. Thus if we employ the weights w_{kj} of Table 6.5, which corresponds to the use of the repeated Simpson's rule in combination with the $\frac{3}{8}$th rule, we see, when we take limits in (6.69) as $h \to 0$ with rh fixed, that

$$\tau_r(h) = O(h^4) \quad \text{as} \quad h \to 0 \,, \quad rh \text{ fixed}$$

and $f(x)-\tilde{f}(x) = O(h^4)$ provided that starting values are accurate to within $O(h^3)$.

It is not surprising to discover that the local truncation error for the equations, giving $\tilde{f}(h),\tilde{f}(2h),\ldots,\tilde{f}((q-1)h)$ need only be of order h^{p-1} to 'sustain' $O(h^p)$ accuracy in the calculation of subsequent values. Indeed, if we consider the linear equation (6.66) we observe that for all reasonable choices of weights $w_{k,j}$,

$$w_{k,j} = O(h) \quad (j = 0,1,2,\ldots,(q-1)) \ ,$$

and since the contributions from $\tilde{f}(jh)$ $(j = 0,1,2,\ldots,q-1)$ are multiplied by the factors w_{kj}, errors in q function values each of which are $O(h^{p-1})$ affect the computed values $\tilde{f}(kh)$ $(k \geq q)$ by an amount which is $O(h^p)$.

Of course, the weights w_{kj} for large k may correspond to quite accurate formulae, and the errors in the starting values $\{\tilde{f}(kh),\ 0 \leq k \leq q-1\}$ may adversely affect the overall accuracy obtained. This is the case with some of the methods associated with prescriptions for weights w_{kj} which we have given above. The usefulness of the table of weights supplied earlier is not invalidated, however, as we shall show after the following Examples.

Example 6.23.

Consider the weights w_{kj} given in Table 6.3. For a slight increase in simplicity we shall first consider the method applied to the linear Volterra equation to give (6.66), where $K(x,y)$ and $f(x)$ have as many derivatives as required. We choose $\lambda=1$ for notational convenience.

Now we have $\tilde{f}(0) = f(0) = g(0)$ and hence

$$\tilde{f}(h) - \tfrac{1}{2}h\{K(h,0)f(0) + K(h,h)\tilde{f}(h)\} = g(h) \ .$$

On the other hand,

$$f(h) - \tfrac{1}{2}h\{K(h,0)f(0) + K(h,h)f(h)\} = g(h) + \tau_1 \ ,$$

where $\tau_1 = O(h^3)$ as $h \to 0$, so that $(1-\tfrac{1}{2}hK(h,h))\{f(h)-\tilde{f}(h)\} = \tau_1 = O(h^3)$ and, as $h \to 0$,

$$\eta_1 = \tilde{f}(h)-f(h) = O(h^3) \ .$$

Now, from the equation for $\tilde{f}(2h)$ and the result that $\tilde{f}(h) = f(h) + \eta_1$ (where $\eta_1 = O(h^3)$)

$$\{1-\tfrac{1}{3}hK(2h,2h)\}\tilde{f}(2h)-\tfrac{1}{3}h\left[K(2h,0)f(0)+4K(2h,h)\{f(h)+\eta_1\}\right] = g(2h) \quad ,$$

whilst on replacing $\tilde{f}(2h)$ by $f(2h)$, and η_1 by zero, the right-hand side becomes $g(2h) + \tau_2$, where $\tau_2 = O(h^5)$. We obtain, therefore,

$$\{1-\tfrac{1}{3}hK(2h,2h)\}\{f(2h)-\tilde{f}(2h)\} = \tau_2 + \tfrac{4}{3}h \frac{K(2h,h)\tau_1}{\{1-\tfrac{1}{2}hK(h,h)\}} = O(h^4)$$

and $f(2h)-\tilde{f}(2h) = O(h^4)$, as $h \to 0$. We obtain a similar result, $f(kh)-\tilde{f}(kh) = O(h^4)$ as $h \to 0$ with k fixed (for $k = 3,4,5,\ldots$) on repeating the argument. We have *not*, however, established the behaviour of the error in $f(kh)-\tilde{f}(kh)$ as $h \to 0$ with kh fixed. (To do this we would have to adopt a more careful approach, embodied in the proof of Theorem 6.1.)

It is fairly clear that the preceding remarks can be extended to a consideration of the non-linear Volterra equation (6.68), given the existence of the Lipschitz constant L of Theorem 6.1. Our equation for $\tilde{f}(h)$ is

$$\tilde{f}(h) = g(h) + \tfrac{1}{2}h\{F(h,0;f(0)) + F(h,h;\tilde{f}(h))\} \quad ,$$

whereas

$$f(h) = g(h) + \tfrac{1}{2}h\{F(h,0;f(0)) + F(h,h;f(h))\} + \tau_1 \quad ,$$

where $\tau_1 = O(h^3)$ if $F(x,y;f(y))$ is sufficiently differentiable in y. We find that

$$\tilde{f}(h)-f(h) = \tfrac{1}{2}h\{F(h,h;\tilde{f}(h))-F(h,h;f(h))\} - \tau_1 \, .$$

Using the Lipschitz constant L we obtain

$$|\tilde{f}(h)-f(h)| \leq \tfrac{1}{2}hL|\tilde{f}(h)-f(h)| + |\tau_1|$$

and $|\tilde{f}(h)-f(h)| \le |\tau_1|/(1-\frac{1}{2}hL)$ for sufficiently small h. Thus $\tilde{f}(h) = f(h) + \eta_1$, where $\eta_1 = O(h^3)$. A similar analysis for $\tilde{f}(2h)$ gives

$$\tilde{f}(2h)-f(2h) = \tfrac{4}{3}h\{F(2h,h;\tilde{f}(h))-F(2h,h;f(h))\} + $$
$$+ \tfrac{1}{3}h\{F(2h,2h;\tilde{f}(2h))-F(2h,2h;f(2h))\} - \tau_2 ,$$

where $\tau_2 = O(h^5)$. Hence,

$$|\tilde{f}(2h)-f(2h)| \le \tfrac{4}{3}hL|\eta_1| + \tfrac{1}{3}hL|\tilde{f}(2h)-f(2h)| + |\tau_2| ,$$

and it follows that

$$|\tilde{f}(2h)-f(2h)| = O(h^4)$$

since $\eta_1 = O(h^3)$. *

Suppose, now, that we employ a set of weights w_{rj} which correspond to formulae

$$\int_0^{rh} \phi(y)\mathrm{d}y = \sum_{j=0}^{r} w_{rj}\phi(jh) + E_r , \tag{6.71}$$

where E_r is of modestly high order in h as $h \to 0$ with rh fixed. To be specific, we might fix our attention on the use of the Lagrangian form of Gregory's rule with differences of order s, where $s=\min(r,t)$ for some moderately large value of t ($t \ge 3$, say). Then we would obtain, in the notation of Example 6.6,

$$w_{rj} = \Omega_{rj}^{[r]} \; (r \le t) , \quad w_{rj} = \Omega_{rj}^{[t]} \; (r \ge t) \; (j=0,1,2,\ldots,r) \tag{6.72}$$

and

$$E_r = O(h^{s+2}) , \quad (h \to 0 , \quad rh = x) , \tag{6.73}$$

provided (as we shall assume) that the integrand has a sufficient number of derivatives.

Now for $r = 1$ we can do no better than the trapezium rule ($\Omega_{10}^{[1]} = \Omega_{11}^{[1]} = \frac{1}{2}h$) which (as we have seen in Example 6.23) gives $\tilde{f}(h)$ with accuracy $O(h^3)$. Then, with $r = 2$, our choice $\{w_{rj}\}$ with $4\Omega_{20}^{[2]} = 4\Omega_{22}^{[2]} = \Omega_{21}^{[2]} = \frac{4}{3}h$ (that is, Simpson's rule) gives $O(h^4)$ accuracy in $\tilde{f}(2h)$.

Though our subsequent rules are higher-order Gregory rules, the error in $\tilde{f}(h)$ now restricts the accuracy in $\tilde{f}(kh)$ to $O(h^4)$ magnitude, as $h \to 0$ with kh fixed. With sufficiently accurate values $\tilde{f}(h), \tilde{f}(2h), \ldots, \tilde{f}((p-1)h)$, the full accuracy of the Gregory rule (see Example 2.17) would be apparent: if we could find these values with accuracy $O(h^{t+1})$ then we would find that

$$\tilde{f}(kh)-f(kh) = O(h^{t+2}) \quad (h \to 0,\ kh = x) . \tag{6.74}$$

The preceding remarks justify the construction of schemes giving $\tilde{f}(h)$ with higher-order accuracy than $O(h^3)$! Let us establish our basic approach with a simple technique, which we shall develop later into an acceptable scheme.

Let us suppose that we take $t = 3$ in the preceding method, so that we require $\tilde{f}(h)$ with accuracy $O(h^4)$ to maintain overall accuracy. We shall achieve this by using the scheme for the first two stages with step $h_* = \frac{1}{2}h$. Thus we solve (using $f(0) = g(0)$) the equation

$$\tilde{f}(\tfrac{1}{2}h)-\tfrac{1}{4}hF(\tfrac{1}{2}h,\tfrac{1}{2}h;\tilde{f}(\tfrac{1}{2}h)) = g(\tfrac{1}{2}h)+\tfrac{1}{4}hF(\tfrac{1}{2}h,0;f(0)) \tag{6.75}$$

for the value $\tilde{f}(h_*) = \tilde{f}(\frac{1}{2}h)$ and the equation

$$\tilde{f}(h)- \frac{h}{6}F(h,h;\tilde{f}(h)) = g(h)+\frac{h}{6}\{F(h,0;f(0))+4F(h,\tfrac{1}{2}h;\tilde{f}(\tfrac{1}{2}h))\} \tag{6.76}$$

for the value $\tilde{f}(2h_*) = \tilde{f}(h)$. We have indicated, in the previous Example, that $f(2h_*)-\tilde{f}(2h_*) = O(h_*^4)$ and, with $h_* = \frac{1}{2}h$, this gives

$$f(h)-\tilde{f}(h) = O(h^4) . \tag{6.77}$$

which is the accuracy required for $t = 3$. We now employ our scheme to calculate $\tilde{f}(2h), \tilde{f}(3h), \ldots,$ in the normal way with step h, but employing the special value for $\tilde{f}(h)$ obtained in (6.76).

Thus, in particular, with $h_* = \frac{1}{2}h$ we have

$$\tilde{f}(2h) - \tfrac{1}{3}hF(2h,2h;\tilde{f}(2h)) = g(2h) + \tfrac{1}{3}h\{4F(2h,h;\tilde{f}(h)) + F(2h,0;f(0))\}$$

and

$$\tilde{f}(2h) = f(2h) + O(h^5) . \qquad (6.78)$$

Remark. Here it may be noted that if we require $\tilde{f}(h)$ with accuracy $O(h^5)$ then this can be obtained by replacing h by $h^{\#} = \frac{1}{2}h$ in the above procedure. Thus h_* becomes $\frac{1}{2}h^{\#} = \frac{1}{4}h$ and $\tilde{f}(2h)$ becomes $\tilde{f}(2h^{\#}) = \tilde{f}(h)$ to give us an approximation with $\tilde{f}(2h^{\#}) = f(2h^{\#}) + O(h^5)$ (see the following Example).

From what has been stated, it would appear that sufficiently accurate starting values can be obtained by using the basic quadrature schemes with various step-sizes h. It may be necessary to maintain the use of a given step-size over more than one step, as illustrated in the following Example, and to revert to a choice of rule which has occurred earlier in the scheme.

Example 6.24.
We shall indicate a scheme for obtaining values of $\tilde{f}(h), \tilde{f}(2h), \tilde{f}(3h)$ with errors which are $O(h^5)$. Such starting values can be used to compute $\tilde{f}(4h), \tilde{f}(5h), \ldots$ using Gregory's rule incorporating fourth differences to give $O(h^6)$ convergence at a fixed value x of the argument.

The notation

$$\tilde{f}(kh) = \text{Rule } (\tilde{f}(0), \tilde{f}(sh), \ldots, \tilde{f}(k-s)h) , \tilde{f}(kh))$$

will be used to indicate that the rule named is applied with the step sh to the calculation of $\tilde{f}(kh)$. Thus, for example, we interpret $\tilde{f}(8h_0)$= Simpson's rule $(f(0), \tilde{f}(4h_0), \tilde{f}(8h_0))$ to indicate that

$$\tilde{f}(8h_0) - \frac{4h_0}{3}F(8h_0, 8h_0; \tilde{f}(8h_0)) = g(8h_0) +$$

$$+ \frac{4h_0}{3}\{F(8h_0,0;f(0) + 4F(8h_0,4h_0;\tilde{f}(4h_0))\} \quad .$$

The scheme proposed is now as follows. Take $h_0 = \frac{1}{4}h$, and set

$$\tilde{f}(0) = f(0) \equiv g(0)$$

$$\tilde{f}(h_0) = \text{trapezium rule} \quad (\tilde{f}(0),\tilde{f}(h_0))$$
$$(\text{error} = O(h_0^3))$$

$$\tilde{f}(2h_0) = \text{Simpson's rule} \quad (\tilde{f}(0),\tilde{f}(h_0),\tilde{f}(2h_0))$$
$$(\text{error} = O(h_0^4))$$

$$\tilde{f}(4h_0) = \text{Simpson's rule} \quad (\tilde{f}(0),\tilde{f}(2h_0),\tilde{f}(4h_0))$$
$$(\text{error} = O(h_0^5))$$

$$\tilde{f}(8h_0) = \text{Simpson's rule} \quad (\tilde{f}(0),\tilde{f}(4h_0),\tilde{f}(8h_0))$$
$$(\text{error} = O(h_0^5))$$

$$\tilde{f}(12h_0) = \tfrac{3}{8}\text{th rule} \quad (\tilde{f}(0),\tilde{f}(4h_0),\tilde{f}(8h_0),\tilde{f}(12h_0))$$
$$(\text{error} = O(h_0^5)) \quad .$$

Since $h_0 = \frac{1}{4}h$, the scheme gives $\tilde{f}(h),\tilde{f}(2h),\tilde{f}(3h)$ with the required accuracy. The scheme can be extended (taking h_0 to be a smaller fraction of h) to give $O(h^6)$ accuracy, or better. Such a scheme will involve, say, higher-order Newton-Cotes formulae, or Gregory rules. *

We have seen how high-order starting values might be obtained, but the schemes suggested above suffer from the defect that the quadrature formulae employed are closed. This is no great handicap unless a scheme involving many stages (for high-order accuracy) is being used. In this case the required accuracy can always be obtained by replacing each closed formula by a suitable predictor-corrector pair using the closed formula as corrector, and iterating a predetermined number of times to obtain the required order of accuracy.

Consider, for example, the use of the trapezium rule for the nonlinear equation (6.68), to give $\tilde{f}(h)$ with accuracy $O(h^3)$. We may replace the exact solution of the closed formula with the pair of formulae

$$\tilde{f}^{[0]}(h) = g(h) + hF(h,0;f(0)) \quad , \tag{6.79}$$

$$\tilde{f}(h) = g(h) + \tfrac{1}{2}h\{F(h,0;f(0)) + F(h,h;\tilde{f}^{[0]}(h))\} \quad . \tag{6.80}$$

Since $\tilde{f}^{[0]}(h) = f(h) + O(h^2)$, we see that

$$\tilde{f}(h) = g(h)+\tfrac{1}{2}h\{F(h,0;f(0))+F(h,h;f(h))\}+$$

$$+ \tfrac{1}{2}h\left[(\partial/\partial f)F(h,h;f(h))\{\tilde{f}^{[0]}(h)-f(h)\}+o(\tilde{f}^{[0]}(h)-f(h))\right]$$

and hence

$$\tilde{f}(h) = f(h) + O(h^3) \ , \tag{6.81}$$

assuming sufficient smoothness.

Example 6.25.
Let us suppose that Simpson's rule is used with some step h_* which is a fixed multiple of h , and we employ a value $\tilde{f}(h_*) \simeq f(h_*) + O(h^3)$. The value $\tilde{f}(2h_*)$ can be obtained with accuracy $O(h^4)$ if we set

$$\tilde{f}^{[0]}(2h_*) = g(2h_*) + 2h_*F(2h_*,h_*;\tilde{f}(h_*)) \ ,$$

$$\tilde{f}^{[1]}(2h_*) = \left[g(2h_*) + \tfrac{1}{3}h_*\{F(2h_*,0;f(0)) + 4F(2h_*,h_*;f(h_*))\}\right] +$$

$$+ \tfrac{1}{3}h_*F(2h_*,2h_*;\tilde{f}^{[0]}(2h_*)) \ ,$$

and set $\tilde{f}(2h_*) = \tilde{f}^{[1]}(2h_*)$. If $\tilde{f}(h_*)$ has accuracy $O(h^4)$ we can instead set

$$\tilde{f}(2h_*) = \left[g(2h_*) + \tfrac{1}{3}h_*\{F(2h_*,0;f(0)) + 4F(2h_*,h_*;\tilde{f}(h_*))\}\right] +$$

$$+ \tfrac{1}{3}h_*F(2h_*,2h_*;\tilde{f}^{[1]}(2h_*))$$

to obtain an approximation with accuracy $O(h^5)$. A similar scheme can be devised for the $\tfrac{3}{8}$th rule. *

At this point we may recognize the similarity of the formulae

(6.79)-(6.80) with a simple low-order Runge-Kutta method encountered in initial-value problems for ordinary differential equations. Schemes based on the ideas indicated above, replacing closed formulae by fixed combinations of a predictor and a corrector, result in schemes either identical with or similar to basic formulae in the Runge-Kutta schemes for Volterra integral equations (section 6.6). Such schemes have been determined by Pouzet (1960) and Beltyukov (1966). Many starting schemes to be found in the literature are of a similar type, and various steps of the Runge-Kutta formulae can themselves be used as starting procedures. What distinguishes a genuine Runge-Kutta formula from ad hoc schemes devised by the use of quadrature formulae is that the Runge-Kutta schemes are designed to be in a certain sense optimal (giving the highest-order accuracy for a certain number of evaluations of $F(x,y;v)$) when used as starting procedures. The block-by-block methods which we describe later (in section 6.7) are of a different type but these are self-starting and can be used to provide starting procedures for other methods.

We shall conclude this section by giving, as Examples, some starting procedures which have appeared in the literature and which have been constructed along similar lines to those given above.

Example 6.26.

Day (1968) provides a starting method for computing $O(h^4)$ approximations for $f(h), f(2h)$, and $f(3h)$. His suggestion is that these values might be used to start a 'Gregory-method'.

We set $\tilde{f}(0) = f(0) = g(0)$ and

$\tilde{f}_0(h) = g(h) + hF(h,0;f(0))$

(rectangle rule, error = $O(h^2)$)

$\tilde{f}_1(h) = g(h) + \frac{1}{2}h\{F(h,0;f(0)) + F(h,h;\tilde{f}_0(h))\}$

(trapezium rule, error = $O(h^3)$)

$\tilde{f}_0(\frac{1}{2}h) = \frac{1}{2}\{f(0) + \tilde{f}_1(h)\}$

(linear interpolation, error = $O(h^2)$)

$\tilde{f}_1(\frac{1}{2}h) = g(\frac{1}{2}h) + \frac{1}{4}h\{F(\frac{1}{2}h,0;f(0)) + F(\frac{1}{2}h,\frac{1}{2}h;\tilde{f}_0(\frac{1}{2}h)\}$

(trapezium rule, error = $O(h^3)$)

$\tilde{f}(h) = g(h) + \frac{1}{6}h\{F(h,0;f(0)) + 4F(h,\frac{1}{2}h;\tilde{f}_1(\frac{1}{2}h)) + F(h,h;\tilde{f}_1(h))\}$

(Simpson's rule, error = $O(h^4)$)

$$\tilde{f}_0(2h) = g(2h) + 2hF(2h,h;\tilde{f}(h))$$
(mid-point rule, error $= O(h^3)$)
$$\tilde{f}(2h) = g(2h) + \tfrac{1}{3}h\{F(2h,0;f(0)) + 4F(2h,h;\tilde{f}_1(h)) + F(2h,2h;\tilde{f}_0(2h))\}$$
(Simpson's rule, error $= O(h^4)$)

$$\tilde{f}_0(3h) = g(3h) + \tfrac{3}{2}h\{F(3h,h;\tilde{f}(h)) + F(3h,2h;\tilde{f}(2h))\}$$
(2-point open Newton-Cotes, error $= O(h^3)$)
$$\tilde{f}(3h) = g(3h) + \tfrac{3}{8}h\{F(3h,0;f(0)) + 3F(3h,h;\tilde{f}(h)) + $$
$$+ 3F(3h,2h;\tilde{f}(2h)) + F(3h,3h;\tilde{f}_0(3h))\}$$
($\frac{3}{8}$th rule, error $= O(h^4)$) .

(The comments of the form 'error $= O(h^p)$' denote that the left-hand term approximates the appropriate value $f(rh)$ with the order of accuracy stated.)

Observe that the half-open rule displayed earlier for the approximation of $\int_0^{3h}\phi(y)\,dy$ could be used to eliminate the calculation of $\tilde{f}_0(3h)$ and to give a value approximating $f(3h)$ with the required order of accuracy. *

<u>Example 6.27</u>.

Campbell and Day (1970) supply starting procedures which give starting values for a predictor-corrector method in which the corrector is a Gregory formula. The basis of their formulae is the 4-point Lobatto rule, which gives

$$\tilde{f}(h) = g(h) + \frac{1}{12}h\{F(h,0;f(0)) + 5F(h,h_1;\tilde{f}(h_1)) +$$
$$+ 5F(h,h_2;\tilde{f}(h_2)) + F(h,h;\tilde{f}(h))\} ,$$

where $h_1 = \frac{1}{10}h(5-\sqrt{5})$, $h_2 = \frac{1}{10}h(5+\sqrt{5})$. The values $\tilde{f}(h_i)$ $(i = 1,2)$ are computed using the 3-point Radau quadrature rule which gives

$$\tilde{f}(h_i) = g(h_i) + \frac{1}{36}h_i\{4F(h_i,0;f(0)) + (16+\sqrt{6})F(h_i,h_{i,1};\tilde{f}(h_{i,1})) +$$
$$+ (16-\sqrt{6})F(h_i,h_{i,2};\tilde{f}(h_{i,2}))\} ,$$

where $h_{i,1} = \frac{1}{10}(6-\sqrt{6})h_i$ and $h_{i,2} = \frac{1}{10}(6+\sqrt{6})h_i$. The formulae require values of $\tilde{f}(h_{i,1})$ and $\tilde{f}(h_{i,2})$ which are obtained as follows. We set

$$\tilde{f}_0(\tfrac{1}{3}h_{i,j}) = g(\tfrac{1}{3}h_{i,j}) + h_{i,j}F(\tfrac{1}{3}h_{i,j},0;f(0))$$

(rectangular rule)

$$\tilde{f}(\tfrac{1}{3}h_{i,j}) = g(\tfrac{1}{3}h_{i,j}) + \tfrac{1}{6}h_{i,j}\{F(\tfrac{1}{3}h_{i,j},0;f(0)) + F(\tfrac{1}{3}h_{i,j},\tfrac{1}{3}h_{i,j};\tilde{f}_0(\tfrac{1}{3}h_{i,j}))\}$$

(trapezium rule)

$$\tilde{f}(\tfrac{1}{2}h_{i,j}) = g(\tfrac{1}{2}h_{i,j}) + \tfrac{1}{8}h_{i,j}\{F(\tfrac{1}{2}h_{i,j},0;f(0)) + 3F(\tfrac{1}{2}h_{i,j},\tfrac{1}{3}h_{i,j};\tilde{f}(\tfrac{1}{3}h_{i,j}))\}$$

(Radau 2-point rule)

$$\tilde{f}(\tfrac{2}{3}h_{i,j}) = g(\tfrac{2}{3}h_{i,j}) + \tfrac{2}{3}h_{i,j}F(\tfrac{2}{3}h_{i,j},\tfrac{1}{3}h_{i,j};\tilde{f}(\tfrac{1}{3}h_{i,j}))$$

(mid-point rule)

$$\tilde{f}_0(h_{i,j}) = g(h_{i,j}) + \tfrac{1}{4}h_{i,j}\{F(h_{i,j},0;f(0)) + 3F(h_{i,j},\tfrac{2}{3}h_{i,j};\tilde{f}(\tfrac{2}{3}h_{i,j}))\}\ ,$$

(Radau 2-point rule)

and

$$\tilde{f}(h_{i,j}) = g(h_{i,j}) + \tfrac{1}{6}h_{i,j}\{F(h_{i,j},0;f(0)) + 4F(h_{i,j},\tfrac{1}{2}h_{i,j};\tilde{f}(\tfrac{1}{2}h_{i,j})) + F(h_{i,j},h_{i,j};\tilde{f}_0(h_{i,j}))\}$$

(correcting $\tilde{f}_0(h_{i,j})$ by using Simpson's rule). *

Observe that the arguments for using starting procedures of the type described above generally rely on the order of the error as $h \to 0$, and may be invalid for large h. (The same criticism can be levelled at much of the work on Volterra integral equations.)

Remark. In the previous Example we have seen use being made of Lobatto rules. Jain and Sharma (1967) employ Lobatto rules in a different fashion.

The computation of starting values in general involves but a minor part of the computational effort in the numerical solution of a Volterra integral equation. It would appear rare to require more than the first (say) eight values $\tilde{f}(kh)$ $(k = 1,2,3,\ldots,8)$ to be computed by a special starting procedure, the precise number depending on the order of accuracy required. We tend to the view that in this situation efficiency of a starting procedure (measured by the number of evaluations of $F(\ \ )$, say) is of less importance than accuracy and computational simplicity. Thus the repeated use of simple quadrature rules which occur later in a computational scheme may, if the calculations are elegantly arranged, be more attractive than an entirely distinct package of efficient but 'unusual' quadrature rules employed only in the starting procedures. Unfortunately, no general comparison has been made of the accuracy and suitability of various starting methods.

We shall now turn, in the following section, to a discussion of Runge-Kutta formulae. Such formulae can (as mentioned earlier) be employed as starting methods; their prime importance may be for use in this rôle. (The use of extended Runge-Kutta methods for the step-by-step solution of Volterra equations appears to be favoured by the French school, whilst certain other authors regard the methods as inefficient when used for this purpose.)

Example 6.28.

Laudet and Oulès (1960) give a method of Runge-Kutta type for calculating $\tilde{f}(h)$ with accuracy $O(h^5)$. They use this in a quadrature method with the weights in Table 6.5 to compute $\tilde{f}(2h), \tilde{f}(3h), \tilde{f}(4h)$. They employ a predictor-corrector technique, setting $\tilde{f}^{[0]}((n+1)h) = \tilde{f}(nh)$, and appear to iterate to convergence. These authors give as a numerical example the equation

$$f(x) - \int_0^x \left[x \exp\{y(x-2y)\} + \exp(-2y^2)\right]\{f(y)\}^2 dy = 1-x ,$$

with a solution $f(x) = \exp x^2$. Taking $h = \frac{1}{10}$ and solving over $[0,1]$, their approximate values have a maximum error 2×10^{-6} . *

6.6. *Runge-Kutta-type methods for Volterra equations*

Methods of Runge-Kutta type for the Volterra equation

$$f(x) - \int_0^x F(x,y;f(y))\,dy = g(x) \ , \ x \geq 0, \qquad (6.82)$$

can be determined from the work of Pouzet (1960, 1963), Oulès (1960), Aparo (1959), and Beltyukov (1965). These methods are various analogues of the methods of Runge-Kutta type for the solution of initial-value problems in ordinary differential equations, of the form $f'(x)=\Phi(x,f(x))$. The methods can be used to provide starting values for other methods or they can be extended to provide an explicit step-by-step method for the numerical solution of (6.82).

The discussion in the literature is frequently related to the Volterra equation in its canonical form, namely,

$$f(x) = \int_a^x G(x,y;f(y))\,dy \qquad (6.83)$$

(in which there is no free term $g(x)$ outside the integral sign) and the ALGOL procedures published by Pouzet (1970) are for the canonical form. The adaption of the methods to equations of the form (6.82) is readily achieved (see section 6.1 for the conversion of (6.82) to canonical form).

Example 6.29.
Recall that an 'explicit' Runge-Kutta type method for the solution of the equation $\psi'(x) = \Phi(x;\psi(x))$ with $\psi(0) = \psi_0$ is given by setting $\tilde{\psi}(0)=\psi_0$ and for $n = 0,1,2,\ldots$,

$$\tilde{\psi}((n+1)h) = \tilde{\psi}(nh) + \sum_{r=0}^{p-1} \omega_r \tilde{\Phi}_r^{[n]} \ ,$$

where (for $j = 0,1,2,\ldots,(p-1)$) ,

$$\tilde{\Phi}_j^{[n]} = h\Phi((n+\theta_j)h;\tilde{\psi}(nh) + \sum_{r=0}^{j-1} A_{j,r}\tilde{\Phi}_r^{[n]})$$

and the parameters $\theta_j, A_{j,r}$, and ω_j are determined by the particular method. It is convenient if $0 = \theta_0 \leq \theta_1 \leq \ldots \leq \theta_p = 1$ and we can regard $\tilde{\Phi}_r^{[n]}$ as an approximation to $h\psi'((n+\theta_r)h)$. Setting $\omega_r = A_{p,r}$, the value of

$$\tilde{\psi}(nh) + \sum_{r=0}^{j-1} A_{j,r}\tilde{\Phi}_r^{[n]} \qquad (\text{for } j = 1,2,\ldots, p)$$

can be regarded as an approximation to $\psi((n+\theta_j)h)$.

Clearly,

$$\tilde{\psi}((n+1)h) = \psi(0) + \sum_{k=0}^{n} \sum_{j=0}^{p-1} \omega_j\tilde{\Phi}_j^{[k]} \quad ,$$

which is a discretized version of the expression

$$\psi((n+1)h) = \psi(0) + \sum_{k=0}^{n} \int_{kh}^{(k+1)h} \psi'(t)dt \ . \quad *$$

A class of Runge-Kutta methods (due to Pouzet (1960)) can be derived from Runge-Kutta methods for ordinary differential equations. When these methods are applied to the eqn (6.82), we obtain the following formulae.

We write $y_n = nh$ and for $n = 0,1,2,\ldots$ and $q = 1,2,3,\ldots,p$ we set

$$\tilde{f}(y_n+\theta_q h) = \tilde{\phi}_n(y_n+\theta_q h)+h \sum_{s=0}^{q-1} A_{q,s}F(y_n+\theta_q h,y_n+\theta_s h;\tilde{f}(y_n+\theta_s h)) \ , \qquad (6.84)$$

where for $n = 1,2,3,\ldots$ and $x = y_n+\theta_r h$ $(0 \leq r \leq p)$

$$\tilde{\phi}_n(x) = g(x)+h \sum_{j=0}^{n-1} \sum_{s=0}^{p-1} A_{p,s}F(x,y_j+\theta_s h;\tilde{f}(y_j+\theta_s h)),$$

$$\tilde{\phi}_0(x) = g(x) \ , \qquad (6.85)$$

$$\tilde{f}(y_0) = g(y_0), \tilde{f}(y_n) \equiv \tilde{f}(y_{n-1} + \theta_p h) \ .$$

The scheme permits the calculation of $\tilde{f}(h), \tilde{f}(2h), \tilde{f}(3h), \ldots$ in sequence, and we shall refer to it as an *extended Runge-Kutta method.* (The use of the formulae to advance one step and calculate $\tilde{f}((n+1)h)$ given $\tilde{f}(0), \ldots, \tilde{f}(nh)$ will be referred to as a *basic Runge-Kutta method.*)

We suppose $\theta_0, \theta_1, \theta_2, \ldots, \theta_p$ are chosen so that

$$\theta_0 = 0 < \theta_1 \leq \theta_2 \leq \cdots \leq \theta_p = 1 ,$$

and the values $\{A_{r,s}\}$ are chosen to ensure a high order of local truncation error. A suitable choice of values of the parameters $\{A_{r,s}, \theta_r\}$ is given below (but the above conditions on $\{\theta_r\}$ can be relaxed).

The above formulae are seen to have an intuitive meaning in the present context if we write, for $q = 0,1,\ldots,p$,

$$\tilde{f}(y_n + \theta_q h) \simeq f(y_n + \theta_q h)$$

(observing that $\theta_p = 1$). We write, for $y_n \leq x \leq y_{n+1}$ with $y_j = jh$,

$$f(x) = \{g(x) + \sum_{j=0}^{n-1} \int_{y_j}^{y_{j+1}} F(x,y;f(y))\,dy\} + \int_{y_n}^{x} F(x,y;f(y))\,dy , \tag{6.86}$$

subsequently approximating each integral of the form $\int_{y_k}^{x} F(x,y;f(y))\,dy$ (where $x = y_k + \theta_q h$) in terms of values $\tilde{f}(y_k + \theta_s h)$ $(s = 0,1,2,\ldots,(q-1))$. Thus the term in braces in eqn (6.86) is approximated by $\tilde{\phi}_n(x)$. (Pouzet (1960) gives a slightly more general expression for $\tilde{\phi}_n(x)$ than that given above.)

An analogue of the fourth-order Runge-Kutta method for ordinary differential equations is given by choosing $p = 4$:

$$\begin{aligned} &\theta_1 = \theta_2 = \tfrac{1}{2},\ \theta_3 = \theta_4 = 1 ,\ \theta_0 = 0, \\ &A_{10} = A_{21} = \tfrac{1}{2},\ A_{20} = A_{30} = A_{31} = 0 , \\ &A_{40} = A_{43} = \tfrac{1}{6} ,\ A_{41} = A_{42} = \tfrac{1}{3} , \\ &\text{and } A_{32} = 1. \end{aligned} \tag{6.87}$$

This choice of parameters appears to be suitable for practical work.

Example 6.30.

The preceding choice of parameters gives a Runge-Kutta method which can be constructed from the use of quadrature formulae along lines indicated in section 6.5, using auxiliary approximations $\tilde{\phi}_n(x)$ of the term

$$\phi_n(x) = g(x) + \int_0^{nh} F(x,y;f(y))dy$$

occurring in braces in eqn (6.86).

Thus, use of the left-hand rectangle rule gives (for $n = 0,1,2,\ldots$)

$$\tilde{f}_0(nh+\tfrac{1}{2}h) = \tilde{\phi}_n(nh+\tfrac{1}{2}h) + \frac{h}{2}F(nh+\tfrac{1}{2}h,nh;\tilde{f}(nh)) \quad ,$$

where $\tilde{\phi}_n(nh+\frac{1}{2}h)$ is obtained using a formula allied to Simpson's rule in composite form:

$$\tilde{\phi}_n(nh+\tfrac{1}{2}h) = g(nh+\tfrac{1}{2}h) + \tfrac{1}{6}h\sum_{j=0}^{n-1}\Big[F(nh+\tfrac{1}{2}h,jh;\tilde{f}(jh)) +$$

$$+ \{2F(nh+\tfrac{1}{2}h,jh+\tfrac{1}{2}h;\tilde{f}_0(jh+\tfrac{1}{2}h)) +$$

$$+ 2F(nh+\tfrac{1}{2}h,jh+\tfrac{1}{2}h;\tilde{f}_1(jh+\tfrac{1}{2}h))\} +$$

$$+ F(nh+\tfrac{1}{2}h,(j+1)h;\tilde{f}_0((j+1)h))\Big] \quad ,$$

and the right-hand rectangle rule is used to obtain

$$\tilde{f}_1(nh+\tfrac{1}{2}h) = \tilde{\phi}_n(nh+\tfrac{1}{2}h) + \frac{h}{2}F(nh+\tfrac{1}{2}h,nh+\tfrac{1}{2}h;\tilde{f}_0(nh+\tfrac{1}{2}h)) \quad .$$

Further, the mid-point rule gives

$$\tilde{f}_0((n+1)h) = \tilde{\phi}_n((n+1)h) + hF((n+1)h,nh+\tfrac{1}{2}h;\tilde{f}_1(nh+\tfrac{1}{2}h)) \quad ,$$

where $\Phi_n((n+1)h)$ is an obvious modification of the expression for $\Phi_n(nh+\frac{1}{2}h)$, and use of a formula based on Simpson's rule (with step $\frac{1}{2}h$) gives $\tilde{f}((n+1)h) = \tilde{f}_1((n+1)h)$ where

$$\tilde{f}_1((n+1)h) = \Phi_n((n+1)h) + \frac{h}{6}[F((n+1)h,nh;\tilde{f}(nh)) +$$

$$+ \{2F((n+1)h,nh+\tfrac{1}{2}h;\tilde{f}_0(nh+\tfrac{1}{2}h)) +$$

$$+ 2F((n+1)h,nh+\tfrac{1}{2}h;\tilde{f}_1(nh+\tfrac{1}{2}h))\} +$$

$$+ F((n+1)h,(n+1)h;\tilde{f}_0((n+1)h))] \quad .$$

It would appear that, on writing

$$\tilde{f}(nh+\theta_1 h) = \tilde{f}_0(nh+\tfrac{1}{2}h), \tilde{f}(nh+\theta_3 h) = \tilde{f}_0(nh+h) \ ,$$

$$\tilde{f}(nh+\theta_2 h) = \tilde{f}_1(nh+\tfrac{1}{2}h), \tilde{f}(nh+\theta_4 h) = \tilde{f}_1(nh+h) \quad ,$$

the algorithm is identifiable as a Runge-Kutta method. Analysis along the lines of the preceding section (section 6.5) would tend to indicate errors $O(h^2)$ in $\tilde{f}_0(nh+\frac{1}{2}h)$, and in $\tilde{f}_1(nh+\frac{1}{2}h)$ resulting in $O(h^3)$ errors in $\tilde{f}_1((n+1)h)$. *The actual error in* $\tilde{f}_1((n+1)h)$ *is* $O(h^4)$. The secret of the Runge-Kutta approach lies, in this instance, in a construction of $\tilde{f}_0(nh+\frac{1}{2}h)$ and $\tilde{f}_1(nh+\frac{1}{2}h)$ in such a way that (*inter alia*)

$$\tilde{f}_0(nh+\tfrac{1}{2}h) = f(nh+\tfrac{1}{2}h) + \alpha h^2 + O(h^3)$$

and

$$\tilde{f}_1(nh+\tfrac{1}{2}h) = f(nh+\tfrac{1}{2}h) - \alpha h^2 + O(h^3) \ .$$

Then

$$F(x,nh+\tfrac{1}{2}h;\tilde{f}_0(nh+\tfrac{1}{2}h)) + F(x,nh+\tfrac{1}{2}h;\tilde{f}_1(nh+\tfrac{1}{2}h))$$

$$= 2F(x,nh+\tfrac{1}{2}h;f(nh+\tfrac{1}{2}h)) + O(h^3) \quad (n = 0,1,2,\ldots) \ ,$$

and the versions of the repeated Simpson's rule which are employed in fact yield $O(h^4)$ accuracy over a fixed interval $[0,kh]$. *

The Runge-Kutta formulae given above are in the form applicable to the approximate solution of the equation

$$f(x) = \int_0^x F(x,y;f(y))\,dy + g(x) .$$

The formulae are readily adapted to the case where the lower limit of integration is a non-zero constant a, and we seek the approximate solution $\tilde{f}(a+h)$. (This is achieved by writing the integral equation as an equation for $\phi(x) = f(x+a)$, as indicated in section 6.1.) Having made this point let us emphasize that Runge-Kutta methods can be conceived as single-step formulae for $\tilde{f}(a+h)$ which, along with certain 'auxiliary approximations', gives a method for generating the sequence of values $\tilde{f}(2h),\tilde{f}(3h),\tilde{f}(4h),\ldots$. This is seen if we **set $a=nh$,** and for $nh \le x \le (n+1)h$,

$$f(x) = \phi_n(x) + \psi_n(x) , \tag{6.88}$$

where

$$\phi_n(x) = g(x) + \int_0^{nh} F(x,y;f(y))\,dy$$

and

$$\psi_n(x) = \int_{nh}^x F(x,y;f(y))\,dy .$$

On substituting for $f(y)$ in the latter equation we obtain, with $F_n^*(x,y;v) = F(x,y;\phi_n(y)+v)$,

$$\psi_n(x) = \int_{nh}^x F(x,y;\phi_n(y)+\psi_n(y))\,dy = \int_{nh}^x F_n^*(x,y;\psi_n(y))\,dy. \tag{6.89}$$

A single step of the Runge-Kutta method can be used to compute $\tilde{\psi}_n((n+1)h) \simeq \psi_n((n+1)h)$. (We require 'auxiliary approximations' for values of $\phi_r(x)$ required to determine certain values of $F_n^*(x,y;\psi_n(y))$.) If we can also approximate $\phi_n((n+1)h)$ we may set

$$\tilde{f}((n+1)h) = \tilde{\psi}_n((n+1)h) + \tilde{\phi}_n((n+1)h) .$$

The formulae in (6.85) for giving values $\tilde{\phi}_n(x)$ are associated with the parameters $\{A_{rs},\theta_r\}$ of the Runge-Kutta formulae, but other approximations may be more suitable.

Both Pouzet (1963, 1970) and Beltyukov (1965) offer alternative approximations to the term

$$\phi_n(x) = g(x) + \sum_{j=1}^{n}\int_{y_{j-1}}^{y_j} F(x,y;f(y))\,dy$$

in (6.86), obtaining such approximations by the use of any convenient quadrature rule of appropriate accuracy and employing previously calculated values (including intermediate values) of $\tilde{f}(x)$.

Example 6.31.

Beltyukov (1965) gives various approximations of Runge-Kutta type, amongst them being the following.

For the calculation of $\tilde{f}(h)$ set

$$\tilde{f}(h) = \tfrac{1}{4}(k_2+3k_1) + g(h) ,$$

where

$$k_0 = hF(\tfrac{1}{3}h,0;g(0)) ,$$

$$k_1 = hF(h,\tfrac{2}{3}h;\tfrac{2}{3}k_0 + g(\tfrac{2}{3}h)) ,$$

$$k_2 = hF(h,0;k_1-k_0 + g(0)) .$$

Then $f(h)-\tilde{f}(h) = O(h^4)$.

To compute $\tilde{f}(rh)$ we can replace h by rh and repeat the procedure, a process which is acceptable for *small* values of rh . A technique for calculating $\tilde{f}((n+1)h)$, when $\tilde{f}(h),\ldots,\tilde{f}(nh)$ have been found, is obtained on writing

$$\phi_n(x) = \int_0^{nh} F(x,y;f(y))\,dy + g(x) ,$$

so that $f(x) = \phi_n(x) + \psi_n(x)$, where, as in (6.89),

$$\psi_n(x) = \int_{nh}^{x} F(x,y;f(y))dy = \int_{nh}^{x} F(x,y;\phi_n(y)+\psi_n(y))dy$$

$$= \int_{nh}^{x} F^*(x,y;\psi_n(y))dy \ , \quad \text{say,}$$

where $F^*(x,y;z) = F(x,y;z+\phi_n(y))$.

The determination of approximate values of $\phi_n((n+1)h)$ and $\psi_n((n+1)h)$ gives approximate values of $f((n+1)h)$. We may write

$$\tilde{\phi}_n(x) = \sum_{j=0}^{n} w_{nj} F(x,jh;\tilde{f}(jh)) + g(x) \ ,$$

where $\{w_{nj}\}$ is any convenient set of weights of a formula of sufficient accuracy - the Gregory formulae including sufficient difference terms is mentioned by Beltyukov as a possibility. Clearly, n must be sufficiently large ($n > r$, say) for the use of such formulae, but until this is the case the first step of the Runge-Kutta method can be applied with the steps $2h, 3h, .., rh$. Finally, the calculation of $\psi_n((n+1)h)$ is achieved by using a step of the Runge-Kutta method and setting

$$\psi_n((n+1)h) = \tfrac{1}{4}(k_2^{[n]} + 3k_1^{[n]}) \ ,$$

where

$$k_0^{[n]} = hF(nh + \tfrac{1}{3}h, nh; \tilde{\phi}_n(nh)) \ ,$$

$$k_1^{[n]} = hF((n+1)h,\ nh + \tfrac{2}{3}h;\ \tfrac{2}{3}k_0^{[n]} + \tilde{\phi}_n(nh + \tfrac{2}{3}h)),$$

$$k_2^{[n]} = hF((n+1)h, nh; k_1^{[n]} - k_0^{[n]} + \tilde{\phi}_n(nh)) \ .$$

A slightly different method (which is actually the one described by Beltyukov) arises on replacing $\tilde{\phi}_n(nh)$ by $\tilde{f}(nh)$ in the expressions for $k_0^{[n]}$ and $k_2^{[n]}$. The resulting formulae are more economical; the two formulae may have slightly different stability properties. *

The use of the weights A_{rs} and intermediate values $\tilde{f}(rh+\theta_s h)$ to compute $\Phi_n(x)$ appears to be an essential difference between an 'extended Runge-Kutta method' and a method like that of the previous Example which employs Gregory's rule and only the values $\tilde{f}(rh)$. The latter type of method will be referred to as a *mixed Runge-Kutta method*. The work of Garey (1975) is related to such methods, and Baker and Keech (1977) point out that certain extensions of Runge-Kutta methods are unstable.

6.6.1. *Derivation of Runge-Kutta formulae*

In this subsection I shall indicate how basic formulae of Runge-Kutta type can be constructed. Two approaches to this theory will be found in the literature, that of Pouzet (1963) and that of Beltyukov (1966, published in Russian 1965).

Pouzet derives his formulae, for use with integral equations, from corresponding Runge-Kutta formulae for the numerical solution of initial-value problems in ordinary differential equations. Since much work has been performed on the latter topic, there is no need to calculate suitable values of the parameters A_{rs} and θ_s . (The appropriate values can be borrowed from the methods for differential equations which are derived in the literature.) The approach of Beltyukov (1966) is more direct and appears to lead to a more general type of formula.

The approach of Pouzet is illustrated in the following Example. Suppose that a Runge-Kutta formula (of order not exceeding four, according to a suggestion of Pouzet) is given for the approximate solution of

$$\psi'(x) = \Phi(x,\psi(x)) \quad (x > 0)$$

given

$$\psi(0) = \alpha . \tag{6.90}$$

We suppose this basic formula is given in the form

$$\tilde{\psi}(0) = \psi(0) \quad ,$$

$$\tilde{\psi}(\theta_s h) = \tilde{\psi}(0)+h\sum_{r=0}^{s-1} A_{sr}\Phi(\theta_r h,\tilde{\psi}(\theta_r h)) \quad (s = 1,2,3,\ldots(p-1))$$

and, with $\omega_r = A_{p,r}$,

$$\psi(h) = \psi(0) + \sum_{j=0}^{p-1} \omega_j \psi(\theta_j h) \ . \tag{6.91}$$

We assume (Lambert 1973) that $\theta_s = \sum_{r=0}^{s-1} A_{sr}$ $(s = 1,2,3,\ldots,p)$; we have written $A_{p,r} = \omega_r$, and we also suppose that $\theta_0 = 0 \le \theta_1 \le \ldots \le \theta_p = 1$. Such a formula provides a set of parameters A_{rs}, θ_r, and ω_r for use in (6.84) and (6.85). Not all Runge-Kutta formulae for differential equations employ values of θ_r which are naturally ordered numbers, and this restriction by Pouzet seems inessential.

Remark. In the notation of Example 6.29 we may set $n = 0$, to obtain

$$\tilde{\Phi}_0^{[0]} = h\Phi(0,\psi(0)) \ ,$$

$$\psi(\theta_1 h) = \psi(0) + A_{10}\tilde{\Phi}_0^{[0]} \ ,$$

$$\tilde{\Phi}_1^{[0]} = h\Phi(\theta_1 h, \psi(\theta_1 h)) \ ,$$

$$\psi(\theta_2 h) = \psi(0) + A_{20}\tilde{\Phi}_0^{[0]} + A_{21}\tilde{\Phi}_1^{[0]},$$

$$\tilde{\Phi}_2^{[0]} = h\Phi(\theta_2 h, \psi(\theta_2 h)) \ , \ldots,$$

$$\psi(\theta_{p-1} h) = \psi(0) + \sum_{r=0}^{p-2} A_{p-1,r}\tilde{\Phi}_r^{[0]} \ ,$$

$$\tilde{\Phi}_{p-1}^{[0]} = h\Phi(\theta_{p-1} h, \psi(\theta_{p-1} h)) \ ,$$

and finally

$$\psi(h) = \psi(\theta_p h) = \psi(0) + \sum_{r=0}^{p-1} A_{p,r}\tilde{\Phi}_r^{[0]} \tag{6.92}$$

(where $\omega_r = A_{p,r}$ $(r = 0,1,2,\ldots,(p-1))$.

This yields the formulation of the Runge-Kutta methods which is displayed above. Each calculation of a value $\psi(\theta_r h)$ has the appearance of a quadrature rule with jth weight $A_{r,j}$ and corresponding abscissa θ_j used to approximate $\psi(\theta_r h)-\psi(0) = \int_0^{\theta_r h} \psi'(y)\,dy$, where $\psi'(y)$ is obtained from the differential equation on substituting (previously calculated) approximate values of the solution.

Example 6.32.

Let us derive a method for the solution of the linear Volterra equation

$$f(x) - \int_0^x K(x,y)f(y)dy = g(x) .$$

We shall write (formally, see Kershaw (1974), who follows G.F.Miller)

$$K(x,y) = \sum_{j=1}^{\infty} X_j(x)Y_j(y)$$

for $y \leq x$. Then

$$f(x) = g(x) + \sum_{j=1}^{\infty} X_j(x)Z_j(x) \quad ,$$

where

$$Z_j(x) = \int_0^x Y_j(y)f(y)dy .$$

Thus $(d/dx)Z_j(x) = Y_j(x)f(x)$ and $Z_j(0) = 0$. We can apply a Runge-Kutta method to approximate $Z_j(h)$, if we employ

$$\tilde{Z}_j(\theta_s h) \simeq h \sum_{r=0}^{s-1} A_{s,r}Y_j(\theta_r h)f(\theta_r h) \quad , \quad s = 0,1,\ldots,p.$$

The latter expression requires values $f(\theta_r h)$; for these values we substitute values $\tilde{f}(\theta_r h)$, where

$$\tilde{f}(\theta_r h) = g(\theta_r h) + \sum_{j=1}^{\infty} X_j(\theta_r h)\tilde{Z}_j(\theta_r h) \; ,$$

and we then have calculated values satisfying

$$\tilde{Z}_j(\theta_s h) = h \sum_{r=0}^{s-1} A_{s,r}Y_j(\theta_r h)\tilde{f}(\theta_r h) .$$

We eliminate $\tilde{Z}_j(\theta_r h)$ in the expression for $\tilde{f}(\theta_r h)$ using the last equation, and we obtain

$$\tilde{f}(\theta_r h) = g(\theta_r h) + \sum_{j=1}^{\infty} X_j(\theta_r h)\{h \sum_{t=0}^{r-1} A_{r,t} Y_j(\theta_t h)\tilde{f}(\theta_t h)\}$$

$$= g(\theta_r h) + h \sum_{t=0}^{r-1} A_{r,t} K(\theta_r h, \theta_t h)\tilde{f}(\theta_t h) \quad (r = 0,1,2,\ldots,p) .$$

The value of $\tilde{f}(\theta_p h)$ is the required approximation to $\tilde{f}(h)$. We have, formally, extended the Runge-Kutta method for ordinary differential equations to apply to linear Volterra equations of the second kind. *

The approach of Pouzet is a rigorous extension of the ideas portrayed in Example 6.32 to the non-linear Volterra integral equation of the second kind. In the process of extending these ideas, Pouzet (1963) establishes that the Runge-Kutta formulae for Volterra equations have the order of local truncation error to be expected from their application to differential equations.

Beltyukov (1966) adopts a direct approach, displaying the basic philosophy of Runge-Kutta methods in the process. With Beltyukov, let us consider the canonical equation

$$f(x) = \int_0^x G(x,y;f(y))\mathrm{d}y$$

and derive an expression for an approximate value $\tilde{f}(h)$.

If $f'(0), f''(0), f'''(0), \ldots$ could be computed, we might take the Taylor series approximation

$$[f(0)] + hf'(0) + \frac{h^2}{2} f''(0) + \frac{h^3}{3!} f'''(0) + \ldots ,$$

terminated at some power of h, as an approximate value for $\tilde{f}(h)$. Direct differentiation gives (with $f(0) = 0$)

$$f'(x) = G(x,x;f(x)) + \int_0^x G_x(x,y;f(y))\mathrm{d}y , \qquad (6.93)$$

whence $f'(0) = G(0,0;f(0))$ and (writing $G_f(x,y;f(y))$ for $(\partial/\partial f)G(x,y;f(y))$, etc.),

$$f''(x) = G_x(x,x;f(x))+G_y(x,x;f(x))+G_f(x,x;f(x))f'(x) + \\ + G_x(x,x;f(x)) + \int_0^x G_{xx}(x,y;f(y))dy \qquad (6.94)$$

whence

$$f''(0) = 2G_x(0,0;f(0))+G_y(0,0;f(0))+G_f(0,0;f(0))G(0,0;f(0)) .$$

The analysis becomes involved as higher-order derivatives are computed, but Beltyukov gives

$$\begin{aligned} f'''(0) = {} & 3G_{x,x}(0,0;f(0))+3G_{x,y}(0,0;f(0)) + \\ & + G_{y,y}(0,0;f(0))+3G(0,0;f(0))\ G_{x,f}(0,0;f(0))+ \\ & + 2G(0,0;f(0))G_{y,f}(0,0;f(0)) + \\ & + \{G(0,0;f(0))\}^2 G_{f,f}(0,0;f(0)) + \\ & + 2G_x(0,0;f(0))G_f(0,0;f(0)) + \\ & + G_y(0,0;f(0))G_f(0,0;f(0))+G(0,0;f(0))\{G_f(0,0;f(0))\}^2 \end{aligned} \qquad (6.95)$$

and the 25 terms which contribute to $f^{iv}(0)$ can be found also.

If the partial derivatives required above are available we can compute $f(h)$ with accuracy $O(h^4)$, at a cost of ten evaluations of different function values. The basic idea of the Runge-Kutta approach consists of achieving comparable accuracy by repeated evaluation of the function $G(x,y;v)$ (at points which are determined by the computation), maximizing the order of local truncation error for a given number of function evaluations. Let us therefore write (with $f(0) = 0$)

$$\begin{aligned} k_0 &= hG(\alpha_0 h,\beta_0 h;f(0)) , \\ k_1 &= hG(\alpha_1 h,\beta_1 h;A_{10}k_0) , \\ k_2 &= hG(\alpha_2 h,\beta_2 h;A_{21}k_1+A_{20}k_0) , \quad \text{etc.} \end{aligned} \qquad (6.96)$$

Thus

$$k_r = hG(\alpha_r h, \beta_r h; \sum_{s=0}^{r-1} A_{r,s}k_s) \quad (r = 0,1,2,\ldots,(p-1))$$

and we set

$$\tilde{f}(h) = f(0) + \sum_{s=0}^{p-1} \omega_s k_s \,. \tag{6.97}$$

We shall then endeavour to choose constant values for the parameters $\alpha_s, \beta_s, \omega_s, A_{rs}$ (which define the method) so that $\tilde{f}(h) = f(h)+0(h^{\rho})$, where ρ is as large as possible. We require this for all suitably differentiable functions $G(x,y;v)$.

To illustrate the process, consider a choice of parameters defining a formula with $p = 2$. We hope to choose the parameters to give $\rho = 3$.

Now,

$$k_0 = h\{G(0,0;f(0))+\alpha_0 hG_x(0,0;f(0))+\beta_0 hG_y(0,0;f(0))\}+0(h^3)$$

$$\begin{aligned}k_1 &= h[\{G(0,0;f(0))+\alpha_1 hG_x(0,0;f(0))+\beta_1 hG_y(0,0;f(0))\}+\\ &\quad + A_{1,0}k_0 G_f(0,0;f(0))]+0(h^3)\\ &= h[\{G(0,0;f(0))+\alpha_1 hG_x(0,0;f(0))+\\ &\quad + \beta_1 hG_y(0,0;f(0))\}+\\ &\quad + A_{1,0}hG_f(0,0;f(0))G(0,0;f(0))] + 0(h^3)\,.\end{aligned}$$

Thus, collecting together terms in powers of h,

$$\begin{aligned}\tilde{f}(h) &= f(0)+h(\omega_0+\omega_1)G(0,0;f(0)) +\\ &\quad + h^2\{(\alpha_0\omega_0+\alpha_1\omega_1)G_x(0,0;f(0)) +\\ &\quad + (\beta_0\omega_0+\beta_1\omega_1)G_y(0,0;f(0)) +\\ &\quad + \omega_1 A_{1,0}G_f(0,0;f(0))G(0,0;f(0))\} + 0(h^3)\,.\end{aligned} \tag{6.98}$$

On the other hand, the Taylor series for $f(h)$ (employing the expression for $f'(0), f''(0)$ given above) yields

$$\begin{aligned} f(h) = f(0) &+ hG(0,0;f(0)) + \\ &+ \tfrac{1}{2}h^2\{2G_x(0,0;f(0)) + G_y(0,0;f(0)) + \\ &+ G_f(0,0;f(0))G(0,0;f(0))\} + \\ &+ O(h^3) \quad . \end{aligned} \tag{6.99}$$

If $f(h) = \tilde{f}(h) + O(h^3)$ we require that $(\omega_0+\omega_1)G(0,0;f(0))=G(0,0;f(0))$ and that

$$\begin{aligned} &\{(\alpha_0\omega_0+\alpha_1\omega_1)-1\}G_x(0,0;f(0)) + \\ &\quad + \{(\beta_0\omega_0+\beta_1\omega_1)-\tfrac{1}{2}\}G_y(0,0;f(0)) + \\ &\quad + \{(\omega_1 A_{1,0})-\tfrac{1}{2}\}G_f(0,0;f(0)) = 0 \ . \end{aligned} \tag{6.100}$$

To ensure that these equations are satisfied for all suitably differentiable functions $G(x,y;v)$ it is necessary and sufficient to require that

$$\begin{aligned} \omega_0+\omega_1 &= 1 \ , \\ \alpha_0\omega_0+\alpha_1\omega_1 &= 1 \ , \\ \beta_0\omega_0+\beta_1\omega_1 &= \tfrac{1}{2} \ , \end{aligned} \tag{6.101}$$

and

$$\omega_1 A_{1,0} = \tfrac{1}{2} \ .$$

This is a system of 4 non-linear equations in 7 unknowns, and the solution is not unique. Two particular solutions are obtained by taking

$$\omega_0 = \alpha_0 = \beta_0 = 0, \ \alpha_1 = 1 \ , \ A_{1,0} = \beta_1 = \tfrac{1}{2} \ , \ \omega_1 = 1 \ , \tag{6.102}$$

and

$$\alpha_0 = \alpha_1 = 1,\ \beta_1 = A_{1,0} = \tfrac{2}{3},\ \beta_0 = 0,\ \omega_0 = \tfrac{1}{4},\ \omega_1 = \tfrac{3}{4}\ . \qquad (6.103)$$

Example 6.33.
In the case $p = 3$ we obtain a system of 13 equations in 12 unknowns. One solution is given by

$$\alpha_0 = \alpha_2 = 1,\ \alpha_1 = \frac{1}{2},\ \beta_0 = 0,\ \beta_1 = \frac{1}{2},\ \beta_2 = \frac{2}{3}\ ,$$

$$A_{1,0} = \frac{1}{2},\ A_{2,0} = \frac{2}{9},\ A_{2,1} = \frac{4}{9}\ ,\ \omega_0 = \frac{1}{4},\ \omega_1 = 0,\ \omega_2 = \frac{3}{4}\ . \quad *$$

6.7. A block-by-block method

The numerical methods described in preceding sections have been step-by-step methods, in which it has been possible to compute successive values $\tilde{f}(h),\ \tilde{f}(2h), \tilde{f}(3h), \ldots$ one at a time and in sequence. The literature contains a number of descriptions of methods in which a block of values $\tilde{f}(prh), \tilde{f}((pr+1)h),\ \ldots\ \tilde{f}(\{p(r+1)-1\}h)$ is obtained at the rth stage by solving a system of p equations in p unknowns. Such a method is a block-by-block method, and in this section we describe two block-by-block methods due to Linz (1967*a*, 1969*a*) and Weiss (1972). Other block-by-block methods are also indicated.

We consider the equation

$$f(x) - \int_0^x F(x,y;f(y))\,dy = g(x) \quad (x \geq 0)\ , \qquad (6.104)$$

and we seek approximate values of the solution for $0 \leq x \leq X$, with $X = Nph$, where $h > 0$, and N and p are integers. (The basic interval $[0,X]$ is divided into N equal intervals, each of which is further divided into p subintervals of length h.)

For

$$mp < n \leq (m+1)p \quad (m = 0,1,2,\ldots,(N-1))$$

we may write

$$f(nh) - \int_0^{mph} F(nh,y;f(y))dy - \int_{mph}^{nh} F(nh,y;f(y))dy = g(nh) ,$$

and if we employ

$$\int_0^{mph} F(nh,y;f(y))dy \simeq \sum_{j=0}^{mp} w_{n,j} F(nh,jh;f(jh)) \tag{6.105}$$

and

$$\int_{mph}^{nh} F(nh,y;f(y))dy \simeq \sum_{j=mp}^{(m+1)p} w^*_{n,j} F(nh,jh;f(jh)) \tag{6.106}$$

we are led naturally to the approximate equations $\tilde{f}(0) = g(0)$ and

$$\tilde{f}(nh) - \sum_{j=0}^{mp} w_{n,j} F(nh,jh;\tilde{f}(jh)) - \sum_{j=mp}^{(m+1)p} w^*_{n,j} F(nh,jh;\tilde{f}(jh)) = g(nh) \tag{6.107}$$

for $\begin{cases} n = mp+1,\ldots,(m+1)p \\ m = 0,1,2,\ldots,(N-1) \end{cases}$.

(In the case where $m = 0$, the first sum is taken to be zero in (6.107).)

If we set $w^*_{n,j} = 0$ for $j > n$ then the preceding set of equations reverts to the form employed in section 6.2 for the step-by-step approximate solution of the integral equation. What we now propose, however, is that weights $w^*_{n,j}$ be constructed to give an adequate approximation for $\int_{mph}^{nh} F(nh,y;f(y))dy$ in terms of values of the integrand at $y = mph$, $(mp+1)h,\ldots,(m+1)ph$, even though such points lie outside the interval of integration. (It would be possible, if $nh < (m+1)ph$, to set $\Phi(nh,y;f(y)) = F(nh,y;f(y))$ for $y \le nh$, and zero otherwise, and apply a conventional rule to approximate $\int_{mph}^{(m+1)ph} \Phi(nh,y;f(y))dy$, but this is likely to be unsatisfactory in general. We instead prefer to construct the weights $w^*_{n,j}$ directly, to obtain an approximation of a suitable accuracy.)

<u>Example 6.34</u>.

The general procedure for choosing the weights $w_{n,j}$ and $w^*_{n,j}$ is adequately illustrated by considering a specific case, with $p = 2$. If $n = 2k+1$, where k is integer, then $m = k$ in (6.107) and the weights

$w_{n,j}$ can be chosen to be those of a (repeated) Simpson's rule:

$$4w_{n,0} = w_{n,1}(=2w_{n,2}=w_{n,3}=\ldots) = w_{n,2k-1}=4w_{n,2k}=(\tfrac{4}{3})h \ .$$

The associated error is $O(h^4)$ as $h \to 0$ with nh fixed, and this accuracy is maintained if the weights $w^*_{n,j}$ are chosen to give an approximation which is exact if the integrand is a quadratic in y. We therefore choose the weights $w^*_{n,j}$ so that

$$\sum_{j=2k}^{2(k+1)} w^*_{n,j}\phi(jh) = \int_{2kh}^{(2k+1)h} p(y)dy \ ,$$

where $p(x)$ is the quadratic interpolating $\phi(y)$ at $y = 2kh$, $y =(2k+1)h$, and $y = 2(k+1)h$. Thus

$$p(x) = \{\frac{(x-2kh)(x-(2k+1)h)}{2h^2}\}\phi(2(k+1)h) +$$

$$+ \{\frac{(x-2kh)(2(k+1)h -x)}{h^2}\}\phi((2k+1)h) +$$

$$+ \{\frac{(x-(2k+1)h)(x-2(k+1)h)}{2h^2}\}\phi(2kh) \ .$$

Clearly, the weights $w^*_{n,j}$ are the integrals of the terms in braces, and we find that

$$w^*_{2k+1,2k} = \frac{5}{12}h \ ,$$

$$w^*_{2k+1,2k+1} = \frac{2}{3}h \ ,$$

and

$$w^*_{2k+1,2k+2} = \frac{-1}{12}h \ .$$

In the case $n = 2(k+1)$, the situation is simpler. The weights w_{nj} can be chosen as before (setting $w_{2(k+1),j}=w_{2k+1,j}$ $(j=0,1,2,\ldots,2k)$), and since Simpson's rule is exact for quadratics, we can also set

$$4w^*_{2(k+1),2k}= w^*_{2(k+1),2k+1}= 4w^*_{2(k+1),2(k+1)}= \frac{4}{3}h \ .$$

Overall accuracy $O(h^4)$ is preserved. *

The computational scheme arising out of the previous Example has the form

$$\tilde{f}(0) = g(0) \quad ,$$

$$\tilde{f}(h)-\{\tfrac{5h}{12}F(h,0;\tilde{f}(0)) + \tfrac{2h}{3}F(h,h;\tilde{f}(h)) - \tfrac{h}{12}F(h,2h;\tilde{f}(2h))\} = g(h) \quad ,$$

$$\tilde{f}(2h)-\{\tfrac{h}{3}F(2h,0;\tilde{f}(0)) + \tfrac{4h}{3}F(2h,h;\tilde{f}(h)) + \tfrac{h}{3}F(2h,2h;\tilde{f}(2h))\} = g(2h), \text{ etc.},$$

and the solution proceeds in a block-by-block fashion on solving, at each stage, a pair of (in general) non-linear equations. In general, these equations must be solved iteratively. Thus, the equations for $\tilde{f}(h)$ and $\tilde{f}(2h)$ are coupled, and, say, a Newton-Raphson technique for the solution of a system of non-linear equations could be employed to solve these two equations. In the case where the integral equation is linear, the approximate solution involves the solution of a system of linear equations at each stage.

An objection to the scheme in (6.107) is that the calculation of $\tilde{f}(nh)$ involves a system of equations requiring values of $F(nh,y;\tilde{f}(y))$ where $y \geq nh$. In particular, if $F(x,y;f(y)) = K(x,y)f(y)$, the equation for $\tilde{f}((mp+1)h)$ involves $K((mp+1)h,y)$, where $y = (mp+1)h$, $(mp+2)h, \ldots, (m+1)ph$. If we define $K(x,y) = 0$ for $y > x$ then the partial derivatives $(\partial/\partial y)^r K(x,y)$ $(r \geq 0)$ *may be discontinuous* at $x = y$ and the accuracy of the approximation (6.106) will then suffer correspondingly. In general, a smooth extension of $F(x,y;f(y))$ for values of $y > x$ may be difficult to obtain, and setting $F(x,y;f(y)) = 0$ for $y > x$ may give rise to discontinuous derivatives which lower the quality of the approximation.

The previous objection can be overcome, by a suitable modification of the scheme.

<u>Example 6.35.</u>
Linz (1967*a*, 1969*a*) modifies the scheme in Example 6.34 as follows. No difficulty is encountered when discretizing the integral equation where x is an even multiple of h, so the equation where the free term in $\tilde{f}(x)$ is $\tilde{f}(2h), \tilde{f}(4h), \ldots$, respectively, is left unchanged. The non-linear

equation with a free term $\tilde{f}((2k+1)h)$ involves an approximation requiring $F((2k+1)h,\ 2(k+1)h;\ \tilde{f}(2(k+1)h))$ and we revise this equation. For $x = (2k+1)h$ we first employ Simpson's rule with step h to write

$$\int_{2kh}^{(2k+1)h} F((2k+1)h,y;f(y))\,dy \simeq \tfrac{1}{6}h\{F((2k+1)h,2kh;\tilde{f}(2kh)) +$$

$$+ 4F((2k+1)h,(2k+\tfrac{1}{2})h;\tilde{f}((2k+\tfrac{1}{2})h)) +$$

$$+ F((2k+1)h,(2k+1)h;\tilde{f}((2k+1)h))\} \quad ,$$

where the value inserted for $\tilde{f}((2k+\frac{1}{2})h)$ is the value of the quadratic interpolating $\tilde{f}(x)$ at $x = 2kh,\ (2k+1)h,\ 2(k+1)h$. Thus we have

$$\tilde{f}((2k+\tfrac{1}{2})h)) = \tfrac{3}{8}\tilde{f}(2kh) + \tfrac{3}{4}\tilde{f}((2k+1)h) - \tfrac{1}{8}\tilde{f}(2(k+1)h) \ .$$

The resulting approximations are then

$$\tilde{f}(0) = g(0) \ ,$$

$$\tilde{f}((2k+1)h) = g((2k+1)h) +$$

$$+ \tfrac{1}{3}h\{F((2k+1)h,0;\tilde{f}(0)) + 4F((2k+1)h,h;\tilde{f}(h)) +$$

$$+ \ldots + 2F((2k+1)h,\ (2k-2)h;\ \tilde{f}((2k-2)h)) +$$

$$+ 4F((2k+1)h,\ (2k-1)h;\ \tilde{f}((2k-1)h)) +$$

$$+ F((2k+1)h,\ 2kh;\tilde{f}(2kh))\} + \{\tfrac{1}{6}hF((2k+1)h,2kh;\tilde{f}(2kh)) +$$

$$+ \tfrac{2}{3}hF((2k+1)h,(2k+\tfrac{1}{2})h;\tfrac{3}{8}\tilde{f}(2kh)+\tfrac{3}{4}\tilde{f}((2k+1)h)-\tfrac{1}{8}\tilde{f}(2(k+1)h)) +$$

$$+ \tfrac{1}{6}hF((2k+1)h,(2k+1)h;\tilde{f}((2k+1)h))\} \quad , k=0,1,\ldots$$

and

$$\tilde{f}(2(k+1)h) = g(2(k+1)h) + \tfrac{1}{3}h\{F(2(k+1)h,0;\tilde{f}(0)) +$$

$$+ 4F(2(k+1)h,h;\tilde{f}(h)) + \dots +$$

$$+4F(2(k+1)h,(2k+1)h;\tilde{f}((2k+1)h))+F(2(k+1)h,2(k+1)h;\tilde{f}(2(k+1)h))\} \quad .$$

Linz (1967*a*) tabulates maximum errors when this scheme is applied to a number of linear integral equations. For the examples which he considers, the modified method described here is better than the method described in Example 6.34. *

The scheme described in Example 6.35 is clearly more complex (particularly in the case of a non-linear integral equation) than the scheme described in Example 6.34. It would be wise therefore, to dist-inguish cases where one of the techniques is more appropriate than the other.

Example 6.36.
By way of illustration, consider a linear integral equation of the form

$$f(x) - \lambda\int_0^x (x-y+\alpha)^{-2} f(y)\,dy = g(x) \quad (\alpha > 0) \quad .$$

Fox and Goodwin (1953) consider such an equation (studied by Friedlander in 1941) with $\alpha = -\lambda = 2$, $g(x) = (x+2)^{-2}$. We have

$$F(x,y;f(y)) = \lambda(x-y+\alpha)^{-2} f(y)$$

for $y \le x$, and the analytical form for $F(x,y;f(y))$ can also be used for $y > x$.

The suitability of the method indicated in Example 6.34 is determined by the existence and size of the derivatives $(\partial/\partial y)^r F(x,y;f(y))$, for $r = 1,2,3,4$, on an interval $0 \le y \le x+h$, where $0 \le x \le X$. In the present case $F(x,y;f(y))$ and all its derivatives become unbounded at $y = x + \alpha$, and it is necessary to ensure that $h \ll \alpha$ in order to obtain good accuracy. The method will fail if $h = \alpha$.

The suitability of the method indicated in Example 6.35 is also governed by the existence and size of derivatives $(\partial/\partial y)^r F(x,y;f(y))$ (and also on the derivatives of $f(y)$) but now on an interval $0 \le y \le x$, where $0 \le x \le X$. In the present case, the size of α affects the accuracy, but the accuracy is less critically dependent on the relative size of h to α , and the scheme does not break down completely if $h = \alpha$. *

Example 6.37.
The method described in Example 6.35 gives $O(h^4)$ convergence of the approximate values to the value of the solution at a fixed point, given sufficient differentiability. Campbell and Day (1970) have produced a scheme, based on similar principles, which gives $O(h^6)$ accuracy. *

Weiss (1972) gives a description of two systematic versions of the methods indicated above, and these we now outline.

It is clear that the approximation in (6.105) can be obtained naturally by repeating a basic quadrature rule, the basic rule being applied over intervals $[rph,(r+1)ph]$ $(r = 0,1,2,\ldots,(m-1))$. This technique forms part of the schemes constructed by Weiss, but we no longer restrict ourselves to rules which employ equally spaced ordinates. Accordingly, we set $h^* = ph$, and $Nh^* = X$, and we shall generalize the above. Let us choose $0 \le \theta_0 < \theta_1 < \ldots < \theta_p = 1$ and construct quadrature formulae using values of the integrand evaluated at $\theta_0,\theta_1,\ldots,\theta_p$. Such a formula is given by

$$\int_0^1 \phi(x)\,dx \simeq \sum_{k=0}^{p} \omega_k\phi(\theta_k) \quad , \tag{6.108}$$

where $\omega_0,\omega_1,\ldots,\omega_p$ are chosen so that the approximation is exact whenever $\phi(x)$ is a polynomial of degree p . From elementary numerical analysis (by integrating an interpolating polynomial) we know that such a choice is obtained when

$$\omega_k = \int_0^1 l_k(x)\,dx \ , \tag{6.109}$$

where

$$l_k(x) = \prod_{\substack{j\neq k\\ j=0}}^{p} \{(x-\theta_j)/(\theta_k-\theta_j)\} \ . \tag{6.110}$$

Remark. With the choice of weights in (6.109) the quadrature rule (6.108) is a traditional interpolatory rule, derived by integrating the interpolating polynomial

$$P(x) = \sum_{k=0}^{p} l_k(x)\phi(\theta_k) \quad .$$

The composite version of this rule is easily written down, and the error in the m-times repeated rule is $O((\frac{1}{m})^{\rho+1})$, where ρ is the degree of precision of (6.108) and $\phi(x) \in C^{\rho+1}[0,1]$. By construction, $\rho \geq p$, and a suitable choice of $\theta_0, \theta_1, \ldots, \theta_{p-1}$ gives $\rho = 2p$. (If $\frac{1}{2}-\theta_k = \frac{1}{2}+\theta_{p-k}$ $(k = 0,1,2,\ldots,p)$ then $\rho \geq 2[\frac{1}{2}p]+1$.)

The use of (6.108) in a composite form gives the approximation

$$\int_0^{mh^*} F(x,y;\tilde{f}(y))\,dy$$

$$\simeq \sum_{j=0}^{m-1} h^* \sum_{k=0}^{p} \omega_k F(x,jh^*+\theta_k h^*;\tilde{f}(jh^*+\theta_k h^*)) \ , \tag{6.111}$$

and we are led to the equation

$$\tilde{f}(mh^*) - \sum_{j=0}^{m-1} h^* \sum_{k=0}^{p} \omega_k F(mh^*,jh^*+\theta_k h^*;\tilde{f}(jh^*+\theta_k h^*)) = g(mh^*) \quad (m=0,1,2,\ldots,N), \tag{6.112}$$

on discretizing the integral equation (6.104). For a fixed value of mh^*, the local truncation error associated with (6.112) is $O(h^{\rho+1})$, where ρ is the degree of precision of (6.108), provided that $F(mh^*,x;f(x)) \in C^{\rho+1}[0,mh^*]$.

The approximation in (6.112) involves values of $\tilde{f}(jh^*+\theta_k h^*)$ and we require further equations arising from a discretization of (6.104). We therefore seek approximations of the form

$$\int_0^{\theta_j} \phi(y)\,dy \simeq \sum_{k=0}^{p} \omega_k^{[j]} \phi(\theta_k) \quad ,$$

which give rise to an approximation

$$\int_{mh^*}^{(m+\theta_j)h^*} \phi(y)\,dy \simeq h^* \sum_{k=0}^{p} \omega_k^{[j]} \phi(mh^*+\theta_k h^*) \ . \tag{6.113}$$

We obtain a method for constructing the values $\omega_k^{[j]}$ if we ensure that the approximation (6.113) is exact whenever $\phi(y)$ is a polynomial in y of degree $\leq p$. Using the functions in (6.110) we therefore set

$$\omega_k^{[j]} = \int_0^{\theta_j} l_k(y)\,dy \;. \tag{6.114}$$

Remark. The degree of precision (ρ_j^*, say) of the approximation

$$\int_0^{\theta_j} \phi(y)\,dy \simeq \sum_{k=0}^{p} \omega_k^{[j]} \phi(\theta_k)$$

is, with the choice (6.114), at least p . If $\theta_0, \theta_1, \ldots, \theta_{p-1}$ are chosen to give a high-order local truncation error in (6.108) it appears that it may not be simultaneously possible to choose them to give high order of the error in (6.113) $(j = 0,1,2,\ldots,(p-1))$. In general, and assuming sufficient differentiability, the error in (6.113) is $O((h^*)^{p+2})$ as $h^* \to 0$ with fixed mh^* $(j = 0,1,2,\ldots,(p-1))$. (Since $\theta_p = 1$, $\omega_k^{[p]} = \omega_k$ and the error in (6.113) is, in this case, $O((h^*)^{\rho+2})$, where ρ is the degree of precision of (6.108).) *

With the choice of weights $\omega_k^{[j]}$ in (6.114), we may now set

$$\tilde{f}(mh^*+\theta_j h^*) - \sum_{i=0}^{m-1} h^* \sum_{k=0}^{p} \omega_k F(mh^*+\theta_j h^*, ih^*+\theta_k h^*; \tilde{f}(ih^*+\theta_k h^*)) -$$
$$- h^* \sum_{k=0}^{p} \omega_k^{[j]} F(mh^*+\theta_j h^*, mh^*+\theta_k h^*; \tilde{f}(mh^*+\theta_k h^*)) = g(mh^*+\theta_j h^*) \tag{6.115}$$

$$(m = 0,1,2,\ldots,(N-1),\; j = 0,1,\ldots,(p-1),p) \;.$$

(If $\theta_0 = 0$, then taking $m = m_1$ and $j = p$ in eqn (6.115) repeats the equation obtained with $j = 0$ and $m = m_1+1$, since $\theta_p = 1$.) The first sum in eqn (6.115) is taken as zero when $m = 0$.

The set of equations (6.115) give a systematic extension of (6.107), and the local truncation error of (6.115) is (assuming sufficient differentiability) $O((h^*)^{\rho+1})+O((h^*)^{p+2})$ as $h^* \to 0$ with mh^* fixed, where ρ is the degree of precision of (6.108).

If we set $\theta_k = (k/p)$ $(k = 0,1,2,\ldots,p)$, the weights ω_k and $\omega_k^{[j]}$ define a scheme of the form (6.107). It may be observed, consequently, that the scheme based on eqns (6.115) possesses the same feature as a scheme based on (6.107), namely, the use of values of $F(x,y;f(y))$ for $y > x$. Where necessary, we may overcome this possible disadvantage by a revision of the scheme, at the cost of an increase in complexity. Following Weiss, I shall outline a systematic method.

The approximation (6.108) gives, on a change of variables,

$$\int_0^{\theta_j}\phi(y)\,dy \simeq \sum_{k=0}^{p}\theta_j\omega_k\phi(\theta_j\theta_k) ,$$

which in turn gives rise to

$$\int_{mh^*}^{(m+\theta_j)h^*}\phi(y)\,dy \simeq h^*\sum_{k=0}^{p}\omega_k\theta_j\phi(mh^*+\theta_j\theta_k h^*) .$$

The error in this approximation is $O((h^*)^{\rho+2})$ as $h^*\to 0$ (with mh^* fixed), where ρ is the degree of precision of (6.108) $(\rho \geq p)$ and we assume sufficient differentiability. Applying the latter approximation we obtain

$$\int_{mh^*}^{(m+\theta_j)h^*}F(x,y;f(y))\,dy$$

$$\simeq h^*\sum_{k=0}^{p}\theta_j\omega_k F(x,(m+\theta_j\theta_k)h^*;\tilde{f}((m+\theta_j\theta_k)h^*)) .$$

Since we require an approximation in terms of the values $\tilde{f}((m+\theta_j)h^*)$ $(j = 0,1,2,\ldots,p)$, we obtain the values $\tilde{f}((m+\theta_j\theta_k)h^*)$ by polynomial interpolation to these values. Thus interpolation on the interval $[mh^*,(m+1)h^*]$, by means of a polynomial of degree p, gives the approximation

$$\tilde{f}((m+\theta_j\theta_k)h^*) \simeq \sum_{r=0}^{p} l_r(\theta_j\theta_k)\tilde{f}((m+\theta_r)h^*) \quad (j,k = 0,1,2,\ldots,p) , \tag{6.116}$$

using the functions $l_r(x)$ of (6.110). (This is easily verified after some elementary manipulation.) The discretization error in (6.116) is $O((h^*)^{p+1})$, as $h^*\to 0$.

Our general scheme for replacing (6.115) now assumes the form

$$\begin{aligned}\tilde{f}(mh^*+\theta_j h^*)-\sum_{i=0}^{m-1}h^*\sum_{k=0}^{p}\omega_k F(mh^*+\theta_j h^*,ih^*+\theta_k h^*;\tilde{f}(ih^*+\theta_k h^*)) - \\ -h^*\sum_{k=0}^{p}\theta_j\omega_k F(mh^*+\theta_j h^*,mh^*+\theta_j\theta_k h;\sum_{r=0}^{p}l_r(\theta_j\theta_k)\tilde{f}(mh^*+\theta_r h^*))=g(mh^*+\theta_j h^*)\end{aligned} \tag{6.117}$$

$$(m = 0,1,2,\ldots,(N-1),\ j = 0,1,2,\ldots,\ p-1,p).$$

From our earlier remarks we see that the local truncation error of (6.117) is $O((h^*)^{\rho+1}) + O((h^*)^{p+2})$, where ρ is the degree of precision of (6.108), and sufficient differentiability is assumed.

The scheme described in Example 6.35 is a particular case of (6.117), obtained using equally spaced values of θ_k .

Remark. If we consider (6.117) with $m = 0$, and suppose for simplicity that $g(x) = 0$, we have (with $A_{r,k}^{[s]} = l_r(\theta_s\theta_k)$)

$$\tilde{f}(\theta_s h^*) = h^*\theta_s \sum_{k=0}^{p} \omega_k F(\theta_s h^*, \theta_s\theta_k h^*; \sum_{r=0}^{p} A_{r,k}^{[s]}\tilde{f}(\theta_r h^*)) \quad (s = 0,1,2,\ldots,p).$$

Comparing this formula with the Runge-Kutta formulae discussed in section 6.6 we see that there is some motivation for referring to the scheme (like Weiss 1972) as an *implicit Runge-Kutta method*, particularly if $\theta_0,\theta_1,\theta_2,\ldots,\theta_{p-1}$ are chosen to maximize the order of the local truncation error.

Remark. The order of the local truncation error associated with (6.108) is at least $O(h^{p+1})$ as $h \to 0$ with fixed mh , provided $F(x,y;f(y))$ and $f(y)$ are suitably differentiable functions of y . Even when the degree of precision of (6.108) is maximized, the order of the local truncation error is restricted to $O((h^*)^{p+2})$ by the accuracy of the approximation (6.116). For $m > 1$, the order of the local truncation error could be improved by approximating $\tilde{f}((m+\theta_j\theta_k)h^*)$ by means of the value of the interpolating polynomial of appropriate degree agreeing with the values $\tilde{f}(([m-1]+\theta_j)h^*)$, $\tilde{f}((m+\theta_j)h^*)$ $(j = 0,1,2,\ldots,p)$. (The exact degree of the polynomial depends upon whether or not $\theta_0 = 0$, but it is at least $2p$ and the order of local truncation error of (6.119) now rises to $O((h^*)^{\rho+1}) + O((h^*)^{2p+2})$, where $\rho \le 2p$, except in the case $m = 0$.) The computation would require little additional effort. We have not seen this suggestion implemented. *

For an equation of the form

$$f(x) - \lambda\int_0^x f(y)\,dy = g(x) \ ,$$

the methods associated with (6.117) and (6.115) are the same. The block-stability properties (section 6.3) of the methods applied to an equation of this type can be determined readily; the methods already form block-by-block schemes. With ascending values of p we take equally spaced values $\{\theta_i\}$, with $\theta_p=1$, and find the eigenvalues of largest modulus of the corresponding 'amplification matrix' $\underset{\sim}{M}$ of the recurrence, as follows†:

Case (i): $\theta_0 = 0.$

$p = 1$

$$(1 + \tfrac{1}{2}\lambda h)/(1 - \tfrac{1}{2}\lambda h)$$

$p = 2$

$$(1 + \tfrac{1}{2}\lambda h + \tfrac{1}{12}\lambda^2 h^2)/(1 - \tfrac{1}{2}\lambda h + \tfrac{1}{12}\lambda^2 h^2)$$

$p = 3$

$$(1 + \tfrac{1}{2}\lambda h + \tfrac{11}{108}\lambda^2 h^2 + \tfrac{1}{108}\lambda^3 h^3)/(1 - \tfrac{1}{2}\lambda h + \tfrac{11}{108}\lambda^2 h^2 - \tfrac{1}{108}\lambda^3 h^3)\,.$$

Case (ii): $\theta_0 \neq 0.$

$p = 0$

$$1/(1 - \lambda h)$$

$p = 1$

$$(1 + \tfrac{1}{4}\lambda h)/(1 - \tfrac{3}{4}\lambda h + \tfrac{1}{4}\lambda^2 h^2)$$

$p = 2$

$$(1 + \tfrac{1}{3}\lambda h + \tfrac{1}{27}\lambda^2 h^2)/(1 - \tfrac{2}{3}\lambda h + \tfrac{11}{54}\lambda^2 h^2 - \tfrac{1}{27}\lambda^3 h^3)\,.$$

The above expressions for eigenvalues are rational approximations to $\exp \lambda h$. We obtain 'absolute stability' if the modulus of the appropriate eigenvalue is less than unity, and 'relative stability' if it is less than $\exp \lambda h$.

Example 6.38.

The block-by-block method defined by eqn (6.117) was applied to a number

---

† Recall that the block-by-block formulation (and hence $\underset{\sim}{M}$) is not defined uniquely. In some formulations we encounter removable unit eigenvalues. Further details of the stability result appear in Baker and Keech (1975).

of test equations of the form

$$f(x) + \int_0^x K(x,y)f(y)\,dy = g(x) \quad ,$$

where the kernel $K(x,y)$ and the solution $f(x)$ had the form indicated:

(a) $K(x,y) = \cos(x-y)$, $f(x) = 1$

(b) $K(x,y) = \exp(x-y)$, $f(x) = \exp(-x)$

(c) $K(x,y) = \surd(x-y)$, $f(x) = \frac{15}{4}x$

(d) $K(x,y) = \exp\{2(x-y)\}$, $f(x) = \cos x - 2\sin x$

(e) $K(x,y) = \exp(x-y)$, $f(x) = 2x-x^2$

(f) $K(x,y) = \cos(x-y)$, $f(x) = 2\sin x$

(g) $K(x,y) = \sinh(x-y)$, $f(x) = 6(x-x^2)\exp(-x)$

(h) $K(x,y) = \cosh(x-y)$, $f(x) = 1-\frac{1}{2}x^2$,

the inhomogeneous term $g(x)$ being chosen to give the solution indicated. The numerical methods were applied with varying choices of $\theta_r (r = 0,1,2,\ldots,p)$:

Case (i) $\theta_0 = 0,\ \theta_1 = \frac{1}{4},\ \theta_2 = \frac{1}{2}.\ \theta_3 = \frac{3}{4},\ \theta_4 = 1 \quad (p = 4)$,

Case (ii) $\theta_0 = 0,\ \theta_1 = 0{\cdot}2763932,\ \theta_2 = 0{\cdot}7236068,\ \theta_3 = 1 \ (p = 3)$,

Case (iii) $\theta_0 = 0,\ \theta_1 = \frac{1}{3},\ \theta_2 = \frac{2}{3},\ \theta_3 = 1 \quad (p = 3)$

and step $h = 0.075$. The numerical solution was computed on $[0, 1{\cdot}5]$ in each case.

Maximum absolute errors in the computed approximations were:

Case (i): (a) 4×10^{-13}, (b) 5×10^{-12}, (c) 8×10^{-4}, (d) 3×10^{-11}, (e) 4×10^{-12}, (f) 2×10^{-12}, (g) 2×10^{-10}, (h) 1×10^{-12}

Case (ii): (a) 1×10^{-13}, (b) 2×10^{-10}, (c) 9×10^{-4}, (d) 3×10^{-10}, (e) 6×10^{-12}, (f) 4×10^{-10}, (g) 2×10^{-10}, (h) 2×10^{-12}

Case (iii): (a) 3×10^{-9}, (b) 6×10^{-8}, (c) 2×10^{-3},
(d) 9×10^{-7}, (e) 1×10^{-7}, (f) 2×10^{-9},
(g) 1×10^{-6}, (h) 2×10^{-8}.

The relatively bad performance of the methods in producing an accurate solution in (c) is due to the discontinuity in the derivative $(\partial/\partial y)K(x,y)$ at $x = y$.

Since it was clear from the figures above that the numerical method could cope very well with the majority of the test equations, new results were computed with a larger value of h over the extended range $[0,15]$. The following tables give a sample of the behaviour of the errors. Maximum absolute errors in case (ii) with $h = 0{\cdot}75$ were:

(a) 5×10^{-8}; (b) 1×10^{-5}; (c) 2×10^{-1}; (d) 8×10^{1};
(e) 3×10^{-5}; (f) 3×10^{-5}; (g) 8×10^{-4}; (h) 3×10^{-3}.

The figures in (d) display a reasonable accuracy up to $x = 1{\cdot}5$ (maximum error 7×10^{-5}), the maximum error increasing to 4×10^{-2} at $x = 7{\cdot}5$ whereupon the error grows to dominate the computed solution. A comparable (but somewhat smaller) error growth occurs using the θ_i of case (i), for this example. *

6.7.1.

A type of method involving the use of *spline functions* (and, in particular, so-called 'deficient' spline functions mentioned below) has been discussed in the literature. The method reduces in a special case to one of the block-by-block methods described above, as we note in Example 6.39. The integral equation considered here is eqn (6.104), as before.

We choose a step h^* with $Nh^* = X$, and we select $q + 1$ values θ_i with $0 = \theta_0 < \theta_1 < \ldots < \theta_{q-1} < \theta_q = 1$, for use later. The approximation $\tilde{f}(x)$ to the solution $f(x)$ is chosen as a function in $C^{p-q}[0,X]$ which is a polynomial of degree p in each interval $[mh^*, (m+1)h^*]$ $(m = 0,1,2,\ldots)$. The values of p,q are parameters of the method; if $q = 1$ then $\tilde{f}(x)$ is a conventional spline of degree p in the sense of Example 2.4. If $q > 1$, then $\tilde{f}(x)$ is known as a deficient spline, and in the case $q = p$, $\tilde{f}(x)$ is a continuous function which is a polynomial of degree p in each interval $[0,h^*]$, $[h^*,2h^*]$,

$[2h^*,3h^*]$, ...

In the general case we can represent $\tilde{f}(x)$ by setting (with a suitable choice of parameters $a_1^{(r)}, a_2^{(r)}, \ldots, a_q^{(r)}$):

$$\tilde{f}(x) = \sum_{j=0}^{p-q} \frac{1}{j!}(x-rh^*)^j \tilde{f}^{(j)}(rh^*) + \sum_{j=1}^{q} (x-rh^*)^{p-q+j} a_j^{(r)} , \qquad (6.118)$$

for $rh^* \leq x \leq (r+1)h^*$ $(r = 0,1,2,\ldots)$. The required approximation $\tilde{f}(x)$, of this form, is obtained by asking that

$$\tilde{f}^{(j)}(0) = f^{(j)}(0) \quad (j = 0,1,2,\ldots,(p-q)) ,$$

where $f(x)$ is the solution of the given equation, and also requiring (for $m = 0,1,2,\ldots,(N-1)$, $j = 1,2,3,\ldots,q$) that

$$\tilde{f}(mh^*+\theta_j h^*) = g(mh^*+\theta_j h^*) + Q_{m,j}\{F(mh^*+\theta_j h^*,y;\tilde{f}(y))\} , \qquad (6.119)$$

where $Q_{m,j}\{F(mh^*+\theta_j h^*,y;\tilde{f}(y))\}$ is an approximation to

$$\int_0^{mh^*+\theta_j h^*} F(mh^*+\theta_j h^*,y;\tilde{f}(y))dy .$$

Example 6.39.

Suppose that $\theta_j = j/q$ and

$$Q_{m,j}\{F(mh^*+\theta_j h^*,y;\tilde{f}(y))\} =$$

$$= \sum_{k=0}^{r} w_{r,k} F(mh^*+\theta_j h^*,kh^*/q;\tilde{f}(kh^*/q)) ,$$

where the weights $w_{r,j}$ are a set of weights suggested in section 6.1, and $r = mq+j$. Then $\tilde{f}(x)$ is merely a (possibly deficient) spline interpolating the values $\tilde{f}(0),\tilde{f}(h),\tilde{f}(2h)$ computed using the quadrature method with weights $\{w_{r,k}\}$ and a step $h = h^*/q$. *

To define the general method more precisely, the form of $Q_{m,j}\{\quad\}$ must be specified, each type of approximation giving rise to a different method. In particular, suppose that (with $0 \le t_k \le 1$ for $k=0,1,\ldots,n$)

rules of the form $\int_0^{\theta_j} \phi(t)dt \simeq \sum_{k=0}^{n} \omega_k^{[j]}\phi(t_k)$ are known, and set

$\omega_k = \omega_k^{[q]}$. Then we may set, for $Q_{m,j}\{F(mh^*+\theta_j h^*, y; \tilde{f}(y))\}$,

$$h^* \sum_{r=0}^{m} \sum_{k=0}^{n} \gamma_{rk} F(mh^*+\theta_j h^*, (r+t_k)h^*; \tilde{f}((r+t_k)h^*)) \tag{6.120}$$

on taking $\gamma_{rk} = \omega_k$ if $r \le m-1$, and $\omega_k^{[j]}$ if $r = m$. An illustration is

$$\begin{aligned} &h^* \sum_{r=0}^{m-1} \sum_{k=0}^{n} \omega_k F(mh^*+\theta_j h^*, (r+t_k)h^*; \tilde{f}((r+t_k)h^*)) + \\ &\quad + \theta_j h^* \sum_{k=0}^{n} \omega_k F(mh^*+\theta_j h^*, (m+\theta_j t_k)h^*; \tilde{f}((m+\theta_j t_k)h^*)) . \end{aligned} \tag{6.121}$$

With $\Delta\theta_k = \theta_{k+1} - \theta_k$, we can form alternative types of approximation:

$$\begin{aligned} &h^* \sum_{r=0}^{m-1} \sum_{i=0}^{q-1} \sum_{k=0}^{n} \omega_k \Delta\theta_i F((m+\theta_j)h^*, (r+t_k\Delta\theta_i)h^*; \tilde{f}((r+t_k\Delta\theta_i)h^*)) + \\ &\quad + h^* \sum_{i=0}^{j-1} \sum_{k=0}^{n} \Delta\theta_i \omega_k F((m+\theta_j)h^*, (m+t_k\Delta\theta_i)h^*; \tilde{f}((m+t_k\Delta\theta_i)h^*)) . \end{aligned} \tag{6.122}$$

Example 6.40.

Suppose that we take $q = p = n$ and $t_k = \theta_k$ in (6.121). Then the function $\tilde{f}(x)$ is merely a piecewise polynomial interpolant to the values computed using (6.117). (The scheme represented by (6.115) can also be obtained by a suitable choice of $Q_{m,j}\{\ \ \}$ and by taking $q = p$.) *

In the general formulation of the method, the use of an approximation such as (6.120), (6.121), or (6.122), for $Q_{m,j}\{\ \ \}$, and the substitution of representation (6.118) in eqn (6.119), gives rise to an equation for the values $a_k^{(m)}$ $(k = 1,2,3,\ldots,q)$. The solution can therefore proceed as follows. With $m = 0$, eqn (6.119) with $j = 1,2,3,\ldots,q$ gives q equations in q unknowns $a_k^{(0)}$ $(k=1,2,3,\ldots,q)$.

These equations involve the values $\tilde{f}(0), \tilde{f}'(0), \ldots, \tilde{f}^{(p-q)}(0)$ which must be determined from the integral equation. Determination of $a_k^{(0)}$ $(k = 1,2,3,\ldots,q)$ permits the evaluation of $\tilde{f}^{(j)}(h^*)$ $(j = 0,1,2,\ldots,(p-q))$ and, in particular, $\tilde{f}(h^*)$. The substitution of (6.118) in (6.119) with $m = 1$ and $j = 1,2,3,\ldots,q$ gives q equations for q unknowns $a_k^{(1)}$ $(k = 1,2,3,\ldots,q)$ and so on, for $m = 2,3,4,\ldots$.

Example 6.41.

We consider here an application of the foregoing scheme with $p = 2$ and $q = 1$. We have $\theta_0 = 0$ and $\theta_1 = 1$ and we replace h^* by h and consider the linear integral equation

$$f(x) - \lambda\int_0^x K(x,y)f(y)dy = g(x) .$$

In the interval $[0,h]$, $\tilde{f}(x)$ has the form $\tilde{f}(x) = f(0)+xf'(0)+ +a^{(0)}x^2$. We have $f(0) = g(0)$ and $f'(0) = g'(0)+\lambda K(0,0)g(0)$. To obtain $a^{(0)}$ we shall employ an approximation of the form (6.120) and require that

$$f(0)+hf'(0)+h^2a^{(0)} = g(h)+\lambda h\sum_{j=0}^{n}\omega_jK(h,t_jh)\{f(0)+t_jhf'(0)+(t_jh)^2a^{(0)}\} .$$

This equation has a solution $a^{(0)}$ provided that

$$\{1-\lambda h\sum_{j=0}^{n}\omega_jK(h,t_jh)t_j^2\} \neq 0 ,$$

that is, assuming $K(x,y)$ continuous, if h is sufficiently small (since $\Sigma_j\,\omega_j$ is fixed). Now

$$\tilde{f}(h) = f(0)+hf'(0)+a^{(0)}h^2$$

and

$$\tilde{f}'(h) = f'(0)+2a^{(0)}h .$$

In $[h,2h]$ we have

$$\tilde{f}(x) = \tilde{f}(h)+(x-h)\tilde{f}'(h)+a^{(1)}(x-h)^2 ,$$

and we require that

$$\tilde{f}(h)+h\tilde{f}'(h)+a^{(1)}h^2 = g(2h) + \lambda h \sum_{j=0}^{n} \omega_j K(2h,t_j h)\{f(0) + $$

$$+ t_j hf'(0) + (t_j h)^2 a^{(0)}\} +$$

$$+ \lambda h \sum_{j=0}^{n} \omega_j K(2h,h+t_j h)\{\tilde{f}(h)+t_j h\tilde{f}'(h)+(t_j h)^2 a^{(1)}\} \quad .$$

Since $a^{(0)}$ and hence $\tilde{f}(h)$ and $\tilde{f}'(h)$ are known, this equation for $a^{(1)}$ can be solved if h is sufficiently small, and $\tilde{f}(2h)$ and $\tilde{f}'(2h)$ can then be computed. The solution proceeds in a like manner at subsequent steps.

The application of the method to the non-linear equation

$$f(x) - \int_0^x F(x,y;f(y))\mathrm{d}y = g(x)$$

follows similar lines, which should be clear. In particular, we require $\tilde{f}(0) = g(0)$ and $\tilde{f}'(0) = g'(0)+F(0,0;g(0))$ to start the process. The equation for $a^{(0)}$ has the form

$$\tilde{f}(0)+h\tilde{f}'(0)+h^2a^{(0)}=g(h)+ h\sum_{j=0}^{n} \omega_j F(h,t_j h;\tilde{f}(0)+t_j h\tilde{f}'(0)+(t_j h)^2 a^{(0)}) \quad .$$

We obtain a non-linear equation at each stage for $a^{(r)}$, which, if h is sufficiently small, is unique and may be obtained by an iterative method. Thus the predictor-corrector methods have a natural analogue, since $a^{(r)}$ appears linearly in the representation of $\tilde{f}(rh)$. For $r = 0,1,2,\ldots,$ we can construct iterates $\alpha_k^{(r)}$ defined by equations

$$h^2\alpha_{k+1}^{(r)} = \gamma_r-\beta_r h+ h\sum_{j=0}^{n} \omega_j F((r+1)h,(r+t_j)h;\tilde{f}(rh)+t_j h\tilde{f}'(rh)+(t_j h)^2\alpha_k^{(r)}))$$

(where γ_r and β_r can be written down) and $\lim_{k\to\infty} \alpha_k^{(r)} = a^{(r)}$ if h is sufficiently small. (A convenient starting value for the iteration is $\alpha_0^{(r)} = a^{(r-1)}$, if $r \geq 1$; the usual conditions on $F(x,y;v)$ are assumed. The results below arise on iterating to numerical convergence.)

In the following results we indicate the error in approximations computed using $n = 2$, $t_0 = 0$, $t_1 = \frac{1}{2}$, $t_2 = 1$ and $\omega_0 = \omega_2 = \frac{1}{6}$, $\omega_1 = \frac{2}{3}$ (Simpson's rule) and with $h = \frac{2}{5}$. The equations considered are

(a)

$$f(x) - \int_0^x e^{-x}\{f(y)\}^2 dy = \tfrac{1}{2}e^{-x}(1+e^{-2x})$$

with solution $f(x) = e^{-x}$ and

(b)

$$f(x) + \int_0^x (1+x-y)\{f(y)\}^2 dy = 1+x+\tfrac{1}{2}x^2 ,$$

with solution $f(x) = 1$. We tabulate absolute errors:

x	(a)	(b)	x	(a)	(b)
0·4	$3{\cdot}6 \times 10^{-4}$	$1{\cdot}3 \times 10^{-6}$	6·0	$1{\cdot}1 \times 10^{-6}$	$5{\cdot}1 \times 10^{-6}$
0·8	$1{\cdot}3 \times 10^{-4}$	$8{\cdot}0 \times 10^{-7}$	8·0	$1{\cdot}5 \times 10^{-7}$	$2{\cdot}2 \times 10^{-5}$
1·2	$1{\cdot}5 \times 10^{-4}$	$1{\cdot}3 \times 10^{-6}$	10·0	$2{\cdot}0 \times 10^{-8}$	$8{\cdot}1 \times 10^{-5}$
1·6	$7{\cdot}7 \times 10^{-5}$	$1{\cdot}0 \times 10^{-6}$	12·0	$2{\cdot}7 \times 10^{-9}$	$3{\cdot}2 \times 10^{-4}$
2·0	$6{\cdot}4 \times 10^{-5}$	$3{\cdot}9 \times 10^{-7}$	14·0	$3{\cdot}6 \times 10^{-10}$	$1{\cdot}2 \times 10^{-3}$
4·0	$8{\cdot}0 \times 10^{-6}$	$3{\cdot}3 \times 10^{-6}$	16·0	$4{\cdot}9 \times 10^{-11}$	$4{\cdot}6 \times 10^{-3}$. *

<u>Example 6.42</u>.

We consider, now, the application of the general method with $p = 3$ and $q = 2$ to the equation

$$f(x) - \int_0^x F(x,y;f(y))dy = g(x) .$$

Thus, $\tilde{f}(x)$ is a *deficient* cubic spline. (Similar methods with $p = 3$ and $q = 1$, in which $\tilde{f}(x)$ is a conventional cubic spline, are known to be unsatisfactory.) With $q = 2$ we shall take $\theta_0 = 0$, $\theta_1 = \frac{1}{2}$, $\theta_2 = 1$ and we now represent $\frac{1}{2}h^*$ by h. In the interval $[0,h^*] = [0,2h]$ the function $\tilde{f}(x)$ has the representation

$$\tilde{f}(x) = f(0)+xf'(0)+a_1^{(0)}x^2+a_2^{(0)}x^3 \; .$$

With an approximation of the form (6.121) we require of $\tilde{f}(x)$ that (say)

$$f(0)+hf'(0)+a_1^{(0)}h^2+a_2^{(0)}h^3 = g(h) +$$

$$+ h\sum_{j=0}^{n} \omega_j F(h,t_jh;f(0)+t_jhf'(0)+a_1^{(0)}(t_jh)^2+a_2^{(0)}(t_jh)^3) \; ,$$

and

$$f(0)+2hf'(0)+4a_1^{(0)}h^2+8a_2^{(0)}h^3 = g(2h) +$$

$$+ 2h\sum_{j=0}^{n} \omega_j F(2h,2t_jh;f(0)+2t_jhf'(0)+4a_1^{(0)}(t_jh)^2+8a_2^{(0)}(t_jh)^3)$$

and we obtain two non-linear equations in $a_1^{(0)}$ and $a_2^{(0)}$. When such values exist, and in particular if h is sufficiently small, the generalized Newton method can be applied (if the derivatives of $F(x,y;v)$ are known) in order to determine the solutions. In the case of a linear integral equation, the equations in $a_1^{(0)}$ and $a_2^{(0)}$ are linear.

If $a_1^{(0)}$ and $a_2^{(0)}$ have been determined we may write down

$$\tilde{f}(h) = f(0)+hf'(0)+a_1^{(0)}h^2+a_2^{(0)}h^3 \; ,$$

$$\tilde{f}(2h) = f(0)+2hf'(0)+4a_1^{(0)}h^2+8a_2^{(0)}h^3 \; ,$$

$$\tilde{f}'(h) = f'(0)+2ha_1^{(0)}+3a_2^{(0)}h^2 \; ,$$

$$\tilde{f}'(2h) = f'(0)+4ha_1^{(0)}+12a_2^{(0)}h^2 \; ,$$

and solve the equations

$$\tilde{f}(2h)+h\tilde{f}'(2h)+a_1^{(1)}h^2+a_2^{(1)}h^3 = g(3h) +$$

$$+2h\sum_{j=0}^{n} \omega_j F(3h,2t_jh;f(0)+2t_jhf'(0)+4a_1^{(0)}(t_jh)^2+8a_2^{(0)}(t_jh)^3) +$$

$$+ h\sum_{j=0}^{n} \omega_j F(3h,(2+t_j)h;\tilde{f}(2h)+t_jh\tilde{f}'(2h)+a_1^{(1)}(t_jh)^2+a_2^{(1)}(t_jh)^3)$$

and

$$\tilde{f}(2h)+2h\tilde{f}'(2h)+4a_1^{(1)}h^2+8a_2^{(1)}h^3 = g(4h) +$$

$$+ 2h \sum_{i=0}^{1} \sum_{j=0}^{n} \omega_j F(4h,(2i+2t_j)h;\tilde{f}(2ih)+2t_jh\tilde{f}'(2ih)+4a_1^{(i)}(t_jh)^2+8a_2^{(i)}(t_jh)^3)$$

for $a_1^{(1)}$ and $a_2^{(1)}$. We then compute $\tilde{f}(3h)$ and $\tilde{f}(4h)$, and the calculation of $a_1^{(r)}$ and $a_2^{(r)}$ $(r = 2,3,4,\ldots)$ proceeds similarly. *

In Example 6.42 we see an instance of a method in which a set of q parameters is computed at each step, and from these parameters a set of q values of $\tilde{f}(x)$ is computed as part of the computational scheme. The general method described can therefore be regarded as a q-component block-by-block method. Because of the general nature of our description, a wide class of methods is covered, and we must warn the reader that not all cases give rise to satisfactory schemes. Thus if $q = 1$, obvious choices of $Q_{m,j}\{\ \}$ lead to schemes which are *non-convergent* as $h^* \to 0$, when $p \geq 3$. For further reading, refer to Hung (1970), El Tom (1974). (The method discussed by Hung differs, in the case $p \geq 3$, from methods covered here.) See also the remarks of Baker and Keech (to appear), where stability is discussed.

Example 6.43.

In this example we consider the application of the general method using $p = 3$, $q = 2$, $\theta_0 = 0$, $\theta_1 = \frac{1}{2}$, $\theta_2 = 1$ and for $Q_{m,j}\{F((m+\theta_j)h^*, y;\tilde{f}(y))\}$ we take an approximation of the form (6.122). We take as the approximation the sum

$$\tfrac{1}{2}\sum_{k=0}^{n} w_{n,k}F((m+\theta_j)h^*, \tfrac{1}{2}kh;\tilde{f}(\tfrac{1}{2}kh)) \quad (\text{for } j \in \{1,2\}) ,$$

where $h = \frac{1}{2}h^*(=\theta_1 h^*)$, and where $n = 4m+2j$ and the weights $w_{n,k}$ are those in Table 6.3. Thus, $Q_{mj}\{\ \}$ is the approximation to an integral over $[0,(m+\theta_j)h^*]$ obtained with the repeated Simpson's rule using function values at a uniform interval $\frac{1}{4}h^*$, θ_j being either $\frac{1}{2}$ or 1.

The equations to which the method was applied are those discussed in Example 6.41. The corresponding absolute errors obtained with the present method and $h^* = \frac{2}{5}$ are displayed below:

x	(a) (solution $f(x) = e^{-x}$)	(b) (solution $f(x) = 1$)
0·4	$4{\cdot}0 \times 10^{-6}$	$1{\cdot}0 \times 10^{-7}$
0·8	$3{\cdot}6 \times 10^{-6}$	$3{\cdot}5 \times 10^{-7}$
1·2	$2{\cdot}5 \times 10^{-6}$	$1{\cdot}5 \times 10^{-7}$
1·6	$1{\cdot}6 \times 10^{-6}$	$1{\cdot}1 \times 10^{-7}$
2·0	$1{\cdot}0 \times 10^{-6}$	$5{\cdot}0 \times 10^{-8}$
...	...	...
...	...	...
4·0	$1{\cdot}2 \times 10^{-7}$	$1{\cdot}4 \times 10^{-7}$
6·0	$1{\cdot}5 \times 10^{-8}$	$2{\cdot}8 \times 10^{-7}$
8·0	$2{\cdot}1 \times 10^{-9}$	$7{\cdot}8 \times 10^{-7}$
	...	... *

6.8. *Product-integration methods*

Consider the linear Volterra integral equation

$$f(x) - \lambda\int_0^x K(x,y)f(y)\mathrm{d}y = g(x) ,$$

and suppose that we seek the solution for $0 \leq x \leq X$. If the solution $f(x)$ is known to be smooth, we may apply a simple product-integration technique such as was introduced in Chapter 4 and discussed further in Chapter 5.

We require approximations of the form, say,

$$\int_0^x K(x,y)\phi(y)\mathrm{d}y \simeq \sum_{j=0}^{N} \nu_j(x)\phi(jh) \quad (x = h,2h,3h,\ldots,Nh), \tag{6.123}$$

where $Nh = X$. As we have seen, such approximations can generally be constructed in such a way that they are exact when $\phi(x)$ is a continuous function which is piecewise polynomial. In particular the approximation may be constructed to be exact (a) if $\phi(x)$ is linear in each interval $[ih,(i+1)h]$ $(i = 0,1,2,\ldots,(N-1))$ (b) if $\phi(x)$ is quadratic in each interval $[2ih,2(i+1)h]$, and so on.

Let us define, in the notation used for splines, $K_+(x,y) = (x-y)_+^0 K(x,y)$. That is,

$$K_+(x,y) = K(x,y) \text{ if } y \le x, \text{ and } K_+(x,y) = 0 \text{ if } y > x .$$

The approximation (6.123) assumes the form

$$\int_0^X K_+(x,y)\phi(y)\,dy \simeq \sum_{j=0}^{N} \nu_j(x)\phi(jh) , \tag{6.124}$$

which is of the type required in the application of product integration to the approximate solution of Fredholm integral equations. However, we shall now find that $\nu_j(x)$ vanishes if $jh-x$ is sufficiently large.

Consider case (a), and suppose that $kh \le x \le (k+1)h$. Since

$$K_+(x,y) = 0 \quad \text{if} \quad y > x$$

it is clear that $\nu_j(x) = 0$ if $j > (k+1)$. We find

$$\nu_0(x) = \frac{1}{h}\int_0^h K_+(x,y)(h-y)\,dy$$

and, with $kh \le x \le (k+1)h$,

$$\begin{aligned} \nu_j(x) = {} & \frac{1}{h}\int_{jh}^{(j+1)h} K_+(x,y)\{(j+1)h-y\}\,dy \; + \\ & + \frac{1}{h}\int_{(j-1)h}^{jh} K_+(x,y)\{y-(j-1)h\}\,dy \quad (j = 1,2,3,\ldots,k \text{ and } j = k+1). \end{aligned} \tag{6.125}$$

(The first term in (6.125) vanishes in the case $j = k+1$.) The associated rule (6.124) is the generalized trapezium rule, and its error is $O(h^2)$ as $h \to 0$ with fixed x .

If we treat case (b) similarly we obtain the generalized Simpson's rule. We then find that if $kh \le x \le (k+1)h$, then $\nu_j(x) = 0$ when $j > k+2$. Indeed, we find that

$$\nu_0(x) = \frac{1}{2h^2}\int_0^{2h} K_+(x,y)\{(h-y)(2h-y)\}dy$$

and, generally,

$$\nu_{2j+1}(x) = \frac{-1}{h^2}\int_{2jh}^{2(j+1)h} K_+(x,y)\{(y-2jh)(y-2(j+1)h)\}dy\ , \qquad (6.126)$$

whilst

$$\nu_{2j}(x) = \frac{1}{2h^2}\int_{2(j-1)h}^{2jh} K_+(x,y)\{(y-2(j-1)h)(y-(2j-1)h)\}dy\ + \qquad (6.127)$$

$$+\ \frac{1}{2h^2}\int_{2jh}^{2(j+1)h} K_+(x,y)\{(y-(2j+1)h)(y-2(j+1)h)\}dy\ ,$$

where $K_+(x,y) = K(x,y)$ if $y \leq x$, and $K_+(x,y) = 0$ if $x < y$. Thus, the second term on the right in (6.127) is zero if $x = 2jh$. The associated error in (6.123) is $O(h^3)$ as $h \to 0$ with fixed x, provided $\phi(x)$ is sufficiently differentiable.

Another, more simple, approximation is given by the generalized mid-point rule. In the form (6.123) we set $\nu_{2j}(x) = 0$ $(j=0,1,2,\ldots)$ and

$$\nu_{2j+1}(x) = \int_{2jh}^{2(j+1)h} K_+(x,y)dy\ .$$

Then $\nu_{2j+1}(x) = 0$ if $x \leq 2jh$, and the accuracy of (6.123) is $O(h)$, in general.

The application to the linear equation of the approximation (6.123) associated with the (repeated) generalized trapezium rule or the (repeated) generalized mid-point rule gives a system of equations

$$\tilde{f}(ih) - \lambda\sum_{j=0}^{i}\nu_j(ih)\tilde{f}(jh) = g(ih) \qquad (6.128)$$

in which the matrix of coefficients is triangular. These particular product integration formulae therefore give rise to a system of equations which can be solved in a step-by-step fashion. However, the use of the generalized Simpson's rule gives a system of equations

$$\tilde{f}(ih) - \lambda \sum_{j=0}^{i+1} \nu_j(ih)\tilde{f}(jh) = g(ih) \quad (i = 1,2,3,\ldots) \tag{6.129}$$

with $\tilde{f}(0) = g(0)$ in which $\nu_{i+1}(ih) = 0$ if i is even (see eqn (6.126)), but $\nu_{i+1}(ih)$ is non-zero if i is odd (see eqn (6.127)). The system of coefficients in the equations for $\tilde{f}(0), \tilde{f}(h), \tilde{f}(2h)$ thus has the form shown in Table 6.10,

TABLE 6.10

i	$\tilde{f}(0)$	$\tilde{f}(h)$	$\tilde{f}(2h)$	$\tilde{f}(3h)$	$\tilde{f}(4h)$
0	x				
1	x	x	x		
2	x	x	x		
3	x	x	x	x	x
4	x	x	x	x	x
.	.	.	.	.	.
.	.	.	.	.	.
.	.	.	.	.	.

where x denotes a non-zero entry. The use of this product integration method gives a block-by-block method in which ($\tilde{f}(0)$ being known) we solve a system of equations to give $\tilde{f}(h), \tilde{f}(2h)$, another to obtain $\tilde{f}(3h), \tilde{f}(4h)$, and so on.

Consider, now, a general scheme, and suppose that we employ a product integration formula of the form

$$\int_0^x K(x,y)\phi(y)\,dy \simeq \sum_{j=0}^{N} \nu_j(x)\phi(jh)$$

which is exact when $\phi(x)$ is continuous and a polynomial of degree p in each interval $[rph, (r+1)ph]$. The resulting scheme

$$\tilde{f}(kh) - \lambda \sum_{j=0}^{N} \nu_j(kh)\tilde{f}(jh) = g(kh) \tag{6.130}$$

gives rise to a p-component block-by-block method in which we compute $\tilde{f}(h)$, $\tilde{f}(2h)$,..., $\tilde{f}(ph)$ followed by $\tilde{f}((p+1)h), \tilde{f}((p+2)h)$, ... , $\tilde{f}(2ph)$, etc. (Variations in the block structure may be possible.)

Example 6.44.

The earliest suggestion of a systematic application of product-integration techniques to the construction of block-by-block methods for the numerical solution of Volterra equations appears to be that of Young (1954*b*). Wagner (1954) considers the use of the generalized Simpson's rule in the case where $K(x,y)$ is a difference kernel, $k(x-y)$.

Makinson and Young (1960) study test results obtained on applying the block-by-block method indicated by (6.130) in the case $p = 4$. Following the spirit of Makinson and Young (q.v.) we write

$$\tilde{\underset{\sim}{f}}_1 = [\tilde{f}(0),\tilde{f}(h),\ldots,\tilde{f}(4h)]^T \quad ,$$

$$\underset{\sim}{g}_1 = [g(0),g(h),\ldots,g(4h)]^T \quad ,$$

$$\tilde{\underset{\sim}{f}}_2 = [\tilde{f}(4h),\tilde{f}(5h),\ldots,\tilde{f}(8h)]^T \quad , \text{ etc.}$$

and the block-by-block method reduces to the form

$$\tilde{\underset{\sim}{f}}_1 - \underset{\sim}{B}_{11}\tilde{\underset{\sim}{f}}_1 = \underset{\sim}{g}_1 \quad ,$$

$$\tilde{\underset{\sim}{f}}_2 - \underset{\sim}{A}_{21}\tilde{\underset{\sim}{f}}_1 - \underset{\sim}{B}_{22}\tilde{\underset{\sim}{f}}_2 = \underset{\sim}{g}_2 \quad ,$$

and, in general,

$$\tilde{\underset{\sim}{f}}_r - \sum_{s=1}^{r-1} \underset{\sim}{A}_{rs}\tilde{\underset{\sim}{f}}_s - \underset{\sim}{B}_{rr}\tilde{\underset{\sim}{f}}_r = \underset{\sim}{g}_r \quad .$$

Makinson and Young (1960) and Young (1954*b*) study, in particular, the equation

$$f(x) - 2\int_0^x (x-2y)f(y)\,dy = 2x/\sqrt{\pi}$$

which has the solution

$$f(x) = \{2/\sqrt{\pi}\}\int_0^x \exp(-y^2)dy \quad .$$

(For this particular equation $\underset{\sim}{B}_{j,j} = \underset{\sim}{B}_{j-1,j-1} + \underset{\sim}{\Delta}$ and $\underset{\sim}{A}_{i,j} = \underset{\sim}{A}_{i-1,j} + \underset{\sim}{C} = \underset{\sim}{A}_{i,j-1} - 2\underset{\sim}{C}$ where $\underset{\sim}{\Delta}$ and $\underset{\sim}{C}$ are fixed matrices; their entries are displayed by Makinson and Young (1960).) Using a step $h = 0{\cdot}05$, the approximate values of the solution displayed by Makinson and Young have an error not exceeding 8×10^{-9}; the interval of solution $[0,X]$ is $[0,4]$. *

Modifications can be made to the scheme associated with the eqns (6.130) to transform it into a step-by-step method. A natural modification arises when we adapt the techniques developed for the quadrature method. In the following Example we study such a scheme developed for the case $p = 2$. (For larger values of p , more complex schemes would be required.)

Example 6.45.
Consider the case $p = 2$ in the scheme outlined above. If k is an integer and $x = 2kh$, the approximation

$$\int_0^x K(x,y)\phi(y)dy \simeq \sum_{j=0}^{N} \nu_j(x)\phi(jh) \quad (x \le X) \quad ,$$

using the generalized Simpson's rule, reduces to an approximation $\sum_{j=0}^{2k} \nu_j(2kh)\phi(jh)$ involving $\phi(0),\phi(h),\ldots,\phi(2kh)$. On the other hand, if $x = (2k+1)h$ the approximation reduces to $\sum_{j=0}^{2(k+1)} \nu_j((2k+1)h)\phi(jh)$, involving the function value $\phi(2(k+1)h)$ evaluated outside the range of integration. We therefore revise this latter approximation.

In the case $x = (2k+1)h$ we may construct an approximation

$$\int_0^x K(x,y)\phi(y)dy \simeq \sum_{j=0}^{2k+1} \nu_j^*(x)\phi(jh) \quad ,$$

which is exact when $\phi(x)$ is continuous, quadratic in each interval

$[0,2h]$, $[2h,4h],\ldots,[2(k-1)h, 2kh]$ and *linear* in $[2kh,(2k+1)h]$. (The error in such an approximation is $O(h^3)$ as $h \to 0$ with fixed kh.)

Alternatively, and for an expected increase in accuracy (though the order of the error is not generally better) we may construct an approximation

$$\int_0^{(2k+1)h} K((2k+1)h,y)\phi(y)\,dy \simeq \sum_{j=0}^{2k+1} \nu_j^{\dagger}((2k+1)h)\phi(jh) \quad ,$$

which is exact when $\phi(x)$ is continuous, quadratic in each interval $[0,2h]$,$[2h,4h],\ldots,[2(k-2)h,2(k-1)h]$ $(k > 1)$ and *cubic* in the interval $[2(k-1)h,(2k+1)h]$. Let us, in addition, define $\nu_j^{\dagger}(x) = \nu_j(x)$ if $x = 2kh$, where k is an integer. We then obtain the equations

$$\tilde{f}(kh) - \lambda\sum_{j=0}^{k} \nu_j^{\dagger}(kh)\tilde{f}(jh) = g(kh) \quad (k = 2,3,4,\ldots,N)$$

on applying the product-integration formulae. These equations can be solved in a step-by-step fashion, if we compute $\tilde{f}(h)$ by a starting procedure giving at least $O(h^2)$ accuracy and set $\tilde{f}(0) = f(0)$ $(= g(0))$.

Linz (1967*a*, 1969*b*) gives formulae for the weights $\nu_j^{\dagger}(kh)$ in the notation $W_{k,j}$. In the case where $K(x,y) = 1$, the method reduces to the use of the scheme in Table 6.5 with a suitable starting procedure for $f(h)$. More generally we obtain the following expressions for $\nu_j^{\dagger}(kh)$. Adapting the notation of Linz (1969*b*), we set

$$\nu_j^{\dagger}(kh) = 2J(k,j-1,2)-L(k,j-1,2) \quad (j \text{ odd}) \quad ,$$

$$\nu_j^{\dagger}(kh) = \tfrac{1}{2}\{L(k,j-2,2)-J(k,j-2,2)\} +$$

$$+ I(k,j,2)- \tfrac{3}{2}J(k,j,2)+ \tfrac{1}{2}L(k,j,2) \quad (j \text{ even})$$

for even values of k , and odd values of k with $k < j-3$, whilst otherwise we have

$$\nu_j^{\dagger}(jh) = \tfrac{1}{3}J(j,j-3,3) - \tfrac{1}{2}L(j,j-3,3) + \tfrac{1}{6}M(j,j-3,3) \quad ,$$

$$\nu^{\dagger}_{j-1}(jh) = 2L(j,j-3,3) - \tfrac{3}{2}J(j,j-3,3) - \tfrac{1}{2}M(j,j-3,3),$$

$$\nu^{\dagger}_{j-2}(jh) = 3J(j,j-3,3) - \tfrac{5}{2}L(j,j-3,3) + \tfrac{1}{2}M(j,j-3,3) \quad ,$$

$$\nu^{\dagger}_{j-3}(jh) = I(j,j-3,3) - \tfrac{11}{6}J(j,j-3,3)+L(j,j-3,3) -$$

$$- \tfrac{1}{6}M(j,j-3,3) - \tfrac{1}{2}J(j,j-5,2) + \tfrac{1}{2}L(j,j-5,2) \quad ,$$

where $I(p,q,r)$, $J(p,q,r)$, $L(p,q,r)$, and $M(p,q,r)$ are taken to be zero if $q < 0$ or $q \geq p$, and otherwise

$$I(p,q,r) = \int_{qh}^{(q+r)h} K(ph,y)\,dy \quad ,$$

$$J(p,q,r) = \frac{1}{h}\int_{qh}^{(q+r)h} (y-qh)K(ph,y)\,dy \quad ,$$

$$L(p,q,r) = \frac{1}{h^2}\int_{qh}^{(q+r)h} (y-qh)^2K(ph,y)\,dy \quad ,$$

$$M(p,q,r) = \frac{1}{h^3}\int_{qh}^{(q+r)h} (y-qh)^3K(ph,y)\,dy \quad .$$

The method requires a starting procedure to calculate $\tilde{f}(h)$. This value can be obtained with $O(h^3)$ accuracy if we employ the generalized trapezium rule for $\int_0^h K(x,y)\phi(y)\,dy$. *

6.9. *Product-integration techniques for singular linear and non-linear equations*

The construction of formulae of the form (6.123) is generally somewhat difficult, but in the case where $K(x,y)$ is discontinuous (or even badly behaved) the techniques of section 6.8 are invaluable. These techniques may be readily adapted to the case of a (linear or) non-linear integral equation of the form

$$f(x) - \int_0^x K(x,y)\Phi(x,y;f(y))\,dy = g(x) \quad (0 \leq x \leq X) \quad , \qquad (6.131)$$

where $K(x,y)$ is badly behaved and $\Phi(x,y;f(y))$ is well behaved for

$0 \leq y \leq x$, $0 \leq x \leq X$. Typically, $K(x,y)$ may be weakly singular, say

$$K(x,y) = \ln(x-y) \quad (y < x)$$

or

$$K(x,y) = \frac{1}{(x-y)^{\alpha}} \quad (0<\alpha<1) \quad (y < x) \; .$$

Linz (1969*b*) also suggests that the case $K(x,y) = y(x^2-y^2)^{-\frac{1}{2}}$ arises in practice but points out that in the similar case where $K(x,y) = (x^2-y^2)^{-\frac{1}{2}}$, and $\Phi(x,y;f(y)) = f(y)$ the solution $f(x)$ is not unique since it contains an arbitrary component which is any multiple of x .

In this section we shall present formal techniques for the approximate solution of (6.131). A unique solution of (6.131) is assured for $0 \leq x \leq X$ provided that $g(x) \in C[0,X]$, and that $\Phi(x,y;f)$ is continuous in x,y for $0 \leq y \leq x \leq X$ and satisfies the Lipschitz condition

(i) $$|\Phi(x,y;f_1)-\Phi(x,y;f_2)| \leq L|f_1-f_2|$$

uniformly for $0 \leq y \leq x \leq X$, and provided

(ii) $$\int_0^x |K(x,y)|\,dy \leq M < \infty \quad \text{for} \quad 0 \leq x \leq X \, ,$$

provided also (Linz 1969*a*) that for every $\varepsilon > 0$ there exists $\delta > 0$ independent of x,z such that

(iii) $$\int_z^{z+\delta} |K(x,y)|\,dy < \varepsilon \text{ for all } z \in [0,x-\delta],\; x \leq X.$$

(The function $K(x,y) = (x^2-y^2)^{-\frac{1}{2}}$ does not satisfy this last condition.)

In section 6.8 we saw an application of approximations of the form

$$\int_0^x K(x,y)\phi(y)\,dy \simeq \sum_{j=0}^{N} v_j(x)\phi(jh) \quad (0 \leq x \leq Nh=X). \quad (6.132)$$

In our treatment of (6.131) we replace $\phi(y)$ by $\Phi(x,y;f(y))$ and obtain the approximation

$$\int_0^x K(x,y)\Phi(x,y;f(y))\,dy \simeq \sum_{j=0}^{N} \nu_j(x)\Phi(x,jh;f(jh)) \ .$$

The associated numerical method for the solution of (6.131) consists of solving the equations $\tilde{f}(0) = g(0)$ and

$$\tilde{f}(kh) - \sum_{j=0}^{N} \nu_j(kh)\Phi(kh,jh;\tilde{f}(jh)) = g(kh) \ (k=1,2,3,\ldots,N) \qquad (6.133)$$

where $X = Nh$.

In the case where the weights $\nu_j(\ )$ in (6.132) are those of the modified trapezium rule, $\nu_j(kh) = 0$ when $j > k$ and the solution of (6.133) proceeds in a step-by-step fashion. Eqn (6.133) constitutes a non-linear equation for $\tilde{f}(kh)$ and must be solved iteratively. Now $\nu_k(kh) = o(1)$ as $h \to 0$, and a predictor-corrector technique for solving (6.133) can therefore be developed, using (say) an explicit formula similar to (6.133) as predictor; the convergence may be slow, however.

Suppose, now, that the weights $\nu_j(x)$ in (6.132) are those of the generalized Simpson's rule. More generally, suppose these weights are constructed so that the approximation (6.132) is exact whenever $\phi(x)$ is a polynomial of degree p in each interval $[rph,(r+1)ph]$, as indicated above. (If $p = 2$ we obtain the generalized Simpson's rule.) Then eqns (6.133) must be solved in a block-by-block fashion, since, in general,

$$\nu_j(kh) \neq 0 \quad \text{for} \quad j \leq k+r \ , \qquad (6.134)$$

where $r \equiv k \pmod{p}$, although $\nu_j(kh) = 0$ otherwise. Moreover we note (as a consequence of (6.134)) that the approximation (6.133) involves values $\Phi(x,y;\tilde{f}(y))$ in which $y > x$. As in the case of the non-singular equation, there are circumstances in which this can be a disadvantage. It follows that modifications to (6.133) similar to the one described in Example 6.44 can now be particularly useful. As indicated in that Example we construct approximations of the form

$$\int_0^X K(x,y)\phi(y)\,dy \simeq \sum_{j=0}^{N} \nu_j^{\dagger}(x)\phi(jh) ,$$

in which $\nu_j^{\dagger}(kh) = 0$ if $k > j$. We then employ these to produce the equations

$$\tilde{f}(kh) - \sum_{j=0}^{k} \nu_j^{\dagger}(kh)\Phi(kh,jh;\tilde{f}(jh)) = g(kh) \quad (k=1,2,3,\ldots,N) . \qquad (6.135)$$

With $\tilde{f}(0) = g(0)$, eqns (6.135) can be solved in a step-by-step fashion.

The step-by-step methods associated with eqn (6.135) are generally extensions of those step-by-step methods for non-singular equations, discussed in section 6.2, in which weights w_{kj} were formed by applying suitable interpolatory quadrature rules. In (6.135) the weights $\nu_j^{\dagger}(kh)$ may be obtained by defining corresponding product integration formulae. It is possible that, as with the methods of section 6.2, certain methods of constructing the weights $\nu_j^{\dagger}(kh)$ may give unstable discretizations.

The block-by-block methods of section 6.7 have their analogues in the treatment of (6.131) by product-integration techniques. The method discussed in the following Example is in the spirit of the method discussed in Example 6.35.

<u>Example 6.46</u>.

Linz (1969*b*) proposes an alternative scheme for use with the generalized Simpson's rule. This scheme avoids the requirement of values of $\Phi(x,y;f)$ in which $y > x$, but still leads to a block-by-block method. The modified Simpson's rule is applied in a straightforward way to discretize the integral equation (6.131) when $x = kh$ and k is even (see Example 6.45). For odd values of k we now write

$$\tilde{f}(kh) - \sum_{j=0}^{k-1} \nu_j(kh)\Phi(kh,jh;\tilde{f}(jh)) - \{u_{k-1}(kh)\Phi(kh,(k-1)h;\tilde{f}((k-1)h)) +$$

$$+ u_{k-\frac{1}{2}}(kh)\Phi(kh,(k-\tfrac{1}{2})h;\tilde{f}((k-\tfrac{1}{2})h)) + u_k(kh)\Phi(kh,kh;\tilde{f}(kh))\} = g(kh) ,$$

where we write $\tilde{f}((k-\frac{1}{2})h)$ to denote $\frac{3}{8}\tilde{f}((k-1)h)+\frac{3}{4}\tilde{f}(kh) - \frac{1}{8}\tilde{f}((k+1)h)$, (compare with Example 6.35) and the approximation

$$\int_{(k-1)h}^{kh} K(x,y)\phi(y)\,dy \approx u_{k-1}(x)\phi((k-1)h)+u_{k-\frac{1}{2}}(x)\phi((k-\tfrac{1}{2})h)+u_k(x)\phi(kh)$$

is constructed to be exact when $\phi(x)$ is a quadratic. (The required expressions $u_k(x)$ can be deduced from the formulae given in eqns (6.126), (6.127) after replacing h by $\frac{1}{2}h$, where appropriate.) *

For a general block-by-block method analogous to that defined by (6.115), we may proceed as follows. Choosing $0 \le \theta_0 < \theta_1 < \ldots < \theta_p = 1$ we construct formulae

$$\int_{ih^*}^{(i+\theta_j)h^*} K(x,y)\phi(y)\,dy \approx \int_{ih^*}^{(i+\theta_j)h^*} K(x,y) \sum_{r=0}^{p} l_r(y)\phi((i+\theta_r)h^*)\,dy$$

$$= \sum_{r=0}^{p} v_r^{[i,j]}(x)\phi((i+\theta_r)h^*) \quad .$$

We then solve the equations

$$\tilde{f}((m+\theta_j)h^*) - \sum_{i=0}^{m-1}\sum_{r=0}^{p} v_r^{[i,p]}((m+\theta_j)h^*)\Phi((m+\theta_j)h^*,(i+\theta_r)h^*;\tilde{f}((i+\theta_r)h^*)) -$$

$$- \sum_{r=0}^{p} v_r^{[i,j]}((m+\theta_j)h^*)\Phi((m+\theta_j)h^*,(m+\theta_r)h^*;\tilde{f}((m+\theta_r)h^*)) \tag{6.136}$$

$$= g((m+\theta_j)h^*) \quad (m = 0,1,2,\ldots,\ j = 0,1,2,\ldots,p) \quad .$$

<u>Remark</u>. Clearly the kth equation can be written in the form

$$\tilde{f}(u_k) - \sum_{i=0}^{q_k} W_{ki}\Phi(u_k,u_i;\tilde{f}(u_i)) = g(u_k) \quad ,$$

where $u_k = (m+\theta_j)h^*$, $u_0 \le u_1 \le u_2 \le \ldots$ are the ordered abscissae, and $q_k \le (m+1)(p+1)$.

6.10. *Non-singular Volterra equations of the first kind*

We now turn our attention to the equation

$$\int_0^x K(x,y)f(y)\,dy = g(x) \quad (0 \le x \le X) \quad . \tag{6.137}$$

We suppose, here, that $K(x,y)$ is continuous for $0 \le y \le x \le X < \infty$. If we write $K_+(x,y) = K(x,y)$ for $y \le x$, $K_+(x,y) = 0$ for $y > x$, then (6.137) assumes the form of a Fredholm equation of the first kind,

$$\int_0^X K_+(x,y)f(y)\,dy = g(x) \quad (0 \le x \le X) \quad . \tag{6.138}$$

We have already established (see Chapter 5) that it is difficult to treat the Fredholm equation of the first kind. We observed that a solution $f(x)$ may not exist for every choice of $g(x)$, and that, when the solution $f(x)$ does exist, it may not be unique or it may be difficult to determine accurately (owing to the inherent ill-conditioning of the integral equation).

These general conclusions are still valid in the treatment of eqn (6.137). However, as we observed in our discussion of the Fredholm equation of the first kind, the general situation is usually improved upon if $K(x,y)$ is badly behaved (discontinuous, or with a discontinuous derivative $(\partial/\partial y)K(x,y)$) on $x = y$. This affords some possible relief in the case of a Volterra equation (6.137), in particular when

$$K(x,x) \ne 0 \tag{6.139a}$$

or

$$(\partial/\partial y)^r K(x,y)\big]_{y=x} \ne 0 \tag{6.139b}$$

for some moderately small value of r. It remains the case, however, that (6.137) has a solution only for certain functions $g(x)$.

<u>Example 6.47</u>.

Suppose that $K(x,y)$ is continuous for $0 \le y \le x \le X$. Then, setting $x = 0$ in (6.137) we see that if there is a continuous solution $f(x)$ then $g(0) = 0$. *

We saw, in Chapter 1, that when $K_x(x,y) = (\partial/\partial x)K(x,y)$ is continuous for $0 \le y \le x \le X$, eqn (6.137) is equivalent to the equation

$$K(x,x)f(x) + \int_0^x K_x(x,y)f(y)\,dy = g'(x) , \qquad (6.140)$$

provided that $g'(x)$ exists for $0 \le x \le X$.

In the case where $K(x,x) \equiv 0$ this equation is an equation of the first kind, and the equation then possesses a continuous solution only if $g'(0) = 0$ and $g(0) = 0$. In the case where $K(x,x) \neq 0$ for $0 \le x \le X$ in (6.137), we can endeavour to solve, numerically, the equivalent equation of the second kind

$$f(x) + \int_0^x K_0(x,y)f(y)\,dy = g_0(x) \quad (0 \le x \le X) ,$$

where $g_0(x) = g'(x)/K(x,x), K_0(x,y) = K_x(x,y)/K(x,x)$. (There may be some difficulty in evaluating the derivatives required, in a practical problem.) Linz (1967*a*) has proposed a method based on this idea, but we shall also examine some direct methods which are suitable for the numerical solution of (6.137) in the case where $K(x,x) \neq 0$ for $0 \le x \le X$.

If $K(x,x)$ vanishes everywhere in $[0,X]$ then (6.140) reduces to the form of (6.137), and the process of attempting to obtain an equation of the second kind can be repeated. This process eventually succeeds if $(\partial/\partial x)^r K(x,y)]_{y=x} \neq 0$ for $0 \le x \le X$ and some r , but in practice the technique may not be of great help if r is large or the derivatives required are difficult to obtain.

<u>Remark</u>. It may also happen that $K(X_1,X_1) = 0$ for $X_1 < X$, but that $K(x,x) \neq 0$ for $0 \le x \le X_2 < X_1$, and (6.140) can be solved accurately for $0 \le x \le X_2 < X_1$. We may then re-formulate (6.137) as

$$\int_{X_2}^x K(x,y)f(y)\,dy = g(x) - \int_0^{X_2} K(x,y)f(y)\,dy \quad (x \ge X_2) .$$

The second term on the right can be approximated, for $x \ge X_2$, in terms of a computed approximation for $f(y)$ already obtained for $0 \le y \le X_2$ using (6.140).

6.10.1.

When $K(x,x)$ or derivatives $(\partial/\partial x)^r K(x,y)$ are not available, or equations of the form (6.140) are difficult to obtain, consideration should be given to the possibility of treating (6.138), as a Fredholm equation. Schmaedeke (1968) considers regularization methods (see Chapter 5) applied to equations of the form (6.138) with a Volterra kernel. (I incline to the view that for large values of X, this approach may be difficult to implement satisfactorily, though there may be occasions where it is highly successful.)

There are, however, some direct methods which can be applied successfully to the numerical solution of (6.137) when $K(x,x)$ is non-vanishing in $[0,X]$. We shall now investigate the possibility of such methods, assuming that $K(x,y)$ is continuously differentiable with respect to x for $0 \le y \le x \le X$, and assuming that $g(x)$ satisfies the conditions required for the existence of a continuous solution.

6.10.2.

It is tempting to adapt our quadrature methods for Volterra equations of the second kind to deal with equations of the first kind. The methods of section 6.7 can be adapted to this purpose and they generally prove very successful (section 6.10.4) in practice. However, we shall first discuss quadrature methods based on rules of the form

$$\int_0^{kh} \phi(y)\,dy \simeq \sum_{j=0}^{k} w_{k,j}\phi(jh) \quad (k = 1,2,3,\ldots,N) \tag{6.141}$$

(where $Nh = X$). These lead to the approximating system of equations

$$\sum_{j=0}^{k} w_{k,j}K(kh,jh)\tilde{f}(jh) = g(kh) \quad (k = 1,2,3,\ldots,N)\,. \tag{6.142}$$

If $w_{k,k} \neq 0$ $(k = 1,2,3,\ldots,N)$, then eqns (6.142) constitute a system of N equations in the $(N+1)$ unknowns $\tilde{f}(0),\tilde{f}(h),\ldots,\tilde{f}(Nh)$, and cannot be solved. In general we require, in addition to (6.142), an equation determining $\tilde{f}(0)$.

Now $f(0)$ may be determined from (6.140) by

$$f(0) = g'(0)/K(0,0)\,, \tag{6.143}$$

and hence evaluation of (6.143) permits the determination of an exact value for $\tilde{f}(0)$. More generally, it will be necessary to approximate the derivative $g'(0)$ in (6.143) numerically, to obtain a value for $\tilde{f}(0)$.

Example 6.48.
The use of the repeated trapezium rule to give an approximation (6.141), and the use of the correct value (6.143), leads to the equations

$$\tilde{f}(0) = g'(0)/K(0,0) \ ,$$

$$h \sum_{j=0}^{k}{}'' K(kh,jh)\tilde{f}(jh) = g(kh) \quad (k = 1,2,3,\ldots,N) \ ,$$

where $Nh = X$.

We may, alternatively, employ the mid-point rule ($w_{k0} = w_{k2} = w_{k4} = \ldots = 0$ in (6.141), provided that k is even). Eqn (6.142) where $k = 2$ then gives

$$2hK(2h,h)\tilde{f}(h) = g(2h) \ ,$$

and in general we may write

$$2h \sum_{j=0}^{r-1} K(2rh,(2j+1)h)\tilde{f}((2j+1)h) = g(2rh) \quad (r = 1,2,3,..,[\tfrac{1}{2}N]) \ .$$

The preceding equations, for $r = 1,2,3,\ldots,[\frac{1}{2}N]$ (where [ ] denotes the integer part) , give a system of $[\frac{1}{2}N]$ equations in $[\frac{1}{2}N]$ unknowns, and no 'starting' value for $\tilde{f}(0)$ is required in order to compute

$$\tilde{f}(h),\tilde{f}(3h),\tilde{f}(5h),\ldots$$

in a step-by-step fashion (by 'forward substitution'). This scheme is generally considered to be better than the scheme using the trapezium rule. *

It should be emphasized that no starting value (6.143) is required if we employ the mid-point rule or (more generally) any approximation of the form (6.141) in which $w_{k0} = 0$ for all k .

We discovered in our discussion of equations of the second kind that some approximations of the form (6.141) give better approximations than others, and for comparable computational effort. The same is true for equations of the first kind (though apparently for somewhat different reasons).

Example 6.49.

We consider the use of the repeated trapezium rule, as indicated in Example 6.48, to solve the equation

$$\int_0^x f(y)\,dy = g(x) .$$

(For the existence of a solution we require that $g(0) = 0$ and that $g'(x)$ exists; we then have $f(x) = g'(x)$.) In this case we have the equations

$$\tilde{f}(0) = g'(0)$$

$$h \sum_{j=0}^{k}{}'' \tilde{f}(jh) = g(kh) \quad (k = 1,2,3,\ldots,N) \quad ,$$

and the approximate solution can be shown, by induction, to have the form

$$\tilde{f}(rh) = \{2g(rh) + 4\sum_{j=1}^{r-1}(-1)^j g((r-j)h)\}/h + (-1)^r g'(0) .$$

If $h \to 0$ with rh fixed and r always even, $\tilde{f}(rh) = g'(rh) + o(1)$ if $g''(x)$ is continuous. For, since $g(0) = 0$,

$$\tilde{f}(rh) = (2/h^2)\{hg(rh) + 2hg(\overline{r-2}h) + \ldots + hg(0)\} - \frac{2}{h^2}\{2hg(\overline{r-1}h) +$$

$$+ 2hg(\overline{r-3}h) + \ldots + 2hg(h)\} + g'(0) = \tfrac{2}{3}\{g'(rh) - g'(0)\}$$

$$+ \tfrac{1}{3}\{g'(rh) - g'(0)\} + o(1) + g'(0) \quad ,$$

using the error terms for the repeated trapezium and mid-point rules; the result follows.

However, the expression for $\tilde{f}(rh)$ involves all the values $g(h), g(2h), \ldots, g(rh)$ in an expression approximating $g'(rh)$, and it cannot be considered a very satisfactory approximation when we consider the effect of perturbations ε_r in the values $g(rh)$. (Why?)

By way of contrast, the use of the mid-point rule as indicated in Example 6.48 gives values $\tilde{f}(h), \tilde{f}(3h), \tilde{f}(5h), \ldots$ which satisfy the relation

$$\tilde{f}((2r+1)h) = \frac{g(2(r+1)h) - g(2rh)}{2h} .$$

The expression on the right is a local approximation to $g'(rh)$. It should be clear, however, that when $K(x,y)$ is not constant for $0 \leq y \leq x$, the approximation may no longer be a local one.

The order of accuracy is the same with each method, but the use of the mid-point rule requires no starting value and, for the equation considered here, gives a more acceptable approximation to the solution. However, even in the case of the mid-point rule, perturbations in the values $g(2rh)$ are magnified when a small step h is employed. *

In the preceding Example we suggested two quadrature methods which could be considered for practical use on general equations in which $K(x,x) \neq 0$. To test other methods it is often convenient to analyse their performance when applied to the equation $\int_0^x f(y)\,dy = g(x)$. Against this must be weighed the belief, based on reporting of numerical experiments, that certain methods appear successful for particular equations even though they cannot be generally recommended. More work appears to be necessary here.

Example 6.50.

It is possible to analyse the effect of the formula based on the weights in Table 6.2, which correspond to the use of the trapezium rule followed by the (repeated) Simpson's rule, when applied to the equation

$$\int_0^x f(y)\,dy = g(x) ,$$

with $g(0) = 0$. We require the starting value $f(0) = g'(0)$. Differencing odd-numbered and even-numbered equations respectively we find that on employing the operator notation $E\phi(x) = \phi(x+h)$ we may write

$$\tfrac{1}{3}h\{E^2 + 4E + 1\}\,\tilde{f}((r-1)h) = (E^2-1)g((r-1)h) \quad (r=1,2,\ldots) \quad (6.144)$$

with $\tilde{f}(0) = f(0)$ and $\tilde{f}(h) = (2/h)g(h) - f(0)$. Proceeding quite formally we can show that the operators E and $D = (d/dx)$ are related by $E = \exp(hD)$, and by formal manipulation that $(3/h)\{E^2+4E+1\}^{-1}(E^2-1) = D - h^4D^4/180 + O(h^6)$. Thus, we might expect convergence of $\tilde{f}(rh)$ to $f(rh)$, but such an argument is not rigorous (and takes no account of starting values $\tilde{f}(0)$ and $\tilde{f}(h)$, for example). A general line of argument indicates non-convergence, in fact. Whilst it is not a simple matter to discuss convergence in terms of the analytical expression for $\tilde{f}(rh)$, we readily obtain insight into other matters. Thus, if we write $\gamma_0 = \tilde{f}(0)$, $\gamma_1 = \tilde{f}(h)$, and $\gamma_k = 3(E^2-1)g((k-2)h)/h$ for $k\geq 2$, then we find

$$\tilde{f}(kh) = \gamma_k - a_1\gamma_{k-1} + a_2\gamma_{k-2} - a_3\gamma_{k-3} + \cdots \pm a_{k-2}(\gamma_2-\gamma_0) \mp a_{k-1}\gamma_1$$

where $a_0 = 1$, $a_1 = 4$, and $a_{r+2} - 4a_{r+1} + a_r = 0$. Thus,

$$a_r = \left\{\frac{4\tau_1-1}{4\tau_1-2}\right\}\tau_1^r + \left\{\frac{4\tau_2-1}{4\tau_2-2}\right\}\tau_2^r \quad \text{where} \quad \tau_1,\tau_2 = 2\pm\sqrt{3},$$

and $a_r > 0$ grows rapidly with increasing r. The above expression reveals a defect already displayed in Example 6.49, namely the dependence on $g(kh)$, $g((k-1)h)$, ..., $g(h)$ of the value $\tilde{f}(kh)$, but the growth in size of successive values a_r gives increasing contributions from errors present in $\tilde{f}(h)$, for example, and is clearly unstable.

The auxiliary equation associated with eqn (6.144) is of course $t^2 + 4t + 1$, with roots $-\tau_1$ and $-\tau_2$, so that instability might have been anticipated. Lest the reader imagine that the use of the weights in Table 6.3 yield a satisfactory scheme, in contrast with those of Table 6.2 discussed here, we must add that this is not the case. *

The approximation of the form

$$\tilde{f}((2r+1)h) = \{g(2(r+1)h) - g(2rh)\}/\{2h\} \simeq g'((2r+1)h),$$

obtained using the repeated mid-point rule, has modest accuracy and the effect of perturbations is kept reasonably under control by the local nature of the approximation. We would like to control the error propagation whilst improving the order of the discretization error.

Unfortunately, the reported accuracy of high-order quadrature methods of the form (6.142) is not good at all, and Linz (1967*a*) has presented strong arguments for considering that many such high-order methods are non-convergent (or, at least, unsatisfactory - see also Gladwin and Jeltsch (1974)). We shall later see that other types of high-order methods *seem* more satisfactory.

Example 6.51.
Linz (1967*a*) reports on the use of a fourth-order Gregory rule to solve the equation

$$\int_0^x \cos(x-y)f(y)\,dy = \sin x \quad (0 \le x \le 2) .$$

Column (a) gives errors computed with $h = 0\cdot 1$, and column (b) gives errors computed with $h = 0\cdot 05$.

x	(a)	(b)	x	(a)	(b)
1	1×10^{-3}	5×10^{-1}	1·6	2×10^{-1}	1×10^{4}
1·2	8×10^{-3}	2×10^{1}	1·8	1×10^{0}	5×10^{5}
1·4	4×10^{-2}	5×10^{2}	2·0	7×10^{0}	1×10^{7}

A reduction of the step-size gives more rapid error growth, in this example. *

Because practical computation with high-order methods frequently gives results like those reported in Example 6.51, attention has been given to the use of the (repeated) mid-point rule or the (repeated) trapezium rule. We discuss both, but the former is probably the best.

Example 6.52.
The repeated trapezium rule was employed with step $h = 1/2$ to compute an approximate solution to the problem in Example 6.51. The computed errors were $-4\cdot 01\times 10^{-2}$, $1\cdot 42\times 10^{-14}$, $-4\cdot 01\times 10^{-2}$, $1\cdot 42\times 10^{-14}$, $-4\cdot 01\times 10^{-2}$, ... at $x = 0\cdot 5, 1\cdot 0, 1\cdot 5, 2\cdot 0, 2\cdot 5, \ldots$. With $h = \frac{1}{4}$, the computed errors were $-1\cdot 03\times 10^{-2}$, $2\cdot 84\times 10^{-14}$, $-1\cdot 03\times 10^{-2}$, $2\cdot 84\times 10^{-14}$,... at $x = 0\cdot 25, 0\cdot 5, 0\cdot 75, \ldots$. (Exact calculations would produce zeros in place of the figures of order 10^{-14}, but in practice these numbers

grow away from zero.) The maximum absolute errors are 4×10^{-2} and 1×10^{-2} respectively and this suggests that the errors are $O(h^2)$.

Note that an attempt at h^2-extrapolation on the values computed with $h = \frac{1}{2}$, $h = \frac{1}{4}$ does not yield $o(h^2)$ accuracy. Thus, formal h^2-extrapolation for the value of the solution at $x = 0\cdot5$ gives an approximation with error $\frac{1}{3}\{4\times(2\cdot84\times10^{-14}) + 4\cdot01\times10^{-2}\}$ and the corresponding error at $x = 1$ is $\frac{1}{3}\{4\times(-1\cdot03\times10^{-2}) - 1\cdot42\times10^{-14}\}$, so that the approximations obtained with $h = \frac{1}{4}$ are actually better than those obtained by a routine application of an h^2-extrapolation formula. This phenomenon is due to the 'oscillation' in the error, which is a general feature of the method using the trapezium rule. *

Numerical examples of the application of the repeated trapezium rule generally display an 'oscillatory' behaviour of the error, which can be explained by the theory. Jones (1961), Linz (1967*a*) and Kobayasi (1967) remark on this feature. To eliminate the oscillations, various methods of smoothing the values have been proposed.

In one such smoothing method we proceed as follows. We compute the values $\tilde{f}(rh)$ such that

$$\tilde{f}(0) = \frac{g'(0)}{K(0,0)} \quad , \qquad h \sum_{j=0}^{r}{}'' K(rh,jh)\tilde{f}(jh) = g(rh) \quad .$$

Having computed $\tilde{f}(h),\tilde{f}(2h),\tilde{f}(3h),\ldots$ we set

$$\overline{f}(rh) = \tfrac{1}{2}\{\tilde{f}((r-1)h) + \tilde{f}((r+1)h)\} \quad ,$$

and then form

$$\hat{f}(rh) = \tfrac{1}{2}\{\tilde{f}(rh) + \overline{f}(rh)\} \quad (r=1,2,3,\ldots). \qquad (6.145)$$

This method of smoothing was proposed by Jones and is discussed by Linz (1967*a*).

The order of accuracy of the values $\tilde{f}(h),\tilde{f}(3h),\tilde{f}(5h),\ldots$ computed using the (repeated) mid-point rule, and of the values $\hat{f}(rh)$ defined by (6.145) using the (repeated) trapezium rule, is normally $O(h^2)$ as $h \to 0$ with rh fixed. Assuming sufficient differentiability of $K(x,y)f(y)$, we may establish that (for the computed values referred to here)

$$f((2r+1)h)-\tilde{f}((2r+1)h) = \alpha_1((2r+1)h)h^2+O(h^4) \qquad (6.146)$$

when using the repeated mid-point rule, and, for the repeated trapezium rule and eqn(6.145),

$$f(ph)-\hat{f}(ph) = \alpha_2(ph)h^2 + O(h^4) \quad , \qquad (6.147)$$

where $\alpha_1(x)$ and $\alpha_2(x)$ are continuous functions. Relations of this type may be employed to justify the use of h^2-extrapolation to obtain results with $O(h^4)$ accuracy. An h^2-extrapolation method based on the mid-point rule will be described next.†

If we denote the half-step-size used to compute the results by means of a suffix, we might use the (repeated) mid-point rule to compute

$$\tilde{f}_h(h),\tilde{f}_h(3h),\tilde{f}_h(5h),\tilde{f}_h(7h),\ldots$$

using a step $2h$, and

$$\tilde{f}_{\frac{1}{3}h}(\tfrac{1}{3}h),\tilde{f}_{\frac{1}{3}h}(1h),\tilde{f}_{\frac{1}{3}h}(\tfrac{5}{3}h),\tilde{f}_{\frac{1}{3}h}(\tfrac{7}{3}h),f_{\frac{1}{3}h}(3h),\ldots, \text{ using a step } \tfrac{2}{3}h,$$

to obtain a pair of estimates for each value $f(h),f(3h),f(5h),\ldots$. New estimates with $O(h^4)$ errors would be obtained on calculating

$$\tfrac{1}{8}\{9\tilde{f}_{\frac{1}{3}h}((6r+3)\tfrac{h}{3})-\tilde{f}_h((2r+1)h)\} \simeq f((2r+1)h) \quad (r=0,1,2,\ldots). \qquad (6.148)$$

When using the (repeated) trapezium rule we might, similarly, compute

$$\tfrac{1}{3}\{4\hat{f}_{\frac{1}{2}h}(2p\tfrac{h}{2})-\hat{f}_h(ph)\} \simeq f(ph) \quad , \qquad (6.149)$$

using the smoothed values (6.145). I prefer the use of (6.148), as above.

It should be noted that approximations $\tilde{f}(rh)$ obtained using the (repeated) trapezium rule satisfy a relation of the form

---

† Such an extrapolation process is called 'passive' since the extrapolated values are not used in the calculation of subsequent values $\tilde{f}(rh)$.

$$f(rh)-\tilde{f}(rh) = \{\varepsilon(rh)+(-1)^r\delta(rh)\}h^2+O(h^4)$$

and values $\tilde{f}(rh)$ computed with step h and step $\frac{1}{2}h$ cannot be used to eliminate the h^2 term in the error. *It would, however, be possible to use steps h and $\frac{1}{3}h$ to eliminate the h^2 term of the error* by computing

$$\frac{1}{8}\{9\tilde{f}_{\frac{1}{3}h}(3r\frac{h}{3}) - \tilde{f}_h(rh)\} \simeq f(rh) . \tag{6.150}$$

Instead of applying the method of h^2-extrapolation, we might consider using a method of deferred correction. Linz (1967*a*) considers such a possibility, basing his method on finite-difference approximations to the derivative correction terms of the quadrature rule. For his method, the kernel $K(x,y)$ needs to be known outside the range $0 \le y \le x \le X$, and there are various difficulties which suggest that the technique should be applied only with caution.

6.10.3.

An indirect high-order method, for the step-by-step solution of eqn (6.137) using quadrature rules, is obtained if we discretize eqn (6.140). We again require $K(x,x) \neq 0$ for $0 \le x \le X$, and values $\tilde{K}_x(x,y)$ approximating $K_x(x,y) = (\partial/\partial x)K(x,y)$ and $\tilde{g}'(x)$ approximating $g'(x)$ must be calculated either by exact differentiation or by using finite-difference approximations. We suppose that

$$|\tilde{K}_x(ih,jh)-K_x(ih,jh)| = O(h^p) \tag{6.151}$$

as $h \to 0$ (for $0 \le j \le i$) at a fixed point (ih,jh) $(ih \le X)$, and we suppose that

$$|\tilde{g}'(ih)-g'(ih)| = O(h^p) \tag{6.152}$$

as $h \to 0$ at a fixed point $ih \le X$. Then for the approximate solution of eqn (6.140), namely,

$$f(x) + \int_0^x K_0(x,y)f(y)dy = g_0(x) , \tag{6.153}$$

we may use a step-by-step quadrature method for the solution of an equation of second kind and set

$$\tilde{f}(rh)+\sum_{j=0}^{r} w_{rj}\tilde{K}_0(rh,jh)\tilde{f}(jh) = \tilde{g}_0(rh) \tag{6.154}$$

with

$$\tilde{K}_0(x,y) = \tilde{K}_x(x,y)/K(x,x) \simeq K_0(x,y) \tag{6.155}$$

and

$$\tilde{g}_0(x) = \tilde{g}'(x)/K(x,x) \simeq g_0(x) \; . \tag{6.156}$$

In this way we approximate the solution of equation (6.137). If $O(h^p)$ accuracy is normally expected from the quadrature method applied directly to eqn (6.153), this rate of convergence is preserved when $\tilde{K}_0(x,y)$ and $\tilde{g}_0(x)$ satisfy (6.151) and (6.152) respectively.

6.10.4.

The block-by-block methods discussed in section 6.7 for the treatment of equations of the second kind have their analogues in the treatment of the equation of the first kind. Whilst Linz (1967*a*) appears to have decided, initially, that such methods do not prove satisfactory in use, more recent work of de Hoog and Weiss (1973*a*) has established that the methods generally give high-order accuracy under reasonable assumptions. Our own view is that such methods provide quite powerful tools for the solution of Volterra equations of the first kind, provided certain restrictions are satisfied.

A natural analogue of eqns (6.115) for the treatment of the equation

$$\int_0^x K(x,y)f(y)\mathrm{d}y = g(x)$$

arises on making a choice

$$0 \le \theta_0 < \theta_1 < \ldots < \theta_p = 1 \; , \tag{6.157}$$

and assumes the form (in the notation of eqn (6.115)):

$$h^*\sum_{i=0}^{m-1}\sum_{k=0}^{p}\omega_k K(mh^*+\theta_j h^*, ih^*+\theta_k h^*)\tilde{f}(ih^*+\theta_k h^*) + \tag{6.158}$$

$$+ h^*\sum_{k=0}^{p}\omega_k^{[j]}K(mh^*+\theta_j h^*, mh^*+\theta_k h^*)\tilde{f}(mh^*+\theta_k h^*) = g(mh^*+\theta_j h^*)$$

$$(m = 0,1,2,\ldots,\ j = 0,1,2,\ldots,p).$$

In the case where $\theta_0 = 0$, it is sufficient to set $j = 0,1,2,\ldots, p-1$, and the equation corresponding to $j = m = 0$ is replaced by the special starting value

$$\tilde{f}(0) = g'(0)/K(0,0)\ , \tag{6.159}$$

but no such starting value is required in the case where $\theta_0 > 0$.

Since values of $K(x,y)$ for $y > x$ are required in (6.158), we may construct the modified method, which is given by the analogue of equations (6.117), and set

$$h^*\sum_{i=0}^{m-1}\sum_{k=0}^{p}\omega_k K(mh^*+\theta_j h^*, ih^*+\theta_k h^*)\tilde{f}(ih^*+\theta_k h^*) +$$

$$+ h^*\sum_{k=0}^{p}\theta_j\omega_k K(mh^*+\theta_j h^*, mh^*+\theta_j\theta_k h^*)\sum_{r=0}^{p}l_r(\theta_j\theta_k)\tilde{f}(mh^*+\theta_r h^*) =$$

$$= g(mh^*+\theta_j h^*) \quad (m = 0,1,2,\ldots,\ j = 0,1,2,\ldots,p)\ . \tag{6.160}$$

We again employ the starting value (6.159) in the case $\theta_0 = 0$. (The function $l_r(x)$ in eqn (6.160) is defined by eqn (6.110), and is $\prod_{\substack{j\neq r\\ j=0}}^{p}\{(x-\theta_j)/(\theta_r-\theta_j)\}$.

In practice there are some restrictions in employing (6.158) and (6.160). The methods seem unsatisfactory when $K(x,x)$ vanishes, and if $\theta_0 = 0$ the theory suggests that $\theta_1,\theta_2,\ldots,\theta_{p-1}$ should be chosen so that $|\eta| \le 1$ where

$$\eta = \{\prod_{k=0}^{p-1}(1-\theta_k)\}/\{\prod_{k=1}^{p}(-\theta_k)\}\ . \tag{6.161}$$

Here $\theta_0 = 0$, $\theta_p = 1$. The theory seems to indicate that better results may be obtained if $|\eta| < 1$.) No such restrictions appear to be necessary in the case $0 < \theta_0 < \theta_1 < \ldots < \theta_p = 1$.

In the following two Examples, some results of my experience in using (6.160) on some simple test equations are given. For these examples, the exact starting value (6.159) was employed. In all the test equations, the solution $f(x)$ is well behaved.

Example 6.53.

Here we consider the numerical solution of the simple equation

$$\int_0^x (\tfrac{1}{2} + x^4 - y^4) f(y)\,dy = \tfrac{1}{2}x + \tfrac{4}{5}x^5 ,$$

of which the exact solution is $f(x) \equiv 1$. This test equation is considered by Linz (1967*a*) and de Hoog and Weiss (1973*a*). Linz applied a fourth-order block-by-block method (analogous to that described in Example 6.34) to compute the approximate solution of this equation at points $h, 2h, 3h, \ldots$. With $h = 0{\cdot}1$ he reported disastrous instability, which is worse on reducing the step to $h = 0{\cdot}05$ and appears to indicate non-convergence. A further reduction of h does actually give improved results, however (see the thesis of Weiss (1972) for details).

The errors shown in Table 6.11 correspond to the use of $\theta_0 = 0$, $\theta_1 = \frac{1}{3}$, $\theta_2 = \frac{2}{3}$, $\theta_3 = 1$ and $h^* = \frac{3}{20}$ and $h^* = \frac{3}{80}$ respectively in eqn (6.160).

TABLE 6.11

ERRORS

x	Step $h^* = \frac{3}{20}$	Step $h^* = \frac{3}{80}$
0·05	$-1{\cdot}39 \times 10^{-7}$	$2{\cdot}28 \times 10^{-8}$
0·10	$-3{\cdot}89 \times 10^{-6}$	$-1{\cdot}52 \times 10^{-8}$
0·15	$-1{\cdot}79 \times 10^{-5}$	$8{\cdot}57 \times 10^{-11}$
0·20	$5{\cdot}87 \times 10^{-6}$	$2{\cdot}30 \times 10^{-8}$
0·25	$-9{\cdot}77 \times 10^{-6}$	$-1{\cdot}47 \times 10^{-8}$
0·30	$3{\cdot}68 \times 10^{-7}$	$1{\cdot}30 \times 10^{-9}$

ERRORS		
x	Step $h^* = \frac{3}{20}$	Step $h^* = \frac{3}{80}$
0·35	$1{\cdot}90 \times 10^{-7}$	$2{\cdot}46 \times 10^{-8}$
0·40	$-3{\cdot}26 \times 10^{-6}$	$-1{\cdot}22 \times 10^{-8}$
0·45	$-1{\cdot}70 \times 10^{-5}$	$6{\cdot}26 \times 10^{-9}$
0·50	$7{\cdot}47 \times 10^{-6}$	$2{\cdot}99 \times 10^{-8}$
0·55	$-7{\cdot}85 \times 10^{-6}$	$-5{\cdot}22 \times 10^{-9}$
0·60	$5{\cdot}02 \times 10^{-6}$	$1{\cdot}80 \times 10^{-8}$
0·65	$2{\cdot}89 \times 10^{-6}$	$4{\cdot}04 \times 10^{-8}$
0·70	$9{\cdot}52 \times 10^{-7}$	$6{\cdot}46 \times 10^{-9}$
0·75	$-1{\cdot}46 \times 10^{-5}$	$3{\cdot}75 \times 10^{-8}$
0·80	$1{\cdot}30 \times 10^{-5}$	$5{\cdot}36 \times 10^{-8}$
0·85	$-4{\cdot}25 \times 10^{-6}$	$1{\cdot}89 \times 10^{-8}$
0·90	$1{\cdot}77 \times 10^{-5}$	$6{\cdot}48 \times 10^{-8}$
0·95	$2{\cdot}50 \times 10^{-6}$	$6{\cdot}55 \times 10^{-8}$
1·00	$4{\cdot}90 \times 10^{-6}$	$2{\cdot}94 \times 10^{-8}$
1·05	$-2{\cdot}73 \times 10^{-5}$	$1{\cdot}16 \times 10^{-7}$
1·10	$1{\cdot}92 \times 10^{-5}$	$8{\cdot}49 \times 10^{-8}$
1·15	$-9{\cdot}92 \times 10^{-6}$	$5{\cdot}18 \times 10^{-8}$
1·20	$5{\cdot}82 \times 10^{-5}$	$2{\cdot}73 \times 10^{-7}$
1·25	$-2{\cdot}17 \times 10^{-5}$	$1{\cdot}60 \times 10^{-7}$
1·30	$1{\cdot}39 \times 10^{-5}$	$1{\cdot}44 \times 10^{-7}$
1·35	$-1{\cdot}46 \times 10^{-4}$	$9{\cdot}22 \times 10^{-7}$
1·40	$8{\cdot}05 \times 10^{-5}$	$5{\cdot}31 \times 10^{-7}$
1·45	$-4{\cdot}75 \times 10^{-5}$	$5{\cdot}98 \times 10^{-7}$
1·50	$4{\cdot}89 \times 10^{-4}$	$4{\cdot}70 \times 10^{-6}$

TABLE 6.11 .
(concluded)

Whilst the left-hand column is grouped in blocks, the corresponding values on the right do not group into their natural blocks. *

Example 6.54.

The absolute block-stability concepts of section 6.3 can be used to analyse the absolute error propagation when applying (6.158) to the equation

$$\int_0^x f(y)\,dy = g(x) \ .$$

We obtain a natural block-by-block recurrence of the form (6.31), where, if $\theta_0 \neq 0$, $\rho(\underset{\sim}{M}) = 0$ and if $\theta_0 = 0$ the eigenvalues of $\underset{\sim}{M}$ of largest modulus are $= -1, 1, -1, \ldots$ in the cases $p = 1,2,3,\ldots$ respectively, using equally spaced θ_i . This indicates the superior stability properties of the method with $\theta_0 \neq 0$. (We note, in passing, that for the quadrature method using the mid-point rule we have a recurrence (6.31) with $\rho(\underset{\sim}{M}) = 0$ and using the trapezium rule we have a recurrence where the eigenvalue of $\underset{\sim}{M}$ of largest modulus is -1.) Further details of this stability analysis appear in Baker and Keech (1975), Keech (1976).*

Example 6.55.

We consider here some numerical results computed using the block-by-block method associated with eqns (6.160) and an exact starting value $\tilde{f}(0)=f(0)$. We consider a number of simple test equations of the form

$$\int_0^x K(x,y)f(y)\,dy = g(x) \quad ,$$

where

(a): $K(x,y) = \cos(x-y)$, $g(x) = \sin x$, $f(x) = 1$

(b): $K(x,y) = \exp(x-y)$, $g(x) = \sinh x$, $f(x) = \exp(-x)$

(c): $K(x,y) = \exp(2(x-y))$, $g(x) = \sin x$, $f(x) = \cos x - 2\sin x$

(d): $K(x,y) = \exp(x-y)$, $g(x) = x^2$, $f(x) = 2x-x^2$

(e): $K(x,y) = \cos(x-y)$, $g(x) = x\sin x$, $f(x) = 2\sin x$

(f): $K(x,y) = \cosh(x-y)$, $g(x) = x$, $f(x) = 1 - \frac{1}{2}x^2$.

(I also tried two cases where $K(x,x) = 0$, namely $K(x,y) = \sqrt{(x-y)}$ and $K(x,y) = \sinh(x-y)$, and found a catastrophic growth of errors which made nonsense of the computed approximations on the range considered.)

TABLE 6.12

Table of absolute errors ($h^* = 0{\cdot}075$)

Case	x	0·15	0·30	0·45	0·60	0·75	0·90	1·05	1·20	1·35	1·50
(a)	(i)	$1{\cdot}3\times10^{-12}$	$1{\cdot}4\times10^{-12}$	$1{\cdot}9\times10^{-13}$	$4{\cdot}7\times10^{-12}$	$3{\cdot}6\times10^{-12}$	$1{\cdot}1\times10^{-11}$	$1{\cdot}0\times10^{-11}$	$1{\cdot}1\times10^{-11}$	$1{\cdot}7\times10^{-11}$	$7{\cdot}6\times10^{-12}$
	(ii)	$2{\cdot}5\times10^{-13}$	$4{\cdot}8\times10^{-13}$	$8{\cdot}5\times10^{-13}$	$8{\cdot}7\times10^{-13}$	$1{\cdot}6\times10^{-12}$	$1{\cdot}2\times10^{-12}$	$2{\cdot}3\times10^{-12}$	$5{\cdot}2\times10^{-12}$	$2{\cdot}7\times10^{-12}$	$4{\cdot}1\times10^{-12}$
	(iii)	$2{\cdot}2\times10^{-12}$	$4{\cdot}3\times10^{-12}$	$6{\cdot}1\times10^{-12}$	$7{\cdot}1\times10^{-12}$	$7{\cdot}6\times10^{-12}$	$8{\cdot}8\times10^{-12}$	$1{\cdot}1\times10^{-11}$	$1{\cdot}1\times10^{-11}$	$1{\cdot}2\times10^{-11}$	$1{\cdot}4\times10^{-11}$
(b)	(i)	$5{\cdot}2\times10^{-10}$	$9{\cdot}7\times10^{-10}$	$1{\cdot}4\times10^{-9}$	$1{\cdot}7\times10^{-9}$	$2{\cdot}1\times10^{-9}$	$2{\cdot}4\times10^{-9}$	$2{\cdot}6\times10^{-9}$	$2{\cdot}9\times10^{-9}$	$3{\cdot}1\times10^{-9}$	$3{\cdot}3\times10^{-9}$
	(ii)	$4{\cdot}5\times10^{-9}$	$8{\cdot}5\times10^{-9}$	$1{\cdot}2\times10^{-8}$	$1{\cdot}5\times10^{-8}$	$1{\cdot}8\times10^{-8}$	$2{\cdot}1\times10^{-8}$	$2{\cdot}3\times10^{-8}$	$2{\cdot}6\times10^{-8}$	$2{\cdot}8\times10^{-8}$	$3{\cdot}0\times10^{-8}$
	(iii)	$4{\cdot}5\times10^{-8}$	$8{\cdot}8\times10^{-8}$	$1{\cdot}3\times10^{-7}$	$1{\cdot}7\times10^{-7}$	$2{\cdot}1\times10^{-7}$	$2{\cdot}4\times10^{-7}$	$2{\cdot}8\times10^{-7}$	$3{\cdot}2\times10^{-7}$	$3{\cdot}6\times10^{-7}$	$4{\cdot}0\times10^{-7}$
(c)	(i)	$1{\cdot}4\times10^{-9}$	$2{\cdot}8\times10^{-9}$	$4{\cdot}3\times10^{-9}$	$5{\cdot}8\times10^{-9}$	$7{\cdot}3\times10^{-9}$	$8{\cdot}7\times10^{-9}$	$1{\cdot}0\times10^{-8}$	$1{\cdot}1\times10^{-8}$	$1{\cdot}3\times10^{-8}$	$1{\cdot}4\times10^{-8}$
	(ii)	$9{\cdot}8\times10^{-9}$	$2{\cdot}0\times10^{-8}$	$3{\cdot}1\times10^{-8}$	$4{\cdot}2\times10^{-8}$	$5{\cdot}2\times10^{-8}$	$6{\cdot}3\times10^{-8}$	$7{\cdot}3\times10^{-8}$	$8{\cdot}2\times10^{-8}$	$9{\cdot}1\times10^{-8}$	$9{\cdot}9\times10^{-8}$
	(iii)	$3{\cdot}0\times10^{-8}$	$8{\cdot}7\times10^{-8}$	$1{\cdot}7\times10^{-7}$	$3{\cdot}0\times10^{-7}$	$4{\cdot}6\times10^{-7}$	$6{\cdot}6\times10^{-7}$	$9{\cdot}1\times10^{-7}$	$1{\cdot}2\times10^{-6}$	$1{\cdot}6\times10^{-6}$	$2{\cdot}0\times10^{-6}$
(d)	(i)	$5{\cdot}2\times10^{-11}$	$1{\cdot}0\times10^{-10}$	$1{\cdot}5\times10^{-10}$	$2{\cdot}0\times10^{-10}$	$2{\cdot}5\times10^{-10}$	$3{\cdot}0\times10^{-10}$	$3{\cdot}5\times10^{-10}$	$3{\cdot}9\times10^{-10}$	$4{\cdot}3\times10^{-10}$	$4{\cdot}6\times10^{-10}$
	(ii)	$1{\cdot}2\times10^{-12}$	$2{\cdot}4\times10^{-12}$	$3{\cdot}9\times10^{-12}$	$4{\cdot}4\times10^{-12}$	$4{\cdot}6\times10^{-12}$	$1{\cdot}2\times10^{-12}$	$2{\cdot}2\times10^{-12}$	$2{\cdot}4\times10^{-12}$	$1{\cdot}0\times10^{-11}$	$1{\cdot}4\times10^{-11}$
	(iii)	$3{\cdot}6\times10^{-8}$	$7{\cdot}4\times10^{-8}$	$1{\cdot}1\times10^{-7}$	$1{\cdot}5\times10^{-7}$	$1{\cdot}9\times10^{-7}$	$2{\cdot}4\times10^{-7}$	$2{\cdot}8\times10^{-7}$	$3{\cdot}3\times10^{-7}$	$3{\cdot}7\times10^{-7}$	$4{\cdot}3\times10^{-7}$
(e)	(i)	$1{\cdot}2\times10^{-9}$	$2{\cdot}4\times10^{-9}$	$3{\cdot}5\times10^{-9}$	$4{\cdot}6\times10^{-9}$	$5{\cdot}5\times10^{-9}$	$6{\cdot}3\times10^{-9}$	$6{\cdot}9\times10^{-9}$	$7{\cdot}4\times10^{-9}$	$7{\cdot}7\times10^{-9}$	$7{\cdot}8\times10^{-9}$
	(ii)	$7{\cdot}9\times10^{-9}$	$1{\cdot}6\times10^{-8}$	$2{\cdot}3\times10^{-8}$	$3{\cdot}0\times10^{-8}$	$3{\cdot}6\times10^{-8}$	$4{\cdot}1\times10^{-8}$	$4{\cdot}6\times10^{-8}$	$4{\cdot}9\times10^{-8}$	$5{\cdot}1\times10^{-8}$	$5{\cdot}3\times10^{-8}$
	(iii)	$2{\cdot}2\times10^{-8}$	$4{\cdot}3\times10^{-8}$	$6{\cdot}3\times10^{-8}$	$8{\cdot}2\times10^{-8}$	$9{\cdot}9\times10^{-8}$	$1{\cdot}2\times10^{-7}$	$1{\cdot}3\times10^{-7}$	$1{\cdot}4\times10^{-7}$	$1{\cdot}4\times10^{-7}$	$1{\cdot}5\times10^{-7}$
(f)	(i)	$1{\cdot}7\times10^{-11}$	$3{\cdot}3\times10^{-11}$	$4{\cdot}9\times10^{-11}$	$6{\cdot}8\times10^{-11}$	$8{\cdot}2\times10^{-11}$	$1{\cdot}0\times10^{-10}$	$1{\cdot}4\times10^{-10}$	$1{\cdot}5\times10^{-10}$	$1{\cdot}6\times10^{-10}$	$1{\cdot}8\times10^{-10}$
	(ii)	$3{\cdot}8\times10^{-13}$	$4{\cdot}3\times10^{-13}$	$4{\cdot}5\times10^{-13}$	$1{\cdot}1\times10^{-12}$	$7{\cdot}5\times10^{-13}$	$2{\cdot}5\times10^{-12}$	$1{\cdot}4\times10^{-12}$	$3{\cdot}1\times10^{-12}$	$8{\cdot}3\times10^{-13}$	$2{\cdot}5\times10^{-13}$
	(iii)	$2{\cdot}7\times10^{-10}$	$7{\cdot}3\times10^{-10}$	$1{\cdot}4\times10^{-9}$	$2{\cdot}2\times10^{-9}$	$3{\cdot}2\times10^{-9}$	$4{\cdot}3\times10^{-9}$	$5{\cdot}6\times10^{-9}$	$7{\cdot}2\times10^{-9}$	$8{\cdot}8\times10^{-9}$	$1{\cdot}1\times10^{-8}$

Various choices of abscissae θ_k $(k = 0,1,2,\ldots,p)$ were employed:

case (i): $\theta_0 = 0$, $\theta_1 = \frac{1}{4}$, $\theta_2 = \frac{1}{2}$, $\theta_3 = \frac{3}{4}$, $\theta_4 = 1$ $(p = 4)$,

case (ii): $\theta_0 = 0$, $\theta_1 = 0{\cdot}2763932$, $\theta_2 = 1-\theta_1$, $\theta_3 = 1$ $(p = 3)$,

case (iii): $\theta_0 = 0$, $\theta_1 = \frac{1}{3}$, $\theta_2 = \frac{2}{3}$, $\theta_3 = 1$ $(p = 3)$,

with step-size $h^* = 0{\cdot}075$. Results obtained are summarized in Table 6.12. (The choice (ii) gives the Lobatto rule using 4 abscissae in (6.108).)

*

6.11. *Product-integration methods for equations of the first kind with continuous or weakly singular kernels*

Product-integration formulae of the type

$$\int_0^x K(x,y)\phi(y)\,dy \simeq \sum_{j=0}^{N} v_j(x)\phi(jh) \quad (x \le X = Nh) , \tag{6.162}$$

where $K(x,y)$ is either a continuous Volterra kernel, or has a weak singularity, can be employed successfully to treat the Volterra equation of the first kind. The construction of formulae of the type (6.162) has been discussed earlier, in section 6.8, and their use to approximate the equation of the first kind

$$\int_0^x K(x,y)f(y)\,dy = g(x) \quad (0 \le x \le X)$$

leads to a step-by-step or block-by-block method. Many methods of this type are obvious analogues of the methods given in section 6.8 for equations of the second kind.

Thus the use of the generalized trapezium rule, which gives the approximation

$$\int_0^{rh} K(rh,y)\phi(y)\,dy \simeq \sum_{j=0}^{r-1} \frac{1}{h}\int_{jh}^{(j+1)h} K(rh,y)\{(y-jh)\phi((j+1)h) +$$

$$+ \{(j+1)h-y\}\phi(jh)\}\,dy \quad (r = 1,2,3,\ldots,N) ,$$

gives a step-by-step method provided that the starting value $\tilde{f}(0) \simeq f(0)$ is obtained independently. (If $K(x,y)$ is weakly singular, the value $g'(0)/K(0,0)$ has no meaning and cannot be taken as the value of $f(0)$.†)

Example 6.56.

Anderssen and White (1971) apply the generalized trapezium rule to the numerical solution of the test problem

---

† For various types of kernels with weak singularities, an analytical expression can be obtained for the value of $f(0)$. Some examples appear below.

$$\int_0^x \cos(x-y)f(y)\,dy = 1 - \cos x \,.$$

(The exact solution is $f(x) = x$.)

The equations to be solved are

$$\tilde{f}(0) = g'(0)/K(0,0)$$

and

$$\sum_{j=0}^{r} \nu_j(rh)\tilde{f}(jh) = g(rh) \quad (r = 1,2,3,\ldots) \;,$$

where $K(x,y) = \cos(x-y)$ and the expression for $\nu_j(rh)$ is given in eqn (6.125).

The errors in the approximate solution computed on $[0,2]$ with a step $h = 0{\cdot}15$ do not exceed 2×10^{-6}. The results are superior to those computed using the unmodified mid-point rule (section 6.10) followed by an h^2-extrapolation using results obtained with $h = 0{\cdot}3$ and $h = 0{\cdot}1$. The reader may refer to the original paper of Anderssen and White (1971) for fuller details. *

The generalized mid-point rule applied to the equation of the first kind gives a system of equations

$$\sum_{j=0}^{r} \nu_{2j+1}(2(r+1)h)\tilde{f}((2j+1)h) = g(2(r+1)h) \quad (r = 0,1,2,\ldots,[\tfrac{1}{2}(N-1)]) \tag{6.163}$$

where

$$\nu_{2j+1}(x) = \int_{2jh}^{2(j+1)h} K(x,y)\,dy \,. \tag{6.164}$$

Thus eqns (6.163) constitute a system of equations for $\tilde{f}(h), \tilde{f}(3h)$, $\tilde{f}(5h)$, ... which can be solved in a step-by-step fashion, and no starting value is required - a desirable feature.

Product-integration formulae of the type (6.162) are particularly appropriate in the numerical solution of equations of the form

$$\int_0^x \frac{H(x,y)}{(x-y)^{\frac{1}{2}}} f(y)\,dy = g(x) \quad (0 \leq x \leq X) \;,$$

where $H(x,y)$ is continuous for $0 \leq x \leq X$, or, more generally,

$$\int_0^x \frac{H(x,y)}{(x^p-y^p)^\alpha} f(y)\,dy = g(x) \quad (0 \leq x \leq X) , \tag{6.165}$$

where $p = 1$ or 2 and $0 < \alpha < 1$. Equations of the form (6.165) are sometimes called generalized Abel's equations.

Example 6.57.
Atkinson (1971*a*) has considered the approximate solution of the equation

$$\int_0^x \frac{f(y)\,dy}{(x^2-y^2)^{\frac{1}{2}}} = g(x) \quad (0 \leq x \leq X)$$

using product integration. The equation in question is somewhat unusual since the integral operator with kernel $K(x,y) = (x^2-y^2)^{-\frac{1}{2}}$ $(0 \leq y \leq x \leq X)$, and $K(x,y) = 0$ for $y > x$, is not compact on $C[0,X]$. However, the solution of the equation can be expressed as

$$f(x) = \frac{2}{\pi}\frac{d}{dx}\int_0^x \frac{yg(y)}{(x^2-y^2)^{\frac{1}{2}}}\,dy , \quad (X \geq x > 0) ,$$

assuming the existence of the latter term, and $f(0) = (2/\pi)g(0)$.

Use of the generalized trapezium rule to approximate integrals of the form $\int_0^x (x^2-y^2)^{-\frac{1}{2}}\phi(y)\,dy$ produces the system of equations

$$\tilde{f}(0) = (2/\pi)g(0) ,$$

$$\sum_{j=0}^{r} \nu_j(rh)\tilde{f}(jh) = g(rh) \quad (r = 1,2,3,\ldots,N) ,$$

where $Nh = X$, and we can write

$$\nu_j(rh) = \frac{1}{h}\int_{(j-1)h}^{(j+1)h} \frac{\delta_j(y)}{\{(rh)^2-y^2\}^{\frac{1}{2}}}\,dy \quad (1 \leq j \leq r-1) ,$$

with

$$\delta_j(y) = \begin{cases} (1/h)\{y-(j-1)h\} & ((j-1)h \leq y \leq jh) \\ \\ (1/h)\{(j+1)h-y\} & (jh \leq y \leq (j+1)h) \end{cases},$$

whilst

$$\nu_0(rh) = \frac{1}{h}\int_0^h \frac{h-y}{\{(rh)^2-y^2\}^{\frac{1}{2}}}dy$$

and

$$\nu_r(rh) = \frac{1}{h}\int_{(r-1)h}^{rh} \frac{y-(r-1)h}{\{(rh)^2-y^2\}^{\frac{1}{2}}}\,dy\ .$$

The choice $g(x) = \frac{1}{4}\pi x^2$ corresponds to the solution $f(x) = x^2$ and the error in the solution at $x = 5$ computed with varying steps in h is $(h = \frac{1}{10})$ $1{\cdot}6 \times 10^{-3}$, $(h = \frac{1}{20})$ $4{\cdot}1 \times 10^{-4}$, $(h = \frac{1}{40})$ $1{\cdot}0 \times 10^{-4}$, and $(h = \frac{1}{80})$ $2{\cdot}6 \times 10^{-5}$. The results indicate $O(h^2)$ convergence.

For the equation with $g(x) = \frac{1}{2}\pi J_0(x)$ the solution is $f(x)=\cos(x)$. Errors computed at $x = 2(2)20$, using steps $h = \frac{1}{10}$ and $h = \frac{1}{20}$ are:

x	$\frac{1}{10}$	$\frac{1}{20}$
2·0	$2{\cdot}9 \times 10^{-4}$	$7{\cdot}7 \times 10^{-5}$
4·0	$5{\cdot}6 \times 10^{-4}$	$1{\cdot}4 \times 10^{-4}$
6·0	$-7{\cdot}5 \times 10^{-4}$	$-1{\cdot}9 \times 10^{-4}$
8·0	$7{\cdot}1 \times 10^{-5}$	$2{\cdot}2 \times 10^{-5}$
10·0	$6{\cdot}9 \times 10^{-4}$	$1{\cdot}7 \times 10^{-4}$
12·0	$-6{\cdot}5 \times 10^{-4}$	$-1{\cdot}7 \times 10^{-4}$
14·0	$-1{\cdot}5 \times 10^{-4}$	$3{\cdot}5 \times 10^{-5}$
16·0	$7{\cdot}7 \times 10^{-4}$	$2{\cdot}0 \times 10^{-4}$
18·0	$-4{\cdot}9 \times 10^{-4}$	$-1{\cdot}3 \times 10^{-4}$
20·0	$-3{\cdot}6 \times 10^{-4}$	$-8{\cdot}9 \times 10^{-5}$

The error 'growth' appears to be modest. *

Let us consider the equation

$$\int_0^x \frac{H(x,y)}{(x-y)^\alpha} f(y)\,dy = g(x) \quad (0 \le x \le X) \quad , \qquad (6.166)$$

where $0 < \alpha < 1$. If $H(x,x) \neq 0$ $(0 \le x \le X)$, $g(0) = 0$, and $H(x,y)$, $(\partial/\partial x)H(x,y)$ and $g'(x)$ are continuous (for $0 \le y \le x \le X$) , and

$$G(x) = \int_0^x \frac{g(y)}{(x-y)^{1-\alpha}}\,dy$$

is continuously differentiable for $0 \le x \le X$, then eqn (6.166) has a unique solution which is continuous for $0 \le x \le X$. Indeed, under these conditions we obtain

$$\int_0^x L(x,y)f(y)\,dy = G(x) \ ,$$

where $L(x,y) = \int_y^x H(s,y)/\{(x-s)^{1-\alpha}(s-y)^\alpha\}ds$,

on replacing x by s in (6.166), multiplying both sides by $(x-s)_+^{\alpha-1}$, and then integrating with respect to s .

We have thus obtained an equation of the first kind in which the kernel $L(x,y)$ is finite. Indeed,

$$L(x,y) = \int_0^1 \frac{H(y+\tau(x-y),y)}{\tau^\alpha(1-\tau)^{1-\alpha}}\,d\tau$$

so that $\|L(x,y)\|_\infty \le \|H(x,y)\|_\infty \pi/(\sin \pi\alpha)$.

Since $L(x,x) = H(x,x)\pi/(\sin \pi\alpha) \neq 0$ for $0 \le x \le X$, we find

$$f(x) + \int_0^x \frac{(\partial/\partial x)L(x,y)}{L(x,x)} f(y)\,dy = \frac{G'(x)}{L(x,x)} \quad (0 \le x \le X)$$

on differentiating the equation of the first kind.

The latter equation provides the basis of indirect methods for the solution of (6.166) but in the present context it may be used to supply the value $f(0) = G'(0)/L(0,0)$ and to assess the degree of differentiability of the solution $f(x)$.

Remark. If $(d/dx)^{r+1}G(x)$ is continuous for $0 \le x \le X$, and the mixed partial derivatives of $H(x,y)$ of order r and $(\partial/\partial x)^{r+1}H(x,y)$ are continuous for $0 \le y \le x \le X$, then $(d/dx)^{r}f(x)$ is continuous for $0 \le x \le X$. The properties of $G(x)$ can be investigated by considering the expression

$$G'(x) = \Gamma(\alpha) \sum_{j=0}^{r-1} \frac{1}{\Gamma(\alpha+j+1)} g^{(1+j)}(0)x^{\alpha+j} +$$

$$+ \frac{1}{\Gamma(\alpha+r)} \int_0^x g^{(r+1)}(y)\,(x-y)^{\alpha+r-1}dy \quad (0 \le x \le X) \quad ,$$

assuming $g(x) \in C^{r+1}[0,X]$; this follows on repeated integration by parts. (Here $\Gamma(\ )$ is the gamma function .) We may require the use of product integration formulae in the evaluation of $G'(x)$ if $f(x)$ is to be determined numerically from tabular values of $G'(x)$.

As is frequently the case, a mathematical study of the integral equation, which is undertaken before a numerical method is applied, will generally bring benefits. Abel's equations have received quite extensive study in the literature; see Linz (1967*b*) for example.

Example 6.58.
To treat it in the manner indicated above, the equation

$$\gamma(x) = 2\int_x^1 \{y\psi(y)/(y^2-x^2)^{\frac{1}{2}}\}dy \quad (x \le 1)$$

can be written, by means of a change of variables, in the form

$$g(x) = \int_0^x \{f(y)/(x-y)^{\frac{1}{2}}\}dy. \quad *$$

In eqns (6.165) and (6.166) we have cases where the kernel $K(x,y)$ has the form $H(x,y)/(x^p-y^p)^{\alpha}$. Approximations of the form (6.162) may not be easy to obtain in the case where $H(x,y)$ is complicated. In this situation, the product integration techniques discussed in Chapter

5 (see section 5.6) can be applied. In particular, we may construct approximations of the form

$$\int_0^x \frac{1}{(x^p-y^p)^\alpha}\phi(y)dy \simeq \sum_{j=0}^{N} \nu_j(x)\phi(jh) \quad (0 \le x \le X = Nh)$$

and set

$$\sum_{j=0}^{N} \nu_j(rh)\{H(rh,jh)\tilde{f}(jh)\} = g(rh) \quad (r = 1,2,3,\ldots,N) .$$

(A starting value for $\tilde{f}(0)$ is required if $\nu_0(rh) \neq 0$, where $r \in \{1,2,3,\ldots,N\}$.) Such approximations will usually be simple ones if $H(x,y)$ is unknown for $y > x$, and the approximations most frequently encountered in the literature involve use of the generalized trapezium rule or generalized mid-point rule. (These approximations give step-by-step methods for calculating $\tilde{f}(x)$, and do not require values of $H(x,y)$ with $y > x$. The generalized mid-point rule gives a method which does not require a starting value.) In the case of (6.166), the weight function $(x-y)^{-\alpha}$ is a function of $(x-y)$ and the weights $\nu_j(rh)$ are easily obtained.

<u>Example 6.59</u>.

The weights of the generalized mid-point rule give approximations

$$\int_0^{2r\eta} (2r\eta-y)^{-\alpha}\phi(y)\,dy \simeq \sum_{j=0}^{r-1} \nu_{2j+1}(2r\eta)\phi((2j+1)\eta) ,$$

where

$$\nu_{2j+1}(2r\eta) = \int_{2j\eta}^{2(j+1)\eta} (2r\eta-y)^{-\alpha}dy$$

$$= (2\eta)^{1-\alpha}\{(r-j-1)^{1-\alpha}-(r-j)^{1-\alpha}\}/(\alpha-1)$$

is a function of $r-j$.

With $Nh = X$ it is convenient to set $\eta = \frac{1}{2}h$ and the generalized mid-point method applied to (6.166) then gives equations which may be written in the form

$$\sum_{j=0}^{r-1} V_{r-j} H(rh,(j+\tfrac{1}{2})h)\tilde{f}((j+\tfrac{1}{2})h) = g(rh) \quad (r = 1,2,3,\ldots,N)$$

where $V_s = h^{1-\alpha}\{s^{1-\alpha}-(s-1)^{1-\alpha}\}/(1-\alpha)$.

Weiss (1972) presents numerical results computed when applying the method to the equation

$$\int_0^x (x-y)^{-\frac{1}{2}}\exp\{-\tfrac{1}{2}(x-y)\}f(y)\,dy$$

$$= x^{-\frac{1}{2}}\exp\{-\tfrac{1}{2}(1+x)^2/x\} \quad (0 \leq x \leq 1) .$$

Maximum errors $1{\cdot}2 \times 10^{-2}$, $5{\cdot}2 \times 10^{-3}$, $1{\cdot}2 \times 10^{-3}$, and 2×10^{-4} are reported with $h = \frac{1}{10}$, $h = \frac{1}{30}$, $h = \frac{1}{90}$, and $h = \frac{1}{270}$ respectively, and the results suggest that the errors are $O(h^{3/2})$ as $h \to 0$. The reader will find further discussion in Weiss's thesis. *

I conclude this section by noting that collocation methods and Galerkin methods could also be applied, in suitable cases, to Volterra equations of the first kind.

6.12. *Convergence properties of numerical methods for equations of the second kind*

We first consider convergence results for the numerical solution of the equation

$$f(x) - \int_0^x F(x,y;f(y))\,dy = g(x) \quad (0 \leq x \leq X) \qquad (6.167)$$

supposing that $g(x)$ and $F(x,y;v)$ satisfy the conditions (a)-(c) stated in the opening paragraph of section 6.4. In particular we suppose that the Lipschitz condition (6.52) is satisfied, namely,

$$|F(x,y;v_1)-F(x,y;v_2)| \leq L|v_1-v_2| \quad \text{for} \quad 0 \leq y \leq x \leq X .$$

Let us suppose that approximate values $\tilde{f}(0),\tilde{f}(h),\tilde{f}(2h)$, ... are obtained (with $Nh = X$, say) by solving the equations

$$\tilde{f}(0) = g(0)$$

$$\tilde{f}(kh) - \sum_{j=0}^{k} w_{kj} F(kh, jh; \tilde{f}(jh)) = g(kh) \quad (6.168)$$

for $k = q, (q+1), (q+2), \ldots, N$, the values $\tilde{f}(h), \tilde{f}(2h), \ldots \tilde{f}((q-1)h)$ being obtained by a suitable starting method (see section 6.5). We suppose that

$$\tilde{f}(rh) = f(rh) + \xi_r(h) \quad (r = 1, 2, \ldots, (q-1)) , \quad (6.169)$$

and since $g(0)$ may not be evaluated exactly, we can also suppose that (6.169) is valid for $r = 0$ and

$$\xi(h) = \sum_{r=0}^{q-1} |\xi_r(h)| . \quad (6.170)$$

Finally, we suppose that $hWL < 1$ where

$$W \equiv W_N = \max_{0 \le j \le k \le N} |w_{kj}| / h < \infty ,$$

and we define

$$t_r(h) = \int_0^{rh} F(rh, y; f(y)) \mathrm{d}y - \sum_{j=0}^{r} w_{rj} F(rh, jh; f(jh)) , \quad (6.171)$$

and

$$\tau(h) \equiv \tau_N(h) = \max_{q \le r \le N} |t_r(h)| . \quad (6.172)$$

The assumptions made above are those stated in Theorem 6.1, and we shall prove that theorem shortly. Let us first establish the following result.

LEMMA 6.1. *Given that* $hWL \le \rho < 1$, *eqn* (6.168) *has a unique solution* $\tilde{f}(kh)$ *in terms of* $\tilde{f}(0), \tilde{f}(h), \ldots, \tilde{f}((k-1)h)$, *and this solution may be found by a predictor-corrector technique.*

Proof. Consider the iteration $\tilde{f}^{[r+1]}(kh) = \psi_k(\tilde{f}^{[r]}(kh))$, where

$$\psi_k(f) = \gamma_k + w_{k,k}F(kh,kh;f) ,$$

γ_k is the right-hand term of eqn (6.54), and

$$|\psi_k(f_1) - \psi_k(f_2)| \leq hWL|f_1 - f_2| \qquad (6.173)$$

for arbitrary f_1, f_2 . If $\tilde{f}_1, \tilde{f}_2$ are two solutions of the equation $f = \psi_k(f)$ it follows that, if $|\tilde{f}_1 - \tilde{f}_2| \neq 0$,

$$|\tilde{f}_1 - \tilde{f}_2| = |\psi_k(\tilde{f}_1) - \psi_k(\tilde{f}_2)| \leq hWL|\tilde{f}_1 - \tilde{f}_2| < |\tilde{f}_1 - \tilde{f}_2| ,$$

since $hWL \leq \rho < 1$. Hence $|\tilde{f}_1 - \tilde{f}_2| = 0$, and any solution of (6.168) is unique.

Now, for $r \geq 1$ we see, using (6.173), that

$$|\tilde{f}^{[r+1]}(kh) - \tilde{f}^{[r]}(kh)| = |\psi_k(\tilde{f}^{[r]}(kh)) - \psi_k(\tilde{f}^{[r-1]}(kh))|$$

$$\leq \rho|\tilde{f}^{[r]}(kh) - \tilde{f}^{[r-1]}(kh)| ,$$

where $\rho < 1$. It follows that

$$|\tilde{f}^{[r+j]}(kh) - \tilde{f}^{[r+j-1]}(kh)| \leq \rho^j |\tilde{f}^{[r]}(kh) - \tilde{f}^{[r-1]}(kh)| \quad (j = 1,2,3,\ldots);$$

hence

$$|\tilde{f}^{[n]}(kh) - \tilde{f}^{[n-1]}(kh)| \leq \rho^{n-1}|\tilde{f}^{[1]}(kh) - \tilde{f}^{[0]}(kh)|$$

and

$$|\tilde{f}^{[n]}(kh) - \tilde{f}^{[n+m]}(kh)|$$

$$\leq \sum_{j=1}^{m} |\tilde{f}^{[n+j]}(kh) - \tilde{f}^{[n+j-1]}(kh)|$$

$$\leq \{\sum_{j=1}^{m} \rho^j\} |\tilde{f}^{[n]}(kh) - \tilde{f}^{[n-1]}(kh)|$$

$$\leq \rho \frac{\rho^m - 1}{\rho - 1} |\tilde{f}^{[n]}(kh) - \tilde{f}^{[n-1]}(kh)| , \qquad (6.174)$$

where $\rho < 1$. It follows, immediately, that the sequence $\tilde{f}^{[r]}(kh)$ $(r = 0,1,2,\ldots)$ is a Cauchy sequence and hence has a limit point. Since $\psi_k(f)$ is continuous,

$$\lim_{r\to\infty} \tilde{f}^{[r+1]}(kh) = \lim_{r\to\infty} \psi_k(\tilde{f}^{[r]}(kh)) = \psi_k(\lim_{r\to\infty} \tilde{f}^{[r]}(kh))$$

and it follows that $\lim_{r\to\infty} \tilde{f}^{[r]}(kh)$ is the required value $\tilde{f}(kh)$.

Remark. If we allow m to tend to infinity in (6.174) we find that

$$|\tilde{f}^{[n]}(kh)-\tilde{f}(kh)| \leq \{\rho/(1-\rho)\}|\tilde{f}^{[n]}(kh)-\tilde{f}^{[n-1]}(kh)|$$

$$\leq \{\rho^n/(1-\rho\}|\tilde{f}^{[1]}(kh)-\tilde{f}^{[0]}(kh)| \quad (\rho<1). \quad (6.175)$$

In our statement of Theorem 6.1 we gave the result

$$|\tilde{f}(rh)-f(rh)| \leq \frac{\tau_N(h)+hLW\xi(h)}{1-hLW} \exp(\frac{LWrh}{1-hLW}) \quad (r=q,(q+1),\ldots,N). \quad (6.176)$$

We shall establish this result as a corollary of the following lemmas. Note that $W \equiv W_N$ in (6.176).

LEMMA 6.2. *If* $|\zeta_r| \leq A \sum_{i=0}^{r-1} |\zeta_i| + B$ $(r = q,(q+1), \ldots)$ *where* $A > 0$, $B > 0$, *and* $\sum_{i=0}^{q-1} |\zeta_i| \leq \zeta$, *then* $|\zeta_r| \leq (A\zeta+B)(1+A)^{r-q}$ $(r = q,(q+1), \ldots)$.

Proof. The proof is by induction. Clearly, $|\zeta_q| < A\zeta+B$, so the result is true for $r = q$. Suppose that the result is true for $r = q,(q+1), \ldots (m-1)$. Then

$$|\zeta_m| \leq A \sum_{i=0}^{m-1} |\zeta_i|+B \leq A \sum_{i=0}^{q-1} |\zeta_i|+\{A \sum_{i=q}^{m-1} |\zeta_i|+B\}$$

$$\leq A\zeta+\{A(A\zeta+B) \sum_{i=q}^{m-1} (1+A)^{i-q}+B\}$$

$$= (B+A\zeta)+A(A\zeta+B) \frac{(1+A)^{m-q}-1}{A} = (B+A\zeta)(1+A)^{m-q},$$

and the result follows by induction.

LEMMA 6.3. *Suppose that* $A = hL_* \geq 0$ *and* $ph = x \geq 0$. *Then*

$$(1+A)^{p-q} \leq \exp(L_* x) \quad \text{if} \quad p \geq q \ .$$

Proof. $A \geq 0$ *and* $(1+A)^{p-q} \leq (1+A)^p \leq (1+A+\frac{1}{2}A^2+\frac{1}{6}A^3 \ldots)^p$

$$= \{\exp A\}^p = \exp(L_* x) \ .$$

We now give the proof of (6.176).
We have

$$f(rh) - \sum_{j=0}^{r} w_{r,j} F(rh,jh;f(jh))$$

$$= g(rh)+\{\int_0^{rh} F(rh,y;f(y))dy - \sum_{j=0}^{r} w_{r,j} F(rh,jh;f(jh))\} \ , \qquad (6.177)$$

whereas for $r \geq q$ there exist values $\{\tilde{f}(rh)\}$ satisfying

$$\tilde{f}(rh) - \sum_{j=0}^{r} w_{r,j} F(rh,jh;\tilde{f}(jh)) = g(rh) \ , \qquad (6.178)$$

provided that $hWL \leq \rho < 1$.
We write $\xi_r(h) = \tilde{f}(rh)-f(rh)$ and we find, for $r \geq q$,

$$\xi_r(h) - \sum_{j=0}^{r} w_{r,j}\{F(rh,jh;\tilde{f}(jh))-F(rh,jh;f(jh))\} = -t_r(h)$$

in the notation of (6.171). Hence

$$|\xi_r(h)| \leq |t_r(h)| + \sum_{j=0}^{r} |w_{r,j}| \left|F(rh,jh;\tilde{f}(jh))-F(rh,jh;f(jh))\right|$$

$$\leq |t_r(h)| + hWL \sum_{j=0}^{r} |\xi_j(h)| \ .$$

Consequently, when $hWL < 1$ and $q \leq r \leq N$,

$$(1-hWL)|\xi_r(h)| \leq \tau_N(h) + hWL \sum_{j=0}^{r-1} |\xi_j(h)| \ .$$

We now apply Lemmas 6.2, 6.3 with $\xi_r(h) = \zeta_r$, $B = \tau_N(h)/(1-hWL)$ and $A = hWL/(1-hWL)$, where $W \equiv W_N$, and we establish immediately the required result (6.176) $(r = q,(q+1),\ldots,N)$.

As immediate corollaries of the proof given above we have the results following the definition below.

DEFINITION 6.12. *A 'consistent step-by-step quadrature method'* (6.168) *is one in which* $\xi(h) \to 0$ *as* $h \to 0$ *and such that* $\sup_N W_N < \infty$ *and*

$$\lim_{\substack{h\to 0\\ Nh=X}} \sup_{q\le r\le N} \left| \sum_{j=0}^{r} w_{r,j}\Phi(rh,jh) - \int_0^{rh} \Phi(rh,y)\,dy \right| = 0 \qquad (6.179)$$

for all functions $\Phi(x,y)$ *which are continuous for* $0 \le y \le x \le X$.

If a method is consistent, (6.176) holds for all h sufficiently small with W replaced by $\sup_N W_N$, and convergence follows given the conditions in section 6.4:

THEOREM 6.2. (a) *All consistent step-by-step quadrature methods are convergent.* (b) *If* $\tau_N(h) = O(h^p)$ *as* $h \to \infty$ *with* Nh *fixed,* $\xi(h) = O(h^{p-1})$, *and* $\lim_{N\to\infty} W_N < \infty$, *then* $f(x)-\tilde{f}(x) = O(h^p)$ *at each mesh-point* $x \in \{rh \,|\, q \le r \le N\}$.

It may be observed that convergence occurs, in the normal sense that

$$\lim_{\substack{h\to 0\\ Nh=X}} \sup_{0\le r\le N} |\tilde{f}(rh)-f(rh)| = 0,$$

provided $\xi(h) \to 0$, $\lim W_N < \infty$, and (6.179) is satisfied for the *particular* function $\Phi(x,y) = F(x,y;f(y))$. (Continuity of $F(x,y;f(y))$ is in general assumed, along with other conditions of section 6.4.)

All the schemes in section 6.2 which have been described as giving suitable schemes for constructing the weights $w_{r,j}$ do give consistent methods satisfying (6.179). Generally, if $\sum_{j=0}^{r} w_{r,j}\phi(jh)$ is, for every r, a Riemann sum (see Chapter 2) approximating $\int_0^{rh} \phi(y)\,dy$, then (6.179) is satisfied, and $W_N \leq 1$ for all N.

(Remark. Indeed, we then have

$$\sup_{r \leq N} \left| \sum_{j=0}^{r} w_{r,j} \Phi(rh,jh) - \int_0^{rh} \Phi(rh,y)\,dy \right| \leq X \omega_y(\Phi;h), \quad X = Nh \text{ fixed},$$

where

$$\omega_y(\Phi;\delta) = \sup_{\substack{|y_1-y_2| \leq \delta \\ 0 \leq |y_1|,|y_2| \leq x \leq X}} |\Phi(x,y_1) - \Phi(x,y_2)| .$$

By uniform continuity, $\omega_y(\Phi;h) \to 0$ as $h \to 0$, and the method is consistent.)

The choice of a Gregory rule does not always give a Riemann sum, as above, but the choice suggested in Example 6.6 with a *fixed* upper bound on the number of differences taken in the Gregory correction does give a scheme satisfying (6.179). The argument required to establish this is a modification of the argument in the preceding remark, and it can be extended to cover a general class of quadrature methods discussed later in this section.

Remark. In practice we are generally unable to solve eqn (6.168) exactly, because of rounding error and because, when $F(x,y;f)$ is non-linear in f, the predictor-corrector iteration must generally be terminated before the exact solution is obtained. Let us suppose that $\tilde{f}(0) = g(0)$,

$$\tilde{f}(kh) - \sum_{j=0}^{k} w_{k,j} F(kh,jh;\tilde{f}(jh)) = g(kh) ,$$

and that the values $\hat{f}(kh)$ which we compute satisfy the perturbed equation

$$\hat{f}(kh) - \sum_{j=0}^{k} w_{k,j} F(kh,jh;\hat{f}(jh)) = g(kh) + s_k \quad (k = 0,1,2,\ldots,N,)$$

where $Nh = x$, and the sum is taken to be zero if $k = 0$. If we write $\zeta_k = \hat{f}(kh) - \tilde{f}(kh)$ then we obtain (for $k \geq 0$)

$$\zeta_k = \sum_{j=0}^{k} w_{k,j}\{F(kh,jh;\tilde{f}(jh))-F(kh,jh;\hat{f}(jh))\} + s_k ,$$

and hence

$$|\zeta_k| \leq \sum_{j=0}^{k} |w_{k,j}|L|\zeta_j| + |s_k|$$

or, if $hWL \leq \rho < 1$,

$$(1-hWL)|\zeta_k| \leq hWL \sum_{j=0}^{k-1} |\zeta_j| + |s_k| .$$

If we define

$$\sigma_N = \max_{k=0,1,2,\ldots,N} |s_k| ,$$

then applying Lemmas 6.2 and 6.3 (and supposing that $hWL < 1$) we find

$$|\tilde{f}(rh)-\hat{f}(rh)| \leq \frac{\sigma_N}{1-hLW} \exp\left(\frac{LWrh}{1-hLW}\right) \quad (r \leq N) , \tag{6.180}$$

in analogy with eqn (6.176).

The bound on the error $\tilde{f}(rh)-f(rh)$ given in (6.176) depends exponentially upon the value of $x = rh$. In other words, it would be consistent with (6.176) if, for a fixed step h , the errors $\tilde{f}(x)-f(x)$ grow exponentially with x . Fortunately this result can be improved upon, and our confidence in the stability of certain methods can be somewhat restored. A simple result is given in the following Example. Our aim is to show that with *certain* methods, $\tilde{f}(x) = f(x)+h^p\varepsilon(x)+o(h^p)$, where the form of $\varepsilon(x)$ can be determined.

<u>Example 6.60</u>.

Suppose that we employ the trapezium rule with step h and solve the equations

$$\tilde{f}(rh) - h \sum_{j=0}^{r}{}'' F(rh,jh;\tilde{f}(jh)) = g(rh) \quad (r = 1,2,3,\ldots)$$

with $\tilde{f}(0) = g(0)$. Then if $\xi_r(h) = \tilde{f}(rh)-f(rh)$, and $F(x,y;f(y))$ and

$f(x)$ are sufficiently differentiable,

$$\xi_r(h) = h^2\,\varepsilon(rh) + O(h^3)\ .$$

Here, $\varepsilon(x)$ is the solution of the equation

$$\varepsilon(x) - \int_0^x K(x,y)\,\varepsilon(y)\mathrm{d}y = -\nu(x)\ , \qquad (6.181)$$

where

$$K(x,y) = (\partial/\partial f)F(x,y;f(y))$$

and

$$\nu(x) = -\frac{1}{12}\left[\frac{\mathrm{d}}{\mathrm{d}y}F(x,y;f(y))\right]_{y=0}^{y=x}$$

Proof. An outline of the proof of this result will be given without paying full attention to rigour. Consider eqn (6.179) with the choice $w_{rj} = h$ for $r > j > 0$, $w_{rr} = w_{r0} = \frac{1}{2}h$, $q = 1$, and $\tilde{f}(0) = g(0)$. We obtain $\xi_0(h) = 0$ and

$$\xi_r(h) - h\sum_{j=0}^{r}{}''\{K(rh,jh)\xi_j(h) + O(\{\xi_j(h)\}^2)\} = -t_r(h) = O(h^2) \quad (r=1,2,3,\ldots,N)$$

on setting

$$F(rh,jh;\tilde{f}(jh)) - F(rh,jh;f(jh)) = (\tilde{f}(jh)-f(jh))K(rh,jh) +$$

$$+\tfrac{1}{2}(\tilde{f}(jh)-f(jh))^2\left[(\partial/\partial f)^2F(x,y;f)\right]_{f=f(jh)} + \ldots\ .$$

Considering the situation as $h \to 0$ with Nh fixed, we have, from (6.176), the result $\xi_j(h) = O(h^2)$. Then

$$h\sum_{j=0}^{r}{}''O(\{\xi_j(h)\}^2) = O(h^4)$$

for $rh \le X$, and hence

$$\xi_r(h) - h \sum_{j=0}^{r}{}'' K(rh,jh)\,\xi_j(h) = -t_r(h) + O(h^4) \; .$$

Since, indeed,

$$t_r(h)$$

$$= \int_0^{rh} F(rh,y;f(y))\,dy - h \sum_{j=0}^{r}{}'' F(rh,jh;f(jh))$$

$$= \frac{-h^2}{12}\left[\frac{d}{dy} F(rh,y;f(y))\right]_0^{rh} + O(h^4) \; ,$$

we obtain

$$\xi_r(h) - h \sum_{j=0}^{r}{}'' K(rh,jh)\,\xi_j(h) = -h^2\nu(rh) + s_r \; , \qquad (6.182)$$

where $s_r = O(h^4)$. Now if $\varepsilon(x)$ satisfies eqn (6.181), and

$$\tilde{\varepsilon}(rh) - h \sum_{j=0}^{r}{}'' K(rh,jh)\,\tilde{\varepsilon}_j(h) = -\nu(rh) \qquad (6.183)$$

it follows from (6.176) that $\tilde{\varepsilon}(rh) = \varepsilon(rh) + O(h^2)$. It also follows (if we apply the result expressed in (6.180) to compare eqn (6.182) with h^2 times the eqn (6.183)) that

$$|h^2\tilde{\varepsilon}(rh) - \xi_r(h)| = O(\max_k |s_k|) = O(h^4) \; .$$

Thus

$$\xi_r(h) = h^2\varepsilon(rh) + O(h^4) \; .$$

(A similar result governing the growth of rounding error can be obtained, assuming that the square of individual errors is negligible.) *

Let us interpret the result of the preceding Example by considering the effect of a small perturbation in the original integral equation. We have

$$f(x) - \int_0^x F(x,y;f(y))\,dy = g(x) \quad (0 \le x \le X) \; , \qquad (6.184)$$

and we suppose that

$$f_\varepsilon(x) - \int_0^x F(x,y;f_\varepsilon(y))dy = g(x)+\varepsilon s(x) , \qquad (6.185)$$

where $\varepsilon > 0$, $\sup_{0\le x\le X} |s(x)| \le 1$, and ε^2 is negligible compared with ε .

Our aim is to show that when (6.184) and (6.185) are satisfied for $0 \le x \le X$, and $F(x,y;f)$ is sufficiently differentiable,

$$f_\varepsilon(x) = f(x) + d_\varepsilon(x) + e_\varepsilon(x) ,$$

where $d_\varepsilon(x)$ satisfies an integral equation of the form

$$d_\varepsilon(x) - \int_0^x K(x,y)d_\varepsilon(y)dy = \varepsilon s(x) \quad (0 \le x \le X) , \qquad (6.186)$$

and where $K(x,y)$ is related to $F(x,y;v)$ and $f(x)$, and

$$\sup_{0\le x\le X} |e_\varepsilon(x)| = O(\varepsilon^2) .$$

This result will be established in Theorem 6.3.

We begin with a lemma which establishes a weaker result.

LEMMA 6.4. *If* $f(x)$ *and* $f_\varepsilon(x)$ *satisfy the equations above,* $\varepsilon > 0$, *and* $\sup_{0\le x\le X} |s(x)| \le 1$, *then*

$$\sup_{0\le x\le X} |f_\varepsilon(x)-f(x)| = O(\varepsilon)$$

as $\varepsilon \to 0$.

Proof. Since

$$f_\varepsilon(x)-f(x) = \varepsilon s(x) + \int_0^x \{F(x,y;f_\varepsilon(y))-F(x,y;f(y))\}dy ,$$

we have (employing the Lipschitz constant L)

$$|f_\varepsilon(x)-f(x)| \le \varepsilon|s(x)|+L\int_0^x |f_\varepsilon(y)-f(y)|\,dy . \tag{6.187}$$

If we define the function $\delta_\varepsilon(x)$ by the equation

$$\delta_\varepsilon(x) = \varepsilon|s(x)|+L\int_0^x \delta_\varepsilon(y)\,dy \tag{6.188}$$

then eqn (6.188) has the solution

$$\delta_\varepsilon(x) = \varepsilon\{|s(x)|+L\int_0^x e^{L(x-y)}|s(y)|\,dy\}$$

(see Example 6.2) and it is readily seen that $|f_\varepsilon(x)-f(x)| \le \delta_\varepsilon(x)$ Indeed, from (6.187) we see that

$$|f_\varepsilon(x)-f(x)| = \varepsilon|s(x)|+L\int_0^x |f_\varepsilon(y)-f(y)|\,dy-\sigma(x) ,$$

where $\sigma(x) \ge 0$, and on solving this equation

$$|f_\varepsilon(x)-f(x)| = \delta_\varepsilon(x)-\{\sigma(x)+L\int_0^x e^{L(x-y)}\sigma(y)\,dy\}$$

$$\le \delta_\varepsilon(x) .$$

Since

$$\sup_{0\le x\le X} |\delta_\varepsilon(x)| = O(\varepsilon)$$

the lemma is established.

We can now prove the following theorem.

THEOREM 6.3. *If* $f(x)$ *and* $f_\varepsilon(x)$ *satisfy* (6.184) *and* (6.185), *and* $(\partial/\partial f)^2F(x,y;f)$ *is continuous for* $0 \le y \le x \le X$, $|f| < \infty$, *then, as* $\varepsilon \to 0$,

$$f_\varepsilon(x) = f(x)+d_\varepsilon(x)+O(\varepsilon^2)$$

uniformly for $0 \le x \le X$, *where* $d_\varepsilon(x)$ *is defined by eqn* (6.186) *with*

$$K(x,y) = (\partial/\partial f)F(x,y;f)\Big]_{f=f(y)} .$$

Proof. We have

$$F(x,y;f_\varepsilon(y))-F(x,y;f(y)) = [f_\varepsilon(y)-f(y)]K(x,y) + \\ + \tfrac{1}{2}\{f_\varepsilon(y)-f(y)\}^2(\partial/\partial f)^2F(x,y;f)\Big]_{f=f_\theta(y)} ,$$

where

$$f_\theta(y) = \theta f_\varepsilon(y)+(1-\theta)f(y)$$

for some $\theta \in (0,1)$. We therefore obtain, from (6.184) and (6.185), the equation

$$\{f_\varepsilon(x)-f(x)\}-\int_0^x K(x,y)\{f_\varepsilon(y)-f(y)\}dy = \varepsilon s(x)+r_\varepsilon(x) ,$$

where

$$\sup_{0\le x\le X} |r_\varepsilon(x)| = ||r_\varepsilon(x)||_\infty = O(||f_\varepsilon(x)-f(x)||_\infty^2) = O(\varepsilon^2)$$

from Lemma 6.4. If

$$e_\varepsilon(x) - \int_0^x K(x,y)e_\varepsilon(y)dy = r_\varepsilon(x)$$

then clearly

$$\sup_{0\le x\le X} |e_\varepsilon(x)| = O(\varepsilon^2)$$

and

$$f_\varepsilon(x)-f(x) = d_\varepsilon(x) + e_\varepsilon(x) ,$$

where

$$d_\varepsilon(x) - \int_0^x K(x,y)d_\varepsilon(y)dy = \varepsilon s(x) .$$

The preceding analysis is intended to establish that the effect of a perturbation $\varepsilon s(x)$ is to produce a solution $f_\varepsilon(x) \approx f(x)+d_\varepsilon(x)$, where $d_\varepsilon(x)$ satisfies eqn (6.186). The theory is asymptotic since we assume that ε^2 is negligible and (consequently) that $|f_\varepsilon(x)-f(x)|^2$ is negligible for all $x \in [0,X]$.

If we compare the result of Theorem 6.3 with the discussion of Example 6.60, we gain some insight. In the Example we saw that the use of the repeated trapezium rule with step h gives an approximate solution $\tilde{f}(x)$ with

$$\tilde{f}(x)-f(x) = h^2\varepsilon(x) + O(h^4) ,$$

where $\varepsilon(x)$ satisfies eqn (6.186) obtained on setting $\varepsilon s(x) = -\nu(x)$. Theorem 6.3 indicates that perturbations in the integral equation can be expected to produce changes of this type. We may consider that the use of the trapezium rule to obtain a numerical scheme gives a good representation of the integral equation, in the sense that no spurious solutions (solutions not inherent in the equation itself†) dominate the error. Such conclusions are only asymptotic, however, since we may find that the approximate solution $\tilde{f}(x)$ obtained using the trapezium rule satisfies a relation

$$\tilde{f}(x)-f(x) = h^2\varepsilon(x)+h^4\gamma(x)+O(h^6),$$

and though we have said nothing about the terms $h^4\gamma(x) + O(h^6)$, these may be significant.

In section 6.3 we suggested that certain methods gave 'unstable' discretizations of the integral equation which they were meant to solve. For such methods we would expect that the dominant part of the error does not behave like the solution $d_\varepsilon(x)$ of an equation of the form (6.186). Kobayasi (1967) and Linz (1967*a*) and Noble (1969*a*) have examined this question.

The analysis of Noble (1969*a*) provides an extension of the ideas

---

† The precise set of functions which can be expected to arise from a perturbation of the integral equation depends, of course, upon the nature of possible perturbations! In (6.186), $d_\varepsilon(x)$ depends upon the behaviour of $s(x)$.

developed at the end of section 6.3 so that they apply to non-linear Volterra equations. If $\tilde{f}(0),\tilde{f}(h),\ldots,\tilde{f}((q-1)h)$ are computed by starting procedures and

$$\tilde{f}(rh)-\sum_{j=0}^{r} w_{r,j}F(rh,jh;\tilde{f}(jh)) = g(rh) \quad (r = q,(q+1),\ldots,N;Nh=X) \qquad (6.189)$$

whilst

$$f(rh)-\sum_{j=0}^{r} w_{r,j}F(rh,jh;f(jh)) = g(rh)+t_r(h) \quad (r = 1,2,3,\ldots,N)$$

then

$$\{f(rh)-\tilde{f}(rh)\}-\sum_{j=0}^{r} w_{r,j}\{F(rh,jh;f(jh))-F(rh,jh;\tilde{f}(jh))\}$$

$$= t_r(h) \quad (r = q,(q+1),\ldots,N) \,. \qquad (6.190)$$

Let us suppose that

$$\max_{q\le r\le N} |t_r(h)| = \tau_N(h) = O(h^p)$$

and

$$\max_{0\le r\le q-1} |f(rh)-\tilde{f}(rh)| = O(h^{p-1}) \,,$$

and let us assume that the conditions of Theorem 6.2 are satisfied. Then $|f(rh)-\tilde{f}(rh)| = O(h^p)$ as $h \to 0$ with rh fixed and lying in $[0,X]$.

We now see, from (6.190), that

$$[f(rh)-\tilde{f}(rh)] - \sum_{j=0}^{r} w_{r,j}K(rh,jh)[f(jh)-\tilde{f}(jh)]$$

$$= t_r(h)+u_r(h) \quad (r = q,(q+1),\ \ldots,N) \,, \qquad (6.191)$$

where $\max|t_r(h)| = O(h^p)$, and $\max |u_r(h)| = O(h^{2p})$ as $h \to 0$ with Nh fixed.

Let us *suppose* that for a fixed step h we can find a smooth function $\mu(x)$ such that $\mu(rh) = t_r(h)+u_r(h)$. Then eqn (6.191) has the appearance of a discretized form of the equation

$$\varepsilon(x) - \int_0^x K(x,y)\varepsilon(y)dy = \mu(x) \tag{6.192}$$

obtained by writing

$$\tilde{\varepsilon}(rh) - \sum_{j=0}^{r} w_{r,j}K(rh,jh)\tilde{\varepsilon}(jh) = \mu(rh) \quad (r=q,(q+1),\ldots,N) \tag{6.193}$$

and identifying $\tilde{\varepsilon}(x)$ with $f(x)-\tilde{f}(x)$.

We shall not regard eqns (6.189) as an unstable discretization of the integral equation if the dominant term of the error $\tilde{f}(x)-f(x)$ behaves like the solution $\varepsilon(x)$ of an equation of the form (6.192) (since perturbations in the integral equation naturally produce changes satisfying such an equation). However, we saw in section 6.3 that the numerical solution of an equation of the form (6.192) using a set of weights w_{rj} together with certain starting values *can* be unsatisfactory. The method can result in values $\tilde{\varepsilon}(rh)$ governed by (6.193) but containing a 'spurious' component which is unrelated to eqn (6.192). The size of this spurious component relative to the solution of (6.189) will not be discussed here, but this heuristic argument might suggest that a choice of weights w_{rj} which is unsuitable for the treatment of a linear integral equation is also unsuitable for a non-linear equation. With an unsuitable choice, the dominant term of the error $\tilde{f}(rh)-f(rh)$ is not related to an equation of the form (6.192). Such 'parasitic' terms may arise whether or not starting errors are present.

The preceding comments have lacked rigour, and heuristic arguments are sometimes unreliable. However, Kobayasi (1966) has given a rigorous discussion of the behaviour of the dominant component in the error $\tilde{f}(rh)-f(rh)$. We shall examine this theory below. Noble (1969*a*) refers to the work of Kobayasi as 'a remarkable tour-de-force', and the reader may prefer to pursue the natural extension of the part of the discussion in section 6.3.2 due to Noble (1969*a*) which, if slightly less general than the work of Kobayasi, is somewhat simpler to follow.

Our summary of the theory of Kobayasi will therefore omit some of the details of the analysis.

6.12.1.

Kobayasi (1966) is concerned with the behaviour of approximations $\tilde{f}(rh)$ $(q \leq r \leq N, Nh = x)$ satisfying eqns (6.189), the values $\tilde{f}(h)$, $\tilde{f}(2h),\ldots,\tilde{f}((q-1)h)$ being obtained by some starting procedure.

Certain assumptions (or limitations) are imposed concerning the choice of weights w_{rj}; these assumptions are sufficient, however, to permit a study of all the schemes of sections 6.2.1 and 6.2.2. The underlying assumption is to the effect that the 'algorithm' for constructing the weights w_{rj} is determined by a fixed value k and hence by $\rho \equiv r \pmod{k}$. Thus, the 'prescription' for writing down the weights $\{w_{rj}\}$ is to be the same as that for writing down the weights $\{w_{r+k,j}\}$. Kobayasi assumes that $k \leq q$, which can always be achieved by choosing to regard a sufficient number of values as starting values.

We suppose, with Kobayasi, that the approximations

$$\int_0^{rh} \phi(y)\,dy \simeq \sum_{j=0}^{r} w_{r,j}\phi(jh) \quad (r \geq q) \tag{6.194}$$

are derived by constructing approximations

$$\int_0^{k_\rho h} \phi(y)\,dy \simeq h \sum_{j=0}^{k_\rho^*} \alpha_{\rho,j}\phi(jh)\ , \tag{6.195}$$

$$\int_{k_\rho h+skh}^{k_\rho h+(s+1)kh} \phi(y)dy \simeq h \sum_{j=0}^{k} \gamma_j\phi((ks+k_\rho+j)h) \tag{6.196}$$

$$(s = 0,1,2,\ldots,M_\rho-1\ ; M_\rho = (r-\rho)/k)\ ,$$

$$\int_{(r-\rho+k_\rho)h}^{rh} \phi(y)\,dy \simeq h \sum_{j=0}^{k_\rho^{**}} \beta_{\rho j}\phi((r-j)h)\ , \tag{6.197}$$

and summing these contributions. Suitable constraints are imposed on k_ρ, k_ρ^*, and k_ρ^{**}. The value of k referred to above occurs in (6.196). The resulting method is consistent.

<u>Remark</u>. Simple examples occur if $k_\rho \equiv 0$, and the approximation in

(6.195) is zero. The rule (6.194) may then be obtained by applying the repeated form of a closed $(k+1)$-point rule (6.196), repeated over $[0,kmh]$; the remainder of the interval $[0,rh]$ is covered by a rule with weights $\beta_{\rho,j}$. (Compare this with the schemes in Tables 6.3 and 6.5.) Clearly, as r changes the length of the interval $[kmh,rh]$ can grow until its length is kh whereupon it collapses to zero, and with $k_\rho \equiv 0$, M_ρ is increased by one. We give various illustrations in the next Examples.

In the general case we suppose that the weights w_{rj} can be defined for $\rho = 0,1,2,\ldots,(k-1)$, $r \equiv \rho \bmod(k)$ and $r \geq q$ by requiring that

$$\sum_{j=0}^{r} w_{r,j}\phi(jh) = h\sum_{j=0}^{k_\rho^*} \alpha_{\rho,j}\phi(jh)+\sum_{j=0}^{M_\rho-1} h\sum_{i=0}^{k} \gamma_i\phi((kj+i+k_\rho)h)+h\sum_{j=0}^{k_\rho^{**}} \beta_{\rho,j}\phi((r-j)h) . \tag{6.198}$$

Eqn (6.198) is required to hold for $r = q_\rho,(q_\rho+k),(q_\rho+2k),\ldots$ for all bounded functions $\phi(x)$, where

$$q_\rho = \rho+k\left[\frac{q-\rho-1}{k}+1\right] .$$

Here, $[x]$ denotes the greatest integer which does not exceed x, and $k_\rho,k_\rho^*,k_\rho^{**}$ are integers such that $-1 \leq k_\rho^* \leq q_\rho$, $-1 \leq k_\rho^{**} \leq q_\rho$, $0 \leq k_\rho \leq \rho$, and $M_\rho = (r-\rho)/k$. We adopt the conventions that $\sum_{j=0}^{-1} u_j = 0$, and that $|\alpha_{\rho,k_\rho^*}| > 0$, $|\beta_{\rho,k_\rho^{**}}| > 0$, and we suppose that $k \leq q$.

<u>Example 6.61</u>.

Consider the choice of weights given in Table 6.2. (This results on applying the trapezium rule and repeated Simpson's rule.) No starting procedure is required for the method associated with these weights, but to ensure that $q \geq k$ we set $q = \max(k,1)$. The rule (6.196) is Simpson's rule and the weights w_{rj} depend on whether r is even or odd, and hence $k = 2$. When r is even, $r \equiv 0 \bmod(2)$ and $\rho = 0$ so that $k_0^* = k_0^{**} = -1$ and $k_0 = 0$, whilst $q_0 = 2$. When r is odd,

$r \equiv 1 \bmod(2)$ and $\rho = 1$ so that $k_1^{**} = -1$ whilst $k_1^* = k_1 = 1$, $q_1 = 3$. We shall encounter a modification of the scheme based on the weights of Table 6.2, in Example 6.63.

Here, let us consider the scheme associated with the weights in Table 6.3. We again have $k = 2$ in (6.196) and set $q = 2$. The case r even corresponds to $\rho = 0$ and $k_0^* = k_0^{**} = -1$, $k_0 = 0$, and $q_0 = 2$. With r odd, $\rho = 1$, $k_1^* = -1$, $k_1^{**} = 1$, and $q_1 = 3$.

For the schemes associated with the weights of Tables 6.4 and 6.5, the use of the trapezium rule when $r = 1$ is atypical and the formula for computing $\tilde{f}(h)$ should be regarded as a starting procedure; indeed, we set $q = 2$. The methods of obtaining the weights depend on whether r is even or odd (and greater than 1) and $k = 2$. In each case (Table 6.4 and Table 6.5) $k_0 = 0$, $k_0^* = k_0^{**} = -1$, $q_0 = 2$, $q_1 = 3$. In the case of Table 6.4, $k_1 = 3$, $k_1^* = 3$, $k_1^{**} = -1$ and in the case of Table 6.5, $k_1 = 1$, $k_1^* = -1$, and $k_1^{**} = 3$. *

The principal result of Kobayasi is embodied in the following theorem.

THEOREM 6.4. *Suppose that the starting values* $\tilde{f}(0), \tilde{f}(h), \ldots, \tilde{f}((q-1)h)$ *satisfy relations of the form*

$$\xi_r(h) \equiv \tilde{f}(rh) - f(rh) = \hat{\xi}_r h^{q^*} + O(h^{q^*+1}) \quad (r = 0,1,2,\ldots,(q-1)), \qquad (6.199)$$

and the values $\tilde{f}(qh), \tilde{f}((q+1)h), \ldots$ *are defined by eqn* (6.189), *where the weights* w_{rj} *are chosen as assumed. Then, if* $F(x,y;v)$ *and* $g(x)$ *are sufficiently differentiable, there is an integer* $p>0$ *such that*

$$\tilde{f}(rh) - f(rh) = h^p \sum_{j=0}^{k-1} \theta_j^r e_j(rh) + h^{q^*+1} \sum_{j=0}^{k-1} \theta_j^r d_j(rh) + \{O(h^{q^*+2}) + O(h^{p+1})\}$$

$$\text{as } h \to 0,$$

where $rh \in [0,X]$ *is fixed*,

$$\theta_j = \exp 2\pi i j/k \quad (\text{where } i = \sqrt{-1}),$$

and $\tau_N(h) = O(h^p)$ *as* $h \to 0$ *with* $Nh = X$ *fixed. The functions* $e_j(x)$

and $d_j(x)$ *satisfy equations of the form*

$$\sum_{j=0}^{k-1} \theta_j^{\rho} e_j(x) = \sum_{j=0}^{k-1} \lambda_j \theta_j^{k_\rho} \int_0^x K(x,y) e_j(y)\,dy - \psi_\rho(x) \quad (\rho = 0,1,2,\ldots,(k-1)) \quad (6.200)$$

and

$$\sum_{j=0}^{k-1} \theta_j^{\rho} d_j(x) = \sum_{j=0}^{k-1} \lambda_j \theta_j^{k_\rho} \int_0^x K(x,y) d_j(y)\,dy + c_\rho K(x,0) \quad (\rho = 0,1,2,\ldots,(k-1)) \quad (6.201)$$

with $K(x,y) = (\partial/\partial v)F(x,y;v)\Big]_{v=f(y)}$, *and* $\lambda_r = \sum_{j=0}^{k} \gamma_j \theta_r^j / k$ $(r = 0,1,2,\ldots,(k-1))$. (*Here*, $\psi_0(x), \psi_1(x), \ldots, \psi_{k-1}(x)$ *are certain functions determined by the weights* w_{rj} *of* (6.194) *and* $F(x,y;f(y))$, *and* $c_0, c_1, \ldots, c_{k-1}$ *are constants which are similarly dependent.*)

Remark. If, as for Table 6.2, $k_\rho = \rho$, $\rho = 0,1,2,\ldots,(k-1)$ then we find that

$$e_\rho(x) = \lambda_\rho \int_0^x K(x,y) e_\rho(y)\,dy - \hat{\psi}_\rho(x)$$

and

$$d_\rho(x) = \lambda_\rho \int_0^x K(x,y) d_\rho(y)\,dy + \hat{c}_\rho K(x,0) \quad (\rho = 0,1,\ldots,(k-1)),$$

where

$$\hat{\psi}_\rho(x) = \sum_{j=0}^{k-1} \theta_\rho^{-j} \psi_j(x)/k$$

and

$$\hat{c}_\rho = \sum_{j=1}^{k-1} \theta_\rho^{-j} c_j / k .$$

The preceding theorem may be established as a consequence of certain lemmas. To gain insight let us first state a convergence result.

LEMMA 6.5. *Under the conditions of Theorem 6.4,*

$$\lim_{h\to 0}|\tilde{f}(rh)-f(rh)| = 0$$

for fixed $rh \in [0,X]$ *provided that*

$$\sum_{j=0}^{k} \gamma_j = k , \qquad (6.202)$$

for all $h > 0$. (*Here*, $\gamma_0,\gamma_1,\ldots,\gamma_k$ *are defined in eqn* (6.196).)

A proof may be found in the paper of Kobayasi (1966), or deduced from Theorem 6.2(a).

It is clear that the dominant terms in the local truncation error must be obtained in order to establish Theorem 6.4. We have the following result, which we do not prove.

LEMMA 6.6. *Let* $F(x,y;v)$ *and* $g(x)$, *and hence* $f(x)$, *be sufficiently differentiable on* $[0,X]$. *Then if*

$$t_r(h) = \int_0^{rh} F(rh,y;f(y))\,dy - \sum_{j=0}^{r} w_{rj}F(rh,jh;f(jh)) ,$$

where the weights w_{rj} *satisfy the conditions* (6.198) *and* (6.202), *there exist functions* $\tilde{\omega}_\rho(x)$ *and positive integers* p_ρ *such that*

$$t_r(h) = h^{p_\rho}\tilde{\omega}_\rho(rh) + O(h^{p_\rho+1}) \quad (\rho = 0,1,2,\ldots,(k-1))$$

as $h \to 0$ *with* $r \in \{q_\rho,(q_\rho+k),\ldots\}$ *and* rh *fixed.*

(A proof may be established by considering an Euler-MacLaurin formula for the repetition of the rule in (6.196), and the further use of Taylor-series expansions. Details can be found in Kobayasi (1966).)

COROLLARY 6.1. *Suppose that* $p = \min_{0\le\rho\le k-1} p_\rho$. *Then*

$$t_r(h) = h^p\psi_\rho(rh)+O(h^{p+1}) \quad (\rho = 0,1,2,\ldots,(k-1), r = q_\rho,(q_\rho+k),\ldots) ,$$

where $\psi_\rho(x) = 0$ *if* $p_\rho > p$ *and* $\psi_\rho(x) = \tilde{\omega}_\rho(x)$ *if* $p_\rho = p$.

The functions $\psi_\rho(x)$ occurring in this corollary are the functions required in the statement of Theorem 6.4.

Example 6.62.

Consider a formula (Table 6.2) derived from the use of Simpson's rule, repeated where possible over $[0,rh]$ where $r \geq 2$ is even. We assume the continuity of $(\partial/\partial y)^4 F(x,y;f(y))$. Then (see section 2.12.1)

$$t_r(h) = -\frac{1}{180}h^4 [(\partial/\partial y)^3 F(rh,y;f(y))]_{y=0}^{y=rh} + O(h^5)$$

if r is even. Suppose that when $r \geq 1$ is odd we apply the trapezium rule over $[0,h]$ and the (repeated) Simpson's rule over $[h,rh]$, provided $r \geq 1$. Then for odd values of $r \geq 1$

$$t_r(h) = -\frac{1}{12}h^2 [(d/dy)F(rh,y;f(y))]_{y=0}^{y=h} -$$

$$-\frac{1}{180}h^4 [(d/dy)^3 F(rh,y;f(y))]_{y=h}^{y=rh} + O(h^5) .$$

This last expression involves values $(d/dy)^s F(x,y;f(y))$ evaluated at $y = h$ for $s = 1$ and $s = 3$. We write

$$(d/dy)F(rh,y;f(y))]_{y=h} = (d/dy)F(rh,y;f(y))]_{y=0} +$$

$$+ h(d/dy)^2 F(rh,y;f(y))]_{y=0} + O(h^2)$$

(using the Taylor series) and we obtain

$$t_r(h) = -\frac{1}{12}h^3 \times (d/dy)^2 F(rh,y;f(y))]_{y=0} + O(h^4) .$$

Here we have $k = 2$, $\rho \in \{0,1\}$, $q_0 = 2$, and $q_1 = 3$. The only starting value required is $\tilde{f}(h)$. We have $k_\rho^{**} = -1$ for $\rho \in \{0,1\}$, $k_0^* = -1$, and $k_1^* = 1$, where $p_0 = 4$, $\tilde{\omega}_0(x) = -\frac{1}{180}[(d/dy)^3 F(x,y;f(y))]_{y=0}^{y=x}$, $p_1 = 3$, and $\tilde{\omega}_1(x) = -\frac{1}{12} \times (d/dy)^2 F(x,y;f(y))]_{y=0}$. *

LEMMA 6.7. *Under the conditions assumed above,*

$$|\tilde{f}(rh)-f(rh)| = O(h^{p^*})$$

as $h \to 0$ *with* rh *fixed in* $[0,X]$ *where*

$$p^* = \min(p,q^*+1)$$

Remark. This lemma is a strengthening of Lemma 6.5, which is clearly subsumed by the statement of Theorem 6.1. The rate of convergence follows from (6.176).

We now have sufficient preparation to enable us to discuss Kobayasi's proof of Theorem 6.4, which will now be outlined.

The statement of Theorem 6.4 involves the functions $\psi_\rho(x)$ which are those of Corollary 6.1 to Lemma 6.6, and certain constants c_ρ which are given by

$$c_\rho = \sum_{j=0}^{m_\rho} \alpha_{\rho,j}\hat{\xi}_j + \sum_{i=0}^{m^*_\rho}\sum_{j=0}^{m^{**}_\rho} \gamma_j\hat{\xi}_{ik+j+k_\rho} \quad (\rho = 0,1,2,\ldots,k-1), \tag{6.203}$$

where $m_\rho = \min(q-1,k^*_\rho)$, $m^*_\rho = [(q-1-k_\rho)/k]$, where $[x]$ is the integer part of x, and $m^{**}_\rho = q-1-km^*_\rho$.

With the preceding notation, we define the functions $e_\rho(x)$ and $d_\rho(x)$ by eqns (6.200) and (6.201), and set

$$z_r = \{f(rh)-\tilde{f}(rh)\}-h^p\sum_{j=0}^{k-1}\theta^r_j e_j(rh)-h^{q^*+1}\sum_{j=0}^{k-1}\theta^r_j d_j(rh) \quad (r=q,q+1,\ldots) \tag{6.204}$$

and $z_0 = z_1 = \ldots = z_{q-1} = 0$. Our task is to prove that, with $p^* = \min(p, q^*+1)$,

$$z_r = O(h^{p^*+1}) \tag{6.205}$$

as $h \to 0$ with rh fixed. To this end we establish that

$$z_r - \sum_{j=0}^{r} w_{r,j}K(rh,jh)z_j = \sigma_r \quad (r \geq q), \tag{6.206}$$

where $\sigma_r = O(h^{p^*+1})$, as $h \to 0$ with rh fixed, and $K(x,y) =$

$= (\partial/\partial v)F(x,y;v)]_{v=f(y)}$. Hence,

$$|z_r| \leq A \sum_{j=0}^{r-1} |z_j| + B ,$$

where

$$B = \sup|\sigma_r|/|1-w_{rr}K(rh,rh)|, \; A = \sup|w_{rj}K(rh,jh)/(1-w_{rr}K(rh,rh))|$$

and, using Lemmas 6.2 and 6.3 we may deduce that $z_r = O(h^{p^*+1})$ as $h \to 0$ with rh fixed.

We suppose that (6.206) serves as a definition for σ_r . Then from (6.204), after some manipulation, we may show that

$$\sigma_r = \sigma_r^* + \sigma_r' - \sigma_r^{**} - \sigma_r'' , \tag{6.207}$$

where we set

$$\sigma_r^* = \{\tilde{f}(rh)-f(rh)\}- \sum_{j=0}^{r} w_{r,j}K(rh,jh)\{\tilde{f}(jh)-f(jh)\}+O(h^{p^*+1}) , \tag{6.208}$$

$$\sigma_r' = \sum_{j=0}^{q-1} w_{r,j}K(rh,jh)\xi_j(h) , \tag{6.209}$$

$$\sigma_r^{**} = h^p \sum_{j=0}^{k-1} \{\theta_j^r e_j(rh)- \sum_{i=0}^{r} w_{r,i}K(rh,ih)\theta_j^i e_j(ih)\} , \tag{6.210}$$

and

$$\sigma_r'' = h^{q^*+1} \sum_{j=0}^{k-1} \{\theta_j^r d_j(rh)- \sum_{i=0}^{r} w_{r,i}K(rh,ih)\theta_j^i d_j(ih)\} . \tag{6.211}$$

We may show, further, that if $r \equiv \rho \bmod(k) \geq q$

$$\sigma_r^* = -h^p\psi_\rho(rh) + O(h^{p^*+1}) , \tag{6.212}$$

$$\sigma_r' = h^{q^*+1}c_\rho K(rh,0) + O(h^{p^*+1}) , \tag{6.213}$$

and

$$\sigma_r^{**} = -h^p\psi_\rho(rh) + O(h^{p^*+1}) , \tag{6.214}$$

$$\sigma_r'' = h^{q^*+1} c_\rho K(rh,0) + O(h^{p^*+1}) \quad , \tag{6.215}$$

so that, from (6.207), $\sigma_r = O(h^{p^*+1})$. Eqn (6.205) then follows, as indicated above, and the proof of Theorem 6.4 is established.

An outline of the proof of Theorem 6.4 has been given, and we now investigate certain details more closely.

The values z_r $(r=q,(q+1),\ldots)$ are defined by (6.204), and elementary manipulation may be employed to establish (6.207) where σ_r is defined by (6.206). We shall concentrate on the proof of eqns (6.212)-(6.215).

To establish (6.212) we write $\xi_r(h) = \tilde{f}(rh)-f(rh)$ for $r \geq 0$, and from Lemma 6.7 we have $\xi_r(h) = O(h^{p^*})$ (where $p^* = \min(p,q^*+1)$), as $h \to 0$ with rh fixed. Let us write

$$f(rh) - \sum_{j=0}^{r} w_{r,j} F(rh,jh;f(jh))-g(rh) = \hat{\sigma}_r . \tag{6.216}$$

Using Lemma 6.6, and Corollary 6.1, we find that

$$\hat{\sigma}_r = \int_0^{rh} F(rh,y;f(y))\,dy - \sum_{j=0}^{r} w_{r,j} F(rh,jh;f(jh))$$

$$= t_r(h) = h^p \psi_\rho(rh) + O(h^{p+1})$$

Moreover,

$$\tilde{f}(rh) - \sum_{j=0}^{r} w_{r,j} F(rh,jh;\tilde{f}(jh)) = g(rh)$$

and hence

$$\xi_r(h) - \sum_{j=0}^{r} w_{r,j} \{F(rh,jh;\tilde{f}(jh))-F(rh,jh;f(jh))\} = -\hat{\sigma}_r . \tag{6.217}$$

On the other hand

$$F(rh,jh;\tilde{f}(jh))-F(rh,jh;f(jh)) = K(rh,jh)\xi_j(h) + O(\{\xi_j(h)\}^2) \tag{6.218}$$

and $\xi_j(h) = O(h^{q^*})$ $(j = 0,1,2,\ldots,(q-1))$ whilst $\xi_r(h) = O(h^{p^*})$, as $h\to 0$ with fixed rh. Since, for all r and any fixed j, $w_{r,j} = O(h)$, whilst $\Omega<\infty$ exists giving $\sum_{j=0}^{r} |w_{rj}| = \Omega rh + O(h)$ for fixed rh, we find that substituting (6.218) and (6.216) in (6.217) gives (for $p^* \geq 1$)

$$\xi_r(h) - \sum_{j=0}^{r} w_{r,j}K(rh,jh)\xi_j(h) = -h^p\psi_\rho(rh) + O(h^{p^*+1}). \quad (6.219)$$

In view of (6.208), this establishes (6.212).

To establish (6.213) we write

$$\xi_r(h) = \hat{\xi}_r h^{q^*} + O(h^{q^*+1}) \quad (r = 0,1,2,\ldots,(q-1))$$

and

$$\sigma'_r = h^{q^*} \sum_{j=0}^{q-1} w_{r,j}\, K(rh,jh)\hat{\xi}_j + O(h^{q^*+2}),$$

as $h \to 0$ with rh fixed. For $r \geq q + q_\rho$, where $\rho \equiv r \bmod(k)$, it may be established (after some manipulation, see Kobayasi (1966)) that

$$\sum_{j=0}^{q-1} w_{r,j}\hat{\xi}_j = c_\rho h \quad ,$$

where c_ρ is defined above. Thus, (6.213) may be established.

The techniques for establishing (6.214) and (6.215) follow similar lines in each case. We shall consider the former equation only. The expression for σ_r^{**} in (6.210) involves a sum of terms in braces. We examine such terms, and we find, for $\rho \equiv r \bmod(k)$,

$$\theta_j^r e_j(rh) - \sum_{i=0}^{r} w_{r,i}K(rh,ih)\theta_j^i e_j(ih)$$

$$= \theta_j^r e_j(rh) - \sum_{s=0}^{M_\rho-1} h \sum_{i=0}^{k} \gamma_i \theta_j^{ks+i+k\rho} K(rh,(ks+i+k_\rho)h)e_j((ks+i+k_\rho)h) + O(h) \quad (6.220)$$

on writing the contributions from (6.195) and (6.197) in the $O(h)$ term.

Since

$$\theta_j^{ks} = 1, \quad \lambda_\rho = \sum_{i=0}^{k} \gamma_i \theta_\rho^i / k \tag{6.221}$$

the right-hand side of eqn (6.220) becomes

$$\theta_j^\rho e_j(rh) - kh \sum_{s=0}^{M_\rho - 1} \theta_j^{k_\rho} \lambda_j K(rh,(ks+k_\rho)h) e_j((ks+k_\rho)h) + O(h) \tag{6.222}$$

$$= \theta_j^\rho e_j(rh) - \theta_j^{k_\rho} \lambda_j \int_0^{rh} K(rh,y) e_j(y)\,dy + O(h) \quad . \tag{6.223}$$

(Here, $kh \sum_{s=0}^{M_\rho - 1} K(rh,(ks+k_\rho)h) e_j((ks+k_\rho)h)$ approximates $\int_0^{rh} K(rh,y) e_j(y)\,dy$ with $O(h)$ accuracy.) Substituting the right-hand side of (6.223) as a replacement for the terms in braces appearing in (6.210) we obtain

$$\sigma_r^{**} = h^p \sum_{j=0}^{k-1} \{\theta_j^\rho e_j(rh) - \theta_j^{k_\rho} \lambda_j \int_0^{rh} K(rh,y) e_j(y)\,dy\} + O(h^{p+1}) ,$$

and, in view of (6.200), we obtain

$$\sigma_r^{**} = -h^p \psi_\rho(rh) + O(h^{p^*+1}) ,$$

which is the required relation (6.214). In the proof of (6.215) we follow similar lines, and employ (6.201).

Example 6.63.

Kobayasi (1966) considers a particular case in which $k_\rho = \rho$ $(\rho=0,1,2,\ldots,(k-1))$, so that the Remark following Theorem 6.4 applies.

In the case considered, the weights $w_{r,j}$ correspond, if r is even, to the use of the repeated Simpson's rule. For r odd, $r \geq 3$, the repeated Simpson's rule is applied over $[h,rh]$ and the rule

$$\int_0^h F(rh,y;f(y))\,dy \simeq \frac{h}{24}\{9F(rh,0;f(0)) + 19F(rh,h;f(h)) -$$

$$- 5F(rh,2h;f(2h)) + F(rh,3h;f(3h))\}$$

is employed over $[0,h]$. The corresponding values of q_ρ, k_ρ, k_ρ^*, etc., supplied by Kobayasi, are as follows. With $k = 2$, $q_0 = 2$, $q_1 = 3$, $M_0 = \frac{1}{2}r, M_1 = \frac{1}{2}(r-1)$, $k_0 = 0$, $k_1 = 1$, $k_0^* = k_0^{**} = k_1^{**} = -1$, $k_1^* = 3$, $\theta_0 = 1$, $\theta_1 = -1$, $\lambda_0 = 1$, $\lambda_1 = -\frac{1}{3}$, $p = p_0 = p_1 = 4$, and, for r sufficiently large,

$$\psi_0(x) = \psi_1(x) = \frac{-1}{180}\left[(d/dy)^3 F(x,y;f(y))\right]_{y=0}^{y=x} .$$

Kobayasi considers the equation

$$f(x) + \int_0^x (1+x-y)\{f(y)\}^2 dy = 1+x+\tfrac{1}{2}x^2 \quad (x \geq 0) ;$$

he finds that the corresponding functions $e_0(x)$ and $e_1(x)$ vanish, whilst

$$d_\rho(x) + 2\lambda_\rho \int_0^x (1+x-y) d_\rho(y) dy = -2\hat{c}_\rho(1+x) \quad (\rho = 0,1) ,$$

where $\hat{c}_0, \hat{c}_1$ are certain constants determined by the starting errors. Thus

$$d_0(x) = -2\hat{c}_0 e^{-x} \cos x$$

and

$$d_1(x) = -\hat{c}_1\{(1+\tfrac{4}{\sqrt{7}}) \exp(\tfrac{1+\sqrt{7}}{3}x) + (1-\tfrac{4}{\sqrt{7}}) \exp(\tfrac{1-\sqrt{7}}{3}x)\} ,$$

so that the error $\tilde{f}(x)-f(x)$ can be shown to have a component which oscillates and grows rapidly as x is increased (using a fixed step h). The instability in this Example displays entirely the effect of starting errors. *

Example 6.64.

Consider either the scheme in Table 6.3 or the scheme in Table 6.5. In either case we have $k = 2$, $\theta_0 = 1$, $\theta_1 = -1$, and $k_\rho \equiv 0$, $\gamma_0 = \gamma_2 = \frac{1}{3}$, $\gamma_1 = \frac{4}{3}$ and $\lambda_0 = 1$, $\lambda_1 = -\frac{1}{3}$. For suitable functions $\psi_\rho(x)$ $(\rho=0,1)$, depending on the scheme, eqns (6.200) become

$$e_0(x)+e_1(x) = \int_0^x K(x,y)e_0(y)\,dy - \frac{1}{3}\int_0^x K(x,y)e_1(y)\,dy - \psi_0(x)$$

and

$$e_0(x)-e_1(x) = \int_0^x K(x,y)e_0(y)\,dy - \frac{1}{3}\int_0^x K(x,y)e_1(y)\,dy - \psi_1(x) \quad .$$

Thus

$$e_0(x) = \int_0^x K(x,y)e_0(y)\,dy - \tfrac{1}{2}\{\psi_0(x)+\psi_1(x)\} - \frac{1}{3}\int_0^x K(x,y)e_1(y)\,dy$$

and

$$e_1(x) = -\tfrac{1}{2}\{\psi_0(x)-\psi_1(x)\} \quad .$$

Consider, now, the case $F(x,y;v) = \lambda v$ and $g(x) = 1$. Then $K(x,y) = \lambda$ and

$$e_0(x)-\lambda\int_0^x e_0(y)\,dy = -\tfrac{1}{2}\{\psi_0(x)+\psi_1(x)\} + \frac{1}{6}\int_0^x \{\psi_0(y)-\psi_1(y)\}\,dy \ ,$$

whilst

$$e_1(x) = -\tfrac{1}{2}\{\psi_0(x)-\psi_1(x)\} \quad .$$

Here, as in general with these schemes, $e_0(x)$ satisfies an equation of the same type as that satisfied by the error arising from a perturbation of the integral equation. Here, we see that we can expect $e_0(x)$ to be exponentially increasing if $\lambda > 0$ and decreasing if $\lambda < 0$ *but the precise form of* $e_0(x)$ *and* $e_1(x)$ *depends on* $\psi_0(x)$ *and* $\psi_1(x)$.

The effect of $e_0(x)$ and $e_1(x)$ at a given point x can be reduced by making h smaller (but, the errors due to rounding actually satisfy similar equations).

The equations associated with Table 6.4, say, are rather different. We find, instead,

$$e_0(x)+e_1(x) = \int_0^x K(x,y)e_0(y)\,dy - \tfrac{1}{3}\int_0^x K(x,y)e_1(y)\,dy - \psi_0(x) ,$$

$$e_0(x)-e_1(x) = \int_0^x K(x,y)e_0(y)\,dy + \tfrac{1}{3}\int_0^x K(x,y)e_1(y)\,dy - \psi_1(x) ,$$

for appropriate functions $\psi_0(x)$ and $\psi_1(x)$. Thus, for example,

$$e_1(x) + \tfrac{1}{3}\int_0^x K(x,y)e_1(y)\,dy = -\tfrac{1}{2}\{\psi_0(x)-\psi_1(x)\} ,$$

and in the case $F(x,y;v) = \lambda v$ we expect $e_1(x)$ to be exponentially increasing when $\lambda < 0$ and $f(x)$ is exponentially decreasing. *

In the general case, let us refer to Theorem 6.4 and write the system of equations for $e_j(x)$ in the form

$$\underset{\sim}{A}\underset{\sim}{e}(x) = \int_0^x K(x,y)\underset{\sim}{B}\underset{\sim}{e}(y)\,dy - \underset{\sim}{\psi}(x) ,$$

where $A_{r,s} = \theta_s^r$, $B_{r,s} = \lambda_s\theta_s^{k_r}$ $(r,s = 0,1,2,\ldots,(k-1))$ and $\underset{\sim}{e}(x) = [e_0(x),e_1(x),\ldots,e_{k-1}(x)]^T$, etc. Then

$$\underset{\sim}{e}(x) = \int_0^x K(x,y)\underset{\sim}{A}^{-1}\underset{\sim}{B}\underset{\sim}{e}(y)\,dy - \underset{\sim}{A}^{-1}\underset{\sim}{\psi}(x) .$$

If $\underset{\sim}{A}^{-1}\underset{\sim}{B}$ is diagonalizable (as assumed by Noble (1969*a*)) and $\underset{\sim}{X}(\underset{\sim}{A}^{-1}\underset{\sim}{B})\underset{\sim}{X}^{-1} = \underset{\sim}{\text{diag}}\,(\mu_0,\mu_1,\ldots,\mu_{k-1}) = \underset{\sim}{D}$, then we obtain on setting $\underset{\sim}{\varepsilon}(x) = \underset{\sim}{X}\underset{\sim}{e}(x)$ an equation of the form

$$\underset{\sim}{\varepsilon}(x) = \int_0^x K(x,y)\underset{\sim}{D}\underset{\sim}{\varepsilon}(y)\,dy - \underset{\sim}{\nu}(x) ,$$

where $\underset{\sim}{\nu}(x) = \underset{\sim}{X}\underset{\sim}{A}^{-1}\underset{\sim}{\psi}(x)$. In the case $K(x,y) = \lambda$, we have an equation which can be written in component form as

$$\varepsilon_i(x) = \lambda\,\mu_i\int_0^x \varepsilon_i(y)\,dy + \nu_i(x) ,$$

and the behaviour of the components $\varepsilon_i(x)$ of the error can be related, qualitatively, to the behaviour of the true solution $f(x)$, in terms of the values μ_i. It would appear that this analysis is related to the analysis of a relatively stable discretization with *small* values of h. We may seek to find $\mu_0 = 1$ and $\text{sign}(\lambda)\mu_i \le \text{sign}(\lambda)$ $(i=1,2,\ldots(k-1))$.

Example 6.65.
If the rule (6.198) has a repetition factor 1, then $\mu_0 = 1$, and $\mu_1 = \mu_2 = \ldots = \mu_{k-1} = 0$ (Noble 1969*a*). *

6.13. *Convergence of certain block-by-block methods*

In section 6.12 we discussed the properties of a class of step-by-step methods for the approximate solution of Volterra equations of the second kind, and we now consider the block-by-block methods of section 6.7.

It is sufficient to consider the methods defined by (6.115) and (6.117) since the methods discussed in the early part of section 6.7 are special cases of these methods, or variants. We shall concentrate on the method defined by eqn (6.115); the reader may provide the extension of the results to (6.117).

Let us first establish that, under sufficiently strict conditions on h^* and $F(x,y;v)$, eqns (6.115) have a unique solution.

LEMMA 6.8. *Suppose that*

$$|F(x,y;v_1)-F(x,y;v_2)| \le L|v_1-v_2|$$

for $-\infty < v_1, v_2 < \infty$, $0 \le x \le X$, $0 \le y \le x + h^*$, *and* $(p+1)h^*\Omega L < 1$, *where*

$$\Omega = \max_{\substack{0\le k\le p\\ 0\le j\le p}} |\omega_k^{[j]}| \text{ . } \dagger$$

---

† Recall that $\theta_p = 1$ and, hence, $\Omega \ge \max_{0\le k\le p} |\omega_k|$. Ω depends upon $\{\theta_i\}$ but is independent of h^*.

Then eqns (6.115) have a unique solution for $m = 0,1,2,\ldots,(N-1)$, $j = 0,1,2,\ldots, p$ *which can be found by a predictor-corrector technique.*

<u>Proof</u>. Consider the iteration defined by a choice $\tilde{f}^{[0]}(u_{m,k})$ and by

$$\tilde{f}^{[r+1]}(u_{m,j}) = h^* \sum_{k=0}^{p} \omega_k^{[j]} F(u_{m,j},u_{m,k};\tilde{f}^{[r]}(u_{m,k}))+\Gamma_{m,j}$$

$$(j = 0,1,2,\ldots,p;\ r = 0,1,2,\ldots),$$

where we have written $u_{m,j} = (m+\theta_j)h^*$ and

$$\Gamma_{m,j} = g(u_{m,j}) + \sum_{j=0}^{m-1} h^* \sum_{k=0}^{p} \omega_k F(u_{m,j},u_{j,k};\tilde{f}(u_{j,k})) .$$

(The iteration is applied with $m = 0$ and $\Gamma_{0,j} = g(u_{0,j})$ until convergence, whereupon $\Gamma_{1,j}$ can be evaluated and the iteration can be applied with $m = 1$, etc.) We have, on differencing equations in the iteration, and employing the Lipschitz condition,

$$|\tilde{f}^{[r+1]}(u_{m,j})-\tilde{f}^{[r]}(u_{m,j})| = |h^* \sum_{k=0}^{p} \omega_k^{[j]} \{F(u_{m,j},u_{m,k};\tilde{f}^{[r]}(u_{m,k})) - F(u_{m,j},u_{m,k};\tilde{f}^{[r-1]}(u_{m,k}))\}|$$

$$\leq h^* \Omega L \sum_{k=0}^{p} |\tilde{f}^{[r]}(u_{m,k})-\tilde{f}^{[r-1]}(u_{m,k})|$$

$$(j = 0,1,2,\ldots,p;\ r= 1,2,3,\ldots) .$$

We write

$$\tilde{\Phi}_m^{[r]} = [\tilde{f}^{[r]}(u_{m,0}),\tilde{f}^{[r]}(u_{m,1}),\ldots,\tilde{f}^{[r]}(u_{m,p})]^T$$

and have

$$||\tilde{\Phi}_m^{[r+1]} - \tilde{\Phi}_m^{[r]}||_\infty \leq \sigma ||\tilde{\Phi}_m^{[r]}-\tilde{\Phi}_m^{[r-1]}||_\infty \quad (r \geq 1)$$

with

$$\sigma = (p+1)h^*\Omega L < 1 .$$

The proof now follows as an analogue of the proof of Lemma 6.1. We see that the sequence $\{\tilde{\Phi}_m^{[r]}\}$ converges to a limit Φ_m which (from the continuity of $F(x,y;v)$ in v) provides a solution

$$\tilde{\Phi}_m = [\tilde{f}(u_{m,0}),\tilde{f}(u_{m,1}),\ldots,\tilde{f}(u_{m,p})]^T$$

of the mth block of equations. Such a solution is unique. (For if there were two solutions corresponding to vectors $\tilde{\Phi}_{m,0}$ and $\tilde{\Phi}_{m,1}$ then we can readily show that

$$\|\tilde{\Phi}_{m,1}-\tilde{\Phi}_{m,0}\|_\infty \le \sigma\|\tilde{\Phi}_{m,1}-\tilde{\Phi}_{m,0}\|_\infty$$

where $\sigma = (p+1)h^*\Omega L < 1$, and hence $\tilde{\Phi}_{m,0} = \tilde{\Phi}_{m,1}$.)

Remark. The predictor-corrector iteration employed in the preceding proof is a simple one, and its use imposes a limitation on the size of h^*. Various simple modifications to the algorithm suggest themselves, but in practice a Newton iteration seems to be favoured for use with large values of h^*. (There may then be no guarantee that the solution of (6.115) is unique.)

We gave an indication, in section 6.7, of the order of local truncation error associated with (6.115). Let us give a precise statement, in the form of the following lemma.

LEMMA 6.9. *Suppose that* $t_{m,j}(h^*)$ *is defined, for* $m = 0,1,2,\ldots,(N-1)$, $j = 0,1,2,\ldots,(p-1), p$, *by the relation*

$$f((m+\theta_j)h^*) - \sum_{i=0}^{m-1} h^* \sum_{k=0}^{p} \omega_k F((m+\theta_j)h^*,(i+\theta_k)h^*;f((i+\theta_k)h^*)) -$$

$$- h^* \sum_{k=0}^{p} \omega_k^{[j]} F((m+\theta_j)h^*,(m+\vartheta_k)h^*;f((m+\theta_k)h^*))$$

$$= g((m+\theta_j)h^*) + t_{m,j}(h^*) \qquad (6.224)$$

and suppose that the degree of precision of (6.108) *is* ρ . *If* $F(x,y;f(y))$ *has, for each* $x \in [0,X]$, *an* r*th derivative in* y *which is continuous*† *for* $y \in [0,x+\delta]$ (*where* $\delta > 0$ *is a fixed constant*) *then*

$$t_{m,j}(h^*) = O((h^*)^s), \text{ where } s = \min(r, p+2, \rho+1)$$

as $h \to 0$ *with* mh^* *fixed, and* $0 \le j \le p$. *Furthermore,*

$$t_{m,p}(h^*) = O((h^*)^q), \text{ where } q = \min(r, \rho+1)$$

as $h^* \to 0$ *with* mh^* *fixed in* $[0,X]$.

<u>Proof</u>. A proof can be based on the discussion of section 2.16, using the relation (2.39) to analyse the errors $t_{m,p}(h^*)$, and (2.39) and (2.38) to analyse $t_{m,j}(h^*)$ for $0 \le j < p$. Thus, for example, the error in (6.113) is $h\times\{O((h^*)^{\rho+1}) + O((h^*)^r)\}$ $(j = 1,2,3,\ldots,(p-1))$ if $\phi(x) \in C^r[mh^*,(m+1)h^*]$. This follows on using (2.38). Also the error in the composite version of (6.108) over a fixed interval is $O((h^*)^r) + O((h^*)^{\rho+1})$, provided that $\phi(x)$ has a (continuous) rth derivative. Setting $\phi(y) = F(x,y;f(y))$ for various values of $x \in [0, X]$, the required results are obtained.

The preceding lemma can be strengthened in a certain sense. We have the following result, which will be used somewhat later.

LEMMA 6.10. *There exists a continuously differentiable function* $\Phi(x)$ *with* $\Phi(0) = 0$ *and continuous functions* $\Psi_i(x)$ $(i = 0,1,2,\ldots,p)$ *such that*

$$t_{mj}(h^*) \quad \int_0^{(m+\theta_j)h^*} F((m+\theta_j)h^*,y;f(y))dy -$$
$$- \sum_{i=0}^{m-1} h^* \sum_{k=0}^{p} \omega_k F((m+\theta_j)h^*,(i+\theta_k)h^*;f((i+\theta_k)h^*)) -$$

---

† It will suffice if the derivative is merely bounded.

$$- h^* \sum_{k=0}^{p} \omega_k^{[j]} F((m+\theta_j)h^*, (m+\theta_k)h^*; f((m+\theta_k)h^*))$$

$$= (h^*)^{p+1}\Phi((m+\theta_j)h^*) + (h^*)^{p+2}\Psi_j((m+\theta_j)h^*) + O((h^*)^{p+3}) \quad (6.225)$$

as $h^* \to 0$ *with* $(m+1)h^*$ *fixed in* $[0,X]$ *and* $0 \le j \le p$ *, provided that* $F(x,y;f(y))$ *has a continuous derivative of order* $p+3$ *with respect to* y *for* $y \in [0,x+\delta]$ *,* $x \in [0,X]$ *.*

Proof (in outline). A proof may be constructed using eqn (2.23), together with an expression for the values of τ_k which occur in (2.23) (see Baker and Hodgson 1971) to analyse the form of the error in the composite version of (6.108). Taylor's series is then applied so that the form of the error in (6.111) with $x = mh^* + \theta_j h^*$ is expressed in terms of y-derivatives at $y = 0$ and $y = x$. The contribution in $t_{mj}(h^*)$ which results from the application of (6.113) can be investigated using Taylor series expansions about $y = mh^* + \theta_j h^*$. Collecting terms will give the required result. Observe that if $\rho > p$, $\Phi(x)$ vanishes.

Let us write

$$\tau(h^*) \equiv \tau_N(h^*) = \max_{\substack{0 \le m \le N-1 \\ 0 \le j \le p}} |t_{mj}(h^*)| , \quad (6.226)$$

and consider the method defined by (6.115). We may expect to obtain a result similar in form to (6.176), and we shall show that this is possible. We require a slight variation on the condition (6.52), namely that

$$|F(x,y;v_1)-F(x,y;v_2)| \le L|v_1-v_2| \quad (6.227)$$

for $0 \le x \le X$, $0 \le y \le x + \delta$, $|v_i| < \infty$ $(i = 1,2)$, where $\delta > 0$ is constant. We now subtract (6.115) from (6.224) and we obtain, on using the condition (6.227),

$$|\xi_{m,j}| \le \{ \sum_{j=0}^{m-1} \sum_{k=0}^{p} |\omega_k| \, |\xi_{j,k}| + \sum_{k=0}^{p} |\omega_k^{[j]}| \, |\xi_{m,k}| \} h^* L + |t_{m,j}(h^*)| , \quad (6.228)$$

where we have written

$$\xi_{m,j} \equiv \xi_{m,j}(h^*)=f((m+\theta_j)h^*)-\tilde{f}((m+\theta_j)h^*) \quad (m=0,1,2,\ldots,(N-1),\ j=0,1,2,\ldots,p),$$

and we assume that $h^* < \delta$.

We now write

$$\xi_m = \max_{0\le j\le p} |\xi_{m,j}| ,$$

and

$$\Omega = (p+1)\times\max_{0\le j,k\le p} |\omega_k^{[j]}| \ge \max_{0\le k\le p} |\omega_k|(p+1).$$

Ω is a constant and we obtain, from (6.228), the result

$$\xi_m \le h^*\Omega L \sum_{i=0}^{m} \xi_i+\tau(h^*) \quad (m = 0,1,2,\ldots,(N-1)) . \tag{6.229}$$

If $h^*\Omega L < 1$ then

$$(1-h^*\Omega L)\xi_m \le h^*\Omega L \sum_{j=0}^{m-1} \xi_j+\tau(h^*) .$$

Dividing throughout by $(1-h^*\Omega L)$ and applying Lemma 6.2 with

$$A = h^*\Omega L/(1-h^*\Omega L)$$

and

$$B = \tau(h^*)/(1-h^*\Omega L)$$

we obtain

$$|\xi_m| \le \frac{\tau(h^*)+h^*\Omega L\xi_0}{1-h^*\Omega L}\left(1+\frac{h^*\Omega L}{1-h^*\Omega L}\right)^m . \tag{6.230}$$

Thus, applying Lemma 6.3,

$$|\xi_m| \leq \frac{\tau(h^*)+h^*\Omega L\xi_0}{1-h^*\Omega L} \exp\left(\frac{\Omega Lmh^*}{1-h^*\Omega L}\right) \quad (m = 1,2,3,\ldots,(N-1)) , \tag{6.231}$$

in analogy with (6.176).

The bound (6.231) involves $\xi_0 = \max_{0\leq j\leq p} |\xi_{0,j}|$. But in (6.228) we have, on setting $m = 0$,

$$|\xi_{0,j}| \leq h^*L \sum_{k=0}^{p} |\omega_k^{[j]}| \, |\xi_{0,k}| + |t_{0,j}(h^*)|$$

$$\leq h^*L\Omega\xi_0 + |t_{0,j}(h^*)| \quad (j = 0,1,2,\ldots,p)$$

so that $\xi_0 \leq h^*L\Omega\xi_0 + \tau(h^*)$, and hence

$$\xi_0 \leq \tau(h^*)/(1- h^*L\Omega) , \tag{6.232}$$

provided that $h^*L\Omega < 1$. If we combine (6.231) and (6.232) we obtain, with some loss of precision,

$$\max_{0\leq i\leq p} |\tilde{f}((m+\theta_i)h^*) - f((m+\theta_i)h^*)|$$

$$\leq \frac{\tau(h^*)(1+h^*\Omega L)}{1 - h^*L\Omega} \exp\left(\frac{\Omega Lmh^*}{1-h^*\Omega L}\right) \tag{6.233}$$

when h^* is sufficiently small. We deduce the following result.

THEOREM 6.5.

(a) *If* $\lim \tau_N(h^*) = 0$ *as* $h^* \to 0$ *with* $Nh^* = X$, $|\tilde{f}(x)-f(x)| \to 0$ *at every mesh-point* $x \in [0,X]$.

(b) *At every mesh-point* $x \in [0,X]$

$$\tilde{f}(x)-f(x) = O((h^*)^s) ,$$

where s *is given in Lemma* 6.9.

The preceding result is comparable with Theorem 6.2. In view of Lemma 6.10 we would hope to be able to strengthen Theorem 6.2 to give a result similar to Theorem 6.4. Weiss, in his thesis (1972), appears to have obtained such a result.

Weiss (1972) obtains an extension of Lemma 6.10 under additional assumptions. Supposing that the degree of precision of (6.108) is ρ, Weiss (1972) establishes that

$$t_{mj}(h^*) = \sum_{k=0}^{\rho-p} \phi_{kj}((m+\theta_j)h^*)(h^*)^{p+k+1} + O((h^*)^{\rho+2}) , \tag{6.234}$$

and obtains explicit expressions for the functions $\phi_{kj}(x)$, under the assumption that $F(x,y;f(y))$ has a derivative of order $\rho+2$ with respect to y which is continuous for $0 \le y \le x + \delta$, $x \in [0,X]$, where $\delta > 0$, $h^* < \delta$. This permits Weiss to claim a result of the form

$$\tilde{f}((m+\theta_j)h^*) = f((m+\theta_j)h^*) + \sum_{k=0}^{\rho-p} (h^*)^{p+k} e_{k,j}((m+\theta_j)h^*) + O((h^*)^{\rho+2}) ,$$

and recurrence relations for the functions $e_{k,j}(x)$ are given. Weiss then gives the result $e_{k,\rho}(x) = 0$, $0 \le k < \rho-p$, which implies that approximations of the values $f(mh^*)$ $(m = 1,2,3,\ldots,(N-1),N)$ have a generally higher order of accuracy (the error being $O((h^*)^{\rho+1})$ than other computed values. (Such a result is clearly motivation for calling the schemes implicit Runge-Kutta schemes.) Further, it would appear that the function $e_{\rho-p,p}(x)$ satisfies an integral equation of the form (6.186), so that the scheme gives a discretization of the integral equation which is not unstable.

Whilst we have outlined the results claimed by Weiss, the proof depends upon a detailed investigation of the functions $\phi_{k,j}(x)$ occurring in (6.234) and will not be pursued here. Further, it is not yet clear to us what effect the stricter assumptions of Lemma 6.10 would have on the theory, but the construction of a theory analogous with that of Kobayasi ought to be possible.

6.14. *Convergence of product-integration methods*

In this section we investigate the convergence of the product-integration methods of section 6.9 applied to an equation of the form (6.131), namely,

$$f(x) - \int_0^x K(x,y)\Phi(x,y;f(y))dy = g(x) \quad (0 \le x \le X) \tag{6.235}$$

with the conditions on $K(x,y)$ and $\Phi(x,y;v)$ which are stated in section 6.9.

It will be recalled that the methods of section 6.9 are extensions either of the methods of section 6.2 or 6.7, employing product-integration formulae in place of conventional quadrature rules. We shall here establish the convergence of a class of methods. (Though the class is a wide one, it does not include all the methods suggested in our text for reasons of simplicity.) Since our proof is similar to earlier ones we shall not give all the details.

Suppose that

$$\{0 \le u_0 < u_1 < u_2 < \ \ldots\ < u_M = X\}$$

defines a partition of the interval $[0,X]$, and suppose that

$$\max_{1\le i\le M} |u_i - u_{i-1}| \to 0$$

as $h^* \to 0$, where h^* is some parameter associated with the partition. (The simplest case occurs with $u_r = rh$, $h = X/N$ on setting $M = N$ and $h^* = h$. Another case arises on setting $u_r = (m+\theta_j)h^*$, where $r = m(p+1)+j$ $(j = 0,1,2,\ldots,p,\ m = 0,1,2,\ldots,(N-1))$ and $0 < \theta_0 < \theta_1 < \ldots < \theta_p = 1$, with $M = N(p+1)-1$.)

We suppose that the discretization of (6.235) is achieved by writing

$$\tilde{f}(u_r) - \sum_{j=0}^{M} W_{rj}\Phi(u_r,u_j;\tilde{f}(u_j)) = g(u_r) \quad (r = 0,1,2,3,\ldots,M), \tag{6.236}$$

where the matrix $\underset{\sim}{W}$ with entries $W_{r,j}$ is block-lower-triangular and all the diagonal blocks (with the possible exception of the leading one) have the same order q. (We shall suppose that the order of the leading block is q_0.) Thus for a step-by-step method, $W_{r,j} = 0$ if $j > r$, $q = q_0 = 1$, and $\underset{\sim}{W}$ is lower triangular. For a block-by-block method defined by (6.136) with $0 < \theta_0 < \theta_1 < \ldots < \theta_p = 1$, the diagonal blocks are of order $q_0 = q = p+1$; the pattern of non-zero entries

W_{rj} in the case $p = 2$ is indicated in (6.237) by means of a ×.

$$\left.\begin{array}{ccc|ccc|ccc}
\times & \times & \times & \cdot & \cdot & \cdot & \cdot & \cdot & \cdot \\
\times & \times & \times & \cdot & \cdot & \cdot & \cdot & \cdot & \cdot \\
\times & \times & \times & \cdot & \cdot & \cdot & \cdot & \cdot & \cdot \\
\hline
\times & \times & \times & \times & \times & \times & \cdot & \cdot & \cdot \\
\times & \times & \times & \times & \times & \times & \cdot & \cdot & \cdot \\
\times & \times & \times & \times & \times & \times & \cdot & \cdot & \cdot \\
\hline
\times & \times & \times & \times & \times & \times & \times & \times & \times \\
\times & \times & \times & \times & \times & \times & \times & \times & \times \\
\times & \times & \times & \times & \times & \times & \times & \times & \times \\
\hline
\end{array}\right. \qquad (6.237)$$

In the case of a block-by-block method (6.136) with $\theta_0 = 0$, we have $q_0 = p+1$, and $q = p$. (The leading diagonal block has its first row consisting of zero entries.)

We require a hypothesis concerning the behaviour of the weights W_{rj}.

HYPOTHESIS 6.1. *We suppose that for each value of the parameter* h^*, *the corresponding matrix* $\underset{\sim}{W}$ *can be written in partitioned form as a block-lower-triangular matrix as derived above. The submatrices will be denoted* $\underset{\sim}{W}^{[i,j]}$, *and we suppose that the norm of every partitioned submatrix* $\underset{\sim}{W}^{[i,j]}$ *can be made arbitrarily small for all* $h^* \le h_0^*$ *by taking* h_0^* *sufficiently small. Thus, given* $\varepsilon > 0$, *there exists* $h_0^*(\varepsilon)$ *such that* $h^* \le h_0^*$ *implies that* $\sup \|\underset{\sim}{W}^{[i,j]}\|_\infty < \varepsilon$.

Provided that $K(x,y)$ satisfies the conditions imposed in section 6.9, this hypothesis is satisfied when the weights W_{rj} are defined by a scheme associated with eqns (6.136), or one of the schemes suggested for constructing (6.133) or (6.135).

With the preceding hypothesis, the existence of values $\tilde{f}(u_r)$ satisfying (6.236) can be established, for all h^* sufficiently small, by examining a predictor-corrector technique and applying a standard argument, as follows.

For simplicity of presentation suppose that $q_0 = q$, and partition the vector $\tilde{\mathbf{f}} = [\tilde{f}(u_0), \tilde{f}(u_1), \ldots, \tilde{f}(u_M)]^T$ into subvectors $\tilde{\Phi}_0, \tilde{\Phi}_1, \ldots,$ each with q components. The entries in $\mathbf{W}^{[r,r]}$ are then $W_{rq+i,rq+j}$ $(i, j = 0,1,2,\ldots,(q-1))$.

The solution of the equations for the values $\tilde{f}(u_i)$ can be achieved by computing $\tilde{\Phi}_0, \tilde{\Phi}_1, \tilde{\Phi}_2, \ldots$ in turn, each vector $\tilde{\Phi}_r$ being obtained by an iterative method. Thus for the components of the iterate $\tilde{\Phi}_r^{[k+1]}$ we set – with $\tilde{f}^{[0]}(x)$ suitably defined –

$$\tilde{f}^{[k+1]}(u_{rq+i}) = \Gamma_{rq+i} + \sum_{j=0}^{q-1} W_{rq+i,\, rq+j}\Phi(u_{rq+i}, u_{rq+j}; \tilde{f}^{[k]}(u_{rq+j}))$$

$$(i = 0,1,2,\ldots,(q-1)),$$

where Γ_{rq+i} depends upon $g(x)$ and $\tilde{\Phi}_0, \tilde{\Phi}_1, \ldots, \tilde{\Phi}_{r-1}$. Employing the Lipschitz condition for $\Phi(x,y;v)$ and Hypothesis 6.1,

$$\|\tilde{\Phi}_r^{[k+1]} - \tilde{\Phi}_r^{[k]}\|_\infty \leq L\varepsilon \|\tilde{\Phi}_r^{[k]} - \tilde{\Phi}_r^{[k-1]}\|_\infty ,$$

for all $h^* \leq h_0^*(\varepsilon)$ and the sequence $\{\tilde{\Phi}_r^{[k]}\}$ converges to $\tilde{\Phi}_r$ if $L\varepsilon = \chi < 1$.

Let us now consider the convergence of $\tilde{f}(u_r)$ to $f(u_r)$ at each mesh-point u_r as $h^* \to 0$. Analogous to (6.235) we have

$$f(u_r) - \sum_{j=0}^{M} W_{r,j}\Phi(u_r, u_j; f(u_j)) = g(u_r) + t_r(h^*), \tag{6.238}$$

where

$$t_r(h^*) = \int_0^{u_r} K(u_r,y)\Phi(u_r,y;f(y))\,dy - \sum_{j=0}^{M} W_{r,j}\Phi(u_r,u_j;f(u_j)). \tag{6.239}$$

We shall suppose that as $h^* \to 0$, $\max|t_r(h^*)| \to 0$, where the maximum is taken over each mesh-point u_r in $[0,X]$. Now subtract (6.235) from (6.238). We employ the Lipschitz constant L for $\Phi(x,y;v)$ and the sub-matrices $\mathbf{W}^{[i,j]}$ of the partitioned form of $\mathbf{W}$, and we write $\boldsymbol{\zeta}_r = \boldsymbol{\phi}_r - \tilde{\boldsymbol{\Phi}}_r$, where

$$\boldsymbol{\phi}_r = [f(u_{rq}), f(u_{rq+1}), \ldots, f(u_{(r+1)q-1})]^T$$

and

$$\underset{\sim}{t}_r = [t_{rq}(h^*), t_{rq+1}(h^*), \ldots, t_{(r+1)q-1}(h^*)]^T .$$

We then obtain, assuming $q = q_0$,

$$||\underset{\sim}{\zeta}_r||_\infty \leq L \sum_{j=0}^{r} ||\underset{\sim}{W}^{[r,j]}||_\infty \, ||\underset{\sim}{\zeta}_j||_\infty + ||\underset{\sim}{t}_r||_\infty$$

whence, if h^* is sufficiently small

$$||\underset{\sim}{\zeta}_r||_\infty \leq \frac{L \max ||\underset{\sim}{W}^{[r,j]}||_\infty}{1-L\,||\underset{\sim}{W}^{[r,r]}||_\infty} \{\sum_{j=0}^{r-1} ||\underset{\sim}{\zeta}_j||_\infty + ||\underset{\sim}{t}_r||_\infty\}. \qquad (6.240)$$

According to Hypothesis 6.1, if $h^* \leq h_0^*(\varepsilon)$, $\max ||\underset{\sim}{W}^{[r,j]}||_\infty < \varepsilon$, and applying Lemma 6.2 the (uniform) convergence of $||\underset{\sim}{\zeta}_r||_\infty$ to zero is established (since $\max ||\underset{\sim}{t}_r||_\infty \to 0$) as $h^* \to 0$, provided that $||\underset{\sim}{\zeta}_0||_\infty \to 0$. In the case $q = q_0$,

$$||\underset{\sim}{\zeta}_0||_\infty \leq ||\underset{\sim}{t}_0||_\infty / \{1-L\,||\underset{\sim}{W}^{[0,0]}||_\infty\}$$

and the result follows. With a slight adjustment the same argument goes over to the case $q_0 \neq q$. We thus deduce Theorem 6.6.

THEOREM 6.6. *Suppose that $\Phi(x,y;v)$ satisfies a Lipschitz condition in the variable v, and suppose that a solution $f(x)$ of (6.235) exists and is unique, and the approximate scheme satisfies Hypothesis 6.1. If $\max|t_r(h^*)| \to 0$ then $\max|f(u_r)-\tilde{f}(u_r)| \to 0$ as $h^* \to 0$.*

6.15. *Convergence of methods for non-singular equations of the first kind*

The methods for the numerical treatment of Volterra equations of the first kind,

$$\int_0^x K(x,y)f(y)\,dy = g(x) \quad (0 \leq x \leq X) \qquad (6.241)$$

(where $K(x,x) \neq 0$ for $0 \leq x \leq X$) do not all yield readily to a rigorous mathematical analysis of their convergence properties. The detailed analysis which we give in this section will be limited to the simpler theories, and we shall limit ourselves to a *statement* of some results which are rather more difficult to prove. Throughout this section, we shall assume the existence of a solution $f(x)$ of the equation to be solved.

Perhaps simplest to analyse is the mid-point method for the approximate solution of the equation (in which $g(0) = 0$):

$$\int_0^x K(x,y)f(y)\,dy = g(x) ,$$

with $0 \leq x \leq X$. Recall that the use of the mid-point rule gives approximate values $\tilde{f}((r+\frac{1}{2})h)$ ($r = 0,1,2,\ldots,(N-1)$, with $Nh = x$), which satisfy the equations

$$h\sum_{j=0}^{r} K((r+1)h,(j+\tfrac{1}{2})h)\tilde{f}((j+\tfrac{1}{2})h) = g((r+1)h) \quad (r = 0,1,2,\ldots,(N-1)). \tag{6.242}$$

Hence, on differencing successive equations, and using the convention $\sum_{j=0}^{-1} u_j = 0$,

$$hK((r+1)h,(r+\tfrac{1}{2})h)\tilde{f}((r+\tfrac{1}{2})h)+h\sum_{j=0}^{r-1}\Delta_x K(rh,(j+\tfrac{1}{2})h)\tilde{f}((j+\tfrac{1}{2})h)=\Delta g(rh) \tag{6.243}$$
$$(r = 0,1,2,\ldots,(N-1)) ,$$

where $\Delta_x K(\hat{x},\hat{y}) = K(\hat{x}+h,\hat{y})-K(\hat{x},\hat{y})$ and $\Delta g(\hat{x}) = g(\hat{x}+h)-g(\hat{x})$. On the other hand,

$$h\sum_{j=0}^{r} K((r+1)h,(j+\tfrac{1}{2})h)f((j+\tfrac{1}{2})h) = g((r+1)h)+t_{r+1}(h), \tag{6.244}$$

where we now have $t_0(h) = 0$ and for $r=0,1,\ldots,N-1$,

$$t_{r+1}(h)=h\sum_{j=0}^{r} K((r+1)h,(j+\tfrac{1}{2})h)f((j+\tfrac{1}{2})h)-\int_0^{(r+1)h} K((r+1)h,y)f(y)\,dy .$$

Thus,

$$t_{r+1}(h) = -\tfrac{1}{24}h^2\left[(\partial/\partial y)K(\bar{x},y)f(y)\right]_{y=0}^{y=(r+1)h} + O(h^q) \ , \quad \bar{x}=(r+1)h \ ,$$

with $q = 4$ if (say) $(\partial/\partial y)^4 K(x,y)f(y)$ exists and is bounded for $0 \le y \le x$, $0 \le x \le X$. Thus if we difference successive equations in (6.244) we obtain, for $r = 0,1,2,\ldots,(N-1)$,

$$hK((r+1)h,(r+\tfrac{1}{2})h)f((r+\tfrac{1}{2})h)+h\sum_{j=0}^{r-1}\Delta_x K(rh,(j+\tfrac{1}{2})h)f((j+\tfrac{1}{2})h)$$

$$= \Delta g(rh)+\{t_{r+1}(h)-t_r(h)\} \ . \quad (6.245)$$

We now write $\xi_r = f((r+\tfrac{1}{2})h)-\tilde{f}((r+\tfrac{1}{2})h)$, and we obtain, on subtracting (6.243) from (6.245),

$$hK((r+1)h,(r+\tfrac{1}{2})h)\xi_r+h\sum_{j=0}^{r-1}\Delta_x K(rh,(j+\tfrac{1}{2})h)\xi_j=t_{r+1}(h)-t_r(h) \quad (6.246)$$

$$(r = 0,1,2,\ldots,(N-1)) \ .$$

Now we have supposed that $K(x,x)$ is non-vanishing for $0\le x\le X$, and if (as we also suppose) $K(x,y)$ is continuous, we can find a value h_0 such that for all positive $h \le h_0$,

$$|K(x,x-\tfrac{1}{2}h)| \ge \tfrac{1}{2}\inf_{0\le x\le X}|K(x,x)| = \sigma > 0 \ .$$

We further suppose that $(\partial/\partial x)K(x,y)$ is continuous for $0\le y\le x\le X$, so that

$$|\Delta_x K(x,y)| \le h \sup_{0\le y\le x\le X}|(\partial/\partial x)K(x,y)| \ .$$

Consequently, for all positive $h \le h_0$ and all values of r,j occurring in (6.246)

$$\left|\frac{\Delta_x K(rh,(j+\tfrac{1}{2})h)}{K((r+1)h,(r+\tfrac{1}{2})h)}\right| \le \Omega h \ ,$$

where Ω is independent of h,r, and j. It follows, from (6.246), that

$$|\xi_r| \le \Omega h\sum_{j=0}^{r-1}|\xi_j|+|t_{r+1}(h)-t_r(h)|/\sigma h \quad (r = 0,1,2,\ldots,(N-1)) \ .$$

Hence, by Lemmas 6.2 and 6.3, taking $q = 0$ therein,

$$|\xi_r| \le \tau^*(h)(1+\Omega h)^r/\sigma h \le \exp(\Omega rh)\tau^*(h)/\sigma h, \quad (r=0,1,2,\ldots,(N-1)) \tag{6.247}$$

where

$$\tau^*(h) \equiv \tau_N^*(h) = \max_{0\le r\le N-1} |t_{r+1}(h)-t_r(h)| .$$

We must now proceed with some care, since if we bound $\tau^*(h)$ by $2\max_{0<r<N}|t_r(h)|$, $\tau^*(h)$ is shown to be $O(h^2)$ (under sufficient conditions), and the bound (6.247) on $|\xi_r|$ is then $O(h)$. We can, with suitable conditions, establish that $\tau^*(h) = O(h^3)$. Indeed, $t_1(h)=O(h^3)$, and

$$t_{r+1}(h)-t_r(h)$$

$$= \{hK((r+1)h,(r+\tfrac{1}{2})h)f((r+\tfrac{1}{2})h)-\int_{rh}^{(r+1)h} K((r+1)h,y)f(y)\,dy\} +$$

$$+ \sum_{j=0}^{r-1} \{h\Delta_x K(rh,(j+\tfrac{1}{2})h)f((j+\tfrac{1}{2})h)-\int_{jh}^{(j+1)h} \Delta_x K(rh,y)f(y)\,dy\} \tag{6.248}$$

(where $\Delta_x K(\hat{x},\hat{y}) = K(\hat{x}+h,\hat{y})-K(\hat{x},\hat{y})$), for $r = 1,2,\ldots,N-1$.

If $(\partial/\partial y)^2 K(x,y)f(y)$ is continuous for $0 \le y \le x$, the first term in braces in the expression (6.248) for $t_{r+1}(h)-t_r(h)$ is $O(h^3)$. A bound on the modulus of the remaining contributions is, for $r \le N-1$,

$$\frac{1}{24}(rh)h^2 \sup_{\substack{0\le y\le x\\ 0\le x\le X-h}} |(\partial/\partial y)^2\{\Delta_x K(x,y)f(y)\}| . \tag{6.249}$$

But

$$\sup|(\partial/\partial y)^2\{\Delta_x K(x,y)f(y)\}| = \sup|(\partial/\partial y)^2\{K(x+h,y)f(y)-K(x,y)f(y)\}|$$

$$\le \sup|h(\partial^3/\partial x\partial y^2)K(x+\theta h,y)f(y)| ,$$

where the supremum is taken over $0 \le y \le x \le X$ and $0 \le \theta \le 1$ (assuming the existence of the mixed derivative). Thus the term in (6.249) is

$O(h^3)$ as $h \to 0$, uniformly for all $rh \le X$. It now follows that $\tau^*(h) = O(h^3)$.

Since $\tau^*(h) = O(h^3)$, (6.247) establishes that the error of the approximate values computed using the mid-point rule is $O(h^2)$.

THEOREM 6.7. *If* $(\partial/\partial x)(\partial/\partial y)^2 K(x,y)f(y)$ *is continuous for* $0 \le y \le x$, $0 \le x \le X$, *and values* $\tilde{f}((r+\frac{1}{2})h)$ *are obtained using the mid-point rule, then* $\tilde{f}(x)-f(x) = O(h^2)$ *uniformly for mesh-points in* $[0,X]$.

The bound (6.247) is generally pessimistic. Fortunately, we can obtain an improved result.

THEOREM 6.8. *Suppose that, in addition to the assumptions of Theorem 6.7,* $(\partial/\partial y)^4 K(x,y)f(y)$ *exists and is bounded for* $0\le y\le x$, $0\le x\le X$. *Then* $\tilde{f}(x)-f(x) = h^2 e(x)+O(h^3)$ *for* x *of the form* $(r+\frac{1}{2})h$ *lying in* $[0,X]$ *where* $\int_0^x K(x,y)e(y)\,dy = \nu(x)$, *with* $\nu(x)= \frac{1}{24}\left[(\partial/\partial y)K(x,y)f(y)\right]_{y=0}^{y=x}$, *provided that* $(\partial/\partial y)^2K(x,y)e(y)$ *is bounded for* $0\le y\le x\le X$.

Proof. Suppose that $e(x)$ satisfies the stated equation. Then

$$h\sum_{j=0}^{r} K((r+1)h,(j+\tfrac{1}{2})h)\{f((j+\tfrac{1}{2})h)+h^2e((j+\tfrac{1}{2})h)\}$$

$$= g((r+1)h) - \frac{1}{24}h^2\left[(\partial/\partial y)K(x,y)f(y)\right]_{y=0}^{y=(r+1)h} + O(h^4) +$$

$$+ h^2\left\{\int_0^{(r+1)h} K((r+1)h,y)e(y)\,dy + O(h^2)\right\} \qquad (6.250)$$

when $K(x,y)e(y)$ has a bounded second derivative in y. If $\phi(x) = f(x)+h^2e(x)$ and $\varepsilon_j = \tilde{f}((j+\frac{1}{2})h)-\phi((j+\frac{1}{2})h)$, the last equation yields, in view of the equation satisfied by $e(x)$,

$$hK((r+1)h,(r+\tfrac{1}{2})h)\varepsilon_r+h\sum_{j=0}^{r-1}\Delta_x K(rh,(j+\tfrac{1}{2})h)\varepsilon_j = \Delta\hat{t}_r(h) \quad ,$$

in analogy with (6.246), where $\hat{t}_r(h)$ gives the $O(h^4)$ terms in (6.250).

It follows, since $|\Delta\hat{t}_r(h)| \le 2\max|\hat{t}_r(h)|$, that

$$|\tilde{f}((r+\tfrac{1}{2})h)-\phi((r+\tfrac{1}{2})h)| \le 2\exp\Omega rh \max|\hat{t}_r(h)|/\sigma h ,$$

and the right-hand term in the bound is $O(h^3)$.† Thus $\tilde{f}(x) = f(x) + h^2 e(x) + O(h^3)$ at points $x = (r+\frac{1}{2})h$ in $[0,X]$.

Observe that the equation for $e(x)$ can be recast as an equation of the second kind of the form

$$e(x) + \int_0^x K_0(x,y)e(y)\,dy = \nu_0(x) \quad (0 \le x \le X) ,$$

where

$$K_0(x,y) = \{(\partial/\partial x)K(x,y)\}/K(x,x) \text{ and } \nu_0(x) = \nu'(x)/K(x,x) .$$

The conditions of the theorem can therefore be expressed as conditions on $K(x,y)$ and $f(x)$.

The analysis of the mid-point method is fairly simple in comparison with the analysis for the method employing the trapezium rule. We gain some insight into the reasons for this if we consider the next Theorem, which is due to Kobayasi (1967).

THEOREM 6.9. *Suppose that* $\int_0^x K(x,y)f(y)\,dy = g(x)$ (*for* $0 \le x \le X$) *and* $h\sum_{j=0}^{r}{}'' K(rh,jh)\tilde{f}(jh) = g(rh)$ *for* $r = 1,2,3,\ldots,N$ (*with* $Nh = x$), *and* $\tilde{f}(0) = f(0) + O(h^3)$. *If the functions* $\varepsilon(x)$ *and* $\delta(x)$ *are defined by the equation*

$$\int_0^x K(x,y)\varepsilon(y)\,dy = -\frac{1}{12}\Big[(\partial/\partial y)K(x,y)f(y)\Big]_{y=0}^{y=x} \qquad (6.251)$$

and with $(\partial/\partial y)K(x,y) = K_y(x,y)$,

$$K(x,x)\delta'(x) + K_y(x,x)\delta(x) = 0 ,$$

$$\delta(0) = -\varepsilon(0) , \qquad (6.252)$$

† The argument is of a type now familiar to the reader, and follows similar lines to the proof of (6.247).

then $\tilde{f}(rh) = f(rh) + h^2[\varepsilon(rh) + (-1)^r\delta(rh)] + O(h^3)$, *under sufficient differentiability conditions.*

It will be observed that this theorem may be compared with Theorem 6.8, but the results are not directly analogous because the dominant error in results obtained using the trapezium rule contains two components, one of which is oscillatory. Theorem 6.9 is of some importance because it establishes that there are limitations on any attempt to implement a process of h^2-extrapolation. (See, however, the corollary to Theorem 6.10.) Further, we expect instability if either $\varepsilon(x)$ or $\delta(x)$ grow exponentially, and we note that

$$\delta(x) = -\varepsilon(0)\exp\{-\int_0^x \frac{K_y(t,t)}{K(t,t)}dt\} .$$

The proof of Theorem 6.9 requires careful attention to details and will not be pursued here. We instead establish $O(h^2)$ convergence. We use the style of proof developed by Linz (1967*a*), who followed the lines of argument used by Jones (1962).

We have

$$h\sum_{j=0}^{r}{}''K(rh,jh)\tilde{f}(jh) = g(rh) \quad (r = 1,2,3,\ldots,N)$$

with $\tilde{f}(0) = g'(0)/K(0,0) = f(0)$. In comparison,

$$h\sum_{j=0}^{r}{}''K(rh,jh)f(jh) = g(rh) + t_r(h) \quad (r = 1,2,3,\ldots,N) ,$$

where, now,

$$t_r(h) = h\sum_{j=0}^{r}{}''K(rh,jh)f(jh) - \int_0^{rh} K(rh,y)f(y)dy .$$

Assuming sufficient differentiability,

$$t_r(h) = h^2\nu(rh) + O(h^4) ,$$

where

$$\nu(x) = \frac{1}{12}[(\partial/\partial y)K(x,y)f(y)]_{y=0}^{y=x} , \tag{6.253}$$

and clearly $t_r(h) = O(h^2)$, as $h \to 0$ with rh fixed. (However, $t_1(h)$ and $t_2(h)$ are $O(h^3)$.)

If we write $\xi_r = f(rh)-\tilde{f}(rh)$ $(r = 1,2,3,\ldots,N)$ then we find

$$h \sum_{j=0}^{r}{}'' K(rh,jh)\xi_j = t_r(h) \quad (r = 1,2,3,\ldots,N) \tag{6.254}$$

and, hence, ξ_1 and ξ_2 are $O(h^2)$.

Differencing successive equations in (6.254),

$$\frac{h}{2}K((r+1)h,(r+1)h)\xi_{r+1} + \frac{h}{2}K(rh,rh)\xi_r + h\sum_{j=0}^{r}{}'\Delta_x K(rh,jh)\xi_j = \Delta t_r(h) \tag{6.255}$$

$$(r = 1,2,3,\ldots,(N-1))$$

(where $\sum_{j=0}^{r}{}' u_j = \frac{1}{2}u_0 + \sum_{j=1}^{r} u_j$ and $\Delta t_r(h) = t_{r+1}(h)-t_r(h)$, etc.). We wish to show, using (6.255), that $\xi_r = O(h^2)$, as $h\to 0$ with rh fixed. We have, on differencing (6.255) again,

$$\frac{h}{2}\{K((r+1)h,(r+1)h)\xi_{r+1} - K((r-1)h,(r-1)h)\xi_{r-1}\} +$$

$$+ h\Delta_x K(rh,rh)\xi_r + h\sum_{j=0}^{r-1}{}'\Delta_x^2 K((r-1)h,jh)\xi_j = \Delta^2 t_{r-1}(h)$$

$$(r = 2,3,4,\ldots,(N-1)) .$$

Consequently, it can be shown that

$$\xi_{r+1} = \xi_{r-1}(1+hM_r) - hL_r\xi_r - h^2\sum_{j=0}^{r-1}{}' N_{rj}\xi_j +$$

$$+ 2\{\Delta^2 t_{r-1}(h)\}/\{hK((r+1)h,(r+1)h)\} ,$$

where

$$L_r = 2\Delta_x K(rh,rh)/(hK((r+1)h,(r+1)h)),$$

$$M_r = \{K((r-1)h,(r-1)h)-K((r+1)h,(r+1)h)\}/\{hK((r+1)h,(r+1)h)\} ,$$

$$N_{rj} = 2\Delta_x^2 K((r-1)h,jh)/\{h^2K((r+1)h,(r+1)h)\} .$$

Assuming sufficient differentiability of $K(x,y)$ and $f(x)$, $|L_r| \le 2\,||K_x(x,y)||_\infty/\sigma$, where $|K(x,x)| \ge \sigma > 0$ for $0 \le x \le X$, etc., and $|L_r|$, $|M_r|$, $|N_{rj}|$ can be bounded by a constant C independent of r and h , uniformly if $rh \in [0,X]$. Furthermore (compare with the discussion of the size of the term $\tau^*(h)$ in (6.247)), the constant C can be chosen so that $2\{\Delta^2 t_r(h)/h\} \le Ch^3\sigma$. With such a choice of C ,

$$|\xi_{r+1}| \le |\xi_{r-1}|(1+hC)+hC|\xi_r|+h^2C\sum_{j=0}^{r-1}|\xi_j|+Ch^3 \quad (r = 2,3,4,\ldots,(N-1)).$$

Suppose that $|\xi_i| \le \zeta_i$ $(i = 0,1,2)$ where $\zeta_0 < \zeta_1 < \zeta_2 = O(h^2)$ and

$$\zeta_{r+1} = (1+hC)\zeta_r + hC\zeta_r+h^2C\sum_{j=0}^{r}\zeta_j+h^3C \quad (r = 2,3,\ldots,N-1). \qquad (6.256)$$

It is not difficult to show, by induction, that $|\xi_r| \le \zeta_r \le \zeta_{r+1}$. Moreover, we find on differencing (6.256) that

$$\zeta_{r+1}-(2+2hC+h^2C)\zeta_r+(1+2hC)\zeta_{r-1} = 0 \quad (r = 3,4,5,\ldots,(N-1)). \qquad (6.257)$$

The solution of this homogeneous difference equation is given by

$$\zeta_r = c_1t_1^r + c_2t_2^r \quad (r = 2,3,4,\ldots,N) \quad ,$$

where t_1,t_2 are the roots of the auxiliary equation associated with (6.257), and c_1 and c_2 are determined by the values ζ_2,ζ_3 . We find that

$$t_1,t_2 = 1+hC+\tfrac{1}{2}h^2C\pm\tfrac{1}{2}\{(2+2hC+h^2C)^2 -4(1+2hC)\}^{\frac{1}{2}} \quad ,$$

and, by the choice of starting values, $\zeta_2 = O(h^2)$, $\zeta_3 = \zeta_2+O(h^3)$. We find, from a simple argument, that c_1 and c_2 are $O(h^2)$. On the other hand, there exists a constant, M say, such that $|t_i| \le e^{Mh}$ $(i=1,2)$, and hence $t_i^r \le e^{Mrh}$. Thus $\zeta_r \le \Gamma h^2 e^{Mrh}$ for some constant Γ . Since $|\xi_r| \le \zeta_r$, the $O(h^2)$ convergence of ξ_r to zero (as $h \to 0$ with rh fixed in $[0,X]$) is established.

Having established that $\xi_r = O(h^2)$ it is possible to return to eqn (6.254) and employ it to establish, directly, a result which may be regarded as a corollary to Theorem 6.9:

THEOREM 6.10. *In the notation of Theorem 6.9,*

$$\tfrac{1}{2}\{\tilde{f}((r+1)h) + \tilde{f}(rh)\} = \tfrac{1}{2}\{f((r+1)h) + f(rh)\} + h^2\varepsilon(rh)+O(h^3)$$

at each mesh-point rh *in the interior of* $[0,X]$, *assuming sufficient differentiability.*

<u>Proof</u>. We shall establish that $\tfrac{1}{2}(\xi_{r+1}+\xi_r) = h^2\varepsilon(rh)+O(h^3)$, in the notation of eqns (6.251) and (6.254).

We write $\Delta_x K(x,y) = K(x+h,y)-K(x,y)$, $\Delta_y K(x,y) = K(x,y+h)-K(x,y)$, etc., and we obtain, from eqn (6.254),

$$\tfrac{1}{2}hK(rh,rh)(\xi_{r+1}+\xi_r)+\{\tfrac{1}{2}h\Delta_x K(rh,rh)\xi_r\}$$
$$+\{\tfrac{1}{2}h[K((r+1)h,(r+1)h)-K(rh,rh)]\xi_{r+1}\}$$
$$+h\sum_{j=0}^{r}{}''\Delta_x K(rh,jh)\xi_j = \Delta t_r(h) .$$

Now, each of the terms in braces is $O(h^4)$, assuming the existence of $K_x(x,y) = (\partial/\partial x)K(x,y)$ and $(d/dx)K(x,x)$ (say). Further,

$$h\sum_{j=0}^{r}{}''\Delta_x K(rh,jh)\xi_j = \tfrac{1}{2}h\sum_{j=1}^{r}\{\Delta_x K(rh,jh)\xi_j+\Delta_x K(rh,(j-1)h)\xi_{j-1}\}$$
$$= h\sum_{j=1}^{r}\tfrac{1}{2}\Delta_x K(rh,(j-1)h)(\xi_j+\xi_{j-1})$$
$$+\tfrac{1}{2}h\sum_{j=1}^{r}\Delta_x\Delta_y K(rh,(j-1)h)\xi_j ,$$

where $\Delta_y K(\hat{x},\hat{y}) = K(\hat{x},\hat{y}+h) - K(\hat{x},\hat{y})$. Now

$$\Delta_y K(x,y) = h(\partial/\partial y)K(x,y) + O(h^2) ,$$

and

$$\Delta_x \Delta_y K(x,y) = h^2(\partial^2/\partial x \partial y)K(x,y) + O(h^3),$$

assuming sufficient differentiability, whilst $\xi_j = O(h^2)$. Thus on writing $\eta_j = \frac{1}{2}(\xi_j + \xi_{j+1})$, we find

$$hK(rh,rh)\eta_r + h \sum_{j=0}^{r}{}'' \Delta_x K(rh,jh)\eta_j = \Delta t_r(h) + O(h^4) ,$$

uniformly for all r with $rh \in [0,X]$, as $h \to 0$. Now $\Delta t_r(h) = h^2\{\nu((r+1)h) - \nu(rh)\} + O(h^4) = h^3\nu'(rh) + O(h^4)$, $\Delta_x K(rh,jh) = h(\partial/\partial x)K(rh,jh) + O(h^2)$, and $\eta_j = O(h^2)$. We therefore find that

$$K(rh,rh)\eta_r + h \sum_{j=0}^{r}{}'' (\partial/\partial x)K(rh,jh)\eta_j = h^2\nu'(rh) + O(h^3)$$

uniformly for all r with $rh \in [0,X]$.

We now consider the scaled equations

$$\mu_r + h \sum_{j=0}^{r}{}'' \frac{(\partial/\partial x)K(rh,jh)}{K(rh,rh)} \mu_r = \frac{\nu'(rh)}{K(rh,rh)} + O(h)$$

(where $(\partial/\partial x)K(rh,jh) = K_x(rh,jh))$, and we compare this with the equations obtained when the trapezium rule is applied to the equation

$$\varepsilon(x) + \int_0^x \frac{(\partial/\partial x)K(x,y)}{K(x,x)} \varepsilon(y)\,dy = \frac{-\nu'(x)}{K(x,x)} ,$$

which is obtained on differentiating (6.251). It can be shown rigorously that $\mu_r = -\varepsilon(rh) + O(h)$ and hence

$$\eta_r = \frac{1}{2}(\xi_{r+1} + \xi_r) = -h^2\varepsilon(rh) + O(h^3)$$

uniformly at mesh-points rh in $[0,X]$.

As a result of the preceding theorem, we obtain the following corollary.

COROLLARY 6.2. *With sufficient differentiability conditions,*

$$\hat{f}(rh) = f(rh) + h^2\{\varepsilon(rh)+\tfrac{1}{4}f''(rh)\} + O(h^3) ,$$

where

$$\hat{f}(rh) = \tfrac{1}{4}\{\tilde{f}((r-1)h) + 2\tilde{f}(rh)+\tilde{f}((r+1)h)\}$$

at each mesh-point rh *in the interior of* $[0,X]$.

Proof. Since $\tilde{f}((r+1)h)+\tilde{f}(rh) = 2f(rh)+hf'(rh)+\tfrac{1}{2}h^2f''(rh)+2h^2\varepsilon(rh)+O(h^3)$ and $\tilde{f}(rh)+\tilde{f}((r-1)h) = 2f(rh)-hf'(rh)+\tfrac{1}{2}h^2f''(rh)+2h^2\{\varepsilon(rh)-h\varepsilon'(rh)\}+O(h^3)$, the result follows.

The above corollary provides a justification of h^2-extrapolation on the values $\hat{f}(rh)$. Moreover, $\hat{f}(x)$ is a function whose dominant error term depends upon $\varepsilon(x)$ and $f''(x)$, and contains no contribution from the function $\delta(x)$ of Theorem 6.9.

The preceding analysis excludes the effect of rounding error. For a discussion which takes this into account we refer to Kobayasi (1967).

6.16. *Block-by-block methods for non-singular equations of the first kind*

We now consider the convergence of certain block-by-block methods described in section 6.10 for the solution of Volterra equations of the first kind (see section 6.10.4). In this treatment we follow the lines of argument used by de Hoog and Weiss (1973*a*).

We shall now consider the scheme associated with eqns (6.158) and defined by a choice of points $\{\theta_i\}$ $(i = 0,1,2,\ldots,p)$. We first consider the case where $\theta_0 \neq 0$, and

$$0 < \theta_0 < \theta_1 < \ldots < \theta_p = 1 .$$

The scheme is defined by eqns (6.158), that is,

$$h^* \sum_{i=0}^{m-1} \sum_{k=0}^{p} \omega_k K(u_{m,j},u_{i,k})\tilde{f}(u_{i,k})+h^* \sum_{k=0}^{p} \omega_k^{[j]} K(u_{m,j},u_{m,k})\tilde{f}(u_{m,k})=g(u_{m,j})$$

with $u_{m,j} = mh^* + \theta_j h^*$ $(m = 0,1,2,\ldots,(N-1),\ j = 0,1,2,\ldots,p)$, $Nh^* = X$.

Since $\theta_0 > 0$, no starting value is required. If $\tilde{f}(x)$ is replaced in (6.158) by $f(x)$ then it is necessary to replace $g((m+\theta_j)h^*)$ by $g((m+\theta_j)h^*)-t_{m,j}(h^*)$. Here $t_{m,j}(h^*)$ is the local truncation error which may be defined by (6.225) on interpreting $f(x)$ to be the solution of the equation of the first kind (6.241), and on setting $F(x,y;f(y)) = K(x,y)f(y)$. From Lemma 6.10,

$$t_{m,j}(h^*) = (h^*)^{p+1}\Phi((m+\theta_j)h^*)+(h^*)^{p+2}\Psi_j((m+\theta_j)h^*)+O((h^*)^{p+3}) \quad (6.258)$$

if $(\partial/\partial y)^{p+3}K(x,y)f(y)$ is continuous for $y \in [0,x+\delta]$, $x \in [0,X]$, for some $\delta > 0$. Clearly, if we write $\xi_{m,j} = f((m+\theta_j)h^*)-\tilde{f}((m+\theta_j)h^*)$ we find that

$$h^*\sum_{i=0}^{m-1}\sum_{k=0}^{p}\omega_k K((m+\theta_j)h^*,(i+\theta_k)h^*)\xi_{i,k} + h^*\sum_{k=0}^{p}\omega_k^{[j]}K((m+\theta_j)h^*,(m+\theta_k)h^*)\xi_{m,k} = -t_{m,j}(h^*). \quad (6.259)$$

Now set m to $m-1$ and $j = p$ in this equation and subtract the resulting equation from (6.259). We obtain, if $K(x,x) \neq 0$ for $0 \leq x \leq X$,

$$\sum_{k=0}^{p}\omega_k^{[j]}\frac{K((m+\theta_j)h^*,(m+\theta_k)h^*)}{K(mh^*,mh^*)}\xi_{m,k} + h^*\sum_{i=0}^{m-1}\sum_{k=0}^{p}\omega_k\frac{K((m+\theta_j)h^*,(i+\theta_k)h^*)-K(mh^*,(i+\theta_k)h^*)}{h^*K(mh^*,mh^*)}\xi_{i,k}$$

$$= -\left\{\frac{t_{m,j}(h^*)-t_{m-1,p}(h^*)}{h^*K(mh^*,mh^*)}\right\}$$

$$= O((h^*)^{p+1}) \quad (m = 1,2,3,\ldots,N-1),$$

since the function $\Phi(x)$ in (6.258) is differentiable. Since the right-hand side is $O((h^*)^{p+1})$, there is a constant C_1 such that its modulus may be bounded by $C_1(h^*)^{p+1}$ for h^* sufficiently small ($h^* < \delta$).

Now, we suppose that

$$|K(x,x)| \geq \alpha > 0 \text{ , and } |(\partial/\partial x)K(x,y)| \leq M_1$$

for $0 \leq y \leq x \leq X$, so that

$$|\omega_k\{K((m+\theta_j)h^*,(i+\theta_k)h^*)-K(mh^*,(i+\theta_k)h^*)\}/\{h^*K(mh^*,mh^*)\}|$$

$$\leq M_1 \max|\omega_k|/\alpha = C_2 ,$$

say. Consequently, for $j = 0,1,\ldots,p$,

$$\left|\sum_{k=0}^{p} \omega_k^{[j]} \frac{K((m+\theta_j)h^*,(m+\theta_k)h^*)}{K(mh^*,mh^*)} \xi_{m,k}\right| \leq C_2 h^* \sum_{i=0}^{m-1} \sum_{k=0}^{p} |\xi_{i,k}| + C_1(h^*)^{p+1} \quad (m = 1,2,3,\ldots,(N-1)). \tag{6.260}$$

On the other hand, if $(\partial/\partial x)K(x,y)$ and $(\partial/\partial y)K(x,y)$ are continuous,

$$\frac{K((m+\theta_j)h^*,(m+\theta_k)h^*)}{K(mh^*,mh^*)} = 1 + O(h^*) ,$$

so that a constant C_3 exists such that

$$\sum_{k=0}^{p} \left|\omega_k^{[j]} \frac{K(mh^*,mh^*)-K((m+\theta_j)h^*,(m+\theta_k)h^*)}{K(mh^*,mh^*)}\right| \leq C_3 h^* .$$

If we write $\zeta_i = \max_{0\leq k\leq p} |\xi_{ik}|$, then we obtain, from (6.260),

$$\left|\sum_{k=0}^{p} \omega_k^{[j]} \xi_{m,k}\right| \leq C_3 h^* \zeta_m + C_2(p+1)h^* \sum_{i=0}^{m-1} \zeta_i + C_1(h^*)^{p+1} \quad (m = 1,2,3,\ldots,(N-1);\ j = 0,1,\ldots,p). \tag{6.261}$$

To relate the left-hand side to ζ_m we employ the following lemma.

LEMMA 6.11. *For any choice* $0 < \theta_0 < \theta_1 < \ldots < \theta_p = 1$, *the matrix* $\underset{\sim}{W}$ *with* (j,k)*th element* $\omega_k^{[j]}$ $(j,k = 0,1,..,p)$ *is non-singular and if* $\sum_{k=0}^{p} \omega_k^{[j]} \alpha_k = \beta_j$ *for* $j = 0,1,2,\ldots,p$ *then* $\max_{0\le k\le p} |\alpha_k| \le C_4 \max_{0\le k\le p} |\beta_k|$ *for some constant* C_4.

Proof. If $\underset{\sim}{W}$ is singular then $\gamma_0, \gamma_1, \ldots, \gamma_p$ (not all vanishing) can be found such that

$$\sum_{k=0}^{p} \omega_k^{[j]} \gamma_k = 0 \quad (j = 0,1,2,\ldots,p) .$$

Thus if $\gamma(x)$ is the polynomial of degree p with $\gamma(\theta_i) = \gamma_i$ $(i = 0,1,2,\ldots,p)$, it follows that

$$\int_0^{\theta_j} \gamma(x)\,dx = 0 \quad (j = 0,1,2,\ldots,p)$$

and hence

$$\sum_{j=0}^{p} a_j \theta_i^{j+1}/(j+1) = 0 \quad ,$$

where $\gamma(x) = \sum_{j=0}^{p} a_j x^j$. Since the matrix having as its elements θ_i^{j+1} is well known to be non-singular, the only possibility is $a_0 = a_1 = \ldots = a_p = 0$, and hence $\gamma_0 = \gamma_1 = \ldots = \gamma_p = 0$, contrary to assumption. Thus $\underset{\sim}{W}$ is non-singular and we may set $C_4 = \|\underset{\sim}{W}^{-1}\|_\infty$.

It follows from Lemma 6.11 that

$$\zeta_m \le C_4\{C_3 h^* \zeta_m + C_2(p+1)h^* \sum_{i=0}^{m-1} \zeta_i + C_1(h^*)^{p+1}\} .$$

When $C_4 C_3 h^* < 1$,

$$\zeta_m \le \{\frac{C_2(p+1)C_4}{1 - C_4 C_3 h^*}\} h^* \sum_{i=0}^{m-1} \zeta_i + \frac{C_1 C_4}{1 - C_4 C_3 h^*} (h^*)^{p+1} \quad (m = 1,2,3,\ldots,(N-1)). \quad (6.262)$$

On the other hand, it is clear that

$\zeta_0 \le C_5 \max_{0\le j\le p} |t_{0j}(h^*)|/h^*$, and $\max_{0\le j\le p} |t_{0j}(h^*)| = O((h^*)^{p+2})$.

Consequently,

$$\zeta_0 = O((h^*)^{p+1}) \quad , \tag{6.263}$$

and from (6.262) and (6.263), together with Lemma 6.1, we obtain the result that $|\zeta_m| = O((h^*)^{p+1})$ $(m = 0,1,2,\ldots,(N-1))$, uniformly in m as $h^* \to 0$.

We have established the following result.

THEOREM 6.11. *Suppose that the Volterra equation of the first kind has a (unique) solution* $f(x)$ $(0 \le x \le X)$ *which has a continuous derivative of order* $p+3$ *(for* $0 \le x \le X$*) ; suppose that* $(\partial/\partial x)K(x,y)$ *and* $(\partial/\partial y)^{p+3}K(x,y)$ *are continuous for* $0 \le y \le x+\delta$, $x \le X$, *where* $\delta > 0$ *is fixed. Then using* (6.158) *the rate of convergence of* $\tilde{f}(x)$ *to* $f(x)$ *is* $O((h^*)^{p+1})$ *at any fixed mesh-point in* $[0,X]$ *when* $0 < \theta_0 < \theta_1 < \ldots < \theta_p = 1$.

The asymptotic expansion for the local truncation errors $t_{m,j}(h^*)$ in (6.258) can be used to improve this result. We have the following result of Weiss (1972).

THEOREM 6.12. *Subject to the conditions of the preceding theorem,* $\xi_{m,j} = (h^*)^{p+1}e_j((m+\theta_j)h^*) + O((h^*)^{p+2})$, *where, for* $j = 0,1,2,\ldots,p$,

$$K(x,x)\sum_{k=0}^{p}\omega_k^{[j]}e_k(x)+\theta_j\int_0^x(\partial/\partial x)K(x,y)\sum_{k=0}^{p}\omega_k e_k(y)\,dy$$

$$= \quad \theta_j\Phi'(x)+\Psi_j(x)-\Psi_p(x)$$

and

$$\xi_{m,j} = f((m+\theta_j)h^*)-\tilde{f}((m+\theta_j)h^*) \ .$$

<u>Proof</u>. If $f((m+\theta_j)h^*)-(h^*)^{p+1}e_j((m+\theta_j)h^*)$ is substituted for $\tilde{f}((m+\theta_j)h^*)$ in equations defining the approximate solution $\tilde{f}(x)$, the

residual will be found to be $O((h^*)^{p+2})$. Using the lines of proof employed for the preceding theorem, the result follows.

At the time of writing, the existing theories for block-by-block methods defined by (6.158) with $\theta_0 = 0$ are somewhat involved, and the theory for the case $\theta_0 > 0$ is, by comparison, rather simple. Indeed, with $p = 1$ and $\theta_0 = 0$, $\theta_1 = 1$ the method defined by (6.158) amounts to the use of the trapezium rule, and a proof of Theorem 6.9 is, we note, relatively involved.

However, there is a theory, due to Weiss, for the case $\theta_0 = 0$. I shall state one of his results (see Weiss 1972).

THEOREM 6.13. *Suppose that* $K(x,y)$ *is* $p+5$ *times continuously differentiable for* $0 \le y \le x+\delta$, $0 \le x \le X$, *where* $\delta > 0$ *is a constant, and suppose that* $(\partial/\partial x)^{p+6}K(x,y)$ *is continuous for* $0 \le y \le x \le X$ *and* $(d/dx)^{p+6}g(x)$ *is continuous for* $0 \le x \le X$. *Suppose also that* $K(x,x) \ne 0$ *for* $0 \le x \le X$ *and the solution* $f(x)$ *of the equation of the first kind exists. Then the scheme defined by eqns* (6.158) *with* $\theta_0 = 0$ *is convergent if and only if*

$$-1 \le \eta = \{ \prod_{k=0}^{p-1} (1-\theta_k) / \prod_{k=1}^{p}(-\theta_k)\} \le 1$$

and the order of convergence is p *if* $\eta = 1$, *and* $p+1$ *otherwise.*

The dominant component in the error of the computed solution is investigated by Weiss, who concludes that the schemes with $|\eta| < 1$ have satisfactory stability properties, but that schemes with $\eta = 1$ are *less satisfactory*.

We shall now consider product-integration methods for Volterra equations of the first kind.

6.17. *Theory of product-integration methods for first kind equations*

A general and complete theory for product integration methods for Volterra equations of the first kind does not appear to exist at the time of writing. Rigorous theories for classes of equations with weakly singular kernels appear to exist only for the case of the generalized mid-point or generalized trapezium rules (or the generalized Euler method†) applied to (6.165). Whilst the case where $K(x,y) = H(x,y) / (x-y)^{\alpha}$ $(0 < \alpha < 1)$ appears to be tractable, the equation where $K(x,y) = H(x,y)/(x^2-y^2)^{\alpha}$ $(0 \le y \le x)$ has, at present, been satisfactorily treated only in the case where $H(x,y) = 1$ (see Atkinson 1971*a*).

We shall consider here the use of the generalized mid-point rule to give an approximate solution satisfying the equations

$$\sum_{j=0}^{i-1} V_{i-j} H(ih,(j+\tfrac{1}{2})h)\tilde{f}((j+\tfrac{1}{2})h) = g(ih) \quad (i = 1,2,3,\ldots,N), \tag{6.264}$$

where $Nh = X$, and

$$V_r = \int_{(r-1)h}^{rh} t^{-\alpha}\mathrm{d}t =$$

$$= h^{1-\alpha} \{r^{(1-\alpha)}-(r-1)^{(1-\alpha)}\}/(1-\alpha) .$$

We shall consider the convergence of the approximate solution.

If $\tilde{f}(x)$ is replaced by $f(x)$ in the eqn (6.264) we must replace $g(ih)$ by $g(ih)+t_i(h)$, where $t_i(h)$ is the local error and $t_i(h) = \sum_{j=0}^{i-1} \tau_{ij}$, with

$$\tau_{ij} = \int_{jh}^{(j+1)h} \frac{H(ih,(j+\frac{1}{2})h)f((j+\frac{1}{2})h)-H(ih,y)f(y)}{(ih-y)^{\alpha}}\,\mathrm{d}y .$$

If we write $f((j+\frac{1}{2})h)-\tilde{f}((j+\frac{1}{2})h) = \xi_j$ then, clearly,

$$\sum_{j=0}^{i-1} H(ih,(j+\tfrac{1}{2})h)V_{i-j}\xi_j = t_i(h) . \tag{6.265}$$

As in earlier discussions on equations of the first kind, we difference successive equations governing the errors. We here difference

† The generalized Euler method is based on the analogue of the left end-point rectangular rule in product integration.

eqn (6.265) and we now find that

$$H((i+1)h,(i+\tfrac{1}{2})h)V_1\xi_i + \sum_{j=0}^{i-1} H((i+1)h,(j+\tfrac{1}{2})h)(V_{i+1-j}-V_{i-j})\xi_j +$$

$$\sum_{j=0}^{i-1} \{\Delta_x H(ih,(j+\tfrac{1}{2})h)\} V_{i-j}\,\xi_j = \Delta t_i(h) \ . \qquad (6.266)$$

Let us now suppose that $|H(x,x)| \geq 2\sigma > 0$ for $0 \leq x \leq X$, and that $(\partial/\partial x)(\partial/\partial y)^2 H(x,y)$ and $f''(x)$ are continuous for $0 \leq x \leq X$, $0 \leq y \leq x$. (These conditions may be relaxed.) We suppose that $|(\partial/\partial y)H(x,y)| \leq L$ and $|(\partial/\partial x)H(x,y)| \leq L$ for $0 \leq y \leq x \leq X$, and h_0 is sufficiently small that $h_0 L < \sigma$, and $h \leq h_0$. Then $H((i+1)h,(i+\tfrac{1}{2})h) \neq 0$, and

$$\xi_i = \sum_{j=0}^{i-1} A_{i,j}\xi_j + \gamma_i \qquad (6.267)$$

where, from (6.266),

$$A_{i,j} = a_{i-j}\frac{H((i+1)h,(i+\tfrac{1}{2})h)+\{H((i+1)h,(j+\tfrac{1}{2})h)-H((i+1)h,(i+\tfrac{1}{2})h)\}}{H((i+1)h,(i+\tfrac{1}{2})h)} -$$

$$- \{(i-j)^{1-\alpha}-(i-j-1)^{1-\alpha}\}\Delta_x H(ih,(j+\tfrac{1}{2})h)/H((i+1)h,(i+\tfrac{1}{2})h) \ ,$$

with $a_k = \{V_k - V_{k+1}\}/V_1$, and where

$$\gamma_i = (1-\alpha)\Delta t_i(h)/\{H((i+1)h,(i+\tfrac{1}{2})h)h^{1-\alpha}\} \ .$$

It now follows that, for h_0 sufficiently small,

$$|\xi_i| \leq \sum_{j=0}^{i-1} b_{i-j}|\xi_j| + |\gamma_i| \quad (i = 1,2,3,\ldots,(N-1)) \ ,$$

where,

$$0 < b_{i-j} = a_{i-j}(1+2hL(i-j)/\sigma) + (2hL/\sigma)\{(i-j)^{1-\alpha}-(i-j-1)^{1-\alpha}\}$$

when $h \leq h_0$. (This we have assumed.)

The analysis of Weiss (1972), which we follow in outline here, establishes the behaviour of ξ_i from (6.267) together with a knowledge of the behaviour of $|\gamma_i|$. The analysis is unfortunately somewhat involved. Reference may be made to the thesis of Weiss for further details.

We have

$$|\gamma_i| \le \left|\frac{1-\alpha}{H((i+1)h,(i+\tfrac{1}{2})h)\,h^{1-\alpha}} \int_{ih}^{(i+1)h} \frac{1}{((i+1)h-y)^{\alpha}}\{\psi_i(y)-\psi_i((i+\tfrac{1}{2})h)\}dy\right.$$

$$\left. - \sum_{j=0}^{i-1}\int_{jh}^{(j+1)h}\{\frac{1}{(ih-y)^{\alpha}} - \frac{1}{((i+1)h-y)^{\alpha}}\}\{\psi_i(y)-\psi_i((j+\tfrac{1}{2})h)\}dy\right| +$$

$$+ \left|\frac{1-\alpha}{H((i+1)h,(i+\tfrac{1}{2})h)h^{1-\alpha}}\sum_{j=0}^{i-1}\int_{jh}^{(j+1)h}\frac{1}{(ih-y)^{\alpha}}\{\phi_i(y)-\phi_i((j+\tfrac{1}{2})h)\}dy\right| ,$$

where $\psi_i(y) = H((i+1)h,y)f(y)$ and $\phi_i(y) = \Delta_x H(ih,y)f(y)$; and from this it may be deduced (Weiss 1972) that

$$|\gamma_i| \le Chi^{-\alpha} \quad (i = 1,2,3,\ldots,(N-1))$$

for some constant C .

To deduce the behaviour of the values ξ_i we proceed as follows. We shall use the following lemma, which can be established by induction.

LEMMA 6.12. *If* $|\xi_0| \le \gamma_0$ *and*

$$|\xi_i| \le \sum_{j=0}^{i-1}|\beta_{ij}||\xi_j|+|\gamma_i| \quad (i = 1,2,3,\ldots) ,$$

and where

$$\rho_i = 1 - \sum_{j=0}^{i-1}|\beta_{ij}| > 0$$

then $|\xi_i| \le \Gamma$ *if* $|\gamma_0| \le \Gamma$ *and* $|\gamma_i| \le \Gamma\rho_i$ $(i = 1,2,3,\ldots)$.

To apply this lemma to the present discussion, we first choose M giving $\eta = X/M$ such that $2\eta L(1+\alpha)/\sigma \le \frac{1}{2}(1-\alpha)$, and we consider the behaviour of the errors in each interval $[0,\eta)$, $[\eta,2\eta)$, ..., $[(M-1)\eta,X]$ separately. For values of $i = 1,2,3,\ldots,(r-1)$ with $0 < (i+\frac{1}{2})h < \eta$

we have

$$|\xi_0| \le Ch ,$$

$$|\xi_i| \le \sum_{j=0}^{i-1} b_{i-j}|\xi_j| + Chi^{-\alpha} ,$$

and on calculating $\sum_{r=1}^{i} a_r$, $\sum_{r=1}^{i} ra_r$ and $\sum_{r=1}^{i} |r^{1-\alpha}-(r-1)^{1-\alpha}|$ we obtain

$$\sum_{j=0}^{i-1} b_{i-j} \le 1-(1-\alpha)i^{-\alpha}+\tfrac{1}{2}(1-\alpha)\alpha i^{-(1+\alpha)} +$$

$$+ (h/\eta)\{2\eta L(1+\alpha)/\sigma\}\left[i^{1-\alpha}+\{\alpha(1-\alpha)i^{-\alpha}/2(1+\alpha)\}\right] .$$

Since $ih < \eta$ and η has been chosen as indicated above,

$$\sum_{j=0}^{i-1} b_{i-j} \le 1-\tfrac{1}{2}(1-\alpha)^2|1-\alpha/\{2(1+\alpha)\}|i^{-\alpha}$$

for the values of i considered and, then,

$$\rho_i = 1-\sum_{j=0}^{i-1} b_{i-j} \ge C_1 i^{-\alpha} > 0 ,$$

where C_1 is some constant. Applying Lemma 6.12, $|\xi_i| \le (C/C_1)h$ if $(i+\tfrac{1}{2})h \in [0,\eta]$. The analysis now proceeds in a block-by-block fashion, over the finite number (M) of subintervals of length η . Thus for convergence at points in $[\eta,2\eta)$

$$|\xi_{r_1}| \le \mu_0 + Ch$$

$$|\xi_{r_1+i}| \le \sum_{j=0}^{i-1} b_{i-j}|\xi_{r_1+j}| + \mu_i + Chi^{-\alpha} ,$$

where

$$\mu_i = \sum_{j=1}^{r_1} b_{i+j}|\xi_{r_1-j}|$$

and hence $|\mu_i| \leq Dhi^{-\alpha}$ $(i = 1,2,3,\ldots,(r_2-1))$ corresponding to points $(r_1+i+\frac{1}{2})h \in [\eta,2\eta)$. Thus

$$|\xi_{r_1+i}| \leq \sum_{j=0}^{i-1} b_{i-j}|\xi_{r_1+j}| + (D+C)hi^{-\alpha} ,$$

and on applying Lemma 6.12,

$$|\xi_{r_1+i}| \leq \{(D+C)/C_1\}h .$$

Since there are a finite number of subintervals of length η, the process establishes $O(h)$ convergence of the mid-point method at all mesh-points in $[0,X]$.

There are conditions under which an improved rate of convergence can be established. Weiss (1972) gives the following result.

THEOREM 6.14. *If* $f'(x)$ *is Lipschitz-continuous on* $[0,X]$ *and* $f(0) = f'(0) = 0$, *and if* $(\partial/\partial x)H(x,y),(\partial/\partial y)H(x,y),(\partial^2/\partial x\partial y)H(x,y)$, *and* $(\partial^2/\partial y^2)H(x,y)$ *are Lipschitz-continuous with respect to* x *and* y *on* $0 \leq y \leq x \leq X$, *then the generalized mid-point method is convergent of order* $2-\alpha$.

The proof of this theorem is similar to the analysis given here, but more attention must be paid to the size of γ_i . Compare Theorem 6.14 with the experimental results of Example 6.59.

The form of analysis adopted in these proofs is conceptually simple, but the manipulation required is involved (see Weiss (1972) for full details). It is to be hoped that a more powerful theory could be found which would simplify the theory and give greater insight into the performance of various numerical methods for a class of problems which are undoubtedly of some importance. (Since writing this, we have been referred to the thesis of Benson.)

REFERENCES

In the following references, the titles of papers or books appearing in Russian have been translated into English. Where possible, I have referenced English language translations of such entries, rather than Russian originals. A number of titles which have been published since the body of the text was prepared have been included in these references.

AHLBERG, J.H., NILSON, E.N., and WALSH, J.L. (1967). *The theory of splines and their applications*. Academic Press, New York and London.

AITKEN, A.C. (1964). *Determinants and matrices* (9th edn). Oliver and Boyd, Edinburgh.

ALBERT, G.E. (1956). *A general theory of stochastic estimates of the Neumann series for the solutions of certain Fredholm integral equations and related series*. Symposium on Monte Carlo Methods, University of Florida (1954). Wiley, New York and London.

ALDER, B., FERNBACH, S., and ROTENBERG, M. (1963). *Methods in computational physics*, Vol. 1 - Stastical physics. Academic Press, New York and London.

ANDERSON, D.G. (1965). *Iterative procedures for nonlinear integral equations*. J. Ass. comput. Mach. 12,547- 60.

ANDERSSEN, A.S. and WHITE, E.T. (1971). *Improved numerical methods for Volterra integral equations of the first kind*. Comput. J. 14 , 442-3.

ANDERSSEN, R.S. (1976). *Stable procedures for the inversion of Abel's equation*. J. Inst. Math. & its Appl. 17,329-42.

ANSELONE, P.M. (ed.) (1964). *Nonlinear integral equations*. University of Wisconsin Press, Madison.

— (1965). *Convergence and error bounds for approximate solutions of integral and operator equations*. In Error in digital computation (ed. L.B. Rall), Vol. 2, pp. 231-52, Wiley, New York.

— (1967*a*). *Uniform approximation theory for integral equations with discontinuous kernels*. SIAM J. numer. Anal. (Soc. ind. appl. Math.) 4,245-353.

— (1967*b*). *Perturbation of collectively compact operators*. MRC tech. Summ. Rep. 726.

— (1967*c*). *Collectively compact and totally bounded sets of linear operators*. MRC tech. Summ. Rep. 766. (See also J. Math. Mech. 17,613-22 (1968).)

— (1971). *Collectively compact operator approximation theory, and applications to integral equations*. Prentice-Hall, Englewood Cliffs.

— and GONZALEZ-FERNANDEZ, J.M. (1965). *Uniformly convergent approximate solutions of Fredholm's integral equations*. J. math. Analysis Applic. 10,519-36.

— and MOORE, R.H. (1964). *Approximate solutions of integral and operator equations*. J. math. Analysis Applic. 9,268-77.

— and PALMER, T.W. (1967*a*). *Collectively compact sets of linear operators*. MRC tech. Summ. Rep. 740. (See also Pacif. J. Math. 25,417-22 (1968).)

— — (1967*b*). *Spectral analysis of collectively compact strongly convergent operator sequences*. MRC tech.

Summ. Rep. 741. (See also Pacif. J. Math. 25,423-31 (1968).)

- - (1967*c*). *Spectral properties of collectively compact sets of linear operators.* MRC tech. Summ. Rep. 767. (See also J. Math. Mech. 17,853-60 (1968).)

- and RALL, L.B. (1968). *The solution of characteristic value-vector problems by Newton's method.* Num. Math. 11,38-45.

APARO, E. (1959). *Sulla risoluzione numerica delle equazioni integrali di Volterra di seconda specie.* Atti Accad. naz. Lincei RC. 26,183-8.

APOSTOL, T.M. (1957). *Mathematical analysis: a modern approach to advanced calculus.* Addison-Wesley, Reading, Mass.

ARSENIN, V.J. (1965). *The discontinuous solutions of equations of the first kind.* USSR comput. Math. math. Phys. 5,202-9.

ARTHUR, D.W. (1973). *The solution of Fredholm integral equations using spline functions.* J. Inst. Math & its Appl. 11,121-9.

ATKINSON, K.E. (1966). *Extension of the Nyström method for the numerical solution of linear integral equations of the second kind.* Ph.D. Thesis, University of Wisconsin. (See also MRC tech. Summ. Rep. 686.)

- (1967). *The numerical solution of Fredholm integral equations of the second kind.* SIAM J. numer. Anal. (Soc. ind. appl. Math.) 4,337-48.

- (1969). *The numerical solution of integral equations on the half-line.* SIAM J. numer. Anal. (Soc. ind. appl. Math.) 6,375-97.

– (1971*a*). *The numerical solution of an Abel integral equation by a product trapezoidal method.* Report, Computer Centre, The Australian National University , Canberra.

– (1971*b*). *A survey of numerical methods for the solution of Fredholm integral equations of the second kind.* Proceedings of the symposium on the numerical solution of integral equations with physical applications.- Society for Industrial and Applied Mathematics. (See also Atkinson (1976).)

– (1972*a*). *The numerical solution of Fredholm integral equations of the second kind with singular kernels.* Report, Computer Center, Australian National University, Canberra.

– (1972*b*). *The numerical evaluation of the Cauchy transform on a simple closed curve.* SIAM J. numer. Anal. (Soc. ind. appl. Math.) 9, 284-99.

– (1973*a*). *The numerical evaluation of fixed points for completely continuous operators.* SIAM J. numer. Anal. (Soc. ind. appl. Math.) 10, 797-807.

– (1973*b*). *Iterative variants of the Nyström method for the numerical solution of integral equations.* Num. Math. 22 , 17-31.

– (1975). *Convergence rates for approximate eigenvalues of compact integral operators.* SIAM J. numer. Anal. (Soc. ind. appl. Math.) 12, 213-22.

– (1976). *A survey of numerical methods for the solution of Fredholm integral equations of the second kind.* Society for Industrial and Applied Mathematics, Philadelphia.

– (1976). *An automatic program for linear Fredholm integral equations of the second kind.* A.C.M. Trans. math. Software (Ass. comput. Mach.) 2,154-71.

BAKER, C.T.H. (1968). *On the nature of certain quadrature formulas and their errors.* SIAM J. numer. Anal. (Soc. ind. appl. Math.) 5,783-804.

– (1970). *The error in polynomial interpolation.* Num. Math. 15,315-9.

– (1971). *The deferred approach to the limit for eigenvalues of integral equations.* SIAM J. numer. Anal. (Soc. ind. appl. Math.) 8,1-10.

– , FOX, L., WRIGHT, K., and MAYERS, D.F. (1964). *Numerical solution of integral equations of the first kind.* Comput. J. 7,141-8.

– and HODGSON, G.S. (1971). *Asymptotic expansion for integration formulas in one or more dimensions.* SIAM J. numer. Anal. (Soc. ind. appl. Math.) 8,473-80.

– and KEECH, M.S. (1975). *Stability regions for numerical methods for Volterra integral equations.* Numerical Analysis Rep. 12, Dept. of Mathematics, University of Manchester.

– – (1977). *On the instability of a certain Runge-Kutta procedure for a Volterra integral equation.* Numerical Analysis Rep. 21, Dept. of Mathematics, University of Manchester. To appear in A.C.M. Trans.math.Software 1978/79.

– – (to appear) *Stability regions in the numerical treatment of Volterra integral equations.* SIAM J. numer. Anal. (Soc. ind. appl. Math.) 15, 394-417.

– and RADCLIFFE, P.A. (1970). *Error bounds for some Chebyshev methods of approximation and integration.* SIAM J. numer. Anal. (Soc. ind. appl. Math.) 7,317-27.

BAKUSHINSKII , A.B. (1965). *A numerical method for solving Fredholm integral equations of the first kind.* USSR comput. Math. math. Phys. 5,226-33.

– (1968). *Some properties of regularizing algorithms.* USSR comput. Math. math. Phys. 8,254-9.

– and STRAKHOV, V.N. (1968). *The solution of some integral equations of the first kind by the method of successive approximation.* USSR comput. Math. math. Phys. 8,250-6.

BARRODALE, I. and ROBERTS, F.D.K. (1972). *Solution of an overdetermined system of equations in the* l_1*-norm.* Report No. 69, University of British Columbia, Victoria.

– – (1973). *An improved algorithm for discrete* l_1*-linear approximation.* SIAM J. numer. Anal.(Soc. ind. appl. Math.) 10, 843- 8.

– and YOUNG, A. (1970). *Computational experience in solving linear operator equations using the Chebyshev norm.* In Numerical approximations to functions and data, pp.115-42. The Athlone Press, London.

BARTELS, R.H. (1971). *A stabilization of the simplex method.* Num. Math. 16,414-34.

– and GOLUB, G.H. (1969). *The simplex method of linear programming using LU decomposition.* Communs Ass. comput. Mach. 12,266-8; 275-8.

BATEMAN, H. (1922). *Numerical solution of linear integral equations.*

Proc. R. Soc. 100,441-9.

BAUER, F.L., RUTISHAUSER, H., and STIEFEL, E. (1963). *New aspects in numerical quadrature.* Proc. Symp. appl. Math. 15,119-218.

BELTYUKOV, B.A. (1964). *Approximate solution of Fredholm integral equations by Bateman's method.* Uchen. Zap. Irkutskogu ped. Inst. 20,177-84. (In Russian.)

- (1965). *The connection between the coincidence method and the method of substitution of a degenerate kernel.* USSR comput. Math. math. Phys. 5,238-41.

- (1965). *An analogue of the Runge-Kutta methods for the solution of a nonlinear integral equation of the Volterra type.* Different. Equations 1,417-33. (Translation from Russian.)

- (1966). *Solution of nonlinear integral equations by Newton's method.* Different. Equations 2,555-60. (Translation from Russian.)

BENNETT, J.H. (1964). *Integral equation methods for transport problems. Some numerical results.* Num. Math. 6,49-54.

BENSON, M.T. (1973). *Errors in numerical quadrature for certain singular integrands, and the numerical solution of Abel's integral equation.* Ph.D. thesis, Mathematics Department, University of Wisconsin-Madison.

BERG, L. (1955). *Lösungsverfahren für singuläre Integralgleichungen. I.* Math. Nachr. 14, 193-212.

BERNIER, I. (1945). *Les principales méthodes de résolution numérique des équations intégrales de Fredholm et de Volterra.* Annls Radioléct. 1, 311- 8.

BJÖRCK, A. (1967*a*). *Solving linear least squares problems by Gram-Schmidt orthogonalisation.* Nord. Tidskr. Inf.-behandl. 7, 1-21.

— (1967*b*). *Iterative refinement of least squares solutions.* Nord. Tidskr. Inf.-behandl. 7,257-78; 8,8-30.

— and GOLUB, G.H. (1967). *Iterative refinement of least-squares solution by Householder transformations.* Nord. Tidskr. Inf.- behandl. 7,322-37.

BÔCHER, M. (1912). *An introduction of the study of integral equations* (2nd edn). Cambridge University Press, Cambridge.

BOWDLER, H.J., MARTIN, R.S., PETERS, G., and WILKINSON, J.H. (1966). *Solution of real and complex systems of linear equations.* Num. Math. 8,217-34.

— — , REINSCH, C., and WILKINSON, J.H. (1968). *The QR and QL algorithms for symmetric matrices.* Num. Math. 11,292-306.

BOWNDS, J.M. (1976). *On solving weakly-singular Volterra equations of the first kind with Galerkin approximations.* Maths Comput. 30,747-57.

BRAKHAGE, H. (1960). *Über die numerische Behandlung von Integralgleichungen nach der Quadraturformelmethode.* Num. Math. 2,183-96.

— (1961*a*). *Zur Fehlerabschätzung für die numerische Eigenwertbestimmung bei Integralgleichungen.* Num. Math. 3,174- 9.

— (1961*b*). *Bemerkungen zur numerischen Behandlung und Fehlerabschätzung bei singulären Integralgleichungen.* Z. angew. Math. Mech. 41,T12-T14.

BRUNNER, H. (1973*a*). *The solution of Volterra integral equations of the first kind by piecewise polynomials.* J. Inst. Math. & its Appl. 12, 295-302.

- (1973*b*). *On the numerical solution of a class of Abel integral equations.* Proceedings of the 3rd Manitoba Conference on numerical mathematics, pp. 105-22, Utilitas Math., Winnipeg.

- (1974). *Global solution of the generalized Abel equation by implicit interpolation.* Maths Comput. 28, 61-8.

- (1975*a*). *On the approximate solution of the Abel integral equation with discontinuous solution.* Nord. Tidskr. Inf.-behandl. 15, 136-43.

- (1975*b*). *Projection methods for the approximate solution of integral equations of the first kind.* Proceedings of the 5th Manitoba Conference on numerical mathematics, pp. 3-23, Winnipeg.

BÜCKNER, H.F. (1949). *Ein unbeschränkt anwendbares Iterationsverfahren für Fredholmsche Integralgleichungen.* Math. Nachr. 27, 304-13.

- (1950). *Konvergenzuntersuchungen bei einem algebraischen Verfahren zur näherungsweisen Lösung von Integralgleichungen.* Math. Nachr. 3, 358-72.

- (1952). *Die praktische Behandlung von Integralgleichungen.* Springer-Verlag, Berlin.

- (1962). *Numerical methods for integral equations.* In Survey of numerical analysis (ed. J. Todd) pp. 439-67. McGraw-Hill, New York.

BULIRSCH, R. and STOER, J. (1966). *Asymptotic upper and lower bounds for results of extrapolation methods.* Num. Math. 8, 93-101.

BUSINGER, P. and GOLUB, G.H. (1965). *Linear least squares solutions by Householder transformations.* Num. Math. 7, 269-76.

CAMPBELL, G.M. (1970). *On the numerical solution of Volterra integral equations.* Ph.D. Thesis, Pennsylvania State University, U.S.A.

- and DAY, J.T. (1970). *An empirical method for the numerical solution of Volterra integral equations of the first kind.* (Manuscript, Pennsylvania State University, U.S.A.)

- - (1971). *A block-by-block method for the numerical solution of Volterra integral equations.* Nord. Tidskr. Inf.-behandl. 11, 120-4.

CHASE, S.M. and FOSDICK, L.D. (1969). *An algorithm for Filon quadrature.* Communs Ass. comput. Mach. 12, 453-8.

CHENEY, E.W. (1966). *Introduction to approximation theory.* McGraw-Hill, New York.

CLENSHAW, C.W. and CURTIS, A.R. (1960). *A method for numerical integration on an automatic computer.* Num. Math. 2, 197-205.

C.N.R.S. (1970). *Procédures ALGOL en analyse numérique,* Vol. 2, Centre National de la Récherche Scientifique, Paris.

COCHRAN, J.A. (1965). *The existence of eigenvalues for the integral equations of laser theory.* Bell Syst. tech. J. 44, 77-88.

- (1972). *Analysis of linear integral equations.* McGraw-Hill, New York. (Also circulated privately as a manuscript (1971).)

COLDRICK, D.B. (1972). *Methods for the numerical solution of integral equations of the second kind.* Ph.D. Thesis, University of Toronto (also University of Toronto Computer Science Technical Report No.45).

COLLATZ, L. (1940). *Schrittweise Näherungen bei Integralgleichungen und Eigenwertschranken.* Math. Z. 46,692-708.

- (1941). *Einschliessungssatz für die Eigenwerte von Integralgleichungen.* Math. Z. 47,395-8.

- (1966*a*). *The numerical treatment of differential equations* (transl. from German by P. Williams) (3rd edn). Springer-Verlag, Berlin.

- (1966*b*). *Functional analysis and numerical mathematics* (transl. from German by H. Oser). Academic Press, New York.

COURANT, R. and HILBERT, D. (1953). *Methods of mathematical physics* Vol. 1. Interscience, New York.

COX, M.G. (1972). *The numerical evaluation of B-splines.* J.Inst. Maths & its Appl. 10, 134-49.

CROUT, P.D. (1940). *An application of polynomial approximation to the solution of integral equations arising in physical problems.* J. math. Phys. 19,34-92.

- and HILDEBRAND, F. (1941). *A least squares procedure for solving integral equations by approximation.* J. math. Phys. 20,310-35.

CRYER, C.W. (1967). *On the calculation of the largest eigenvalue of an integral equation.* Num. Math. 10,165-76. (See also MRC tech. Summ. Rep. 704.)

CUTKOSKY, R.E. (1951). *A monte Carlo method for solving a class of integral equations*. J. Res. natn. Bur. Stand. B 47,113-15.

DANTZIG, G.B. (1963). *Linear programming and extensions*. Princeton University Press, New Jersey.

DAVIS, A.M. (1970). *Waves in the presence of an infinite dock with gap*. J. Inst. Math. & its Appl. 6, 141-56.

DAVIS, H.T. (1962). *Introduction to nonlinear differential and integral equations*. Dover, New York. (Originally published by United States Atomic Energy Commission (1960).)

DAVIS, P.J. (1965). *Interpolation and approximation*. (2nd edn). Blaisdell, New York and Toronto.

— and RABINOWITZ, P. (1967). *Numerical integration*. Blaisdell, Waltham, Massachussetts.

— — (1975). *Methods of numerical integration*. Academic Press, New York.

DAY, J.T. (1967). *A starting method for solving nonlinear Volterra integral equations*. Maths Comput. 21,179-88.

— (1968). *On the numerical solution of Volterra integral equations*. Nord. Tidskr. Inf.-behandl. 8,134-7.

de BOOR, C.R. (1966). *The method of projections as applied to the numerical solution of two-part boundary value problems*. Ph.D. Thesis, University of Michigan, Ann Arbor, Michigan.

— (1972). *On calculating with B-splines*. J. Approx. Theory 6,50-62.

de FIGUEIREDO, R.J.P., and NETRAVALI, A.N. (1975). *Spline approximation to the solution of the linear Fredholm integral equation of the second kind*. MRC tech. Summ. Rep. 1343.

de HOOG, F. and WEISS, R. (1971). *High order methods for first kind Volterra integral equations*. Report, Computer Centre, The Australian National University.

— — (1973*a*). *High order methods for Volterra integral equations of the first kind*. SIAM J. numer. Anal. (Soc. ind. appl. Math.) 10, 647-60. (Submitted 1972.)

— — (1973*b*). *Asymptotic expansions for product integration*. Maths Comput. 27, 295-306.

— — (1973*c*) *On the solution of Volterra integral equations of the first kind*. Num . Math. 21, 22-32.

— — (1974). *High order methods for a class of Volterra integral equations with weakly singular kernels*. SIAM J. numer. Anal. (Soc. ind. appl. Math.) 11, 1166-80.

— — (1975). *Implicit Runge-Kutta methods for second kind Volterra integral equations*. Num. Math. 23 , 199-214. (Submitted 1972.)

DEJON, B. and WALTHER, A. (1960). *General report on the numerical treatment of integral and integro-differential equations*. <u>In</u> Symposium on the numerical treatment of ordinary differential equations, integral equations, and integro-differential equations, pp. 647-71. Birkhauser-Verlag, Basel.

de la VALLEE POUSSIN, C.J. (1925). *On the approximation of functions of a real variable and on quasi-analytic functions*. Rice Inst. Pamph. XII 2,105.

DELVES, L.M. (1974). *An automatic Ritz-Galerkin procedure for the numerical solution of linear Fredholm integral equations of the second kind.* (Manuscript, University of Liverpool.)

— and WALSH, J.E. (ed.) (1974). *Numerical solution of integral equations* (Proceedings of Liverpool-Manchester Summer School). Clarendon Press, Oxford.

DIXON, V.A. (1974). In *Software for numerical mathematics* (ed. D.J. Evans) Academic Press, London.

DOMBROVSKAJA, I.N. (1964). *The approximate solution of Fredholm integral equations of first kind.* Ural Gos. Univ. Mat. Zap. 4 Tetrad 4, 30-5. (Transl. from Russian by L.B. Rall: MRC tech. Summ. Rep. 652 (1966).)

DOUGLAS, J. (1960). *Mathematical programming and integral equations.* In Symposium on the numerical treatment of ordinary differential equations, integral equations, and integro-differential equations, pp.269-74, Birkhauser-Verlag, Basel.

DUNFORD, N. and SCHWARTZ, J.T. (1957). *Linear operators,* Vol. 1. Interscience, New York.

EDELS, H., HEARNE, K., and YOUNG, A. (1962). *Numerical solution of the Abel integral equation.* J. Math. Phys. 41, 62-75.

EINARSSON, B. (1971). *Numerical solution of Abel's integral equation with spline functions.* FOA 2 Rapport, C2455-11(25). Försvarets Forskningsanstalt, Avdelning, Stockholm.

ELDÉN, L. (1974). *Numerical methods for the regularization of Fredholm integral equations of the first kind.* Linköping University (Sweden) Mathematics Department Report LiH-MAT-R-1974-7.

EL-GENDI, S.E. (1969). *Chebyshev solution of differential, integral, and integro-differential equations.* Comput. J. 12, 282- 7.

ELLIOTT, D. (1960). *The numerical solution of integral equations using Chebyshev polynomials.* J. Aust. math. Soc. 1, 344-56.

- (1963). *A Chebyshev series method for the numerical solution of Fredholm integral equations.* Comput. J. 6,102-11.

- (1965). *Truncation errors in two Chebyshev series approximations.* Maths Comput. 19, 234-48.

- and WARNE, W.G. (1967). *An algorithm for the numerical solution of linear integral equations.* ICC Bull. (Int. Comput. Cent.) 6, 207-24.

EL TOM, M.E.A. (1971). *Application of spline functions to Volterra integral equations.* J. Inst. Math. & its Appl. 8, 354-7.

- (1973). *Numerical solution of Volterra integral equations by spline functions.* Nord. Tidskr. Inf.-behandl. 13, 1-7.

- (1974*a*). *On the numerical stability of spline function approximations to solutions of Volterra integral equations of the second kind.* Nord. Tidskr. Inf.-behandl. 14, 136-43.

- (1974*b*). *On spline function approximations to the solution of Volterra integral equations of the first kind.* Nord. Tidskr. Inf.-behandl. 14,288-97.

- (1976). *Applications of spline functions to systems of Volterra integral equations of the first and second kinds.* J. Inst. Math. & its Appl. 17, 295-310. (Manuscript, CERN Geneva 1974*c*.)

ERDOGAN, F. (1969). *Approximate solutions of systems of singular integral equations.* SIAM J. appl. Math. (Soc. ind. appl. Math.) 17,1041-59.

EVANS, G.C. (1910). *Volterra's integral equation of the second kind with discontinuous kernel.* Trans. Am. math. Soc. 11, 393-413.

FANG WANG HAP (FAN WAN HAP) see HAP, FANG WANG.

FERRAR, W.L. (1958). *Integral claculus.* Clarendon Press, Oxford.

FETTIS, H.G. (1964). *On the numerical solution of equations of the Abel type.* Maths Comput. 18, 491-6.

FORSYTHE, G.E. (1967). *Today's computational methods of linear algebra.* SIAM Rev. (Soc. ind. appl. Math.) 9,489- (Stanford University Computer Science Technical Report No. CS46 (1966)).

- and MOLER, C.B. (1967). *Computer solution of linear algebraic systems.* Prentice-Hall, Englewood Cliffs, New Jersey.

FOSDICK, L.D. (1968). *A special case of the Filon quadrature formula.* Math. Comput. 22,71-81.

FOX, L. (1957). *The numerical solution of two-point boundary value problems in ordinary differential equations.* Clarendon Press, Oxford.

- (ed.) (1962). *Numerical solution of ordinary and partial differential equations.* Pergamon Press, Oxford.

- (1964). *An introduction to numerical linear algebra.* Clarendon Press, Oxford.

- (1967). *Romberg integration for a class of singular integrands.* Comput. J. 10,87-93.

- and GOODWIN, E.T. (1953). *The numerical solution of nonsingular linear integral equations.* Phil. Trans. R. Soc. 245,501-34.

- and MAYERS, D.F. (1969). *Computing methods for scientists and engineers.* Clarendon Press, Oxford.
- and PARKER, I.B. (1968). *Chebyshev polynomials in numerical analysis.* Clarendon Press, Oxford.

FRANCIS, J.G.F. (1961-2). *The QR transformation.* Comput.J. 4,265-71;332-45

FRANKLIN, J.N. (1970). *Well-posed stochastic extensions of ill-posed linear problems.* J. math. Analysis its Appl. 31, 682-716.

- (1974). *On Tikhonov's method for ill-posed problems.* Maths Comput. 28, 889-907.

FRIE, W. (1963). *Zur Auswertung der Abelschen Integralgleichung.* Annln Phys. 10,332-9.

FRIEDLANDER, F.G. (1941). *The reflexion of sound pulses by convex parabolic reflectors.* Proc. camb. phil. Soc. 37,134-49.

FRÖBERG, C.E. (1970). *Introduction to numerical analysis* (2nd edn). Addison-Wesley, Reading, Massachusetts.

GABDULHAEV, B.G. (1968). *Approximate solution of singular integral equations by the method of mechanical quadratures.* Soviet Math. 9,329-32. (Translation from Russian.)

- (1970). *A direct method for solving integral equations.* Am. math. Soc. Transl. Ser. 2, 91,213-23. (Translation from Russian.)

GANADO, A.N. (1971). *Some numerical problems in integral equations.* D. Phil. thesis, University of Oxford, England.

GARABEDIAN, H.L. (ed.) (1965). *Approximation of functions.* Elsevier, Amsterdam.

GAREY, L. (1972). *Predictor-corrector methods for nonlinear Volterra integral equations of the second kind.* Nord. Tidskr. Inf.-behandl. 12, 325-33.

— (1975). *Solving nonlinear second kind Volterra equations by modified increment methods.* SIAM J. numer. Anal. (Soc. ind. appl. Math.) 12 501-8.

GARWICK, J.V. (1952). *On the numerical solution of integral equations.* In Proceedings of the 11th Scandinavian Mathematical Congress, Trondheim (1949), Tanums Forlog, Oslo.

GLADWIN, C.J. (1975). *Numerical solution of Volterra integral equations of the first kind.* Ph.D. thesis, Department of Mathematics, Dalhousie University, Nova Scotia.

— and JELTSCH, R. (1974). *Stability of quadrature methods for first kind Volterra integral equations.* Nord. Tidskr. Inf-behandl. 14, 144-51.

GODUNOV, S.K. and RYABENKI, V.S. (1964). *Theorey of difference schemes* (translated from Russian by E. Godfredsen). North - Holland, Amsterdam.

GOHBERG, I.C. and FELDMAN, I.A. (1965). *On a method of reduction for systems of equations of Wiener-Hopf type.* Soviet Math. 6, 1433-6.

GOLDENSTEIN, L.S. and GOHBERG, I.C. (1960). *On a multidimensional integral equation on a half-space, whose kernel is a function of the difference of the arguments, and on a discrete analogue of this equation.* Soviet Math. 1, 173-6.

GOLUB, G.H. (1965). *Numerical methods for solving least squares problems.* Num. Math. 7, 206-16.

— and KAHAN, W. (1964). *Calculating the singular values and pseudo-inverse of a matrix.* SIAM J. numer. Anal. (Soc. ind. appl. Math.) 2 205-24. (Also as Stanford Technical Report CS. 8.)

- and REINSCH, C. (1970). *Singular value decomposition and least-squares solutions.* Num. Math. 14,403-20. (Also as Stanford Technical Report CS. 133.)

GOOD, I.J. (1957). *On the numerical solution of integral equations.* Mathl Tabl. natn. Res. Coun. Wash. 11, 82-3.

GOURSAT, E. (1964). *A course in mathematical analysis,* Vol. III, Part 2 - Integral equations, calculus of variations. Dover, New York. First published by Gauthier-Villas, Paris (1923). (Translated from French 5th edn (1956).)

GRAM, C. (1964). *Definite integral by Romberg's method.* Nord. Tidskr. Inf.-behandl. 4,54-60.

GREEN, C.D. (1969). *Integral equation methods.* Nelson, London.

GROETSCH, C.W. (1975). *On existence criteria and approximate procedures for integral equations of the first kind.* Maths Comput. 29, 1105-8.

HADLEY, G. (1962). *Linear programming.* Addison-Wesley, Reading Massachusetts.

HALMOS, P.R. (1958). *Finite-dimensional vector spaces* (2nd edn). Van Nostrand, Princeton.

HÄMMERLIN, G. (1960). *Ein Verfahren zur numerischen Behandlung von homogenen Integralgleichungen 2. Art mit Fehlerabschätzung.* Z. angew. Math. Mech. 40,T12-T14 (also *ibid* 42, 439-463).

HAMMING, R.W. (1962). *Numerical methods for scientists and engineers.*

McGraw-Hill, New York.

HANSON, R. (1971). *A numerical method for solving Fredholm integral equations of the first kind, using singular values.* SIAM J. numer. Anal. (Soc. ind. appl. Math.) 8,616-22.

- (1972). *Integral equations of immunology.* Communs Ass. comput. Mach. 15,883-90.

HAP, FANG WANG (1965). *Approximate solution of singular integral equations.* USSR comput. Math. math. Phys. 5,1-19.

HAYES, J.G. (ed.) (1970). *Numerical approximations to functions and data.* The Athlone Press, London.

HAYES, J.K., KAHANER, D.K., and KELLNER, R.G. (1972). *An improved method for conformal mapping.* Maths Comput. 16,327-34.

- - - (1975). *A numerical comparison of integral equations of first and second kinds for conformal mapping.* Maths Comput. 29,512-21.

HENRICI, P. (1962*a*). *Lecture notes on elementary numerical analysis.* Wiley, New York.

- (1962*b*). *Discrete variable methods in ordinary differential equations.* Wiley, New York.

- (1962*c*). *Error propagation for difference methods.* Wiley, New York.

- (1964). *Elements of numerical analysis.* Wiley, New York.

HERBOLD, R.J. (1968). *Consistent quadrature schemes for the numerical solution of boundary value problems by variational techniques.* Doctoral thesis, Case Western Reserve University, Cleveland.

HERR, D.G. (1974). *On a statistical model of Strand and Westwater for the numerical solution of a Fredholm integral equation of the first kind.* J. Ass. comput. Mach. 21,1-5.

— (1976). *On Strand and Westwater's minimum-rms. estimation of the numerical solution of a Fredholm integral equation of the first kind.* SIAM J. numer. Anal. (Soc. ind. appl. Math.) 13 427-31.

HERTLING, J. (1971). *Numerical treatment of singular integral equations by interpolation methods.* Num. Math. 18,101-12.

HILDEBRAND, F.B. (1941). *The approximate solution of singular integral equations arising in engineering practice.* Proc. Am. Acad. Arts Sci. 74,287-95.

— (1965). *Methods of applied mathematics* (2nd edn). Prenctice-Hall, Englewood Cliffs, New Jersey.

HILGERS, J.W. (1974*a*). *Non-iterative methods for solving operator equations of the first kind.* Ph.D. thesis,University of Wisconsin (also MRC tech. Summ. Rep. 1413).

— (1974*b*). *Approximating the optimal regularization parameter.* MRC tech. Summ. Rep. 1472.

— (1976). *On the equivalence of regularization and certain reproducing kernel Hilbert space approaches for solving first kind problems.* SIAM J. numer. Anal. (Soc. ind. appl. Math.) 13,172-84.

H.M.S.O. (1961). *Modern computing methods* (2nd edn.) H.M.S.O., London.

HOLYHEAD, P.A.W. (1974). *Multistep methods for Volterra integral equations of the first kind.* Ph.D. thesis, University of Southampton.

— and McKEE, S. (1976). *Stability and convergence of multistep methods for linear Volterra integral equations of the first kind.* SIAM J. numer. Anal. (Soc. ind. appl. Math.) 13, 269-92.

— — and TAYLOR, P.J. (1975). *Multistep methods for solving linear Volterra integral equations of the first kind.* SIAM J. numer. Anal. (Soc. ind. appl. Math.) 12, 698-711.

HODGSON, G.S. (1969). *Numerical integration.* M.Sc. thesis, University of Manchester.

HOUSEHOLDER, A.S. (1964). *The theory of matrices in numerical analysis.* Blaisdell, New York.

HUDAK, Y.I. (KHUDAK, Y.I.) (1966). *On the regularization of solutions of integral equations of the first kind.* USSR comput. Math. math. Phys. 6, 217-21.

HUNG, HING-SUM (1970). *The numerical solution of differential and integral equations by spline functions.* MRC tech. Summ. Rep. 1053.

IKEBE, Y. (1970). *The Galerkin method for numerical solution of Fredholm integral equations of the second kind.* Report C.N.A.-5, Centre for Numerical Analysis, University of Texas at Austin.

— (1972). *The Galerkin method for the numerical solution of Fredholm integral equations of the second kind.* SIAM Rev. (Soc. ind. appl. Maths.) 14, 465-91.

— LYNN, M.S., and TIMLAKE, W.P. (1969). *The numerical solution of the integral equation formulation of the single interface Neumann problem.* SIAM J. numer. Anal. (Soc. ind. appl. Maths.)

6,334-46.

ISAACSON, E., and KELLER, H.B. (1966). *Analysis of numerical methods.* Wiley, New York.

IVANOV, V.V. (1956). *Approximate solution of singular integral equations when the integral is not taken along a closed contour.* Dokl. Akad. Nauk SSSR 111, 933-6. (In Russian.)

– (1956). *The approximate solution of singular integral equations.* Dokl. Akad. Nauk SSSR 110, 15-18. (In Russian.)

– (1961). *Approximate solution of singular integral equations.* Investigations in modern problems of the theory of functions of a complex variable. Fizmatgiz, Moscow. (In Russian.)

– (1963). *Method of approximate solution of systems of singular integral equations.* USSR Comput. Math. math. Phys. 3, 892-917.

– (1965). *Methods for the solution of singular integral equations.* Akademii Nauk SSSR institute Nauch Informacii. pp. 125-75. (In Russian.)

– (1976). *The theory of approximate methods and their application to the numerical solution of singular integral equations.* (Transl. A. Ideh.) Noordhoff, Leyden.

– and KARAGODOVA, E.A. (1961). *Approximate solution of singular integral equations of convolution type by Galerkin's method.* Ukr. mat. Zh. 13, 28-38. (In Russian.)

JAIN, M.K. and SHARMA, K.D. (1967). *Numerical solution of linear differential equations and Volterra integral equations using Lobatto quadrature formula.* Comput. J. 10, 101-7.

JONES, J.G. (1961). *On the numerical solution of convolution integral equations and systems of such equations.* Maths Comput. 15, 131-42.

JOYCE, D.C. (1971). *Survey of extrapolation processes in numerical analysis.* SIAM Rev. (Soc. ind. appl. Math.) 13, 435-90.

KADNER, H. (1960). *Optimale Kollokation für lineare Integralgleichungen 2. Art.* Z. angew. Math. Mech. 40, T16-T17.

- (1963). *Untersuchungen zu Quadraturformelmethoden für lineare Integralgleichungen 2. Art auf der Grundlage der Kollokation.* Wiss. Z. tech. Hochsch. (Univ.) Dresden, 12 119-20.

- (1967). *Die numerische Behandlung von Integralgleichungen nach der Kollokationsmethode.* Num. Math. 10, 241-61.

KAGIWADA. H. and KALABA, R.E. (1967). *Initial-value methods for the basic boundary value problems and integral equations of radiative transfer.* J. comput. Phys. 3, 322- 9.

- - and SCHUMITZKY, A. (1967). *An initial-value method for general Fredholm integral equations.* RAND Memorandum RM-5307-PR. Rand Corporation.

KAMMERER, W.J. and NASHED, M.Z. (1971). *A generalization of a matrix iterative method of G. Cimmiro to best approximate solution of linear integral equations of the first kind.* Atti Accad. naz. Lincei RC. VIII Ser., 51, 20-5.

KANTOROVITCH, L.V. (1934). *On the approximate computation of some types of definite integrals and other applications of the method of removing singularities.* Mat. Sb. 41, 235-45. (In Russian.)

- (1948). *Functional analysis and applied mathematics.*

Usp. mat. Nauk 3,89-185. (In Russian.)

- (1952). *Functional analysis and applied mathematics.* Rep. natn. Bur. Stand. 1509. (Transl. from Usp. mat. Nauk 3, 89-185.)

- and AKILOV, G.P. (1964). *Functional analysis in normed spaces* (transl. D. Brown). Pergamon, Oxford.

- and KRYLOV, V.I. (1958). *Approximate methods of higher analysis.* Fizmatgiz, Moscow.

- - (1964). *Approximate methods of higher analysis* (transl. from 3rd Russian edn by C.D. Benster). Interscience, New York; Noordhoff, Gröningen.

KAPPENKO, L.N. (1967). *Approximate solution of a singular integral equation by means of Jacobi polynomials.* J. appl. Math. Mech. 30,668-75. (Translation from Russian.)

KASPSHITSKAYA, M.F. and LUCHKA, A.Y. (1968). *The collocation method.* USSR Comput. Math. math. Phys. 8,19-39.

KATO, T. (1966). *Perturbation theory for linear operators.* Springer-Verlag, Berlin.

KEECH, M.S. (1974). *Some aspects of the numerical solution of Volterra integral equations of the second kind.* M.Sc. thesis, University of Manchester.

- (1976). *Semi-explicit, one-step methods in the numerical solution of non-singular Volterra integral equations of the first kind.* Numerical Analysis Rep. 19, Dept. of Mathematics, University of Manchester. (See also BIT, 17, 312-20(1977).)

KELLER, H.B. (1965). *On the accuracy of finite difference approximation to the eigenvalues of differential and integral operators.* Num. Math. 7, 412- 9.

- (1968). *Numerical methods for two-point boundary-value problems.* Blaisdell, Waltham, Massachusetts.

- and WENDROFF, B. (1957). *On the formulation and analysis of numerical methods for time-dependent transport equations.* Communs pure appl. Math. 10, 567-82.

KENT, G. and MAUTZ, J. (1969). *The numerical solution of a Volterra integral equation.* J. comput. Phys. 3, 399-415.

KERSHAW, D. (1963). *A numerical solution of an integral equation satisfied by the velocity distribution around a body of revolution in axial flow.* Report ARL/R1/Math 3.45-ARC23, 470. Published by HMSO (1963) as Ministry of Aviation R M No.3308.

- (1974). In *Numerical solution of integral equations* (eds. L.M.Delves and J.E. Walsh), pp.140-61. Clarendon Press, Oxford.

KHUDAK, Y.I. See HUDAK, Y.I.

KOBAYASI, M. (1966). *On numerical solution of the Volterra integral equations of the second kind by linear multistep methods.* Rep. statist. Applic. Res., Tokyo, 13, 1-21.

- (1967). *On numerical solution of the Volterra integral equations of the first kind by trapezoidal rule.* Rep. statist. Applic. Res., Tokyo, 14, 1-14.

KOLMOGOROV, V. and FOMIN, S. (1957). *Elements of the theory of functions and functional analysis.* Graylock Press, New York.

KOROVKIN, P.P. (1960). *Linear operators and approximation theory.* Hindustan, India. (Translation from Russian.)

KOWALEWSKI, G. (1930). *Integralgleichungen.* de Gruyter, West Berlin and Leipzig.

KRASNOSELSKII, M.A. (1950). *The convergence of the Galerkin method for non-linear equations.* Dokl. Akad. Nauk SSSR 73,1121-4. (In Russian.)

— (1963). *Topological methods in the theory of non-linear equations* (transl. from Russian by A. Armstrong). Pergamon Press, Oxford.

— VAINIKKO, G.M., ZABREIKO, P.P., RUTITSKII, Y.B., and STETSENKO, V.Y. (1972). *Approximate solution of operator equations* (transl. from Russian: D. Loyuish). Noordhoff, Gröningen.

KREIN, M.G. (1962). *Integral equations on a half-line with a kernel depending on the difference of the arguments.* Am. math. Soc. Transl. 22,163-288. (Translation from Russian.)

KREIN, S.G. (ed.) (1972). *Functional analysis* (transl. from Russian by R. Flaherty). Noordhoff, Gröningen.

KRYLOV, V.I. (1962). *Approximate calculation of integrals* (transl. from Russian by A. Stroud). Macmillan, New York.

KSCHWENDT, H. and HEMBD, H. (1971). *Konvergenzbetrachtungen bei einem Lösungsverfahren für Integralgleichungen.* Z. angew. Math. Mech. 51,T25-T27.

KUSSMAUL, R. and WERNER, P. (1967). *Fehlerabschätzungen für ein numerisches Verfahren zur Auflösung linearer Integralgleichungen mit schwachsingularen Kernen.* Computing 3, 22-46.

LAMBERT, J.D. (1973). *Computational methods in ordinary differential equations.* Wiley, London and New York.

LANDWEBER, L. (1951). *An iteration formula for Fredholm integral equations of the first kind.* Amer. J. Math. 73,615-24.

— (1974). *Axisymmetric potential flow in a circular tube.* J. Hydronautics, 8,137-45.

LAPIDUS, L. and SEINFELD, J. (1971). *Numerical solution of ordinary differential equations.* Academic Press, New York.

LAUDET, M. and OULÈS, H. (1960). *Sur l'intégration numérique des équations intégrales du type de Volterra.* In Symposium on the numerical treatment of ordinary differential equations, integral and integro-differential equations pp.117-21. Proceedings of the Rome Symposium organized by the Provisional International Computation Centre. Birkhauser-Verlag, Basel.

LAWSON, C.L. and HANSON, R.J. (1974). *Solving least-squares problems.* Prentice-Hall, Englewood Cliffs.

LAVRENTIEV, M.M. (1967). *Some improperly posed problems of mathematical physics.* Springer-Verlag, Berlin. (Translation from Russian.)

LINZ, P. (1967*a*). *The numerical solution of Volterra integral equations by finite difference methods.* MRC tech. Summ. Rep. 825.

— (1967*b*). *Applications of Abel transforms to the numerical solution of problems in electrostatics and elasticity.* MRC tech. Summ. Rep. 826.

— (1969*a*). *A method for solving nonlinear Volterra integral equations of the second kind.* Math. Comput. 23,595-600.

— (1969*b*). *Numerical methods for Volterra integral equations with* *singular kernels.* SIAM J. numer. Anal. (Soc. ind. appl. Math.) 6,365-74.

- (1970). *On the numerical computation of eigenvalues and eigenvectors of symmetric integral equations.* Maths Comput. 24, 905-9.

- (1971). *Product integration methods for Volterra integral equations of the first kind.* Nord. Tidskr. Inf.-behandl. 11,413-21.

LIUSTERNIK, L.A. and SOBOLEV, V.J. (1961). *Elements of functional analysis.* Ungar, New York. (Translation from Russian.)

LOGAN, J.E. (1976). *The approximate solution of Volterra integral equations of the second kind.* Ph.D. thesis, University of Iowa.

LONGMAN, I.M. (1958). *On the numerical evaluation of Cauchy principal values of integrals.* Mathl Tabl. natn. Res. Coun. Wash. 12, 205-7.

LONSETH, A.T. (1947). *The propagation of error in linear problems.* Trans. Am. math. Soc. 62,193-212.

- (1954). *Approximate solutions of Fredholm-type integral equations.* Bull. Am. math. Soc. 60,415-30.

LOVITT, W.V. (1924). *Linear integral equations.* McGraw-Hill, New York. Reissued by Dover, New York (1950).

LUCHKA, A.Y. (1965). *The method of averaging functional corrections-theory and applications.* Academic Press, New York. (Translation from Russian.)

LUKE, Y.L. (1963). *On the approximate solution of integral equations of the convolution type.* Midwest Research Institute, Report ARL 63-33.

LYNESS, J.N. (1965*a*). *Symmetric integration rules for hypercubes. I - Error coefficients.* Maths Comput. 19,260-76.

— (1965*b*). *Symmetric integration rules for hypercubes. II - Rule projection and rule extension.* Maths Comput. 19, 394-407.

— (1965*c*). *Symmetric integration rules for hypercubes III - Construction of integration rules using null rules.* Maths Comput. 19,625-37.

MAKINSON, G.J. and YOUNG, A. (1960). *The stability of solutions of differential and integral equations.* In Symposium on the numerical treatment of ordinary differential equations, integral and integro-differential equations, pp.499-509. Proceedings of the Rome Symposium organized by the Provisional International Computational Centre. Birkhauser-Verlag, Basel.

MARSH, T. and WADSWORTH, M. (1976). *An iterative method for the solution of Fredholm integral equations of the second kind.* J. Inst. Math. & its Appl. 18,57-66.

MAYERS, D.F. (1962). In *Numerical solution of ordinary and partial differential equations* (ed. L. Fox). Pergamon Press, Oxford.

MIKHLIN, S.G. (1964*a*). *Integral equations and their applications to some problems of mechanics, mathematical physics and engineering* (2nd edn). (Transl. from Russian by A.H. Armstrong.) Pergamon Press, Oxford. (1st edn 1957.)

— (1964*b*). *Variational methods in mathematical physics* (transl. from Russian: T. Boddington). Pergamon Press, Oxford.

- (1971). *The numerical performance of variational methods* (transl. from Russian by R.S. Anderssen). Noordhoff, Gröningen.

- and SMOLITSKY, K.L. (1967). *Approximate methods for solution of differential and integral equations.* American Elsevier, New York. (Translation from Russian.)

MILLER, G.F. (1956). *A note on the numerical solution of certain non-linear integral equations.* Proc. R. Soc. A 236, 529-34.

MINERBO, G.N. and LEVY, M.E. (1969). *Inversion of Abel's integral equation by means of orthogonal polynomials.* SIAM J. numer. Anal. (Soc. ind. appl. Math.) 6,598-616.

MOLER, C.B. and STEWART, G.W. (1973). *An algorithm for generalized matrix eigenvalue problems.* SIAM J. numer. Anal. (Soc. ind. appl. Math.) 10,241-56.

MORRIS, J.L. (ed.) (1969). *Conference on the numerical solution of differential equations.* Lecture notes in mathematics, No. 109. Springer-Verlag, Berlin.

MORSE, P.M. and FESHBACH, H. (1953). *Methods of theoretical physics,* Part 1. McGraw-Hill, New York.

MUSKHELISHVILI, N.I. (1953). *Singular integral equations* (transl. from Russian by J.R.M. Radok). Noordhoff, Gröningen.

MYSOVSKIH, I.P. (1956). *Estimation of error arising in the solution of an integral equation by the method of mechanical quadratures.* Vest. leningr. gos. Univ. 11,66-72. (In Russian.)

- (1957). *Computation of the eigenvalues of integral equations by means of iterated kernels.* Dokl. Akad. Nauk SSSR 115,45-8. (In Russian.)

– (1961). *Estimate of the error involved in the numerical solution of a linear integral equation.* Soviet Math. 2, 1268-70.

– (1963). *An error bound for the numerical solution of a non-linear integral equation.* Soviet Math. 4, 1603-7.

– (1964*a*). *On error bounds for approximate methods of estimation of eigenvalues of Hermitian kernels.* Am. math. Soc. Transl. 35, 237-50. (Translation from Russian.)

– (1964*b*). *On error bounds for eigenvalues calculated by replacing the kernel by an approximating kernel.* Am. math. Soc. Transl. 35, 251-62. (Translation from Russian.)

NAG (1973). *Numerical algorithms group: MINI-manual to the program library.* Oxford University Computing Laboratory.

NASHED, M.Z. (1974). *On moment-discretization and least-squares solutions of linear integral equations of the first kind.* MRC tech. Summ. Rep. 1371.

– (ed.) (1976). *Generalized inverses and applications.* Academic Press, New York.

– and WAHBA, G. (1974*a*). *Convergence rates of approximate least-squares solutions of linear integral and operator equations of the first kind.* Maths Comput. 28, 69-80.

– – (1974*b*). *Generalized inverses in reproducing kernel spaces: An approach to regularization of linear operator equations.* SIAM J. math. Anal. (Soc. ind. appl. Math.) 5, 974-87.

– – (1974*c*). *Regularization and approximation of linear operator equations in reproducing kernel spaces.* Bull.

Amer. math. Soc. 80 1213-8.

NATANSON, I.P. (1961). *Constructive theory of functions.* AEC-tr-4503 (i,ii), U.S. Atomic Energy Commission. (Translation from Russian.)

– (1965). *Constructive function theory,* Vol. I (1964), Vols. II, III (1965). Ungar, New York. (Translation from Russian.)

NIKOLSKII, S.M. (1966). *Quadrature formulae* (transl. from Russian by J.B. Sykes). A.E.R.E. Transl. 1055. (Original in Russian (1958).)

NILSSON, K. (1971). *A method of solving an integral equation and improving the solution with Richardson extrapolation.* Report NA 7147 Computer Science Department, Royal Institute of Technology, Stockholm.

NOBLE, B. (1964). *The numerical solution of non-linear integral equations and related topics.* In Nonlinear integral equations (ed. P.M. Anselone), pp.215- 8, University of Wisconsin Press, Madison.

– (1969*a*). *Instability when solving Volterra integral equations of the second kind by multistep methods.* In Conference on the numerical solution of differntial equations, pp.23- 39. Lecture notes in Mathematics, No.109. Springer-Verlag, Berlin.

– (1969*b*). *Applied linear algebra.* Prentice-Hall, Englewood Cliffs, New Jersey.

– (1971*a*). *A bibliography on methods for solving integral equations.* Author listing. MRC tech. Summ. Rep. 1176.

– (1971*b*) -- Subject listing. MRC tech. Summ. Rep. 1177.

– (1977) *The numerical solution of integral equations.* In

The state of the art in numerical analysis (D.A.H.Jacobs, editor) Academic Press New York .

NYSTRÖM, E.J. (1928). *Über die praktische Auflösung von linearen Integralgleichungen mit Anwendungen auf Randwertaufgaben der Potentialtheorie.* Commentat.physico-math. 4,1-52.

— (1929). *Über die praktische Auflösung von Integralgleichungen.* Commentat.physico-math. 5,1-22.

— (1930). *Über die praktische Auflösung von Integralgleichungen mit Anwendungen auf Randwertaufgaben.* Acta Math. Stockh. 54,185-204.

ORTEGA, J.M. (1972). *Numerical analysis:a second course.* Academic Press, New York.

— and RHEINBOLDT, W.C.(eds.)(1970). *Studies in numerical analysis.* Vol. 2: Numerical solutions of non-linear problems. Society for Industrial and Applied Mathematics, Philadelphia.

— — (1971). *Iterative solution of non-linear equations in several variables.* Academic Press, New York.

OULÈS, H. (1960). *Sur l'intégration numérique de l'équation intégrale de Volterra de seconde éspèce.* C.r.hebd. Seanc. Acad. Sci., Paris, 250, 1433-5.

PEASE, M.C. 3rd (1965). *Methods of matrix algebra.* Academic Press, New York.

PETERS, G. and WILKINSON, J.H. (1970*a*). *The least squares problem and pseudoinverses.* Comput. J.

— — (1970*b*). *Eigenvectors of real and complex matrices by LR and QR triangularizations.* Num. Math. 16, 181-204.

PHILLIPS, D.L. (1962). *A technique for the numerical solution of certain integral equations of the first kind.* J. Ass. comput. Mach. 9, 84-96.

PHILLIPS, G.M. (1970). *Analysis of numerical iterative methods for solving integral and integrodifferential equations.* Comput. J. 13,297-300.

PHILLIPS, J.L. (1969). *Collocation as a projection method for solving integral and other operator equations.* Ph.D. thesis, Purdue University, Lafayette, Indiana.

\- (1972). *The use of collocation as a projection method for solving linear operator equations.* SIAM J. numer. Anal. (Soc. ind. appl. Math.) 9,14-27.

PICC (1960). *Symposium on the numerical treatment of ordinary differential equations, integral and integro-differential equations.* Proceedings of the Rome Symposium organized by the Provisional International Computation Centre. Birkhauser-Verlag, Basel.

POGORZELSKI, W. (1966). *Integral equations and their applications.* Pergamon Press, Oxford. (Translation from Polish.)

POLSKII, N.I. (1962). *Projection methods in applied mathematics.* Soviet Math. 3,488-91.

POUZET, P. (1960). *Méthode d'intégration numérique des équations intégrales et intégro-différentielles du type de Volterra de seconde espèce. Formules de Runge-Kutta.* In Symposium on the numerical treatment of ordinary differential equations, integral and integro-differential equations, pp.362-368. Proceedings of the Rome Symposium organized by the Provisional International Computation Centre. Birkhauser-Verlag, Basel.

\- (1962). *Étude, en vue de leur traitement numérique d'équations intégrales et intégro-différentielles du type de Volterra pour des problèmes de conditions initiales.* Thesis, University of Strassbourg.

— (1963). *Étude en vue de leur traitement numérique des équations intégrales de type Volterra.* Revue Fr. Trait. Inf. 6 79-112.

— (1970). In *Procédures Algol en analyse numérique*, pp.201-4. Centre National de la Recherche Scientifique.

POWELL, M.J.D. (1967). *On the maximum errors of polynomial approximation defined by interpolation and least-squares criteria.* Comput. J. 9,404-7.

RABINOWITZ, P. (1968). *Applications of linear programming to numerical analysis.* SIAM Rev. (Soc. ind. appl. Math.) 10,121-59.

RAKOTCH, E. (1975). *Numerical solution for eigenvalues and eigenfunctions of a Hermitian kernel and an error estimate.* Maths Comput. 29,794-805.

— (1976). *Numerical solution with large matrices of Fredholm's integral equation.* SIAM J. numer. Anal. (Soc. ind. appl. Math.) 13,1-7.

— and STEINBERG, J. (1970). *Error estimate for the numerical solution of Fredholm's integral equation.* Num. Math. 15, 320-8.

RALL, L.B. (ed.) (1965). *Error in digital computation*, Vol. 2. Wiley, New York.

— (1965). *Numerical integration and the solution of integral equations by the use of Riemann sums.* SIAM Rev. (Soc. ind. appl. Math.) 7,55-64.

- (1969). *Computational solution of nonlinear operator equations.* Wiley, New York.

RALSTON, A. (1965). *A first course in numerical analysis.* McGraw-Hill, New York.

REISZ, F. and Sz.NAGY, B. (1955). *Functional analysis.* Ungar, New York.

RIBIÈRE, G. (1967). *Regularization d'operateurs.* Revue Fr. Inf. Recherche Opérationelle 1, 57-79.

RICE, J.R. (1969). *The approximation of functions,* Vol. 2: Advanced topics. Addison-Wesley, Reading, Massachusetts.

RIVLIN, T.J. (1969). *An introduction to the approximation of functions.* Blaisdell, Waltham, Massachusetts.

ROARK, A.L. (1971). *On the eigenproblem for convolution integral equations.* Num. Math. 17, 54-61.

- and SHAMPINE, L.F. (1965). *On a paper of Roark and Wing.* Num. Math. 7, 394-5.

- and WING, G.M. (1965). *A method for computing the eigenvalues of certain integral equations.* Num. Math. 7, 159-70.

SALA, I. (1963). *On the numerical solution of certain boundary value problems and eigenvalue problems of the 2nd and 4th order with the aid of integral equations.* Acta Polytech. scand. Series (*d*) 9, 1-24.

SCHLESSINGER, S. (1957). *Approximating eigenvalues and eigenfunctions of symmetric kernels.* SIAM J. appl.Math. (Soc. ind. appl. Math.) 5, 1-14.

SCHMAEDEKE, W.W. (1968). *Approximate solutions of Volterra integral equations of the first kind.* J. math. Analysis Applic. 23, 604-13.

SCRATON, R.E. (1969). *The solution of integral equations in Chebyshev series.* Maths Comput. 23, 837-44.

SHINBROT, M. (1958). *A generalization of Latta's method for the solution of integral equations.* Q. appl. Math. 16, 415-21.

— (1969). *On the range of general Wiener-Hopf operators.* J. Math. Mech. 18, 587-601.

— (1970). *The solution of some integral equations of Wiener-Hopf type.* Q. appl. Math. 28, 15-36.

SIMMONS, G.F. (1963). *Introduction to topology and modern analysis.* McGraw-Hill, New York.

SINGLETON, R.C. (1967). *On computing the Fast Fourier Transform.* Communs Ass. comput. Mach. 10, 647-54.

SMIRNOV, V.I. (1964). *Course of higher mathematics,* Vol. 4. Pergamon Press, Oxford.

SMITHIES, F. (1962). *Integral equations* (2nd edn). Cambridge University Press, Cambridge.

SOFRONOV, I.D. (1956). *On approximate solution of singular integral equations.* Dokl. Akad. Nauk SSSR 111, 37-9. (In Russian.)

SPENCE, A. (1974). *The numerical solution of the integral equation eigenvalue problem.* D.Phil. thesis, University of Oxford, England.

— (1975). *On the convergence of the Nyström method for the integral equation eigenvalue problem.* Num . Math. 25, 57-66.

— (1976). *Error bounds and estimates for eigenvalues of integral equations.* School of Math. Res. Rep: Math/NA/2, University of Bath (U.K.).

SPOHN, D. (1965). *Sur les formules a pas liés dans l'intégration numérique des équations intégrales du type de Volterra.* In Quatrième Congrès de traitement de l'information, Versailles 1964, pp.349-56. Dunod, Paris.

SQUIRE, W. (1969). *Numerical solution of linear Volterra equations of the first kind.* Aerospace Engineering Report TR-15, West Virginia University.

— (1970). *Integration for engineers and scientists.* American Elsevier, New York.

STENGER, F. (1972). *The approximate solution of Wiener-Hopf integral equations.* J. math. Analysis Applic. 37, 687-724.

— (1973). *An algorithm for the approximate solution of Wiener-Hopf integral equations.* Communs Ass. comput. Mach. 16, 708-10.

STETTER, H.J. (1965). *Asymptotic expansions for the error of discretization algorithms for non-linear functional equations.* Num. Math. 7 18-31.

— (1973). *Analysis of discretization methods for ordinary differential equations.* Springer-Verlag, Berlin.

STEINBERG, J. (1970). *Numerical solution of Volterra integral equations.* Num. Math. 19, 212-17.

STIBBS, D.W.N. and WEIR, R.E. (1959). *On the H-functions for isotropic scattering.* Mon. Not. R. astr. Soc. 119, 512-25.

STRAND, O.N. (1974). *Theory and methods related to the singular-function expansion and Landweber's iteration for integral equations of the first-kind.* SIAM J. numer. Anal. (Soc. ind. appl. Math.) 11, 798-825.

— and WESTWATER, E.R. (1968*a*). *Statistical estimation of the numerical solution of a Fredholm integral equation of the first kind.* J. Ass. comput. Mach. 15, 100-14.

— — (1968*b*). *Minimum-RMS estimation of the numerical solution of a Fredholm integral equation of the first kind.* SIAM J. numer. Anal. (Soc. ind. appl. Math.) 5, 287-95.

STROUD, A.H. (1965). *Error estimates for Romberg quadrature.* SIAM J. numer. Anal. (Soc. ind. appl. Math.) 2, 480-8.

— (1966). *Estimating quadrature errors for functions with low continuity.* SIAM J. numer. Anal. (Soc. ind. appl. Math.) 3, 420-4.

— (1971). *Approximate calculation of multiple integrals.* Prentice-Hall, Englewood Cliffs, New Jersey.

— and SECREST, D. (1966). *Gaussian quadrature formulas.* Prentice-Hall, Englewood Cliffs, New Jersey.

SYMM, G.T. (1966). *An integral equation method in conformal mapping.* Num. Math. 9, 250-8.

SZEGŐ, G. (1959). *Orthogonal polynomials.* American Mathematical Society, New York.

TAYLOR, A.E. (1958). *Introduction to functional analysis.* Wiley, New York.

THACHER, H.C. Jr. (1964). *An efficient composite formula for multi-dimensional quadrature.* Communs Ass. comput. Mach. 7,23-5.

THOMAS, J.A. (1971). *The numerical solution of certain eigenvalue problems in integral equations and ordinary differential equations.* M.Sc. thesis, University of Manchester.

THOMAS, K.S. (1975). *On the approximate solution of operator equations.* Num. Math. 23,231-9.

THOMPSON, G.T. (1957). *On Bateman's method for solving linear integral equations.* J. Ass. comput. Mach. 4,314-28.

TIKHONOV, A.N. (TIHONOV, A.N.) (1963*a*). *On the solution of ill-posed problems and the method of regularization.* Soviet Math. 4, 1035-8.

\- (1963*b*). *Regularization of ill-posed problems.* Soviet Math. 4,1624-7.

\- (1964). *Solution of nonlinear integral equations of the first kind.* Soviet Math. 5,835-8.

\- and GLASKO, V.B. (1964). *The approximate solution of Fredholm integral equations of the first kind.* USSR comput. Math. math. Phys. 4,236-7. (Translations from Russian.)

TODD, J. (ed.) (1962). *Survey of numerical analysis.* McGraw-Hill, New York.

\- and WARSCHAWSKI, S.E. (1955). *On the solution of the Lichtenstein-Gershgorin integral equation in conformal mapping. II. Computational experiments.* In Experiments in the computation of conformal maps. National Bureau of Standards Applied Mathematics Series, Vol. 42,31-44.

TRAUB, J. (1964). *Iterative methods for the solution of equations.* Prentice-Hall, Englewood Cliffs.

TREFFTZ, E. (1933). *Über Fehlerabschätzung bei Berechnung von Eigenwerten.* Math. Annln 5, 595-604.

TRICOMI, F.G. (1957). *Integral equations.* Interscience, New York.

TWOMEY, S. (1963). *On the numerical solution of Fredholm integral equations by the inversion of the linear system produced by quadrature.* J. Ass. comput. Mach. 10,97-101.

— (1965). *The application of numerical filtering to the solution of integral equations encountered in indirect sensing measurements.* J. Franklin Inst. 279,95-109.

VAINIKKO, G.M. (1964). *Asymptotic evaluation of the error of projection methods in the problem of eigenvalues.* USSR comput. Math. math. Phys. 4,9-36.

— (1965). *Evaluation of the error of the Bubnov-Galerkin method in an eigenvalue problem.* USSR Comput. Math. math. Phys. 5,1-31.

— (1967). *Galerkin's perturbation method and the general theory of approximate methods for non-linear equations.* USSR Comput. Math. math. Phys. 7,1-41.

— and DEMENTEVA, A.M. (1968). *The rate of convergence of a method of mechancial quadrature in an eigenvalue problem.* USSR Comput. Math. math. Phys. 8,226-34.

VARAH, J.M. (1973). *On the numerical solution of ill-conditioned linear systems with applications to ill-posed problems.* SIAM J. numer. Anal. (Soc. ind. appl. Math.) 10,257-67.

VARGA, R.S. (1962). *Matrix iterative analysis.* Prentice-Hall, Englewood Cliffs.

– (1970). *Accurate numerical methods for nonlinear boundary value problems.* SIAM Fall Meeting, 1968. In Studies in numerical analysis (ed. J.M. Ortega and W.C. Rheinboldt), Vol. 2, Numerical solutions of non-linear problems. Society for Industrial and Applied Mathematics, Philadelphia.

VEKUA, N.P. (1967). *Systems of singular integral equations.* (transl. from Russian by A. Gibbs and G. Simmons). Noordhoff, Grönigen.

VLADIMIROV, V.S. (1956). *On the application of the Monte Carlo method for obtaining the lowest characteristic number and the corresponding eigenfunction for a linear integral equation.* Theory Probab. Applic. 1, 101-16. (Translated from Russian.)

WACHSPRESS, E.L. (1966). *Iterative solution of elliptic systems.* Prentice-Hall, Englewood Cliffs.

WAGNER, C. (1951). *On the solution of Fredholm integral equations of second kind by iteration.* J. math. Phys. 30, 23-30.

– (1952). *On the numerical evaluation of Fredholm integral equations with the aid of the Liouville-Neumann series.* J. math. Phys. 30, 232-4.

– (1954). *On the numerical solution of Volterra integral equations.* J. math. Phys. 32, 289-301.

WAHBA, G. (1969). *On the approximate solution of Fredholm integral equations of the first kind.* MRC tech. Summ. Rep. 990.

– (1973). *Convergence rates for certain approximate solutions to Fredholm integral equations of the first kind.* J. Approx. Theory 7, 167-85.

— (1975). *Practical approximate solutions to linear operator equations when the data are noisy.* Dept. Statistics Tech. Rep. 430, University of Wisconsin - Madison. (To appear SIAM J. numer. Anal.)

— (1976). *A survey of some smoothing problems and the method of generalized cross-validation for solving them.* Dept. Statistics tech. Rep. 457, University of Wisconsin - Madison. (To appear in Proceedings of a Conference on the Applications of Statistics, Dayton, Ohio, June 1976.)

WALSH, J.E. (ed.) (1966). *Numerical analysis: an introduction.* Academic Press, London.

WANG, J.Y. (1976*a*). *On the discretisation error of the weighted Simpson rule.* Nord. Tidskr. Inf.-behandl. 16,205-14.

— (1976*b*). *On the numerical computation of eigenvalues and eigenfunctions of compact integral operators using spline functions.* J. Inst. Math. & its Appl. 18,177-88.

WEISS, R. (1972). *Numerical procedures for Volterra integral equations.* Ph.D. thesis, Computer Centre, Australian National University, Canberra.

WELSCH, J.H. (1966*a*). *Algorithm 280: Abscissas and weights for Gregory quadrature* [D1]. Communs Ass. comput. Mach. 9,271.

— (1966*b*). *Algorithm 281: Abscissas and weights for Romberg quadrature* [D1]. Communs Ass. comput. Mach. 9,271.

WIELANDT, H. (1956). *Error bounds for eigenvalues of symmetric integral equations.* Proc. Symp. appl. Math. 6,261-82.

WILKINSON, J.H. (1963). *Rounding errors in algebraic processes.* H.M.S.O. London.

— (1965). *The algebraic eigenvalue problem.* Clarendon Press, Oxford.

— and REINSCH, C. (1971). *Linear algebra: handbook for automatic computation,* Vol. 2. Springer-Verlag, Berlin.

WING, G.M. (1965). *On a method for obtaining bounds on the eigenvalues of certain integral equations.* J. math. Analysis Applic. 11, 160-75.

— (1967). *On certain Fredholm integral equations reducible to initial value problems.* SIAM Rev. (Soc. ind. appl. Math.) 4, 655-70.

WOLFE, M.A. (1969). *The numerical solution of non-singular integral and integro-differential equations by iteration with Chebyshev series.* Comput. J. 12, 193-6.

YOUNG, A. (1954*a*). *Approximate product integration.* Proc. R. Soc. A 224, 552-61.

— (1954*b*). *The application of product-integration to the numerical solution of integral equations.* Proc. R. Soc. A 224, 561-73.

YOSIDA, K. (1965). *Functional analysis.* Springer-Verlag, Berlin.

ZAANEN, A.C. (1953). *Linear analysis: measure and integral, Banach and Hilbert space, linear integral equations.* Interscience, New York.

— (1964). *Linear analysis* (4th printing). North-Holland, Amsterdam and Noordhoff, Gröningen.

ZABREYKO, P.P. KOSHELEV, A.I., KRASNOSEL'SKII, M.A., MIKHLIN, S.G.,

... , RAKOVSHCHIK, L.S. and STET'SENKO, V. Ya. (1975). *Integral equations - a reference text.* (Translated from Russian and edited by T.O. Shaposhnikova, R.S. Anderssen, and S.G.Mikhlin.) Noordhoff International, Leyden.

ZAHAR-ITKIN,M.(1964). *On the growth of eigenvalues of a linear integral equation.* Soviet Math. 5, 1348-51.

— (1966). *Growth of eigenvalues of a linear integral equation.* Vest. mosk. gos. Univ. Ser. 1, 21 3-19. (In Russian.)

For further reading, and additional references, consult

BRUNNER, H. (1977). *The approximate solution of integral equations by projection methods based on collocation.* Lecture notes, Mathematics Department, University of Trondheim (ISBN 82-7151 - 022 - 3)

JASWON, M.A. and SYMM, G.T. (1977). *Integral equation methods in potential theory and electrostatics.* Academic, London.

SCHNEIDER, C. (1977). *Beiträge zur numerischen Behandlung schwach-singulärer Fredholmscher Integralgleichungen zweiter Art.* Thesis, Johannes Gutenberg-Universität, Mainz.

INDEX

Abel's equation 69, 917, 980
adjoint operator 26
allow 241
angle 21
Arzela-Ascoli theorem 49
A-stability 810
asymptotic behaviour 112
bound 147, 177, 293, 365, 466, 473

Banach space 45
Banach-Steinhaus theorem 55
basic functions 87
Bateman's method 388
Bernoulli numbers 114
best approximation 93, 95, 100, 103
block-by-block methods 798, 864, 952, 974
boundary-value problem 57, 675
bounded operator 25
bounded set 48
B-splines 91, 92

Cauchy criterion 45
Cauchy kernel 583
Cauchy principal value 69, 637
Cauchy sequence: see Fundamental sequence 45
Cauchy-Schwarz inequality 21
characteristic function 5
characteristic value 4
Chebyshev polynomials 94
Chebyshev set: see Haar set 88
closed sequence 47, 102
closure 101
collectively compact set of linear operators 617
collocation 206, 340, 396, 487, 510, 564, 567, 701
compact operator 48, 596
compact set 47
completely continuous operator 48
completeness 20, 45, 46, 47, 101
complete sequence 46
conditioning (condition number) 155, 160, 168, 230, 297, 421, 427, 650
conformal mapping 67
continuity: Lipschitz- 93
: modulus of 93, 125, 595
continuous operator 48
contraction mapping 51
convergence: essential 19
in mean 19
in norm 19
relatively uniform 19
uniform 19
convergence of Fourier series 99
convergence theory 124, 234, 293, 325, 336, 340, 432, (438), 457, 473, 477, 479, 491, 509, 596, 719, 922-985
convolution kernel 57
Courant's method 169
cubature 144

deferred approach to the limit 115, 151, 178, 293, 362, 467, 907
deferred correction 188, (295), 368, 473
degree of precision 107
difference correction: see deferred correction 188
difference equations 764, 770, 792
differences 117
differential equations 56
Dirichlet problem 68
discontinuities 526
discretization parameter 148
dock problem 567
dominant part 76, 583

eigenfunction 5, 11, 167
generalized 29
left 30, 168
right 30
eigenproblem 5, 26, 28, 167, 563
eigensystem (total) 31
eigenvalue 5, 11, 167
multiple 28
problem 4, 11, 167
eigenvector 11, 158
generalized 159
El-gendi's method 390, 403
Elliott's method 402
equation of the first kind 3, 40, 635, 896, 963

of the second kind 4, 33, 352, 520, 589, 759, 922
of the second kind, not uniquely solvable 570
of the second kind, intrinsically singular 69, 577
equicontinuity 49, 56
error bounds 99, 128, 269, 293, 322, 329, 461, 512, 837
Euler-Maclaurin formula 113, 301, 468
expansion methods 205, 316, 389, 481, 509, 699, 717
exponent (of weakly singular kernel) 68
extrapolation 87

finite differences 117
Fourier-Chebyshev sum 99
Fourier coefficients 98, 142
Fourier-Legendre series 101
Fourier series 97
generalized 100
Fourier sum 99
Fréchet derivative 720, 721
Fredholm alternative 15, 34
Fredholm integral equation 8, 167, 352, 519, 585
linear 10, 167, 352, 521, 635
non-linear 77, 685, 825
of first kind 40, 635, 896, 979
of second kind 33, 352, 520, 759, 922
quasi- 584
functional 574
function space 21
fundamental sequence 45

Galerkin method 214, 330, 406, 496, 509, 559, 566, 707
Galerkin perturbation method 726
Gaussian elimination 154
with pivoting 154
Gaussian quadrature 133
generalized integration rules 135, 139, 198, 346, 380, 479, 539, 546, 651, 688, 885, 915, 959, 979
mid-point rule 135
parabolic rule 135, 136
Simpson's rule 136
trapezium rule 135

Geometric series theorem 51
Green's function 58
generalized 59
Gregory's rules 117, 120, 188, 366, 777.

Haar set (Haar system) 88, 96
Hämmerlin's method 202
Hammerstein equation 78, 688
H-equation 685, 693, 701, 749
Hermitian kernel 12, 220, 229, 269, 276, 285, 287, 316, 322, 330, 374, 566, 637
Hessenberg form 158
Hilbert kernel 583
Hölder's inequality 129
Hilbert space 49

ill-conditioning 155, 162, 168, 297, 421, 643, 788, 790
ill-posed problem 635, 640, 641
index 76, 579, 583
induced metric 44
inequalities 21
infinite limits of integration 577
infinite system (of equations) 391
initial-value problem 56
inner product 20
instability 702, 791, 823, 901, 903
integral equation 1, 2
coupled 10, 522, 820
Fredholm 8, 167, 352, 519
Hammerstein 78, 688
linear 167, 352, 521, 759
nonlinear 77, 685, 825, 892, 922
Urysohn 78, 685
Volterra 7, 755
integral operator 17
integro-differential equation 2, 64, 670
interpolation 87
osculating 89
piecewise polynomial 90
polynomial 88
iterated kernel 27, 71, 565
iterative improvement 155
iterative refinement 155

Jackson's theorem 93
Jacobian matrix 164

Kantorovitch theorem 83

kernel: adjoint 26
centro-symmetric 355
continuous 8
convolution 57
degenerate 32, 169, 231, 236, 385, 429, 435
difference 57
discontinuous 7, 526
finite rank (see degenerate) 32
Fredholm 8
Hermitian (see Hermitian kernel) 12
iterated 24, 27, 71
logarithmic 68
normal 27
self-adjoint 27
symmetric (see also Hermitian) 12, 173, 710
transposed 30
Volterra 8
weakly-singular 68
knots (of splines) 92

Lagrangian interpolation 88
Latta's method 639
least-squares problem 95, 103, 219, 349, 507, 655
Lebesgue constant 94, 100
Lebesgue integrals 46
Legendre polynomials 95
linear independence 28
linear integral equations: classification 3
linear programming 162, 416, 655
linear space 18
Lipschitz condition 93
Lipschitz constant 93
Love's equation 358

measure (of quadrature rule) 125, 241, 268, 444
metric 44
mid-point rule 107
multiplicity of eigenvalue 26, 158
algebraic 26, 158
geometric 26, 158

Neumann series 36
Newton's method 80, 163, 164, 686, 720, 722
modified 84, 165

nonlinear integral equations 77, 685, 825, 892, 922
norm 18, 19
$L^p(L_p\text{-})$ 19, 21
mean-square (L_2-) 21
operator 21, 23, 24
uniform (L_∞-) 19
normal equations 156
normed linear space 19
numerical methods (basic) 152
Nyström extension 170, 176, 252, 267, 280, 357, 364, 440, 444, 448, 452, 455, 467
Nyström method 170, 356

operator 17
adjoint 26
bounded 25
compact (= completely continuous) 48
completely continuous 48
continuous 48
linear 17
nonlinear 720
projection 264
order notation 112
orthogonal functions 12

partition 123
Peano constants 130, 289, 464
Peano's theory 128
pivoting 154
power method 158
predictor-corrector method 827
product (weighted) integration 131, 198, 346, 539, 546, 616, 885, 915, 959, 979.
product rules 145
projection method 509, 596, 719
operator 27, 264, 509

QR algorithm 158
quadrature formula (rule) 106
methods 169, 231, 356, 432, 645, 726, 825, 896, 922, 963
modified 193, 375, 477, 534, 589, 603
quadrature operator 280, 453
quadrature rule 106
quasi-Fredholm equation 584, 587
QZ algorithm 158

radiative transfer (equation of) 685, 693, 701, 749
range 48
rank of a matrix 56
Rayleigh quotient 220, 330, 349
Rayleigh-Ritz method 222, 316, 330, 567
Rayleigh-Ritz-Galerkin methods 330, 406, 496
rectangle rule 107
regular value 15, 26
regularization 644, 655
relatively uniform convergence 19
repetition factor 823
resolvent equation 37
resolvent kernel 36
  Neumann series 36
resolvent operator 37
resolvent set 49
Riemann integral 19
Riemann sum 123
Ritz-Galerkin type method 496
Romberg scheme 115, 124, 153
Runge-Kutta method 165, 849

Schwartz inequality 21
Scraton's method 393
secant method 164
self-adjoint 27
sequences: U- and T- 151
sequential compactness (relative) 47
Simpson's rule 107
singular functions 43, 643
  system 43
  values 44, 156
  vectors 156
singular integral equation 69, 577
singular value decomposition 156, 643
spectral radius 13, 25
spectrum 5, 49
splines (polynomial-) 91, 402, 877
  B- 91
  cubic 91
stability 788
  A- 810
subordinate norm 21

Taylor series 89, 129
theorem: Arzela-Ascoli 49, 54
  Banach-Steinhaus 55
  Geometric series 51
  Jackson's 93
  Kantorovitch 84
  Peano's 129
  Taylor's 129
three-eighths rule 107
trapezium rule 107

uniform boundedness principle 55
Urysohn equation 78, 685

variational method (see also Rayleigh-Ritz-Galerkin methods) 717
Volterra equation 7
  of the first kind 8, 896, 915, 963
  of the second kind 8, 759, 922
Volterra kernel 8
  continuous 8

weak singularity 68, 526, 534, 539, 556, 563, 589, 637, 892, 915, 959, 979
weight function 131
width (of partition) 124
Wiener-Hopf equations 578